**Handbook of Defence
Electronics and Optronics**

Handbook of Defence Electronics and Optronics

Fundamentals, Technologies
and Systems

Anil K. Maini

Consultant, Defence Technologies
Formerly, Director, Laser Science and Technology Centre,
New Delhi, India

Registered Offices
John Wiley & Sons, Inc., 111 River Street, Hoboken, NJ 07030, USA
John Wiley & Sons Ltd, The Atrium, Southern Gate, Chichester, West Sussex, PO19 8SQ, UK

Editorial Office
The Atrium, Southern Gate, Chichester, West Sussex, PO19 8SQ, UK

For details of our global editorial offices, customer services, and more information about Wiley products visit us at www.wiley.com.

Wiley also publishes its books in a variety of electronic formats and by print-on-demand. Some content that appears in standard print versions of this book may not be available in other formats.

Library of Congress Cataloging-in-Publication data applied for

ISBN: 9781119184706

Cover design by Wiley
Cover image: © enot-poloskun/GettyImages

Set in 10/12pt Warnock by SPi Global, Pondicherry, India

Printed in Great Britain by TJ International Ltd, Padstow, Cornwall

10 9 8 7 6 5 4 3 2 1

"A wee bit of heaven drifted down from above.... a handful of happiness,
a heart full of love for sure
Precious, priceless and lovable...the gift of life,
So sacred and pure"
Dedicated with love and blessings
To my new born bundle of joy, my grandson Kiaan

Anil K. Maini

Contents

Preface

Defence Electronics and Optronics, today, is a complete subject in itself. It includes, in its vast domain a wide range of subjects, radar systems, communication satellites, electronic warfare, directed-energy weapons, precision guided munitions, laser systems, optronic sensors, nuclear weapons, space warfare, and so on. Since the early 1940s, during the era of World War II, electronics and optronics have penetrated almost every conceivable area of application of both the tactical battlefield and the strategic domain. The applications have grown at a very fast rate, not only in already existing domains; newer application areas for defence electronics and optronics are finding favour with the Armed Forces. The enormity of the subject of *Defence Electronics and Optronics* and the interest it currently holds internationally in terms of ever increasing usage for a variety of scenarios, and also the kind of interest shown by the Armed Forces and investments being made towards their research and development, underlines the importance of a book that addresses all these topics.

This book comprehensively covers the subject of *Defence Electronics*; covering all topics related to *Defence Electronics* and *Defence Optronics*. The book begins with *Military Communications* in Chapter 1. The opening chapter focuses on communication techniques and systems; antennas and propagation modes; optical communications, including both free space communication, fibreoptic communication; emerging concepts such as software defined radio, net-centric warfare and C^{4}ISR, and some representative military communications equipment for the whole range of applications. *Radar Fundamentals* in the second chapter and *Military Radars* in the third follow that. Chapter 2 presents a detailed description of fundamentals such as radar's operational parameters, radar range equation, radar transmitters and receivers, radar antennas and different types of radar based on principle of operation such as continuous wave (CW) radar, FM-CW radar, pulse Doppler radar, moving target indicator (MTI) radar, tracking radar, pulse compression radar, synthetic aperture radar, over-the-horizon radar (OTHR), monostatic and bistatic radar, surveillance radar and laser radar. Chapter 3 comprehensively covers common military radar systems including target detection, surveillance and tracking radars, fire control radars, ground penetration radars and weapon locating radars. The emphasis in this chapter is on the salient features and applications of major international radar systems in these categories with an overview of involved technologies.

The fourth and fifth chapters cover *Satellite Technology* and *Military Satellites*, respectively. Chapter 4 covers satellite orbits and trajectories, in-orbit operations, satellite hardware and components of a satellite network. Chapter 5 presents an overview of military communication satellites, reconnaissance satellites, SIGINT satellites, early warning satellites, weather forecasting satellites, navigation satellites, and related topics.

Electronic Warfare, covered in the sixth chapter, is the next major topic covered in the book. The chapter extensively covers both electronic warfare and electro-optic warfare systems. Electronic warfare systems' classification, involved technologies and systems are comprehensively described first in this chapter. Major topics covered under electronic warfare systems include different categories of electronic warfare systems, electronic support measures (ESM) such as signal intelligence, radiation intelligence and telemetry intelligence; passive and active electronic countermeasures (ECM) such as chaff, decoys and various types of jammers and electronic counter countermeasures (ECCM). Stealth technologies are also discussed in the chapter. Salient features of major international electronic warfare systems and their deployment scenarios is another highlight of this chapter. The next major topic discussed in this chapter relates to electro-optic countermeasures (EOCM). EOCM systems play an important role in the present-day warfare due to widespread use of lasers and other electro-optic systems. Both passive as well as active electro-optic countermeasures are discussed in this chapter with particular emphasis on laser warning and countermeasures and missile approach warning sensor and infrared countermeasures. Active protection systems are briefly discussed towards the end of the chapter.

Laser technology, optoelectronics and their military applications are discussed next in Chapters 7–10. This section begins with Chapter 7 on *Laser Fundamentals* covering operational basics of lasers and related concepts, laser parameters and their measurement techniques and different types of lasers, mainly including solid state, gas and semiconductor lasers. This is followed up by a comprehensive description of electronics that goes along with a laser optics module to make it a laser source or laser system in *Laser Electronics*. Chapter 8 begins with a brief overview of basic building blocks of laser electronics before it moves on to comprehensive treatment of electronics for solid state, gas and semiconductor laser sources.

Optronic sensors are used in a wide range of military applications both as a part of an overall system and also as individual devices. Chapter 9 on *Photo Sensors and Related Devices* begins with an overview of photo sensors covering important types, major performance specifications and application circuits. These photo sensors are generally used in laser range finders and related devices, laser seekers, laser warning sensors, LIDAR receivers, LADAR sensors, and so on. This is followed by discussion on sensor systems such as night vision devices (NVD), thermal imaging (TI) sensors, CCD and CMOS sensors, FLIR (forward looking infrared) sensors and navigation sensors, including ring laser and fibreoptic gyroscopes.

Having discussed lasers in terms of their operational basics, different types and the electronics that goes with them to make usable laser systems, Chapter 10 discusses tactical military applications of lasers and related devices in *Military Laser Systems*. Major laser systems discussed in this chapter include laser aiming devices, laser range finders and target designators; laser based sensor systems, including laser proximity sensors, laser bathymetry sensors, laser based explosive detection sensors, LADAR sensors and LIDAR sensors.

Precision guided munitions including both radar guided munitions and electro-optically guided munitions are discussed in Chapter 11, titled *Precision Guided Munitions*. The chapter begins with an introduction to different guidance techniques followed by detailed discussion of radar guided munitions, laser guided munitions, infrared guided missiles and GPS/INS guided weapons. Advantages and limitations of different categories of precision guided munitions along with salient features of some common international weapon systems in these categories are in focus in this chapter.

A category of weapon systems that has rapidly evolved in the last decade, transforming itself from laboratory prototypes to field deployable systems, is the class of directed-energy weapon (DEW) systems. DEW systems have been projected to replace kinetic energy weapons for tactical applications in not-too-distant future and strategic applications by 2030. In the concluding chapter of the book on *Directed-Energy Weapons*, after a brief introduction to history of origin of the DEW concept, different categories of DEW systems are discussed. These include particle beam weapons, high-power microwaves, laser-based DEWs and laser-induced plasma channel weapons. The focus is, however, on the two major categories of DEW systems, namely, high-power microwaves (HPM) and laser DEW systems. Merits and demerits of these systems, the involved hardware, major international systems and their application potential are covered in the chapter.

This book is the only one of its kind on the subject of defence electronics and defence optronics that amalgamates the whole gamut of topics in this area. Major topics exhaustively covered in the book include the operational fundamentals of radar, military radar systems, operational fundamentals of satellites, military satellites, electronic countermeasures and counter countermeasures, electro-optic countermeasures, laser fundamentals, laser electronics, tactical military laser systems, radar and electro-optically guided and GPS/INS guided precision strike weapons, fibreoptic and free space laser communication, optronic sensors including photo sensors, LIDAR and LADAR sensors, spectroscopic and interferometric sensors, proximity sensors, bathymetry sensors, particle beam weapons, laser induced plasma channel (LIPC) weapons, less-lethal laser systems including laser dazzlers, laser ordnance disposal systems and lethal directed-energy laser weapons, including chemical, solid state and fibre-based DEW systems, high-power microwaves and E-bombs. The book covers each of the topics in their entirety, from fundamentals to advanced concepts, military systems and related technologies, thereby, leading the reader logically from the operational basics of military systems to involved technologies and battlefield deployment and applications. Each of the topics is discussed keeping in view the military applications. The book also gives an overview of important military systems in different categories along with their application potential. Current status of various military technologies and systems and future trends are also discussed. An Illustrated Glossary at the end of each chapter summarizes important terms, definitions and concepts. A comprehensive bibliography at the end of each chapter will particularly interest researchers.

It is intended to be a reference book for engineers and scientists working in R&D centres, the defence industry and academic institutes engaged in research, development and use of defence electronics and optronics systems. The book also fulfils the requirements of a text book for graduate level students and a reference book for researchers and for industry and military professionals. It is also intended for a wide cross-section of professionals working in the Armed Forces. The book is also intended to be a useful reference for defence experts and strategic planners. I hope that the book will be well received by the readers. Suggestions from readers to make the book more useful in future editions would be highly appreciated.

Anil K. Maini

1

Military Communications

There is a host of technologies that are in use in the state-of-the-art communications equipment used by the Armed Forces world over. Be it the land-based systems or systems in use at sea, in air or space, military communications equipment embraces many technologies. No one technology dominates military communications systems; instead, a number of technologies are used to provide secure and reliable communications. Different generations of communications equipment have been in use by the Armed Forces for various applications over the last 100 years or so. Improvements seen in each new generation of communications equipment have been largely driven by the development of better hardware, including improved components, more sophisticated circuits and more precise manufacturing. The opening chapter begins with discussion on the fundamental topics of communication such as communication techniques and systems; antennas and propagation modes; optical communications including both free-space communication, fibreoptic communication and laser communication, particularly for underwater applications. This is followed by detailed description of emerging concepts employed in the current generation of communications equipment such as software-defined radio, net-centric warfare and C^4ISR. Some representative military communications equipment for the whole range of applications are briefly discussed towards the end.

1.1 Introduction to Military Communications

Military communications technologies are complex and wide ranging. Development of new technologies and advances in existing technologies has led to different generations of communications equipment. Each generation of equipment has leveraged enhanced life and performance of components and emergence of a range of new components due to technological advances. Extended operating time of portable radios used by the Armed Forces in the battlefield due to availability of new battery technologies is one such example. Some of the major concerns faced by military planners relate to improving security and reliability of communications. Another concern relates to integration, which means achieving interoperability among a wide range of communications systems and technologies.

Features and capabilities of communications equipment both for commercial and military usage are undergoing revolutionary changes leading to availability of new generation of sophisticated communications devices and equipment enabling faster, more secure, less costly and more flexible communications. As outlined in the previous paragraph, security and interoperability are the two major concerns. While security-related issues have been resolved a large extent,

Handbook of Defence Electronics and Optronics: Fundamentals, Technologies and Systems, First Edition. Anil K. Maini.
© 2018 John Wiley & Sons Ltd. Published 2018 by John Wiley & Sons Ltd.

Figure 1.1 LG's V10 smart phone. (*Source:* LG Electronics, https://creativecommons.org/licenses/by/2.0/deed. en.CC BY 2.0.)

integration of contrasting communications technologies (or in other words interoperability of different technologies and equipment) is one of the most important challenges facing military technology developers.

Modern radio and networking technologies such as smart phones, tablets, high-speed networks and other sophisticated technologies offer many new opportunities, though they too pose challenges vis-à-vis security and interoperability issues. Very few communication devices have seen such rapid growth and usage and consequential benefits as the smart phones and tablets. Smart phones with touch screen interfaces, internet access and an operating system capable of executing downloaded apps perform many of the functions of a computer. A tablet too is a portable PC with a form factor slightly larger than that of a smart phone. Both can fit into the cargo pocket of a soldier's uniform. Smart phone and tablet apps have given troops the ability to perform a range of tasks anytime anywhere and allowed commanders to instantly distribute essential documents directly to the troops. Network and device security concerns had earlier hindered widespread deployment of smart phones in the Armed Forces and with the availability of new generation smart phones, such as those using Google's Android 6.0 Marshmallow OS, these concerns have been addressed. This has even brought smart phones onto classified networks enabling soldiers access secret level mission command computer systems. Reportedly, the US Government has certified some smart phones, such as the LG G5 using Android OS version 6.0.1 and the V10 using Android OS version 5.1.1 (Figure 1.1), for use in environments where security is the top concern.

Keeping pace with smart phone and other commercial radio innovations, the next major military communications relevant technology evolving quite rapidly is that of *Ground Mobile Radio* (GMR). GMR of the future will focus on two main approaches, namely *Soldier Radio Waveform* (SRW) and *Wideband Networking Waveform*. SRW is an open-standard voice and data waveform used to extend wideband battlefield networks to the tactical edge. It is designed as a mobile ad-hoc waveform and it functions as a router within a wireless network. It is used to transmit vital information over long distances and elevated terrains including mountains and other natural or

Figure 1.2 AN/PRC-154 JTRS Rifleman Radio.

manmade obstructions, and allows communication without a fixed infrastructure such as cellular tower or satellite. The WNW is the next-generation high throughput military waveform, developed under the Joint Tactical Radio System (JTRS) Ground Mobile Radio (GMR) program. It uses the Orthogonal Frequency Division Multiplexing (OFDM) Physical Layer. With its mobile ad-hoc networking (MANET) capabilities, the waveform is designed to work well in both urban landscape as well as a terrain-constrained environment, since it can locate specific network nodes and determine the best path for transmitting information. Combination of these two technologies allows secure networked communications among platoon, squad and team level soldiers. It will also facilitate communication with combat commanders via satellite. The *JTRS-HMS* (Joint Tactical Radio System Handheld Manpack Small form fit) *Rifleman RadioType AN/PRC-154* (Figure 1.2) developed by Thales and General Dynamics, designed to deliver networking connectivity to front line troops and capable of transmitting voice and data simultaneously via SRW (Soldier Radio Waveform), is one example of a GMR accepted for military use. JTRS also interfaces with smart phones. A vehicle-mounted software-defined radio system for ground mobile communications is the one being developed under the Mid-Tier Networking Vehicular Radio (MNVR) programme of the U.S. Army based on the Falcon family of wide band tactical radios. The Harris Corporation has developed the AN/VRC-118 (V)1 under this programme (Figure 1.3).

Another significant technological development has been in the field of wireless networking such as the *Mesh Networks* including *Mobile Ad-hoc Networks* (MANETs). These networking technologies are potentially capable of supporting both JTRS as well as smart phones. Also, these networking technologies provide high-bandwidth networking capabilities for handheld radios, ground and airborne vehicle communications, and security and tactical wireless sensors such as those used to monitor wireless security cameras positioned around critical infrastructure. MANETs can be networked to interconnect multiple mobile phones within a specified coverage area offering greater bandwidth and better connectivity. One application of the MANET is its use by convoys and other team-oriented missions to remain in constant communication with their movement spread over a large terrain. Another application of mesh networks is their use for control and coordination of unmanned ground vehicles. These remotely controlled unmanned vehicles following predefined paths may be used as targets by fighter aircraft pilots during training exercises in the same manner as Pilotless Target Aircraft (PTA) used by Air-Defence ground forces for training purposes.

Figure 1.3 AN/VRC-118 (V)1 MNVR. (*Source:* Courtesy of Harris Corporation.)

Satellite communication too plays an important role in military communications. Though smart phones and other cutting edge communications technologies have impacted on the utility of satellites for military communications, satellite communication continues to remain relevant with its potential of providing ubiquitous satellite coverage to terrestrial communications systems including smart phones. It would be worthwhile mentioning here that, other than the communications services, military satellites are extensively used for intelligence gathering, weather forecasting, early warning and providing navigation and timing data. Software Reprogrammable Payload (SRP), a satellite-rooted technology with its down-to-earth communication potential, is an adaptation of a small radio receiver designed for space applications into a full-fledged radio frequency system initially targeted for UAS (Unmanned Airborne System) communications. SRP is nothing but an airborne SDR (Software-Defined Radio) that facilitates beyond line-of-sight communications. The SRP development program is a joint effort between the Office of Naval Research (ONR), Naval Research Lab (NRL) and Marine Corps Aviation. SRP is a flexible, reconfigurable while-in-operation software-defined radio designed to meet current and future requirements of Unmanned Aircraft System (UAS) communications by Marine Corps. It is currently targeted at the American unmanned aerial vehicle AAI Shadow. The ability to reconfigure SRP's function in operation ensures that marines are able to share data, access capabilities and effectively command while they engage the adversary. SRP, configured around a software-defined radio platform, is designed to perform multiple functions, which include UHF communications relay with interference mitigation, UHF IP router capability, an automated identification system, single channel ground and airborne radio systems and so on. SRP has an open architecture very similar to that of JTRS and is interoperable with it.

Another communication technology that can become a potential game changer in military communications is that of *Cognitive Radio* for reasons of being inherently interoperable, having higher compatibility, reduced interference and enhanced security. The concept of cognitive radio addresses the problem of spectrum congestion that causes acute scarcity of spectrum space. It uses computer intelligence to automatically adapt to band conditions and user requirements.

Cognitive radio in fact refers to an array of technologies that allow radios self-reconfiguration in terms of operating mode selection, optimal power output and dynamic spectrum access for interference management. Cognitive radios have the ability to monitor, sense and detect the conditions of their operating environment, and dynamically reconfigure their own characteristics to best match those conditions. Due to the dynamic access feature, cognitive radio applies situation-aware access to available bands to choose the right radio band for the right purpose. Cognitive technologies including the dynamic spectrum access are being increasingly incorporated into communication devices and technologies such as smart phones, ground mobile radios, mesh networks and other emerging military communications technologies. Cognitive technology developed by XG Technology Inc. that uses six algorithms to evaluate spectrum conditions has already been tested by the US Army.

Many new communication technologies are being developed and maturing. In future, there will be a focus on adoption of more easily developed and deployable technologies due to shrinking military budgets. It will also drive them towards looking at commercial communication technologies, which will further lead to a more collaborative approach and greater focus on communication technologies with multiple users.

1.2 Communication Techniques

In the previous section, we briefly discussed different current and emerging military communication devices and technologies. These are discussed in detail in the latter part of the chapter. Some representative military communication equipment for a range of application scenarios is discussed towards the end of the chapter. Keeping in view the target readers, before we get down to discussing specific military communication technologies and equipment, it would be worthwhile discussing fundamental topics of communication as that would provide a better understanding of more advanced topics.

1.2.1 Types of Information Signals

When it comes to transmitting information over an RF communication link, be it a terrestrial link or a satellite link, it is essentially voice, data or video. A communication link therefore handles three types of signals; namely voice signals like those generated in telephony, radio broadcast and the audio portion of a TV broadcast, data signals produced in computer-to-computer communications and video signals like those generated in a TV broadcast or video conferencing. Each of these signals is called a base band signal. The base band signal is subjected to some kind of processing known as base band processing to convert the signal to a form suitable for transmission. Band limiting of speech signals to 3000 Hz in telephony and use of coding techniques in case of digital signal transmission are examples of base band processing. The transformed base band signal then modulates a high-frequency carrier so that it is suitable for propagation over the chosen transmission link. The demodulator on the receiver end recovers the base band signal from the received modulated signal. The three types of information signals are briefly described in the following paragraphs.

1.2.1.1 Voice Signals

Though the human ear is sensitive to a frequency range of 20 Hz–20 kHz, the frequency range of a speech signal is less than this. For the purpose of telephony, the speech signal is band limited to an upper limit of 3400 Hz during transmission. The quality of received analogue voice signal has been specified by CCITT to give a worst-case base band signal-to-noise ratio of 50 dB. Here, the signal is considered to be a standard test tone and maximum allowable base band signal noise power is 10 nW. Other than signal bandwidth and signal-to-noise ratio, another important parameter that characterizes voice signal is its dynamic range. Speech or voice signal is characterized to have a large dynamic range of 50 dB.

In the case of digital transmission, the quality of recovered speech signal depends upon the number of bits transmitted per second and the bit error rate (BER). The BER to give good speech quality is considered to be 10^{-4}; that is, 1 bit error in 10 kB though a BER of 10^{-5} or better is common.

1.2.1.2 Data Signals

Data signals refer to a digitized version of a large variety of information services including voice telephony, video and computer generated information exchange. It is indeed the most commonly used vehicle for information transfer due to its ability to combine on to a single transmission support the data generated by a number of individual services, which is of great significance when it comes to transmitting multimedia traffic integrating voice, video and data.

Again, it is the system bandwidth that determines how fast the data can be sent in a given period of time expressed in bits/second. Obviously, the bigger the size of file to be transferred in a given time, the faster the required data transfer rate or greater the required bandwidth. Transmission of video signal requires a much larger data transmission rate (or bandwidth) than that required by transmission of a graphics file. A graphics file requires a much a larger data transfer rate than that required by a text file. The desired data rate may vary from a few tens of kb/s to tens of Mb/s for various information services. However, data compression techniques allow transmission signals at a rate much lower than that theoretically needed to do so.

1.2.1.3 Video Signals

The frequency range or bandwidth of a video signal produced as a result of television quality picture information depends upon the size of the smallest picture information, referred to as a pixel. The greater the number of pixels, the higher the signal bandwidth. As an example, in the 625-line, 50 Hz TV standard where each picture frame having 625 lines is split into two fields of 312.5 lines and the video signal is produced as a result of scanning 50 fields per second in an interlaced scanning mode, assuming a worst-case picture pattern where pixels alternate from black to white to generate one cycle of video output, the highest video frequency is given by eqn. 1.1.

$$f = \frac{(aN/2)}{t_h} \tag{1.1}$$

where N = Number of lines per frame
t_h = Time period for scanning one horizontal line

For a 625-line, 50 Hz system, it turns out to be 6.5 MHz. This calculation, however, does not take into account the lines suppressed during line and frame synchronization. For actual picture transmission, the chosen bandwidth is 5 MHz for a 625-line, 50 Hz system and 4.2 MHz for a 525-line, 60 Hz system. And it does not seem to have any detrimental effect on picture quality.

1.2.2 Amplitude Modulation

In *Amplitude Modulation*, the instantaneous amplitude of the carrier signal varies directly as the instantaneous amplitude of the modulating signal. The frequency of the carrier signal

Figure 1.4 Amplitude modulation.

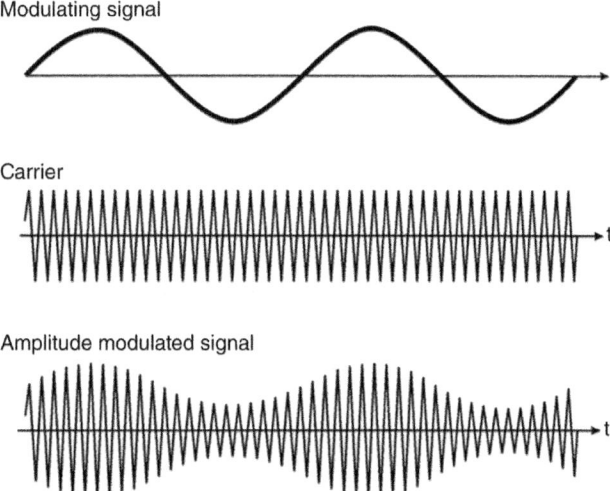

Modulating signal

Carrier

Amplitude modulated signal

remains constant. Figure 1.4 shows the modulating signal, carrier signal and modulated signal in the case of a single tone modulating signal.

If the modulating signal and the carrier signal are expressed, respectively, by $v_m = V_m \cos\omega_m t$ and $v_c = V_c \cos\omega_c t$, then the modulated signal can be expressed mathematically by eqn. 1.2.

$$v(t) = V_c\left(1 + m\cos\omega_m t\right)\cos\omega_c t \tag{1.2}$$

Where $m = $ Modulation index $= V_m/V_c$

When more than one sinusoidal or cosinosoidal signals with different amplitudes amplitude modulate a carrier, the overall modulation index in that case is given by eqn. 1.3.

$$m = \sqrt{\left(m_1{}^2 + m_2{}^2 + m_3{}^2 + \ldots\ldots\ldots\right)} \tag{1.3}$$

Where m_1, m_2, m_3 are modulation indices corresponding to individual signals.

Percentage of modulation or *depth of modulation* is given by $m \times 100$ and for depth of modulation equal to 100%, $m = 1$ or $V_m = V_c$.

1.2.2.1   Frequency Spectrum of the AM Signal

Expanding the expression for the modulated signal given in eqn. 1.2, we get

$$v(t) = V_c\cos\omega_c t + \left(\frac{mV_c}{2}\right)\cos\left(\omega_c - \omega_m\right)t + \left(\frac{mV_c}{2}\right)\cos\left(\omega_c + \omega_m\right)t \tag{1.4}$$

The frequency spectrum of an amplitude modulated signal in case of a single frequency modulating signal thus contains three frequency components; namely the carrier frequency component (ω_c), the sum component ($\omega_c + \omega_m$) and the difference frequency component ($\omega_c - \omega_m$). The sum component represents the upper side band and the difference component the lower side band. Figure 1.5 shows the frequency spectrum.

It should be mentioned here that, in actual practice, the modulating signal is not a single frequency tone. In fact, it is a complex signal. This complex signal can always be represented mathematically in terms of sinusoidal and cosinosoidal components. Thus if a given modulating signal is equivalently represented as a sum of, say, three components (ω_{m1}, ω_{m2} and ω_{m3}),

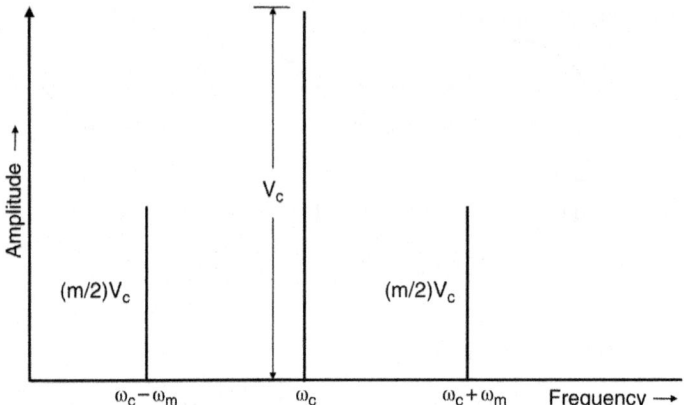

Figure 1.5 Frequency spectrum of the AM signal.

then the frequency spectrum of the AM signal, when such a complex signal amplitude modulates a carrier, contains the frequency components (ω_c), $(\omega_c + \omega_{m1})$, $(\omega_c - \omega_{m1})$, $(\omega_c + \omega_{m2})$, $(\omega_c - \omega_{m2})$, $(\omega_c + \omega_{m3})$ and $(\omega_c - \omega_{m3})$.

1.2.2.2 Power in the AM Signal

The total power (P_t) in an AM signal is related to the unmodulated carrier power (P_c) by eqn. 1.5.

$$P_t = P_c \left[1 + \left(\frac{m^2}{2} \right) \right] = P_c + P_c \left(\frac{m^2}{4} \right) + P_c \left(\frac{m^2}{4} \right) \tag{1.5}$$

Where $(P_c m^2/4)$ is the power in either of the two side bands; that is, upper and lower side bands. For 100% depth of modulation for which $m = 1$, total power in an AM signal is $(3P_c/2)$ and power in each of the two side bands is $(P_c/4)$ with the total side band power equal to $(P_c/2)$. These expressions indicate that, even for 100% depth of modulation, power contained in the sidebands, which contains the actual information to be transmitted, is only one-third of the total power in the AM signal.

Power content of different parts of the AM signal can also be expressed in terms of peak amplitude of unmodulated carrier signal (V_c) by eqn. 1.6.

$$\text{Total power in AM signal, } P_t = \frac{V_c^2}{2} + m\frac{V_c^2}{8} + m\frac{V_c^2}{8} \tag{1.6}$$

Power in either of the two side bands $= m\dfrac{V_c^2}{8}$

1.2.2.3 Noise in the AM Signal

We shall now examine the noise performance when an AM signal is contaminated with noise. Let us assume S, C and N are the signal, carrier and noise power levels, respectively. Let us also assume that the receiver has a bandwidth of B. In the case of a conventional double side band system, it equals $2f_m$ where f_m is the highest modulating frequency. If N_b is the noise power at the output of the demodulator, then

$N_b = AN$, where A is the scaling factor for the demodulator

Now, signal power in each of the side band frequencies is one-quarter of the carrier power as explained in the earlier paragraphs. That is, $S_L = S_U = C/4$

Also, $S_{bL} = S_{bU} = AC/4$

Where S_L = Signal power in lower side band frequency before demodulation.

S_U = Signal power in upper side band frequency before demodulation.
S_{bL} = Signal power in lower side band frequency after demodulation.
S_{bU} = Signal power in upper side band frequency after demodulation.

Since both lower and upper side band frequencies are identical before and after demodulation, they will add coherently in the demodulator to produce a total base band power S_b given by eqn. 1.7.

$$S_b = 2\left(S_{bL} + S_{bU}\right) = 2\left[\left(\frac{AC}{4}\right) + \left(\frac{AC}{4}\right)\right] = AC \tag{1.7}$$

Combining the expressions for S_b and N_b, we get the following relationship between S_b/N_b and (C/N).

$$\frac{S_b}{N_b} = \frac{C}{N} \tag{1.8}$$

Where $N = N_o B$ with N_o being noise power spectral density in W/Hz and B being the receiver bandwidth.

However, this relationship is only valid for a modulation index of unity. The generalized expression for modulation index of m is given by eqn. 1.9.

$$\frac{S_b}{N_b} = m^2\left(\frac{C}{N}\right) \tag{1.9}$$

So far, we have been talking about a single frequency modulating signal. In the case where the modulating signal is a band of frequencies, we would get a lower and an upper side band and we shall get a frequency spectrum such as the one shown in Figure 1.6. Incidentally, the spectrum shown represents a case where the modulating signal is the base band signal of telephony ranging from 300 to 3400 Hz.

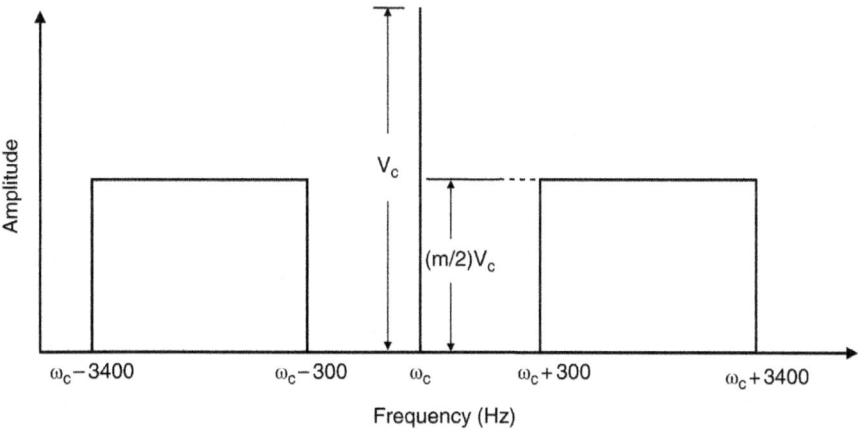

Figure 1.6 Spectrum of the AM signal for a multi-frequency modulating signal.

1.2.2.4 Different Forms of Amplitude Modulation

We have seen in the preceding paragraphs that the process of amplitude modulation produces two side bands, each of which contains the complete base band signal information. Also, the carrier contains no base band signal information. Therefore, if one of the side bands was suppressed and only one side band transmitted, it would make no difference to the information content of the modulated signal. In addition, it would have the advantage of requiring only one-half of the bandwidth required as compared to the conventional double side band signal. Also, if the carrier were also suppressed before transmission, it would lead to a significant saving in the required transmitted power for a given power in the information carrying signal. That is why the single side band suppressed carrier mode of amplitude modulation is very popular. In the following paragraphs, we shall briefly outline some of the practical forms of amplitude modulation systems.

1.2.2.4.1 A3E System

This is the *standard AM* system used for broadcasting. It uses the double side band with full carrier. The standard AM signal can be generated by adding a large carrier signal to the Double Side Band Suppressed Carrier (DSBSC) or simply the DSB signal. The DSBSC signal in turn can be generated by multiplying the modulating signal $m(t)$ and the carrier ($\cos\omega_c t$). Figure 1.7 shows the arrangement for generating the DSBSC signal.

Demodulation of the standard AM signal is very simple and is implemented by using what is known as an envelope detection technique. In a standard AM signal, when the amplitude of the unmodulated carrier signal is very large, the amplitude of modulated carrier is proportional to the modulating signal. Demodulation in this case simply reduces to detection of envelope of modulated carrier regardless of the exact frequency or phase of the carrier. Figure 1.8 shows the envelope detector circuit used for demodulating the standard AM signal. Capacitor C filters out the high-frequency carrier variations.

Demodulation of DSBSC signal is carried out by multiplying the modulated signal by a locally generated carrier signal and then passing the product signal through a low pass filter.

Figure 1.7 Generation of the DSBSC signal.

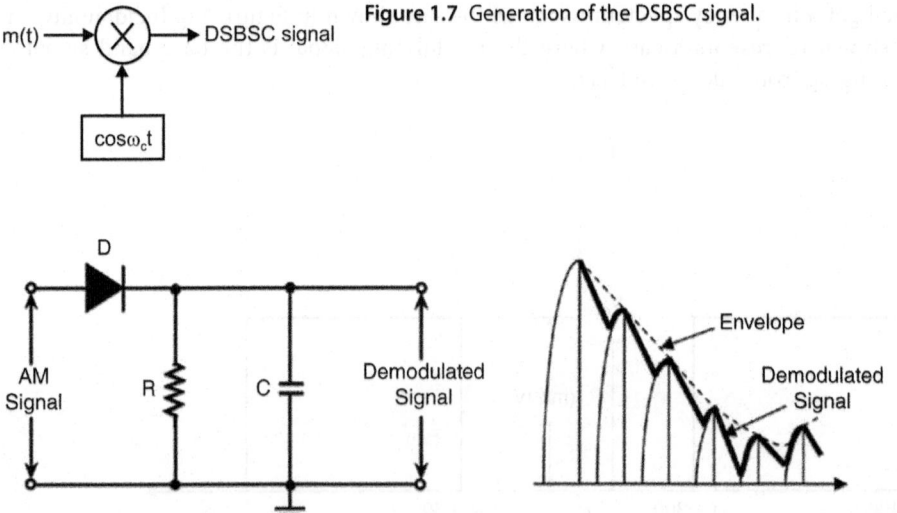

Figure 1.8 Envelop detector for demodulating standard AM signal.

1.2.2.4.2 H3E System

This is the Single Side Band, Full Carrier system (SSBFC). H3E transmission could be used with A3E receivers with distortion not exceeding 5%. One method to generate an SSB signal is to first generate a DSB signal and then suppress one of the side bands by the process of filtering. This method, known as the Frequency Discrimination method, is illustrated in Figure 1.9. In practice, this approach poses some difficulty because the filter needs to have sharp cut-off characteristics.

Another method for generating an SSB signal is the phase shift method. Figure 1.10 shows the basic block-schematic arrangement. The blocks labelled – $\pi/2$ are phase shifters that add a lagging phase shift of $\pi/2$ to every frequency component of the signal applied at the input to the block. The output block can either be an adder or a subtractor. If $m(t)$ is the modulating signal and $m'(t)$ is the modulating signal delayed in phase by $\pi/2$, then the SSB signal produced at the output can be represented by eqn. 1.10.

$$x_{SSB}(t) = m(t)\cos\omega_c t \pm m'(t)\sin\omega_c t \tag{1.10}$$

The output with a + sign is produced when the output block is an adder and with – when the output block is a subtractor.

The difference signal represents the upper side band SSB signal while the sum represents the lower side band SSB signal. For instance, if $m(t)$ is taken as $\cos\omega_m t$, then $m'(t)$ would be $\sin\omega_m t$. The SSB signal in case of the minus sign would then be

$$\cos\omega_m t.\cos\omega_c t - \sin\omega_m t.\sin\omega_c t = \cos(\omega_m + \omega_c)t \tag{1.11}$$

In case of the plus sign, it would be

$$\cos\omega_m t.\cos\omega_c t + \sin\omega_m t.\sin\omega_c t = \cos(\omega_c - \omega_m)t \tag{1.12}$$

Figure 1.9 Frequency discrimination method for generating an SSBFC signal.

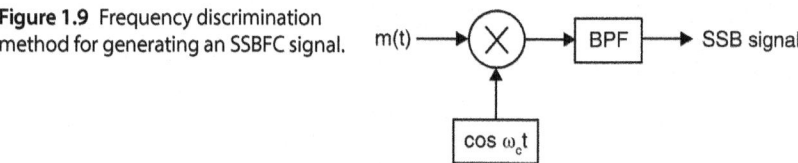

Figure 1.10 Phase shift method for generating an SSBFC signal.

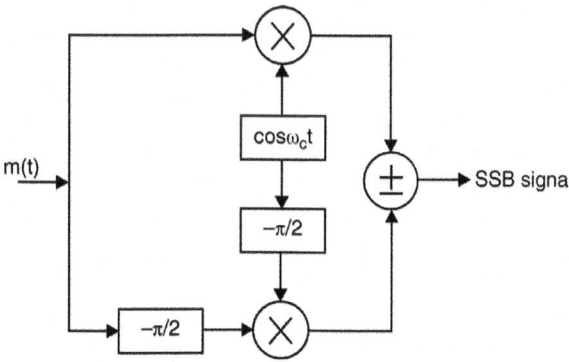

1.2.2.4.3 R3E System

This is the *Single Side Band Reduced Carrier* system, also called the *pilot carrier* system. Re-insertion of a carrier with a greatly reduced amplitude before transmission aims to facilitate receiver tuning and demodulation. This reduced carrier amplitude is 16 or 26 dB below the value it would have had it not been suppressed in the first place. This attenuated carrier signal, while retaining the advantage of saving in power, provides a reference signal to help demodulation in the receiver.

1.2.2.4.4 J3E System

This is the *Single Side Band Suppressed Carrier (SSBSC) system*. This system is usually referred to as SSB, in which a carrier is suppressed by at least 45 dB in the transmitter. It was not popular initially due to the requirement of high receiver stability. However, with the advent of synthesizer-driven receivers, it has now become the standard form of radio communication.

Generation of SSB signals was briefly described in the earlier paragraphs. Suppression of carrier in an AM signal is achieved in the building block known as the *Balanced Modulator*. Figure 1.11 shows the typical circuit implemented using FETs. The modulating signal is applied in push-pull to a pair of identical FETs as shown and as a result, the modulating signals appearing at the gates of the two FETs are 180° out of phase. The carrier signal, as is evident from the circuit, is applied to the two gates in phase. The modulated output currents of the two FETs produced as a result of their respective gate signals are combined in the centre-tapped primary of the output transformer. If the two halves of the circuit are perfectly symmetrical, it can be proved with the help of simple mathematics that the carrier signal frequency will be completely cancelled in the modulated output and the output would contain only the modulating frequency, sum frequency and difference frequency components. The modulating frequency component can be removed from the output by tuning the output transformer. Demodulation of SSBSC signals can be implemented by using a coherent detector scheme as outlined in case of demodulation of DSBSC signal in earlier paragraphs. Figure 1.12 shows the arrangement.

1.2.2.4.5 B8E System

This system uses two independent side bands with the carrier either attenuated or suppressed. This form of amplitude modulation is also known as Independent Side Band (ISB) transmission and is usually employed for point-to-point radio telephony.

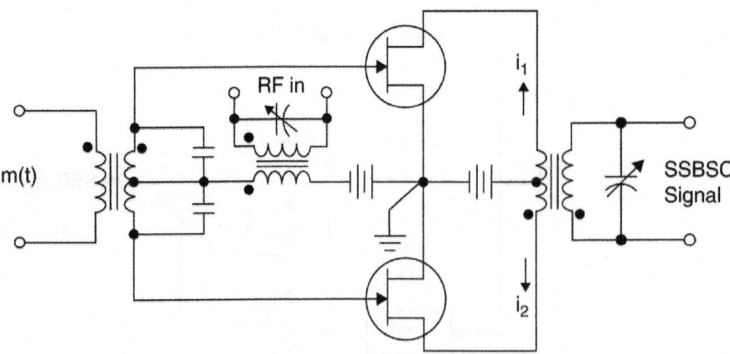

Figure 1.11 Balanced modulator.

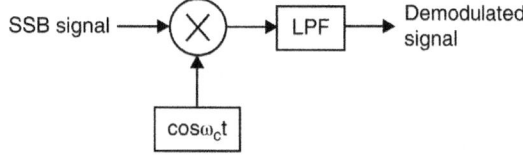

Figure 1.12 Coherent detector for demodulation of a SSBSC signal.

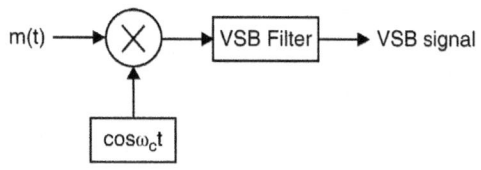

Figure 1.13 Generation of a VSB signal.

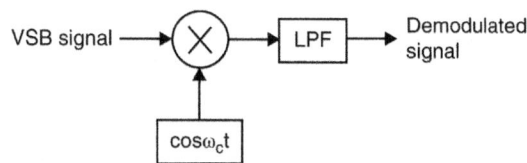

Figure 1.14 Demodulation of a VSB signal.

1.2.2.4.6 C3F System

Vestigial Side Band (VSB) transmission is the other name for this system. It is used for transmission of video signal in commercial television broadcasting. It is a compromise between SSB and DSB modulation systems in which a vestige or part of the unwanted side band is also transmitted usually with a full carrier along with the other side band. The typical bandwidth required to transmit a VSB signal is about 1.25 times that of an SSB signal. VSB transmission is used in commercial TV broadcasting to conserve bandwidth.

VSB signal can be generated by passing a DSB signal through an appropriate side band shaping filter as shown in Figure 1.13. The demodulation scheme for the VSB signal is shown in Figure 1.14.

1.2.3 Frequency Modulation

In *Frequency Modulation*, the instantaneous frequency of the modulation signal varies directly as the instantaneous amplitude of the modulating or base band signal. The rate at which these frequency variations take place is of course proportional to the modulating frequency. If the modulating signal is expressed by $v_m = V_m \cos\omega_m t$, then instantaneous frequency, f, of an FM signal is mathematically expressed by eqn. 1.13.

$$f = f_c\left(1 + KV_m \cos\omega_m t\right) \tag{1.13}$$

Where f_c = unmodulated carrier frequency

V_m = Peak amplitude of modulating signal
ω_m = Modulating frequency
K = Constant of proportionality

The instantaneous frequency is at a maximum when $\cos\omega_m t = 1$ and minimum when $\cos\omega_m t = -1$. This gives:

$$f_{\max} = f_c\left(1 + KV_m\right) \text{ and } f_{\min} = f_c\left(1 - KV_m\right) \tag{1.14}$$

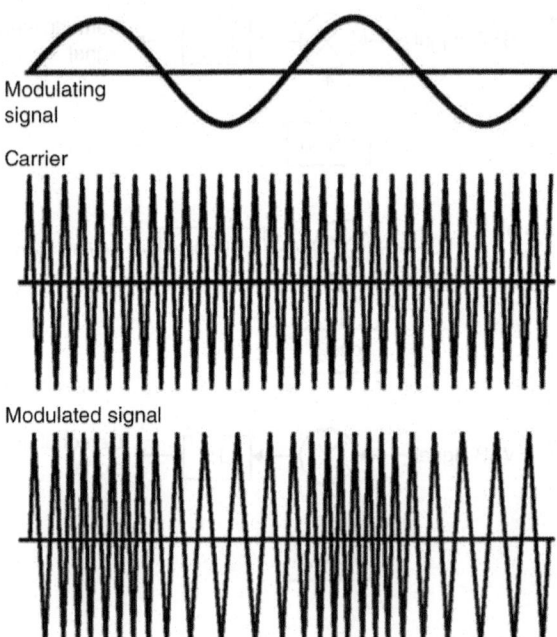

Figure 1.15 Frequency modulation.

Modulating signal

Carrier

Modulated signal

Frequency deviation, δ, is one of the important parameters of an FM signal and is given by $(f_{max} - f_c)$ or $(f_c - f_{min})$. This gives

$$\text{Frequency deviation}, \delta = KV_m f_c \tag{1.15}$$

Figure 1.15 shows the modulating signal (taken as a single tone signal in this case), the unmodulated carrier and the modulated signal. An FM signal can be mathematically represented by eqn. 1.16.

$$v(t) = A\sin\left[\omega_c t + \left(\frac{\delta}{f_m}\right)\sin\omega_m t\right] = A\sin\left[\omega_c t + m_f \sin\omega_m t\right] \tag{1.16}$$

Where, $m_f = Modulation\ index = \delta/f_m$

A is the amplitude of the modulated signal that in turn is equal to the amplitude of the carrier signal.

Depth of modulation in the case of an FM signal is defined as the ratio of frequency deviation, δ, to maximum allowable frequency deviation. Maximum allowable frequency deviation is different for different services and is also different for different standards, even for a given type of service using this form of modulation. For instance, maximum allowable frequency deviation for commercial FM radio broadcast is 75 kHz. It is 50 kHz for the FM signal of TV sound in CCIR standards and 25 kHz for FM signal of TV sound in FCC standards.

1.2.3.1 Frequency Spectrum of the FM Signal

We have seen that an FM signal involves the sine of a sine. The solution of this expression involves the use of Bessel Functions. The expression for the FM signal can be rewritten as:

$$v(t) = A \begin{bmatrix} J_0(m_f)\sin\omega_c t + J_1(m_f)\{\sin(\omega_c + \omega_m)t - \sin(\omega_c - \omega_m)t\} \\ +J_2(m_f)\{\sin(\omega_c + 2\omega_m)t - \sin(\omega_c - 2\omega_m)t\} \\ +J_3(m_f)\{\sin(\omega_c + 3\omega_m)t - \sin(\omega_c - 3\omega_m)t\} + \ldots\ldots\ldots \end{bmatrix} \tag{1.17}$$

Thus, the spectrum of an FM signal contains the carrier frequency component and an apparently infinite number of side bands. In general, $J_n(m_f)$ is the Bessel function of the first kind and nth order. It is evident from this expression it is the value of m_f and the value of the Bessel functions that will ultimately decide the number of side bands having significant amplitude and therefore the bandwidth. The following observations can be made from eqn. 1.17.

1.2.3.2 Narrow Band and Wide Band FM

An FM signal, whether it is a *Narrow Band FM signal* or a *Wide Band FM signal*, is decided by its bandwidth and in turn by its modulation index. For a modulation index m_f much less than 1, the signal is considered the narrow band FM signal. It can be shown that for an m_f less than 0.2, 98% of the normalized total signal power is contained within the bandwidth. Bandwidth for narrow band FM is given by eqn. 1.18.

$$Bandwidth = 2(m_f + 1)\omega_m \approx 2\omega_m \text{ for } m_f \ll 1 \tag{1.18}$$

where ω_m is the sinusoidal modulating frequency.

In case of FM signal with an arbitrary modulating signal $m(t)$ band limited to (ω_M), we define another parameter, called the Deviation Ratio (D) as $D = (Maximum\ frequency\ deviation)/(Bandwidth\ of\ m(t))$. The deviation ratio, D, has the same significance for arbitrary modulation as the modulation index m_f for sinusoidal modulation. The bandwidth in this case is given by eqn. 1.19.

$$Bandwidth = 2(D+1)\omega_M \tag{1.19}$$

This expression for bandwidth is generally referred to as Carson's rule. In the case of $D \ll 1$, the FM signal is considered a narrow band signal and the bandwidth is given by eqn. 1.20.

$$Bandwidth = 2(D+1)\omega_M \approx 2\omega_M \tag{1.20}$$

In the case where $m_f \gg 1$ (for sinusoidal modulation) or $D \gg 1$ (for arbitrary modulation signal band limited to ω_M, the FM signal is termed the wide band FM and the bandwidth in this case is given by eqn. 1.21.

$$Bandwidth = 2m_f\omega_m \text{ or } 2D\omega_M \tag{1.21}$$

1.2.3.3 Noise in the FM Signal

As we shall see in the following paragraphs, frequency modulation is far less affected by presence of noise compared to the effect of noise on an amplitude modulated signal. Whenever a noise voltage with peak amplitude (V_n) is present along with a carrier voltage of peak amplitude (V_c), the noise voltage amplitude modulates the carrier with a modulation index equal to (V_n/V_c). It also phase modulates the carrier with a phase deviation equal to $[\sin^{-1}(V_n/V_c)]$. This expression for phase deviation results when a single frequency noise voltage is considered vectorially and the noise voltage vector is superimposed on the carrier voltage vector. In case of voice communication, an FM receiver is not affected by the amplitude change as it can be removed in the receiver in the limiter circuit. Also, an AM receiver will not be affected by the phase change. It is therefore the effect of phase change on the FM receiver and the effect of amplitude change on the AM receiver that can be used as the yardstick for determining the noise performance of the two modulation techniques. Two very important aspects that need to be addressed when we set out to compare the two communication techniques vis-à-vis their noise performance are the effects of modulation index and the signal-to-noise ratio at the receiver input. Without going into detailed analysis of effects of modulation index and signal-to-noise ratio, we can summarize that an FM system offers a better performance than an AM system provided that (1) the modulation index is greater than unity, (2) the amplitude of carrier is greater than maximum noise peak amplitudes and (3) the receiver is insensitive to amplitude variations.

1.2.3.3.1 Pre-Emphasis and De-Emphasis

Noise has a greater effect on the higher modulating frequencies than it has on lower ones. This is because of the fact that FM results in smaller values of phase deviation at the higher modulating frequencies, whereas the phase deviation due to white noise is constant for all frequencies. Due to this, *S/N* deteriorates at higher modulating frequencies. If the higher modulating frequencies above a certain cut-off frequency were boosted at the transmitter prior to modulation according to a certain known curve and then reduced at the receiver in the same fashion after the demodulator, a definite improvement in noise immunity would result. The process of boosting the higher modulating frequencies at the transmitter and then reducing them in the receiver are, respectively, known as pre-emphasis and de-emphasis. Figure 1.16 shows the pre-emphasis and de-emphasis curves.

Having briefly discussed noise performance of an FM system, it would be worthwhile presenting the mathematical expression that could be used to compute the base band signal-to-noise ratio at the output of the demodulator. Without getting into intricate mathematics, we can write the following expression for base band signal-to-noise ratio (S_b/N_b).

$$\left(\frac{S_b}{N_b}\right) = 3\left(\frac{f_d}{f_m}\right)^2 \left(\frac{B}{2f_m}\right)\left(\frac{C}{N}\right) \tag{1.22}$$

Where f_d = Frequency deviation

f_m = Highest modulating frequency
B = Receiver bandwidth
C = Carrier power at receiver input
N = Noise power (kTB) in bandwidth B.

The expression 1.22 does not take into account the improvement due to use of pre-emphasis and de-emphasis. In that case the expression gets modified to eqn. 1.23.

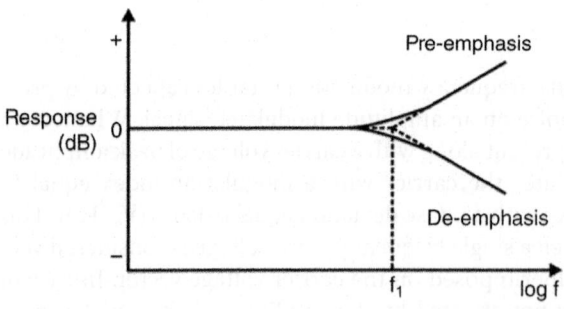

Figure 1.16 Pre-emphasis and de-emphasis curves.

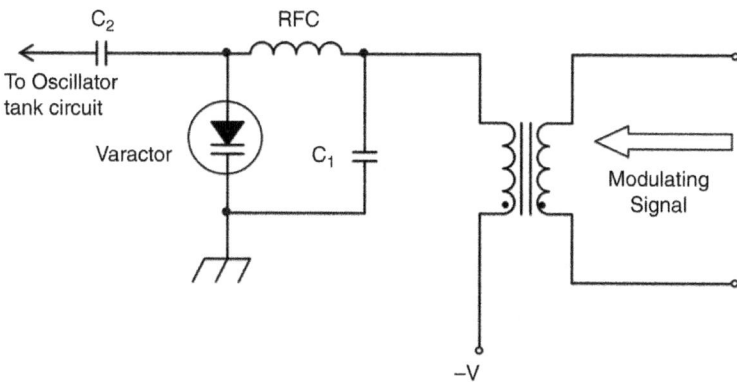

Figure 1.17 LC oscillator-based direct method of FM signal generation.

$$\left(\frac{S_b}{N_b}\right)=\left(\frac{f_d}{f_1}\right)^2\left(\frac{B}{f_m}\right)\left(\frac{C}{N}\right)$$

(1.23)

where f_1 = Cut-off frequency for the pre-emphasis/de-emphasis curve.

1.2.3.4 Generation of FM Signals

Though there are many possible schemes that can be used to generate an FM signal, all of them depend simply on varying the frequency of an oscillator circuit in accordance with the modulating signal input. One of the possible methods is based on the use of a varactor (a voltage variable capacitor) as a part of the tuned circuit of an L-C oscillator. The resonant frequency of this oscillator will not vary directly with the amplitude of the modulating frequency as it is inversely proportional to square root of the capacitance. However, if the frequency deviation is kept small, the resulting FM signal is quite linear. Figure 1.17 shows the typical arrangement when the modulating signal is an audio signal. This is also known as the *direct method* of generating an FM signal as in this case, the modulating signal directly controls the carrier frequency.

Another direct method scheme that can be used for generation of an FM signal is the *reactance modulator*. In this, the reactance offered by a three-terminal active device such as an FET or a bipolar transistor forms a part of the tuned circuit of the oscillator. The reactance in this case is made to vary in accordance with the modulating signal applied to the relevant terminal of the active device. For example, in case of FET, the drain-source reactance can be shown to be proportional to the transconductance of the device, which in turn can be made to depend on the bias voltage at its gate terminal. The main advantage of using the reactance modulator is that large frequency deviations are possible and thus less frequency multiplication is required. One of the major disadvantages of both these direct method schemes is that carrier frequency tends to drift and therefore additional circuitry is required for frequency stabilization. The problem of frequency drift is overcome in crystal controlled oscillator schemes.

While crystal control provides a very stable operating frequency, the exact frequency of oscillation in this case mainly depends upon the crystal characteristics and to a very small extent on the external circuit. For example, a capacitor connected across the crystal can be used to change its frequency typically from 0.001 to 0.005%. The frequency change may be

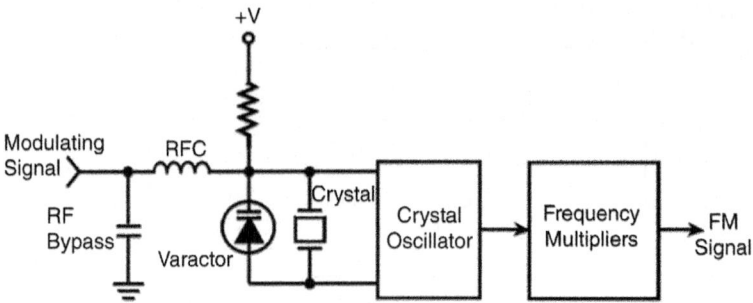

Figure 1.18 Crystal control oscillator-based scheme for FM signal generation.

linear only up to a change of 0.001%. Thus, a crystal oscillator can be frequency modulated over a very small range by a parallel varactor. The frequency deviation possible with such a scheme is usually too small to be used directly. The frequency deviation in this case is then increased by using frequency multipliers as shown in Figure 1.18.

Another approach that eliminates the requirement of extensive chains of frequency multipliers in direct crystal controlled systems is an *indirect method* where frequency deviation is not introduced at the source of RF carrier signal; that is, the oscillator. The oscillator in this case is crystal controlled to get the desired stability of the unmodulated carrier frequency and the frequency deviation is introduced at a later stage. The modulating signal phase modulates the RF carrier signal produced by the crystal controlled oscillator. Since frequency is nothing but rate of change of phase, phase modulation of the carrier has the associated frequency modulation. Introduction of a leading phase shift would lead to an increase in the RF carrier frequency and a lagging phase shift results in a reduced RF carrier frequency. Thus, if the phase of the RF carrier is shifted by the modulating signal in a proper way, the result is a frequency modulated signal. Since phase modulation also produces little frequency deviation, a frequency multiplier chain is required in this case too.

1.2.3.5 Detection of FM Signals

Detection of an FM signal involves the use of some kind of a frequency discriminator circuit that can generate an electrical output directly proportional to the frequency deviation from the unmodulated RF carrier frequency. The simplest of the possible circuits would be the *balanced slope detector* that makes use of two resonant circuits; one off-tuned to one side of the unmodulated RF carrier frequency and the other off-tuned to the other side of it. Figure 1.19 shows the basic circuit. When the input to this circuit is at the unmodulated carrier frequency, the two off-tuned slope detectors (or the resonant circuits) produce equal amplitude but out-of-phase outputs across them. The two signals after passing through their respective diodes produce equal amplitude opposing DC outputs that combine to produce a zero or near-zero output. When the received signal frequency is towards either side of the centre frequency, one output has higher amplitude than the other to produce a net DC output across the load. The polarity of the output produced depends on which side of the centre frequency the received signal is. Such a detector circuit, however, does not find application for voice communication because of its poor linearity of response.

Another class of FM detectors, known as *quadrature detectors*, use a combination of two quadrature signals, that is, two signals 90° out of phase, to get the frequency discrimination property. One of the two signals is the FM signal to be detected and its quadrature counterpart is generated by using either a capacitor or an inductor. The principle of operation of quadrature

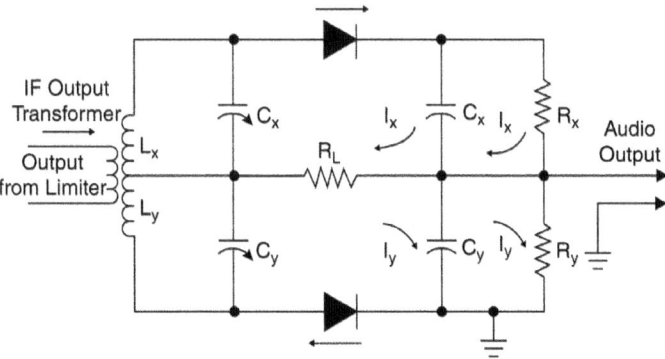

Figure 1.19 Basic circuit of balanced slope detector.

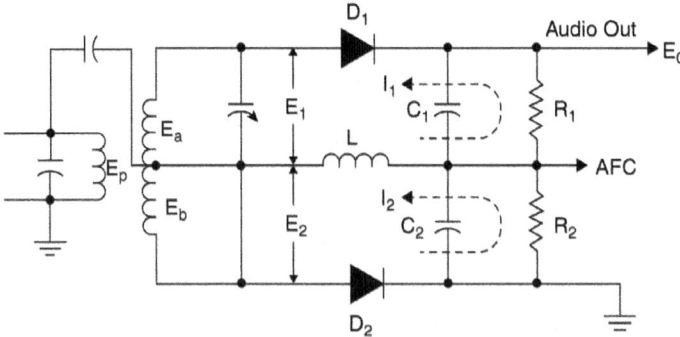

Figure 1.20 Foster–Seeley frequency discriminator.

detector forms the basis of two most commonly used FM detectors namely the Foster–Seeley FM Discriminator and the Ratio Detector.

In the *Foster–Seeley Frequency Discriminator* circuit of Figure 1.20, the two Quadrature signals are provided by the primary signal E_p as appearing at the centre tap of secondary and E_b. We can appreciate that E_a and E_b are 180° out of phase and also that E_p available at the centre tap of the secondary is 90° out of phase with the total secondary signal. Signals E_1 and E_2, appearing across the two halves of the secondary, have equal amplitudes when the received signal is at the unmodulated carrier frequency as shown in the phasor diagram. E_1 and E_2 cause rectified currents I_1 and I_2 to flow in the opposite directions with the result that voltage across R_1 and R_2 are equal and opposite. The detected voltage is zero for $R_1 = R_2$. The conditions when the received signal frequency deviates from the unmodulated carrier frequency value are also shown in the phasor diagrams. In case of frequency deviation, there is a net output voltage whose amplitude and polarity depends upon the amplitude and sense of frequency deviation.

Another commonly used FM detector circuit is the *ratio detector*. This circuit has the advantage that it is insensitive to short term amplitude fluctuations in the carrier and therefore does not require an additional limiter circuit. The circuit configuration, as can be seen from Figure 1.21, is similar to the one given in case of Foster–Seeley discriminator circuit, except for a couple of changes. These are a reversal of diode connections and addition of a large capacitor C_3. The time constant $[(R_1 + R_2)C_3]$ is much larger than the time period of even the lowest modulating frequency of interest. The detected signal in this case appears across the C_1–C_2 junction. The sum output across R_1–R_2 and hence across C_1–C_2 remains constant for a given

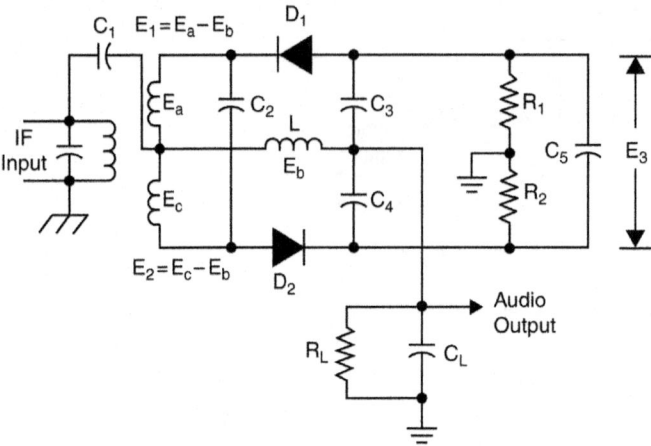

Figure 1.21 Ratio detector.

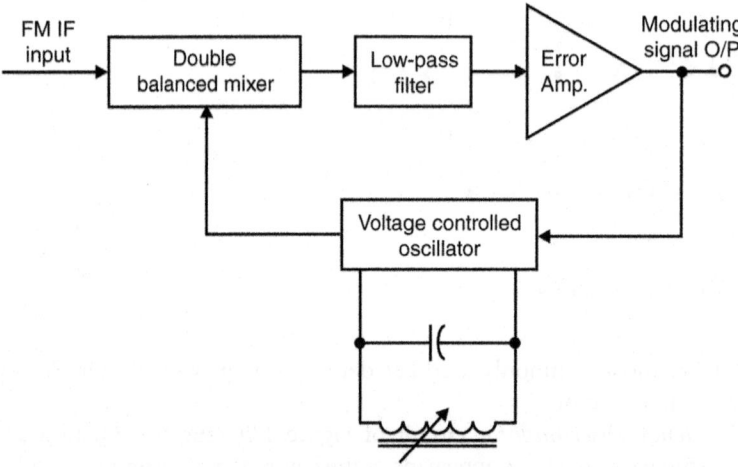

Figure 1.22 PLL-based FM detector.

carrier level, and also is insensitive to rapid fluctuations in carrier level. However, if the carrier level changes very slowly C_3 would charge/discharge to the new carrier level. The detected signal therefore is not only proportional to the frequency deviation, it also depends upon average carrier level.

Yet another form of FM detector is the one implemented using a phase locked loop (PLL). A PLL, as we know, has a phase detector (usually a double balanced mixer), a low pass filter and an error amplifier in the forward path and a voltage controlled oscillator (VCO) in the feedback path. The detected output appears at the output of error amplifier as shown in Figure 1.22. A PLL-based FM detector functions as follows.

The FM signal is applied to the input of the phase detector. The VCO is tuned to a nominal frequency equal to unmodulated carrier frequency. The phase detector produces an error voltage depending upon frequency and phase difference between the VCO output and instantaneous frequency of input FM signal. As the input frequency deviates from the centre frequency, the error voltage produced as result of frequency difference after passing through the low pass filter and error amplifier drives the control input of the VCO to keep its output frequency always in lock with the instantaneous frequency of the input FM signal. As a result, the error amplifier always represents the detected output. The double balanced mixer nature of phase detector suppresses any carrier level changes and therefore the PLL-based FM detector requires no additional limiter circuit.

A comparison of the three types of FM detectors would reveal that the Foster–Seeley type FM discriminator offers excellent linearity of response, is easy to balance and the detected output depends only on frequency deviation. But it needs high gain RF and IF stages to ensure the limiting action. The ratio detector circuit on the other hand requires no additional limiter circuit; detected output depends both on frequency deviation as well as on average carrier level. However, it is difficult to balance. The PLL-based FM detector offers excellent reproduction of modulating signal, is easy to balance and has low cost and high reliability.

1.2.4 Pulse Communication Systems

Pulse communication systems differ from continuous-wave communication systems in the sense that the message signal or intelligence to be transmitted is not supplied continuously as in case of AM or FM. In turn, it is sampled at regular intervals and it is the sampled data that is transmitted. All pulse communication systems fall into either of the two categories; namely, analogue systems and digital systems. Analogue and digital communication systems differ in the mode of transmission of sampled information. In case of analogue communication systems, the representation of sampled amplitude may be infinitely variable whereas in digital communication systems, a code representing the sampled amplitude to the nearest predetermined level is transmitted.

1.2.5 Analogue Pulse Communication Systems

Important techniques that fall in the category of analogue pulse communication systems include:

1) Pulse Amplitude Modulation
2) Pulse Width (or Duration) Modulation
3) Pulse Position Modulation

1.2.5.1 Pulse Amplitude Modulation
In the case of *Pulse Amplitude Modulation* (PAM), the signal is sampled at regular intervals and the amplitude of each sample, which is a pulse, is proportional to the amplitude of the modulating signal at the time instant of sampling. The samples, as shown in Figure 1.23, can have either a positive or negative polarity. In a single-polarity PAM, a fixed DC level can be added to the signal as shown in Figure 1.23(c). These samples can then be transmitted either by a cable or used to modulate a carrier for wireless transmission. Frequency modulation is usually employed for the purpose and the system is known as PAM-FM.

1.2.5.2 Pulse Width Modulation
In case of *Pulse Width Modulation* (PWM), as shown in Figure 1.24, the starting time of the sampled pulses and their amplitude is fixed. The width of each pulse is made proportional to the amplitude of the signal at the sampling time instant.

(a)

(b)

(c)

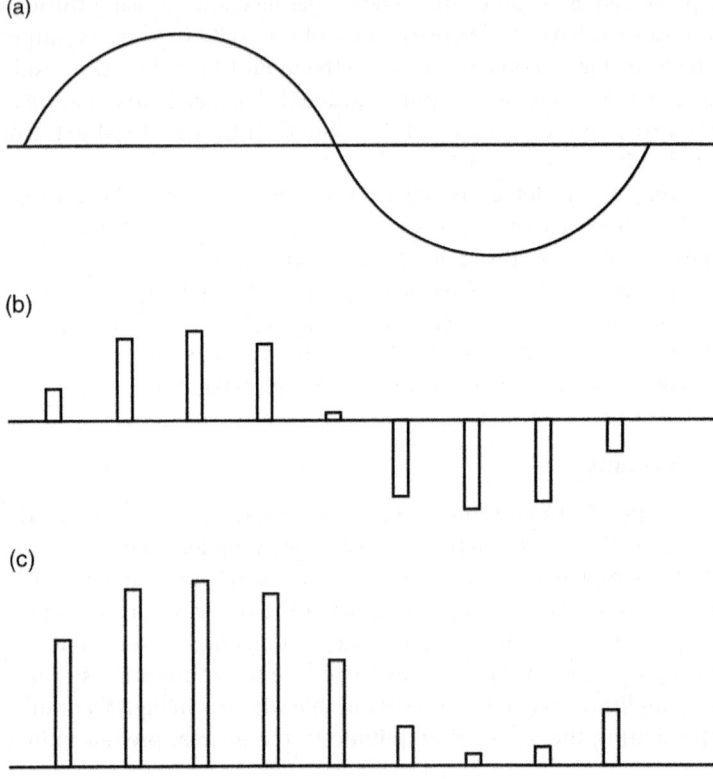

Figure 1.23 Pulse amplitude modulation: (a) modulating signal, (b) double polarity PM signal and (c) single polarity PAM signal.

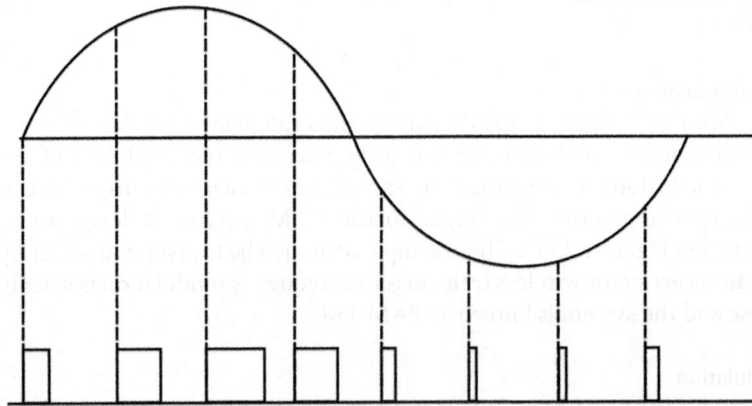

Figure 1.24 Pulse width modulation.

1.2.5.3 Pulse Position Modulation

In case of *Pulse Position Modulation* (PPM), the amplitude and width of sampled pulses is maintained as constant and the position of each pulse with respect to the position of a recurrent reference pulse varies as a function of instantaneous sampled amplitude of the modulating signal. In this case, the transmitter sends synchronizing pulses to operate timing circuits in the receiver.

A pulse position modulated signal can be generated from a pulse width modulated signal. In a PWM signal, as we know, the position of leading edges is fixed, whereas that of trailing edges depends upon the width of pulse, which in turn is proportional to amplitude of modulating signal at the time instant of sampling. Quite obviously, the trailing edges constitute the pulse position modulated signal. The sequence of trailing edges can be obtained by differentiating the PWM signal and then clipping the leading edges as shown in Figure 1.25.

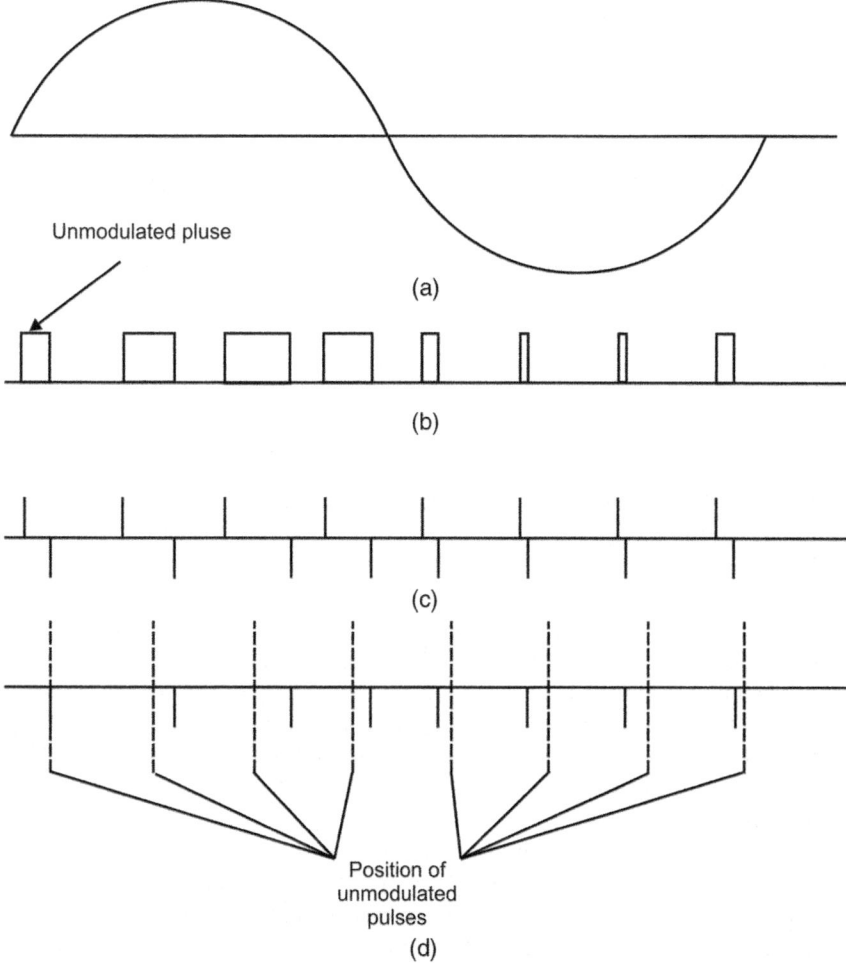

Figure 1.25 Pulse position modulation: (a) modulating signal, (b) pulse width modulated signal, (c) differentiated pulse width modulated signal and (d) pulse position modulated signal.

Pulse width modulation and pulse position modulation both fall in the category of Pulse Time Modulation (PTM).

1.2.6 Digital Pulse Communication Systems

Digital pulse communication techniques differ from the analogue pulse communication techniques described in the previous paragraphs in the sense that, in the case of analogue pulse modulation, the sampling process transforms the modulating signal into a train of pulses with each pulse in the pulse train representing the sampled amplitude at that instant of time. This is one of the characteristic features of the pulse, such as amplitude in the case of PAM, width in the case of PWM and position of leading or trailing edges in the case of PPM, which is varied in accordance with the amplitude of the modulating signal. What is important to note here is that the characteristic parameter of the pulse, which is amplitude or width or position, is infinitely variable. As an illustration, if in case of pulse width modulation, every volt of modulating signal amplitude corresponded to 1 μs of pulse width, then 5.23 and 5.24 V amplitudes would be represented by 5.23 and 5.24 μs, respectively. Further, there could be any number of amplitudes between 5.23 and 5.24 V. It is not the same in the case of digital pulse communication techniques, to be discussed in the paragraphs that follow. In the case of digital pulse communication techniques, each sampled amplitude is transmitted by a digital code representing the nearest predetermined level.

Important techniques that fall in the category of digital pulse communication systems include:

1) Pulse Code Modulation (PCM)
2) Differential PCM
3) Delta Modulation
4) Adaptive Delta Modulation.

1.2.6.1 Pulse Code Modulation

In *Pulse Code Modulation* (PCM), the peak-to-peak amplitude range of the modulating signal is divided into a number of standard levels, which in case of binary system is an integral power of 2. The amplitude of the signal to be sent at any sampling instant is the nearest standard level. For example, if at a particular sampling instant, the signal amplitude is 3.2 V, it will not be sent as a 3.2 V pulse as might have been in case of PAM or a 3.2 μs wide pulse as in case of PWM; instead it will be sent as the digit-3 if 3 V is the nearest standard amplitude. And in a case where the signal range has been divided into 128 levels, it will be transmitted as 0000011. The coded waveform would be like that shown in Figure 1.26(a). This process is known as *quantizing*. In fact, a supervisory pulse is also added with each code group to facilitate reception. Thus, the number of bits for 2^n chosen standard levels per code group is $n + 1$. Figure 1.26(b) illustrates the quantizing process in PCM.

As is evident from Figure 1.26(b), the quantizing process distorts the signal. This distortion is referred to as quantization noise, which is random in nature as the error in the signal's amplitude and that actually sent after quantization is random. Maximum error can be as high as half of the sampling interval, which means if the number of levels used were 16, it would be 1/32 of the total signal amplitude range. It may be mentioned here that it would be unfair to say that a PCM system with 16 standard levels will necessarily have a signal-to-quantizing noise ratio of 32:1; because neither the signal nor the quantizing noise will always have its maximum value. Signal-to-noise ratio depends upon many other factors too and also its dependence on the number of quantizing levels is statistical in nature. Nevertheless, an

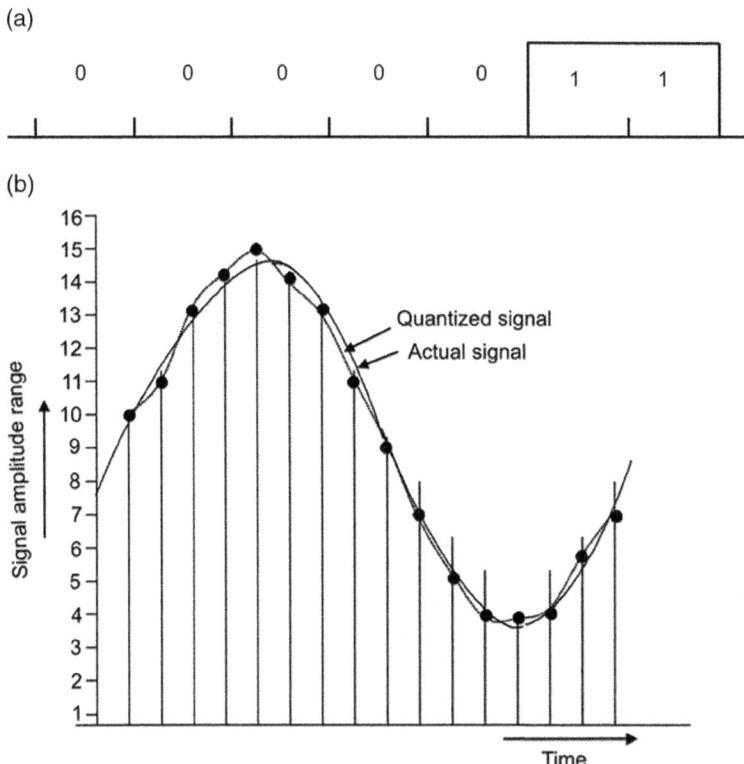

Figure 1.26 The quantization process: (a) coding waveform and (b) actual and quantized signals.

increase in number of standard levels does lead to an increase in signal-to-noise ratio. In practice, for speech signals, 128 levels are considered adequate. Also, the greater the number of levels, the greater the number of bits to be transmitted and therefore the higher the required bandwidth. In binary PCM where a binary system of representation is used for encoding various sampled amplitudes, the number of bits to be transmitted per second would be given by nf_s where $n = \log_2 L$, L = number of standard levels. Also, $f_s \geq f_m$ where f_m = message signal bandwidth.

Assuming that PCM signal is a low pass signal of bandwidth, f_{PCM}, then the required minimum sampling rate would be $2f_{PCM}$.

$$\text{Therefore, } 2f_{PCM} = nf_s \text{ or } f_{PCM} = \left(\frac{n}{2}\right)f_s \tag{1.24}$$

Generating a PCM signal is a complex process. The message signal is usually sampled and first converted into a PAM signal, which is then quantized and encoded. The encoded signal can then be transmitted either directly via a cable or used to modulate a carrier using analogue or digital modulation techniques. PCM-AM is quite common.

1.2.6.2 Differential PCM
Differential PCM is similar to conventional PCM. The difference between the two lies in the fact that, in differential PCM, each word or code group indicates a difference in amplitude

(a)

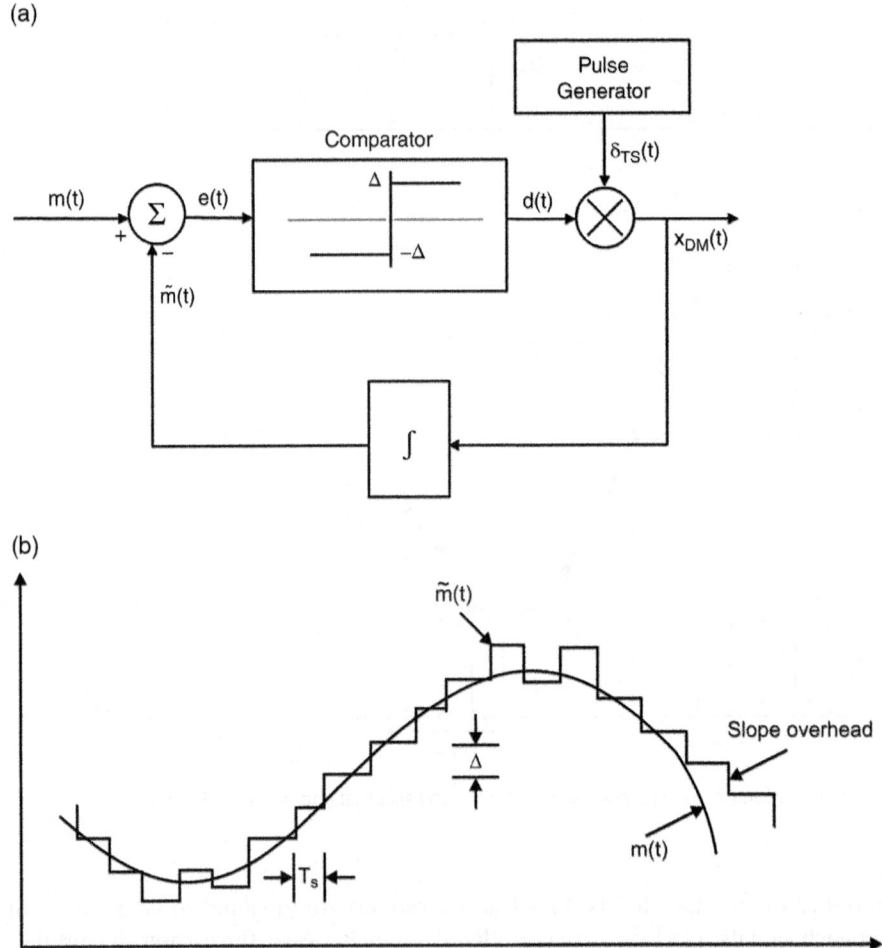

(b)

Figure 1.27 (a) Delta modulator and (b) output waveform.

(positive or negative) between the current sample and the one immediately preceding. Thus, it is not the absolute but the relative value that is indicated. The bandwidth required as a consequence is lower compared to that required in the case of normal PCM.

1.2.6.3 Delta Modulation

Delta modulation has various forms. In one of the simplest forms, only one bit is transmitted per sample just to indicate whether the amplitude of the current sample is greater or smaller than the amplitude of the immediately preceding sample. It has extremely simple encoding and decoding processes but then it may result in tremendous quantizing noise in case of rapidly varying signals.

Figure 1.27(a) shows a simple delta modulator system. The message signal $m(t)$ is added to a reference signal with the polarity shown. The reference signal is integral of the delta modulated signal. The error signal $e(t)$ produced is fed to a comparator. The output of the comparator is $(+\Delta)$ for $e(t) > 0$ and $(-\Delta)$ for $e(t) < 0$. The output of the delta modulator is a series of impulses with the polarity of each impulse depending upon the sign of $e(t)$ at the sampling time instance. Integration of delta modulated output $x_{DM}(t)$ is a staircase approximation of message signal $m(t)$, as shown in Figure 1.27(b).

The delta modulated signal can be demodulated by integrating the modulated signal to get the staircase approximation and then passing it through a low pass filter. The smaller the step size Δ, the better the reproduction of the message signal. However, small step size must be accompanied by a higher sampling rate if the slope overload phenomenon is to be avoided. In fact, to avoid slope overload and associated signal distortion the following condition should be satisfied.

$$\frac{\Delta}{T_s} = \left[\frac{dm(t)}{d(t)}\right]_{max} \tag{1.25}$$

Where T_s = time between successive sampling time instants.

1.2.6.4 Adaptive Delta Modulator

This is a type of delta modulator. In delta modulation, the dynamic range of amplitude of message signal $m(t)$ is very small due to threshold and overload effects. This problem is overcome in an adaptive delta modulator. In adaptive delta modulation, the step size Δ is varied according to the level of message signal. The step size is increased as the slope of the message signal increases to avoid overload. The step size is reduced to reduce the threshold level and hence the quantizing noise when the message signal slope is small. In case of adaptive delta modulation, however, the receiver also needs to be adaptive. The step size at the receiver should also be made to change to match the changes in step size at the transmitter.

1.2.7 Sampling Theorem

During our discussion on digital pulse communication techniques such as pulse code modulation, delta modulation and so on, we have noticed that the three essential processes of such a system are sampling, quantizing and encoding. *Sampling* is the process in which a continuous time signal is sampled at discrete instants of time and its amplitude at those discrete instants of time are measured. *Quantization* is the process by which the sampled amplitudes are represented in the form of a finite set of levels. The *encoding* process designates each quantized level by a code. Digital transmission of analogue signals has been made possible by sampling the continuous time signal at a certain minimum rate, which is dictated by what we call *sampling theorem*.

Sampling theorem states that a band limited signal with the highest frequency component as f_M Hz can be recovered completely from a set of samples taken at the rate of f_s samples per second provided that $f_s \geq 2 f_M$. This theorem is also known as the *Uniform Sampling Theorem* for base band or low pass signals. The minimum sampling rate of $2f_M$ samples per second is called the *Nyquist rate* and its reciprocal is the *Nyquist interval*. For sampling of band pass signals, lower sampling rates can sometimes be used.

The sampling theorem for band pass signals states that if a band pass message signal has a bandwidth of f_B and an upper frequency limit of f_u, then the signal can be recovered from the sampled signal by band pass filtering if $f_s = (2 f_u / k)$ where k is the largest integer not exceeding (f_u / f_B).

1.2.8 Shannon–Hartley Theorem

The *Shannon-Hartley theorem* describes the capacity of a noisy channel (assuming that the noise is random). According to this theorem,

$$C = B\log_2\left[1+\left(\frac{S}{N}\right)\right] \text{bits/second} \tag{1.26}$$

Where C = Channel capacity in bits/second

B = Channel bandwidth in Hz
S/N = Signal-to-noise ratio at the channel output or receiver input.

The Shannon–Hartley theorem underlines the fundamental importance of bandwidth and signal-to-noise ratio in communication. It also shows that, for a given channel capacity, we can exchange increased bandwidth for decreased signal power. It may be mentioned that increasing the channel bandwidth by a certain factor does not increase the channel capacity by the same factor in a noisy channel, as would be suggested by Shannon–Hartley theorem apparently. This is because increasing the bandwidth also increases noise, thus decreasing S/N ratio. However, channel capacity does increase with increase in bandwidth; the increase will not be in the same proportion.

1.2.9 Digital Modulation Techniques

Base band digital signals have significant power content in the lower part of the frequency spectrum. Because of this, these signals can be conveniently transmitted over a pair of wires or coaxial cables. At the same time, for the same reason, it is not possible to have efficient wireless transmission of base band signals as it would require prohibitively large antennas, which would not be a practical or feasible proposition. Therefore, if base band digital signals are to be transmitted over a wireless communication link, they should first modulate a CW high-frequency carrier. Three well-known techniques available for the purpose include:

1) Amplitude Shift Keying (ASK)
2) Frequency-Shift Keying (FSK)
3) Phase Shift Keying (PSK)

Each one of these is described in the following paragraphs. Of the three techniques used for digital carrier modulation, PSK and its various derivatives such as Differential PSK (DPSK), Quadrature PSK (QPSK) and Offset Quadrature PSK (O-QPSK) are the most commonly used ones; more so for satellite communications because of certain advantages they offer over others. PSK is therefore described in a little more detail.

1.2.9.1 Amplitude Shift Keying (ASK)
In the simplest form of ASK, the carrier signal is switched ON and OFF depending upon whether a 1 or 0 is to be transmitted (Figure 1.28). For obvious reasons, this form of ASK is also known as *ON-OFF Keying* (OOK). The signal in this case is represented by:

$$X_c(t) = A\sin \omega_c t \text{ for bit 1 and } X_c(t) = 0 \text{ for bit 0.}$$

ON-OFF keying has the disadvantage that appearance of any noise during transmission of bit 0 can be misinterpreted as data. This problem can be overcome by switching the amplitude of

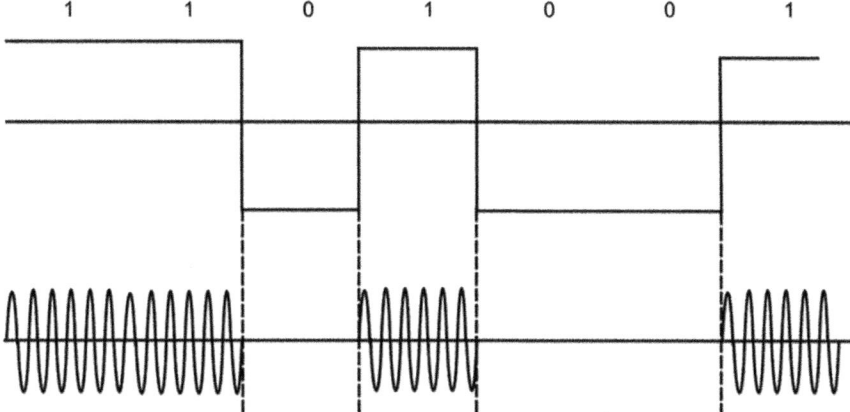

Figure 1.28 Amplitude shift keying.

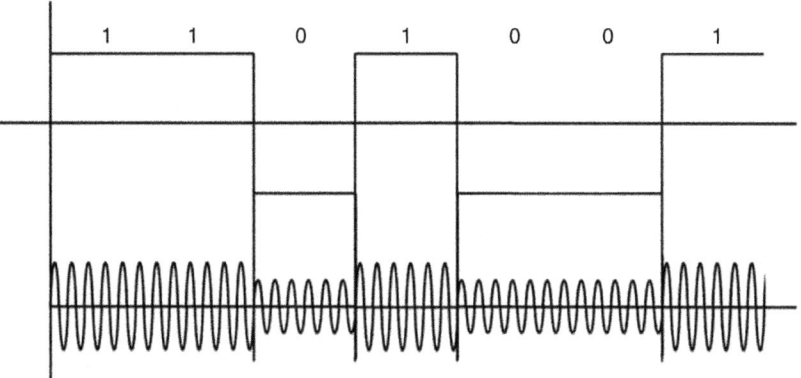

Figure 1.29 Two-amplitude shift keying.

the carrier between two amplitudes, one representing a 1 and the other representing a 0 as shown in Figure 1.29. Again, the carrier can be suppressed to have maximum power in information carrying signals and also one of the side bands can be filtered out to conserve the bandwidth.

1.2.9.2 Frequency Shift Keying (FSK)

In FSK, it is the frequency of the carrier signal that is switched between two values, one representing bit 1 and the other representing bit 0, as shown in Figure 1.30. The modulated carrier signal in this case is represented by:

$$X_c(t) = A\sin\omega_{c1}t \text{ for bit 1 and}$$
$$X_c(t) = A\sin\omega_{c2}t \text{ for bit 0.}$$

In the case of FSK, when modulation rate increases, the difference between the two chosen frequencies to represent a 1 and a 0 also needs to be higher. Keeping in view the restriction in available bandwidth, it would not be possible to achieve bit transmission rate beyond a certain value.

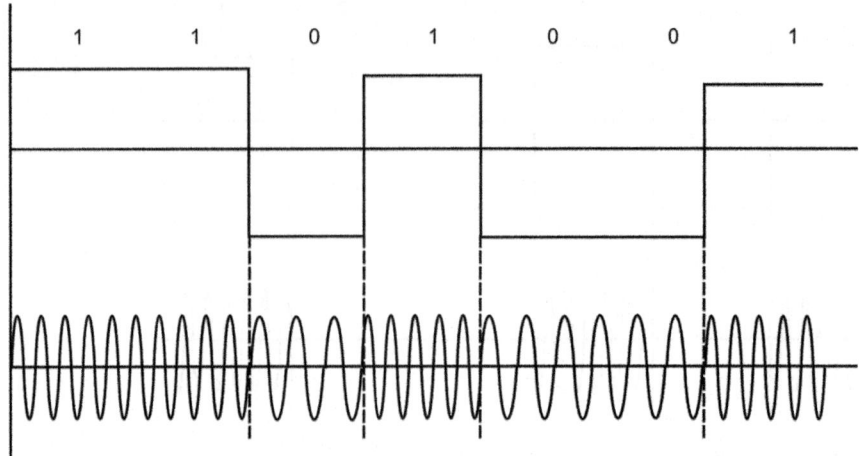

Figure 1.30 Frequency shift keying.

1.2.9.3 Phase Shift Keying (PSK)

In PSK, the phase of the carrier is discretely varied with respect to either a reference phase or to the phase of the immediately preceding signal element in accordance with the data being transmitted. For example, when encoding bits, the phase shift could be 0° for encoding a bit 0 and 180° for encoding a bit 1, as shown in Figure 1.31. The phase shift could have been −90° for encoding a bit 0 and + 90° for encoding a bit 1. The essence is that representations for 0 and 1 are a total of 180° apart. Such PSK systems in which the carrier can assume only two different phase angles are known as Binary Phase Shift Keying (BPSK) systems. We can appreciate that in BPSK systems each phase change carries one bit of information. This, in other words, means that bit rate equals the modulation rate. Now if the number of recognizable phase angles were increased to four, then two bits of information could be encoded into each signal element.

Coming back to BPSK, the carrier signals used to represent 0 and 1 bits could be expressed as:

$$X_{c0}(t) = A\cos(\omega_c t + \theta_0)$$
$$X_{c1}(t) = A\cos(\omega_c t + \theta_1)$$

Since the phase difference between two carrier signals is 180°; that is, $\theta_1 = \theta_0 + 180°$
Then, $X_{c0}(t) = A\cos(\omega_c t + \theta_0)$ and $X_{c1}(t) = -A\cos(\omega_c t + \theta_0)$.

1.2.9.4 Differential Phase Shift Keying (DPSK)

Another form of PSK is *DPSK*. Instead of an instantaneous phase of carrier determining which bit is transmitted, it is the change in phase that carries message intelligence. In this system, one logic level (say 1) represents a change in phase of carrier and the other logic level (i.e., 0) represents a no change in phase. In other words, if a digit changes in the bit stream from 0 to 1 or 1 to 0, a 1 is transmitted in the form of change in phase of carrier signal. And in case, there is no change, a 0 is transmitted in the form of no phase change in carrier.

The BPSK signal is detected using a coherent demodulator where a locally generated carrier component is extracted from received carrier by a PLL circuit. This locally generated carrier assists in the product demodulation process where the product of the carrier and the received

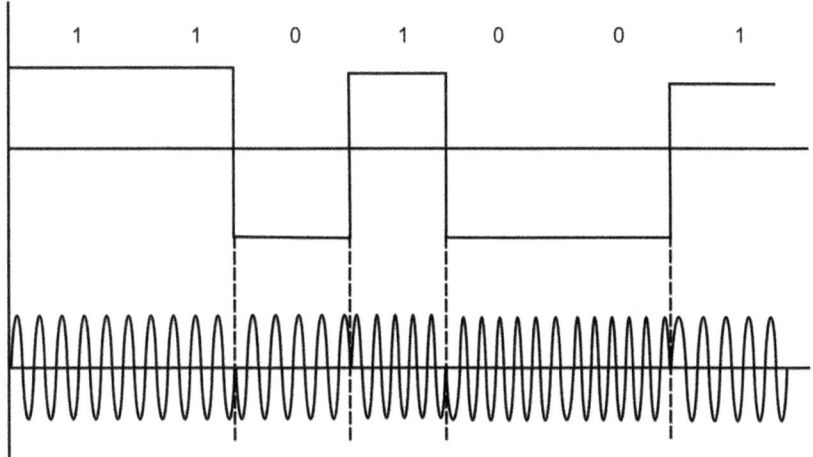

Figure 1.31 Binary phase shift keying.

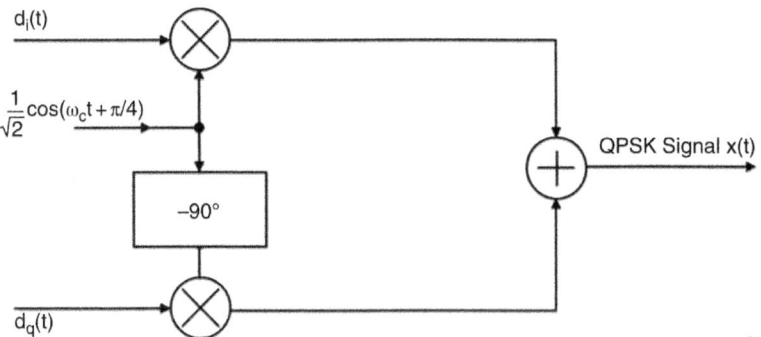

Figure 1.32 Block-schematic arrangement of a typical QPSK modulator.

modulated signals generate the demodulated output. There could be difficulty in successfully identifying the correct phase of regenerated signal for demodulation. DPSK takes care of this ambiguity to a large extent.

1.2.9.5 Quadrature Phase Shift Keying (QPSK)

QPSK is the most commonly used of all forms of PSK. A QPSK modulator is nothing but two BPSK modulators operating in quadrature. The input bit stream (d_0, d_1, d_2, d_3, d_4...) representing the message signal is split into two bit streams, one with, say, even numbered bits (d_2, d_4...) and the other with odd numbered bits (d_1, d_3, d_5,...). Also, in QPSK, if each pulse in the input bit stream has a duration of T seconds, then each pulse in the even/odd numbered bit streams has a pulse duration of $2T$ seconds. Figure 1.32 shows the block-schematic arrangement of a typical QPSK modulator. One of the bit streams $d_i(t)$ feeds the in-phase modulator while the other bit stream $d_q(t)$ feeds the quadrature modulator. The modulator output can be written as eqn. 1.27.

$$x(t) = \frac{1}{\sqrt{2}}\left[d_i(t)\cos(\omega_c t + \pi/4)\right] + \frac{1}{\sqrt{2}}\left[d_q(t)\sin(\omega_c t + \pi/4)\right] \qquad (1.27)$$

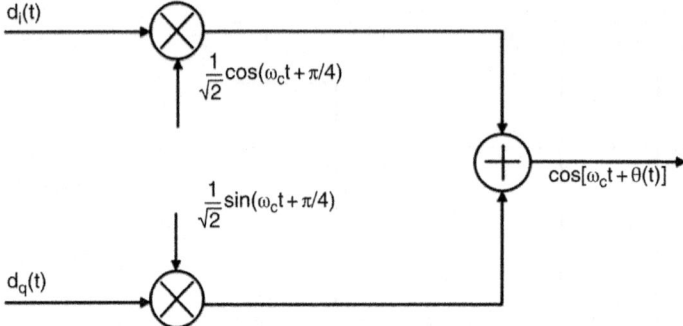

Figure 1.33 Conceptual diagram of QPSK.

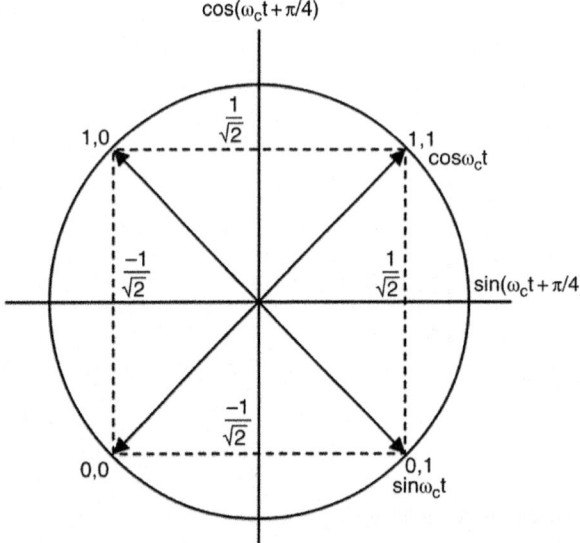

Figure 1.34 QPSK phase diagram.

This expression can also be written in a simplified way as $x(t) = \cos[\omega_c t + \theta(t)]$.

The in-phase bit stream represented by $d_i(t)$ modulates the cosine function and has the effect of shifting the phase of the function by 0 or π radians. This is equivalent to BPSK. The other pulse stream represented by $d_q(t)$ modulates the sine function thus producing another BPSK like output that is orthogonal to the one produced by $d_i(t)$. The vector sum of the two produces a QPSK signal given by $x(t) = \cos[\omega_c t + \theta(t)]$. Figure 1.33 illustrates it further.

$\theta(t)$ will have any of the four values of 0°, 90°, 180° and 270° depending upon the status of pair of bits having one bit from $d_i(t)$ bit stream and the other from the $d_q(t)$ bit stream. Four possible combinations are 00, 01, 10 and 11. Figure 1.34 illustrates the process further. It shows all possible four phase states. The in-phase bits operate on the vertical axis at phase states of 90° and 270° whereas the quadrature-phase channel operates on horizontal axis at phase states of 0° and 180°. The vector sum of the two produces each of the four phase states as shown. As mentioned earlier, phase state of QPSK modulator output depends upon a pair of bits. The phase states in the present case would be 0°, 90°, 180° and 270° for input combinations 11, 10, 00 and 01, respectively.

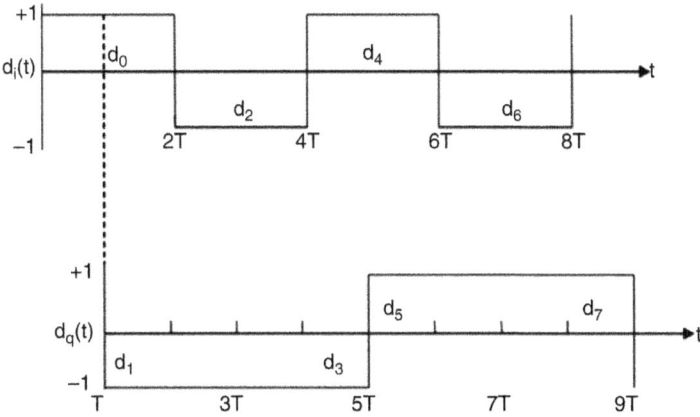

Figure 1.35 Offset QPSK waveform.

Since each symbol in the case of QPSK comprises of two bits, symbol transmission rate is half of bit transmission rate of BPSK, the bandwidth requirement is halved. The power spectrum for QPSK is same as that for BPSK.

1.2.9.6   Offset QPSK

The *Offset QPSK* is similar to QPSK with the difference that the alignment of the odd/even streams is shifted by an offset equal to T seconds as shown in Figure 1.35. In the case of QPSK, as explained earlier, carrier phase change can occur every $2T$ seconds. If neither of the two streams changes sign, the carrier phase remains unaltered. If only one of them changes sign, carrier phase undergoes a change of +90° or −90° and both change sign, the carrier phase undergoes a change of 180°. In such a situation, the QPSK signal no longer has a constant envelope if the QPSK signal is filtered to remove spectral side lobes. If such a QPSK signal is passed through a nonlinear amplifier, the amplitude variations could cause spectral spreading to restore unwanted side lobes, which in turn could lead to interference problems. Offset QPSK overcomes this problem. Due to staggering of in-phase and quadrature-phase bit streams, the possibility of the carrier phase changing state by 180° is eliminated as only one bit stream can change state at any time instant of transition. A phase change of +90° or −90° does cause a small drop in the envelope but it does not fall to near zero as in the case for QPSK for 180° phase change.

1.2.10   Multiplexing Techniques

Multiplexing techniques are used to combine several message signals into a single composite message so that they can be transmitted over a common channel. The multiplexing technique ensures that the different message signals in the composite signal do not interfere with each other and that they can be conveniently separated out at the receiver end. The two basic multiplexing techniques in use include:

1) Frequency Division Multiplexing (FDM)
2) Time Division Multiplexing (TDM)

While *FDM* is used with signals that employ analogue modulation techniques, *TDM* is used with digital modulation techniques where the signals to be transmitted are in the form of a bit stream. The two techniques are briefly described in the following paragraphs.

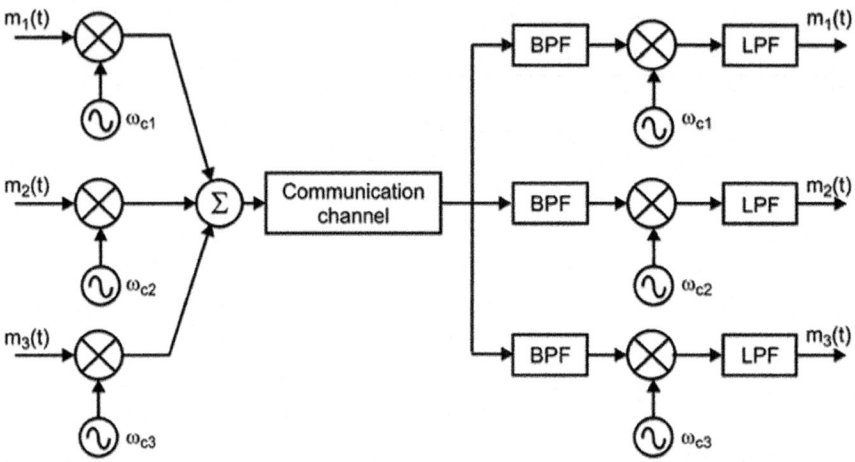

Figure 1.36 Frequency division multiplexing.

1.2.10.1 Frequency Division Multiplexing

In the case of *Frequency Division Multiplexing* (FDM), different message signals are separated from each other in frequency. Figure 1.36 illustrates the concept of FDM showing simultaneous transmission of three message signals over a common communication channel. As is clear from the block-schematic arrangement shown, each of the three message signals modulates a different carrier. The most commonly used modulation technique is the single side band (SSB) modulation. Any type of modulation can be used as long as we ensure that the carrier spacing is sufficient to avoid a spectral overlap. On the receiving side, band pass filters separate out the signals, which are then coherently demodulated as shown. The composite signal formed by combining different message signals after they have modulated their respective carrier signals may be used to modulate another high-frequency carrier before it is transmitted over the common link. In that case, these individual carrier signals are known as sub-carrier signals.

FDM is used in telephony, commercial radio broadcast (both AM and FM), television broadcast, communication networks and telemetry. In the case of commercial AM broadcast, the carrier frequencies for different signals are spaced 10 kHz apart. This separation is definitely not adequate if we consider a high-fidelity voice signal with a spectral coverage of 50 Hz–15 kHz. Because of this reason, AM broadcast stations using adjacent carrier frequencies are usually geographically far apart to minimize interference. In case of FM broadcast, the carrier frequencies are 200 kHz apart. In the case of long-distance telephony, 600 or more voice channels, each with a spectral band of 200 Hz–3.2 kHz, can be transmitted over a coaxial or microwave link using SSB modulation and a carrier frequency separation of 4 kHz.

1.2.10.2 Time Division Multiplexing

Time Division Multiplexing (TDM) is used for simultaneous transmission of more than one pulsed signals over a common communication channel. Figure 1.37 illustrates the concept. Multiple pulsed signals are fed to a type of electronic switching circuitry called the *commutator* in the figure. All the message signals, which have been sampled at least at the Nyquist rate (the sampling is usually done at 1.1 times the Nyquist rate to avoid aliasing problem), are fed to the commutator. The commutator interleaves different samples from different sampled message signals so as form a composite interleaved signal. This composite signal is then transmitted

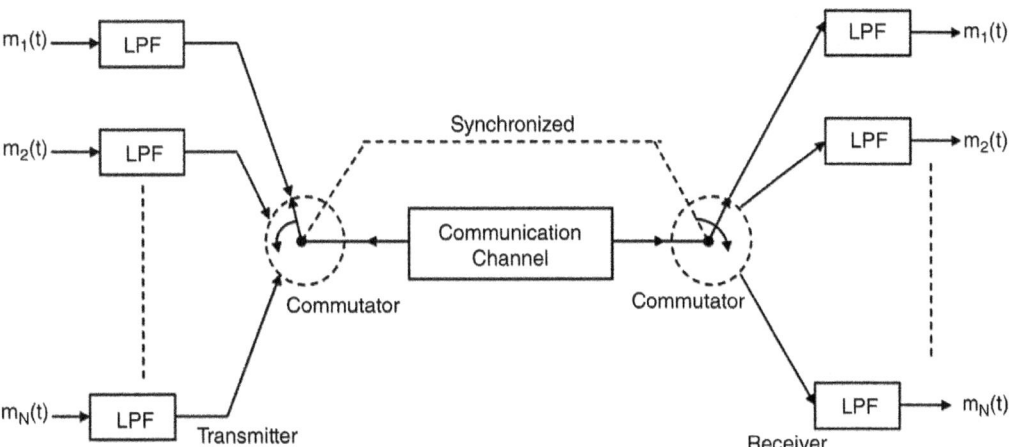

Figure 1.37 Time division multiplexing.

over the link. In the case where all message signals have same bandwidth, one commutation cycle will contain one sample from each of the messages. But in case signals have different bandwidths, then one would need to transmit more number of samples per second of the signals with a larger bandwidth. As an illustration, if there are three message signals with respective sampling rates of 2.4, 2.4 and 4.8 kHz, then each cycle of commutation will have one sample each from the first two messages and two samples from the third message.

At the receiving end, the composite signal is de-multiplexed using a similar electronic switching circuitry that is synchronized with the one used at the transmitter. TDM is widely used in telephony, telemetry, radio broadcast and data processing. If T is the sampling time interval of the time multiplexed signal of n different signals each with a sampling interval of (T_s), then $T = T_s/n$. Also, if the time multiplexed signal is considered a low pass signal with a bandwidth of f_{TDM} and f_m is the bandwidth of individual signals, then $f_{TDM} = nf_m$.

1.3   Communication Transmitters and Receivers

Having discussed different analogue and digital modulation and the corresponding demodulation techniques, in the following paragraphs we shall briefly describe (with the help of block-schematic arrangements) different types of transmitters and receivers that use these modulation and demodulation schemes.

1.3.1   Elements of the Communication System

The basic communication system – irrespective of whether it is an analogue or digital communication system, microwave communication system or the one operating at relatively lower carrier frequency, point-to-point communication or broadcast communication, the message signal to be transmitted is audio, video or data – has three essential elements, namely the *Transmitter*, the *Transmission* or *Communication Channel* and the *Receiver*. Extending it further, the communication systems would have *Source of Message Signal* and *Input Transducer* preceding the transmitter and *Output Transducer* following the receiver. Figure 1.38 shows the block-schematic arrangement of a generalized

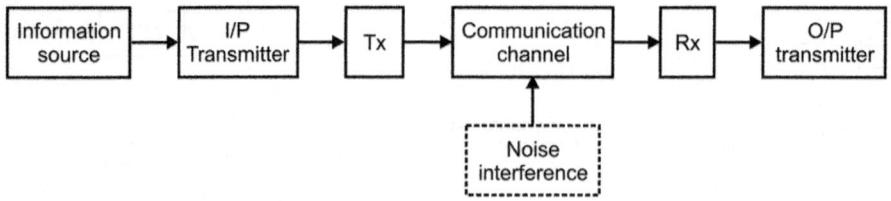

Figure 1.38 Block-schematic arrangement of a generalized communication system.

communication system. The block marked Noise/Interference represents the noise and interference that gets introduced into the transmitted signal as it propagates through the communication channel.

The *source of information* or message signal generates the information or message signal to be transmitted. It could be audio, video or data. The *input transducer* is used only in cases where the message signal to be transmitted is not electrical in nature. The transducer converts the non-electrical message signal into a time varying electrical signal. A microphone used to convert sound waves into an equivalent electrical signal is an example of transducer. The function of the *transmitter* is to process the electrical signal from different aspects. The primary function of transmitter is modulation, which is the process of changing one of the properties (amplitude, frequency or phase) of the high-frequency carrier signal in accordance with the message signal. The message signal may need to be processed before modulation. Processing may involve filtering such as restricting the range of audio frequencies in an AM radio broadcast transmitter, signal amplification and so on. The term *channel* means the medium through which the message travels from the transmitter to the receiver. In other words, we can say that the function of the channel is to provide a physical connection between the transmitter and the receiver.

Communication channels provide the physical medium for transmission of signals from one point to another. Communication channels are either point-to-point channels such as microwave links, wired lines, optical fibres or broadcast channels where a single transmitter signal reaches a large number of receivers; an example being the one provided by a satellite in geostationary orbit covering about one-third of Earth's surface. As the signal propagates through the communication channel, the signal gets distorted due to noise. Though noise may creep into the signal at any point, it adversely affects the signal the most while it is propagating through the channel. The *receiver* receives the noise-affected signal and reproduces the message signal from the distorted modulated signal by a process called demodulation or detection. The *output transducer* restores the demodulated electrical signal representing the message to its original form.

1.3.2 Classification of Transmitters

Radio transmitters are generally classified on the basis of operating carrier frequency, type of modulation techniques used by the transmitter and the nature of service provided by the transmitter. On the basis of operating frequency, they are classified as Medium Frequency (MF) transmitters, High-Frequency (HF) transmitters, Very High Frequency (VHF) transmitters and Ultra-High Frequency (UHF) transmitters. On the basis of type of modulation, they are classified as Amplitude Modulated (AM) transmitters, Frequency Modulated (FM) transmitters, Phase Modulated (PM) transmitters, Amplitude Shift Keying (ASK) transmitters,

Frequency-Shift Keying (FSK) transmitters and Phase Shift Keying (PSK) transmitters and different variants. According to the nature of service or function provided by the transmitter, we have radio telegraph transmitters, television transmitters, radar transmitters, navigation transmitters and so on.

MF transmitters operate over the 300 kHz–3 MHz frequency band. MF transmitters are used for AM radio broadcast services that employ the 535–1705 kHz band, maritime and aircraft navigation, amateur radio, cordless phones and so on. *HF transmitters* that operate over the HF band (3–30 MHz) and use the ionosphere as a means of electromagnetic wave propagation (commonly known as sky wave propagation) are particularly suitable for long-distance communication across intercontinental distances. The band is extensively used by international shortwave broadcasting stations as the entire short waveband of frequencies (2.310–25.820 MHz) falls within the HF band. Other examples of HF transmitter use include aviation communication, such as HF radios used on board aircraft to provide effective communication over long-distance oceanic and transpolar routes, weather stations and amateur radio and citizens' band services. *Amateur radio* relates to use of radio frequency spectrum for voice, text, image and data communication for non-commercial purposes such as wireless experimentation, radio sport, training, emergency services and so on. It is allocated one band in the MF (Medium Frequency) band and several bands in the HF band. The *Citizens' Band* (CB) *Radio Service* that is allocated 40 channels in the 26.965–27.405 MHz frequency band is a private, two-way, short-distance voice communications service for the personal or business activities of the general public. It may also be used for voice paging. *VHF transmitters* operate over the VHF band (30–300 MHz) and use space wave line-of-sight propagation. The radio horizon is slightly longer than the geometric line-of-sight horizon due to bending of electromagnetic waves by the atmosphere. VHF transmitters are commonly used for FM radio broadcasting employing the 87.5–108 MHz frequency band with a few exceptions such as 76–90 MHz used in Japan and 65–74 MHz used in the Eastern Bloc of the erstwhile Soviet Republic, television broadcasting, two-way land mobile radio systems for various private, business, military and emergency services, long-range data communication, amateur radio and marine communications. Air traffic control communications and air navigation systems are other applications. *UHF transmitters* operate over the UHF-band (300 MHz–3 GHz). UHF transmitters find applications in a large number of communication services, which include television broadcasting, cell phone communication, cordless phones, satellite communication, WiFi and Bluetooth services and so on.

On the basis of type of modulation used, transmitters are classified as AM transmitters, FM transmitters and PM transmitters, ASK transmitters, FSK transmitters and different variants of PSK transmitters. *AM transmitters* employ amplitude modulation. These transmitters are used in medium wave (MW) and short wave (SW) frequency bands for AM broadcast. Based on the transmitted power levels, AM transmitters use a high-level modulation scheme where the transmitted power needs to be of the order of kilowatts and low-level modulation where only a few watts of power need to be transmitted. *FM transmitters* employ frequency modulation. One of the most common applications of FM transmitters is in FM broadcasting, which is nothing but radio broadcasting using frequency modulation. FM broadcasting first began in 1945 and is now used worldwide to provide high-fidelity voice and music over broadcast radio. FM broadcasting employs the VHF band of 30–300 MHz. FM broadcast range is limited by optical visibility. Range doesn't increase linearly with transmitted power. An Effective Radiated Power (ERP) of 30 W would give a coverage range of about 15 km and an ERP of 300 W would probably give a range of 45 km, provided it is not limited by optical visibility. In addition to FM broadcasting, other important applications of FM transmitters include telemetry, radar, two-way radio systems and medical diagnostics. *PM transmitters* employ phase modulation. Phase modulation and frequency modulation are closely linked together (frequency is rate of change of phase) and it is often used in many transmitters and receivers such as two-way radios and mobile radio communications including maritime mobile radio communications.

Like the AM, FM and PM transmitters that use analogue modulation techniques, we have transmitters such as ASK, FSK and PSK transmitters employing digital modulation techniques. *ASK modulation* is commonly used in LED transmitters for transmission of digital data through optical fibres. Other applications of ASK transmitters include remote control operations and security systems. *FSK transmitters* are also used for remote control and security systems. Integrated circuits that can generate different forms of ASK and FSK signals are commercially available for these applications. ASK and FSK transmitters operate in the VHF/UHF bands. *PSK transmitters* employ any of the different forms of phase shift keying, which include Phase Shift Keying (PSK), Binary Phase Shift Keying (BPSK), Quadrature-Phase Shift Keying (QPSK), 8 Point Phase Shift Keying (8 PSK), 16 Point Phase Shift Keying (16 PSK), Quadrature Amplitude Modulation (QAM), 16 Point Quadrature Amplitude Modulation (16 QAM) and 64 Point Quadrature Amplitude Modulation (64 QAM). PSK is particularly suited to data communications. PSK in its different forms is extensively used for wireless LANs, Radio Frequency Identification (RFID) and Bluetooth communication.

Based on the function performed by transmitter, they are classified into Television Transmitters, Radar Transmitters, Radio Control Transmitters, Navigation Transmitters and so on. *Television transmitters* use AM transmitters for transmission of picture information and FM transmitters for transmission of sound. *Radar transmitters* may be one of two types; namely the Continuous-Wave Radar Transmitter and Pulsed Radar Transmitter. *Radio Control Transmitters* are used for remote control operations. Navigation Transmitters and Receivers are used as navigational aids for sea and air navigation.

1.3.3 Continuous-Wave (CW) Transmitter

The CW transmitter makes use of a radio communication technique that uses an undamped continuous-wave signal of constant amplitude and frequency. The signal is obviously a sinusoidal waveform. The CW transmitter is principally used for radiotelegraphy; that is, for the transmission of short or long pulses of RF energy to form the dots and dashes of the Morse code characters. A significant advantage of CW transmission is its narrow bandwidth, which not only reduces output power requirement on the part of the transmitter but also makes the communication immune to noise and interference.

Figure 1.39 shows the simplified block-schematic arrangement of a CW transmitter. It comprises an RF oscillator that generates the sinusoidal waveform at the desired frequency; a buffer or preamplifier to amplify the oscillator output to a level sufficient to drive the power amplifier; a power amplifier that amplifies the RF oscillations to power level to get the desired range; a device to turn the RF output on and off, a process known as keying, in accordance with the intelligence to be transmitted and a power supply to operate different sections of the transmitter and an antenna to radiate keyed RF signal. The buffer also serves the purpose of isolating the amplifier stages from the oscillator. Variations in source voltage or/and changes in amplifier due to keying would vary the load on the oscillator and hence change the frequency. A buffer stage ensures that the oscillator sees a constant load.

There is currently no commercial traffic using Morse code characters, but it is still popular with amateur radio operators. Another application area where information is transmitted using Morse code characters is in non-directional beacons used for air navigation.

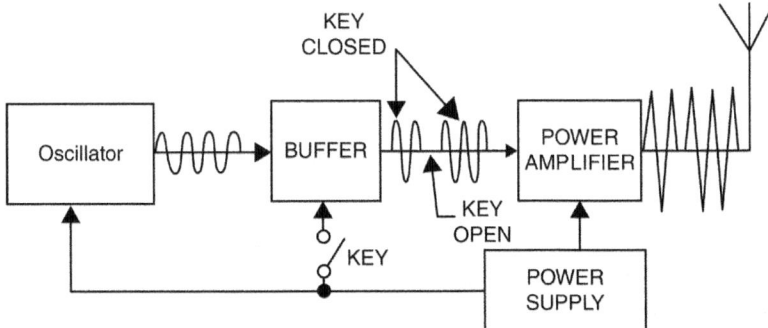

Figure 1.39 Block-schematic arrangement of a CW transmitter used for wireless telegraphy.

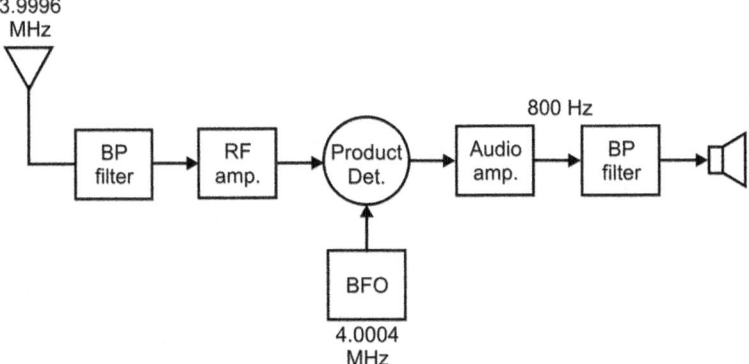

Figure 1.40 CW receiver used in wireless telegraphy.

1.3.4 CW Receiver

A CW receiver is used for detection of continuous-wave communication signals. Figure 1.40 shows a simplified block-schematic arrangement of a *direct conversion receiver* that can be used for detection of CW signals such as those transmitted in wireless telegraphy using Morse code. The incoming RF signal is passed through a *Band Pass Filter* (BPF) with its centre frequency equal to the transmitted carrier frequency. The band pass filter ensures that the signal fed to the *RF amplifier*, and subsequently to the *product detector*, is very close in frequency to the carrier frequency and that unwanted frequencies are eliminated. The incoming signal after passing through the band pass filter is mixed with the output of a *Beat Frequency Oscillator* (BFO) in the product detector. The product detector converts the incoming signal from RF frequencies to audio frequencies. It is simply another mixer whose output is the difference between the BFO frequency and the received carrier RF frequency. The BFO has a frequency very close to that of the RF input so that the output signal of the product detector is in the audio range. For CW, the difference between the input frequency and the BFO frequency is kept to about 800 Hz. This 800 Hz tone is then amplified in an *audio amplifier* and fed to a *loudspeaker*. This is the tone we hear when we listen to CW. Since the BFO can mix with two different frequencies, one above and one below the received carrier frequency to get the same output tone, for this reason the input RF signal is band pass filtered so as to contain only a narrow range of frequencies. This keeps the audio output cleaner with fewer undesired signals.

1.3.5 Amplitude Modulated (AM) Transmitter

As outlined in an earlier paragraph, AM transmitters are broadly classified as those employing *low-level modulation* and those employing *high-level modulation*. While detailing the amplitude modulation technique, we discussed different forms of amplitude modulation including Double Side Band (DSB) with full carrier, also known as Standard AM or A3E, Double Side Band with Suppressed Carrier (DSBSC), Single Side Band (SSB) with full carrier known as H3E, Single Side Band with Reduced Carrier (R3E) and Single Side Band with Suppressed Carrier (SSBSC) known as J3E. In the following paragraphs, we shall discuss AM transmitters employing low-level and high-level modulation schemes, which will be followed by a brief description of standard AM broadcast transmitters and SSB transmitters.

1.3.5.1 Low-Level AM Transmitter

In the case of low-level modulation, amplitude modulation of the carrier signal takes place at a much lower power level of the carrier signal and the modulated signal is then amplified in a chain of linear amplifiers to raise the power required for transmission. Figure 1.41 shows a block-schematic arrangement of a low-level amplitude modulated transmitter. The *RF oscillator* generates the carrier signal of desired frequency. The *buffer* is nothing but an amplifier stage required to raise power level of carrier signal before it is fed to *modulator*. This ensures that we don't draw too much power from the oscillator that could lead to change in its frequency. The modulating signal, which is an audio signal in case of AM broadcast, is also amplified to the desired level before it is fed to the modulator. The modulator is also a kind of amplifier operating in the nonlinear region. As its output, it produces a double side band full carrier signal. The modulated signal is then amplified in the *driver amplifier* and the *power amplifier* to raise the power to desired level as required for transmission by antenna. It should be mentioned that all amplifiers following the modulator must be linear. That is, they are either Class A amplifiers or Class B push-pull amplifiers. The advantage of low-level amplitude modulation is that not much audio power is required. The disadvantage is reduced efficiency due to use of linear amplifiers. Low-level amplitude modulation is usually used in low power transmitters such as those in Walkie-Talkies.

1.3.5.2 High-Level Amplitude Modulated Transmitter

In the case of high-level modulation, the carrier signal is modulated at a much higher power level equal to the power to be transmitted. The carrier signal here is amplified in a chain of Class C amplifiers to raise the power to the level required for transmission. Figure 1.42 shows a block-schematic arrangement of a high-level amplitude modulated transmitter. The RF oscillator generates the carrier signal of the desired frequency. The buffer is nothing but an

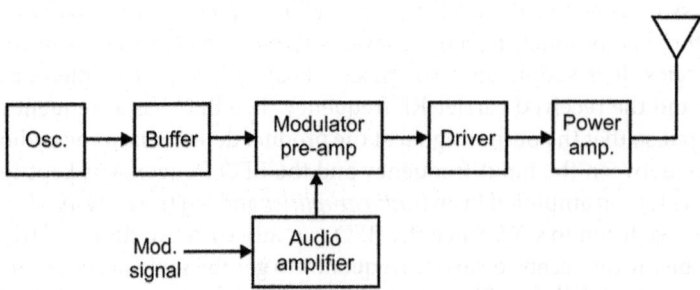

Figure 1.41 Block-schematic of a low-level amplitude modulated transmitter.

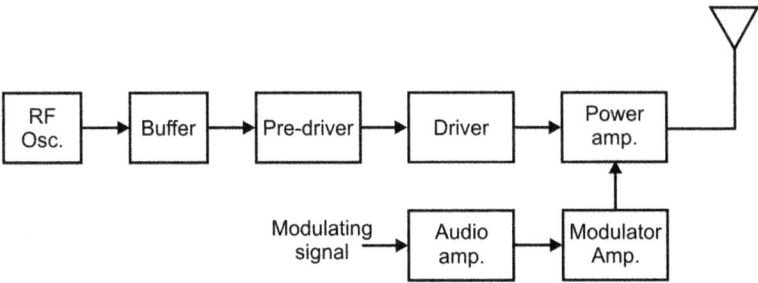

Figure 1.42 Block-schematic of a high-level amplitude modulated transmitter.

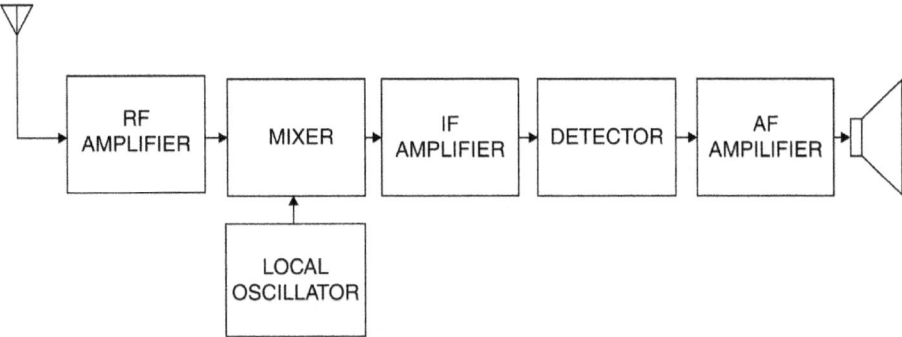

Figure 1.43 Block-schematic arrangement of a superheterodyne AM broadcast receiver.

amplifier stage required to raise the power level of carrier signal before it is fed to modulator and serves the same purpose as the one in case of low-level modulation. The modulating signal, which is an audio signal in case of AM broadcast, is also amplified to the desired level before it is fed to the power amplifier where the modulation takes place. As outlined before, all amplifiers used to amplify the carrier signal can be high efficiency Class C amplifiers as they have to amplify only a pure sinusoidal signal. This is the key advantage of high level modulation. The disadvantage is that the modulating signal needs to be amplified to a high level to be able to modulate a high-power carrier. High level modulation is generally used in high power AM transmitters used at broadcast stations and also in high power amateur transmitters operating in the AM mode only.

1.3.6 AM Receiver

The *superheterodyne receiver* is the receiver of choice in modern AM receivers. Figure 1.43 shows block-schematic arrangement of a superheterodyne AM receiver. The RF amplifier amplifies the entire band of frequencies, such as the broadcast band in the case of an AM broadcast transmitter or the amateur band in case of an amateur radio transmitter, received by the antenna. The amplified RF signal is fed to the mixer, which is a nonlinear device. The mixer is also fed with the signal from a local oscillator. The local oscillator frequency is kept higher than the received RF signal frequency by an amount equal to intermediate frequency. The intermediate frequency is 455 kHz in case of an AM broadcast receiver. While tuning in different radio stations operating at different carrier frequencies, the resonant frequencies of tuned circuits of RF amplifier and local oscillator are simultaneously varied to maintain this constant difference of 455 kHz. The signal at the output of the mixer is also an amplitude modulated signal with its side band information intact, except for the fact that the carrier frequency is now 455 kHz. In fact, in an AM broadcast receiver, it will be 455 kHz irrespective of the station tuned in to. An intermediate frequency amplifier comprises a cascade arrangement of several stages of amplification. Different stages of IF amplification are sharply tuned to 455 kHz with the desired bandwidth. Most of signal amplification occurs in IF amplifiers. The signal is then demodulated. The amplified modulated signal at intermediate frequency is demodulated in the envelope detector. It is a diode detector with a low pass filter to filter out the high-frequency carrier. The audio amplifier stage amplifies the demodulated signal before it is fed to the loudspeaker. The audio amplifier is a two-stage amplifier comprising a preamplifier and a power amplifier.

One of the problems of superheterodyne receivers arises from the image frequency. Image frequency is the signal frequency that would produce the same intermediate frequency as the desired signal frequency. For example, if the desired signal frequency is 1600 kHz, then the local oscillator frequency would be 2055 kHz. The image frequency in this case is 2510 kHz and also produces a 455 kHz signal frequency at the output of the mixer. Therefore, it is likely to interfere with the desired frequency. The image frequency signal, if lying within the broadcast band, is substantially attenuated in the tuning stage of the RF amplifier. Also, radio broadcast stations operating in the same area are assigned frequencies to avoid such a situation.

Some receivers such as amateur receivers have two intermediate frequencies. These are 10.7 MHz (first IF) and 455 kHz (second IF). While the first IF is produced by a variable frequency local oscillator; generation of second IF employs fixed frequency local oscillator. Use of two intermediate frequencies enhances selectivity further.

Sensitivity, selectivity, fidelity and noise performance are the important characteristics of an AM broadcast receiver. *Sensitivity* is a measure of a receiver's ability to produce the desired *S/N* ratio for weak received signals. It is measured as received signal strength in microvolts to produce a signal-to-noise ratio of 10 dB. *Selectivity* is the ability of the receiver to reject unwanted signals at the input of receiver. The *fidelity* of a receiver is its ability to accurately reproduce at its output the signal that appears at its input. A broader receiver bandwidth produces high fidelity. This is in contrast to the requirement of narrower bandwidth for better selectivity. Most receivers are a compromise between good selectivity and high fidelity. *Noise* limits the usable input signal of the receiver. A noise level of 0 dB is ideal.

1.3.7 Single Side Band (SSB) Transmitter

SSB modulation as outlined earlier is a type of amplitude modulation in which one of the side bands, either the upper side band or lower side band, is suppressed and only one side band is transmitted. SSB modulation and SSBSC (Single Side Band Suppressed Carrier) modulation techniques offer more efficient use of transmitter bandwidth and power. Figure 1.44 shows the block-schematic arrangement of an SSB transmitter. The modulator combines the modulating signal input, which is an audio signal, and the carrier signal input to produce the modulated signal with carrier and two side bands. The filter selects the desired side band and suppresses the other one. The filter pass band can also be so chosen as to suppress the carrier along with one of the side bands. In most cases, SSB generators operate at very low frequencies as compared to the normally transmitted frequencies. The filter output is translated to the desired frequency in the mixer stage. The mixer is fed with the filter output and the carrier generator output after its frequency is multiplied in the frequency multiplier. The output from the mixer is fed to a linear power amplifier to raise the signal power to the desired level for transmission.

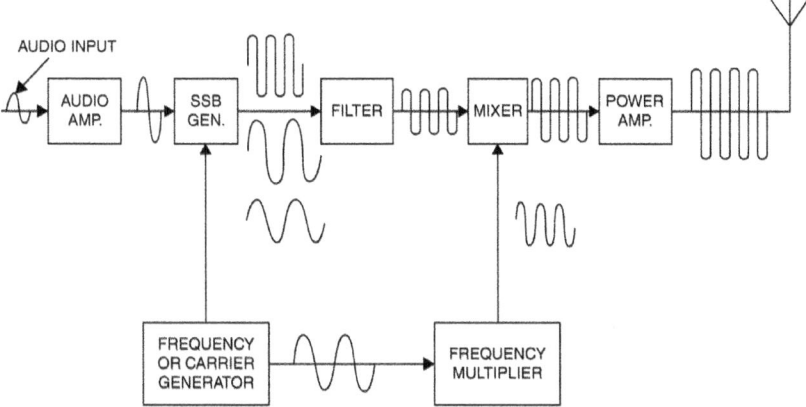

Figure 1.44 Block-schematic arrangement of an SSB transmitter.

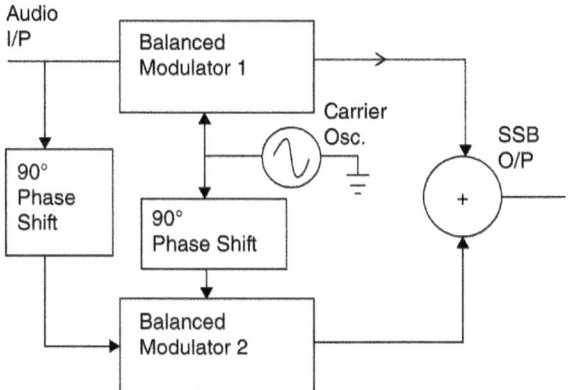

Figure 1.45 Block-schematic of an SSBSC transmitter using the phase shift method.

Another method of generating the SSB signal is the phase shift method. Figure 1.45 shows the block-schematic arrangement of an SSB transmitter based on the phase shift method. The phase shift technique uses two balanced modulators. One of the balanced modulators is fed with modulating signal, which is the audio signal, and the high-frequency carrier signal. The other balanced modulator is fed with modulating signal and the carrier signal with both phases shifted by 90°. Both balanced modulators produce the product of the two signals fed at the inputs. If the carrier and modulating signals are respectively expressed by $V_c \cos \omega_c t$ and $V_m \cos \omega_m t$, then the output of first balanced modulator will be $1/2[\cos(\omega_c - \omega_m)t - \cos(\omega_c + \omega_m)t]$. The output of second balanced modulator that is fed with phase shifted signals is given by $1/2[\cos(\omega_c - \omega_m)t + \cos(\omega_c + \omega_m)t]$. The two signals present at the outputs of balanced modulators are summed up to produce the difference frequency component only. That is, the output contains only the lower side band and the carrier is also suppressed. The output is therefore an SSBSC signal.

1.3.8   SSB Receiver

An SSB receiver is also a superheterodyne receiver, discussed in Section 1.3.6, comprising an RF amplifier, a mixer along with a local oscillator, intermediate frequency amplifier stage, a

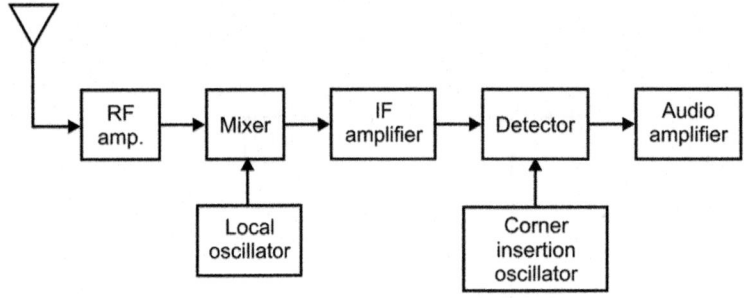

Figure 1.46 Block-schematic of an SSB receiver.

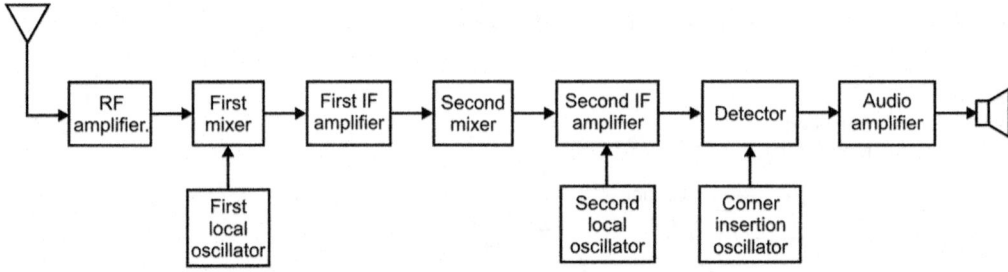

Figure 1.47 Block-schematic of a double conversion SSB receiver.

detector, an audio amplifier stage and a loudspeaker. Figure 1.46 shows the simplified block-schematic of an SSBSC receiver. One difference between the standard AM broadcast receiver and the SSBSC receiver is that the carrier signal needs to be reinserted at the detector for the detection process to occur. This may be done by using another oscillator called a *carrier insertion oscillator*. In a case where a pilot carrier is transmitted, it may be separated or removed from the signal at the output of IF amplifier using a filter, amplified and then reinserted at the detector. The receiver will also have AGC (Automatic Gain Control) and AFC (Automatic Frequency Control) circuits, not shown in the block diagram. The AGC circuit controls gain of RF and IF amplifiers and AFC controls the frequency of local oscillator. Functions of different sections have been explained in the case of an AM broadcast receiver.

Double conversion superheterodyne receiver configuration is also used for SSB receivers. In this case, the AFC, again not shown in the diagram for the sake of simplicity, controls the frequency of second local oscillator. Figure 1.47 shows the block-schematic of a double conversion SSB receiver.

1.3.9 Frequency Modulated (FM) Transmitter

In the case of a frequency modulated signal, the carrier frequency is varied in accordance with the amplitude of modulating signal with the amplitude of carrier signal remaining nearly constant. Figure 1.48 shows the block-schematic arrangement of an FM transmitter that employs a direct method of generation of FM signals. The *carrier oscillator* generates a stable sinusoidal signal at the centre frequency, also called the rest frequency, of the carrier oscillator. It is the frequency of the carrier oscillator when no modulating signal is applied to it. The *buffer*

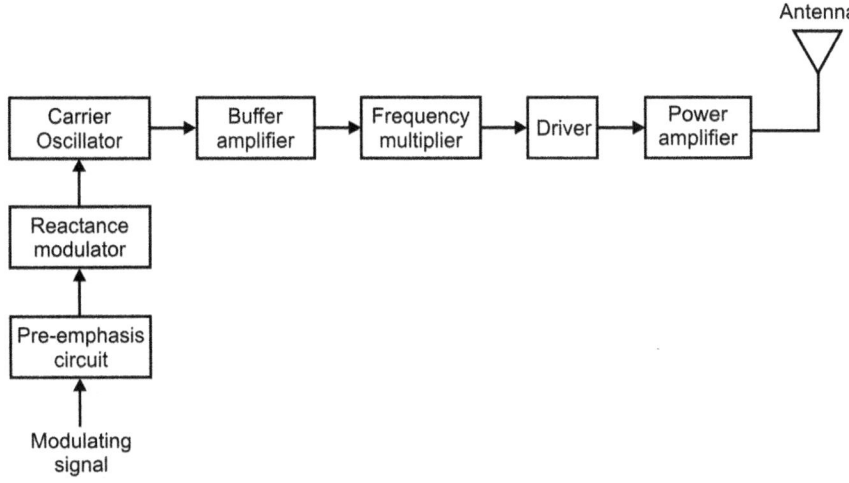

Figure 1.48 Block-schematic arrangement of an FM transmitter.

amplifier acts as a constant high impedance load on the oscillator to help stabilize the oscillator frequency. The buffer amplifier may have a small gain. The *pre-emphasis circuit* is used to boost the amplitudes of higher frequencies, generally 2–15 kHz, in the modulating signal. This is done to increase signal-to-noise ratio. It increases intelligibility and fidelity. The *reactance modulator* changes the carrier oscillator frequency on application of a modulating signal. The greater the peak-to-peak amplitude of the modulating signal, the larger the frequency deviation of the carrier frequency from its rest value. A reactance modulator is a circuit whose reactance, capacitive or inductive, is connected across the resonant circuit of the carrier oscillator. The reactance and hence the oscillator frequency is made to vary by the changing amplitude of the modulating signal. A *frequency multiplier* multiplies the frequency and is used to increase frequency deviation. Frequency multipliers are tuned-input, tuned-output RF amplifiers in which the output resonant circuit is tuned to a multiple of the input frequency. Multiplication factors of 2, 3 and 4× are common. The output of frequency multiplier feeds the driver amplifier, which in turn feeds the power amplifier to raise the RF power to a level desired for transmission. The power amplifier output feeds the transmitting antenna through an impedance matching network.

In the direct method of generating an FM signal, frequency stability of the carrier oscillator is a concern. This shortcoming is overcome in the *indirect method* that attempts to generate FM signal from a phase modulated signal. FM and PM signals are interrelated: one cannot exist without the other. In the indirect method, also known as *Armstrong's method*, the information signal is first integrated and then used to phase modulate a crystal controlled oscillator thereby achieving exceptionally high-frequency stability of the centre frequency. The modulated signal is a narrow band FM signal. A frequency multiplier is used to transform it to a wide band FM signal.

1.3.10   FM Receiver

Figure 1.49 shows a block-schematic arrangement of a superheterodyne FM receiver. An *RF amplifier* amplifies the signal intercepted by the antenna. The amplified signal is fed to the *mixer*. The second input to the mixer is from the *local oscillator*. The intermediate frequency signal is amplified by a chain of *IF amplifiers*. The *limiter* following the IF stage removes noise

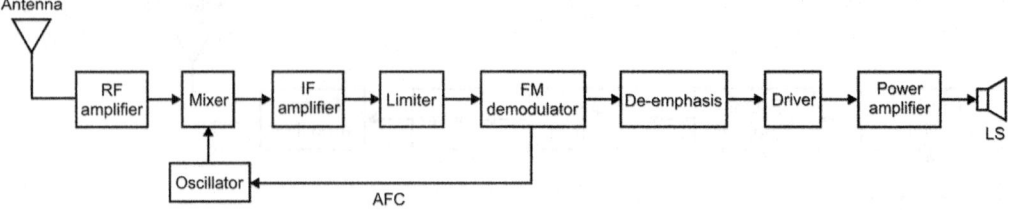

Figure 1.49 Block-schematic of an FM receiver.

from the received modulated signal and produces a constant amplitude FM signal at an intermediate frequency. The output of the limiter circuit is applied to the *FM demodulator* circuit. FM demodulator circuits such as the Foster–Seeley discriminator and ratio detector have been discussed in Section 1.2.3. A phase locked loop also makes a very good FM demodulator. In a PLL-based FM demodulator, the incoming FM signal can be fed into the reference input and the VCO drive voltage used to provide the demodulated output. Then there is the quadrature FM detector. A quadrature FM detector comprises a phase shift network, a mixer and a low pass filter. The FM signal and the phase shifted FM signal are applied to the mixer. The mixer output feeds a low pass filter. The output of low pass filter is the demodulated signal. The demodulated output is *de-emphasized* to attenuate the high-frequency components. This is done to bring them back to their original amplitudes as these are boosted prior to transmission. The de-emphasized signal is fed to the *audio amplifier* stage comprising a driver amplifier and a power amplifier. A limiter circuit is required only in the case where an FM discriminator is used as an FM demodulator. In the case of a ratio detector, a limiter circuit is not required as the ratio detector limits the amplitude of the signal. An *automatic frequency control* (AFC) circuit is required to have a stable local oscillator frequency.

1.3.11 Phase Modulated (PM) Transmitter and Receiver

Frequency modulated and phase modulated signals are interrelated. Therefore, a PM transmitter and an FM transmitter have similar structures. The difference between the two lies in the type of modulator used. While frequency modulators encode information by changing the frequency of the carrier wave in accordance with amplitude variations of information signal, phase modulators do so by changing the phase of the carrier wave again in accordance with the amplitude variations of information signal. The PM receiver and FM receiver also have similar structures, except for the fact that a PM receiver doesn't have a de-emphasis stage following the demodulator. Because of this, in the case of a PM transmitter, no pre-emphasis stage is needed to boost the high-frequency components of the modulating signal.

1.3.12 Amplitude Shift Keying (ASK) Transmitter

Various digital modulation techniques have been discussed in terms of their operational principle, merits and demerits earlier in Section 1.2.9. In this sub-section and the following Section 1.3 sub-sections, we shall discuss transmitters and receivers for ASK, FSK and different forms of PSK. As outlined earlier, in the case of ASK, the high-frequency sinusoidal carrier signal is given two or more discrete amplitude levels depending upon the number of levels adopted by the digital message. In the case of Binary ASK the message sequence has two levels, one of which is typically zero. The data rate is a sub-multiple of the carrier frequency. One of the disadvantages is that it doesn't have a constant envelope, which makes processing such as power amplification more difficult. Another disadvantage is that ASK is sensitive to atmospheric nose, distortions and propagation conditions. However, demodulation is relatively easier. ASK digital modulation scheme is used to transmit digital data over optical fibres, point-to-point military communication applications and so on. Figure 1.50 shows a simplified block-schematic of a binary ASK modulator. The diagram is self-explanatory. The sinusoidal carrier is applied to the input of a controlled electronic switch. The message sequence is fed to the control input of the switch. Logic 1 closes the switch thereby passing the carrier signal on to the output. The switch remains open for logic 0, thereby blocking the carrier from appearing at the output.

1.3.13 ASK Receiver

Figure 1.51 shows use of a simple product detector for demodulation of a binary ASK signal. The product detector multiplies the modulated signal and locally generated carrier signal. The product detector works like a frequency mixer. The output of product detector contains the message signal frequency component and twice the carrier frequency component. The output of low pass filter is the message signal. In the case of the schematic in Figure 1.51, frequency and phase of the locally generated carrier signal must be matched to those in the case of a transmitted carrier signal. A more sophisticated receiver makes use of two product detectors. The modulated signal is fed to one of the inputs of both product detectors. While the other input to one of the product detectors is the carrier signal, the other input of the second product detector is fed with a carrier signal that is phase shifted by 90°. The outputs of the two product detectors are combined to produce a demodulated output.

1.3.14 Frequency Shift Keying (FSK) Transmitter and Receiver

FSK is a digital modulation technique in which digital information is transmitted through discrete frequency changes of a carrier wave. The simplest FSK is binary FSK (BFSK) that uses a pair of discrete frequencies to transmit binary information. In the case of BFSK, two frequencies

Figure 1.50 Block-schematic of a binary amplitude shift keying (BASK) modulator.

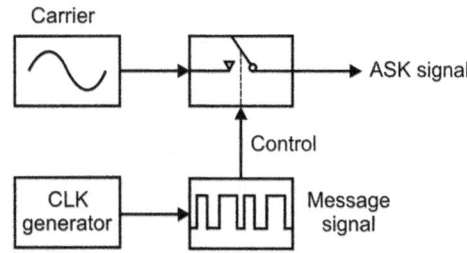

Figure 1.51 Simplified block-schematic arrangement of a BASK demodulator.

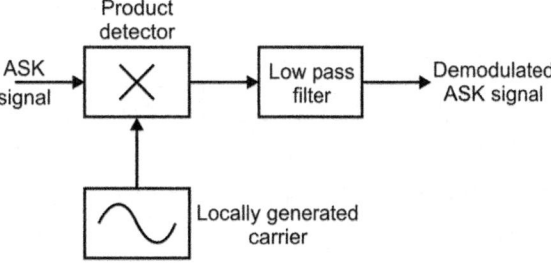

used to transmit 1 (MARK) and 0 (SPACE) are equal to integral multiples of bit rate. Figure 1.52 represents the basic block-schematic arrangement of a BFSK transmitter. The binary sequence to be encoded is applied to two mixers; directly to one of them and through an inverter to the other. The two mixers are fed at the other input by carrier waves of two discrete frequencies. The outputs of two mixers are combined to produce the BFSK output. As is self-explanatory, when 1 is being transmitted, the output is the carrier wave at frequency f_1 and when a 0 is being transmitted, the output is the carrier wave at frequency f_2.

There are two broad categories of FSK demodulators, namely the FM detector type demodulators and filter-type demodulators. In the case of FM detector type FSK demodulators, the FSK signal is treated as an FM signal with binary modulation. The FSK signal is fed to a band pass filter to remove out-of-band interference and then to a limiter to remove AM interference. The amplitude limited signal is then demodulated using an FM detector such as the Foster–Seeley FM discriminator or ratio detector. The demodulated signal is low pass filtered to remove interference at frequencies above the baud rate. The decision-making circuit finally converts all positive voltages into binary 1 s and all negative voltages into binary 0 s. Figure 1.53 shows the simplified block-schematic, which is self-explanatory. This category of FSK demodulators, though simple to implement, are non-optimal.

In the category of filter-type FSK demodulators, we have the non-coherent FSK demodulators and the coherent FSK demodulators. Figure 1.54 shows a simplified block-schematic of a non-coherent filter-type FSK demodulator. The information bits from an FSK signal are demodulated by subtracting the amplitude of the detected SPACE component from the amplitude of the detected MARK component. When the difference is above a certain threshold level, a MARK is assumed to be sent. If the difference is less than the threshold, a SPACE is assumed to be sent. The two-band pass filters are centred on the Mark and Space frequencies, respectively.

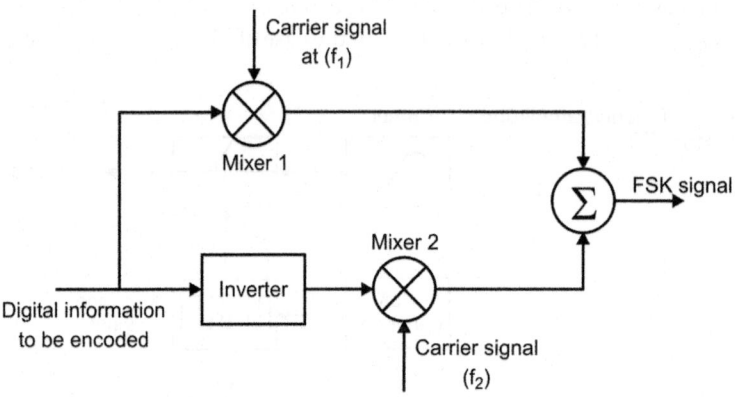

Figure 1.52 Generation of a BFSK signal.

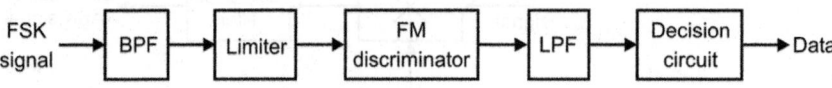

Figure 1.53 FM detector-type FSK demodulator.

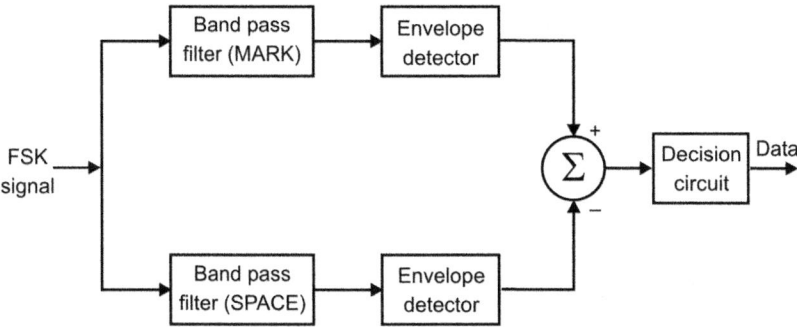

Figure 1.54 Noncoherent filter type FSK demodulator.

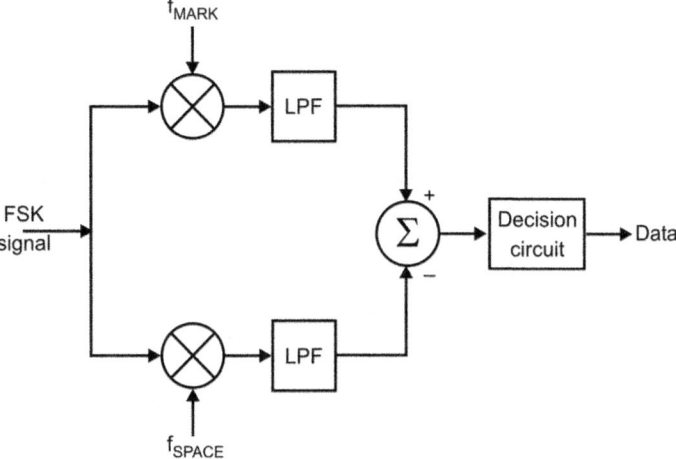

Figure 1.55 Coherent filter type FSK demodulator.

Figure 1.55 shows a simplified block-schematic of a coherent filter-type FSK demodulator. In the FSK demodulator of Figure 1.54, the detectors operate on base band signals. With the base band approach, the input signal is mixed by locally generated coherent MARK and SPACE carrier frequencies. The two base band signals present at the outputs of mixers are then filtered by identical low pass filters. The low pass filtered MARK and SPACE signals are combined and fed to the decision circuit. The output of decision circuit is the data. The base band approach is easily adaptable to moving MARK and SPACE frequencies without needing to change data filters. However, different low pass filters are needed when the baud rate changes.

1.3.15 Phase Shift Keying (PSK) Transmitters and Receivers

PSK is a digital modulation scheme that encodes data by changing the phase of a carrier wave. There are a number of PSK modulation schemes such as BPSK, DPSK, Quadrature-Phase Shift Keying (QPSK) and Offset QPSK and so on. These digital modulation schemes have been discussed in Section 1.2.9 previously. Different PSK modulation schemes use a finite number

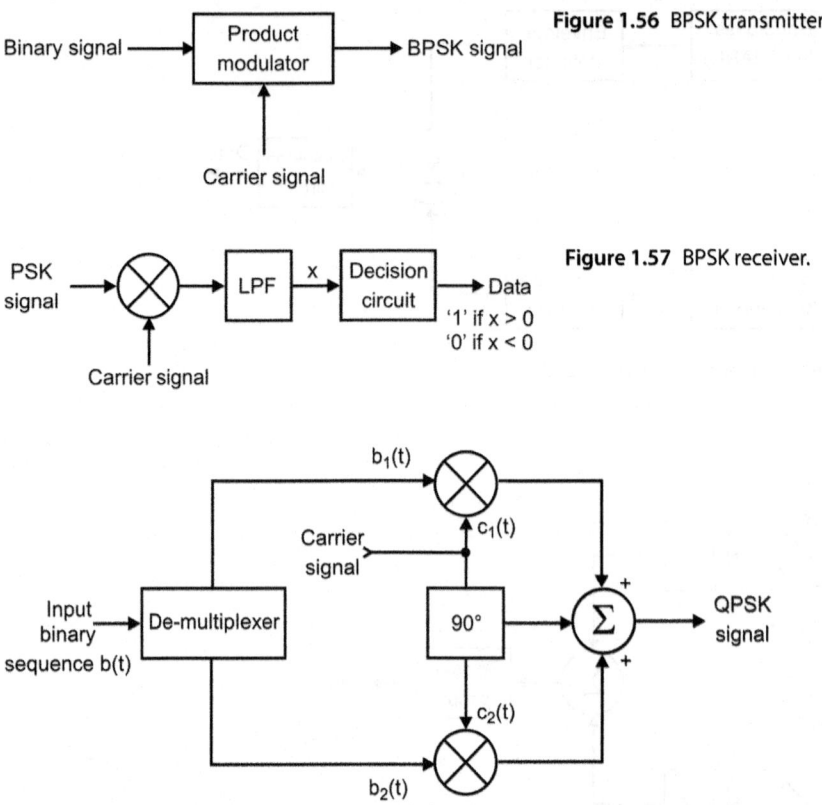

Binary signal → | Product modulator | → BPSK signal

Carrier signal

Figure 1.56 BPSK transmitter.

PSK signal → ⊗ → LPF → x → | Decision circuit | → Data

Carrier signal

'1' if x > 0
'0' if x < 0

Figure 1.57 BPSK receiver.

$b_1(t)$

Carrier signal

$c_1(t)$

Input binary sequence b(t) → De-multiplexer → 90° → Σ → QPSK signal

$c_2(t)$

$b_2(t)$

Figure 1.58 Block-schematic of a QPSK transmitter.

of phases, each assigned a unique pattern of binary digits. Each pattern of bits representing a certain phase forms a symbol. On the receiver side, the demodulator determines the phase of the received signal and maps it back to the symbol it represents to recover the transmitted data.

In the case of BPSK, a carrier signal can assume only two values of phase. Each phase value therefore carries one bit of information and the bit rate is same as the modulation rate. Figure 1.56 shows a simplistic BPSK transmitter. In the figure, E_b is transmitted signal energy per bit; T_b is time interval between adjacent bits, n is a fixed integer and $(f_c = n/T_b)$ is the carrier frequency.

Figure 1.57 shows the basic BPSK receiver. The received BPSK signal is multiplied with a locally generated coherent reference carrier signal. The output of the mixer is low pass filtered and fed to a decision circuit where it is compared with a threshold of 0 V. The receiver produces 1 for the filter output greater than zero and a 0 for filter output less than zero.

In QPSK, also known as quaternary PSK, 4-PSK or 4-QAM, the information is encoded in phase states of the carrier, which takes on one of four equally spaced values such as $\pi/4$, $3\pi/4$, $5\pi/4$ and $7\pi/4$. Figure 1.58 shows a simplified block-schematic of a QPSK transmitter. The input binary sequence $b(t)$ is de-multiplexed into two separate binary sequences comprising odd and even numbered input bits denoted by $b_1(t)$ and $b_2(t)$. $b_1(t)$ and $b_2(t)$ are used to modulate two carrier signals that are in phase quadrature. The two binary PSK signals produced as a result are added to produce the desired QPSK signal.

Figure 1.59 shows a block-schematic of a QPSK receiver. The QPSK signal is applied to a pair of correlators, each comprising a cascaded arrangement of a mixer and a low pass filter.

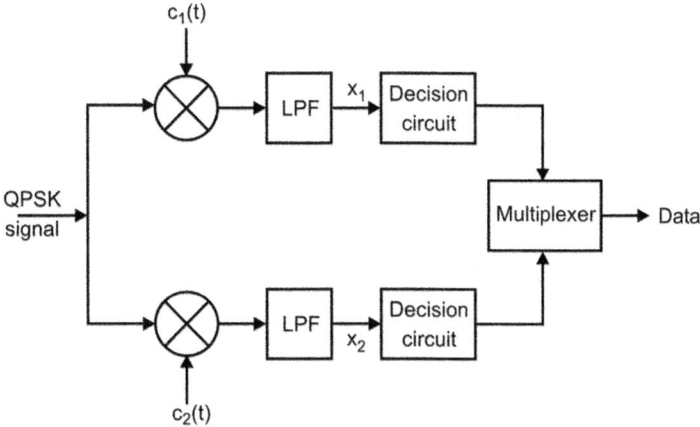

Figure 1.59 Block-schematic of a QPSK receiver.

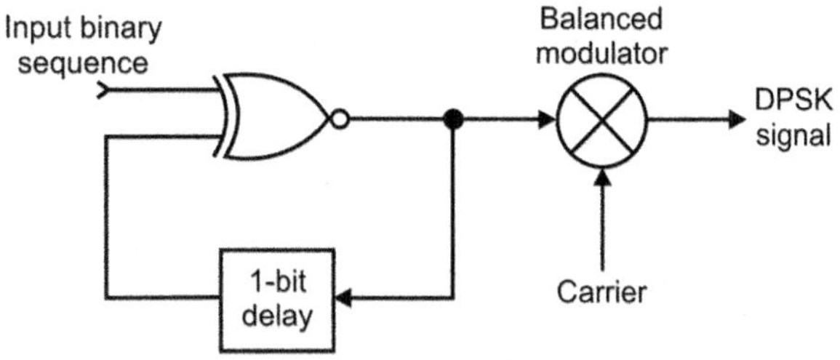

Figure 1.60 DPSK modulator.

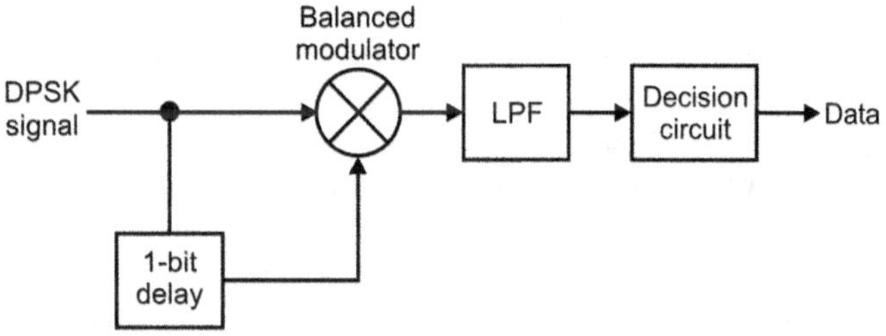

Figure 1.61 DPSK demodulator.

The two mixers are also fed with locally generated carrier signals $c_1(t)$ and $c_2(t)$. The outputs x_1 and x_2 of two low pass filters are each compared with a threshold of 0 V. The decision circuit takes the decision in favour of 1 for x (x_1 and x_2) greater than 0 V and 0 for x (x_1 and x_2) less than 0 V. These two channels are combined in a multiplexer to get the original binary output.

DPSK, a common form of phase modulation encodes data by changing the phase of the carrier wave. In the case of DPSK, as outlined in Section 1.2.9.4, the information is encoded by change in phase. While logic state 1 is represented by a change in phase of the carrier, logic level 0 is represented by no change in phase. Figure 1.60 illustrates the generation of DPSK signal. The binary sequence to be encoded is applied to one of the inputs of an X-NOR gate. The second input to the X-NOR gate is the output fed back via a 1-bit delay circuit. The resulting bit stream is applied to the balanced modulator to produce a DPSK signal. In the DPSK demodulator (Figure 1.61), the DPSK signal is passed to the balanced modulator directly and also through a 1-bit delay circuit. The resulting signal feeds a low pass filter. The output of the low pass filter is applied to a decision circuit, which produces binary data.

1.4 Antennas, Transmission Media and Propagation Modes

Antennas and transmission lines constitute a vital interface between the transmitter output and free space, and also between the propagating medium and receiver input. Transmission lines and waveguides constitute transmission media. Transmission media are primarily used in communication systems to carry signals from transmitter output to the input of transmitting antenna and from receiving antenna to the input of receiver. A waveguide does the same job at microwave frequencies as the transmission lines usually do at relatively lower radio frequencies. The mode of propagation of electromagnetic waves through the propagating medium is largely governed by the operating frequency. In this section, we shall discuss principles and applications of transmission lines, waveguides and antennas. Different propagation modes are briefly touched upon towards the end of the section.

1.4.1 Transmission Line Fundamentals

1.4.1.1 Transmission Line Equivalent Circuit

As outlined before, transmission lines are used in communication systems to carry signals from the transmitter output to the input of the transmitting antenna and from the receiving antenna to the input of the receiver. They are also used to for other applications such as impedance matching. Principle of operation of a transmission line can be best understood with the help of its electrical equivalent circuit. Figure 1.62 shows the lumped component equivalent network of a radio frequency (RF) transmission line supporting a transverse electromagnetic (TEM) mode. In the equivalent network shown, R and L are the equivalent series resistance and equivalent series inductance, respectively, per unit length of the line. G and C are equivalent shunt conductance and equivalent shunt capacitance, respectively, per unit length of the line. In an ideal lossless transmission line, $R = G = 0$. The incremental length here is chosen to be the one that is much smaller than the wavelength of the propagating signal. It should be mentioned here that transmission lines support two types of modes at microwave frequencies: (1) TEM and (2) non-TEM modes of propagation. The four basic parameters characterizing a TEM mode are the characteristic impedance Z_0, the phase velocity n_p, the attenuation constant a and the peak power handling capability P_{max}. In case of transmission lines supporting a non-TEM mode, these four parameters also depend upon the type of supported mode in addition to depending upon the geometrical features and material properties of the transmission line.

1.4.1.2 Transmission Line Losses

The three major sources of losses in RF transmission lines include: (1) copper losses (also referred to as I^2R losses), (2) dielectric losses and (3) radiation losses. Copper loss or I^2R loss is due to the resistance associated with the conductors constituting the transmission line. This loss appears in the form of heat. This loss is frequency dependent and increases with an increase in frequency. Dielectric loss also appears in the form of heat and increases with an increase in

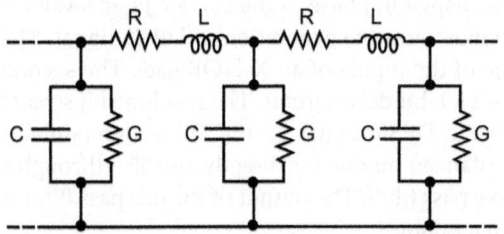

Figure 1.62 Equivalent circuit of a transmission line.

frequency. This loss is due to leakage through the dielectric. Radiation loss is due to radiation of RF power to free space or nearby circuits. Although transmission lines are not lossless, for all practical purposes they can be assumed to be so.

1.4.1.3 Transmission Line Propagation Modes

Two types of modes propagating in transmission lines are: (1) the TEM mode and (2) non-TEM modes. In the TEM mode (also called the principal mode), the electric and magnetic field vectors are perpendicular to one another and transverse to the direction of propagation of the signal. The TEM mode has no cut-off frequency. Besides the fundamental TEM mode or principal TEM mode, transmission lines can also support various non-TEM higher-order modes referred to as TE_{mn} (transverse electric) and TM_{mn} (transverse magnetic) modes. In the case of TE modes, there is no electric field component in the direction of propagation, and for TM modes, there is no magnetic field component in the direction of propagation. The subscript m signifies the number of full-period variations of the radial component of the field in the angular direction and n denotes the number of half-period variations of the angular component of the field in the radial direction.

1.4.1.4 Transmission Line Parameters

Important transmission line parameters include: (1) characteristic impedance, (2) propagation constant, (3) reflection coefficient, (4) standing wave ratio, (5) input impedance, (g) return loss and (7) mismatch loss.

Characteristic impedance of a transmission line is its input impedance if it was infinitely long. Refer to the transmission line equivalent circuit of Figure 1.62. It can be proved with simple mathematics that the characteristic impedance of this line is given by $Z_0 = \sqrt{\dfrac{(R + j\omega L)}{(G + j\omega C)}}$.

Where R = distributed resistance per unit length

L = distributed inductance per unit length
G = distributed shunt conductance per unit length
C = distributed shunt capacitance per unit length.

In a lossless transmission line, $R = 0$ and $G = 0$, so the expression for characteristic impedance line reduces to $Z_0 = \sqrt{\dfrac{L}{C}}$.

Characteristic impedance, as is clear from its definition and the relevant mathematical expression, is characteristic of the line and is independent of the length of the line. As all practical transmission lines are going to be of finite length, the significance of this parameter arises from the fact that if a finite line is terminated in a load impedance equal to the characteristic impedance of the line, its input impedance in that case will also equal the characteristic impedance.

The *propagation constant, γ*, is a measure of the attenuation and the phase shift of the incident waves travelling from the source to the load end of the transmission line. The propagation constant, for practical purposes, is a complex quantity having a real part known as attenuation constant α and an imaginary part called as phase shift constant β. The propagation of a wave along a transmission line can be mathematically expressed as $\gamma = \alpha + j\beta = \sqrt{(R + j\omega L)(G + j\omega C)}$. For a lossless line, $\gamma = j\omega\sqrt{LC}$, which gives $\alpha = 0$ and $\beta = \omega\sqrt{LC}$.

When a transmission line is terminated in load impedance that is not equal to its characteristic impedance, part of the signal energy sent down the line is reflected back. The ratio of the reflected signal amplitude to the incident one is defined as the *reflection coefficient*. It may be expressed as a magnitude only and denoted by ρ or as a complex value having both a magnitude and a phase and denoted by (Γ) with $\rho = |\Gamma|$. Reflection coefficient is expressed as $\rho = |\Gamma| = |(Z_L - Z_0)/(Z_L + Z_0)|$.

Whenever a signal travelling along a transmission line comes across a discontinuity or whenever the line is terminated in a load other than the characteristic impedance of the line, a part of the whole incident energy is reflected back. Under such circumstances, we have two counter-propagating waves in the transmission line. At all those points, where the waves are in phase, they add producing a signal maximum, and at all those points where they are out of phase, they produce a signal minimum. Thus, we have points of signal maxima and signal minima along the line except for the case where there is no discontinuity and where the line is terminated in its characteristic impedance. VSWR, an abbreviation for *voltage standing wave ratio*, is the ratio of E_{max} to E_{min}. It is a measure of the mismatch at the discontinuity. VSWR of unity implies a zero reflection coefficient and thus a perfect match. VSWR of infinity implies a unity reflection coefficient and thus a complete mismatch. VSWR is expressed as $VSWR = (1 + \rho)/(1 - \rho)$. Also, $\rho = (VSWR - 1)/(VSWR + 1)$.

Input impedance of a section of transmission line of length (l), characteristic impedance (Z_0) and terminated in a load impedance Z_L is expressed as $Z_{in} = Z_0[(Z_0 + Z_L \tanh \gamma l)/Z_L + Z_0 \tanh \gamma l]$. In the case of an ideal transmission line, the expression reduces to $Z_{in} = Z_0[(Z_0 + Z_L \tan \beta l)/Z_L + Z_0 \tan \beta l]$. Further, the expressions for Z_{in} for short circuited and open circuited lines, respectively, would be $Z_{in} = jZ_0 \tan \beta l$ and $Z_{in} = -jZ_0 \cot \beta l$. Another interesting observation is that, for lines whose length is an odd integral multiple of $\lambda/4$, impedance inversion takes place from load to the source and such a line can then be used as an impedance transformer. It can be verified that for such a line, $Z_{in}.Z_L = Z_0^2$.

The *return loss* signifies the total round trip loss of the signal and is defined as the ratio of the incident power to the reflected power at a point on the transmission line. It is expressed in decibels. Return loss, $L_r = -20 \log \rho$, where ρ is the magnitude of reflection coefficient.

Mismatch loss is the loss due to reflection from a mismatch. It is defined as the ratio of incident power to the difference of incident and reflected power expressed in decibels.

Mismatch loss, $L_m = -10 \log(1 - \rho^2)$.

1.4.2 Types of Transmission Lines

The two commonly used types of transmission lines on radio frequencies include *open-wire lines*, also known as parallel wire lines and the *coaxial lines*. In the category of open-wire lines, the two-wire balanced configuration, whose cross-section is shown in Figure 1.63(a), is more common. A coaxial transmission line [Figure 1.63(b)] comprises a conducting shell and a solid tape or a braided conductor surrounding an isolated concentric inner conductor. The inner conductor is either solid or stranded. Open-wire lines suffer from radiation losses and cross-talk. Radiation losses become prohibitively large at microwave frequencies. Coaxial lines, however, have much better shielding properties and therefore much lower radiation losses.

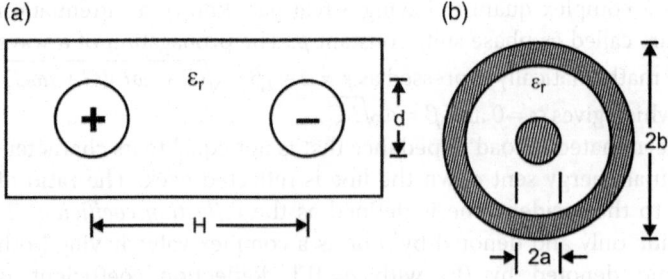

Figure 1.63 (a) Open wire line and (b) coaxial line.

Coaxial lines are, however, unbalanced lines. TEM is the dominant mode. Referring to the two-wire balanced transmission line of Figure 1.63(a), characteristic impedance of the line is given by $Z_0 = (260/\sqrt{\varepsilon_r})\log_{10}(2H/d)$ Ω. Referring to the coaxial line in Figure 1.63(b), characteristic impedance of the line is given by $Z_0 = (138/\sqrt{\varepsilon_r})\log_{10}(b/a)$ Ω.

1.4.3 Impedance Matching using Transmission Lines

Impedance matching is an important requirement in microwave circuit design in order to ensure that there is maximum transfer of power from source to load, that amplitude and phase imbalances are reduced in power distribution networks and that power loss in feed lines is minimized. Use of a transmission line to provide an impedance match involves a transmission line section of characteristic impedance Z_0 and length l, which depend upon the nature of impedances to be matched. The transmission line section used for matching is connected in either of the different possible configurations again depending upon the matching requirement. A typical matching problem, in practice, involves matching complex impedance, which could be either input or output impedance of a device, to real impedance. The commonly used configurations include use of stubs and quarter-wave transformers. In stub matching, again there is a single stub matching technique and a double stub matching technique. Single stub matching technique uses either a shunt stub or a series stub. Various techniques outlined here are briefly described in the following paragraphs.

1.4.3.1 Single Stub Matching

A stub is basically a shorted or open section of a transmission line used in conjunction with transmission lines to provide impedance match and cancel out reflections if any. As the shorted and open transmission line sections present pure reactance, their introduction does not absorb any power. Figure 1.64 illustrates the use of a single stub, a shunt stub as it is connected across the main transmission line, to provide an impedance match. Here, a transmission line having characteristic impedance of Z_0 is shown terminated in a complex load admittance of $(g_L + jb_L)$. As a first step, we locate a point on the transmission line where the normalized admittance is $(1 + jb_L)$. It may be mentioned here that $(g_L + jb_L)$ is also the normalized load. In the second step, put a stub across the transmission line at that point with the stub designed to offer a susceptance of $-jb_L$. Thus, the transmission line with a characteristic impedance of Z_0 gets matched to a complex load. It is usual practice to use shorted stubs rather than open ones as it is impossible to get a perfect open. An open stub if used will always be terminated in free-space impedance. Figure 1.65 shows the use of series stub. Again, in the first step, we locate a point on the transmission line where the normalized impedance looking towards the load end is $(1 + jX)$. At that point, a stub is added with the stub offering a normalized reactance of $-jX$. As is clear from Figure 1.65, the feed line needs to be cut for insertion of series stub. This technique is therefore not commonly used as it is difficult to fabricate in coaxial and strip lines.

Figure 1.64 Single stub matching using a shunt stub.

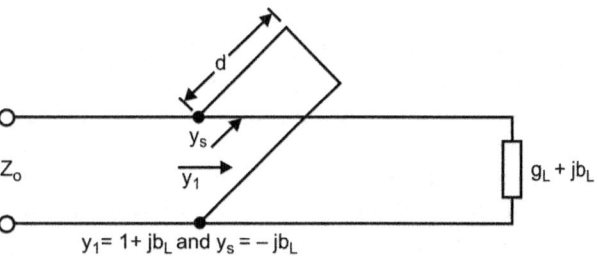

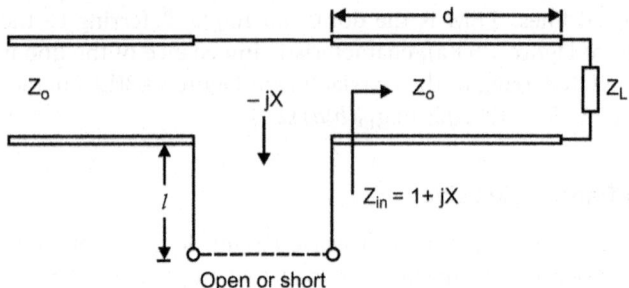

Figure 1.65 Single stub matching using a series stub.

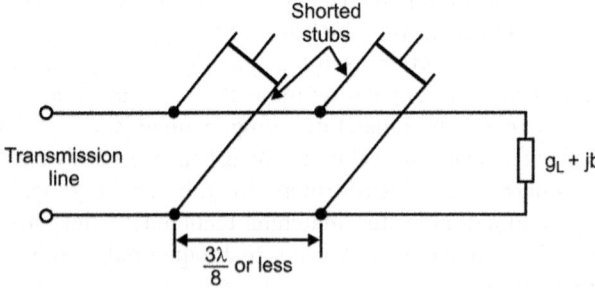

Figure 1.66 Double stub matching.

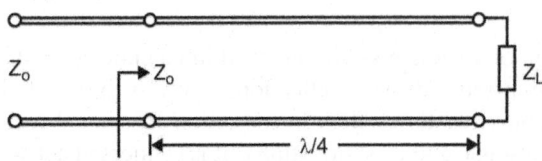

Figure 1.67 Quarter wave transformer matching for a resistive load.

1.4.3.2 Double Stub Matching

With the single stub matching of the type discussed in Section 1.4.3.1, it is sometimes impractical to put the stub at the intended location, more so in coaxial lines and waveguides. In such cases, a double stub matching technique is preferred. In double stub matching, as shown in Figure 1.66, the two stubs are put across the main line at fixed points spaced $3\lambda/8$ or even closer. These stubs have adjustable shorting plungers that can be adjusted to cancel out most of the reflections.

1.4.3.3 Quarter-Wave Transformer

A quarter-wave transformer ($\lambda/4$ long line) can be used to match both real as well as complex load impedance to a transmission line. If the main line characteristic impedance is Z_0 and it is to be matched to a load with a resistance R_L, the characteristic impedance of the quarter-wave section required for matching would be $\sqrt{R_L.Z_0}$. Figure 1.67 shows the interconnection. If the load impedance is complex, say $(R_L + jX_L)$, it should first be converted into real impedance by means of an additional length l of a line to cancel out the reactive component. If R_L is the real impedance looking into the input of this additional length towards the load end, then the characteristic impedances of the quarter-wave line section are given by $Z_T = \sqrt{R_L.Z_0}$.

The interconnections are shown in Figure 1.68. The reactive part of the load can also be tuned out by using a stub as shown in Figure 1.69. In this case, the complex load impedance can be made to present the real impedance $R_L{}'$ to the quarter-wave line section by means of an $(n\lambda/8)$ length of a line having characteristic impedance equal to the magnitude of the load impedance.

Figure 1.68 Quarter wave transformer matching for a complex load.

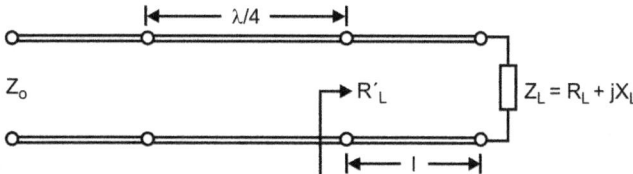

Figure 1.69 Tuning out the reactive part of a complex load.

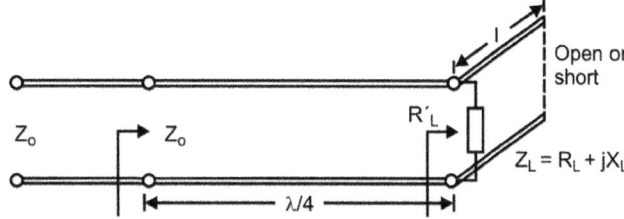

1.4.4 Waveguide Fundamentals

A waveguide does the same job with microwaves that the transmission lines usually do at relatively lower frequencies. At microwaves, it is more convenient to talk in terms of electric and magnetic fields propagating in the transmission medium rather than voltages and currents we are familiar with in the case of transmission lines. At relatively lower frequencies, extending up to say 100 MHz or so, the AC circuit theory is very well developed and almost all electronic functions can be implemented with available lumped components such as resistors, capacitors, inductors and so on, interconnected with wires to form a circuit. This approach of the AC circuit theory, which is nothing but an approximation to the field theory explained by Maxwell's equations and the wave equation, however, breaks down as we operate at higher frequencies exceeding 1 GHz or so. At those frequencies, it is more relevant to talk in terms of electromagnetic fields. The reason for this is very simple. At microwave frequencies, where the corresponding wavelengths are typically few tens of centimetres or lower, the size of lumped circuit elements and interconnecting wires becomes comparable to the wavelength and they behave like antennas. Because of this, the electromagnetic energy instead of remaining confined to the circuit gets radiated. The waveguide is the transmission medium of choice at higher frequencies.

1.4.4.1 Types of Waveguides

A waveguide is nothing but a conducting tube through which energy is transmitted in the form of electromagnetic waves. The waveguide can be considered to be a boundary that confines the waves to the space enclosed by boundary walls. An ideal waveguide would perform this task without any loss of energy or any distortion of the propagating wave. Actual waveguides, however, only approximate to this ideal condition. The waveguide can assume any shape theoretically, but the analysis of irregularly shaped guides becomes very difficult. Two popular types of waveguides are rectangular and circular waveguides, and again out of the two, the former is more extensively used. Less commonly used waveguide types include elliptical, cylindrical, and irregular waveguides. Figure 1.70(a) and (b) shows the outlines of rectangular and circular waveguides, respectively. A rectangular waveguide is characterized by its wide dimension (a) and narrow dimension (b) whereas a circular waveguide is characterized by its internal diameter (d).

(a)   (b)

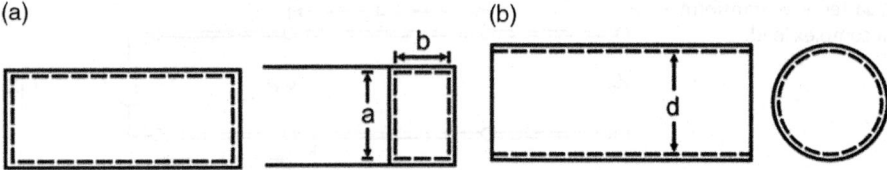

Figure 1.70 Types of waveguides. (a) Rectangular waveguide and (b) circular waveguide.

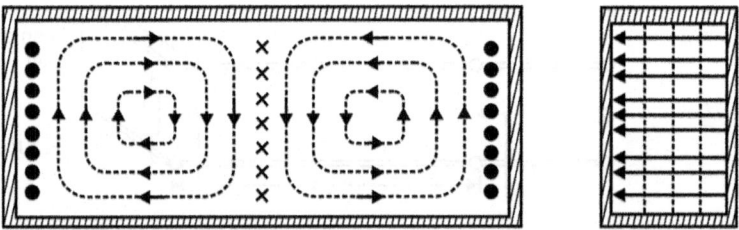

Figure 1.71 Waveguide modes.

1.4.4.2 Waveguide Modes

There will, in general, be infinite number of possible electric and magnetic field configurations inside the waveguide if there was no upper limit for the frequency of the signal to be transmitted. Each of these field configurations is called a mode. There are two types of modes: transverse magnetic (TM) modes and transverse electric (TE) modes. In TM modes, magnetic lines are entirely transverse to the direction of propagation of the electromagnetic wave. The electric field has a component in that direction. In TE mode, the electric field lines are entirely transverse to the direction of propagation whereas magnetic field has a component along the direction of propagation. Various propagation modes, both TM and TE, are designated by two subscripts. The first subscript indicates the number of half-wave variations of the electric field in the wide dimension of the waveguide whereas the second subscript indicates the number of half-wave variations along the narrow dimension of the waveguide. For instance, in TE10 mode, which is the simplest mode, there is only one-half-wave variation of electric field along the wide dimension and there is no electric field variation along the narrow dimension. Refer to Figure 1.71. It may be mentioned that this subscript notation is only for rectangular waveguides. In circular waveguides, the subscripts are there but they do not carry the same meaning as they do in case of rectangular waveguides.

The *dominant mode* propagating in a waveguide is one which has the highest cut-off wavelength for a waveguide of given dimensions. The cut-off wavelength of a waveguide is the highest signal wavelength that can propagate in a given waveguide. It will be seen that TE10 mode is the dominant mode in rectangular waveguides. Now, if we choose the guide dimensions in such a way that the signal wavelength is less than the cut-off wavelength for TE10 mode and greater than the cut-off wavelength at all other modes, which is easily achievable, we can ensure that only TE10 mode propagates. That is why TE10 mode is called the dominant mode. Even if a higher mode gets excited due to a discontinuity in the waveguide; it would soon die out as the guide would not support that mode. TE11 mode has the highest cut-off wavelength in a circular waveguide and we can always choose a diameter so that only TE11 mode propagates. This should then be the dominant mode in circular guides. However, due to the unsymmetrical nature of this mode, as shown in Figure 1.72(a), and due to the symmetrical nature of a circular guide, this mode is not the most popular because a bend or a discontinuity

(a) (b) (c)

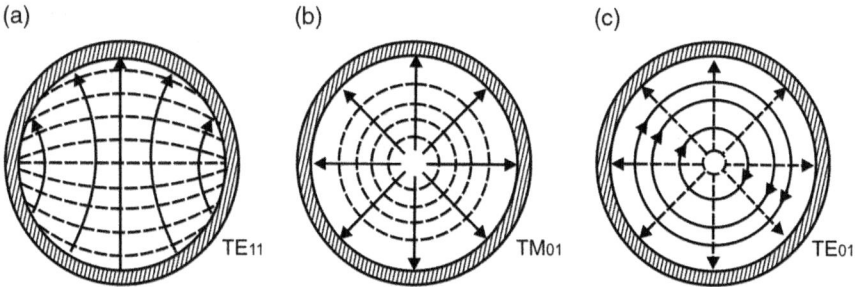

Figure 1.72 Circular waveguide modes. (a) TE11 mode, (b) TM01 mode and (c) TE01 mode.

in the guide might twist the mode leading to propagation with the wrong polarization. TM01 and TE01 modes, however, are symmetrical as shown in Figure 1.72(b) and (c), respectively. TM01 mode is used where symmetry is important whereas TE01 is used for long-distance waveguide runs as it has the least attenuation of all the commonly used modes in circular waveguides. Also, its attenuation decreases as frequency increases and is thus useful at higher microwave frequencies.

1.4.4.3 Waveguide Parameters

Important waveguide parameters include; (1) cut-off wavelength, (2) guide wavelength, (3) group and phase velocities and (4) characteristic wave impedance. Each of these parameters is briefly described in the following paragraphs.

As already outlined, there are a number of possible electric field and magnetic field configurations (called modes) that can exist in a waveguide. The modes that can exist and sustain in a waveguide are a function of waveguide dimensions and the frequency of the propagating signal. Each mode has a cut-off wavelength, that is, for a particular mode to sustain the wavelength corresponding to the signal frequency must be less than the cut-off wavelength for that mode. The *cut-off wavelength* for rectangular guides for both TE_{mn} and TM_{mn} is given by $\lambda_c = 2/\sqrt{(m/a)^2 + (n/b)^2}$ where a is the wide dimension and b is the narrow dimension of the waveguide. The cut-off wavelength of a circular waveguide with an internal diameter d is given by $\lambda_c = (\pi d/K_r)$, where K_r is the solution of a Bessel function equation. The values of K_r for the TE_{01}, TE_{11}, TE_{21}, TE_{02}, TE_{12} and TE_{22} modes are 3.83, 1.84, 3.05, 7.02, 5.33 and 6.71, respectively. The values of K_r for TM_{01}, TM_{11}, TM_{21}, TM_{02}, TM_{12} and TM_{22} modes are 2.4, 3.83, 5.14, 5.52, 7.02 and 8.42, respectively.

Guide wavelength, that is, the wavelength of the travelling wave propagating inside the waveguide, is always different from the free-space wavelength λ. The guide wavelength λ_g, the cut-off wavelength λ_c and the free-space wavelength λ are interrelated by $\lambda_g = \lambda/\sqrt{[1-(\lambda/\lambda_c)^2]}$.

The velocity of propagation in a waveguide is the product of guide wavelength λ_g and frequency f. As $\lambda_g > 1$, it appears as if *phase velocity* v_p is greater than the speed of light. This appears to contradict the law that no signal can be transmitted faster than the speed of light. Also, in waveguides, it is found that intelligence or modulation does not travel at a velocity (v_p). When a modulated carrier travels through a waveguide, the modulation envelope travels with a velocity far lower than that of the carrier and even lower than the speed of light. The velocity of modulation envelope is called *group velocity* v_g. As v_g is less than v_p, the modulation keeps slipping backwards with respect to the carrier as the modulated signal travels in a waveguide. In an air-filled or hollow waveguide, the phase and group velocities are related to speed of light by $v_p = (\lambda_g/\lambda)c$ and $v_g = (\lambda/\lambda_g)c$, which gives $v_p v_g = c^2$.

Characteristic impedance is another important waveguide parameter. The generalized expression for the characteristic impedance Z_0 of waveguide for TE modes is given by $Z_0 = 377 \times (\sqrt{\mu/\varepsilon}) \times (b/a) \times (\lambda_g/\lambda)$. The generalized expression for the characteristic impedance Z_0 of waveguide for TM modes is given by $Z_0 = 377 \times (\sqrt{\mu/\varepsilon}) \times (b/a) \times (\lambda/\lambda_g)$. For rectangular waveguides, a is the wide dimension and b is the narrow dimension. For circular waveguides, $a = b$.

1.4.5 Antenna Fundamentals

An antenna is a structure that transforms guided electromagnetic waves into free-space electromagnetic waves and vice versa. The guided electromagnetic waves look more appropriate when the feeder connecting the output of the transmitter and the antenna or the input of the receiver and the antenna is a waveguide, which is generally true when we talk about microwave frequencies and microwave antennas. In case of other antennas such as those at high frequency (HF) and very high frequency (VHF), the term *guided electromagnetic waves* mentioned previously would be interpreted as a guided electromagnetic signal in the form of current and voltage. Sometimes, an antenna is considered a system that comprises everything connected between the transmitter output or the receiver input and free space. This includes, in addition to the component that radiates other components such as the feeder line, balancing transformers and so on. An antenna is a reciprocal device, that is, its directional pattern as receiving antenna is identical to its directional pattern when the same is used as a transmitting antenna provided; of course, it does not employ unilateral and nonlinear devices such as some ferrites. Also, reciprocity applies, provided the transmission medium is isotropic and the antennas remain in place with only transmit and receive functions interchanged. Antenna reciprocity also does not imply that antenna current distribution is the same on transmission as it is on reception.

1.4.5.1 Radiation Mechanism

When a radio frequency (RF) signal is applied to the antenna input, there is current and voltage distribution on the antenna that lead to the existence of an electric and a magnetic field. The electric field reaches its maximum coincident with the peak value of the voltage waveform. If the frequency of the applied RF input is very high, the electric field does not collapse to zero as the voltage goes to zero. A large electric field is still present. During the next cycle, when the electric field builds up again, the previously sustained electric field gets repelled from the newly developed field. This phenomenon is repeated again and again and we get a series of detached electric fields moving outwards from the antenna. According to laws of electromagnetic induction, a changing electric field produces a magnetic field and a changing magnetic field produces an electric field. It can be noticed that when the electric field is at its maximum, its rate of change is zero and when the electric field is zero, its rate of change is maximum. This implies that the magnetic field's maximum and zero points correspond to the electric field's zero and maximum points, respectively. That is, the electric and magnetic fields are at right angles to each other and so are the detached electric and magnetic fields. The two fields add vectorially to give one field that travels in a direction perpendicular to the plane carrying mutually perpendicular electric and magnetic fields.

1.4.5.2 Characteristic Parameters

Important basic characteristic parameters of antennas relevant to all types of antennas used in various types of communication systems, radar systems and satellites include; (1) antenna reciprocity, (2) directive gain, (3) power gain, (4) effective isotropic radiated power, (5) directional pattern, (6) beam width, (7) bandwidth, (8) polarization, (9) impedance and (10) aperture.

An antenna is a *reciprocal device*. That is, its directional pattern as a receiving antenna is identical to its directional pattern when the same is used for transmitting antenna; provided that, of course, it doesn't use any unilateral and nonlinear devices such as some ferrites. Also, reciprocity applies provided that the transmission medium is isotropic and the antennas remain in place with only transmit and receive functions interchanged. Antenna reciprocity also doesn't imply that antenna current distribution is the same on transmission as it is on reception.

The *directive gain* in a given direction is defined as the ratio of the power density of the radiated electromagnetic energy in that direction to the power density in the same direction and at the same distance due to an isotropic radiator with both antennas radiating the same total power. Directive gain is always specified for a given direction and would have maximum value in the direction of maximum radiation. This maximum directive gain is termed the *directivity* and is usually expressed in decibels (dB). An antenna with a directivity of 20 dB would produce a power density at a given distance in the direction of maximum radiation when radiating a certain total power that would be 100 times the power density resulting from an isotropic radiator at the same point when radiating the same total power. The generalized expressions for directive gain are given by $4\pi P(\theta,\phi)/P_R$. Directivity is expressed by $4\pi[P(\theta,\phi)]_{max}/P_R$. $P(\theta,\phi)$ = Power radiated in the direction (θ,ϕ), P_R is isotropic radiated power and $[P(\theta,\phi)]_{max}$ is the maximum radiated power of a directive antenna.

Power gain is defined as the ratio of the power density at a given distance in the direction of maximum radiation intensity to the power density at the same distance because of an isotropic radiator of the same total power fed to the two antennas. The definition of power gain is similar to that of directive gain or directivity, except that it is not the power radiated by the antenna but the power fed to the antenna that is considered while computing the gain. It takes into account the antenna losses and thus is of greater practical importance. The generalized expression for power gain is given by: power gain = $4\pi[P(\theta,\phi)]_{max}/P_{in}$. P_{in} = Input power = $P_R + P_L$; P_L = power loss and P_R = isotropic radiated power.

The *Effective Isotropic Radiated Power* (EIRP) is the more appropriate figure of merit of the antenna. It is given by the product of transmitter power and antenna gain. An antenna with a power gain of 40 dB and a transmitter power of 1000 W would mean an EIRP of 10 MW. That is, 10 MW of transmitter power, when fed to an isotropic radiator, would be as effective in the desired direction as 1000 W of power fed to a directional antenna with a power gain of 40 dB in the desired direction.

The antenna *directional pattern* or *radiation pattern* is a normalized distribution plot of electromagnetic energy in a three-dimensional (3D) angular space. The parameters to be plotted could be radiation intensity, which is the power per unit solid angle or the power density. A more commonly used representation of directional pattern is the two-dimensional (2D) plot. There are, again, various types of 2D plots. One of the types is the polar plot of radiation intensity or power density shown in Figure 1.73 that we are all quite familiar with. Another is the principal plane elevation pattern as shown in Figure 1.74. This is the pattern drawn by sectioning the 3D pattern with a vertical plane through the peak of the beam and a zero azimuth angle. A similar pattern called the *principal plane azimuth pattern* could be drawn by sectioning the 3D pattern through the peak of the beam and a zero elevation angle. Though the 2D patterns obtained by sectioning with planes other than the principal planes (called cardinal planes) can also be drawn, the azimuth and elevation patterns usually suffice in most of the cases. The main beam of the pattern is called the main lobe and the beams in directions other than the direction of maximum radiation are called sidelobes. High sidelobe levels, with a few exceptions, are always undesirable. The sidelobe level of an antenna pattern is usually specified in terms of relative sidelobe level, which is the peak level of the highest sidelobe relative to the peak of the main lobe. For instance, a relative sidelobe level of 20 dB means that the peak power density in the side lobe is 1/100th of the peak power density in the main lobe.

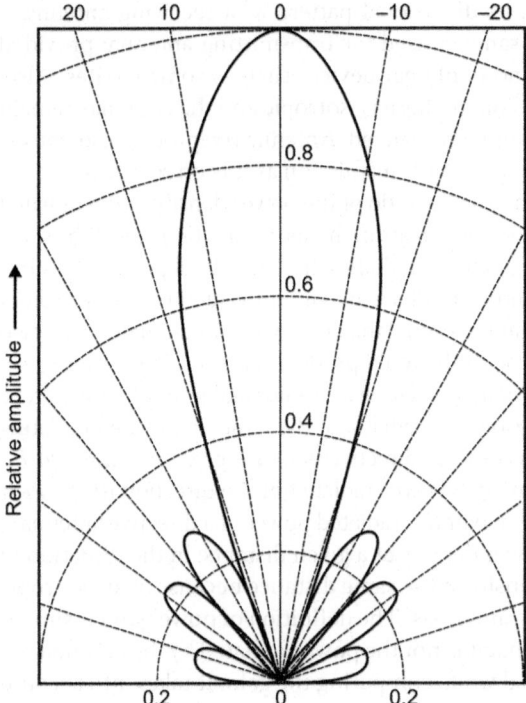

Figure 1.73 Polar plot.

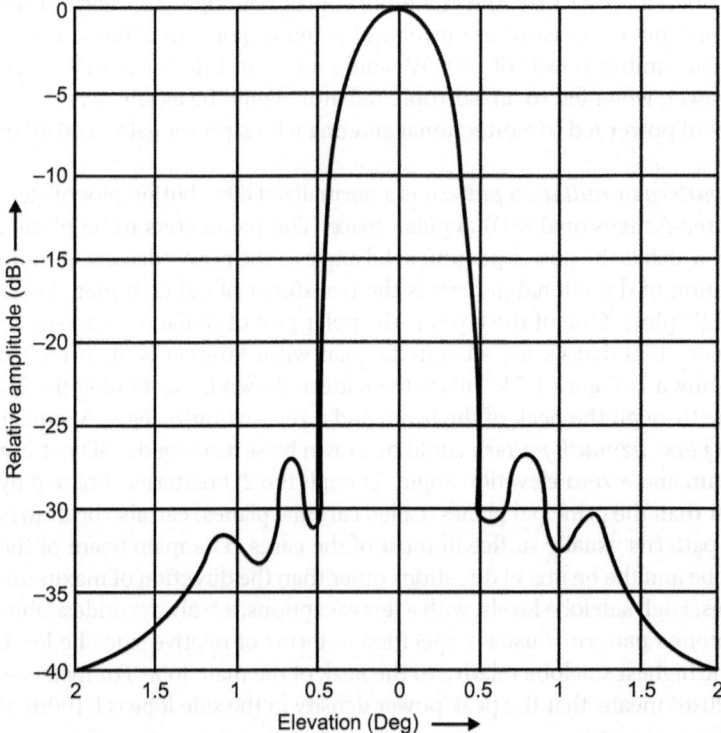

Figure 1.74 Principle plane elevation pattern.

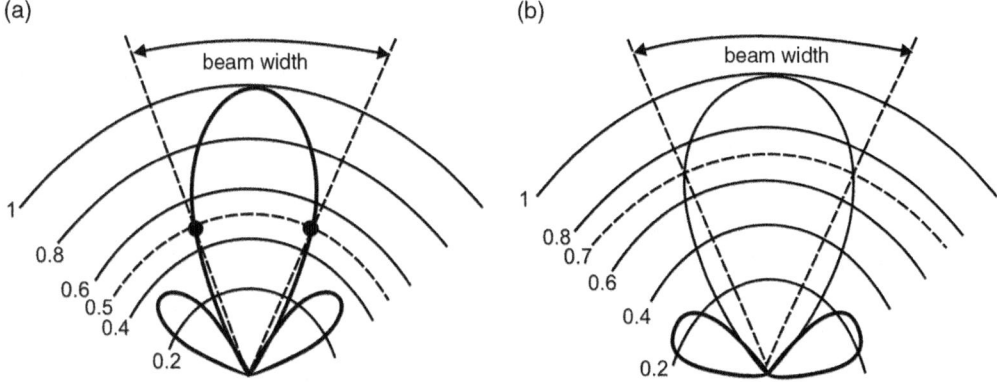

Figure 1.75 Beam width. (a) Power density radiation pattern and (b) field intensity radiation pattern.

The common types of antenna radiation patterns include the (1) omnidirectional (azimuth plane), beam, (2) pencil beam, (3) fan beam and (4) shaped beam. The *omnidirectional beam* is commonly used in communication and broadcast applications for obvious reasons. The azimuth plane pattern is circular and the elevation pattern has some directivity to increase the gain in horizontal directions. A *pencil beam* is a highly directive pattern whose main lobe is confined to within a cone of a small solid angle and it is circularly symmetric about the direction of maximum intensity. A *fan beam* is narrow in one direction and wide in the other. A typical application of such a pattern would be in search or surveillance radars in which the wider dimension would be vertical and the beam is scanned in azimuth. The last application would be in height-finding radar where the wider dimension is in the horizontal plane and the beam is scanned in elevation. There are applications that impose beam-shaping requirements on the antenna. One such requirement, for instance, is to have a narrow beam in azimuth and a shaped beam in the elevation such as in case of air search radar.

Beam width gives the angular characteristics of radiation pattern. It is taken as the angular separation either between the half power points on its power density radiation pattern [Figure 1.75(a)] or between 3 dB points on the field intensity radiation pattern [Figure 1.75(b)]. It is measured in degrees and with reference to the main lobe. Antennas also have 6-dB beam widths and null-to-null beam widths. Null-to-null beam width is the width of the response between the minima surrounding the main lobe and is approximately twice the 3-dB beam width for most antenna responses. The parameter is particularly relevant to the antennas producing narrow beams such as those in tracking radars. An antenna's power gain $G(\theta,\phi)$ is related to its beam width parameters by $G(\theta,\phi) = 4\pi/\Omega$. Ω is the solid angle (in steradians) $\Delta\theta.\Delta\phi$. $\Delta\theta$ is the beam width in the azimuth direction (in radians) and $\Delta\phi$ is the beam width in the elevation direction (in radians).

Antenna bandwidth is in general the operating frequency range over which the antenna gives a certain specified performance. Antenna bandwidth is always defined with reference to a certain parameter such as gain or input impedance or standing wave ratio (SWR). It is generally taken as the frequency range around the nominal centre frequency over which power gain falls to half of the maximum value. When referenced to the SWR, one may specify a 2:1 SWR bandwidth and so on. The lower the operating frequency, the narrower the bandwidth. It follows from the rule that in case of a resonant circuit, for a given quality factor Q, the bandwidth is directly proportional to the centre frequency.

Antenna polarization, whether it is transmitting or receiving, is the direction of electric field vector with respect to the ground. While receiving, it is considered for the orientation of electromagnetic wave that the antenna responds best to. From antenna reciprocity, we can say that the antenna would respond most optimally to an electromagnetic wave that would have the same polarization as that of the transmitted wave radiated from the same antenna. It is a normal practice to consider the antenna itself as being polarized. The polarization of the antenna is the same as that of the electromagnetic wave it radiates or best responds to. The polarization of an antenna can be classified into two broad categories: linear polarization and elliptical polarization. *Linear polarization* could be either horizontal polarization or vertical polarization. *Circular polarization* is a special case of *elliptical polarization*. In linear polarization, the electric vector lies in a plane. If the plane is horizontal, it is horizontally polarized and if the plane is vertical, it is vertically polarized. An inclined plane leads to what may be referred to as *slant polarization*. Slant polarization is a general case of linear polarization having both horizontal and vertical components. It is called linear polarization because the direction of the resultant E vector is constant with respect to time. In the generalized case of a linearly polarized wave, the two mutually perpendicular components of the E vector are in phase. When the two components of the E vector are not in phase, it can be verified that the tip of the resultant traverses an ellipse as the RF signal goes through one complete cycle. This is called *elliptical polarization*. This polarization could have right-hand sense or left-hand sense depending upon whether the E vector moves clockwise or anticlockwise when viewed as a wave receding from the observation point in the direction of propagation. Elliptical polarization has two orthogonal linearly polarized components. When the magnitudes of these components become equal and the phase difference between the two becomes 90°, polarization becomes circular polarization. Again, we have either right-hand circular polarization (RHCP) or left-hand circular polarization (LHCP).

Cross polarization is the component that is orthogonal to the desired polarization. For instance, a horizontally polarized antenna may also radiate vertical polarization in some directions of propagation or a vertically polarized antenna may radiate horizontal polarization in some directions. The other example could be that of an RHCP antenna also radiating LHCP and an LHCP antenna also radiating RHCP. A well-designed antenna should have a cross-polarized component at least 20 dB below the desired polarization in the direction of the main lobe and 5–10 dB below the desired polarization in the direction of side lobes.

If the received electromagnetic wave is of a polarization different from the one the antenna is designed for, a polarization loss results. This loss in decibels in the case of linear polarization is given by Polarization loss $= 20\log(1/\cos\phi)$. Here, ϕ is the angle between the polarization of the received wave and that of the antenna.

The *antenna impedance* at a given point in the antenna is given by the ratio of voltage to current at that point. As the magnitude of voltage and current vary along the antenna length, the impedance also varies being minimum at the point of the voltage node or minima, such as the centre point of a half-wave dipole and maximum at the point of the current node, such as the centre point of a full-wavelength long antenna. The input impedance of an antenna is of considerable importance to engineers as it is desirable to supply the maximum amount of transmitter power to the antenna. For this, the characteristic impedance of the feeder line must match the antenna input impedance at the chosen feed point. The antenna impedance is resistive if it is resonant at the operating frequency. The antenna resistance further comprises of two components, namely, the radiation resistance R_r and the R_d loss resistance. Radiation resistance is basically the resistance that, if the antenna is terminated, would dissipate the same power as that being radiated by the antenna. It is given by the radiated power divided by square of feed current. The loss resistance is contributed to by factors such as eddy current losses in metallic objects lying in the vicinity of induction field of antenna, losses in imperfect dielectrics, corona effects and so on. Antenna efficiency is defined as $\eta = R_r/(R_r + R_d)$.

The antenna aperture is the physical area of the antenna projected on a plane perpendicular to the direction of the main beam or the main lobe. In the case where the main beam axis is parallel to the principal axis of the antenna, it is the same as the physical aperture of the antenna itself. For a given antenna aperture A, the directive gain of the antenna at an operating wavelength of 1 is given by $(4\pi A/\lambda^2)$. This expression is valid only when the aperture A is uniformly illuminated. Typical antennas are not uniformly illuminated and have a tapered illumination, the maximum being at the centre and lower towards the edges. This is done to reduce the sidelobe level. Because of this nonuniform illumination, the antenna gain falls from its maximum value of $(4\pi A/\lambda^2)$. This is where the term effective aperture A_e of the antenna comes into the picture. It is that aperture area that, when uniformly illuminated, gives the same gain as that offered by a nonuniformly illuminated antenna of aperture A. Thus, the gain of a practical antenna is given by $(4\pi A_e/\lambda^2)$. Here A and A_e are interrelated by $A_e = \eta A$ and η is the aperture efficiency (or effectiveness).

1.4.6   Types of Antennas

In the paragraphs to follow, we shall briefly describe the operational aspects of major types of antennas. Different types of antennas include the following.

1) Hertz antenna
2) Marconi antenna
3) Dipole antenna
4) Yagi-Uda antenna
5) Rhombus antenna
6) Reflector antenna
7) Lens antenna
8) Horn antenna
9) Helical antenna
10) Log periodic antenna
11) Phased array antenna
12) Microstrip antenna.

1.4.6.1   Hertz, Dipole and Marconi Antennas

A *Hertz antenna* is a straight length of a conductor that is a half-wave long. It may be placed vertically to produce vertically polarized waves [Figure 1.76(a)] or in horizontal position to produce horizontally polarized waves [Figure 1.76(b)]. A *dipole antenna* is also a straight radiator usually fed at the centre and producing the maximum of radiation in a place perpendicular to the antenna axis. A dipole that is a half-wavelength long is called a half-wave dipole [Figure 1.76(c)].

The vertical antenna that is a quarter-wave long and is fed against an infinitely large perfectly conducting plane is called a quarter-wave monopole or *Marconi antenna*. It has the same radiation characteristics above the plane as the half-wave dipole antenna in free space. The Marconi antenna has an edge over the Hertz antenna when it is to be used as a transmitting antenna at low frequencies, as its length is half of the required length of Hertz antenna for a given transmission frequency. Also, a Marconi antenna produces vertically polarized waves, ideally suited for transmission and propagation of relatively lower frequency RF signals. The radiation resistance of the half- and quarter-wave monopoles can be determined to be equal to 73 Ω and 36.5 Ω, respectively. The generalized expressions for the impedance are given $0.609\eta I_{RMS}^2/2\pi$ (monopole) and $0.609\eta I_{RMS}^2/\pi$ (dipole). Here, I_{RMS} is the RMS value of antenna current and η is the characteristic impedance of medium = 377 Ω for free space.

A modification of the half-wave dipole is the folded dipole suitable for TV reception purposes. A folded dipole [Figure 1.76(d)] comprises two half-wave dipoles connected at the ends and one of them fed at the centre. It may be constructed by folding a full-wave long conductor. The second element gets its excitation from the field produced by the driven element. The folded dipole electrically behaves in the same fashion as a straight dipole, physical construction being the only difference. Addition of this second element increases the input impedance of the antenna, which is given by $Z_{in} \times$ [(Cross-sectional area of all conductors)/Cross-sectional area of driven element].

1.4.6.2 Yagi-Uda Antenna

A *Yagi-Uda antenna* comprises a half-wave dipole with parasitic elements to enhance the directionality of the radiation pattern. It is the most commonly used antenna type for HF and VHF communications. The simplest Yagi antenna would be a three-element array with a centre-fed half-wave dipole as the driven element, one parasitic element smaller in length than the driven element by about 4% called the director is placed in front of the driven element and another parasitic element longer in length than the driven element by about 5%, called the reflector, is placed behind the driven element (Figure 1.77). The director enhances the directivity of the radiation pattern and the reflector suppresses the radiation in the backwards direction; that is, when used as a receiving antenna it does not receive from that direction thus improving the front-to-back ratio. The director–dipole spacing is approximately 0.12λ whereas the reflector–dipole spacing is 0.2λ.

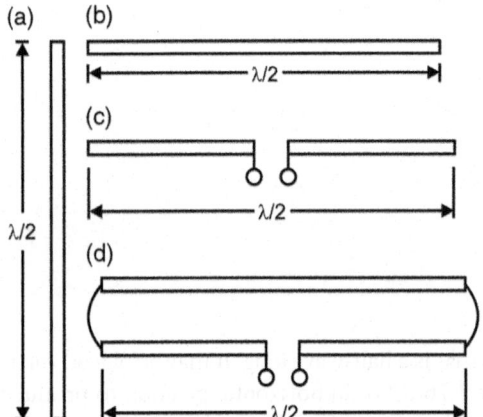

Figure 1.76 (a) Vertical Hertz antenna. (b) Horizontal Hertz antenna. (c) Dipole antenna. (d) Folded dipole.

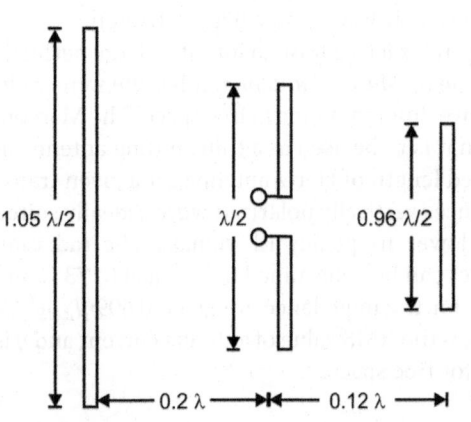

Figure 1.77 Yagi-Uda antenna.

1.4.6.3 V-Antenna and Rhombic Antenna

These are long-wire antennas. In a *V-antenna*, the conductors are arranged to form a V-shape and they are fed in phase opposition at the apex [Figure 1.78(a)]. Such an arrangement produces a high gain bidirectional pattern as shown in Figure 1.78(b). If the antenna is to be used as a wideband antenna, the apex angle is a compromise between an optimum for the lowest and the highest frequencies in terms of number of half wavelengths in each leg. In a *rhombic antenna*, conductors are arranged to form a rhombus. It is a combination of two long-wire V-antennas [Figure 1.78(c)]. In this case too, the length of the legs and the apex angle control the shape and directivity of the pattern. The gain of a rhombic antenna, whose individual legs are of the same lengths as those of a V-antenna, will be approximately double. The resonant rhombic antenna produces a bidirectional radiation pattern as shown in Figure 1.78(d).

1.4.6.4 Reflector Antennas

A *reflector antenna* is made in different types, shapes and configurations depending upon the shape of the reflector and the type of feed mechanism. It is by far the most commonly used antenna type in all those applications that require high gain and directivity. High gain and a highly directional radiation pattern, which are antenna parameters that are essentially the same, are the characteristics typical of both terrestrial and satellite-based communication links, radar systems, direction-finding systems and so on. While communicating in the UHF and microwave frequency bands, it is important to have narrow beam width to avoid interference with other transmissions. In a radar system such as tracking radar, accuracy and resolution of measurement of angular information are equally important. Angular resolution, which is the ability to discriminate between two targets located close to each other, again depends upon the narrowness of the beam width. The narrower the beam, the higher the angular resolution. Now the gain or the directivity of the antenna is directly proportional to the size of the antenna. The antenna dimensions need to be much larger than the operating wavelength for achieving high directivity, a requirement that would not be practicable at relatively lower frequencies. At UHF and above it does become practicable. For example, at 10 GHz, $\lambda = 3$ cm with a 3-m diameter dish would give a dimension that is 100 times the operating wavelength. Of course, there is a small overlap region between VHF (30 – 300 MHz) and UHF (300 – 3000 MHz) and some of the antenna types to be used for higher-end VHF and lower-end UHF are common.

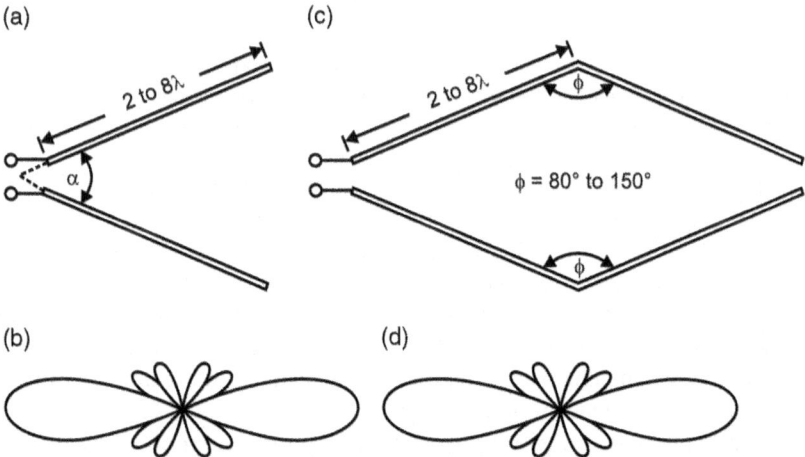

Figure 1.78 (a) V-antenna. (b) Directional pattern of a V-antenna. (c) Rhombic antenna. (d) Directional pattern of a rhombic antenna.

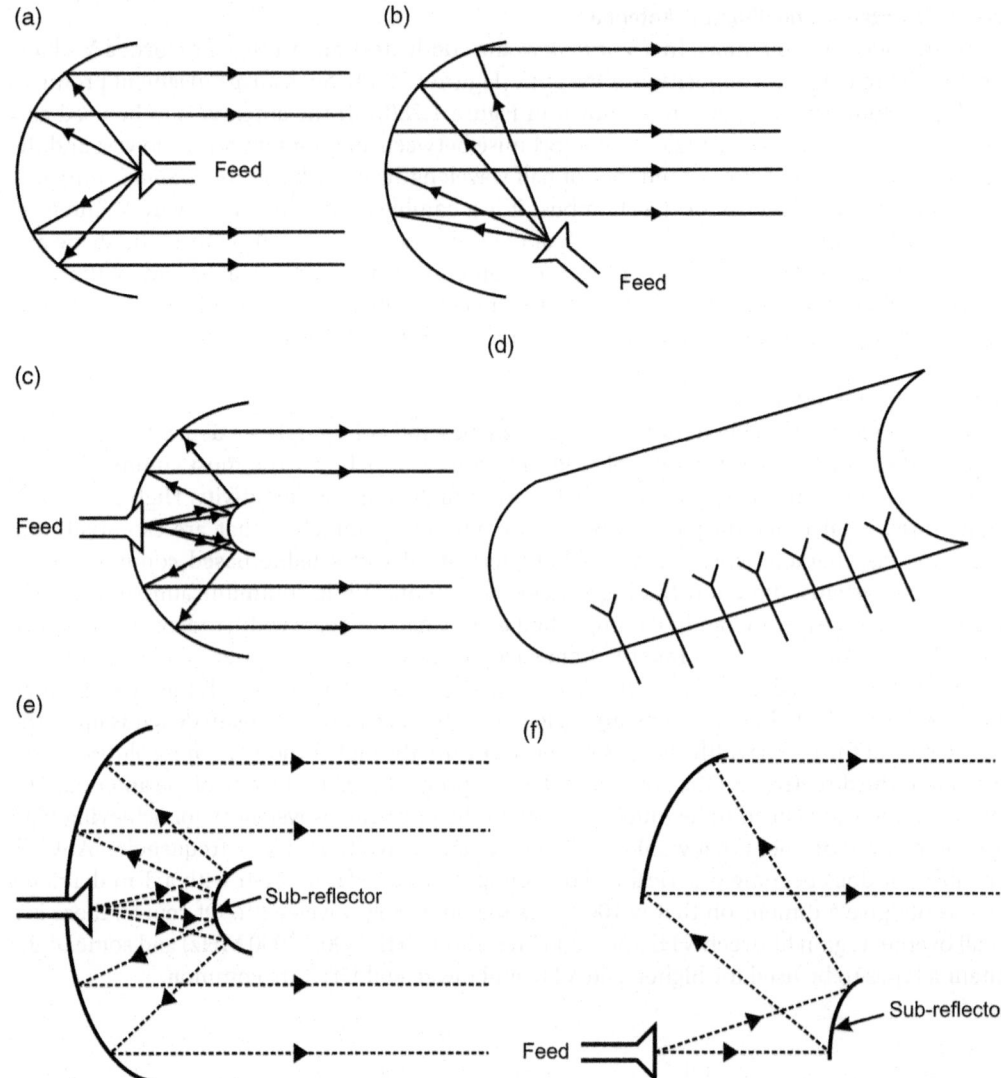

Figure 1.79 Reflector antennas: (a) Focal point fed parabolic antenna. (b) Offset-fed sectioned parabolic antenna. (c) Cassegrain fed antenna. (d) Array-fed cylindrical reflectors. (e) Symmetrical Cassegrain antenna. (f) Offset Cassegrain antenna.

A reflector antenna, in essence, comprises a reflector and a feed antenna. As mentioned earlier, depending upon the shape of the reflector and the feed mechanism, there are different types of reflector antennas suitable for different applications. The reflector is usually parabolic, also called parabolic reflector, or a section of parabolic or cylindrical reflectors. A cylindrical reflector has a parabolic surface in one direction only. The feed mechanisms include the feed antenna placed at the focal point of the parabolic reflector or the feed antenna placed off the focal point. Another common feed mechanism is the Cassegrain feed. Cylindrical reflectors are fed by an array of feed antennas. The feed antenna is usually a dipole or a horn. These antennas are thus available in many types and configurations, some of the more commonly used ones include; (1) a focal point fed parabolic reflector [Figure 1.79(a)], (2) an offset fed sectioned parabolic reflector [Figure 1.79(b)], (3) a Cassegrain fed reflector [Figure 1.79(c)] and (4) an array fed cylindrical reflector [Figure 1.79(d)].

Power gain of a focal point fed parabolic reflector antenna is given by $(4\pi A_e/\lambda^2)$. Here, A_e is the effective aperture area and λ is the operating wavelength. If D and h are the mouth diameter of the reflector and aperture efficiency, respectively, then power gain is given by $(\pi\eta^2 D/\lambda^2)$. The 3-dB beam width of such an antenna is given by $70(\eta/D)$.

If the feed antenna beam width is excessive, it causes a spill over producing an undesired antenna response in that direction. And if it is too small, only a portion of the reflector is illuminated with the result that antenna produces a wider beam and a consequent lower gain. Focal length is another important design parameter. A long focal length reflector antenna would produce more error at the feed than that produced by a short focal length reflector. However, focal length cannot be increased arbitrarily as long focal length reflectors need a larger support structure for the feed and hence a greater aperture blockage.

The directional pattern of the feed determines the illumination of the reflector. The angle subtended by the feed antenna at the edges of the reflector is given by $4\tan^{-1}[1/(4f/D)]$. According to a rule of thumb, the 3 dB beam width should be equal to 0.9 times the subtended angle.

Feed together with its support is one of the major causes of aperture blockage, which is further one of the major causes of sidelobes. In applications where the feed antenna is rather large so as to block a portion of the reflector aperture with significant effects on the radiated beam in terms of increased sidelobe content, an offset fed parabolic reflector antenna is one of the solutions. The shortcomings of the focal point fed parabolic reflector antenna, such as aperture blockage and lack of control over main reflector illumination, can also be overcome by adding a secondary reflector. The contour of the secondary reflector determines the distribution of power along the main reflector thereby giving control over both amplitude and phase in the aperture. The Cassegrain antenna derived from telescope designs is the most commonly used antenna using multiple reflectors. The feed antenna illuminates the secondary reflector, which is a hyperboloid. One of the foci of the secondary reflector and the focus of the main reflector are coincident. The feed antenna is placed on the other focus of the secondary reflector. The reflection from the secondary reflector illuminates the main reflector. Figure 1.79(e) shows the arrangement. Symmetrical Cassegrain systems usually produce a large aperture blockage, which can be minimized by choosing the diameter of the secondary reflector equal to that of the feed. Blockage can be completely eliminated by offsetting both the feed as well as the secondary reflector as shown in Figure 1.79(f). Such an antenna is capable of providing a very low sidelobe level.

A cylindrical parabolic antenna uses a reflector that is a parabolic surface only in one direction and is not curved in the other. It is fed by an array of feed antennas, which gives it much better control over reflector illumination. Electronic steering of the output beam is also more convenient in an array fed cylindrical antenna. Symmetrical parabolic cylindrical reflectors, however, suffer from a large aperture blockage. A cylindrical reflector fed from an offset placed multiple element line source offers excellent performance.

1.4.6.5 Lens Antenna

Like reflector antennas such as parabolic reflectors, lens antennas are another example of application of rules of optics to microwave antennas. In the case of former, it is the laws of reflection; the lens antennas depend on refraction phenomenon for their operation. Lens antennas are made of dielectric material and Figure 1.80(a) explains the principle of operation. A point source of operation is placed at the focal point of the lens. Due to the curvature of the lens, rays close to the edges are refracted more than the rays close to the centre. This explains why the rays get collimated and become parallel to the lens axis after passing through the lens, though they are inclined in the space between the lens and the point source. Similarly, on reception, the rays arriving parallel to the lens axis get focused onto the focal point where the feed antenna is placed. Another way of explaining the operation of a lens antenna is as follows. Refer to Figure 1.80(b). Spherical waves emitted by the point source get transformed to plane

(a)

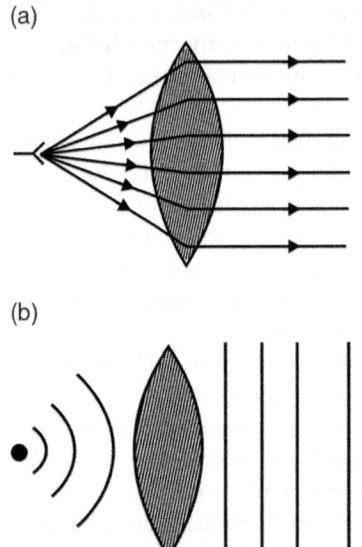

(b)

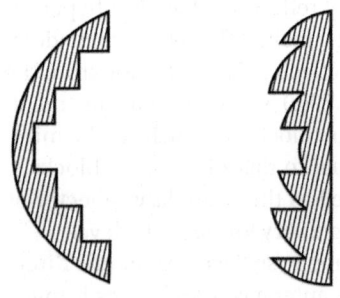

Figure 1.80 Lens antenna: (a) principle of operation using a ray diagram and (b) principle of operation using a wave front.

Figure 1.81 Zoned lenses.

waves during transmission and the reverse process occurs during reception. This happens as the spherical waves travelling closer to the centre are slowed down more than those travelling away from the centre. Same reasoning explains the reverse process during reception. Also, for efficient operation of a lens antenna, the thickness of the lens antenna at the centre should be much larger than the operating wavelength. This makes a lens antenna less attractive at operating frequencies less than 10 GHz. Even at frequencies around 10 GHz, there are serious thickness and weight issues, which are overcome by using what are known as Fresnel or zoned lenses. Two types of zoned lenses are shown in Figure 1.81. Zoning not only overcomes the weight problem, it also absorbs less energy. A thicker lens would absorb a higher proportion of the radiation. The thickness t of each step in a zoned lens is related to wavelength to ensure that phase difference between the rays passing through the centre and those passing through adjacent section is 2π radians or an integral multiple of it. A zone lens, because of its thickness being related to operating wavelength, has a small operational frequency range.

1.4.6.6 Horn Antennas

If the abrupt discontinuity of a waveguide is transformed to a more gradual one, we get the *horn antenna*. It may be mentioned here that a transmission line or a waveguide open circuited at the load end would radiate electromagnetic energy into the atmosphere very inefficiently, mainly due to the impedance mismatch between the transmission line or waveguide and the atmosphere. Making the discontinuity more gradual only improves the impedance match and thereby the coupling of electromagnetic energy to the atmosphere. There are various types of horn antennas such as the *sectoral horn* [Figure 1.82(a)] where the flare is only on one side, the *rectangular pyramidal horn* [Figure 1.82(b)] where the flare is on both sides and the *conical horn* [Figure 1.82(c)], which is a natural extension of a circular waveguide.

The important design parameters of a horn antenna include flare length and flare angle [Figure 1.82(d)]. If the flare angle is too small, the antenna has low directivity and also emitted electromagnetic waves are spherical and not planar. Too large a flare angle also leads to loss of directivity due to diffraction effects. Horns could have simple straight flares or exponential flares. These are commonly used as feed antennas for reflector type antennas. When more demanding antenna performance is desired in terms of polarization diversity, low sidelobe level, high radiation efficiency and so on, the feeds also become more complex. Segmented, finned and multimode horns may be used. Some combinations of horn antennas and parabolic reflectors such as Cass-Horn and Hog-Horn antennas have gain and beam width specifications matching those of parabolic reflectors of comparable dimensions.

Figure 1.82 (a) Sectoral horn. (b) Rectangular pyramidal horn. (c) Conical horn. (d) Flare length and flare angle.

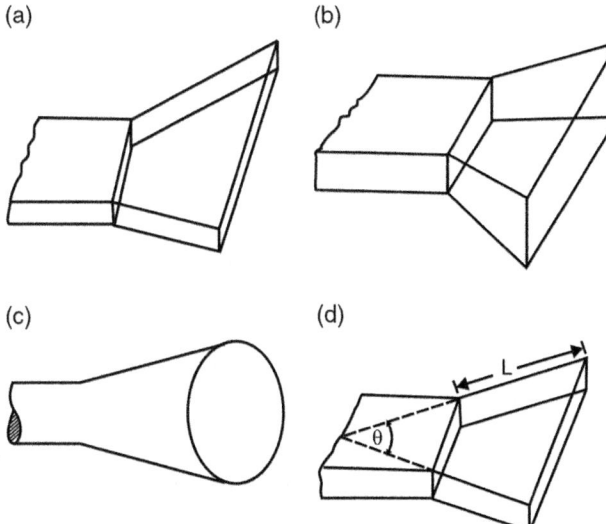

Figure 1.83 The helical antenna. (a) Helical antenna. (b) photograph of a representative helical antenna.

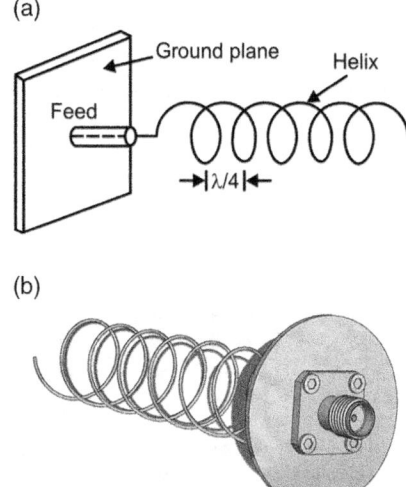

1.4.6.7 Helical Antenna

A *helical antenna* is a broadband VHF and UHF antenna. Most of its applications are due to the circularly polarized waves it produces. VHF and UHF propagation undergoes a random change in its polarization as it propagates through the atmosphere due to various factors like Earth's magnetic field, ionization of different regions of atmosphere and so on, Faraday rotation being the main cause. The propagation gets more severely affected in case of trans-ionospheric communications such as those involving satellites. Circular polarization is, to a great extent, immune to these polarization changes. On the other hand, horizontally polarized waves would not be received at all if its polarization was rotated by 90° and became vertically polarized. Figure 1.83(a) shows a typical helical antenna. Figure 1.83(b) shows a photograph of a representative helical antenna. The ground plane is a wire mesh. The antenna has two operating modes with the one producing a circularly or elliptically polarized broadside pattern with the emitted wave perpendicular to the helical axis and the other producing a circularly polarized end fire pattern with emitted wave along the helical axis. For the first mode, the helix circumference is much smaller than operating wavelength whereas for the second mode, it is approximately equal to operating wavelength. The second mode is the more common of the two.

1.4.6.8 Log Periodic Antenna

The *log periodic antenna* is another broadband VHF and UHF antenna capable of providing enormous bandwidth. It is a driven array and is made in a very large variety of shapes and configurations. One of the most commonly used types is a driven array of dipoles as shown in Figure 1.84(a). The array is driven by a feeder line that is transposed between adjacent elements so that feed to a given element is 180° out of phase with that to the adjacent elements. The lengths of the dipoles and the inter-dipole spacing is governed by the relation $R_1/R_2 = R_2/R_3 = R_3/R_4 \ldots\ldots\ldots\ldots\ldots\ldots = L_1/L_2 = L_2/L_3 = L_3/L_4 \ldots\ldots\ldots\ldots\ldots = k$(constant). Also, the typical values for the convergence angle and constant angle k are 30° and 0.7, respectively. The lowest and highest frequencies of operation are respectively determined by the longest and shortest dipoles. The cut-off frequencies are the ones for which the length is $\lambda/2$. Straight dipoles are usually used for the UHF-band and the dipoles are bent like V-antennas as shown in Figure 1.84(b) for operation in the VHF band.

1.4.6.9 Phased Array Antennas

A *phased array antenna*, or more appropriately a phase steered array antenna, is where the radiated beam (or the axis of the main lobe of the radiated beam) can be steered by feeding the

(a)

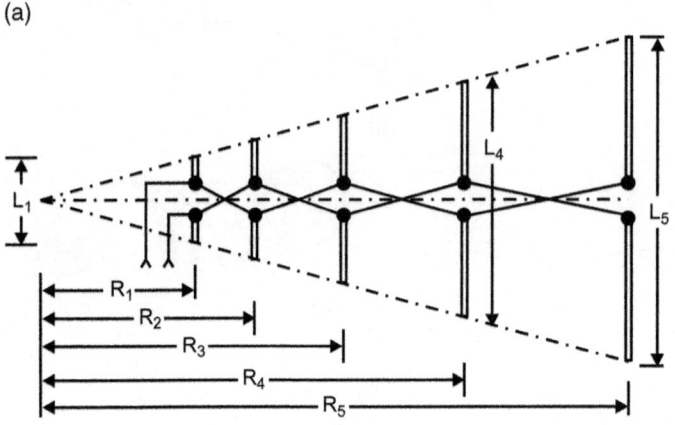

(b)

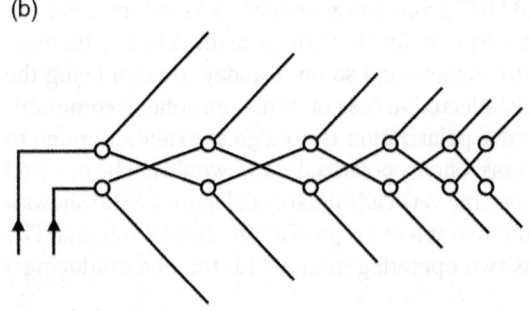

Figure 1.84 (a) Using straight dipoles. (b) Using bent dipoles.

elements of the array with signals having a certain fixed phase difference between adjacent elements of the array during transmission. On reception, they work exactly the same way and instead of splitting the signals among elements, the elemental signals are summed. Receive steering uses the same phase angles as transmit steering from antenna reciprocity principle. Phased array antennas are extensively used in different types of radars including those used for surveillance, tracking, air defence and so on and don't have much relevance to military communications.

1.4.6.10 Microstrip Antennas

A *microstrip* consists of a thin strip sitting on a dielectric that rests on a ground plane. A microstrip when used as a transmission line has a tendency to radiate from irregularities and sharp corners, which indicates that such a component could possibly be used as an antenna. Microstrip antennas radiate efficiently as devices on microstrip printed circuit boards and the microstrip antenna arrays consist of microstrip elements, feed mechanisms, phasing networks and any other microstrip devices. The most commonly used microstrip antenna element is a rectangular element photo-etched from one of the sides of a double-sided printed circuit board with the other side used as a ground plane as shown in Figure 1.85. The element is fed from a coaxial feed. The length L here is the most critical device dimension and is slightly less than half the operating wavelength in the dielectric substrate material. Length L is expressed by $L = \left(0.49\lambda_0 / \sqrt{\varepsilon_r}\right)$. Here ε_r is the relative dielectric strength of printed circuit substrate material. The thickness t is in the order of 0.01λ. The selected value of thickness is based on the desired bandwidth and commercially available thickness. The width W must be less than a wavelength in the dielectric substrate material so that higher-order modes are not excited. However, this is not the constraint if multiple feeds are used to eliminate higher-order modes. Width W decides the input impedance of the antenna element. The expected bandwidth (in MHz) can be computed from $(50 \times f \times t)$ where f is operating frequency in GHz and t is thickness in cm. The input impedance can be computed from $60\lambda_0 / W$ (for $\lambda/2$ element) and $120\lambda_0 / W$ (for $\lambda/4$ element).

1.4.7 Propagation Modes

The subject of electromagnetic wave propagation is of immense importance to all those associated with radio communications, be it two-way radio communications links, point-to-point radio communications or radio broadcasting or be it mobile radio communications, as electromagnetic wave propagation is significantly affected by the media it travels through in terms of the quality of received signal. It is therefore important to know the electromagnetic wave propagation characteristics likely to be encountered. There is a number of different mechanisms by which electromagnetic waves travel through media and the resultant signal may comprise a combination of several signals that have travelled by different paths. These signals may either add constructively or combine destructively, thereby causing an increase in signal strength in some places and complete loss of signal in others. Also, signals travelling via different paths may be delayed causing distortion of the resultant signal. In the following sections, we shall discuss the different regions of the atmosphere that are of great importance to the propagation

Figure 1.85 Microstrip antenna.

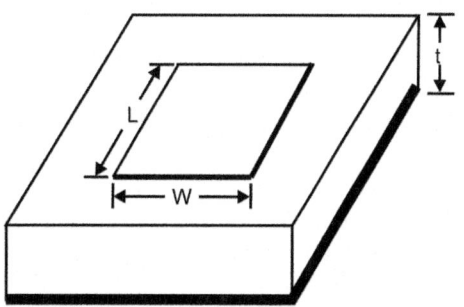

of electromagnetic waves followed by a brief discussion on different modes of electromagnetic wave propagation.

1.4.7.1 Different Regions of the Atmosphere

The different regions of the atmosphere based on their properties may be classified as the *troposphere*, a region extending to altitudes of about 10 km above the surface of Earth; the *stratosphere* extending from 10 to 50 km and the *mesosphere*, located between 50 and 80 km above Earth's surface and ionosphere extending from 60 to about 1000 km. There are other classifications depending upon which properties one is interested in. From the viewpoint of radio propagation, the two main regions of interest are the troposphere and ionosphere.

The *troposphere* is the lowest region of the atmosphere. What we term 'weather' occurs in this region with low clouds occurring at altitudes of up to 2 km, medium clouds occurring at altitudes of up to 4 km and the highest clouds occurring at 10 km or so. Within the troposphere, the temperature steadily falls with height. The troposphere plays an important role in radio wave propagation, particularly in VHF and UHF frequency ranges. Tropospheric propagation refers to the lower atmosphere of the Earth causing bending, scattering and/or reflection of electromagnetic waves, thereby sometimes enhancing their usable communication range by letting them propagate over the horizon, but also compounding interference-related problems. The tendency of electromagnetic waves to bend towards the surface of Earth occurs due to varying index of refraction in the troposphere, which is further due to varying propagation velocity at different altitudes in the troposphere. Different propagation velocities are due to different air density values. The refractive index is the highest near the surface of Earth and decreases with increase in altitude. This produces a tendency for electromagnetic waves to bend towards the surface of Earth. The troposphere scatters electromagnetic waves over a vast range. The effect is more pronounced at UHF and microwave frequencies. An electromagnetic wave beamed slightly above the horizon can get scattered up to several miles. This makes over-the-horizon communications possible. A related effect is that of *ducting*. Under certain specific conditions of the troposphere, when a cool air mass is overlain by a warm air mass, the electromagnetic waves striking the boundary at a near grazing angle from beneath can propagate over hundreds and sometimes to thousands of miles due to waves alternately bouncing off the frontal boundary and the Earth's surface.

The *ionosphere* extends from 60 to 1000 km above the surface of Earth. It is the layer of the Earth's atmosphere that is ionized by solar and cosmic radiation. It is formed by ionization of the Earth's atmosphere by high energy from the Sun and from cosmic rays. The radiation interacts with the molecules to produce free electrons and positive ions. It is found that the level of free electrons varies throughout the ionosphere and, as a result of this, electromagnetic waves are affected more in certain regions than others. These regions are often known as layers, which are given designations D, E and F1, F2.

The *D layer* is the lowest in altitude of all the regions extending from 60 to 90 km. The D layer is present during the day only when radiation is being received from the Sun. After sunset, in the absence of solar radiation to retain ionization levels and because of density of molecules at this altitude, free electrons and ions quickly recombine resulting in the vanishing of the D layer. The D layer has the effect of attenuating radio signals passing through it. The attenuation decreases with increase in frequency. For example, the medium wave broadcast band may not be heard in the regions beyond the ground wave coverage during the day. The same may be heard at further-off distances during night in the absence of the D layer when the signals are reflected from higher layers of ionosphere. The E layer extends from 100 to 125 km. The *E layer* reflects radio signals while they undergo some attenuation. The E layer too significantly reduces in strength after sunset due to recombination of free electrons and positive ions, though some ionization remains. The *F layer* is above both the D and E layers. It is the most important region for long-distance HF communications. During the day it often splits into two regions known as the F1 and F2 layers with the F1 layer being the lower in altitude of the two. At night, the two layers combine to give one layer called the F2 layer. The characteristics of the F layer are significantly affected by the time of day, season and the state of the Sun. Summer and winter time altitudes of F1 and F2 layers are 300 km and 400 km (summer) and 200 km and 300 km (winter). Night time altitude of the F layer may be 250 – 300 km. These are only approximate values. In the case of the F layer too, during night time, the ionization density reduces like in the D and E layers; the process of recombination of free electrons and positive ions that causes reduction in ionization strength is, however, much slower in the case of the F layer due to lower air density at higher altitudes. As a consequence of this, the F layer is able to support night time radio communications.

1.4.7.2   Modes of Propagation

Electromagnetic waves, after being radiated by transmitting antennas, may be divided into various parts. One part travels along the surface of Earth and is known as the *ground wave*. The other part moves upwards and is known as the *sky wave* (Figure 1.86). The ground wave further comprises the *surface wave* and *space wave*. The surface wave travels in contact with the surface of Earth. The space wave travels in the space just above the surface of Earth. It is composed of the *direct wave* and *reflected wave*. To summarize, the ground wave comprises the surface wave, direct wave and reflected wave (Figure 1.87).

Ground waves propagate along the boundary between the Earth's surface and the atmosphere. In the case of a vertical antenna, ground waves leave the antenna in all directions. In the case of horizontal antennas, they leave the antenna mainly from the broad side. Since the

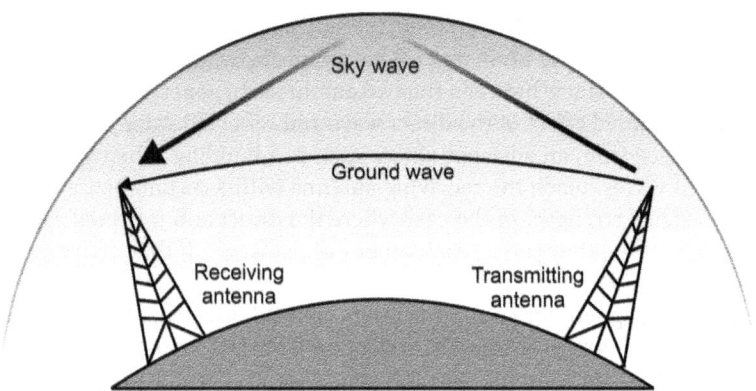

Figure 1.86 Different modes of propagation.

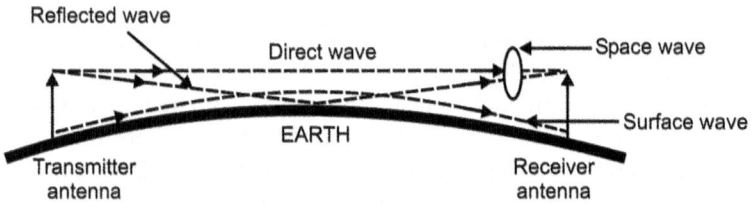

Figure 1.87 Components of a ground wave.

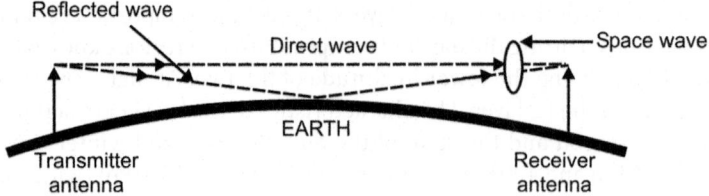

Figure 1.88 Space wave propagation.

ground is a good conductor, ground waves are always vertically polarized as horizontally polarized ground waves would get shorted out by the conductivity of the ground. Ground waves are attenuated as they propagate along Earth's surface as the Earth is not a perfect conductor. The attenuation is more pronounced at higher frequencies, which limits the usefulness of the ground wave propagation mechanism to frequencies to below 3 MHz. It is, however, the preferred propagation type for long-distance communication for frequencies below 3 MHz. Since sea water has higher conductivity, ground waves can propagate over longer distances over sea.

Space waves propagate directly between the transmitting antenna and receiving antenna, which necessitates there is a line-of-sight path between the two (Figure 1.88). The maximum geometric line-of-sight distance between the two antennas depends upon the heights of them and is mathematically expressed by $D \text{ (in km)} = 3.57 \times \left[\sqrt{H_t} + \sqrt{H_r} \right]$ where H_t and H_r, respectively, are the heights of transmitting and receiving antennas in metres. In practice, the radio waves don't travel in straight lines. Instead, due to refractive effects of atmospheric layers, they experience bending towards the surface of Earth, thereby effectively increasing the radio horizon distances by a factor approximately equal to 1.15 (increase of 15%) under normal weather conditions. As a result, this expression gets modified to $D \text{ (in km)} = 4.12 \times \left[\sqrt{H_t} + \sqrt{H_r} \right]$. Space waves also have the characteristic of bouncing off hard objects, which can lead to total blockage of the signal if there is an obstacle between the two antennas. In general, what reaches the receiving antenna is the combined effect of the direct wave and reflected wave. The reflection could be from Earth's surface from an adjacent object such as a building. The two waves, that is the direct and reflected waves, reach the receiving antenna with a certain phase difference leading to reduction in signal strength. In the case where the direct and reflected signals become out of phase with each other, they cancel each other out. However, if they arrive at the receiving antenna in phase, they add up.

Sky waves make use of the ionosphere for communication. Ionospheric propagation allows radio signals to be received at much longer distances. Reflection from the ionosphere is called a 'hop'. The radio signal after reflection from the ionosphere travels back towards Earth and is received by the receiving antenna. The radio signal may bounce off the Earth's surface again and get reflected from the ionosphere again only to be received on Earth's surface at a further distance. There can be several reflections between Earth's surface and ionosphere in what is known as multi-hop transmission (Figure 1.89).

The preferred mode of propagation is determined by the frequencies involved and the distance to be covered. Ground wave propagation, which is mainly dominated by the surface wave, is useful only for short-distance communication. Surface waves attenuate rapidly as they propagate due to Earth being a good conductor and there comes a point where the signal strength of the surface wave becomes too weak to be received and detected. As outlined earlier, higher frequencies suffer greater attenuation. Thus, a higher frequency wave travels considerably smaller distances than a lower frequency wave. Therefore, ground wave propagation is effective for short-distance communication at low frequencies. At higher frequencies, the only

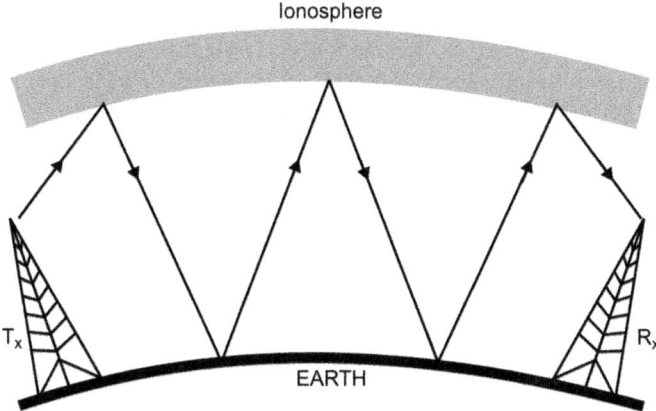

Ionosphere

T_x

R_x

EARTH

Figure 1.89 Sky wave propagation.

alternative is to use space waves. To avoid a cancellation effect, their path difference may get considerably increased by increased heights of transmitting and receiving antennas. For long-distance communication, sky wave propagation is the preferred mode. In this, high-frequency signals get reflected from the ionosphere and reach the Earth. In the vicinity of the point where the waves reach after reflection, the signal is considerably stronger.

1.5 Optical Communication

Communication technology has experienced a continual development to higher and higher carrier frequencies, starting from a few hundred kilohertz in Marconi's time to several hundred terahertz since we employed lasers in fibre systems. The main driving force was that the usable bandwidth and the consequently the transmission capacity increased in direct proportion to carrier frequency. Another asset comes into play in free-space point-to-point links. The minimum divergence obtainable with a freely propagating beam of electromagnetic waves scales proportional to the wavelength. The jump from microwaves to light waves therefore means a reduction in beam width by orders of magnitude, even if we used transmit antennas of much smaller diameter. The reduced beam width does not only imply increased intensity at the receiver site but also reduced cross-talk between closely operating links and less chance for eavesdropping. Space communication, as employed in satellite-to-satellite links, is traditionally performed using microwaves. For more than 25 years, however, laser systems are being investigated as an alternative. One hopes that mass, power consumption and size of an optical transceiver module will be smaller than that of a microwave transceiver. Also, fuel consumption for satellite attitude control when quickly redirecting antennas should be less for optical antennas. On the other hand, a new set of problems would need to be addressed in connection with the extreme requirements for pointing, acquiring and tracking the narrow-width laser beams.

1.5.1 Advantages and Limitations

Optical communication is a communication technology that uses light propagating in the communication medium to transmit data for telecommunications or computer networking. The communication media could be free space, which means air, outer space, vacuum or something similar, water as is the case in underwater communication or an optical transmission line such as optical fibre cable.

Key advantages of using optical communication, fibreoptic/free space or underwater include high achievable data transmission rates, low bit error rates, immunity to electromagnetic interference, full duplex operation, higher communication security and no necessity for a Fresnel zone. Also, the light beam can be very narrow, which makes it hard to intercept.

Key disadvantages include beam dispersion particularly in free space and underwater communication applications, signal attenuation due to atmospheric absorption and adverse weather conditions, scintillation and signal swamping when the Sun goes exactly behind the transmitter. These factors cause an attenuated receiver signal and lead to higher bit error ratio (BER). To overcome these issues, designers have found some solutions such as multi-beam or multipath architectures, which use more than one sender and more than one receiver. Some state-of-the-art devices also have larger fade margin (extra power, reserved for rain, smog, fog). To keep an eye-safe environment, good free-space optical communication systems have a limited laser power density and support laser classes 1 or 1M. Attenuation due to atmospheric conditions, which are exponential in nature, limits practical range of free-space optical communication (FSO) devices to several kilometres. In the following paragraphs two common modes of optical communication are described at length; namely, free-space communication and fibreoptic communication.

1.5.2 Free-Space Communication

Free-space optical communication is a communication technology that makes use of light as the carrier to transmit intelligence. Free space here means air, outer space, vacuum or something similar. The other common form of optical communication in use is optical fibre communication that makes use of an optical transmission line such as optical fibre cable. The technology is useful where physical connections are impractical due to high costs or other considerations. In addition, there is a host of other advantages that come along with use of optical communication. These have already been outlined in the previous paragraph. Figure 1.90 shows the basic block-schematic arrangement of a free-space optical communication link. The diagram is self-explanatory.

Practical free-space point-to-point optical links are usually implemented by using infrared laser light, although low-data-rate communication over short distances is possible using LEDs. Maximum range for terrestrial links is in the order of 2–3 km, but the stability and quality of the link is highly dependent on atmospheric factors such as rain, fog, dust and heat.

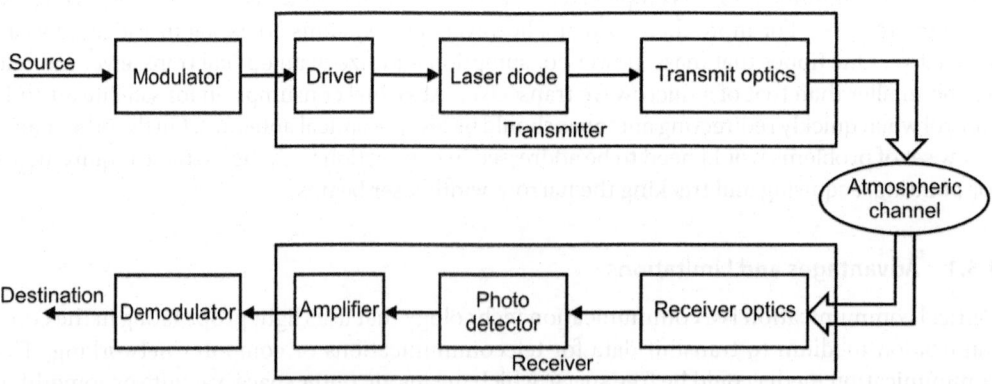

Figure 1.90 Block-schematic arrangement of a free-space optical communication link.

Data transmission rates approaching 1 Gbps have been demonstrated in the case of free-space optical communication, though not for terrestrial links.

Preferred wavelengths for free-space optical communication are 850 and 1550 nm. Operation in 3–5 µm and 8–14 µm bands has also been used due to excellent atmospheric transmission characteristics in them. Selection of optimum wavelength for free-space communication depends upon many factors, which include required transmission distance, eye-safety considerations, availability of components, cost and so on. Recent studies have revealed that operation in the 3–5 µm and 8–14 µm bands does not offer a significant advantage as compared to 850 and 1550 nm bands to counteract scattering losses. Also, availability of sources and detectors is limited in mid-infrared and far-infrared bands. Another advantage of 1550 nm comes in its eye-safety feature. Regulatory agencies allow approximately 100 times higher laser power levels for 1550 nm compared to 850 nm. In general, choice of a specific wavelength is not so important as long as it is not strongly absorbed in the atmosphere.

In outer space, the communication range of free-space optical communication is currently in the order of several thousand kilometres, but can be extended to cover interplanetary distances of millions of kilometres using optical telescopes as beam expanders. Use of optical communication technology involving detection and emission of laser light by space probes has been done several times in the past. A two-way distance record for communication was set by the Mercury laser altimeter instrument aboard the *Messenger* (an acronym for MErcury Surface, Space ENvironment, GEochemistry and Ranging) spacecraft. This infrared diode-pumped neodymium laser, designed as a laser altimeter for a Mercury orbit mission and known as Mercury Laser Altimeter (MLA), set a record of two-way communication across a distance of 24 million km, as the craft neared Earth on a fly-by in May, 2005.

The space-based free-space optical communication concept has been around for many years and particularly in the last few years, significant advances have been made for the concept to fructify in both civilian and government funded non-classified and classified applications. The primary market of free-space optical communication today is that of Inter-Satellite Links (ISL). There is scope in providing space-earth optical communication links in spite of the discouraging atmosphere related issues. For example, lot of R&D activity is known to be funded to develop a satellite to submarine optical communication link, which mainly interests military strategists. In such a link of course, problems are encountered not only from atmospheric issues but also from issues concerning propagation of lasers through turbulent waters of the oceans.

Coming back to the primary application of free-space optical communication links in today's world, that is, inter-satellite links (ISLs), it may be mentioned here that inter-satellite communications are mainly used for networking of a satellite constellation. The involved data rates could vary from hundreds of Mbps to several Gbps. ISLs are in use for all types of satellite orbits including low earth orbits (LEO), medium earth orbits (MEO), geosynchronous earth orbits (GEO) and even highly elliptical orbits (HEO). Though there are currently operational satellite constellations employing RF inter-satellite links, examples being the Iridium satellite system and NASA's TDRSS (Tracking and Data Relay Satellite System), the future definitely belongs to optical ISLs. This is supported by the fact that most of the commercial satellite constellations being announced now will use optical ISLs. The SILEX optical communication system is another example. SILEX payload embarked on the European Space Agency's *Artemis* (Advanced Relay and Technology Mission Satellite) spacecraft and also on the French Earth observation satellite SPOT-4. It uses GaAlAs laser diodes as the source and is used to transmit data at 50 Mbps from a low earth orbit to geostationary orbit.

The TSAT (Transformational Satellite System) is yet another example of a contemporary satellite constellation employing laser intersatellite links. The system is designed to provide a

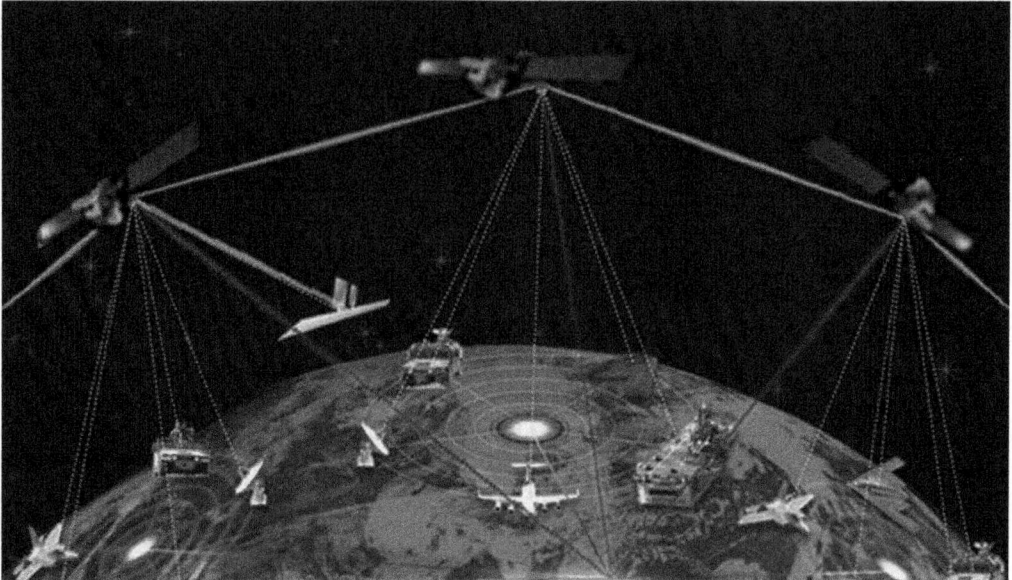

Figure 1.91 The TSAT satellite constellation.

protected, secure Internet-like communication system that integrates space, air, ground and sea networks. The TSAT programme is composed of three segments and a systems engineering and integration function. The Space Segment will consist of five satellites in geosynchronous orbits interconnected by high-data rate laser cross links. TSAT will use internet-like technology to connect war fighters all over the world in a global information network with unprecedented carrying capacity, accessibility, reliability and immunity to jamming, eavesdropping and nuclear effects. It is the backbone of twenty-first-century net-centric warfare and is projected to revolutionize military communications. Figure 1.91 shows a photograph of the TSAT satellite constellation.

1.5.3 Fibreoptic Communication

Fibreoptic communication is also a form of optical communication and it differs from the free-space communication in respect to transmission medium, which in this case is a fibreoptic cable rather than free-space. While the advantages of optical communication discussed in the previous paragraphs in the case of free-space communication equally hold good for fibreoptic communication, some of the key limitations of free-space optical communication encountered due to atmospheric propagation issues are overcome in fibreoptic communication. In practice, laser-based communication today is dominated by fibreoptic transmission. Earlier, the life times of semiconductor diode lasers and the fibre losses were too high to make laser-based fibreoptic communication an attractive alternative to other forms of communication. With advances in both semiconductor diode laser and fibre technologies, these shortcomings have been overcome. State-of-the-art semiconductor diode lasers have life times of greater than 10^7 hours and fibre loss is as small as a small fraction of dB/km. Today, fibreoptic communication links are a reality for both intra-city and trunk telephone lines, video data links and computer-to-computer communications.

The semiconductor diode laser is the natural choice for fibreoptic communication as these are suitably small and have a configuration for efficient coupling into the small-diameter core

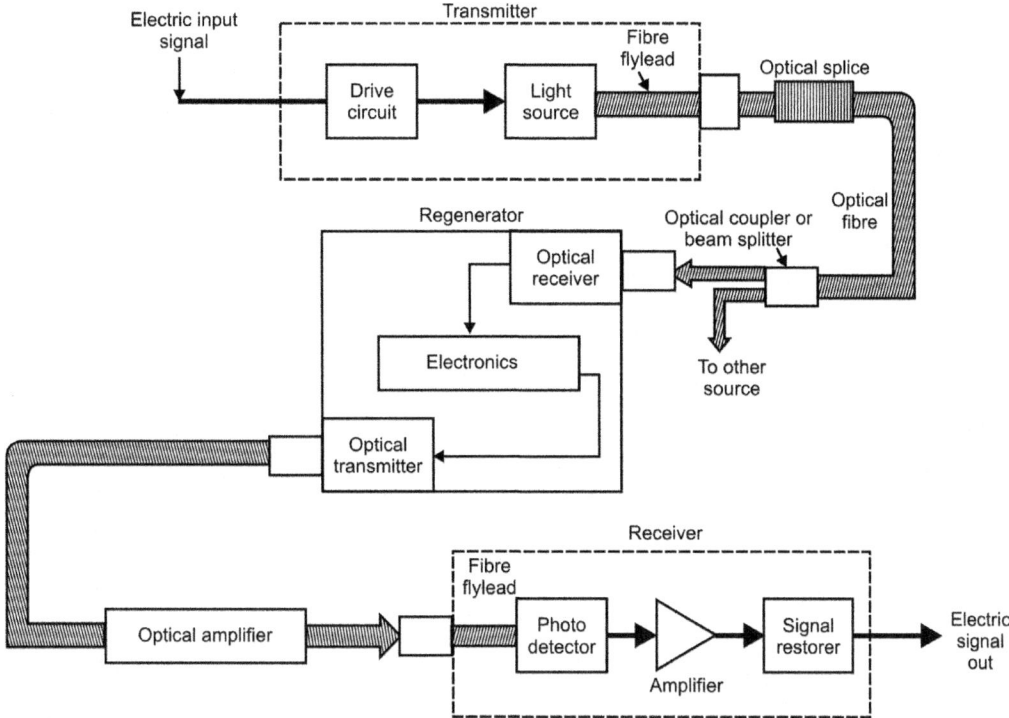

Figure 1.92 Basic block-schematic of a fibreoptic communication link.

of an optical fibre cable. Semiconductor diode lasers operating at CW power levels of a few milliwatts are suitable for fibreoptic communication. These lasers can be easily modulated by drive current modulation up to frequencies in the gigahertz range.

Laser wavelengths in use for fibreoptic communication are 0.85, 1.3 and 1.55 μm. The first practical fibreoptic communication systems employed 0.85 μm as it matched the available AlGaAs lasers. With advances in fibre technology and the opening up of lower loss windows first at 1.3 μm and then at 1.55 μm, these became the preferred wavelengths for long-distance, high performance systems. Corresponding semiconductor diode laser and detector types for these operational wavelengths are AlGaAs and silicon for 0.85 μm, InGaAsP and InGaAs or Germanium for 1.3 μm and InGaAsP and InGaAs for 1.55 μm. Typical fibre losses at these wavelengths are 2 dB/km at 0.85 μm, 0.5 dB/km at 1.3 μm and 0.2 dB at 1.55 μm.

Figure 1.92 shows a typical fibreoptic communication link. The laser is pulse code modulated with intelligence to be transmitted through drive current modulation and is coupled into the fibre. On the receiver side, laser light is detected and the intelligence signal is recovered. Optical amplifiers are used to reinforce the signal strength every few km of fibre cable length to counter signal degradation caused by various loss mechanisms such as absorption, scattering and modal and chromatic dispersion. To summarize, the process of fibreoptic communication involves generating the optical signal modulated with intelligence to be transmitted using a transmitter, relaying the signal along the fibre making sure that the signal does not become too distorted or weak, receiving the optical signal and converting it back into an electrical signal representing the original intelligence signal.

In terms of choice of different components for fibreoptic communication, preferred types of laser sources and detectors for the commonly used wavelength bands were briefly discussed in an earlier paragraph. To summarize, fibres with loss figures of 2 dB/km at 0.85 μm, 0.5 dB/km at 1.3 μm and 0.2 dB/km at 1.55 μm are available. Also, fibres with bandwidth-distance products approaching 3000 MHz/km are available. Semiconductor diode lasers with 10 mW power levels and modulation rates approaching 10 GHz with a life time of 10^7 hours are also available. LEDs providing 0.1 mW into the fibre with a modulation rate of 200 MHz and life time in the order of 10^6–10^7 are also available. PIN type photo diodes with responsivity and NEP figures of 0.5 A/W and 10^{-12} W/√Hz, respectively, are used. Also, more sensitive APDs with responsivity and NEP figures of 80–100 A/W and 10^{-14} W/√Hz are commercially available for the purpose. Couplers and splices with insertion loss in the range of 0.1–0.5 dB are available for use. Splices are usually added to the link to allow fibre cable repair in case of need.

We have seen four generations of fibreoptic communication and currently we are in the fifth generation. This first-generation system operated at 0.85 μm wavelength at a bit rate of 45 Mbps and repeater spacing of up to 10 km. In April 1977, General Telephone and Electronics sent the first live telephone traffic using fibreoptics at a 6 Mbps data rate in Long Beach, California. The second generation of fibreoptic communication operated at 1.3 μm and used InGaAsP semiconductor lasers. These fibreoptic systems were initially limited by dispersion of multimode fibres. The advent of single-mode fibres in 1981 significantly improved system performance. Third-generation fibreoptic systems operated at 1.55 μm. The difficulty faced earlier in terms of pulse spreading at 1.55 μm was largely overcome by using dispersion-shifted fibres designed to have minimal dispersion at 1.55 μm or by limiting the laser spectrum to a single longitudinal mode. The features of third-generation systems allowed commercial fibreoptic systems to operate at 2.5 Gbit/s with repeater spacing in excess of 100 km. The fourth generation of fibreoptic communication systems used optical amplification to reduce the need for repeaters. It also explored use of wavelength-division multiplexing to increase data capacity. These features brought revolutionary improvements to the performance of fibreoptic systems. In fact, since 1992, the data rate has doubled every six months until it reached a figure of 10 Tbps after 2000. In 2006, a bit rate of 14 Tbps was reached over a single 160 km line using optical amplifiers.

In the fifth generation of fibreoptic communications systems, the focus is on extending the wavelength range over which a WDM system can operate. The conventional wavelength window, known as the C band, covers the wavelength range 1.53–1.57 μm. The dry fibre promises an extension of that range to 1.30–1.65 μm. Other features of fifth generation fibreoptic systems include the concept of optical solitons, which involves use of pulses of a specific shape that helps them preserve their shape by counteracting the effects of dispersion with the nonlinear effects of the fibre.

1.6 Software-Defined Radio

There are diverse areas in which communications devices and systems are put to use by a large cross-section of users in civilian and military sectors. Also there are many ways by which one would like communicate, which principally include voice, video and data communications. Others include broadcast messaging, command and control communications, emergency response communications and so on. Conventionally, it wouldn't be feasible for a generic hardware platform to fit all these application scenarios. Software-Defined Radio (SDR) allows a common platform to be used across a number of areas. The basic concept of the SDR is that the functions to be performed by the radio can be totally configured or defined by the software. In addition to defining radio configuration, there is also the possibility that it can be reconfigured in case its scope of operation is changed or it is to be employed for another role or as the existing standards get the upgrades. The concept of SDR is equally applicable to both military as well as commercial sectors. Joint Tactical Radio System (JTRS), a radio intended for military applications and briefly discussed in an earlier section in the chapter, has been a major initiative in the military domain. JTRS allowed the use of a single hardware platform to communicate in different application scenarios by simply reloading or reconfiguring the software required for the intended application. One of the common applications applicable to the commercial world is the ease with which frequently occurring upgrades of standards can be incorporated, such as at cellular base stations by using a generic platform. These changes such as those from *Universal Mobile Telecommunications System* (UMTS), a third-generation mobile cellular system based on the GSM (*Global System for Mobile*) communication standard to *High-Speed Packet Access* (HSPA) and on to *Long Term Evolution* (LTE) technology, a standard used for wireless transmission of high-speed data for mobile phones and data terminals, could be incorporated simply by uploading new software and reconfiguring it without any hardware changes irrespective of the fact that different operating frequencies and modulation schemes are used.

There are two main categories of radio using software, which include the software controlled radio and SDR. The SDR Forum, working in collaboration with the Institute of Electrical and Electronic Engineers (IEEE) P1900.1 group defines the two categories of radio as follows. *Software controlled radio* is the one in which software is used to control some or all of the physical layer functions of the radio that are fixed within the radio. That is, the functions performed by the radio are not software defined or reconfigurable; instead the software is used to only control the predefined functions. In the case of software-defined radio, some or the entire physical layer functions are software-defined. Software in this case can be used to alter the specifications and functions of the radio. The SDR is configured around a generic hardware platform comprising digital signal processors as well as general-purpose processors to implement transmit and receive radio functions. The hardware platform is used to operate in different application scenarios and offer a range of performance specifications as defined by the software. The software is used to reconfigure the hardware platform for various applications.

1.6.1 Different Tiers of SDR

Different tiers of SDR in essence define the different levels of SDR that may exist. It may be mentioned here that it may not always be practically feasible for an SDR to have all possible features in one radio. Different tiers describe the level of software definability of the radio in terms of what is configurable and what is not. Different tiers include Tier 0, Tier 1, Tier 2, Tier 3 and Tier 4. Each one of them is briefly described in the following paragraph.

Tier 0 is the level assigned to anon-configurable hardware radio. That is, no changes can be made to the radio by software. Tier 1 is a software controlled radio where limited functions are controllable. In the case of Tier 1 SDR, control functionality is implemented in software, but change of attributes such as type of modulation and operating frequency band cannot be implemented without changing hardware. These may be power levels, interconnections and so on, but not mode or frequency. Tier 2 SDR is more of a software controlled radio (SCR) though there is significant proportion of the radio that is software configurable. Tier 2 SDR is capable of covering a wide frequency range and executes software to provide variety of modulation techniques, wideband or narrow band operation, and communications security functions. The RF front-end still remains hardware based and nonreconfigurable. Tier 3 SDR is the ideal SDR. It possesses all of capabilities of software-defined radio, but eliminates analogue amplification and heterodyne mixing prior to A/D conversion and after D/A conversion. It could be said to have full programmability. Tier 4 is the ultimate software radio (USR) and is a stage further on from the Ideal Software Radio (ISR). Not only does this form of software-defined radio have full programmability, but it is also able to support a broad range of functions and frequencies at the same time. It is the ideal software-defined radio on a chip that requires no external antenna and doesn't have any frequency restrictions. It can perform a wide range of adaptive services.

1.6.2 Advantages of SDR

The key advantages of SDR originate mainly from its capability to reconfigure its common hardware platform using software and the waveform portability.

1) The concept of SDR allows implementation of a family of radios using a common hardware architecture, thereby facilitating quick introduction of many a new product into the market. Software re-usage across a range of radio products reduces costs and remote re-programming enables fixing bugs in radios while they are in service. For the service providers, the use of a common hardware platform for multiple applications significantly reduces logistical support and operating expenditures. Remote software downloads can be used to increase capacity.

2) SDR waveform portability when ensured by incorporating certain steps at the early stages of design lends interoperability. In addition to the capability of an SDR to reconfigure itself, another major advantage is that of waveform portability. Capability of reusing waveforms for various applications leads to huge savings in cost. It also mitigates obsolescence as the existing waveforms get transferred to newer platforms with the development of technology.

1.6.3 SDR Hardware Architecture

There are different tiers or levels of SDR, as outlined previously. The complexity of the hardware platform and associated control and management software to an extent varies with the level the radio belongs to. Irrespective of the hardware platform and software complexity, there are certain basic functional blocks that are present in the architecture of SDR. Figure 1.93 shows the block-schematic arrangement of an ideal SDR highlighting the basic functional blocks.

The first basic functional block is that of *RF amplification*, which includes power amplification of the signal present at the output of digital-to-analogue converter (DAC) on the transmit side and low noise amplification of received antenna signal on the receive side. A power amplifier raises the level of the RF signal to the required power suitable for transmission and

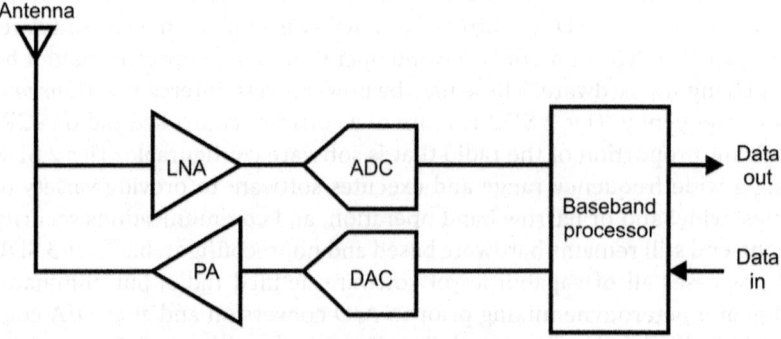

Figure 1.93 Block-schematic arrangement of an ideal SDR.

the low noise amplifier amplifies the received antenna signal before it is further passed onto the analogue-to-converter (ADC). An ideal SDR digitally codes and modulates the data that's going to be communicated in a baseband processor before transmitting it. Also, the digital output of the ADC is processed in a baseband processor to recover the originally transmitted signal. In the case of an ideal SDR, the antenna connects directly to the low noise amplifier (LNA) and the ADC or the power amplifier and the DAC. It is the baseband processor that handles all radio functions. In many designs, some analogue processing may be required, which would typically involve converting the signal to and from the final radio frequency. Some intermediate frequency processing may also be present.

A digital-to-analogue converter (DAC) on the transmit side and analogue-to-digital converter (ADC) on the receive side constitute *digital conversion* building block. It is at this stage that the signal is converted between the digital and analogue formats. While on transmit side digital conversion, the maximum frequency and the required power level are the some of the key issues to be addressed, on the receive side, the maximum frequency and number of bits to give the required quantization noise are of great importance.

The *baseband processor* is at the very centre of SDR. It performs all radio functions including filtering, up/down conversion, modulation/demodulation and digital baseband. One of the key issues of the baseband processor is the amount of processing power required. The greater the processing power, higher would be the current consumption, which in turn requires additional cooling. This is particularly important if size were a limitation.

Most wireless activity is above the VHF and UHF frequency band and well into the microwave region. Though it may be feasible to realize an ideal SDR at relatively lower frequencies, it is not so at frequencies normally encountered in the case of practical SDRs. As a consequence of this, modern SDRs need to use mixers in the front-end to perform analogue up-conversion and down conversion. Figure 1.94 shows the block-schematic arrangement of a practical SDR.

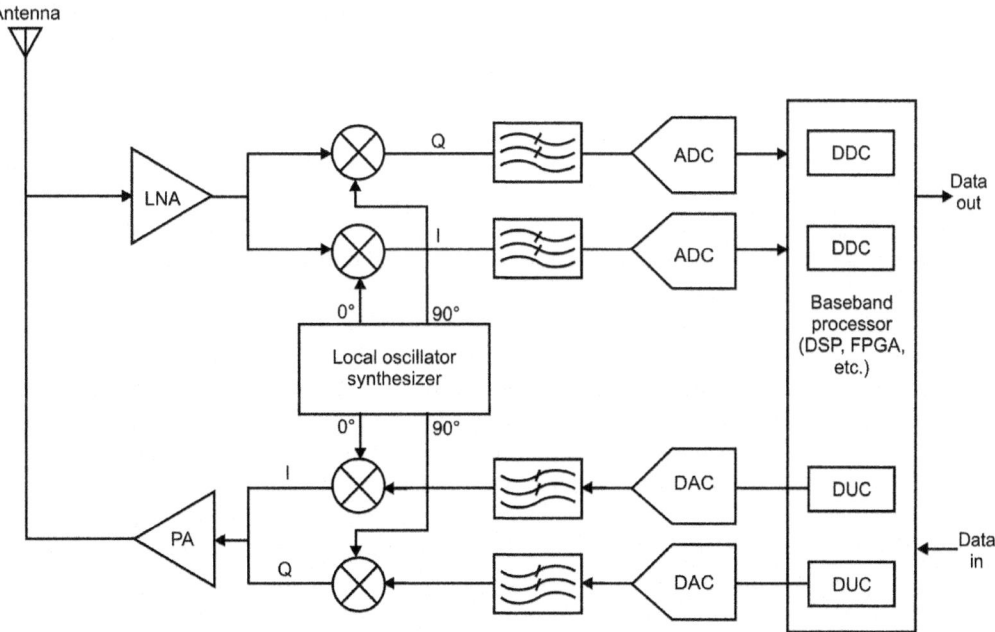

Figure 1.94 Block-schematic arrangement of a practical, modern SDR.

A mixer following the low noise amplifier in the receive channel downconverts the received RF signal to an intermediate frequency (IF) that can be handled by today's ADCs. In the transmission channel, an up-convert mixer up-converts the DAC signal to the final transmission frequency before it is fed to the power amplifier for subsequent transmission. As is evident from the block diagram, I/Q mixer format has been used to preserve phase and frequency information contained in most digital modulation schemes.

Digital Up Conversion (DUC) is employed in the transmission channel to boost the lower set of frequencies to an intermediate frequency that is closer to transmit frequency. Digital Down Conversion (DDC) is commonly used in the receive channel after the ADC to further lower the data rate so that memory requirements may be relaxed and processing speeds are more moderate. Ideally, we would have liked to have ADC and DAC to be fast enough to eliminate these conversions so that all baseband processing is performed digitally. But that is not the case as yet. Though these devices come as individual ICs, their functions also can be implemented in the baseband processor. The baseband processors may be fast standard processors like those found in PCs, laptops, programmable DSPs or FPGAs. Modern SDRs typically use both a DSP and an FPGA, with the processing duties divided up as appropriate to the capabilities of each.

1.6.4 SDR Security

While the emergence of SDRs has added many new features such as re-configurability, waveform portability and interoperability to both military and commercial radios, it has also triggered new sets of security problems not encountered by conventional radios. It is important to understand security requirements of a communication network in general before we look at the security risks that affect these networks in an adverse manner. SDRs and cognitive radios should have capabilities to address various security risks that can undermine the following security requirements. Different security requirements that are applicable to almost all communication networks including SDRs and cognitive radios encompass a *controlled access to resources* allowing one to access information or resources only when authorized for the same; robustness to provide the specified communication services; *confidentiality and integrity* that ensures confidentiality of stored and communicated data while guaranteeing integrity of a system and also that of stored and communicated data; *compliance to a regulatory framework*; *accountability or non-repudiation* implying that the system entities own the responsibility for any of their performed actions and system capability to establish and *verifying the claimed identity* of any player in the network.

Different types of security threats are briefly described as follows. These threats are broadly classified as threats that are common to both conventional radios and SDRs and those that are specific to SDRs. *Insertion of malicious software* is a type of threat similar to mobile malware and relates to introduction of malicious software on an SDR. *Alteration or destruction of configuration data* is a threat that identifies the alteration or destruction of configuration data leading to corruption or removal of this data from the SDR platform and needed by it to perform its intended functions. *Artificial consumption of resources* is a type of threat that identifies the abnormal increase in processing or memory resources of the SDR platform and may be caused by various things such as insertion of malicious software, alteration or destruction of configuration data or even physical failure. *Alteration or destruction of waveform code* is a type of threat that relates to alteration or destruction of the waveform code required to support a radio access technology or air interface, thereby affecting one or more radio waveforms but not the SDR itself. *Alteration or destruction of a real-time operating system* alters or destroys components or the real-time operating system (RTOS). Alteration or destruction of the software framework is a threat relating to alteration or destruction of elements of the software framework and middleware, which support the waveforms and applications. *Alteration or destruction of user data* relates to alteration or destruction of user data. *Software failure* relates to failure of any of the components of the real-time operating system, the software framework, waveforms or applications. *Hardware failure* is a type of threat identifying a generic hardware failure in the hardware platform. *Extraction of configuration data, waveform data* and *user data* are eavesdropping threats where the attacker collects configuration data, waveform data and user data, respectively, which can be used in subsequent attacks. Yet another type of security threat relates to *download and activation of a malicious software waveform* on the SDR platform. *Unauthorized use of SDR services* is a security breach, where a waveform or application can access or use services of the SDR platform without proper access level. *Data repudiation* or rejection relates to the possibility of repudiating or rejecting access or provision of data and services.

1.7 Network-Centric Warfare

Network-centric warfare (NCW) is a military doctrine that uses networking of sensors, planners and decision makers and shooting platforms to create enhanced shared awareness of the battle space. Shared awareness of battle space for the armed forces increases synergy for command and control, which leads to superior decision making and an ability to coordinate complex military operations spread over a wide geographical coverage to give an overwhelming military advantage. Coordinated efforts and synchronized operations result in greater inflicted lethality on the adversary forces and increased survivability of own forces.

There are slight variations in how network-centric warfare is known and interpreted in different countries. The United States uses the term *Network-Centric Warfare* (NCW), which is characterized by the ability of geographically dispersed forces to create a high level of shared battle space awareness that can be exploited through various network-centric operations to execute commander's intent. In the United Kingdom, the term *Networked Enabled Capability* (NEC) is used. The NEC concept is based on use of three elements namely *sensors* to gather information, *network* to fuse and disseminate sensor data and *strike assets* to deliver the military effect. In Australia, the term *Network Enabled Operations* (NEO) was initially used. The concept was derived its power from effectively linking different elements to conduct military operations. Under the NEO concept, platforms are treated as nodes of a network. They all collect, share and access information, which is used to create a common, real-time battlespace picture. This allows a greater level of situational awareness, coordination and offensive potential than is currently the case. The use of the term NEO has been discontinued and the US terminology of NCW has been adopted instead.

1.7.1 OODA Loop

In order to understand the role of different components of NCW in improving the overall efficacy of military engagement, we must first understand the complete sequence of events that must take place in a military engagement. This is best described by the US Air Force strategist John Boyd's *OODA* (*Observation-Orientation-Decision-Action*) loop. The OODA loop is fundamental to all military operations at both tactical and strategic levels as the adversary must be observed to gather information, the attacker must orient himself as per the situation or context, take decision and then act accordingly. With everything else being equal, the player with faster OODA loop would have a distinctive edge over his opponent by preempting or blocking any move that the opponent with relatively sluggish OODA loop intends to execute.

The OODA loop as outlined here has four components. It is imperative that all four components are accelerated to achieve an overall higher operational tempo. While the first three of the four components relate to information processing and therefore largely governed by networking capabilities, the fourth component is the kinetic component of the loop and is associated with movement and application of fire power. The first three components of observation, orientation and decision are all about gathering and distributing information, analysing it and then deciding how to act upon it. Networking accelerates the observation and orientation phases thereby facilitating the decision phase.

However, efficient networking capability only is not the solution to all problems. Overall combat effectiveness is linked to the kinetic phase of the OODA loop. NCW scholars have tried to use the well-established laws such as the Metcalfe's law and Amdahl's law from commercial domains to military operations. While Metcalfe's law presents a best possible scenario for distribution of information gathered by platform sensors by stating that utility of a network increases as a function of square of number of nodes or platforms in the network, Amdahl's law explains the limits to the gains achieved in networked systems arising from decision and action phases of the OODA loop. According to Amdahl's law, increasing the number of assets in the system increases the achieved effect at best only by the number of assets added as the real improvement is constrained by queuing effects observed in positioning of the assets or platforms that perform engagements. Another important aspect of NCW operations is that the networking does permit a significant improvement in operational tempo in all those cases where the targeting information is lacking such as in close air support operations against highly mobile ground targets. On the contrary, networking would have very little impact in cases where the targeting information is well-known or where the operation is constrained by number of platforms in use.

1.7.2 Advantages and Shortcomings

The salient features of the network-centric warfare doctrine as outlined in the previous paragraphs highlight the advantages it brings to the modern battlefield. To summarize, key advantages expected from application of the NCW doctrine to military operations include the following.

1) The NCW doctrine with its networking capabilities allows deployment of relatively smaller size units with fewer platforms and supplies without the need for a tight formation. This enables accomplishment of a given mission with greater efficacy and lower cost. Also, networking allows use of new tactics. It gives to the troops far greater battle space awareness and ability to track movements of fellow soldiers when they are spread over a large coverage area in small independent units. Swarm tactics involving the use of a decentralized force against the adversary with focus on mobility, communication, coordination and unit autonomy used by the US Armed Forces during Operation Iraqi Freedom is an example. Another example of use of networking and communications capabilities has been during the US Operation Enduring Freedom in Afghanistan when Special Forces on the ground could execute a coordinated operation between them and the air support through use of data and voice links provided by communications satellites thereby enabling attack aircraft destroy targets laser designated by ground forces.

2) The networking capability of NCW allows individual units benefit from the experiences of experts located far away from war scene in case situation demands.

3) The NCW doctrine exploits the shared battle space awareness of the Armed Forces to achieve strategicand tactical objectives through close coordination not only in a specific theatre of operations, but also of dispersed forces on a global level.

4) Implementation of the NCW doctrine significantly reduces sensor-to-shooter time by allowing soldiers in the field conduct an on-site analysis of raw intelligence available from sensors on platforms to facilitate quick action.

Though NCW is essential to modern-day military operations, it has issues that must be addressed to take full advantage of this military doctrine. Technological innovations come with both pros and cons. Implementation of new concepts in warfare can create a new set of vulnerabilities an adversary can seek to exploit. NCW is no exception. Some of the key problem areas include the following.

1) NCW's heavy reliance on technology and infrastructure runs the risk of crippling military operations in the event of failure of technology or incapacitation of infrastructure through attack by adversary; should there be no alternative warfare strategy in a non-network-centric operational scenario. Widespread use of GPS jammers capable of blocking GPS signals that are central to navigation and precision guidance capabilities strengthen the point.
2) Another flaw in NCW, if one may say so, is the premise that machine intelligence and analysis is superior and therefore can be used to replace the soldier in the loop, though there is no viable proof to substantiate the claim. There are numerous instances of massive communication, information, security and processing failures of commercial computer networks. While such failures to a certain degree may be acceptable in case of commercial networks, the same would not be true for networks dictating military operations. Information and networking alone are therefore no substitutes for combat manoeuvre and the massing of Armed Forces. NCW is akin to a chess game where situational awareness alone is not enough to win you the game, making a move by instead anticipating the enemy's next move is the key.

1.8   C4ISR

C4ISR is an acronym for *Command, Control, Communications, Computers, Intelligence, Surveillance and Reconnaissance.* The C4ISR concept encompasses all systems, technologies and procedures that are used to gather and disseminate information. C4ISR systems allow military commanders to understand their operational environment, identify mission-critical factors and control their assets. It is a highly multi-disciplinary concept involving application of command and control, communications, intelligence, surveillance and reconnaissance systems and technologies. Command and control functions involve planning, directing, coordinating and controlling operations for accomplishment of intended mission. These functions are performed through an arrangement of personnel, equipment, communications and facilities. Creation of situational awareness is an essential element of command and control systems, which they provide by fusing data streams from multiple sensors and feeding the decision support software. Other software components as applied to C4ISR systems include programmes and algorithms used to ensure interoperability among disparate communication systems, encryption algorithms to ensure security of communications, communications networking protocols and inertial navigation. Threat warning systems and electronic countermeasures such as jamming techniques and counter-countermeasures such as decoys are also included in the command and control domain. Communications and computing technologies process and transport information and are the enabling technologies that support command and control (C2) and intelligence, surveillance and reconnaissance (ISR) functions. ISR functions involve use of remote optical, infrared and radar sensors placed on platforms such as satellites and unmanned vehicles.

1.8.1 Command and Control

Command and control, communications and information infrastructure, data fusion and information management systems, radar, infrared and optical sensors, electronic warfare, cyber security and space-based surveillance constitute important elements of C4ISR. Command and control systems give the Armed Forces the decisive advantage by enabling concurrent and not serial planning and decision making, thereby keeping Armed Forces ahead of their adversaries in accelerated operational environments. The *Theatre Battle Management Core System* (TBMCS) developed by Lockheed–Martin is one such command control system. It is the primary system for planning and executing the joint air campaign, coordinating and directing flying operations from a wide range of airborne platforms such as F-16 fighters, refuelling tankers, helicopters, unmanned aerial vehicles and even cruise missiles. TOPLITE from Rafael, a highly stabilized, multi-role, multi-sensor optronic payload, is a day/night surveillance and targeting system configured for naval, air and ground surveillance and targeting systems. It is designed for a wide range of missions, from law enforcement observation through surveying and fire-control to missile targeting.

1.8.2 Communications

Next-generation communication networks need to be secure, resilient and adaptable. The Universal Communications Platform (UCP) from Lockheed–Martin is one such system. The UCP integrates any radio communications systems using existing communications infrastructure to form a cost-effective, dedicated network for complete communications including voice, video and data. The Warfighter Information Network–Tactical (WIN-T) by the same company is a tactical tele-communications system for the army. The WIN-T network provides mobile, secure and seamless C4ISR capabilities capable of supporting multimedia tactical information systems. C4I-CONNECT from Rafael, a voice communication system developed for Air Force is another example. The SEA-COM system again from Rafael is a state-of-the-art communications suite for naval platforms.

1.8.3 Intelligence, Surveillance and Reconnaissance

ISR systems, as outlined earlier, use thousands of sensors housed on manned and unmanned land-based, airborne, sea-based and space-based military platforms. Data from sensors is processed, analysed, fused to generate and disseminate mission-critical information. *P-3 Orion* from Lockheed–Martin is one such ISR platform. It is a premier multi-mission maritime long-endurance aircraft that performs air, surface and subsurface patrol and reconnaissance tasks over extended periods. SEA SPOTTER, infrared Stabilized Stare and the Track (IRS^2T) system, developed by Rafael, enable situational awareness and the automatic, passive detection of sea skimming missiles, air-to-surface smart bombs, fast strike aircraft, helicopters, marine vessels and rubber boats.

1.8.4 Cyber Security and EW Systems

Sophisticated *cyber solutions* are used to defend networks and systems from advanced persistent threats. Electronic warfare systems include *Naval EW systems* are used to intercept signals that identify both imminent and potential threats and even protect vessels from anti-ship missile attacks, *Ground EW Systems* such as the *Symphony Counter – Improvised Explosive Device (C – IED) Defeat* system and *Airborne EW Systems* such as the *Electronic Support Measures (ESM), Radar Warning Receivers (RWR), COMINT (Communications Intelligence)* and *ELINT (Electronic Intelligence)* systems.

C-PEARL from Rafael is a compact, lightweight state-of-the-art ESM system enabling automatic detection and identification of threats in complex electromagnetic environments. C-PEARL has military, naval and homeland security applications.

1.9 Representative Military Communications Equipment

In the following paragraphs, we shall briefly discuss salient features and capabilities of some representative devices and systems under the categories of smart phones, radios including ground mobile radios, SDRs and cognitive radios and C4ISR systems including command and control, communications and ISR systems. Representative EW systems are covered in Chapter 6 on *Electronic Warfare*.

1.9.1 Smart Phones

Smart phones and tablets are becoming increasingly popular with the Armed Forces, particularly as the associated security concerns have been addressed to a large extent. An increasing number of smart phones certified for military use are now commercially available. Some common types include *LG G5* using Android OS version 6.0.1 and *LG V10* using Android OS version 5.1.1, Motorola's *AME 1000 secure mobile telephony system* configured around *ES400 Enterprise smart phone*, *Solarin* from Sirin Labs and the *Boeing Black Smartphone*.

The United States' *National Information Assurance Partnership* (NIAP) for compliance in meeting international security standards in corporate environments has certified model G5 and V10 from LG Electronics for military use. It may be mentioned here that NIAP employs a Common Criteria Evaluation and Validation Scheme for certification for security conformance and is recognized by the governments of 25 Common Criteria member countries including, Canada, France, Germany, India, Japan, South Korea and the UK. LG V5 and G10 smart phones feature LG's *GATE (Guarded Access to Enterprise)* technology for enhanced platform, network, and application security.

The LG G5 employs GSM/HSPA/LTE/CDMA networking technology and is powered by Qualcomm MSM8996 Snapdragon 820 octa-core chip set that runs on Android OS, v6.0.1 (Marshmallow). It comes with a 5.3-inch 1440×2560 pixel touch screen display, 4 GB of RAM and 32 GB of internal storage. Solarin packs a 23.8-megapixel primary camera on the rear and an 8-megapixel front camera. Connectivity options include WiFi, GPS, Bluetooth, NFC, radio, an infrared port and USB. The phone includes fingerprint technology, accelerometer, gyro, proximity, compass, barometer and colour spectrum sensors. The LG V10 employs GSM/HSPA/CDMA/EVDO/LTE networking technology and is powered by a Qualcomm MSM8992 Snapdragon 808Hexacore chip set that runs on Android OS, v5.1.1 (Lollipop), upgradable to v6.0 (Marshmallow). It comes with a 5.7-inch 1440×2560 pixel touchscreen display, 4 GB of RAM and 32/64 GB of internal storage. It has a 16-megapixel primary camera and 5-megapixel duo secondary camera. Connectivity options include WiFi, GPS, Bluetooth, NFC, radio, infrared port and USB. The phone includes fingerprint technology, accelerometer, gyro, proximity, compass, barometer and colour spectrum sensors.

Motorola's *AME-1000 Secure Mobile Telephony Solution* combines hardware- and software-based security that meets federal information security standards and also provides National Security Agency (NSA)-approved Suite B encrypted voice communications. The AME-1000 is configured around the ES400 enterprise smart phone (Figure 1.95). It comprises an ES400 with a CRYPTR micro encryption module, Apriva Voice software and an Apriva gateway infrastructure. It employs GSM/HSPA networking technology and is configured around the Qualcomm

Figure 1.95 Motorola's ES400 smart phone.

MSM7627 chip set that runs on the Microsoft Windows Mobile 6.5.3 Professional operating system (OS). It has a 3.0-inch 480 × 640 pixel resolution touch screen display. Connectivity options include WiFi, Bluetooth, radio, GPS and USB. It includes fingerprint technology and accelerometer sensors.

Solarin from Sirin Labs (Figure 1.96) is yet another smart phone designed for military use and other applications where security and privacy are highly desirable. It was launched in May 2016. Reportedly, it is equipped with the most advanced security features and incorporates privacy technology currently unavailable in other secure smart phones. The smart phone includes military grade encryption embedded in its hardware and also has built-in security software. In addition, it comes with a physical security switch, which can be used to put the phone into a shielded mode for making secure phone calls. This special switch at the back flips it into a cyber-secure mode, thereby allowing only outgoing voice calls and securely encrypted messaging.

Solarin is powered by an octa-core Qualcomm Snapdragon 810 processor that runs on the Android 5.1 operating system (OS). It comes with a 5.50-inch 1440 × 2560 pixel touchscreen display, 4 GB of RAM and 128 GB of internal storage. Solarin packs a 23.8-megapixel primary camera on the rear and an 8-megapixel front camera. The SirinSolarin is a single SIM (GSM) smart phone that accepts a Nano-SIM. Connectivity options include WiFi, GPS, Bluetooth, NFC and 4G. The phone includes a proximity sensor, ambient light sensor, accelerometer and gyroscope.

Figure 1.96 Solarin smart phone.

Figure 1.97 Boeing Black smart phone.

The *Boeing Black* smart phone (Figure 1.97) with its security and modularity features allows use of the same mobile platform across a range of missions and configurations. The smart phone employs specific hardware and software architecture for various configurations depending on the intended mission. Security features include Embedded FIPS 140–2 Key Storage, Hardware Inhibits, Trusted Modules and Configurable OS Security Policies. Major device specifications include a 4.3-inch qHD display with 540×960 pixel resolution, Bluetooth v2.1 + EDR enabled connectivity, a dual 1.2 GHz ARM Cortex-A9 processor, dual SIM support that allows users to switch between government and commercial networks, micro-SD card slot and Android OS with enhanced software security configuration allowing users to configure the device for maximum mission productivity and security.

1.9.2 Tactical Radios

Tactical radios are widely used to deliver secure, reliable and mission-critical information to the intended recipients on the battlefield. Tactical handheld and manpack radios for individual soldiers and radios mounted on ground vehicles and aerial platforms such as fixed or

Figure 1.98 JTRS-HMS AN/PRC-155 Manpack Radio.

rotary-wing aircraft and unmanned aircraft systems deliver high-speed multiband voice, data and video to provide enhanced situational awareness enabling soldiers to switch wave-forms and networks on-the-move as per the requirements of the intended mission. A large number of tactical radios including handheld radios, manpack radios, HF radios, VHF radios, UHF radios, VHF/UHF radios and multiband radios to name a few are available from leading international companies such as Thales, General Dynamics, Motorola and the Harris Corporation. Some representative tactical radios are briefly covered as regards their salient features in the following paragraphs. These include the *JTRS-HMS (Joint Tactical Radio System – Handheld Manpack Small form fit) Rifleman Radio* Type *AN/PRC-154* and *Manpack radio Type AN/PRC-155* by Thales and General Dynamics, the *AN/VRC-118 (V) 1* developed under MNVR (Mid-tier Networking Vehicular Radio) US Army programme by the Harris Corporation, the *Falcon-III AN/PRC-152A* wideband networking handheld radio and *Falcon-III RF-7850A-MR* multichannel networking radio from the Harris Corporation, the *FlexNet-One* wideband vehicular software-defined radio from Thales and the *SRX-2200* combat radio from Motorola.

JTRS-HMS is a family of software-programmable and hardware configurable digital radios designed to provide increased flexibility, adaptability and interoperability to support military communications requirements of diverse missions of the Army, Marine Corps, Navy and Air Force. The JTRS-HMS family includes essentially any portable ground radio unit not mounted on vehicles. Different variants of JTRS family of tactical radios include the *Handheld Rifleman Radio Type AN/PRC-154* and *Two-channel Manpack Radio Type AN/PRC-155* (Figure 1.98) and forms suitable for integration on platforms requiring a small form fit radio. The AN/PRC-154 Rifleman Radio is an individual soldier's radio that is used to simultaneously trans-mit voice and data utilizing the UHF-band (225–450 MHz) and L band (1250–1390 MHz, 1750–1850 MHz) and the Soldier Radio Waveform (SRW). It brings secure squad-level communications to the soldier by encrypting unclassified information (NSA, Type-2). The AN/PRC-155 Manpack Radio is two-channel networking software-defined radio operating in the 30–2200 MHz (banded) frequency range. The PRC-155 Manpack is the only tactical radio to demonstrate the successful use of all three networking waveforms including Soldier Radio Waveform (SRW), Wideband Networking Waveform (WNW) and Mobile User Objective System (MUOS) waveform. The SRW delivers secure networked voice and data communications for individual soldiers. The WNW provides the backbone function needed to seamlessly transport large amounts of data across the tactical network. The MUOS is a military satellite communications system that enables secure, mobile networked communica-tions. The MUOS waveform enables satellite communications and leverages advanced wireless technology that is similar to terrestrial cellular communications. The AN/PRC-155 Manpack delivers secure communications using NSA Type-1 encryption of classified information. With two channels, the Manpack can run different waveforms simultaneously, eliminating the need for more than one radio at any location.

AN/VRC-118 (V)1 was developed by the Harris Corporation under the US Army's Mid-tier Networking Vehicular Radio (MNVR) programme. This programme provides software-programmable digital radios to support army tactical communications requirements from company through to brigade. AN/VRC-118 (V) 1 provides a dynamic, self-forming and self-healing wireless network for both mobile and stationary forces. The radio utilizes two high-bandwidth waveforms namely the SRW and the WNW, which increases overall connectivity and network capability by operating as a node. Mission-critical information gets routed from sender to receiver using the best possible route through various hops from one AN/VRC-118(V)1 MNVR to another. The MNVR also provides a terrestrial data path if or when SATCOM is denied. It can also exchange mission-critical data with Army Aviation platforms through an Airborne Maritime Fixed station (AMF) Small Airborne Networking Radio (SANR). It is interoperable with Falcon, SINCGARS (Single Channel Ground and Airborne Radio System) and other legacy radio systems. SINCGARS is a new family of VHF-FM combat net radios that can operate in a hostile environment by means of its Electronic Counter Countermeasure (ECCM) features. It provides the primary means of command and control for Infantry, Armour and Artillery Units.

Falcon-III AN/PRC-152A by the Harris Corporation is handheld wideband networking radio that provides simultaneous voice and high-speed networked data with NSA-certified security using the Harris Sierra-II encryption module. The radio delivers secure, IP-based mobile ad-hoc networking to dismounted frontline operators, thereby putting enhanced connectivity in the hands of warfighters. Falcon-III AN/PRC-152A seamlessly connects dismounted and upper echelon networks to provide situational awareness and data-on-demand for faster decision making. The *Falcon-III RF-7850A-MR* multichannel networking radio, also from the Harris Corporation, extends the tactical network into the aerial tier. The radio facilitates ground-to-air communications between command posts and the frontline forces by simultaneously delivering mission-critical voice, high-speed IP-networked data and full motion video. The radio is airborne certified for fixed and rotary winged aircraft. It is fully interoperable with legacy waveforms and all radios in the Falcon family. It is easily integrable on a range of military platforms to support multiple missions, waveforms and modes of operation.

The *FlexNet-One* (Figure 1.99) from Thales is a compact vehicular V/UHF (30–512 MHz) software-defined radio equipped with features such as mobile ad-hoc networking that is

Figure 1.99 FlexNet-One.

Figure 1.100 SRX-2200 combat radio.

self-healing and self-organizing allowing automatic routing upon mobile nodes, open architecture that is fully compliant with Software Communications Architecture (SCA) to ensure waveform portability and accommodate customized requirements and upgraded functionality, interoperability that allows the radio to be immediately reconfigured to provide interoperability with the PR4G standard that makes it form/fit compatible with the worldwide operated Thales PR4G vehicular tactical radio and other standard and national waveforms, easy integration with IP networks and applications and embedded or external high level encryption including customer specific encryption.

Motorola's *SRX 2200 Combat Radio* (Figure 1.100) operates in the 700/800 MHz (763–776 MHz, 851–870 MHz), VHF (136–174 MHz) and UHF Range-1 (380–470 MHz). It is designed specifically for tactical and base personnel, equipped with host of battlefield-tested and military-trusted features. Salient features of the SRX 2200 include embedded Individual Location Information (ILI), night vision goggle compatibility, tactical inhibition, Federal Information Processing Standard (FIPS) 140–2 Level-3 validated encryption capability for secure voice and data communications and radio-to-radio text messaging allowing deployment in most sensitive operations, forwards and backwards compatibility with all Motorola mission-critical radios including the most recent Project-25 (P-25) standards of interoperability and its compliance to rigorous MIL-810 environmental specifications and IP67 submersion specifications.

Cognitive radio is a radio communications technology that uses knowledge about environment, internal state and any predefined objectives to decide the relevant operational parameters thereby allowing the most efficient use of radio spectrum for the prevailing conditions. Cognitive radios have the ability to monitor, sense and detect the conditions of their operating environment, and dynamically reconfigure their own characteristics to best match those conditions. A cognitive radio is able to look at the spectrum, detect available or free frequencies and then implement the best form of communication for the required conditions by selecting the frequency band, the type of modulation and power levels most suited to the requirements, prevailing conditions and the geographic regulatory requirements. The xMAX system by xG Technology Inc. is one such cognitive radio network solution that provides end-to-end Internet Protocol (IP) network solution incorporating xG's patented cognitive radio technologies to deliver the first fully mobile Voice over Internet Protocol (VoIP) and broadband network that also supports any smart phone, tablet and other commercial WiFi or IP-enabled devices.

1.9.3 C4ISR Systems

C4ISR systems discussed in this section include the *Theatre Battle Management Core System* (TBMCS) developed by Lockheed–Martin and *TOPLITE* from Rafael (Command and Control), the *Universal Communications Platform* (UCP) and *Warfighter Information Network–Tactical* (WIN-T) from Lockheed–Martin, *C4I-CONNECT* and *SEA-COM system* from Rafael (Communications) and *P-3 Orion* from Lockheed–Martin and *SEA SPOTTER* developed by Rafael (ISR).

Developed and fielded by Lockheed–Martin, the *Theatre Battle Management Core System* (TBMCS), a set of hardware and software application tools, is the primary air warfare tool used for planning and execution of theatre air battle plan for intelligence and operations personnel at force and unit levels. It is the workhorse engine of the Air Operations Centre (AOC) for coordinating and directing flying operations from a wide range of airborne assets such as fighter jets, refuelling tankers, helicopters, unmanned aerial vehicles and even cruise missiles. TBMCS feeds real-time, decision-quality information to a cross-section of users from the Joint Forces Air Commander to the staff at an Air Operations Centre and further on to pilots and weapons control officers on the battlefield. TBMCS horizontally integrates numerous stove-piped systems, often developed in isolation with disregard to how it might be integrated with future technologies, to provide mission-critical information in real time to users across the battlefield accelerating the decision cycle to enable forces to act faster and decisively. Developed and fielded by Lockheed–Martin, TBMCS is resident at Air Operations Centres (AOCs) of the Air Force and Joint Command Air Operations Centres and Navy ships around the globe.

Physical elements of TBMCS include workstations running the core *Contingency Theatre Automated Planning System* (CTAPS), *Combat Intelligence System* (CIS), *Wing Command and Control System* (WCCS) and *Command and Control Information Processing System (C2IPS) software applications* at various nodes of the Theatre Air Control System (TACS). There are other elements such as *Air Force Mission Support System* (AFMSS), *Navy/Marine Joint Maritime Command and Control Information Processing System* (JMCIS), *Army/Marine Advanced Field Artillery Tactical Data System* (AFATDS) that are not considered core systems. These systems are also equally important in the integrated system for command and control of joint air operations. Each element of the system exchanges information with the *Joint Forces Air Component Commander* (JFACC) and the staff through automated links to TBMCS. TBMCS functionality supports the command and control of joint air operations regardless of the JFACC's Service affiliation.

TOPLITE Electro-Optic System (EOS) from Rafael (Figure 1.101) is a multirole, multi-sensor optronic payload that can be configured for military, airborne, naval and homeland security applications. It is a derivative of Rafael's state-of-the-art LITENING targeting and navigation pod. TOPLITE EOS systems support defence and homeland security applications ranging from missile targeting and guidance for precision-guided weapons to fire-control and law enforcement observation. It is used for day/night target observation, detection, identification and recognition of targets in adverse weather conditions for guidance of precision-guided weapons. Different subsystems of TOPLITE payload include 4-axis Gimbals (360° continuous or ±165° azimuth coverage, +85° to −35° or −85° to +35° elevation coverage, up to 90°/s slew rate, up to 100°/s² LOS acceleration and better than 20 micro-radian LOS stabilization on a manoeuvring helicopter flight), third-generation (3–5 μm) Focal Plane Array (FPA) with 320 × 240 pixel resolution (TOPLITE-II) and 640 × 480 pixel resolution (TOPLITE-III) or second generation and third-generation 8–12 micron FPA, B/W or colour CCD, eye-safe laser rangefinder, Laser designator (optional), NVG (Night Vision Goggle) compatible 0.808 μm Laser Marker and Advanced correlation tracker. TOPLITE systems are currently operational

Figure 1.101 TOPLITE EOS system. (*Source:* Courtesy of Konflikty.pl.)

on land-based, airborne and sea-based platforms in the United States, Israel, Australia and many other countries across the worldfor weapons guidance and control and in surveillance, detection and identification systems.

The *Universal Communications Platform* (UCP) from Lockheed–Martin integrates fixed and mobile radio systems thereby transforming any radio system into a fully IP-based network. With the new generation of IP communications technologies and protocols, the UCP enables nearly any type of existing radio system to perform with state-of-the-art IP features. It enables interoperability with other communications and data-related systems, and allows monitoring, control and dispatch from any location with a network connection and a smart phone, laptop, Personal Computer (PC) or a Personal Digital Assistant (PDA). The communications platform has been designed on an open-architecture framework, which allows the UCP to be easily and seamlessly integrated and deployed to a wide range of government and civilian applications, including first responder, law enforcement and counterterrorism. The *Tactical Deployable Unit* (TDU) version of UCP integrates land mobile radio units, LTE, Tactical Communications and WiFi in a seamless operation. It is a self-contained, portable system that includes fully capable communications for operation centres, mobile units and the Department of Defence and Department of Homeland Security applications. It supports a large number of tactical radios including the Thales MBITR/JEM, Harris PRC 117F/150/152, General Dynamics URC-200, PSC-5, Collins GRC-171, Harris R-2368, Harris RF-590A, Harris RF-1310A, Harris RF-5800M-HH, Harris RF-5800H-MP and Harris RF-310M-HH; Ground Mobile radios including the Motorola P25, Tetra, Harris P25, Open Sky, EF Johnson P25, TDMA, Thales P25 and Thales Liberty; Cellular Radios including the GSM 2/3G, CDMA and Satellite Radios including BGAN, Iridium and Thuraya. TDU enables first responders to set up for operation at any place and at any time in less than 15 minutes. *Network Interface Unit* (NIU), *Radio Control Unit* (RCU) and *Multi-Radio Unit* (MRU) are the other UCP products from Lockheed–Martin. NIU provides an IP-based gateway for communications interoperability. It allows up to four separate audio devices to be accessed remotely through a LAN or WAN using a common IP language and supports any device that supports two- or four-wire analogue audio. A Radio Control Unit (RCU) enables IP-based remote control radios worldwide. It provides access and

Figure 1.102 WIN-T Increment 2 communication node.

control of embedded radios through an IP Network either in conjunction with or independent of NIU and MRU products. This allows all radios connected to the RCU to be remotely controllable from any client device including PC, laptop, PDA, cellular smart phones, SIP phones, iPhones and iPads. MRU connects multiple radio transceivers. The unit works in conjunction with NIU and RCU to provide IP access and control of up to four separate portable radio transceivers. It can be configured and tuned remotely with the most commonly used web browsers via the web-based interface.

The *Warfighter Information Network – Tactical* (WIN- T) developed by *General Dynamics Mission systems* is a telecommunications network backbone that delivers voice and data communication services without the need for any fixed infrastructure. *WIN-T Increment 1* provides voice and data communications at the halt. It began to be fielded in 2004 and is currently in use by the US Army, National Guard and Army Reserves. It was used to support combat missions during *Operation Enduring Freedom* and *Operation Iraqi Freedom* where it provided a high-speed, interoperable voice and data communications network down to the battalion level. *WIN-T Increment 2* provides connectivity to the soldiers on the move. It does so through tactical communication nodes that provide mobile communications infrastructure on the battlefield, thereby maintaining network connectivity even after the fixed infrastructure has been removed. Figure 1.102 shows the mobile communications infrastructure on a vehicle. WIN-T Increment 2 is a completely ad-hoc, self-forming network that enables commanders and select staff to manoeuvre anywhere on the battlefield and maintain connectivity to the network. This makes them far less vulnerable to attack as there is no need for them to stop and set up communications. The 10th Mountain Division of the US Army reportedly was the first to get equipped with this capability in July 2013 when deployed in Afghanistan. *WIN-T Increment 3* is the research and development component of the WIN-T programme and is aimed at improving the capability of all increments of the network. The objective of WIN-T Increment 3 is that the network keeps pace with advances in technology and that the entire WIN-T portfolio remains cyber secure with ongoing upgrades and Type-1 encryption for the network. Another objective is to expand the reach of the network so as to be able to support a highly dispersed force over isolated areas.

Figure 1.103 The P-3 Orion. (*Source:* Courtesy of the US Navy).

C4I CONNECT from Rafael is a Radio/Voice-over-IP (RoIP/VoIP) communication system leveraging the most advanced industry standard technologies such as IP, VoIP and RoIP. It is designed for the specific needs of battlefield command and control (C2) applications. A distinctive feature of the C4I CONNECT is its seamless interoperability, which provides seamless access to various types of UHF, VHF, HF and SATCOM radios and to any legacy telephony system. Other key features include built-in IP-PBX capability that supports advanced telephony features such as call transfer, caller ID and call waiting indication; an embedded conference bridge with capabilities to run an unlimited number of broadcast or conference calls; secure voice communication that ensures that all voice sessions are protected and only authorized users are allowed to participate in a session; user/officer mobility that allows the user to be located anywhere in the country and still be able to participate in a mission taking place elsewhere and a built-in digital recording system that records all voice, data and video sessions over the tactical network.

The *SEA-COM system* developed by Rafael is a state-of-the-art, IP-based communication suite for naval platforms such as ships and submarines. SEA-COM fully integrates internal communication including voice, data and video and external communication systems including HF, V/UHF and SATCOM. The SEA-COM communication suite maintains continuity of operations in the harsh environmental conditions of the sea due to use of a redundant IP network and an automatic reconfiguration management system.

The P-3 Orion (Figure 1.103) from Lockheed–Martin is a multi-mission maritime aircraft designed to perform air, surface and subsurface patrol and reconnaissance tasks over extended periods of time. It is a long-endurance aircraft capable of performing designated tasks far from support facilities. P-3 Orion is fitted with sophisticated detection equipment including Infrared and long-range electro-optical cameras and a special imaging radar that allow it to monitor activity from a comfortable distance. The P-3 Orion can carry a variety of weapons in its large internal weapons bay and a number of external weapons hard points. The P-3 airborne platform is used for a variety of missions including submarine hunting, over-land peacekeeping and surveillance operations, protection of shipping lanes, prevention of illegal immigration, anti-terrorism missions, providing warning of potential threats to ground convoys and so on. Lockheed–Martin's Mid-Life Upgrade (MLU) programme ensures that P-3 Orion continues to evolve and remains mission ready for decades to come. P-3 Orion has provided support for *Operation Unified Assistance*, a humanitarian mission of the US military in the wake of Tsunami that struck South-East Asia in December 2004; *Hurricane Katrina* that struck the Gulf Coast of the United States in August 2005; the ongoing *Operation Atlanta* that started in December 2008, a counter-piracy military operation in the Gulf of Aden and the *BP Horizon oil rig disaster*, also known as the *Deep Water Horizon Oil Spill* that started in the Gulf of Mexico in April 2010 and capped in July 2010. No other aircraft is better suited for these missions and Lockheed–Martin's P-3 Mid-Life Upgrade Programme will help ensure the P-3 is mission ready for decades to come. It is mainly used by the US Navy, NASA, Canadian Armed Forces, Royal Australian Air Force and Royal New Zealand Air Force.

The *Sea Spotter* from Rafael is a third-generation infrared staring and tracking system that is capable of automatically locating both surface and airborne targets including surface-to-surface missiles, supersonic and subsonic sea skimming missiles, combat aircraft, gliding bombs, ARM weapons, helicopters, ships, rubber boats and small target vessels such as submarine periscopes. Based on infrared sensors, Sea Spotter is a completely passive system, which makes it invisible to the adversary's sensors unlike traditional radar sensors that give away a vessel's location. It has high probability of detection and an extremely low false alarm rate due to use of a continuously staring infrared sensor coupled with unique image-processing algorithms like 'track before detect' and 'multiple target tracking'. This is unlike the Infrared Search and Track (IRST) systems of the previous generation that used scanning sensors.

Illustrated Glossary

A3E System This is the Standard AM system used for broadcasting. It uses double side band with a full carrier.

Adaptive Delta Modulation This is a type of delta modulator. In delta, the dynamic range of amplitude of message signal $m(t)$ is very small due to threshold and overload effects. In adaptive delta modulation, the step size (Δ) is varied according to the level of message signal. The step size is increased as the slope of the message signal increases to avoid overload. The step size is reduced to reduce the threshold level and hence the quantizing noise when the message signal slope is small.

AME-1000 A secure mobile telephony solution from Motorola. It combines hardware and software-based security that meets federal information security standards and also provides National Security Agency (NSA)-approved Suite B encrypted voice communications. AME-1000 is configured around ES400 enterprise smart phone and comprises ES400 with a CRYPTR micro encryption module, Apriva Voice software and an Apriva gateway infrastructure.

Amplitude Modulated (AM) Transmitters AM transmitters employ amplitude modulation. These transmitters are used in medium-wave (MW) and short-wave (SW) frequency bands for AM broadcast. Based on the transmitted power levels, AM transmitters use high level modulation scheme where the transmitted power needs to be of the order of kilowatts and low-level modulation where only a few watts of power needs to be transmitted.

Amplitude Modulation This is the analogue modulation technique in which the instantaneous amplitude of the carrier signal varies directly as the instantaneous amplitude of the modulating signal. The frequency of the carrier signal remains constant.

Amplitude Shift Keying (ASK) This is a digital modulation technique. In the simplest form of ASK, the carrier signal is switched ON and OFF depending upon whether a 1 or 0 is to be transmitted. This is also known as ON-OFF keying. In another form of ASK, the amplitude of the carrier is switched between two different amplitudes.

Amplitude Shift Keying (ASK) Transmitters ASK transmitters use ASK modulation commonly used in LED transmitters for transmission of digital data through optical fibres. ASK transmitters operate in the VHF/UHF-band. Other applications of ASK transmitters include remote control operations and security systems.

Antenna An antenna is a structure that transforms guided electromagnetic waves into free-space electromagnetic waves and vice versa.

AN/VRC-118 (V) 1 A vehicular radio developed by the Harris Corporation under the Mid-tier Networking Vehicular Radio (MNVR) programme of the US Army. The MNVR programme provides software-programmable digital radios to support Army tactical communications requirements from company through brigade. AN/VRC-118 (V) 1 provides a dynamic, self-forming and self-healing wireless network for both mobile and stationary forces.

B8E System This system uses two independent side bands with carrier either attenuated or suppressed. This form of amplitude modulation is also known as independent side band (ISB) transmission and is usually employed for point-to-point radio telephony.

Bandwidth (Antenna) Antenna bandwidth is, in general, the operating frequency range over which the antenna gives a certain specified performance. Antenna bandwidth is always defined with reference to a certain parameter such as gain or input impedance or standing wave ratio (SWR).

Beam Width (Antenna) This gives the angular characteristics of radiation pattern. It is taken as the angular separation either between the half power points on its power density radiation pattern or between 3 dB points on the field intensity radiation pattern

Boeing Black This is a smart phone from Boeing. With its security and modularity features, it allows use of the same mobile platform across a range of missions and configurations. The Boeing Black smart phone employs specific hardware and software architecture for various configurations depending on the intended mission.

C3F System Also known as Vestigial Side Band transmission. This is used for transmission of video signal in commercial television broadcasting. It is a compromise between SSB and DSB modulation systems in which a vestige or part of the unwanted side band is also transmitted usually with a full carrier along with the other side band. The typical bandwidth required to transmit a VSB signal is about 1.25 times that of an SSB signal.

C4I CONNECT The C4I CONNECT from Rafael is a Radio/Voice-over-IP (RoIP/VoIP) communication system leveraging the most advanced industry standard technologies such as IP, VoIP and RoIP. It is designed for the specific needs of battlefield command and control (C2) applications.

C4ISR An acronym for Command, Control, Communications, Computers, Intelligence, Surveillance and Reconnaissance. C4ISR concept encompasses all systems, technologies and procedures that are used to gather and disseminate information. C4ISR systems allow military commanders to understand their operational environment, identify mission-critical factors and control their assets.

Characteristic Impedance (Transmission Line) This is its input impedance if it was infinitely long. Characteristic impedance is characteristic of the line and is independent of the length of the line.

Characteristic Impedance (Waveguide) This is another important waveguide parameter. The generalized expression for the characteristics impedance Z_0 of waveguide for TE modes is given

by $Z_0 = 377 \times \sqrt{(\mu/\varepsilon)} \times (b/a) \times (\lambda_g/\lambda)$. The generalized expression for the characteristic impedance Z_0 of waveguide for TM modes is given by $Z_0 = 377 \times \sqrt{(\mu/\varepsilon)} \times (b/a) \times (\lambda/\lambda_g)$. For rectangular waveguides, a is the wide dimension and b is the narrow dimension. For circular
waveguides, $a = b$.

Citizens Band (CB) The CB Radio Service that is allocated 40 channels in 26.965–27.405 MHz
frequency band is a private, two-way, short-distance voice communications service for
personal or business activities of the general public and may also be used for voice paging.

Cognitive Radio This refers to an array of technologies that allow radios self-reconfiguration
in terms of operating mode selection, optimal power output and dynamic spectrum access
for interference management. Cognitive radios have the ability to monitor, sense and detect
the conditions of their operating environment, and dynamically reconfigure their own characteristics to best match those conditions.

Continuous-Wave (CW) Receiver The CW receiver is used for detection of continuous-
wave communication signals. A direct conversion receiver used for detection of CW signals,
such as those transmitted in wireless telegraphy using Morse code, is a CW receiver.

Continuous-Wave (CW) Transmitter The CW transmitter makes use of a radio communication technique that uses an undamped continuous-wave signal of constant amplitude and
frequency. The CW transmitter is principally used for radiotelegraphy; that is, for the transmission of short or long pulses of RF energy to form the dots and dashes of the Morse code
characters.

Cross-Polarization (Antenna) This is the component that is orthogonal to the desired
polarization.

Cut-off Wavelength (Waveguide) This is the highest wavelength that can propagate through
the waveguide. Cut-off wavelength for rectangular guides for both TE$_{mn}$ and TM$_{mn}$ is given
by $\lambda_c = 2/\sqrt{(m/a)^2 + (n/b)^2}$ where a is the wide dimension and b is the narrow dimension of
the waveguide. The cut-off wavelength of a circular waveguide with an internal diameter d is
given by $\lambda_c = \pi d / K_r$ where K_r is the solution of a Bessel function equation.

Delta Modulation This has various forms. In one of the simplest forms, only one bit is transmitted per sample just to indicate whether the amplitude of the current sample is greater or
smaller than the amplitude of the immediately preceding sample.

Differential PCM This is similar to conventional PCM. The difference between the two
lies in the fact that in differential PCM, each word or code group indicates difference
in amplitude (positive or negative) between the current sample and the immediately
preceding one.

Differential Phase Shift Keying This is another form of PSK. In this, instead of instantaneous
phase of carrier determining which bit is transmitted, it is the change in phase that carries
message intelligence. In this system, one logic level (say 1) represents a change in phase of
carrier and the other logic level (i.e., 0) represents a no change in phase.

Dipole Antenna This is also a straight radiator usually fed at the centre and producing maximum of radiation in a place perpendicular to the antenna axis.

Directional Pattern (Antenna) The antenna Directional Pattern or Radiation Pattern is a
normalized plot of distribution of electromagnetic energy in a three-dimensional (3D) angular space. The parameters to be plotted could be radiation intensity, which is the power per
unit solid angle or the power density.

Directive Gain (Antenna) Directive gain in a given direction is defined as the ratio of the
power density of the radiated electromagnetic energy in that direction to the power density
in the same direction and at the same distance due to an isotropic radiator with both antennas radiating the same total power.

Dominant Mode (Waveguide) Dominant mode propagating in a waveguide is one which has the highest cut-off wavelength for a waveguide of given dimensions.

Effective Isotropic Radiated Power (EIRP) This given by product of transmitter power and antenna gain. An antenna with a power gain of 40 dB and a transmitter power of 1000 W would have an EIRP of 10 MW. That is, 10 MW of transmitter power when fed to an isotropic radiator would be as effective in the desired direction as 1000 W when fed to a directional antenna with 40 dB power gain in desired direction.

Falcon-III AN/PRC-152A This is a handheld wideband networking radio from the Harris Corporation. It provides simultaneous voice and high-speed networked data with NSA-certified security using the Harris Sierra-II encryption module.

Falcon-III RF-7850A-MR This is a multichannel networking radio from Harris Corporation. It extends the tactical network into the aerial tier. The radio facilitates ground-to-air communications between command posts and the frontline forces by simultaneously delivering mission-critical voice, high-speed IP-networked data and full motion video.

Fidelity (Communication Receiver) This refers to its ability to accurately reproduce at its output the signal that appears at its input. Broader receiver bandwidth produces high fidelity. This is in contrast to the requirement of narrower bandwidth for better selectivity.

FlexNet-One This is a compact vehicular software-defined radio from Thales.

Frequency Division Multiplexing In the case of FDM, different message signals are separated from each other in frequency. FDM is used in telephony, commercial radio broadcast (both AM and FM), television broadcast, communication networks and telemetry.

Frequency Modulated (FM) Transmitters These employ frequency modulation. One of the most common applications of FM transmitters is in FM broadcasting, which is nothing but radio broadcasting using frequency modulation. FM broadcasting first began in 1945 and is now used worldwide to provide high-fidelity voice and music over broadcast radio. FM broadcasting employs the VHF band at 30–300 MHz.

Frequency Modulation This is the analogue modulation technique in which the instantaneous frequency of the modulation signal varies directly as the instantaneous amplitude of the modulating or base band signal. The rate at which these frequency variations take place is of course proportional to the modulating frequency.

Frequency-Shift Keying FSK is the frequency of the carrier signal that is switched between two values, one representing bit '1' and the other representing bit '0'.

Frequency-Shift Keying (FSK) Transmitters FSK transmitters employ FSK modulation and operate in the VHF/UHF-band. They are generally used for remote control and security systems available for these applications.

Ground Wave This is that part of propagating electromagnetic wave that travels along the surface of Earth. The ground wave further comprises the surface wave and space wave. A surface wave travels in contact with the surface of Earth. A space wave travels in the space just above the surface of Earth. It is composed of a direct wave and reflected wave. Space waves propagate directly between transmitting antenna and receiving antenna, which necessitates there is a line-of-sight path between the two.

Guide Wavelength (Waveguide) This is the wavelength of the travelling wave propagating inside the waveguide. The guide wavelength λ_g, the cut-off wavelength λ_c and the free-space wavelength (λ) are interrelated by $\lambda_g = \lambda / \sqrt{1 - (\lambda/\lambda_c)^2}$.

H3E System This is the single side band, full carrier AM system.

Helical Antenna This is a broadband VHF/UHF antenna. Most of its applications are due to the circularly polarized waves it produces.

Hertz Antenna This is a straight length of a conductor that is half-wave long. It may be placed vertically to produce vertically polarized waves or in horizontal position to produce horizontally polarized waves.

High-Frequency (HF) Transmitters HF transmitters operate over the HF band (3–30 MHz) and use ionosphere as means of electromagnetic wave propagation that is commonly known as sky wave propagation are particularly suitable for long-distance communication across intercontinental distances. The band is extensively used by international short wave broadcasting stations as the entire short wave band of frequencies (2.310–25.820 MHz) falls within the HF band.

High Level Amplitude Modulated Transmitter This employs high level modulation. In a case of high level modulation, the carrier signal is modulated at a much higher power level equal to the power to be transmitted. The carrier signal here is amplified in a chain of Class C amplifiers to raise the power to the level required for transmission.

Horn Antenna This is a simple development of the waveguide transmission line. It is essentially a section of waveguide where the open end is flared to provide a transition to the areas of free space. The horn antenna is used in the transmission and reception of RF microwave signals. The antenna is normally used in conjunction with waveguide feeds.

Impedance (Antenna) Antenna impedance at a given point in the antenna is given by the ratio of voltage to current at that point. As the magnitude of voltage and current vary along the antenna length, the impedance also varies being minimum at the point of voltage node or minima such as the centre point of half-wave dipole and maximum at the point of current node such as the centre point of full-wave length long antenna.

Ionosphere This is a region of atmosphere extending from 60 km to about 1000 km. The ionosphere further comprises of different layers designated as D, E, F1 and F2 layers.

J3E System This is the Single Side Band Suppressed Carrier system. This is the system usually referred to as SSB, in which carrier is suppressed by at least 45 dB in the transmitter.

JTRS-HMS This is a family of software-programmable and hardware configurable digital radios designed to provide increased flexibility, adaptability and interoperability to support military communications requirements of diverse missions of the Army, Marine Corps, Navy and Air Force. The JTRS-HMS family includes essentially any portable ground radio unit not mounted on vehicles. Different variants of JTRS family of tactical radios include the Handheld Rifleman Radio Type AN/PRC-154 and Two-channel Manpack Radio Type AN/PRC-155 and form suitable for integration on platforms requiring a small form fit radio.

LG G5 This is a smart phone from LG Electronics. It uses Android OS version 6.0.1 and is certified by National Information Assurance Partnership (NIAP) for military use.

LG V10 This is smart phone from LG Electronics. It uses Android OS version 5.1.1 and is certified by National Information Assurance Partnership (NIAP) for military use.

Log Periodic Antenna This is a broadband VHF/UHF antenna capable of providing enormous bandwidth.

Low-Level Amplitude Modulated Transmitter This employs low-level modulation. In the case of low-level amplitude modulation, amplitude modulation of carrier signal takes place at a much lower power level of the carrier signal and the modulated signal is then amplified in a chain of linear amplifiers to raise the power required for transmission.

Marconi Antenna This is the vertical antenna that is a quarter-wave long and is fed against an infinitely large perfectly conducting plane.

Medium Frequency (MF) Transmitters MF transmitters operate over the 300 kHz–3 MHz frequency band and are used for AM radio broadcast services that employ (535–1705) kHz band, maritime and aircraft navigation, amateur radio, cordless phones and so on.

Mesosphere This is a region of atmosphere located between 50 and 80 km above Earth's surface.

Microstrip Antenna This consists of a thin strip sitting on a dielectric that rests on a ground plane.

Mid-tier Networking Vehicular Radio Programme (MNVR) The MNVR programme aims to provide software-defined, multichannel networking radios for a wide variety of Army tactical vehicles to meet the Army's requirement for a mid-tier wideband networking capability. It provides self-forming and self-healing communication networks from the brigade to the platoon level throughout the full range of military operations.

Mismatch Loss (Transmission Line) This is the loss due to reflection from a mismatch. It is defined as the ratio of incident power to the difference of incident and reflected power expressed in decibels.

Mobile Ad-Hoc Network (MANET) MANET is a type of ad-hoc network that can change locations and configure itself on the fly. MANETs can be networked to interconnect multiple mobile phones within a specified coverage area offering greater bandwidth and better connectivity.

Network-Centric Warfare (NCW) NCW is a military doctrine that uses networking of sensors, planners and decision makers and shooting platforms to create enhanced shared awareness of the battle space.

Offset QPSK This is similar to QPSK with the difference that the alignment of the odd/even streams is shifted by an offset equal to T seconds, which is the pulse duration of the input bit stream.

OODA Loop This describes sequence of events that must take place in a military engagement. The OODA loop is fundamental to all military operations at both tactical and strategic levels as the adversary must be observed to gather information, the attacker must orient himself as per the situation or context, take decision and then act accordingly.

P-3 Orion The P-3 Orion from Lockheed–Martin is a multi-mission maritime aircraft designed to perform air, surface and subsurface patrol and reconnaissance tasks over extended periods of time.

Phased Array Antenna A Phased Array Antenna or, more appropriately, a phase steered array antenna is the one where the radiated beam (or the axis of the main lobe of the radiated beam) can be steered by feeding the elements of the array with signals having a certain fixed phase difference between adjacent elements of the array during transmission. On reception, they work exactly the same way and instead of splitting the signals among elements, the elemental signals are summed.

Phase Modulated (PM) Transmitters PM transmitters employ phase modulation. Phase modulation and frequency modulation are closely linked together (frequency is rate of change of phase) and is often used in many transmitters and receivers such as two-way radios and mobile radio communications including maritime mobile radio communications.

Phase Shift Keying (PSK) In PSK, the phase of the carrier is discretely varied with respect to either a reference phase or to the phase of the immediately preceding signal element in accordance with the data being transmitted. For example, when encoding bits, the phase shift could be 0° for encoding a bit 0 and 180° for encoding a bit 1. The phase shift could have been −90° for encoding a bit 0 and + 90° for encoding a bit 1.

Phase Shift Keying (PSK) Transmitters PSK transmitters employ any of the different forms of phase shift keying, which include Phase Shift Keying (PSK), Binary Phase Shift Keying (BPSK), Quadrature-Phase Shift Keying (QPSK), 8-Point Phase Shift Keying (8-PSK), 16-Point Phase Shift Keying (16-PSK), Quadrature Amplitude Modulation (QAM), 16-Point Quadrature Amplitude Modulation (16-QAM) and 64-Point Quadrature Amplitude Modulation (64-QAM). PSK is particularly suited to data communications. PSK in its different forms is extensively used for wireless LANs, Radio Frequency Identification (RFID) and Bluetooth communication.

Polarization (Antenna) Antenna polarization is the direction of electric field vector with reference to ground in the radiated electromagnetic wave while transmitting and the orientation of the electromagnetic wave again in terms of the direction of electric field vector the antenna responds best to while receiving.

Power Gain (Antenna) This is defined as the ratio of the power density at a given distance in the direction of maximum radiation intensity to the power density at the same distance due to an isotropic radiator for the same total power fed to the two antennas.

Propagation Constant (Transmission Line) The propagation constant, γ, is a measure of the attenuation and the phase shift of the incident waves travelling from the source to the load end of the transmission line. Propagation constant, for practical reasons, is a complex quantity with a real part known as the attenuation constant, α, and an imaginary part called the phase shift constant, β.

Pulse Amplitude Modulation (PAM) In PAM, the signal is sampled at regular intervals and the amplitude of each sample, which is a pulse, is proportional to the amplitude of the modulating signal at the time instant of sampling.

Pulse Code Modulation (PCM) In PCM, the peak-to-peak amplitude range of the modulating signal is divided into a number of standard levels, which in case of binary system is an integral power of 2. The amplitude of the signal to be sent at any sampling instant is the nearest standard level. This nearest standard level is then encoded into a group of pulses. The number of pulses (n) used to encode a sample in binary PCM equals $\log_2 L$ where L is number of standard levels.

Pulse Position Modulation (PPM) In the case of PPM, the amplitude and width of sampled pulses is maintained as constant and the position of each pulse with respect to the position of a recurrent reference pulse varies as a function of instantaneous sampled amplitude of the modulating signal.

Pulse Width Modulation (PWM) In the case of PWM the starting time of the sampled pulses and their amplitude is fixed. The width of each pulse is made proportional to the amplitude of the signal at the sampling time instant.

Quadrature-Phase Shift Keying (QPSK) QPSK is the most commonly used of all forms of PSK. A QPSK modulator is nothing but two BPSK modulators operating in Quadrature. The input bit stream (d_0, d_1, d_2, d_3, d_4....) representing the message signal is split into two bit streams, one with, say, even numbered bits (d_0, d_2, d_4....) and the other with odd numbered bits (d_1, d_3, d_5.......). The two bit streams modulate the carrier signals, which have a phase difference of 90° between them. The vector sum of the output of two modulators constitutes the QPSK output.

Quarter-Wave Transformer (Transmission Line) This is $\lambda/4$ long line. It can be used to match both real as well as complex load impedance to a transmission line.

R3E System This is the single side band reduced carrier type AM system also called the pilot carrier system.

Reflection Coefficient (Transmission Line) When a transmission line is terminated in load impedance which is not equal to its characteristic impedance, part of the signal energy sent down the line is reflected back. The ratio of the reflected signal amplitude to the incident one is defined as the reflection coefficient. It may be expressed as a magnitude only and denoted by Γ or as a complex value having both a magnitude and a phase and denoted by ρ, with $\rho = |\Gamma|$.

Reflector Antenna This comprises of a reflector and a feed antenna. Depending upon the shape of the reflector and the feed mechanism, there are different types of reflector antennas suitable for different applications. The reflector is usually a parabolic, also called a parabolic reflector, or a section of parabolic or cylindrical reflectors. A cylindrical reflector has a parabolic surface in one direction only. The feed mechanisms include the feed antenna placed at the focal point of the parabolic reflector or the feed antenna placed off the focal point. Another common feed mechanism is the Cassegrain feed. Cylindrical reflectors are fed by an array of feed antennas. The feed antenna is usually a dipole or a horn.

Return Loss (Transmission Line) This signifies the total round trip loss of the signal and is defined as the ratio of the incident power to the reflected power at a point on the transmission line.

Rhombic Antenna This is a long-wire antenna. In a rhombic antenna, conductors are arranged to form a rhombus. It is a combination of two long-wire V-antennas.

Sampling Theorem This states that a band limited signal with the highest frequency component as f_M Hz can be recovered completely from a set of samples taken at the rate of f_s samples per second provided that $f_s \geq 2f_M$.

SEA-COM This system developed by Rafael is state-of-the-art, IP-based communication suite for naval platforms such as ships and submarines. SEA-COM fully integrates internal communication including voice, data and video and external communication systems including HF, V/UHF and SATCOM.

Sea Spotter Sea Spotter from Rafael is a third-generation infrared staring and tracking system that is capable of automatically locating both surface and airborne targets including surface-to-surface missiles, supersonic and subsonic sea skimming missiles, combat aircraft, gliding bombs, ARM weapons, helicopters, ships, rubber boats and small target vessels such as a submarine periscope.

Selectivity (Communication Receiver) This is the ability of a receiver to reject unwanted signals at the input of receiver.

Sensitivity (Communication Receiver) This is a measure of a receiver's ability to produce the desired signal-to-noise (S/N) ratio for weak received signals. It is measured as received signal strength in microvolts to produce a signal-to-noise ratio of 10 dB.

Shannon–Hartley Theorem Shannon–Hartley theorem describes the capacity of a noisy channel (assuming that the noise is random). According to this theorem,
$$C = B\log_2[1+(S/N)] \text{ bits/second}$$
The Shannon–Hartley theorem underlines the fundamental importance of bandwidth and S/R ratio in communication. It also shows that for a given channel capacity, we can exchange increased bandwidth for decreased signal power.

Single Side Band (SSB) Receiver The SSB is also a superheterodyne receiver comprising an RF amplifier, a mixer along with a local oscillator, intermediate frequency amplifier stage, a detector, an audio amplifier stage and a loudspeaker. One difference between the standard AM broadcast receiver and the SSBSC receiver is that the carrier signal needs to be reinserted at the detector for the detection process to occur.

Single Side Band (SSB) Transmitter An SSB transmitter employs single side band modulation, a type of amplitude modulation in which one of the side bands, either upper side band or lower side band, is suppressed and only one side band is transmitted. SSB modulation and SSBSC (Single Side Band Suppressed Carrier) modulation techniques offer more efficient use of transmitter bandwidth and power.

Sky Wave These make use of the ionosphere for communication. Ionospheric propagation allows radio signals to be received at much longer distances.

Smart Phone This is a mobile phone that is capable of performing many of the functions of a computer with its touch screen interface, internet access and an operating system capable of running downloaded applications.

Software-Defined Radio (SDR) SDR is the radio in which some or all of the physical layer functions are software defined. Software in this case can be used to alter the specifications and functions of the radio. The SDR is configured around a generic hardware platform comprising digital signal processors as well as general-purpose processors to implement transmit and receive radio functions.

Software Reprogrammable Payload (SRP) SRP is a satellite-rooted technology with its down-to-earth communication potential. It is an adaptation of a small radio receiver designed for space applications into a full-fledged radio frequency system initially targeted for UAS (Unmanned Airborne System) communications.

SOLARIN This is a smart phone developed by Sirin Labs designed for military use and other applications where security and privacy are highly desirable. The smart phone includes military grade encryption embedded in its hardware and also has built-in security software. In addition, it comes with a physical security switch, which can be used to put the phone into a shielded mode for making secure phone calls.

Soldier Radio Waveform (SRW) SRW is an open-standard voice and data waveform used to extend wideband battlefield networks to the tactical edge. It is designed as a mobile ad-hoc waveform and it functions as a router within a wireless network. It is used to transmit vital information over long distances and elevated terrains including mountains and other natural or manmade obstructions and allows communication without a fixed infrastructure.

SRX-2200 This is a combat radio from Motorola. It operates in the 700/800 MHz (763–776 MHz, 851–870 MHz), VHF (136–174 MHz) and UHF Range-1 (380–470 MHz). It is designed specifically for tactical and base personnel, equipped with host of battlefield-tested and military-trusted features.

Stratosphere This is a region of atmosphere extending from 10 to 50 km.

Stub (Transmission Line) Basically, this is a shorted or open section of a transmission line used in conjunction with transmission lines to provide impedance match and cancel out reflections.

Superheterodyne Receiver This is one of the most commonly used types of receiver in a variety of applications from broadcast receivers to two-way radio communications links as well as many mobile radio communications systems. It uses frequency mixing to convert the received radio frequency (RF) signal to a fixed intermediate frequency (IF) which can be more conveniently processed than the original carrier frequency. Basic building blocks of a superheterodyne receiver include RF amplifier, mixer, local oscillator, IF amplifier, suitable detector, audio amplifier and loudspeaker.

Tablet This is a wireless, portable personal computer with a touch-screen interface with a form factor typically smaller than a notebook computer but larger than a smart phone.

Theatre Battle Management Core System (TBMCS) The TBMCS, developed and fielded by Lockheed–Martin, is a set of hardware and software application tools. It is the primary air warfare tool used for planning and execution of theatre air battle plan for intelligence and operations personnel at force and unit levels.

Time Division Multiplexing (TDM) TDM is used for simultaneous transmission of more than one pulsed signals over a common communication channel and different message signals are separated from each other in time. TDM is widely used in telephony, telemetry, radio broadcast and data processing.

TOPLITE This is an electro-optic system from Rafael. It is a multirole, multi-sensor optronic payload that can be configured for military, airborne, naval and homeland security applications. It is a derivative of Rafael's state-of-the-art LITENING targeting and navigation pod.

Transmission Line This is a communication medium used in communication systems to carry signals from transmitter output to the input of transmitting antenna and from receiving antenna to the input of receiver. They are also used to for other applications such as impedance matching.

Troposphere This is a region of atmosphere extending to altitudes of about 10 km above the surface of Earth.

Ultra-High Frequency (UHF) Transmitters UHF transmitters operate over the UHF-band (300 MHz–3 GHz). UHF transmitters find applications in a large number of communication services, which includes television broadcasting, cell phone communication, cordless phones, satellite communication, WiFi and Bluetooth services and so on.

Universal Communications Platform (UCP) The UCP from Lockheed–Martin integrates fixed and mobile radio systems, thereby transforming any radio system into a fully IP-based network to meet the demanding requirements of rapidly changing tactical and emergency situations.

V-antenna This is a long-wire antenna. In a V-antenna, the conductors are arranged to form a V-shape and they are fed in phase opposition at the apex.

Very High Frequency (VHF) Transmitters VHF transmitters operate over the VHF band (30–300 MHz) and use space wave line-of-sight propagation. The radio horizon is slightly longer than the geometric line-of-sight horizon due to bending of electromagnetic waves by the atmosphere. VHF transmitters are commonly used for FM radio broadcasting employing 87.5–108 MHz frequency band with few exceptions such as 76–90 MHz used in Japan and 65–74 MHz used in Eastern Bloc of the former Soviet Republic, television broadcasting, two-way land mobile radio systems for various private, business, military and emergency services, long-range data communication, amateur radio and marine communications. Air traffic control communications and air navigation systems are other applications.

Warfighter Information Network – Tactical (WIN-T) This is a telecommunication network backbone developed by General Dynamics Mission systems. It delivers voice and data communication services without the need for any fixed infrastructure. WIN-T Increment 1 provides voice and data communications at the halt. WIN-T Increment 2 provides connectivity to the soldiers on the move. WIN-T Increment 3 is the research and development component of the WIN-T programme and is aimed at improving the capability of all increments of the network.

Waveguide This is nothing but a conducting tube through which energy is transmitted in the form of electromagnetic waves. The waveguide can be considered to be a boundary that confines the waves to the space enclosed by boundary walls.

Wideband Networking Waveform (WNW) WNW is the next-generation high throughput military waveform, developed under the Joint Tactical Radio System (JTRS) Ground Mobile Radio (GMR) programme. It uses the Orthogonal Frequency Division Multiplexing (OFDM) Physical Layer. With its mobile ad-hoc networking (MANET) capabilities, the waveform is designed to work well in both urban landscape as well as a terrain-constrained environment, since it can locate specific network nodes and determine the best path for transmitting information.

xMAX system This is a cognitive radio network solution from xG Technology Inc. It provides end-to-end Internet Protocol (IP) network solution incorporating xG's patented cognitive radio technologies to deliver the first fully mobile Voice over Internet Protocol (VoIP) and broadband network that also supports any smart phone, tablet and other commercial WiFi or IP-enabled devices.

Yagi-Uda Antenna This comprises of a half-wave dipole with parasitic elements to enhance the directionality of the radiation pattern.

Bibliography

1 Calcutt, D. and Tetley, L. (1994) *Satellite Communications: Principles and Application*, Edward Arnold, a member of the Hodder Headline Group, London.

2 Elbert, B.R. *Introduction to Satellite Communication*, Altech House, Boston, MA.

3 Gedney, R.T., Schertler, R. and Gargione, F. (2000) *The Advanced Communication Technology*, SciTech., New Jersey.

4 Kadish, J.E. (2000) *Satellite Communications Fundamentals*, Artech House, Boston, MA.

5 Kennedy, G. and Davis, B. (1992) *Electronic Communication Systems*, McGraw-Hill Education, Europe.

6 Maral, G. and Bousquet, M. (2002) *Satellite Communication Systems: Systems, Techniques and Technology*, John Wiley & Sons, Ltd, Chichester.

7 Richharia, M. (1999) *Satellite Communication Systems*, Macmillan Press Ltd.

8 Stallings, W. (2004) *Wireless Communications and Networks*, Prentice-Hall, Englewood Cliffs, NJ.

9 Tomasi, W. (1997) *Advanced Electronic Communications Systems*, Prentice-Hall, Englewood Cliffs, NJ.

10 Tomasi, W. (2003) *Electronic Communications System*, Prentice-Hall, Englewood Cliffs, NJ.

11 Elmasry, G.F. (2012) *Tactical Wireless Communications and Networks: Design Concept and Challenges*, Wiley-Blackwell.

12 Garstka, J.J., Stein, F.P. and Alberts, D.S. (2016) *Network-Centric Warfare – Developing and Leveraging Information Superiority*, ALPHA Editions.

13 Mir, N.F. (2006) *Computer and Communication Networks*, Prentice-Hall.

14 Williamson, J. *(Annual Book) Jane's Military Communications*, HIS Global Inc.

15 Gething, M.J., Ebbutt, G., Williamson, J. and Fuller, M. *(Annual Book), Jane's C4ISR & Mission Systems: Maritime*, HIS Global Inc.

16 Ebbutt, G., Gething, M.J., Streetly, M., Williamson, J. and Downs, E. *(Annual Book), Jane's C4ISR & Mission Systems: Air*, HIS Global Inc.

17 Ebbutt, G. and Gething, M.J. *(Annual Book), Jane's C4ISR & Mission Systems: Land*, HIS Global Inc.

18 Sanders, C. (2013), *Applied Network Security Monitoring: Collection Detection and Analysis*, Syngress.

19 Loo, J., Mauri, J.L. and Ortiz, J.H. (2012), *Mobile Ad Hoc Networks: Current Status and Future Trends*, CRC Press.

2

Radar Fundamentals

Radar is an acronym for *Radio Detection and Ranging*, which tends to suggest that it is a piece of equipment that can be used to detect and locate a target. Though this does continue to be the primary function of radar, modern radar does much more than just detection and ranging. It is used to determine the velocity of moving targets and also find out many more characteristics about the target such as its size, shape and other physical features including, for example, the type and number of engines used on an aircraft. Radar is extensively used in many civilian and military applications. Radar has been and will continue to be an essential capability for militaries worldwide. This chapter gives a comprehensive treatment to the radar fundamentals covering a wide cross-section of topics including basic radar functions, related performance parameters, radar range equation, radar waveforms, radar transmitters, receivers and displays, radar antennas and types of radar. Different radar types, based on their operational principles and discussed in this chapter, include continuous-wave radar, FM-CW radar, moving target indicator (MTI), pulse Doppler radar, tracking radar, synthetic aperture radar, pulse-compression radar, monostatic and bistatic radar, over-the-horizon radar, primary and secondary surveillance radar and laser radar. Application of these radar types for specific military roles is covered in the following Chapter 3 on *Military Radars*.

2.1 Introduction to Radar

Radar is a standalone active system with its own transmitter and receiver. It is primarily used for detecting the presence and finding the exact location of a far-off target. It does so by transmitting electromagnetic energy, in the form of short bursts in most of the cases, and then detecting the echo signal returned by the target. The range of the intended target is computed from the time that elapses between the transmission of energy and reception of echo (Figure 2.1). The location of the target can be determined from the angle/direction of arrival of the echo signal by using a scanning antenna, preferably transmitting a very narrow width beam. As mentioned in the introductory paragraph, radar today does much more than just detecting a target and finding its location. Radar can be used to determine the velocity of a moving target, track a moving target and even determine some of the physical features of the target. Of course, no single radar type can be used to perform all the functions. There are different types best suited to different applications. In addition, radar is a principal source of navigational aid to aircraft and ships. It forms a vital part of an overall weapon guidance or a fire-control system. Behind most of the radar functions lies its capability to detect a target and find its range and velocity.

Handbook of Defence Electronics and Optronics: Fundamentals, Technologies and Systems, First Edition. Anil K. Maini.
© 2018 John Wiley & Sons Ltd. Published 2018 by John Wiley & Sons Ltd.

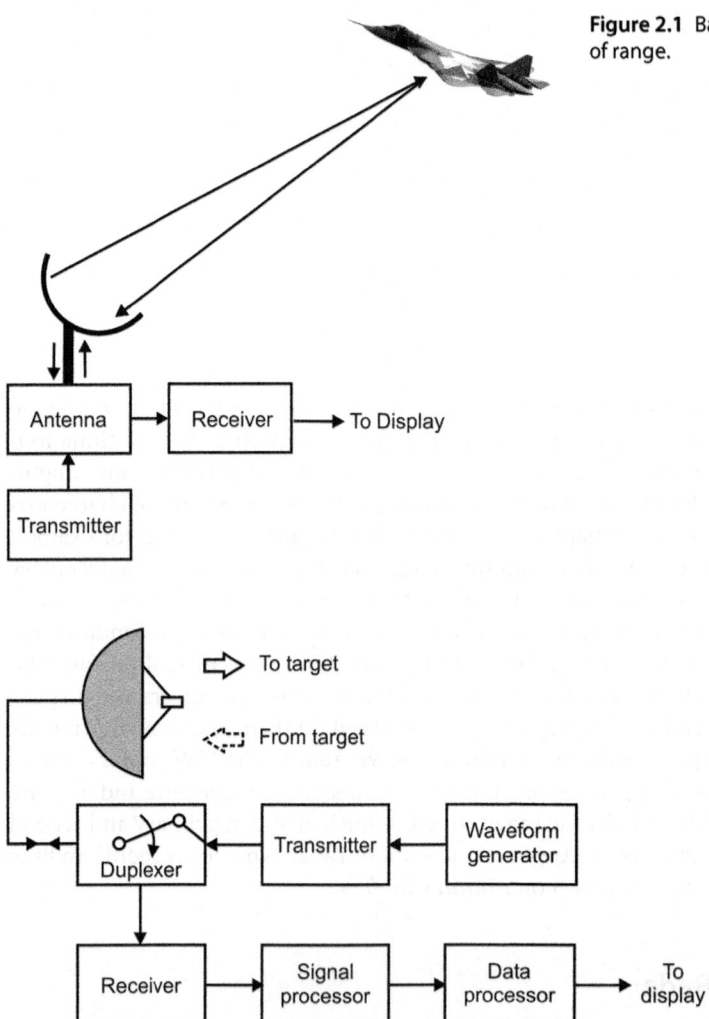

Figure 2.1 Basic radar: measurement of range.

Figure 2.2 Basic components of a radar system.

2.1.1 Basic Radar System

The basic components of a radar system are shown in the block-schematic arrangement of Figure 2.2. The radar signal waveform as generated by the waveform generator modulates a high-frequency carrier and the modulated signal is raised to the desired power level in the transmitter portion. The transmitter could be a power amplifier employing any of the micro-wave tube amplifiers such as Klystron, Travelling Wave Tube (TWT), Crossed Field Amplifier (CFA) or even a solid-state device. The radar waveform is generated at a low power level, which makes it far easier to generate different types of waveforms required for different radars. The most common radar waveform is a repetitive train of short pulses. CW is employed to determine the radial velocity of the moving target from Doppler frequency shift; FM-CW is used where it is desired to measure range with a CW waveform. Pulse-compression waveforms are used when it is desired to have the advantage of longer range capability of a long pulse and higher range resolution capability of a shorter pulse. These waveforms are shown in Figure 2.3.

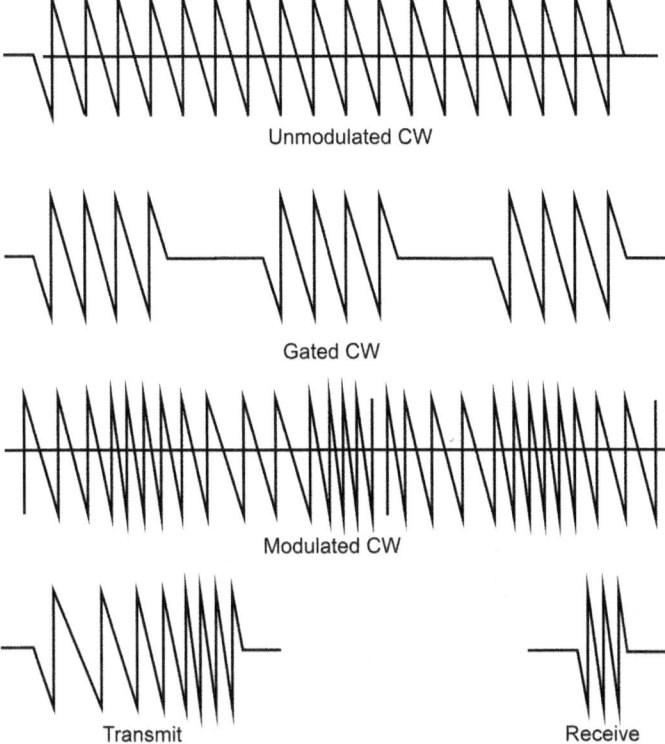

Figure 2.3 Radar waveforms.

The average output power requirement of radar could be as small as a few tens of milliwatts for very short-range radars to several megawatts for Over-The-Horizon-Radar (OTHR).

The *duplexer* allows the same antenna to be used for both transmission as well as reception. It acts as a switch disconnecting the receiver from the antenna during the time the relatively much higher power transmitter is ON to protect the receiver from getting damaged. On reception, the weak received signal is routed to the receiver by the duplexer. The duplexer usually makes use of gas filled transmit/receive tubes that are basically sections of transmission line filled with a low breakdown voltage gas [Figure 2.4(a)]. These tubes get fired due to presence of high power to direct the transmitter output to antenna. After the transmitter signal is radiated, these tubes de-ionize or recover quickly to direct any received signal to the receiver input. A circulator is sometimes used to provide further isolation between transmitter and receiver [Figure 2.4(b)]. A circulator as a component can also be used as a duplexer [Figure 2.4(c)]. The circulator duplexer contains a high power RF circulator comprising of signal couplers and phase shifters such that a signal entering one port has a low attenuation path only to the next port in a particular direction. The low attenuation paths in the circulator shown in Figure 2.4(c) are 1–2, 2–3, 3–4 and 4–1. All other paths are high attenuation paths.

The *antenna* acts as an interface between the radar transmitter output and free space. Mechanically steered parabolic reflector antennas and electronically steered antenna arrays are commonly used.

The echo signal received by the antenna is directed to the receiver input. The receiver is usually of the superheterodyne type. The receiver filters out-of-band interference. It also amplifies the desired signal to a level adequate for operating subsequent circuits.

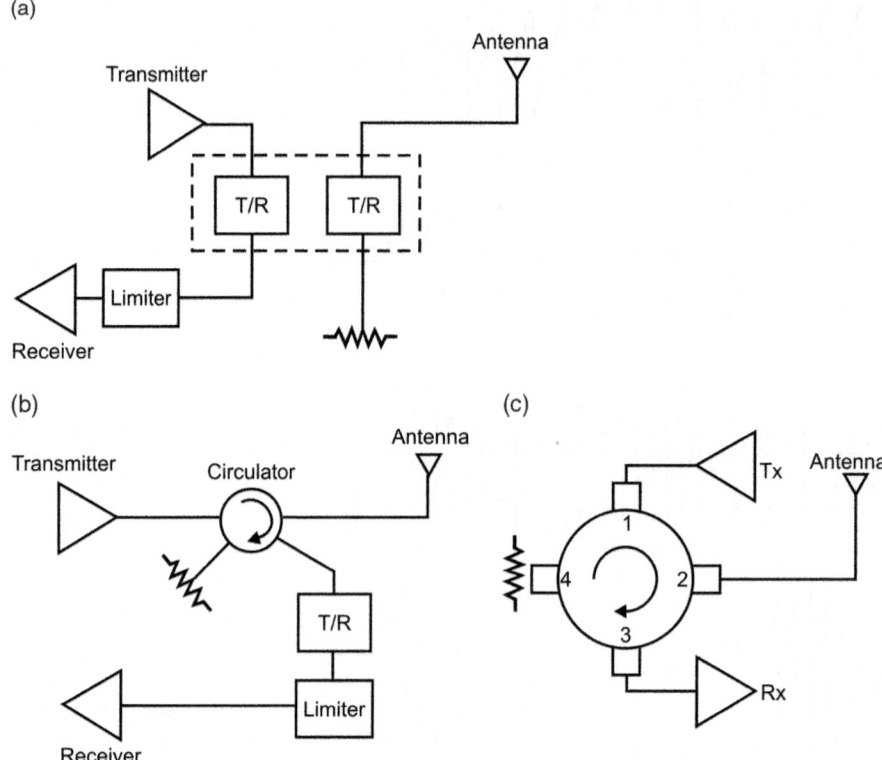

Figure 2.4 Use of T/R tube and circulator for radar transmitter-receiver isolation. (a) The duplexer usually makes use of gas filled transmit/receive tubes that are basically sections of transmission line filled with a low breakdown voltage gas. (b) A circulator is sometimes used to provide further isolation between transmitter and receiver. (c) A circulator as a component can also be used as a duplexer.

The purpose of *signal processing* is to reject the undesired signals such as clutter and enhance the desired signals due to the targets. It is done prior to the section that makes the decision as to whether the target is present and in case of target being present, extracts the information such as range, Doppler and so on.

Data processing refers to the processing done after the detection decision has been made. Functions like automatic tracking, target recognition are examples of data processing in a radar system.

The *display* puts the processed information in a form usable by radar operators and others wanting to use the information such as air traffic controllers, weapon system operators and so on. The operation of radar and the sequence of events that take place from start to finish can be summarized in case of typical pulsed radar as follows.

The transmitter generates a repetitive pulse train with each pulse having a burst of RF signal. The pulse parameters, of course, vary with the type of radar and the mode in which it is operating. The duplexer routes the pulsed electromagnetic energy to the transmitting antenna which concentrates the energy fed to its input into a narrow beam in the direction of the intended target. At the same time, a time base is initiated coinciding with the transmission time instant of the pulse. The electromagnetic wave propagates through the atmosphere. This wave gets reflected from the target due to difference in the impedance characteristics of the targets. The impedance offered by the atmosphere (or more precisely, the free space) to the propagating electromagnetic wave is 377 Ω and any discontinuity encountered causes the wave to get reflected. The amount of reflection depends upon the characteristics of the target. The target reflects the wave in all directions and the portion of the reflected energy travelling in the direction of the radar constitutes the echo or the backscatter.

It may be mentioned here that the reflection also occurs from ground and sea surfaces, atmospheric conditions like clouds, turbulence and so on. These reflections occurring in the direction of the radar constitute *clutter*. The backscatter energy travels back to the radar and a portion of it along with a portion of the clutter is intercepted by the radar's receiving antenna, which in the present case is same as the transmitting antenna. The amount of the backscatter energy intercepted by the antenna depends upon the capture area of the antenna. The received signal is routed to the receiver by the duplexer. The signal that contains both the desired echo as well as the interfering signals and noise gets processed in the receiver.

The processed information is then subjected to the detection threshold comparison and if the signal is larger than the detection threshold, detection is said to occur. If the detection is caused by the desired target, a target is said to be present and if the same occurs due to interfering signals, detection is a false alarm. The detection threshold in fact is so chosen as to minimize the probability of false alarm. Another detection error is the one when the radar fails to detect an existent target due to the target echo signal being weak and being not able to cross the detection threshold. When detection occurs, that is, when the processed signal crosses the detection threshold, the time base initiated at the start is strobed and the round trip propagation time measured to determine the target range. The antenna's position encoders are also strobed to determine the angle-of-arrival of the echo at the time of detection. If the target is a mobile one, its radial velocity information is contained in the Doppler shift, which can be used to determine the target velocity.

2.1.2   Radar Classification

Radars can be classified on the basis of: (1) operational frequency band, (2) transmit wave shape and spectrum, (3) PRF class and (4) intended mission and mode.

2.1.2.1   Operational Frequency Band

Radars typically operate in frequency range of a few tens of MHz to few tens of GHz. Radars operating up to about 30 MHz make use of ionospheric reflection as a means to detect targets lying beyond the radar horizon. Over-The-Horizon-Radar (OTHR) belongs in this category.

Very long-range early warning radars are found in the VHF and UHF bands (30 MHz to 1 GHz).

L band (D Band in the new designation) radars operating in the 1–2 GHz frequency band are usually long-range military radars and air traffic control radars.

S band (E/F band in the new designation) radars operating in the 2–4 GHz band are usually the medium-range ground-based and shipboard search radars and air traffic control radars.

C band (G Band in the new designation) radars operating in the 4–8 GHz frequency band are usually search and fire-control radars of moderate range, weather detection radars and metric instrumentation radars.

X band (I/J band in the new designation) radars operating in the 8–12.5 GHz frequency band are mostly airborne multimode radars.

Ku, K and Ka bands (J, K and L bands in the new designation) operating in the 12.5–18 GHz frequency band (Ku), 18–26.5 GHz frequency band (K) and 26.5–40 GHz frequency band (Ka) are used for short-range applications due to severe atmospheric attenuation in these bands. These include short-range terrain avoidance and terrain following radars and space-based radars.

Radars operating in the infrared and visible bands (Laser radars) are mainly used as *Rangefinders* and *Designators*.

2.1.2.2 Transmit Wave Shape and Spectrum

Based on the transmit wave shape and spectrum, radars are classified as *unmodulated CW radar* capable of finding target velocity only, *modulated CW radar* capable of finding both range and velocity, *gated CW pulsed radar* and *modulated pulsed radar*. FM-CW radars belong to the class of modulated CW radars. Pulse Doppler radar is a popular type belonging to the category of gated CW pulse radars. Pulse-compression radar falls in the category of modulated pulse radar.

2.1.2.3 PRF Class

Based on PRF class, we have Low PRF radars including *Moving Target Indicator* (MTI) and *Moving Target Detector* (MTD), High PRF radars such as *Pulse Doppler radar* and Medium PRF radars.

2.1.2.4 Intended Mission and Mode

Radars with surface-to-surface mission are usually short-range radars that do not use Doppler. Radars with surface-to-air missions include surveillance radar, early warning radar, weather radar, fire-control radar, metric instrumentation radar, OTHR and so on. Radars with air-to-air missions are tracking radars, track-while-scan radars, airborne weather radars and radar designators. Radars with air-to-surface mission include terrain following and avoidance radars, synthetic aperture radars (SAR), ground mapping radars, radar altimeter and so on.

2.2 Basic Radar Functions

The basic functions that radar can perform include target detection, identifying target location in range and angular position and determining target velocity. The radar performs these tasks provided that the target echo signals after signal processing are sufficiently stronger than the interfering signals like noise generated in the receiver; clutter that is unwanted signal echo due to reflections from land, sea, clouds and so on; a jamming signal, which is an intentional interference; electromagnetic interference (EMI), which is an accidental interference from friendly sources such as communication systems, other radars and spillover, which is due to leakage from transmitter into receiver occurring mainly in CW radars. It may also be mentioned that not all radars are capable of measuring all these listed parameters.

2.2.1 Target Detection

Detection is the process of determining whether or not a target is present. There are four possible conditions of detection. If a target is present and detection also occurs, the result is considered correct [Figure 2.5(a)]. Similarly, if there is no target and the radar display also shows no detection condition, the result is again correct [Figure 2.5(b)]. If the target is present and radar fails to show it on display, an error is said to occur [Figure 2.5(c)]. But if the target is absent and radar shows detection, it is a different form of error referred to as a false alarm [Figure 2.5(d)]. Both of the last two conditions are error conditions; the one of false alarm is usually considered far more serious and undesirable. Such a tricky situation usually occurs because the target echo and interference signals have more or less the same shape after they have been processed in the receiver and only way to discriminate between the two is by amplitude comparison. The radar can often confuse between a weak target and a strong interference residue. For this reason, detection can only be described by probabilities; the *probability of detection* and *probability of false alarm*.

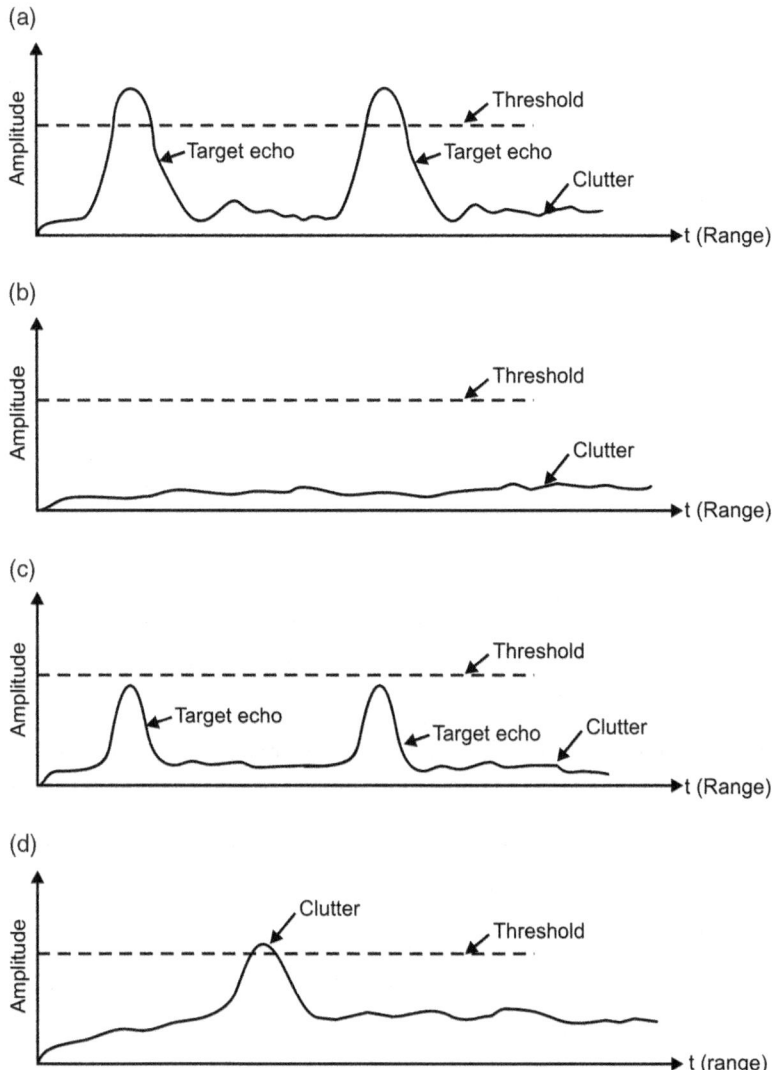

Figure 2.5 Target detection. Parts (a)–(d) show the four possible conditions of detection.

2.2.1.1 Probability of Detection and Probability of False Alarm

Probability of detection (P_d) is the probability that for any given evaluation of signal plus interference and a threshold, the result will be detection if the target is present. *Probability of false alarm* (P_{fa}) is the probability that interference alone will exceed the threshold for a compound test.

There are many other terms related to the two probabilities. These include False Alarm Number (FNA), Probability of Signal (P_s), Probability of Noise (P_n), False Alarm Time (FAT) and False Alarm Rate (FAR). *False Alarm Number* (FAN) is the reciprocal of probability of false alarm. P_s is the probability that on any given single test of signal plus interference and threshold, the result will be a threshold crossing if the target is present. P_d differs from P_s in the sense that the former is the result of many consecutive signal echoes processed together. P_n is the probability that noise alone will cross the threshold for a single test. P_{fa} is related to the P_n in the same way as P_d is to P_s. FAT is the mean time between noise threshold crossings.

Different terms are computed from the following equations. Probability of false alarm is given by eqn. 2.1.

$$P_{fa} = 1/\left[(FAT) \times B\right] \tag{2.1}$$

B = System bandwidth at the point of test

FAR is the average number of false alarms per second and is given by the reciprocal of FAT. That is,

$$FAR = 1/FAT \tag{2.2}$$

Also,

$$FAR = P_{fa} \times B \tag{2.3}$$

2.2.2 Target Location

The target location is expressed in terms of its range, azimuth angle and elevation angle (Figure 2.6). Range is the shortest distance of the target from the radar regardless of direction. Azimuth angle is the angle between the antenna beam's projection on the local horizontal and some reference. The azimuth reference in case of land-based radars is usually the true north. Ship-borne radars usually reference the ship's head, which is a line parallel to the ship's roll axis. Airborne radars reference the roll axis on the local horizontal plane. Elevation angle is the angle between radar antenna's beam axis and the local horizontal. Local horizontal in case of land-based radars is the plane passing through antenna's centre of radiation and perpendicular to Earth's radius passing through the same point. For ship- and airborne radars, it is also the plane containing the vehicle's pitch and roll axes.

2.2.2.1 Ranging

Ranging is based on the principle of measuring the time delay between the transmission of a pulse of electromagnetic energy by the radar and the detection of the received echo (Figure 2.7). The product of measured time difference and the velocity of propagation of electromagnetic waves is equal to twice the target range. Time difference is measured from the centre of the transmitted pulse to the centre of the received echo in what is known as the centroid ranging. In another case used less often called leading edge ranging, it is measured

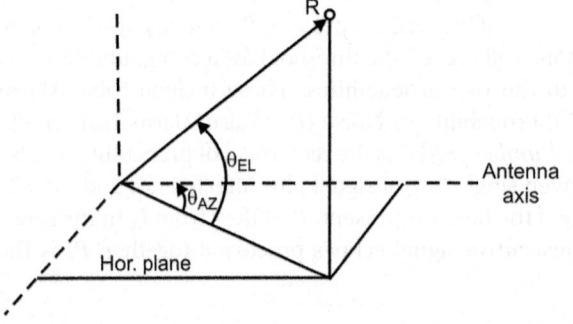

Figure 2.6 Target location.

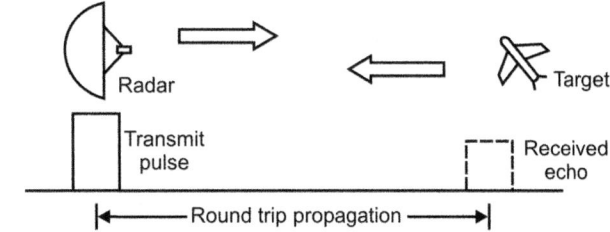

Figure 2.7 Ranging.

from leading edge of the transmitted pulse to the leading edge of the received echo pulse. The target range can be computed from eqn. 2.4.

$$R = \left[(c \times t)/2 \right] \tag{2.4}$$

R = Target range
t = Round trip propagation time
c = Velocity of propagation.

2.2.2.2 Range Ambiguity

Range ambiguity occurs when multiple target positions produce the same reported information and the radar is unable to distinguish between the two in terms of range. It occurs when the received echoes are attributed to wrong transmitted pulses. Refer to Figure 2.8(a). The transmit pulses T_1, T_2, T_3 and so on produce echoes E_1, E_2, E_3, respectively. Most radar systems, for a given received echo, take the last transmitted pulse as the reference for computing the round trip propagation time and hence the range. If there are no other transmissions between a transmitted pulse and the corresponding received echo, this assumption is correct and the target is called a Range Zone 1 target and is ranged correctly. Now, refer to Figure 2.8(b). Here, the echo for a given transmitted pulse is received after the next adjacent transmitted pulse has been radiated with the result that the radar, on the assumption of the Range Zone1 target, measures the range as R_a and not R leading to ambiguity. This is a Range Zone 2 target and this fact could be used to determine the true range R from apparent range R_a. To sum up, any target for which the range zone is not known is ambiguous in range. Since radars initially assume the targets to be Range Zone 1, there are no ambiguities when the round trip propagation time is less than the time period of the transmitted pulse train.

Figure 2.8 Range ambiguity: (a) Range Zone-1 target ranged correctly and (b) Range Zone-2 target ambiguously.

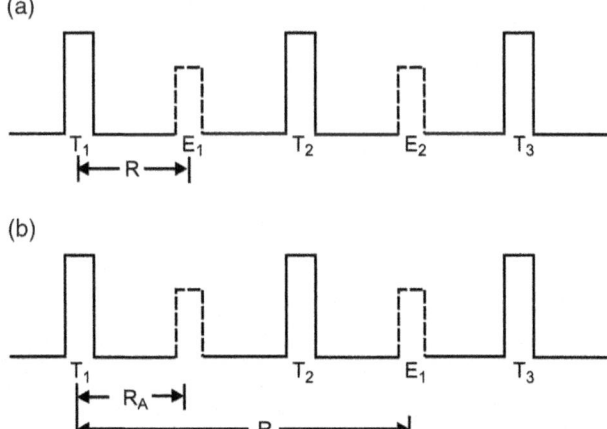

2.2.2.3 Range Quantization and Granularity

Range results from the digital nature of the radar's signal processing, data processing and display sections. Due to this, data occurs in discrete values. The smallest output data change is termed as granularity. Such radars that handle target positions digitally can only report a target's range in discrete steps. If the smallest output data change corresponds to a range of 100 m, radar would give range information in steps of 100 m. A target with a true range of say 5.23 km will be displayed as having a range of 5.2 km and the one with a range of 5.27 km will be ranged as 5.3 km. It must not be confused with range resolution, which is the smallest range difference between the closely located targets that the radar can detect as two separate targets. Range resolution and other related topics are discussed in detail in the next section.

2.2.2.4 Angular Position

Angular position is determined by measuring angular coordinates; that is, azimuth and elevation angles. These can be determined from the knowledge of the direction of arrival of the echo pulse or in other words, the direction of antenna's main beam at the time of detection. This of course assumes that the detection signal originated from the direction of antenna's main beam. Apparent ambiguity can occur if the signal enters through the side lobe.

2.2.3 Target Velocity

When the target is moving with respect to radar, the frequency of received echo is different from the transmitted frequency by an amount equal to the Doppler shift whose magnitude and sense are, respectively, proportional to the magnitude of the radial component of the relative velocity between the target and the radar and the sense of this velocity component. The Doppler shift is positive or the received frequency is higher when the target is moving towards the radar and negative or the received frequency is lower when the target is moving away from the radar. The target's radial velocity information is extracted from this Doppler frequency shift. Doppler shift f_d is given by eqn. 2.5.

$$f_d = 2V_r / \lambda \tag{2.5}$$

V_r = Radial velocity difference between target and radar
λ = Operating wavelength.

If θ is the angle between the target/radar velocity vector and the radar antenna axis (Figure 2.9), then

$$f_d = \left[(2V_r / \lambda) \cos\theta \right] \tag{2.6}$$

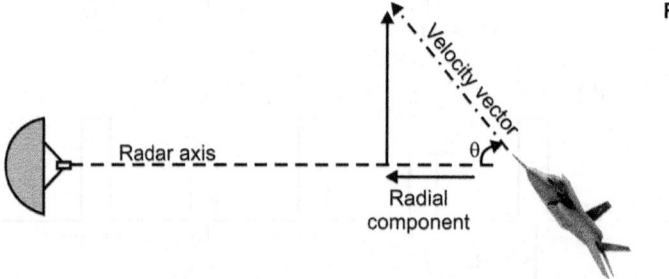

Figure 2.9 Target velocity.

Radar axis

Velocity vector

θ

Radial component

Further, if θ_h and θ_v are the horizontal and vertical angles, respectively, between the radar's axis and the target's velocity vector, then

$$\cos\theta = \cos\theta_h \cos\theta_v \tag{2.7}$$

2.2.3.1 Doppler Ambiguity

Doppler ambiguities exist in all those radar systems where the signals are processed digitally including pulse Doppler radars, digitally processed CW radars and so on. In case of digital processing, the spectra of transmit and receive signals are discrete in nature. In case of a radar making use of Doppler principle, the received spectrum will be offset in frequency from the transmit frequency by Doppler shift. For extracting true information, it is therefore necessary to know the transmit spectral lines that caused the Doppler shifted received echo. Doppler ambiguity will occur if it is not known as to which transmit spectral line caused a particular spectral line. As the data is sampled at a rate equal to the pulse repetition rate of the transmit pulses and the Nyquist criterion must be met with, the unambiguous Doppler frequency for a given pulse repetition frequency is given by:

$$\Delta f = \pm[PRF/2] \tag{2.8}$$

Δf = Doppler shift = $2V_r/\lambda$

If the target's maximum radial velocity is $\pm V_r$, then the minimum value of PRF, which is unambiguous in both magnitude and sense of velocity, is given by eqn. 2.9.

$$(PRF)_{min} = 4V_r/\lambda \tag{2.9}$$

If it is desired to avoid ambiguity only in the Doppler magnitude, then

$$(PRF)_{min} = 2V_r/\lambda \tag{2.10}$$

It is indeed a practice in some pulse Doppler radars which rely on detections in multiple PRFs during the time on target to resolve ambiguity in Doppler sense. To sum up, we can say that low PRF radars are unambiguous in range but highly ambiguous in Doppler whereas high PRF radars are unambiguous in Doppler but highly ambiguous in range.

2.2.3.2 Doppler Granularity

Doppler granularity does the same to Doppler shift information as range granularity does to range information. It is the result of target echoes being processed in a discrete set of band widths called Doppler bins. The result is that the information on Doppler frequency shift and hence the target velocity is also in discrete form. If each Doppler bin is 100 Hz wide, the Doppler shift information will be displayed in steps of 100 Hz.

2.3 Accuracy and Resolution

Accuracy and *resolution* are the two most important parameters concerning the radar system performance. While accuracy tells about the precision with which a radar can measure individual parameters like range, Doppler, angular coordinates and so on, Resolution defines the ability of the radar to effectively handle multiple targets. Again, we could talk about range resolution, Doppler resolution, angular resolution and so on. The two parameters are discussed in detail in the following paragraphs.

2.3.1 Accuracy

Accuracy of measurement of a given parameter is the deviation of the measured value from its true value. For instance, in case of target range, an accuracy of ±10 m for a target at 1 km would mean that the measured range could be anywhere between 990 and 1010 m. Accuracy of radar parameters is usually expressed in absolute terms. Accuracy of different parameters such as range, Doppler and angular position would be expressed in metres, Hz and degrees, respectively.

There are two types of errors that lead to measurement inaccuracy. These are: (1) bias errors and (2) noise errors. Bias errors refer to continuous deviation or offset from the true value and are usually due to factors like equipment miscalibration, servo lags and so on. Noise errors refer to random uncertainties in the measured parameter and are primarily due to the echo signal getting contaminated by noise and other interfering signals. These errors are expressed as a standard deviation of random uncertainties.

Though, broadly speaking, the mechanisms that spoil the accuracy in case of major radar parameters like range, angular position and Doppler are the same, there are one or two changes in each of them. The factors affecting range accuracy, Doppler accuracy and tracking accuracy are briefly described in the following paragraphs.

2.3.2 Range Accuracy

Range Accuracy primarily depends upon two factors; namely the accuracy with which the radar determines the time of arrival of the echo pulse and the accuracy of measurement of time delay between the transmitted signal and the received echo. The error in the measurement of time of arrival is caused primarily by noise and other forms of interference contaminating the echo. The time of arrival is measured at the centroid of the echo pulse, though in some applications the leading edge of the envelope is also used. Due to noise and interference, the signal centroid occurs at a time other than that of the actual centre of the echo pulse. The range error δR due to a time of arrival error of $\delta\tau$ can be computed from eqn. 2.11.

$$\delta R = V_p \times (\delta\tau/2) \tag{2.11}$$

V_p = Velocity of propagation.

The other predominant source of error affecting range accuracy is the time delay error caused by erroneous time bases and miscalibration of systems. Range error due to this factor is given by eqn. 2.12.

$$\delta R = c \times (\delta T_p/2) \tag{2.12}$$

δT_p = total round trip propagation error.

2.3.3 Angular Position Accuracy

Then angular accuracy with which a radar can establish the angular coordinates of a target include the antenna pattern beam width, echo signal-to-noise ratio, scintillation (target amplitude fluctuations) and glint (target phase fluctuations) in case of a range finding radar and in addition the servo system noise in case of a tracking radar. A narrower beam width, higher S/N ratio, lower scintillation and glint and reduced servo noise are definitely the desired features. Precise determination of angular coordinates of a target is very vital to tracking radar. So, with

Figure 2.10 Angular position accuracy: (a) error versus range in the case of conical scanning and sequential lobing and (b) error versus range in the case of monopulse tracking.

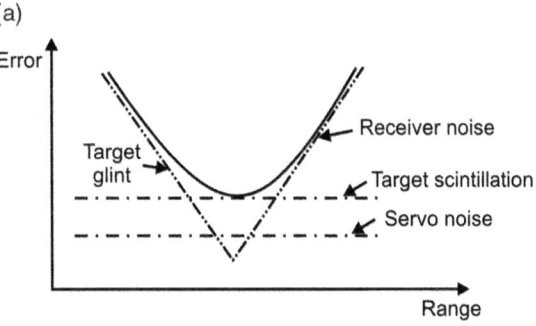

(a)

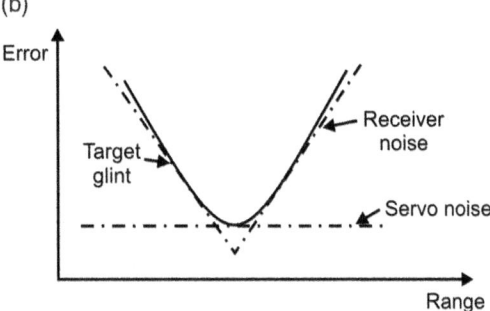

(b)

reference to tracking radar, Figure 2.10(a) and (b) shows the angular error as a function of target range in case of conical scan and lobing and monopulse tracking radars. The contribution of various error contributing factors outlined previously is also presented in these figures. As is obvious from the given curves, monopulse radar is not affected by scintillation. The error dependence on beam width and *S/N* ratio is expressed by eqn. 2.13.

$$\delta\theta = \frac{\theta_3}{K_1\sqrt{2\tau B(S/N)}} \tag{2.13}$$

θ_3 = 3 dB beam width
τ = Pulse width
B = Equivalent noise bandwidth
S/N = Power signal-to-noise ratio
K_1 = Constant whose value depends upon the tracking technique used (whether conical scan or monopulse).

Angular error due to glint is given by eqn. 2.14.

$$\delta\theta_g = K_2/R \tag{2.14}$$

K_2 = Constant whose value depends upon target span and glint parameters.

Scintillation track errors refer to the errors caused by amplitude fluctuations and thus only those radars that depend for their operation on sequential target amplitude variations to generate tracking information are affected by scintillation phenomenon. It is particularly severe at low elevation angles.

Another factor that can produce angular error is multipath propagation of signal; that is, the signal is travelling from radar to target and back following more than one propagation paths. One effective solution to this problem is to use a very narrow elevation beam so as to prevent reflection points on ground and water surfaces from being in the main response of the antenna.

2.3.4 Resolution

Measurement resolution of a certain radar parameter is the ability of the radar to measure that parameter for multiple targets. For example, high range resolution in radar would allow the same to measure range of two closely located targets in range. It also refers to the ability of radar to detect multiple features on the same target. For example different parts of an aircraft may return different Doppler information back to the radar due to some relative motion between them and this Doppler information could be used by the radar to detect certain physical features of the target. Radar targets could be resolved in four dimensions; that is, range, azimuth cross range, elevation cross range and Doppler.

2.3.5 Range Resolution

Range resolution is the ability of the radar to detect multiple targets separated in radial range but with the same angular position (Figure 2.11). It depends upon the processed pulse width which in other words is indicative of RF signal bandwidth. Except for radars that use special techniques to alter the received pulse width during processing as is the case with pulse-compression radar, the processed pulse width is the same as the transmitted pulse width. The narrower the pulse width, the closer the two targets can be in range and still be detected as two separate targets rather than one big target. This phenomenon is illustrated in Figure 2.12. Range resolution is expressed by eqn. 2.15.

$$\Delta R = (c\tau/2) = (c/2B) \tag{2.15}$$

τ = Processed pulse width
B = Bandwidth

Effective bandwidth of any processed signal is approximately reciprocal of its pulse width. To get a feel of the range resolution, it would be worth stating here that a wide processed echo pulse would mean a range resolution of 150 m; that is, targets separated in range by a distance equal to or greater than 150 m would get resolved.

As is clear from this description, a higher range resolution means a shorter pulse width that in conventional pulsed radar would also mean a shorter transmitted pulse width. We have also seen earlier that a longer pulse is required for a higher ranging capability, which implies that conventional pulsed radar cannot achieve both a higher range as well as higher range

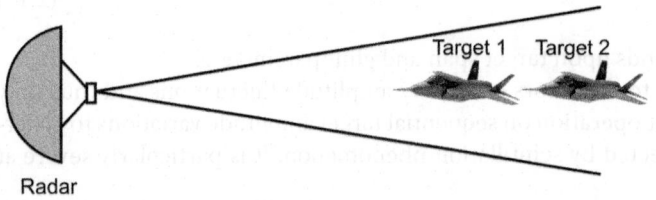

Figure 2.11 Range resolution.

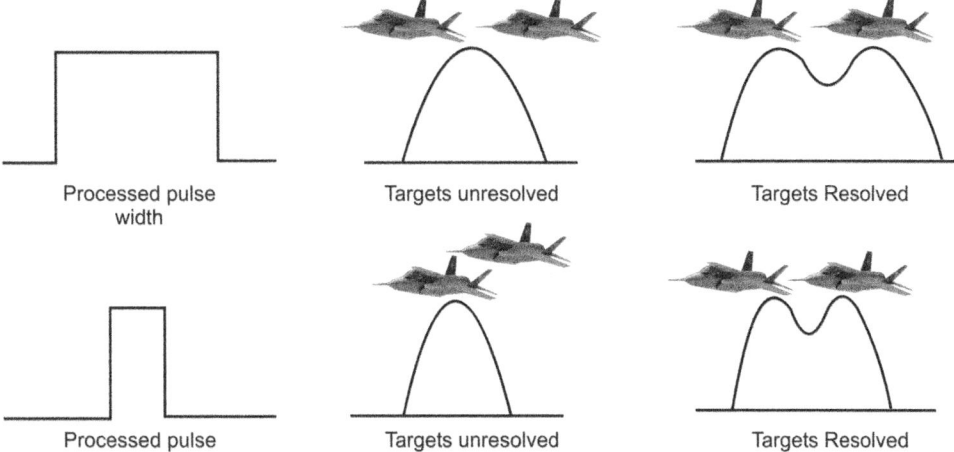

Figure 2.12 Pulse width dependence of range resolution.

resolution capabilities as the requirements for the two are opposing. Since from the application viewpoint, a higher range capability is inescapable, a longer transmitted pulse is unavoidable. In that case, one of the methods to enhance resolution is to compress the received pulse during processing. This is implemented in pulse-compression radar. The operational mechanism of pulse-compression radar will be discussed in the latter part of the chapter.

2.3.6 Cross-Range Resolution

Cross-range resolution is the resolution in a direction perpendicular to the antenna axis. We have the azimuth cross range and the elevation cross range (Figure 2.13).

The cross-range resolution depends upon the antenna beam width. Narrower beam width has higher resolution and thus can resolve more closely spaced targets in the cross range. This is illustrated in Figure 2.14. Cross-range resolution is expressed by eqn. 2.16.

$$\Delta R\left(\text{cross range}\right) = \left(R \times \theta_3\right) = \left[\left(R \times \lambda\right)/D_{\text{eff}}\right] \tag{2.16}$$

A higher cross-range resolution would mean an antenna with a much large aperture. A technique to enhance cross-range resolution even with an antenna of a practical size is used in a Synthetic Aperture Radar (SAR) where the antenna is moved across the target and the echo signals received with the antenna in different positions. These signals are processed in such a way as to achieve the same result that would have been achieved if the antenna had an aperture equal to the distance moved by the antenna (Figure 2.15).

Figure 2.13 Cross-range resolution.

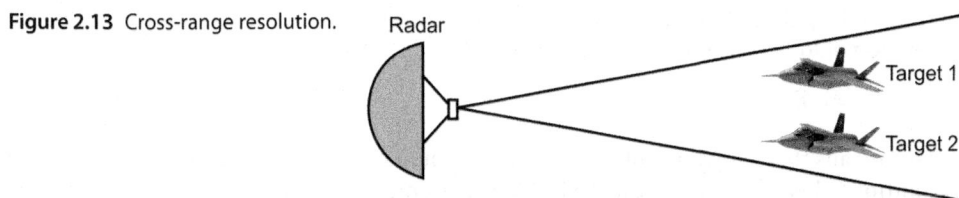

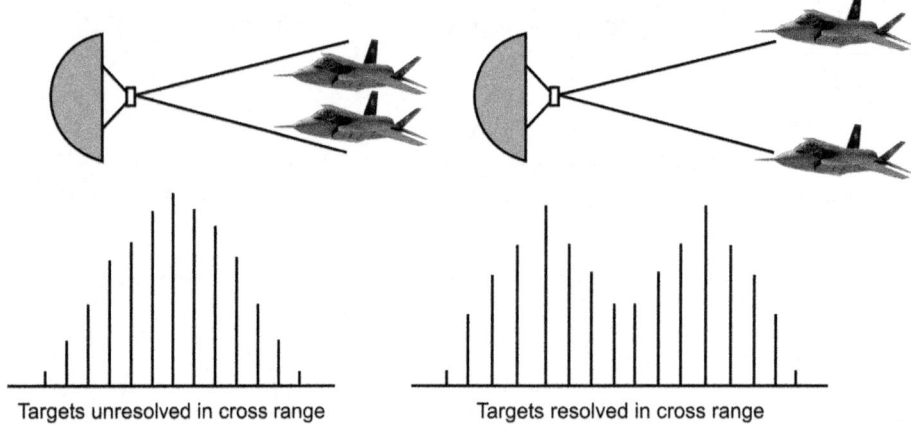

Targets unresolved in cross range | Targets resolved in cross range

Figure 2.14 Pulse width dependence of cross-range resolution.

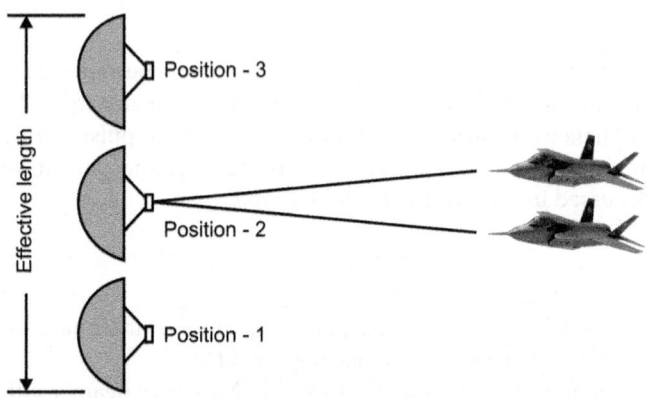

Figure 2.15 Synthetic aperture radar concept to increase cross-range resolution.

2.3.7 Doppler Resolution

Doppler resolution is the ability of the radar to detect and differentiate between the targets at the same range and angular position but having different radial velocities. Higher Doppler resolution is particularly useful in identifying physical characteristics of the target such as an aircraft by resolving the net target movement and movement of some other parts of the targets such as rotating blades of the engine on the basis of Doppler. As the moving targets produce a Doppler shift proportional to their relative radial velocity with respect to the radar, the Doppler shifts produced by the two targets will be different. This difference between the two Doppler frequencies that can be resolved is the reciprocal of the look or dwell time. For instance, if the look time is 1 s, the targets producing a Doppler difference of 1 Hz can only be resolved. On the other hand, a look time of 10 s will allow Doppler difference as small as 100 mHz to be resolved. Similarly, if the look time is 0.1 s, the targets with a radial velocity difference that produces a minimum Doppler difference of 10 Hz will only be resolved. Resolvable Doppler difference as a function of look time is expressed by eqn. 2.17.

$$\Delta f_d = 1/T \tag{2.17}$$

This means that a longer look or dwell time would be required to achieve a higher Doppler resolution. This goes against the desirable requirements of range processing where a smaller look time is preferred so as to ensure that the target range does not change much as the range information is being processed.

2.4 Radar Cross-Section

Radar cross-section (RCS) of target is an indicator of its susceptibility to detection by radar. The lower the radar cross-section, the higher the immunity of the target to detection by the radar. On the contrary, a higher RCS makes it easier to detect and identify an object. It is a measure of the extent the target reflects radar signal in the direction of the radar for a given radar signal strength intercepted by the target. The RCS of target is a significant parameter as it directly affects the survivability of the target in a hostile environment. For example, reduction in the RCS of an aircraft reduces the required jammer power to achieve the same effectiveness; reduces the burnthrough range for same jammer power and reduces the detection range if jamming were not considered. In the following paragraphs different aspects of target radar cross-sections are discussed including the RCS concept, factors determining RCS, RCS measurement and RCS reduction mechanisms.

2.4.1 RCS Concept

RCS of a target, denoted by σ, is measured as a ratio of the transmitted radar signal power back-scattered from the target per unit solid angle in the direction of radar to the radar signal power intercepted by the target. Conceptually, RCS is measured by comparing strength of reflected signal from the target to the reflected signal from a perfectly smooth conducting metal sphere with a frontal or projected area of 1 m^2 (Figure 2.16). RCS is measured in m^2 and is therefore the projected area of an isotropically radiating perfectly conducting sphere that would reflect the same power in the direction of radar as the one that is actually reflected by the target for a given incident power. RCS is also measured in dBsm (or dBm^2), which is decibels relative to 1 m^2. RCS in dBsm or dBm^2 is expressed as 10 log σ where σ is RCS in m^2.

A sphere is used for comparison while computing RCS as a sphere projects the same area irrespective of its orientation. Also, RCS of a sphere is independent of frequency provided that the operating wavelength is much smaller than both the range as well as the radius of the sphere. Most structures including a sphere exhibit different RCS dependence on operating frequency.

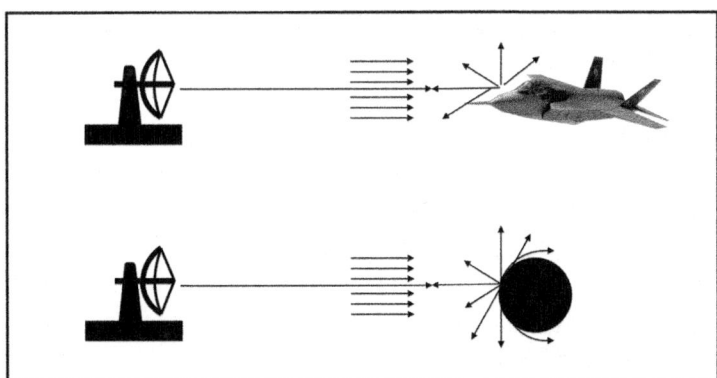

Figure 2.16 Concept of RCS.

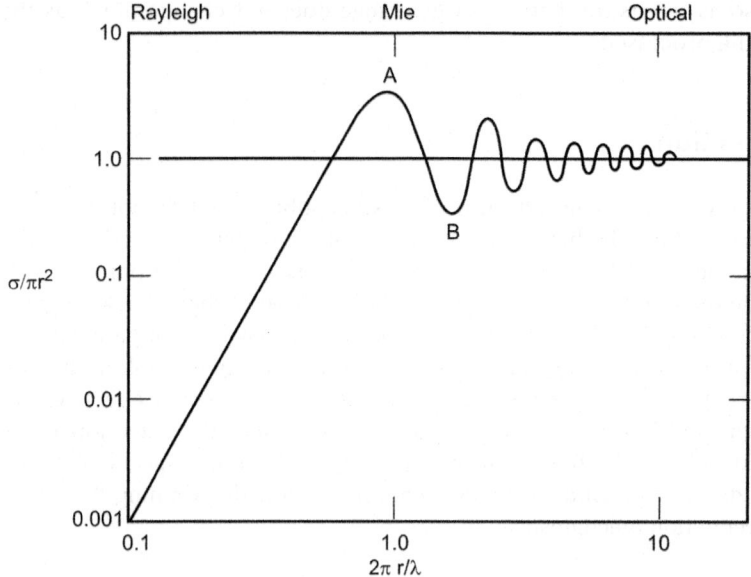

Figure 2.17 RCS of a metal sphere.

There are three identifiable frequency regimes; namely the low frequency or Rayleigh region, resonance or Mie region and high-frequency or optical region. Rayleigh scattering is the predominant source of scattering when target dimensions are much lower than the radar wavelength. This region occurs at lower frequencies and in this region RCS is proportional to the fourth power of the frequency. In the case of Mie scattering, target dimensions and the radar wavelength are of the same order. The Mie region corresponds to medium frequencies. The RCS oscillates in the resonance region. In the optical region corresponding to high frequencies, target dimensions are very large compared to the radar wavelength. In this region RCS is roughly the same size as the real area of target.

Refer to Figure 2.17. In the Rayleigh region, RCS (σ) of a sphere is given by eqn. 2.18.

$$\sigma = \left(\pi r^2 \right) \times \left[7.11 \times \left(kr \right)^4 \right] \tag{2.18}$$

where $k = 2\pi / \lambda$

In the Mie region, σ of a sphere is given by eqns. 2.19 and 2.20.

$$\sigma_{MAX} = 4\pi r^2 \tag{2.19}$$

$$\sigma_{MIN} = 0.26\pi r^2 \tag{2.20}$$

In the optical region, σ of a sphere is given by eqn. 2.21.

$$\sigma = \pi r^2 \tag{2.21}$$

RCS of some other basic shapes are expressed by eqns. 2.22 (for a flat plate), 2.23 (for a cylinder) and 2.24 (for a dihedral corner reflector).

$$\sigma \left(\text{flat plate} \right) = 4\pi a^2 b^2 / \lambda^2 \tag{2.22}$$

a, b = Width and height of the plate
λ = Operating wavelength

$$\sigma \text{ (cylinder)} = 4\pi ab^2 / \lambda \tag{2.23}$$

a, b = Width and height of the plates making the corner reflector
λ = Operating wavelength

$$\sigma \text{ (diheral corner reflector)} = 8\pi a^2 b^2 / \lambda^2 \tag{2.24}$$

a, b = Width and height of the plates making the corner reflector
λ = Operating wavelength.

Back to the concept of radar cross-section, the conceptual definition of RCS considers the fact that not all of the radiated energy falls on the target. RCS may be viewed as the product of three factors namely the projected cross-section, reflectivity and directivity. RCS is the product of the three factors. We have discussed the projected cross-section. Reflectivity is the fraction of the radar signal power intercepted by the target that is scattered by the target. Directivity is given by the ratio of the backscattered power into the radar's direction to the power that would have been backscattered considering an isotropic radiation; that is, the one that scatters uniformly in all directions.

2.4.2 Factors Determining RCS

Radar cross-section of target is influenced by both radar signal parameters such as operating wavelength and polarization as well as target characteristics such as size, shape, orientation and so on. Important factors that influence radar cross-section include the following.

- Target size and shape and surface roughness
- Target material composition with reference to its electromagnetic properties
- Relative size of target in relation to operating wavelength
- Target orientation.

2.4.2.1 Target Size and Shape

RCS is directly proportional to target size. Larger the target, greater is its RCS value. In addition to the absolute size of the target, its shape also influences the RCS. Different shapes present different incident angles to the radar signal. Radar waves that make large angles of incidence are reflected away from the direction of the radar and therefore contribute to reducing the RCS. Very large incidence angles produce equally large angles of reflection leading to forward scattering. This makes the target stealthier. For example, the F-117A Nighthawk fighter aircraft by Lockheed–Martin is designed to have flat and large angled surfaces, which significantly contributes to its having a low RCS. The RCS of F-117A stealth fighter is estimated to be between 10 cm² and 100 cm². Similarly, air frame shaping such as alignment of planform edges and fixed-geometry S-ducts (or serpentine inlets) that prevent line-of-sight of the engine faces from any exterior view in F-22 Raptor fighter aircraft are important factors that give it an extremely low RCS of 1 cm².

Indentations present in relief in a surface such as those arising from open bomb bays, ordnance pylons, joints between constructed sections and engine intakes are potent corner reflectors contributing toward increase in RCS form many orientations. It is more so as it is impractical to coat these surfaces with radar absorbent materials.

2.4.2.2 Target Material

The materials used in the construction of the target and also the materials used to coat the surfaces play a significant role in determining the RCS. There are materials such as metals that are strong reflectors of radar waves. Even a thin layer of metal coating makes the object a strong reflector of radar energy. Chaff that is often made of metallized plastic or glass is a good example. There are materials such as wood, plastic, and fibre glass that are less reflective. Use of radar absorbent materials significantly reduces the RCS. The F-117A nighthawk stealth fighter and the bomber B-2 are well-known examples of stealth technology that minimize RCS by using both aerodynamic geometry and radar absorbent materials applied to the surface of their fuselages. Radar absorbent materials minimize the reflection of radar waves thereby reducing the RCS. There are two broad categories of radar absorbing materials; namely *impedance matching absorbers* and *resonant absorbers.* There are other absorbing material configurations that have features of both the classifications.

Impedance matching absorbers are based on minimizing reflection at the interface of the propagating medium such as free space or air and the material surface. As is evident from the expression of reflection coefficient, the reflection coefficient becomes zero when material offers an impedance equal the impedance of the propagating medium. In the case of radar waves propagating through free space, it occurs for a material impedance of 377 Ω. In practical impedance matching absorber materials, the minimized reflection coefficient depends upon the magnitude of impedance step at the interface of propagating and material media. Three types of impedance matching radar absorbing materials are in common use. These are pyramidal, tapered and matched absorbers. Impedance matching absorbers tend to be bulky and heavy as for significantly reducing reflection, the material needs to be one or more wavelengths thick.

Resonant absorbers are based on the principle of destructive interference between the two reflected waves; first, that is reflected from the first interface of the propagating medium and the material medium and second, that is reflected from the metal backing. At the first interface, the radar wave is partly reflected and partly transmitted. The one that is reflected undergoes a phase reversal of π radians. The one that is transmitted travels through the length of material absorber and gets reflected from metal backing again undergoing a phase reversal of π radians. If the optical path difference of the transmitted wave is odd multiple of $\lambda/2$; the two waves will destructively interfere at the surface. Total reflected radar signal strength becomes zero when the magnitudes of the two reflected waves are equal. Resonant absorbers are also called tuned absorbers or quarter wavelength absorbers. Dallenbach layers, Salisbury screen and Jaumann layers are common resonant absorbers.

In addition to physical shape and composite materials making up the target, RCS also depends upon various subsystems such as antennas and other sensors mounted on the target platform. These subsystems may be designed to meet low RCS requirements in addition to meeting subsystems' performance requirements. In some cases, these subsystems may increase the RCS of an otherwise low RCS platform. One such example is the use of reciprocal high gain antenna operating in the frequency band that also includes the radar operating frequency. If the antenna beam pointed towards the radar, it will increase the RCS.

2.4.2.3 Operating Wavelength

RCS is strong function of operating frequency or wavelength. The three frequency regimes, namely the low frequency or Rayleigh regime, mid-frequency or Mie regime and high-frequency or optical regime, were described in an earlier paragraph. The three regimes, respectively, correspond to the conditions of $(2\pi r/\lambda) < 1$ (Rayleigh regime), $(2\pi r/\lambda) \cong 1$ (Mie regime) and $(2\pi r/\lambda) > 1$ (optical regime). RCS is a function of relative size of target with respect to operating wavelength. As outlined earlier, the RCS is approximately equal to the real area of the target when the target size is much smaller than the operating wavelength. For a target size roughly equal to the operating wavelength, the RCS may be greater or smaller than the real area depending upon operating wavelength before it approaches the real value in the optical region. For a target size much less than the operating wavelength, RCS varies as λ^{-4}.

2.4.2.4 Target Orientation

Target orientation with respect to radar line-of-sight strongly influences the RCS. For example, a fighter aircraft presents a much larger area when viewed from the side than when it is viewed from the front. The fact that military targets such as fighter aircraft have many reflecting elements and shapes and also that targets move relative to radar line-of-sight, relative orientations of various reflecting elements and shapes on the target structure make RCS dependence on target orientation a very complex phenomenon.

2.4.3 Radar Cross-Sections of Typical Targets

Typical values of radar cross-sections of some common military targets are 0.5 m^2 (missiles), 2–3 m^2 (small combat aircraft), 5–6 m^2 (large combat aircraft), 100 m^2 (cargo aircraft), less than 0.1 m^2 (stealth aircraft) and 3000–1 000 000 m^2 (ships). RCS of stealth fighter aircraft is several orders of magnitude lower than the RCS of conventional fighter aircraft. For example, RCS of the F-15 fighter aircraft is 25 m^2 while in the case of F-22 Raptor, it is 0.0001 m^2. Also, RCS of the B-52 strategic bomber and B-2 strategic stealth bomber, respectively, are 100 m^2 and 0.1 m^2. As another example, while the MiG-29 and MiG-29K have an RCS of about 5 m^2 and 1 – 1.5 m^2, respectively, the stealthier MiG-35 has an RCS of 0.3 m^2. The extent to which use of stealth technology is able to reduce the radar cross-section of these gigantic structures can be gauged from the fact that an insect and a bird typically have radar cross-sections of 0.00001 m^2 and 0.01 m^2. A human being has an RCS of 1 m^2. These figures are similar to those observed in the case of state-of-the-art stealth fighter aircraft.

Modern stealthy cruise missiles may have RCSs as small as 0.1 m^2. A low RCS value drastically reduces maximum detection range from missile defences allowing minimal time for intercept. As an example, the US airborne warning and control system (AWACS) radar system designed to detect typical non-stealthy cruise missiles at a range of at least 227 km could detect stealthy cruise missiles only when they were within 108 km from air defences. This resulted in a significant reduction in the time available to the air-defence systems such as surface-to-air missile (SAM) battery to engage and destroy the target.

Naval surface vessels have relatively much larger RCSs compared to land-based and airborne targets. Stealthy ships employ stealth technology to make sure that they are harder to detect by radar, visual, sonar and infrared sensors. Some well-known examples of naval surface vessels employing stealth technology include the *Admiral Gorshkov-class frigate* of Russia, the French *La Fayette-class frigate*, the Chinese PLA Navy's *Type 022 missile boat*, *Type 054A frigate*, *Type 052C, 052D* and *055 destroyers*, German *MEKO ships*, the Indian *Shivalik-class frigate* and *Kolkata-class destroyer*, the Singaporean *Formidable-class frigate*, the British *Type 45 destroyer* and the US Navy's *Zumwalt-class destroyer*.

Table 2.1 summarizes the RCSs of some common military targets.

2.4.4 RCS Measurement

The RCS of a target is a vital parameter for the design of the whole gamut of static and mobile military targets including aircraft designed for different roles, surface-to-surface, surface-to-air, air-to-surface and air-to-air missiles, helicopters, unmanned aerial vehicles, different

Table 2.1 RCS of military targets.

Target	RCS (in m^2)	RCS (in dBm^2)
F-4 Phantom-II	6	7.8
F-15	10–15	10–11.75
F-16	5	7
F-16C	1.2	0.8
F-18	1	0
F-22A Raptor	0.0002–0.0005	−37–33
F-35A Lightning II	0.0015	−28.2
F-117A Nighthawk	0.001–0.01	−30–20
B-52	100	20
B-2	0.0001	−40
B-1A Excalibur	10	10
B-1B Lancer	1	0
Su-27	10–15	10–11.75
Su-30 MKI	10	10
Su-35	2	3
Su-47	0.1	−10
Su-T50	0.01	−20
MiG-29	5	7
MiG-29K	1–1.5	0–1.7
MiG-35	0.3	−5.2
MiG-44	0.1	−10
Rafale	0.1–0.2	−10–7
Typhoon	0.5	−3
LCA	1.5	1.7
Mirage-2000	1–2	0–3
Tornado	8	9
JAS-39	0.5	−3
Tomahawk SLCM	0.5	−3
Tomahawk ALCM	0.05	−13
Artillery Shell	0.0001	−40
227 mm MRLS Rocket	0.018	−17.45
T-33 Shooting Star	10	10
Tank	6–9	7.8–9.5
Truck	6–10	7.8–10
B-70A Valkyrie	40	16
Mortar Bomb	0.01	−20

classes of naval vessels including patrol boats, frigate and destroyer class ships and aircraft carriers, ground military vehicles, combat platforms, launchers, airport buildings and other important strategic installations. While it may be a simple exercise to estimate the RCS of simple geometrical configurations, theoretical estimation of RCS of actual military targets is a very complex exercise and therefore impractical. It is therefore important to carry out actual measurement of RCS for given target size and measurement specifications such as operational range, frequency, aspect angles and so on.

2.4.4.1 Principle of RCS Measurement

Measurement of a RCS basically involves illuminating the test target with the radar signal at different viewing angles, collecting radar signal back scattered from the target and then comparing it with radar signal back scattered from a calibrated target. The test target and the calibrated target suitable interface with each other are mounted on a positioning platform enabling different target orientations for the purpose of measurement. The target needs to be placed far enough so that the incident wave is an acceptably plane wave. The test setup basically comprises of suitable instrumentation radar, a positioning system for targets, test and calibrated targets, a low background environment with far field behaviour and a suitable data acquisition and control system. Figure 2.18 shows the basic block-schematic arrangement of the RCS measurement setup, which is self-explanatory.

Different approaches adopted for the measurement of a radar cross-section include the following:

- Continuous-wave (CW) RCS measurement
- Stepped CW RCS measurement
- Gated CW RCS measurement
- Frequency modulated CW RCS measurement
- Pulsed RCS measurement.

In the case of *CW RCS measurement*, a tuned narrowband receiver is used to record the CW signal at the desired frequency. This technique uses vector subtraction between measurements made with the target and background and background alone to identify correct vector signal of the target component. If the subtraction were not perfect; this approach would not be able to resolve small scatterers. Stepped CW measurement overcomes this shortcoming. In *stepped CW measurement*, instead of transmitting a single frequency, a band of frequencies is transmitted. By using time domain gating, problems due to imperfect subtraction can be overcome by gating the desired target zone. Another advantage of this approach is that it can lead

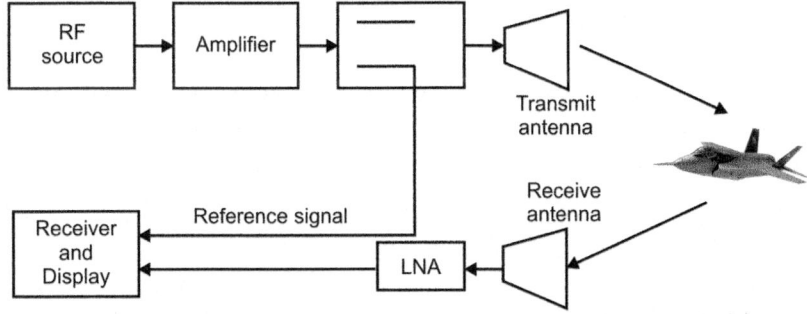

Figure 2.18 Basic block-schematic arrangement of an RCS measurement setup.

to ISAR (Inverse Synthetic Aperture Radar) imaging. *Frequency modulated CW RCS measurement* is a faster implementation of stepped CW RCS measurement where radar is specially designed to perform measurements in a single shot. In the case of *gated CW RCS measurement*, a pulsed CW waveform is generated by chopping the CW signal in the transmit chain. This helps in cutting down the target-only zone from clutter. The approach has the advantages of both pulsed and CW measurements. A modification of gated CW RCS measurement employs stepped CW or FM-CW signals within the pulse envelope. In the case of pulsed CW, which is a special implementation of a gated CW, the generated signal is a pulse rather than chopped CW.

2.4.4.2 Types of RCS Measurement

There are two broad classifications of RCS measurement based on transmit and receive location and the type of desired information about the target. In the first type, we have the monostatic and bistatic RCS and in the second type, we measure either the gross RCS or carry out diagnostic imaging to identify prominent scatters or the hot spots on the target.

In the case of *monostatic RCS measurement*, transmit and receive observation points are co-located. The target in this case is observed by the same radar. Transmit and receive observation points in the case of *bistatic RCS measurement* are spatially separated. The target in this case is illuminated by a radar station that is different from the one that is used to observe the target. *Gross RCS* is the far field RCS of the target measured from different angles as a function of frequency and polarization. Gross RCS is measured by collecting reflectivity data at the specified values of frequency and transmit and receive polarization with and without target in the required annular sector. The measurements are calibrated with reference to standard targets. *Diagnostic imaging* makes use of microwave imaging to determine prominent scattering centres

2.4.5 RCS Reduction and Enhancement

As outlined in an earlier paragraph, reduction in RCS enhances target survivability, reduces detection range of radar allowing target aircraft penetrate deeper into enemy territory without being visible to radar, reduces radar burnthrough range for a given jammer power and also reduces required jammer power for same effectiveness.

The *radar-to-target detection range* without considering the jammer is proportional to one-quarter the power of the radar cross-section of target. That is, $R \propto \sigma^{1/4}$. A 16-fold reduction in RCS reduces the detection range to half. The radar range equation is discussed in detail in Section 2.6. The radar burnthrough range is the detection range where the target return signal can first be detected through the jamming signal. That is, it is the range corresponding to cross-over point where the target return signal equal the received jamming signal. Radar burnthrough range is proportional to the square root of the RCS of a target. That is, $R_{BT} \propto \sigma^{1/2}$. A four-fold reduction in RCS reduces burnthrough range to half of its original value. Reduction in RCS also reduces the required *jammer power* in the same proportion to maintain a certain specified effectiveness. This is evident from the equations for the received signal power in the two-way range equation and received jamming signal in the one-way range equation.

We have discussed various factors influencing radar cross-section of target in an earlier section. These factors can be exploited to minimize the RCS. Common ways of reducing the RCS of a target include physical target shaping, using radar absorbing materials on target surface and using active elements on the surface. A single layer coating is effective only over a narrow bandwidth. In the case of wider bandwidths, it is necessary to use multiple layers of composite materials. Active elements work on the principle of phase cancellation in the desired direction. Target shaping relies on changing the direction of scattered energy from angular region of interest to an unimportant region. Considering the importance of aerodynamic shape of airborne targets, a trade-off needs to be worked out while deciding on target shaping. Static targets though are more amenable to target shaping.

While in most cases all efforts are made to minimize the RCS, there are some applications that require enhancement or accentuation of RCS. These include targets such as training aircraft and artificial airborne targets such as pilotless target aircraft. Training aircraft need to be continuously tracked and enhancement of RCS makes tracking more reliable. Similarly, pilotless target aircraft used for evaluation of missile systems are also tracked by radars. The radar cross-section of a small pilotless drone may be augmented to give it radar cross-section of a much larger aircraft. Augmentation of RCS achieves reliable tracking. Common methods to enhance the RCS include use of Luneburg lenses, corner reflectors and transponders with amplifiers.

The *Luneburg lens* is a passive RCS augmentation device. It is used to increase the radar reflectivity of a target without the use of additional energy. Luneburg lens is usually composed of concentric dielectric shells. Radar energy incident on one of the faces of the lens is focused at a point on the rear conductive surface of the lens, which then reflects radar energy back to the source. A *corner reflector* like a Luneburg lens is also a passive retro-reflector that reflects the incident radar energy back in the direction of the source. Unlike Luneburg lenses and corner reflectors that are passive augmenters of RCS, a *transponder* is an active augmentation device that works on the principle of capturing a portion of radar signal illuminating the target, amplifying the signal and transmitting it back to the radar.

2.5   Radar Clutter

Radar clutter is nothing but unwanted echoes. These undesired echoes could originate from a number of sources such as land or sea surfaces, insects, animals or birds, weather conditions like rain or atmospheric turbulences, objects deployed as countermeasures like chaff and decoys and so on. Clutter may be divided into three broad categories, including *surface clutter* originating from objects on land and sea surfaces; *volume clutter* produced by chaff and weather conditions such as rain and clutter originating from point targets such as birds, animals, vehicles and structures. The term 'clutter' to an extent is application specific. Clutter in one application may be a genuine target in another. For example, for radar tracking a land target such as tank to guide a missile to hit the target, scattering from vegetation on land surface or from weather conditions such as rain would be clutter. On the other hand, for airborne remote sensing radar, reflection of radar energy from natural vegetation is the primary target. Also, backscattering from atmospheric particles and turbulences would be a genuine signal for weather radar.

2.5.1   Surface Clutter

Surface clutter includes both ground clutter and sea clutter. The magnitude of clutter, that is, the magnitude of undesired radar signal backscattered in the direction of radar depends upon the nature of material composition, surface roughness and the angle the radar beam makes with the surface in azimuth and elevation directions. The backscattered radar energy is also a function of radar signal wavelength and polarization. The reflected signal is the phasor sum of reflections from a large number of individual scatterers. These individual sources of scatter may be static such as in the case of buildings, tree trunks and so on, or moving like in the case of rain drops, leaves or ripples on the sea surface. Individual sources of clutter vary spatially and temporally.

There are three major forms of scattering the clutter is generally classified into. These are specular, retro and diffuse reflections. In the case of a *specular* (or mirror like) *reflection* that occurs when the surface is smooth as compared to wavelength of operation, most of the incident energy is reflected away from the radar for angles away from normal to the surface. As a result there is negligible clutter. This is illustrated in Figure 2.19. In the case of *retro-reflection* depicted in Figure 2.20, reflection occurs at multiple reflecting surfaces. Most of radar energy is reflected back in the direction of radar producing significant clutter strength. *Diffuse reflection* occurs off microscopically rough surfaces. In this case, radar energy is reflected in all directions, a small portion of which is in the direction of radar constituting clutter. The magnitude of clutter in this case is fluctuating. Figure 2.21 shows diffuse reflection off a rough surface.

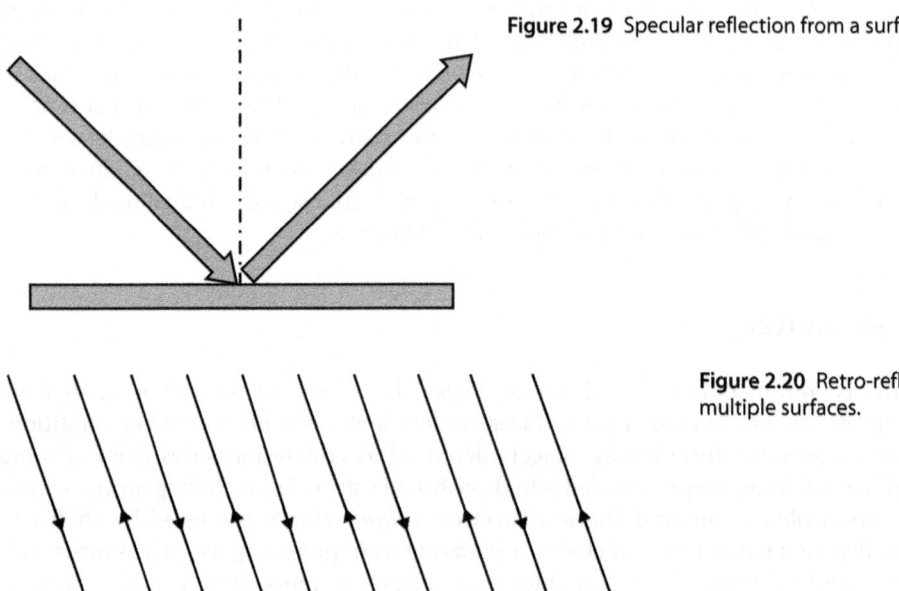

Figure 2.19 Specular reflection from a surface.

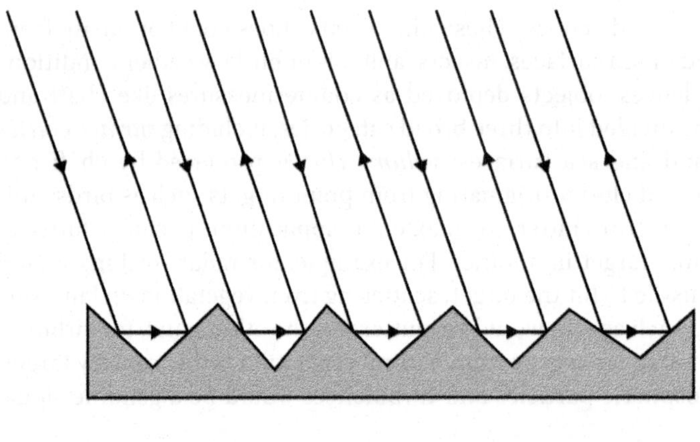

Figure 2.20 Retro-reflection from multiple surfaces.

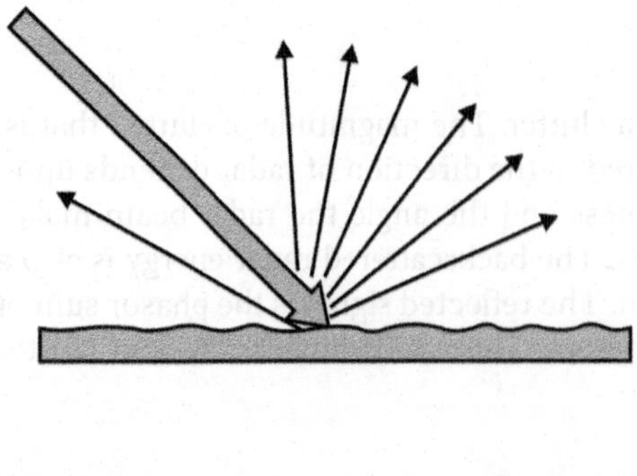

Figure 2.21 Diffuse reflection.

2.5.1.1 Constant γ Model for Surface Clutter

The constant γ model is generally used to express mean reflectivity in the case of surface clutter including ground as well as sea clutter. Mean reflectivity is more relevant due to statistical nature of clutter. According to this model, mean reflectivity σ_{MEAN} is expressed by eqn. 2.25.

$$\sigma_{MEAN} = \gamma \sin\psi \qquad (2.25)$$

σ_{MEAN} = Mean reflectivity (in m^2/m^2)
γ = Scattering effectiveness parameter
ψ = Grazing angle made by radar beam with surface (in radians).

Scattering effectiveness parameter γ is around -5 dB for urban and mountainous regions, in the range of -10–15 dB for land covered with crops, bushes and trees and -20 dB for flat land such as deserts, land covered with grass and marsh. Figure 2.22 shows variations of σ_{MEAN} as a function of grazing angle for different types of surfaces. The modelled value of mean reflectivity is independent of operational wavelength and polarization. Also, the measured value of mean reflectivity is less than the modelled value for low values of grazing angle due to effects of propagation factor and is higher than the value predicted by the model at higher angles due to the contribution from quasi-specular reflections.

Application of the γ-model to sea clutter requires that the γ-dependence on wind direction and operational wavelength is also taken into account. γ-dependence on wind speed and direction as

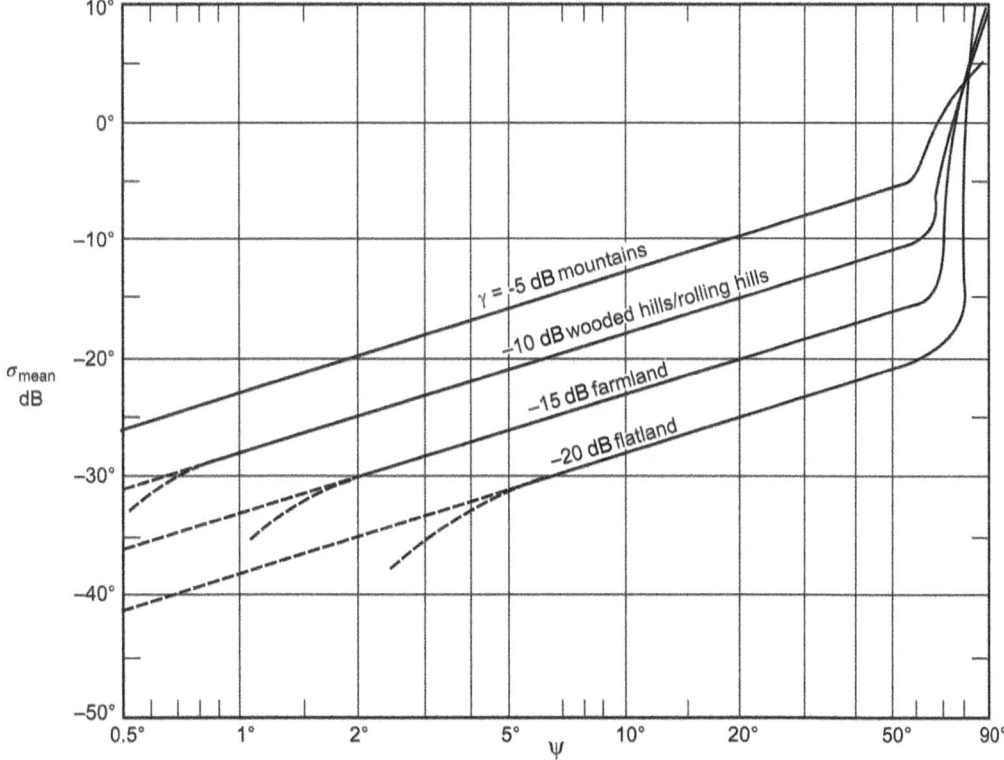

Figure 2.22 Mean reflectivity versus grazing angle for different types of surfaces.

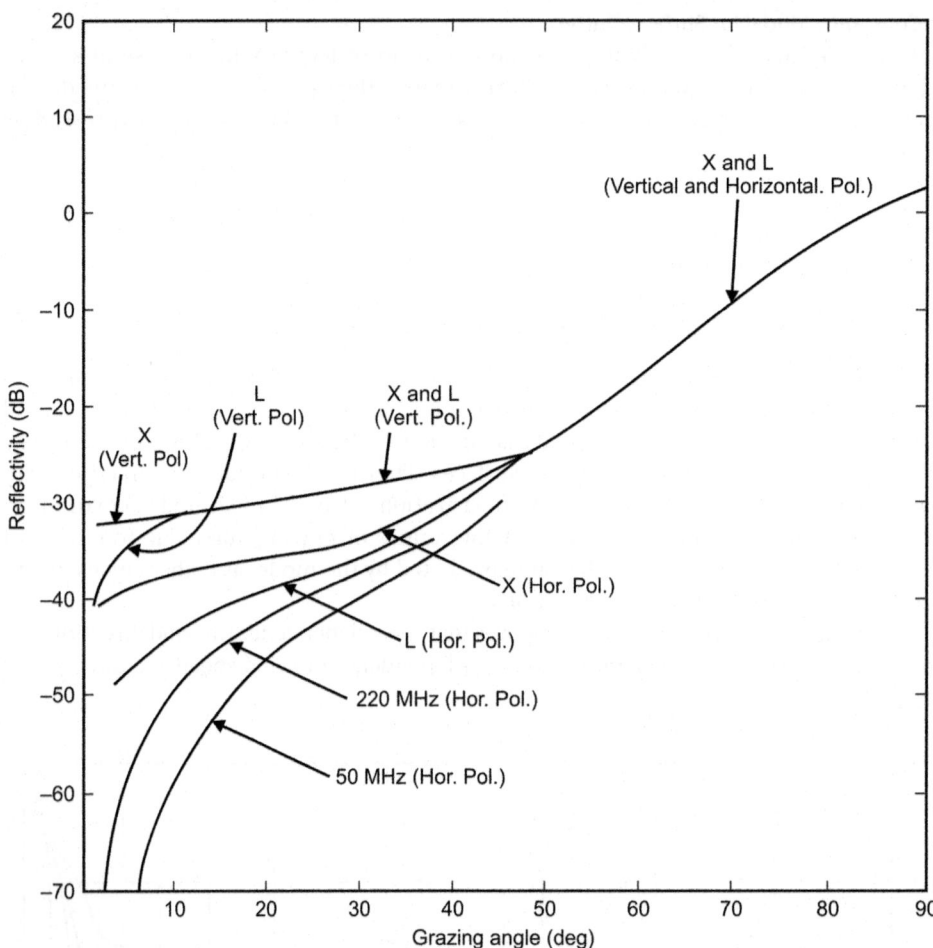

Figure 2.23 Sea clutter reflectivity versus grazing angle for different wavelength bands and polarizations.

expressed by the Beaufort wind scale K_B and operational wavelength λ is given by eqn. 2.26. There is also a polarization-dependent component at low grazing angles.

$$10\log\gamma = 6K_B - 10\log\lambda - 64 \tag{2.26}$$

Beaufort scale associates a different number with different types of wind. For example, Beaufort numbers for light breeze, strong breeze and a violent storm are, respectively, 4–6, 22–27 and 56–63. Figure 2.23 depicts the sea clutter reflectivity as a function of grazing angle for different operational wavelength bands and polarization.

2.5.1.2 Computing Surface Clutter

The radar cross-section of surface clutter is given by the product of mean reflectivity and surface area and is expressed by eqn. 2.27. This is with the assumption that the target is close to Earth's surface such that the Earth and the target are in the same range resolution cell.

$$Surface\quad clutter \ = \sigma_{MEAN} \times A \tag{2.27}$$

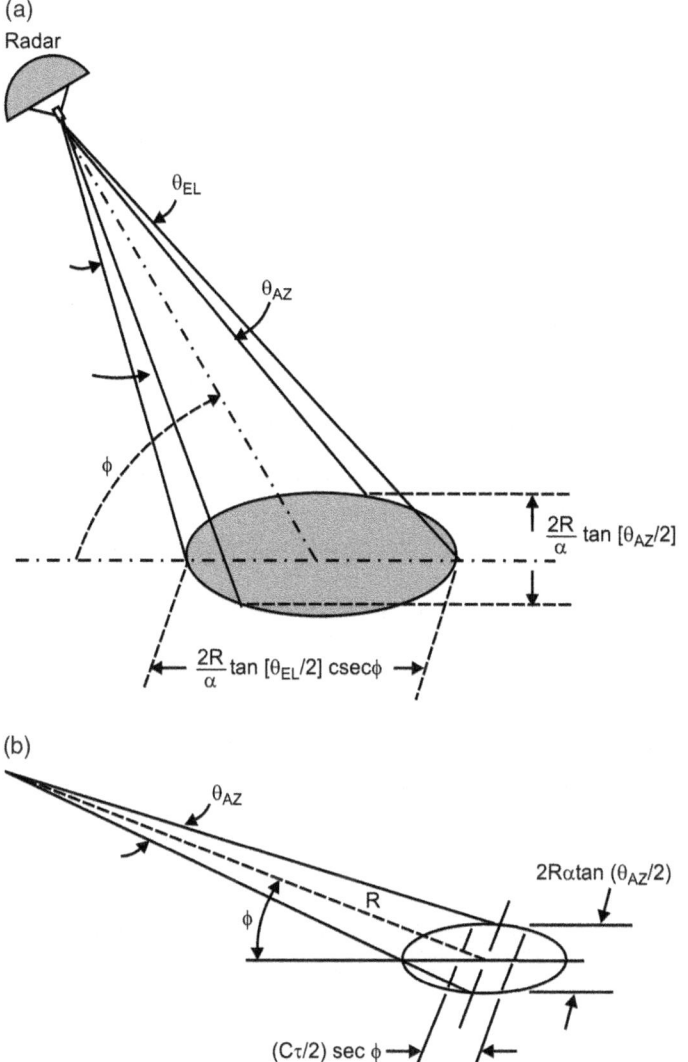

Figure 2.24 (a) Beam width limited illumination. (b) Pulse width limited illumination.

There are two common cases of surface clutter, namely the beam width limited surface clutter applicable to high-angle illumination of surface and pulse width limited surface clutter applicable to low angle illumination. In the case of beam width limited clutter characterized by high-angle illumination and depicted in Figure 2.24(a), the illuminated area is given by the area of the elliptical footprint and is computed as follows.

If a and b are the semi-minor and semi-major axes of the elliptical illumination on the surface, then the surface area is given by eqn. 2.28.

$$A = \pi ab \tag{2.28}$$

$$a = \left(\frac{R}{\alpha}\right) \times \left(\frac{\tan\theta_{AZ}}{2}\right)$$

$$b = \left(\frac{R}{\alpha}\right) \times \left(\frac{\tan\theta_{EL}}{2}\right) c\sec\phi$$

Substituting the values of a and b

$$A = \left(\pi \frac{R^2}{\alpha^2} \right) \times \left(\frac{\tan\theta_{AZ}}{2} \right) \times \left(\frac{\tan\theta_{EL}}{2} \right) c \sec\phi \qquad (2.29)$$

A = Illuminated surface area (in m^2)
ϕ = Grazing angle made by radar beam with surface (in radians)
α = Beam shape factor = 1.33 for a Gaussian beam
θ_{AZ} = Full angle 3B beam width in azimuth direction (in radians)
θ_{EL} = Full angle 3 dB beam width in elevation direction (in radians)
R = Slant range to the surface (in m).

This mode can be also used to compute the surface clutter from trees and walls by applying appropriate geometry and assuming that the tree surface is impenetrable by radar signal.

In the case of pulse width limited model applicable to low angle surface irradiation depicted in Figure 2.24(b), the projected pulse width is much shorter than the length of the elliptical footprint due to elevation beam width. In this case, the illuminated area is given by the area of a rectangle, one of whose sides is the length of elliptical footprint in azimuth direction and the other is the pulse length measured along the surface in the elevation direction.

Length of elliptical footprint in the azimuth direction $= \left(\frac{2R}{\alpha} \right) \times \left(\frac{\tan\theta_{AZ}}{2} \right)$

Pulse length measured along the surface $= \left(\frac{c\tau}{2} \right) \times \sec\phi$

Illuminated area is given by eqn. 2.30.

$$A = \left(\frac{2R}{\alpha} \right) \times \left(\frac{\tan\theta_{AZ}}{2} \right) \times \left(\frac{c\tau}{2} \right) \sec\phi = \left(\frac{Rc\tau}{\alpha} \right) \times \left(\frac{\tan\theta_{AZ}}{2} \right) \times \sec\phi \qquad (2.30)$$

c = Velocity of electromagnetic waves = 3×10^8 m/s
τ = Transmitted pulse width (in s)

Grazing angles ϕ in both cases are given by eqns. 2.31 and 2.32.

$$\tan\phi > \left(\frac{2\pi R}{\alpha} \right) \times \left(\frac{\tan\theta_{EL}}{2} \right) \times \left(\frac{c\tau}{2} \right) \qquad (2.31)$$

$$\tan\phi < \left(\frac{2\pi R}{\alpha} \right) \times \left(\frac{\tan\theta_{EL}}{2} \right) \times \left(\frac{c\tau}{2} \right) \qquad (2.32)$$

The illuminated surface areas can be used to compute the signal-to-clutter ratio in the two cases. It can be seen that the signal-to-clutter ratio in the beam width limited case varies inversely as a square of slant range, R. In the pulse width limited case, it varies inversely as R.

2.5.2 Volume Clutter

Weather phenomena such as rain, dust, hail, snow and so on, and electronic countermeasures such as chaff are common examples of volume scatter. Volume scatter formulation is appropriate in case of weather phenomena as there is always some penetration of radar energy in these cases. A parameter analogous to mean reflectivity, σ_{MEAN}, defined earlier in case of surface clutter is considered for volume clutter. It is defined as the RCS per unit illuminated

Figure 2.25 Computation of volume scatter.

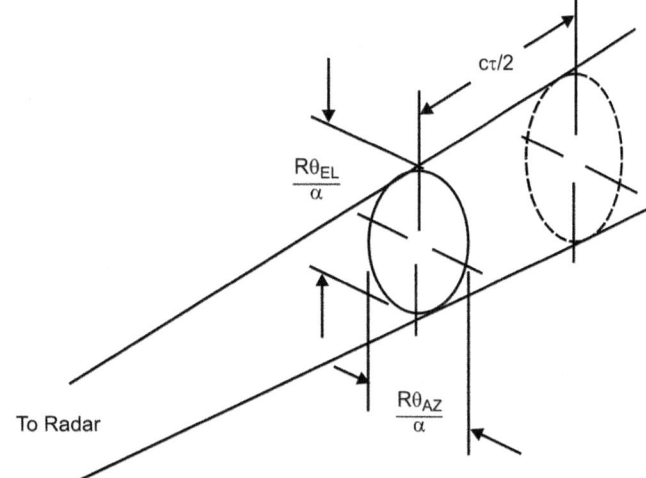

volume and is given by ratio of RCS of illuminated volume (in m^2) to illuminated volume (in m^3). It is denoted by η and measured in m^{-1}.

Radar cross-section in the case of volume clutter $= \eta \times V$ (2.33)

η = Volume reflectivity (in m^{-1})
V = Illuminated volume.

Illuminated volume can be computed by referring to Figure 2.25. It is equal to the volume of an elliptical cylinder whose height and the two diameters are given by eqns 2.34, 2.35 and 2.36.

$$\text{Height} = \left(c\tau/2 \right) \tag{2.34}$$

$$\text{Diameter in azimuth} = \left(R \times \theta_{AZ}/\alpha \right) \tag{2.35}$$

$$\text{Diameter in elevation} = \left(R \times \theta_{EL}/\alpha \right) \tag{2.36}$$

Volume V is given by eqn. 2.37

$$V = \left(\pi R^2 \times \theta_{AZ} \times \theta_{EL} \times c \times \tau/8\alpha^2 \right) \tag{2.37}$$

RCS of volume clutter is then computed from the product ηV assuming that there is minimal attenuation of radar signal over the length of the cell. This may be true where rain or foliage penetration is good. This expression may not hold well in cases of dry or dense green foliage where the attenuation is not insignificant. In such cases, the RCS is computed from eqn. 2.38.

$$\sigma = \eta V e^{-c\tau\alpha} \tag{2.38}$$

$$\sigma_{dB} = \ln\left(\eta V \right) - c\tau\alpha \tag{2.39}$$

When the radar is limited by volume clutter, signal-to-clutter ratio varies inversely as a square of slant range.

2.5.3 Clutter from Point Objects

Reflections from birds, insects, human beings, buildings and so on are examples of clutter due to point targets. The clutter due to birds is often quite serious as they can fly up to speed of 50 knots, which makes their rejection difficult by Doppler or MTI (Moving Target Indicator) processing. Radar echoes from birds are a moving clutter that can cause serious degradation of radar performance. The problem is accentuated during migratory season. Most civil air traffic control and air surveillance radars use sensitivity time control to turn down receiver gain at short ranges to minimize clutter from nearby birds and insects. Military radar cannot do so as their intended targets can also have RCSs as small as that of a bird. The RCS of birds is strongly dependent upon resonance effects. The RCS of a specified bird varies with frequency band of operation. For a given frequency, it is observed to be a strong function of its mass. For frequency bands of interest to radar, the RCS may vary from 0.1 cm^2 for small birds to 500 cm^2 for large birds.

RCS of insects is appreciable if their length exceeds $\lambda/3$. Also, RCS of insects when viewed broadside is about 10–1000 times larger than viewed head on. RCS of small and large insects are typically 0.01 cm^2 and 10 cm^2.

2.6 Radar Range Equation

The radar range equation relates the radar's detection range to various radar and target parameters. These parameters include the transmitted power, transmit antenna gain, radar cross-section of the target, receive antenna aperture, minimum detectable power at the receiver input and various loss factors. The range equation has been derived from first principles step-by-step in the following paragraphs. A brief description of different parameters entering the range equation has also been given along with different steps of derivation of equation, in particular emphasizing the significance of these parameters vis-à-vis the maximum detection range of the radar.

The first step is to determine the effective radiated power (ERP) in the direction of the main beam of the radar's transmitting antenna. It is the product of power delivered to the antenna input and the antenna gain and is expressed by eqn. 2.40.

$$ERP = P_t.G_t \tag{2.40}$$

P_t = Transmitter power at antenna input
G_t = Transmitting antenna gain.

Antenna gain is the effectiveness with which it concentrates the power fed to its input in a preferred direction. An isotropic radiator when fed with power equal to ERP will produce the same power density at a given range as that produced by a highly directional antenna when fed with a power equal to ERP divided by its gain. As the emitted electromagnetic energy propagates in the forward direction, its power density, which is the power per unit area of beam cross-section goes on reducing. Now area of the beam at any given range is the area of sphere whose radius equals the range in case of an isotropic radiator. In case of an antenna with gain, it is the area of the sphere with radius equal to range divided by antenna gain. The power density at a range R_t can be computed from eqn. 2.41.

$$\text{Power density at the target} = \left[ERP/4\pi R_t^2 \right] \text{ or } \left[P_t/\left(4\pi R_t^2/G_t\right)\right] = \left[P_t.G_t/4\pi R_t^2 \right] \text{W/m}^2 \tag{2.41}$$

We should remember that if a power (P_t) fed to an antenna produces a certain power density at a given range, then a power $ERP(= P_tG_t)$ fed to the isotropic radiator would produce same power density at that point.

A part of this energy is reflected from the target depending upon reflection characteristics of the target. It may be mentioned here that reflection of energy occurs if the energy propagating in free space with an impedance of 377 Ω encounters a discontinuity in impedance in the form of target. The part of reflected energy travelling toward the radar is called *backscatter* and it is this backscatter that we are interested in. There are various target related parameters that, together with the incident power density, determine the backscatter. These parameters are combined together and expressed in the form of target's RCS σ.

The RCS of a target is defined as the ratio of its effective isotropically reflected power to the illumination or incident power density. It may be remembered that only the power reflected towards the radar is a part of RCS.

The RCS is made up of three components; namely the area of the target, target reflectivity at the polarization of the receiving antenna of radar and directionality of reflected power. The target's RCS fluctuates in both amplitude and phase and the two types of fluctuations are respectively known as *scintillation* and *glint*. While scintillation causes variations in the received echo power, the glint affects the wave fronts echoing from the target.

Coming back to the radar range equation, the effective power reflected from the target in the direction of radar can be expressed by eqn. 2.42.

$$P_{rt} = \frac{P_t.G_t}{4\pi R_t^2}\sigma \tag{2.42}$$

P_{rt} = Backscattered power
σ = Target's radar cross-section (RCS).

P_{rt}, the backscatter propagates towards the radar and the power density at the radar's receiving antenna due to backscatter P_{dr} can be expressed by eqn. 2.43.

$$P_{dr} = \frac{P_{rt}}{4\pi R_r^2} \tag{2.43}$$

R_r = Range from target to radar receiving antenna
Substituting the value of Pt, we get

$$P_{dr} = \frac{P_t G_t \sigma}{\left(4\pi\right)^2 R_t^2 R_r^2} \tag{2.44}$$

In case of monostatic radar (a radar with co-located transmitter and receiver), $R_t = R_r = R$.

Therefore, $$P_{dr} = \frac{P_t G_t \sigma}{\left(4\pi\right)^2 R^4} \tag{2.45}$$

If A_e is the effective aperture area of the antenna, then
Received power,

$$P_r = \frac{P_t G_t \sigma A_e}{\left(4\pi\right)^2 R^4} \tag{2.46}$$

Effective area and gain are interrelated by eqn. 2.47.

$$G = 4\pi A_e / \lambda^2 \tag{2.47}$$

λ = Operating wavelength

Equation 2.47 can now be expressed by eqn. 2.48.

This gives

$$P_r = \frac{P_t G_t \sigma G_r \lambda^2}{(4\pi)^3 R^4} \tag{2.48}$$

G_r = Gain of receive antenna

For common transmit and receive antenna, $G_t = G_r = G$

The expression for P_r can therefore be written as eqn. 2.49.

$$P_r = \frac{P_t G^2 \sigma \lambda^2}{(4\pi)^3 R^4} \tag{2.49}$$

The received *S/R* power ratio determines whether or not sufficient signal is present to detect the target. The noise power at the input to the receiver is given by eqn. 2.50.

$$P_n = KT_0 BF \tag{2.50}$$

K = Boltzmann constant (1.38×10^{-23} W–s/°K)

T_0 = Absolute temperature

B = Noise bandwidth

F = Noise figure.

The *S/N* power ratio at the receiver input can be written as eqn. 2.51.

$$S/N = \frac{P_r}{P_n} \tag{2.51}$$

Or

$$P_r = kT_0 BF(S/N) \tag{2.52}$$

If $S_{\min}$ is the minimum detectable signal power, then the maximum radar range can be expressed by eqn. 2.53.

$$(R_{\max})^4 = (P_t \times G^2 \times \lambda^2 \times \sigma) / \left[(4\pi)^3 \times S_{\min} \right] \tag{2.53}$$

Since $G = \dfrac{4\pi A_e}{\lambda^2}$ or $G^2 = \dfrac{(4\pi)^2 A_e^2}{\lambda^4}$

Therefore,

$$(R_{\max})^4 = \left[\frac{P_t A_e^2 \sigma}{4\pi \lambda^2 S_{\min}} \right] \tag{2.54}$$

The minimum detectable signal $S_{\min}$ is usually 10–20 dB stronger than the noise at a point in the receiver where the detection decision is made. The minimum detectable signal can be expressed as a signal-to-noise ratio, required for reliable detection, times the receiver noise. Also $S_{\min}$ is a statistical quantity and must be described in terms of probability of detection and probability of false alarm. Equation 2.54, which can be expressed in various forms, is called the radar range equation. In addition to its utility for prediction of range, the equation also forms a good basis for preliminary system design by allowing the designer to appreciate the effect of various radar and target parameters on range and thus optimize them for best performance. Equation 2.54, however, assumes that there are no propagation losses as the signal propagates from radar to target and from target to radar and also that the target lies in the beam maxima. The propagation losses are due to absorption, diffraction, certain types of refraction effects and so on. The losses are expressed in the form of pattern propagation factors F_t and F_r. F_t is defined as the ratio of field strength E at the target position to that which would exist at the same distance from radar in free space and in the direction of antenna beam maxima. F_r is similarly defined. This modifies the range equation to eqn. 2.55.

$$\left(R_{\max}\right)^4 = \left[\frac{P_t . G^2 \lambda^2 \sigma F_t^2 F_r^2}{\left(4\pi\right)^3 S_{\min}}\right] \tag{2.55}$$

Another loss element is the loss of power in the transmission line connecting the transmitter output to the transmitting antenna input. In that case P_t is to be replaced by P_t/L_t where L_t is the ratio of transmitter output power to the power actually delivered to the antenna. For a lossless transmission line, $L_t = 1$. There can be other similar loss factors. All these loss factors are multiplicative and can be expressed as a single loss factor L, which gives another form of range equation that takes into account the loss factor as eqn. 2.56.

$$R_{\max} = \left[\frac{P_t G^2 \lambda^2 \sigma F_t^2 F_r^2}{\left(4\pi\right)^3 S_{\min} L}\right]^{1/4} \tag{2.56}$$

2.6.1 Evaluation of Range Parameters

The factors that determine the detection range of radar discussed in the preceding paragraphs are highly interdependent. The definitions of most of these factors are arbitrary. Therefore, while evaluating these factors, it is important that the definitions chosen are mutually compatible.

P_t is the transmitted power called peak transmitted power. In pulse radars, it is the average power during the pulse. If τ is the pulse width and $W(t)$ the instantaneous power, then P_t is expressed by eqn. 2.57.

$$P_t = \frac{1}{\tau} \int_{-\tau/2}^{\tau/2} W(t)\,dt \tag{2.57}$$

τ = Pulse width

It is clear from the range equation that the *transmitted power* needs to be increased 16 times in case it is desired to double the range with other parameters remaining unchanged. This means that it will not be feasible to increase ranging capability of the radar by increasing the transmitted power without any limit.

Antenna gain is the other parameter that can be used to enhance the ranging capability of the radar. If the same antenna is used for both transmission and reception, then maximum detection range is directly proportional to square root of antenna gain and is given by eqn. 2.58.

$$R_{max} \propto G^{1/2} \tag{2.58}$$

This shows that the antenna gain needs to be increased by a factor of four only in order to double the maximum range.

Radar cross-section of the target is another parameter that affects the range. This is a factor that is beyond the control of the radar, though it has been effectively utilized by aircrafts to avoid detection by radar. The modern fighter aircrafts are so designed as to minimize their radar cross-section by using what is known as stealth technology. Stealth technology is discussed in Chapter 6 on *Electronic Warfare.*

Presence of λ (or f) in the radar range equation makes it clear that range equation is frequency dependent. The frequency dependence of the range is a complex phenomenon. It is so because a number of other parameters are also frequency dependent. For example, most antennas have frequency dependent gain. The radar range is also affected by the receiver bandwidth and the receiver signal-to-noise ratio. A more sensitive radar receiver would have a higher ranging capability. To sum up, it is not practical to play with various range determining parameters outlined before without considering their effects on the other desired performance specifications of the radar system. One also needs to weigh the effectiveness of a certain parameter to enhance the range against its implications on other performance specifications. Higher antenna gain and a more sensitive radar receiver, for instance, seem to be much better options than increasing the transmitted power to increase the range.

2.7 Radar Waveforms

A large variety of transmit waveforms is employed in radar systems depending upon the type of radar and its intended function. Waveforms other than the conventional gated CW and the CW are often used by the radar to enable it to perform either additional functions or the existing functions with enhanced capability. For example, a CW radar capable of measuring target speeds only could be made to measure the target range also by modulating the transmitted signal using frequency modulation such as in FM-CW radar. Also, parameters like range resolution and Doppler resolution could be enhanced by using special waveforms. Waveform characteristics like spectrum, spectral bandwidth, auto-correlation function and so on can be used to determine the suitability of a given waveform. For example, the Fourier transform of the waveform's auto-correlation function determines its power spectrum. Also, narrow auto-correlation indicates good range resolution capability. Some common radar waveforms along with their characteristic features are briefly described in the following paragraphs.

2.7.1 Continuous Wave (CW)

The basic CW is a single frequency unmodulated signal. Such a waveform can be used only for measuring target speeds as it is good only for recovering Doppler shifts. Since the transmitted signal is devoid of any time dependent variation in any of the signal characteristics, it cannot be used for range measurement. CW is often frequency modulated such as in FM-CW radar to enable it measure both range as well as Doppler. The modulation used could be either FM or PM. Again, in FM-CW, the modulating signal could be a sine or a triangle [Figure 2.26(a) and (b)]. The figures are self-explanatory.

Figure 2.26 Continuous wave: (a) sine wave modulation and (b) triangular modulation.

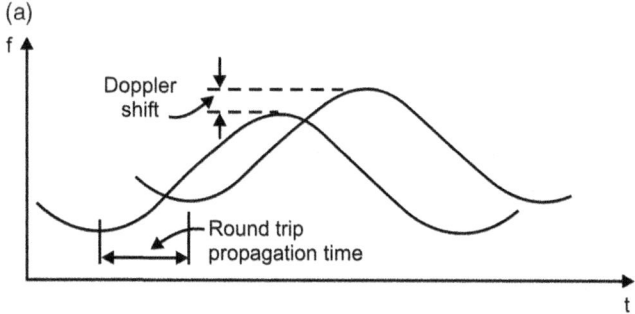

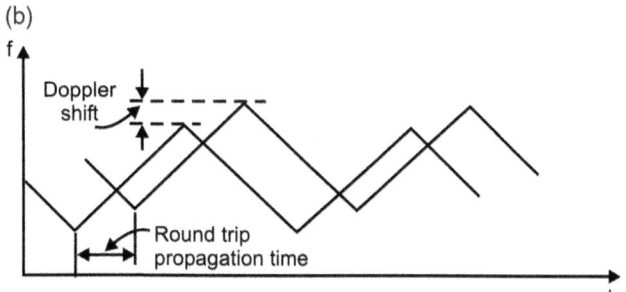

Figure 2.27 Single frequency CW.

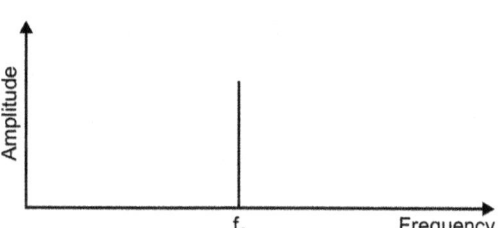

The spectrum of an unmodulated CW is ideally a single frequency (Figure 2.27). In case of modulated CW, the spectrum depends upon the type of modulation, modulating signal shape and frequency, frequency deviation caused by the modulating signal and so on. If the modulating signal is a sinusoid, the spectrum can be found in the same way as we analyse an FM wave. The bandwidth can be determined from eqn. 2.59.

$$B = 2\left(M_f + 1\right)f_m \tag{2.59}$$

M_f = Modulation index and f_m = modulating frequency.

In case of triangular modulation, the spectrum is more complex. The Fourier series expansion of a triangular waveform is nothing but the sum of sinusoids of fundamental (the modulating frequency) and its odd harmonics. Each component in this case will have its own set of side bands.

2.7.2 Gated CW Pulsed Waveform

Gated CW pulsed waveform is commonly used in pulsed radars. It is basically a sinusoidal signal gated to a rectangular time envelope, the frequency versus time plot of which is shown in Figure 2.28. This type of waveform can be expressed by eqn. 2.60.

$$\left[u(0) - u(\tau)\right].A\cos\left[\phi(t) + \phi_0\right] \tag{2.60}$$

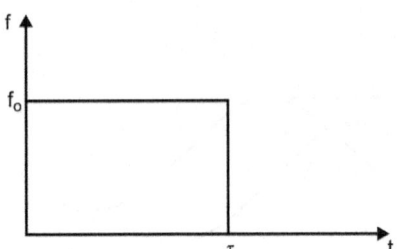

Figure 2.28 Gated CW pulse: frequency versus time plot.

(a)

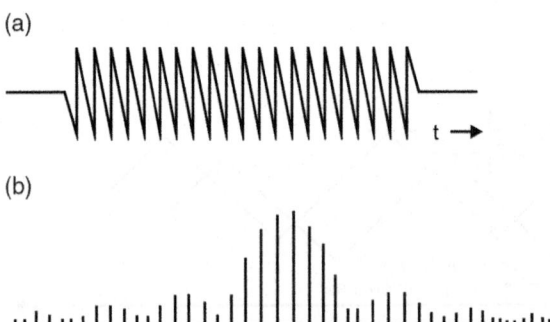

(b)

Figure 2.29 Gated CW pulse: (a) time domain response and (b) Frequency domain response.

Where $u(0)$ and $u(\tau)$ are unit step functions occurring at time $t = 0$ and $t = \tau$, respectively.

A = Peak amplitude of the waveform
$\phi(t)$ = The waveform's phase function
ϕ_0 = Absolute phase of the wave, usually defined as zero.

The waveform phase can be computed from eqn. 2.61.

$$\phi(t) = \int \omega(t)\,dt \tag{2.61}$$

ω = Radian frequency = $2\pi f_0$
This gives $\phi(t) = 2\pi f_0 t$
The frequency domain response is given by eqn. 2.62.

$$\sin\left[\pi\left(f - f_0\right)\tau\right] / \pi\left(f - f_0\right)\tau \tag{2.62}$$

f_0 = Frequency of sinusoid being gated
τ = Gating pulse width.

The time domain and frequency domain responses are shown in Figure 2.29(a) and (b).
Now $(\sin x / x) = 0$ gives $x = \pm n\pi$
This implies that the frequency domain response will have nulls located at frequencies given by eqn. 2.63.

$$\pi\left(f - f_0\right).\tau = \pm n\pi \tag{2.63}$$

This gives $f = [f_0 \pm n/\tau]$

Therefore, nulls are located at values of f that are away from f_0 by a quantum equal to integral multiple of reciprocal of τ. The main lobe has a width equal to $2/t$. The matched bandwidth equals $1/\tau$. Different spectral lilies (each line representing an individual sinusoid) are separated by the frequency equal to PRF.

2.7.3 Linear Frequency Modulated (LFM) Gated Pulse

In the LFM gated pulse also known as 'chirp', the frequency is swept linearly across the pulse. The frequency versus time graph is as shown in Figure 2.30(a). The graph shown is for a down-chirp where the frequency is swept from a higher to a lower value. In the up-chirp shown in Figure 2.30(b), the frequency is swept from a lower to a higher value. The advantage of having modulation in the transmitted pulse is that it is possible to achieve large bandwidth even with long transmitted pulse widths. In an unmodulated pulse discussed earlier, the bandwidth is approximately equal to reciprocal of the pulse width and the only way to enhance bandwidth is by reducing the pulse width. The chirp and other modulated waveforms achieve their bandwidth through modulation and not the pulse width. If the frequency is swept from f_1 to f_2, then frequency difference df is given by eqn. 2.64.

$$df/dt = (f_2 - f_1)/\tau \tag{2.64}$$

$f_1 > f_2$ for down-chirp and $f_1 < f_2$ for up-chirp

Instantaneous frequency $f(t)$ as a function of time is given by eqn. 2.65.

$$f(t) = \int \left[(f_2 - f_1)/\tau \right].dt = \left[(f_2 - f_1)/\tau \right].t + K \tag{2.65}$$

K = Constant of integration

$$K = f_1$$

This gives:

$$f(t) = \left[(f_2 - f_1)/\tau \right].t + f_1$$

Phase function $\phi(\tau)$ can be expressed by eqn. 2.66.

$$\phi(t) = \pi.f(t).t^2 + 2\pi.f_1(t) + \phi_0 \tag{2.66}$$

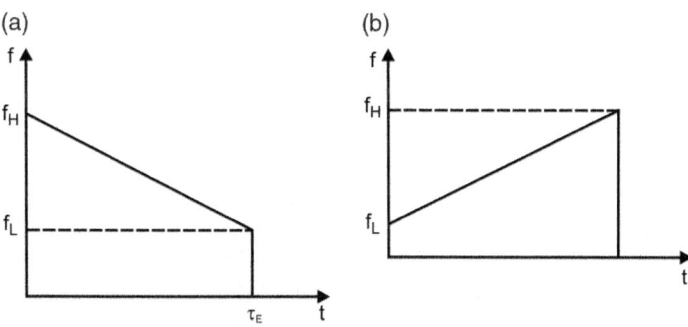

Figure 2.30 Linear frequency modulated gated pulse frequency versus time plot: (a) down-chirp and (b) up-chirp.

(a)

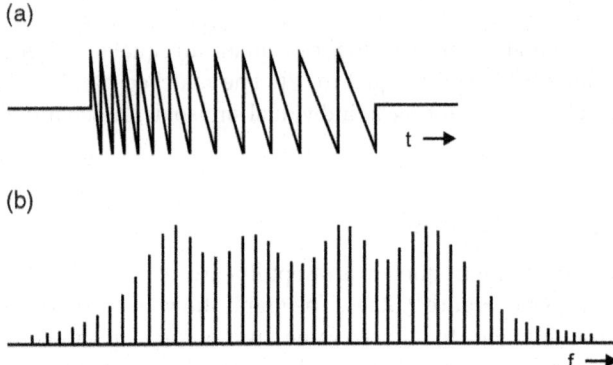

(b)

Figure 2.31 Linear frequency modulated gated pulse: (a) time domain response and (b) frequency domain response.

The time domain and frequency domain responses are shown in Figure 2.31(a) and (b), respectively. The centre of spectrum f_0 is given by eqn. 2.67.

$$f_0 = (f_1 + f_2)/2 \tag{2.67}$$

The matched filter bandwidth is independent of the pulse width and is equal to sweep bandwidth. The reciprocal of sweep bandwidth approximately equals the compressed pulse width. The spectral lines are apart by PRF.

One example of use of this type of waveform is the pulse-compression radar. The transmitted pulse has a relatively large pulse width with the RF swept across the width by employing either linear FM or nonlinear FM (to be discussed next). The receiver processes the received echo pulse in such a way as to compress the pulse with the compressed pulse width approximately equal to reciprocal of swept bandwidth. This feature enables the pulse-compression radar to have both a higher ranging as well as higher range resolution capability.

2.7.4 Nonlinear FM Gated Pulse

In this, the frequency is swept across the pulse width non-linearly. Various nonlinear functions have been used for the purpose. One such typical frequency versus time relationship is shown in Figure 2.32. It is a non-symmetric quadratic waveform. A symmetric quadratic is also used. The waveform shown in Figure 2.31 is expressed by eqn. 2.68.

$$f(t) = \left[\left\{ u\left(-\frac{\tau}{2}\right) - u(0) \right\} \times \frac{4(f_1 - f_0)^2}{\tau^2} \cdot t^2 + f_0 \right] + \left[\left\{ u(0) - u(\tau/2) \right\} \times \frac{4(f_2 - f_0)^2}{\tau^2} t^2 + f_0 \right]$$

$$f_0 = \frac{f_1 - f_2}{2} \tag{2.68}$$

Functions other than quadratics are also used in nonlinear FM gated pulse waveforms. The Taylor function is a typical example. The spectrum of nonlinear FM, as expected, depends upon frequency versus time characteristics. The time domain and frequency domain responses are shown in Figure 2.33(a) and (b), respectively.

Figure 2.32 Nonlinear FM gated pulse.

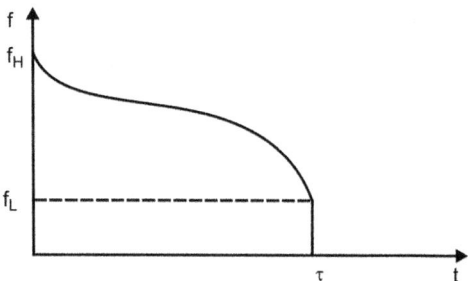

Figure 2.33 Nonlinear FM gated pulse:
(a) time domain response and (b) frequency
domain response.

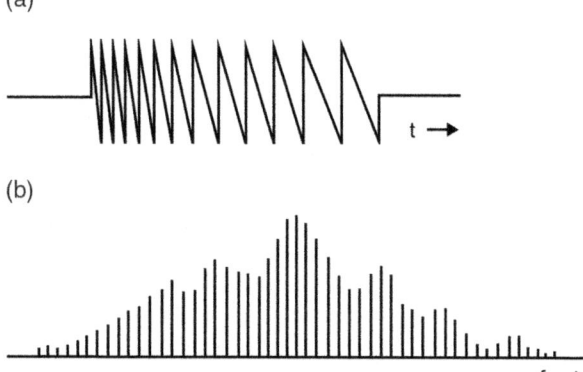

Figure 2.34 V-FM gated pulse.

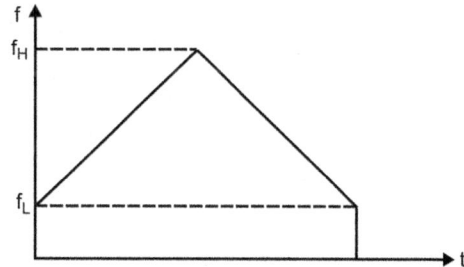

The envelope and centre of the spectrum are determined by frequency versus time characteristics of the modulation. The centre of spectrum is not necessarily in the centre. Again, compressed pulse width is reciprocal of the swept bandwidth and different spectral lines are PRF apart.

2.7.5 V-FM Gated Pulse

This type of pulse has a frequency sweep in one direction up to one-half of the pulse width and then in the opposite direction across the remaining half (Figure 2.34). The sweep is not necessarily linear but is usually so. The spectrum is similar to the one shown in case of linear FM if the sweep employed is linear which usually the case is. The centre of the spectrum is the mean of the lowest and highest frequencies, the compressed pulse width is reciprocal of the swept bandwidth and different spectral lines are PRF apart.

2.7.6 Phase Coded Waveforms

Phase coding is another method of achieving the desired characteristics of enhanced range resolution from long transmitted pulse widths, immunity to jamming signals and so on. In this, the RF being gated is a single frequency sinusoid which has been divided to sub-pulses with the number of sub-pulses depending upon the length of the code sequence. The phase of the sub-pulses is switched usually between 0° and 180° depending upon the logic status of the bit sequence. Figure 2.35 shows the use of a 13-bit Barker code to do phase coding. Here, 1 represents 0° and 0 represents 180° 2, 3, 4, 5, 7, 11 and 13-bit Barker codes are available (Table 2.2). As is seen from the table, there are only seven known Barker codes and nine known sequences which limit the signal security and make it prone to jamming. The main advantage of Barker code lies in the fact that they have simple and well behaved auto-correlation functions. Their auto-correlation functions have only three absolute values; that is, 0, 1 and N where N is the number of bits in the sequence. This property makes the pulse-compression process very simple. The compressed pulse width equals the width of the sub-pulse.

Another way of phase coding is by using pseudo random sequences where 1s and 0s appear in a reproducible random order. The advantage here is that there is no limit to the number of pseudo random codes available for use and also there is no limit to the number of bits in the sequence, which is limited only by transmitted energy, signal bandwidth and minimum range considerations. This technique is far more secure as one could transmit different codes on different pulses or different group of pulses. $Sub-pulse \ \ width \ = \tau/N$, which gives $Bandwidth \ = N/\tau$. The spectrum of the phase coded waveform is highly code dependent. The centre of the spectrum is the frequency of the sinusoid being phase coded.

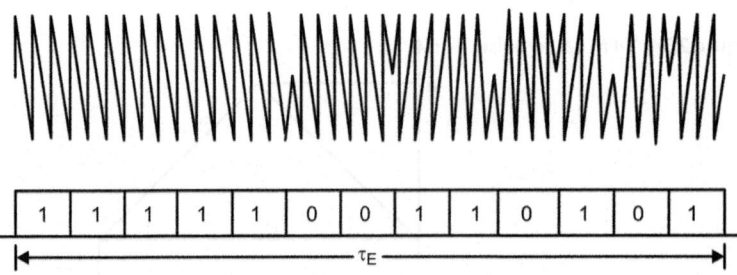

| 1 | 1 | 1 | 1 | 1 | 0 | 0 | 1 | 1 | 0 | 1 | 0 | 1 |

τ_E

Figure 2.35 Phase coded waveforms using 13-bit Barker code.

Table 2.2 Barker codes.

No. of Bits	Known Sequences
2	11, 10
3	110
4	1110, 1101
5	11101
7	1110010
11	11100010010
13	1111100110101

2.8 Radar Transmitters

Radar transmitter is the portion of the radar system that generates the RF signal of the desired shape and spectrum at the required power level to illuminate the target. Its output feeds the radar antenna via a transmission line. The antenna acts as an interface between the transmitter output and the propagation medium. There are three general categories of radar transmitters namely *coherent transmitters, quasi-coherent transmitters* (also known as *coherent-on-receive transmitters*) and *non-coherent transmitters*. Each of these types is briefly described next.

2.8.1 Coherent Transmitters

Coherent transmitters are those in which the phase of the transmitted signal is derived from stable internal sources. The signal phase in this case is constant and predictable. This type of radar can measure the Doppler shift of any target because stability of phase and its prior knowledge enables the radar compare the phase of the Doppler shifted echo with its corresponding illumination pulse to recover the Doppler information. Figure 2.36 shows the phases of the illumination signal and the echo signal for a target whose radial velocity component causes a phase change of 30° between two consecutive received pulses.

Figure 2.37 shows a generic block diagram of a coherent transmitter. The *waveform generator* generates the waveform to be transmitted. The waveform in a coherent transmitter is generated by making use of a coherent oscillator (abbreviated to COHO) and a stable local oscillator (abbreviated to STALO). The two oscillators interact to form all the required sinusoids. The STALO generates a frequency equal to the difference between the transmitted frequency and the receiver's intermediate frequency. The COHO operates at the intermediate frequency.

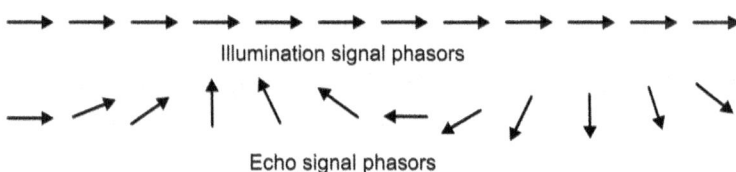

Figure 2.36 Coherent transmitter. Predictable phase of transmit signal.

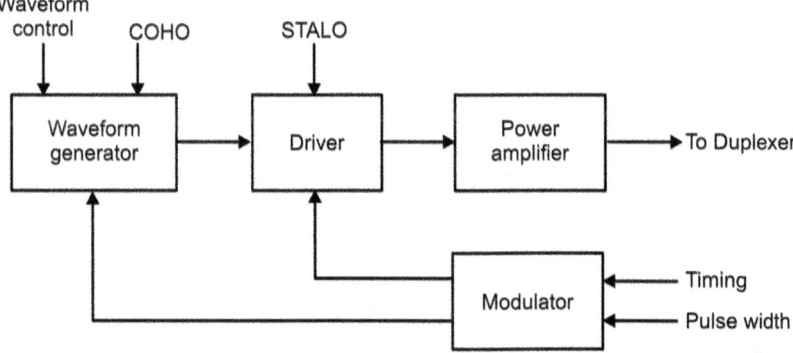

Figure 2.37 Generic block diagram representation of a coherent transmitter.

The sum of COHO and STALO outputs generate the illumination frequency. This is only one of the schemes. In another implementation, a single frequency standard is used to synthesize the COHO, STALO and TRANSMIT frequency.

The *driver block* translates the signal generated by the waveform generator at COHO frequency to the radar's transmit frequency and is amplified to a level suitable for operating the power amplifier. Any frequency agility to be incorporated into the transmit signal is introduced in this block. The signal is raised to the desired power level in this block. In case of transmitter employing power oscillator, the power amplifier is usually absent.

The *modulator block* controls different components of the transmitter. It is linked to the waveform generator, the driver and the power amplifier. In case of low-level modulation, the modulator provides control commands to various components whereas in case of high level modulation setup, it provides DC operating voltages and currents for the power amplifying devices.

2.8.2 Quasi-Coherent (Coherent-on-Receive) Transmitter

Quasi-coherent systems have unpredictable transmit signal phases. These phases are actually measured and phase information retained for use as internal reference. The system always has the phase information on only the latest transmit pulse and every time the transmitter sends out a new pulse, previous references are forgotten. The phase relationships in the case of a coherent-on-receive system are depicted in Figure 2.38. As is clear from the diagram, the transmit phases are random but since the system is capable of remembering the phase of the last transmitted pulse, the phase difference information between a given transmitted pulse and the corresponding received echo pulse is the same as it would be in case of a coherent system. Again, the frequency generation scheme uses two internal oscillators; that is, COHO running at receiver's intermediate frequency and STALO that serves as the receiver's local oscillator. The frequency of STALO is the difference between the transmitted frequency and the COHO frequency. Figure 2.39 shows the frequency generation scheme. The AFC ensures that the STALO operates at the correct frequency and the COHO lock circuit phase locks the COHO output to each transmit pulse. Obviously, every time the transmitter fires a new pulse with an unpredictable phase, COHO is related to this phase and the previous information is lost. Because of this, radars employing a coherent-on-receive transmitter can be used to extract Doppler information only if PRF is relatively low.

Figure 2.40 shows the simplified block diagram of a coherent-on-receive transmitter. The transmitter uses a power oscillator to generate the output at required power level rather than a stable master oscillator at a lower power level followed by a power amplifier as is the case in a coherent transmitter.

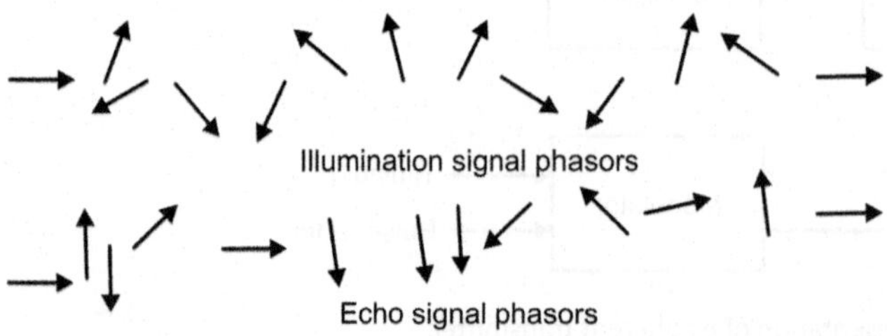

Figure 2.38 Phase relationship in a quasi-coherent transmitter.

Illumination signal phasors

Echo signal phasors

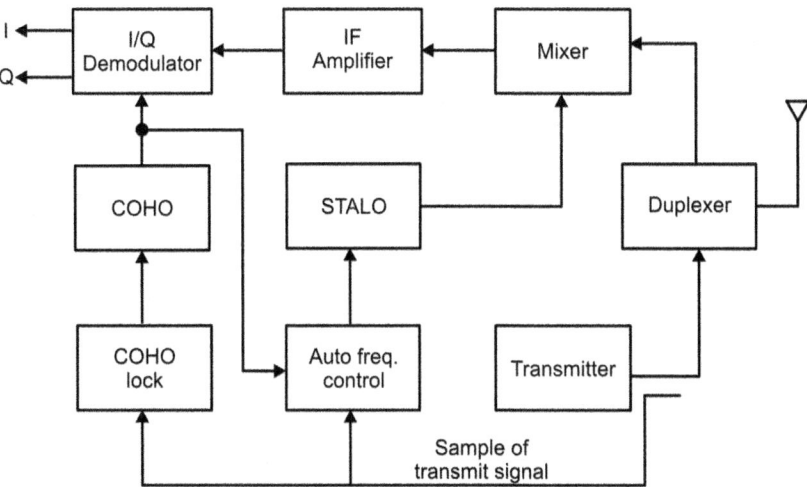

Figure 2.39 Frequency generation scheme in a quasi-coherent transmitter.

Figure 2.40 Simplified block diagram of a quasi-coherent transmitter.

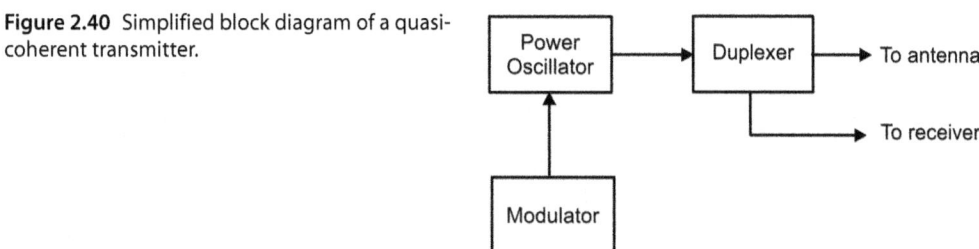

2.8.3   Non-Coherent Transmitters

The transmitter block diagram is similar to the one shown in the case of a coherent-on-receive system. The frequency generation scheme uses no COHO. The receiver local oscillator (STALO in case of previous systems) may or may not be locked to the transmit frequency. A typical frequency generation scheme for a non-coherent transmitter is shown in Figure 2.41. Non-coherent radars are not suited to recovering Doppler information.

2.8.4   Transmitter Parameters

The important transmitter parameters include pulse width, pulse repetition frequency, peak power, average power, duty cycle, pulse energy, efficiency and transmitter stability.

2.8.4.1   Pulse Width

Pulse width is the length of the time the target illumination is on during each transmission. In a radar system that does not use pulse-compression, the received pulse width is only slightly different from the transmitted pulse width and for all practical purposes, can be considered equal to the transmitted pulse width. In case of pulse-compression radar, the received pulse width is widely different from the transmitted pulse width and is significantly smaller. The ratio of the expanded pulse width to the compressed pulse width in the pulse-compression radar is termed the *compression ratio*.

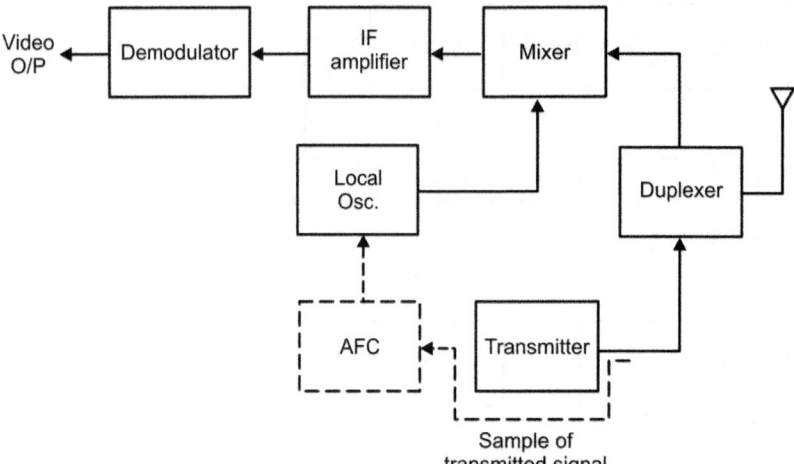

Figure 2.41 Frequency generation scheme of a non-coherent transmitter.

2.8.4.2 Pulse Repetition Frequency

The number of illumination pulses transmitted per second is the *Pulse Repetition Frequency* (PRF). The PRF is constant in some radar systems and agile in some others. This PRF agility could be from pulse to pulse, from one group of pulses to another or from scan-to-scan. PRF agility has many uses including prevention of blind speeds in an MTI, resolution of range and Doppler ambiguities and building into the radar system immunity to jamming signals.

The PRF of radar is a very important parameter. In fact, radars are even classified as low, medium and high PRF radars on the basis of the pulse repetition frequency. A low PRF radar is mainly for ranging as it is highly unambiguous in range. It is also used in radars where true Doppler shift need not be known as such a system would be highly ambiguous in Doppler. MTI (Moving Target Indicator) is an example. A typical low PRF for a moderate range system is 500 PPS. Such a system would have an unambiguous range up to 300 km and a Doppler shift measuring capability of at the most equal to 250 Hz. High PRF radar, on the other hand is unambiguous in Doppler but highly ambiguous in range. Pulse Doppler radar is an example. A typical high PRF for X band radar is 300 000 PPS. This radar would have a Doppler measuring capability of up to 150 kHz but its maximum unambiguous range is only 500 m. Medium PRF is ambiguous in both range as well as Doppler but not as severe as a low PRF system for Doppler or a high PRF system for range. A typical medium PRF is 15 000 PPS. The reciprocal of PRF is the Pulse Repetition Interval (PRI).

2.8.4.3 Peak Power

Peak power is the RMS signal power during the time the transmitter is on.

2.8.4.4 Average Power

Average power is the power transmitted by the radar averaged over a long time. For rectangular envelopes, average power is the product of the peak power and the duty cycle.

2.8.4.5 Duty Cycle

Duty cycle is the fraction of the total time the transmitter is on. It is the product of the transmitted pulse width and the PRF. For variable PRFs, the duty cycle can be expressed as the ratio of the total transmitter on time per PRF cycle to the total time in one PRF cycle.

$$\text{Duty} = \frac{\text{Total transmitter on time per } PRF \text{ cycle}}{\text{Total time in one } PRF \text{ cycle}}$$

For pulse envelopes other than rectangular, duty cycle is calculated from the integral of the envelope over one pulse cycle. Another term related to the duty cycle is the *duty cycle correction factor*. It is the duty cycle expressed in negative decibels. That is,

$$\text{Duty cycle correction factor} = -10\log(Duty \quad Cycle)$$

Duty cycle correction factor when added to the average power in decibels gives the peak power in decibels.

2.8.4.6 Pulse Energy

Pulse energy is the energy in the transmitted pulse. It is the product of peak power and the pulse width or the product of average power and the pulse repetition interval. *Look energy* is the total energy transmitted during one data gathering cycle. It is the product of pulse energy times number of pulses in one look or dwell period.

2.8.4.7 Transmitter Stability

Transmitter stability is another important parameter particularly important for good Doppler performance. A stable transmitter generates spectrally pure waves having only the intended modulations. The transmitter needs to have both amplitude and phase stability. Amplitude instability may lead to a stationary clutter producing spurious targets with apparent Doppler shifts. As an illustration, if the transmitted waveform gets amplitude modulated by, say, 100 Hz due to a ripple frequency of 50 Hz of single phase power supply, the clutter would produce two spurious moving targets having Doppler shifts at ±100 Hz. Phase instability manifests itself in the form of stationary clutter appearing like a moving target.

2.9   Radar Receivers

A radar receiver is almost invariably of the superheterodyne type, the basic block-schematic of which is shown in Figure 2.42. The crystal video receiver (Figure 2.43) and the homodyne receiver (Figure 2.44) have very limited use for radar applications.

The *RF section* amplifies the received echo along with the interference contaminating the signal. Its other functions are to filter out unwanted signals especially at the image frequencies and to attenuate the signals that are strong enough to saturate the subsequent stages in the receiver. In fact, the RF section amplifies signal pulse noise to a level where the noise generated in the subsequent stages does not significantly contribute to the S/N ratio. In a nutshell, RF amplifiers introduce as a little noise as possible but at the same time amplify that noise along with the signal enough so as to swamp the noise generated in the later stages.

The *mixer* along with the local oscillator, which is STALO in this case, converts the received signal at the transmit frequency or Doppler shifted transmit frequency in case of moving targets, down to the Intermediate Frequency (IF) which is the COHO frequency. One of the disadvantages here is that any input frequency that is equal to the desired input frequency plus twice the intermediate frequency will also be handled by the mixer in the same way and produce an interfering signal at the mixer output at the receiver IF. The undesired input must therefore be rejected before it reaches the mixer. One way to solve this is to use a higher

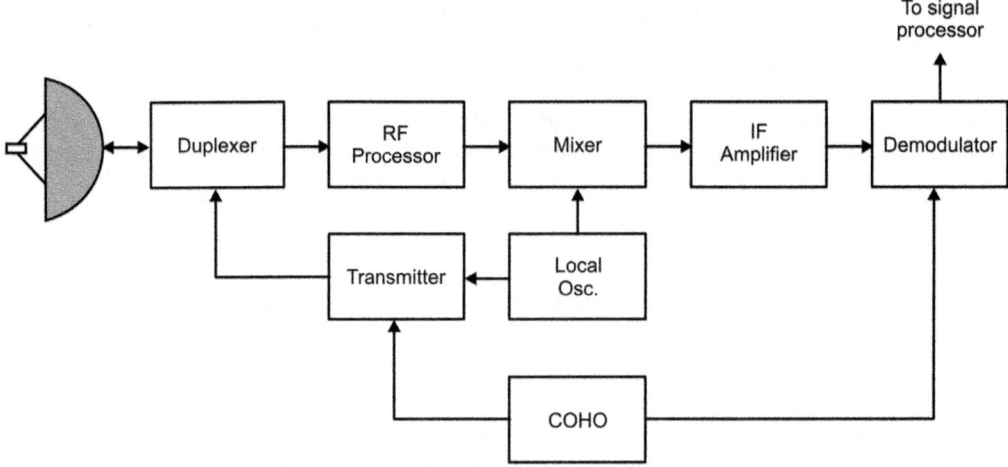

Figure 2.42 Super-heterodyne receiver.

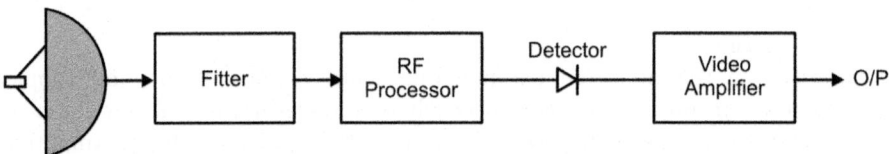

Figure 2.43 Crystal video receiver.

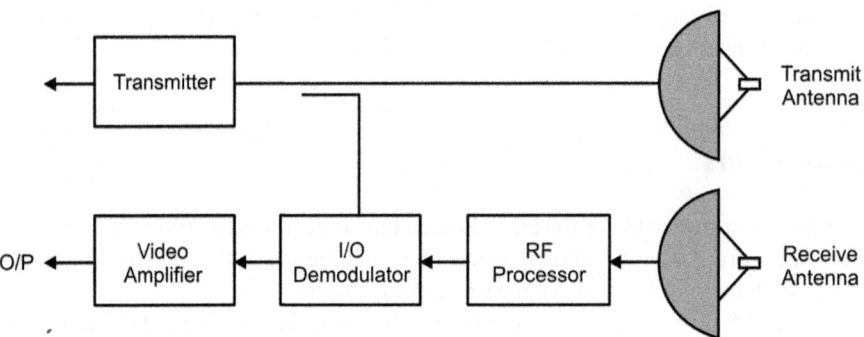

Figure 2.44 Homodyne receiver.

IF so that the image frequency is far away from signal frequency and can therefore be effectively filtered out in the RF section. The other technique to overcome this image frequency problem, particularly when it is close to the signal frequency and cannot be filtered out effectively in the RF section, is to use a special type of mixer called an *image reject mixer*, which is basically a combination of two mixers producing an IF output corresponding to signal and Image frequency at different ports. Details are beyond the scope of the text.

Higher IFs would be preferred as they lead to larger separation between the desired signal frequency and unwanted image frequency. But a higher IF has an associated drawback in that amplification and filtering, the primary functions of the IF section, become difficult at higher frequencies.

The *local oscillator* provides the RF power needed by the mixer for frequency conversion. Different devices used for local oscillators include the Reflex klystron, Gunn diode, crystal oscillator followed by multiplier chain and frequency synthesizer. Coherent receivers almost invariably use either the crystal type or the synthesizer type local oscillators.

The *IF section* is the place where bulk of receiver amplification and selectivity (or filtering) occurs. The gain of IF section may be as high as 120 dB, which in fact necessitates that the gain stages are highly stable. Two types of filtering processes are executed. The first is the channel-select filtering that processes both signal plus interference within the signal bandwidth and therefore rejects out-of-band interference. The other is matched filtering. A matched filter processes signal plus residual interference in such a way as to allow maximum of signal and reject maximum of interference so as to maximize signal-to-interference ratio.

The *demodulator* recovers the information at the base band from signal plus interference at IF. The available demodulation schemes include envelope detection, synchronous detection and I/Q demodulation. The first recovers only the amplitude of signal plus interference. It disregards the signal phase and is therefore unusable for recovering Doppler information that resides there. Synchronous detection recovers both amplitude and phase. Also, in the case of synchronous demodulation, signals that are in phase with the receiver COHO frequency (i.e., IF), which is one of the inputs to the demodulator, are only recovered. Signals in phase quadrature are lost. The I/Q demodulator (Figure 2.45) gives complete information on magnitude and phase. The amplitude and phase in this case are given by eqn. 2.69.

$$Amplitude = \sqrt{I^2 + Q^2}, \ Phase = \tan^{-1}\left(\frac{Q}{I}\right) \tag{2.69}$$

$V_s \cos(2\pi f_c t + \phi_s)$ is the signal at output of the IF section

f_c = COHO frequency
ϕ_s = Signal phase
V_s = Peak amplitude of signal

The output of demodulator feeds the signal processor discussed in the next section.

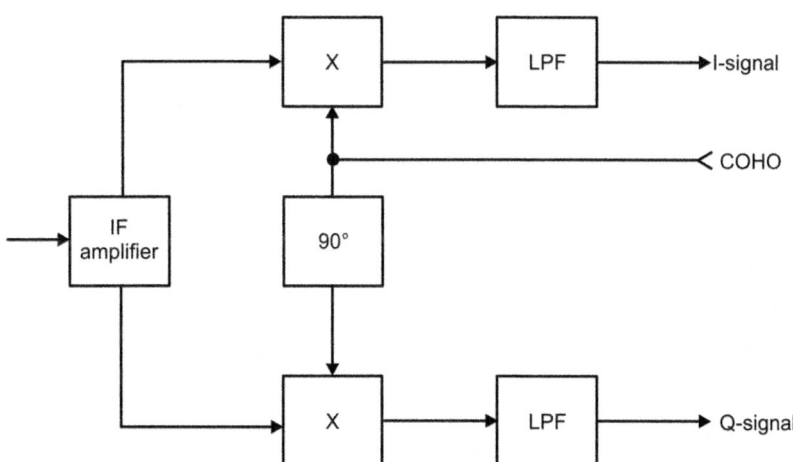

Figure 2.45 I/Q demodulator.

2.9.1 Receiver Parameters

The key parameters of a radar receiver include gain, sensitivity, linearity and dynamic range.

2.9.1.1 Gain and Gain Control

A receiver's gain is defined as the ratio of receiver output power to that at its input.
   That is,

$$G = \left[\text{Output Power/Input Power}\right] = P_o / P_i \tag{2.70}$$

Gain of a radar receiver is invariably controlled. For reasons that would be obvious in the discussion to follow. The technique used to do that depends upon the type of radar and its intended purpose. The available and commonly used techniques include *Manual Gain Control, Automatic Gain Control (AGC), Sensitivity Time Control* and *Instantaneous AGC*.

In *manual gain control*, the receiver gain is constant but a changeable constant selectable usually by the operator. The output from the receiver will also vary with changes in input as the gain is fixed for a given setting. The situation is illustrated in Figure 2.46. Figure 2.46(a) shows the input to the receiver for three different looks at the same target. Figure 2.46(b) shows the corresponding outputs.

In *AGC*, the receiver gain is decided by the received signal strength itself with the gain being inversely proportional to the received signal strength. Such a gain control mode holds the receiver output constant despite variations in the input signal strength (Figure 2.47). The AGC circuit bandwidth decides as to how fast the echo signal amplitude can fluctuate with AGC producing a constant output. This type of gain control is used in radars that view only one target at a time such as in single target track mode.

Sensitivity time control looks into the aspect of signal amplitude variation as a function of range. Due to the fact that the received signal power is inversely proportional to $(\text{Range})^4$,

(a)

(b)

Figure 2.46 Manual gain control: (a) inputs and (b) corresponding outputs.

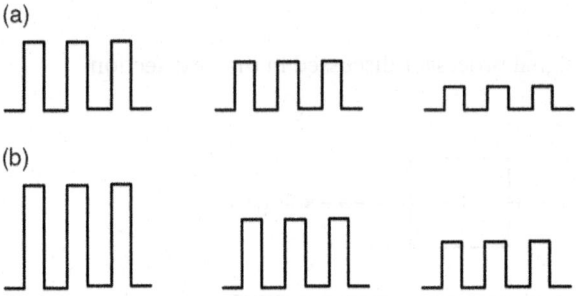

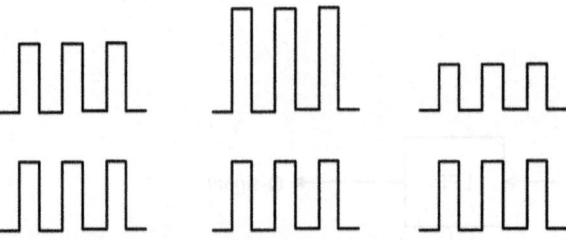

Figure 2.47 Automatic gain control.

the power would reduce to 1/16 of its original value for only doubling of range. Sensitivity time control varies the receiver gain with time beginning with the lowest value at the time of transmitter sending out a pulse. Sensitivity time control is primarily used in search radar. This technique is not suitable for CW radars and radars employing medium and high PRFs. In medium to high PRF radars, echo arrival times are independent of the last transmitted pulse.

Instantaneous AGC causes the receiver output power to be proportional to the rate of change of input power. It is particularly effective when the received signal is accompanied by extended large amounts of interference, such as extended clutter. While AGC controls receiver gain by averaging over many returns and thus taking several tens of milliseconds, instantaneous AGC acts by averaging over a few range resolution intervals involving only a few microseconds.

2.9.1.2 Sensitivity and Noise

Sensitivity of a receiver is in general the minimum signal level it can detect and also perform the intended function. In communication receivers, it is usually expressed in absolute terms such as microvolts into 50 Ω or dBm. In radar receivers, it is also sometimes expressed in terms of signal-to-noise or signal-to-clutter ratio. Sensitivity of a receiver is very intimately linked to receiver noise, which again is expressed in various forms such as noise factor/noise figure, noise temperature, minimum discernible signal and so on.

The most predominant source of receiver noise is the *thermal noise* that for all practical purposes can be assumed to be a *white noise*. White noise is where the noise power generated per unit bandwidth is the same at all frequencies. Now the RMS noise voltage across the terminals of an open circuited resistor R is given by eqn. 2.71.

$$V_n = \sqrt{4KTB_nR} \tag{2.71}$$

B_n = Equivalent noise bandwidth

The equivalent noise bandwidth is defined as the width of a rectangular filter whose peak amplitude response equals the mid-band response of actual filter and whose area equals that under the actual filter (Figure 2.48). It is expressed by eqn. 2.72.

$$B_n = \frac{\int\limits_{-\infty}^{\infty} |H(f)|^2 \, df}{|H(f_o)|^2} \tag{2.72}$$

The noise power that would be delivered by R into a matched termination (which is again R) is given by eqn. 2.73.

$$P_n = kTB_n \tag{2.73}$$

Figure 2.48 Equivalent noise bandwidth.

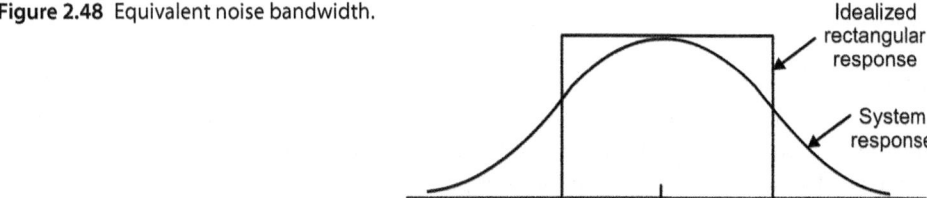

The *minimum discernible signal* is another parameter that can be used to express the receiver sensitivity. It is the minimum signal that can be discerned from noise, usually by an operator. With an expert operator, minimum discernible signal occurs at a signal-to-noise ratio of unity.

Noise factor/noise figure is a measure of thermal noise generated in a receiver as compared to noise produced by an ideal receiver. Noise figure is noise factor expressed in decibels.

Noise factor is,

$$F = [S/N]_i / [S/N]_o \tag{2.74}$$

The ideal noise factor would be unity implying that $[S/N]$ at the input to the receiver is same as $[S/N]$ at the output. The noise power at the receiver output is therefore given by eqn. 2.75.

$$P_{no} = KT_o B_n FG \tag{2.75}$$

G = Gain

Another parameter used to express receiver noise is the *equivalent noise temperature*. The equivalent noise temperature of a system in general is the temperature at which an ideal system produces the same thermal noise power as the total noise power being produced by the actual system. The equivalent temperature is not a measurable quantity. It only indicates the noise power above $KT_o B$ added by the receiver. With reference to the receiver, if F is the receiver noise factor, the equivalent noise temperature can be expressed by eqn. 2.76.

$$TE = T_o [F - 1] \tag{2.76}$$

2.9.1.3 Linearity

A system is considered *linear* when the system gain is not a function of amplitude of the input signal. The output in this case is directly proportional to the input. Also, if the input is the sum of multiple inputs, the output is the same sum of multiple inputs amplified by a constant equal to system gain. This also implies that no new signal components are introduced during the process of amplification. If the system is nonlinear, the output, in addition to the gain value, also depends upon the input signal amplitude. And if this system is fed with multiple inputs, intermodulation components are also produced at the output. Nonlinearity has many associated problems. First, intermodulation products interfere with desired echoes. Strong interfering signals which otherwise do not fall within the range and Doppler bins are likely to leak into those bins and affect processing. In a typical case, a nonlinearity amplified clutter in an MTI may be shown as a moving target.

Nonlinear amplification is helpful sometimes but it should be resorted to only in those applications where nonlinear effects are not important. A nonlinear receiver such as a long receiver enables the operator to view a weak signal in the presence of a strong interference by compressing the output dynamic range. But this would be possible only if the two targets fall in different range bins.

2.9.1.4 Dynamic Range

Dynamic range is defined as the ratio of the strongest to the weakest signal that the system can handle without any degradation in other system parameters. As an illustration, if a radar receiver offers same performance for an input signal power as small as 10 μW and as large as 100 mW, then its dynamic range would be 40 dB.

2.9.2   Signal Processor

In a radar system, the signal processor follows the receiver, the demodulator section of the receiver to be more precise. Signal processor's main function is to process the demodulated signal in such a way as to enhance the signal and suppress the noise as much as possible. In other words, it gives a preferential treatment to the echo signal compared to the interfering signals. It does so basically by dividing the signal space into various segments called bins in one or more dimensions like range, Doppler and so on. Having done that, the improvement in signal-to-noise ratio can be achieved either by concentrating the signal into a single bin and spreading the interference equally in all bins or by concentrating the signal into one bin and interference into another bin.

Figure 2.49 shows a typical block-schematic of a radar signal processor. The *analogue-to-digital converter* does signal translation from analogue form to the digital form followed by *matched filtering*. In the case of pulse-compression radar, pulse-compression also takes place in this block. This is followed by signal filter, which is a periodic band reject filter used to reject clutter and with a response of the type shown in Figure 2.50. It needs to be periodic because the clutter that it is supposed to reject appears at multiple of PRF due to an aliasing effect. *Spectrum analysis* segregates the signal on the basis of Doppler shift. It not only allows measurement of target speeds, any residual clutter is also suppressed due to the fact that Doppler shift due to clutter would be different from that due to targets. Any other random noise also gets spread equally through all Doppler frequencies thus reducing the effect of noise when it competes with the Doppler signal of a given target. Threshold detection is finally established usually following the principle of Constant False Alarm Rate (CFAR).

2.9.2.1   Signal Processing Parameters

Important signal processing parameters include bandwidth, process gain and jamming margin. Each of these is described briefly in the following paragraphs.

2.9.2.1.1   Bandwidth

Bandwidth, with reference to a radar receiver, is particularly relevant to the signal processing section that follows the demodulator. We have the information bandwidth, which is the bandwidth of the information carried by the received signal, and the signal bandwidth. The signal

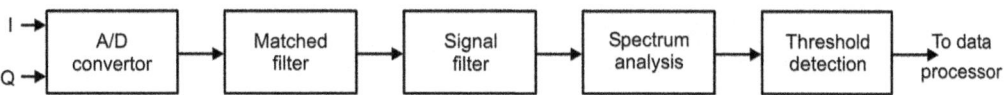

Figure 2.49 Block diagram of a signal processor.

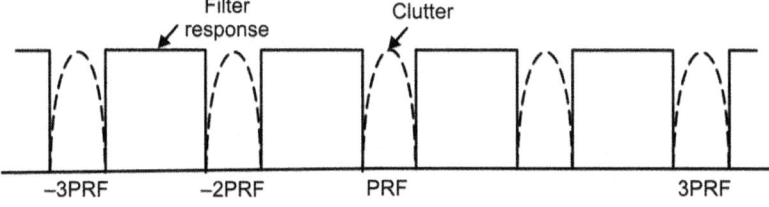

Figure 2.50 Signal filter response.

bandwidth B, its signal-to-noise ratio S/N and the rate of information or information capacity, C, are interrelated by the famous Shannon equation given by eqn. 2.77.

$$C = B\log_2\left[1+\left(S/N\right)\right] \tag{2.77}$$

One interpretation of this equation is that, by having a large S/N ratio, we can afford to transmit information at a rate far exceeding its signal bandwidth. Another is that if the information rate is kept deliberately much lower than the bandwidth, then this low rate information can be efficiently recovered even under the conditions of S/N ratio much less than unity. For an S/N ratio far less than unity, information capacity is expressed as eqn. 2.78.

$$C = 1.44 \times B \times \left(S/N\right) \tag{2.78}$$

This implies that if $C = B/3$, an S/N ratio of 0.23 would also allow signal recovery. This latter interpretation is more relevant to radar and is a typical situation encountered in radar.

2.9.2.1.2 Process Gain
Process gain tells us about the effectiveness of the signal processor in improving the signal-to-processor ratio of the signal. It is defined as the ratio of signal-to-interference ratio at the input of the receiver to signal-to-interference ratio at the output and is given by eqn. 2.79.

$$\text{Process Gain} = \frac{\left(S/I\right)_i}{\left(S/I\right)_o} \tag{2.79}$$

Process gain is also expressed as ratio of signal bandwidth to information bandwidth.

If the interference being considered is the clutter, as usually is the case in an MTI, then the same parameter is referred to as the *MTI Improvement Factor*.

$$\text{Improvement Factor} = \frac{\left(\text{Signal/Clutter}\right)_i}{\left(\text{Signal/Clutter}\right)_o} \tag{2.80}$$

2.9.2.1.3 Jamming Margin
The *jamming margin* is defined as the ratio of interference to signal at the input of the signal processor, which produces a minimum detectable signal at the output. The jamming margin is related to the process gain by eqn. 2.81.

$$\text{Jamming Margin} = \frac{\text{Process Gain}}{\left(S/I\right)_{\min}} \tag{2.81}$$

$[(S/I)_o]_{\min}$ = Minimum (S/N) at the output of the processor for proper detection for a given probability of detection.

The jamming margin is known as sub-clutter visibility if the interference is the clutter. Sub-clutter visibility is defined as ratio of the ratio of the improvement factor to the minimum MTI output signal-to-clutter ratio (SCR) required for proper detection for a given probability of detection.

$$\text{Sub-Clutter Visibility} = \frac{I}{\left[\left(S/C\right)_o\right]_{MIN}}$$

$[(S/C)_o]_{\min}$ = Minimum output signal-to-clutter ratio
I = Improvement factor.

2.10   Radar Displays

Various types of displays are used with radar systems. Some of them are common while others are used for specific applications. Some of the more commonly used radar displays include the A-Scope or A-Scan, B-Scope, F-Scope and Plan Position Indicator (PPI). Each one of these is briefly described next.

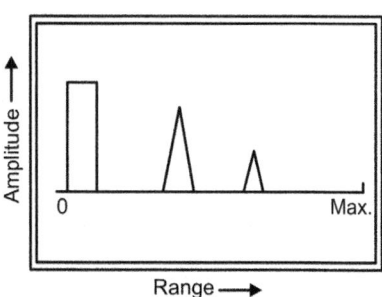

Figure 2.51 A-Scope.

2.10.1   A-Scope

The A-scope represents an oscilloscope like display where the horizontal coordinate represents the range (in terms of time) and the vertical coordinate represents the target echo amplitude (Figure 2.51). It is the most commonly used display. Horizontal sweep is triggered every time a pulse is transmitted providing a reference point. The end of sweep, that is, the right extreme of the display represents the maximum range capability of the radar. The echo signal causes a deflection in the vertical direction. The separation between the starting reference and the echo deflection represents the target range. The deflection is either linearly or logarithmically proportional to target amplitude.

A slight variation of A-scope is the A/R scope. Here, any desired segment of time base can be expanded (Figure 2.52). It is commonly used in tracking radars. Yet another variation of A-scope popular with tracking radars is the R-scope (Figure 2.53). In this, a limited range segment around the centre, that is adjustable, is displayed. The range segment is usually the tracking range gate.

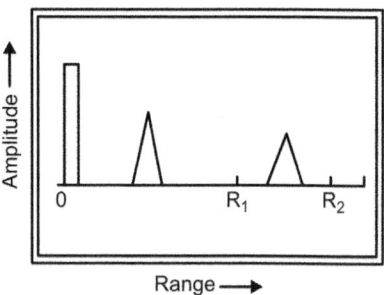

Figure 2.52 A/R Scope.

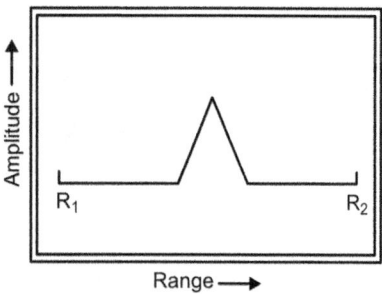

Figure 2.53 R-Scope.

2.10.2   B-Scope

This is an intensity modulated display with horizontal and vertical axes respectively representing azimuth angle and range (Figure 2.54). The entire lower edge of the display is the radar location. It is commonly used in airborne radar particularly when the aircraft is on an intercept mission. It shows the true range. The cross-range dimension however gets distorted on this display. Even if two targets are at a constant cross range, they appear at different separations at different ranges.

In another operational mode of a B-scope called the B-prime scope, the vertical axis represents a target's radial velocity rather than its range (Figure 2.55). The velocity is zero along a horizontal line in the centre.

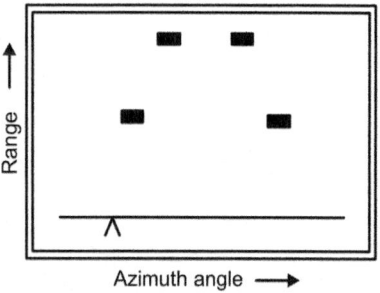

Figure 2.54 B-Scope.

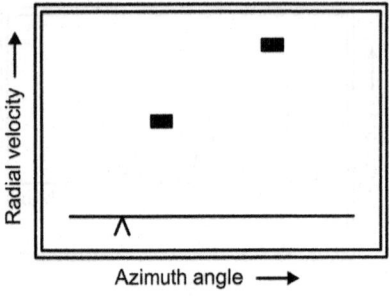

Figure 2.55 B-Prime Scope.

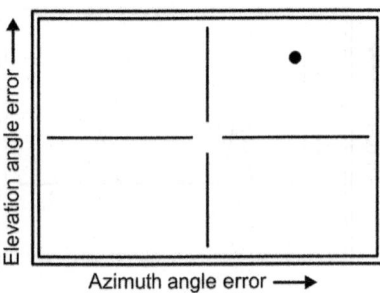

Figure 2.56 F-Scope.

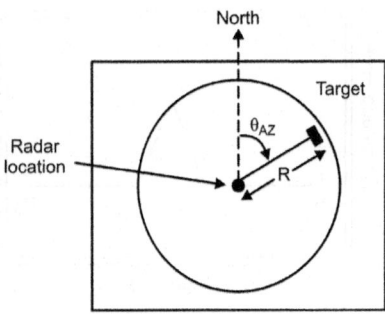

Figure 2.57 Plan Position Indicator (PPI).

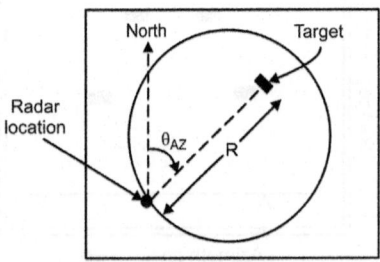

Figure 2.58 Offset or Sector PPI.

Targets above this line are those that are closing on to the radar and those below this line are those that are moving away.

2.10.3 F-Scope

Horizontal and vertical axes of an F-scope display represent azimuth and elevation track error, respectively (Figure 2.56). The centre of the display indicates the antenna's beam axis location. The blip's displacement from the centre indicates target's position with respect to the antenna beam axis.

2.10.4 Plan Position Indicator (PPI)

This is an intensity modulated map like circular display that gives target location in polar coordinates (Figure 2.57). The radar location is in the centre of the display. The target range is represented by the radial distance from the centre and the target's azimuth angle is given by the angle from the top of the display, usually north, clockwise. In some types of PPI display called Offset or Sector PPI, the radar location is offset from the centre of the display (Figure 2.58). This is commonly used in search radars.

2.11 Radar Antennas

The antenna structure acts as an interface that couples the transmitter output (in the case of the transmitting antenna) and receiver input (in the case of the receiving antenna) to free space. It performs the function of transmitting high-frequency electrical currents fed to it from the output of transmitter into electromagnetic waves in a transmitting antenna and that of intercepting electromagnetic waves and generating equivalent electrical signal in the case of a receiving antenna. In other words, it converts guided electromagnetic waves into free-space electromagnetic waves and vice versa. A radar antenna needs to generate different beam characteristics in terms of beam shape, beam width, sidelobe level and directivity to steer the beam in different directions to meet the diverse requirements of the intended application. Antenna technology has gone through several changes commensurate with the evolution of complex radar systems for various military applications. Antennas have been discussed at length in Chapter 1 on *Military Communications.* Important antenna characteristics and parameters including those of relevance to radar antennas, and different types of antennas including those used in radar systems were discussed in Sections 1.4.5 and 1.4.6 of that chapter.

2.12 Types of Radar

There is a large variety of radar systems categorized on the basis of operational frequency band, transmitted waveform shape and spectrum, PRF class and intended mission. Radar classification based on these parameters was discussed in the earlier part of the chapter. In the paragraphs to follow radar systems are discussed, including different types of CW and pulse radar systems. The basic radars in the CW radar category are unmodulated and modulated CW radars. The moving target indicator (MTI) and pulse Doppler radars are the basic pulse radar systems. In addition to these basic radars, specific radars such as pulse-compression radar, synthetic aperture radar, and over-the-horizon radar systems are also discussed.

2.12.1 Continuous-Wave (CW) Radar

CW radar (unmodulated CW radar) transmits an unmodulated CW sinusoidal signal and looks for Doppler shift in the received echo, which is again CW, to measure target velocity. Obviously, unmodulated CW radar is fundamentally incapable of measuring target range though the same would be possible by using some sort of coding in the transmitted waveform. FM-CW radar is discussed a little later in the instance that has both Doppler as well as ranging capability. CW radar has a number of advantages such as having very little spread in the transmitted spectrum, peak power being only slightly greater than the average power and its ability to handle Doppler of targets at any range and having any conceivable velocity without any ambiguity. As we shall see in the latter part of the chapter, this feature is achieved in MTI and Pulse Doppler radars at the cost of system complexity. One of the main disadvantages of unmodulated CW radar is the direct leakage of transmitter and its associated noise into the receiver.

Figure 2.59 shows a block-schematic of simple unmodulated CW radar. The transmitter is an unmodulated CW oscillator operating on a single frequency. A small part of this signal is coupled to the mixer with a directional coupler. The CW echo is routed to one of the

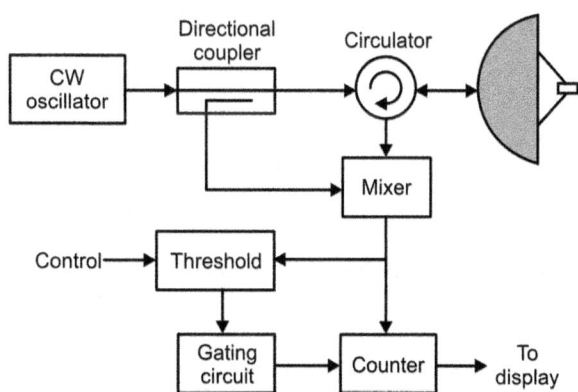

Figure 2.59 Block-schematic of an unmodulated CW radar.

input ports of mixer with microwave circulator. The mixer output is the Doppler shift signal. And if this signal is stronger than a certain threshold, it is gated to a counter which measures this frequency. The Doppler frequency is then transformed to velocity information and displayed.

A CW radar of this type must detect weak echoes in the presence of strong transmit signal. That is why, a single antenna system as shown in Figure 2.59 is used only in very low power (1–100 mW) systems. For high power CW radars, the transmitter leakage is so severe that it necessitates the use of separate transmit and receive antennas. Police radars and radar proximity fuses are examples of simple low power unmodulated radar sensors.

2.12.2 FM-CW Radar

FM-CW radar overcomes the inherent shortcoming of in-capability to measure target range of unmodulated CW radar. CW radar can be made capable of measuring range by using some kind of modulation or coding. In FM-CW radar, the transmitted signal is a frequency modulated CW signal. The FM signal is generated by using either a sinusoidal or a triangular modulating signal. Another way of achieving ranging capability with CW radars is by using phase coding. The maximum range capability on all the three cases previously would be determined by the period of modulation.

2.12.2.1 FM-CW Radar with Sinusoidal Modulation
The modulating signal here is a sinusoidal signal that leads to a sinusoidal frequency versus time characteristics of the transmitted signal. The range is determined by the instantaneous phase shift between transmit and receive signals. Figure 2.60(a) shows the frequency versus

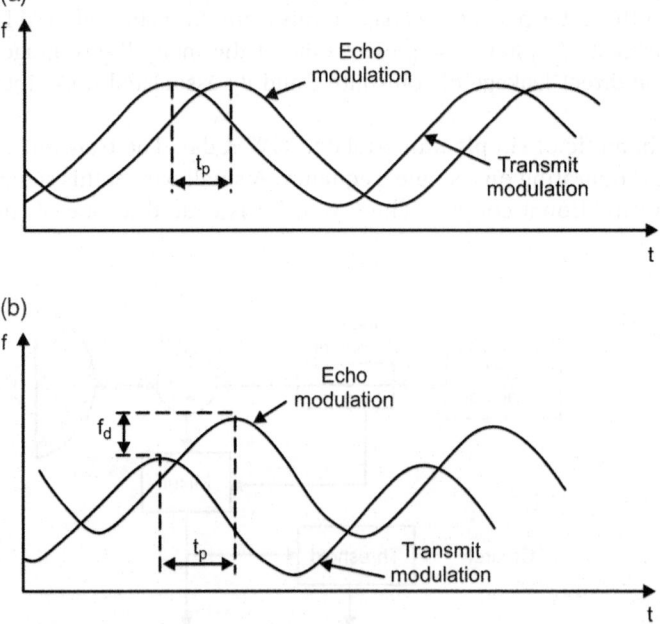

Figure 2.60 Frequency versus time characteristics in FM-CW radar with sinusoidal modulation. (a) In the absence of Doppler shift. (b) With Doppler shift.

time characteristics of transmit and echo signals when there is no Doppler shift. The phase shift Δf corresponds to a two-way propagation time of T_p given by eqn. 2.82.

$$T_p = \left[\Delta\phi/360 \cdot f_m \right] \tag{2.82}$$

$\Delta\phi$ = Phase difference in degrees
f_m = Modulating frequency

If the target is moving, the echo signal will also have a Doppler shift [Figure 2.60(b)]. The Doppler shift f_d is indicative target's radial velocity. The Doppler is extracted by comparing DC values of the demodulated transmitted and echo signals. Sinusoidal FM-CW radar has a disadvantage in that it is not capable of resolving multiple targets. If there is more than one target, each demodulated echo will be sinusoid of modulating frequency having a phase shift and DC offset corresponding to its own target range and velocity. The demodulated echo from multiple targets is then the sum of demodulated echoes from individual targets with the net result that the system perceives it is a single target with range and Doppler, which is an average of all. Sinusoidal FM is therefore used only in case of single target track radars.

2.12.2.2 FM-CW Radar with Triangular Modulation

The transmitted signal in this case, is modulated by a triangular signal that leads to an up-sweep portion where the frequency increases from a low value f_L to a high value f_H and a down-sweep portion where the frequency decreases from a high value f_H back to a low value f_L. In the case of FM-CW radar with triangular modulation, it is the instantaneous frequency difference between the transmitted signal and the received echo that determines the range. Figure 2.61(a) shows the frequency versus time characteristics of transmit and echo signals in case of a target

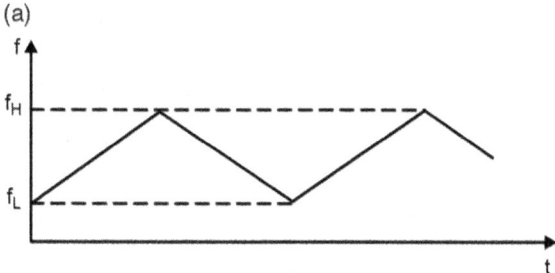

(a)

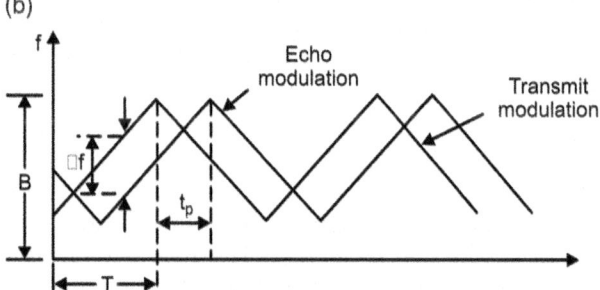

(b)

Figure 2.61 Frequency versus time characteristics in FM-CW radar with triangular modulation. (a) Frequency versus time characteristics of a transmitted signal. (b) Frequency versus time characteristics of an echo signal in the absence of Doppler shift.

producing a zero Doppler shift. The instantaneous frequency difference in this case is constant throughout the up-sweep and down-sweep with the difference that transmitted frequency is greater during the up-sweep and it is the opposite during the down-sweep. The range can be computed from eqn. 2.83.

$$R = \frac{CT}{2B} \cdot \Delta f$$
$$B = f_H - f_L$$

(2.83)

Δf = Instantaneous frequency difference
T = Round trip propagation time

In case the target produces a Doppler shift f_d [Figure 2.61(b)], then the instantaneous frequency difference during the up-sweep and down-sweep is, respectively, given by eqns 2.84 and 2.85.

$$\text{Instantaneous frequency difference} (\text{up-sweep}) = \left[-\Delta f + f_D \right]$$

(2.84)

$$\text{Instantaneous frequency difference} (\text{down-sweep}) = \left[\Delta f + f_d \right]$$

(2.85)

Δf = (Received signal frequency − Transmit signal frequency) due to range delay only
Now, $f(US) = -\Delta f + f_d$ and $f(DS) = \Delta f + f_d$

$$\text{Therefore,} f_d = \left[f(DS) + f(US) \right] / 2 \text{ and } \Delta f = \left[f(DS) - f(US) \right] / 2$$

(2.86)

Thus both Doppler and range information can be determined by measuring instantaneous frequency difference during up-sweep and down-sweep. While using these expressions, due consideration should be given to the sign of the instantaneous frequencies; that is, up-sweep difference frequency [$f(US)$] and down-sweep difference frequency [$f(DS)$] with the former to be given a negative sign and the latter obviously the positive as the two have opposite sense. To simplify the calculations, if we only take the magnitudes of the two difference frequencies, then the average of the two gives Δf and half of the difference between the two gives f_d.

Such radar is also capable of resolving more than one target. In this case, individual targets produce their own up-sweep and down-sweep frequency differences that can be sorted out by spectrum analysis of the composite received signal.

Since Doppler shift is a function of the transmit frequency, the Doppler shift may itself change significantly during sweep period. In order to avoid this problem, sweep bandwidths in most FM-CW radars are much narrow as compared to carrier frequency. Typically, it is a few MHz for a carrier frequency of several GHz.

2.12.2.3 Phase Coded FM-CW Radar

The transmit signal is CW with phase modulation. Range in this case is given by the time difference between transmission of a code segment and reception of the same (Figure 2.62). The time difference is measured using a correlation technique. The received signal is correlated with delayed transmit signal. The delay that corresponds to correlation function maximum is the propagation time. Range in ambiguity in this case requires that the selected code has only one peak in its auto-correlation function over the range of possible propagation times.

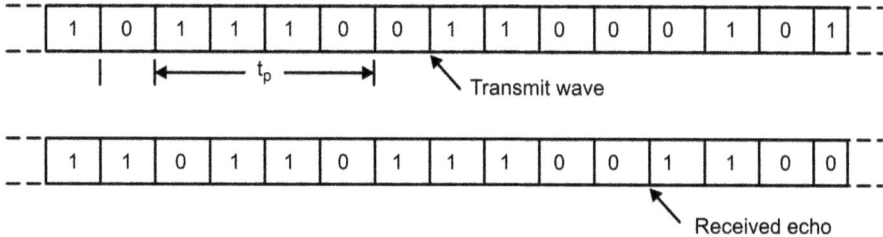

Figure 2.62 Phase coded FM-CW radar.

2.12.3 Moving Target Indicator (MTI)

The *MTI* radar system very effectively handles moving targets such as aircraft and is capable of measuring their range and radial velocity component in the presence of strong clutter due to stationary and even slow moving undesired objects such as buildings, clouds, rain and so on. Again, it is based on the Doppler shift imparted to the transmit signal by the moving target to determine the target's radial velocity component. The range, of course, is measured from the time lapse between the transmit signal and the received echo. The Doppler shift is not measured exactly in the same way as it is in case of CW radar where the process is more or less straightforward. An MTI, being a pulse system, relies on the phase difference between the transmitted signal and the corresponding echo to compute the Doppler. This phase difference for successive transmit pulses of RF energy and their corresponding echoes changes in case of moving targets at a rate equal to the Doppler frequency shift. The phase difference, however, remains the same in case of stationary targets and changes at a very small rate in case of slow moving targets so as to be easily distinguishable from the phase difference information produced by relatively much faster desired targets. The principle of operation of echo signal processing is shown in the block-schematic of Figure 2.63. Each echo from a given range gate is subtracted coherently from a delayed version of the previous echo from the range gate. If the target is stationary, both echoes would produce the same phase difference and there would be complete cancellation provided the noise is absent. If the echo has changed phase slightly due to target motion, the cancellation would be incomplete. For a target in uniform motion, there would be a constant change in phase from pulse to pulse and there is no cancellation.

As is clear from this description, there is a need to maintain phase coherence because it is in the phase difference between the transmit signal and received echo for successive pulses where the target's Doppler information resides. Based on the manner in which this phase coherence is ensured, there are two commonly used MTI configurations. These are Coherent MTI and COHO-STALO type MTI. The latter is basically a coherent-on-receive system. The basics of these two configurations have been discussed in an earlier part of the chapter. Figure 2.64

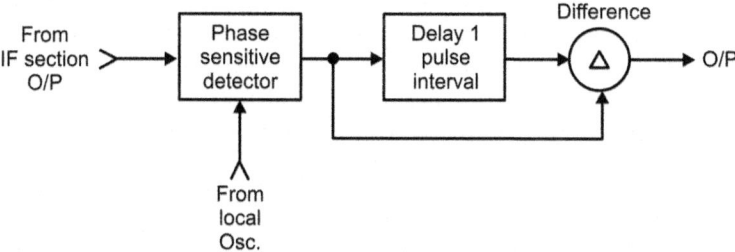

Figure 2.63 Echo signal processing in MTI.

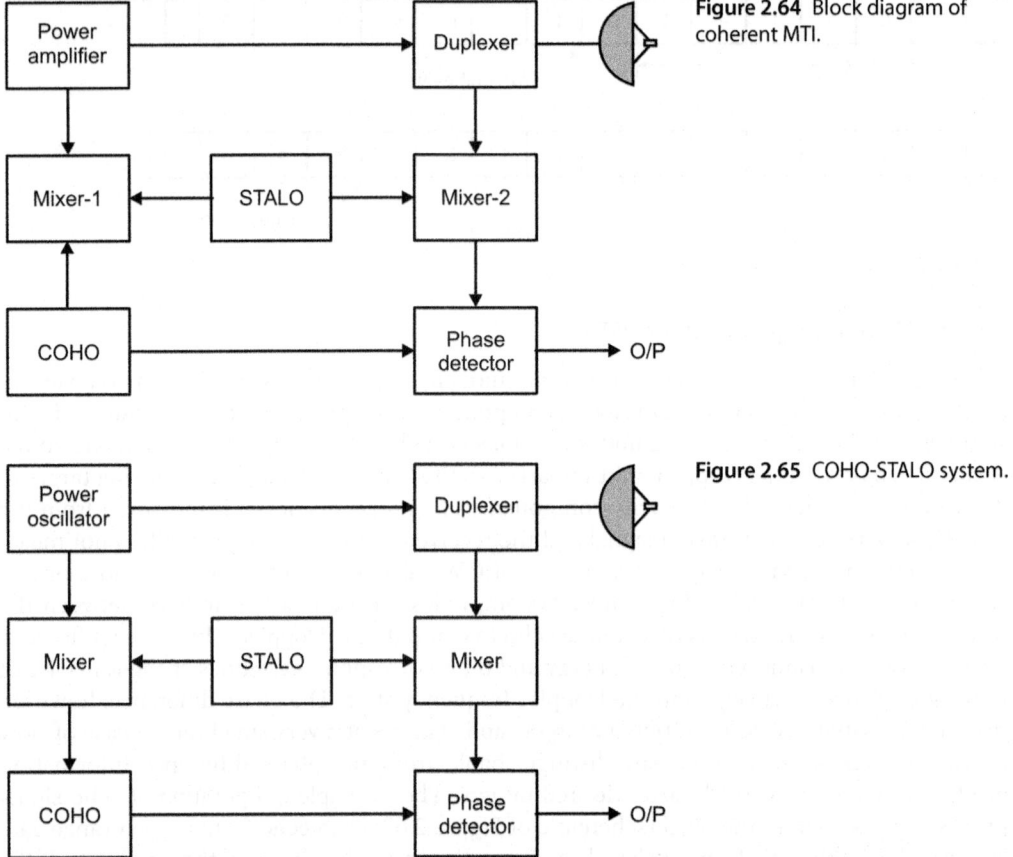

Figure 2.64 Block diagram of coherent MTI.

Figure 2.65 COHO-STALO system.

shows the block-schematic arrangement of a coherent MTI system. Here, the transmitter is a power amplifier with its RF source controlled by a crystal reference. Both the local oscillator as well reference oscillator frequencies are synthesized from a stable reference. The transmitted frequency is the sum of the two produced by Mixer 1.

The reference oscillator output is also at the receiver intermediate frequency (IF). The received signal is routed to Mixer 2 whose other input is from the local oscillator. The mixer output at the receiver IF, which is same as the reference oscillator frequency, is fed to the phase sensitive detector where it is phase sensitively detected. The output is fed to the signal processor section similar to the one shown in Figure 2.63 earlier.

In the COHO-STALO system shown in Figure 2.65, the transmitter is a power oscillator, usually a magnetron oscillator. In this case, the receiver has an extremely stable local oscillator called STALO. A sample of the transmitted RF output power at low levels is mixed with STALO output and locking pulse thus generated then triggers another oscillator called COHO whose output becomes the coherent reference. It is clear that it is the difference output of Mixer 1 here that triggers the COHO. The rest of the system is self-explanatory.

Figure 2.66 shows the photograph of TPS-703 tactical mid-range air surveillance radar from the Northrop–Grumman Corporation. The radar provides MTI and Moving Target Detection (MTD) processing to full range, which facilitates cancellation of both fixed and moving clutter. While the full-range MTI also ensures detection and tracking of low-altitude targets at all

Figure 2.66 TPS-703 tactical mid-range air surveillance radar.

ranges in heavy clutter conditions, MTD suppresses moving clutter and ensures optimal target detection and tracking even in heavy rain and chaff.

2.12.3.1 Blind Speeds in MTI

As mentioned earlier, the computation of Doppler shift in case of pulsed radar is not as simple and straightforward as it is in case of CW radar. In a CW radar, Doppler shift is recovered by measuring the frequency displacement of the echo spectrum from the transmit spectrum [Figure 2.67(a)]. The process is simple because the transmit spectrum is a single line. In pulsed systems, the transmit spectrum comprises of an infinite number of spectral lines separated by the PRF. In such a case, if the target does not produce any Doppler shift, the situation is shown in Figure 2.67(b). This is the case when the target is either stationary or is moving at the same rate as the radar or its velocity vector is perpendicular to the radar axis. The echo and the transmit signal are at the same frequency. If the target's radial component of velocity is such that the Doppler shift produced is less than half of the PRF, the Nyquist criterion is satisfied and the Doppler information is extracted without any ambiguity. The situation is depicted in Figure 2.67(c). It may be mentioned here that frequency of a sampled wave is recovered as the smallest frequency span from the received spectrum line to the closest transmit spectral line. The sampling rate is the same as the PRF. How, if the target's radial component of velocity vector is such that the Doppler shift produced as a result is more than half of the sampling rate, that is, PRF, as shown in Figure 2.67(d), the Doppler measurement is now ambiguous as the Doppler measurement still reads the Doppler shift as the location of the received spectral line to the nearest transmit line. The ambiguity in Doppler measurement arising out of under-sampling also called *aliasing* leads to the reported or apparent Doppler shift being different from the true Doppler shift. The true and apparent Doppler shifts are interrelated by eqn. 2.87.

$$f_A = \left[\left(f_d \text{ MOD } PRF \right) - PRF \right] \text{MOD } PRF \text{ or } f_A = \left[\left(f_d \text{ MOD } PRF \right) + PRF \right] \text{MOD } PRF \qquad (2.87)$$

whichever gives a smaller absolute value.

The MOD operator is the remainder of the division process. For instance, A MOD B is the remainder of the division of A by B. As an illustration, for $f_d = 1000$ Hz and PRF = 400 Hz, f_d MOD PRF would be 200. Similarly, PRF MOD f_d would be 400 as 400/1000 gives a remainder of 400 only.

Another very serious problem is that of *blind Doppler* or *blind speeds*. When the Doppler shift equals an integer multiple of PRF, the moving target echo signal's spectral lines coincide with the spectral lines of the transmit signal and so are the spectral lines of the stationary target echoes [Figure 2.67(e)]. Another way of saying the same thing would be that when the target's radial velocity component is such that it travels a distance of $n.\lambda/2$ (where n is an integer) along

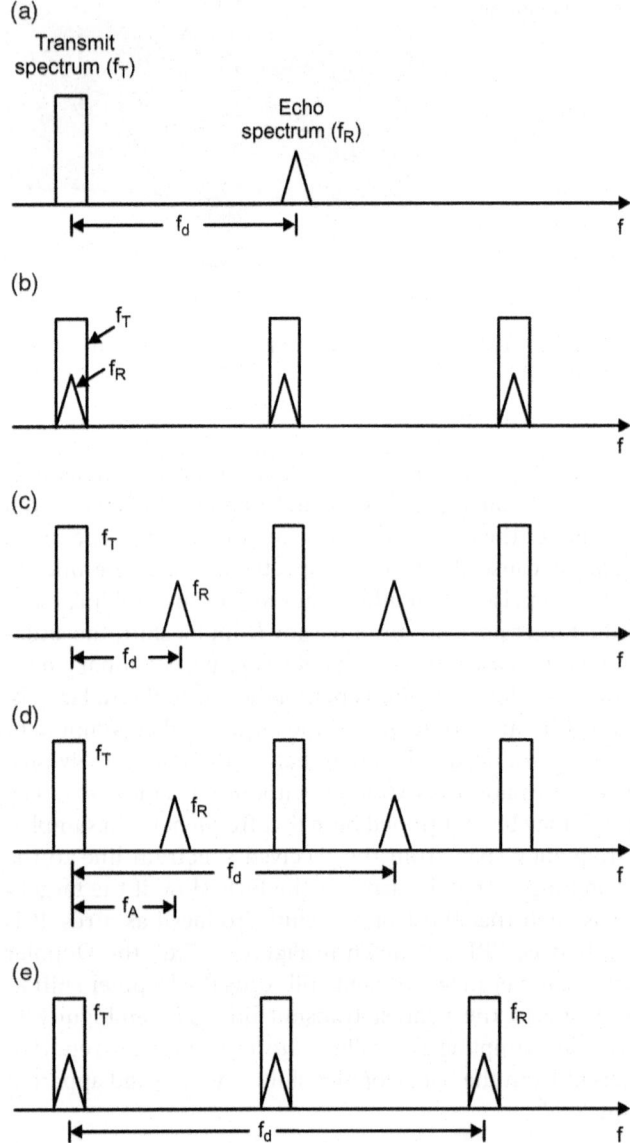

Figure 2.67 Blind speeds in MTI. (a) Doppler shift, (b) no Doppler shift, (c) Doppler shift is less than half the PRF, (d) Doppler shift is more than half the PRF and (e) Doppler shift equals the PRF.

the radar axis during the time between successive transmit pulses, then the phase difference between the corresponding echo pulses would be $2n\pi$ radians, which is equivalent to no phase change or a stationary target. Though use of Doppler filters effectively attenuates echoes at zero Doppler shifts and at integer multiples of PRF for clutter rejection, a moving target producing these Doppler shifts cannot be detected. Such Doppler shifts and the associated radial velocity components are called blind Doppler and blind speeds respectively. The blind shifts and blind speeds can be computed from eqn. 2.88.

$$f_B = n.PRF \tag{2.88}$$

$V_B = [(n \times c \times PRF)/2f]$

$n = \pm 1, \pm 2, \pm 3, \pm 4$

f = Operating frequency.

One of the effective solutions to the problem of blind speeds lies in PRF staggering. Blind Doppler shifts are the Doppler shifts equal to PRF or its integer multiples. For a fixed PRF, the Doppler phase is sampled at the same point in each cycle. If the PRF was varied, the Doppler phase would be sampled at different points in different cycles with the result that coincidence of Doppler shift spectral lines with the transmit signal spectral lines can be avoided thus facilitating recovery of Doppler information. The PRF stagger can be pulse to pulse or look-to-look or even scan-to-scan.

2.12.4 Pulse Doppler Radar

The principle of operation of a Pulse Doppler Radar is similar to that of an MTI radar in the sense that both make use of Doppler shift caused by reflection of transmit signal from a moving target. Low PRF radars designed to avoid range ambiguity problem and based on Doppler shift to extract target speed information have become to be known as MTI radars to distinguish them from pulse Doppler radars that are usually high PRF systems. Medium PRF pulse Doppler radars are also prevalent. To establish connectivity with what we have already discussed, the main difficulty with MTI systems is that the Doppler sampling rate, which equals the radar PRF, is too low for the speed of the modern aircraft with the result that Doppler shift information from most of the realistic targets is highly under-sampled. The Nyquist criterion is not met and this essentially leads to aliasing problem. Aliasing further leads to ambiguous estimates of target speed. Occurrence of blind speeds, where the target appears stationary and unresolvable against background clutter, is still worse.

Pulse Doppler radar, being a high PRF radar, offers solution to such types of Doppler ambiguities. The PRF is classified as high when it is fast enough to sufficiently sample the highest possible Doppler shift induced by the moving target. In fact, it has to be at least twice the Doppler shift. Assuming that targets with maximum closing velocities are to be accounted for, typical PRF for an X band radar may be in the range of 300 kHz or so. Such a high pulse repetition rate however does not allow unambiguous recovery of target range. The maximum unambiguous range for a PRF of 300 kHz, for instance, would only be 500 m, which for all practical purpose would be as good as nothing.

Pulse Doppler radars are generally divided into two broad PRF categories namely medium PRF and high PRF. While a high PRF pulse Doppler radar, as mentioned before, is ambiguous in range and unambiguous in velocity, medium PRF radar is ambiguous in both range and velocity and these ambiguities need to be resolved during processing. This gives one an impression that medium PRF Pulse Doppler radar due to its range and Doppler ambiguities would have a limited use. In reality, it is not so. With radar signal processing technology making significant advances, it has become possible to resolve these ambiguities. Along with pulse-compression, it allows the designers to build detection (requiring high-energy, long transmitted pulses and high PRF), Doppler (requiring high PRF) and Range Resolution (requiring short processed pulses) capabilities. A typical medium PRF is in the range of 10–20 kHz. X band radar operating at, say, 15 kHz would have a maximum unambiguous range of about 10 km and a maximum unambiguous closing speed of 0.3 Mach.

A pulse Doppler radar employs a coherent radar system architecture where the transmit signal and receiver local oscillator are synchronized to a highly stable reference. Coherent, coherent-on-receive and non-coherent types of radar architectures have been discussed in detail in the earlier part of this chapter.

Figure 2.68 HARD-3D radar.

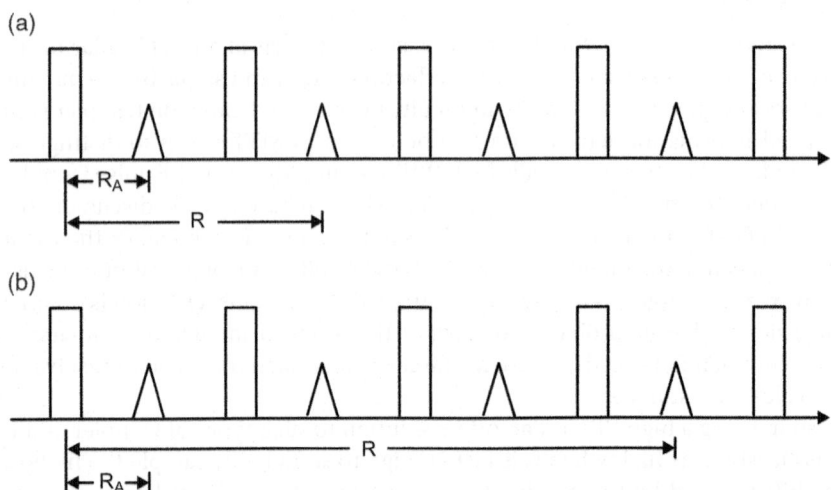

Figure 2.69 True and apparent ranges: (a) zone-2 target and (b) zone-4 target.

Figure 2.68 shows a photograph of three-dimensional short-range air-defence search and acquisition radar (Type HARD 3D) of the pulse Doppler type. Mounted on a tracked vehicle for high mobility, the radar operates in the X band. It is optimized for detection of helicopters and other low-level air threats. It has features like low power output that gives it a good anti-jamming performance, pulse-to-pulse frequency agility to give it enhanced ECCM capability and pulse compression.

2.12.4.1 True and Apparent Ranges

The *apparent range*, which is the range corresponding to the time difference between the received echo pulse and the last transmitted pulse, is different from the *true range* if the target is not a zone 1 target. Figure 2.69(a) shows the true and apparent propagation times for a zone 2 target while Figure 2.69(b) shows the same for a zone 4 target. The two propagation times are interrelated by eqn. 2.89.

$$T_A = T_P \, \text{MOD} \, PRI \tag{2.89}$$

PRI = Pulse Repetition Interval
T_A = Apparent propagation time
T_P = True propagation time

The apparent and true target ranges are interrelated by

$$R_A = R \operatorname{MOD}\left[(c \times PRI)/2\right]$$

Equation 2.89 allows us to find apparent propagation time or the apparent range from known magnitude of true propagation time and range respectively. In practice, it is necessary to find the true values from the apparent values. The relevant expressions are given by eqns 2.90 and 2.91.

$$T_P = T_A \times (N_R - 1) \times PRI \tag{2.90}$$

$$R = R_A + (c/2)\left[(N_R - 1)/PRI\right] \tag{2.91}$$

The targets interpreted as being at incorrect ranges due to ambiguity are called *range ghosts* and the process of determining true range from apparent range is called *range de-ghosting*. The equations above also reveal that for targets that outside range zone-1 the apparent range is a function of radar PRF. This phenomenon is used as the basis for range de-ghosting in most radars. Thus, radars use multiple PRFs to determine true range of ambiguous targets. The maximum range that a given PRF can determine unambiguously corresponds to a PRF value given by the highest common factor (HCF) of different PRFs used by radar.

Occurrence of a *blind range* is yet another form of range ambiguity when the radar fails to detect the target though the target is very much there and is sending back to the radar a fairly strong echo signal. A blind range occurs when the time of occurrence of a received echo pulse coincides with the time of occurrence of a transmit pulse. Blind ranges can be computed from eqn. 2.92.

$$RB = N_R \times c \times (PRI/2) \tag{2.92}$$

N_R is an integer greater than zero. It equals n for an n-zone target.

2.12.5   Tracking Radar

The primary function of tracking radar, as the name suggests, is the automatic tracking of moving targets. It is usually a ground-based system used to track the airborne targets. The tracking radar antenna sends out a very narrow beam whose width could be anywhere between fraction of a degree to a degree or so in both azimuth and elevation to get the desired resolution for tracking purpose. One can, however, visualize that it would be necessary to acquire the target with search radar having a beam of relatively much large width before a track action is initiated. In the track mode, whenever the target tends to move away from the radar beam axis, an error signal is generated, which in the closed loop is used to steer the radar antenna either mechanically or electronically to keep the target always illuminated by the radar beam.

2.12.5.1   Track Modes

Tracking could be carried out using Range (called *Range Tracking*), Doppler (called *Doppler* or *Velocity Tracking*) and Angular (called *Angle Tracking*) information. This allows the radar to follow the motion of a target in azimuth and elevation (due to angle tracking), range (due to range tracking) and Doppler (due to Doppler tracking). However, not all radars track in all dimensions. Different track modes include the following.

1) Single Target Track (SIT)
2) Spotlight Track
3) Multi-Target Track
4) Track-While-Scan (TWS).

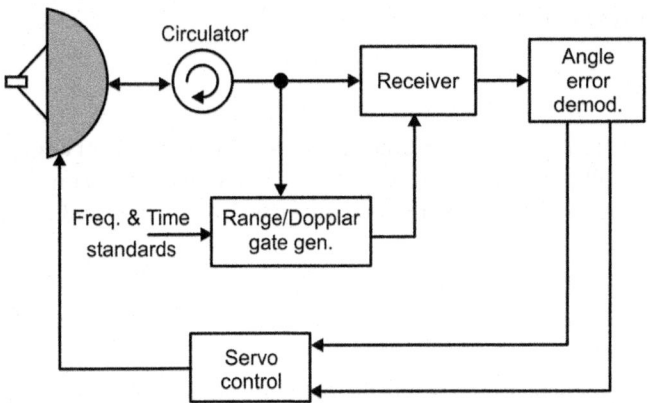

Figure 2.70 Block-schematic of a tracking radar.

In the *SIT*, the radar tracks a single target. It is continuously dedicated to a single moving target. Such a radar samples the target information at the radar PRF. Single target trackers are capable of tracking targets with great accuracy. In the *spotlight track*, the radar sequentially dwells upon various targets spending a certain specified time on each target. It is not as accurate as the single target track due to the fact that a given target is likely to undergo a change in its coordinates during the time between two successive dwell periods. A *multi-target track* mode is capable of simultaneously tracking multiple targets with an accuracy matching that of a single target track. In a *TWS* system, the radar samples the position of several targets once per scan and then with the help of certain extrapolation algorithms, estimates the position of targets between samples. TWS is truly search radar's operational mode. It is not essentially a tracking operation because for true multi-target tracking, each target must be sampled at the Nyquist rates corresponding to the radar servo loop and target manoeuvring bandwidths. The required sampling rate may typically be 10–20 samples per second for each target. On the other hand, in a TWS process, the target may be sampled once every 10–15 s.

Figure 2.70 shows a basic block-schematic arrangement of tracking radar. Most tracking radars use angular information as the basis for tracking operation. But for accurate tracking, it is important that the radar concentrates on one target at a time. If there is more than one target in the radar antenna's beam, techniques should be used to ignore other returns from other targets. Range gating (which is a part of the range tracker) and Doppler gating (which is a part of a Doppler tracker) can be used for the purpose. Range and Doppler trackers are described in subsequent paragraphs. In fact, time and frequency control for range and Doppler gating is done in range and Doppler trackers, respectively. The angular error signals for the desired target to be tracked are developed in the error demodulator block, which is also controlled by the Range/Doppler gate generation block and then fed back to the steerable antenna into a closed loop for tracking.

2.12.5.2 Tracking Radar Types

Tracking radars are classified based on the methodology used to develop angular errors. The commonly used tracking methodologies include the following.

1) Lobe Switching
2) Sequential Lobing
3) Conical Scan
4) Amplitude Comparison Monopulse
5) Phase Comparison Monopulse.

2.12.5.2.1 Lobe Switching

In the *lobe switching* tracking technique, the antenna beam is rapidly switched between two positions around the antenna axis in a single plane as shown in Figure 2.71. The amplitudes of the echoes from the target to be tracked for the two lobe positions are compared. The difference between the two amplitudes indicates the location of the target with reference to the antenna axis. When the target is on the axis, the difference is zero as the echo amplitudes for the two lobe positions are identical. To summarize, the amplitude and sense of the difference signal can be used to generate the correction signal, which with the help of servo loop control can be used to move the antenna so as to bring the target on the antenna axis. The lobe switching technique has the disadvantage that it loses its effectiveness if the target cross-section changed between different returns in one scan.

2.12.5.2.2 Sequential Lobing

In *sequential lobing*, a squinted radar beam (a radar beam whose axis has been shifted slightly off the antenna axis) is sequentially placed in discrete angular positions, usually four, around the antenna axis (Figure 2.72). The angular information on the target is determined by processing several target echoes. The track error information is contained in the target signal amplitude variations. The squinting and squinted beam switching between different positions

Figure 2.71 Lobe switching.

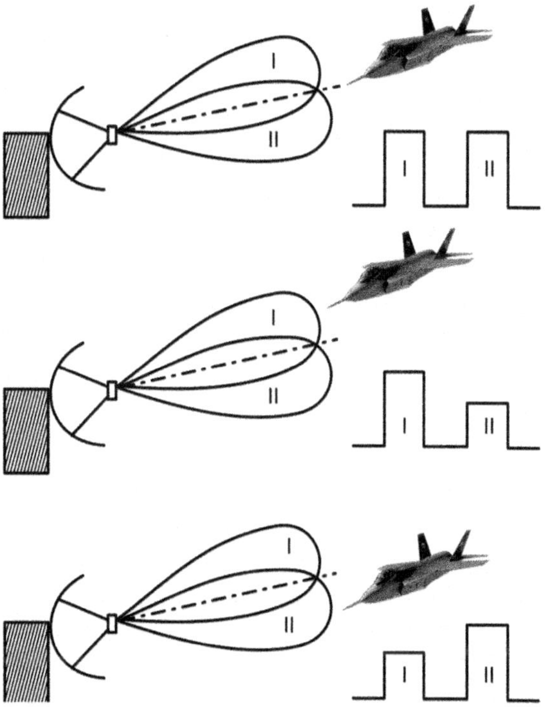

Figure 2.72 Sequential lobing.

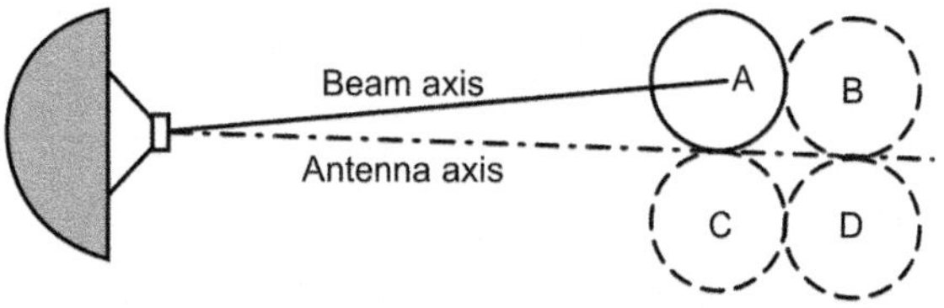

is done electronically in modern radars using this tracking methodology. Since the beams can be switched very rapidly using electronic means, the transmitted beam is usually not scanned. The lobing is done on receive-only. Also, virtually any scanning pattern can be used. The scan pattern can be changed from scan-to-scan. It is because of this reason that this type of tracking radar is less affected by amplitude modulated jamming.

2.12.5.2.3 Conical Scanning

This is similar to sequential lobing described above except for the difference that in the case of *conical scanning*, the squinted beam is scanned rapidly and continuously in a circular path around the axis (Figure 2.73). If the target to be tracked is off the antenna axis, the amplitude of the target echo signal varies with the antenna's scan position. The tracking system senses these amplitude variations as a function of scan position to determine target's angular coordinates. The error information is then used to steer the antenna axis so as to coincide with the target location. For true tracking, the scan frequency must be such that Nyquist criterion for the sampling rate is met. In pulsed radar, there must at least be four pulses per scan; two for generating azimuth error signal and two for generating elevation error signal. This implies that the maximum scan frequency can be one-quarter of the radar PRF. The actual scan frequency also depends upon the scan mechanism and is smaller of the two values set by maximum scan rate dictated by Nyquist criterion and the scan mechanism capability. The antenna beam is squinted and scanned either mechanically by offsetting the feed and rotating it or electronically with the help of phase shifters. Mechanical scans are usually much slower than electronic scans. Typical scan rates are 30–40 scans per second.

Figure 2.74 shows a block-schematic arrangement of a conical scanning system. The scan reference is a sinusoidal signal varying at the scan rate and phase locked to the scan signal. The functions of the other building blocks are self-explanatory.

The conical scan tracking technique is highly vulnerable to amplitude modulated jamming particularly the gain inversion jamming. In gain inversion jamming, the target carrying the jammer receives the radar's transmitted signal. The jammer demodulates the amplitude variation,

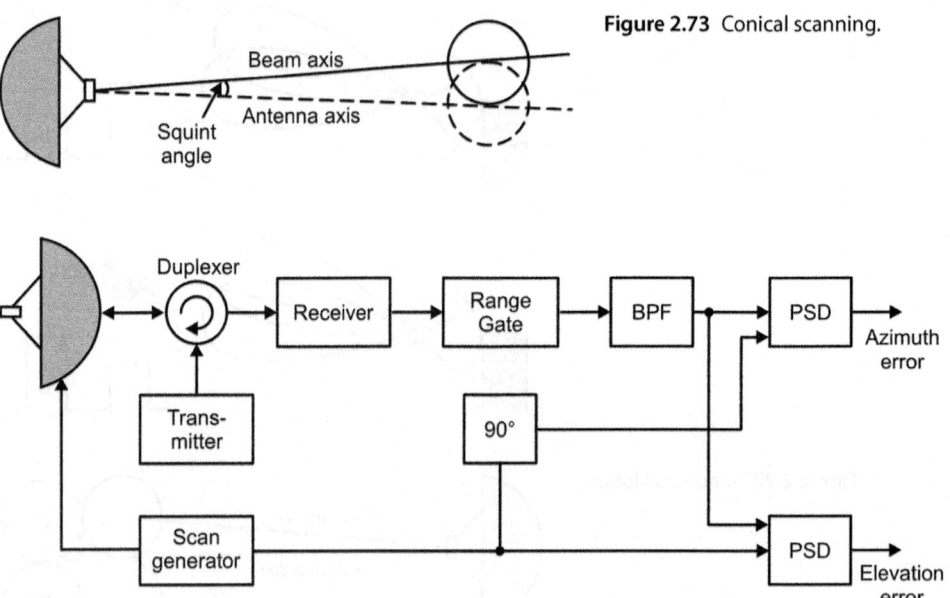

Figure 2.73 Conical scanning.

Figure 2.74 Block diagram of a conical scanning system.

inverts it and then sends it back towards the radar. The radar starts tracking the jammer signal if the jammer signal is much stronger than the echo signal. Since the jammer transmits inverted angle information, the radar will never be able to track the target.

2.12.5.2.4 Amplitude Comparison Monopulse Tracking

One of the major disadvantages of sequential techniques including *lobe switching, sequential lobing* and *conical scan* is that their tracking accuracy gets severely affected if the target's radar cross-section changes during the time when the beam is being switched or scanned, as the case may be, to get the desired number of samples. In addition, these techniques also suffer from their vulnerability to AM jamming. Monopulse tracking such as amplitude comparison and phase comparison overcomes these shortcomings by generating all the required angle error information from one pulse only.

The basic principle of operation of *amplitude comparison monopulse tracking* can be explained with the help of Figure 2.75. Figure 2.75(a) shows a radar antenna and the received wave front when the target is on the antenna axis. In this case, the received wave front will be focused onto a spot on antenna axis as shown. If the antenna used four feeds placed symmetrically around the focal point representing four quadrants A, B, C and D as shown, then the amount of energy falling on each feed would be the same as shown in Figure 2.76(a). Now, if the target is located off axis as shown in Figure 2.75(b), then different feeds would receive different energies depending

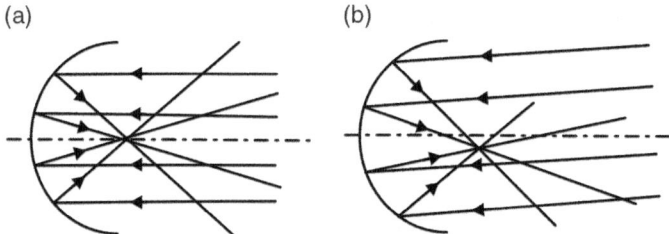

Figure 2.75 Amplitude comparison monopulse tracking: (a) target on-axis and (b) target off-axis.

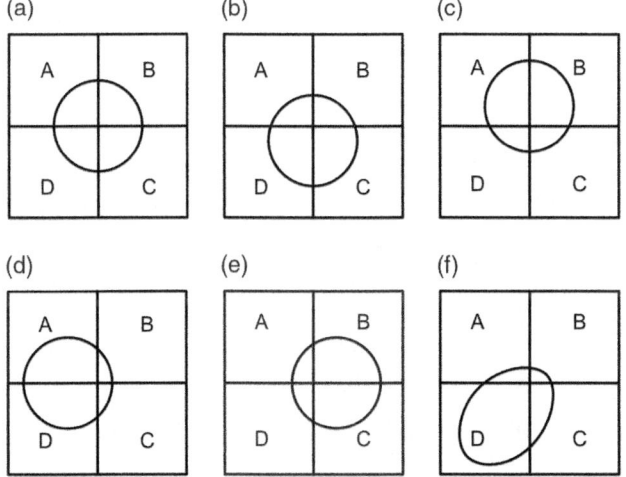

Figure 2.76 Received energy as a function of angle: (a) target on-axis, (b) target off-axis, (c) target is below the antenna axis but has the same azimuthal location as the antenna axis, (d) target is to the right of the antenna axis, (e) target is to the left of the antenna axis and (f) target is above and to the right of the antenna axis.

upon magnitude and sense of this off-axis angle. For instance, if the target is above the axis and having the same azimuthal location as the antenna axis, then the received energy would be distributed as shown in Figure 2.76(b). Figure 2.76(c) gives the received pattern when the target is below the antenna axis but has the same azimuthal location as that of the antenna axis. Figure 2.76(d) and (e) depict the condition when the target is to the right and left of the antenna axis, respectively. with the target having same elevation location as the antenna axis in both cases. Figure 2.76(f) shows a condition where the target is above and to the right of the antenna axis.

The amplitude of the received echo pulse at the outputs of various feeds can be appropriately processed to determine azimuth and elevation error signals along with some other useful error signals. One such processing method is shown in Figure 2.77. The azimuth and elevation error signals in this case are, respectively, given by eqns. 2.93 and 2.94.

$$\text{Azimuth error signal} = (A+D)-(B+C) \tag{2.93}$$

$$\text{Elevation error signal} = (A+B)-(C+D) \tag{2.94}$$

$(A+B+C+D)$ gives the sum channel.

It is not necessary to pair A, D and B, C only as shown in Figure 2.97. One could pair A, B and C, D also as shown in Figure 2.78. In that case, the azimuth error would be difference of first

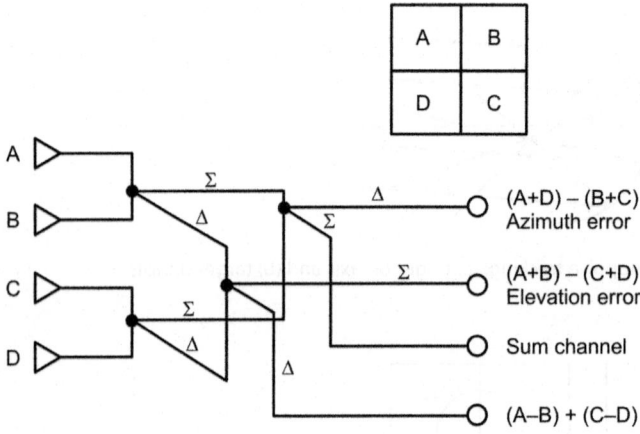

Figure 2.77 Error signal processing in amplitude comparison monopulse tracking.

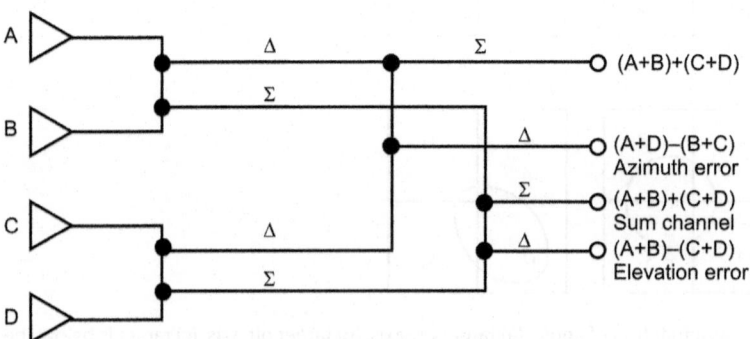

Figure 2.78 Error signal processing paired as *A*, *B* and *C*, *D*.

differences rather than difference of first sums and the elevation error would be difference of the first sums rather than the sum of the first differences.

In the amplitude comparison monopulse tracking technique, it is important that signals arriving at various feeds are in phase. This is not a problem when using reflector antennas with feeds that are physically small, usually a few wavelengths across. In case of arrays where the antenna surface is very large, signals arriving from different off-axis angles present different phases to different segments into which the array has been divided. These phases need to be equalized before error signals are developed.

Amplitude comparison monopulse is somewhat more complex to implement with array antennas. Now for tracking, one needs to transmit the transmit beam and four squinted beams for error detection, which could be conveniently implemented by using a four- or five-horn reflector feed. In case of an array, the array is divided into four quadrants. Signals from each of the four quadrants feed a beam forming network comprising of phase shifters necessary to produce that much squinted beam. These four beams are then fed to monopulse comparator to generate sum, azimuth and elevation error signals required for tracking.

2.12.5.2.5   Phase Comparison Monopulse Tracking

In *phase comparison monopulse tracking*, it is the phase difference between the received signals in different antenna elements that contains information on angle errors. In all, at least two antenna elements are required each for azimuth and elevation error detection. When the target is on axis [Figure 2.79(a)]; the magnitude of phase difference would be zero. If it is off-axis, then magnitude and sense of the phase difference would determine the magnitude and sense of the off-axis angle [Figure 2.79(b)]. This technique is very sensitive; that is, the phase difference produced per unit angular error increases if the elements are wide apart. But if they are too far apart, an off-axis signal may produce identical phases at the antenna elements (Figure 2.80). This gives rise to ambiguity. A practical system could have two pairs of antenna elements each for azimuth and elevation. The outer pair gives the desired sensitivity while the inner pair resolves ambiguity.

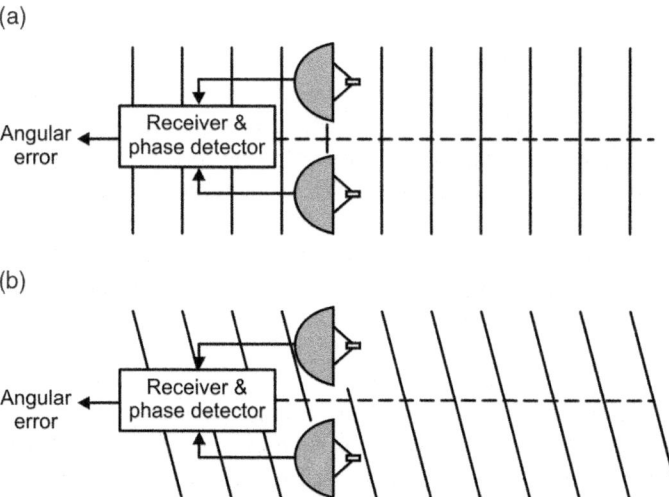

Figure 2.79 Phase comparison monopulse tracking.

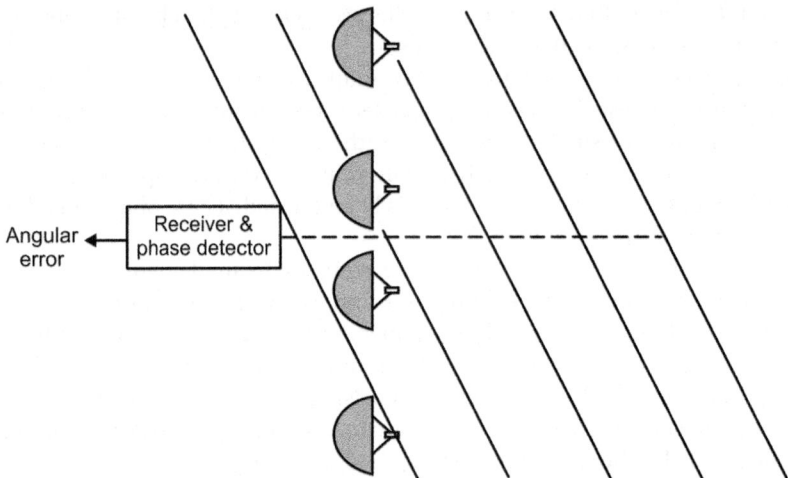

Figure 2.80 Phase comparison monopulse tracking with antenna elements wide apart.

2.12.5.3 Range Tracking

Range tracking is the process of tracking a moving target based on its range coordinates. Even though the commonly used tracking methodology in tracking radars is angle tracking, a range tracker forms a part of the angle tracker also. A range tracker in that case continuously measures the target range and based on the range data generates a range (or time) gate so that the target is at the centre of the gate. Range tracking thus provides an effective means of distinguishing the desired target to be tracked by using, may be, angular means from other targets within radar beam. Doppler tracking discussed in the next section is another.

The first step in any tracker, range or otherwise, is target acquisition, which provides an idea about the target coordinates so that a radar beam can be pointed in that direction. A range tracker could do the job of target acquisition very well. Typically, the range tracker divides the minimum to maximum range into small range increments and as the antenna scans a given region, it examines each of the range increments in a given direction simultaneously for presence of target. The antenna is made to scan slowly enough for the target to remain within the radar beam width as different range increments are being examined in a given direction.

A range tracker is a closed loop system. The error corresponding to deviation of target's range location from the centre of the range gate is sensed and fed back to the range gate generating circuitry to reposition the gate in such a way that the target is at the centre. The commonly used technique of sensing range tracking error is that of using split gate comprising of an 'Early Gate' and a 'Late Gate' as shown in Figure 2.81. When the target is at the centre, the area under the echo pulse when early gate is open is same as the area under the pulse when the late gate is open. If the signals under the two gates are integrated and a difference taken, it would be zero. If the target is off-centre, one signal will be greater than the other. The magnitude and sense of the difference signal can be used to reposition the gate.

2.12.5.4 Velocity Tracking

Velocity tracking is a process that makes use of Doppler shift information. It is Doppler tracking error using split filter error detection. The track error is represented by the difference between target IF and receiver's normal IF. The error after filtering is used to change the receiver local oscillator frequency until Doppler shifted signal is nominal IF. Figure 2.82 shows the block-schematic arrangement in a velocity tracker.

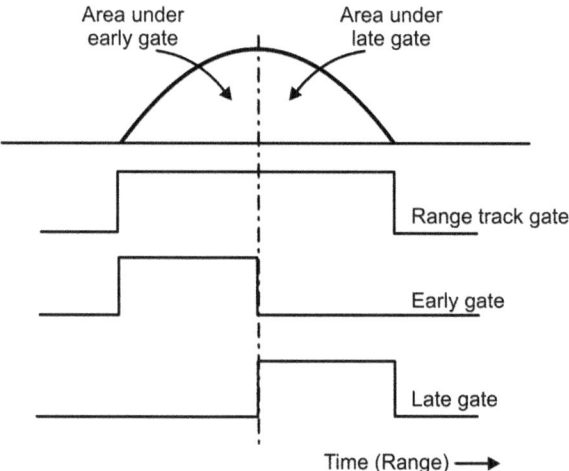

Figure 2.81 Split gate method of sensing range tracking error.

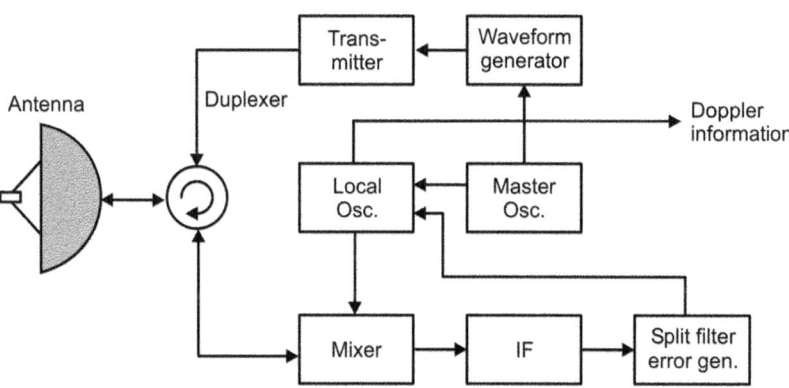

Figure 2.82 Block diagram of a velocity tracker.

2.12.6 Pulse-Compression Radar

It is always desirable to have long transmitted pulse in radar to enhance its detection capability. A long transmit pulse, for a given FRF, increases the average power and reduces the ratio of peak power to average power. In other words, for a given average power requirement depending upon the required ranging capability, the peak power stress would be less in the case of radar asked to transmit a longer pulse. On the contrary, the range resolution capability of the radar is inversely proportional to pulse width. A narrower processed pulse is desirable for achieving higher range resolution. In the conventional pulsed radar where the transmit signal is gated CW pulse of a fixed frequency, transmit and received pulse widths are approximately the same and therefore a higher detection capability could be achieved only at the expense of poorer range resolution. We could have the best of both worlds if somehow we could transmit a long pulse and have a narrow processed pulse. This is exactly what is done in *pulse-compression radar*. Pulse-compression radar transmits a wide pulse called *expanded pulse* to achieve higher average power and detection capability and processes the received pulse to a

(a)   (b)

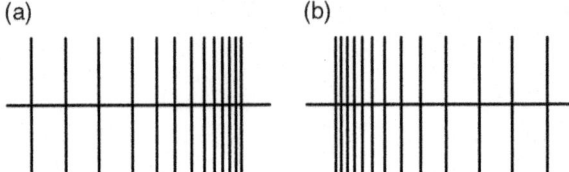

Figure 2.83 Pulse compression techniques: (a) up-chirp and (b) down-chirp.

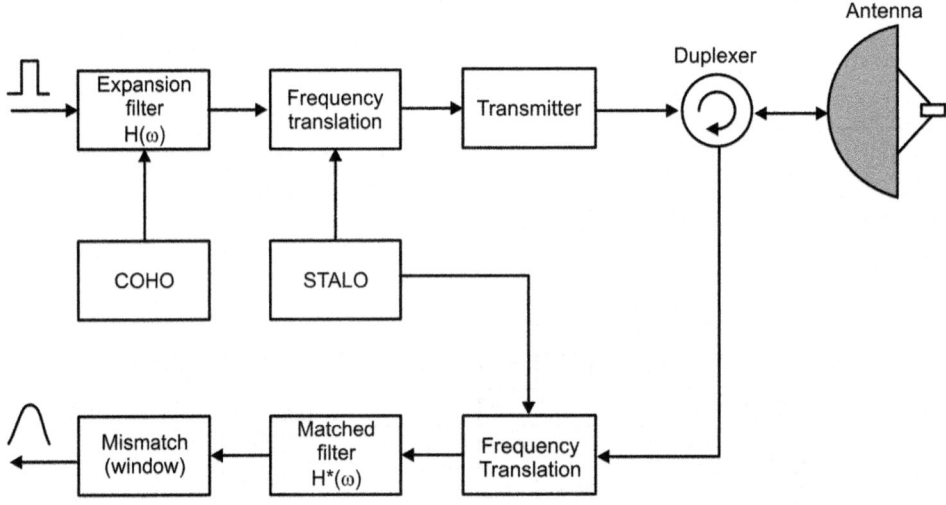

Figure 2.84 Block diagram of pulse compression radar.

narrow pulse called the *compressed pulse*. The compressed pulse width is approximately equal to reciprocal of the signal bandwidth in the expanded pulse.

There are two basic techniques commonly used for pulse-compression. One is analogue in nature and the other is digital. In analogue pulse-compression, the expanded transmit pulse, instead of being a single frequency burst, has a frequency modulation across the pulse. It could be an up-chirp where the frequency linearly varies from a low value to a high value [Figure 2.83(a)] or a down-chirp where it varies from an initial high value to a final low value [Figure 2.83(b)]. In the digital scheme, some kind of phase coding is given to the RF across the pulse. The received echoes are compressed either by using a matched filter concept where the compression filter in the receiver is matched to the transmit wave or by the process of correlation where a delayed version of transmit signal is correlated with the received echo. Analogue pulse-compression is done using matched filter concept while correlation technique adapts better to digital pulse-compression.

Figure 2.84 shows the basic block-schematic arrangement of a pulse-compression radar using the matched filter concept. The diagram is self-explanatory except for a couple of points. The compression filter response is complex conjugate of that of expression filter. From the property that *If $H(\omega)$ is Fourier transform of $h(t)$, then $H^*(\omega)$ is the Fourier transform of $h(-t)$*, any phase change introduced into the signal by $H(\omega)$, such as linear FM, is undone in $H^*(\omega)$. So, if expansion filter is fed with a short pulse with RF at receiver IF (COHO), its output is the expanded pulse with modulated RF. The received echo at the input of $H^*(\omega)$ is again an expanded pulse with RF equal to Doppler shifted IF whereas its output is a compressed pulse whose

(a) (b)

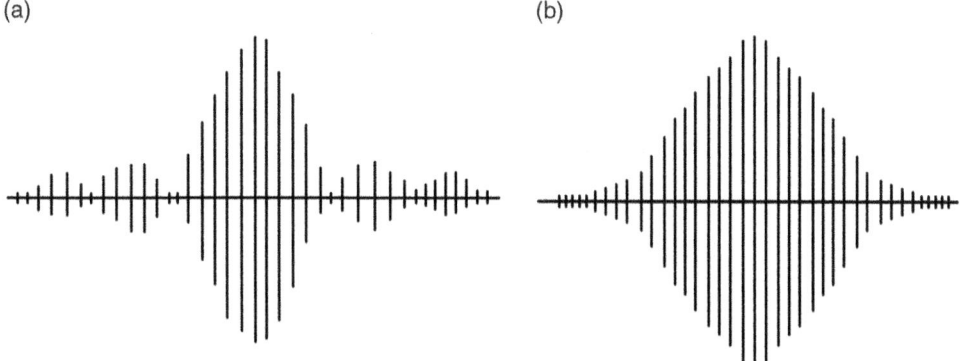

Figure 2.85 Compressed pulse: (a) with leakage in time region and (b) reduced leakage using mismatch.

Figure 2.86 Digital pulse compression.

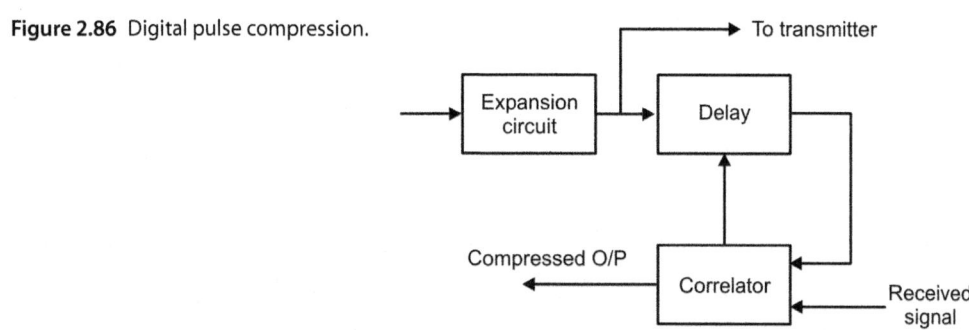

width is approximately equal to the short pulse at the input of the expansion filter. The RF of the compressed pulse equals the Doppler shifted IF.

The compressed wave in practice is not exactly the same as the COHO pulse that was expanded. Compressed pulse in fact leaks into time region other than that occupied by echo [Figure 2.85(a)]. This leads to low-amplitude desired echoes getting hidden by large interfering signals. This necessitates introduction of a mismatch (similar to a window) to reduce leakage [Figure 2.85(b)]. But in the process, the compressed pulse width increases a bit. This can be compensated for by starting with a pulse a bit shorter than the desired compressed pulse width.

Figure 2.86 shows the schematic arrangement for achieving pulse-compression in radar employing digital pulse-compression. A delayed copy of the transmit wave is correlated with received echo to compress the pulse. In principle, it is equivalent to the matched filter concept.

To achieve analogue pulse expansion and compression, the commonly used device is the SAW device. By proper design of inter-digital transducers, desired expansion filter character-istics can be achieved. The compressed pulse is the matched version of it. Figure 2.87(a) shows a typical SAW based expansion filter producing an up-chirp expanded output. A matched compression filter is shown in Figure 2.87(b). Detailed discussion of SAW devices is beyond the scope of this text. In case of digital pulse-compression, a variety of codes are available. Barker code bit sequences and pseudo random bit sequences are the commonly used ones. Details are again beyond the scope of the text.

Figure 2.88 shows a photograph of the ship-borne surveillance radar operating in the E/F band (Type TIGER DRBV-15). It employs pulse-compression techniques and is designed to

(a)

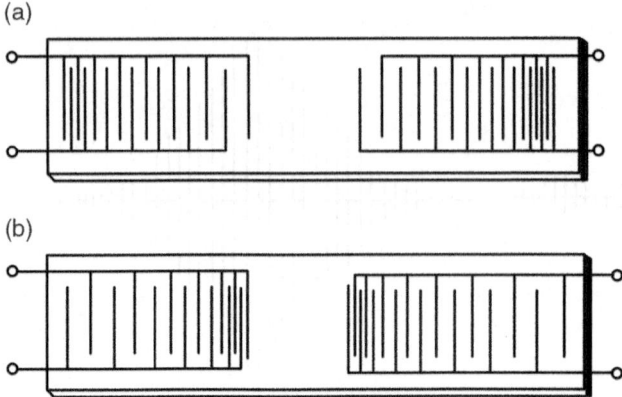

(b)

Figure 2.87 SAW-based filter: (a) expansion and (b) compression.

Figure 2.88 Pulse compression radar type TIGER DRBV. (*Source:* Courtesy of Jörg Wat.)

perform the functions of air surveillance, surface surveillance, anti-missile surveillance and target designation for weapon systems in a severe clutter and jamming environment.

2.12.7 Synthetic Aperture Radar

In an earlier part of the chapter, the way in which the physical size of a real antenna (physically realizable antenna) limits the minimum achievable beam width was mentioned. The minimum achievable beam width may not be adequately small to yield cross-range resolution that would be desirable for some specific applications such as terrain mapping, imaging and so on.

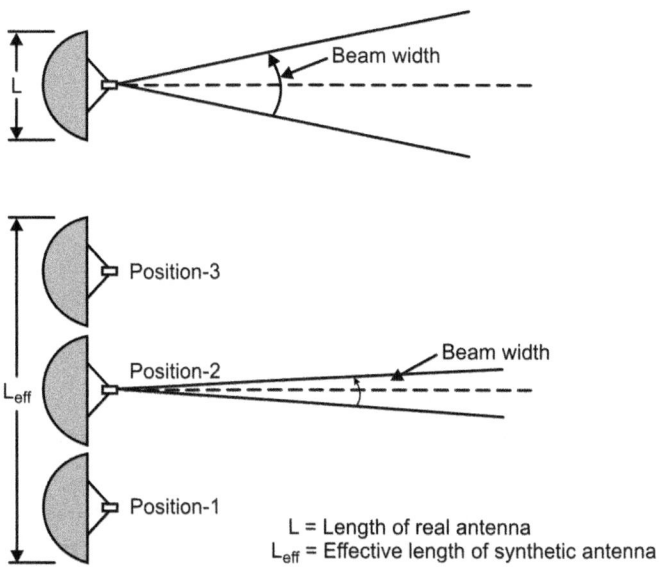

Figure 2.89 Principle of synthetic aperture radar.

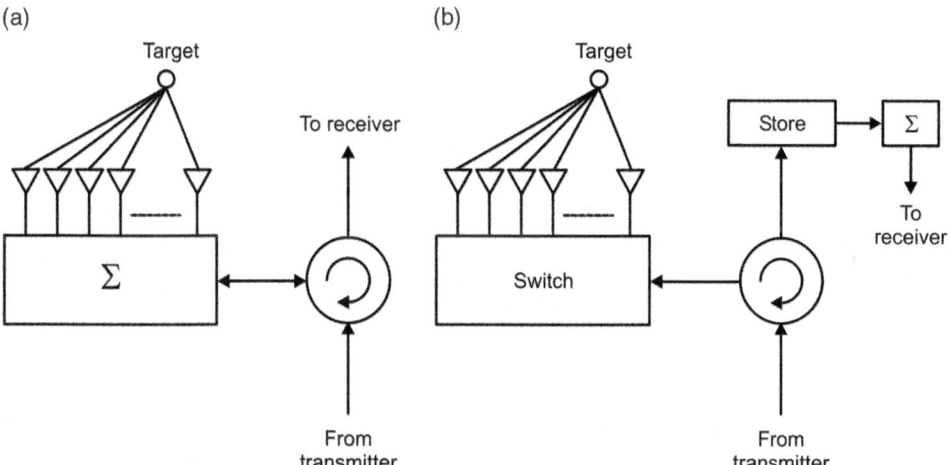

Figure 2.90 Process of achieving narrow beam width: (a) a large physical antenna and (b) a synthesized antenna.

Synthetic Aperture Radar (SAR) synthesizes a large antenna by moving a real antenna through different locations across the volume of interest. The synthesized length then equals the distance moved by the real antenna (Figure 2.89). The antenna moves through different locations and a pulse transmitted from each location. The received echoes from different locations are stored and then processed in such a way that the antenna behaves as if it was as long as the distance travelled. If there was an antenna physically as long as the synthesized antenna, the incremental elements of the synthesized antenna would be formed simultaneously here and the narrow beam would be formed by summing outputs from all those elements. In the case of synthesized antenna, these incremental elements would be formed sequentially. The outputs from these elements are stored until the full array is formed. The outputs are then summed to form the narrow beam. The two situations are depicted in Figure 2.90(a) and (b).

The cross-range resolution of synthetic aperture radar is given by eqn. 2.95.

$$\text{Cross-Range resolution, } X_s = \left[\lambda.R/2.L_{eff} \right] \tag{2.95}$$

Also, $L_{eff} = R\lambda/D_{eff}$

$$\text{Therefore, } X_s = \left[\lambda.R/2\right].\left[D_{eff}/R.\lambda \right] = D_{eff}/2 \tag{2.96}$$

L_{eff} = Effective length of synthesized antenna
D_{eff} = Effective length of real antenna

Equation 2.95 for cross-range resolution implies the following:

1) The cross-range resolution is independent of range.
2) Cross-range resolution is not a function of operating wavelength.
3) A smaller real antenna makes a better synthetic antenna.

The first statement is true because effective length of synthetic antenna increases with range. So, any degradation in cross-range resolution with range would get compensated by increase in L_{eff}. If synthetic antenna has a larger beam width due to increase in operating wavelength, the real antenna would also have increased beam width which in turn means increased L_{eff}. Smaller real antenna means large beam width and a large L_{eff}, Synthetic aperture radars are used for high-resolution ground mapping from airborne platforms.

2.12.8 Inverse Synthetic Aperture Radar (ISAR)

ISAR is an extension of the concept of synthetic aperture radar. In ISAR, the radar is usually stationary (it could also be moving) and it is the target motion, or rotation to be more precise, that is used to synthesize a large antenna. It takes advantage of the differential Doppler shift that exists between the echoes from different objects in the same range bin when the target is rotating to enhance the cross-range resolution. These are commonly used for imaging tactical targets from airborne, shipboard and land-based platforms.

2.12.9 Over-the-Horizon Radar (OTHR)

The maximum ranging capability of microwave radar gets limited by the radar horizon even if there is adequate transmitted power to receive a detectable signal. The shortcoming could be overcome to some extent by elevating the height of the transmitting antenna as shown in Figure 2.91. But this is certainly not the solution when we are talking about ranges of the order of thousands of

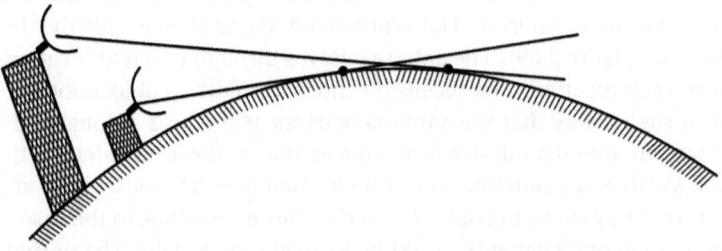

Figure 2.91 Extending range by increasing antenna height.

kilometres. *OTHR*, as the name suggests, can look at the targets that lie beyond the radar horizon. This is made possible by making use of propagation modes other line-of-sight propagation used by microwave radar. These include ground wave propagation and sky wave propagation. The radar is operated in the high-frequency (HF) band of the electromagnetic spectrum. The frequency of operation therefore lies between 3 and 30 MHz.

Two types of OTHR have been developed known as the *surface wave* (or *ground wave*) radar and the *sky wave radar* making use of the surface wave and ionospheric propagation modes, respectively. Surface wave OTHR uses the surface propagation mode to look over the immediate horizon. It is predominantly monostatic in nature. *Monostatic radar has a co-located transmitter and receiver.* Typically, such an OTHR has a maximum ranging capability of 400 km or so.

Sky wave OTHR makes use of ionosphere to transmit electromagnetic waves to very long distances beyond the radar horizon. It is bistatic in nature. *Bistatic radar* has its transmitter and receiver widely separated from each other. The radar beam is beamed up towards the sky. It gets reflected from the ionosphere to illuminate the target. The radar signal as reflected from the target reaches the radar receiver again via the ionosphere (Figure 2.92). The radar transmitter and receiver in an OTHR are typically separated by about 100 km. As is clear, OTHR can look at targets close to the radar site. Typically, it can detect targets between 1000 and 4000 km. This type of OTHR is particularly suited to the defence and remote ocean sensing needs of countries that have very extensive geographical areas such as USA, Russia, Australia and so on. OTHR performance, due to the fact that it makes use of ionospheric propagation, suffers from sun spot activity, day/night cycle, seasonal variations and so on.

Targets of interest for an OTHR are the same as they are for the microwave radar, which includes aircraft, missiles, ships, other strategic locations and so on. In addition, due to the fact that the wavelengths used are of the same order as that of the ocean gravity waves, this type of radar has also been used with great advantage in providing information on the wave height directional spectrum and consequently by inference on surface winds and ocean currents. It can also be effectively used for observing various forms of high-altitude atmospheric ionization phenomenon such as those due to meteors, aurora, missile launches and so on. Figure 2.93 shows transmit [Figure 2.93(a)] and receive [Figure 2.93(b)] sides of an OTHR (Type AN/ FPS-118). The system is capable of providing electronic surveillance of aircrafts at extended

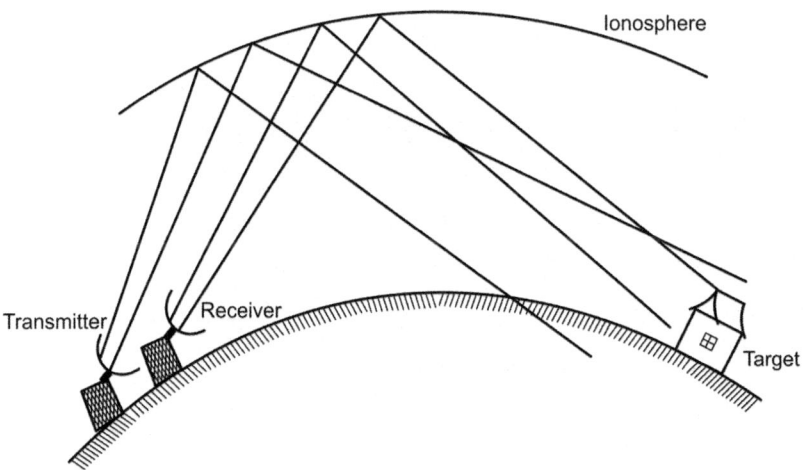

Figure 2.92 Principle of sky wave OTHR.

(a)

(b)

Figure 2.93 AN/FPS-18 radar: (a) the transmitter side and (b) the receiver side. (*Source:* Courtesy of Robert Hicks, NPS.)

Figure 2.94 Basic monostatic radar system.

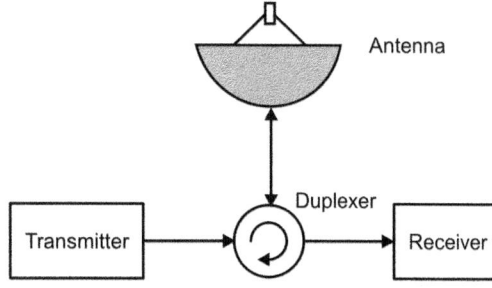

Figure 2.95 Basic bistatic radar system.

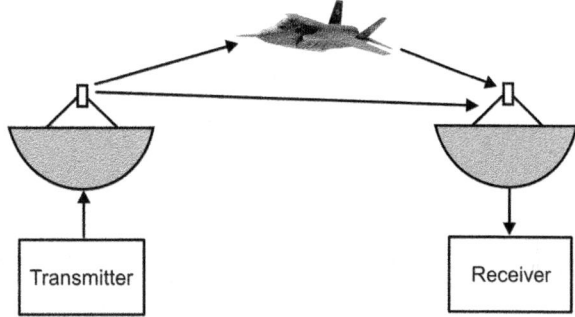

ranges of 800–2880 km. Operating in the HF band between 5 and 28 MHz where the radar energy is reflected by the ionosphere, this system is designed for over-the-horizon detection and tracking of aircraft and cruise missiles flying at any altitude. Transmit and receive sites are separated by about 150–200 km.

OVERSEER is yet another radar system that uses OTHR concept. This radar developed by GEC-Marconi is a ship-borne tracking radar, which gives continuous surface vessel surveillance with ranges up to 370 km from the shore. It operates in 4–12 MHz band with a peak radiated power output of 40 KW.

2.12.10 Monostatic and Bistatic Radars

In the basic *monostatic radar* system, the radar transmitter and receiver are co-located and usually a single antenna is used for both transmission and reception. A duplexer is used to separate transmit and receive functions both in time and power amplitude domains. Figure 2.94 shows the basic monostatic radar system.

In a *bistatic radar* system, separate antennas are used for transmission and reception purpose and the two antennas may sometimes be hundreds of kilometres apart as is the case in OTHR. Figure 2.95 shows the basic setup. In a typical setup, the receiver receives both directly from the transmitter as well as from the target. The two propagation times as well as azimuth and elevation angle measurements at the receiving site give the target location.

2.12.11 Primary and Secondary Surveillance Radar

A *primary radar* is the one we have been discussing up until now. It makes use of reflection from the target to determine various target related parameters and characteristics. A *secondary surveillance radar* differs from primary in the sense that it does not make use of reflection of

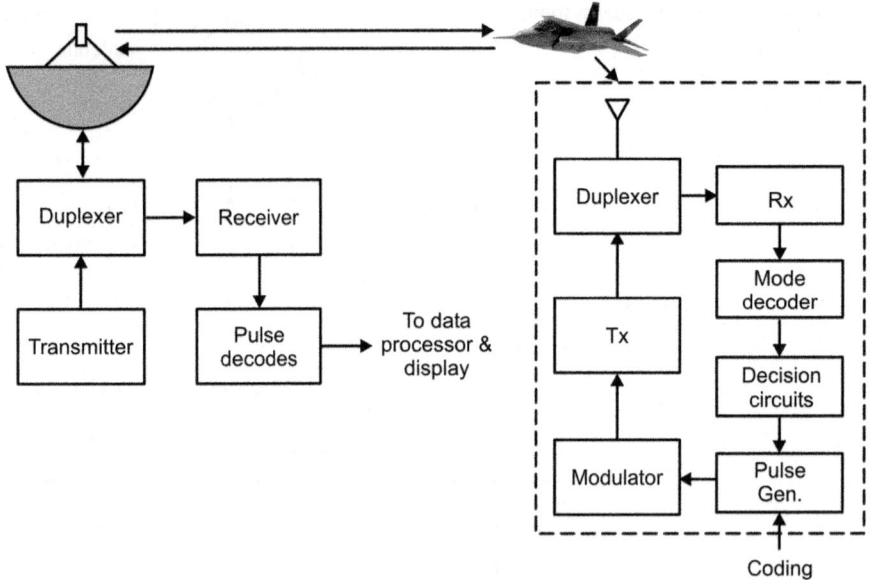

Figure 2.96 Block diagram of a secondary surveillance radar.

the transmitted energy. In a secondary surveillance radar system, pulse transmission from ground is received at the target, usually an aircraft, detected, and decoded in the transponder. The aircraft transponder then transmits coded pulses back to ground after a certain known delay. Thus identification of the aircraft can be established on ground. A radar system of this kind serves an extremely useful purpose in identifying friendly and enemy aircraft. IFF system (Identification Friend or Foe) is another name for such systems. Figure 2.96 shows a typical block-schematic arrangement of an SSR system along with a typical block representation of an aircraft transponder.

2.12.11.1 Performance Comparison of Primary and Secondary Surveillance Systems

1) In terms of positional data, primary radar is characterized by comparatively better resolution, whereas the SSR system has higher range accuracy potential.
2) Identification of target capability is inherent in an SSR system, whereas the same has to be inferred from the position data in case of the primary radar system.
3) In a primary radar system, the received signal is a function of equivalent echoing area and thus the target size and shape. In an SSR system, the signals from even small aircraft are as strong from large aircraft.
4) SSR system is completely free from the constraints of ground clutter as the ground station is not meant to receive the transmitted frequency and reflected energy if any is totally rejected. On the other hand, primary radar has to include clutter rejection systems. This can sometimes inhibit target detection in high-clutter regions.
5) An SSR system has a very high degree of dependence on the target cooperation and is completely useless if the target does not cooperate. That is, targets must have properly working transponders for an SSR system to work. Primary radar functions as long as there is a target present within its detection range.
6) Range capability of an SSR system is the compound of two virtually independent inverse square law functions, one for the interrogation range and the other for the reply range. Primary radar range, when a single antenna is used, is an inverse fourth power law function.

2.12.12 Laser Radar

Laser radar uses a laser beam instead of microwaves. In other words, the transmitted electromagnetic energy lies in the optical spectrum in laser radars whereas in microwave radars, it is in the microwave region. The frequencies involved are therefore very high, ranging from 30 to 300 THz and the corresponding wavelengths ranging from 10 to 1 μm in practical systems. This results in tremendous increase in angular resolution. Also, as the operating frequency increases, so does the available absolute bandwidth and thus the time or range resolution. It is feasible to generate laser pulses as narrow as a picosecond. Another advantage of laser radars is their immunity to jamming when compared to microwave counterparts. Laser radars can be used for all those applications where microwave radar has been used. These include target detection and ranging, target characterization, imaging and so on.

One of the most common functions of radar is the determination of target range. Laser radar counterpart of microwave radar that performs this function is popularly known as a *laser rangefinder.* A conventional laser rangefinder uses direct or incoherent detection. The term laser radar is usually associated with systems that transmit modulated CW signal and use coherent detection techniques. Both these types are briefly described in the following paragraphs.

2.12.12.1 Coherent Laser Radar

Laser radar is functionally identical to convectional microwave radar except for the fact that the transmitted electromagnetic signal is in the optical band. All those functions that can be performed by microwave radar can be better performed by laser radar except for some shortcomings, which will be discussed toward the end of this section. Like its microwave counterpart, laser radar can be used for range finding, determining target velocity, target tracking and even determining important physical features of the target not possible to the same level of resolution and detail with a microwave radar.

Figure 2.97 shows a block-schematic representation of coherent laser radar. The laser transmits a signal that is reflected from the target and collected by the radar receiver. The diagram shown is that of the monostatic system in which the transmitter and receiver share common optics. The Transmit-to-Receive (T/R) switch allows the use of common optics,

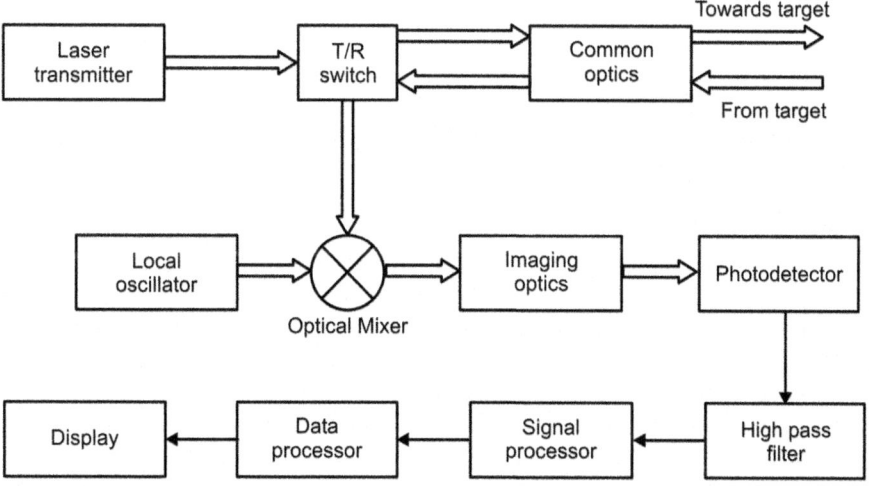

Figure 2.97 Block diagram of a coherent laser radar.

comprising a beam expander and scanning optics, during transmission and reception. The received optical signal is coherently detected in the optical mixer whose output is focused on to the photosensitive detector by the imaging optics. The photosensitive detector generates an electrical signal that is processed to extract the desired information about the target. In case of a bistatic system, the transmitting and receiving optics are separate.

2.12.12.2 Advantages and Disadvantages

The superiority of a laser radar over a microwave radar originates from the higher resolution and narrower beam width offered by the former. Both these features stem from the much higher carrier frequency of the laser radar. Narrower beam width implies higher angular resolution. These features allow much higher accuracy measurements of target range, velocity and angular position.

The narrow beam width of laser radars is ideally suited to target tracking applications. Surveillance function becomes difficult with this radar. In order for laser radar to perform the function of surveillance radar, the repetition rate of the laser radar needs to be very high so that large search volumes usually associated with surveillance radar can be interrogated within the prescribed time. The other alternative to overcome this problem is to use multiple beams. The higher resolution of laser radar allows recognition and identification of certain unique target features such as target shape, size, velocity, spin, vibration and so on. These high-resolution radars are also called *imaging radars*.

High frequencies associated with laser radars permit detection of smaller objects. This is made possible by the fact that laser radar output wavelengths are much smaller than small sized practical objects we can think of. In other words, the laser radar cross-section of a given object would be much larger than the microwave radar cross-section of the same thing. In fact, airborne aerosols and rain too have significantly larger laser radar cross-sections to allow range and velocity measurements of aerosols, which is very important for a number of meteorological functions.

The major shortcoming of laser radar or for that matter any laser device is its inability to penetrate weather. The atmosphere affects optical radiation far more severely than it affects microwave radars. The atmosphere absorbs and scatters optical radiation very strongly though the severity of effects also depends upon operating wavelength. In spite of this, laser rangefinders and laser radars are being increasingly exploited for various battlefield applications due to other superior features, which also include their high degree of immunity to jamming.

2.12.13 Millimetre-Wave Radar

Millimetre-Wave (MMW) radars operate in the millimetric wave bond, usually considered as 30–300 GHz with operating frequencies obviously higher than those used in microwave radars but much lower than those in laser radars. As a result, these radars benefit from the use of the higher frequencies and have superior range and angular resolution performance than the microwave radars. The atmospheric effects on millimetric wave propagation are not as severe as they are in case of laser radars. Thus, to an extent, these radars combine the advantages of both microwave and laser radars less affected by atmosphere (microwave radar) and higher range and angular resolution (laser radar).

These radars have becomes particularly attractive due to development of better and cost-effective components in this band. Also millimetric wave components are smaller and lighter than their microwave counterparts with the result that MMW radars lighter and smaller. MMW radars usually operate at 35, 94, 140 and 200 GHz as the atmosphere offers minimum attenuation at these frequencies.

Major applications of MMW radars include missile seekers, terminal guidance, airborne ground mapping, ground-to-ground surveillance and other similar applications where narrower beam width, increased antenna gain and smaller physical size are an advantage and where increased propagation losses vis-à-vis microwave frequencies are tolerable.

Illustrated Glossary

Antenna The antenna performs the function of transmitting high-frequency electrical currents fed to it from the output of transmitter into electromagnetic waves in a transmitting antenna and that of intercepting electromagnetic waves and generating equivalent electrical signal in the case of a receiving antenna. In other words, it converts guided electromagnetic waves into free-space electromagnetic waves and vice versa.

Apparent Range The apparent range corresponds to the time difference between the received echo pulse and the last transmitted pulse. It is different from the true range when the round trip time of the transmitted pulse is greater than the time period between two successive transmitted pulses.

A-Scope A type of radar display similar to an oscilloscope-like display with the horizontal axis representing range in terms of time and the vertical axis representing the echo signal amplitude.

Backscatter This is the part of reflected energy from the target travelling towards the radar.

Beam Width (Antenna) The angular separation between the half power points on the power density radiation pattern.

Bistatic Radar In a bistatic radar system, separate antennas are used for transmission and reception. The antennas may or may not be co-located. Over-The-Horizon-Radar is an example of a bistatic radar system.

Blind Speeds Blind speeds are those radial speeds of the moving target that produce a Doppler shift that is an integral multiple of the pulse repetition frequency of the transmitted pulse train. Radar fails to detect such speeds and to the radar it appears as if the target were stationary.

B-Scope A type of radar display. It is an intensity modulated display with horizontal and vertical axes representing azimuth angle and range, respectively.

Coherent Radar Transmitter Coherent radar transmitters are those in which the phase of the transmitted signal is derived from stable internal sources with the result that signal phase is constant and predictable.

Conical Scan A type of tracking methodology in which the antenna is rapidly scanned in a circular path around the antenna axis.

Cross-Polarization (Antenna) This is the component orthogonal to the desired polarization.

Cross-Range Resolution The resolution in a direction perpendicular to antenna axis. We have azimuth cross-range and elevation cross-range resolutions.

CW Radar This refers to an unmodulated continuous-wave radar in which transmitted signal is an unmodulated sinusoidal signal. It measures target velocity from the Doppler shift.

Doppler Ambiguity This is a phenomenon leading to inaccurate measurement of Doppler shift. It occurs when it is not known as to which transmit spectral line caused a particular received spectral line.

Doppler Granularity This occurs due to target echoes being processed in a discrete set of bandwidths called Doppler bins with the result that information of Doppler shift and hence the target velocity is also in discrete form.

Doppler Resolution This is the ability of the radar to detect and differentiate between the largest at the same range and angular position but having different radial velocities.

Duplexer This allows the same radar antenna to be used for both transmission as well as reception.

FM-CW Radar In the case of FM-CW radar, the transmitted signal is a frequency modulated CW signal. Such a radar is capable of measuring both target range and velocity.

F-Scope A type of radar display in which horizontal and vertical axes respectively represent azimuth and elevation track error. Centre of display indicates antenna's beam location.

Inverse Synthetic Aperture Radar (ISAR) This is an extension of the concept of synthetic aperture radar. In ISAR, the radar is usually stationary and it is the target motion that is used to synthesize a large antenna.

Laser Radar This is a type of radar in which the transmitted electromagnetic energy is in the optical region of the electromagnetic spectrum. Laser radar offers extremely high range and angular resolution and immunity to jamming but is severely affected by atmosphere.

Lens Antenna This is an antenna made from dielectric material. It depends upon refraction phenomenon for operation.

Lobe Switching This is a type of tracking methodology in which the antenna beam is rapidly switched between two positions in the same plane around antenna axis.

Millimetric Wave (MMW) Radar MMW radar operates in the millimetric wave band of 30–300 GHz. With a higher operating frequency and consequent higher bandwidth, the radar offers superior range and angular resolution performance when compared to microwave radar. MMW radar is also less affected by atmosphere than laser radar.

Monopulse Radar This is a type of tracking radar that generates the required tracking error signals from a single pulse.

Monostatic Radar This is a type of radar system in which transmitter and receiver are co-located and usually a single antenna is used for both transmission and reception.

Moving Target Indicator (MTI) This is a type of radar that can very effectively measure the radial velocity and range of a moving target even in the presence of a strong clutter from stationary and slow moving undesired objects.

Over-The-Horizon-Radar (OTHR) This is a type of radar that can look at targets lying beyond the radar horizon. It operates in HF band (3–30 MHz) and uses propagation modes other than line-of-sight propagation.

PPI Stands for Plan Position Indicator. This is an intensity modulated map-like display that gives target locations in polar coordinates. Radar location is at the centre of the display.

Primary Radar This is radar that makes use of electromagnetic energy reflected from the target to determine target related parameters. Except for secondary surveillance radar, all other radar systems can be classified as primary radars.

Pulse-Compression Radar This is a type of radar that offers higher range detection capability by transmitting expanded or elongated pulse and also higher range resolution capability by processing a compressed pulse during reception.

Pulse Doppler Radar This is a Doppler radar operating at higher pulse repetition frequency. Its operational principle is similar to that of MTI except that the latter operates at low PRF.

Quasi-Coherent Radar Transmitter Also known as coherent-on-receive, these systems have unpredictable transmit signal phases. The system always has the phase information on only the latest transmit pulse and every time the transmitter sends out a new pulse, the previous reference is lost.

Radar An acronym for *radio detection and ranging*. Modern radar, however, does much more than detection and ranging. It is used to determine target velocity and many more characteristic features of the target.

Radar Cross-Section of Target (RCS) RCS is defined as the ratio of its effective isotropically reflected power to the incident power density. It has three components; namely the area of the target, directionality of reflected power and target reflectivity at the polarization of the radar's receiving antenna.

Range Accuracy The deviation of measured range value from the true value. It primarily depends upon the accuracy with which radar determines the time of arrival of echo pulse and the accuracy with which time interval between the transmitted pulse and the received echo is measured.

Range Ambiguity This leads to inaccurate measurement of range. It occurs when multiple target positions produce reported information and the radar is unable to distinguish between the two in terms of range. It occurs when received echo pulses are attributed to wrong transmitted pulses.

Range Equation This relates the radar's detection range to various radar and target parameters. These parameters include; transmitted power, transmit antenna gain, radar cross-section of the target, receive antenna aperture, minimum detectable signal by radar receiver and various loss factors. The range equation explicitly brings out the significance of various parameters with respect to the maximum detection range of the radar.

Range Granularity This results from the digital nature of the radar's signal processing, data processing and display sections due to which such radar can report the target range only in discrete steps.

Range Quantization Another name for range granularity.

Range Resolution This is the ability of the radar to detect multiple targets separated in radial range but having same angular position.

Range Tracking This is the process of tracking a target using range coordinates. A range tracker continuously measures the target range and based on range data generates a range or time gate so that the target is at the centre of the gate.

Secondary Surveillance Radar This is a radar system that doesn't make use of reflected energy from the target. It is basically an IFF (Identification Friend and Foe) system in which the transmitted energy is received by the target such as aircraft where it is decoded and responded with thereby identifying itself.

Sequential Lobing This is a type of tracking methodology in which a squinted radar beam having radar beam axis slightly off the antenna axis is sequentially placed in usually four positions around an antenna's axis. The angular information on the target is determined by processing several target echoes.

Synthetic Aperture Radar (SAR) A type of radar in which a relatively small antenna synthesizes a very large antenna thereby achieving a high angular resolution by moving the antenna across the volume of interest.

Tracking Radar A type of radar designed for tracking moving targets. Such a radar transmits a very narrow beam with beam width anywhere between fraction of a degree in both azimuth and elevation.

Velocity Tracking This is the process of tracking moving targets using Doppler information.

Bibliography

1 Skolnik, M.I. (2008), *Radar Handbook*, McGraw-Hill Education.
2 Skolnik, M.I. (2002), *Introduction to Radar Systems*, McGraw-Hill Education.
3 Richards, M.A., Scheer, J.A. and Holm, W.A. (2010), *Principles of Modern Radar-Basic Principles*, SciTech Publishing.

4 Levanon, N. (1988), *Radar Principles*, Wiley-Interscience.

5 Richards, M.A. (2005), *Fundamentals of Radar Signal Processing*, McGraw Education.

6 Charvat, G.L. (2014), *Small and Short-Range Radar Systems*, CRC Press.

7 Griffiths, H.D. (2014), *Stimson's Introduction to Airborne Radar*, SciTech Publishing.

8 Knott, E.F., Shaeffer, J.F. and Tuley, M.T. (2004), *Radar Cross-Section*, SciTech Publishing.

9 Barton, D.K. (1972: Volume 1, 1974: Volume 2, 1975: Volume 3, 1975: Volume 4, 1975: Volume 5, 1977: Volume 6, 1978: Volume 7), *Radars Volumes 1–7*, Artech House.

10 Jenn, D.C. (2005), *Radar and Laser Cross-Section Engineering*, AIAA.

11 Eaves, J.L. and Reedy, E.K. (2011), *Principles of Modern Radar*, Springer.

12 Tuzlukov, V. (2012), *Signal Processing in Radar Systems*, CRC Press.

13 Jeffrey, T.W. (2008), *Phased Array Radar Design*, SciTech Publishing.

14 Barton D.K. (2012), *Radar Equations for Modern Radar*, Artech House.

15 Toomay, J.C. (2004), *Radar Principles for Non-specialist*, SciTech Publishing.

16 Faulconbridge, I. (2002), *Radar Fundamentals*, Argos Press.

17 Peebles, P.Z. (1998), *Radar Principles*, Wiley-Interscience.

3

Military Radars

Radar fundamentals were comprehensively covered in the previous chapter. Major topics discussed at length in the previous chapter included radar technology fundamentals, the radar functions put to use in a wide range of civilian and military applications and the different types of radar systems. The present chapter discusses specific military applications of radar systems. While the previous chapter covered different types of radar systems based on their operational principle, the present chapter discusses specific military radar systems based on their intended function. Major military applications of radar systems covered here include surveillance-based radar systems such as ground and area surveillance radars, air surveillance radar and ground penetration radar and tracking based radar systems such as weapon locating radar, ballistic missile defence radar, mortar or shell-tracking radar and so on. In addition to these, radars for navigation and weather forecasting are also discussed.

3.1 Military Applications of Radar Systems

Though radar systems are extensively used in a wide range of civilian applications in the areas of science, meteorology and air traffic control, use of radars by law enforcement agencies and military applications outnumber all other radar applications. Major radar systems in use by the armed forces and law enforcement agencies include the police radar used for detecting traffic rule violations, surveillance-based radar systems including battlefield surveillance radar, ground penetration radar, air surveillance radar and tracking based applications such as air-defence radar, weapon locating radar and ballistic missile defence radar. Military radars are also used for navigation, weather forecasting and Identification Friend or Foe (IFF).

3.1.1 Surveillance-Based Applications

Surveillance radar sensors are used to monitor activity surrounding critical and/or strategic assets such as military installations, border crossings, airports, ports and harbours, nuclear research and nuclear power generation facilities, missile and satellite launch stations, oil refineries, ammunition storage depots and so on. Surveillance functions may include intended targets underneath ground level, on ground level or in the air space surrounding the critical asset. There are primary radars and secondary surveillance radars. Both types were discussed at length in Chapter 2. While primary radar systems measure only the range and bearing of intended targets by detecting the transmitted radio frequency signal reflected off the target, secondary surveillance radar (SSR) relies on targets equipped with a radar transponder that

Handbook of Defence Electronics and Optronics: Fundamentals, Technologies and Systems, First Edition. Anil K. Maini.
© 2018 John Wiley & Sons Ltd. Published 2018 by John Wiley & Sons Ltd.

Figure 3.1 Battlefield Surveillance Radar-Short Range (BFSR-SR). (*Source:* Arun Vishwakarma, https://commons.wikimedia.org/wiki/ File:BFSR-SR_with_thermal_imager.JPG. CC BY 2.5 IN.)

replies to each interrogation signal by transmitting a response containing encoded data. Air traffic control (ATC) radar is an example of secondary surveillance radar system. ATC radar not only measures the range and bearing of the aircraft, it also requests additional information from the aircraft itself such as its identity and altitude. The IFF system is another example of an SSR system. Common surveillance-based military radar systems include ground (or area) surveillance radar, air surveillance radar and ground penetration radar (GPR).

State-of-the-art *ground* (or *area*) *surveillance radar* scans track movements of targets such as an individual walking or crawling towards a facility with precision, speed and reliability. Such radars typically have ranges of several hundred metres to over 10 km. Battlefield surveillance radar is the most commonly used application of ground surveillance. These radars are generally suit case sized tripod-mounted portable systems. Those with longer ranges are mounted on a vehicular platform. The Battlefield Surveillance Radar-Short-Range (BFSR-SR) of the DRDO, India (Figure 3.1) is an example of man-portable battlefield surveillance radar that can search a specified sector, simultaneously perform track while scan on multiple targets and carry out Doppler-based classification of various ground surface targets. There are hundreds of other ground surveillance radars with similar or enhanced features available from major international manufacturers of Defence and security equipment. Some representative systems are briefly described in the following pages.

Military application of *air surveillance radar* primarily involves monitoring the airspace to detect hostile aircraft and directing defensive measures against them. Conventional air surveillance radar called two-dimensional (2D) radar measures the location of a target in two dimensions including range and azimuth. Air surveillance radar capable of determining the elevation angle in addition to the target range and azimuth angle is known as three-dimensional (3D) radar.

The elevation angle allows computation of target height. The 3D air surveillance radar measures range in a conventional manner but has an antenna that is mechanically or electronically rotated about a vertical axis to obtain a target's azimuth angle and has either fixed multiple beams in elevation or a scanned pencil beam to measure its elevation angle. There are other types of radar such as the electronically scanned phased arrays and tracking radars that measure the target location in three dimensions. It is essential for air surveillance radar to be able to look around the corners to provide better coverage and capability to detect ground-hugging airborne targets. Over-the-horizon-radar (OTHR) exploits certain features of Earth's atmosphere enabling it detect low-flying aircraft over ranges of thousands of kilometres. Air surveillance radars are generally located on elevated platforms to maximize coverage area. Coverage area and capability to detect ground-hugging aircraft can be further enhanced by mounting radar on an airborne platform. The Airborne Warning and Control System (AWACS) is one such example.

State-of-the-art air surveillance radars are designed to detect, locate, track and classify a wide range of targets including traditional fixed and rotary-wing aircraft, non-traditional targets like ultra-lights, Para Gliders and Unmanned Aerial Vehicles (UAVs: also referred to as drones), ballistic missiles and even birds, thereby providing early warning, situational awareness and tactical ballistic missile surveillance and defence. Radars used by air traffic controllers for both approach phase surveillance and en-route surveillance are also examples of air surveillance radars. These radar systems are offered by international giants including Lockheed–Martin, Northrop–Grumman, the Thales Group, NEC Corporation and SRC, Inc. AN/TPS-59, AN/TPS-117 and AN/TPS-77 long-range air surveillance radar from Lockheed–Martin; AN/TPS-78 advanced air surveillance radar and E-3 Sentry AWACS aircraft from Northrop–Grumman; LSTAR family of air surveillance radars from SRC, Inc.; NS-100 surface and air surveillance radar and SMART-S MK2 3D medium to long-range surveillance radar from the Thales Group are some representative state-of-the-art radar systems designed for air surveillance missions.

AN/TPS-59 (Figure 3.2) is an active electronically scanned array (AESA), three-dimensional, transportable long-range ground-based air surveillance radar developed for the US Marine Corps to provide a mobile long-range surveillance capability and designed tactical ballistic missile defence. The radar uses active beam steering in elevation and mechanical steering in azimuth. AN/TPS-117 is dual-use air traffic monitoring and air-defence radar. AN/TPS-77 is long-range air surveillance radar designed to be a multi-mission radar. It incorporates the best features of both AN/TPS-59 and AN/TPS-117 to create an economical, high performing radar that can be transported via C-130, C-17, truck, rail or helicopter. Situational awareness of potential hostile targets and of friendly forces has always been considered a key component in obtaining and sustaining military superiority over adversaries. Airborne surveillance radar that can maintain situational awareness of potential targets and friendly aircraft over hundreds of square miles of airspace in any direction became a reality with the introduction of the AWACS. The E-3 Sentry from Northrop–Grumman is an Airborne Warning and Control System (AWACS) aircraft that provides all-weather surveillance, command, control and communications needed by commanders of air tactical forces. The E-3 Sentry is a modified Boeing 707/320 commercial airframe with a rotating radar dome mounted just above the aircraft fuselage. The E-3 Sentry has proven its efficacy in wartime operations such as Operation Desert Storm and, more recently, Operation Enduring Freedom as well as ongoing peacekeeping and humanitarian efforts. The Northrop–Grumman AN/TPS-78 is among the latest generation of highly mobile, state-of-the-art radars made possible by advances in high power transistor technology and designed to operate in some of the harshest, most demanding environments. The radar has proven particularly adept at detecting small targets in areas of difficult land and sea clutter, making it well suited for use in mountainous and coastal regions. The AN/TPS-78 is available

Figure 3.2 AN/TPS-59 radar. (*Source:* Courtesy of Roman Yurek, US Marine Corps.)

in two versions including the AN/TPS-78 long-range radar and the mid-range radar known as the TPS-703. The *LSTAR* family of air surveillance radars provides 360 degree, 3D electronic scanning capabilities for detecting and tracking the most difficult airborne targets including fixed and rotary-wing aircraft, Unmanned Aerial Vehicles referred to as drones, paragliders and so on. The *LSTAR* systems are ideally suited to border air surveillance, protection of critical infrastructure and local airspace management. The NS-100 from the Thales Group is a dual-axis, multi-beam air and ground surveillance radar (see Figure 3.3). The NS-100 offers enhanced surveillance capability with its new multi-mission capabilities, which include swarm defence, anti-piracy, UAV control and weapon support for active missiles.

Ground penetrating radar (GPR) is also surveillance-based radar that makes use of radar pulses to image subsurface structures. It has a wide range of civilian and military applications. GPR applications are limited only by the imagination and availability of suitable instrumentation. While on one hand, GPR is used for measuring snow and ice thickness and quality of ski slope management and avalanche prediction, locating buried utilities, forensic investigations and archaeological digs, in the military domain it is considered an effective tool for mine site evaluation and locating buried landmines.

Subsurface landmines are serious threat not only in wartime but also in the decades that follow. The landscapes of countries like Angola, Cambodia, the former Yugoslavia and even the deserts of Libya are still contaminated with leftovers from wars fought previously. Ground penetrating radar signals are not completely absorbed or reflected off air-ground surface boundary and can penetrate into soil provided the moisture content is not too high. Therefore, GPR can be used to detect and subsequently destroy these mines.

Figure 3.3 NS-100 air and ground surveillance radar.

Figure 3.4 LMX-200 GPR. (*Source:* Courtesy of Sensors and Software, Inc.).

LMX-100 and LMX-200 (Figure 3.4) from M/s Sensors and Software, Inc. are two GPR systems designed for applications, which include clearing areas of unknown underground obstructions before excavation, verifying as-built records and historical plans to confirm the present and precise location of buried infrastructure and developing detailed maps of underground infrastructure for design purposes. The SIR family of GPRs (SIR-20, SIR-30, SIR-3000 and SIR-4000) from M/s Geophysical Survey Systems, Inc. are compact lightweight systems suited to application areas, which include concrete inspection, utility location, geological investigation, archaeological surveys, forensic investigations, detection of buried land mines, mining and turf analysis and so on. *StructureScan* and *UtilityScan* families of GPRs are the other systems from GSSI for various applications. NIITEK, which is one of the three companies under Chemring Sensors and Electronic systems, produces a number of GPR systems designed for detection of landmines, improvised explosive devices, and other buried explosive hazards. These include GROUNDSHARK, MINESHARK, Minestalker, HMDS (Husky Mounted Detection System) series and R-VISOR.

3.1.2 Tracking Radar-Based Applications

Tracking radar detects and follows the intended targets so as to determine their trajectory, a function that is put to use in a wide range of civilian and military applications. One such widely used application of tracking radar is for air traffic control. Air traffic controllers rely on systems installed both at airports as well as at strategic spots on the ground beneath air traffic lanes for effective air traffic control extending to hundreds of kilometres. Tracking radars installed at airports are generally short-range radars that are intended to track aeroplanes, vehicles and even individuals on the surface in and around the airport. There is a large number of military applications that rely for their functioning on tracking radars. Armed forces use tracking radars to keep track of friendly and enemy platforms, which include land-based vehicles such as tanks, airborne targets such as aircraft, unmanned aerial vehicles, missiles, rockets and ships. Radar is used to monitor enemy targets to determine if they represent an immediate threat. In case of an imminent threat, radar may track the target and then use the track information to employ suitable defensive or offensive countermeasures such as using guided missiles or aircraft to intercept the enemy targets. Another important application of tracking radars is in removal of space debris. Space debris comprises used rocket stages and leftovers from completed missions, fragmented and inactive satellites and asteroids. Tracking radar may be used to track the space debris to determine if it poses any threat to major space assets such as space stations. The spacecraft may be manoeuvred to get out of the way in the case of any possibility of collision. Common military radars employing tracking radar concept or a combination of tracking and surveillance concepts include *fire-control radar, weapon locating radar* also called *counter-battery radar* or *shell tracking radar*.

Fire-control radar is a tracking radar specifically designed for integration with air-defence weapon systems. The radar component of the platform measures the coordinates of the intended target or targets in terms of their azimuth, elevation, height, range and velocity, which may be used to determine the target trajectory and to predict its future position. These radars provide continuous position data on a single or multiple targets enabling the associated guns or guided weapons to be directed and locked on to targets. The AN/APG-78 Longbow fire-control radar (Figure 3.5) is one such radar that is integrated with radar-guided air-to-surface, fire-and-forget type Longbow Hellfire missile on Apache attack helicopters enabling them to detect, classify and prioritize ground targets day or night, in adverse weather conditions, and then attack those targets with pinpoint accuracy.

The RDY-3 from the Thales Group is multirole airborne fire-control radar with its air-to-air function offering all aspect look-up/look-down detection; its air-to-sea function providing all the modes required for BVR (Beyond Visual Range) attack of surface ships and the air-to-ground function providing all the modes required for day and night all-weather operations. INDIRA series of 2D radars used to search and track low-flying cruise missiles, helicopters and aircraft and RAJENDRA 3D medium-range fire-control radar used with AKASH surface-to-air missiles are land-based fire-control radars developed by Defence Research and Development Organization (DRDO), India. The CEROS-200 from the SAAB Group is a fire-control radar designed for naval platforms. It is capable of tracking multiple supersonic missiles as well as

surface targets extremely close to the ship. It is interfaced with missile or gun systems to provide defence against threats including advanced sea skimming missiles or asymmetric surface threats in littoral environments. These are only some representative examples and a small fraction of hundreds of other land-based, airborne and ship-based fire-control radars in use by the Armed Forces all over the world.

Weapon Locating Radar (WLR) is primarily designed to locate hostile guns, rocket launchers and artillery units by detecting and then tracking incoming artillery shells, rockets and mortar rounds. Information on the point of origin of enemy fire is then used by counter-battery fire. Weapon locating radar is also known by names of *shell-tracking radar* and *counter-battery radar* for obvious reasons. In its secondary role, WLR can use its shell-tracking ability to track and observe the fall of shot from friendly weapon platforms. The information thus obtained can be used to provide corrections to own fire.

AN/TPQ-37 manufactured by Thales Raytheon Systems is one such mobile weapon locating radar. The radar uses a combination of radar techniques, computer-controlled signal processing and automatic height correction to detect, verify and track the projectiles in flight. The track information is then used to extrapolate both the firing position and the impact point. COBRA (Counter-Battery Radar), a collaborative long-range battlefield radar programme between France, Germany and Turkey, is an advanced land-based weapon locating system that comprises high performance radar, advanced processing and an integrated command, control and communication system. The COBRA system is designed to locate mortars, rocket launchers and artillery batteries and to provide information for counter-fire. It can also be used to monitor breaches of ceasefire when deployed in a peacekeeping role. The AN/TPQ-50 Lightweight Counter Mortar Radar (LCMR) is a man-portable and HMMWV 1152A-mountable lightweight radar system (Figure 3.6) from SRC, Inc. It is used to locate rocket, artillery and mortar rounds. The radar is capable of accurately determining *points of origin* and *points of impact* of incoming rounds over a range of 10 km. The MAMBA weapon

Figure 3.6 AN/TPQ-50 LCMR. (*Source:* Courtesy of SRC, Inc.)

Figure 3.7 BEL WLR. (*Source:* Sniperz11, https://commons.wikimedia.org/wiki/File:WLR_Mockup.jpg. CC BY 2.5 IN.)

locating radar from UK, ARTHUR (Artillery Hunting Radar) from Sweden, SLC-2 and BL-904 WLR and Fire Finder Radars from China and BEL WLR (joint development project undertaken by DRDO India and Bharat Electronics Limited (BEL) are some other representative examples of weapon locating radar systems. The design and performance of the BEL WLR (Figure 3.7) is similar to the US AN/TPQ-37 radar.

3.1.3   Multifunction Radar

Multifunction radar is primarily designed to perform the functions of target surveillance and tracking as a single unit, and is configured around active electronically scanned array (AESA) radar. Multifunction radars are used on land-based, seaborne and airborne platforms. Actual functions performed by multifunction radars used on different platforms are application specific. Shipborne or land-based multifunction radar generally performs the functions of surface search, short, medium and long-range search, multiple target tracking, weapon control and control of remotely piloted aerial vehicles. Multifunction radars on airborne platforms, in addition to surveillance and tracking based functions, may also take over the functions of navigation or terrain mapping. During surveillance, the objective of multifunction radar is to search a given volume of space within a given time frame with probability of detection and false alarm as determined by operational requirement. The surveillance function has an additional requirement of specified positional accuracy of detection in case its output feeds track while scan process. Multifunction radar has a small reaction time as compared to conventional surveillance radar as the former can rapidly confirm detections. This function known as *plot confirmation* minimizes the reaction time of multifunction radar. Track initiation follows plot confirmation. The purpose of track initiation process is to establish a track of sufficient position and velocity accuracy so that it can be handed over to track maintenance function.

An important attribute of multifunction radars are flexible enough to be easily reconfigured to perform different functions as per the operational requirement. Reconfiguration is possible due to flexibility in changing the parameters such as search volume, selected waveforms, signal processing and control algorithms. The flexibility of altering important system parameters also allows the user to rapidly reconfigure the radar for certain unforeseen operational scenarios.

A large number of multifunction radars are currently in use and under development for land-based, sea-based, and airborne platforms. HERAKLES from Thales Group representing state-of-the-art in naval radar is one such multifunction radar designed to be the sole radar onboard high-value ships. The radar concurrently performs all the functions involved in the establishment of a 3D air and surface picture, missile detection and deployment of anti-surface and air-defence weapons. The SAMPSON radar from BAE Systems and fitted to Royal Navy's Type 45 Destroyers is another multifunction radar. SAMPSON multifunction radar is at the core of the Sea Viper naval air-defence system. It is capable of simultaneously detecting and tracking hundreds of targets and is compatible with both active and semi-active homing missile systems. It provides surveillance and dedicated tracking in a single system, enabling the Type 45 platforms to defend itself and other ships in its company from attack. Yet another multifunction radar designed for naval vessels is the AN/SPY-3 manufactured by Raytheon. It is an active electronically scanned array radar designed to meet all horizon search and fire-control requirements. It can detect the most advanced low-observable anti-ship cruise missile threats and support fire-control illumination requirements for compatible missiles. The RDY-3 from the Thales Group is advanced multifunction radar that features all the functions required on a multirole aircraft of the twenty-first-century. It is a day/night, all-weather, airborne multirole fire-control radar delivering long-range targeting and firing and simultaneous multi-shoot capabilities. The radar is available with different antenna sizes that allow to be used for modernization of light and medium multirole fighters such as the Mirage V, Mirage F1, Mirage 2000, Mig-29 and LCA/Tejas. The ELM-2248 MF-STAR from Israel Aerospace Industries is multifunction radar that features surveillance, track and guidance radar functions in one radar (Figure 3.8). The radar employs multi-beam and pulse Doppler techniques as well as robust ECCM techniques to extract fast, low RCS targets from complex clutter and jamming environments and is compatible with active and semi-active homing missile systems.

Figure 3.8 ELM-2248 MF-STAR. (*Source:* Indian Navy, https://commons.wikimedia.org/wiki/File:ELM_2248_MF-STAR_radar_onboard_INS_Kolkata_(D63)_of_the_Indian_Navy.png. CC BY 2.5 IN.)

3.2 Ground (or Area) Surveillance Radar Systems

Ground (or area) surveillance radars are used in a variety of battlefield and security-related applications. Some of the well-established application areas include their use in urban warfare manoeuvres, covert surveillance, counterterrorism, maritime surveillance, border patrol and security, monitoring activity surrounding critical and strategic assets and tactical battlefield applications. In most applications, the ground surveillance radar performs the function of detection, classification and recognition of targets on ground level or the air space close to the ground level. While in the case of protection of critical and/or strategic assets, the radar may be used to monitor any kind of suspicious activity in the area surrounding the asset; in border patrol application, the ground surveillance radar may be used to track the movements of individual walking or even crawling personnel. In a tactical battlefield scenario, the area of activity of ground surveillance radar extends well beyond the border fence where it is used to monitor concentration of enemy troops and vehicles, carry out classification of enemy vehicles and track the movement of personnel and vehicles. In a tactical battlefield scenario, situational awareness is the key to battlefield dominance. The field commander who knows about the enemy's location, concentration and the types of forces and fighting platforms being deployed enjoys a great tactical advantage.

Whether it is protection of a critical asset such as an airport or a nuclear facility or guarding the border, ground surveillance radar provides effective surveillance that extends deep beyond the perimeter or border fence enabling security troops maintain constant and uninterrupted vigilance and track suspicious movements to warn of evolving threats. These radar sensors may also be linked directly to remotely located firing posts to engage any confirmed potential threats. The efficacy of ground surveillance radars is often augmented by use of panoramic electro-optic scanners. In the following paragraphs are discussed some representative ground surveillance radar systems. Different radar systems are briefly described as regards their salient features, operational parameters and application areas they are particularly suited to.

3.2.1 Design Considerations

Given the application domain of ground surveillance radars; *size and weight* is an important requirement, more so for tactical battlefield applications where the radar must be small enough to be deployed with ease. At the same time, the radar needs to be powerful enough to perform the intended role of detection and classification of targets.

Choice of radio frequency and antenna type are the other key issues. Many of the ground surveillance radars operate in the X band (8.5–10.68 GHz), Ku band (13.4–14/15.7–17.7 GHz) and K band (24.05–24.25 GHz). The higher the operating frequency of radar, the sharper the resolution of the image that the radar generates. A higher operating frequency leads to larger number of echoes from the target, which further leads to a greater detail in the picture seen by the radar.

Antenna design is another important consideration. Active Electronically Scanned Array (AESA) antennas are being increasingly employed in ground surveillance radars. The AESA antenna houses a multitude of Transmit/Receive (T/R) modules on its surface. Each of these T/R modules generates its own RF pulse and processes the characteristics of the returned echo. Use of multiple T/R modules allows the radar perform different tasks simultaneously. Another advantage of AESA is that the serviceability of radar is not affected due to failure of one T/R module.

3.2.2 Representative Ground Surveillance Radar Systems

There are hundreds of ground surveillance radars available from major international companies to provide short, medium and long-range surveillance for a variety of application scenarios, some of which have been highlighted in earlier paragraphs. It is not feasible to cover each one of those ground surveillance radars. Some representative systems are briefly described in the following paragraphs. Ground surveillance radars briefly discussed in this section include; (1) Battlefield Surveillance Radar – Short-Range (BFSR-SR) by DRDO, India, (2) the Ground Observer-12 (GO-12) ground surveillance radar by the Thales Group, (3) the EL/M-2129 Movement Detection and Security Radar (MDSR) by IAI Elta Systems Ltd, (4) the AN/PPS-5 Ground Surveillance Radar, (5) SpotterRF radars from SpotterRF, (6) the Blighter-200 series and B400 series radars from Blighter Surveillance Systems, (7) the Scanter-1002 from Terma A/S(h) SNAR-10M1 Mobile Ground Surveillance Radar from Russia, (8) the BUR Ground Surveillance Radar by EADS and (9) the GR-20, GR-40 and GR-05 Ground Surveillance Radars from Belgian Advanced Technology Systems.

3.2.2.1 Battlefield Surveillance Radar – Short-Range (BFSR-SR)

Battlefield Surveillance Radar – Short-Range (BFSR-SR), developed by the DRDO (Defence Research and Development Organization), India, and manufactured by Bharat Electronics Ltd, is a man-portable battery-powered short-range track-while-scan 2D surveillance and acquisition radar capable of searching a specified sector for multiple ground surface targets to provide day/night all-weather surveillance. BFSR-SR is configured around a coherent pulse Doppler radar sensor. The radar sensor is characterized by a low probability of intercept due to its low peak transmitted power and use of digital pulse-compression technology. Also, digital pulse-compression technology provides adaptive RF power management based on depth of surveillance thus enabling the radar to resolve closely spaced targets.

Major constituent parts of BFSR-SR include an antenna/electronics assembly, power pack, integrated control and display unit (CDU) and a mounting tripod. Targets are classified automatically using Doppler target classification. Output Doppler return is provided to the operator through headphones for the Doppler audio to aid in manual classification. The radar sensor head that operates over 21 channels in the J band (frequency of operation: 10–20 GHz) of NATO designation (related bands in the IEEE designation: X and Ku bands) consists of a planar antenna, transmit, receive and processing modules within a single block. The radar employs a solid-state transmitter module, a superheterodyne type receiver, antenna made up of microchip patch array antennas, a single-board FPGA (Field Programmable Gate array) implementation signal processor, a PPI (Plan Position Indicator) or a B-Scope display and a two-wire ISDN protocol used as an interface between a radar sensor head and control and display unit. The radar has inbuilt software to detect, track and classify targets diversity of moving targets such as a single crawling or walking person, a group of walking people, moving light and heavy vehicles at various distances. It comprises an inbuilt interface for automatic transfer of target data and Doppler audio to remote locations and integrated image sensors, which enhance the ability to identify and classify the targets. Radar ranges for different types of targets are specified as 500 m (crawling person), 2 km (walking person), 5 km (group of people) and 8 km (light vehicles) and 10 km (heavy vehicles) with an accuracy 20 m (RMS) and resolution of 50 m. Other important performance specifications of radar include azimuth scan sector of 30–180° with resolution of better than 4°, elevation coverage of −60–+15°, track-while-scan of 50 targets and instrumented range of 18 km.

Other important features of the radar include an inbuilt Global Position System (GPS) for self-location of radar and Digital Magnetic Compass (DMC) for North alignment, thermal imager integration, built-in training simulator and provision for networking multiple radar sensors for wider area coverage. Potential applications include border surveillance, battlefield surveillance, intelligence gathering, protection of sensitive sites, protection of industrial facilities, power plants and prevention of infiltration and illegal immigration. The radar is in use in India (Indian Army and Border Security Force), Indonesia and Sudan. The radar is under the trial and evaluation phase in a number of other countries.

3.2.2.2 Ground Observer-12 (GO-12) Ground Surveillance Radar

Ground Observer-12 (GO-12) is lightweight, compact ground surveillance radar (see Figure 3.9. The radar sensor is configured around a pulse Doppler radar operating in the Ku band. With its true manpack configuration, high scan rates, easy system integration and networking capability, the radar is suitable for a wide range of applications including tactical battlefield surveillance, border patrol and coast and site surveillance both by military/paramilitary forces or civilian users.

The GO-12 radar set comprises of radar sensor, laptop MMI, power pack and a tripod stand. It operates in the Ku band (12–18 GHz) with ≥440 MHz bandwidth. Due to its standard Ethernet/ASTERIX interface, GO-12 can easily be integrated into systems and operated by client-server principle via any IP network from any standard PC. Salient features of the radar include multiple operational modes (manual acquisition, surveillance, single target tracking and fire registration), track-while-scan for 40 targets, multiple sectors, Doppler tone based automatic target classification, recording/playback function and so on. Key performance specifications include coverage up to 27 km, scan speeds of 12°/s to 32°/s, typical detection ranges of 7–9 km (single person), 10–12 km (light vehicle), 14–17 km (heavy vehicle) and 13–15 km (helicopter). These detection ranges are typically for probability of detection of 90% and probability of false alarm of 10^{-6}. The system has specified MTBF of >4000 h for tactical use and >20 000 h for the sensor on a fixed site.

3.2.2.3 EL/M-2129 Movement Detection and Security Radar (MDSR)

Israel Aerospace Industries (IAI) offers a family of ground surveillance radars for protection of sensitive installations, borders and coastline. Some of these radars include EL/M-2226

Figure 3.9 Ground Observer-12 ground surveillance radar.

Advanced Coastal Surveillance Radar, EL/M-2107 Miniature Movement Detection Digital Radar, EL/M-2127 short-range miniature ground surveillance radar, EL/M-2129 Medium-Range Movement Detection Radar and EL/M-2140NG Long-Range Movement Detection Radar. EL/M-2127 is short-range miniature ground surveillance radar for detection of movement of persons and vehicles up to 2 km. It is used for protection of sensitive sites and ambushes. The EL/M-2107 is a high-resolution advanced miniature radar for security applications designed for automatic detection and classification of moving objects inside a guarded area. A cluster of these radars can be installed and connected to the control station to cover the area to be protected and generate an audio-visual alarm whenever an intrusion is identified inside the guarded area. The EL/M-2107 is interfaceable to a day and night electro-optical sensor that will automatically be directed towards the detected moving target for identification purposes. The EL/M-2226 is state-of-the-art coastal surveillance radar optimized for the detection and tracking of smart surface targets even under extremely adverse sea conditions providing a reliable tactical situation display.

EL/M-2129 is an X band coherent pulse-compression ground-based radar that detects and monitors moving targets such as walking persons, moving vehicles and flying objects inside the areas of interest at ranges of up to 30 km. The radar can be used either in standalone mode or integrated into systems providing battlefield surveillance, tactical intelligence and field artillery correction by detecting shell impact location. The radar is available in two variants with transmitter powers of 5 W and 25 W. Detection ranges of 25 W radar are 10 km (walking person), 16 km (helicopters), 17 km (light vehicles), 30 km (heavy vehicles) and 12 km (155 mm artillery shell). Range accuracy in all these cases is 25 m. It has operator selectable sector search of 10–350°. The radar can track more than 100 targets. The radar can be used on three possible installations including fixed installation such as on a tower or top of a building, mobile installation such as on a light vehicle and portable installation on a tripod stand.

EL/M-2140 NG like EL/M-2129 is also ground surveillance radar for military and paramilitary applications. The radar also monitors and detects movements inside the area of interest. It is capable of automatically detecting moving targets such as walking persons, vehicles and low-flying objects such as low-flying aircraft, hovering helicopters and gliders. The operational range in this case is up to 60 km.

3.2.2.4 AN/PPS-5 Ground Surveillance Radar

The AN/PPS-5 Ground Surveillance Radar is a lightweight, man-portable, battery-powered, ground- surveillance radar that has been around since the Vietnam War. AN/PPS-5 is a pulse Doppler radar that is designed to detect, locate, identify and track moving personnel and vehicles day or night under virtually all-weather conditions. The radar displays targets in a multimodal manner, both aurally and visually. The visual display is a Plan Position Indicator (PPI), and the aural indicator produces tones corresponding to target velocity. The system can operate in an automatic sector scanning mode or in a manual searchlighting mode. The radar comprises of two major assemblies including the tripod-mounted radar sensor and radar control indicator connected by a remote cable. Communications are provided by combat net radio (CNR). Different variants of the radar that have evolved over the years include AN/PPS-5, AN/PPS-5B, MSTAR, ARINE, AN/PPS-5C and AN/PPS-5D. AN/PPS-5 operates in the 16–16.5 GHz (J band). Detection ranges are 5 km (personnel) and 10 km (vehicles) with a range accuracy of 20 m. The AN/PPS-5B is an improved version of AN/PPS-5 with its maximum range extended to 20 km. AN/PPS-5C (Figure 3.10) operates in the 10–20 GHz (J band). Its maximum range is specified as 24 km. Detection ranges are 7 km (walking person), 16 km (light vehicle), 23 km (heavy vehicle) and 12 km (artillery fall of shot) with a range accuracy of about 8 m (RMS). AN/PPS-5D is produced by Syracuse Research Inc. In AN/PPS-5D, the analogue detection system of AN/PPS-5B is replaced by a digital detection system. AN/PPS-5D is connected to a ruggedized laptop computer that shows target detection on real-time mapping software.

Different variants of AN/PPS-5 were employed during Operation Desert Storm and were in service with the Dutch and British armies during the Bosnia hostilities in 1996. They were also

Figure 3.10 AN/PPS-5C ground surveillance radar.

deployed in Saudi Arabia in 1990–1991 with forward artillery observers of the Royal Artillery as part of the British contribution to operations Desert Shield and Desert Storm. The radar was also used during the conflicts in Bosnia, Afghanistan and Iraq.

3.2.2.5 Spotter RF Radar

The SpotterRF series of radars are ultra-lightweight and compact radars designed and particularly suited for detection and tracking of potential threats before they are on the fence line in a perimeter surveillance application. The family of SpotterRF surveillance radars such as Model nos. CK2, C20, C40, C100 and C400 are available for perimeter security of critical infrastructure with coverage areas up to 300 acres. SpotterRF also offers radars for battlefield surveillance. The M600D radar is an example.

Spotter CK2 (Figure 3.11) is a low-cost security radar that has a range of 100 m, field-of-view of 20° to provide coverage area of 1750 m^2. The 100 mW radar transmitter operates at 24 GHz. It detects and pin points the precise GPS coordinates of threats in all-weather conditions. Spotter C20 (Figure 3.12) provides a 130 m range with a beam width of 120° covering an area of 16 187 m^2. Three such radars can be integrated to provide full 360° coverage (Figure 3.13). The radar features auto-camera slew-to-cue capability, smart phone and tablet integration and GPS coordinates output every second. Spotter C40 can track any moving target in its line-of-sight up to a range of 350 m in all-weather conditions including presence of fog, rain, snow and so on. With its field-of-view of 90°, it provides coverage of 82,467 m^2. C40-EXT has a range of 450 m and a coverage area of 1, 32, 536 m^2. With a weight of less than 1.5 pounds (~3 kg), a C40 radar sensor can be easily placed anywhere on existing structures, poles, fences, trees and so on. It connects easily to existing Ethernet networks with Power over Ethernet switches and is ideally suited to perimeter security of oil wells, petrochemical plants, utility substations, power plants, cell phone towers, mines and so on.

The Spotter C550 radar is based on the same technology as used in the C40 radar but with enhanced performance in terms of range and coverage area. C550 covers 550 000 m^2 with ranges of up to 850 m on a walking person. C550-EXT has an extended range of 950 m. Due to its larger coverage, it is particularly useful for security of larger areas such as borders,

Figure 3.11 CK2 security radar. (*Source:* Courtesy of SpotterRF.)

Figure 3.12 Spotter C20 radar. (*Source:* Courtesy of SpotterRF.)

Figure 3.13 Three C20 radars to give 360° coverage. (*Source:* Courtesy of SpotterRF.)

ports, docks and wharfs, dams, bridges and water processing plants. The Spotter C950 (Figure 3.14) is the newest and most powerful commercial radar produced by SpotterRF. One C950 can detect a walking target up to 1350 m and a moving vehicle up to 2 km. With a field-of-view of 90°, single sensor provides a coverage area of 1.2 million m². It has all the characteristic features of C550 and like C550 is well suited to protection of larger areas. M600D is military radar with a range of 1000 m. with sensor field-of-view of 90°, it provides coverage area of 607 028 m².

Figure 3.14 SpotterRF 950 radar. (*Source:* Courtesy of SpotterRF.)

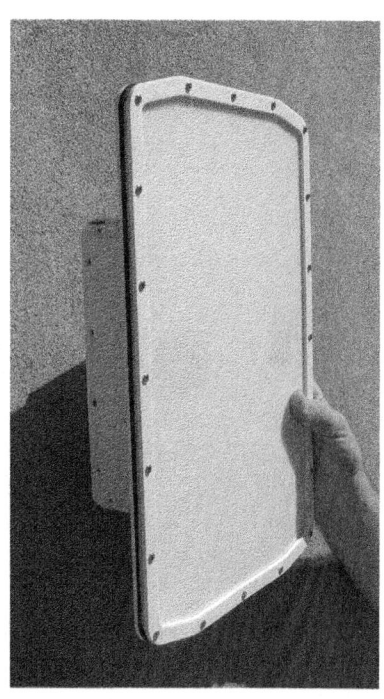

3.2.2.6 Blighter-B200 and B400 Series Radars

The Blighter B200 series and B400 series are ground surveillance radars from Blighter Surveillance Systems, capable of detecting potential threats including crawling and walking personnel and moving vehicles and suited to fixed, mobile and portable applications. The Blighter-B202 Mk-2 radar is one such radar in the B200 series. It is compact and lightweight medium-range ground surveillance radar capable of scanning and detecting moving persons including crawlers and vehicles up to range of 8 km. The radar employs a unique combination of FM-CW and Doppler processing technology to offer unsurpassed ground clutter cancellation with the ability to detect incredibly slow movement. Low transmitted power makes the radar safe for human operation and also electronically covert. Radar's wide elevation beam of 20° ensures that it can detect far-off and close by targets simultaneously even when it is mounted on a tower or hill-top to maximize long-range detection performance. Detected targets are reported via a standard Ethernet TCP/IP network connection, allowing target recognition and identification through the automatic slew-to-cue of optional cameras or thermal imagers. The radar outputs accurate positional information, latitude and longitude coordinates, range and bearing. The radar is built to withstand harsh environmental conditions and operate in all-weather conditions on 24/7 basis. A built-in precipitation filter suppresses false alarms caused due to rain or snow. The radar operates in the Ku band and employs fully electronic scanning in azimuth with azimuth scan angle of 80° or 90° using passive electronically scanned array. Detection ranges are 1.5 km (crawling person with RCS of 0.1 m^2), 3.3 km (walking person with RCS of 1 m^2), 8 km (moving vehicles with RCS of 30 m^2), maximum targets per scan of 700 and one false alarm per day and minimum detectable target radial velocity of 370 m/h. The radar also supports multi-radar configuration.

The Blighter B400 series radars (Figure 3.15) scan and detect moving vehicles and persons and offer enhanced detection range as compared to B200 series radars. These have all the technology related features of the B200 series radars including E-scan, low power FM-CW Doppler

Figure 3.15 Blighter B400 series radar (Courtesy: Blighter Surveillance systems).

technology, wide elevation beam and multi-radar configuration support. Detection ranges are 4.6 km (crawling person with RCS of 0.1 m^2), 11 km (walking person with RCS of 1 m^2), 20 km (moving vehicle with RCS of 30 m^2), 25 km (moving large vehicle with RCS of 100 m^2) and 32 km (large moving vessel with RCS of 1000 m^2). Specifications of number of targets per scan, false alarm rate and minimum detectable radial velocity are same as in the case of B200 series radars. B400 series radars can be used in multi-radar configurations to provide wider azimuth coverage. Single radar provides azimuth coverage of 90°. Two-, three- and four-radar configurations can be used to provide azimuth coverage of 180, 270 and 360, respectively, as shown in Figure 3.16(a)–(d). The two-radar configuration [Figure 3.16(b)] has one main radar unit and one auxiliary radar unit. The three-radar configuration [Figure 3.16(c)] has one main radar unit and two auxiliary radar units while the four-radar configuration [Figure 3.16(d)] has three auxiliary radar units beside one main radar unit.

3.2.2.7 Scanter-1002 Radar
The Scanter-1002 from Terma (Figure 3.17) is a short-to-medium range ground surveillance radar designed for protection of critical infrastructure. The radar operates in the Ku band and comprises a radar sensor employing high-resolution Doppler technology, processing hardware, rotating mechanics and an IP connection. The radar is designed for detection fast and slow moving threats such as humans, vehicles, boats and other small moving objects providing uninterrupted full 360° coverage up to 452 km^2 on a 24/7 basis in all weather conditions. With 360° surveillance and high update rates, the radar and its embedded tracker capable of simultaneously tracking more than 100 targets provide constant information on potential threats along with behaviour analysis of the detected targets by determining course, speed and heading to maximize the time available for assessment and possible intervention. Detection ranges are 3.7 km (walking person with RCS of 1 m^2), 7.3 km (light vehicles with RCS of 10 m^2) and 12 km (heavy vehicles with RCS of 100 m^2). Range and angular accuracies are <5 m and <2°, respectively. The minimum detection range is 20 m. The radar can be used either as autonomous radar or as part of an integrated network of sensors. It conveniently interfaces

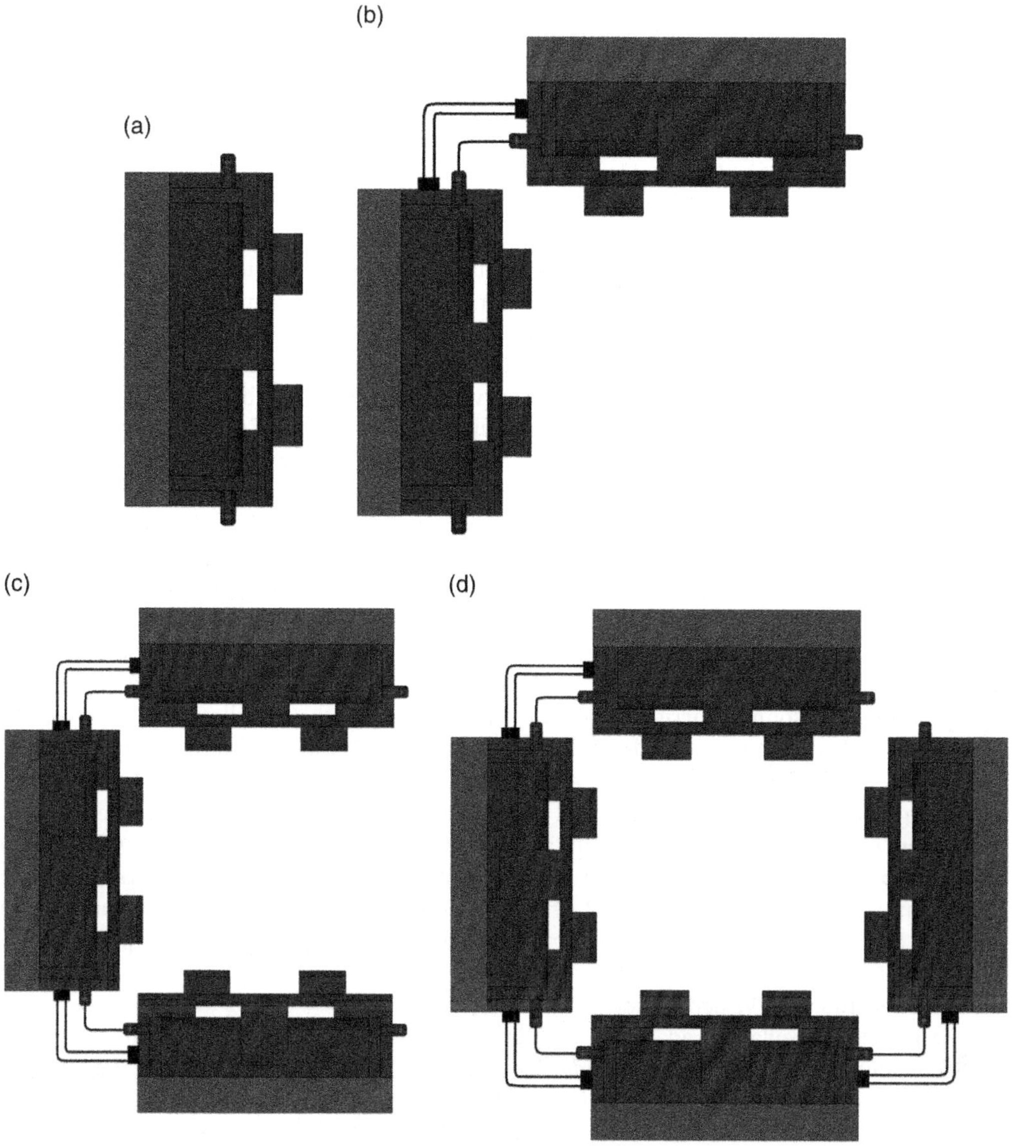

(a)

(b)

(c)

(d)

Figure 3.16 Multi-radar configuration. (a) Single. (b) Two. (c) Three. (d) Four.

Figure 3.17 Scanter-1002 ground surveillance radar.

with other sensors such as video cameras and infrared sensors. For example, the surveillance and tracking capabilities of the radar can be enhanced by interfacing the radar with a suitable pan-tilt zoom camera that slews automatically to the radar-detected threat for immediate identification and tracking. Also, in combination with the camera, timely and snap identification of a potential target can be done with minimal operator intervention.

The Scanter-2100 series and -2200 series from Terma are medium-range 2D fully coherent pulse-compression radars designed for surface surveillance. Both radars operate in the X band (9.25–9.45 GHz). The design of both radars is based on the technology gains from the Scanter-2001 and Scanter-5000 series coastal surveillance radars and the Scanter-1002 ground surveillance radar. These radars are particularly suitable for detection of targets from small ships to large vessels, surveillance of rivers, in-shore waters and VTS (Vessel Traffic Services) areas.

3.2.2.8 SNAR-10M1 Mobile Ground Surveillance Radar

The SNAR-10M1 (Russian name: 1RL232–2M and NATO code name: Big Fred) is a modernized version of the SNAR-10 battlefield surveillance radar and was unveiled by NPO Strela during Defence Exhibition Oboron Expo 2014 in Russia. It is mobile battlefield surveillance radar built to detect and locate moving ground and sea-surface targets. The radar is capable of detecting and locating single moving ground targets (single person, group of vehicle, light vehicle, Armoured Personnel Carrier, tank), moving convoys (rocket, artillery and mechanized units), sea-surface targets (motor boat, ship, assault landing craft etc.) and low-flying targets (helicopter, sports plane, UAV) for ranges up to 40 km. The radar can also detect and locate to a very high accuracy bursts of shells and mortar bombs, which allows radar to be used for adjustment of own artillery fire. Autonomous navigation facilities with GLONASS/GPS, digital terrain map, intercommunication and external communication facilities and data transmission equipment are the standard equipment of SNAR-10M1 radar set. The radar is mounted on a standard MT-LB multipurpose tracked armoured vehicle [Figures 3.18(a) and (b)].

3.2.2.9 BUR Ground Surveillance Radar

The battlefield surveillance radar with German Army designation *Bodenuberwachungsradar* (BUR) meaning *Ground Overwatch Radar* has been developed by Cassidian, the defence and security division of EADS. BUR radar is intended to close the gap earlier experienced by the German Armed Forces in the area of ground surveillance. The radar is proposed to be used from 2012 to 2037. The radar is capable of detecting and tracking movements of potential targets on ground, over water and in the air space close to the ground with speed, precision, and reliability. It tracks personnel and vehicular movements on ground up to a radius of 40 km. Use of Active Electronically Scanned Array (AESA) technology and resultant delay free electronic beam scanning allows the radar perform multiple reconnaissance tasks such as simultaneous ground and air scans thus achieving much greater performance level as compared to mechanically scanned radars. Reportedly, it can do the job of three conventional radars. BUR radar is integrated on DINGO-2 wheeled vehicle with high tactical mobility (Figure 3.19). DINGO-2 vehicle's innovative safety concept offers its passengers the highest possible protection against anti-tank and anti-personnel mines, IEDs, small arms and fragments from artillery and mortar ammunition. The vehicle is also equipped with a protection kit against NBC agents.

3.2.2.10 GR-20, GR-40 and GR-05 Ground Surveillance Radars

GR-20 radar, developed by Belgian Advanced Technology Systems (BATS), is coherent pulse-compression ground-based radar ground surveillance radar operating in the X band and designed to be used by military, paramilitary and security agencies for surveillance and protection applications. Major application areas include protection of critical infrastructure, border surveillance and protection, harbour surveillance and artillery fire correction. Artillery fire

(a)

(b)

Figure 3.18 (a) NAR-10M1 surveillance radar mounted on an MT-LB vehicle. (b) Close up view. (Part (a) *Source:* Vitaly V. Kuzmin, http://www.vitalykuzmin.net/Copyright-policy.CC BY-SA 4.0.)

correction function is provided by detection of shell impact location. The radar automatically detects moving potential threats such as walking persons, vehicles, gliders, helicopters and so on over ranges up to 30 km. Two variants of radar with transmitted output power of 5 W and 25 W are available. Detection ranges in the case of 5 W radar are 8 km (walking person), 15 km (light vehicle), 15 km (helicopter), 24 km (heavy vehicle) and 10 km (artillery shell). The same in case of 25 W radar are 10, 16, 17, 30 and 12 km in the same order. Specified range accuracy in all cases is 25 m. Search sector and azimuth accuracy in both variants are 10–350° and 0.5°, respectively. Both variants of GR-20 can simultaneously track more than 100 targets.

Figure 3.19 BUR ground surveillance radar. (*Source:* Sonaz, https://commons.wikimedia. org/wiki/File:ATF_Dingo_2_mit_B%C3%9CR. jpg. CC BY 3.0.)

The BATS' GR-40 ground surveillance radar is coherent pulse-compression radar operating in the X band and capable of detecting moving targets including walking personnel, vehicles and aerial targets such as hovering helicopters, gliders and low-flying aircraft up to range of 60 km. Specific detection ranges are 15 km (pedestrian), 25 km (helicopter) and 60 km (heavy vehicle) with range accuracy of 50 m. Artillery shell impact is detectable to greater than 15 km with accuracy of 25 m. The radar is capable of tracking more than 100 targets.

The BATS' GR-05 tactical ground surveillance radar, which is an extension of the GR-27 radar family, is capable of detecting and tracking intruders' movements in protected and selected zone of interest. It uses an X band pulse Doppler radar sensor that provides full 360° coverage in all-weather conditions on 24/7 basis. The radar detects and tracks movements of pedestrians and vehicles up to ranges of 5 and 8 km, respectively, and can track while scan more than 200 targets. Other important specifications of GR-20 radar include range accuracy of 1 m, best range resolution of 4 m, azimuth accuracy of 0.5°, minimum detectable range of 50 m and azimuth beam width of 10° and track update rate of ±5 s.

3.3 Air Surveillance Radar Systems

As outlined briefly in Section 3.1.1 while introducing surveillance-based applications of radar, air surveillance radars are used to carry out surveillance of the air space surrounding a critical asset and are designed to detect, locate, track and classify a wide range of both friendly and hostile targets. Typical targets include both conventional targets such as different categories of aircraft and unconventional targets including unmanned aerial vehicles (UAV), drones, ultra-lights and ballistic missiles. Air traffic control of both civilian and military aircraft constitutes another important application of air surveillance radars. We have 2D and 3D air surveillance radars. While the former provide information on target range and target position with respect to the radar location in terms of azimuth angle, the latter also gives altitude information in addition to range and azimuth. While surveillance radars used for detection of hostile airborne targets are primary surveillance radars, air traffic control radars are an integrated setup of primary and secondary surveillance radars. Modern air surveillance radars range from small, portable, low power systems to larger and more powerful platform-mounted systems. These radars can be configured for fixed site installation or for a transportable application using shipping cases or even for a platform-mounted mobile radar application. All modern air surveillance radars like the ground surveillance radars are designed to be network ready for easy integration into command and control (C2) systems. Also, these radars can be conveniently interfaced with other types of sensor systems such as electro-optic sensors.

3.3.1 Airport Surveillance Radar

Airport surveillance radar is the key constituent of air traffic control and management function, which includes collision avoidance between different aircraft, prevention of obstructions on the ground and maintenance of orderly flow of traffic. The radar is used to detect aircraft and send detailed positional information to the air traffic control system. The air traffic controllers use the information to safely and efficiently guide the aircraft from gate to gate. Different phases of air traffic control include the ground control, air traffic control during landing and take-off operations and en-route surveillance when the aircrafts are at medium to high-altitude. Though surveillance is most widely provided by primary and secondary radars; there are newer technologies such as Global Positioning System (GPS) based Automatic Dependent Surveillance (ADS) and multilateration are also in use. It is important to note that no single solution yields the most optimal results in all scenarios though one may outweigh another for a given situation. A combination of different technologies such as radar-based sensors, multilateration, and automatic dependent systems would only produce the best results.

Airport surveillance systems use an integrated system of primary and secondary surveillance radars. The primary surveillance radar (PSR) component is mainly used during the landing and take-off phase and sometimes for en-route surveillance. Primary radar provides independent surveillance and detects all aircraft regardless of type of equipment onboard the aircraft under surveillance (Figure 3.20). It detects the position of the aircraft but does not identify the aircraft. For landing and take-off phase traffic management, it is mainly used around airports. The radar is also used at locations in certain countries for en-route surveillance. Major advantages of primary surveillance radar technology for air traffic control include capability to detect non-cooperative aircraft, high level of data integrity and low infrastructure costs. Key limitations include the following. The target aircrafts cannot be identified; their altitude cannot be determined; requirement of higher transmitted power limits range and the latency is high and update rate is low.

The secondary surveillance radar (SSR) is also used for approach (landing and take-off) phase as well as for en-route surveillance. The SSR also determines the altitude and identity of aircraft in addition to detecting aircraft's position. In order for the SSR to determine additional information, the aircraft needs to be fitted with a transponder onboard. The transponder responds to the interrogation signal beamed toward to the aircraft by the SSR with a coded message (Figure 3.21). The coded message contains information on aircraft's identity, altitude and some additional information depending upon the type of transponder onboard the aircraft. The SSR then transmits all this information to the air traffic control station where it is displayed as an aircraft label. Key advantages of secondary surveillance radar technology in air traffic control are its capability to identify the aircraft and also know its altitude. Range is also

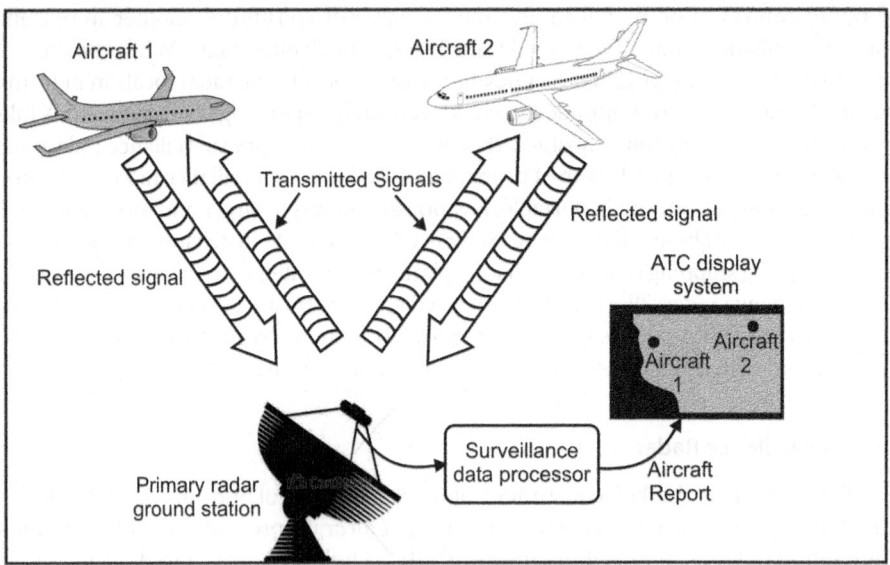

Figure 3.20 Primary surveillance radar for air traffic control.

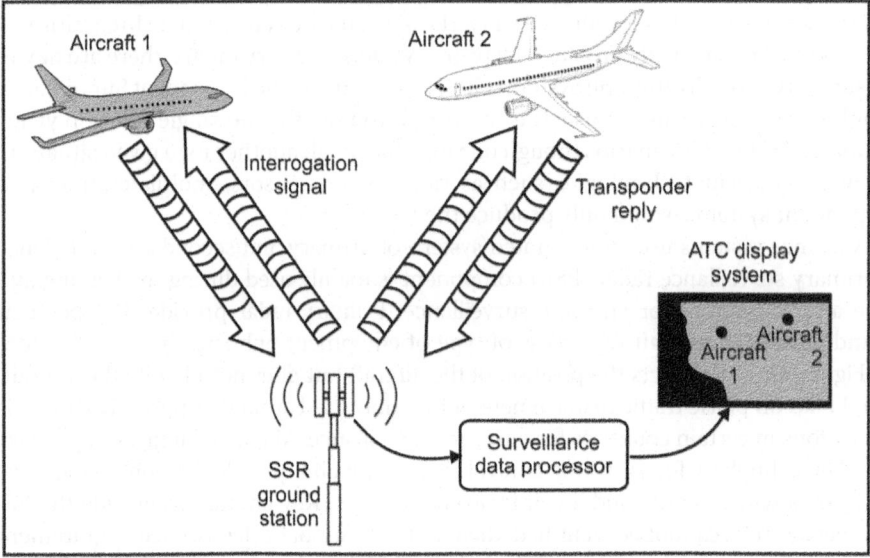

Figure 3.21 Secondary surveillance radar for air traffic control.

relatively higher due to interrogation and response signals travelling only one way distance. High latency and low update rate are some of the key limitations.

3.3.2 Multilateration System

Multilateration is used for ground, terminal approach and en-route surveillance. The system not only detects the aircraft position but also identifies the aircraft and can receive additional information. Multilateration system makes use of several beacons that receive signals

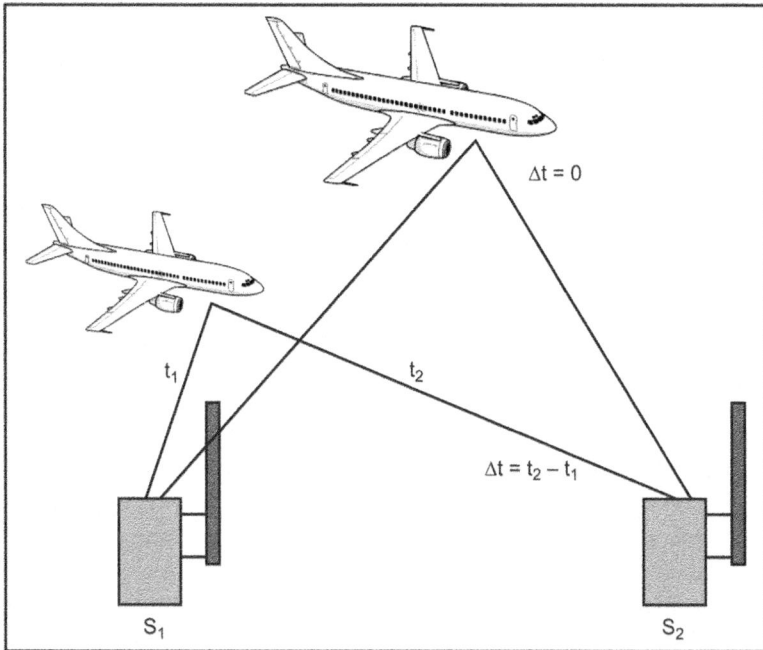

Figure 3.22 Multilateration: The TDOA technique.

transmitted by the aircraft transponder. The signals are either unsolicited ones known as squitters or are responses to the interrogation signals sent out by multilateration or secondary surveillance radar systems. The received signals are further processed to determine aircraft position, identity, and other information. The operational principle of multilateration system is based on Time Difference of Arrival (TDOA) technique. Principle of TDOA technique is illustrated in Figure 3.22. The TDOA technique takes advantage of the fact that signals travelling from a given target to a set of spatially separated sensor locations will take different times. The time difference of arrival of the signal at different sensors translates into a corresponding range difference between the target and the sensors. With reference to Figure 3.22, a time difference of $\Delta = (t_2 - t_1)$ would imply a range difference of $c\Delta$ where c is the velocity of electromagnetic waves. This measured time difference of Δ constrains the position of the aircraft to a certain set of points that would lie on a hyperbola. In the case of three-dimensional space, the set of points would lie on a surface called a hyperboloid. Detection by the three sensors would produce two hyperboloids and the target position is then constrained to the curve given by intersection of two hyperboloids. Addition of a fourth sensor allows unambiguous determination of target position.

Multilateration system offers a number of compelling advantages, which makes it suitable for difficult environments, complex aerospace and congested airports. Other advantages are their low infrastructure and life cycle costs, which make it particularly attractive where the high life cycle costs associated with radar systems are either not justifiable or affordable. The sensors are compact, purely passive and have minimal requirements for power and network connectivity. The MAGS wide area multilateration system from the Thales Group is a common wide area multilateration system designed to meet the requirements of cooperative surveillance for airport surface, terminal manoeuvring area and en-route coverage.

3.3.3 Automatic Dependent Surveillance

In the case of Automatic Dependent Surveillance (ADS), the aircraft broadcasts to the ground stations its position and other information that it determines with the help of Global Navigation Satellite System (GNSS) and other resources on board the aircraft. The information received by the ground stations is processed and then sent to Air Traffic Control (ATC) centre (Figure 3.23). We have two variants of Automatic Dependent Surveillance known as Automatic Dependent Broadcast-Broadcast (ADS-B) and Automatic Dependent Broadcast-Contract (ADS-C). In the case of ADS-B, the broadcasts can be processed by any receiving unit, which means that it can be used for both ground as well as airborne air traffic control applications. ADS-B equipped aircraft broadcast on its position and other information such as altitude, speed and identity occurs once every second without any intervention from ground systems. In the case of ADS-C, the Air Traffic Control (ATC) centre has a contract with the aircraft by virtue of which it asks the aircraft to send information on position and other parameters such as altitude, speed, expected route and meteorological data at regular intervals through point-to-point communications with a ground station. It also means that only those ATC centres that have a contract can receive the broadcasts. ADS-C is an effective system for surveillance of areas such as oceanic or desert areas where other surveillance systems are either impossible or impractical.

In air traffic control and management application, to meet the diverse requirements of complex approach areas, oceanic and mountainous regions and desert terrains, optimal results are obtained when several surveillance technologies are combined together. The important surveillance technologies in use for air traffic control and management have been briefly discussed in previous paragraphs. The TopSky-Tracking system by Thales Group is one such multi-sensor tracking system that fuses data pertaining to a single aircraft from multiple sensors including primary and secondary surveillance radar systems, multilateration system and Automatic Dependent Surveillance-Broadcast into a single surveillance track thereby taking advantage of the best contribution from each surveillance sensor with complete disregard to their respective drawbacks.

3.3.4 Representative Air Surveillance Radar Systems

Air surveillance radars discussed in this section include; (1) AN/TPS-59, (2) AN/TPS-117, (3) AN/TPS-77 (long-range air surveillance radars from Lockheed–Martin), (4) AN/TPS-78 advanced air surveillance radar from Northrop–Grumman, (5) E-3 Sentry AWACS aircraft from Northrop–Grumman, (6) LSTAR family of air surveillance radars from SRC, Inc., (7) NS-100

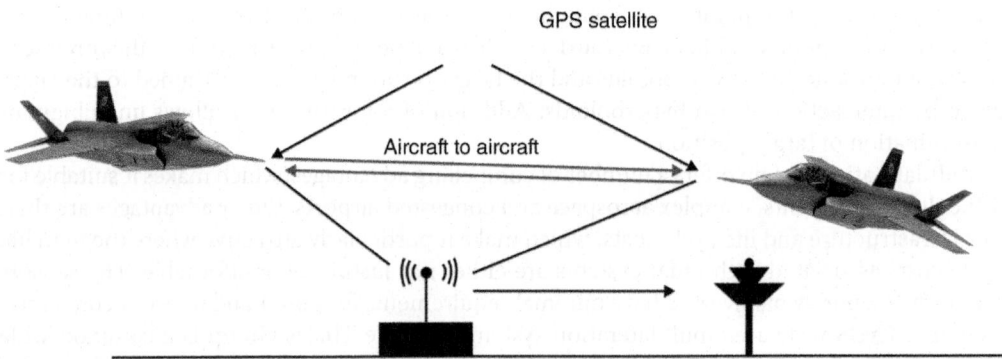

Figure 3.23 Automatic Dependent Surveillance-Broadcast (ADS-B).

surface and air surveillance radar, (8) SMART-S MK2 3D medium to long-range surveillance radar, (9) MAGS wide area multilateration system all three from the Thales Group, (10) Airport Surveillance Radar ASR-11 from Raytheon and finally (11) 3D Central Acquisition Radar (3D-CAR) from DRDO, India.

3.3.4.1   AN/TPS-59 Long-Range Air Surveillance Radar

The AN/TPS-59 by Lockheed–Martin (Figure 3.24) is a solid-state three-dimensional long-range air surveillance radar operating in the D/L band (1215–1400 MHz) and capable of detecting and tracking air breathing targets within 555 km (300 nautical miles) and tactical ballistic missiles up to a maximum range of 740 km (400 nautical miles). Azimuth, elevation and height coverage, respectively, are 360°, 0–19° (air breathing targets) and 0–30.5 km (air breathing targets). Elevation and height coverage specifications in the case of ballistic missile targets, respectively, are 0–60° and 0–305 km. Range, azimuth and height accuracy specifications are 30 m (at 150 km), 300 m (at 185 km) and 3 mrad (185 km). Other major operational specifications of the radar include peak transmitted RF power of 46 kW, average RF power of 11 kW, antenna aperture size of 49.2 m^2 and dual scan rate of 6/12 RPM. Scanning in azimuth direction is achieved by mechanical rotation of the antenna. The radar employs a linear phase-controlled array of transmitters and receivers for scanning in the elevation direction. Though primarily three-dimensional radar, the radar can also operate in the two-dimensional mode in case its general-purpose computer fails. The radar is connected to a Tactical Air Operations Module (TAOM) via a fibreoptic cable for passing information on air breathing targets and tactical ballistic missiles.

Figure 3.24 AN/TPS-59 long range air surveillance radar. (*Source:* Courtesy of Roman Yurek, U.S. Marine Corps.)

The AN/TPS-59 radar system was fielded in 1985. The radar was upgraded in 1998 to AN/TPS-59 (Version 3) by adding tactical ballistic missile detection and tracking capability. The improved radar is capable of detecting and tracking multiple theatre ballistic missiles to provide information on point of origin and point of contact as a part of theatre missile defence. The TPS-59 has successfully demonstrated its capabilities against various short-range ballistic missiles and cruise missiles, including successful intercepts of these targets. The radar is in use by the US Marine Corps, Egypt, Bahrain and Taiwan. The radar has been used in the past in Operation Desert Storm (also known as the Gulf War) during 1990–1991, Operation Iraqi Freedom (also known as the Iraq War) during 2003–2011 and Operation Enduring Freedom starting in 2001 to the present. Operation Enduring Freedom is the code name given by Government of the United States of America to describe war on global terrorism with locations in different countries including Afghanistan, Philippines, the Trans-Sahara and Caribbean and Central America.

3.3.4.2 AN/TPS-117 and AN/FPS-117 Long-Range Air Surveillance Radar

The AN-TPS-117 also belongs to the Lockheed–Martin's family of ground-based medium and long-range air surveillance radars, which also includes AN-TPS-59 and AN/TPS-77. It is a mobile, miniaturized variant of AN/FPS-117 long-range air surveillance radar in use in a large number of countries including Australia, Belgium, Brazil, Croatia, Denmark, Estonia, Germany, Hungary, Iceland, Indonesia, Iraq, Italy, Jordan, Kuwait, Latvia, Pakistan, Romania, Saudi Arabia, Singapore, South Korea, Taiwan, Thailand, Greece and the UK.

The radar operates in the D/L band (1215–1400 MHz). While AN/TPS-59 radar has the US Marine Corps as its primary user, AN/TPS-117 has been developed for the US Air Force. Like other radars in the family, it uses mechanical steering for azimuth scanning and active electronic scanning with a linear phased array of solid-state transmitters. The radar can be deployed in both fixed and transportable configurations. AN/TPS-117 and other radars in the family specialize in early warning situational awareness and tactical and ballistic missile defence. Major performance specifications of the radar include the following: Peak and average RF power levels of 24.6 kW and 4.5 kW, respectively; antenna aperture size of 52.6 m^2; azimuth coverage of 360°; elevation coverage of −6–+20°; maximum range and height specifications of 470 and 30.5 km, respectively; range, azimuth and height accuracy figures of better than 50 m, 0.18° at 250 km and 762 m at 250 km. Other important operational specifications include availability, critical MTBF (Mean Time between Failures) and MTTR (Mean Time to Repair) of 99.5%, greater than 2000 h and less than 45 min.

3.3.4.3 AN/TPS-77 Long-Range Surveillance Radar

AN/TPS-77 (Figure 3.25) is the latest radar in the TPS-59, TPS-117 and TPS-77 family of three-dimensional all solid-state air surveillance radars. The mobile AN/TPS-77 radar shares about 80–90% Line Replaceable Units (LRUs) and maintenance activity with the AN/FPS-117 fixed position radar system. The radar is configurable on a single C-130 aircraft or two medium trucks. The radar operates in the D/L band (1215–1400 MHz) and employs mechanical steering for azimuth scan and the phased array concept for elevation scan. The radar's planar array antenna comprising 34 all solid-state transmitters and receivers directly coupled to their antenna elements produces a pencil beam and is rotated to provide full 360° azimuth coverage. The pencil beam is electronically scanned to provide specified elevation coverage. It may be mentioned here that the L band is the preferred band for land-based long-range air surveillance radars due to its superior long-range detection performance. Also, full monopulse processing in both azimuth and elevation allows accurate position determination with every target hit. The radar has the design flexibility to meet the emerging theatre ballistic missile threat; a capability

Figure 3.25 AN/TPS-77 long range air surveillance radar. (*Source:* Nick-D, https://commons.wikimedia.org/wiki/File:RAAF_AN-FPS-117_radar_in_2007.jpg. CC BY-SA 3.0.)

that has been proven with AN/TPS-59 Version-3 radar is use with the US Marine Corps. The radar is inherently immune to electromagnetic interference, though a full suite of electronic protection features is also available.

Major performance specifications of the radar include the following: peak and average RF power levels of 19.9 kW and 3.6 kW, respectively; antenna aperture size of 27.1 m^2; azimuth coverage of 360°, Elevation coverage of −6−+20°, maximum range and height specifications of 470 and 30.5 km, respectively; range, azimuth and height accuracy figures of better than 50 m, 0.25° at 250 km and 915 m at 250 km. Other important operational specifications include availability, critical MTBF (Mean Time between Failures) and MTTR (Mean Time to Repair) of 99.5%, greater than 2000 h and less than 45 min.

3.3.4.4 AN/TPS-78 Long-Range Air Surveillance Radar

AN/TPS-78 (Figure 3.26) is the latest in the AN/TPS-70 family of air surveillance radars from Northrop–Grumman. The radar employs advanced solid-state radar technologies with operationally proven designs such as those used earlier in AN/FPS-130 long-range air-defence radar and the ASR-12 air traffic control radar. The radar is available in two variants, namely the AN/TPS-78 long-range radar and TPS-703 mid-range radar. The two radars share all major subsystems and Line Replaceable Units (LRU). Also, the two radars have more than 90% software commonality.

The radar employs a stacked-beam architecture rather than pencil beam architecture to provide full time full volume coverage of all targets at all azimuth positions. The AN/TPS-78 detects targets at all altitudes, including low-altitude targets in heavy ground and sea clutter

Figure 3.26 AN/TPS-78 long range air surveillance radar.

and the radar suffers no degradation in detection performance in the presence of long-range clutter. Though the stacked-beam architecture may appear to be less immune to jamming as compared to radar employing pencil beam that is electronically scanned in elevation; AN/TPS-78 radar overcomes this with its superior ECCM (Electronic Counter-Countermeasures) features such as frequency agility, low antenna sidelobes, CFAR (Constant False Alarm Rate), JATS (Jamming Analysis and Transmission Selection) and so on. Constant false alarm rate schemes refer to the use of adaptive algorithm that varies the detection threshold as a function of sensed environment. Frequency agility, which refers to radar's ability to rapidly change its transmission frequency in a pseudorandom manner so as to maintain a narrow instantaneous bandwidth over a wide operating bandwidth, is a very efficient technique to make the radar less vulnerable to jamming signal. JATS system implements self-adaptive frequency agility by controlling the transmission frequency in real time by analysing the jamming spectrum. The radar's low antenna side lobes maximize the transmitted energy on the target rather than broadcasting it as a beacon thereby preventing the disclosure of radar's location to the adversary and making it far less prone to attack by anti-radiation missiles. Also, low receive side lobes and narrow bandwidth improve radar's performance in jamming environment. In addition, a relatively smaller antenna gives it a small visual profile.

Another important feature of AN/TPS-78 radar is its full-range MTI (Moving Target Indicator) and MTD (Moving Target Detection) processing that cancels both fixed and moving clutter. While full-range MTI ensures detection and tracking of low-altitude targets at all ranges in heavy clutter conditions; MTD allows optimal target detection and tracking, even in heavy rain and chaff by suppressing moving clutter. The radar is configurable on both fixed as well as mobile platforms. AN/TPS-78 radar is in operation with the United States Air Force, Colombian Air Force, Royal Thai Air Force, Mexican Army and the Armed Forces of many other countries. Major performance specifications of AN/TPS-78 and TPS-703 are summarized in Table 3.1.

Table 3.1 Performance specifications of AN/TPS-78 and TPS-703.

Characteristic	AN/TPS-78	TPS-703
Instrumented Range	240 nmi (445 km)	75 nmi (140 km)
Operating frequency band	2.8–3.1 GHz	2.8–3.1 GHz
Height Coverage	100 K/500 K Ft (30.5/152.4 km)	100K Ft (30.5 km)
Elevation Coverage	0–20°	0–40°
Antenna size	5.5 m × 2.5 m	5.5 m × 2.0 m
Mean Time Between Critical Failure (MTBCF)	>2000 h	>2000 h
Moving Target Indicator (MTI)	≥50 dB Full Range	≥50 dB Full Range
Moving Target Detection (MTD)	Rain/Chaff Performance	Rain/Chaff Performance
Cooling	Air	Air
Operating Temperature	−30°C to + 55°C	−30°C to + 55°C

Figure 3.27 E-3 Sentry AWACS (Courtesy: Northrop–Grumman).

3.3.4.5 E3 Sentry AWACS

Surveillance radar on an airborne platform is the key to providing situational awareness of the friendly forces and potential hostile targets in the tactical battlefield scenario in achieving and sustaining military superiority over adversaries by increasing effectiveness of the friendly forces engaged in a variety of missions. The E-3 Sentry by Northrop–Grumman (Figure 3.27) is an Airborne Warning and Control System (AWACS) aircraft comprising of a very sophisticated and powerful surveillance radar system and command, control and communications equipment to provide all-weather surveillance to the tactical air forces covering more than 500 000 km^2 of air space around the aircraft. The S band (E/F-band) surveillance radar is mounted atop the fuselage of the aircraft in a rotating dome. The dome is rotated to provide 360° of azimuth coverage. Electronic scanning is used in elevation to determine altitude. It uses high pulse repetition frequency pulse Doppler radar waveforms to discriminate between aircraft targets and clutter returns. Also, low sidelobe antenna beam gives optimal performance in all terrains including urban and mountainous terrains. AWACS has been operational since 1977 and is projected to be in service beyond 2035. In order to meet the evolving challenges, AWACS mission effectiveness is being constantly enhanced through modernization and sustainment programmes. Radar System Improvement Programme (RSIP) is the most significant of these programmes. It is aimed at enhancing the operational capability of the AWACS radar against the growing threats posed by smaller targets, cruise missiles, and electronic countermeasures. Some of the important features introduced in AWACS through RSIP include advanced pulse Doppler waveforms, pulse-compression, and new processing algorithms increasing the detection and tracking range of the radar two-fold. Range and angular (azimuth and elevation) resolution specifications have also improved by factors of 6:1 and 2:1, respectively. Reliability and maintainability aspects of the radar have also improved thereby increasing radar availability and reducing time to repair. The enhanced radar is also more immune to electronic attack.

RSIP has also enabled multiple radar modes giving it operational flexibility. These include *pulse Doppler non-elevation scan (PDNES)* mode, *pulse Doppler elevation scan (PDES)* mode, *beyond-the-horizon (BTH)* mode, *interleaved* mode, *maritime* mode, and *passive* mode. The PDNES mode provides surveillance of aircraft down to the surface by using pulse Doppler radar, with Doppler filters and a sharply defined antenna beam. PDES mode is similar to PDNES mode except that target elevation is derived from electronic vertical scan. BTH mode uses pulse radar that provides surveillance over extended range where ground clutter is in horizon shadow. In the case of the maritime mode, a very short pulse is used to decrease the sea clutter patch for detection of large and small surface ships in various sea states. In the interleaved mode of operation, PDES and BTH modes can be used simultaneously with either portion active or passive. Also, maritime and PDNES modes can be used simultaneously. Use of passive mode is effective while operating in jammed environment. In this mode, radar transmission is shut in selected subsectors while receivers continue to receive and process data.

Northrop–Grumman's AWACS has proven amply demonstrated its potential as the premier air warfare surveillance, command and control aircraft both in wartime operations as well as on-going peacekeeping humanitarian missions. It was first used during cold war days when it was used to keep constant vigil in the skies over central Europe and the Far East. Subsequently, it was used in Operation Desert Storm where it provided radar surveillance and control for more than 120 000 coalition sorties. Following the Operation Desert Storm, AWACS was used as a critical component of Operation Northern Watch to keep vigil in Southwest Asia to enforce UN Security Council resolutions. E-3 Sentry AWACS was extensively used in Operation Allied Force during Balkans campaign and also during Operation Enduring Freedom that began in 2001 and is continuing till present.

3.3.4.6 LSTARM Family of Surveillance Radars

SRC's LSTAR family of lightweight, surveillance and target acquisition radars with their 3D electronic scanning capabilities are designed to detect and track difficult airborne targets such as unmanned aircraft systems (unmanned aerial vehicles, remotely piloted vehicles, drones), fixed and rotary-wing aircraft, ultra-lights, paragliders and hand gliders. The LSTAR family has two variants, namely LSTAR (V)2 and LSTAR (V)3. Typical application areas of both

Figure 3.28 LSTAR (V)3 air surveillance radar. (*Source:* Courtesy of SRC, Inc.)

variants include border air surveillance, local air space management and protection of critical infrastructure and vital assets. LSTAR radars are designed to specifically address clutter problems induced by wind turbines. LSTAR is also used as a part of ground-based sense and avoid (GBSAA) system that facilitates operation of unmanned aircraft systems' flights in the domestic airspace without a chase plane or a ground observer. LSTAR radars in the GBSAA system are used to monitor air traffic and provide the necessary information to the UAS operator for maintaining separation between the UAS and the other airborne traffic. Both radars are transportable and have flexible mounting options, low maintenance and life cycle costs and high mean time between failures (MTBF). LSTAR (V)2 is tripod mounted or tower mounted. LSTAR (V)3 is mountable on tripod or pedestal, roof-top, tower or a vehicular platform. Figure 3.28 shows LSTAR (V) 3 on a vehicular platform. Both radars can be networked and can be integrated with other sensors such as electro-optic sensors. Major performance features and specifications are summarized in Table 3.2.

3.3.4.7 NS-100 Surface and Air Surveillance Radar
The NS-100 from the Thales Group (Figure 3.29) is a 3D surveillance S band (E/F-band) AESA (Active Electronically Scanned Array) radar, Due to its highly modular and scalable architecture, the radar is suitable for a wide range of naval ships. The radar has a minimum range of 150 m and maximum instrumented range of 200 km. With its full digital beam forming with dual-axis multi-beam processing using AESA technology, it is capable of detecting a wide variety of airborne targets such as fighter jets, unmanned aerial vehicles, high diving missiles, sea skimmers and helicopters putting different requirements on the radar in one single-mode. NS-100 is a multi-sensor integrated platform. Integration of multiple sensors enhances overall system capability and addresses footprint issues. The radar is designed to integrate IFF interrogator transponder, Scout Mk3 FM-CW antenna, Infrared camera, AIS (Automatic Identification System) and ADS-B (Automatic Dependent Surveillance – Broadcast) antennas and receiver.

Table 3.2 Specifications of LSTAR (V)2 and LSTAR (V)3 Radars.

Features/Specifications	LSTAR (V)2	LSTAR (V)3
Azimuth Coverage	360°	360°
Elevation Coverage	0–30°	0–30°
Instrumented Range (km)	40	50 km
System Weight (kg)	68 (Transportable) 114 (Rugged enclosure)	Less than 227 kg
Operating Temperature Range	–	−32–+49°C
Power Requirement (W)	1200	3000
Key Benefits	Low false alarm rate, Low life cycle cost, Full remote and unattended operation, Can be integrated with visible/IR cameras or secondary surveillance radars	Low false alarm rate, Low life cycle cost, Full remote and unattended operation, Can be integrated with visible/IR cameras or secondary surveillance radars

Figure 3.29 NS-100 surface and air surveillance radar.

Typical applications include 2D surface surveillance, 3D air surveillance and weapon support, surface gun fire support, jammer surveillance and IFF interrogation support.

3.3.4.8 SMART-S MK2 3D Surveillance Radar

The SMART-S Mk2 from the Thales Group is a modern medium- to long-range S band (E/F-band) naval air and surface surveillance radar that with its multi-beam concept is capable of detecting difficult targets including stealth targets even in cluttered environments. This feature supports long missions and operation in littoral environment. The radar is being increasingly considered as the main air and surface surveillance radar in a one radar concept for naval platforms such as light

Figure 3.30 SMART-S Mk2 radar on ABSALON class frigate. (*Source:* Courtesy of Konflikty.pl.)

frigates, corvettes and landing platform docks (LPD). Furthermore, SMART-S Mk2 is designed to match the full performance of surface-to-air missiles (SAM), such as the Evolved Sea Sparrow Missile (ESSM). The radar has two operating modes including medium-range mod up to 150 km at 27 RPM and the long-range up to 250 km at 13.5 RPM. Other major specifications of the radar include maximum instrumented range of 250 km, minimum range of 150 m, maximum elevation coverage of 70°, tracking 3D capacity of 500 targets and range, bearing and elevation accuracy figures of <20 m, <5 mrad and <10 mrad, respectively. Typical applications include 2D surface surveillance, 3D air surveillance with fast target alerts, surface gun fire support, jammer surveillance and IFF interrogation support. Detection performance of the radar is typically characterized by its capability to detect small missiles up to 50 km and marine patrol aircraft up to 200 km.

A large number of SMART-S Mk2 3D surveillance radars are either in use or being considered for use by naval forces all over the world. Some of the naval platforms and the user services of the radar include the ABSALON-class support ships of the Royal Danish Navy (Figure 3.30), Sigma-class corvette of the Netherlands, MILGEM corvette and G-class frigate of the Turkish Navy, Barbaros-class frigates also of the Turkish Navy, Halifax-class frigates of the Royal Canadian Navy, Khareef-class corvette of the Royal Navy of Oman, Almirante-Padilla class frigates of the Colombian Navy and Incheon-class frigates of the Republic of Korea Navy. Other naval vessels that are being equipped with SMART-S Mk2 radar include Second Generation Patrol Vessel called Littoral Combat Ships (LCS) of the Royal Malaysian Navy, C28A corvette or frigate of the Algerian National Navy, ORP Slazak (formerly Gawron-class corvette) of the Polish Navy and ANZAC-class frigates operated by the Royal Australian Navy and Royal New Zealand Navy.

3.3.4.9 MAGS Wide Area Multilateration System
MAGS is a wide area multilateration (WAM) system designed and built by Thales Air Systems with the primary objective of providing high precision and high update rate secondary surveillance to Air Traffic Controllers. The system can operate with high reliability in the most stringent environments and is flexible and scalable enough to be tailored to meet requirements of different customers. MAGS is designed to meet the surface, precision approach monitoring and en-route cooperative surveillance needs and provide simultaneous multilateration and

ADS-B surveillance. Thales' WAM system is in use at the Frankfurt airport. Precision Approach Monitor (PAM) at the Frankfurt airport has been integrated with the WAM system of Thales. The new system offers a faster update rate and updates aircraft information every second on a radar screen, compared to the previous system that updated information only every 5 s, thereby significantly enhancing airport's safety standards. This also allows the system to quickly identify and correct variations if any from the allocated routings.

3.3.4.10 ASR-11 Digital Airport Surveillance Radar

The Airport Surveillance Radar, Model 11 (ASR-11) from the Raytheon Corporation is an ajoint Federal Aviation Administration (FAA)/Department of Defense (DOD) procurement programme with the United States Air Force (USAF) assuming overall lead responsibility. The ASR-11 digital solid-state surveillance radar provides to the airport terminal area primary surveillance coverage up to 116 km (60 nautical miles) and secondary surveillance up to 232 km (120 nautical miles). The ASR-11 uses advanced signal processing to provide improved target and weather processing. The ASR-11 surveillance radar is replacing current ageing analogue radars such as ASR-7, ASR-8, and counterpart military radars at the nation's civilian airports as well as at numerous DOD airfields. Replacement of ageing radars with the ASR-11 is expected to improve reliability, provide additional weather data, reduce life cycle costs and improve performance. The ASR-11 will interface with the new *Standard Terminal Automation Replacement System* (STARS) and other digital automation systems to provide air traffic controllers with a state-of-the-art aircraft and weather detection system in the airport environment. This has been made possible due to fully digital surveillance and weather data outputs from ASR-11. The ASR-11 with its capability to produce simultaneous outputs in several formats makes it compatible with FAA and DOD non-digital automation systems.

3.3.4.11 Central Acquisition Radar (3D-CAR)

The 3D Central Acquisition Radar (3D CAR) developed by the Defence Research and Development Organization (DRDO), India is the state-of-the-art multifunction medium-range surveillance radar capable of simultaneously tracking multiple targets such as fighter jets and missiles travelling at supersonic speeds of up to 3000 m/s (approximately 3 Mach). The radar has an operating range of up to 170 km and an altitude of up to 15 km. The 3D CAR has its origin in a collaborative programme between DRDO and Polish Institute of Technology (PIT) to develop a family of mobile S band 3D radars. The cooperation was mainly in the areas of development of planar array and general architecture. The radar is configured on two high mobility Tatra vehicles called *Radar Sensor Vehicle* (RSV) and the *Data Centre Vehicle* (DCV) and a third Tatra vehicle houses the power unit in order to meet the operational and battlefield mobility requirements. The transmitter, the receiver and the antenna along with its hoisting and rotating mechanism are housed on the RSV. The operator's console and control post-provisions are housed in a shelter located on the DCV. Tatra platform is a modified heavy duty truck built by Bharat Earth Movers Limited (BEML), an Indian public sector enterprise. The radar is equipped with ECCM (Electronic Counter-Countermeasures) features like frequency agility and jammer analysis. 3D CAR is also integrated with a secondary surveillance radar IFF system to discriminate between friendly and hostile aircraft.

The 3D CAR has further been developed indigenously into two special variants, namely ROHINI radar (Figure 3.31) for the Indian Air Force and REVATHI radar for the Indian Navy. These replace the original joint development items, such as the planar array antenna, with new locally developed ones that are more capable than the original design. A 3D Tactical Control Radar (TCR) has also been developed for the Indian Army. ROHINI is medium-range 3D surveillance radar operating in the S band (E/F-band). The radar is capable of Track-While-Scan

Figure 3.31 ROHINI radar.

of airborne targets up to a range of 150 km subject to line-of-sight clearance and radar horizon and an altitude of 15 km. The radar can detect low-altitude targets as well as supersonic aircraft and missiles flying at velocities up to 3 Mach. It has the ECCM and IFF features of 3D CAR. Radar antenna is mechanically rotated to provide azimuth coverage of 360°. It employs seven beams to provide elevation coverage of 30°. The radar employs multi-beam coverage in the receive mode to provide required discrimination in elevation.

REVATHI is medium-range 3D surveillance radar to be fitted in the ASW (Anti-Submarine Warfare) Corvette class of ships to detect air and sea-surface targets up to a range of 200 km. REVATHI radar too employs multi-beam coverage in the receive mode, eight beams in this case, to achieve required discrimination and granularity in elevation data and also the elevation coverage of 50°. Again, the antenna is rotated mechanically to achieve azimuth coverage of 360°.

The 3D TCR is also state-of-the-art medium-range surveillance and tracking radar developed for the Indian Army for detection and identification of airborne targets. The radar is used to transmit target data to the air-defence weapon system for exercising required control. The radar is capable of Track-While-Scan of airborne targets up to a range of 90 km for fighter aircraft and 65 km for unmanned aerial vehicles subject to line-of-sight clearance and radar horizon. The radar is so designed as to be directly usable in the battlefield. The radar is configured on two Tatra vehicles; the *Radar Sensor Vehicle* (RSV) housing the electronics subsystems and the antenna and its hoisting and rotating mechanism and the *Power Source Vehicle* (PSV) housing the generators and the UPS (Uninterrupted Power Supply). Again, the antenna is mechanically rotated to achieve azimuth coverage of 360°. The elevation coverage is 50° up to an altitude of 10 km provided by eight beams in the receive mode. Multi-beam coverage in the receive mode also provides necessary discrimination in elevation data.

3.4 Ground Penetrating Radar Systems

Ground penetrating radar (GPR) is an effective non-invasive and non-destructive method of imaging the subsurface and is widely used for host of civilian and military applications, which include engineering and construction industry, archaeological investigation, and forensic investigation, geological and environmental science, utility detection and concrete inspection, detection of buried mines, unexploded ordnances and so on.

3.4.1 Operational Principle

GPR uses the same principle as the conventional radar in the sense that it also derives the desired information about the target from the fraction of transmitted electromagnetic energy backscattered from the target. The difference lies in the location and nature of target under surveillance. While in the case of conventional radar, we may be looking at a land-based or an airborne target; in the case of GPR, we are intending to detect subsurface objects or anomalies. Also, the target in the case of former could be an aerial one such as a fighter aircraft, an unmanned aerial vehicle, a mortar or a missile, a land-based one including Main Battle Tank (MBT) or an Infantry Combat Vehicle (ICV) or even a human intruder. In the case of GPR, the target could be a buried structure, a subsurface landmine, or unexploded ordnance. It could even be some kind of anomaly in the structure.

Ground penetrating radar employs electromagnetic waves in the frequency range of 10–5000 MHz to map structures and features buried underneath the ground surface and manmade structures. The GPR sends out a short pulse of electromagnetic energy in the chosen frequency into the subsurface (ground or engineered structure) under investigation and records the amplitude or strength of the backscattered signal. There is a backscattered or reflected signal whenever the transmitted electromagnetic energy pulse encounters an interface with different electrical conduction properties or dielectric permittivity from the surrounding material. The amplitude of the reflected signal is determined by the contrast in the permittivity and conductivity values of the material and the anomaly or obstacle. Higher the contrast, greater will be the amplitude of reflected signal. While a small or large fraction of the transmitted energy is reflected, the rest of it keeps travelling through the material until it dissipates. The rate of attenuation again depends upon properties of the material. Materials with a high dielectric constant slow down the travelling wave and therefore it doesn't penetrate very far. Materials with high electrical conductivity attenuate the signal very rapidly. Metals therefore don't allow signals to pass through. Metals are considered as complete reflectors and therefore materials beneath a metal sheet or fine metal mesh are not visible to the radar. Table 3.3 shows electrical properties of different geological media.

The GPR measures the travel time between the transmitted signal and the signal reflected from the discontinuity, if any, in the subsurface region. In fact, radar transmits repetitive pulses coupled to the ground surface and records the corresponding reflected pulses. As the antenna is moved across the surface, a continuous profile record of the two-way travel time is generated. Travel time information is converted into depth information from the knowledge of signal velocity. Typically, radar is able to measure the depth of discontinuities in subsurface soil to

Table 3.3 Electrical properties of different geological media.

Material	Dielectric constant	Conductivity (ms/m)	Velocity (m/ns)	Attenuation (dB/m)
Air	1	0	0.3	0
Distilled water	80	0.01	0.033	0.002
Fresh water	80	0.5	0.033	0.1
Sea water	80	30 000	0.01	1000
Dry sand	3–5	0.01	0.15	0.01
Saturated sand	20–30	0.1–1.0	0.06	0.03–0.3
Limestone	4–8	0.5–2	0.12	0.4–1
Shale	5–15	1–100	0.09	1–100
Silt	5–30	1–100	0.07	1–100
Clay	4–40	2–1000	0.06	1–300
Granite	4–6	0.01–1	0.13	0.01–1
Salt (dry)	5–6	0.01–1	0.13	0.01–1
Ice	3–4	0.01	0.16	0.01

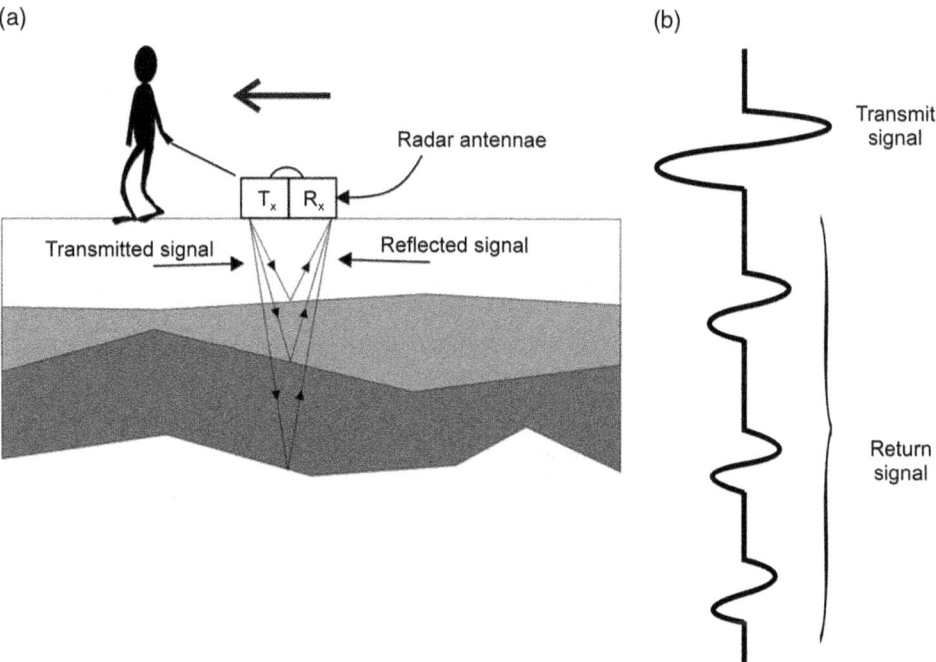

(a)

(b)

Figure 3.32 Operational principle of GPR. (a) Single pulse transmitted into a subsurface region. (b) Corresponding signal strength versus depth plot.

within few centimetres depending upon electrical and magnetic properties of the earth medium, the depth and size of target and frequency of operation. Figure 3.32(b) shows the signal strength versus depth plot for a single pulse transmitted into a subsurface region as shown in Figure 3.32(a). The data collected in the field using GPR is subjected to rigorous processing to filter out interfering signals and enhance the visibility of desired signals. The processed data leads to what is called a *radargram* that represents a two-dimensional cross-sectional image of the subsurface region. GPR technique is best suited to dry sandy soil with little salt content. Heavy clay based soils are difficult to penetrate with GPR. Penetration depth is severely comprised in clay soils where it may be limited to a few feet or even less. The GPR technique can be used through fresh water though it doesn't work in the presence of salt. Also, GPR works very well through ice and snow.

3.4.2 Design Considerations

The efficacy of a GPR measurement for a given application is mainly governed by selection of operating frequency, time window, sampling interval, spacing between discrete GPR measurements or station spacing, choice of antenna and spacing between transmit and receive antenna.

Choice of optimum *operating frequency* is a trade-off between spatial resolution, depth of penetration and system portability. Higher operating frequency yields better spatial resolution but only at the cost of reduced penetration depth. Table 3.4 gives typical resolution and penetration figures for different operating frequencies. Often, it is better to trade off resolution for penetration. Higher resolution would be a waste if the target cannot be detected. Spatial resolution is given by half of the operating wavelength. Signal velocity in a medium of dielectric constant K is given by $(3 \times 10^8 / \sqrt{K})$ m/s. The operating frequency may be computed to a reasonably good estimate for a given value of spatial resolution X from eqn. 3.1.

$$f (\text{in MHz}) = 150 / X \sqrt{K} \qquad (3.1)$$

K = Dielectric constant of most material
   The *time window* is estimated from eqn. 3.2.

$$\text{Time window} (\text{in ns}) = (2D/v) \qquad (3.2)$$

D = Maximum expected depth (in m)
v = Minimum expected signal velocity (m/ns)

The time window may be taken to be about 30% greater than the value computed from eqn. 3.2 to allow for uncertainties in depth and velocity variations.

Table 3.4 Penetration and resolution figures for different operating frequencies.

Operating Frequency (MHz)	Penetration in clean dry sandy soil (m)	Penetration in dense wet clay soil (m)	Smallest object visible to GPR (m)
100	18	6	Tunnel @ 18 m depth 0.6 m pipe @ 6 m depth
250	12	4	0.9 m pipe @ 12 m depth 0.15 m pipe @ 4 m depth
500	4.5	1.8	0.1 m pipe @ 4 m depth 0.5 cm hose @ 1.8 m depth
1000	2	0.9	0.5 cm hose @ 0.9 m depth Wire mesh, shallow
2000	0.6	0.15	Monofilament fishing line

Sampling time interval is another important design criterion. The desired sampling interval or sampling frequency is dictated by the Nyquist rate, which states that the sampling frequency should at least be twice the high-frequency signal in the record. In other words, sampling time interval should at the most be equal to half of the period of the highest frequency signal. In most GPR systems, bandwidth to centre frequency ratio is about 1, which implies that the highest frequency is about 1.5 times the centre frequency. Taking a safety margin of two, the sampling time interval may be computed from eqn. 3.3.

$$t = 1000/6f \tag{3.3}$$

f = Centre frequency in MHz
t = Time is time in ns

Using expression 3.3, the sampling time intervals for centre frequencies of 10 MHz and 1000 MHz are computed as 16.7 ns and 0.17 ns, respectively.

Yet another important design criterion is the time interval between two discrete radar measurements spatially called *station spacing*. The optimum station spacing is related to the centre frequency and the relative permittivity of the medium and may be computed from eqn. 3.4.

$$t_x = 75/f\sqrt{K} \tag{3.4}$$

f = Centre frequency (in MHz)
K = Relative permittivity of medium

The chosen station spacing is such that Nyquist sampling intervals are not exceeded to ensure that ground response is not spatially aliased. Exceeding Nyquist sampling interval will not adequately define steeply dipping reflectors though it reduces data volume and survey time. Also, nothing is gained from spatial oversampling.

Choice of antenna in GPR is a trade-off between several factors, which include plan resolution, antenna size, signal processing methodology employed and ability to penetrate the material. A high gain antenna is generally desirable to achieve acceptable plan resolution. To achieve high gain, antenna size needs to be greater than the wavelength of the lowest transmitted signal frequency. To reduce antenna and still have the desired high gain necessitates the carrier frequency to be high. Penetration depth reduces with increase in frequency. Also, the transmit-receive antenna needs to be well isolated from the effect of the material lest unwanted signals with characteristics similar to the target mask the target return.

Most GPR systems are bistatic. That is, they use separate antennas from transmission and reception. Though the two antennas are usually packaged as a single module and there is no provision for varying spacing between the two, the ability to vary spacing between transmit and receive antennas can be used advantage in maximizing target coupling. The ability to vary the antenna spacing can be a powerful aid in optimizing the system for specific types of target detection. Optimum antenna separation S may be estimated from eqn. 3.5.

$$S = (2 \times Depth)/\sqrt{(K-1)} \tag{3.5}$$

K = Relative permittivity of medium.

Higher separation helps in the case of flat lying planar targets. As a rule of thumb, in lesser known survey areas, S is taken as 20% higher than the target depth.

3.4.3 Representative GPR Systems

GPR systems discussed in this section include the LMX-100 and LMX-200 from M/S Sensors & Software, Inc., SIR family of GPRs (SIR-20, SIR-30, SIR-3000 and SIR-4000) from Geophysical Survey Systems, Inc. (GSSI), StructureScan and UtilityScan systems also from GSSI and GROUNDSHARK, MINESHARK, Minestalker series, HMDS (Husky Mounted Detection System) series and R-VISOR, all from NIITEK.

3.4.3.1 LMX-100 and LMX-200 Ground Penetration Radar Systems

The *LMX-100 Locate & Mark System* from M/S Sensors & Software is ground penetrating radar particularly suitable for locating and marking underground both metallic and non-metallic objects including metallic pipes and cables, non-metallic objects such as plastic pipes, conduits, concrete ducts and dielectric fibreoptic cables. LMX-100 is designed to operate in a wide range of environmental conditions. The radar provides a balance between conflicting requirements of larger depth of penetration and higher resolution. Salient features of LMX-100 include high-resolution ultra-wideband GPR sensor that is ground coupled for maximum signal penetration, rugged field-proof display and multi-language selectable menu with more than 10 languages, lightweight and rugged fibreglass cart that eliminates interference from metallic structures and a fully enclosed odometer enabling precision data collection even in poor terrain. Major specifications of LMX-100 include a maximum penetration depth of 8 m, dynamic depth range of 0.8–8 m, spatial interval of 5 cm, real-time in-field analysis of depth estimates, image storage capacity of >10 000 graphic data images depending on an external flash memory up to 64 GB and operating temperature range of –40–+50°C (for the sensor) and –10–+50°C (for the display).

The LMX-200 GPR system, like the LMX-100, is also designed to locate and mark unground metallic and difficult-to-detect non-metallic utilities. The radar can locate and mark both shallow and deep unground targets including non-metallic PVC and asbestos cement pipes, concrete storm and sewer systems, underground storage tanks and drainage tiles, septic system components and non-utility structures such as vaults, foundation walls and concrete pads. Compared to the LMX-100 GPR, LMX-200 has advanced GPS and underground mapping capabilities. Some of the advanced features of LMX-200 include a high-resolution touch screen Digital Video Logger (DVL) that has high contrast and is sunlight visible, Ultra-Wide Band (UWB) GPR technology and advanced signal enhancement technology. Internal and optional external GPS for geo-referencing of targets, online report generation capability and connection to WiFi for wireless transmission of reports to remote locations via email. Salient features and major performance specifications of LMX-100 and LMX-200 are summarized in Table 3.5.

3.4.3.2 SIR Family of GPRs

Some of the representative GPR systems in the SIR family of GPRs from M/S Geophysical Survey Systems, Inc. and discussed in this section include Dual-Channel GPR SIR-20, Multichannel (configurable two, four and eight channels) SIR-30, Single Channel SIR-3000 with capability of in-the-field 3D imaging and SIR-4000 capable of operating with analogue and digital antennas.

The GSSI SIR-20 is a versatile dual-channel GPR control unit particularly suited to applications such as road structure assessment, bridge deck inspection, rail bed inspection, concrete inspection, archaeology, geological investigation and mining. The GPR is coupled with a rugged Panasonic ToughBook PC that provides a familiar windows environment for collecting, storing, processing, and transferring data. There are application specific processing functions that enable the user to present the interpreted GPR results in practical, useful formats.

Table 3.5 Features and specifications of LMX-100 and LMX-200.

Feature/Specification	LMX-100	LMX-200
Dynamic Depth Range (m)	0.8–8.0	0.8–8.0
Spatial Interval (cm)	5	5
Depth Estimates	Real Time Analysis	Real Time Analysis
Signal Enhancement	DynaQ Stacking, Spatial Filtering	Dyna-Q Stacking/Dyna-T, Spatial Filtering
Image Storage Capacity	>10 000 Graphic Data Images Depending on external flash memory (up to 64 GB)	320 km of line data in internal memory
Dimensions: Height × Length × Width (cm)	115 × 100 × 70	115 × 100 × 70
Weight (kg)	22	22
Power	1.25 A @ 12 VDC	1.25 A @ 12 VDC
Environmental Qualification	IP 65 −40°C to +50°C (Sensor) −10°C to +50°C (Display)	IP 65 −40°C to +50°C (Sensor) −10°C to +50°C (Display)
Other Features	Ultra Wide Band GPR Technology, Fully Enclosed Odometer, Field-Proof Display, Multi-language Menu	Ultra Wide Band GPR Technology, Internal and Optional External GPS, Wi-Fi Connection, On Site Report Generation, High Resolution Touch Screen DVL

The SIR–20 is compatible with all GSSI antennas, allowing the user to address the full range of GPR applications outlined before. The GPR is GPS compatible. Though ideally suited to vehicle-mounted applications, it is also configurable on a cart-based system or mobile on-site. Major specifications are outlined in Table 3.6, which summarizes a comparison of SIR-20, SIR-30, SIR-3000 and SIR-4000 systems.

The GSSI SIR-30 Radar Control Unit is an advanced multichannel radar control unit. Like SIR-20, it is compatible with most GSSI antennas and supports multiple mounting configurations. Radar is characterized by high-speed GPR data collection at more than 5792 scans/second with four channels. The system records data from one to four channels simultaneously with provision for two four-channel systems to be configured as an eight-channel system. Typical applications include road structure assessment, utility designation, bridge deck inspection, rail bed inspection, airport runway inspection, detection of cavities and detection of clean/fouled ballasts. Table 3.6 includes the major specifications of the SIR-30 radar control unit.

The *GSSISIR-3000 GPR Control Unit* is a portable and lightweight system designed for single-user operation. Its advanced signal processing and display capabilities allow in-the-field 3D imaging on a high-resolution colour display that is visible over a wide range of light conditions. It is compatible with most GSSI antennas. A Windows-based user interface, large internal data storage and a removal flash memory are the other features that make it a versatile sensor. Typical applications include utility location, concrete inspection, geological investigation, archaeological surveys, forensics, mining and bridge deck inspection. Table 3.6 includes the major performance specifications of the SIR-3000 GPR control unit.

Table 3.6 Specification comparison of the SIR-20, SIR-30, SIR-3000 and SIR-4000 GPR systems.

Specification	SIR-20	SIR-30	SIR-3000	SIR-4000
Antenna Support	Compatible with all GSSI antennas	Compatible with most GSSI antennas	Compatible with most GSSI antennas	Compatible with all GSSI antennas
Number of Channels	Records data from 1 or 2 hardware channels simultaneously; 1 to 4 data channels selectable	Records data from 1 to 4 hardware channels simultaneously; Can be configured as 8-channel system	1	Records data from 1 single-frequency antenna or 1 dual-frequency antenna
Display Type	13.3 inch colour LCD display on ToughBook laptop		Enhanced 8.4: TFT, 800x600 resolution, 64K colours	Enhanced 10.4" LED display with internal high brightness, Active matrix 1024 768 resolution and 32-bit colour
Display Mode	Line scan and O-scope, 3D	Line scan and O-scope	Line scan and O-scope, 3D	Linescan, Linescan plus O-scope, Wiggle trace Full 3D, 256 colour bins are used to represent the amplitude and polarity of the signal
Data Format	RADAN (.dzt)	RADAN (.dzt)	RADAN (.dzt)	RADAN (.dzt)
Data Storage	80 GB internal memory, Optional external memory	Internal memory: 500 GB (4-channel), 250 GB (2-channel)	Internal memory: 2 GB Flash memory card, Compact Flash port: Accepts CF memory up to 8 GB (using FAT 16 file format)	32 GB
Scan Interval Samples PerScan	User selectable 256, 512, 1024, 2048, 4096, 8192	User selectable 256, 512, 1024, 2048, 4096, 8192, 16 384	User selectable 256, 512, 1024, 2048, 4096, 8192	User-selectable, up to 400 scans/s 256, 512, 1024, 2048, 4096, 8192, 16 384
Transmit Range	Up to 500 kHz (International only)	Up to 800 KHz (International), US/Canada and CE rates depend on antenna model	Up to 100 kHz	Up to 800 KHz (International), US/Canada and CE rates depend on antenna model
Operating Modes	Free run, survey wheel, point mode	Continuous (time) or survey wheel (distance triggered)	Free run, survey wheel, point mode	Continuous (time) or survey wheel (distance triggered) or point mode

Available Filters (Processing)	Low and High Pass, Infinite Impulse Response (IIR), Finite Impulse Response (FIR), Boxcar and Triangular filter	Infinite Impulse Response (IIR) - Low and High Pass, vertical and horizontal Finite Impulse Response (FIR) - Low and High Pass, vertical and horizontal	Vertical: Low Pass and High Pass IIR and FIR Horizontal: Stacking, Background Removal	Infinite Impulse Response (IIR) - Low and High Pass, vertical and horizontal Finite Impulse Response (FIR) - Low and High Pass, vertical and horizontal, Migration, Surface Position Tracking, Signal Noise Floor Tracking, Adaptive Background Removal
Input Power	12 VDC, 18Ah (External)	260W max (120W typical) at 95-250VAC 50/60Hz or +10VDC to +28VDC	10.8 VDC Internal	–
Dimensions	466 × 395 × 174 mm	450 × 330 × 130 mm	315 × 220 × 105 mm	360 × 250 × 70 mm
Weight	12.2 kg with laptop	8.4 kg	4.1 kg including Battery	4.53 kg with Battery
Operating Temperature Range	–10°C to +40°C	–10°C to +50°C	–10°C to +40°C	–20°C to +40°C

Figure 3.33 SIR-4000 GPR Control Unit.

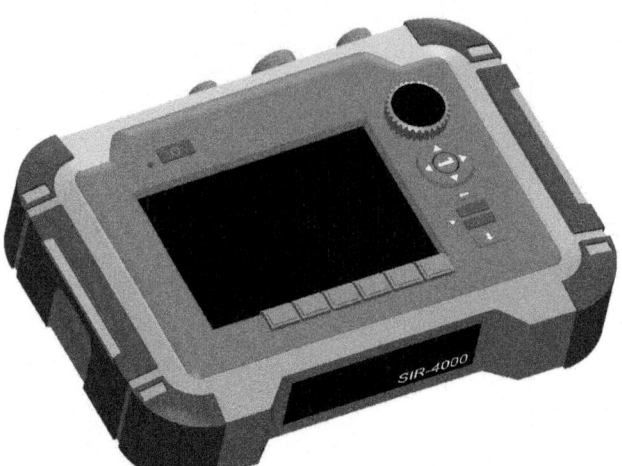

The GSSI SIR-4000 (Figure 3.33) is a single channel GPR control unit designed to operate with analogue antennas as well as new generation digital antennas. The sensor records data from one single frequency analogue antenna or one dual frequency digital antenna. SIR-4000 offers several deployment options for in-field accessibility and unique data collection modules including Quick 3D, UtilityScan and StructureScan and Expert Mode for efficient data collection and visualization. Windows 8 based interface, GPS integration, large data storage and convenient file transfer options are the other important features. Typical applications include utility designation, concrete inspection, archaeological surveys, forensics, mining, environmental assessment and geology. Table 3.6 includes the major performance specifications of the SIR-4000 GPR control unit.

GSSI designs and manufactures GPR antennas covering a wide frequency range from 15 MHz to 2600 MHz suiting various application areas outlined in the previous paragraphs in this section while describing some of the GSSI GPR control units. The centre frequency, penetration depth and typical applications of different GPR antennas offered by GSSI are summarized in Table 3.7.

3.4.3.3 Utility Scan Series GPR Systems

The UtilityScan series of ground penetrating radar systems from GSSI offers three variants including *UtilityScan*, *UtilityScan-DF* and *UtilityScan-LT*. The UtilityScan family of GPRs are configurable and have the flexibility to be used for a wide range of utility detection and location applications. Typical uses include detection and location of metallic and non-metallic utilities, archaeological surveys, forensic investigations, road inspection, bridge deck assessment, geological investigation, damage prevention and environmental remediation. *UtilityScan GPR* solution employs SIR-3000 radar control unit and is compatible with multiple antenna options in the frequency range of 2600–270 MHz (with rugged survey cart) and 2600–400 MHz (with compact and standard survey carts) offering penetration depth of 0–6 m and 0–4 m, respectively, for 400 MHz and 270 MHz. The *UtilityScan-DF* uses a customized Panasonic Toughpad FZ-G1 radar control unit and a dual frequency (300 MHz and 800 MHz) digital antenna offering maximum penetration depth up to 5 m. With its survey speed of 10 km/h, it offers fast and efficient data collection. The easy-to-use touch screen interface allows viewing of shallow and deep targets. UtilityScan-LT is the low-cost version of industry standard UtilityScan. It employs a 400 MHz antenna offering a maximum penetration depth of 4 m.

Table 3.7 GSSI GPR antennas.

S. No.	Antenna type	Centre frequency (MHz)	Penetration depth (m)	Application
1.	Multiple low frequency antenna Model: 3200 MLF	15–80	0–50	Deployed in discrete measurements or continuous profile data collection modes
2.	Monostatic antenna Model: 3207AP	100	2–15	Deep sub-surface investigation
3.	Bistatic antenna Model: 3207F	100	1–30	Deep sub-surface investigation
4.	Antenna Model: 5106/5106A	200	0–9	Geotechnical and environmental applications
5.	Antenna Model: 50270S	270	0–6	Utility mapping and shallow engineering
6.	Dual frequency digital antenna Model: D50300/800	300/800	7/4	Utility, archaeological and environmental surveys
7.	Antenna Model: 3101D/3101A	900	0–1	Concrete assessment, void detection, shallow pipe location
8.	General purpose antenna Model: 51600S	1600	0–0.5	Inspection of concrete structures, bridge deck condition assessment
9.	Horn antenna Model: 41000S	1000	0–0.9	Pavement thickness monitoring, road condition assessment
10.	Horn antenna Model: 42000S	2000	0–0.75	Pavement thickness monitoring, road condition assessment
11.	Antenna Model: 62000	2000	0–0.4	Ability to reach tightly spaced areas such as corners, against walls and around corners
12.	Antenna Model: 52600S	2600	0–0.4	Inspection of concrete structures

3.4.3.4 Structure Scan Series GPR Systems

Different variants of StructureScan family of GPRs include *StructureScan Standard, StructureScan Optical, StructureScan Mini, StructureScan Mini-HR* and *StructureScan Mini-XT*. StructureScan Standard is configured around an SIR-3000 radar control unit and can be used with a 1600 MHz or 2600 MHz antenna providing penetration depth of 0.45 m. The GPR provides a non-destructive means of accurately inspecting concrete structures including location of embedment inside concrete structures prior to drilling, cutting or coring thereby allowing concrete and construction professionals identify conduits and avoid dangerous and costly hits. Typical uses include concrete and structure inspection, void detection and location, measurement of slab thickness and so on. 2 GB of internal memory, high-resolution (800 × 600) colour display, RS-232, compact flash memory, USB Master and Slave ports and a Windows-based user interface are the other features. *StructureScan Optical* too is configured around SIR-3000 radar control unit and 1600 MHz and 2600 MHz antennas. It is equipped with an optical barcode reader and patented Smart Pad technology that facilitates easy wall and floor scanning. Other features are similar to those of StructureScan Standard.

The *StructureScan Mini* is a portable handheld easy-to-transport GPR designed for concrete inspection and analysis offering a lightweight alternative to other GPR systems. With its 1600 MHz

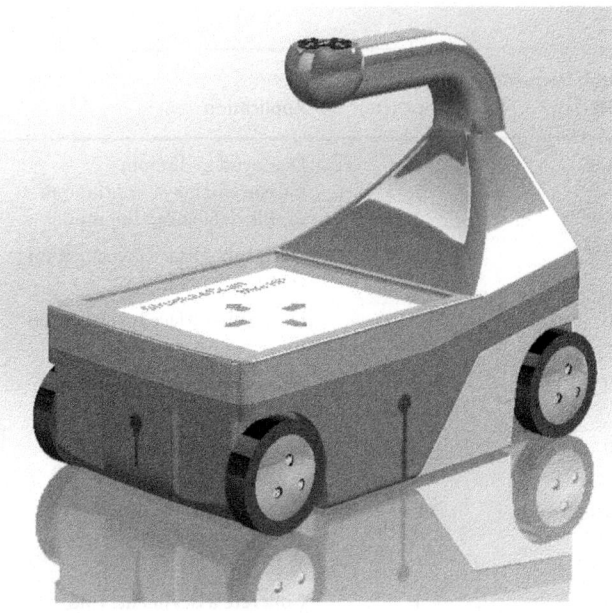

Figure 3.34 StructureScan Mini-HR GPR.

antenna, it offers a perfect blend of high-resolution and penetration depth. It is available with 2D or 3D capabilities. *StructureScan Mini-HR* (Figure 3.34) with its 2600 MHz antenna offers higher resolution than *StructureScan Mini* allowing it to delineate smaller targets with better horizontal and vertical resolution. StructureScan Mini-HR is also available in two models including the 2D system for real-time location of targets and 3D system for an X-ray like image. *StructureScan Mini-XT* is the newest member of StructureScan Mini family of all-in-one handheld GPR systems. With its 2700 MHz antenna, it offers superior target resolution than the other radars in this family. The antenna that is positioned about 8 mm off the surface allows radar use over rough concrete surfaces. Also, it offers excellent near-surface resolution without compromising ability to resolve deeper targets. The radar is particularly suited to determining position and depth of metallic and non-metallic objects in concrete structures.

3.4.3.5 GROUNDSHARK Dual Sensor System

The GROUNDSHARK by NIITEK is a high performance handheld dual sensor system capable of detecting, locating and visualizing buried high metallic, low metallic, non-ferrous, nonmagnetic and non-metallic objects in challenging soil conditions such as changing and uneven terrain. Visualization of subsurface here includes ground/asphalt/cement layers. The system's customizable software yields superior false alarm rate (FAR) and probability of detection performance. The radar system comprises of a two-channel GPR with a co-located single frequency continuous-wave electromagnetic induction based metal detector with distinct/separable tones based on target detection via metal detector or GPR. The system employs modular architecture with field-replaceable components including field-replaceable high-resolution (800 × 480) visual display upgradable with compatible night vision goggles. The system also has an extensive Built-In Test (BIT) with operator notification of system malfunctions.

3.4.3.6 MINESHARK Dual Sensor System

NIITEK's MINESHARK is a handheld dual sensor system capable of detection and subsurface visualization of buried anti-personnel (AP) and anti-tank (AT) landmines, improvised explosive devices (IED) and other buried explosive hazards. It detects, discriminates and identifies metal targets including single strand wires and carbon rods in challenging soil conditions such as changing and uneven terrain. The system's customizable software yields superior false alarm rate (FAR) and probability of detection performance. The dual sensor system comprises of NIITEK's combat-proven dual-channel ground penetrating radar (GPR) and Minelab's advanced single frequency, continuous-wave electromagnetic induction based metal detector with distinct/separable tones based on target detection via metal detector or GPR. MINESHARK's dual sensor configuration is particularly suited to countermine operations, battle area clearance, explosive ordnance disposal (EOD) spot inspection and IED detection. MINESHARK is capable of detection, discrimination and identification of metal targets in challenging soil conditions including single strand wires and carbon rods. MINESHARK's automatic detection and recognition software facilitates determination of the type of buried threat. Optional GPS and compatible night vision goggles are also available for system upgradation.

3.4.3.7 Minestalker Series

The Minestalker Series of systems (Figure 3.35) is designed to be used for detecting and marking buried landmines and other explosive hazards. It can be remotely controlled to carry out demining operation, thereby preventing loss of life. Minestalker uses NIITEK's high performance front mounted VISOR-3200 ground penetrating radar (GPR). Minestalker combines advanced real-time automatic target recognition (ATR) algorithms, optional integrated metal detection and user-friendly software in a ruggedized package. It is configurable on a variety of vehicular platforms and can also be integrated with a wide array of robotic systems. High probability of detection, low false alarm rate, real-time GPR imagery with 3D visualization, physical marking and marking by GPS coordinates, capability to store 100 km of scanned data on local drive, optional see-deep metal detector array and low-cost navigation options are some of the other important features.

Figure 3.35 Minestalker series GPR. (*Source:* Courtesy of NIITEK/Chemring Sensors & Electronic Systems.)

3.4.3.8 Husky Mounted Detection System (HMDS) Series

The HMDS is configured around VISOR ground penetrating radar and is capable of automatic detection, recognition and precision marking of buried threats including metallic and non-metallic mines, improvised explosive devices (IEDs) and other explosive hazards (Figure 3.36). HMDS has been integrated on to HUSKY Mark III Single Occupant Vehicle-Mounted Mine Detection System (VMMD) or HUSKY Mark IV Dual Occupant VMMD. The HMDS is a kit that integrates with the Husky vehicle and has four ground penetrating radar panels. The system is also readily adaptable to robotic platforms. Major constituent parts of the HMDS include GPR sensor with high probability of detection and low false alarm rate, Remote Visualization (R-VIS) capability and Built-in Test (BIT); Radar Mounting and Positioning System (RMPS), Control and Display Unit (CDU) comprising sun-readable LCD, resistive touch screen and integrated numeric keypad; a Navigation Subsystem with inertial and satellite-based corrections and GPS coordinates; Marking Subsystem and Emergency Stop (E-Stop) Switches that can halt power to hydraulics and Cabin Control Unit (CCU). Optional Remote Visualization (R-VIS) feature allows the HMDS operator to transmit real-time data to a secondary display located in a mine protected vehicle for threat analysis. The United States and its allies have successfully used HMDS in Afghanistan. HMDS is in use for detection and marking of IEDs and other explosive hazards in the United States (Army and Marine Corps), Canada, Australia, Spain, Turkey, Italy, Angola, Cambodia and other nations thereby saving human lives on a daily basis.

Figure 3.36 Husky Mounted Detection System. (*Source:* Courtesy of NIITEK/Chemring Sensors & Electronic Systems.)

Figure 3.37 R-VISOR GPR. (*Source:* Courtesy of NIITEK/Chemring Sensors & Electronic Systems.)

3.4.3.9 R-VISOR GPR System

NIITEK's R-VISOR or Robot Visor (Figure 3.37) is a dual sensor comprising high performance ground penetrating radar and a sophisticated metal detector. The ground penetrating radar and the metal detector work in tandem to offer the user ability to precisely mark and visualize buried improvised explosive devices (IEDs), anti-personnel and anti-tank landmines. R-VISOR is integrated on the TALON-class robotic vehicle and is readily adaptable to other robotic platforms such as the Turkish ASELSAN robot. With its automated detection feature, the system is particularly suited to route clearance operations and/or rapidly clearing small areas of interest, pathways, landing zones and mine fields with operators at safe standoff distances.

3.5 Weapon Locating Radar Systems

Weapon locating radar, also known as *gun locating radar* and *counter-battery radar*, provides a significant tactical advantage in the battlefield by detecting the location or point of origin of hostile fire including artillery shells, mortars, unguided rockets and so on, facilitating the friendly forces to launch a counter-fire with pin-point accuracy and thereby forcing the adversary to deploy massively disproportionate fire power. Weapon locating radar is always attached to an artillery battery or their support group. It uses a number of radar techniques, advanced signal processing, and automatic height correction to detect, authenticate and track the projectiles in flight. Subsequent to this, it constructs new data points outside a discrete set of known data points for both determination of the firing location and impact point. The detection, location and tracking operations of intended targets are executed by advanced algorithms and state-of-the-art-hardware. Weapon locating radar typically detects the point of origin of the adversary fire immediately after it is launched and long before it hits the friendly forces. Detection of adversary's firing location is so rapid that the coordinates of the firing weapon are with the operator before the enemy projectile lands. While the primary function of weapon locating radar is to detect and track incoming artillery rounds, mortar and rockets and locate their launchers, in its secondary role it can also track and observe the impact point of the round from friendly guns and provide fire corrections to counter-battery fire. Many weapon locating radars, Raytheon's AN/TPQ-37 radars, for example, are also able to compute the likely point of impact of incoming hostile projectile. This information can be used to warn the friendly troops targeted by the incoming projectiles. Given that time-of-flight of the hostile projectile is typically under a minute, it can be effective only if the targeted troops are in the vicinity of the weapon locating radar. Otherwise, one would require an automated and fast communications between the radar location and location of troops.

Most state-of-the-art weapon locating radars allow networking of operation of multiple radar sensors in tandem for better accuracy and more information. Also, they can simultaneously track multiple targets fired at low or high angles and at all aspect angles. For example, in the case of BEL (Bharat Electronics Limited) Weapon Locating Radar (WLR), up to 99 weapon locations can be tracked and stored at any time. They are capable of operating in high density fire environment and in severely cluttered and jamming conditions. Modern weapon locating radars also have remote operation feature, which allows the operators command the radar sensor from a safe standoff distance thereby making them far less vulnerable if the radar were targeted by the adversary's fire. Another feature allows the data to be transmitted to a command centre for better situational awareness at the higher echelons of the command hierarchy.

Most weapon locating radars use the C band (4–8 GHz) or S band (2–4 GHz). Some of the radar systems have used the X band (8–12 GHz) and Ku band (12–18 GHz) also. These frequency bands are as per the IEEE designation. The higher the operating frequency, greater the accuracy for small size targets. However, higher frequencies suffer greater attenuation. The detection range of weapon locating radar is governed by the radar cross-section of the intended target. Typical values of RCS for different weapon locating radar relevant targets include 0.01 m (mortar bomb), 0.001 m (artillery shell), 0.009 m (light rocket such as 122 mm rocket) and 0.018 m (heavy rocket such as 227 mm rocket). Modern radars have detection range of greater than 50 km for mortar bombs and rockets. This is around 30 km in the case of artillery shells. Achievable detection range is subject to the condition that the trajectory is high enough to be seen by the radar at these ranges. Since, for good accuracy, one needs to track a reasonable length of the trajectory close to the firing weapon, long-range detection doesn't guarantee good accuracy. Accuracy of determination of location of firing weapon is typically specified by a parameter called CEP (Circular Error Probable). Modern radars have a CEP of about 0.3–0.4% of range. CEP is the radius of the circle centred on the actual location, the boundary of which includes 50% of the measured locations.

3.5.1 Operational Principle

All modern weapon locating radars employ phased array scanning technique to scan a sector of air space. The radar usually scans one quadrant; that is, a 90° sector to look for any hostile targets. The radar in this case can electronically scan up to ±45° from its mean bearing. To achieve 360° coverage from a given position, the radar array can be rapidly rotated by 135° on either side to quickly change the scanning sector in response to threats. Upon detecting an incoming round, the radar automatically acquires and classifies the threat and initiates a track sequence, while continuing to search for new targets. The radar needs to track the intended target for sufficient time to record a segment of the trajectory. The required track time depends upon the radar type and is typically 5–8 s for AN/TPQ-37 Fire Finder Radar. The track data is processed in real time with a state-of-the-art digital signal processor. The radar along with the digital terrain data bases extrapolates the round's point-of-origin. The computed point-of-origin is then reported to the radar operator, thus allowing friendly artillery to direct counter-battery fire towards the enemy artillery.

Weapon locating radar has two modes of operation namely the *friendly mode* and the *hostile mode*. In the friendly mode, it tracks the projectile fired from the friendly fire location and extrapolates the impact or burst location from the track data. Also in this case, the radar tracks the projectile in the descending leg of the trajectory. In the case of hostile mode, it detects and tracks the projectile fired from the hostile firing location during the ascending phase of the trajectory and extrapolates the hostile firing location and the impact or burst predict point from the analysis of track data. Both modes are similar except for those differences previously mentioned and some others explained in the following paragraphs. Different steps involved in determining the hostile firing location include (1) establishing the search fence, (2) verifying the search fence penetration, (3) validating the trajectory, (4) tracking the projectile and (5) extrapolating the firing location and also the prediction of impact point from the track data.

The three-dimensional space within which the target can possibly be detected and tracked is defined by the minimum and maximum detection ranges of the radar, the azimuth search sector called the search fence and the vertical scan angle. The search fence is typically 1600 mils (90°) or ± 800 mils (±45°) around the radar's azimuth of orientation in the hostile mode of operation. A narrowed search fence may be employed if the tactical situation so demands. In the case of friendly mode of operation, the search fence is much narrower, typically 440 mils instead of 1600 mils. The vertical scan angle extends from the search fence to the maximum scan elevation of the radar. In order that the radar is able to determine the firing location and also predict the impact point, it is imperative that there is sufficient amount of vertical scan available known as track volume at the points where an object passes through the detection area. Typically, radars require a minimum of 50 mils of track volume to track a round for long enough to be able to determinethe intended parameters.

Aspect angle and speed of the object are the other important parameters determining the efficacy of weapon locating radar. The aspect angle, which is the angle measured from radar antenna to the target path of the object, must be greater than 1600 mils (90°). This ensures that the intended target is travelling towards the radar. Also, the velocity of the intended target must be within the minimum and maximum velocity thresholds for the specific radar.

Target detection, verification and location process takes place as follows. When the target enters the search fence established by the radar by sensing a series of beams conforming to the terrain, the radar determines target's speed, azimuth, elevation and range. From this information, it predicts the target's next location. It then sends out verification beams to establish whether the target has ballistic trajectory. If the trajectory is non-ballistic it is ignored by radar. Having verified the ballistic trajectory, the radar initiates track sequence. Track data is used to mathematically extrapolate a predicted launch and impact point. The radar stops tracking the target if any of the following conditions exist: (1) the intended objective has been met, (2) there is a certain specified number of sequential misses depending upon the radar: this number is five for AN/TPQ-37 and (3) the predicted azimuth and elevation positions for the next track update is outside radar's limits of search sector and search elevation, respectively. Figure 3.38 illustrates the operational scenario of a weapon locating radar. The diagram is self-explanatory.

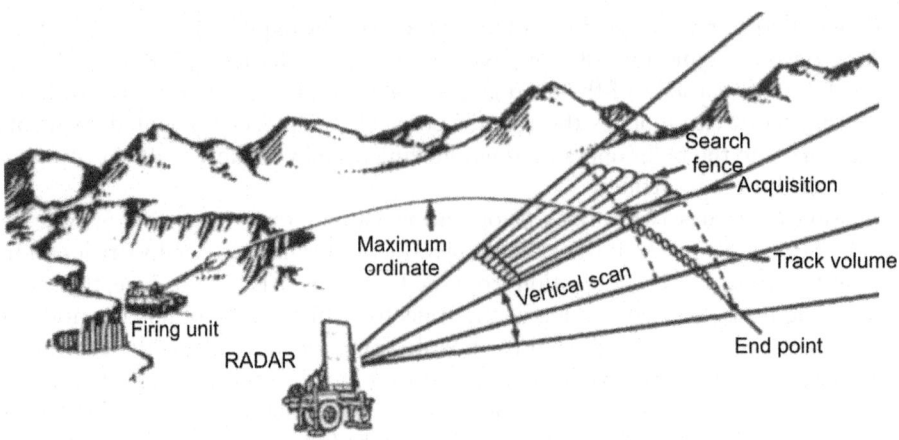

Figure 3.38 Operational scenario of a weapon locating radar.

3.5.2 Representative Weapon Locating Radar Systems

Weapon locating radar systems discussed in this section include; (1) AN/TPQ-37 from Thales Raytheon Systems, (2) AN/TPQ-50 from SRC, Inc., (3) COBRA (Counter-Battery Radar), (4) ARTHUR (Artillery Hunting Radar) from Sweden, (5) MAMBA from the UK, (6) SLC-2 and (7) BL-904 fire finding radars from China and (8) BEL WLR from India.

3.5.2.1 AN/TPQ-37 Fire Finder Radar

The AN/TPQ-37 by Thales Raytheon Systems (Figure 3.39) is mobile electronically steered S band (2–4 GHz) long-range weapon locating radar, also called the Fire Finder radar. It is capable of detecting and tracking artillery, mortar and rocket fire from single or multiple positioned hostile weapon platforms to determine the point-of-origin of attack thereby enabling friendly forces to direct an effective counter-fire. It is also capable of detecting and tracking short-range tactical missiles. In a secondary role, it also determines the impact point of the rounds from friendly launchers to provide fire corrections to counter-battery fire. It can also predict the impact point of hostile fire. The radar has 15 different operating frequencies in the S band, a 90° azimuth scan sector and detection range of 3–50 km depending upon type of round. The detection range depends upon the radar cross-section of the target and is much greater in the case of missiles than what it is for artillery rounds or mortars or rockets. Other important operational features include first round detection capability, capability to perform high-burst, datum plane and impact registrations and permanent storage of up to 99 targets. The radar system comprises of Antenna-Transceiver Group, Operations Control Group mounted on a M-35 series truck, a 60 kW and a 400 Hz Generator Set mounted on a 5-tonne truck. the antenna-transceiver group comprises of antenna, transmitter, receiver and associated electronics mounted on M-1048 trailer, a 6-tonne four wheeled flatbed cargo trailer towed by the 5-tonne truck hosting the Generator Set.

AN/TPQ-37 radar has evolved over the years through different versions of the radar, which include AN/TPQ-37 (V) 6, AN/TPQ-37 (V) 7/8 and the most recent AN/TPQ-37 (V) 9. AN/TPQ-37 (V) 6 is a long-range weapon locating radar capable of locating artillery shells and rockets within the accuracy of weapons systems. The system though accurate lacks mobility and also needs to be dismantled into constituent parts if it were to be airlifted in any military transport aircraft smaller than the C-5 heavy cargo transport aircraft. The radar was extensively used during Operation Desert Storm. Most of the shortcomings of (V) 6 radar

Figure 3.39 AN/TPQ-37 (RMI) weapon locating radar. (*Source:* Courtesy of US Army.)

were subsequently taken care of in the development of AN/TPQ-37(V) 7 and (V) 8 radars. Addition of a mobility package, in which a C-130 roll-on/roll-off fixed to the radar systemallows the antenna subsystem to be rolled off its trailer directly onto the ramp of C-130 transport aircraft. Other additions and modifications implemented in (V) 7 and (V) 8 included the following: a Medium Tracked Suspension system (MTSS) to address manoeuvrability issues encountered during Operation Desert Storm, the S-250 military shelter, which makes it possible to transport and operate a wide variety of communications and electronic equipment to the field as a mobile unit, was moved to a M1097 Heavy High Mobility Multipurpose Wheeled Vehicle (HMMWV), Modular Azimuth Positioning System (MAPS) was installed for self-survey capability and memory keep-alive circuits. Also, the old cooler was replaced by a new redesigned cooler.

AN/TPQ-37 (V) 9 also known as AN/TPQ-37 RMI (Reliability Maintainability Improvement) is the most recent radar in the family (Figure 3.39). The radar incorporates many advanced technologies to allow an already combat-proven weapon locating radar remain as the mainstay well into the twenty-first-century. The new radar has an upgraded radar transmitter incorporating 12 modern air-cooled power amplifier modules (PAMs), a high power RF combiner and a fully automated transmitter control unit thereby enhancing life and reliability, continued radar operation under fault conditions and reduced maintenance time. It has and an upgraded radar processor offering improved reliability and power consumption performance, combining the latest VME-64x architecture and full high/low temperature performance with proven

Figure 3.40 AN/TPQ-47 weapon locating radar. (*Source:* Courtesy of GlobalSecurity.org.)

AN/TPQ-37 operational and maintenance software programs. All AN/TPQ-37 (V) 9 radars (RMI upgrades) have improved system availability by 90%, a significant reduction in life cycle cost, extended system supportability and useful life beyond 2018 and retain all proven AN/TPQ-37 performance capabilities.

Further efforts in the process of evolution of AN/TPQ-37 family of weapon locating radars have led to development of the AN/TPQ-37 P^{3}I (Block II) radar also known as AN/TPQ-47 radar (Figure 3.40). The AN/TPQ-47 provides rapid target location and classification at greater ranges with improved accuracy. Target detection and location ranges in AN/TPQ-47 radar are 18 km (for light mortars), 30 km (for heavy mortars), 60 km (for artillery and light rockets) and 100 km (for heavy rockets). It can also detect tactical ballistic missiles up to 300 km in range. The radar can be transported in a single C-130 transport aircraft or multiple CH-47 helicopter lifts. The radar with its significantly enhanced performance capabilities is projected to revolutionize the way the military conduct weapon locating missions in future.

3.5.2.2 AN/TPQ-50 Counter-Fire Radar

The AN/TPQ-50 manufactured by SRC, Inc. is an L band (1–2 GHz) lightweight counter-fire radar capable of detecting and tracking rockets, mortars and artillery shells up to 10 km with an accuracy of 50 m as specified by circular error probable (CEP). The radar is mountable on towers, tripods or light vehicles such as the High Mobility Multipurpose Wheeled Vehicle (HMMWV). The HMWWV-mounted version is shown in Figure 3.41. The radar is capable of locating Points Of Origin (POO) as well as Points Of impact (POI), which it accomplishes by detecting and tracking the round and extrapolating the track data to determine the POO and POI with desired accuracy. It determines the POO with better accuracy for the projectiles following flatter trajectories or those coming from low quadrant elevations thereby facilitating

Figure 3.41 AN/TPQ-50 counter fire radar. (*Source:* Courtesy of Major Linda Wade, US Army.)

a more precise counter-fire response. The radar provides an azimuth coverage of the full 360°, which allows it to simultaneously detect and track rounds fired from different locations within its coverage area of about 315 square km defined by its detection radius of 10 km. The elevation coverage is 0–30°. The radar can also be configured for a focused coverage of smaller areas with higher update rates. The information on POO can be reported back to integrated control and command centre for initiating a counter-fire response. The radar has been tested against stringent US military standards, which include MIL-STD-461E, MIL-STD-464A and MIL-STD-810G. The radar also is digitally interoperable with Advanced Field Artillery Tactical Data system (AFATDS, which is the Fire Support Command and Control (C2) system employed by the US Army and Marine Corps units to provide automated support for planning, coordinating, controlling and executing fires and effects and Forward Area Air-Defence and Control (FAADC2) that collects, processes and disseminates real-time target tracking and cuing information to all short-range air-defence weapons and provides command and control for the counter-rocket, artillery and mortar system-of-systems.

3.5.2.3 COBRA Counter-Battery Radar

COBRA is the result of a collaborative programme between Germany, France and Turkey to develop long-range battlefield radar. The radar comprises of a high performance radar sensor, advanced processing and an integrated command, control and communication system (Figure 3.42). COBRA is a singularly effective force on the battlefield capable of detection, location and classification of hostile rockets launchers, mortars and artillery shells and prediction of point-of-impact (POI) of shells. The radar adjusts and registers friendly fire, determines jamming data, and communicates with battle forces. The radar is used to support artillery units in a wide range of battlefield scenarios; for example, in a peacekeeping role, it can be used to

Figure 3.42 COBRA counter battery radar. (*Source:* Bukvoed, https://de.wikipedia.org/wiki/Datei:Shilem-beyt-hatotchan-4.jpg. CC BY-SA 3.0.)

supervise a ceasefire and thereby monitor any breaches of ceasefire. It also supports peace-keeping measures during an intervention in defence of a nation where return fire is required and a large spread of forces occurs. And of course, it acts as a major force multiplier during major conflicts by defending friendly artillery batteries.

The radar operates in the C band (4–8 GHz) with agility in waveform and frequency, has detection coverage of 90° out to a range of 40 km corresponding to 1200 km^2 and is capable of locating and acquiring up to greater than 240 batteries in less than 2 min The radar employs a fully-active phased array antenna with 2870 T/R modules containing MMICs providing electronic scanning both in azimuth and elevation. COBRA, due to its advanced features, is one of the most widely used weapon locating radars in the world today. More than 600 COBRA systems are used by the US and its Allies.

3.5.2.4 ARTHUR – Artillery Hunting Radar

ARTHUR, an acronym for Artillery Hunting Radar is a mobile weapon locating radar system originally developed by *Ericsson Microwave Systems* (now *SAAB Microwave Systems*) in collabo-ration with the Swedish and the Norwegian Armed Forces, particularly the Defence Material Administration and Norwegian Army Material Command (Figure 3.43). The radar operates in the C band (4–8 GHz) and is partly based on the same technology as the GIRAFFE and employs a passive electronically scanned phased array antenna for both azimuth and elevation scanning

Figure 3.43 ARTHUR artillery hunting radar. (*Source:* Defense Command, https://commons.wikimedia.org/wiki/File:ARTHUR_(radar)_Danish.jpg. CC BY-SA 4.0.)

that enables location of the launch site with high accuracy immediately after detection of a projectile. *GIRAFFE* is the family designation of a series of C band pulse Doppler surveillance radars and combat control centres from Ericsson for mobile and static short- or medium-range Command, Control, Communications and Intelligence (C^3I) air-defence systems. The ARTHUR radar searches for flying artillery shells, tracks their trajectories for several seconds and based on the track information received by radar, it extrapolates the track data to compute not only the location of the sites firing those shells or mortars or rockets, but also determines the point of impact. The system is capable of monitoring up to 100 targets a minute in the search mode and can simultaneously track up to eight flying shells in the track mode. The information on the location of firing weapon site is transferred to the friendly firing units including Air Force assets. There are three operational modes in which ARTHUR weapon locating radar can be operated in. These include the *weapon locating mode*, the *fire-control mode* and the *sense and warn mode*. In the case of weapon locating mode, the radar determines the location of the guns, mortars, or rocket launchers. In the case of fire-control mode, the radar predicts the impact point of friendly shells and the information is used to adjust the fire of own artillery on to target coordinates to improve precision of friendly fire. In sense and warn mode, extrapolation of points of impact of incoming fire is used to trigger timely warning to own troops. The radar can operate either as a standalone medium-range weapons locating radar or as a long-range weapon locating radar with a coordinated configuration of 2–4 radars facilitating a constant surveillance of an area of interest. The radar can be easily transported by C-130 transport aircraft or it can be hung under a heavy duty helicopter such as a Chinook helicopter.

The ARTHUR weapon locating radar system has undergone upgradations from time to time. The original ARTHUR radar system also called ARTHUR MOD A could locate guns at 15–20 km and 120 mm mortars at 30–35 km with a circular error probability in the 0.45% of range. The first upgraded version of ARTHUR called ARTHUR MOD B was capable of locating guns at 20–25 km and 120 mm mortars at 35–40 km with a circular probable error of 0.35% of range. A further upgrade of ARTHUR known as ARTHUR MOD C uses a larger antenna. The radar detects guns at 30 km, mortars at 55 km and rockets at 50–60 km depending on their size. It does so with CEP 0.2% of range for guns and rockets and 0.1% for mortars. ARTHUR is or has been in service in the Armed Forces of Sweden, Norway, Canada (leased), Italy (leased for use in Iraq 2003/2004 and then furnished since 2013 in five units), the Czech Republic, Greece, Spain, Singapore, Malaysia and the UK (leased until MAMBA becomes available). South Korea reportedly purchased an upgraded version of ARTHUR in 2007 with a follow-up order in 2011.

3.5.2.5 MAMBA Radar

MAMBA (Figure 3.44) is an acronym for Mobile Artillery Monitoring Battlefield Asset. This is artillery hunting radar similar to ARTHUR radar. In fact, ARTHUR MOD C radar evolved out of MAMBA radar. It is currently deployed with the 5th Regiment Royal Artillery of the British Army and was deployed operationally for the first time in April 2002. MAMBA is capable of automatically detecting, locating and classifying artillery shells, rockets and mortars. It is capable of processing up to 100 rockets, mortars or artillery shells and provides the operator with eight POO grids, thereby offering a capability of locating eight targets simultaneously. All acquired data is automatically transmitted to a combat control centre for necessary action including counter-battery fire operationor improving precision of friendly fire. The radar system also has its own basic command, control and communications (C^3) system for direct control of counter-battery fire. The detection range of MAMBA radar is 20 km for howitzers and 30 km for rockets with a Circular Error Probability (CEP) of around 30 m at the maximum

Figure 3.44 MAMBA radar. (*Source:* Photo: Graeme Main/MOD.)

range. The system is mounted on an all-terrain Alvis Hagglunds BV206 tracked vehicle and is easily transportable by aircraft or helicopters. It is currently deployed in Afghanistan to provide constant 24 hour C-IDF cover. The UK Ministry of Defence (MoD) is evaluating options for the Future Weapon Locating Radar (FWLR) to replace both COBRA and MAMBA radars, the assessment phase of which has already begun.

The FWLR is expected to have the capability to detect, locate and classify simultaneous indirect fire events to a probability of minimum 85% and most desirably 100% within at least a 90° arc with a single system. It should generate POO and POI for these in desirably ≤5 s to provide a CEP of ≤ 50 m to a distance of 30 km minimum and most desirably 100 km. The radar should be capable of operating day and night and in all weather conditions in the specified climatic conditions, with a time into and out of action of less than 2 min and for more than 12 h continuously.

3.5.2.6   SLC-2 Fire Finding Radar

The Chinese SLC-2 Fire Finding Radar's development has been based on the well-established and widely deployed AN/TPQ-37 radar. In fact, SLC-2 radar development was driven by the shortcomings of AN/TPQ-37 experienced by Chinese during its use by them. Type-373 radar was developed first to address the shortcomings of AN/TPQ-37 radar. One of those shortcomings was that it was less effective against projectiles with flat trajectory. Another shortcoming was its lower reliability and tendency to malfunction when deployed in environments with high humidity and high level of rainfall experienced in southern China, high salinity of coastal regions, high altitudes in south-western China and daily high-temperature differences observed in north-western China. Yet another problem was that, when two artillery batteries separated by more than 200 m fired simultaneously, though AN/TPQ-37 provided accurate coordinates for distance, the position coordinates were not determined with less accuracy. The SLC-2, which is an improvement on the Type-373 radar, is a fully solid-state, highly digitized version of Type 373 radar employing planar active phased array antenna technology instead of passive phased array antenna technology used in Type 373 radar. SLC-2 radar is designed to detect and locate hostile artillery shells, unguided rockets and ground-to-ground missile launchers immediately after the round is fired. It also supports friendly artillery by providing location data of the enemy launchers for guiding counter-battery fire and improves precision of friendly fire. The radar can also be used to detect and track low-flying targets such as light aircraft, helicopters and Remotely Piloted Vehicles (RPVs).

Important SLC-2 radar parameters and features include S band (2–4 GHz) operation, a detection range of 35 km for artillery shells and 50 km for rockets, accuracy of 35 m for ranges less than 10 km and 0.35% of the range for ranges greater than 10 km, peak power of 45 kW, active phased array antenna with electronic scanning both in azimuth and elevation, track-while-scan, sophisticated computer-controlled digital signal processing, comprehensive online or offline BITE, automatic/manual height correction with digital/video map and electronic counter-countermeasures (ECCM). SLC-2 is in use by Pakistan and Bangladesh in addition to the Chinese Armed Forces.

3.5.2.7   BL-904 Artillery Locating Radar

The Artillery Locating Radar Type BL-904 was developed by the North Industries Group Corporation (NORINCO) of China as a part of the PLZ-45 155 mm self-propelled howitzer system. The radar is similar to AN/TPQ-36 weapon locating radar, as claimed by the manufacturer, in both appearance and performance. The radar operates in the C band (4–8 GHz) and employs phased array radar technology. It is capable of detecting and tracking up to eight targets (artillery shells, mortar bombs, howitzer projectiles and rockets) and is equipped with

Figure 3.45 BL-904 Artillery Locating Radar.

GPS that enables it transmit target coordinates to the battery command centre in real time. The radar can also be used totrack friendly artillery fire, compute the impact error of the friendly artillery round and provide correction to improve precision of friendly artillery fire. The entire system is configured on two 6 × 6 trucks (Figure 3.45). One of them carries the radar censor including the phased array radar antenna and the other hosts the command cabin. And the system power supply. The command cabin contains the operation and control panel, data processing equipment, displays and so on.

The radar has azimuth sector scan of 180° and elevation scan of –5–+12°. It offers two azimuth scan modes. In the wide scan mode, the detection ranges are 15 km (82 mm mortar), 16 km (122 mm howitzer) and 18 km (155 mm howitzer). In the narrow scan mode, the detection ranges are 20 km (122 mm howitzer), 25 km (155 mm howitzer) and 30 km (273 mm rocket).

3.5.2.8 BEL Weapon Locating Radar

Weapon Locating Radar (WLR) (Figure 3.46) is a joint project undertaken by the Electronics and Radar Development Establishment (LRDE) of the Defence Research and Development Organization (DRDO) of India and Bharat Electronics Limited (BEL). In the primary role, the BEL WLR, like the other weapon locating radars including the ones discussed in the preceding paragraphs, is used to detect and track artillery shells, mortar bombs and unguided rockets to determine the location of hostile guns firing them and transmit the data to the counter-fire installation for retaliatory strike before the hostile weapon is redeployed. In the secondary role, the WLR can track the trajectories of friendly fire and predict the impact point to provide corrections to own fire. The radar data can also be displayed at a convenient remote screen to protect operators from any targeted attacks on the radar.

BEL WLR uses passive phased array radar technology and is similar to AN/TPQ-37 in design and performance. The radar's phased array design and algorithms permits the WLR tooperate effectively even under severe clutter and high density fire environment. The radar has an azimuth scan sector of ±45° and the radar antenna is slewable up to ±135° to change the azimuth scan sector to provide full 360° coverage. The radar is configured on two vehicles including the radar vehicle and power source-come-BITE vehicle. The radar

Figure 3.46 BEL WLR.

vehicle contains the antenna and electronics and power source-come-BITE vehicle contains two diesel generator sets and radar target simulator.

The radar operates in the C band (4–8 GHz). The instrumented range is 50 km. The detection range is 2–20 km (mortar bombs >81 mm), 2–30 km (>105 mm guns) and 4–40 km (unguided rockets). The azimuth scan sector as mentioned earlier is ±45° with a slewability of ±135°. The elevation scan sector is −5–+75°. It can simultaneously track a maximum of seven targets fired at both low and high firing angles and can track and store a maximum of 99 weapon locations at any time. Digital map storage capability is 100 × 100 km. Probability of detection and false alarm rate respectively are 0.9 and 10^{-6}. Multiple radars can be networked and operated in tandem to provide more information with enhanced accuracy. It can be operated in rigorous environmental conditions including operating temperature range of −20–+55°C, storage temperature of −40–+70°C, damp heat of 95% RH at 40°C and operational altitude of 4900 m. The radar is also designed to meet international standards on shock and vibration and Electromagnetic Interference (EMI)/Electromagnetic Compatibility (EMC). The radar is being further improved to have longer ranges and to build compact versions for better operation in mountainous terrains.

3.6   Fire-Control Radar Systems

Fire-control radar, as outlined in an earlier paragraph, is a tracking radar that is used usually in conjunction with an air-defence weapon system. Its job is to determine the trajectory of the intended hostile target or targets, which it does by measuring different coordinates, and then predict its future course of travel. The radar provides continuous position data on a single or multiple targets enabling the associated weapon system such as guns or guided weapons to be directed and locked on to targets. The fire-control radar rapidly searches, detects, locates, classifies, prioritizes and tracks multiple hostile targets and provides positional information on the intended target to the sensors enabling rapid weapon system lock-on to the target. The AN/APG-78 Longbow fire-control radar integrated with radar-guided air-to-surface, fire-and-forget type Longbow Hellfire missile on Apache attack helicopters, RDY-3 from the Thales Group multirole airborne fire-control radar, the INDIRA series of 2D radars and RAJENDRA 3D medium-range land-based fire-control radar used with AKASH surface-to-air missiles developed by the Defence Research and Development Organization (DRDO), India, CEROS-200 fire-control radar from the SAAB Group designed for naval platforms and interfaced with missile or gun systems to provide defence against threats including advanced sea skimming missiles or asymmetric surface threats in littoral environments are only some representative examples of land-based, airborne and ship-based fire-control radars in use by the Armed Forces all over the world.

Fire-control radars need to detect and locate target before they can initiate a track sequence. Locating the target involves a search function. The search and track functions have conflicting antenna beam width and shape requirements. While search function requires the antenna to transmit a wide fan-shaped beam, track function requires a pencil beam. This makes using fire-control radar impractical for locating the targets as a narrow beam can easily miss the targets. In a practical system, the search radar determines the target coordinates such as range, bearing and elevation and then hands over the information to the fire-control radar. Some fire-control radars have built-in search and track radars while others rely on separate search radars.

3.6.1 Representative Fire-Control Radar Systems

Representative fire-control radar systems discussed in this section include the AN/APG-78 Longbow fire-control radar, RDY-3 fire-control radar, RAJENDRA 3D fire-control radar, CEROS-200 fire-control radar, GRIFO-S fire-control radar, ELM-2032 multimode airborne fire-control radar and SAMPSON multifunction radar.

3.6.1.1 AN/APG-78 Longbow Fire-Control Radar

The AN/APG-78 radar forms part of the Longbow fire-and-forget anti-armour system that is fitted to AH-64D Apache battlefield reconnaissance and attack helicopters. The AN/APG-78 Longbow radar's weapon system is the AGM-114L air-to-surface radar-guided fire-and-forget type Hellfire missile. The radar is mounted on top of the helicopter's main rotor mast (Figure 3.47). The Longbow system is manufactured by a Joint Venture of Lockheed–Martin and Northrop–Grumman and is a modification of the Apache helicopter that consists of an upgraded airframe, newly developed radar and the Longbow Hellfire missile. The Hellfire missile has a millimetre wave seeker that allows the missile to perform in full fire-and-forget mode over a range of 0.5–8 km. The missile guidance is configured around a millimetre wave seeker coupled with inertial guidance. The missile is capable of locking on before or after launch and has been extensively tested for homing on to intended targets in adverse weather conditions and battlefield obscurants such as countermeasures. The radar provides high performance with very low probability of intercept. AN/APG-78 Longbow fire-control radar is a Ka band (26.5–40 GHz) radar operating at a centre frequency of 35 GHz. The Longbow radar provides rapid target area search, automatic target detection, classification and prioritization of fixed and moving targets at maximum standoff range and missile fire-and-forget capabilities within a 55 km^2 area. The Longbow fire-control radar also incorporates an integrated radar frequency interferometer (designated as the AN/APR-48A Radio Frequency Interferometer – RFI) for passive location and identification of radar-emitting hostile threats to cue the AN/APG-78 radar, while the Longbow system as a whole identifies and ranks targets in priority order for

Figure 3.47 AN/APG-78 fire control radar. (*Source:* Hunini, https://commons.wikimedia.org/wiki/File:JGSDF_AH-64D%EF%BC%8874506%EF%BC%89_APG-78_Longbow_millimeter-wave_fire-control_radar.JPG. CC BY-SA 3.0.)

attack by RF Hellfire missiles. In addition to its fire-control functions, AN/APG-78 also has terrain profiling and air overwatch capabilities.

Other than the United States, the countries that are using AN/APG-78 Longbow radar or have expressed keen interest to procure the radar include the United Kingdom, the Netherlands, the United Arab Emirates, Kuwait, Israel, Egypt, Saudi Arabia, South Korea and Taiwan. The UK has operated Apaches with Longbow radar in Afghanistan where the radar has helped in improving situational awareness while allowing Apache pilots to act as aerial coordinators as well as conduct command and control operations.

3.6.1.2 RDY-3 Fire-Control Radar

The Thales RDY-3 is a multirole fire-control radar designed to meet all requirements of air-to-air, air-to-ground and air-to-surface missions with long-range targeting, firing and simultaneous multi-shoot capabilities. The radar is capable of operating in day/night and all weatherconditions. Other important features of the radar include advanced ECCM (Electronic Counter-Countermeasures), modular architecture, use of COTS components for obsolescence protection, compatibility with different antenna sizes, advanced signal processing techniques, efficient BITE, very low false alarm rate, ease of maintenance and low life cycle cost. The modular architecture of radar comprises for Line Replaceable Units (LRUs), which include antenna unit, processing unit, transmitter and exciter/receiver. Availability of different antenna sizes allows modernization of light and medium multirole fighters such as different variants of Mirage (Mirage 2000, Mirage F1 and Mirage V), Mig-29 and LCA Tejas. The RDY-3 radar is also being considered for upgrading aircrafts to provide better air-to-air and air-to-ground capabilities. One such exercise relates to upgradation of Mirage-2000s by Indian Air Force who have entered into a contract with Thales, France. Under the contract, BEL (Bharat Electronics Limited), India jointly with Thales, France will upgrade Mirage-2000s to bring them to the full Mirage 2000v5 Mk 2 standard. The upgrade would also include a new night-vision-compatible all-digital cockpit and improved electronic warfare systems.

RDY-3 offers air-to-air, air-to-ground and air-to-sea functional modes. Air-to-air mode provides better search domain management and situational awareness. This reduces pilot work load and allows ranking and sorting of highest priority targets. Other features of air-to-air mode include all aspect look-up/look-down detection, automatic waveform management and antenna scanning, simultaneous multi-target fire-control with multi-target automatic lock-on and track-while-scan and IFF interrogation capability. The air-to-ground mode features air-to-ground ranging and other operational modes such as navigation and surveillance in all-weather conditions day and night conditions. The radar also provides target acquisition and very low-altitude penetration with enhanced terrain avoidance, high-resolution SAR (Synthetic Aperture Radar) imagery and moving target indication (MTI) and tracking. In the air-to-sea mode, the RDY-3 radar can engage multiple targets and TWS (track-while-scan) simultaneously, target's RCS (Radar Cross-Section) assessment and also provide ISAR (Inverse SAR) imagery as an option.

3.6.1.3 RAJENDRA 3D Fire-Control Radar

RAJENDRA, developed by the Defence Research and Development Organization (DRDO) of India is multifunction, 3D passive electronically scanned phased array radar constituting the primary sensor at the flight level for AKASH surface-to-air missile (SAM) weapon system (Figure 3.48). The radar is capable of simultaneously searching, tracking and engaging multiple targets and providing command and guidance to missiles. While acquiring hostile targets, the sequence involves search, confirm, track, interrogate targets and assign and lock-on launchers. While engaging targets, the sequence is launch, acquire, track and guide missiles. The radar has a large number of integrated phase shifters including 4000 phase shifters operating in the 4–8 GHz band in a surveillance antenna array and 1000 phase shifters operating in an 8–20 GHz in engagement antenna array to carry out electronic beam steering. In addition, it

Figure 3.48 RAJENDRA radar.

has a 16-element IFF array. The IFF feature allows the radar to distinguish between friendly and hostile targets. Each AKASH battery has four AKASH launchers with each launcher having three missiles. Each AKASH battery is attached to RAJENDRA radar. The radar can guide two of the missiles in each launcher, which means it can guide a maximum of eight missiles in flight. Each AKASH battery may have battery surveillance radar. Four AKASH batteries constitute a group and each group has one Central Acquisition Radar (CAR) that acts as an early warning sensor. RAJENDRA radar is capable of acquiring aircraft targets either independently or handed over from group control centre via the 3D CAR or from the battery surveillance radar. The radar antenna has azimuth scan sector of ±45°. The antenna array is mounted on a rotating turnstile to provide full 360° coverage. The elevation coverage is 30°. The radar has a tracking range of 60 km for aircraft targets at medium height. The radar is mounted on a T-72 chassis. The antenna array can be folded flat when the vehicle is in motion. The radar is in use by the Indian Armed Forces including the Indian Air Force and Indian Army.

3.6.1.4   CEROS-200 Fire-Control Radar

CEROS-200 (Figure 3.49) is a radar and optronic tracking system specifically designed to provide protection to naval platforms against modern threats including sea skimming missiles and asymmetric surface threats in littoral environments when interfaced to gun or missile systems. The radar is capable of operating in all-weather conditionsand providing specified capabilities over both long distances as well as extremely close to the platform. Other important features of the radar include automatic target detection and lock-on, frequency agility, capability to operate in cluttered and jamming environment, high acquisition speed and great tracking precision combined with the ability to track any target in any weather situation. One of the most significant features of the radar is the CHASE algorithm, a patented method for processing of the complex radar target return signal to eliminate multipath effects while detecting and tracking very low-flying targets such as sea skimming missiles. Use of CHASE algorithm produces better than 0.2 mrad tracking accuracy while operating in calm sea conditions and better than

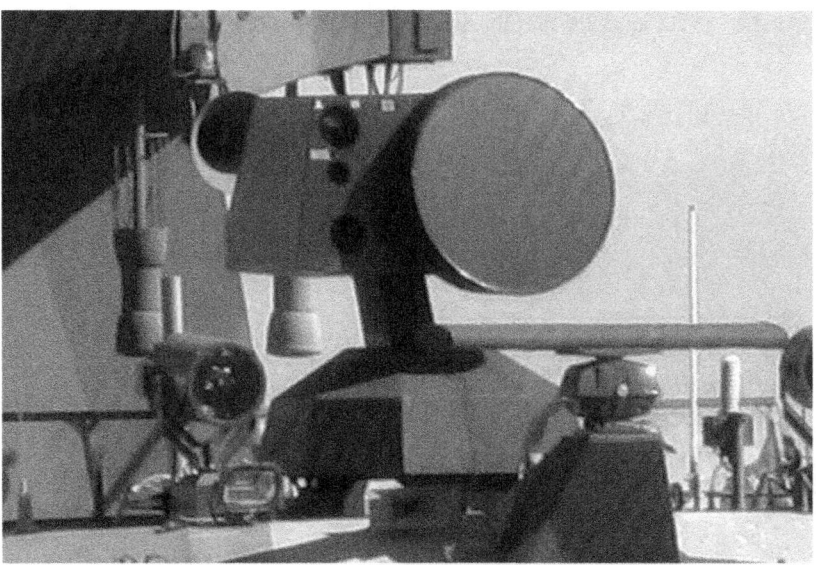

Figure 3.49 CEROS-200 radar. (*Source:* MusicToDieTo, https://commons.wikimedia.org/wiki/File:Saab_Ceros_200.jpg. CC BY-SA 3.0.)

0.4 mrad tracking accuracy while operating in rough sea environment. With these tracking accuracy figures, the radar when combined with SAAB's gun fire-control provides unparalleled performance for gun engagements. The radar is highly resistant to jamming environments. It does so by selecting a suitable frequency agility pattern between 32 pulse bursts and pulse Doppler signal processing and four pulse bursts and MTI (moving target indication) processing or pulse-to-pulse agility. The ECCM (Electronic Counter-Countermeasures) capability of the radar comes from its very low antenna side lobes, large number transmit frequencies, random selection of frequency, very large bandwidth, lock-on jam and track on jam.

CEROS-200 radar is interfaced with SAAB's 9LV fire-control system (FCS) or combat management system (CMS) for precise control of naval guns and surface-to-air missiles (SAM). The 9LV family offers complete C^{4}I for all types of naval platforms, including frigates, patrol vessels, large ships and submarines. 9LV offers exceptional operational capabilities to the command team ranging from FCS solutions to complex Combat Management Systems. The radar combined with 9LV provides rapid and reliable defence against any threat in any environment and provides an extremely fast and accurate sensor-to-shooter cycle.

3.6.1.5 GRIFO-S Fire-Control Radar

The GRIFO-S Radar as shown in Figure 3.50 (from Selex ES Limited, a Finmeccanica company), is the fourth generation of X band (8–12 GHz) coherent pulse Doppler multimode fire-control radars incorporating a comprehensive suite of air-to air and air-to-surface and missions and navigation modes, modular architecture comprising configurable number of compact line replaceable units (LRUs), comprehensive ECCM suite, fully coherent air-cooled TWT (Travelling Wave Tube) based transmitter with wide frequency agility, monopulse flat plate slotted array antenna tailored to installation with guard horns and IFF dipoles and dual-channel receiver with advanced digital signal and data processing using adaptive pulse-compression technology. The high level of modularity comprising configurable LRUs makes it easy for integration on various platforms, such as JF-17, F-16, Mirage V, Mirage 2000, MiG 23, MiG 25 and MiG 29. It has the capability to interface with an existing-field-proven set of radar modes such as the sensor fusion with IRST, shows compatibility with state-of-the-art semi-active and active missiles including AIM-9L-M-X and PYTHON-4 and BVR (Beyond Visual Range) missiles

Figure 3.50 GRIFO-S radar.

such as AIM-120 AMRAAM, MICA and Derby, demonstrates high reliability, capability of long-range detection and ranging in all scenarios including look-up and look-down detection and head-on and tail target aspects, high-resolution imaging (using sub-metric SAR) and identification capability, wide scan sector and multi-target tracking. Major operational specifications include operation in the X band, scan sector coverage of ±60° in both azimuth and elevation, scan rate in the range of 100–400°/s, capability to track-while-scan 10 targets, SAR resolution of better than 1 m and look-up detection range of greater than 50 NM (80 km)

3.6.1.6 ELM-2032 Multimode Fire-Control Radar

ELM-2032 is multi-mission airborne radar designed to enhance air-to-air, air-to-surface and air-to-sea capabilities of fighter aircraft. The radar incorporates advanced technologies that greatly enhance surveillance and fire-control capabilities in air-to-air, air-to-ground and air-to-sea missions. Modular hardware architecture, software control and flexible avionics make it easy for the radar to be installed on a range of aircraft such as F-4, F-5, Sea Harrier, MiG-21, Mirage, F-16 and Kfir C-10 and customized as per the user's requirements. The radar has also been selected for upgrade programmes of South Korean F/A-50 (strike version of the T-50 trainer) and Indian Tejas Light Combat Aircraft. In the air-to-air missions, the radar provides long-range target detection and tracking for weapon delivery. In close combat operations, it also allows automatic target acquisition. Air-to-air target detection and tracking range of radar is 80 NM (150 km). In the air-to-ground missions, the radar in addition to providing air-to-ground ranging, it also provides very high-resolution mapping through SAR (Synthetic Aperture Radar) and surface target detection and tracking over RBM (Real Beam Map), DBS (Doppler Beam Sharpening) and SAR maps. Air-to-ground mapping and target detection is possible up to a range of 80 NM (150 km). In air-to-sea missions, the radar provides long-range target detection and tracking along with target classification using range signatures and Inverse Synthetic Aperture Radar (ISAR). The detection and tracking range in this case is 160 NM (300 km).

3.6.1.7 SAMPSON Multifunction Radar

The SAMPSON multifunction radar (Figure 3.51) manufactured by BAE Systems Maritime is an active electronically scanned array (AESA) radar that constitutes the heart of Sea Viper naval air-defence system. The radar is fitted to the Royal Navy's fleet of Type-45 destroyers. The radar operates in the S band (2–4 GHz) and has an operational range of 400 km. The radar is also being considered to provide credible ballistic missile defence amid ever increasing threat of ballistic missile attacks. The radar supports point and area defence against aerial threats in heavily cluttered environment. The radar is compatible with both semi-active and active homing missiles to provide midcourse guidance and supports fully automatic operation to provide rapid response. The system maintainability is very high due to easy fault diagnosis using built-in test facilities, absence of high voltages, high power microwave parts and associated cooling systems, which further enhances the operational availability.

Different operational modes of radar include long and medium-range search mode, surface picture search mode, high-speed horizon search, high-angle search and track mode and multiple target tracking and multiple channel fire-control. In its anti-air warfare capability, the radar can simultaneously detect and track hundreds of aerial targets. It may be mentioned here that a long-range volume search takes a lot of radar's resources, which is likely to affect the radar performance in performing other tasks if the volume search function were to be combined with other functions. It is because of this reason that multifunction radar is often combined with separate volume search radar. This would also allow use of two different operating frequencies, one for the multifunction radar and another for volume search radar. Use of different frequencies can be used to advantage to optimize the performance of both functions.

Figure 3.51 Sampson multi-function radar on a Type-45 destroyer. (*Source:* Hpeterswald, https://commons.wikimedia.org/wiki/File:HMS_Daring_SAMPSON_is_a_multi-function_AESA_radar.jpg.)

3.7 Space-Based Radar Systems

Radar on a space-borne platform is surveillance, reconnaissance and target detection and acquisition system capable of simultaneously supporting a wide variety of missions covering tactical, operational and strategic levels. Some of these include target detection and tracking, wide area surveillance, battlefield management, command and control and attack operations. When it comes to detection and identification of slow moving ground targets and a variety of other imaging, sensing and communications missions, radars mounted on space-based platforms have certain distinctive advantages over those on airborne platforms. These include the following.

1) Space-based radars provide a coverage area that is much larger than what is possible with an airborne radar system. Space-based radars in fact have the potential of providing nearly global coverage on a continuous basis. Space-based radar, because of its height can cover a very large area on Earth. By adjusting the speed andorbit parameters of the space platform, it is possible to scan different parts of the earth periodically to collect data. Space-based radar can be remotely controlled to provide surveillance with little human intervention. As a result, targets of interest can be identified and tracked in greaterdetail. Also, high-resolution images can be obtained.
2) Space-based radars are capable of providing information about geographic areas inaccessible to aircraft and allow reconnaissance of a particular geographic area for a longer period of time.

3) Radar can operate in all-weather conditions. Radars can see through clouds, can operate at night, and are not dependent on energy emitted or reflected by the target. This is, however, applicable to radars based on any other platform.

4) Formation flight of multiple satellites carrying radars offers single pass synthetic aperture radar imaging with practically unlimited aperture size, digital terrain elevation data collection, precision geolocation and high-data rate secure communications. The United States Air Force Research Laboratory's (AFRL) planned flight experiment, dubbed the TechSat-21 and scheduled for launch in 2006, to demonstrate a formation of three lightweight, high performance microsatellites operating together as a *virtual satellite* with a single, large radar antenna aperture is one such example. The flight experiment was intended to demonstrate the ability of multiple small satellites flying in formation to perform missions traditionally carried out by single, larger satellites and capability to rapidly change formation based on mission requirements. The technological issues involved therein turned out to be far more challenging than originally thought. The project was cancelled in 2003 reportedly due to numerous cost overruns.

The space-based radar systems in fact would move most of the reconnaissance and intelligence gathering functions performed by airborne platforms such Airborne Warning and Control System (AWACS) discussed earlier in the chapter, Joint STARS (Surveillance, Targeting and Attack Radar System) and Rivet Joint surveillance aircraft into space. Rivet Joint is the United States Air Force (USAF) surveillance aircraft employing RC-135 V/W aircraft equipped with an extensive array of sophisticated intelligence gathering equipment (Figure 3.52) enabling military specialists to monitor the electronic activity of adversaries. The aircraft was widely used in Operation Desert Storm. The Joint STARS is a joint development

Figure 3.52 RC-135 Rivet Joint Surveillance Aircraft. (*Source:* Courtesy of US Airforce.)

project of the United States Air Force (USAF) and Army. The aircraft configured around Boeing 707 aircraft (Figure 3.53) is equipped with radar and electronics systems to provide an airborne, standoff range, surveillance and target acquisition radar and command and control centre. Figure 3.54 shows an artist's view of inside of the aircraft. The aircraft has been successfully used during Operation Desert Storm in 1991, Operation Iraqi Freedom in 2003 and

Figure 3.53 Joint STARS aircraft. (*Source:* Courtesy of US Airforce.)

Figure 3.54 Artist's concept of crew working inside Joint STARS aircraft. (*Source:* Courtesy of US Airforce.)

also during peacekeeping missions in Bosnia-Herzegovina and during the Kosovo crisis. E-10 MC2A multirole aircraft configured around Boeing 767–400ER aircraft from Northrop–Grumman is yet another airborne platform equipped with radar and intelligence gathering systems. The aircraft is proposed to replace Boeing 707 based E-3 Sentry, E-8 Joint STARS and RC-135 Rivet Joint. The aircraft development has followed a spiral model with Multi-Platform Radar Technology Insertion Programme (MP-RTIP) to provide joint cruise missile defence capability in spiral-1, AWACS capabilities in spiral-2 and SIGINT/ELINT capability in spiral-3. The objective of moving these functions conventionally performed by these aircraft would be to provide nearly continuous worldwide coverage and to reduce the deployment time and the logistical costs associated with the airborne systems.

In terms of technologies involved and the deployment aspects, space-based radars are substantially different from radars used on airborne platforms. Because of size, weight and power restrictions imposed by the satellite launch vehicles and the satellite platform itself, extreme thermal, radiation and vibration conditions the payload is subjected during launch and exposure to harsh space environment make the space-based radar development and production far more challenging than other types of radar systems including those on airborne platforms. The distances involved in the case of space-based radars are much longer than encountered in the case of airborne radars. Also, there are challenges associated with deployment of antennas large enough to detect moving targets. A virtual antenna array comprising of multiple satellites sharing information is one possible solution but that is also not without any technical hurdles. In addition, there are challenges associated with range fold over phenomenon and the Doppler shift induced by motion of space platform and the motion of Earth around its own axis. Sufficient advances have made to find solutions to these problems. Detailed discussion on these is beyond the scope of this text.

3.7.1 Representative Space-Based Radar Systems

A number of remote sensing satellite missions have carried radar systems as the main payload to be used for various applications ranging from monitoring of Earth's resources to disaster management. Some of these include RADARSAT of Canada, RISAT of India, RORSAT of Russia, SEASAT, TerraSAR-X and TanDEM-X of Germany, EL/M-2070 TecSAR of Israel, SAR-LUPE and SARah of Germany are some of the prominent candidates in the category of radar imaging and reconnaissance satellites.

RADARSAT is Canada's first advanced Earth observation satellite project owned and operated by Canadian Space Agency (CSA) (Figure 3.55). RADARSAT satellites focus on the use of radar sensors to provide unique information about the Earth's surface through most weather conditions and even darkness, which can be used in monitoring the environment and managing the Earth's natural resources. Currently the RADARSAT series comprise of RADARSAT-1 and -2 satellites with RADARSAT constellation comprising of three satellites planned for launch in 2018. Both RADARSAT-1 and -2 use Synthetic Aperture Radar (SAR) operating in the C band as the payload. The radar constellation will open up new application areas while continuing to provide C band radar data to RADARSAT-2 users. RISAT is a series of radar imaging reconnaissance satellites built by Indian Space Research Organization (ISRO). In RISAT series, RISAT-1 and RISAT-2 were launched, respectively, in 2012 and 2009. RISAT-1 was the second RISAT satellite to be launched after RISAT-2. RISAT-1 carries on board a C band (5.35 GHz) Synthetic Aperture Radar (SAR) that provides Earth observation day and night and in all-weather conditions. RISAT-2 carries an X band SAR operating at 9.59 GHz and manufactured by Israel Aerospace Industries. The payload data are downlinked at a rate of 620 Mbit/s. The radar is designed to monitor Indian borders as a part of anti-infiltration and anti-terrorist operations. Applications also include tracking hostile ships posing threat to national security.

Figure 3.55 RISAT-2 radar imaging satellite.

The RORSAT (Radar Ocean Reconnaissance Satellite) series of Soviet reconnaissance satellites launched between 1967 and 1988 is another satellite using active radar as the main payload. The satellite was used to monitor NATO and merchant vessels. SEASAT was yet another remote sensing satellite that carried on board radar altimeter to measure the height of the spacecraft above the ocean surface and synthetic aperture radar (SAR) to monitor global surface wave field and polar sea ice conditions. TerraSAR-X and TanDEM-X radar imaging satellites from Germany use synthetic aperture radar operating in the X band to provide high-resolution images with resolution up to 1 m. TerraSAR-X, in addition to its civilian applications such as topographic mapping, environmental applications, rapid emergency response, land cover and land use mapping, is also used in defence and security-related applications including border control through detection of paths (changes), fences and moving objects. TanDEM-X satellite is similar to TerraSAR-X satellite. In fact, it is the name of the satellite mission flying the two such satellites in a formation flight with typical inter-satellite distance of 250–500 m (Figure 3.56). The main objective of this twin satellite constellation is the generation of global digital elevation models trade-marked WorldDEM. The Israeli reconnaissance satellite EL/M-2070 TecSAR, also known by the names of Tech SAR Polaris and Ofek-8, is a state-of-the-art radar imaging satellite that employs an X band synthetic aperture radar capable of imaging with a resolution up to 10 cm. TecSAR has greatly enhanced Israel's intelligence gathering capability and therefore is of great strategic importance.

SAR-LUPE is Germany's reconnaissance satellite operated by the German Armed Forces (Bundeswehr) and used for military applications. The satellite uses X band synthetic aperture radar capable of providing high-resolution images day and night in all-weather conditions with resolutions up to 50 cm. SAR-LUPE programme is in fact a constellation of five satellites placed in three 500 km orbits in planes that are approximately 60° apart. SAR-LUPE satellites will be replaced in future by SARah satellites with enhanced capabilities during 2017–2019. The SARah-series will comprise three radar satellites and one optical satellite (Figure 3.57). The SARah system's radar satellites comprise one active phased array-antenna satellite built by EADS Astrium and two passive reflector antenna satellites provided by OHB-System. Two passive-antenna SARah satellites will fly in formation with the active-antenna spacecraft. The SARah system is planned to provide a higher resolution than the predecessor SAR-LUPE.

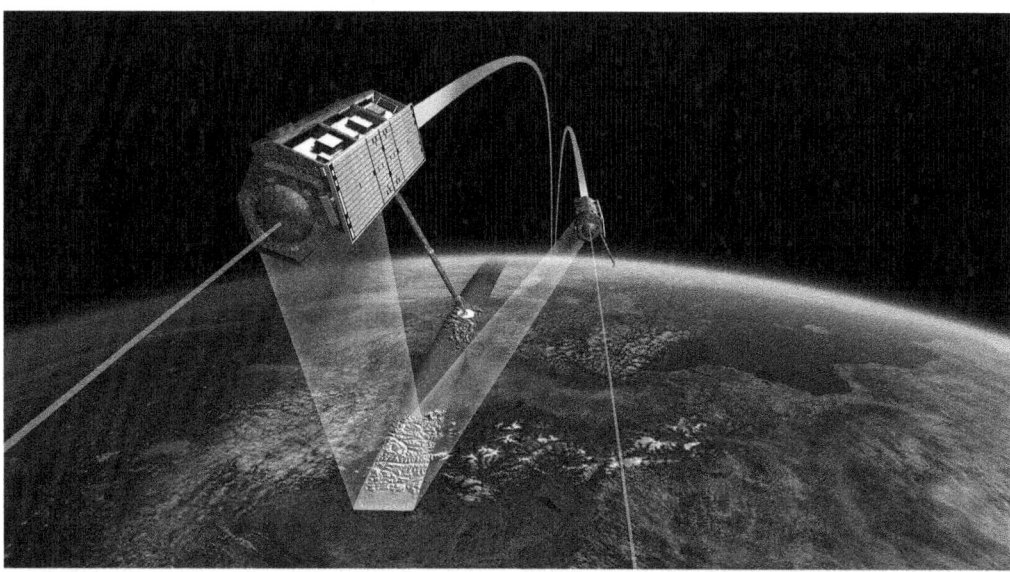

Figure 3.56 TanDEM-X in formation flight. (*Source:* DLR, https://commons.wikimedia.org/wiki/File:TDX-Formation.jpg. CC BY 3.0.)

Figure 3.57 Artist's rendition of the SARah reconnaissance satellite.

The *Space-Based Radar* (SBR) programme proposed by the United States Department of Defense is aimed at providing on demand near-continuous surveillance and reconnaissance for battle space characterization for global and theatre users including military strategists and combatant commanders. SBR system is proposed as a constellation of active radar satellites that would allow detection and tracking from space of ground targets such as land vehicles, sea-surface targets such as ocean going vessels and airborne targets such as aircraft. The target data would then be relayed to regional and national command centres and also to airborne command posts such as E-10 MC2A.The SBR system would enable the military have a penetrative insight into denied areas of interest on a non-intrusive basis without risking resources or rendering own personnel vulnerable not currently possible with existing capabilities.

The SBR system would combine Surface Moving Target identification (SMTI), Synthetic Aperture Radar (SAR) and Digital Terrain and Elevation Data (DTED) to detect, track and target stationary or moving combatants almost anywhere on Earth in near real time. The SMTI capability, for example, would allow tracking of truck size vehicle of about 10 m^2 radar cross-section at 2800 km. The Surface Moving Target Indication (SMTI) and Synthetic Aperture Radar (SAR) imaging capabilities of the SBR system could majorly contribute to tracking mobile platforms and to locating and monitoring ballistic missile installations of interest.

The SBR system as outlined earlier would be a constellation of satellites, which could be some combination of Low Earth Orbit (LEO) satellites at a nominal altitude of 1000 km and Medium Earth Orbit (MEO) at a nominal altitude of 10 000 km. LEO satellites could support both the SMTI and SAR imaging requirements with reasonable combinations of power and aperture. Fewer MEO satellites would be needed to support SMTI but would require relatively longer processing times to provide the SAR images of desired quality. LEO systems are characterized as 1X, 2X and 3X. For the 1X system, the power-aperture (power in kW x aperture in m^2) is 35 kWm^2. A LEO constellation would require about 21 satellites to provide persistent global access. With less than a 21-satellite constellation, a combination of space-based system and other assets would be required to provide persistent access to areas of interest at times when needed. It may be mentioned here that selecting and tracking specific targets of interest requires a capability to change rapidly between SMTI and SAR imaging capabilities. While SAR imaging is required to identify the objects of interest before it has started moving; SMTI tracks the movement of objects.If the object is intermittently halting and moving, SMTI alone cannot do the job. Both SAR imaging and SMTI would be required to know whether the new movement is that of the same object. The need for the same becomes further more important and critical in the case of missile launchers.

A Formal Request for Proposal (RFP) was reportedly issued in 2003 for the concept development effort of the SBR programme. The Raytheon Company was awarded a contract in 2003 to define, analyse, design and demonstrate an SBR pre-prototype payload consisting of an electronic scanned array and an onboard processing component. In 2004, Lockheed–Martin and Northrop–Grumman were taken on board as the concept development prime contractors and awarded study contracts. The SBR programme was renamed Space Radar (SR) after programme was restructured in 2005. As of 2005, SR was intended to provide synthetic aperture radar mapping and surface moving target indication capabilities in all-weather conditions. The initial plans are for a constellation of nine satellites.

3.8 Police Radar

Police radars are speed detectors where the microwave signal transmitted by the radar bouncing off the moving vehicle is Doppler shifted in frequency depending upon the relative speed between the radar and the vehicle. Figure 3.58 illustrates the operational principle. The beat

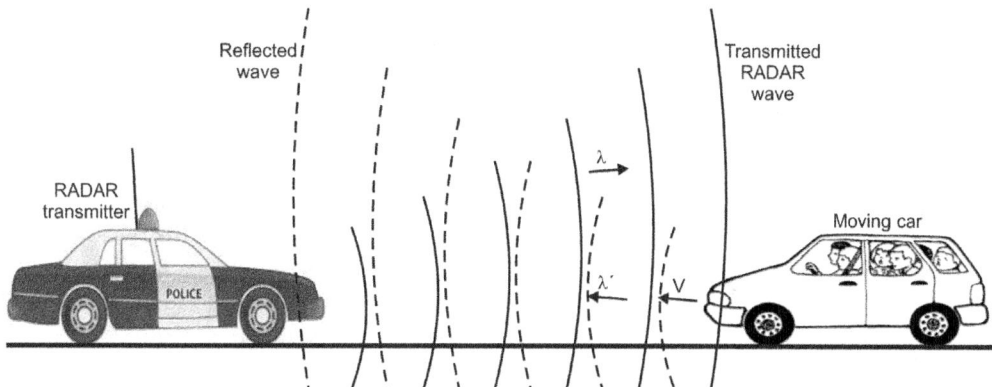

Figure 3.58 Operational principle of police radar.

frequency between the transmitted frequency and the received frequency is proportional to the relative speed and therefore can be used to extract information on the speed of the moving vehicle. If the radar is stationary, which it is in most cases, then the beat frequency (Δf) is given by eqn. 3.6. The beat frequency is positive when the vehicle is moving towards the radar.

$$\Delta f = \left(f' - f \right) = \left(2v/c \right) \times f \tag{3.6}$$

f = Frequency of transmitted radar signal
f' = Frequency of reflected radar signal
v = speed of moving vehicle
c = Speed of electromagnetic waves in free space

It is immaterial whether the radar is stationary or moving. The beat frequency always is representative of relative speed between the radar and the vehicle. Also, it makes no difference with most speed radars whether the car is moving towards or away from the radar. One only needs to know the beat frequency and the speed of the police cruiser carrying radar in order to calculate the target's speed. However, various errors can affect the accuracy of the reading. The most significant source of error is the cosine error. The cosine error occurs when the radar beam strikes the moving vehicle at an angle with the direction of the moving vehicle. As a result of this, the radar measures a speed that is lower than the actual speed by a factor equal to $\cos\theta$ where θ is the angle. The larger the angle, greater will be the error. This error may, however, be compensated for by inputting the angle. Another source of error is frequency drift due to temperature and voltage fluctuations, which can also be minimized through compensating circuitry.

Police radars are operated in two modes namely *Continuously-ON (CO) Mode* and the *Instant-ON Mode*. In the CO mode of operation, the radar is continuously transmitting RF signal without any interruption. The disadvantage of this mode of operation is that such radar can be detected long distance by using suitable radar signal detectors. This allows mischievous elements to avoid detection by police radars while exceeding the speed limit. This shortcoming, however, can be overcome to great extent by operating the radar from a covert position such as hiding themselves among heavy foliage and pointing the radar beam across the road at an angle and not directly at the approaching vehicles. Such trap configurations are not easily detectable even by best of radar signal detectors. The Instant-ON radar gun is usually triggered by the operator on seeing an approaching vehicle thereby denying the vehicle driver any opportunity to detect the radar signal and evade detection. Newer technologies such as use of narrower, polarized and ultra-low powered beams are being incorporated to make sure that they are harder to detect. These radars are also equipped with cameras for visual confirmation by the operator. A wider output beam makes detection of radar transmission easier for radar detector and also provides ample reaction time to the driver to slow down while vehicles moving ahead are being targeted.

Police radars operate on one of the following three frequency bands, namely X and operating at the 10.525 GHz, K band operating at 24.150 GHz and Ka band operating in the frequency range of 33.4–34.4 GHz. The older versions operated on the X band. In fact, they were the most predominantly used police radars till the mid-1970s. K band radar guns were first used in 1976 onwards. Most of them have been replaced by either K band or Ka band radars. Police radar guns began using Ka band frequencies from 1987 with the introduction of photo radar operating at 34.3 GHz followed by Stalker operating in 34.2–35.2 GHz in 1991 and the BEE 36A operating in 33.4–34.4 GHz in 1992. The key advantage of K band and more so the Ka band radar guns with respect to X band radar guns is their much smaller size, which gives them numerous mounting options in addition to many other features. Reduction in size is partially due to use of advanced technologies and partially due to use of smaller wavelength leading to reduction in size of components. Some of the features available in the smaller Ka band radar guns include detachable computer and display unit that minimizes equipment in the cockpit area of patrol vehicle, handheld remote control, front and rear antenna connection ports, communication port and software for opposing direction moving operational mode. These radar guns are interfaceable to Mobile Video Recorders or Mobile Data Computers, which further helps in meeting the objective of downsizing the equipment in the cockpit area of patrol vehicle. Also, spectral analysis of signals received from different moving vehicles made possible due to use of digital signal processing allows differentiation of multiple targets.

3.8.1 Representative Police Radar Systems

There are a large number of international companies manufacturing Police Radar systems including both mounted radar systems and handheld radar guns designed for law enforcement applications. These radars offer a range of performance specifications in terms of operational range, mounting options, display features and so on. It is not possible to discuss police radars manufactured by all these companies. Stalker Radar and MPH Industries are the two better known manufacturers of police radars. Kustom Signals Inc. also offers a range of police radar systems. In the following paragraphs are briefly discussed radars from these three companies. These would include SE-1 and SE-2 Stalker PHODAR, Handheld Stalker-I and Stalker-II radars, Stalker-2X, Stalker DSR, Stalker DUAL and Stalker PATROL dashboard-mounted radars (all from Stalker Radar); BEE-III and Ranger EZ directional radars, PYTHON, PYTHON-S and ENFORCER mounted speed radars and Speed gun PRO and Z-45 handheld radars (all from MPH Industries) and TALON and FALCON-HR handheld radars, RAPTOR-RP1 and Eagle-II dashboard-mounted radars (all from Kustom Signals Inc.)

3.8.1.1 Stalker Radar Systems

Stalker PHODAR is a sophisticated 3D tracking radar system with an integrated high-resolution camera and designed for automated photo/video traffic violation enforcement for collection and ticketing. While Stalker PHODAR SE-1 employs a 5 megapixel camera for both high definition video and still images; Stalker PHODAR SE-2 employs a dual camera system with 24 megapixel camera used for still pictures and 2 megapixel camera used for high definition video. Both systems can monitor up to 32 vehicles travelling in both directions on a multi-lane roadway. The camera system generates high-quality photos and video including vehicle indication,

Figure 3.59 Stalker-II radar gun.

number plate, driver facial characteristics and road plan. Also, both systems can be operated either using external touch screen monitor or through a TCP/IP remote connection. Stalker PHODAR offers flexible mounting options. There are three possible ways of traffic monitoring, including monitoring from the side of roadway in which case it is usually tripod or utility pole mounted, from the centre of the roadway with gantry or bridge mounting and from a parked vehicle on the side of roadway.

Stalker-I SDR is handheld stationary direction sensing radar configured around the Ka band Doppler radar operating at 34.7 GHz. Operational specifications and salient features include backlit LCD with three-speed windows, including target speed, lock/fast peed and expansion window, backlit LCD for night time operation, detachable/rechargeable sealed battery handle with a 7.2 V, 2000 mAh lithium-ion battery, use of advanced DSP technology, faster speed tracking, speed range of 8–321.9 km/h (Standard) and 24–321.9 km/h (menu selectable option), speed measurement accuracy of +2/−3 km/h and auto-self-test that is performed every 10 minutes while transmitting. Stalker-II uses advanced DSP technology to offer the performance of vehicle-mounted radar in a handheld package. It is available in two models, namely *Stalker-II SDR* (Stalker-II Stationary Direction Sensing Radar) as shown in Figure 3.59) and Stalker-II MDR (Stalker-II Moving Direction Sensing Radar). Common specifications and features of Stalker-II SDR and Stalker-II MDR include faster speed tracking, track-thru lock, three-window display featuring LCD display windows for Strongest Target, Faster Target and Patrol Speed (in moving mode) with direction arrows that indicate the direction of travel for both the strongest and faster targets, backlit LCD for night time operation and remote control option. Both versions use Ka band Doppler radar operating at 34.7 GHz, perform auto-self-test every 10 minutes while transmitting and 7.2 V/2000 mAh Li-ion battery. Stalker-II MDR stationary speed range is 8–322 km/h (Standard) and 24–321.8 km/h (option menu selectable). Moving speed range is defined in terms of patrol speed that is selectable by a P.S. 5/20 key, opposite lane target speed and same lane target speed. Select 5 in the patrol window for acquisition of

8–145 km/h and 20 for acquisition of 32–145 km/h. Patrol speed once locked tracks up to 241 km/h. Opposite lane speed is 322 km/h maximum closing, 32–314 km/h for 8 km/h patrol speed and 56–209 km/h for 112 km/h patrol speed. Same lane target speed is ±70% of patrol speed within 8 km/h of patrol speed. Same lane patrol speed must be greater than 25.7 km/h. Speed accuracy is +2/−3 km/h stationary and ±3 km/h moving. In the case of Stalker-II SDR, speed range is 8–322 km/h (Standard) and 240–322 km/h (option menu selectable). Speed accuracy is +2/−3 km/h.

The Stalker 2X is a dashboard-mounted police radar. It comprises two independent Ka band radar units operating on a single five-window display that allows traffic on four stationary target zones or two moving target zones to be monitored simultaneously. The audio Doppler tone while operating in opposite direction represents target's actual speed and not the closure speed. Stalker DSR is also dashboard-mounted police radar. The radar is capable of displaying speeds of both faster and the strongest targets simultaneously. The operator can display or lock both strong and fast targets while in stationary mode or same lane mode or opposite lane mode. The audio Doppler tone, as in the case of Stalker-2X while operating in the opposite lane, directly correlates to the target speed and not the closure speed irrespective of patrol speed. Stalker DUAL is a dual antenna moving/stationary Doppler radar operating in the Ka band. It is the industry standard in moving police radar offering longest range in moving police radar that includes automatic same direction operation. With its direction sensing capability, the radar determines whether the targets in the same direction are closing in or going away from the radar allowing the radar to determine same direction and opposing direction speeds with the same efficacy. Another salient feature of radar is that it displays the strongest and the fastest targets simultaneously on separate display windows. Also, it is easier to correlate the Doppler audio with target speed as the radar compensates for patrol speed variations when generating Doppler audio. Stalker PATROL is also dual antenna moving/stationary K band Doppler radar operating at 24.150 GHz. It offers quality and reliability of Stalker radars at an affordable price.

3.8.1.2 Police Radars from MPH Industries

Three categories of police radars from MPH Industries include Directional Radars such as BEE-III and Ranger EZ, Mounted Speed Radars including PYTHON, PYTHON-FS and ENFORCER and the Handheld Radar Guns, namely Speedgun-Pro and Z-45. BEE-III is one of the most compact directional moving radars. It is multi-piece radar with detachable display unit and wireless remote control. The radar operates in Ka band. Salient features of the radar include patented *Automatic Same Direction* (ASD) technology that enables radar to decide with 100% accuracy whether the target is approaching or receding and also allows radar in stationary mode to select a lane and measure traffic in that lane while completely ignoring traffic in the other lane: the POP mode that allows radar to measure speed of potential speed violators without setting off single radar detector and city and highway modes that minimize shadowing and combining. Major operational specifications include the following. The target range is about 1.6 km for average size vehicle. In stationary mode, target speed range is 24–320 km/h. In same direction moving mode, patrol speed range is 19–128 km/h in city mode and 32–128 km/h for highway mode and target speed range is ±70% of patrol speed and not within 4.8 km/h of patrol speed. In the opposite direction moving mode, patrol speed range is 19–128 km/h in city mode and 32–128 km/h for highway mode and target speed range is 24–320 km/h closing speed. The operational frequency is 33.8 GHz±100 MHz. *Ranger EZ* (Figure 3.60) like the BEE-III is also a directional radar with detachable display unit and wireless remote control. In addition to having all features of BEE-III, it also measures the target's range that allows determination of which vehicle is being measured by the radar, thereby giving

Figure 3.60 Ranger-EZ radar. (*Source:* Courtesy of mPH Industries.)

positive target identifications and preventing mistakes. Ranger EZ radar is also equipped safety-alert mechanism that allows the officers time to move out of danger in case of inattentive vehicles failing to slow down in response to your lights. The safety-alert mechanism called SafetyZone traffic alert makes traffic stops far safer.

PYTHON is mounted moving radar system that can operate from both stationary as well as moving patrol vehicles. It is available in X, K and Ka band versions. Its advanced digital signal processing technology allows simultaneous monitoring of all vehicles in radar beam. The basic PYTHON radar is equipped with stationary mode and opposite direction moving mode software. *PYTHON-FS* also has the fastest vehicle and same direction moving modes added to it. Both PYTHON and PYTHON-FS have a *city mode* and *highway mode* that help in minimizing patrol speed errors at low speed in the case of city mode and high speed of the highway mode. Also both radars have a speedometer interface that takes the approximate information on patrol speed and feeds it to radar to help it determine correct patrol speed. Target range is about 1.6 km in case of vehicles of average size and varies with vehicle size, terrain, weather and traffic conditions. *ENFORCER*, like the BEE-III radar, is also a compact multi-piece radar with detachable display unit. It operates in the Ka band and is also equipped with POP technology that measures traffic speeds without alerting radar detectors. The radar has same direction and fastest vehicle modes, selectable city and highway modes.

Speedgun-Pro is handheld police radar that can also be mounted on a patrol vehicle. The handheld version can be powered either from the vehicle battery or from internal rechargeable batteries. The radar measures targets in only one selected direction or in both directions in which case the radar tells the operator the direction the vehicle is moving in. Availability of POP technology allows measurement of target speeds without activating radar detectors. The fastest vehicle mode allows tracking fast moving vehicles before they slow down on noticing the patrol vehicle. *Z-45* is handheld stationary traffic radar operating in the K band. The operating frequency is 24.150±0.1 GHz. Two operating modes allow track-through-lock and simultaneous tracking of fastest and strongest targets. Target speed range is 11–240 km/h with an accuracy of ±2 km/h. Target range is about 1 km for an average size vehicle.

3.8.1.3 Police Radars from Kustom Signals Inc.

Two categories of police radars from Kustom Signals Inc. include handheld radars (TALON and FALCON-HR) and dashboard-mounted radars (RAPTOR-RP1 and Eagle-II). The *TALON series* of police radars are high performance Ka band Doppler radars that can be operated in stationary and moving modes (Figure 3.61). The radar operates at 35.5±0.1 GHz. Both hand-held and dashboard-mounted variants are available. Important features include use of digital signal processing technology that allows tracking multiple targets, VSS interface available with dashboard-mounted option allows automatic mode switching between stationary and moving modes eliminating patrol shadowing and combining errors, track-through-lock allows tracking history after target lock, backlit display for nigh time operation and smart patrol search that rejects inaccurate patrol speeds. Speed measurement accuracy is ±1 km/h in stationary mode and +2/−3 km/h in moving mode. *FALCON-HR* is K band directional radar operating at 24.150±0.1 GHz. It has all features of TALON series radar such as VSS interface, track-through-lock, digital signal processing for tracking multiple targets and smart patrol search and so on. Target speed range in stationary mode is 8–321 km/h with accuracy of ±1 km/h. In moving mode, it is 8–193 km/h with accuracy of +2/−3 km/h. The speed range in case of opposite direction target is 32–321 km/h and in case of same direction target, it is 5 km/h to 65% of patrol speed.

RAPTOR RP-1 is mounted radar designed for in-car and motorcycle operations and has options for single or dual K or Ka band antennas. In addition to innovative features such as smart patrol search, track-through-lock and VSS interface also available with TALON and FALCON-HR radars, it also has DuraTrak target tracking bars supports tracking history increasing target identification confidence. Speed measurement accuracy is ±1 km/h in stationary mode and +1/−2 km/h in moving mode. *EAGLE-II* family radars are dashboard-mounted radars with options for single or dual Ka band antennas. Three variants of the radar are available. These include Eagle-II, Golden Eagle-II and Directional Golden Eagle-II. All variants have VSS Interface and Smart Patrol Search features. Golden Eagle-II and Directional Golden Eagle-II also have separable display, same direction and fastest vehicle modes. Directional Golden Eagle radar has approaching and receding directional modes (Figure 3.62).

Figure 3.61 TALON series radar.

Figure 3.62 Eagle-II family radar.

Illustrated Glossary

Airport Surveillance Radar Airport surveillance radar is the key constituent of air traffic control and management function, which includes collision avoidance between different aircraft, prevention of obstructions on the ground and maintenance of orderly flow of traffic. The radar is used to detect aircraft and send detailed positional information to the air traffic control system.

Air Surveillance Radar Air surveillance radars are used to carry out surveillance of the air space surrounding a critical asset and are designed to detect, locate, track and classify a wide range of both friendly and hostile targets. Typical targets include both conventional targets such as different categories of aircraft and unconventional targets including unmanned aerial vehicles (UAV), drones, ultra-lights and ballistic missiles.

AN/APG-78 Fire-Control Radar AN/APG-78 Longbow fire-control radar is a Ka band (26.5–40 GHz) radar operating at a centre frequency of 35 GHz. The Longbow radar provides rapid target area search, automatic target detection, classification and prioritization of fixed and moving targetsat maximum standoff range and missile fire-and-forget capabilities within a 55 km^2 area. The AN/APG-78 radar forms part of the Longbow fire-and-forget anti-armour system that is fitted to AH-64D Apache battlefield reconnaissance and attack helicopters.

AN/PPS-5 Radar This is pulse Doppler radar that is designed to detect, locate, identify and track moving personnel and vehicles day or night under virtually all-weather conditions.

AN/TPQ-37 Fire Finder Radar AN/TPQ-37 by Thales Raytheon systems is a mobile electronically steered S band (2–4 GHz) long-range weapon locating radar capable of detecting and tracking artillery, mortar and rocket fire from single or multiple positioned hostile weapon platforms to determine the point-of-origin of attack thereby enabling friendly forces to direct an effective counter-fire. It is also capable of detecting and tracking short-range tactical missiles. In the secondary role, it also determines the impact point of the rounds from friendly launchers to provide fire corrections to counter-battery fire. It can also predict the impact point of hostile fire.

AN/TPQ-50 Counter-Fire Radar AN/TPQ-50, manufactured by SRC, Inc., is an L band (1–2 GHz) lightweight counter-fire radar capable of detecting and tracking rockets, mortars and artillery shells up to 10 km with an accuracy of 50 m as specified by circular error probable (CEP). The radar is mountable on towers, tripods or light vehicles such as a High Mobility Multipurpose Wheeled Vehicle (HMMWV).

AN/TPS-117 Radar AN-TPS-117 also belongs to the Lockheed–Martin's family of ground-based medium and long-range air surveillance radars, which also includes AN-TPS-59 and AN/TPS-77. It is a mobile, miniaturized variant of AN/FPS-117 long-range air surveillance radar. The radar operates in the D/L band (1215–1400 MHz).

AN/TPS-59 Radar AN/TPS-59 developed by Lockheed–Martin is solid-state three-dimensional long-range air surveillance radar operating in D/L band (1215–1400 MHz) and capable of detecting and tracking air breathing targets within 555 km (300 nautical miles) and tactical ballistic missiles up to a maximum range of 740 km (400 nautical miles).

AN/TPS-77 Radar AN/TPS-77 is the latest radar in the TPS-59, TPS-117 and TPS-77 family of three-dimensional all solid-state air surveillance radars. The mobile AN/TPS-77 radar shares about 80–90% Line Replaceable Units (LRUs) and maintenance activity with the AN/FPS-117 fixed position radar system. The radar is configurable on a single C-130 aircraft or two medium trucks. The radar operates in D/L band (1215–1400 MHz) and employs mechanical steering for azimuth scan and phased array concept for elevation scan.

AN/TPS-78 Radar AN/TPS-78 is the latest in the AN/TPS-70 family of air surveillance radars from Northrop–Grumman. The radar employs advanced solid-state radar technologies with operationally proven designs such as those used earlier in AN/FPS-130 long-range air-defence radar and the ASR-12 air traffic control radar. The radar is available in two variants namely the AN/TPS-78 long-range radar and TPS-703 mid-range radar.

ARTHUR Weapon Locating Radar ARTHUR, an acronym for Artillery Hunting Radar, is a mobile weapon locating radar system originally developed by *Ericsson Microwave Systems* (now *SAAB Microwave Systems*) in collaboration with the Swedish and the Norwegian armed Forces. The radar operates in the C band (4–8 GHz) and is partly based on the same technology as the GIRAFFE and employs a passive electronically scanned phased array antenna for both azimuth and elevation scanning that enables location of the launch site with high accuracy immediately after detection of a projectile.

ASR-11 Radar ASR-11 is Airport Surveillance Radar from Raytheon Corporation. It is a joint Federal Aviation Administration (FAA)/Department of Defense (DOD) procurement programme with the United States Air Force (USAF) assuming overall lead responsibility. The ASR-11 digital solid-state surveillance radar provides to the airport terminal area primary surveillance coverage up to 116 km (60 nautical miles) and secondary surveillance up to 232 km (120 nautical miles). The ASR-11 uses advanced signal processing to provide improved target and weather processing.

Automatic Dependent Surveillance (ADS) In the case of ADS, the aircraft broadcasts to the ground stations its position and other information that it determines with the help of the Global Navigation Satellite System (GNSS) and other resources on board the aircraft. The information received by the ground stations is processed and then sent to the Air Traffic Control (ATC) centre

Battlefield Surveillance Radar – Short-Range (BFSR-SR) BFSR-SR, developed by DRDO (Defence Research and Development Organization), India and manufactured by Bharat Electronics Limited, is man-portable battery-powered short-range track-while-scan 2D surveillance and acquisition radar capable of searching a specified sector for multiple ground surface targets to provide day/night all-weather surveillance.

BEE-III Radar BEE-III is compact directional moving police radar from MPH Industries. It is multi-piece radar with detachable display unit and wireless remote control. The radar operates in the Ka band.

BEL Weapon Locating Radar (WLR) BEL weapon locating radar in the primary role is used to detect and track artillery shells, mortar bombs and unguided rockets to determine the location of hostile guns firing them and transmit the data to the counter-fire installation for retaliatory strike before the hostile weapon is redeployed. In the secondary role, the WLR can track the trajectories of friendly fire and predict the impact point to provide corrections to own fire. It is a joint project undertaken by Electronics and Radar Development Establishment (LRDE) of Defence Research and Development Organization (DRDO) of India and Bharat Electronics Limited (BEL).

BL-904 Artillery Locating Radar This is capable of detecting and tracking up to eight targets (artillery shells, mortar bombs, howitzer projectiles and rockets) and is equipped with GPS that enables it transmit target coordinates to the battery command centre in real time. The radar can also be used to track friendly artillery fire, compute the impact error of the friendly artillery round and provide correction to improve precision of friendly artillery fire. It was developed by North Industries Group Corporation (NORINCO) of China as a part of PLZ-45 155 mm self-propelled howitzer system. The radar is similar to AN/TPQ-36 weapon locating radar, as claimed by the manufacturer, in both appearance and performance. The radar operates in the C band (4–8 GHz) and employs phased array radar technology.

Blighter B200 Series Radars This series comprises ground surveillance radars from Blighter Surveillance Systems, capable of detecting potential threats including crawling and walking personnel and moving vehicles and suited to fixed, mobile and portable applications.

Blighter B400 Series Radars The Blighter B400 series are also ground surveillance radars from Blighter Surveillance Systems, capable of detecting potential threats including crawling and walking personnel and moving vehicles and suited to fixed, mobile and portable applications.

BUR Ground Surveillance Radar BUR is the battlefield surveillance radar with German Army designation Bodenuberwachungsradar (BUR) meaning Ground Overwatch Radar. It was developed by Cassidian, the defence and security division of EADS. The radar is capable of detecting and tracking movements of potential targets on ground, over water and in the air space close to the ground with speed, precision and reliability. It tracks personnel and vehicular movements on ground up to a radius of 40 km.

Central Acquisition Radar (3D-CAR) The 3D Central Acquisition Radar (3D CAR) developed by the Defence Research and Development Organization (DRDO), India is multifunction medium-range surveillance radar capable of simultaneously tracking multiple targets such as fighter jets and missiles travelling at supersonic speeds of up to 3000 m/s (approximately 3 Mach). The radar has an operating range of up to 170 km and an altitude of up to 15 km.

CEROS-200 Radar This is a radar and optronic tracking system specifically designed to provide protection to naval platforms against modern threats including sea skimming missiles and asymmetric surface threats in littoral environments when interfaced to gun or missile systems. The radar is capable of operating in all-weather conditions and providing specified capabilities over both long distances as well as extremely close to the platform.

COBRA Counter-Battery Radar This is counter-battery radar capable of detection, location and classification of hostile rockets launchers, mortars, and artillery shells and prediction of point-of-impact (POI) of shells. The radar comprises of a high performance radar sensor, advanced processing and an integrated command, control and communication system. COBRA is the result of a collaborative programme between Germany, France and Turkey to develop long-range battlefield radar.

E-3 Sentry E-3 Sentry, by Northrop–Grumman, is an Airborne Warning and Control System (AWACS) aircraft comprising of a very sophisticated and powerful surveillance radar system and command, control and communications equipment to provide all-weather surveillance to the tactical air forces covering more than 500 000 km^2 of air space around the aircraft.

EAGLE-II Series Radars The EAGLE-II family of police radars from Kustom Signals Inc. are dashboard-mounted radars with options for single or dual Ka band antennas. Three variants of the radar are available. These include Eagle-II, Golden Eagle-II and Directional Golden Eagle-II. All variants have VSS Interface and Smart Patrol Search features. Golden Eagle-II and Directional Golden Eagle-II also have separable display, same direction and fastest vehicle modes. Directional Golden Eagle radar has approaching and receding directional modes.

ELM-2032 Multifunction Radar This is multi-mission airborne radar designed to enhance air-to-air, air-to-surface and air-to-sea capabilities of fighter aircrafts. The radar incorporates advanced technologies that greatly enhance surveillance and fire-control capabilities in air-to-air, air-to-ground and air-to-sea missions.

EL/M-2070 TecSAR This, also known by the names of Tech SAR Polaris and Ofek-8, is a radar imaging satellite that employs an X band synthetic aperture radar capable of imaging with a resolution up to 10 cm.

EL/M-2107 Radar This is miniature high-resolution movement detection digital radar from Israel Aerospace Industries designed for security applications designed for automatic detection and classification of moving objects inside a guarded area. A cluster of these radars can be installed and connected to the control station to cover the area to be protected and generate an audio-visual alarm whenever an intrusion is identified inside the guarded area.

EL/M-2127 Radar This is a short-range miniature ground surveillance radar from Israel Aerospace Industries designed for detection of movement of persons and vehicles up to 2 km. It is used for protection of sensitive sites and ambushes.

EL/M-2129 Radar This is a medium-range movement detection radar from Israel Aerospace Industries. It is an X band coherent pulse-compression ground-based radar that detects and monitors moving targets such as walking people, moving vehicles and flying objects inside the areas of interest at ranges of up to 30 km.

EL/M-2140NG Radar This is long-range movement detection radar from Israel Aerospace Industries.

EL/M-2140 NG, like EL/M-2129, is also a ground surveillance radar for military and para-military applications.

EL/M-2226 Radar This is an advanced coastal surveillance radar from Israel Aerospace Industries optimized for the detection and tracking of smart surface targets even under extremely adverse sea conditions providing a reliable tactical situation display.

Fire-Control Radar This is tracking radar that is used usually in conjunction with an air-Defence weapon system and is used to determine the trajectory of the intended hostile target or targets, which it does by measuring different coordinates and then predicts its future course of travel.

ENFORCER ENFORCER, like the BEE-III radar, is also compact multi-piece radar with detachable display unit. It operates in the Ka band. It is also equipped with POP technology that measures traffic speeds without alerting radar detectors. The radar has same direction and fastest vehicle modes, selectable city and highway modes.

FALCON-HR Radar FALCON-HR from Kustom Signals Inc. is a K band directional police radar operating at 24.150±0.1 GHz. It has all features of TALON series radar such as VSS interface, track-through-lock, digital signal processing for tracking multiple targets and smart patrol search and so on.

GIRAFFE Surveillance Radar This is the family designation of a series of the C band pulse Doppler surveillance radars and combat control centres from Ericsson for mobile and static short- or medium-range command, control, communications and intelligence (C^3I) air-defence systems.

GR-05 Radar GR-05 tactical ground surveillance radar, developed by Belgian Advanced Technology Systems (BATS), is an extension of the GR-27 radar family. It is capable of detecting and tracking intruders' movements in protected and selected zone of interest.

GR-20 Radar GR-20 radar, developed by BATS, is coherent pulse-compression ground-based radar ground surveillance radar operating in the X band and designed to be used by military, paramilitary and security agencies for surveillance and protection applications.

GR-40 radar GR-40 ground surveillance radar, developed by BATS, is coherent pulse-compression radar operating in the X band and capable of detecting moving targets including walking personnel, vehicles and aerial targets such as hovering helicopters, gliders and low-flying aircraft up to a range of 60 km.

GRIFO-S Radar The GRIFO-S Radar from Selex ES Limited, a Finmeccanica company, is a fourth generation of X band (8–12 GHz) coherent pulse Doppler multimode fire-control radar incorporating comprehensive suite of air-to air and air-to-surface and missions and navigation modes.

Ground Observer-12 (GO-12) This is a lightweight, compact ground surveillance radar configured around a pulse Doppler radar operating in the Ku band. The radar is suitable for a wide range of applications including tactical battlefield surveillance, border patrol, and coast and site surveillance both by military/paramilitary forces or civilian users.

Ground Penetrating Radar (GPR): GPR is a non-invasive and non-destructive method of imaging the subsurface. It is widely used for host of civilian and military applications, which include engineering and construction industry, archaeological investigation and forensic investigation, geological and environmental science, utility detection and concrete inspection, detection of buried mines and unexploded ordnances and so on.

GROUNDSHARK GPR The GROUNDSHARK by NIITEK is a high performance handheld dual sensor system capable of detecting, locating and visualizing buried high metallic, low metallic, non-ferrous, nonmagnetic and non-metallic objects in challenging soil conditions such as changing and uneven terrain.

Ground Surveillance Radar (GSR) This type of radar performs the function of detection, classification and recognition of targets on ground level or the air space close to the ground level. Some of the well-established application areas include their use in urban warfare manoeuvres, covert surveillance, counterterrorism, maritime surveillance, border patrol and security, monitoring activity surrounding critical and strategic assets and tactical battlefield applications.

Husky Mounted Detection System (HMDS) The HMDS is configured around VISOR ground penetrating radar. It is capable of automatic detection, recognition and precision marking of buried threats including metallic and non-metallic mines, improvised explosive devices (IEDs) and other explosive hazards.

LMX-100 GPR The LMX-100 Locate & Mark System from M/S Sensors & Software is a ground penetrating radar particularly suitable for locating and marking underground both metallic and non-metallic objects including metallic pipes and cables, non-metallic objects such as plastic pipes, conduits, concrete ducts and dielectric fibreoptic cables. LMX-100 is designed to operate in a wide range of environmental conditions.

LMX-200 GPR The LMX-200 GPR system, like the LMX-100, is also designed to locate and mark unground metallic and difficult-to-detect non-metallic utilities. The radar can locate and mark both shallow and deep unground targets including non-metallic PVC and asbestos cement pipes, concrete storm and sewer systems, underground storage tanks and drainage tiles, septic system components and non-utility structures such as vaults, foundation walls and concrete pads. LMX-200 has advanced GPS and underground mapping capabilities.

LSTAR Radar The LSTAR family of surveillance and target acquisition radars have 3D electronic scanning capabilities and are designed to detect and track difficult airborne targets such as unmanned aircraft systems (unmanned aerial vehicles, remotely piloted vehicles and drones), fixed and rotary-wing aircraft, ultra-lights, paragliders and hand gliders. LSTAR family has two variants, namely LSTAR (V)2 and LSTAR (V)3.

MAGS MAGS is a wide area multilateration (WAM) system designed and built by Thales Air Systems with the primary objective of providing high precision and high update rate secondary surveillance to Air Traffic Controllers.

MAMBA Artillery Hunting Radar MAMBA is an acronym for Mobile Artillery Monitoring Battlefield Asset. This is artillery hunting radar similar to the ARTHUR radar. It is capable of processing up to 100 rockets, mortars or artillery shells and provides the operator with eight Point-Of-Origin (POO) grids thereby offering a capability of locating eight targets simultaneously.

MINESHARK GPR MINESHARK by NIITEK is a handheld dual sensor system capable of detection and subsurface visualization of buried anti-personnel (AP) and anti-tank (AT) landmines, improvised explosive devices (IED) and other buried explosive hazards.

MINESTALKER Series GPR This series of GPR systems from NIITEK is designed to be used for detecting and marking buried landmines and other explosive hazards.

Multifunction Radar This is primarily designed to perform as a single unit the functions of target surveillance and tracking and is configured around active electronically scanned array (AESA) radar. Actual functions performed by multifunction radars used on different platforms are application specific. Shipborne or land-based multifunction radar generally performs the functions of surface search, short, medium, and long-range search, multiple target tracking, weapon location and control of remotely piloted aerial vehicles. Multifunction radars on airborne platforms, in addition to surveillance and tracking based functions, may also take over the functions of navigation or terrain mapping.

Multilateration System This system makes use of several beacons that receive signals transmitted by the aircraft transponder. It is used for ground, terminal approach and en-route surveillance. The system not only detects the aircraft position but also identifies the aircraft and can receive additional information.

NS-100 Radar The NS-100 from the Thales Group is a 3D surveillance S band (E/F-band) AESA (Active Electronically Scanned Array) radar, Due to its highly modular and scalable architecture, the radar is suitable for a wide range of naval ships. The radar has a minimum range of 150 m and maximum instrumented range of 200 km.

Ofek-8 See EL/M-2070 TecSAR.

Police Radar Police radars are speed detectors where the microwave signal transmitted by the radar bouncing off the moving vehicle is Doppler shifted in frequency depending upon the relative speed between the radar and the vehicle.

PYTHON Radar PYTHON is mounted moving police radar system that can operate from both stationary as well as moving patrol vehicles. It is available in X, K and Ka band versions.

PYTHON-FS Radar PYTHON-FS is police radar from the PYTHON series by MPH Industries. It also has the fastest vehicle and same direction moving modes added to it.

RADARSAT RADARSAT is Canada's advanced Earth observation satellite project owned and operated by Canadian Space Agency (CSA) thatuses radar sensors to provide information about the Earth's surface through most weather conditions and even darkness. This feature can be used in monitoring the environment and managing the Earth's natural resources.

RAJENDRA Radar RAJENDRA, developed by Defence Research and Development Organization (DRDO) of India, is a multifunction, 3D passive electronically scanned phased array radar constituting the primary sensor at the flight level for AKASH surface-to-air missile (SAM) weapon system. The radar is capable of simultaneously searching, tracking and engaging multiple targets and providing command and guidance to missiles.

Ranger EZ Radar A Ranger EZ such as the BEE-III is a directional radar with detachable display unit and wireless remote control. In addition to having all the features of BEE-III, it also measures a target's range, which allows determination of which vehicle is being measured by the radar thereby giving positive target identifications and preventing mistakes. Ranger EZ radar is also equipped safety-alert mechanism that allows the officers time to move out of danger in case of inattentive vehicles failing to slow down in response to your lights.

RAPTOR RP-1 Radar RAPTOR RP-1 from Kustom Signals Inc. is mounted radar designed for in-car and motorcycle operations and has options for single or dual K or Ka band antennas. In addition to features such as smart patrol search, track-through-lock and VSS interface also available with TALON and FALCON-HR radars, it also has DuraTrak target tracking bars supports tracking history increasing target identification confidence.

RDY-3 Fire-Control Radar The RDY-3 fire-control radar from Thales is a multirole fire-control radar designed to meet all requirements of air-to-air, air-to-ground and air-to-surface missions with long-range targeting, firing and simultaneous multi-shoot capabilities. RDY-3 offers air-to-air, air-to-ground and air-to-sea functional modes.

REVATHI Radar REVATHI is a medium-range 3D surveillance radar to be fitted in ASW (Anti-Submarine Warfare) Corvette class of ships to detect air and sea-surface targets up to a range of 200 km.

RISAT RISAT is a series of radar imaging reconnaissance satellites built by Indian Space Research Organisation (ISRO).

ROHINI Radar ROHINI is medium-range 3D surveillance radar operating in the S band (E/F band). The radar is capable of Track-While-Scan of airborne targets up to a range of 150 km subject to line-of-sight clearance and radar horizon and an altitude of 15 km. The radar can detect low-altitude targets as well as supersonic aircraft and missiles flying at velocities up to 3 Mach.

RORSAT RORSAT (Radar Ocean Reconnaissance Satellite) is a series of Soviet reconnaissance satellites launched between 1967 and 1988. They use active radar as the main payload. The satellite was used to monitor NATO and merchant vessels.

R-VISOR GPR R-VISOR or Robot Visor is a dual sensor comprising of high performance ground penetrating radar and a sophisticated metal detector. The ground penetrating radar and the metal detector work in tandem to offer the user ability to precisely mark and visualize buried improvised explosive devices (IEDs), anti-personnel and anti-tank landmines.

SAMPSON Multifunction Radar The SAMPSON multifunction radar manufactured by BAE Systems Maritime is active electronically scanned array (AESA) radar that constitutes the heart of Sea Viper naval air-defence system. The radar is fitted to the Royal Navy's fleet of Type 45 destroyers. The radar operates in the S band (2–4 GHz) and has an operational range of 400 km.

SAR-LUPE SAR-LUPE is Germany's reconnaissance satellite operated by the German Armed Forces (Bundeswehr) and used for military applications. The satellite uses X band synthetic aperture radar capable of providing high-resolution images day and night in all-weather conditions with resolutions up to 50 cm.

Scanter-1002 Radar The Scanter-1002 radar from Terma is a short-to-medium range ground surveillance radar designed for protection of critical infrastructure.

Scanter-2100 Series Radars The Scanter-2100 series radars from Terma are medium-range 2D fully coherent pulse-compression radars designed for surface surveillance.

Scanter-2200 Series Radars The Scanter-2200 series radars from Terma are also medium-range 2D fully coherent pulse-compression radars designed for surface surveillance.

SEASAT This is a remote sensing satellite that carries on board a radar altimeter to measure the height of spacecraft above the ocean surface and synthetic aperture radar (SAR) to monitor global surface wave fields and polar sea ice conditions.

SIR-20 GPR This is a versatile dual-channel GPR control unit particularly suited to applications such as road structure assessment, bridge deck inspection, rail bed inspection, concrete inspection, archaeology, geological investigation and mining.

SIR-30 GPR The SIR-30 Radar Control Unit is an advanced multichannel radar control unit. Like SIR-20, it is compatible with most GSSI antennas and supports multiple mounting configurations. Typical applications include road structure assessment, utility designation, bridge deck inspection, rail bed inspection, airport runway inspection, detection of cavities and detection of clean/fouled ballasts.

SIR-3000 GPR The SIR-3000 GPR Control Unit is a portable and lightweight system designed for single-user operation. Its advanced signal processing and display capabilities allow in-the-field 3D imaging on a high-resolution colour display that is visible over a wide range of light conditions. It is compatible with most GSSI antennas.

SIR-4000 GPR The SIR-4000 is a single channel GPR control unit designed to operate with analogue antennas as well as new generation digital antennas.

SLC-2 Radar This Chinese radar is designed to detect and locate hostile artillery shells, unguided rockets and ground-to-ground missile launchers immediately after the round is fired. It also supports friendly artillery by providing location data of the enemy launchers for guiding counter-battery fire and improves precision of friendly fire. The radar can also be used to detect and track low-flying targets such as light aircraft, helicopters and Remotely Piloted Vehicles (RPVs). SLC-2 Fire Finding Radar's development is based on the well-established and widely deployed AN/TPQ-37 radar.

SMART-S Mk2 Radar SMART Mk2 is a modern medium to long-range S band (E/F band) naval air and surface surveillance radar from the Thales group. With its multi-beam concept, it is capable of detecting difficult targets including stealth targets even in cluttered environments.

SNAR-10M1 Radar SNAR-10M1 radar (Russian name: 1RL232–2M and NATO code name: Big Fred) is a modernized version of SNAR-10 battlefield surveillance radar that was unveiled by NPO strela during Defence Exhibition Oboronexpo 2014 in Russia. It is a mobile battlefield surveillance radar built to detect and locate moving ground and sea-surface targets.

Space-Based Radar (SBR) SBR or radar on a space-borne platform is surveillance, reconnaissance and target detection and acquisition system capable of simultaneously supporting a wide variety of missions covering tactical, operational and strategic levels. Some of these include target detection and tracking, wide area surveillance, battlefield management, command and control and attack operations.

Space-Based Radar (SBR) Programme The SBR programme proposed by the United States Department of Defense is aimed at providing on demand near-continuous surveillance and reconnaissance for battle space characterization for global and theatre users including military strategists and combatant commanders. The SBR system is proposed as a constellation of active radar satellites that would allow detection and tracking from space of ground targets such as land vehicles, sea-surface targets, such as ocean going vessels and airborne targets, such as aircraft.

Speedgun-Pro Radar This is a handheld police radar from MPH Industries. It can also be mounted on a patrol vehicle.

SpotterRF Radars These are ultra-lightweight and compact radars from SpotterRF and particularly suited for detection and tracking of potential threats before they are on the fence line in a perimeter surveillance application.

Stalker-I SDR This is a handheld stationary direction sensing police radar configured around Ka band Doppler radar operating at 34.7 GHz.

Stalker-II Radar A police radar that uses advanced DSP technology to offer the performance of vehicle-mounted radar in a handheld package. It is available in two models, namely *Stalker-II SDR* (Stalker-II Stationary Direction Sensing Radar) and Stalker-II MDR (Stalker-II Moving Direction Sensing Radar).

Stalker 2X Radar This is a dashboard-mounted police radar. It comprises of two independent Ka band radar units operating on a single five-window display that allows traffic on four stationary target zones or two moving target zones to be monitored simultaneously.

Stalker DUAL This is a dual antenna moving/stationary Doppler radar operating in the Ka band. It offers long-range in moving police radar that includes automatic same direction operation.

Stalker PATROL This is a dual antenna moving/stationary K band Doppler radar operating at 24.150 GHz.

Stalker PHODAR A 3D tracking radar system with an integrated high-resolution camera and designed for automated photo/video traffic violation enforcement for collection and ticketing.

StructureScan GPR This is configured around the SIR-3000 radar control unit. The GPR provides non-destructive means of accurately inspecting concrete structures including location of embedment inside concrete structures prior to drilling, cutting or coring thereby allowing concrete and construction professionals identify conduits and avoid dangerous and costly hits. Different variants of StructureScan family of GPRs include StructureScan Standard, StructureScanOptical, StructureScan Mini, StructureScan Mini-HR and StructureScan Mini-XT. Typical uses include concrete and structure inspection, void detection and location, measurement of slab thickness and so on.

Tactical Control Radar (TCR) TCR is a state-of-the-art 3D medium-range surveillance and tracking radar developed for the Indian Army for detection and identification of airborne targets.

TALON Series Police Radars This series of police radars from Kustom Signals Inc. are high performance Ka band Doppler radars that can be operated in stationary and moving modes. The radar operates at 35.5±0.1 GHz. Both handheld and dashboard-mounted variants are available.

TanDEM-X A radar imaging satellite similar to TerraSAR-X satellite.

TecSAR Polaris See EL/M-2070 TecSAR.

TerraSAR-X A radar imaging satellite from Germany that uses synthetic aperture radar operating in the X band to provide high-resolution images with resolution up to 1 m. TerraSAR-X in addition to its civilian applications such as topographic mapping, environmental applications, rapid emergency response, land cover and land use mapping, is also used in defence and security-related applications including border control through detection of paths (changes), fences and moving objects.

TPS-703 Radar A mid-range variant of AN/TPS-78 air-defence radar.

UtilityScan GPR These GPRs are configurable and have the flexibility to be used for a wide range of utility detection and location applications. Typical uses include detection and location of metallic and non-metallic utilities, archaeological surveys, forensic investigations, road inspection, bridge deck assessment, geological investigation, damage prevention and environmental remediation. The UtilityScan series of ground penetrating radar systems from GSSI offers three variants including UtilityScan, UtilityScan-DF and UtilityScan-LT.

Weapon Locating Radar Weapon Locating Radar, also known as Gun Locating Radar and Counter-Battery Radar, is used for detecting the location or point of origin of hostile fire including artillery shells, mortars, unguided rockets and so on. facilitating the friendly forces to launch a counter-fire with pin point accuracy and thereby forcing the adversary to deploy massively disproportionate fire power. Weapon locating radar is always attached to an artillery battery or their support group.

Z-45 Traffic Radar The Z-45 radar from MPH Industries is handheld stationary traffic radar operating in K band. Two operating modes allow track-through-lock and simultaneous tracking of fastest and strongest targets.

Bibliography

1 Richards, M.A., Scheer, J.A. and Holm, W.A. (2010), *Principles of Modern Radar-Basic Principles*, SciTech Publishing.

2 Charvat, G.L. (2014), *Small and Short-Range Radar Systems*, CRC Press.

3 Griffiths, H.D. (2014), *Stimson's Introduction to Airborne Radar*, SciTech Publishing.

4 Knott, E.F., Shaeffer, J.F. and Tuley, M.T. (2004), *Radar, Cross-Section*, SciTech Publishing.

5 Barton, D.K. (1972: Volume 1, 1974: Volume 2, 1975: Volume 3, 1975: Volume 4, 1975: Volume 5, 1977: Volume 6, 1978: Volume 7), *Radars, Volumes* 1 to 7, Artech House.

6 Jeffrey T.W. (2008), *Phased Array Radar Design*, SciTech Publishing.

7 Barton D.K. (2012), *Radar Equations for Modern Radar*, Artech House.

8 Toomay J.C, (2004), *Radar Principles for Non-Specialist*, SciTech Publishing.

9 General Books LLC (2010), *Weapon Locating Radar*, General Books LLC.

10 Persico, R. (2014), *Introduction to Ground Penetrating Radar: Inverse Scattering and Data Processing*, Wiley-IEEE Press.

11 Conyers, L.B. (1997), *Ground Penetrating Radar: A Primer for Archaeologist*, AltaMira Press, USA.

12 Jol, H.M. (2009), *Ground Penetrating Radar: Theory and Applications*, Elsevier Science.

13 Daniels, D.J. (2004), *Ground Penetrating, Volume* 1, IET, 2004-Technology and Engineering.

14 Macfadzean, R.H.M. (2000), *Surface-Based Air-Defence System Analysis*, Artech House.

15 Stevens, M. (1988), *Secondary Surveillance Radar*, Artech House.

16 Blackman, S. (1999), *Design and Analysis of Modern Tracking Systems*, Artech House.

17 Langford, L. (1998), *Understanding Police Traffic Radar and Lidar*, Law Enforcement Services.

18 Sawicki, D. (2015), *Police Radar Basics*, CreateSpace Independent Publishing Platform.

19 Cantafio, L.J. (1989), *Space-Based Radar Handbook*, Artech House.

20 Pillai, S. and Yong Li, K. (2008), *Space-Based Radar: Theory and Applications*, McGraw-Hill Education.

21 Himed, B., Yong Li, K. and Pillai, S.U. (2006), *Remote Sensing Using Space-Based Radar*, Springer.

4

Satellite Technology

Since the launch of Sputnik-1 in 1957, over 9000 satellites have been launched for a variety of applications. The horizon of satellite applications has extended far beyond providing interconti-nental communication services and satellite television. Other than communication and television broadcast related applications, some of the most significant and talked about applications of satellites are in the fields of remote sensing and Earth observation. Atmospheric monitoring and space exploration are the other major frontiers where satellite usage has been exploited a great deal. And then there are host of military applications, which include secure communications, navigation, surveillance and reconnaissance making satellites a key component of military strategy and national defence. Satellite technology and their military applications are discussed in this chapter and the one following this. The present chapter comprehensively describes satellite technology fundamentals. Beginning with a brief outline on basic principles of orbiting satellites, the chapter goes on to discuss satellite launch and in-orbit operations. This is followed by a description of different subsystems of satellite such as satellite propulsion, thermal control, power supply, attitude and orbit control, telemetry, tracking and command, payload and antenna sub-systems. Multiple access techniques including frequency division multiple access, time division multiple access, code division multiple access and space division multiple access techniques are discussed next. This is further followed up by detailed account of satellite link design and networking concepts. Satellite applications for different military roles are covered in Chapter 5 on *Military Satellites*.

4.1 Basic Principles of Orbiting Satellites

The motion of natural and artificial satellites around Earth is governed by two forces. One of them is the centripetal force directed towards the centre of the Earth due to gravitational force of attraction of the Earth and the other is the centrifugal force that acts outwards from the centre of the Earth (Figure 4.1). It may be mentioned here that centrifugal force is the force exerted during circular motion, by the moving object upon the other object around which it is moving. In case of a satellite orbiting Earth, the satellite exerts centrifugal force. However, the force causing the circular motion is centripetal force. In the absence of this centripetal force, the satellite would have continued to move in a straight line at a constant speed after injection. The centripetal force directed at right angles to the satellite's velocity towards the centre of Earth transforms the straight line motion in circular or elliptical one, depending upon the satel-lite velocity. Centripetal force further leads to a corresponding acceleration called centripetal acceleration as it causing a change in the direction of the satellite's velocity vector. The centrifugal

Handbook of Defence Electronics and Optronics: Fundamentals, Technologies and Systems, First Edition. Anil K. Maini.
© 2018 John Wiley & Sons Ltd. Published 2018 by John Wiley & Sons Ltd.

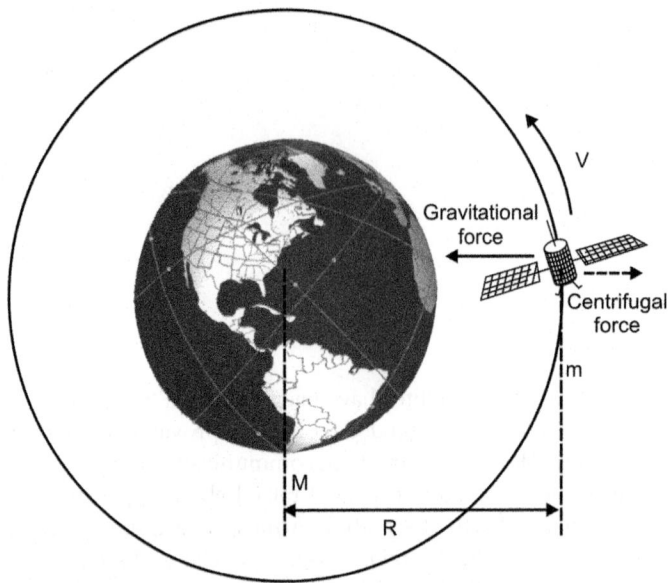

Figure 4.1 Gravitational and centrifugal forces acting on the bodies orbiting Earth.

force is simply the reaction force exerted by the satellite in a direction opposite to that of the centripetal force. This is in accordance with Newton's third law of motion, which states that for every action there is an equal and opposite reaction. This implies that there is a centrifugal acceleration acting outwards from the centre of the Earth due to the centripetal acceleration acting towards the planet's core. The only radial force acting on the satellite orbiting Earth is the centripetal force. The centrifugal force is not acting on the satellite, it is only a reaction force exerted by the satellite. The two forces can be explained from Newton's law of gravitation and Newton's second law of motion respectively as outlined in the following paragraphs.

4.1.1 Newton's Laws

Newton's laws relevant to satellite motion include Newton's law of gravitation and Newton's second law of motion.

4.1.1.1 Newton's Law of Gravitation

According to Newton's law of gravitation, every particle irrespective of its mass attracts every other particle with a gravitational force whose magnitude is directly proportional to the product of the masses of the two particles and inversely proportional to the square of the distance between them. It is given by eqn. 4.1.

$$F = \frac{Gm_1 m_2}{r^2} \tag{4.1}$$

m_1, m_2 = Masses of the two particles
r = Distance between the two particles
G = Gravitational constant = 6.67×10^{-11} m^3/kg.s^2.

The force with which a particle of mass m_1 attracts the particle with mass m_2 equals the force with which a particle of mass m_2 attracts the particle with mass m_1. The forces are equal in magnitude but opposite in direction (Figure 4.2). The acceleration, which force per unit mass,

Figure 4.2 Newton's law of gravitation.

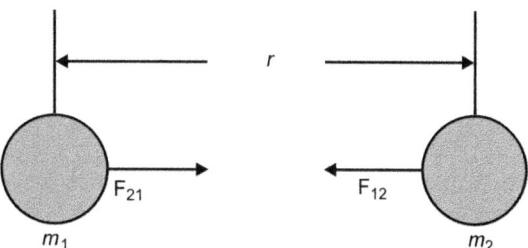

experienced by the two particles, however, depends on their masses. A larger mass experiences lesser acceleration. Newton also explained that, although the law is strictly applied to particles, it is applicable to real objects as long as the size of the particles is small as compared to distance between them. He also explained that a uniform spherical shell of matter would behave as if the entire mass of it were concentrated at its centre.

4.1.1.2 Newton's Second Law of Motion

According to Newton's second law of motion, force equals product of mass and acceleration. In case of a satellite orbiting Earth, if the orbiting velocity is v, then the acceleration, called centripetal acceleration, experienced by the satellite at a distance of r from the centre of the Earth would be v^2/r. And if the mass of satellite is m, it would experience a reaction force of mv^2/r. This is the centrifugal force directed outwards from the centre of the Earth and for a satellite it is equal in magnitude to the gravitational force.

If the satellite orbited the Earth with a uniform velocity v, which would be the case when the satellite orbit is circular, then the equating the two forces mentioned previously would lead to an expression for the orbital velocity v as given in eqn. 4.2.

$$\frac{Gm_1m_2}{r^2} = \frac{m_2v^2}{r} \tag{4.2}$$

$$v = \sqrt{\frac{Gm_1}{r}} = \sqrt{\frac{\mu}{r}} \tag{4.3}$$

m_1 = Mass of the Earth
m_2 = Mass of the satellite
μ = G.m_1 = 3.986013 × 10^5 km^3/s^2 = 3.986013 × 1014 N-m^2/kg.
The orbital period in such a case can be computed from eqn. 4.4.

$$T = \frac{2\pi r^{3/2}}{\sqrt{\mu}} \tag{4.4}$$

In case of an elliptical orbit, the forces governing the motion of the satellite are also the same. The velocity at any point on an elliptical orbit at a distance d from the centre of Earth is given by the formula expressed by eqn. 4.5.

$$v = \sqrt{\mu\left(\frac{2}{d} - \frac{1}{a}\right)} \tag{4.5}$$

a = semi-major axis of the elliptical orbit.

The orbital period in case of an elliptical orbit is given by eqn. 4.6.

$$T = \frac{2\pi a^{3/2}}{\sqrt{\mu}}$$ (4.6)

The movement of a satellite in an orbit is governed by three Kepler's laws.

4.1.2 Kepler's Laws

Johannes Kepler gave a set of three empirical expressions that explained planetary motion. These laws were later vindicated when Newton gave the law of gravitation. Though given for planetary motion, these laws are equally valid for the motion of natural and artificial satellites around Earth or for any body revolving around other body. Here, we shall discuss these laws with reference to the motion of artificial satellites around Earth.

4.1.2.1 Kepler's First Law

The orbit of a satellite around Earth is elliptical with the centre of Earth lying at one of the foci of the ellipse (Figure 4.3). The elliptical orbit is characterized by its semi-major axis a and eccentricity e. Eccentricity is the ratio of distance between the centre of the ellipse and either of its foci ($= a \times e$) to the semi-major axis of the ellipse, a. A circular orbit is a special case of an elliptical orbit where the foci merge together to give a single central point and the eccentricity becomes zero. Other important parameters of an elliptical satellite orbit include its apogee (farthest point of orbit from Earth's centre) and perigee (Nearest point on the orbit from Earth's centre) distances. These are described in subsequent paragraphs.

For any elliptical motion, law of conservation of energy is valid at all points on the orbit. Law of conservation of energy states that energy is neither created nor destroyed, it can only be transformed from one form to another. In context of satellites, it means that the sum of the kinetic and the potential energy of the satellite always remain constant. The value of this constant is equal to $[(-Gm_1m_2)/2a]$.

$m_1 =$ Mass of the Earth
$m_2 =$ Mass of the satellite
$a =$ Semi-major axis of the orbit

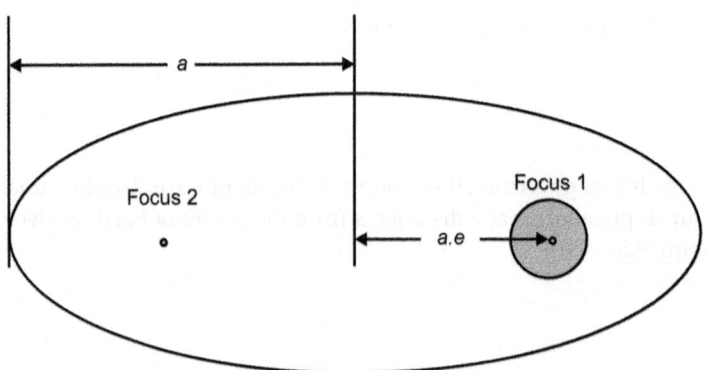

Figure 4.3 Kepler's first law.

Kinetic and potential energy of a satellite at any point at a distance r from the centre of Earth are given by eqns 4.7 and 4.8.

$$\text{Kinetic Energy} = \frac{1}{2}(m_2 v^2)$$ (4.7)

$$\text{Potential Energy} = \left(-\frac{Gm_1 m_2}{r}\right)$$ (4.8)

$$\frac{1}{2}(m_2 v^2) - \left(\frac{Gm_1 m_2}{r}\right) = -\left(\frac{Gm_1 m_2}{2a}\right)$$

$$v^2 = Gm_1\left(\frac{2}{r} - \frac{1}{a}\right)$$

$$v = \sqrt{\mu\left(\frac{2}{r} - \frac{1}{a}\right)}$$ (4.9)

4.1.2.2 Kepler's Second Law

The line joining the satellite and the centre of Earth sweeps out equal areas in the plane of the orbit in equal times (Figure 4.4). That is, the rate dA/dt at which it sweeps area A is constant. Rate of change of the swept out area is given by eqn. 4.10.

$$\frac{dA}{dt} = \frac{\text{Angular Momentum of the satellite}}{2m}$$ (4.10)

m is the mass of the satellite

Hence, Kepler's second law is also equivalent to the law of conservation of momentum, which implies that the angular momentum of the orbiting satellite given by the product of radius vector and component of linear momentum perpendicular to the radius vector is constant at all points on the orbit.

Angular Momentum of the satellite of mass m is given by $m.r^2.\omega$, where, ω is the angular velocity of the radius vector. This further implies that the product $(mr^2\omega = m.\omega r.r) = (m.v'.r)$ remains constant. v' is the component of satellite's velocity v in the direction perpendicular to the radius vector and is expressed as $v.\cos\gamma$, where γ is the angle between the direction of

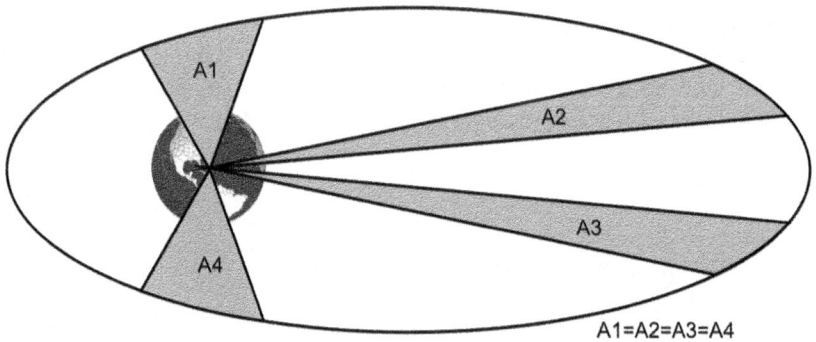

Figure 4.4 Kepler's second law.

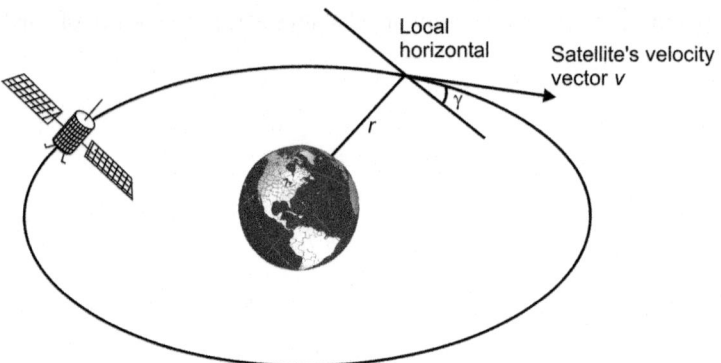

Figure 4.5 A satellite's position at any given time.

motion and the local horizontal, which is the plane perpendicular to radius vector r (Figure 4.5). This leads us to the conclusion that product $r.v. \cos\gamma$ is constant. The product reduces to $r.v$ in the case of circular orbits and also at apogee and perigee points in case of elliptical orbits due to angle γ becoming zero. It is interesting to note here that velocity component v' is inversely proportional to distance r. Qualitatively, this implies that the satellite is at its lowest speed at the apogee point and the highest speed at the perigee point.

In other words, for any satellite in an elliptical orbit, the dot product of its velocity vector and the radius vector at all points is constant. Hence,

$$v_p r_p = v_a r_a = vr\cos\gamma \tag{4.11}$$

v_p = Velocity at the perigee point
r_p = Perigee distance
v_a = Velocity at the apogee point
r_a = Apogee distance
v = Satellite velocity at any point in the orbit
r = Distance of the Point
γ = Angle between the direction of motion and the local horizontal.

4.1.2.3 Kepler's Third Law

According to the third law, also known as the law of periods, the square of time period of any satellite is proportional to the cube of semi-major axis of its elliptical orbit. The expression for the time period can be derived as follows:

Let us assume a circular orbit with radius r. Remember that a circular orbit is only a special case of an elliptical orbit with both semi-major axis and semi-minor axis equal the radius. Equating the gravitational force with the centrifugal force, we get eqn. 4.12.

$$\frac{Gm_1m_2}{r^2} = \frac{m_2v^2}{r} \tag{4.12}$$

Replacing v with $\omega^2 r$ in the eqn. 4.12, we get eqn. 4.13.

$$\frac{Gm_1m_2}{r^2} = \frac{m_2\omega^2r^2}{r} = m_2\omega^2r \tag{4.13}$$

This gives $\omega^2 = \dfrac{Gm_1}{r^3}$. Substituting $\omega = \dfrac{2\pi}{T}$, we get eqn. 4.14.

$$T^2 = \left(\frac{4\pi^2}{Gm_1}\right) r^3 \tag{4.14}$$

The same can also be written as eqn. 4.15.

$$T = \left(\frac{2\pi}{\sqrt{\mu}}\right) r^{3/2} \tag{4.15}$$

Equation 4.15 holds good for elliptical orbits provided we replace r with semi-major axis a. This gives us the expression for the time period of an elliptical orbit as given in eqn. 4.16

$$T = \left(\frac{2\pi}{\sqrt{\mu}}\right) a^{3/2} \tag{4.16}$$

4.1.3   Orbital Parameters

The satellite orbit, which in general is elliptical, is characterized by a number of parameters. These not only include the geometrical parameters of the orbit but also parameters that tell us about its orientation with respect to the Earth. We shall discuss the following orbital elements and parameters in the following paragraphs:

1) Ascending and descending nodes
2) Equinoxes
3) Solstices
4) Apogee
5) Perigee
6) Eccentricity
7) Semi-major axis
8) Right ascension of the ascending node
9) Inclination
10) Argument of the perigee
11) True anomaly of the satellite
12) Angles defining the direction of the satellite.

4.1.3.1   Ascending and Descending Nodes

The satellite orbit cuts the equatorial plane at two points, the first, called the *descending node* (N_1), where the satellite passes from the northern hemisphere to the southern hemisphere and the second, called the *ascending node* (N_2), where the satellite passes from the southern hemisphere to the northern hemisphere (Figure 4.6).

4.1.3.2   Equinoxes

Inclination of the equatorial plane of Earth with respect to the direction of the Sun, defined by the angle formed by the line joining the centre of Earth and sun with the Earth's equatorial plane follows a sinusoidal variation and completes one cycle of sinusoidal variation

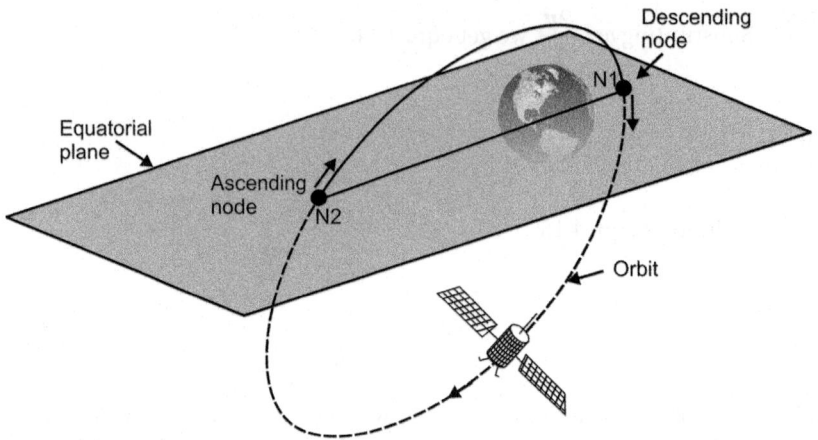

Figure 4.6 Ascending and descending nodes.

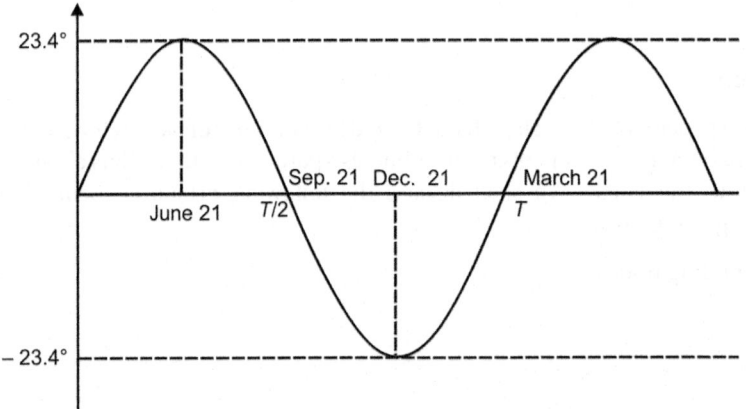

Figure 4.7 Yearly variation of angular inclination of the Earth with the Sun.

over a period of 365 days (Figure 4.7). The sinusoidal variation of angle of inclination is defined by eqn. 4.17.

$$\text{Inclination Angle (in degree)} = 23.4\sin(2\pi t/T) \tag{4.17}$$

$T = 365$ days

This expression tells us that the inclination angle is zero for $t = T/2$ and T. This is observed to occur around the 21 March, called the *Spring Equinox* and around the 21 September, called the *Autumn Equinox*. The two equinoxes are understandably spaced 6 months apart. During equinoxes, as we can see, the equatorial plane of the Earth will be aligned with the direction of the Sun. Also, the line of intersection of the Earth's equatorial plane and Earth's orbital plane that passes through the centre of the Earth is known as line of equinoxes. The direction of this line with respect to the direction of the Sun on the 21 March determines a point at infinity called the *Vernal Equinox* (*V*) (Figure 4.8).

Figure 4.8 The vernal equinox.

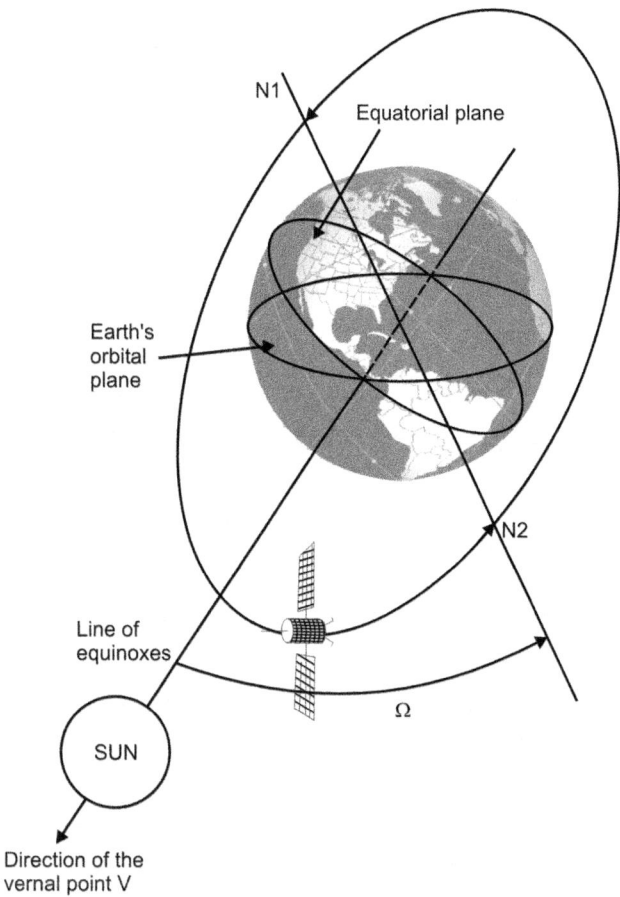

4.1.3.3 Solstices

Solstices are the times when the inclination angle is at its maximum; that is, 23.4°. These too occur twice during a year around the 21 June, called the *Summer Solstice*, and around the 21 December, called the *Winter Solstice*.

4.1.3.4 Apogee

Apogee is the point on the satellite orbit that is at farthest distance from the centre of Earth (Figure 4.9). Apogee distance can be computed from the known values of orbit eccentricity e and the semi-major axis a from eqn. 4.18.

$$\text{Apogee distance} = a \times (1 + e) \tag{4.18}$$

The apogee distance can also be computed from the known values of perigee distance and velocity V_p at the perigee from eqn. 4.19.

$$V_p = \sqrt{\frac{2\mu}{\text{Perigee distance}}} - \sqrt{\frac{2\mu}{(\text{Perigee distance} + \text{Apogee distance})}} \tag{4.19}$$

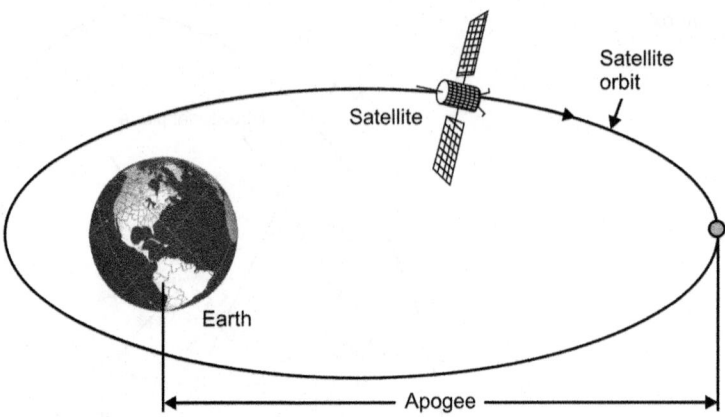

Figure 4.9 Apogee.

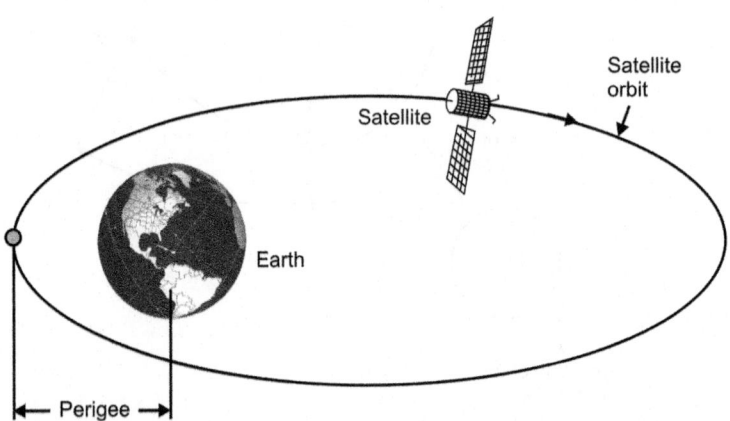

Figure 4.10 Perigee.

Where $V_p = V \times [d \cos\gamma / (\text{Perigee distance})]$ with V being velocity of the satellite at a point distant d from the centre of Earth.

4.1.3.5 Perigee

Perigee is the point on the orbit that is nearest to the centre of Earth (Figure 4.10). Perigee distance can be computed from the known values of orbit eccentricity e and the semi-major axis a from eqn. 4.20.

$$\text{Perigee distance} = a \times (1 - e) \tag{4.20}$$

4.1.3.6 Eccentricity

Orbit eccentricity e is the ratio of the distance between the centre of the ellipse and the centre of Earth to the semi-major axis of the ellipse. It can be computed from any of the following expressions given in eqns 4.21, 4.22 and 4.22.

$$e = \frac{(Apogee - Perigee)}{(Apogee + Perigee)} \tag{4.21}$$

$$e = \frac{(Apogee - Perigee)}{2a} \tag{4.22}$$

$$e = \frac{\sqrt{(a^2 - b^2)}}{a} \tag{4.23}$$

a and b are semi-major and semi-minor axes, respectively.

4.1.3.7 Semi-Major Axis

This is a geometrical parameter of the elliptical orbit. It can, however, be computed from known values of apogee and perigee distances from eqn. 4.24.

$$a = \frac{(Apogee + Perigee)}{2} \tag{4.24}$$

4.1.3.8 Right Ascension of the Ascending Node

Right ascension of the ascending node tells about the orientation of the line of nodes, which is the line joining the ascending and descending nodes, with respect to the direction of vernal equinox. It is expressed as an angle Ω measured from vernal equinox toward the line of nodes in the direction of rotation of the Earth (Figure 4.11). The angle could be anywhere from 0° to 360°.

Acquisition of the correct angle of right ascension of ascending node Ω is important to ensure that the satellite orbits in the given plane. This can be achieved by choosing an appropriate injection time depending upon the longitude. Angle Ω can be computed as a difference of two angles; one, angle (α) between the direction of vernal equinox and the longitude of the injection point and two, angle β between the line of nodes and the longitude of injection point as shown in Figure 4.12. β can be computed from eqn. 4.25.

$$\sin \beta = \frac{(\cos i . \sin l)}{\cos l . \sin i} \tag{4.25}$$

Where i is orbit inclination and l is the latitude at the injection point.

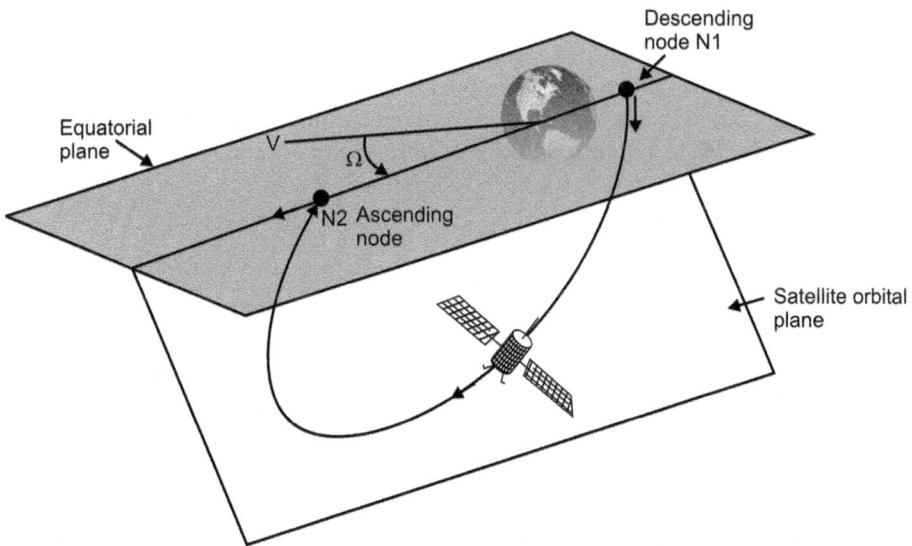

Figure 4.11 Right ascension of ascending node.

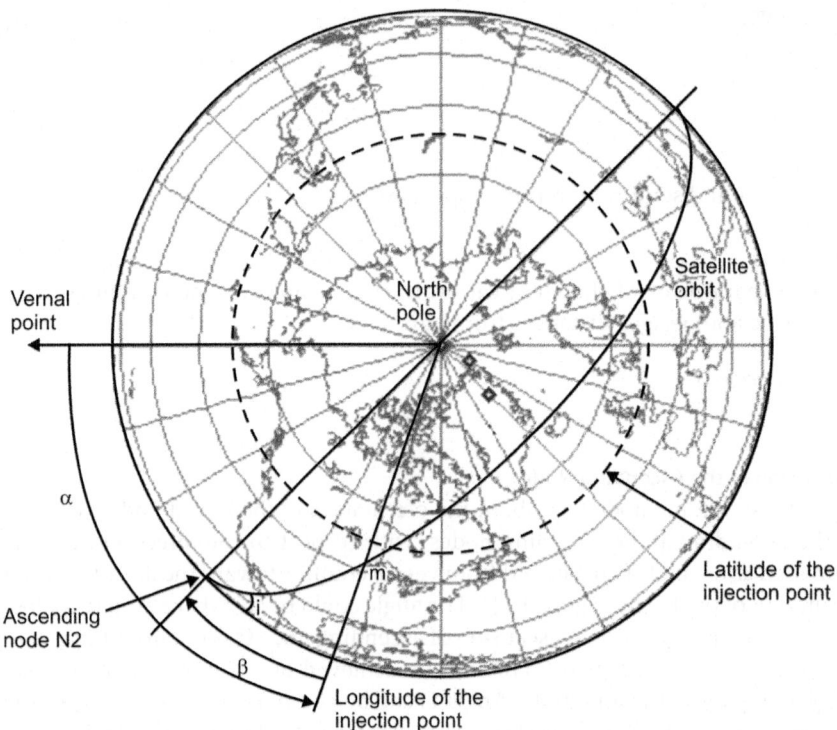

Figure 4.12 Computation of the right ascension of ascending node.

4.1.3.9 Inclination

Inclination is the angle that the orbital plane of the satellite makes with the equatorial plane. It is measured as follows. The line of nodes divides both the equatorial plane as well as the orbital plane into two halves. Inclination is measured as the angle between that half of the orbital plane containing the trajectory of the satellite from the descending node to the ascending node to that half of the equatorial plane containing trajectory of a point on equator from n_1 to n_2 where n_1 and n_2 are, respectively, the points vertically below the descending and ascending nodes (Figure 4.13). Inclination angle can be determined from the latitude l at the injection point and the angle A_z between the projection of satellite's velocity vector on the local horizontal and north. It is given by eqn. 4.26.

$$\cos i = \sin A_z \cos l \tag{4.26}$$

4.1.3.10 Argument of Perigee

This parameter defines the location of the major axis of the satellite orbit. It is measured as angle ω between the line joining the perigee and the centre of Earth and the line of nodes from ascending to the descending node in the same direction as that of the satellite orbit (Figure 4.14).

4.1.3.11 True Anomaly of the Satellite

This parameter is used to indicate the position of the satellite in its orbit. This is done by defining an angle θ, calledthe true anomaly of the satellite, formed by the line joining the perigee and centre of Earth with the line joining the satellite and centre of Earth (Figure 4.15).

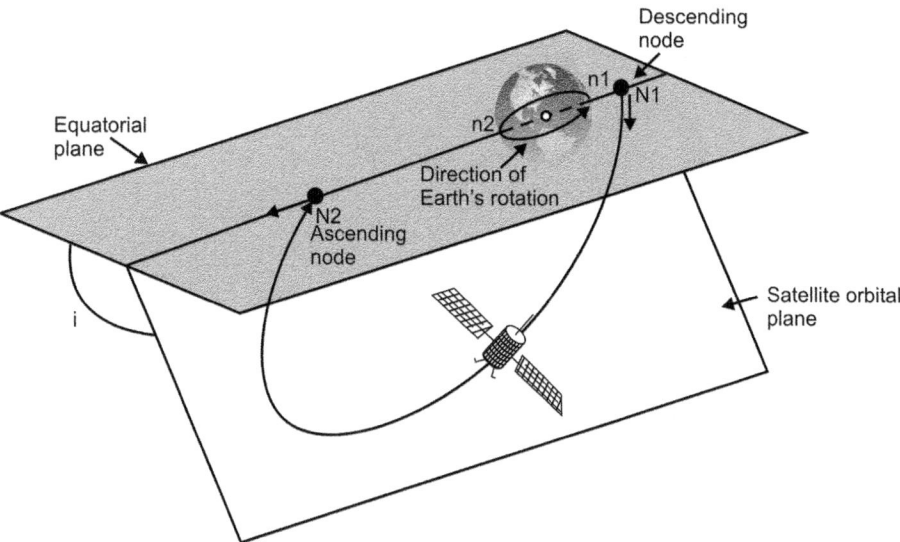

Figure 4.13 Angle of inclination.

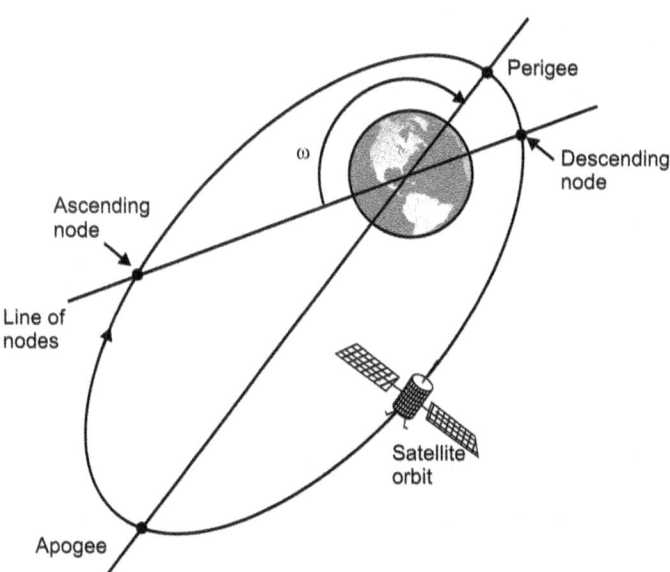

Figure 4.14 Argument of perigee.

4.1.3.12   Angles Defining the Direction of a Satellite

The direction of the satellite is defined by two angles; first, by angle γ between the direction of the satellite's velocity vector and its projection in the local horizontal and second, by angle A_z between the north and the projection of satellite's velocity vector on the local horizontal (Figure 4.16).

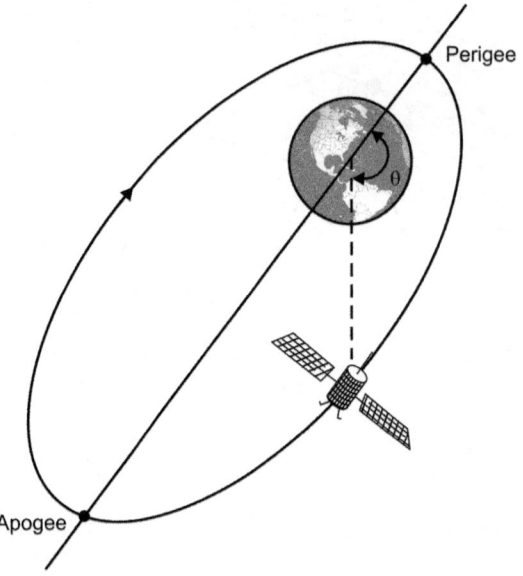

Figure 4.15 True anomaly of a satellite.

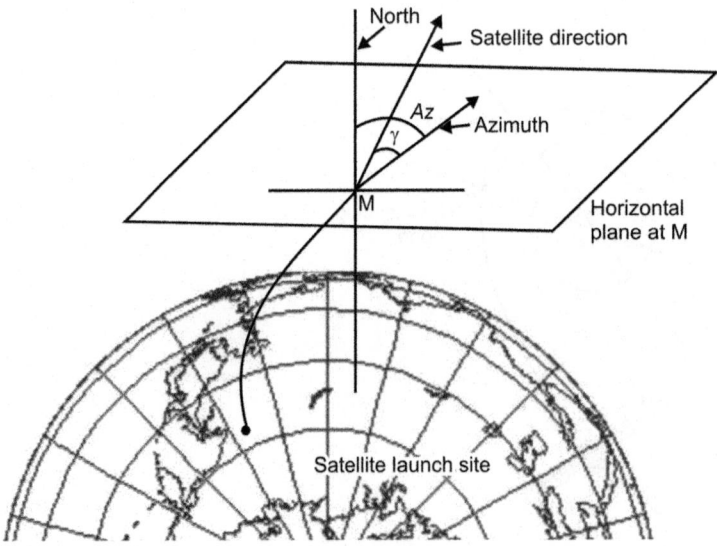

Figure 4.16 Angles defining the direction of a satellite.

4.1.4 Injection Velocity and Satellite Trajectory

The horizontal velocity with which a satellite is injected into space by the launch vehicle has a direct bearing on the satellite trajectory. The phenomenon is best explained in terms of the three cosmic velocities. The general expression for the velocity of the satellite at the perigee point, assuming an elliptical orbit, is given by eqn. 4.27.

$$v = \sqrt{\left(\frac{2\mu}{r}\right) - \left(\frac{2\mu}{R+r}\right)} \tag{4.27}$$

R = Apogee distance
r = Perigee distance
$\mu = GM$ = Constant

The *first cosmic velocity* v_1 is the one at which apogee and perigee distances are equal, that is, $R = r$ and the orbit is circular. Expression 4.27 then reduces to eqn. 4.28.

$$v_1 = \sqrt{\left(\frac{\mu}{r}\right)} \tag{4.28}$$

Thus, irrespective of the distance r of satellite from centre of Earth, if the injection velocity is equal to the first cosmic velocity also sometimes called first orbital velocity, the satellite follows a circular orbit and moves with a uniform velocity equal to $\sqrt{(\mu/r)}$. A simple calculation will tell us that for satellite at 35 786 km above the surface of Earth, the first cosmic velocity turns out to be 3.075 km/s and the orbital period is 23 hours, 56 minutes, which is equal to a time period of one sidereal day – the time taken by Earth to complete one full rotation around its axis with reference to distant stars. This confirms why a geostationary satellite needs to be at a height of 35 786 km above the surface of Earth. Different types of orbits are discussed at length in the following pages.

If the injection velocity happens to be less than the first cosmic velocity, the satellite follows a ballistic trajectory and falls back to Earth. In fact, in this case, the orbit is elliptical and the injection point is at apogee and not the perigee. If the perigee lies in the atmosphere or exists only virtually below the surface of Earth, the satellite accomplishes a ballistic flight and falls back to Earth.

For injection velocity greater than the first cosmic velocity and less than the second cosmic velocity, that is $v > \sqrt{(\mu/r)}$ and $v < \sqrt{(2\mu/r)}$, the orbit is elliptical and eccentric. The orbit eccentricity is between 0 and 1. The injection point, in this case is the perigee and the apogee distance attained in the resultant elliptical orbit depends upon the injection velocity. The higher the injection velocity, greater is the apogee distance. The apogee distance can also be computed from the known value of injection velocity, which is also the velocity at the perigee point as the perigee coincides with injection point, and velocity v at any other point in the orbit distant d from centre of Earth from eqn. 4.29.

$$V_p = \sqrt{\left(\frac{2\mu}{r}\right) - \left(\frac{2\mu}{R+r}\right)} = \frac{(vd\cos\gamma)}{r} \tag{4.29}$$

When the injection velocity equals $\sqrt{(2\mu/r)}$, the apogee distance R becomes infinite and the orbit takes the shape of a parabola and the orbit eccentricity is 1. This is the *second cosmic velocity* v_2. At this velocity, the satellite escapes the Earth's gravitational pull. For injection velocity greater than second cosmic velocity, the trajectory is hyperbolic within the Solar System and the orbit eccentricity is greater than 1.

If the injection velocity is increased further, we reach a stage where the satellite succeeds in escaping from the Solar System. This is known as the *third cosmic velocity* and is related to the motion of planet Earth around the sun. The third cosmic velocity is mathematically expressed by eqn. 4.30.

$$V_p = \sqrt{\left(\frac{2\mu}{r}\right) - v_t^2 \left(3 - 2\sqrt{2}\right)} \tag{4.30}$$

v_t is speed of Earth's revolution around the Sun.

Beyond the third cosmic velocity, we have the region of hyperbolic flights outside the Solar System. Coming back to elliptical orbits, greater is the injection velocity from the first cosmic velocity, greater is the apogee distance. This is evident from the generalized expression for the velocity of the satellite in elliptical orbits. The velocity at the perigee point is given by eqn. 4.31.

$$V_p = \sqrt{\left(\frac{2\mu}{r}\right) - \left(\frac{2\mu}{R+r}\right)} \tag{4.31}$$

For a given perigee distance r, a higher velocity at the perigee point, which is also the injection velocity, necessitates that apogee distance R is greater. For a given perigee distance r, the injection velocity and corresponding apogee distance are related by eqn. 4.32.

$$\left(\frac{v_2}{v_1}\right)^2 = \frac{\left(1 + r/R_1\right)}{\left(1 + r/R_2\right)} \tag{4.32}$$

4.1.5 Types of Satellite Orbits

Satellite orbits can be classified on the basis of (1) orientation of the orbital plane, (2) eccentricity and (3) distance from Earth.

4.1.5.1 Orientation of the Orbital Plane

The orbital plane of the satellite can have various orientations with respect to the equatorial plane of the Earth. The angle between the two planes is called the angle of inclination of the satellite. On this basis, the orbits can be classified as equatorial orbits, polar orbits and inclined orbits.

In the case of an *Equatorial Orbit*, the angle of inclination is zero, that is, the orbital plane coincides with the equatorial plane (Figure 4.17). A satellite in the equatorial orbit has latitude of 0°. For angle of inclination equal to 90°, the satellite is said to be in *polar orbit* (Figure 4.18). For an angle of inclination between 0° and 180°, the orbit is said to be an *inclined orbit*.

For inclination between 0° and 90°, the satellite travels in the same direction as the direction of rotation of the Earth. The orbit in this case is referred to as direct or *prograde orbit* (Figure 4.19). For inclination between 90° and 180°, the satellite orbits in a direction opposite to the direction of rotation of Earth and the orbit in this case is called a *retrograde orbit* (Figure 4.20).

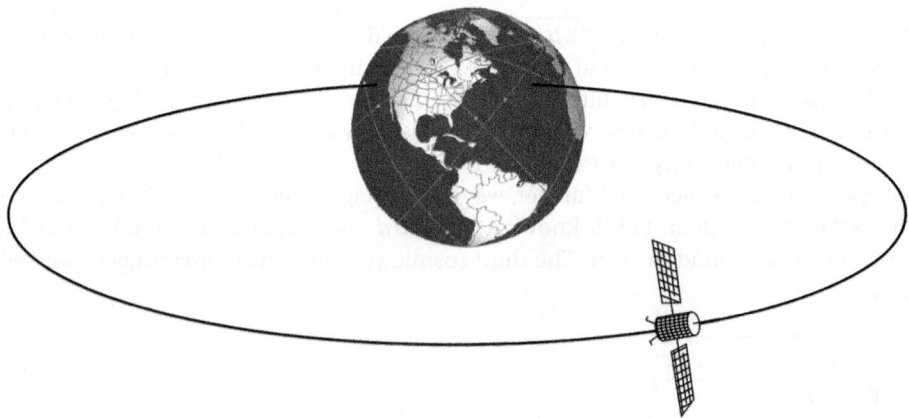

Figure 4.17 Equatorial orbit

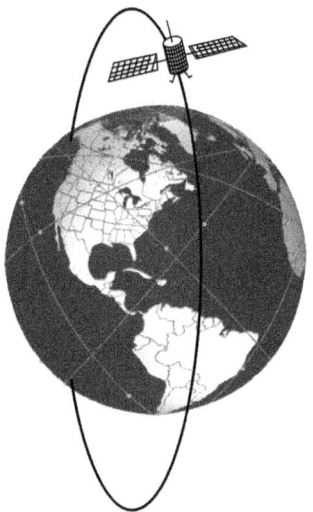

Figure 4.18 Polar orbit.

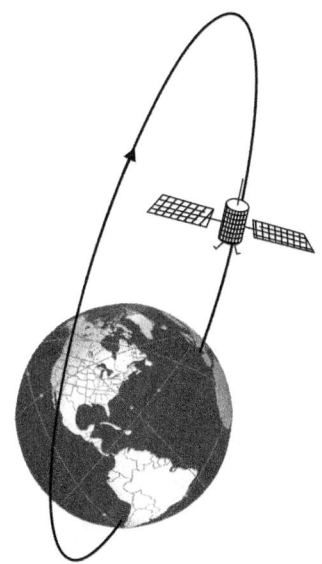

Figure 4.19 Prograde orbit.

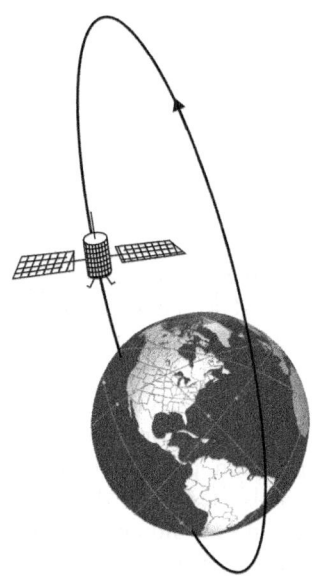

Figure 4.20 Retrograde orbit.

4.1.5.2 Eccentricity of Orbit

On the basis of eccentricity, the orbits are classified as *elliptical* [Figure 4.21(a)] and *circular* [Figure 4.21(b)] orbits. When the orbit eccentricity lies between 0 and 1, the orbit is elliptical with the centre of Earth lying at one of the foci of the ellipse. When the eccentricity is zero, the orbit becomes circular. It's worth mentionung here that all circular orbits are eccentric to some extent. As an example, INSAT-3B, an Indian satellite in the INSAT-series providing communication and meteorological services, has an eccentricity figure of 0.0002526. Eccentricity figures for GOES-9 and METEOSAT-7, both offering Earth observation and environmental monitoring services, are 0.0004233 and 0.0002526, respectively.

4.1.5.3 Molniya Orbit

Highly eccentric, inclined and elliptical orbits are used to cover higher latitudes, which are otherwise not covered by geostationary orbits. A practical example of this type of orbit is the Molniya Orbit (Figure 4.22). It is a widely used satellite orbit, used by Russia and other countries of erstwhile Soviet Union for providing communication services. Typical eccentricity and orbit inclination figures for Molniya orbit are 0.75 and 65°, respectively. The apogee and perigee points are about 40 000 and 400 km, respectively, from the surface of Earth.

The Molniya orbit serves the purpose of a geosynchronous orbit for high latitude regions. It has a 12-hour orbit and a satellite in this orbit spends about 8 hours above a particular high latitude station before diving down to a low-level perigee at equally high southern latitude. Usually, three satellites at different phases of the same Molniya orbit are capable of providing uninterrupted service.

(a)   (b)

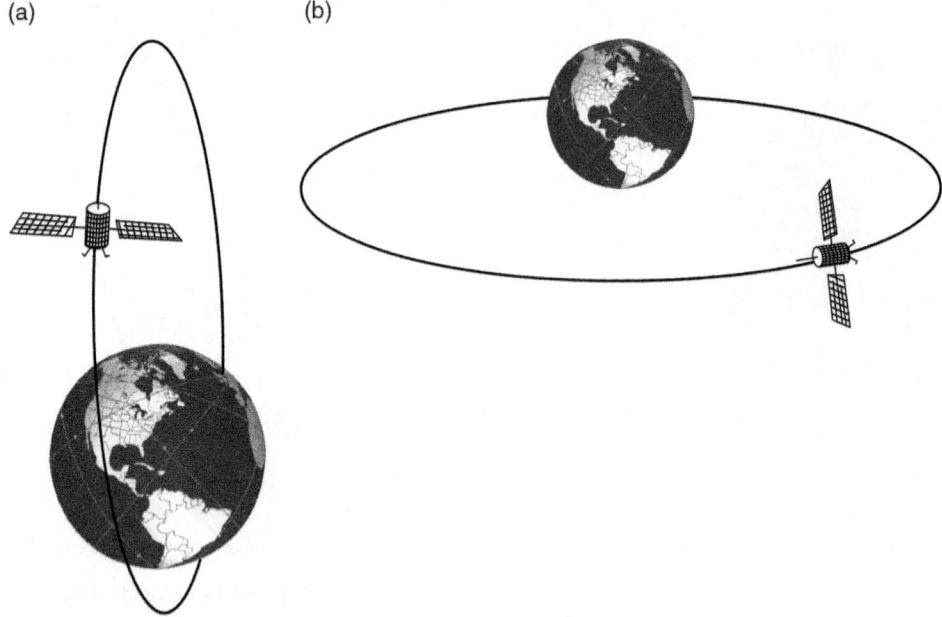

Figure 4.21 (a) Elliptical orbit. (b) Circular orbit.

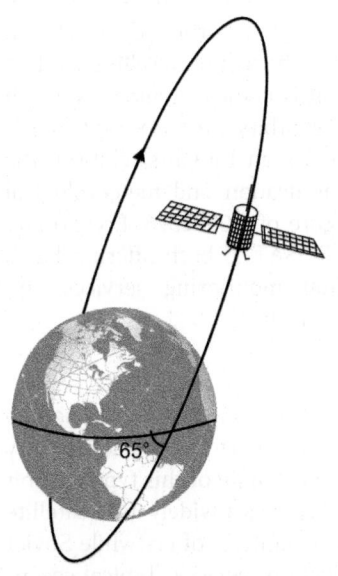

Figure 4.22 Molniya orbit.

4.1.5.4 Distance from Earth

Again depending upon intended mission, satellites may be placed in orbits at varying distances from the surface of Earth. Depending upon the distance, these are classified as Low Earth Orbits (LEO), Medium Earth Orbits (MEO) and Geostationary Orbits (GEO), as shown in Figure 4.23.

Satellites in LEO circle the Earth at a height of around 160 km to 500 km above the surface of Earth. These satellites being closer to the surface of Earth have much shorter orbital periods and smaller signal propagation delays. A lower propagation delay makes them highly suitable for communication applications. Due to lower propagation paths, the power required for signal transmission is also less with the result that the satellites are of small physical size and are inexpensive to build. But due to shorter orbital period of the order of an hour and a half or so, these satellites remain over a particular ground station for a short time. Hence, several of these satellites are needed for 24-hour coverage. One important application of LEO satellites for communication is the project Iridium, which is a global communication system conceived by Motorola that makes use of satellites placed in low Earth orbits (Figure 4.24). A total of 66 satellites are arranged in a distributed architecture with each satellite carrying (1/66) of the total system capacity. The system is intended to provide a variety of telecommunication services on the global level. The project is named *Iridium*, as, early on, the constellation was proposed to have 77 satellites and the atomic number of iridium is 77.

Figure 4.23 LEO, MEO and GEO orbits.

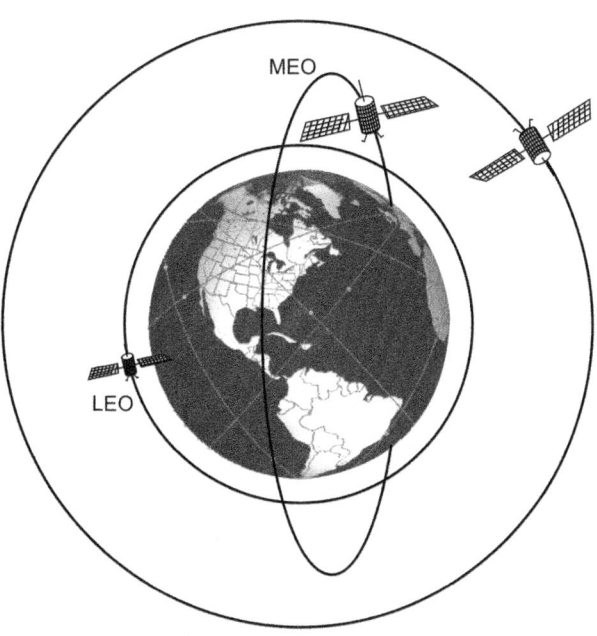

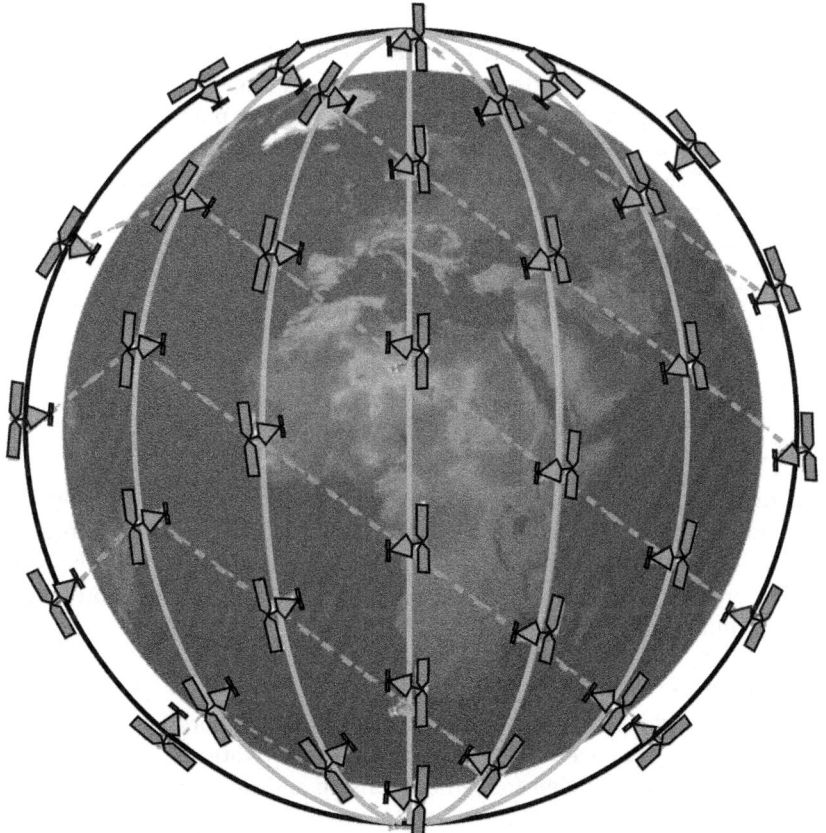

Figure 4.24 The Iridium constellation of satellites.

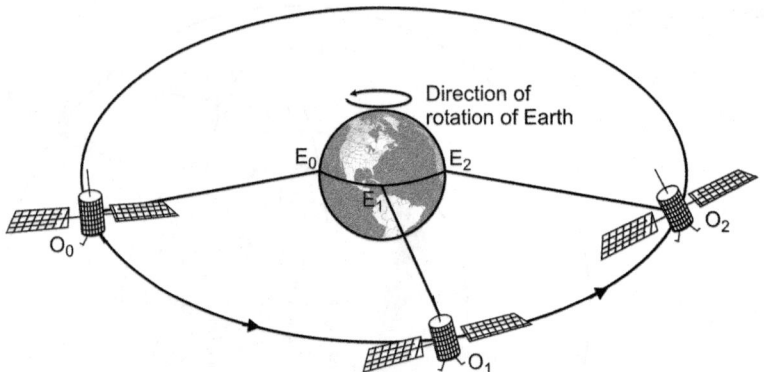

Figure 4.25 GEO satellites appear stationary with respect to a point on Earth.

Other applications where LEO satellites can be put to use are surveillance, weather forecasting, remote sensing and scientific studies.

MEO satellites orbit at a distance of approximately 10 000 km to 20 000 km above the surface of Earth. These have an orbital period of 6–12 h. These satellites stay in sight over a particular region of the Earth for a longer time. Transmission distance and propagation delays are greater than those for LEO satellites. These orbits are generally polar in nature and are mainly used for communication and navigation applications.

A GEO is such that the orbital velocity, which is a function of the distance of the satellite from the surface of Earth, equals the speed of Earth's rotation. If such an orbit were in the plane of equator and circular, it would remain stationary with respect to a given point on the equator. These orbits are the GEOs. For the satellite to have such an orbital velocity, it needs to be at a height of about 36 000 km, 35 786 km to be precise, above the surface of Earth. To be more precise and technical, in order to remain above the same point on the Earth's surface, a geostationary satellite must fulfil the following conditions.

1) It must have a constant latitude, which is possible only at 0° latitude.
2) Orbit inclination should be zero.
3) It should have a constant longitude and thus have a uniform angular velocity, which is possible when the orbit is circular.
4) The orbital period should be equal to 23 hours, 56 minutes for 360°, which implies that the satellite must orbit at a height of 35,786 km above the surface of Earth.

In case the orbit is such, then as the satellite moves from a position O1 to O2 in its orbit, a point vertically below on the equator moves with the same angular velocity and moves from E1 to E2 as shown in Figure 4.25. Satellites in geostationary orbits play an essential role in relaying communication and TV broadcast signals around the globe. They also perform meteorological and military surveillance functions very effectively.

4.1.5.5 Sun-Synchronous Orbit
Another type of satellite orbit, which could have been categorized as LEO orbit on the basis of distance from the surface of Earth, needs a special mention and treatment because of its particular importance to satellites intended for remote sensing and military reconnaissance applications.

A sun-synchronous orbit, also known as a helio-synchronous orbit, is one that lies in a plane that maintains a fixed angle with respect to the Earth–Sun direction. In other words, the orbital

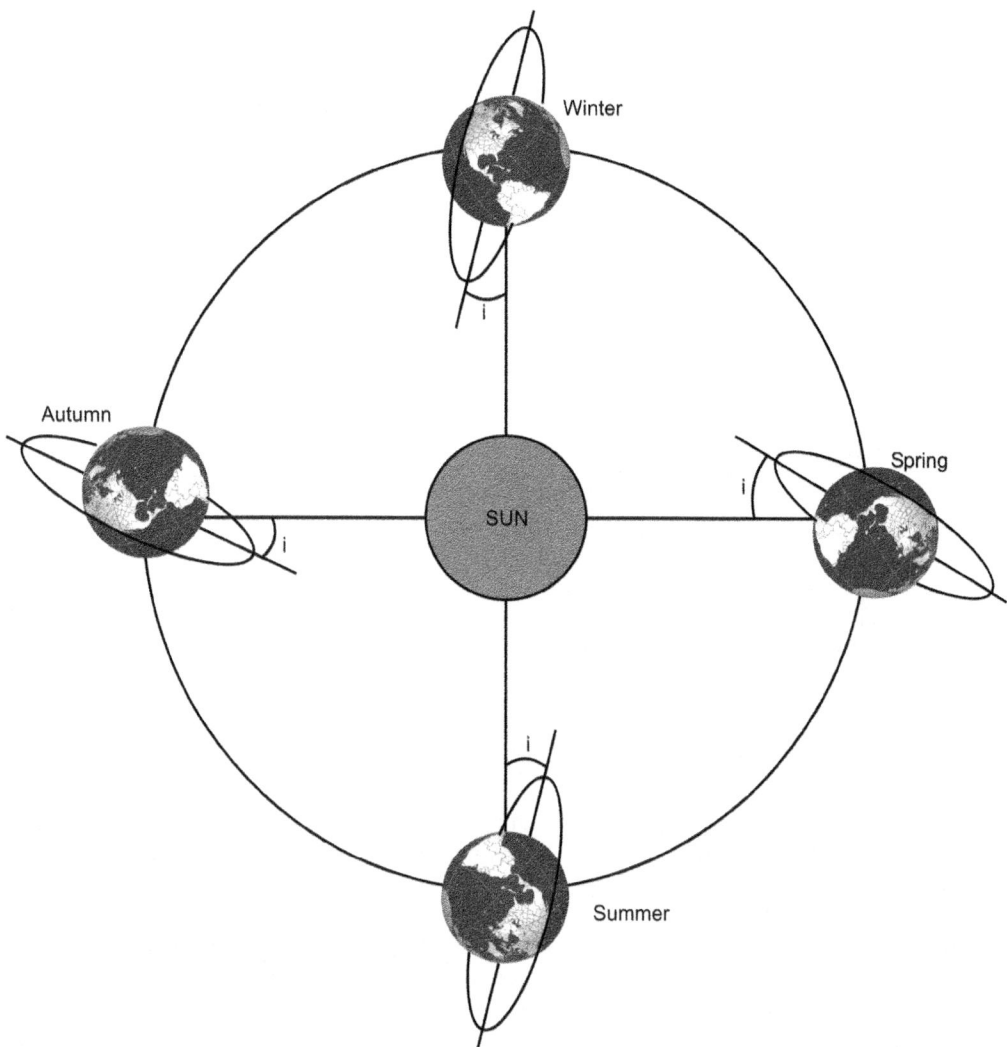

Figure 4.26 Sun-synchronous orbit.

plane has a fixed orientation with respect to the Earth–Sun direction and the angle between the orbital plane and the Earth–Sun line remains constant throughout the year as shown in Figure 4.26. Satellites in sun-synchronous orbits are particularly suited to applications like passive remote sensing, meteorological and atmospheric studies.

As a result of this property, sun-synchronous orbits ensure that:

1) The satellite passes over a given location on the Earth every time at the same local solar time, thereby guaranteeing almost the same illumination conditions, varying only with seasons.
2) The satellite ensures coverage of the whole surface of Earth, being quasi-polar in nature.

Every time a sun-synchronous satellite completes one revolution around Earth, it traverses a thin strip on the surface of Earth. During the next revolution immediately following this, it traverses another strip shifted westwards and the process of shift continues with successive revolutions as shown in Figure 4.27. Depending upon the orbital parameters and speed of

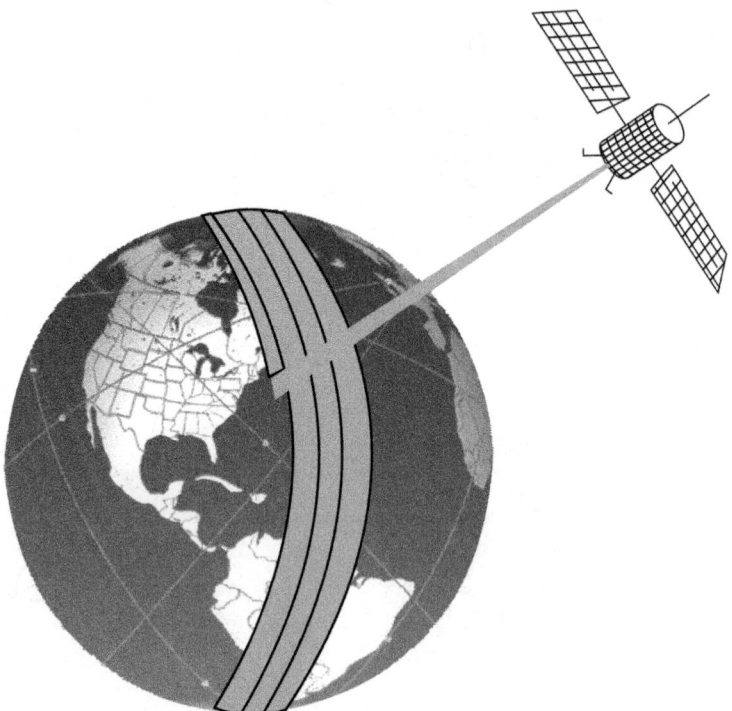

Figure 4.27 Earth coverage of Sun-synchronous satellites.

rotation of Earth, after a making a certain number of revolutions around Earth, it comes back close to the first strip that it had traversed. It may not exactly overlap the first strip, as the mean distance between the two strips called, *Tracking Interval*, may not be integral multiple of the equatorial perimeter. But one can certainly calculate the number of revolutions required before the satellite repeats the same strip sequence. This is called one complete orbital cycle, which is basically the time that elapses before the satellite revisits a given location. To be more precise, by orbital cycle, what we mean is the whole number of orbital revolutions which a satellite must describe in order to be once again flying in the same direction over the same point on the Earth's surface. LANDSAT-1, 2 and 3 satellites for instance have an orbital cycle of 18 days. IRS-1A and IRS-P2 from India have, respectively, an orbital cycle of 22 and 24 days.

We shall illustrate the concept of an orbital cycle further with the help of an example. Let us take the case of LANDSAT-1, 2, 3 satellites with an orbital cycle of 18 days. During the course of a given revolution – let us say revolution 'n' – the satellite crosses the equatorial plane at the descending node above a certain point on the surface of Earth. During the next adjacent pass or revolution, the track shifts to the west by a distance known as the tracking interval. The tracking interval depends upon the speed of Earth's rotation and the nodal precession of the orbit concerned for one revolution. Fourteen revolutions later, which take about 24 hours and a few minutes, we find that the track is located slightly to the west of the first revolution designated revolution 'n'. Due to cumulative effect of these small drifts after every 14 revolutions, we find that the satellite over flies the same point on Earth in the same direction once every 251 revolutions. The figure of 251 comes from $(18 \times 14 - 1)$ thus explaining an 18-day orbital cycle for LANDSAT-1, 2 and 3 satellites.

Another parameter of relevance to satellites in Sun-synchronous orbit is the crossing time for descending and ascending nodes. Satellites of LANDSAT-series, SPOT-series, IRS-series move along their descending trajectory above the sunward face and their ascending trajectory at night. Descending node-crossing times for LANDSAT-1, 2, 3 satellites is 0930 h and for LANDSAT-4 and 5, it is 0937 h. In case of IRS-series and SPOT-series satellites, it is 1030 h.

4.2 Satellite Launch and In-Orbit Operations

Fundamental issues such as laws governing motion of artificial satellites around Earth, different orbital parameters, types of orbits and their suitability for a given application and so on, related to orbital dynamics were addressed in the previous section. The next obvious step is to understand the launch requirements to acquire the desired orbit, which should then lead us to various in-orbit operations such as orbit stabilization, orbit correction and station-keeping that are necessary for keeping the satellite in the desired orbit. The present section addresses all these issues and also other related issues like Earth coverage, eclipses, pointing towards a given satellite from Earth.

4.2.1 Acquiring the Desired Orbit

In order to ensure that the satellite acquires the desired orbit, that is, the orbit with desired values of orbital elements/parameters such as orbital plane, apogee and perigee distances and so on, it is important that right conditions are established at the satellite injection point. For instance, in order to ensure that the satellite orbits within a given plane; the satellite must be injected at a certain specific time, depending upon the longitude of the injection point, at which the line of nodes makes the required angle with the direction of vernal equinox. Put in simple words, for a given orbital plane, the line of nodes will have a specific angle with vernal equinox. To acquire this angle, the satellite must be injected at the desired time depending upon the longitude of the injection point.

4.2.1.1 Parameters Defining Satellite Orbit
Satellite orbit is completely defined by the following parameters.

1) Right ascension of ascending node
2) Inclination angle
3) Position of major axis of the orbit
4) Shape of elliptical orbit
5) Position of satellite in its orbit

4.2.1.1.1 Right Ascension of the Ascending Node
The angle Ω defining the right ascension of the ascending node is basically the difference between two angles θ_1 and θ_2 where θ_1 is the angle made by the longitude of the injection point at the time of launch with the direction of the vernal equinox and θ_2 is the angle made by the longitude of the injection point at the time of launch with the line of nodes as shown in Figure 4.28. θ_2 can be computed from eqn. 4.33.

$$\sin\theta_2 = \frac{\cos i \sin L}{\cos L \sin i} \tag{4.33}$$

i = Angle of inclination
L = Longitude of injection point

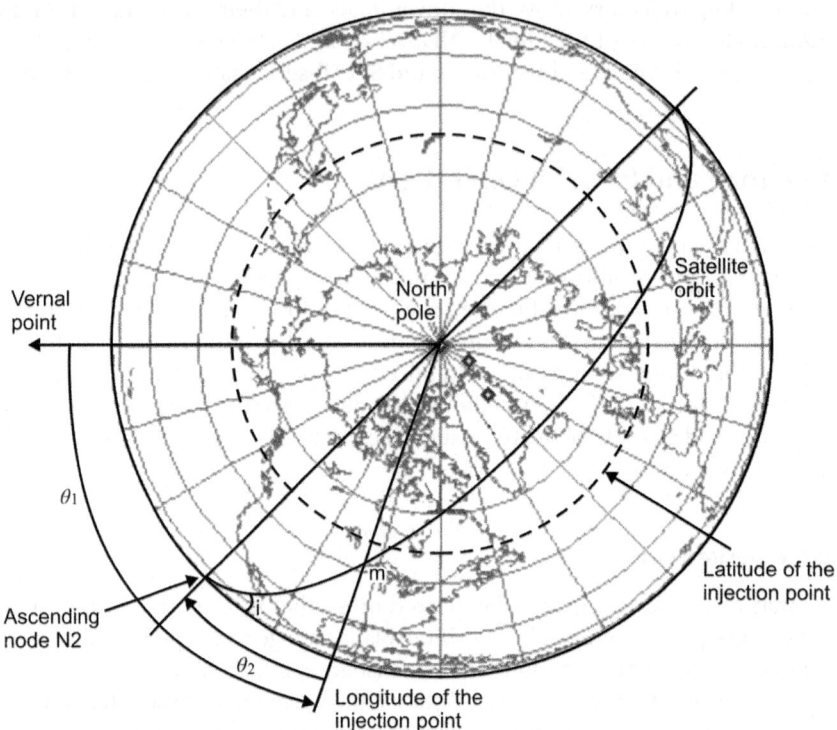

Figure 4.28 Right ascension of the ascending node.

Thus for a known angle of inclination, the time of launch and hence the longitude of the injection point can be so chosen as to get the desired angle Ω.

4.2.1.1.2 Inclination Angle
The angle of inclination of the orbital plane can be determined from the known values of angle of azimuth A_z and the latitude of the injection point using eqn. 4.34.

$$\cos i = \sin A_z \cos l \qquad (4.34)$$

It is obvious from eqn. 4.34 that for the angle of inclination to be zero, the right-hand side of the expression must equal unity. This is possible only for $A_z = 90°$ and $l = 0°$. Following other inferences can be drawn from expression 4.34: (1) For $A_z = 90°$, $i = I$ and (2) For $A_z < 90°$, $\sin A_z < 1$ and therefore $i > I$.

Thus we can conclude that the satellite will tend to orbit in a plane that is inclined to the equatorial plane at an angle equal to or greater than the latitude of the injection point. This will be evident if we look at some of the prominent satellite launch sites and the corresponding latitudes of those locations. The launch site Kourou in French Guyana is at latitude of 5.2°N. When we discuss the typical launch sequences in the latter part of this chapter, we shall see that the orbital plane acquired by the satellite launched from Kourou immediately after injection is 7°, which is later corrected to get an inclination of zero or near zero through manoeuvres. The latitude of another prominent launch site Baikonur in Russia is 45.9°. Study of the launch sequence for satellites launched from this base tells that the satellite acquires an initial orbital inclination of 51°. Both these examples confirm the axiom or the concept outlined in an earlier paragraph. We shall find a similar result if we look at other prominent launch sites at Cape Canaveral, Sriharikota, Vandenberg and Xichang.

4.2.1.1.3 *Position of Major Axis of the Orbit*

Position of major axis of the satellite orbit is defined by argument of perigee ω. Refer to Figure 4.29. If the injection point happens to be different from the perigee point, then ω is the difference of two angles ϕ and θ as shown in Figure 4.29: ϕ can be computed from eqn. 4.35.

$$\sin\phi = \frac{\sin l}{\sin i} \tag{4.35}$$

and θ can be computed from eqn. 4.36.

$$\cos\theta = \frac{(dV^2\cos^2\gamma - \mu)}{(e.\mu)} \tag{4.36}$$

The terms have their usual meaning and are indicated in Figure 4.29.

In case of the injection point being the same as the perigee point (Figure 4.30), $\gamma = 0°, \theta = 0°$. This gives eqn. 4.37.

$$\sin\omega = \sin\phi = \frac{\sin l}{\sin i} \tag{4.37}$$

4.2.1.1.4 *Shape of Elliptical Orbit*

Shape of orbit is defined by the orbit eccentricity e, semi-major axis a, apogee distance r_a and perigee distance r_p. The elliptical orbit can be completely defined by either a and e or by r_a and r_p. The orbit is usually defined by apogee and perigee distances. The perigee distance can be computed from known values of distance d from the focus (which in this case is centre of Earth) at any point in the orbit and the angle γ that the velocity vector V makes with the local horizontal. r_p is computed from eqn. 4.38.

$$r_p = \frac{d^2V^2\cos^2\gamma}{\mu(1+e)} \tag{4.38}$$

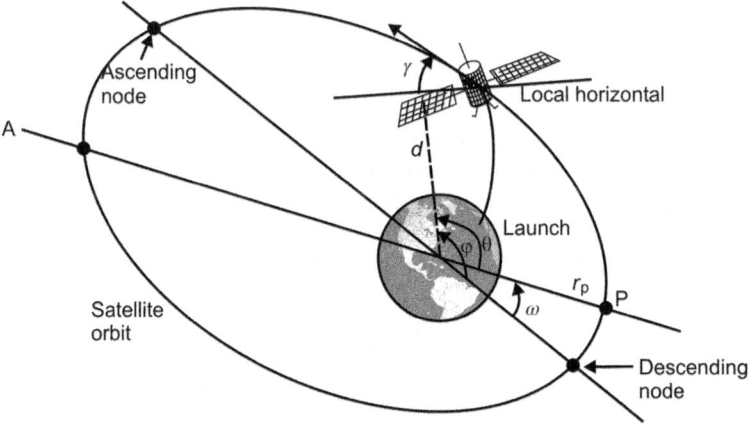

Figure 4.29 Argument of perigee.

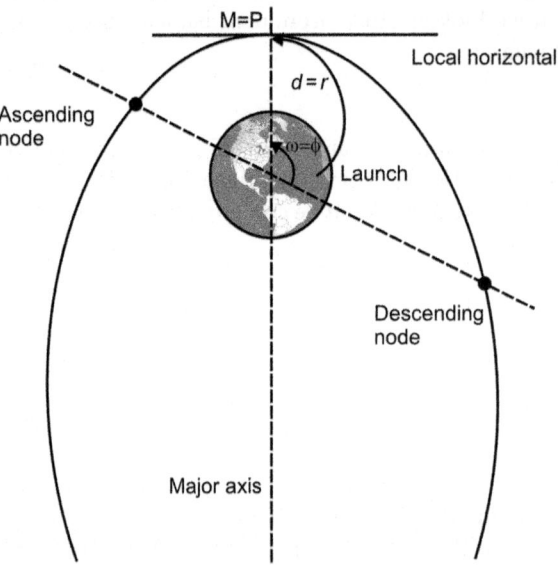

The apogee distance r_a can be computed from eqn. 4.39.

$$r_a = \frac{d^2 V^2 \cos^2 \gamma}{\mu(1-e)} \tag{4.39}$$

4.2.1.1.5 Position of Satellite in Its Orbit

Position of satellite in its orbit can be defined by a time parameter t, which is the time that elapsed after a time instant t_0 when the satellite last passed through a reference point. The reference point is usually the perigee. The time t that has elapsed after the satellite last passed through perigee point can be computed from eqn. 4.40.

$$t = \left(\frac{T}{2\pi}\right)(u - e \sin u) \tag{4.40}$$

T is orbital period and angle u is eccentric anomaly of the current location of the satellite (Figure 4.31).

4.2.1.2 Modifying Orbital Parameters

In the preceding paragraphs, we have discussed the parameters that define an orbit. We have also discussed the factors each one of these parameters is dependent on with the help of relevant mathematical expressions. But when it comes to acquiring a desired orbit, this is perhaps not enough. The discussion would be complete only if we knew how each one of these parameters could be modified if needed with the help of commands transmitted to the satellite from ground control station. In the paragraphs to follow, we shall look at this part of the discussion.

4.2.1.2.1 Right Ascension of Ascending Node

This parameter, as mentioned before, defines the orientation of the orbital plane. Due to Earth's equatorial bulge, there is nodal precession, that is, change in the orientation of the orbital plane with time as the Earth revolves around the sun. We shall see that this perturbation to the orbital

Figure 4.31 Position of a satellite in its orbit.

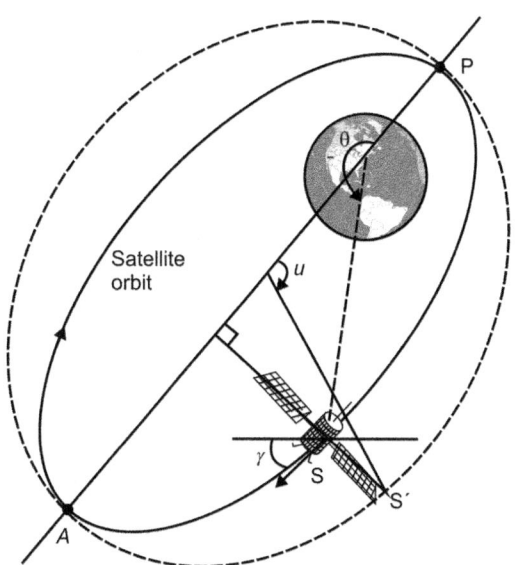

plane depends upon the orbit inclination angle i and also the apogee and perigee distances. The rotational perturbation experienced by orbital plane (in degrees) during one orbit period is given by eqn. 4.41.

$$\Delta\Omega = -0.58 \times \left[\frac{D}{(r_a + r_p)}\right]^2 \times \left[\frac{1}{(1-e^2)}\right]^2 \times \cos i \tag{4.41}$$

D = Diameter of Earth

In case of a circular orbit, $e = 0$, which gives $r_a = r_p$

It is clear from expression 4.41 that the perturbation is zero only in case of inclination angle being 90°. Also, the farther the inclination angle from 90° and shorter the apogee and perigee distances, larger is the perturbation. Any experimental manoeuvre that changes the orientation is a very expensive affair. It is always preferred to depend upon natural perturbation for the purpose.

Also, when orbit inclination angle $i < 90°$, Ω is negative. That is, orbital plane rotates in a direction opposite to the direction of rotation of Earth (Figure 4.32). When orbit inclination angle $i > 90°$, $\Delta\Omega$ is positive. That is, the orbital plane rotates in the same direction as the direction of rotation of Earth (Figure 4.33).

4.2.1.2.2 *Inclination Angle*

Though there is a natural perturbation experienced by the orbital plane inclination to the equatorial plane due to the Sun-Moon attraction phenomenon, this change is small enough to be considered negligible. A small change Δi to the inclination angle i can be affected externally by applying a thrust Δv to the velocity vector V at an angle of $90° + \Delta i/2$ to the direction of the satellite as shown in Figure 4.34. Δv is given by eqn. 4.42. The thrust is applied at either of the two nodes.

$$\Delta v = 2V \sin\left(\frac{\Delta i}{2}\right) \tag{4.42}$$

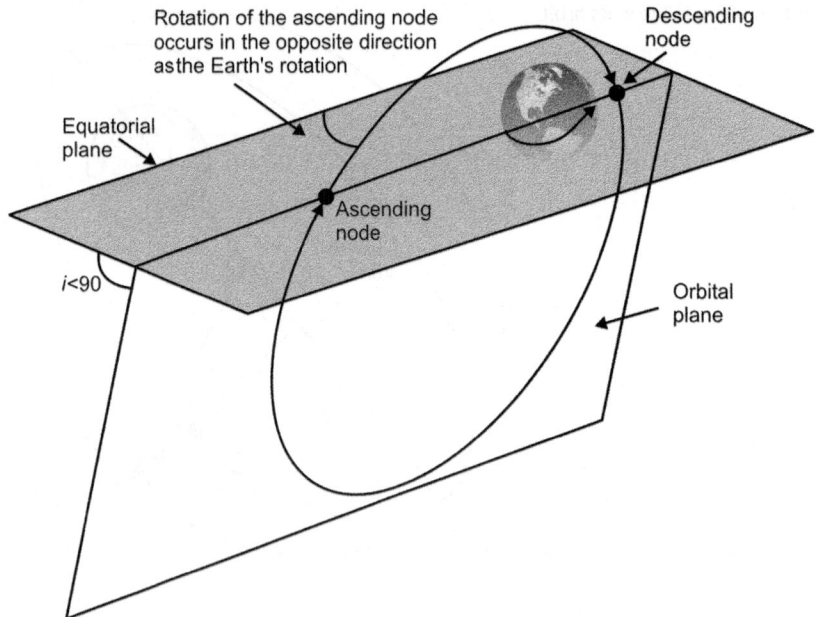

Figure 4.32 Rotation of the orbital plane when the inclination angle is less than 90°.

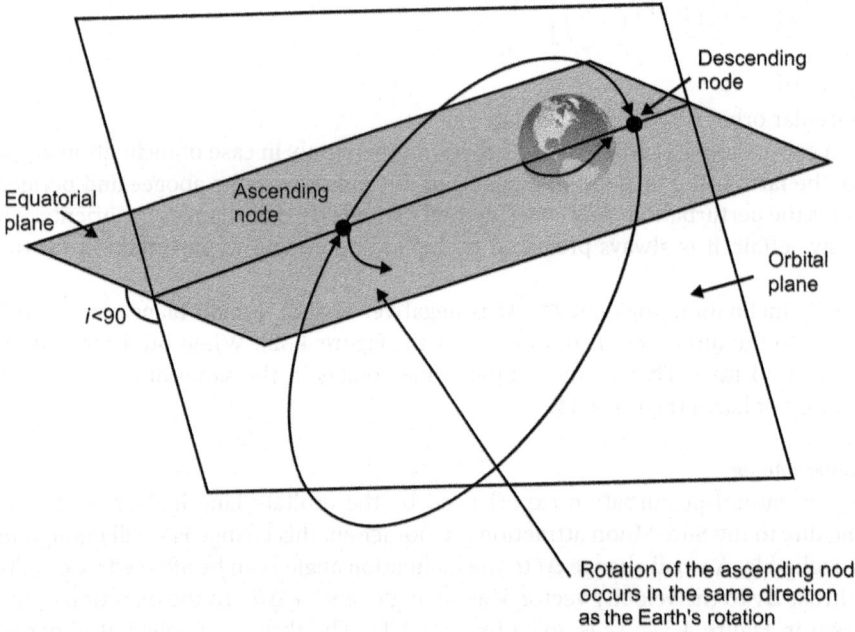

Figure 4.33 Rotation of the orbital plane when the inclination angle is greater than 90°.

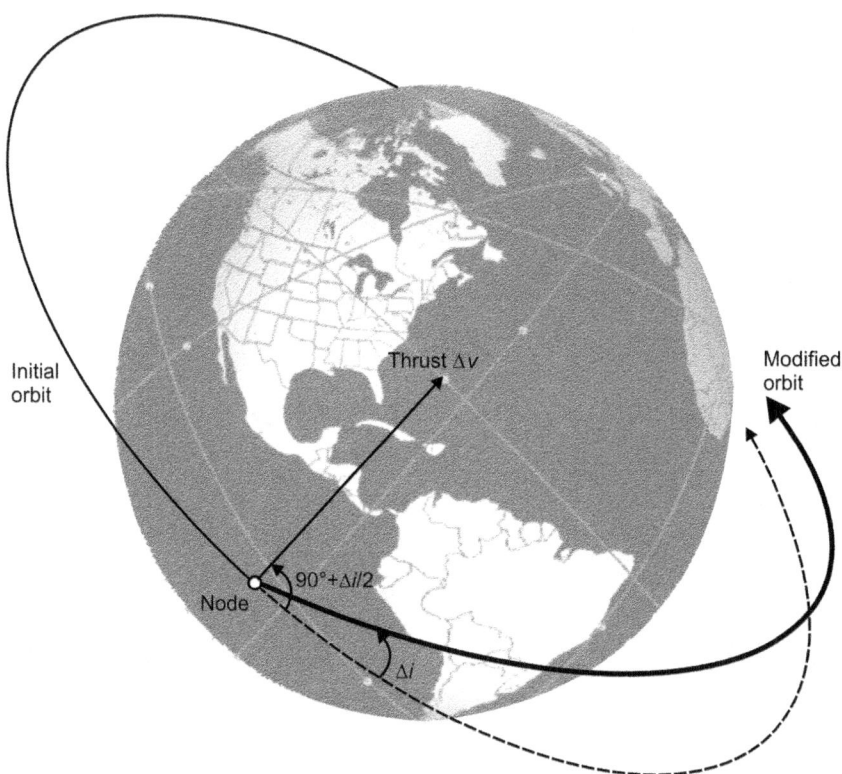

Figure 4.34 Changing the inclination angle.

4.2.1.2.3 Position of Major Axis of the Orbit

Position of the major axis, as mentioned in earlier paragraphs, is defined by the parameter called the argument of perigee, ω. This parameter too, like the right ascension of the ascending node, undergoes natural perturbation due to equatorial bulge of Earth. The phenomenon is known as apsidal precession. Farther the angle of inclination from (63° 26'), larger is the rotation of the perigee. Also closer the satellite to the centre of Earth, larger is the rotation. Moreover, rotation of perigee occurs in the opposite direction to the satellite motion if the inclination is greater than 63°26' and in the same direction of the satellite motion if the inclination angle is less than 63°26'. This is illustrated in Figure 4.35(a) and (b).The rotation experienced by ω in degrees due to natural perturbation in one orbit is given by eqn. 4.43.

$$\Delta\omega = 0.29 \times \left[\frac{\left(4 - 5\sin^2 i\right)}{\left\{\frac{1}{\left(1-e^2\right)}\right\}^2 \times \left\{\frac{D}{\left(r_a + r_p\right)}\right\}^2} \right] \qquad (4.43)$$

In case of a circular orbit, $e = 0$, $r_a = r_p$

In order to affect an intentional change $\Delta\omega$, a thrust Δv is applied at a point in the orbit where the line joining this point and the centre of Earth makes an angle $\Delta\omega/2$ with the major

(a)

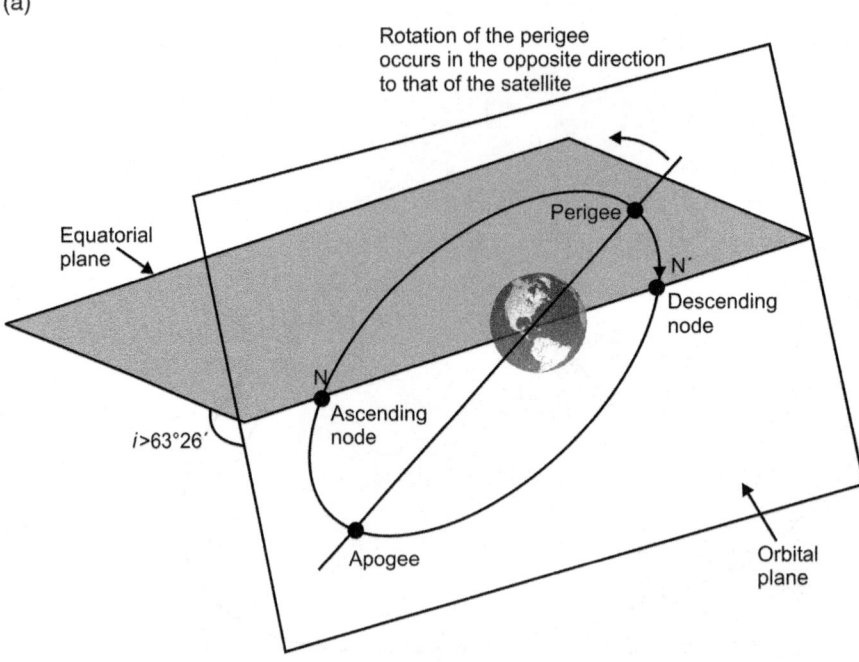

(b)

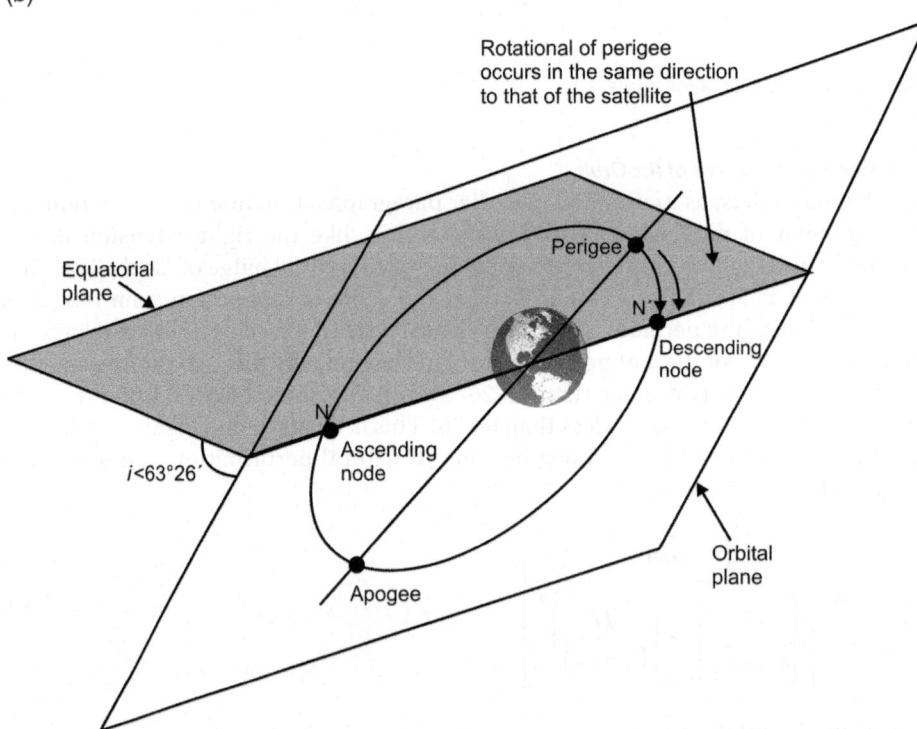

Figure 4.35 (a) Rotation of perigee when the inclination angle is greater than 63°26. (b) Rotation of perigee when the inclination angle is less than 63°26.

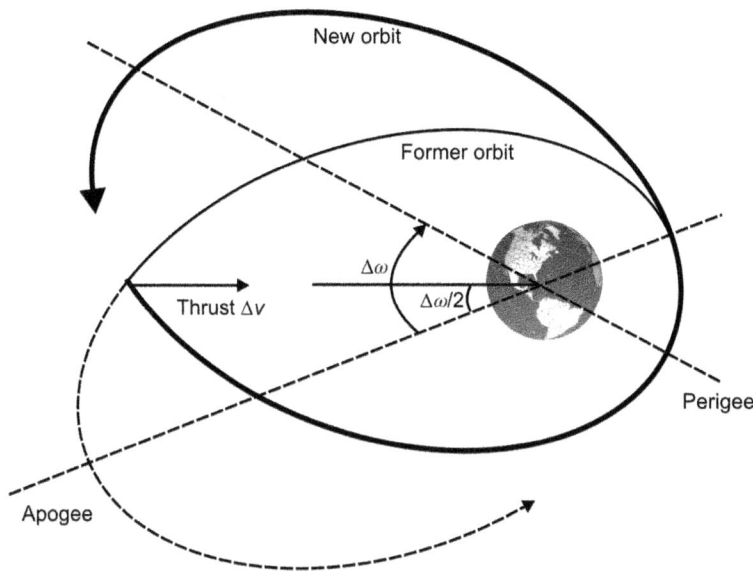

Figure 4.36 Changing the argument of perigee.

axis. The thrust is applied in a direction towards the centre of Earth as shown in Figure 4.36. The thrust Δv can be computed from eqn. 4.44.

$$\Delta v = 2\sqrt{\left(\frac{\mu}{r_p}\right)} \times \left[\frac{e}{\sqrt{(1+e)}}\right] \times \sin\left(\frac{\omega}{2}\right) \tag{4.44}$$

4.2.1.2.4 Shape of an Elliptical Orbit

Apogee and perigee distances fully specify the shape of the orbit as the other parameters like eccentricity e and semi-major axis a can be computed from this pair of distances. Apogee distance gets affected; in fact, it reduces due to atmospheric drag and to a lesser extent by the solar radiation pressure. In fact, every elliptical orbit tends to become circular with a radius equal to the perigee distance with time. It is interesting to note that the probable lifetime of a satellite in a circular orbit at a height of 200 km is only a few days and that of a satellite at a height of 800 km is about 100 years. For a geostationary satellite, it would be more than million years.

Coming back to modifying the apogee distance, it can be intentionally increased or decreased by applying a thrust Δv at the perigee point in the direction of motion of satellite or opposite to direction of motion of satellite, respectively (Figure 4.37). Δr_a can be computed from eqn. 4.45.

$$\Delta v = \frac{\Delta r_a \mu}{\left[V_p\left(r_a+r_p\right)^2\right]} \tag{4.45}$$

V_p is the velocity at the perigee

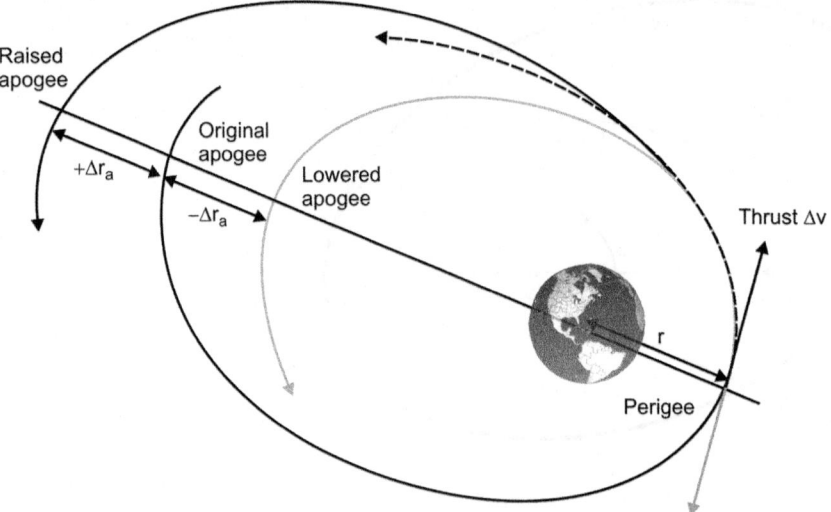

Figure 4.37 Changing the apogee distance.

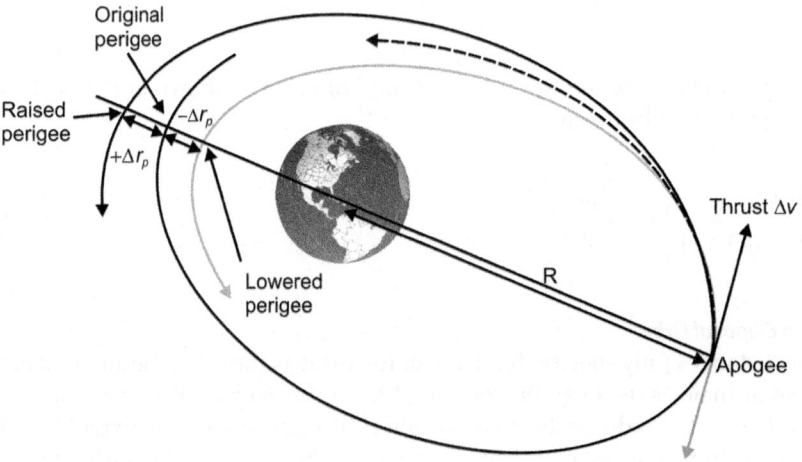

Figure 4.38 Changing the perigee distance.

The perigee distance can similarly be increased or decreased by applying a thrust Δv at the apogee point in the direction of motion of satellite or opposite to direction of motion of satellite, respectively (Figure 4.38). Δr_p can be computed from eqn. 4.46.

$$\Delta v = \frac{\Delta r_p \mu}{\left[V_a \left(r_a + r_p \right)^2 \right]} \tag{4.46}$$

V_a is velocity at the apogee point

So far, we have talked about modifying apogee and perigee distances. A very economical manoeuvre that can be used to change the radius of a circular orbit is the Hohmann transfer that uses an elliptical trajectory tangential to both the current and modified circular orbit. The process requires application of two thrusts Δv and $\Delta v'$. The thrust is applied in the direction

Figure 4.39 Increasing the radius of the circular orbit.

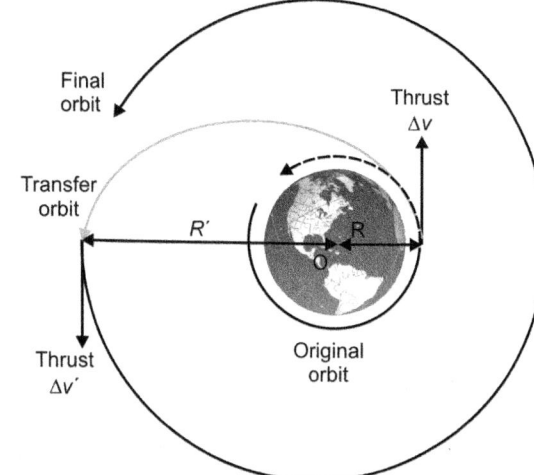

Figure 4.40 Decreasing the radius of the circular orbit.

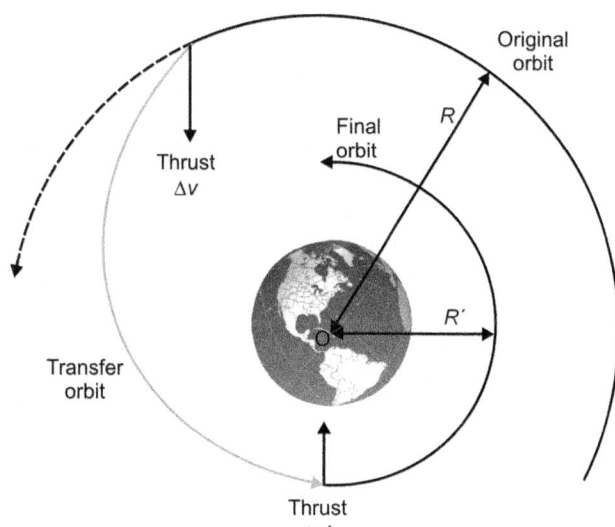

of motion of satellite if the orbit radius is to be increased (Figure 4.39) and opposite to the direction of motion of the satellite if the orbit radius is to be decreased (Figure 4.40). The two thrusts can be computed from eqns. 4.47 and 4.48.

$$\Delta v = \sqrt{\left(\frac{2\mu R'}{\{R(R+R')\}}\right)} - \sqrt{\left(\frac{\mu}{R}\right)} \tag{4.47}$$

$$\Delta v' = \sqrt{\left(\frac{\mu}{R'}\right)} - \sqrt{\left(\frac{2\mu R}{\{R'(R+R')\}}\right)} \tag{4.48}$$

R = Radius of initial orbit and R' = Radius of final orbit

4.2.1.2.5   Position of a Satellite in Its Orbit

The position of a satellite was defined earlier in terms of its passage time at the perigee point. Modifying apogee and perigee distances also change the time of satellite's passage at the perigee point due to change in velocity of the satellite. A similar manoeuvre can be used to modify the longitudinal location of a geostationary satellite.

4.2.2   Satellite Launch Sequence

In the following paragraphs, we shall discuss typical launch sequences employed worldwide for putting the satellites in the geostationary orbit as putting a satellite in a lower circular or elliptical orbit is only a step towards achieving the geostationary orbit.

4.2.2.1   Types of Launch Sequence

There are two broad categories of launch sequence; one that is employed by expendable launch vehicles such as ARIANE of European Space Agency and ATLAS CENTAUR and THOR DELTA of USA and the other that is employed by reusable launch vehicle such as SPACE SHUTTLE of USA and the BURAN of Russia. Irrespective of whether a satellite is launched by a reusable launch vehicle like SPACE SHUTTLE or an expendable vehicle like ARIANE, the satellite heading for a geostationary orbit is first placed in a transfer orbit. The transfer orbit is elliptical in shape with its perigee at an altitude between 200 and 300 km and apogee at the geostationary altitude. In some cases, the launch vehicle injects the satellite directly into a transfer orbit of this type. Following this, an apogee manoeuvre circularizes the orbit at geostationary altitude. The last step is then to correct the orbit for its inclination. This type of launch sequence is illustrated in Figure 4.41.

In the second case, the satellite is first injected into a low Earth circular orbit. In the second step, the low Earth circular orbit is transformed into an elliptical transfer orbit with a perigee manoeuvre. Circularization of the transfer orbit and then correction of orbit inclination follow this. This type of sequence is illustrated in Figure 4.42.

In the following paragraphs, we shall discuss some typical launch sequences observed in case of various launch vehicles used to deploy geostationary satellites from some of the prominent launch sites all over the world. The cases presented here include launch of geostationary satellites

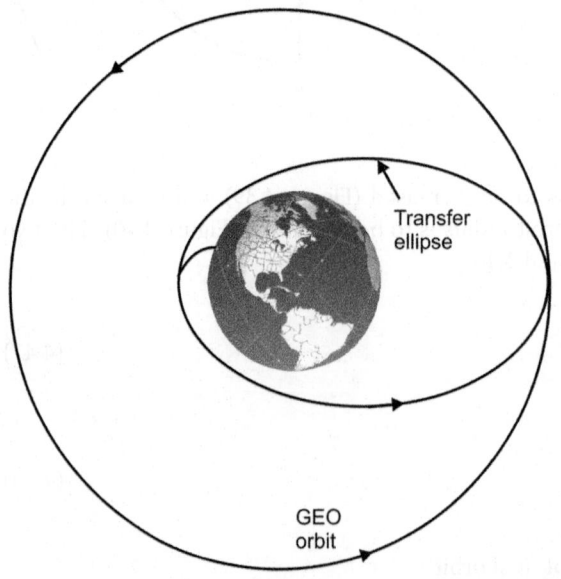

Figure 4.41 Possible geostationary satellite launch sequence: first type.

Figure 4.42 Possible geostationary satellite launch sequence: second type.

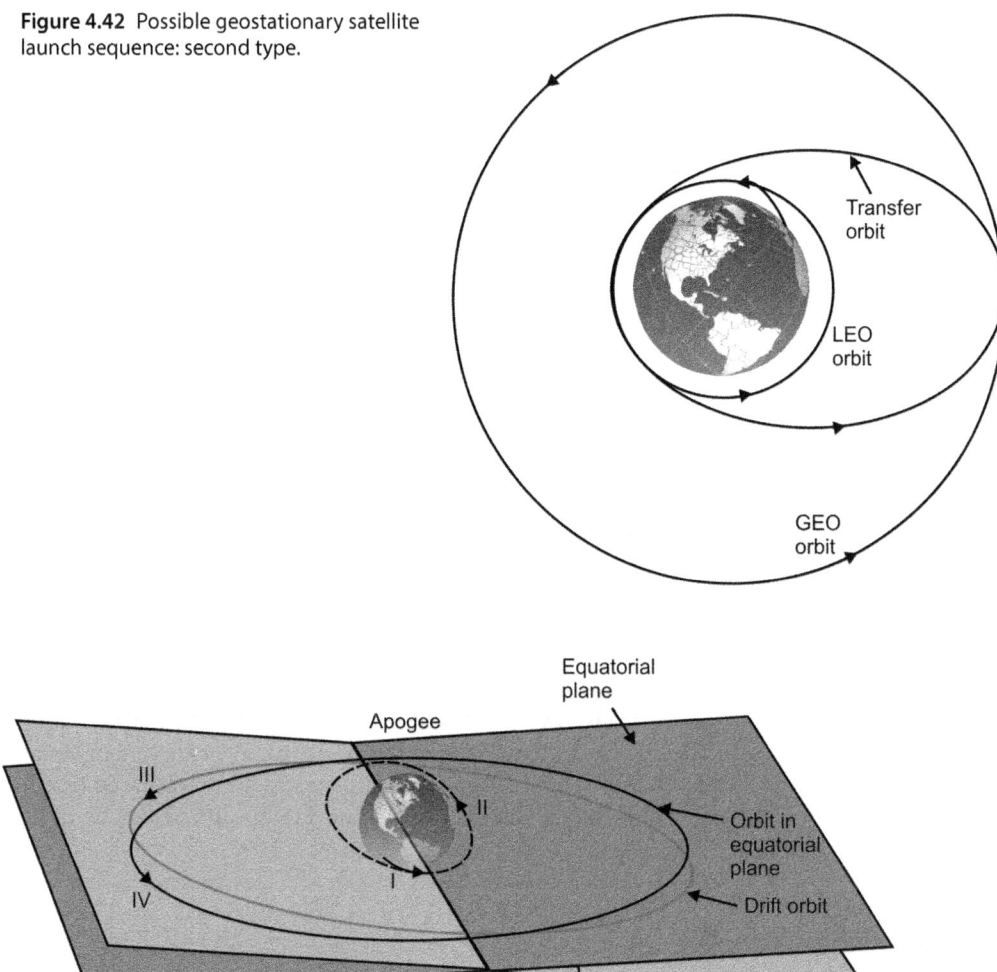

Figure 4.43 Typical launch of a geostationary satellite from Kourou.

from Kourou in French Guiana, Cape Canaveral situated towards the eastern coast of Latin America and Baikonur in Russia. We shall also describe a typical space shuttle launch. The two types of launch sequence briefly described in the preceding lines shall be amply evident in the real life illustrations that follow.

4.2.2.2 Launch from Kourou

We shall illustrate a typical ARIANE launch of a geostationary satellite from Kourou in French Guiana. Different steps involved in the entire process are:

1) The launch vehicle takes the satellite to a point, which is intended to be the perigee of the transfer orbit, at a height of about 200 km above the surface of Earth. The satellite along with its apogee boost motor is injected before the launch vehicle crosses the equatorial plane as shown in Figure 4.43. The injection velocity is such that the injected satellite attains an eccentric elliptical orbit with an apogee at about 36 000 km. The orbit is inclined at about 7°, which is expected, as the latitude of the launch site is 5.2°.

2) In the second step, after the satellite has completed several revolutions in the transfer orbit, apogee boost motor is fired during the passage of satellite at apogee point. The resulting thrust gradually circularizes the orbit. The orbit now is a circular orbit with an altitude of 36 000 km.
3) Further thrust applied at apogee point bring the inclination to 0° thus making the orbit a true circular and equatorial orbit
4) The last step is to attain correct longitude and attitude. This is also achieved by applying thrust either tangential or normal to the orbit.

4.2.2.3 Launch from Cape Canaveral
Different steps involved in the process of launching a geostationary satellite from Cape Canaveral are:

1) The launch vehicle takes the satellite to a point, which is intended to be the perigee of the transfer orbit, at a height of about 300 km above the surface of Earth and injects the satellite first into a circular orbit called the parking orbit. The orbit is inclined at an angle of 28.5° with the equatorial plane as shown in Figure 4.44. Reasons for this inclination angle have been explained earlier.
2) In the second step, a perigee manoeuvre associated with firing of perigee boost motor transforms the circular parking orbit to an eccentric elliptical transfer orbit with perigee and apogee distances of 300 and 36 000 km, respectively.
3) In the third step, an apogee manoeuvre similar to the one used in case of Kourou launch circularizes the transfer orbit. Till now, the orbit inclination is 28.5°. In another apogee manoeuvre, orbit inclination is brought to 0°. The thrust required in this manoeuvre is obviously much larger than that required in case inclination correction in Kourou launch.
4) In the fourth and last step, several small manoeuvres are used to put the satellite in desired longitudinal position.

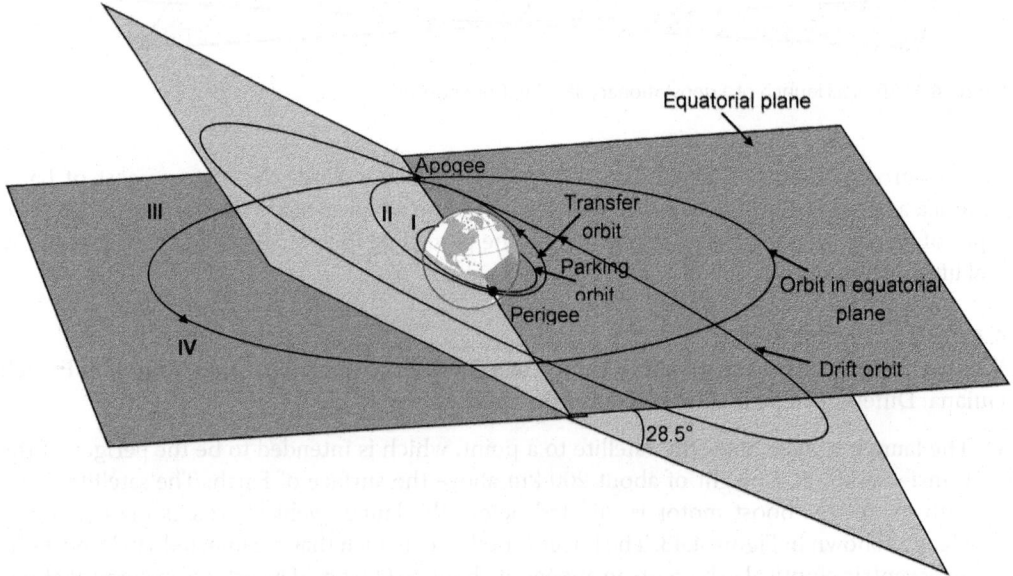

Figure 4.44 Typical launch of a geostationary satellite from Cape Canaveral.

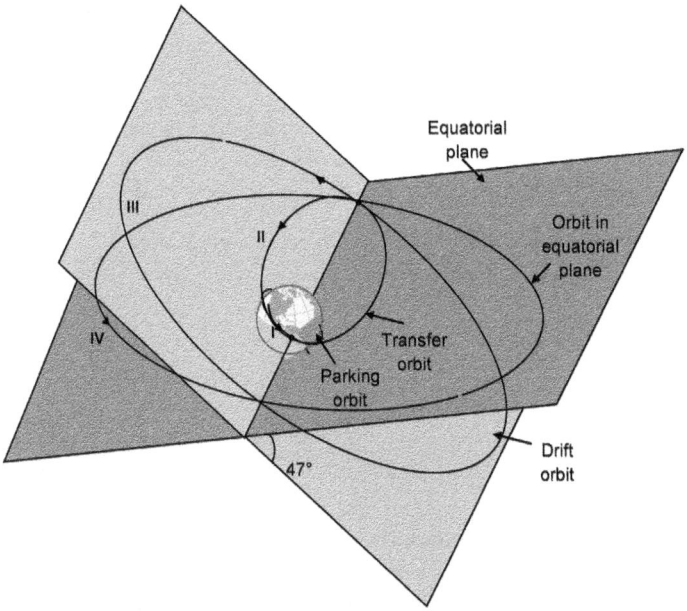

Figure 4.45 Typical launch of a geostationary satellite from Baikonur.

4.2.2.4 Launch from Baikonur

The launch procedure for a geostationary launch from Baikonur is similar to the one described in the launch case from Cape Canaveral. The different steps involved in the process of launch of a geostationary satellite from Baikonur are:

1) The launch vehicle injects the satellite in a circular orbit with an altitude of 200 km and an inclination of 51°.
2) In the second step, during the first passage of the satellite through the intended perigee, a manoeuvre puts the satellite in the transfer orbit with an apogee of little above 36 000 km. The orbit inclination now is 47°.
3) In the third step, the transfer orbit is circularized and inclination corrected at the descending node.
4) In the fourth and last step, satellite drifts to its final longitudinal position. Different steps are depicted in Figure 4.45.

4.2.2.5 Space Shuttle Launch

The space shuttle launch is also similar to launch procedure described in case of launch from Cape Canaveral and Baikonur. The space shuttle injects the satellite along with its perigee motor in a circular parking orbit at an altitude of 100 km. After several revolutions, the perigee motor is fired by signals from an on-board timing mechanism and the thrust generated thereby puts the satellite in a transfer orbit with an apogee at geostationary height. The perigee motor is then jettisoned. The orbit is then circularized with an apogee manoeuvre.

4.2.3 Orbit Perturbations

The satellite, once placed in its orbit, experiences various perturbing torques. These include gravitational forces from other bodies like solar and lunar attraction, magnetic field interaction,

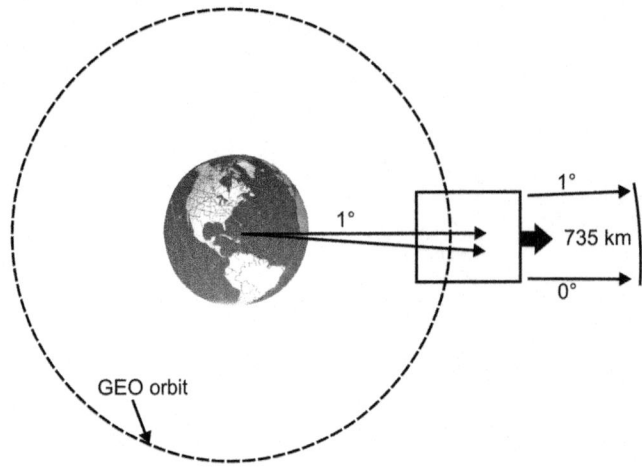

Figure 4.46 Drift of a geostationary satellite.

solar radiation pressure and so on. Due to these factors, the satellite orbit tends to drift and its orientation also changes. The satellite's position thus needs to be controlled both in the east–west as well as the north–south directions. The east–west location needs to be maintained to prevent radio frequency (RF) interference from neighbouring satellites. It may be mentioned here that in the case of a geostationary satellite, a 1° drift in the east or west direction is equivalent to a drift of about 735 km along the orbit (Figure 4.46). The north–south orientation has to be maintained to have proper satellite inclination. The Earth is not a perfect sphere and is flattened at the poles. The equatorial diameter is about 20–40 km more than the average polar diameter. Also, the equatorial radius of the Earth is not constant. In addition, the average density of Earth is not uniform. All of this results in a non-uniform gravitational field around the Earth that, in turn, results in variation in gravitational force acting on the satellite due to the Earth. The effect of variation in the gravitational field of the Earth on the satellite is more predominant for geostationary satellites than for satellites orbiting in low Earth orbits as in the case of these satellites the rapid change in the position of the satellite with respect to the Earth's surface will lead to the averaging out of the perturbing forces. In the case of a geostationary satellite, these forces result in an acceleration or deceleration component that varies with the longitudinal location of the satellite.

In addition to the variation in the gravitational field of the Earth, the satellite is also subjected to the gravitational pulls of the Sun and the Moon. The Earth's orbit around the Sun is an ellipse whose plane is inclined at an angle of 7° with respect to the equatorial plane of the sun. The Earth is tilted around 23° away from the normal to the ecliptic. The Moon revolves around the Earth with an inclination of around 5° to the equatorial plane of the Earth. Hence, the satellite in orbit is subjected to a variety of out-of-plane forces which change the inclination of the satellite's orbit. The gravitational pulls of the Earth, Sun and Moon have negligible effects of the satellites orbiting in LEO orbits, where the effect of atmospheric drag is more predominant. As the perturbed orbit is not an ellipse anymore, the satellite does not return to the same point in space after one revolution. The time elapsed between the successive perigee passages is referred to as anomalistic period. The anomalistic period T_A is given by eqn. 4.49.

$$t_A = \frac{2\pi}{\omega_{mod}} \tag{4.49}$$

Where

$$\omega_{\text{mod}} = \omega_0 \times \left[1 + \left\{ \frac{K\left(1 - 1.5\sin^2 i\right)}{a^2\left(1 - e^2\right)^{3/2}} \right\} \right]$$

ω_0 is the angular velocity for spherical Earth, $K = 66\,063.1704$ km^2, a is the semi-major axis, e is the eccentricity and $i = \cos^{-1} WZ$, WZ is the z-axis component of the orbit normal.

The attitude and orbit control system maintains the satellite's position and its orientation and keeps the antenna pointed correctly in the desired direction (boresighted to the centre of the coverage area of the satellite). The orbit control is performed by firing thrusters in the desired direction or by releasing jets of gas. It is also referred to as station-keeping. Thrusters and gas jets are used to correct the longitudinal drifts (in-plane changes) and the inclination changes (out-of-plane changes). It may be mentioned that the manoeuvres required for correcting longitudinal drifts (referred to as the north–south manoeuvre) require a much larger velocity increment as compared to the manoeuvres required for correcting the inclination changes (referred to as the east–west manoeuvre). Hence, generally a different set of thrusters or gas jets is used for north–south and east–west manoeuvres.

4.2.4 Satellite Stabilization

Commonly employed techniques for satellite attitude control include:

1) Spin Stabilization
2) Three-axis or body stabilization.

4.2.4.1 Spin Stabilization

In a spin-stabilized satellite, the satellite body is spun at a rate between 30 and 100 RPM about an axis perpendicular to the orbital plane (Figure 4.47). Like a spinning top, the rotating body offers inertial stiffness, which prevents the satellite from drifting from its desired orientation.

Spin-stabilized satellites are generally cylindrical in shape. For stability, the satellite should be spun about its major axis having maximum moment of inertia. To maintain stability, the moment of inertia about the desired spin-axis should at least be 10% greater than the moment of inertia about the transverse axis.

Figure 4.47 Spin stabilized satellite.

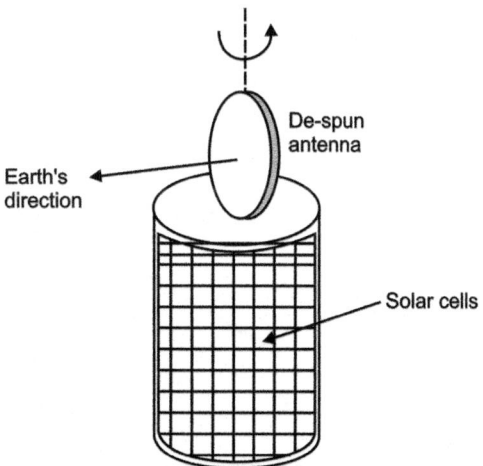

There are two types of spinning configurations employed in spin-stabilized satellites. These include the simple spinner configuration and the dual spinner configuration. In simple spinner configuration, the satellite payload and other subsystems are placed in the spinning section, while the antenna and the feed are placed in the de-spun platform. The de-spun platform is spun in a direction opposite to that of spinning satellite body. In dual spinner configuration, the entire payload along with the antenna and the feed is placed on the de-spun platform and the other subsystems are located on the spinning body. Modern spin-stabilized satellites almost invariably employ a dual spinner configuration. It may be mentioned here that mounting of an antenna system on the de-spun platform in both configurations ensures a constant pointing direction of the antennas. In both configurations, solar cells are mounted on the cylindrical body of the satellite. INTELSAT-1 to INTELSAT-4, INTELSAT-6 and TIROS-1 are some of the popular spin-stabilized satellites.

4.2.4.2 Three-Axis or Body Stabilization

In the case of three-axis stabilization also known as body stabilization, stabilization is achieved by controlling the movement of the satellite along the three axes; that is, yaw, pitch and roll with respect to a reference (Figure 4.48).

The system uses reaction wheels or momentum wheels to correct orbit perturbations. The stability of the three-axis system is provided by the active control system, which applies small corrective forces on the wheels to correct the undesirable changes in the satellite orbit.

Most three-axis stabilized satellites use momentum wheels. The basic control technique used here is to speed up or slow down the momentum wheel depending upon the direction in which the satellite is perturbed. The satellite rotates in a direction opposite to that of the speed change of the wheel. For example, an increase in speed of the wheel in clockwise direction will make the satellite to rotate in counter clockwise direction. Momentum wheels rotate in one direction and can be twisted by a gimbal motor to provide the required dynamic force on the satellite.

An alternative approach is to use reaction wheels. Three reaction wheels are used, one for each axis. They can be rotated in either direction depending upon the active correction force. The satellite body is generally box shaped for three-axis stabilized satellites. Antennas are mounted on the Earth facing side and on the lateral sides adjacent to it. These satellites use flat

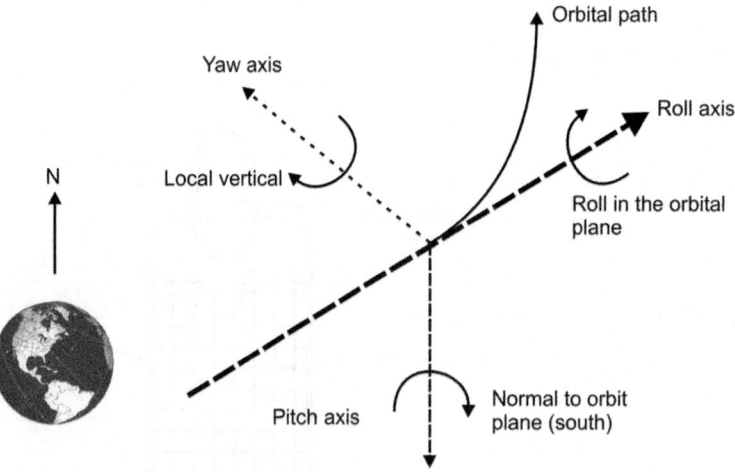

Figure 4.48 Three-axis stabilization.

solar panels mounted above and below the satellite body in such a way that they always point towards the Sun, which is an obvious requirement. Some popular satellites belonging to the category of three-axis stabilized satellites include the INTELSAT-5, INTELSAT-7, INTELSAT-8, GOES-8 and GOES-9, TIROS-N and INSAT-series of satellites. A brief comparison of spin-stabilized and three-axis stabilized satellites is presented next.

1) In comparison to spin-stabilized satellites, three-axis stabilized satellites have more power generation capability and more additional mounting area available for complex antennas structures.
2) Spin-stabilized satellites are simpler in design and less expensive than three-axis stabilized satellites.
3) Three-axis stabilized satellites have the disadvantage that extendible solar array used in these satellites is unable to provide power when the satellite is in transfer orbit, as the array is still stored inside the satellite during this time.

4.2.4.3 Station-Keeping

Station-keeping is the process of maintenance of a satellite's orbit against different factors that can cause temporal drift. Satellites need to have their orbits adjusted from time to time because the satellite even though initially placed in the correct orbit can undergo a progressive drift due to some natural forces such as minor gravitational perturbations due to the Sun and Moon, solar radiation pressure, Earth being an imperfect sphere and so on. The orbital adjustments are usually made by releasing jets of gas or by firing small rockets tied to the body of the satellite.

In case of spin-stabilized satellites station-keeping in the north–south direction is maintained by firing thrusters parallel to the spin-axis in a continuous mode. The east–west station-keeping is obtained by firing thrusters perpendicular to the spin-axis. In case of three-axis stabilization station-keeping is achieved by firing thrusters in the east–west or the north–south direction in a continuous mode.

4.2.5 Satellite Eclipses

With reference to satellites, an eclipse is said to occur when sunlight fails to reach the satellite's solar panel due to an obstruction from a celestial body. The major and most frequent source of eclipse is due to the satellite coming in the shadow of the Earth (Figure 4.49). This is known as a solar eclipse. The eclipse is total, that is, the satellite fails to receive any light whatsoever if it passes through the umbra, which is the dark central region of the shadow, and receives very little light if it passes through the penumbra, which is the lighter region surrounding the umbra (Figure 4.50).

The eclipse occurs as the equatorial plane is inclined at a constant angle of about 23.5° to the ecliptic plane, which is the plane of the Earth's orbit extended to infinity. The eclipse is seen on 42 nights during spring and an equal number of nights during autumn. The effect is the worst

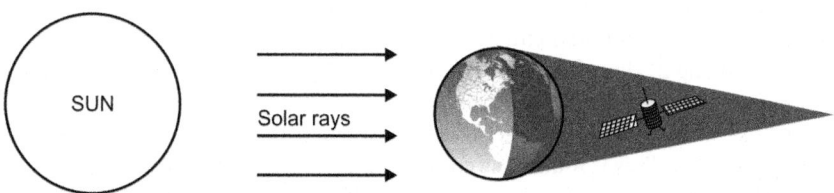

Figure 4.49 Solar eclipse.

Figure 4.50 Umbra and penumbra.

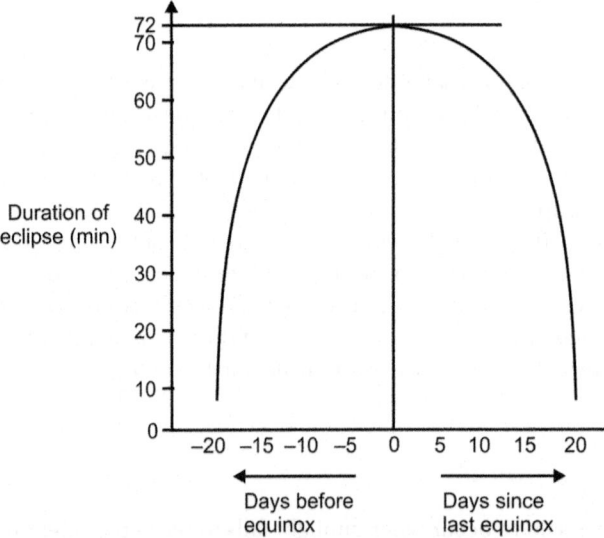

Figure 4.51 Duration of eclipse before and after equinox.

during equinoxes and it lasts for about 72 minutes. Equinox, as explained earlier, is the point in time when the sun crosses the equator making the day and night times equal. The spring and autumn equinoxes, respectively, occur around the 21 March and 21 September. During the equinoxes in March and September, the satellite, Earth and the Sun are aligned at midnight as per local time and the satellite spends about 72 minutes in total darkness. From 21 days before and 21 days after the equinoxes, the satellite crosses the umbral cone each day for some time thereby receiving only a part of solar light for that time. During the rest of year, the geostationary satellite orbit passes either above or below the umbral cone. It is at maximum distance at the time of solstices, above the umbral cone at the time of the summer solstice (20–21 June) and below it at the time of the winter solstice (20–21 December). Hence, duration of eclipse increases from zero to about 72 minutes starting 21 days before the equinox and then decreases from 72 minutes to zero during 21 days following the equinox. Duration of an eclipse on a given day around the equinox can be seen from the graph of Figure 4.51.

Another type of eclipse known as a lunar eclipse occurs when the Moon's shadow passes across the satellite (Figure 4.52). This is much less common and occurs once in 29 years. In fact, for all practical purposes, when we talk about an eclipse with respect to satellites, we refer to a solar eclipse. Talking of solar eclipse, failure of sunlight to reach the satellite interrupts the battery recharging process. The satellite is depleted of its electrical power capacity. It does not

Figure 4.52 Lunar eclipse.

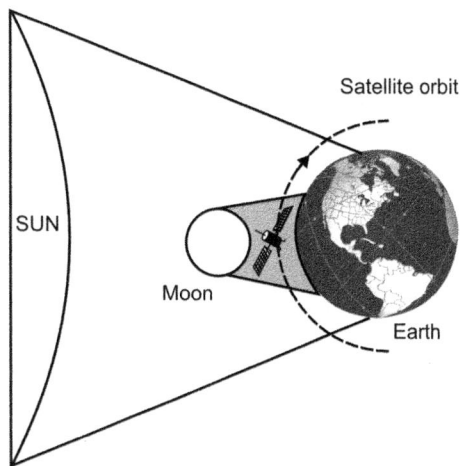

significantly affect low power satellites, which can usually continue their operation with the backup power. The high power satellites, however, shut off for all but essential services.

4.2.6 Look Angles of a Satellite

The look angles of a satellite refer to the coordinates to which an Earth station must be pointed in order to communicate with the satellite and are expressed in terms of azimuth and elevation angles. In case an Earth station is within the footprint or coverage area of a geostationary satellite, it can communicate with the satellite by simply pointing its antenna towards it. The process of accurately pointing the Earth station antenna toward the satellite can be accomplished if the azimuth and elevation angles of the Earth station location are known. Also, elevation angle affects the slant range; that is, line-of-sight distance between the Earth station and the satellite.

The *azimuth angle* (A) of an Earth station is defined as the angle produced by line of inter-section of local horizontal plane and the plane passing through the Earth station, the satellite and the centre of Earth with the true north. We can visualize that this line of intersection between the two previously mentioned planes would be one of the many possible tangents that can be drawn at the point of location of Earth station. Depending upon location of Earth station and the sub-satellite point, which is the point of intersection of line joining satellite and centre of Earth and the equator, the azimuth angle can be computed as under.

For an Earth station in the northern hemisphere,
$A = 180° − A'$............ when the Earth station is west of the satellite
$A = 180° + A'$............ when the Earth station is east of the satellite

For an Earth station in the northern hemisphere,
$A = A'$.................. when the Earth station is west of the satellite
$A = 360° − A'$............ when the Earth station is east of the satellite

A' can be computed from:

$$A' = \tan^{-1}\left[\frac{\tan|\theta_s - \theta_L|}{\sin\theta_l}\right]$$

θ_s = Satellite longitude
θ_L = Earth station longitude
θ_l = Earth station latitude.

The Earth station *elevation angle E* is the angle between the line of intersection of local horizontal plane and the plane passing through the Earth station, the satellite and the centre of Earth and the line joining Earth station and satellite. Figure 4.53(a) and (b) shows elevation angle for two different satellite and Earth station positions. It can be computed from eqn. 4.50.

(a)

(b)

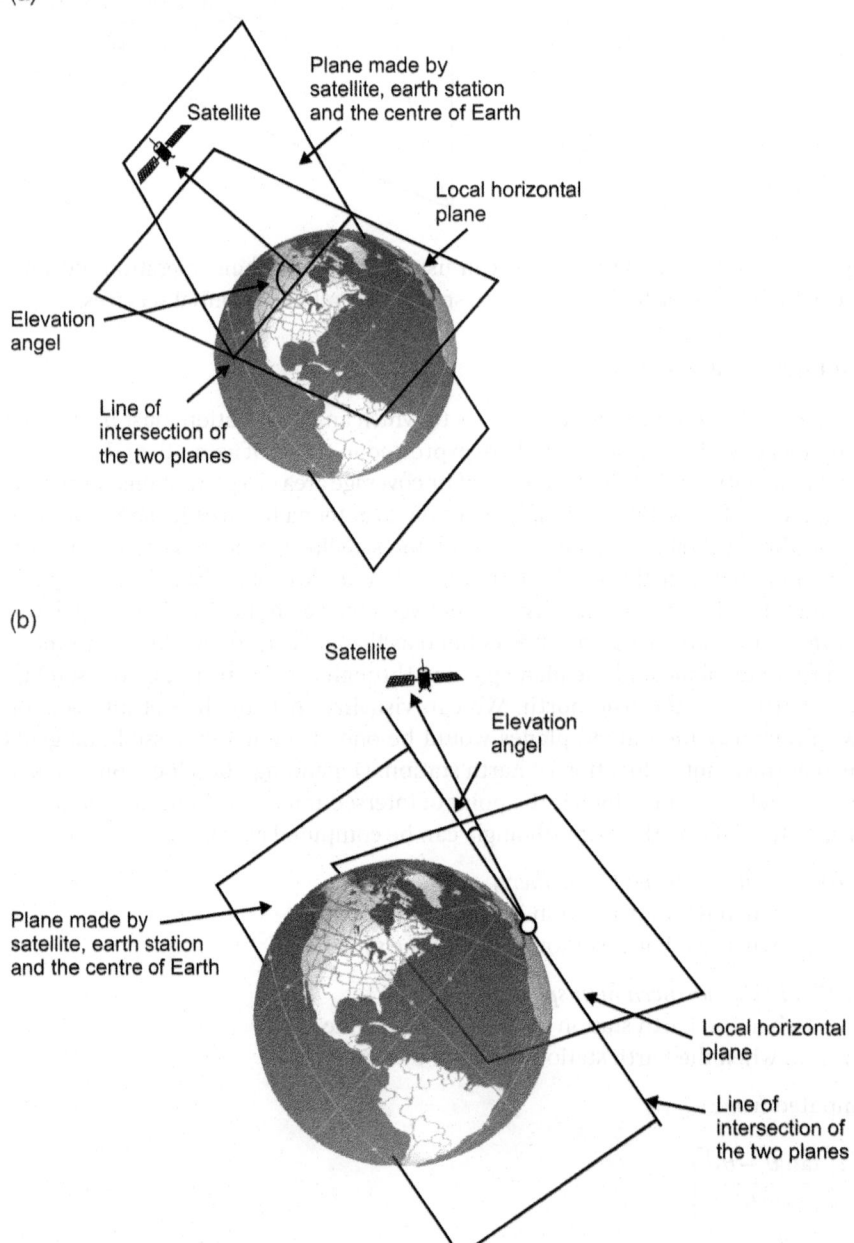

Figure 4.53 Elevation angle.

$$E = \tan^{-1}\left[\frac{\{r - R\cos\theta_I \cos|\theta_S - \theta_L|\}}{R\sin\cos^{-1}\left(\cos\theta_I \cos(\theta_S - \theta_L)\right)}\right] - \cos^{-1}\left(\cos\theta_I \cos|\theta_S - \theta_L|\right) \tag{4.50}$$

r = orbital radius
R = Earth's radius.

Elevation angle E has a direct bearing on slant range. The smaller the angle of elevation of the Earth station, the larger the slant range and the coverage angle. Slant range and coverage angle can be computed from eqns 4.51 and 4.52. As is evident from expression 4.50, a zero angle of elevation leads to maximum coverage angle. A larger slant range means longer propagation delay time and a greater impairment of signal quality, as it has to travel a greater distance through the Earth's atmosphere.

Slant range,

$$D = R^2 + (R+H)^2 - 2R(R+H)\sin\left[E + \sin^{-1}\left\{\frac{R}{(R+H)}\right\}\cos E\right] \tag{4.51}$$

Coverage angle,

$$2\alpha = 2\sin^{-1}\left\{\frac{R}{(R+H)}\right\}\cos E \tag{4.52}$$

R = Radius of Earth
E = Angle of elevation
H = Height of the satellite above the surface of Earth.

4.2.6.1 Computing the Line-of-Sight Distance between Two Satellites

The line-of-sight distance between two satellites placed in same circular orbit can be computed from triangle ABC formed by the points of location of two satellites and centre of Earth (Figure 4.54).

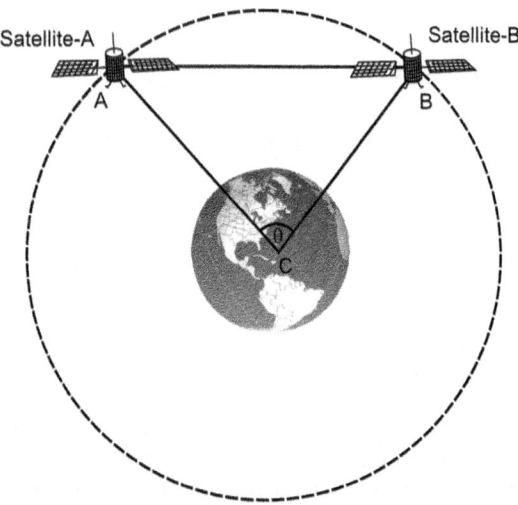

Figure 4.54 Line-of-sight distance between two satellites.

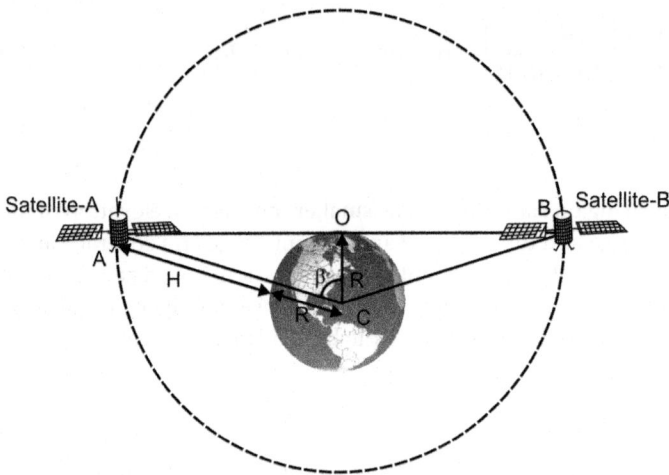

Figure 4.55 Maximum line-of-sight distance between two satellites.

Line-of-sight distance AB in this case is given by eqn. 4.53.

$$AB = \sqrt{\left(AC^2 + BC^2 - 2AC.BC.\cos\theta\right)} \qquad (4.53)$$

Please also note that angle θ is the angular separation of the longitudes of the two satellites. For example, if the two satellites were located at 30°E and 60°E, θ would be equal to 30°. And if the two locations were 30°W and 60°E, then in that case θ would be 90°. Maximum line-of-sight distance between these two satellites would occur when the satellites are so placed that the line joining the two becomes tangent to the Earth's surface as shown in Figure 4.55. In this case, maximum line-of-sight distance (AB) equals OA + OB, which further equals 2OA or 2OB as OA = OB. If R is the radius of Earth and H is the height of satellites above the surface of Earth, then

$$OA = AC\sin\theta = (R+H)\sin\theta$$

Now,

$$\theta = \cos^{-1}\left[\frac{R}{(R+H)}\right] \qquad (4.54)$$

Therefore,

$$OA = (R+H)\sin\left[\cos^{-1}\left\{\frac{R}{(R+H)}\right\}\right] \qquad (4.55)$$

$$\text{Maximum line-of-sight distance} = 2(R+H)\sin\left[\cos^{-1}\left\{\frac{R}{(R+H)}\right\}\right] \qquad (4.56)$$

4.2.7 Earth Coverage

Earth coverage, also known as the *footprint*, is the surface area of the Earth that can possibly be covered by a given satellite. In the discussion to follow, we shall look at the effect of satellite altitude on the Earth coverage provided by the satellite.

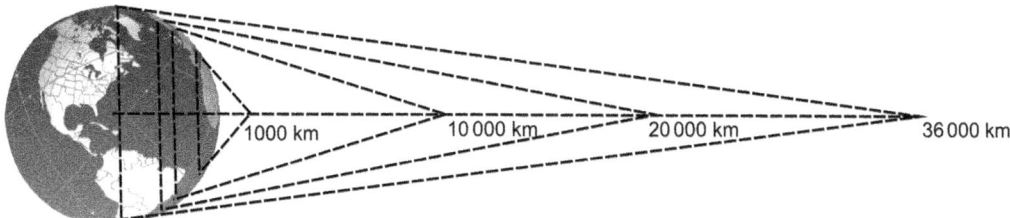

Figure 4.56 Satellite altitude and Earth coverage area

Refer to Figure 4.56. It is evident that the coverage area increases with the height of the satellite above the surface of Earth. It varies from something like 1.5% of Earth's surface for a low Earth satellite orbit at 200 km to about 42% of Earth's surface area for a satellite at geostationary height of 36 000 km. Coverage angle, as mentioned before, can be computed from eqn. 4.57.

Coverage angle,

$$2\alpha = 2\sin^{-1}\left\{\frac{R}{(R+H)}\right\}\cos E \tag{4.57}$$

For maximum possible coverage, $E = 0°$. The expression reduces to eqn. 4.58.

Coverage angle,

$$2\alpha = 2\sin^{-1}\left\{\frac{R}{(R+H)}\right\} \tag{4.58}$$

The coverage angle can be computed to be approximately 150° for a satellite at 200 km and 17° for a satellite at geostationary height of 36 000 km.

4.3 Satellite Hardware

In this section, we shall briefly discuss the different subsystems a typical satellite is made up of and shall address issues like the major function performed by each one of these subsystems along with a brief on their operational considerations.

4.3.1 Satellite Subsystems

Irrespective of the intended application, be it a communications satellite, a weather satellite or even an Earth observation satellite, different subsystems comprising a typical satellite include the following.

1) Mechanical structure
2) Propulsion
3) Thermal control
4) Power supply
5) Tracking, telemetry and command
6) Attitude and orbit control
7) Payload
8) Antennas.

The *structural subsystem* provides the framework for mounting other subsystems of the satellite and also an interface between the satellite and the launch vehicle.

The *propulsion sub-system* is used to provide the thrusts required to impart the necessary velocity changes to execute all the manoeuvres during the lifetime of the satellite. This includes major manoeuvres required to move the satellite from its transfer orbit to geostationary orbit in case of geostationary satellites and also the smaller manoeuvres needed throughout the life span of the satellite, such as those required for station-keeping.

The *thermal control subsystem* is essential to maintain the satellite platform within its operating temperature limits for the type of equipment on board the satellite. It also ensures a reasonable temperature distribution throughout the satellite structure, which is essential to retain dimensional stability and maintain the alignment of certain critical equipment.

The primary function of the *power supply subsystem* is to collect the solar energy, transform it to electrical power with the help of arrays of solar cells and distribute electrical power to other components and subsystems of the satellite. In addition, a satellite also has batteries, which provide standby electrical power during eclipse periods, other emergency situations and also during the launch phase of the satellite when the solar arrays are not yet functional.

The *Telemetry, Tracking and Command (TT&C) subsystem* monitors and controls the satellite right from the lift-off stage to the end of its operational life in space. The tracking part of the subsystem determines the position of the spacecraft and follows its travel using angle, range and velocity information. The telemetry part gathers information on the health of various subsystems of the satellite encodes this information and then transmits the same. The command element receives and executes remote control commands to effect changes to the platform functions, configuration, position and velocity.

The *attitude and orbit control subsystem* performs two primary functions. It controls the orbital path, which is required to ensure that the satellite is in the correct location in space to provide the intended services. It also provides attitude control, which is essential to prevent the satellite from tumbling in space and also to ensure that the antennas remain pointed at a fixed point on the Earth's surface.

The *payload subsystem* is that part of the satellite that carries the desired instrumentation required for performing its intended function and is therefore the most important subsystem of any satellite. The nature of the payload on any satellite depends upon its mission. The basic payload in case of a communication satellite is the transponder, which acts as a receiver, amplifier and transmitter. In case of a weather forecasting satellite, a radiometer is the most important payload. High-resolution cameras, multi-spectral scanners and thematic mappers are the main payloads on board a remote sensing satellite. Scientific satellites have a variety of payloads depending upon the mission. These include telescopes, spectrographs, plasma detectors, magnetometers, spectrometers and so on.

Antennas are used for both receiving signals from ground stations as well as for transmitting signals toward them. There are a variety of antennas available for use on board a satellite. The final choice depends mainly upon the frequency of operation and required gain. Typical antenna types used on satellites include horn antennas, centre fed and offset fed parabolic reflectors and lens antennas.

4.3.2 Mechanical Structure

The mechanical structure weighs between 7 and 10% of the total mass of the satellite at launch. It performs three main functions, namely:

1) It links the satellite to the launcher and thus acts as an interface between the two.
2) It acts as a support for all the electronic equipment carried by the satellite.
3) It serves as a protective screen against energetic radiation, dust and micrometeorites in space.

The mechanical structure that holds the satellite and links it to the launcher is very important. Some of the important design considerations that need to be addressed while designing the mechanical structure of the satellite are briefly described in the following paragraphs.

4.3.2.1 Design Considerations

1) The cost of launching a satellite is a function of its mass. As a result of this, cost of launching one is very high, more so in case of a geostationary satellite. One of the most basic requirements is therefore lightness of its mechanical structure. All efforts are made therefore to reduce the structural mass of the satellite to the minimum. This is achieved by using materials that are light yet very strong. Some of the materials used in the structure include aluminium alloys, magnesium, titanium, beryllium, Kevlar fibres and more commonly the composite materials. Kevlar is a trademark name for a fibre developed and manufactured by the company DuPont. It is characterized by high strength and stiffness and yet low weight and low density. In addition to the lightness of the structure, the choice of material is also governed by many other material properties. The design of the structural subsystem relies heavily on the results of a large number of computer simulations where the structural design is subjected to stresses and strains similar to those likely to be encountered by the satellite during the mission.
2) The structural subsystem design should be such that it can withstand mechanical accelerations and vibrations, which are particularly severe during the launch phase. Therefore the material should be such that it can dampen vibrations. Kevlar has these properties.
3) The satellite structure is subjected to thermal cycles throughout its lifetime. It is subjected to large differences in temperature as the sun is periodically eclipsed by the Earth. The temperatures are typically several hundred degrees Celsius on the side facing the sun and several tens of degrees below zero degrees Celsius on the shaded side. Designers keep this in mind while choosing material for structural subsystem.
4) The space environment generates many other potentially dangerous effects. The satellite must be protected from collision with micrometeorites, space junk and charged particles floating in space. The material used to cover the outside of the satellite should also be resistant to puncture by these fast travelling particles.
5) Structural subsystem also plays an important role in ensuring reliable operation in space of certain processes such as separation of satellite from the launcher, deployment and orientation of solar panels, precise pointing of satellite antennas, operation of rotating parts and so on.

4.3.3 Propulsion Subsystem

The *propulsion subsystem* is used to provide the thrusts required to impart the necessary velocity changes to execute all the manoeuvres during the lifetime of the satellite. This includes major manoeuvres required to move the satellite from its transfer orbit to geostationary orbit in the case of geostationary satellites and also the smaller station-keeping manoeuvres needed throughout the life span of the satellite.

As we shall see in the following paragraphs, most of the onboard fuel, about 95%, is required for east–west station-keeping manoeuvres and only about 5% of the fuel is required for north–south or latitudinal manoeuvres. A small quantity of fuel is retained for the end of satellite's life so that it can be moved out of the orbit by a few kilometres at the end of its life span.

4.3.3.1 Basic Principle

The propulsion system works on the principle of Newton's third law, according to which 'for every action, there is an equal and opposite reaction'. The propulsion system uses the principle of expelling mass at some velocity in one direction to produce thrust in the opposite direction. Depending upon the type of propellant used and mechanism of producing the required thrust, there are three types of propulsion systems in use. These are solid fuel propulsion, liquid-fuel propulsion and electric and ion propulsion. In the case of *solid* and *liquid propulsion* systems, ejection of mass at a high-speed involves generation of a high-pressure gas by high-temperature decomposition of propellants. The high-pressure gas is then accelerated to supersonic velocities in a diverging-converging nozzle. In case of *ion propulsion*, thrust is produced by accelerating charged plasma of an ionized elemental gas such as xenon in a highly intense electrical field.

4.3.3.2 Thrust Force and Specific Impulse

Irrespective of the type, the performance of a propulsion system is measured by the two parameters *thrust force* and *specific impulse*. *Thrust force* is measured in Newton or pounds (force). *Specific impulse* I_{SP} is defined as the impulse, which is the product of force and time, imparted during a time dt by unit weight of the propellant consumed during this time. It is mathematically expressed by eqn. 4.59.

$$\text{Specific Impulse } (I_{sp}) = \frac{Fdt}{gdM} = \left[\frac{F}{g(dM/dt)} \right] \tag{4.59}$$

F = Thrust force
dM = Mass of propellant consumed in time dt
g = Terrestrial gravitational constant = 9.807 m/s^2.

The denominator in expression 4.59 is nothing but the rate at which propellant weight is consumed. Specific impulse thus can also be defined as the thrust force produced per unit weight of propellant consumed per second.

Thrust force F produced when a mass dM is ejected at a velocity v with respect to the satellite can be computed from the velocity increment dv that it imparts to the satellite of mass M using the law of conservation of momentum. According to law of conservation of momentum,
$Mdv = vdM$, which can also be written as:

$$M\frac{dv}{dt} = v\frac{dM}{dt} = F \tag{4.60}$$

If we substitute for F in the expression for specific impulse given here, we get
$I_{sp} = v/g$, which confirms how a specific impulse is expressed in seconds.

Specific impulse is also sometimes expressed in Newton-second/kg or lb (force)-second/lb (mass). The three are interrelated as:

$$I_{sp} \text{ (in seconds)} = \frac{I_{sp}[\text{in lb (force)-second}]}{\text{lb (mass)}} = \frac{I_{sp}[\text{Newton-second/kg}]}{9.807}$$

The significance of specific impulse lies in the fact that it tells us about the mass of the propellant necessary to provide a certain velocity increment to the satellite of a known initial mass. Either of the following expressions given in eqns 4.61 and 4.62 can be used for the purpose.

$$m = M_i \times \left[1 - \exp\left(-\frac{\Delta V}{g \times I_{sp}} \right) \right] \tag{4.61}$$

$$m = M_f \times \left[\exp\left(\frac{\Delta V}{g \times I_{sp}} \right) - 1 \right] \tag{4.62}$$

$$M_i = M_f + m$$

Also, $\Delta V = v \times \log\left(\dfrac{M_i}{M_f}\right)$

Another relevant parameter of interest is the time of operation T of the thrust force F. It is given by eqn. 4.63.

$$T = \frac{g \times m \times I_{sp}}{F} \tag{4.63}$$

This means that when a mass m of a propellant with a specific impulse specification I_{SP} is consumed to produce a thrust force F, then the operational time is given by the ratio of product of specific impulse and weight of propellant consumed to the thrust force produced.

4.3.3.3 Solid Fuel Propulsion

Figure 4.57 shows the cross-section of a typical solid fuel rocket motor. The system comprises of a case usually made of titanium, at the exit of which is attached a nozzle assembly made up of carbon composites. High strength fibres have also been used as the case material. The case is filled with a relatively hard, rubbery, combustible mixture of fuel, oxidizer and binder. The combustible mixture is ignited by a pyrotechnic device in the motor case as shown in the figure. Two igniters are usually employed as to have redundancy. The combustible mixture when ignited burns very rapidly producing an intense thrust, which could be as high as 10^6 N. This type of rocket motor produces a specific impulse of about 300 s. Because of high magnitudes of thrust force and specific impulse, a solid fuel rocket motor is particularly suitable for major orbital manoeuvres such as apogee or perigee kick operations. Either a single solid motor can be integrated into the spacecraft in which case the empty case remains within the spacecraft throughout its lifetime or they can be attached to the bottom of the spacecraft and discarded after use. Although a bi-propellant liquid-fuel rocket motor, to be described in the following paragraphs, also produces thrust force and specific impulse of the same order and is widely

Figure 4.57 Cross-section of a typical solid fuel rocket motor.

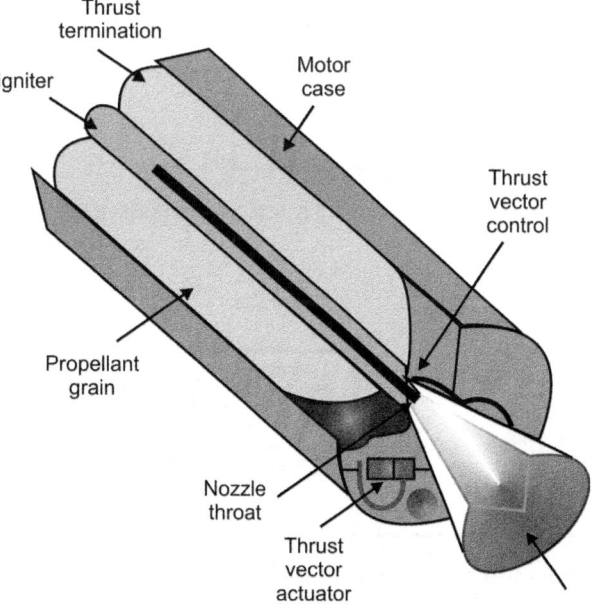

used on geostationary satellites for executing major orbital manoeuvres, a solid rocket motor continues to be used for the same purpose due to its simplicity and efficiency.

4.3.3.4 Liquid Propulsion

Liquid propellant motors can be further classified as *monopropellant motors* and *bi-propellant motors*. Monopropellant motors use a single combustible propellant like hydrazine, which on contact with a catalyst decomposes into its constituents. The decomposition process releases energy resulting in a high-pressure gas at the nozzle. Figure 4.58 shows the cross-section of typical monopropellant liquid-fuel motor.

The specific impulse of a typical monopropellant motor is 200 s and the thrust generated is in the range of 0.05–0.25 N. These motors are used in relatively smaller orbital manoeuvres such as station-keeping manoeuvres where low levels of thrust are required. The performance in terms of thrust produced of a monopropellant can be achieved by using an Electrically Heated Thruster (EHT) in which the propellant is heated using an electrical winding.

A bi-propellant liquid-fuel propulsion system uses separate fuel and oxidizer. The fuel and the oxidizer are stored in separate tanks and are brought together only in the combustion chamber. Figure 4.59 shows a simplified cross-sectional view of a bi-propellant thruster. Having separate fuel and oxidizer produces a greater thrust for the same weight of fuel. This allows a bi-propellant system to yield a longer life. Conversely, it saves on the fuel for a given life. Some of the commonly used fuel-oxidizer combinations include kerosene – liquid oxygen, liquid hydrogen – oxygen and hydrazine or its derivatives – nitrogen tetroxide. The specific impulse can be as high as 300–400 s and the thrust produced can go up to 10^6 N. These systems are particularly suited to major orbital changes requiring large amount of thrust.

Earlier propulsion systems had independent propulsion systems for various thrust requirements such as one for apogee injection, another for orbit control and another for attitude control. A recent trend is to have a common propellant tank system for multiple thrusters. Such a system is called a Unified Propulsion System (UPS). One such unified propulsion system from EADS

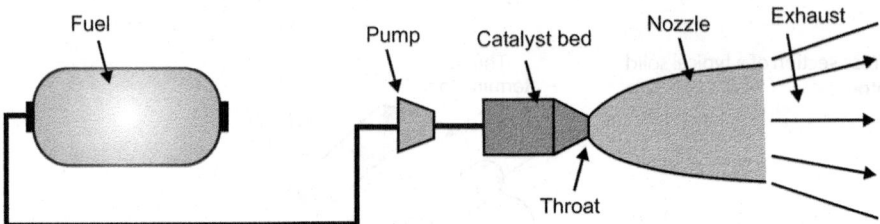

Figure 4.58 Cross-section of a typical monopropellant liquid fuel motor.

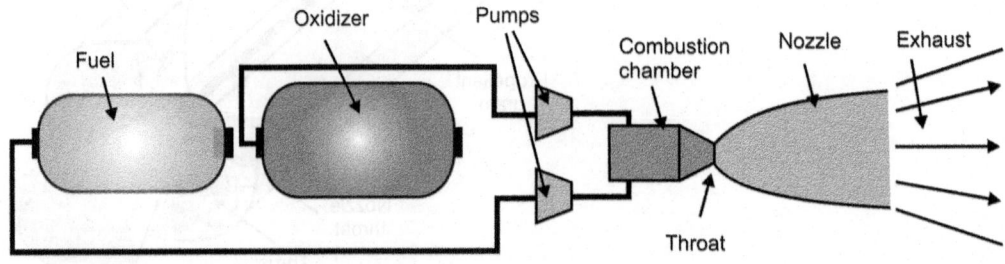

Figure 4.59 Cross-sectional view of a bi-propellant thruster.

Space Transportation is used on METEOSAT. This system feeds two liquid apogee engines of 400 N thrust each and six relatively smaller thrusters of 10 N each.

4.3.3.5 Electric and Ion Propulsion

All electrical propulsion systems use electrical power derived from the solar panels to provide most of the thrust. Some of the common names include ARCJET propulsion, Pulsed Plasma Thruster (PPT), Hall thruster and Ion propulsion.

An *ARCJET thruster* produces a high-energy exhaust plume by electrical heating (using electrical power in the range of 1–20 kW) of ammonia used as fuel. It uses a nozzle to control the plume. It is capable of producing specific impulse in the range of 500–800 s and a thrust that is an order of magnitude smaller than that produced by a monopropellant hydrazine thruster.

A *Pulsed Plasma Thruster (PPT)* uses a Teflon propellant and an electrical power in the range of 100 to 200 W. The magnitude of thrust produced in this case is much smaller than even the ARCJET thruster and is three orders of magnitude smaller than that produced by a monopropellant hydrazine thruster. The magnitude of thrust can be adjusted by varying the pulse rate. This type of thruster is capable of producing a specific impulse in excess of 1000 s.

A *Hall thruster* supports an electric discharge between two electrodes placed in a low-pressure propellant gas. A radial magnetic field generates an electric current due to the Hall Effect. The electric current interacts with the magnetic field to produce a force on the propellant gas in the downstream direction. This type of thruster developed in Russia is capable of producing a thrust magnitude in the order of 1600 s.

In case of *ion propulsion*, thrust is produced by accelerating charged plasma of an ionized elemental gas such as xenon in a highly intense electrical field. It carries very little fuel and in turn relies on acceleration of charged plasma to a high velocity. An ion thruster is capable of producing a specific impulse of the order of 3000 s at an electrical power of about 1 kW. Thrust magnitude is quite low, which necessitates that the thruster is operated over extended periods of time to achieve the required velocity increments.

Figure 4.60 shows the basic arrangement of an ion thruster, which is self-explanatory. A high value of specific impulse coupled with low thrust magnitude makes it ideally suited to attitude

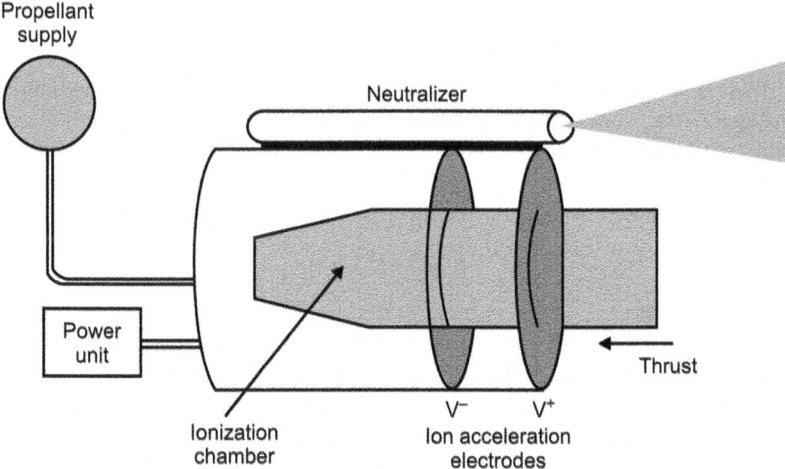

Figure 4.60 Basic arrangement of an ion thruster.

control and station-keeping operations. As a result, this type of thruster is increasingly popular with geostationary satellites where station-keeping requirements remain throughout the lifetime of the satellite. Such a thruster can lead to a lot of savings on station-keeping fuel.

An ion thruster was first flown aboard the PanAmSat-5 launched successfully in 1997. Another popular and proven Ion propulsion system is the RITA (Radio frequency Ion Thruster Assembly) propulsion system. Ion thrusters particularly suit station-keeping and attitude control operations though they can be used for orbital transfers between LEO, MEO and GEO. Compared to classical propellants, ion thrusters do not pollute the space environment as they are driven by environmental xenon gas. Their minimal fuel consumption and long life make them ideally suited for research flights and deep space missions too.

4.3.4 Thermal Control Subsystem

As mentioned earlier, the primary objective of the thermal control system is to ensure that each and every subsystem on board the satellite is not subjected to a temperature that falls outside its safe operating temperature range. Different pieces of equipment may have different normal operating temperature ranges. As an illustration, majority of electronic equipment has an operating temperature range of $-10- +40°C$, the batteries used on board the satellite usually have an operating temperature range of $0- +20°C$, the solar cells have a relatively much wider permissible operating range of $-190- +60°C$ and energy dissipating components such as power amplifiers have an operational temperature limits of $-10- +80°C$. The thermal control system maintains the satellite platform within its operating temperature limits for the type of equipment on board. The thermal control subsystem also ensures the desired temperature distribution throughout the satellite structure, which is essential to retain dimensional stability and to maintain the alignment of certain critical equipment.

4.3.4.1 Sources of Thermal Inequilibrium

There are both internal and external sources that cause changes in the temperature. There are fundamentally three sources of radiation external to the spacecraft. The first and the foremost is the radiation from the Sun. The Sun is equivalent to a perfect black body radiating at an absolute temperature of 5760 K. About 40% of this radiant energy is in the visible spectrum and about 50% of it is in the infrared. This radiant energy produces a flux of about 1370 W/m^2 at the Earth's orbit. Earth and its atmosphere constitute the second source of radiation. Earth, along with its atmosphere, also acts like a black body radiating at 250 K. This radiation is predominantly in the infrared and produces a radiant flux of about 150 W/m^2 in the low Earth satellite orbits. However, flux due to this radiation at geostationary orbit distances is negligible. The third source is the space itself that acts like a thermal sink at 0 K.

Internal to the satellite platform, there are a large number of sources of heat generation as no piece of equipment is 100% efficient. For example, a TWT amplifier with a typical power rating of 200–250 W may have an efficiency of about 40% thus dissipating heat power to the tune of 150 W per amplifier.

4.3.4.2 Mechanism of Heat Transfer

In order to ensure that each and every subsystem on board the satellite operates within its prescribed temperature limits, there is a need to have some mechanism of heat transfer to and from these subsystems. There are three modes of heat transfer that can be used to remove heat from or add heat to a system. These are *conduction* (mechanism of heat transfer through a solid), *convection* (mechanism of heat transfer through a fluid) and *radiation* (mechanism of heat transfer through a vacuum).

On Earth, convection is the dominant mode of heat transfer. Here, heat transfer through radiation is often not significant. Radiation is, however, the dominant mode of heat transfers in space. The convection mode of heat transfer does exist on the satellite platform to a lesser extent where it is used to redistribute heat rather then remove or add any. On a satellite platform, all heat removal or addition must therefore be done through radiation.

4.3.4.3 Types of Thermal Control

Thermal control systems are either *passive* or *active*. *Passive* techniques include having multi-layer insulation surfaces, which either reflect or absorb radiation that is produced internally or generated by an external source. They have no moving parts or electrical power input. These techniques include a good layout plan for the equipment, careful selection of materials for structure, thermal blankets, coatings, reflectors, insulators, heat sinks and so on.

It may be mentioned here that the external conditions are widely different when the satellite is facing the Sun to what they are during the eclipse periods. Also some satellites will always be in an orbit where one side of their body is always facing the Sun and the other facing the colder side of space. In order to achieve thermal regulation, the satellite is shaded as much as possible from changes of radiation from the suSn by using highly reflective coatings and other forms of thermal insulation called thermal blanketing. Thermal blankets, usually golden in colour (gold is a good IR reflector), are used cover the satellite to shade the satellite from excessive heating due to sunlight or to retain internal satellite heat to prevent too much cooling. *Optical solar reflectors* are also used on some satellites for the same purpose.

Active techniques are usually employed to cope with sudden changes in temperature of relatively larger magnitude such as those encountered during an eclipse when the temperature falls considerably. Active systems include remote heat pipes, controlled heaters and mechanical refrigerators. The heaters and refrigerators are controlled either by sensors on board or activated by ground commands.

Heat pipes are highly effective in transferring heat from one location to another. A typical heat pipe of space craft quality has an effective thermal conductivity several thousands of times that of copper. Heat pipes in fact are a fundamental aspect of satellite thermal and structural subsystem design as in the space environment radiation and conduction are the sole means of heat transfer. Whereever possible, heat producing components such as power amplifiers are mounted on the inside of the outside wall. The excess heat is transferred to the outside through thermally conducting heat pipes.

Thermal designs of spin- and three-axis stabilized satellites are different from each other. In a spin-stabilized satellite, the main equipment is mounted within the rotating drum. The rotation of the drum enables every piece of equipment to receive some solar energy and to maintain the temperature to around 20–25°C for most of the time, except during an eclipse. To minimize the effect of eclipse, the drum is isolated from the equipment by heat blankets. The temperature fluctuation is more in case of three-axis stabilized satellites, as their orientation remains fixed with respect to Earth. Therefore, an insulation blanket is placed around the satellite to maintain the temperature of the equipment inside within desirable limits.

4.3.5 Power Supply Subsystem

The *power supply subsystem* generates, stores, controls and distributes electrical power to other subsystems on board the satellite platform. The electrical power needs of a satellite depend upon the intended mission of the spacecraft and the payload that it carries along with it to carry out the mission objectives. The power requirement could vary from a few hundreds of watts to tens of kilowatts.

4.3.5.1 Types of Power Systems

Although power systems for satellite applications have been developed based on use of solar energy, chemical energy and nuclear energy, the solar energy driven power systems are undoubtedly the favourite and are the most commonly used ones. This is due to abundance of mostly uninterrupted solar energy available in the space environment. Here, we are referring to the use of photon energy in the solar radiation. The radiant flux available at the Earth's orbit is about 1370 W/m^2.

There are power systems known as *heat generators* that make use of heat energy in the solar radiation to generate electricity. A parabolic dish of mirrors reflects heat energy of solar radiation through a boiler, which in turn feeds a generator thus converting solar energy into electrical power. This mode of generating power is completely renewable and efficient if the satellite remains exposed to solar radiation. It can also be used in conjunction with rechargeable batteries. This source of power is, however, very large and heavy and is thus appropriate only for large satellites.

Batteries, store electricity in the form of chemical energy, are invariably used together with solar energy driven electrical power generators to meet the uninterrupted electrical power requirement of the satellite. They are never used as the sole medium of supplying the electrical power needs of the satellite. The batteries used here are rechargeable batteries that are charged during the period when solar radiation is falling on the satellite. During the periods of eclipse when solar radiation fails to reach the satellite, the batteries supply electrical power to the satellite.

Nuclear fission is currently the commonly used technique of generating nuclear energy and eventually may be replaced by nuclear fusion technology when the latter is perfected. In nuclear fission, heavy nucleus of an atom is made to split into two fragments of roughly equivalent masses releasing large amount of energy in the process. On satellites, nuclear power is generated in Radio Isotopic Thermoelectric Generators (RTG). The advantage of nuclear power vis-à-vis its use on satellites is that it is practically limitless and it won't run out before the satellite becomes useless for other reasons. The disadvantage is the danger of radioactive spread over the Earth in the event of the rocket used to launch the satellite exploding before it escapes the Earth's atmosphere. Nuclear power is not used in Earth orbiting satellites because when its orbit decays, the satellite falls back to the Earth and burns up in the atmosphere spreading radioactive particles over the Earth. Nuclear power is effective in case of satellites intended for space exploration as these satellites may go deep into space too far from the sun for any solar energy driven power system to be effective. So, nuclear power though seen to be advantageous for interplanetary spacecraft is not exploited for commercial satellites because of the cost and possible environmental hazards.

4.3.5.2 Solar Energy Driven Power System

The major components of a solar power system are the *solar panels* (of which the solar cell is the basic element), rechargeable batteries, battery chargers with inbuilt controllers, regulators and inverters to generate various DC and AC voltages required by various subsystems. Figure 4.61 shows the basic block-schematic arrangement of a regulated bus power supply system. The diagram is self-explanatory. Voltage regulation during the sunlight and eclipse periods is achieved by decoupling the battery from the bus by means of a *Battery Discharge Regulator* (BDR) as shown in the figure. During the sunlight condition, the voltage of the solar generator and also the bus is maintained at constant amplitude with the voltage regulator connected across the solar generator. During the eclipse periods, a battery provides power to the bus at a constant voltage again and the voltage is maintained constant by means of a BDR.

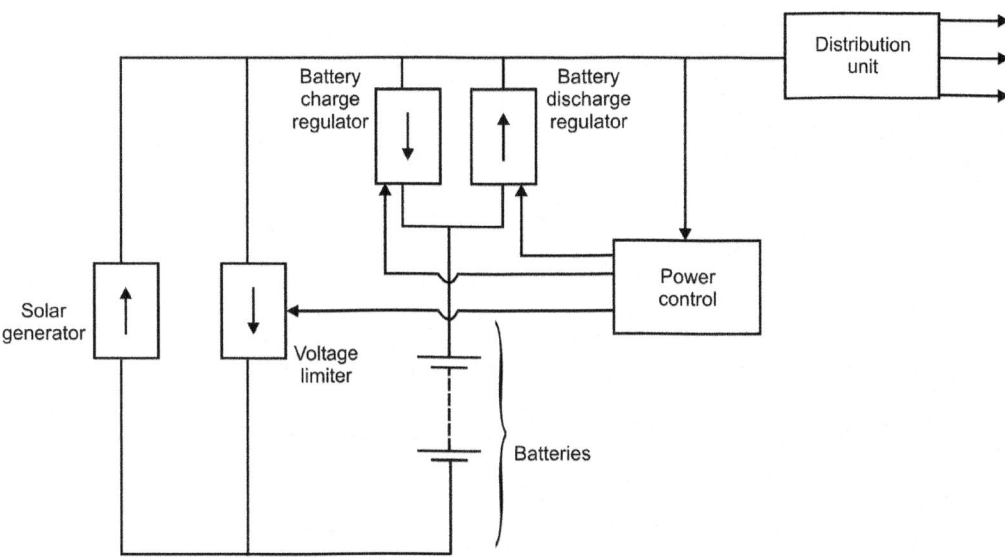

Figure 4.61 Basic block-schematic arrangement of a regulated bus power supply system.

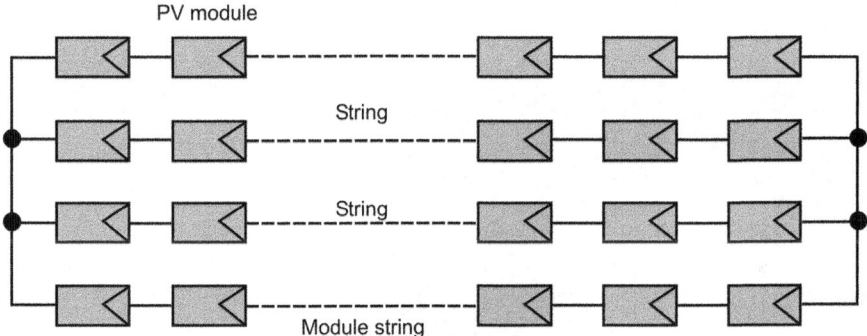

Figure 4.62 Series-parallel arrangement of solar cells.

4.3.5.3 Solar Panels

A solar panel is nothing but a series and parallel connection of a large number of solar cells. Figure 4.62 shows this series-parallel arrangement of solar cells. The voltage output and the current delivering capability of an individual solar cell are very small for it to be any use as an electrical power input to any satellite subsystem. As an example, a typical solar cell would produce 500 mV output with a load current capability of about 150 mA. The series-parallel arrangement is to get the desired output voltage with required power delivery capability. A large surface area is therefore needed in order to produce the required amount of power. The need for large solar panels must, however, be balanced with the need for the entire satellite to be relatively small.

The three-axis body stabilized satellites use flat solar panels (Figure 4.63) whereas spin-stabilized satellites use cylindrical solar panels (Figure 4.64). Both types have their own advantages and disadvantages. In case of three-axis stabilized satellites, the flat solar panels can be rotated to intercept maximum solar energy to produce maximum electricity. However, as the solar panels always face the Sun, they operate at relatively higher temperatures and

Figure 4.63 Flat solar panel on a three-axis stabilized satellite. (*Source:* Courtesy of NASA.)

Figure 4.64 Cylindrical solar panels used on spin-stabilized satellites. (*Source:* Courtesy of NASA.)

thus reduced efficiency as compared to solar panels on spin-stabilized satellites where the cells can cool down when in shadow.

On the other hand, in case of spin-stabilized satellites, such as INTELSAT-VI series satellites, only one-third of the solar cells are facing the Sun at a time and hence more number of cells is needed to get the desired power, which in turn leads to increase in the mass of the satellite. This disadvantage is however partially offset by a reduction in mass of relatively simpler thermal control system and attitude control system in case of spin-stabilized satellites. It may be mentioned here that in case of newer satellites requiring more power, the balance may tilt in favour of three-axis stabilized satellites.

4.3.5.4 Batteries

Batteries are used on board the satellite to meet the power requirements when the same cannot be covered by solar panels as is the case during eclipse periods. Rechargeable batteries are almost invariably used for the purpose. These are charged during the period when solar radiation is available to the satellite's solar panels and then employed during eclipse periods or to meet short term peak power requirement. Batteries are also used during the launch phase, before the solar panels are deployed.

Choice of the right battery technology for a given satellite mission is governed by various factors. These include Frequency of use, Magnitude of load and depth-of-discharge. Generally, fewer cycles of use and less charge demanded of each cycle lead to a longer battery life. The choice of battery technology is closely related to the satellite orbit. Batteries used on board low Earth orbit satellites encounter much larger number of charge/discharge cycles compared to batteries on board geostationary satellites. LEO satellites have orbital period of the order of 100 min and the eclipse period is 30–40 min per orbit. For GEO satellites the orbital period is 24 h and the eclipse duration is 0–72 min during equinoxes. Batteries on LEO satellites are therefore allowed a lower depth-of-discharge. On the other hand, batteries on geostationary satellites can be subjected to a greater depth-of-discharge. One of the major points during battery design is that their capacity is highly dependent on the temperature. As an example, the nickel metal hydride battery has a maximum capacity between the operating temperature of 10–15°C and its capacity decreases at a rate of 1A-h/°C outside this range.

Commonly used batteries are the nickel-cadmium, nickel metal hydride and nickel-hydrogen batteries. These have specific energy specifications of 20–30 Wh/kg (in case of Ni-Cd batteries) and 35–55 Wh/kg (in case of Ni-H and Ni-MH batteries). Small satellites in low Earth orbits mostly employ nickel-cadmium batteries. Nickel-hydrogen batteries are slowly replacing these because of their higher specific energy and longer life expectancy. Currently, GEO satellites mostly employ nickel-hydrogen batteries. Lithium-ion batteries with a specific energy specification of 70–110 Wh/kg are the batteries of the future.

4.3.6 Attitude and Orbit Control Subsystem

The *attitude and orbit control subsystem* performs twin functions of controlling the orbital path, required to ensure that the satellite is in the correct location in space to provide the intended services and to provide attitude control, which is essential to prevent the satellite from tumbling in to space. In addition, it also ensures that the antennas remain pointed at a fixed point on the Earth's surface. The requirements on the attitude and orbit control differ during the launch phase and the operational phase of the satellite.

Orbit control is required to correct for the effects of perturbation forces. These perturbation forces may alter one or more of the orbital parameters. An orbit control subsystem provides correction to these undesired changes. This is usually done by firing thrusters. During the launch phase, the orbit control system is used to affect some of the major orbit manoeuvres and to move the satellite to the desired location.

Attitude of a satellite, or for that matter any space vehicle, is its orientation as determined by the relationship between its axes (yaw, pitch and roll) and some reference plane. The *attitude control* subsystem is used to maintain a certain attitude of the satellite both when it is moving in its orbit and also during its launch phase. Two types of attitude control systems are in common use namely *spin stabilization* and *three-axis stabilization*. During the launch phase, the attitude control system maintains the correct attitude of the satellite so that the satellite is able to maintain link with the ground Earth station and controls its orientation such that the satellite is in the correct direction for an orbital manoeuvre. When the satellite is in orbit, the attitude control system maintains the antenna of the satellite being pointed accurately in the desired direction. The precision with which the attitude needs to be controlled depends on the satellite antenna beam width. Spot beams and shaped beams require more precise attitude control as compared to Earth coverage or regional coverage antennas.

Attitude control systems can be either passive or active. Passive systems maintain the satellite attitude by obtaining equilibrium at the desired orientation. There is no feedback mechanism to check the orientation of the satellite. Active control maintains the satellite attitude by sensing its orientation along the three axes and forming corrections based on these measurements. The basic active attitude control subsystem has three components; one that senses the current attitude of the platform, another that computes the deviations in the current attitude from the desired attitude and a third that controls and corrects the computed errors.

Sensors are used to determine the position of the satellite axis with respect to specified reference directions (commonly used reference directions are the Earth, Sun or a star). Earth sensors sense infrared emissions from the Earth and are used for maintaining the roll and the pitch axis. Sun and the star sensors are generally used to measure the error in the yaw axis. The error between the current attitude and the desired attitude is computed and a correction torque is generated in proportion to the sensed error.

4.3.7 Tracking, Telemetry and Command (TT&C) Subsystem

The TT&C subsystem monitors and controls the satellite right from the lift-off stage to the end of its operational life in space. The tracking part of the subsystem determines the position of the spacecraft and follows its travel using angle, range and velocity information. The telemetry part gathers information on the health of various subsystems of the satellite. It encodes this information and then transmits the same towards the Earth station. The command element receives and executes remote control commands from the control centre on Earth to effect changes to the platform functions, configuration, position and velocity. The TT&C subsystem is therefore very important not only during orbital injection and positioning phase; it also plays an important role throughout the operational life of the satellite.

Figure 4.65 shows the block-schematic arrangement of the basic TT&C subsystem. Tracking, as mentioned earlier, is used to determine the orbital parameters of the satellite on a regular basis. This helps in maintaining the satellite in the desired orbit and in providing look-angle information to the Earth stations. Angle tracking can for instance be used to determine the azimuth and elevation angle from the Earth station. Time interval measurement technique can be used for the purpose of ranging by sending a signal via command link and getting a return via telemetry link. The rate of change of range can be determined either by measuring the phase shift of return signal as compared to that of transmitted signal or by using a pseudorandom code modulation and the correlation between transmitted and received signals.

During the orbital injection and positioning phase, telemetry link is primarily used by the tracking system to establish a satellite-to-Earth station communications channel. After the satellite is put in the desired slot in its intended orbit, its primary function is to monitor the health of various other subsystems on board the satellite. It gathers data from a variety of sensors and then transmits that data to the Earth station. The data includes a variety of electrical and non-electrical parameters. The sensor output could be analogue or digital. Wherever necessary, analogue output is digitized. With the modulation signal as digital, various signals are multiplexed using TDM technique. Since the bit rates involved in telemetry signals are low, it allows a smaller receiver bandwidth at the Earth station to be used with a good signal-to-noise ratio.

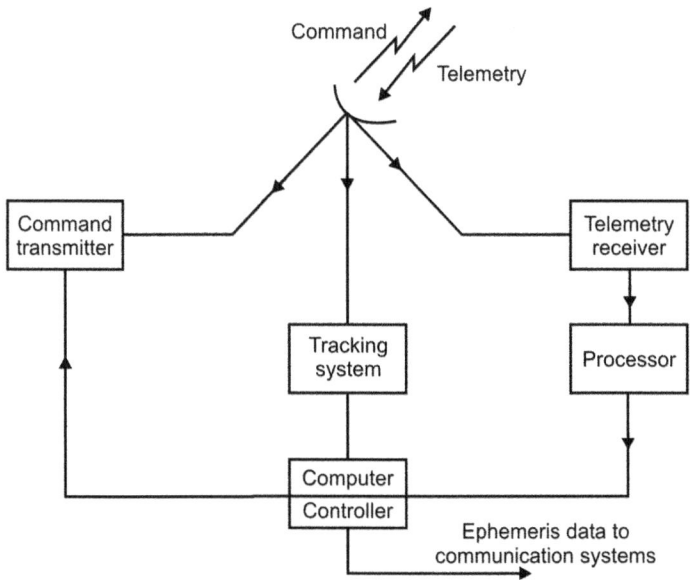

Figure 4.65 Block-schematic arrangement of the basic TT&C subsystem. (*Source:* Courtesy of NASA.)

The command element is used to receive, verify and execute remote control commands from the satellite control centre. The functions performed by the command element include controlling certain functions during orbital injection and positioning phase including firing of apogee boost motor, extending solar panels and so on during the launch phase. When in orbit, it is used to control certain onboard equipment status including transponder switching, antenna pointing control, battery reconditioning and so on. The control commands received by the command element on the satellite are first stored on the satellite and then retransmitted back to the control station via telemetry link for verification. After the commands are verified on ground, a command execution signal is then sent to the satellite to initiate the intended action.

Two well-established and better known integrated TT&C networks used worldwide for telemetry, tracking and command operations of satellites include the ESTRACK network of the European Space Agency (ESA) and the ISTRAC network of the ISRO (Indian Space Research Organization).

4.3.8 Payload

Payload is the most important subsystem of any satellite. Payload can be considered to be the brain of the satellite that performs the intended function of the satellite. The payload carried by a satellite depends upon the mission requirements. The basic payload in case of a *communication satellite* for instance is a *transponder*, which acts as a receiver/amplifier/transmitter. A transponder can be considered to be a microwave relay channel that also performs the function of frequency translation from the up-link frequency to the relatively lower down-link frequency. Thus, a transponder as shown in Figure 4.66 is a combination of elements such as sensitive high gain antennas for transmit-receive functions, a subsystem of repeaters, filters, frequency shifters, low noise amplifiers, frequency mixers and power amplifiers. Satellites employ the L, S, C, X, Ku and Ka microwave frequency bands for communication

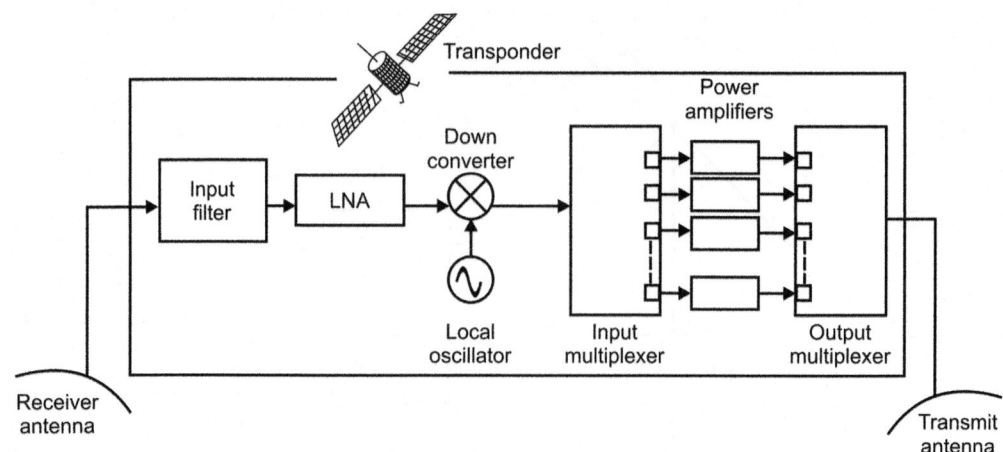

Figure 4.66 Transponder.

purposes with the Ka band being the latest entry into satellite communication bands. Due to the low atmospheric absorption at the L (2 GHz/1 G Hz), S (4 GHz/2 GHz) and C (6 GHz/4 GHz) bands, they were first employed for satellite broadcasting. The C band is the most popular band and is used for providing domestic and international telephone services. Due to advances in the technology of microwave devices, high-frequency Ku and Ka bands are also being extensively used. These high-frequency bands have advantages of higher bandwidth and reduced antenna size. This has led to the revolutionary development in the field of DTH (direct-to-home) services enabling the individual home users to receive TV and broadcast services using antenna sizes as small as 30–50 cm.

In the case of *weather forecasting satellites*, a radiometer is the most important payload. A radiometer is used as a camera and has a set of detectors sensitive to the radiation in the visible, near-IR and far IR bands. Visible images show the amount of sunlight being reflected from the Earth or clouds whereas the IR images provide information on the temperature of the cloud cover or the Earth's surface. The meteorological payload onboard INSAT-3 series satellites for instance includes a very high-resolution radiometer (VHRR) with 2 km resolution in the visible and 8 km resolution in the IR and water vapour channels, respectively, and a CCD camera in the visible (0.63–0.69 μm), near-IR (0.77–0.86 μm) and short-wave IR (1.55–1.70 μm) bands with a 1 km resolution. The payload on board the METEOSAT weather forecasting satellite includes a very high-resolution radiometer. Depending upon the mode of operation radiometers are classified into imagers and sounders.

High-Resolution Visible (HRV) camera, multi-spectral scanners and thematic mappers are the main payloads on board an *Earth observation satellite*. Light and heat reflected and emitted from land and oceans, which contain specific information of the various living and non-living things, are picked up by these sensors. The images produced are then digitized and transmitted to the Earth stations, where they are processed to give the required information.

Scientific satellites have varied payloads depending on their mission. Satellites observing the stars consist of telescope to collect light from stars, and spectrographs operating over a wide range of ultraviolet wavelengths from 120 to 320 nm. Satellites for planetary exploration have varied equipment such as a plasma detector to study solar winds and radiation belts, a magnetometer to investigate the possible magnetic field around the planet, a gamma spectrometer to determine radioactivity of surface rocks, a neutral mass spectroscope, ion mass spectroscope and so on.

4.3.9   Antenna Subsystem

The *antenna subsystem* is one of the most critical components of the spacecraft design because of several well-founded reasons. Some of these are the following.

1) The antenna or antennas (as there is invariably more than one antenna) on board the spacecraft cannot be prohibitively large as large antennas are difficult to mount.
2) Large antennas also cause structural problems as they need to be folded inside the launch vehicle during launch and orbital injection phase and are deployed only subsequent to satellite reaching the desired orbit.
3) The need for having a large antenna arises from the relationship between antenna size and its gain. If the antenna could be as large as desired, we would not have to worry about generating so much power on board the satellite to achieve the required power density at the Earth station antenna.
4) All satellites need a variety of antennas. These include an omnidirectional antenna, which is an isotropic radiator, a Global or Earth coverage antenna, zone coverage antenna and antennas that produce spot beams. In addition, antennas producing spot beams may have a fixed orientation with respect to Earth or may be designed to be steered by remote commands.

An *omnidirectional antenna* is used for TT&C operations during the phase when the satellite has been injected into its parking orbit till it reaches its final position. Unless the high gain directional antennas are fully deployed and oriented properly, the omnidirectional antenna is the only practicable means of establishing a communication channel for tracking, telemetry and command operations.

A *Global or Earth coverage antenna* has a beam width of 17.34°, which is the angle subtended by Earth at a geostationary satellite as shown in Figure 4.67. Any beam width lower than that would have a smaller coverage area and a beam width larger than that would lead to loss of power.

The *zone antenna* offers a coverage that is smaller than that of an Earth coverage antenna. Such an antenna ensures that the desired areas on the Earth are within the satellite's foot print. *Spot beam antennas* concentrate power into a much narrower beam by using large sized reflector antennas and thus illuminate a much smaller area on the Earth. Figure 4.68, which shows a photograph of the INTELSAT-IV satellite, would give you a fairly good idea of the

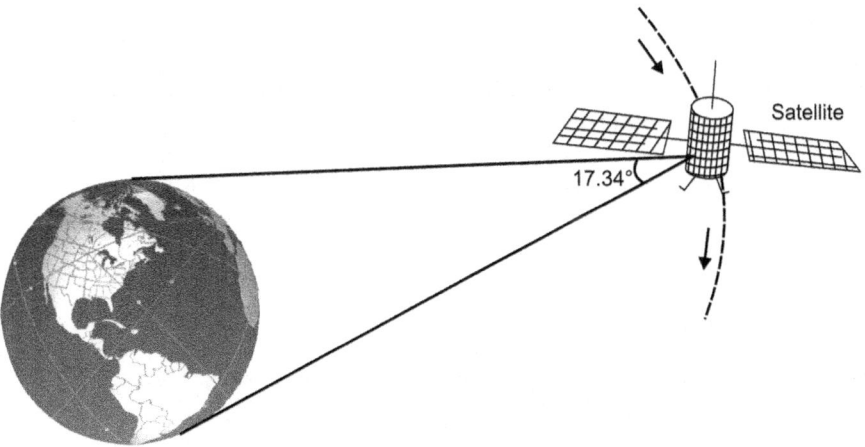

Figure 4.67   Beam width of a global antenna.

Figure 4.68 INTELSAT-4. (*Source: Courtesy of NASA.*)

variety of antennas used on board the spacecraft. As is evident from the photograph, it carries an omnidirectional antenna for TT&C operations, a horn antenna for Earth coverage and two reflector antennas for producing spot beams.

4.4 Multiple Access Techniques

Multiple access means access to a given facility or resource by multiple users. In the context of satellite communication, the facility is the transponder and the multiple users are various terrestrial terminals under the footprint of the satellite. The transponder provides the communication channel/s that receives the signals beamed at it via the up-link and then retransmits the same back to the Earth for intended users via the down-link. Multiple users are geographically dispersed and certain specific techniquesare used to allow them a simultaneous access to the satellite's transponder.

4.4.1 Transponder Assignment Modes

The commonly used transponder assignment modes include the following.

1) Pre-assigned multiple access
2) Demand assigned multiple access
3) Random multiple access.

In case of *Pre-Assigned Multiple Access* (PAMA), the transponder is assigned either permanently for satellite's full lifetime or at least for long durations. The pre-assignment may be that of a certain frequency band, time slot or a code. When it is used infrequently, a link setup with pre-assigned channels is not only costly to the user, the link utilization is also not optimal.

Demand Assigned Multiple Access (DAMA) allows multiple users to share a common link with each user only required putting up a request to the control station or agency for the same as and when it requires using the link. The channel link is only completed as required and a channel frequency is assigned from the available frequencies within the transponder band-width. It is very cost effective for small users who have to pay for the using the transponder capacity only for the time it was actually used.

In case of *Random Multiple Access*, access to the link or the transponder is by contention. A user transmits the messages without knowing the status of messages from other users. Due to random nature of transmissions, data from multiple users may collide. In the case where a collision occurs, it is detected and the data is retransmitted. Retransmission is carried out with random time delays and sometimes may have to be done several times. In such a situation, when all stations are entirely independent there is a great possibility that the messages that collided would be separated out in time on retransmission.

4.4.2 Frequency Division Multiple Access (FDMA)

This is the earliest and still one of the most commonly employed forms of multiple access techniques for communications via satellite. In the case of FDMA, different Earth stations are able to access the total available bandwidth of satellite transponder by virtue of their different carrier frequencies, thus avoiding interference among multiple signals. Figure 4.69 shows the typical arrangement for carrier frequencies for a C band transponder for both up-link and down-link modes. The transponder receives transmissions around 6 GHz and retransmits them around 4 GHz. Figure 4.69 shows the case of a satellite with 12 transponders with each transponder having a bandwidth of 36 MHz and a guard band of 4 MHz between adjacent transponders to avoid interference.

Each of the Earth stations within the satellite's footprint transmits one or more carriers at different centre frequencies. Each carrier is assigned a small guard band, as mentioned previously, to avoid overlapping of adjacent carriers. The satellite transponder receives all carrier frequencies within its bandwidth, does the necessary frequency translation and amplification and then retransmits them back toward Earth. Figure 4.70 illustrates the basic concept of FDMA in satellite communications. Different Earth stations are capable of selecting carriers containing messages of their interest. Two FDMA techniques are in operation today. One of them is multi-channel per carrier (MCPC) where the Earth station frequency multiplexes several channels into one carrier base band assembly, which then frequency modulates an RF carrier and transmits it to FDMA satellite transponder. In the other technique, called Single Channel per Carrier (SCPC), each signal channel modulates a separate RF carrier and is then transmitted to a FDMA transponder. The modulation technique used here could either be frequency modulation (FM) in case of analogue transmission or phase shift keying (PSK) for digital transmission.

Major advantages of FDMA include the simplicity of Earth station equipment and the fact that no complex timing and synchronizing techniques are required. Disadvantages include likelihood of intermodulation problems with its adverse effect on signal-to-noise ratio. The intermodulation products result mainly from the nonlinear characteristics of the TWTA of the transponder, which is required to amplify a large number of carrier frequencies. The problem is further compounded when the TWTA is made to operate near saturation so as to be able to

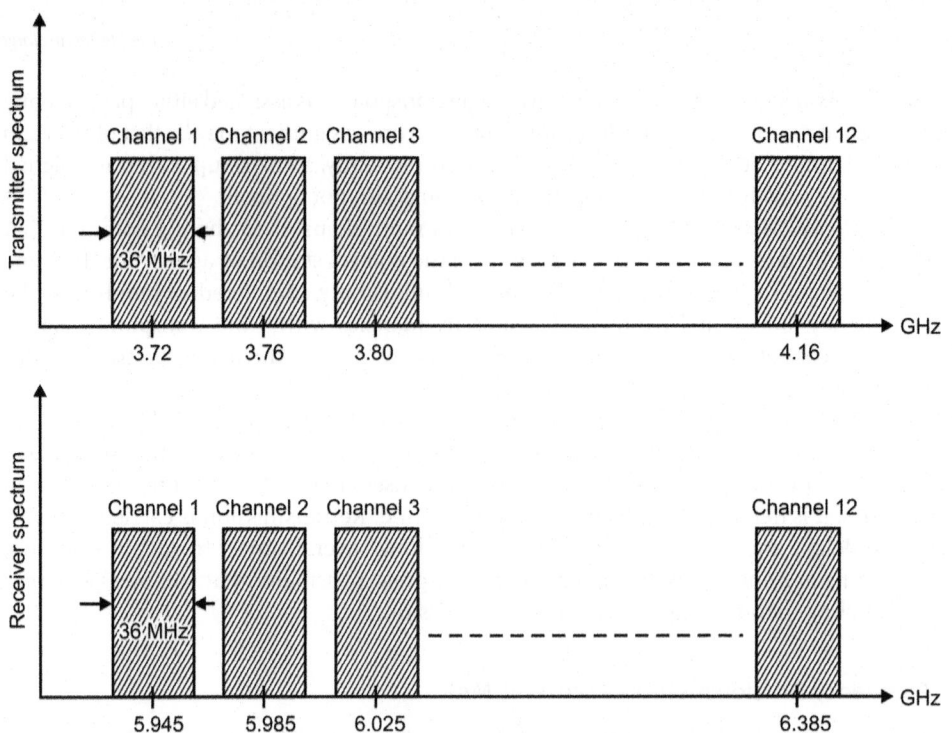

Figure 4.69 Carrier frequencies for a C-band transponder for both the up-link and down-link modes.

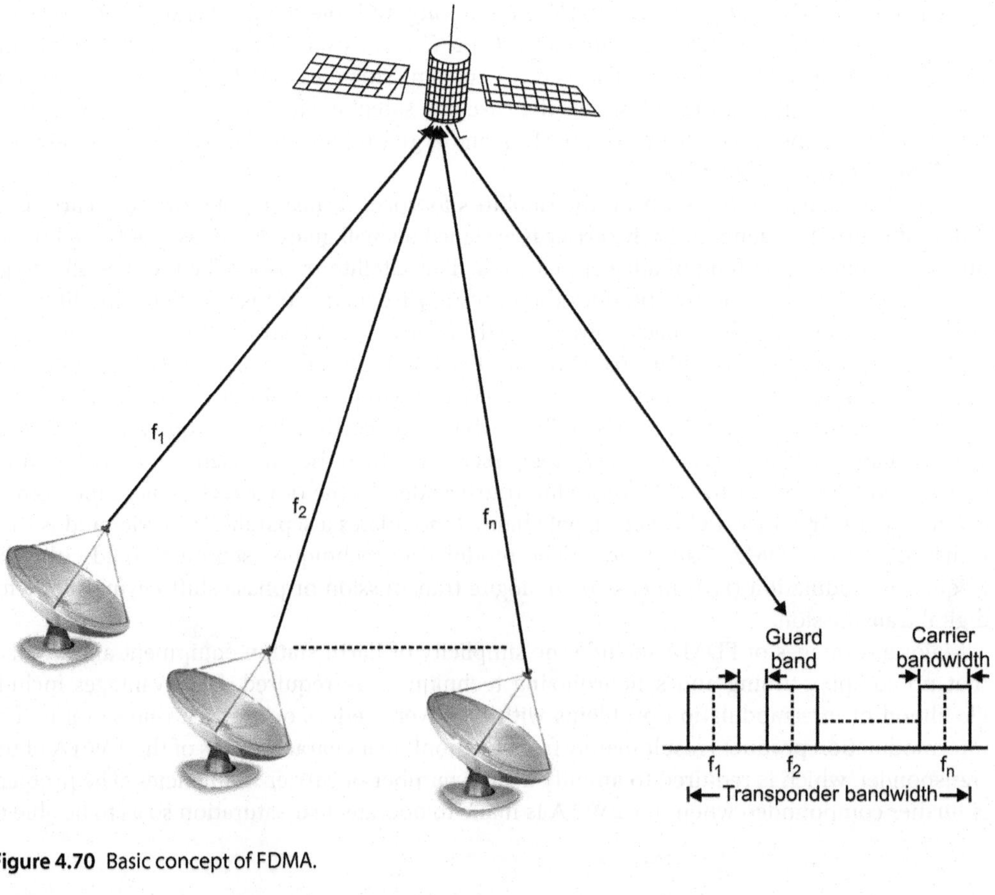

Figure 4.70 Basic concept of FDMA.

supply certain minimum carrier power to reduce down-link noise and the fact that TWTA when operated near saturation exhibits higher nonlinearity.

Common forms of FDMA systems include SCPC/FM/FDMA, SCPC/PSK/FDMA, MCPC/FDM/FM/FDMA and MCPC/PCM-TDM/PSK/FDMA. SCPC and MCPC, respectively, stand for Single Channel Per Carrier and Multiple Channel Per Carrier. FDM and TDM, respectively, stand for Frequency Division Multiplexing and Time Division Multiplexing. PSK stands for Phase Shift Keying. In *SCPC/FM/FDMA* systems, each signal channel modulates a separate RF carrier and the modulation system used here is frequency modulation. The modulated signal is then transmitted to FDMA transponder. The transponder bandwidth is subdivided in such a way that each base band signal channel is allocated a separate transponder subdivision and an individual carrier. The *SCPC/PSK/FDMA* system is the digital form of SCPC system in which the PSK modulation technique is used. In *MCPC/FDM/FM/FDMA* systems, multiple base band signals are grouped together by using frequency division multiplexing to form FDM base band signals. The FDM base band signals are used to frequency modulate pre-assigned carriers and are then transmitted to satellite. The FDMA transponder receives multiple carriers, carries out frequency translation and then separates out individual carriers with the help of appropriate filters. Multiple carriers are then multiplexed and transmitted back to Earth over down-link. The receiving station extracts the channels assigned to that station. In a *MCPC/PCM-TDM/PSK/FDMA system*, multiple base band signals are first digitally encoded using PCM technique and then grouped together to form a common base band assembly using time division multiplexing. This time division multiplexed bit stream then modulates a common RF carrier using phase shift keying as the carrier modulation technique. The modulated signal is transmitted to the satellite, which uses FDMA to handle multiple carriers.

4.4.3 Time Division Multiple Access (TDMA)

TDMA is a technique in which different Earth stations in the satellite footprint making use of satellite transponder use a single carrier on a time division basis. Different Earth stations transmit traffic bursts in a period time frame called a TDMA frame. Over the length of a burst, each Earth station has the entire transponder bandwidth at its disposal. The traffic bursts from different Earth stations are synchronized so that all bursts arriving at the transponder are closely spaced but do not overlap. The transponder works on a single burst at a time and retransmits back to the Earth sequence of bursts. All Earth stations can receive the entire sequence and extract the signal of their interest. Figure 4.71 illustrates the basic concept of TDMA. Disadvantages of TDMA include requirement of complex and expensive Earth station equipment and stringent timing and synchronization requirements. TDMA is suitable for digital transmission only.

4.4.3.1 TDMA Frame Structure

As mentioned before, in a TDMA network, each of the multiple Earth stations accessing a given satellite transponder transmits one or more data bursts. The satellite thus receives at its input a set of bursts from a large number of Earth stations. This set of bursts from various Earth stations is called a TDMA frame. Figure 4.72 shows a typical TDMA frame structure.

As is evident from the frame structure, the start of a frame contains a reference burst transmitted from a reference station in the network. The reference burst is followed by traffic bursts from various Earth stations with guard time between various traffic bursts from different stations. The traffic bursts are synchronized to the reference burst to fix their timing reference. Different parts of the TDMA frame structure are briefly described in the following paragraphs.

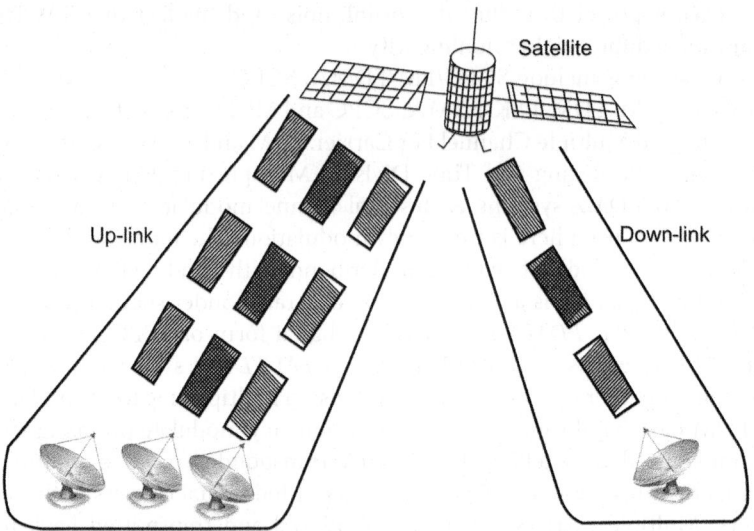

Figure 4.71 Basic concept of TDMA.

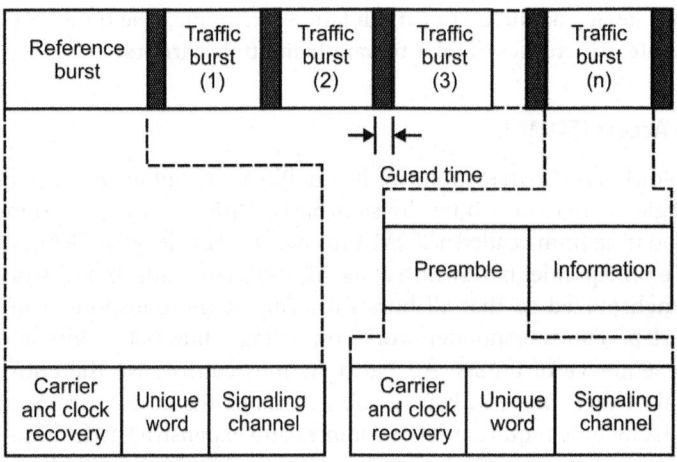

Figure 4.72 Typical TDMA frame structure.

4.4.3.1.1 *Reference Burst*

The reference burst is usually a combination of two reference bursts (RB-1) and (RB-2). The primary reference burst, which can be either RB-1 or RB-2, is transmitted by one of the stations, called primary reference station, in the network. The secondary reference burst, which is RB-1 if the primary reference burst is RB-2 and RB-2 if the primary reference burst is RB-1, is transmitted by another station, called the secondary reference station, in the network. The reference burst automatically switches over to the secondary reference burst in the event of primary reference station failure to provide reference burst to TDMA network. The reference burst does not carry any traffic information and is used to provide timing references to various stations accessing the TDMA transponder.

4.4.3.1.2   Traffic Burst

Different stations accessing the satellite transponder may transmit one or more traffic bursts per TDMA frame and position them anywhere in the frame according to a burst time plan that coordinates traffic between various stations. The timing reference for the location of the traffic burst is taken from time of occurrence of primary reference burst. With this reference, a station can locate and then extract the traffic bursts or portions of traffic bursts intended for it. The reference burst also provides to the station timing references for transmitting its traffic bursts so as to ensure that they arrive at the satellite transponder within their designated positions in the TDMA frame.

4.4.3.1.3   Guard Time

Different bursts are separated from each other by a short guard time, which ensures that bursts from different stations accessing the satellite transponder do not overlap. This guard time should be long enough to allow for differences in transmit timing inaccuracies and also for differences in range rate variation of the satellite.

4.4.3.2   TDMA Burst Structure

A TDMA frame consists of reference and traffic bursts separated by guard time. The traffic burst has two main parts namely the information carrying portion and another sequence of bits preceding the information data called the preamble. The purpose of the preamble sequence of bits is to synchronize the burst and to carry management and control information. The preamble usually consists of three adjacent parts, namely (1) the carrier and clock recovery sequence, (2) the unique word and (3) signalling channel. The reference burst carries no traffic data and contains only the preamble.

4.4.3.2.1   Carrier and Clock Recovery Sequence

Different Earth stations have slight differences in frequency and bit rate. Therefore, the receiving stations must be able to accurately establish the frequency and bit rate of each burst. This is achieved with the help of carrier and clock recovery sequence bits. The length of this sequence usually depends on the carrier-to-noise ratio at the input of demodulator and carrier frequency uncertainty. A higher carrier-to-noise ratio and a lower carrier frequency uncertainty require a smaller bit sequence for carrier and clock recovery and vice versa.

4.4.3.2.2   Unique Word

The unique word is again a sequence of bits that follows the carrier and clock recovery sequence of bits in the preamble. In the reference burst, this bit sequence allows the Earth station to locate the position of traffic burst in the received TDMA frame. The unique word bit sequence in the traffic burst provides timing reference on the occurrence of the traffic burst and also provides timing marker to allow the Earth stations extract their part of traffic burst. The timing marker allows the identification of the start and finish of a message in the burst and helps correct decoding. For obvious reasons, the unique word should have a high probability of detection. For instance, when the unique word of a traffic burst is missed, the entire traffic burst is lost. To achieve this, unique word is sequence of 1s and 0s selected to exhibit good correlation properties to enhance detection.

4.4.3.2.3   Signalling Channel

A signalling channel is used to carry out system management and control functions. A signalling channel of reference burst has three channels, namely (1) an *order wire channel* used to pass instructions to and from Earth stations, (2) a *management channel* transmitted by reference stations to all traffic stations carrying frame management instructions such as changes in burst time plan that coordinates traffic between different stations and (3) a *transmit timing channel* that carries acquisition and synchronization information to different traffic stations enabling them to adjust their transmit burst timing so that similar bursts from different stations reach the satellite transponder within the correct time slot in the TDMA frame.

Signalling channel of the traffic burst also has an order wire channel, which performs the same function as it does in case of reference burst. It also has a service channel, which performs functions like carrying traffic station's status to the reference station, carrying information such as the high bit error rate or unique word loss alarms and so on to other traffic stations.

4.4.3.2.4 Traffic Information
Traffic burst immediately follows the preamble in the TDMA frame structure. Each station in the TDMA network can transmit and receive many traffic bursts and sub-bursts per frame. The length of each sub-burst, which represents information on a certain channel, depends upon the type of service and the number of channels being supported in the traffic burst. For instance, while transmitting a PCM voice channel that is equivalent to data rate of 64 kbps, each sub-burst for this channel would be 64 bits long if the frame time available for the purpose was 1 ms.

4.4.4 Code Division Multiple Access (CDMA)

In case of CDMA, the entire bandwidth of the transponder is used simultaneously by multiple Earth stations at all times. Code division multiple access technique therefore allows multiple Earth stations to access the same carrier frequency and bandwidth at the same time. Each transmitter spreads its signal over the entire bandwidth, which is much wider than that required by the signal otherwise. One of the techniques to do this is to multiply the information signal, which has a relatively lower bit rate, by a pseudo random bit sequence with a much higher bit rate. Interference between multiple channels is avoided as each transmitter uses a unique pseudo random code sequence. Receiving stations recover the desired information by using a matched decoder that works on the same unique code sequence used during transmission.

Let us assume that the message signal is a PCM bit stream. Each message bit is combined with a predetermined code bit sequence. This predetermined code sequence of bits is usually a pseudorandom noise (PN) signal. The bit rate of the PN sequence is kept much higher than the bit rate of the message signal. This spreads the message signal over the entire available bandwidth of the transponder. It is because of this reason that this technique of multiple access is often referred to as Spread Spectrum Multiple Access (SSMA). Spread spectrum operation enables the signal to be transmitted across a frequency band that is much wider than the minimum bandwidth required for transmission of message signal otherwise. The PN sequence bits are often referred to as 'chips' and their transmission rate as 'chip rate'. The receiver is able to retrieve the message addressed to it by using a replica of the pseudorandom sequence used at the transmitter, which is synchronized with the transmitted sequence.

CDMA uses direct sequence (DS) techniques to achieve the multiple access capability. In this, each of the N users is allocated its own PN code sequence. PN code sequences fall in the category of orthogonal codes. Cross-correlation of two orthogonal codes is zero, while their auto-correlation is unity. This forms the basis of each of the N stations being able to extract its intended message signal from a bit sequence that looks like white noise.

4.4.5 Space Division Multiple Access (SDMA)

So far, we have discussed multiple access techniques that allow multiple Earth stations to access given transponder/s capacity without causing any interference among them. In the case of the FDMA technique, different Earth stations are able to access the total available bandwidth in satellite transponder/s by virtue of their different carrier frequencies thus avoiding interference among multiple signals. Here each Earth station is allocated only a part of the total available transponder bandwidth. In the case of the TDMA technique, different Earth stations in satellite's footprint make use of transponder by using a single carrier on a time division basis. In this case, a transponder's entire bandwidth is available to each Earth station on a time shared basis. In the case of the CDMA, the entire bandwidth of the transponder is used simultaneously by multiple Earth stations at all times. Each transmitter spreads its signal over the entire transponder bandwidth. One of the methods to do so is by multiplying the information signal by a unique pseudo random bit sequence. Others include frequency hopping and time hopping techniques. Interference is avoided as each transmitter uses a unique code sequence. Receiving stations recover the desired information by using a matched decoder that works on the same unique code sequence as used during transmission.

SDMA, as outlined in the beginning of the chapter, is a technique that primarily allows frequency reuse where adjacent Earth stations within the footprint of the satellite can use the same carrier transmission frequency and still avoid co-channel interference by using orthogonal antenna beam polarization. Also transmissions from/to a satellite to/from multiple Earth stations can use the same carrier frequency by using narrow antenna beam patterns. As also mentioned earlier, in an overall satellite link, SDMA is usually achieved in conjunction with other types of multiple access techniques such as FDMA, TDMA and CDMA.

4.4.5.1 Frequency Reuse in SDMA

Frequency reuse, as outlined before, is the key feature and the underlying concept of SDMA. In the face of continually increasing demands on the frequency spectrum, it becomes important that frequency bands assigned to satellite communications are efficiently utilized. One of the ways of achieving this is to reuse all or part of frequency band available for the purpose. Another way could be employment of efficient user access methods. Yet another approach could be use of efficient modulation, encoding and compression techniques so as to pack more information into available bandwidth.

Restricting ourselves to frequency reuse, the two methods in common use today for the purpose are *beam separation* and *beam polarization*. Beam separation is based on the fact that if two beams are so shaped that they illuminate two different regions on the surface of Earth without overlapping, then one could use the same frequency band for the two without causing any mutual interference. One could do so by using two different antennas or a single antenna with two feeds as shown in Figure 4.73.

Figure 4.73 Frequency reuse using beam separation.

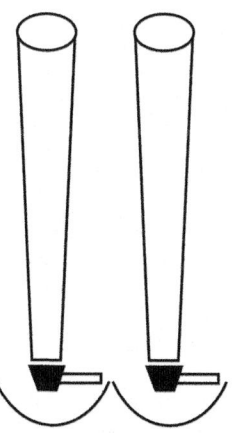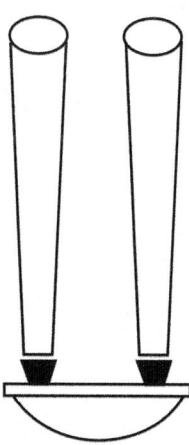

Beam polarization, on the other hand, relies on the principle of using two orthogonally polarized electromagnetic waves to transmit and receive using the same frequency band with no mutual interference between the two. Orthogonal polarizations commonly used include horizontal and vertical polarization or right-hand circular and left-hand circular polarization.

Both the techniques have the capability of doubling the transmission capacity individually and when used in tandem, can increase the capacity by four times. SDMA is seldom used in isolation. It is usually used in conjunction with other types of multiple access techniques discussed earlier including FDMA, TDMA and CDMA. In the following paragraphs, we shall briefly discuss the employment of SDMA separately with each one of these multiple access techniques.

4.4.5.2 SDMA/FDMA System

Figure 4.74 shows a typical block-schematic arrangement of SDMA/FDMA system in which the satellite uses fixed links to route an incoming up-link signal as received by a receiving antenna to a particular down-link transmitter antenna. As is clear from the diagram, the satellite uses multiple antennas to produce multiple beams. The transmitting antenna-receiving antenna combination defines the source and destination Earth stations.

The desired fixed links can be set onboard the satellite by using some form of a switch, which can be selected only occasionally when the satellite needs to be reconfigured. The links could also be configured alternatively by switching the filters with a switch matrix operated by a command link. It may once again be mentioned here that satellite switches are changed only occasionally when the satellite is to be reconfigured.

4.4.5.3 SDMA/TDMA System

This system uses switching matrix to form up-link/down-link beam pairs. In conjunction with TDMA, the system allows TDMA traffic from the up-link beams to be switched to the down-link beams during the course of a TDMA frame. The link between a certain source-destination combination exists at a specified time for the burst duration within the TDMA frame. As an example, signal on beam-1 may be routed to beam-3 during say first 40 μs of a 2 ms TDMA frame and then routed to beam-n during the next 40 μs slot. The process continues until every

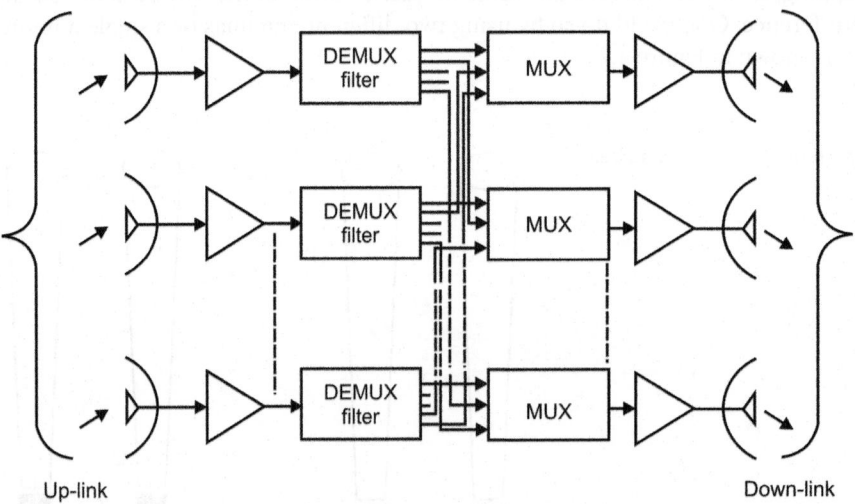

Up-link   Down-link

Figure 4.74 Typical block-schematic arrangement of an SDMA/FDMA system.

Figure 4.75 Typical transponder arrangement for an SDMA/SS/TDMA system.

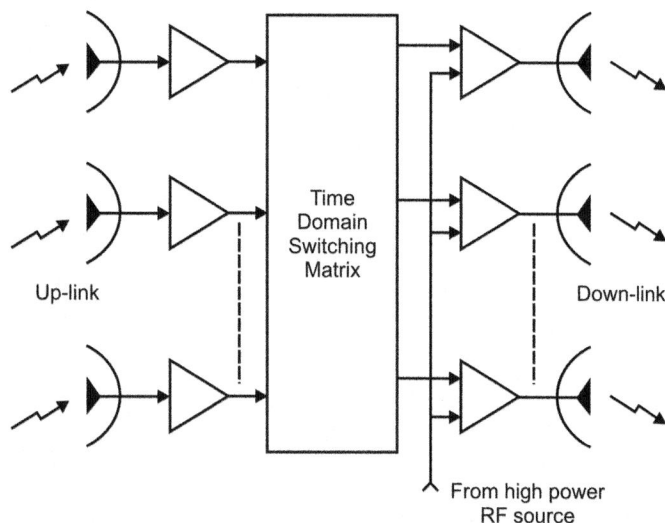

Up-link

Time
Domain
Switching
Matrix

Down-link

From high power
RF source

connection for the traffic pattern has been completed. Figure 4.75 shows a typical transponder arrangement for such an SDMA/SS/TDMA system. SS here stands for 'satellite switched'.

4.4.5.4 SDMA/CDMA System

CDMA provides multiple access to the satellite. The satellite receives an up-link CDMA bit stream, decodes it to determine the destination address and then routes it to the desired down-link. The bit stream is usually re-timed, regenerated and stored in onboard processors before it is retransmitted. This implies that down-link configuration need not be the same as the up-link configuration, thus allowing each link to be optimized.

4.5 Satellite Link Design

In this section, we shall look at the important parameters that govern the design of a satellite communication link. The section begins with a brief introduction to the various parameters that characterize a satellite link or influence its design followed up by detailed description of each one.

4.5.1 Transmission Equation

The transmission equation relates the received power level at the destination, which could be the Earth station or the satellite in the case of a satellite communication link, to the transmitted RF power, the operating frequency and the transmitter-receiver distance. It is fundamental to the design of not only a satellite communication link but any radio communication link because the quality of the information delivered to the destination is governed by the level of the signal power received. The reason for this is that it is the received carrier-to-noise ratio that is going to decide the quality of information delivered and for given noise contribution from various sources both internal and external to the system, the level of received power is therefore vital to the design of communication link. An estimation of received power level in a satellite communication link is made in the following paragraphs.

Let us assume that the transmitter radiates a power P_T watts with an antenna having a gain G_T as compared to isotropic radiation level. The power flux density due to the radiated power in the direction of antenna bore sight at a distance d meters can be expressed by eqn. 4.64.

$$\text{Power flux density} = \frac{P_T G_T}{4\pi d^2} \text{ (in watts/m}^2\text{)} \tag{4.64}$$

The product $P_T G_T$ is nothing but the Effective Isotropic Radiated Power (EIRP). Also, if the radiating aperture A_T of the transmitting antenna is large as compared to λ^2 where λ is the operating wavelength, then G_T equals $4\pi A_T/\lambda^2$. If A_R is the aperture of the receiving antenna, then the received power P_R at the receiver at a distance d from the transmitter can be expressed by eqn. 4.65.

$$P_R = \left(\frac{P_T G_T}{4\pi d^2}\right) A_R \tag{4.65}$$

A_R is related to receive antenna gain by $G_R = (4\pi A_R/\lambda^2)$
The expression for received power gets modified to eqn. 4.66.

$$P_R = \left(\frac{P_T G_T G_R \lambda^2}{\left(4\pi d\right)^2}\right) = \left(\frac{P_T G_T G_R}{\left(4\pi d/\lambda\right)^2}\right) \tag{4.66}$$

The term $4\pi d/\lambda^2$ represents the free-space path loss L_P. Expression 4.66 is also known by the name of the Friis transmission equation. The received power can be expressed in decibels by eqn. 4.67.

$$P_R \text{ (in dB)} = 10\log P_T + 10\log G_T + 10\log G_R - 20\log\left(4\pi d/\lambda\right) \tag{4.67}$$

$$P_R \text{ (in dB)} = EIRP \text{ (in dB)} + G_R \text{ (in dB)} - L_P \text{ (in dB)} \tag{4.68}$$

Equation 4.68 can be modified to include other losses, if any, such as losses due to atmospheric attenuation, antenna losses and so on. For example, if L_A, L_{TX} and L_{RX} are, respectively, the losses due to atmospheric attenuation, transmitting antenna and receiving antenna, then eqn. 4.68 can be rewritten as eqn. 4.69.

$$P_R \text{ (in dB)} = EIRP + G_T - L_P - L_A - L_{TX} - L_{RX} \tag{4.69}$$

4.5.2 Satellite Link Parameters

Important parameters that influence the design of a satellite communication link include the following.

1) Choice of operating frequency
2) Propagation considerations
3) Noise considerations
4) Interference-related problems.

4.5.2.1 Choice of Operating Frequency
The choice of frequency band out of those allocated by the International Telecommunications Union (ITU) for satellite communication services such as the fixed satellite service (FSS), broadcast satellite service (BSS) and mobile satellite service (MSS) is mostly governed by

factors like propagation considerations, coexistence with other services, interference-related issues, technology status, economic considerations and so on. While it may be more economical to use lower frequency bands, there would be interference-related problems as a large number of terrestrial microwave links use frequencies within these bands. Also, lower frequency bands would offer lower bandwidths and hence reduced transmission capacity. Higher frequency bands offer higher bandwidths but suffer from the disadvantage of severe rain induced attenuation, particularly above 10 GHz. Also, above 10 GHz, rain can have the effect of reducing isolation between orthogonally polarized signals in a frequency reuse system. It may be mentioned here that for frequencies less than 10 GHz and elevation angles greater than 5°, atmospheric attenuation is more or less insignificant.

4.5.2.2 Propagation Considerations

The nature of propagation of electromagnetic waves or signals through atmospheric part of an Earth station-satellite link has a significant bearing on the link design. From the viewpoint of the transmitted or received signal, it is mainly the operating frequency and to a lesser extent the polarization that would decide how severe the effect of atmosphere is going to be. From the viewpoint of atmosphere, it is first few tens of kilometres constituting troposphere and then the ionosphere extending from about 80 to 1000 km that do the damage. The effect of atmosphere on the signal is mainly in the form of attenuation caused by atmospheric scattering and scintillation and depolarization caused by rain in troposphere and Faraday rotation in ionosphere. While rain induced attenuation is very severe for frequencies above 10 GHz, polarization changes due to Faraday rotation are severe at lower frequencies and almost insignificant beyond 10 GHz. In fact, atmospheric attenuation is the lowest in the 3–10 GHz window. That is why it is the preferred and most widely used one for satellite communications.

4.5.2.3 Noise Considerations

In both analogue and digital satellite communication systems, the quality of signal received at the earth station is strongly dependent upon the carrier-to-noise ratio of the satellite link. The satellite link comprises of an up-link, the satellite channel and a down-link. The quality of signal received therefore depends upon how strong the signal is as it leaves the originating Earth station and how the satellite receives it. On the down-link, it depends upon how strongly the satellite can re-transmit the signal and then how the destination Earth station receives it. Because of geostationary distances involved, the signals received by the satellite over the up-link and received by the Earth station over the down-link are very weak. Satellite communication systems are therefore particularly susceptible to noise because of their inherent low received power levels. In fact, neither the absolute value of the signal nor that of the noise should be seen in isolation for gauging the effectiveness of the satellite communication link. If the received signal is sufficiently weak compared to the noise level, it may become impossible to detect the signal. Even if the signal is detectable, steps should be taken within the system to reduce the noise to an acceptable level lest it impairs the quality of the signal received.

The sources of noise include natural and manmade sources and also the noise generated in the Earth station and satellite equipment. While manmade noise mainly arises from electrical equipment and is almost insignificant above 1 GHz, the natural sources of noise include solar radiation, sky noise, background noise contributed by Earth, galactic noise due to electromagnetic waves emanating from radio stars in the galaxy and the atmospheric noise caused by lightning flashes and absorption by oxygen and water vapour molecules followed by re-emission of radiation. Sky noise and solar noise can be avoided by proper orientation and directionality of antennas. Galactic noise is insignificant above 1 GHz. Noise due to lightning flashes is also negligible at satellite frequencies.

4.5.2.4 Interference-Related Problems

Major sources of interference include interference between satellite links and terrestrial microwave links sharing the same operational frequency band, interference between two satellites sharing the same frequency band, interference between two Earth stations accessing different satellites operating in the same frequency band, interference arising out of cross-polarization in frequency reuse systems, adjacent channel interference inherent to FDMA systems and interference due to the intermodulation phenomenon.

Interference between satellite links and terrestrial links could further be of two types, first where terrestrial link transmission interferes with reception at a satellite Earth station and the other where transmission from an Earth station interferes with terrestrial link reception. The level of inter-satellite and inter-earth station interference is mainly governed by factors like pointing accuracy of antennas, width of transmit and receive beams, inter-satellite spacing in the orbit of two co-located satellites and so on. Cross-polarization interference is caused by coupling of energy from one polarization state to the other polarization state when a frequency reuse system employs orthogonal linear polarizations (horizontal and vertical polarization) or orthogonal circular polarization (left- and right-hand circular polarization). This coupling of energy occurs due to finite value of cross-polarization discrimination of earth station and satellite antennas and also due to de-polarization caused by rain. Adjacent channel interference arises out of overlapping amplitude characteristics of channel filters. Intermodulation interference is caused by the intermodulation products produced in the satellite transponder when multiple carriers are amplified in the high power amplifier that has both amplitude as well as phase nonlinearity.

4.5.3 Frequency Considerations

The choice of operating frequency for a satellite communication service is mainly governed by factors like propagation considerations, co-existence with other services, noise considerations and interference-related issues. Due to the requirement for co-existence with other services and the fact that there will always be competition to use the most optimum frequency band by various agencies with an eye on getting the best link performance, there is always a need for a mechanism for allocation and coordination of the frequency spectrum on an international level.

4.5.3.1 Frequency Allocation and Coordination

Satellite communication employs electromagnetic waves for transmission of information between Earth and space. The bands of interest for satellite communications lie above 100 MHz including the VHF, UHF, L, S, C, X, Ku and Ka bands. Higher frequencies are employed for satellite communication as the frequencies below 100 MHz are either reflected by the ionosphere or they suffer varying degrees of bending from their original paths due to the refraction by the ionosphere. Initially, the satellite communication was mainly concentrated in the C band (6/4 GHz) as it offered fewest propagation as well as attenuation problems. But due to the overcrowding in the C band and the advances made in the field of satellite technology, which enables it to deal with the propagation problems in the higher frequency bands, newer bands like the Ku and Ka bands are now being employed for commercial as well as military satellite applications. Moreover, the use of these higher frequencies gives satellites an edge over terrestrial networks in terms of the bandwidth offered by a satellite communication system.

As the frequency spectrum is limited, it is evident the frequency bands are allocated in such a manner to ensure their rational and efficient use. The International Telecommunication Union (ITU), formed in 1865, is a specialized institution that ensures the proper allocation of frequency bands as well as the orbital positions of the satellite in the GEO orbit. The ITU carries out these regulatory activities through its four permanent organs namely the General Secretariat, International Frequency Registration Board (IFRB), International Radio Consultative Committee (CCIR) and the International Telegraph and Telephone Consultative Committee (CCITT). The ITU organizes international radio conferences such as the World Administrative Radio Conferences (WARC) and other regional conferences to issue guidelines for frequency allocation for various services. These allocations are made either on an exclusive or shared basis and can be put into effect worldwide or limited to a region. The frequency bands that are allocated internationally are in turn reallocated by the national government bodies of the individual countries. The ITU has divided the frequency allocations for different services into various categories: primary, secondary, planned, shared and so on. In *primary* frequency allocation, a service has an exclusive right of operation. In *secondary* frequency allocation a service is not guaranteed interference protection and neither is permitted to cause interference to services with primary status. WARC divided the globe into three regions for the purpose of frequency allocations. They are:

Region 1: Including Europe, Africa, USSR and Mongolia
Region 2: Including North and South America, Hawaii and Greenland
Region 3: Including Australia, New Zealand and those parts of Asia and the Pacific not included in regions 1 and 2.

It has also classified the various satellite services into the following categories including Fixed Satellite Services, Inter-Satellite links, Mobile Satellite Services comprising land mobile services, maritime mobile services and aeronautical mobile services, broadcasting services, Earth exploration services, space research activities, meteorological activities, space operation, amateur radio services, radio determination, radio navigation, aeronautical radio navigation and maritime radio navigation. Among these applications Fixed Satellite Services (FSS), Mobile Satellite Services (MSS) and Broadcast Satellite Services (BSS) are the principle communication related applications of satellites.

FSS refer to the two-way communication between two Earth stations at fixed locations via a satellite. It supports majority of commercial applications including satellite telephony, satellite television and data transmission services. FSS primarily uses two frequency bands: the C band (6/4 GHz), which provides lower power transmission over a wide geographic area requiring large antennas for reception and the Ku band (14/11 GHz, 14/12 GHz), which offers higher transmission power over smaller geographical area enabling reception by small receiving antennas. All of the C band and much of the Ku band has been allocated for international and domestic FSS applications. The X band (8/7 GHz) is used for providing fixed services to government and military users. EHF, UHF and SHF band transponders are also used in military satellites.

BSS refers to the satellite services that can be received at many unspecified locations by relatively simple receive-only Earth stations. These Earth stations can either be community Earth stations serving various distribution networks or located in homes for Direct-to-Home transmission. The Ku Band (18/12 GHz) and Ka band (30/20 GHz) bands are mainly used for BSS applications like television broadcasting and DTH applications.

MSS refer to the reception by receivers that are in motion like ships, cars, lorries and so on. Increasingly, MSS networks are providing relay communication services to portable handheld terminals. The L Band (2/1 GHz) and S Band (4/2.5 GHz) are mainly employed for MSS services as at these lower microwave frequencies, broader beams are transmitted from the satellite enabling the reception by antennas even if they are not pointed towards the satellite. This in turn makes these frequencies attractive for mobile and personal communications. The lower frequencies in the VHF and UHF bands are mainly employed for messaging and positioning applications as these applications require smaller bandwidths. Some LEO constellations are using these frequencies to provide the previously mentioned applications Current trends in satellite communication indicate the opening of new higher frequency bands for various applications like the Q band (33–46 GHz) and V band (46–56 GHz) are being considered for use in FSS, BSS and inter-satellite communication applications. Another important concept is the sharing of the same frequency bands between GEO and new systems for personal satellite communications using non-GEO orbits.

4.5.4 Propagation Considerations

The nature of propagation of electromagnetic waves through atmosphere has a significant bearing on the satellite link design. As we shall see in the paragraphs to follow, it is first few tens of kilometres constituting the troposphere and then the ionosphere extending from about 80–1000 km that do the major damage. The effect of atmosphere on the signal is mainly in the form of attenuation caused by atmospheric scattering and scintillation, depolarization caused by rain in troposphere and Faraday rotation in ionosphere.

Propagation losses are another two types, namely those that are more or less constant and therefore predictable and those that are random in nature and therefore unpredictable. While free-space loss belongs to the first category, attenuation caused due to rain is unpredictable to a large extent. The second category of losses can only be estimated statistically. The combined effect of these two types of propagation losses is to reduce the received signal strength. Due to the random nature of some types of losses, the received signal strength fluctuates with time and may even reduce to a level below the minimum acceptable limit for as long a period as an hour in 24 hours during the period of severe fading. This is amply illustrated in the graph of Figure 4.76.

4.5.4.1 Free-Space Loss

Free-space loss is the loss of signal strength only due to distance from the transmitter. While free space is a theoretical concept of space devoid of all matter, in the present context, it implies remoteness from all material objects or forms of matter that could influence propagation of electromagnetic waves. In the absence of any material source of attenuation of electromagnetic signal therefore, the radiated electromagnetic power diminishes as inverse square of the distance from the transmitter, which implies that the power received by an antenna of 1-m^2 cross-section shall be $P_t/4\pi R^2$ where P_t is the transmitted power and R is the distance of the receiving antenna from the transmitter. In the case of up-link, the Earth station antenna becomes the transmitter and satellite transponder is the receiver. It is the opposite in the down-link case.

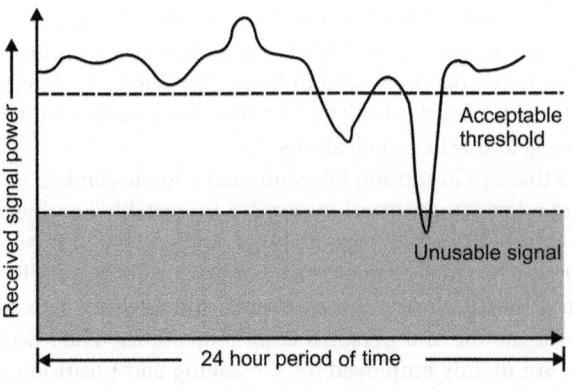

Figure 4.76 Fading phenomenon.

The free-space path loss component can be computed from eqn. 4.70

$$\text{Free space path loss} = \left(\frac{4\pi R}{\lambda}\right)^2 = 20\log\left(\frac{4\pi R}{\lambda}\right) dB \qquad (4.70)$$

λ = operating wavelength
Also, $\lambda = (c/f)$
c = velocity of electromagnetic waves in free space
f = operating frequency.

If c is taken in km/s and f in MHz, then free-space path loss can also be computed from eqn. 4.71.

$$\text{Free space path loss} = (32.5 + 20\log R + 20\log f) dB \qquad (4.71)$$

4.5.4.2 Gaseous Absorption

Electromagnetic energy gets absorbed and converted into heat due to gaseous absorption as it passes through the troposphere. The absorption is primarily due to presence of molecular oxygen and uncondensed water vapour and has been observed to be not so significant as to cause problems in the frequency range of 1–15 GHz. Of the other gases, atmospheric nitrogen does not have a peak while carbon dioxide has shown a peak around 300 GHz. The presence of free electrons in the atmosphere also causes absorption due to collision of electromagnetic waves with electrons. However, electron absorption is significant only at frequencies less than 500 MHz. There are specific frequency bands where the absorption is maximum, near total. The first absorption band is caused due to resonance phenomenon in water vapour and it occurs at 22.2 GHz. The second band is caused by a similar phenomenon in oxygen and it occurs around 60 GHz. These bands are therefore not employed for either up-links or down-links. However, these can be used for inter-satellite links. It may also be mentioned that absorption at any frequency is a function of temperature, pressure, relative humidity and elevation angle.

Absorption increases with decrease in elevation angle E due to increase in transmission path. Absorption at any elevation angle E less than 90° can be computed by multiplying the 90° elevation absorption figure by $(1/\sin E)$. There are two transmission windows in which absorption is either insignificant or have a local minimum. The first window is in the frequency range of 500 MHz to 10 GHz and the second is around 30 GHz. This explains the wide use of 6/4 GHz band. The increasing interest in 30/20 GHz band is due to the second window that shows a local minimum around 30 GHz.

4.5.4.3 Attenuation Due to Rain

After the free-space path loss, rain is the next major factor contributing to loss of electromagnetic energy caused by absorption and scattering of electromagnetic energy by rain drops. While loss of electromagnetic energy due to gaseous absorption tends to remain reasonably constant and predictable, loss due to precipitation in the form of rain, fog, clouds, snow and so on is variable and far less predictable. However, losses can be estimated so as to allow the satellite links to be designed with adequate link margin where ever necessary.

Loss due to rain increases with increase in frequency and reduction in elevation angle. There is not much to worry from rain attenuation for C band satellite links. Attenuation becomes significant above 10 GHz and therefore when a satellite link is planned to operate beyond 10 GHz, an estimate of rain-caused attenuation is made by making extensive measurements at several locations in the coverage area of the satellite system.

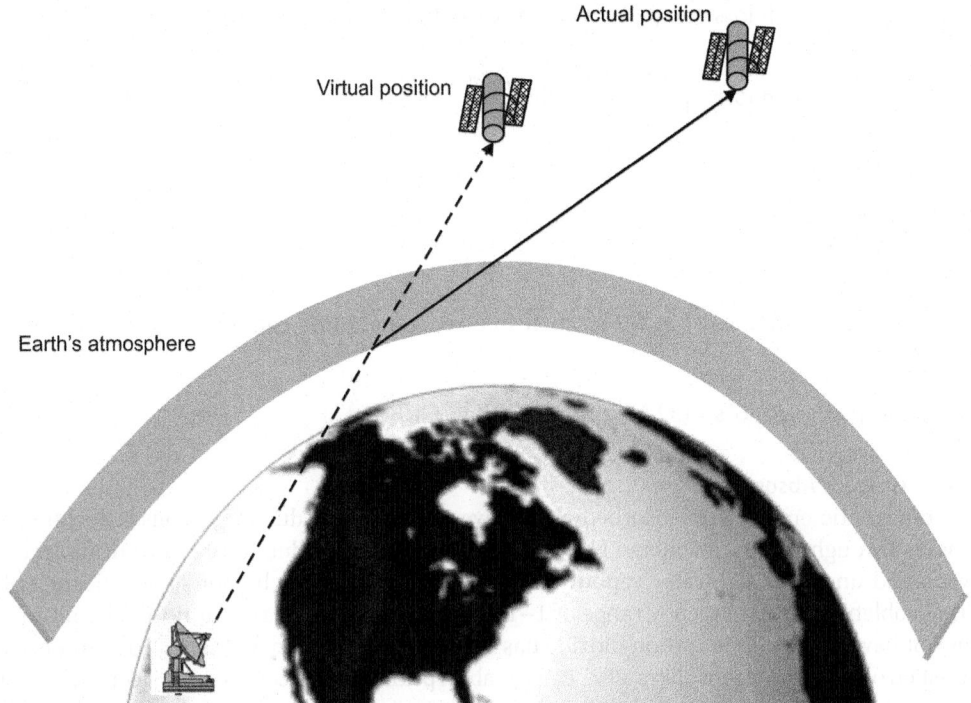

Figure 4.77 Bending of electromagnetic waves caused by refraction.

In addition to attenuation of electromagnetic energy, another detrimental effect caused by rain is the *depolarization*. This reduces the cross-polarization isolation. This effect has been observed to be more severe on circularly polarized waves than the linearly polarized waves. Depolarization of electromagnetic waves due to rain is caused by flattening of supposedly spherical rain drops affected by atmospheric drag. Depolarization is not particularly harmful at C band and lower parts of the Ku band. It is severe at frequencies beyond 15 GHz.

4.5.4.4 Signal Fading Due to Refraction

Refraction is the phenomenon of bending of electromagnetic waves as they pass through different layers of atmosphere. The effect is caused by troposphere, which is the lower portion of the atmosphere. The result of bending of electromagnetic waves is depicted in Figure 4.77. It leads to a virtual position for the satellite slightly above the true position of the satellite. The random nature of bending due to discontinuities and fluctuations caused by unstable atmospheric conditions like temperature inversions, clouds and fog produces signal fading, which leads to loss of signal strength. The effect is more severe for terrestrial links and not too worrisome for satellite links as the amount of bending caused in a satellite link is small relative to the beam width of satellite and Earth station antennas.

4.5.4.5 Ionosphere-Related Effects

The ionosphere, as we all know, is an ionized region in space extending from about 80–1000 km formed by the interaction of solar radiation with different constituent gases in the atmosphere. Electromagnetic waves travelling through the ionosphere are affected in more than one way; some more predominant than the other from the viewpoint of satellite communications. The effects that are of concern and need to be paid attention to include *polarization rotation*, called the *Faraday effect*, and *scintillation*, which is nothing but rapid fluctuation of signal amplitude, phase, polarization or angle-of-arrival. The ionosphere also affects the propagating electromagnetic waves in many other ways such as absorption, causing propagation delay, dispersion and so on. However, these effects are negligible at the frequencies of main interest for satellite communications except for very small time periods during intense solar activity such as solar flares. Also, most ionosphere effects including those of primary interest like polarization rotation and scintillation decrease with increase in frequency with a $1/f^2$ dependence. The two major effects are briefly described in the following paragraphs.

4.5.4.5.1 Polarization Rotation (Faraday Effect)

When an electromagnetic wave passes through a region of high electron content, the plane of polarization of the wave gets rotated due to interaction of the electromagnetic wave with the Earth's magnetic field. The angle through which the plane of polarization rotates is directly proportional to the total electron content of the ionized region such as ionosphere and inversely proportional to square of operating frequency. It also depends upon the state of ionosphere, time of the day, solar activity, the direction of incident wave and so on. Directions of polarization rotation are opposite for transmit and receive signals.

The Faraday effect is more or less predictable and therefore can be compensated for by adjusting the polarization of the receive antenna. Circular polarization is virtually unaffected by Faraday rotation and therefore its impact can be minimized by using circular polarization.

If $\Delta \Psi$ is the polarization mismatch angle caused by Faraday rotation, then this mismatch produces an attenuation of the copular signal given by eqn. 4.72.

$$\text{Attentuation (in dB)} = -20\log(\cos \Delta \psi) \tag{4.72}$$

The mismatch also produces a cross-polarized component, which reduces the cross-polarization discrimination (XPD) given by eqn. 4.73.

$$\text{XPD} = -20\log(\tan \Delta \psi) \tag{4.73}$$

4.5.4.5.2 Scintillation

As outlined before, scintillation is nothing but rapid fluctuation of signal amplitude, phase, polarization or angle-of-arrival. In the ionosphere, scintillation occurs due to small scale refractive index variations caused by local ion concentration. It mainly occurs in the F-region of the ionosphere due to higher electron content in that region. Pockets of local ion concentration produce discontinuities, which cause refraction. As a result, the signal reaches the receiving antenna via two paths, the direct path and the refracted path as illustrated in Figure 4.78. Multipath signals can lead to both signal enhancement as well as signal cancellation depending upon the phase relationship with which they arrive at the receiving antenna. The resultant signal is a vector addition of the direct and refracted signal. In the extreme case, when the strength of the refracted signal is comparable to that of the direct signal, cancellation can occur when the relative phase difference between the two is 180°. On the other hand, an instantaneous recombination of the two signals in phase can lead to signal amplification up to 6 dB.

Scintillation effect is inversely proportional to the square of operating frequency and is predominant at lower microwave frequencies, typically below 4 GHz. Scintillation, however, increases during periods of high solar activity and other extreme conditions such as occurrence of magnetic storms. Also, maximum scintillation is observed in the region of ±25° around equator. Under such adverse conditions, scintillation can cause problems at the 6/4 GHz band too. At the Ku band and beyond, however, the effect is negligible. Also unlike scintillation caused by the

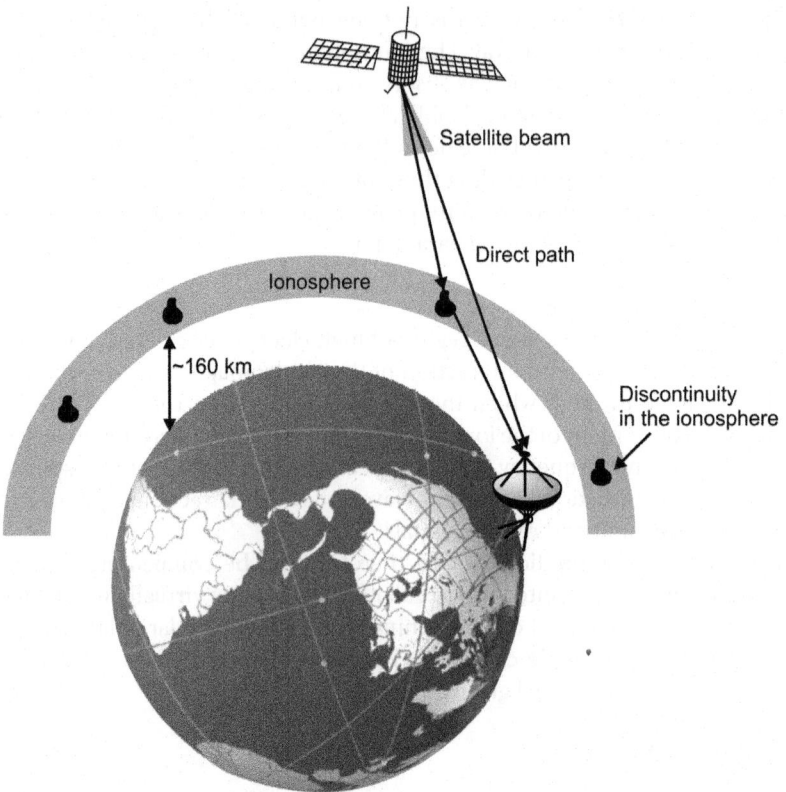

Figure 4.78 Ionospheric scintillation.

troposphere, ionospheric scintillation is independent of the elevation angle. In regions where scintillation is expected to cause problems, adequate link margin in the form of additional transmitted power should catered for to take care of the effect and maintain link reliability.

4.5.4.6 Fading Due to Multipath Signals

Ionospheric scintillation results in a multipath phenomenon where the indirect signal is produced as a result of refraction caused by pockets of ion concentration in the F-region of ionosphere. Multipath signals also result from reflection and scattering from obstacles such as buildings, trees, hills and other manmade and natural objects. Such multipath signals cause problems particularly to mobile satellite terminals where the situation keeps changing with time. In a typical case, a mobile terminal could receive the direct signal and another signal reflected off the highway and the relative phase difference between the two signals could produce either a signal enhancement or fading. In this case, however, the situation remains more or less the same and does not change with time as long as the satellite remains in the same position with respect to the vehicle. On the other hand, reflection off a building or a neighbouring hill produces a time varying fading as the vehicle moves with respect to the point of reflection.

Thus, situation in case of mobile satellite service (MSS) terminals is far more severe and uncertain than it is in case of fixed receiving terminals because of the following reasons. The mobile station-satellite path profile keeps changing continuously with the movement of the mobile terminal. Also, mobile terminals usually employ broadband receiving antennas, which do not provide adequate discrimination against signals received via indirect paths.

4.5.5 Noise Considerations

Satellite communication systems are particularly susceptible to noise because of their inherent low received power levels as the signals received by the satellite over the up-link and received by the Earth station over the down-link are very weak due to involvement of geostationary distances in most cases. Sources of noise include natural and manmade sources and also the noise generated inside the earth station and satellite equipment as outlined during introductory paragraphs in the beginning of the chapter. From the viewpoint of satellite communications, the natural and manmade sources of noise can either be taken care of or are negligible. It is mainly the noise generated in the equipment that needs to be primarily paid attention to. In the paragraphs to follow, we shall briefly discuss various parameters that can be used to describe the noise performance of various building blocks individually and also as a system, which is a cascaded arrangement of those building blocks.

4.5.5.1 Thermal Noise

Thermal noise is generated in any resistor or resistive component of any impedance due to random motion of molecules, atoms and electrons. It is called thermal noise as the temperature of a body is the statistical RMS value of the velocity of motion of these particles. It is also called white noise as, due to randomness of motion of particles, the noise power is evenly spread over the entire frequency spectrum. It is also known as Johnson noise. Noise output power of resistor is given by eqn. 4.74.

$$P_n \propto TB = kTB \tag{4.74}$$

$T =$ Absolute temperature
$B =$ Bandwidth of interest
$k =$ Boltzmann's constant $= 1.38 \times 10^{-23}$ J/K
$P_n =$ Noise power output of a resistor.

Thermal noise power at room temperature in decibels can also be computed from eqn. 4.75

$$P_n \left(\text{in dBm}\right) = -174 + 10 \log B \tag{4.75}$$

where $T = 290$ K
If the resistor is considered as a noise generator with an equivalent noise voltage equal to V_n, then this noise generator will transfer maximum noise power to a matched load that is given by eqn. 4.76.

$$P_n = \frac{V_n^{\,2}}{4R} \tag{4.76}$$

Expression for noise voltage and noise current are given by eqns 4.77 and 4.78.

$$V_n = \sqrt{4kTRB} \tag{4.77}$$

$$I_n = \sqrt{\frac{4kTB}{R}} \tag{4.78}$$

Another term that is usually defined in this context is the noise power spectral density given by eqn. 4.79.

$$P_{no} = kT \ \text{W/Hz} \tag{4.79}$$

This implies that noise power spectral density increases with physical temperature of the device. Also, thermal noise generated in a device can be reduced by reducing its physical temperature or the bandwidth over which the noise is measured or both.

4.5.5.2 Noise Figure

The *noise figure* (F) of a device can be defined as the ratio of its signal-to-noise power at the input to the signal-to-noise power at the output. It is given by eqn. 4.80.

$$\text{Noise Figure} = \frac{(S_i/N_i)}{(S_o/N_o)} = \frac{N_o}{\left[(S_o/S_i)N_i\right]} = \frac{N_o}{GN_i} \tag{4.80}$$

S_i = Available signal power at the input
N_i = Available noise power at the input
S_o = Available signal power at the output
N_o = Available noise power at the output (in a noiseless device)
G = Power gain over the specified bandwidth

Now, $N_i = kT_iB$ where T_i is the ambient temperature in Kelvin.

This gives Noise Figure $= \dfrac{N_o}{GkT_iB}$

The actual amplifier, however, introduces some noise, which gets added to the output noise power. If the noise power introduced is ΔN, then

$$F = \frac{(GkT_iB + \Delta N)}{GkT_iB} = 1 + \frac{\Delta N}{GkT_iB} \tag{4.81}$$

The noise figure is thus ratio of actual output noise to that which would remain if the device itself did not introduce any noise. In case of a noiseless device, $\Delta N = 0$, which gives $F = 1$. Thus the noise figure in case of an ideal device is unity.

4.5.5.3 Noise Temperature

Yet another way of expressing noise performance of a device is in terms of its equivalent noise temperature, T_e. It is the temperature of a resistance that would generate the same noise power at the output of an ideal (i.e. noiseless) device as that produced at its output by an actual device when terminated at its input by a noiseless resistance; that is, a resistance at absolute zero temperature.

Now, noise generated by a device, $\Delta N = GkT_eB$, which when substituted in the expression for the noise figure mentioned in Section 4.5.5.2 gives an expression for the noise figure as eqn. 4.82.

$$F = 1 + (T_e/T_i) \text{ or } T_e = T_i(F - 1) \tag{4.82}$$

If L = loss factor, then the gain G for this attenuator can be expressed as $G = 1/L$. The expression for the total noise power at the output of the attenuator can be written as eqn. 4.83.

$$\text{Total noise output power} = GkT_iB + GkT_eB = \frac{(kT_iB + kT_eB)}{L} \tag{4.83}$$

T_e = Effective noise temperature of the attenuator.

If the attenuator is considered to be at the same temperature T_i as that of the source resistance it is fed from, then

Total noise output power $= kT_iB$

This gives, $\dfrac{(kT_iB + kT_eB)}{L} = kT_iB$

Or $\dfrac{(T_i + T_e)}{L} = T_i$, which gives

$$T_e = T_i(L-1) \tag{4.84}$$

This expression gives the effective noise temperature of noise source at temperature T_i, followed by a resistive attenuator having a loss factor L.

4.5.5.4 Noise Figure and Noise Temperature of Cascaded Stages

So far, we have discussed about the noise figure and noise temperature specifications of individual building blocks. A system such as a receiver would have a large number of individual building blocks connected in series and it is important to determine the overall noise performance of this cascaded arrangement. In the following paragraphs we shall derive expressions for noise figure and effective noise temperature for a cascaded arrangement of multiple stages.

Let us consider a cascaded arrangement of n stages with respective individual gains as G_1, G_2, G_3... G_n and input noise temperature parameters as T_1, T_2, T_n ... T_n. The equivalent noise temperature of cascaded arrangement can be computed from eqn. 4.85.

$$T_{en} = T_1 + \frac{T_2}{G_1} + \frac{T_3}{G_1G_2} + \frac{T_4}{G_1G_2G_3} + \ldots\ldots\ldots + \frac{T_n}{G_1G_2G_3\ldots\ldots\ldots G_{n-1}} \tag{4.85}$$

The same expression can also be written in terms of noise figure specifications of individual stages as eqn. 4.86.

$$F = F_1 + \frac{(F_2-1)}{G_1} + \frac{(F_3-1)}{G_1G_2} + \frac{(F_4-1)}{G_1G_2G_3} + \ldots\ldots\ldots + \frac{(F_n-1)}{G_1G_2G_3\ldots\ldots\ldots G_{n-1}} \tag{4.86}$$

The expressions for noise figure and effective noise temperature highlight the significance of the first stage. As is evident from these expressions, the noise performance of the overall system is largely governed by the noise performance of the first stage. That is why it is important to have the first stage with as low noise as possible.

4.5.5.5 Antenna Noise Temperature

Antenna noise temperature is a measurement of noise entering the receiver via the antenna. The antenna picks up noise radiated by various manmade and natural sources within its directional pattern. Various sources of noise include noise generated by electrical equipment that is almost insignificant above 1 GHz, the noise emanating from natural sources of noise including solar radiation, sky noise, background noise contributed to by Earth, galactic noise due to electromagnetic waves emanating from radio stars in the galaxy and the atmospheric noise caused by lightning flashes and absorption by oxygen and water vapour molecules followed by re-emission of radiation.

Noise from these sources could enter the receiver both through the main lobe as well as through the side lobes of its directional pattern. Thus the noise output from a receiving antenna is a function of the direction in which it is pointing, its directional pattern and state of its environment. The noise performance of the antenna, as mentioned before, can be expressed in terms of a noise temperature called antenna noise temperature. If the antenna noise temperature is T_A K, it implies that the noise power output of the antenna is equal to the thermal noise power generated in a resistor at a temperature of T_A K.

The noise temperature of the antenna can be computed by integrating contributions of all the radiating bodies whose radiation lies within the directional pattern of the antenna. It is given by eqn. 4.87.

$$T_A = \frac{1}{4\pi} \iint G(\theta,\phi) T_b(\theta,\phi) \sin\theta \, d\theta \, d\phi \qquad (4.87)$$

ϑ = Azimuth angle
Φ = Elevation angle
$G(\vartheta, \Phi)$ = Antenna gain in the ϑ and Φ directions
$T_b(\vartheta, \Phi)$ = Brightness temperature in ϑ and Φ directions.

There are two possible situations to be considered here. One is that of the satellite antenna when we are referring to the up-link and the other is that of earth station antenna when we are referring to a down-link. In case of satellite antenna (up-link scenario), the main sources of noise are the Earth and outer space. Again, the noise contribution of the Earth depends upon the orbital position of the satellite and the antenna beam width. In case, the satellite antenna's beam width is more or less equal to angle-of-view of the Earth from the satellite, which is 17.5° for a geostationary satellite, the antenna noise temperature depends upon frequency of operation and orbital position. In the case where the beam width is smaller as that of a spot beam, the noise temperature depends upon the frequency of operation and the area being covered on Earth.

In case of earth station antenna (down-link scenario), the main sources of noise that contribute to antenna noise temperature include the sky noise and the ground noise. The sky noise is primarily due to sources such as radiation from the Sun and Moon and the absorption by oxygen and water vapour in the atmosphere accompanied by emission. The noise from other sources such as cosmic noise originating from hot gases of stars and inter-stellar matter, galactic noise due to electromagnetic waves emanating from radio stars in the galaxy is negligible at frequencies above 1 GHz. Here again we have two distinctly different conditions; one that of clear sky devoid of any meteorological formations and the other that of sky with meteorological formations such as clouds, rain and so on. In the clear sky conditions, the noise contribution is from sky noise and the ground noise. The sky noise enters the system mainly through the main lobe of the antenna's directional pattern and the ground noise enters the system mainly through the side lobes and only partly through the main lobe particularly at low elevation angles.

In addition to the sky noise component, there are also a number of high intensity point sources in the sky noise. These sources subtend an angle that is only a few arcs of a minute and therefore they matter only if the Earth station antenna is high directional and the radiation source lies within the antenna beam width.

Also, sky noise contribution increases significantly when heavenly bodies like the sun and the moon get aligned with the earth station satellite path. The increase in noise temperature is a function of operating frequency and the size of antenna. The average noise temperature contribution due to the sun can be approximated as $12000 f^{-0.75}$; f is the operating frequency in GHz. It may increase by a factor of 100–10 000 during the periods of high solar activity.

4.5.5.6 Overall System Noise Temperature

The overall system noise temperature, which is a cascaded arrangement of antenna, feeder and the receiver, as shown in Figure 4.79, can be expressed at two points; one at the output of antenna, that is, the input of feeder, and two at the input of the receiver.

Figure 4.79 Computation of system noise temperature.

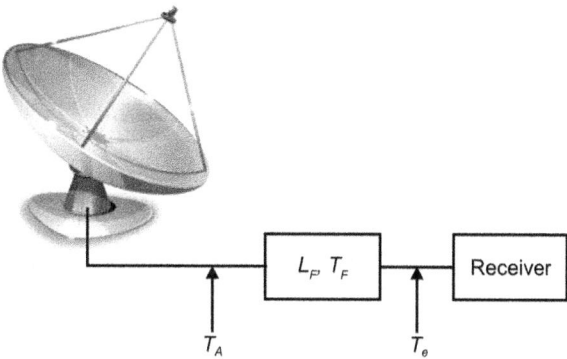

The expression for the system noise temperature with reference to output of antenna can be written as eqn. 4.88.

$$\text{System noise temperature at antenna output} = T_A + T_F(L_F - 1) + T_e L_F \qquad (4.88)$$

T_A = Antenna noise temperature
T_F = Thermodynamic temperature of feeder often taken as ambient temperature
L_F = Attenuation factor of feeder
T_e = Effective input noise temperature of receiver.

System noise temperature when referred to receiver input can be computed from eqn. 4.89.

$$\text{System noise temperature at receiver input} = T_A/L_F + \left[T_F(L_F - 1)\right]/L_F + T_e \qquad (4.89)$$

Expression 4.89 for the noise temperature takes into account the noise generated by antenna and feeder together with receiver noise. The two expressions for the noise temperature in 4.88 and 4.89 highlight another very important point. We notice that the noise temperature at the antenna output is larger than the noise temperature at the receiver input by a factor of L_F. This underlines the importance of having minimum of losses before the first RF stage of the receiver.

4.5.6 Interference-Related Issues

Major sources of interference include the following.

1) Intermodulation distortion
2) Interference between satellite and terrestrial link sharing same frequency band
3) Interference between two satellites sharing same frequency band
4) Interference between two Earth stations accessing different satellites operating in the same frequency band
5) Interference arising out of cross-polarization in frequency reuse systems
6) Adjacent channel interference inherent to FDMA systems.

Each one them is briefly described in the following paragraphs.

4.5.6.1 Intermodulation Distortion

Intermodulation interference is caused due to generation of intermodulation products within the satellite transponder as a result of amplification of multiple carriers in the power amplifier, which is invariably a TWT amplifier. Generation of intermodulation products is due to both amplitude nonlinearity and phase nonlinearity. The characteristics of TWT amplifier are linear only up to a certain low input drive level and becomes increasingly nonlinear as the output power approaches saturation. Intermodulation products can be avoided by operating the amplifier in the linear region by reducing or backing off the input drive. Reduced input drive leads to a reduced output power. This results in a down-link limited system that is forced to operate at a reduced capacity.

Intermodulation distortion is a serious problem when the transponder is made to handle two or more carrier signals. That is why satellite links that use frequency division multiple access technique are particularly prone to this type of interference. On the other hand, single-carrier-per-transponder TDMA system is becoming increasingly popular as in this case the satellite TWT amplifier can be operated at or close saturation level without any risk of generating intermodulation products. This maximizes the EIRP for the down-link.

Another intermodulation interference-related problem associated with FDMA system is that the Earth station needs to exercise a greater control over the transmitted power so as to minimize the overdrive of the satellite transponder and the consequent increase in intermodulation interference. Intermodulation considerations also apply to Earth stations transmitting multiple carriers, which forces the amplifiers at the Earth station to remain under-utilized.

4.5.6.2 Interference between Satellite and Terrestrial Links

Satellite and terrestrial microwave communication links cause interference to each other when they share a common frequency band such as the 6/4 GHz frequency band, which is allocated to both satellite as well as terrestrial microwave links. An Earth station receiving at 4 GHz is susceptible to interference from a 4 GHz transmission from a terrestrial network. Similarly, an Earth station transmitting at 6 GHz is a source of interference for a terrestrial station receiving at 6 GHz. The level of mutual interference between the two is a function of a number of parameters including carrier power, carrier power spectral density and the frequency offset between the two carriers.

The level of interference caused by a terrestrial transmission to a satellite signal reception would depend upon the spectral density of the terrestrial interfering signal and the bandwidth of satellite signal received by the earth station. As an example, for a broadband satellite signal, the whole of interfering carrier power may be applicable whereas for a narrowband satellite signal, the interfering carrier power gets reduced by a factor equal to ratio of the total carrier power and the carrier power included in the narrow bandwidth.

Interference from a narrowband satellite transmission to a terrestrial microwave system can be reduced by using a frequency offset between the satellite and terrestrial carriers. The amount of interference depends upon the frequency difference between the interfering carrier frequency and the terrestrial carrier frequency. The interference reduction factor in this case can be determined by convolving the power spectral densities of the interfering carrier and the terrestrial carrier.

4.5.6.3 Interference Due to Adjacent Satellites

This type of interference is caused by the presence of side lobes in addition to the desired main lobe in the radiation pattern of the Earth station antenna. If the angular separation between two adjacent satellite systems is not too large, it is quite possible that the power radiated through side lobes of the antenna's radiation pattern, whose main lobe is directed towards the intended satellite, interferes with the received signal of the adjacent satellite system. Similarly, transmission from an adjacent satellite can interfere with the reception of an Earth station through the side lobes of its receiving antenna's radiation pattern.

This type of interference phenomenon is illustrated in Figure 4.80. Satellite A and satellite B are two adjacent satellite systems. The transmitting earth station of satellite A on its up-link, in addition to directing its radiated power towards the intended satellite through the main lobe of its transmitting antenna's radiation pattern, also sends some power though unintentionally

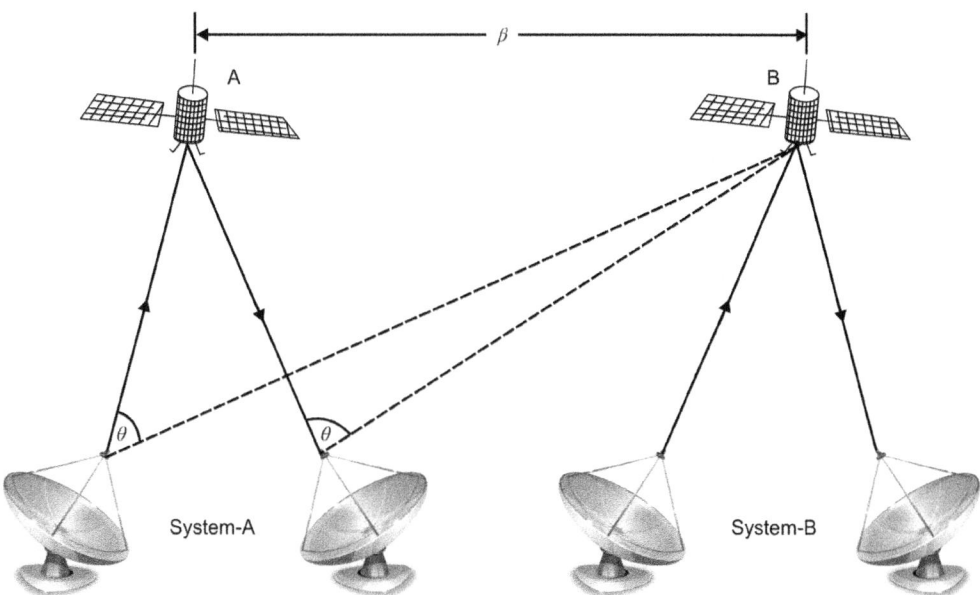

Figure 4.80 Interference due to adjacent satellites.

towards satellite B through the side lobe. The desired and undesired paths are respectively shown by solid and dotted lines. In the figure shown, θ is the angular separation between two satellites as viewed by the earth stations and β is the angular separation between the satellites as viewed from centre of Earth. That is, β is nothing but difference in longitudinal positions of the two satellites. Coming back to the problem of interference, transmission from satellite B on its down-link, in addition to being received by its intended earth station shown by a solid line again, also finds its way to the receiving antenna of the undesired earth station through the side lobe shown by dotted line. Quite obviously, this would happen if the off-axis angle of the radiation pattern of the earth station antenna is equal to or more than the angular separation ϑ between the adjacent satellites.

ϑ and β are interrelated by eqn. 4.90.

$$\theta = \cos^{-1}\left[\frac{\left\{d_A^{\ 2} + d_B^{\ 2} - 2r^2\left(1 - \cos\beta\right)\right\}}{2d_A d_B}\right]1 \qquad (4.90)$$

d_A = Slant range of satellite-A
d_B = Slant range of satellite-B
r = Geostationary orbit radius.

For a known value of ϑ, one can compute the worst-case acceptable value of the off-axis angle of the antenna's radiation pattern. Similarly, for a given radiation pattern and known off-axis angle, one can find out the minimum required angular separation between the two adjacent satellites for them to co-exist without causing interference to each other.

4.5.6.4 Cross-Polarization Interference

Cross-polarization interference occurs in frequency reuse satellite systems. It occurs due to coupling of energy from one polarization state to the other orthogonally polarized state in communications systems that employ orthogonal linear polarizations (horizontal and vertical polarization) and orthogonal circular polarizations (right- and left-hand circular). The coupling of energy of one polarization state to the other polarization state takes place due to finite cross-polarization discrimination of the Earth station and satellite antennas and also by depolarization caused by rain, particularly at frequencies above 10 GHz.

Cross-polarization discrimination is defined as the ratio of power received by the antenna in principal polarization to that received in orthogonal polarization from the same incident signal. A cross-polarization discrimination figure of 30–40 dB along the antenna axis is considered very well. The combined effect of finite values of cross-polarization discrimination for the Earth station and satellite antennas can be expressed in the form of net minimum cross-polarization discrimination for the overall link as given by eqn. 4.91.

$$X = \frac{1}{2}\left[\left(X_e\right)^{-1} + \left(X_s\right)^{-1}\right]^{-1} \tag{4.91}$$

X represents the worst-case carrier-to-cross-polarization interference ratio. It can be taken as an additional source of interference while computing overall carrier-to-noise plus interference ratio.

4.5.6.5 Adjacent Channel Interference

Adjacent channel interference occurs when the transponder bandwidth is simultaneously shared by multiple carriers having closely spaced centre frequencies within the transponder bandwidth. When the satellite transmits back to earth stations lying within its footprint, different carriers are filtered by the receiver so that each Earth station gets its intended signal. Filtering would have easier to realize had there been a large guard band between adjacent channels, which is not practically feasible as that would lead to inefficient use of transponder bandwidth. The net result is that a part of power of the carrier in the channel adjacent to the desired one also gets captured by the receiver due to overlapping amplitude characteristics of channel filters. And this becomes a source of noise.

4.5.7 Antenna Gain-to-Noise Temperature (*G/T*) Ratio

The *G/T* ratio, usually defined with respect to the Earth station receiving antenna, is an indicator of the sensitivity of the antenna to the down-link carrier signal from the satellite. It is the figure of-merit used to indicate the performance of the Earth station antenna and low noise amplifier combine in receiving weak carrier signals. *G* is the receive antenna gain usually referred to the input of the low noise amplifier. It equals the receive antenna gain as computed for instance from $G = \eta(4\pi A_e / \lambda^2)$minus the power loss in the waveguide connecting the output of the antenna to the input of the low noise amplifier. *T* is the Earth station effective noise temperature also referred to the input of low noise amplifier. It may be mentioned here that the value of the *G/T* ratio is invariant irrespective of the reference point chosen to do so. We choose input of a low noise amplifier because it is the point where its contribution is clearly shown.

The system noise temperature at the output of antenna is given by eqn. 4.92.

$$\text{System noise temperature at antenna output} = T_A + T_F\left(L_F - 1\right) + T_e L_F \tag{4.92}$$

T_A = Antenna noise temperature
T_F = Thermodynamic temperature of feeder often taken as ambient temperature
L_F = Attenuation factor of feeder
T_e = Effective input noise temperature of receiver

And system noise temperature when referred to input of low noise amplifier is given by eqn. 4.93.

$$\text{System noise temperature at receiver input} = T_A/L_F + \left[T_F \left(L_F - 1 \right) \right]/L_F + T_e \tag{4.93}$$

T_e can further be expressed by eqn. 4.94.

$$T_{en} = T_1 + \frac{T_2}{G_1} + \frac{T_3}{G_1 G_2} \tag{4.94}$$

Here, T_1, T_2, T_3 are noise temperatures of different stages in the receiver beginning with low noise amplifier and G_1, G_2, G_3 are the corresponding gain values.

From the known values for G and T, the G/T ratio can be computed. To conclude the discussion on the G/T ratio, we can make the following observations.

1) Higher the antenna gain and lower the loss of the feeder connecting output of antenna to input of low noise amplifier, higher would be the G/T ratio.
2) Lower the noise temperature of the low noise amplifier, higher would be the G/T ratio.
3) Higher the gain of the low noise amplifier, lower will be the noise contribution of successive stages in the receiver and higher will be the G/T ratio.

4.5.8 Link Budget

Link budget is a way of analysing and predicting the performance of a microwave communication link for given values of vital link parameters that contribute to either signal gain or signal loss. It is algebraic sum of all gains and losses expressed in decibels as we travel from transmitter to the receiver. The final value thus obtained provides us with the means of knowing the available signal strength at the receiver and therefore also knowing how strong the received signal is with respect to the minimum acceptable level called the threshold level. The difference between the actual value and the threshold is known as the *link margin*. The higher the link margin, the better the quality of service. The link budget thus is a tool that can be used for optimizing various link parameters so as get the desired performance.

The concept of link budget can be illustrated further with the help of a one-way microwave communication link schematic shown in Figure 4.81. Various parameters of interest in this case are as follows.

1) Transmitter power, P_T
2) Power loss in the waveguide connecting the transmitter output to antenna input, L_T

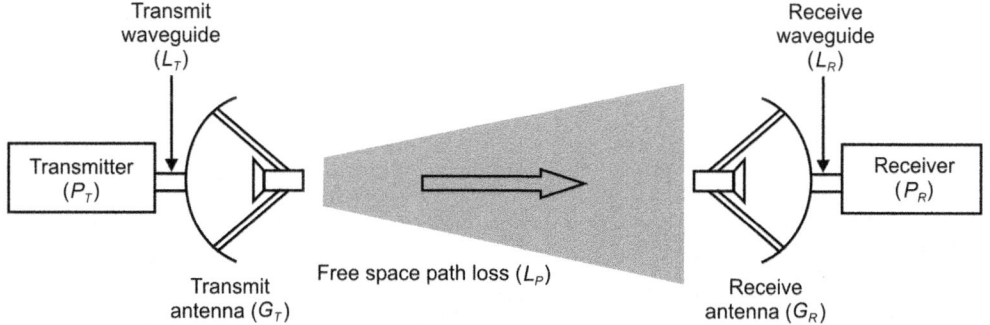

Figure 4.81 Link budget.

3) Transmitting antenna gain, G_T
4) Free-space path loss, L_P
5) Attenuation due to rain, clouds, fog and so on, A
6) Receive antenna gain, G_R
7) Power loss in the waveguide connecting the receive antenna output to receiver input, L_R
8) Received signal power, P_R

The power balance equation describing the link budget in this case is given by eqn. 4.95.

$$P_T - L_T + G_T - L_P - A + G_R - L_R = P_R \tag{4.95}$$

With reference to the satellite link, we can write such an equation for both up-link as well as down-link. Figure 4.82 shows the schematic arrangement for a satellite link indicating various parameters that typically contribute to link budget. For the satellite link shown in Figure 4.82, the up-link and down-link equations can be written as eqn. 4.96 and 4.97.

$$P_{TU} + G_{TU} - L_{PU} - A_U + G_{RU} = P_{RU} \dots\dots\dots\dots\text{up-link} \tag{4.96}$$

$$P_{TD} + G_{TD} - L_{PD} - A_D + G_{RD} = P_{RD} \dots\dots\dots\dots\text{down-link} \tag{4.97}$$

As an example, let us look at the link budget of a typical Ku band satellite-to-DTH receiver down-link. Typical values for various parameters are as follows.

1) Transmit power, $P_T = 25$ dBW
2) Transmit waveguide loss, $L_T = 1$ dB
3) Transmit antenna gain, $G_T = 30$ dB
4) Free-space path loss, $L_P = 205$ dB
5) Receive antenna gain, G_R (for 50 cm diameter dish) $= 39.35$ dB
6) Receive waveguide loss, $L_R = 0.5$ dB

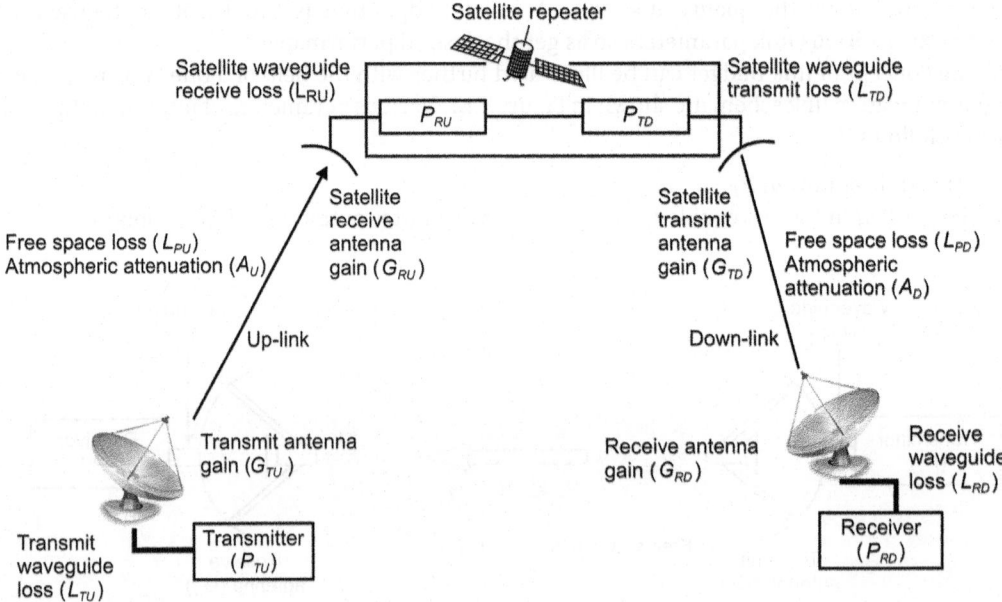

Figure 4.82 Satellite link budget analysis.

Received signal power can be computed from the above data to be equal to (–112.15 dB).

Now if we take the receiver system noise temperature to be 140 K and the receiver bandwidth to be 27 MHz, which are typical values for the link under consideration, then the receiver noise power is would be:

$$N = kTB = 1.38 \times 10^{-23} \times 140 \times 27 \times 10^6 \, W = 5216 \times 10^{-17} \, W = -132.8 \, dB$$

Therefore, received carrier-to-noise (C/N) ratio for this link would be 20.65 dB. This figure can be used to determine the link margin and hence the quality of service provided by the link under clear sky conditions. It can also be used to assess the deterioration in the quality of service in case of hostile atmospheric conditions.

4.6 Networking Concepts

Satellites today provide a wide range of applications and services, which include traditional telephony, radio and television broadcast services and broadband and internet services. The ultimate objective of satellite networks is to provide these services, which are continuously evolving both in terms of type of services as well as the quality of service, to users irrespective of their geographical location and particularly to those in remote locations that are not conveniently accessible to terrestrial networks. Given the enormity and requirements of satellite applications and services in a modern satellite network, one is likely to find use of a single satellite and multiple satellites with inter-satellite links or multi-hop transmission, bench top user terminals and hand-held terminals, fixed Earth stations and transportable Earth stations, satellites only networks and satellite networks integrated with terrestrial networks. With the integration of audio, video and broadband internet services and due to the need to support a wide range of applications and services to users wherever they are, internetworking with terrestrial networks and protocols is also an important part of satellite networks. The core topics of satellite networking briefly discussed in this section include networking technologies, networking protocols, network topologies, and internetworking with terrestrial networks, inter-satellite links and network security.

4.6.1 Network Characteristics

The fundamental characteristics that can be used for the purpose of comparing various available networking options and solutions include availability, reliability, security, speed or throughput, scalability, topology and cost.

4.6.1.1 Availability

Network availability refers to the ability of the network to respond to the users' requests to access it. Availability is the probability that the network will be available to users as and when required. It is typically measured as a percentage of the total time the network is required to be available. One way of computing availability, known as the *availability ratio*, is to take the ratio of the time in minutes the network was actually available during a year to the total number of minutes in a year. The time period considered for quantifying availability is very important. A given percentage availability when measured over a 1-year period is indicative of a much higher degree of availability than when it is measured for a period of 1 month. Another method of expressing availability, particularly with reference to satellite networks, is the *mean time between outages* (MTBO), which is defined as the average duration of a time interval during which the connection is available from the service perspective. Consecutive intervals of available time during which the user attempts to use are concatenated. There are different classes of availability. The standard method of defining classes of availability is to express availability on a scale of nines, such as two nines, three nines and so on. Two nines, three nines and four nines mean availability of 99, 99.9 and 99.999%, respectively.

4.6.1.2 Reliability

The *reliability* of a network is a measure of the reliability of the different components of the network and their interconnections. It is generally measured as the *mean time between failures* (MTBF) or *mean time to repair* (MTTR). For a reliable network, MTBF should be as high as possible and MTTR should be as low as possible. In other words, the fraction of time during which the network is down or unable to support a connection should be as low as possible and once a connection is established the probability of it being terminated due to either failure of a network component or inadequate performance should be as low as possible. The reliability of the network may be measured by its ability to transmit data with an acceptable error rate, which can be specified for each component link of the network. The bit error test (BERT) or the block bit error rate test (BLERT) may be used to determine the reliability of different component parts of the network and hence the reliability of the total network.

4.6.1.3 Security

Networks are the backbone of all communications (audio, video and data) that take place between government agencies, large enterprises, small and medium business houses and individuals. The objective of *network security* is to monitor and prevent unauthorized access, eavesdropping, misuse, modification or denial of use of a network and its resources to authorized users. A network security system typically relies on layers of protection that consists of multiple components, including network monitoring and security software in addition to hardware and appliances. All components work together to increase the overall security of the computer network. The terms 'network security' and 'information security' are often used interchangeably. Network security is generally taken as providing protection at the boundaries of an organization by keeping out intruders or hackers with the help of network security tools such as firewalls, virtual private networks (VPNs), data encryption, and biometric devices and so on. Information security on the other hand explicitly focuses on protecting data resources from malware attack or simple mistakes by people within an organization by use of data loss prevention techniques. Network security is jeopardized by a range of threats including viruses, Trojan horse programmes, vandals, attacks, data interception and social engineering.

4.6.1.4 Throughput

Network throughput is defined as the average rate at which data is transferred through the network in a given time. It is measured in bits per second (bps) and also sometimes in data packets per second or data packets transferred in a given time slot. The throughput is typically less than the transmission speed of the individual links. In a network with multiple links, the overall throughput is equal to or less than the lowest throughput value. The *theoretical maximum throughput* can be considered synonymous with the digital bandwidth capacity and the *effective throughput* allows the subscriber to access data. The latter is always less than the former. For example, a communication link using 9600 bps modems may yield an effective throughput of 6000 bps as there are always some extra bits in the overall bit stream used for functions such as error checking and the information bits are always less than the total number of bits being transmitted. Throughput may be measured as *transfer rate of information bits* (TRIB) by taking the ratio of the number of information bits transferred to the time required for the information bits to be transferred. The throughput of a network is constrained by factors such as network protocols and different components comprising the network, such as switches and routers, type of cabling, whether the cabling is fibreoptic or Ethernet and so on. It is further constrained by the specifications of network adapters used on client systems in the case of wireless networks.

4.6.1.5   Scalability

Scalability is the capability of a network to accommodate increase data rate requirements and number of users. It also indicates how well a network can adapt itself to new applications and replacement of old components by new components with enhanced features. In other words, it is the ability of the network to handle increased quantum of work and to increase total throughput under an increased load when resources are added. In the case of a scalable network, performance improves in direct proportion to the capacity added to the network. Modularity is an important ingredient of all such networks.

4.6.1.6   Topology

Topology is the way different components of the network are connected (*physical topology*) and the logical way data passes through the network from one component or device to the next irrespective of the physical structure of the network (*logical topology*). Common physical topologies include star, bus, mesh, tree and ring. Logical topology defines how the devices communicate across the physical topologies. Two common types of logical topologies are *shared media topology* and *token based topology*. In a *shared media topology* various devices on the network have unrestricted access to the physical media and it suffers from the problem of collisions. Ethernet is an example of shared media logical topology. In *token based logical topology* the sending device uses a token to send a packet of data. The token travels on the network with the data packet. After the data packet is received by the intended machine, the token travels back to the sender. It is taken off the network by the sender and a new empty token is sent out to be used by the next device. Token based networks do not have the collision problem of shared media topology but suffer from latency instead. Token based logical topology is best configured in physical ring topology.

4.6.1.7   Cost

Cost is the expenditure incurred in setting up and maintaining the network.

4.6.2   Network Topologies

Topology of a communication network is the schematic description that defines the network geometry, showing how different nodes are connected to each other and the paths the signals follow from node to node. While the way in which different nodes are connected is more precisely known as physical topology, the nature of the signal flow paths represents logical topology. Common physical topologies include bus, star, ring, mesh and tree topologies. Token based topology and shared media topology are the common logical topologies. The following paragraphs briefly describe different network topologies.

4.6.2.1   Bus Topology

In the case of bus topology all devices to be connected to the network are connected to a central cable that acts as the backbone of the network with the help of interface connectors. This is the simplest of all network topologies, in which all work stations communicate with the other devices through a common cable known as a bus. A network device intended to communicate with another device on the network sends out a broadcast message on the cable. The message is seen by all the devices on the network but is received and processed by the intended recipient only. Terminators are used at the ends of the cable to prevent bouncing of signals. The network

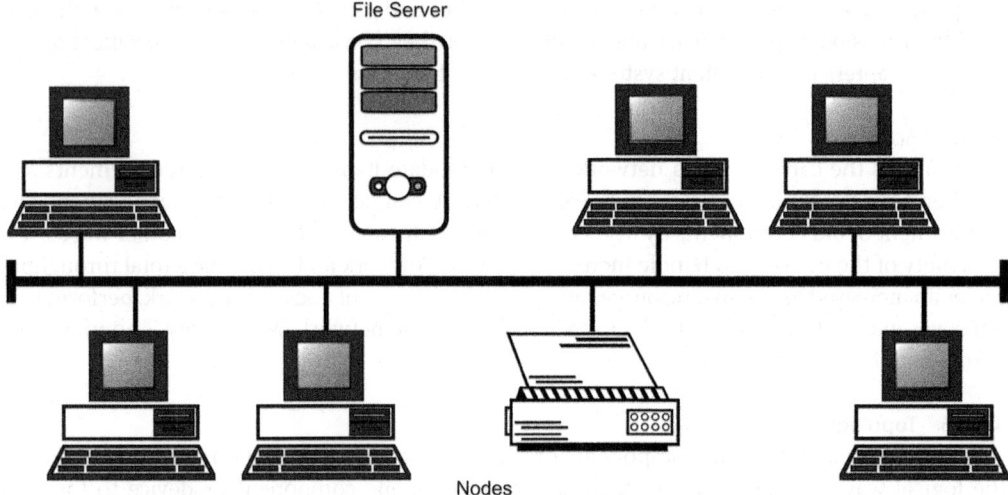

Figure 4.83 Bus topology.

may be extended using barrel connectors. Figure 4.83 shows the basic schematic bus topology. Bus topology is low cost for given a number of devices connected to the network, it is easy to set up and expand, it uses the least cable length and it is well suited to small networks such as LANSs. The disadvantages include relatively high maintenance cost, reducing efficiency with increase in number of devices connected to the network, mandatory requirement of terminations and low security. It is also not suitable for networks with heavy traffic.

4.6.2.2 Star Topology
In star topology, all work stations are connected to a central device with a point-to-point connection, unlike bus topology in which different devices are wired to a common cable that acts as a shared medium. The central device could be a hub, router or switch. The hub may act as a repeater or a signal booster depending on the nature of the central device. In addition to acting as a junction of the network the hub also controls and manages the entire network. The central device can communicate with the hubs of other networks. Different devices in the star network are connected to the central device using unshielded twisted pair Ethernet cables. Each work station on the network communicates with the other workstations through the hub. Figure 4.84 shows the basic schematic diagram of a star network. This network has the advantages of better performance compared to bus topology depending on the capability of the central hub and the ease with which new devices can be added or existing devices can be removed from the network without affecting the rest of the network. The network is easy to monitor and troubleshoot, and the failure of one node does not affect the rest of the network. The network, however, has some shortcomings. The whole of the network fails with failure of the central hub and expansion capability also is governed by the capacity of the central hub.

4.6.2.3 Ring Topology
In ring topology, all workstations are connected to one another in a closed loop, as illustrated in the basic schematic diagram of Figure 4.85. Each workstation is directly connected to one workstation on either of the two sides, as shown in the diagram. Data communication in the case of ring topology takes place with the help of a *token*. Data travels around the network in one direction. The sourcing workstation sends out the data along with token information. Every successive node examines the token to determine if the data were meant for it. If yes, data are received and

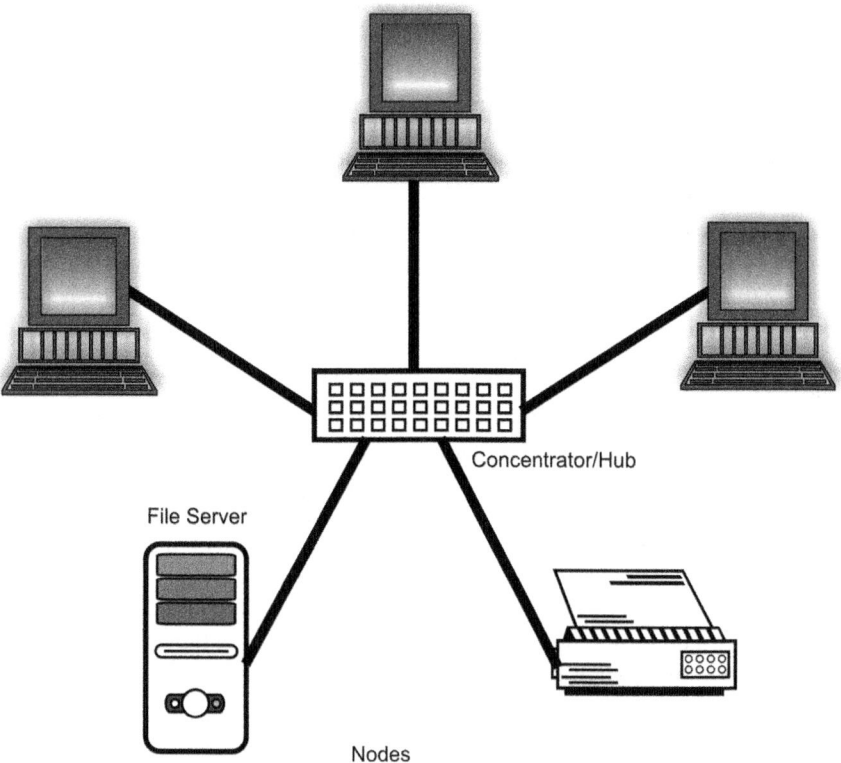

Figure 4.84 Star topology.

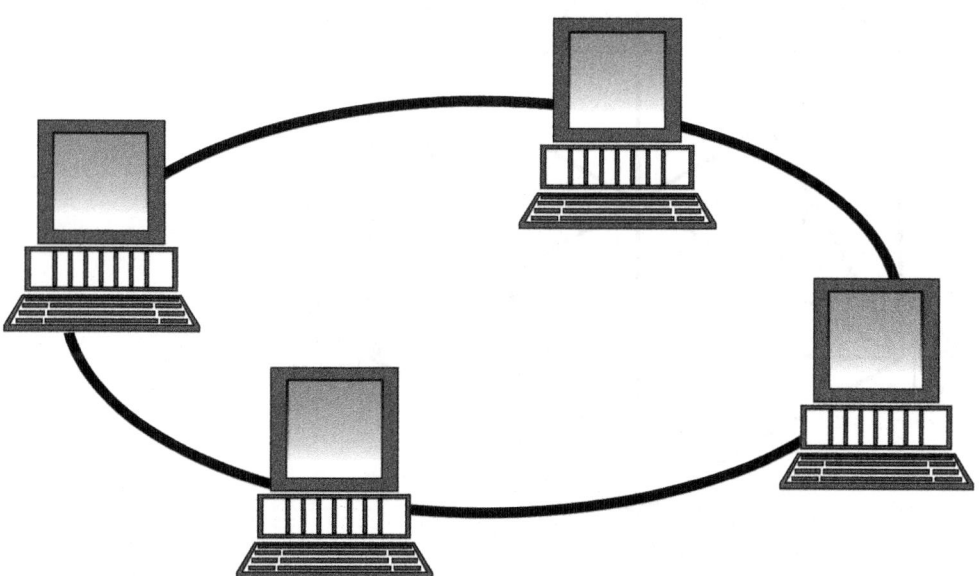

Figure 4.85 Ring topology.

an empty token is passed onto the network. If no, data along with token are passed onto next node. The process continues until the data reach their intended destination. Only those nodes equipped with token can send data. Others have to wait until an empty token reaches them.

In ring topology, data flows in one direction at high-speed and the probability of collision is low as each node gets to send data only when it receives an empty token. Network performance is not affected by the addition of new devices. Each device has equal access to network resources. The topology offers better performance compared to bus topology in the case of increased load on the network. However, it is relatively slow compared to star topology as the data has to pass through all nodes between the source and destination nodes. The network is affected by a fault in any of the nodes. A failure in the cable or any device breaks the loop, bringing down the entire network. The topology is relatively more costly to implement compared to bus and star topologies.

4.6.2.4 Mesh Topology

In mesh topology, each network node is connected to every other node. This is true mesh topology. However, mesh topology suffers from a large number of redundant connections and consequently is more expensive to implement. It is not generally used for computer networks and is preferred for wireless networks. The shortcomings of full or true mesh topology are overcome in the case of partial mesh topology in which some of the nodes are connected, as for true mesh topology, while others are connected to one or two nodes only. In other words, different nodes are either directly or indirectly connected to every other node. This reduces redundancy and the cost of the network. Figure 4.86 shows the basic schematic arrangement of typical true mesh and partial mesh topologies. Mesh topology allows simultaneous transmission from different nodes.

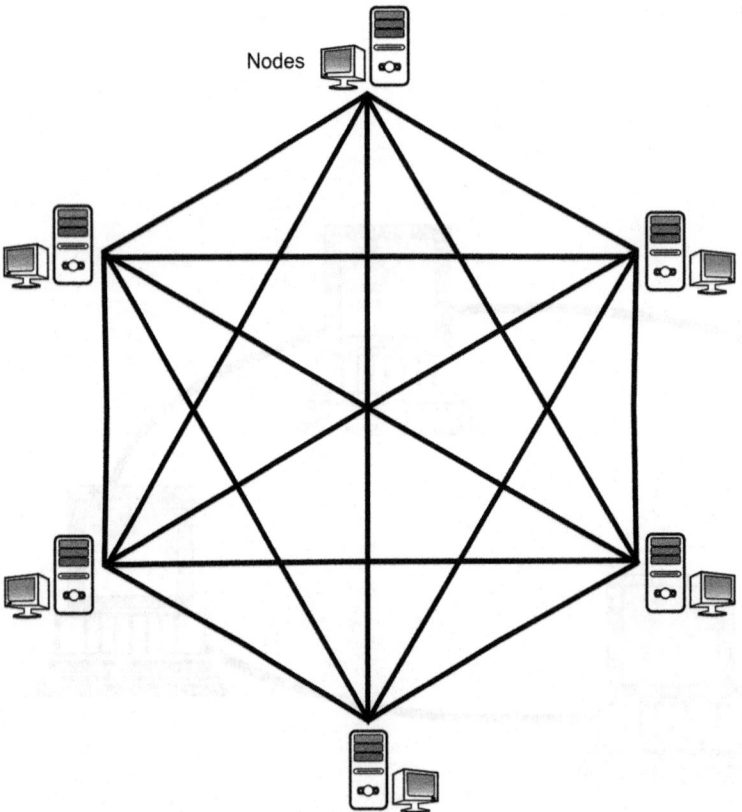

Figure 4.86 Mesh topology.

Also, data transmission is not affected as a result of failure of a particular node. The topology is amenable to expansion and modification without affecting existing nodes. The topology also has several disadvantages such as high cost, high redundancy and is difficult to maintain and administer.

4.6.2.5 Tree Topology

Tree topology combines the features of star and bus topologies. In fact, tree topology is an expanded star topology in which multiple star networks are interconnected by a common bus. Figure 4.87 shows the basic schematic diagram of tree topology. It is known as tree topology for obvious reasons as the common bus represents the stem of the tree and the multiple star networks connected to the bus act like branches. The Ethernet protocol is commonly used in tree topology.

Tree topology allows more efficient network expansion than what is achievable with star or bus topologies individually. While star network expansion is limited by the capacity of the central hub, there is a limit to the maximum number of devices that can be connected to a bus network due to the broadcast traffic the network would generate. Hybrid topology of bus and

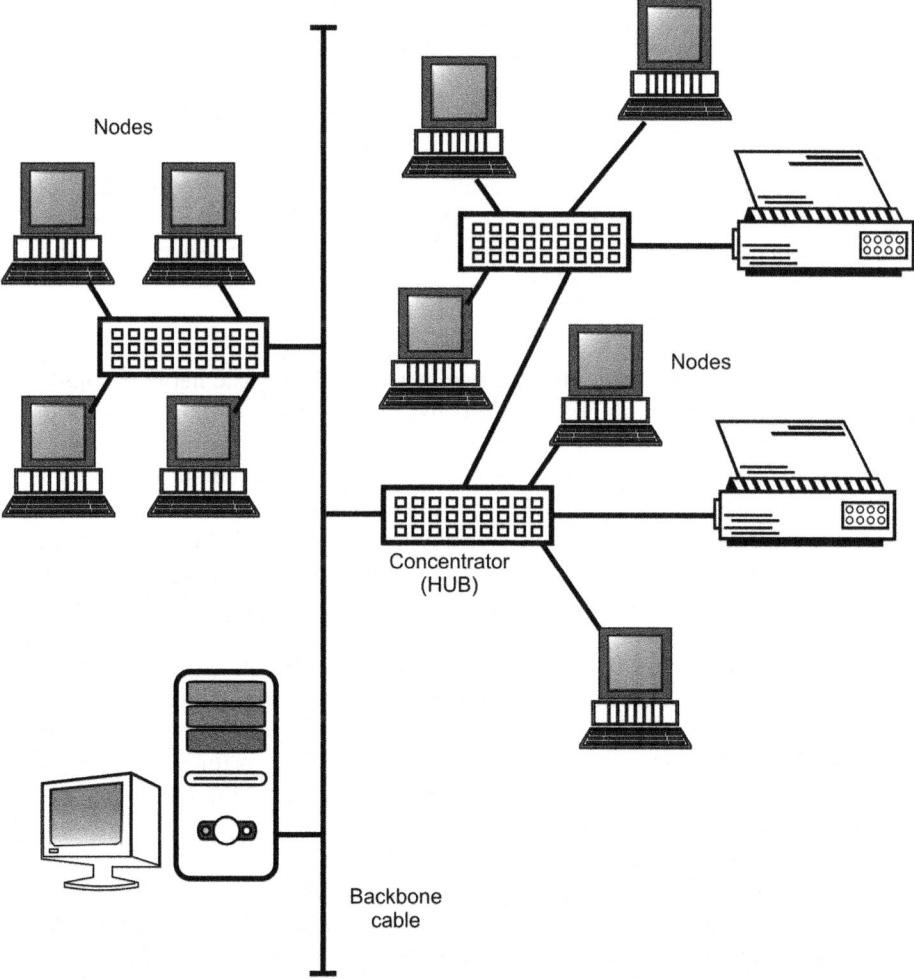

Figure 4.87 Tree topology.

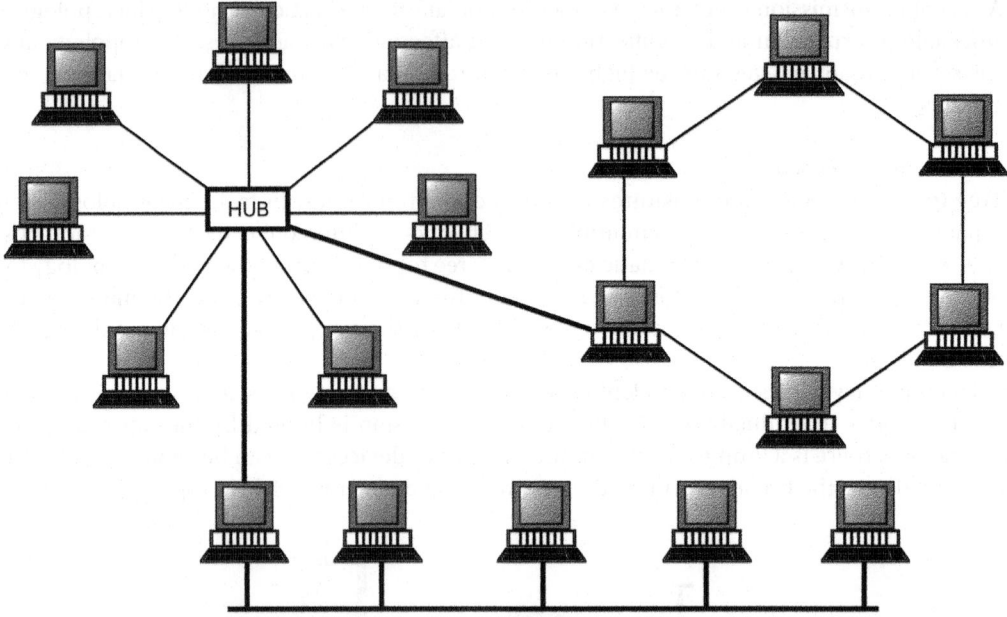

Figure 4.88 Hybrid topology.

star networks overcomes these problems. It also allows easier error detection and correction, and damage to any one segment does not affect other segments of the network. The disadvantages include heavy dependence of tree topology on the central bus and the network scalability being dictated by the type of common cable.

4.6.2.6 Hybrid Topology

Hybrid topology is the result of integration of two or more of the basic network topologies discussed in the previous sections. The resultant hybrid topology combines the good and bad features of all constituent topologies. There is one specific hybrid topology, but different hybrid topologies may be configured depending on application requirement. For example, if the network of one office uses a star topology and that of another a bus topology and if the two networks were to be integrated in a wide area network, a star-bus hybrid topology would be selected. Another common hybrid topology is star-ring topology. Figure 4.88 shows the basic schematic diagram of a hybrid topology that is a combination of star, bus and ring topologies.

Hybrid network topologies are *scalable*, as the network size can be conveniently enhanced without affecting the network architecture; *reliable*, as troubleshooting is relatively easier and the faulty part can be isolated without affecting the network and *flexible*, as they allow design of a network customized for the requirement. Hybridization also allows the good points of the constituent topologies to be maximized and the weak points to be minimized.

4.6.3 Network Technologies

This section describes the technologies used to exchange information between different nodes. A common method of differentiating network technologies is on the basis of the path followed for the flow of information between the communicating devices. There are two broad categories of networking methods, namely circuit switching and packet switching, with the consequent networks known as *circuit switched networks* and *packet switched networks*. These are briefly described in the following paragraphs.

4.6.3.1 Circuit Switched Networks

In a *circuit switched network*, a dedicated physical path is established between the communicating nodes through the network before the start of actual communication. This dedicated path is held for the duration of the communication session. The circuit functions as if the two nodes were physically connected by an electrical circuit. The circuit may be a fixed one that is always present or may be created on a requirement basis. It may be mentioned here that even though alternative paths may exist between the two communicating nodes or devices, the communication in the case of circuit switched networks takes place only over the identified path. The dedicated circuit cannot be used by other callers until the circuit is released and a new connection is set up. Also, the communication channel or circuit remains unavailable for use to other callers even if there is no communication actually taking place. Figure 4.89 illustrates the concept of a circuit switched network. The dedicated circuit for communication between the two devices X and Y is illustrated in the figure. As is evident from the schematic representation of the circuit switched network, communication between the two devices takes place over this dedicated circuit only even though there are several alternative paths available. A telephone system such as the *public switched telephone network k*(PSTN) is a classic example of a circuit switched network.

The circuit switched concept can also be used between two communicating nodes for the transfer of information other than voice. It should not be thought to be a technique used only for connecting analogue and digital voice circuits.

4.6.3.2 Packet Switched Networks

In *packet switched network* technology, there is no dedicated communication path or circuit identified prior to the start of communication for the entire duration of the communication between the source and the destination. The data in this case is broken down into small pieces called packets and then sent over the network based on the destination address contained within the packet. Packets of data from a given communication may take any number of different paths within the network while travelling from source to destination. However, the communication circuit is dedicated for the duration of packet transmission to that packet alone

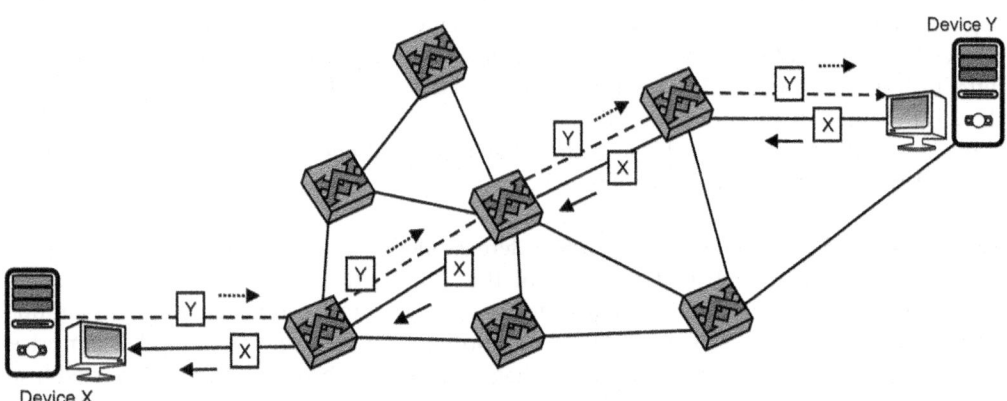

Figure 4.89 Circuit switched network.

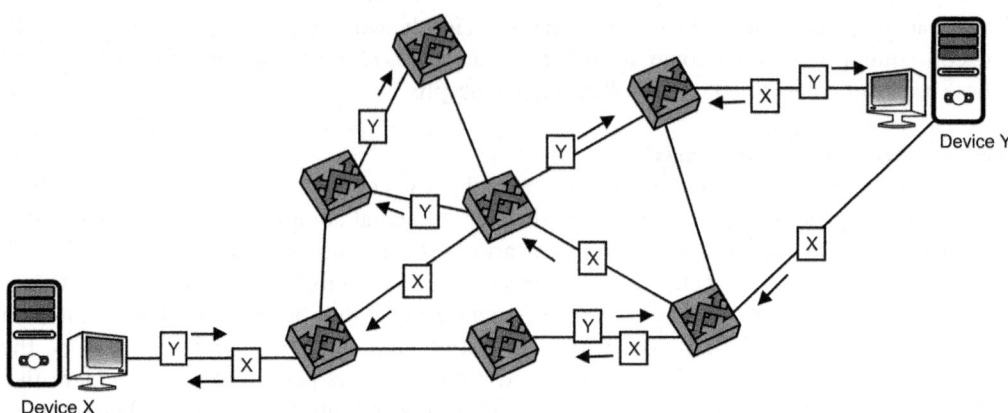

Device Y

Device X

Figure 4.90 Packet switched network.

and is not interrupted to transmit other packets. The data is reassembled into their original form at the receiving end. Figure 4.90 illustrates the concept of a packet switched network.

Packet switching is used in transmitting data over the internet and often over a LAN. The availability of multiple paths allowing the same line to be used for multiple communications simultaneously leads to improved efficiency, particularly when a large volume of traffic is to be handled. The packet switched mode is the signal transmission technology used in all internet communications.

4.6.4 Network Protocols

A networking protocol defines the standard used for communication between different devices connected to a network, such as a LAN, intranet or the internet. Networking protocols include mechanisms for identification of network devices to make connections as well as formatting rules that specify how data is packaged into messages to be sent and received by different devices. To ensure a high performance and reliable network communication, some networking protocols also have message acknowledgement and data compression features built into them.

4.6.4.1 Common Networking Protocols

Hundreds of different networking protocols have been developed, each designed for specific application requirements. Each protocol has its own method of formatting data before it is sent or processing data after it is received. Data compression and error correction techniques also differ for various protocols. It is not feasible to discuss each of these hundreds of protocols, but the most commonly used ones are discussed in the following paragraphs. These include the OSI model, IP, TCP, HTTP, FTP, asynchronous transfer mode (ATM), simple mail transfer protocol and user datagram protocol (UDP).

4.6.4.2 The Open Systems Interconnect (OSI) Reference Model

The OSI reference model is at the core of the OSI standard developed in 1984 by the International Organization for Standardization (ISO), an international federation of national standards organizations representing 130 countries. The OSI reference model comprises seven layers that define the different stages that data must go through when transported from one device to another in a network. Based on the nature of activities performed by different layers of the model the seven layers of the OSI reference model are divided into two sets: the

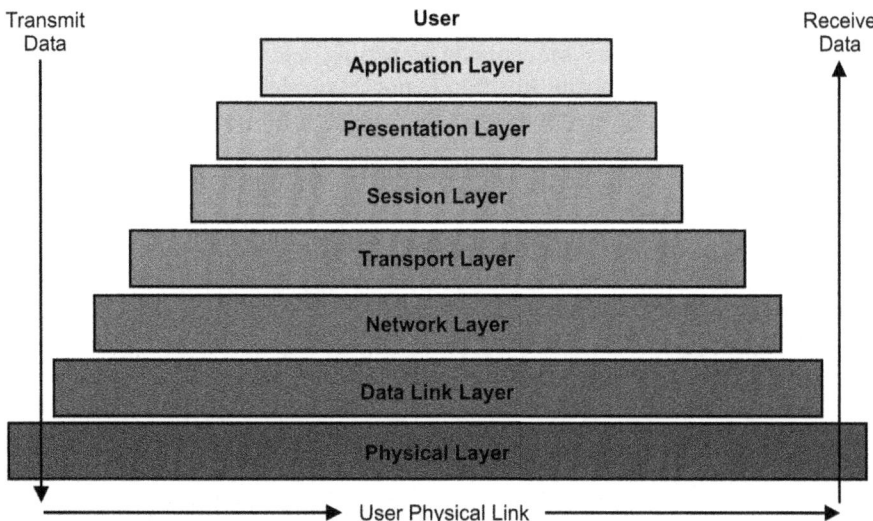

Transmit
Data

User

Receive
Data

Application Layer

Presentation Layer

Session Layer

Transport Layer

Network Layer

Data Link Layer

Physical Layer

User Physical Link

Figure 4.91 Seven layers of the OSI reference model.

transport set comprising layers 1–4 and the *application set* comprising layers 5–7. While the upper layers of the OSI model (layers 5, 6 and 7) manage the application level functions and represent the software that implements network services such as encryption and connection management, the protocols that operate at lower layers control end-to-end transport of data between devices and implement hardware oriented functions such as routing, addressing and flow control. In the OSI reference model, when the two devices communicate with each other, the flow of information begins with the topmost layer on the sending end. The information flows down the stack of layers to the bottom most layer at the sending end. It then traverses the network connection to the bottom most layer of the receiving end, from where it moves up the stack to the topmost layer of the model. Figure 4.91 illustrates the flow of information in the OSI reference model. It may be mentioned here that when the communication is via a network of intermediate systems, only the lower three layers of the OSI protocols are used in the intermediate systems. As stated earlier, the OSI reference model has seven layers. The seven layers of the OSI reference model, beginning with the top most layer, are physical, data link, network, transport, session, presentation and application layers. The following paragraphs give brief descriptions of each of these seven layers.

The *physical layer* (layer 1) transmits bits from one node to another and regulates the transmission of a bit stream over a physical medium. This layer defines how the cable is attached to the network adapter and what transmission technique is used to send data over the cable. The physical layer of the OSI model controls the signalling and transfer of raw bits onto the physical medium. It also describes the electrical/optical, mechanical and functional interfaces to the physical medium. The physical layer performs the following major functions. It defines the physical and electrical specifications of the interface with the physical transmission medium, protocol to establish and terminate a connection between two directly connected nodes, protocol for flow control and conversion of raw data into corresponding signals to be transmitted over the communication medium.

The *data link layer* (layer 2) provides error-free transfer of data frames from one node to another over the physical layer, allowing layers above it to assume virtually error-free transmission over the link. It performs its function by establishing and terminating the logical link between the two nodes, telling the transmitting node to back-off in the absence of any frame buffers, transmitting/receiving frames sequentially, detecting and recovering from errors occurring in the physical layer by retransmitting frames not acknowledged, creating and recognizing frame boundaries, checking received frames for integrity and providing media access management.

The *network layer* (layer 3) controls the internetwork communication and determines the route from the source computer to the destination computer. It performs its functions by routing frames among networks, controlling traffic by instructing a sending station to hold back frame transmission if the router's buffer were filled up and doing frame fragmentation where the router's maximum transmission unit size is less than the frame size. The network layer's two key responsibilities include providing a unique address that identifies the host and the network the host exists on (a function called *logical addressing*) and finding the best path to the destination network and routing data accordingly (a function called *routing*). Two of the most common network layer protocols include IP and internetwork packet exchange (IPX).

The *transport layer* (layer 4) handles error recognition and recovery. It ensures error-free and sequential delivery of messages, and prevents losses and duplications. When necessary, it repackages long messages into small packets at the sending end and rebuilds the data packets into the original message at the receiving end. It also sends receipt acknowledgements at the receiving end. Transport layer communication falls under two categories, namely connection oriented, which requires a connection with specific agreed-upon parameters to be established before data is sent, and connectionless, which requires no connection for data to be sent. The TCP/IP suite incorporates two transport layer protocols: TCP for connection oriented communication and UDP for connectionless communication.

The *session layer* (layer 5) is entrusted with the task of establishing, maintaining and ultimately terminating sessions between devices. It performs the functions that allow devices to communicate over the network, performing security, name recognition, logging and so on. If a session is broken, the session layer attempts to recover it. Sessions communication falls under one of three categories: full duplex, which is simultaneous two-way communication; half duplex, which is two-way communication but not simultaneous and simplex, which is one-way communication. The session layer often has to rely on lower layer protocols for session management as many contemporary protocol suites such as TCP/IP do not implement session layer protocols.

The *presentation layer* (layer 6) translates data from the application layer into a network format while sending information and vice versa while receiving information. Standards have been developed for the formatting of data types such as audio (MIDI, MP3 and WAV), text (RTF, ASCII and EBCDIC), video (MPEG, AVI and MOV) and images (GIF, TIF and JPG). This layer also manages security issues by providing services such as data encryption and compression. The presentation layer translates data from a format used by the application layer into a common format at the sending station and translates data from the common format to a format known to the application layer at the receiving station. In other words, it ensures that the data from the sending application can be understood by the receiving application.

The *application layer* (layer 7) provides the interface between the user application such as an email client or a web browser and the network. The application layer performs the functions of identifying communication partners, determining resource availability and synchronizing communication. Application layer protocols include FTP (via an FTP client), HTTP (via a web browser), POP3 and SMTP (via an email client) and Telnet. Application protocols reside at the application layer. The user interacts with the application, which in turn interacts with the application protocol.

4.6.4.3   Internet Protocol (IP)

IP was developed in the 1970s and is the primary network protocol used on the Internet. It is generally used together with TCP on the internet and many other networks and is referred to as TCP/IP. It is a mechanism for transfer of data packets between computers by allowing computers to be connected by a variety of physical media, including modems, Ethernet cabling, fibreoptics, and radio and satellite links.

Each machine on the internet has a unique number with which it can be identified. This unique number is called its *IP address*. The two common versions of IP include IPv4 and IPv6. IP addresses in IPv4 have a length of four bytes (32 bits). In the newer IPv6, IP addresses are 16 bytes (128 bits) long.

IP addresses are normally expressed in a decimal format as a dotted decimal number comprising four numbers called octets. Each octet can be represented by an 8-bit binary number, which means 32 bits in all for the IP address. This further implies a total of 232 = 4 294 967,296 (approximately 4.3 billion) possible unique values. Octets are used to create classes of IP addresses that can be assigned to a particular entity based on size and requirement. The octets are split into *net* and *host* sections. The net section is used to identify the network the computer belongs to and always contains the first octet. The host section is used to identify the actual computer on the network and always contains the last octet. There are five classes of IP ranges: Class A, Class B, Class C, Class D and Class E. Each class has a range of valid IP addresses. Data on an IP network are organized into packets. Each IP packet includes a header that specifies source, destination and other information about the data other than the message data itself. The IP suite enables these data packets to be transferred between computers based on the IP addresses contained in the data packet header.

4.6.4.4   Transmission Control Protocol (TCP)

TCP is one of the main protocols in the IP suite and is used along with IP to send data packets between computers over the Internet. While IP takes care of handling the actual delivery of the data, TCP takes care of keeping track of the individual data packets that a message is divided into for efficient routing through the Internet. It enables the sending and receiving hosts to establish a connection to exchange data streams and also ensures that the packets are delivered in the same order in which they were sent. TCP is responsible for ensuring the division of data into packets at the sending end and reassembly of data packets into the original message at the receiving end. The role of TCP/IP in data exchange between two computers can be explained with the help of the example of a web server sending an HTML file. The TCP programme layer in the server divides the file into different data packets, which are then numbered. The data packets are forwarded to the IP programme layer one at a time. Though all data packets have the same destination IP address, different data packets may follow different routes through the network to reach the intended destination. At the receiving end computer, TCP reassembles the individual data packets after they have all arrived and forward them as a single file. TCP is a connection oriented protocol, that is, it establishes and maintains the connection until the data to be exchanged by application programmes at the two ends has been completed. TCP works at *transport layer* (layer 4) of the OSI model.

4.6.4.5   Hyper Text Transfer Protocol (HTTP)

HTTP is an application layer protocol built on top of TCP. HTTP utilizes TCP port 80 by default, though other ports such as 8080 can also be used. It is the underlying protocol used by the WWW. HTTP defines how messages are formatted and transmitted. It provides a standard for communication of web browsers and servers. Web browsers (HTTP clients) and servers communicate via HTTP request and response messages. For example, when a URL is entered in the browser an HTTP command is sent to the web server directing it to fetch and transmit the requested web page. The three main HTTP message types are GET, POST and HEAD.

HTML is the other standard controlling the working of the WWW. HTML covers how web pages are formatted and displayed. Currently HTTP version 1.1, which is an improvement over HTTP version 1.0, is in widespread use. One of the shortcomings of HTTP is that it is a stateless protocol, that is, each command is executed independent of knowledge of the previous commands, which makes it difficult to implement websites that react intelligently to user input. This shortcoming is being addressed by a number of new technologies such as ActiveX, JavaScript and Cookies.

4.6.4.6 File Transfer Protocol (FTP)

FTP is a standard based on IP used for transferring files between computers on the Internet. It is an application protocol that makes use of TCP/IP protocols like HTTP, used to transfer displayable web pages and related files, and SMTP, which transfers email. FTP is also used to refer to the process of copying files using FTP technology. Transfer of files using FTP technology takes place as follows. As the first step an FTP client programme initiates a connection to a remote computer running FTP server software. To connect to an FTP server, a username and password as set by the administrator of the server are required. Clients identify an FTP server either by its IP address or by its host name. After the client is connected to the server, copies of files can be sent and/or received singly or in groups. Publically available files can be accessed using the user name 'Anonymous'. Though most network operating systems include simple FTP clients; many alternative third party FTP clients with enhanced performance features are available. FTP supports ASCII (plain text) and binary modes of data transfer, which can be set in the FTP client. The transferred file will not be usable by the intended recipient if it were, for instance, a binary file transferred while in text mode.

4.6.4.7 Simple Mail Transfer Control (SMTP)

SMTP is a TCP/IP protocol used to transfer email messages between servers. It is usually used along with either of the two other protocols, namely POP3 and IMAP, which are used by messaging clients for retrieval of email messages. POP3 and IMAP allow the recipients' saved messages in their server mailbox to be downloaded periodically from the server. This helps SMTP overcome its limited ability to queue messages at the receiving end. In other words, while SMTP is typically used for sending emails, either POP3 or IMAP is used for receiving emails. SMTP usually operates over internet port 25. A large number of mail servers now support *extended simple mail transfer protocol* (ESMPT). ESMTP also allows multimedia files to be delivered as email. X.400, which is an alternative SMTP, is widely used in Europe. SMTP is reliable and simple. In a typical SMTP transaction, a server identifies itself and announces the type of operation it is intending to perform. Once the operation has been authorized by the other server the message is sent. If there is something wrong, such as the wrong address, the receiving server will respond with an appropriate error message.

4.6.4.8 Use Datagram Protocol (UDP)

UDP was introduced in 1980 and is one of the oldest protocols in use. It is an alternative to TCP. Together with IP it is referred to as UDP/IP. UDP does not have some of the desirable features of TCP. While TCP performs the task of dividing the message into data packets at one end and reassembling them at the other end, UDP does not provide this service. Application programmes that use UDP therefore must make sure that the entire message has arrived and is in the right order. Also, checksums that protect data from tampering or getting corrupted during transmission are mandatory in TCP but optional in UDP. However, like TCP, UDP also operates on the transport layer (layer 4) of the OSI reference model. One message unit in the case of UDP network traffic is called a datagram and comprises a header section and a data section. The header section comprises four fields: source port number, destination port number, datagram size and checksum. Each field has a length of two bytes, which implies that the header information is contained in eight bytes. The size of datagrams varies depending on the operating environment but can be a maximum of 65 535 bytes.

4.6.4.9 Asynchronous Transfer Mode (ATM)

ATM is a high-speed networking protocol that supports both voice and data communications. Faster processing and switching speeds are possible as the protocol is designed to be easily implemented by hardware rather than software. It uses asynchronous time division multiplexing. It encodes data into small, fixed-sized cells of 53 bytes length comprising 48 bytes of data and five bytes of header information. ATM operates at the data link layer (layer 2) of the OSI model and transmits data over a physical medium such as fibre or twisted pair cable. It differs from data link technologies such as Ethernet in the sense that it uses fixed length data packets as opposed to variable length packets in Ethernet. It does not utilize routing and hardware devices such as ATM switches are used to establish a point-to-point connection between the end points for flow of data directly from source to destination.

The performance of ATM is often expressed in the form of *optical carrier* (OC) levels written as 'OC-xxx'. OC levels are a set of signalling rates for transmitting digital signals on optical fibres. These are designed for transmission over synchronous optical networks (SONETs) and are also applicable to ATM networks. The base rate is 51.84 Mbps (OC-1). ATM uses some OC levels. The pre-specified bit rates are either 155.520 Mbps (OC-3) or 622.080 Mbps (OC-12). Performance levels as high as 10 Gbps (OC-192) are technically achievable. Along with SONET and several other technologies, ATM is a key component of broadband ISDN (BISDN).

4.6.5 Satellite Constellations

A large number of satellites have been launched individually and in groups for a variety of applications, including communications and other purposes such as remote sensing, meteorology, navigation and so on. While individual satellites in geostationary orbits provide a relatively large fixed footprint on a round the clock basis, those in lower orbits such as low Earth orbits provide a small footprint at any given time that is repeated with a certain periodicity. In both cases, individual satellites do not provide round the clock global or near global coverage. This problem is overcome by having satellite constellations. A satellite constellation is a group of satellites with their operation so synchronized as to provide coordinated ground coverage on a round the clock basis. This implies that the footprints of different satellites in the constellation sufficiently overlap to provide uninterrupted global or near global coverage on a 24×7 basis. There are a number of operational satellite constellations and many more are in the pipeline for a variety of applications including voice and data communication, satellite radio, messaging and navigation. Beginning with the requirements of constellation geometry, major satellite constellations are briefly described in the following paragraphs.

4.6.5.1 Constellation Geometry

There are a number of different constellation geometries to satisfy the intended mission requirements. Three most important orbital parameters governing satellite constellation geometry are *altitude*, *inclination* and *eccentricity*. Orbit altitude is chosen on the basis of both physical and geometric considerations, which include coverage area, time of satellite visibility and revisit periodicity, signal propagation delay, signal power and avoidance of Van Allen radiation belts. Orbit inclination is the second important parameter of a satellite constellation. The choice of orbital inclination is governed by the requirement of global coverage and the minimum angle of elevation, for example higher inclination provides more coverage to polar regions. Similarly, an inclination of around 45° allows coverage of temperate zones and populated regions of the Earth. Orbit eccentricity determines the shape of the orbit, which in turn may affect the dwell time of a satellite in the constellation over a certain specific area on the ground. For example, for a satellite in an eccentric elliptical orbit, the dwell time of the satellite can be maximized over the region of interest by adjusting the position of the apogee.

Satellite constellations are usually designed with the satellites in the constellations having similar orbits, eccentricity and inclination with the advantage that any perturbations affect each satellite in more or less the same manner. This also helps in preserving the constellation geometry and thereby minimizing station-keeping and fuel usage requirements. Also, sufficient separation is maintained between adjacent satellites in the same orbital plane to avoid interference and prevent collision.

There are two major types of satellite constellation: the *polar constellation* and the *Walker constellation*. The Walker constellation has an associated notation proposed by John Walker according to which the constellation is represented by i: $t/p/f$. In this notation, i is the orbital inclination, t is the total number of satellites, p is the number of equally spaced planes and f is the relative spacing between satellites in adjacent planes. Furthermore, there is the Walker Delta constellation and the near-polar Walker Star constellation. While the Galileo navigation system belongs to the former category, the Iridium satellite constellation employs near-polar Walker Star geometry (Figure 4.92). Both polar and Walker constellations are designed to provide global coverage or near global coverage with the minimum number of satellites. However, each constellation has its advantages and disadvantages. The polar constellation provides global coverage including the polar region. On the other hand, the Walker constellation provides coverage only to areas below certain latitude, which for the Globalstar constellation is ±70°.

4.6.5.2 Major Satellite Constellations

Satellite constellations are in use and are being explored for a variety of applications, which include voice communication (e.g. the Iridium and Globalstar constellations), satellite radio (e.g. Sirius XM Radio), broadband networking (e.g. the Teledesic and SkyBridge constellations), messaging (e.g. the Orbcomm constellation) and navigation (e.g. the global positioning system (GPS), global navigation satellite system (GLONASS), Galileo constellations). In addition to these examples of satellite constellations there are many more satellite constellations for different categories of applications. Some of the major satellite constellations are briefly described in the following paragraphs.

4.6.5.2.1 Iridium Satellite Constellation

The *Iridium satellite constellation* is a global satellite network designed to provide voice communication, data, fax and paging services independent of a user's location in the world and of the availability of traditional telecommunication networks. The space segment of the constellation comprises 66 active satellites and 14 spare backup satellites orbiting at an altitude of 780 km in six orbital planes. The ground segment comprises gateways and a system control segment. Iridium subscriber products include phones and pagers that allow users to have access to either a compatible cellular telephone network or the Iridium network.

4.6.5.2.2 Globalstar Constellation

The US Globalstar satellite constellation provides global voice, data, fax and messaging services. The constellation comprises 48 satellites and an additional four in-orbit spares orbiting at an altitude of 1410 km and an inclination of 52°. The satellites are placed in eight orbital planes with six satellites in each plane (Figure 4.93). The constellation provides service on Earth between 70°N latitude and 70°S latitude covering the USA and 120 other countries.

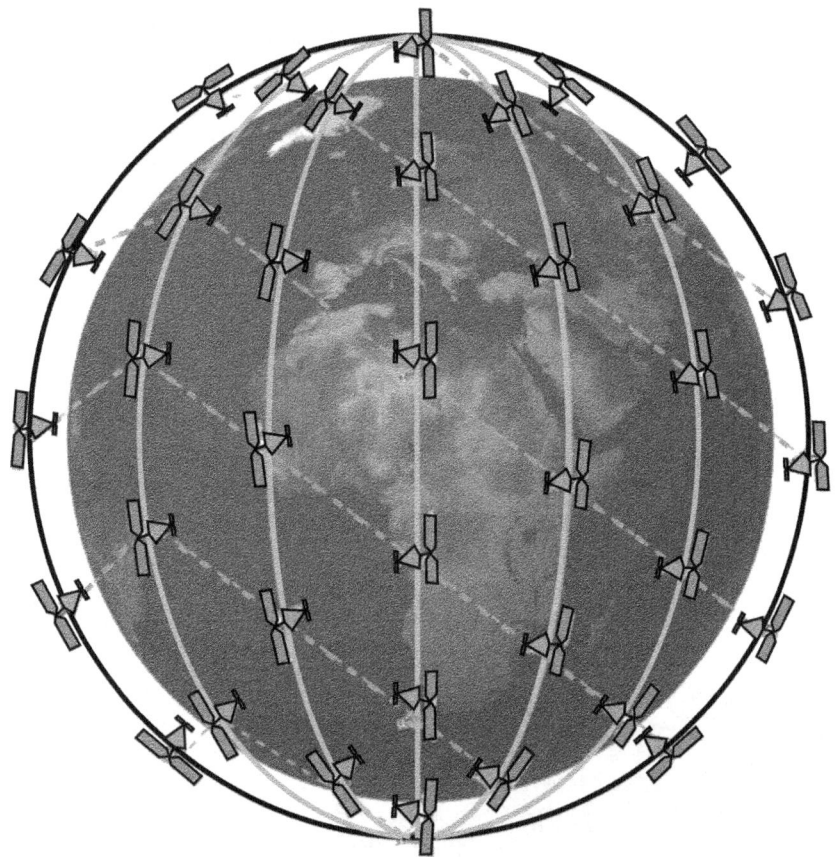

Figure 4.92 Iridium satellite constellation.

Figure 4.93 Globalstar satellite constellation.

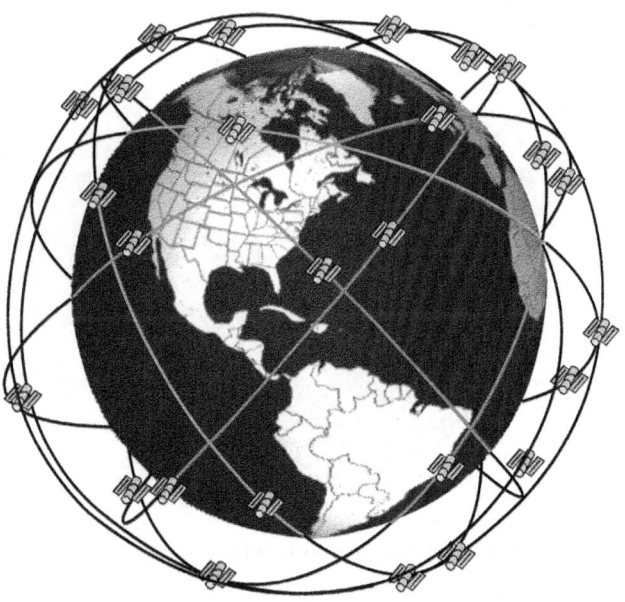

4.6.5.2.3 Sirius XM Radio

Sirius XM Radio was formed by the merger of XM Satellite Radio with Sirius Satellite Radio in July 2008. The company provides two satellite radio services, Sirius Satellite Radio and XM Satellite Radio, in the USA. An affiliate company called XM Canada provides these services in Canada. Currently, the company provides music, sports, news and entertainment channels to listeners. The satellites used for the Sirius radio service are the Radiosat series comprising Radiosat-1 to Radiosat-4, to avoid confusion with Sweden's fleet of Sirius satellites.

4.6.5.2.4 Teledesic Satellite Constellation

The Teledesic satellite constellation is a worldwide satellite network comprising 924 low Earth orbit satellites orbiting at an altitude of 700 km and placed in 21 orbital planes inclined at 98.2° with an adjacent plane separation of 9.5° (Figure 4.94). The constellation offers a wide range of services, including multimedia conferencing, video conferencing, voice communication, video telephony and distance learning, and provides seamless coverage to 100% of the population on Earth on a round the clock basis. The constellation has a peak load capacity of 2 000 000 simultaneous full duplex 16 Kbps connections, with a service quality comparable to today's terrestrial communication systems.

Figure 4.94 Teledesic satellite constellation.

4.6.5.2.5 SkyBridge Satellite Constellation

SkyBridge is a constellation of 80 low Earth satellites divided into two symmetrical Walker sub-constellations of 40 satellites each. The satellites orbit at an altitude of 1457 km. Each satellite provides a 3000 km radius of coverage divided into fixed spot beams of 350 km radius. The SkyBridge satellite constellation is designed to provide the communication infrastructure for a full range of broadband services, including interactive multimedia communication, high-speed data communications and internet access. All traffic management and routing in the SkyBridge constellation is handled on the ground, with no direct links between satellites. The gateway handles interconnections with local servers and terrestrial networks. Being a low Earth orbit satellite constellation the propagation delay is of the order of 20 ms, similar to what it is in the case of landline broadband systems. This allows applications currently used for existing broadband networks to be seamlessly transmitted via SkyBridge.

4.6.5.2.6 Orbcomm Satellite Constellation

Orbcomm is a satellite constellation of 32 satellites located in four orbital planes A, B, C and D, with each plane having eight satellites. Planes A, B and C are inclined at 45° to the equator and the satellites in these planes orbit at an altitude of 825 km. Successive satellites in each of the planes are 45° apart. Plane D is inclined at 0° and the satellites in this orbital plane also orbit at an altitude of 825 km. There are two supplemental orbital planes, Plane F and Plane G, that contain two satellites each and have an altitude of 780 km. Planes F and G are inclined at 70° and 80°, respectively. The two satellites in each of these two planes are 180° apart. Orbcomm has a licence to launch up to 48 satellites. The Orbcomm system comprises a network control centre (NCC) to manage the operation of the overall system and three operational segments: the space segment of the satellites, the ground segment consisting of gateway Earth stations, and the control centre and subscriber segment. A fully deployed constellation is capable of providing a near real-time wireless data communication service worldwide.

4.6.5.2.7 Global Positioning System (GPS)

GPS is satellite-based US global navigation system. The GPS constellation comprises 28 satellites and ground support facilities to provide three-dimensional position, velocity and timing information to users around the world on a 24/7 basis. Of the 28 satellites, 24 satellites are active satellites and remaining four satellites are used as in-orbit spares. The satellites orbit in circular medium Earth orbits (MEO) at an altitude of 20 200 km, inclined at 55° to the equator. Placing them in geostationary orbits would reduce the required number of satellites, but would not provide god polar coverage. The present constellation makes it possible for 4–10 satellites to be visible to all receivers anywhere in the world. Figure 4.95 shows a space segment of GPS.

4.6.5.2.8 Global Navigation Satellite System (GLONASS)

GLONASS is a Russian satellite-based navigation system. The constellation comprises 21 active satellites plus three in-orbit spares in circular medium Earth orbits at a nominal altitude of 19 100 km. The satellites are arranged in three orbital planes inclined at an angle of 64.8°. Each plane comprises eight satellites displaced at 45° with respect to each other. GLONASS provides better coverage compared to the GPS system at higher latitude sites. Figure 4.96 shows a space segment of the GLONASS system.

4.6.5.2.9 Galileo Satellite Navigation System

Galileo is a global satellite navigation system being built by the European Union and the European Space Agency. It is designed to provide European countries with a fully autonomous

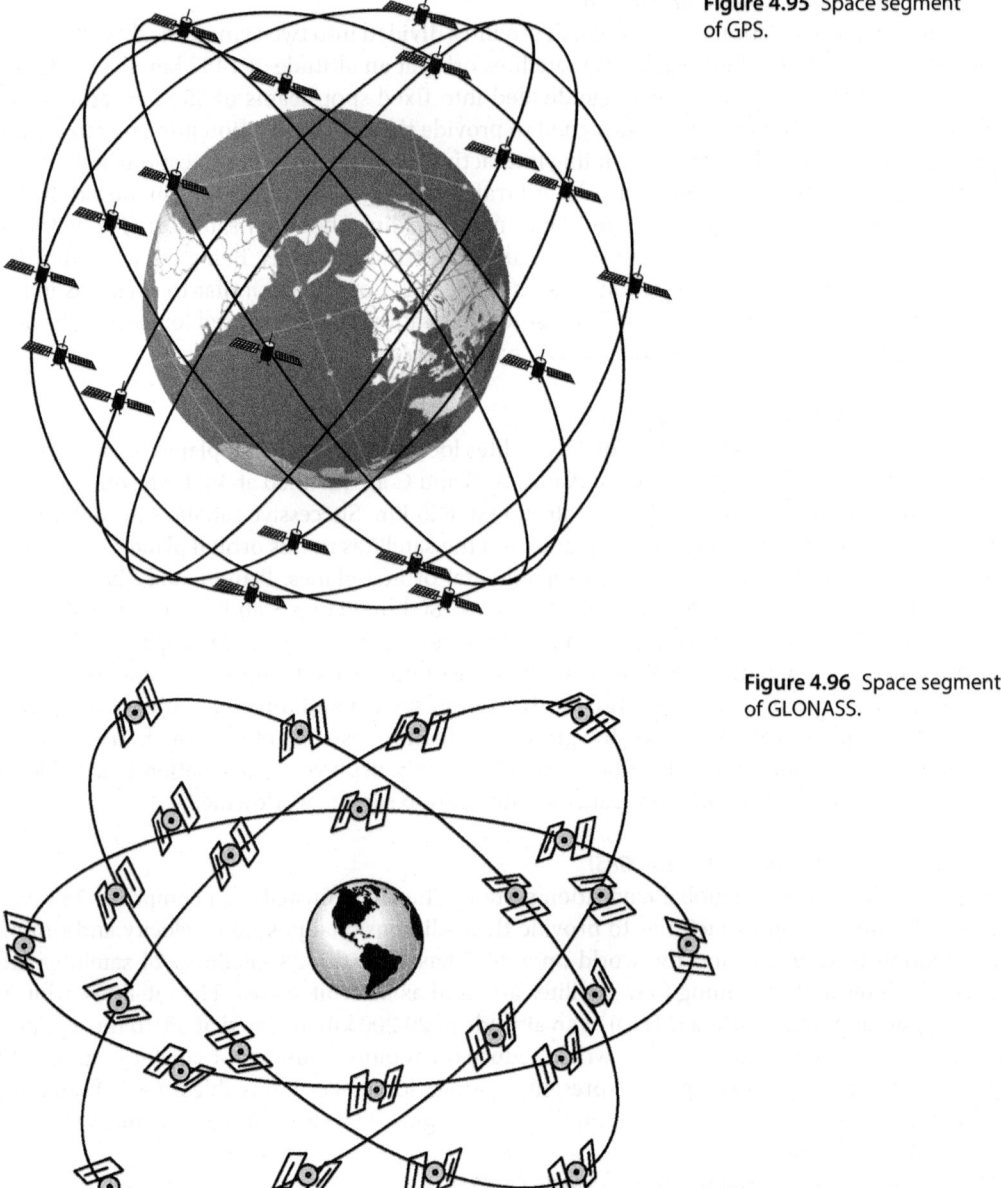

Figure 4.95 Space segment of GPS.

Figure 4.96 Space segment of GLONASS.

and reliable satellite-based positioning, navigation and timing capability independent of the US GPS and Russian GLONASS. Though the Galileo system is intended to be independent of GPS and GLONASS, it is fully interoperable with them thereby making it a fully integrated new element in global satellite navigation systems.

4.6.6 Internetworking with Terrestrial Networks

The techniques of internetworking of similar types of networks, such as internetworking of terrestrial networks, are well-established. Internetworking of different types of networks, such as internetworking of a satellite network with a terrestrial network, encounters multiple problems including those due to different transmission media, different data formats, different transmission speeds and different protocols. Satellite internetworking with other types of networks such as terrestrial networks involves any of the three lower layers of the OSI reference model, namely the physical layer, the data link layer and the network layer. Repeaters, bridges, switches and routers are the commonly used interface elements. Repeaters operate at the physical layer, bridges operate at the data link layer and switches can work at any of the three lower layers, that is, the physical layer, the data link layer or the network layer.

4.6.6.1 Repeaters, Bridges, Switches and Routers

One of the issues that need to be addressed while internetworking a satellite network with a terrestrial network is related to use of different physical media and protocols. Internetworking at the physical layer is at bit level. As the physical layer protocol functions are simple, internetworking between satellite networks and terrestrial networks is relatively easy. Terrestrial networks have much higher data transmission rates compared to satellite networks, and this issue and other issues related to the use of different physical media and protocols need to be addressed by the interconnecting devices. Repeaters, bridges, switches and routers are commonly used network interconnection devices.

A *repeater* is a network device used to regenerate or replicate data signals. It relays data signals between sub-networks using different physical media and protocols. A repeater cannot do intelligent routing, which is performed by bridges and routers.

A *bridge* operates at the data link layer of the OSI model. It serves as the internetworking unit between the satellite network and the terrestrial network. A bridge examines the incoming traffic from the satellite network and decides whether to forward it or discard it. In taking this decision the bridge may look at the source and destination addresses and even the frame size. If the frames are to be forwarded they are formatted according to the protocol of the terrestrial network. The reverse process of handling data flow from a terrestrial network to a satellite network, though similar, is more complicated. The satellite network in its interface with terrestrial networks has to deal with a large number of different types of networks and protocol translations.

Network *switches* are multiport bridges that are used to link multiple computers in a network. Network switches inspect data packets as they are received, determine the source and destination network device of each packet, and forward them to the intended device only. In this way a network switch conserves network bandwidth. Network switches operate at any of the three lower layers depending on the nature of the network.

A *router* is a network device used to connect multiple networks, either LANs or a LAN with an internet service provider's (ISP) network. A router forwards data packets along the networks using headers and routing tables to determine the best paths for forwarding them. Routers are located at the gateways where two or more networks are connected. Routers can be used to interconnect with heterogeneous terrestrial networks. In this case, all user terminals use IP. Routers operate at the network layer of the OSI model.

4.6.6.2 Protocol Translation, Stacking and Tunnelling

Protocol translation, stacking and tunnelling are the three commonly used techniques in internetworking with heterogeneous networks. Protocol translation is implemented through *network address translation* (NAT) and *port translation*. NAT is the translation of an IP address used within one network to a different IP address known within another network.

Protocol stacking refers to a group of protocols that work together to allow software or hardware function. TCP/IP that uses four lower layers of the OSI model is an example.

Protocol tunnelling, also known as *port forwarding*, allows two of the same type of network to communicate with each other through other networks. The tunnelling technique can be used to carry data across an incompatible delivery network or provide a secure path through a network that is perhaps not trustworthy. One application of tunnelling is the transmission of data meant for use within a corporate network through a public network with routing nodes in the public network unaware of the fact that data belongs to the private network. Tunnelling would allow the use of the internet to convey data on behalf of private networks. There are a number of tunnelling protocols in use. Two of the better-known protocols are *point-to-point tunnelling protocol* (PTPP), developed by Microsoft, and *generic routing encapsulation* (GRE) developed by Cisco Systems.

4.6.6.3 Quality of Service (QoS)

QoS is referred to as the collection of networking technologies and techniques to guarantee delivery of a certain performance level. With reference to QoS, elements of performance include network availability or uptime, latency or delay, bandwidth or throughput and error rate. In order to achieve the desired QoS, it is important that data transmission rates, error rates and other characteristics are monitored and measured, improved and to an extent guaranteed in advance. Traffic shaping techniques such as packet prioritization, application classification and queuing at congestion points can be used to improve the QoS. One such protocol that allows expedited delivery of data packets passing through a gateway host is the internet's resource reservation protocol.

There are three fundamental elements in the implementation of QoS, which are QoS identification and marking techniques to coordinate end-to-end quality of service between network elements, tools such as queuing, scheduling and traffic shaping to ensure QoS within individual network elements, and management and accounting functions to administer and control end-to-end traffic across the network.

Illustrated Glossary

Antenna These are used for both receiving signals from ground stations as well as for transmitting signals toward them. There are a variety of antennas available for use on board a satellite. The ultimate choice depends upon the frequency of operation and required gain.

Antenna Aperture This is the physical area of the antenna projected on a plane perpendicular to the direction of the main beam or the main lobe. In the case a of main beam axis being parallel to the principle axis of the antenna, this is the same as the physical aperture of the antenna itself.

Antenna Gain-to-Noise Temperature (G/T) Ratio This is usually defined with respect to the Earth station receiving antenna and is an indicator of the sensitivity of the antenna to the down-link carrier signal from the satellite. It is the figure of-merit used to indicate the performance of the Earth station antenna and low noise amplifier combined to receive weak carrier signals.

Antenna Noise Temperature This is a measure of noise entering the receiver via the antenna. The noise temperature of the antenna can be computed by integrating contributions of all the radiating bodies whose radiation lies within the directional pattern of the antenna.

Apogee Point on a satellite orbit farthest from the centre of the Earth. The apogee distance is the distance of the apogee point from the centre of the Earth.

Argument of Perigee This parameter defines the location of the major axis of the satellite orbit. It is measured as the angle between the line joining the perigee and the centre of the

Earth and the line of nodes from the ascending to the descending node in the same direction as that of the satellite orbit.

Ascending Node The point where the satellite orbit cuts the Earth's equatorial plane, when it passes from the southern hemisphere to the northern hemisphere.

Asynchronous Transfer Mode (ATM) This is a high-speed networking protocol that supports both voice and data communications.

Attitude and Orbit Control This subsystem performs two primary functions. It controls the orbital path, which is required to ensure that the satellite is in the correct location in space to provide the intended services. It also provides attitude control, which is essential to prevent the satellite from tumbling in space and also to ensure that the antennas remain pointed at a fixed point on the Earth's surface.

Availability This refers to the ability of the network to respond to the requests of users intending to access the network.

Azimuth Angle – Earth Station The azimuth angle of an Earth station is the angle produced by the line of intersection of the local horizontal plane and a plane passing through the satellite, Earth station and the centre of the Earth with true north.

Beam Width Defined with respect to the antenna, this is the angular separation between the half power points on the power density radiation pattern.

Bridge A bridge serves as the internetworking unit between the satellite network and the terrestrial network. It examines the incoming traffic from the satellite network and decides on whether to forward it or discard it.

Broadcast Satellite Services (BSS) This refers to the satellite services that can be received at many unspecified locations by relatively simple receive-only Earth stations.

Bus Topology In this network topology, all devices to be connected to the network are connected to a central cable that acts as the backbone of the network with the help of interface connectors.

Centrifugal Force The force acting outwards from the centre of the Earth on any body orbiting it.

Centripetal Force A force that is directed toward the centre of the Earth due to the gravitational force of attraction of Earth.

Circuit Switched Network In the case of a circuit switched network, a dedicated physical path is established between the communicating nodes through the network before the start of actual communication. This dedicated path is held for the duration of the communication session.

Code Division Multiple Access (CDMA) A multiple access technique in which the entire bandwidth of the transponder is used simultaneously by multiple Earth stations at all times.

Cross-Polarization This is the component that is orthogonal to the desired polarization.

Cross-Polarization Discrimination This is defined as the ratio of power received by the antenna in principal polarization to that received in orthogonal polarization from the same incident signal.

Cross-Polarization Interference This occurs in frequency reuse satellite systems. It occurs due to coupling of energy from one polarization state to the other orthogonally polarized state in communications systems that employ orthogonal linear polarizations (horizontal and vertical polarization) and orthogonal circular polarizations (right- and left-hand circular). The coupling of energy from one polarization state to the other takes place due to finite cross-polarization discrimination of the Earth station and satellite antennas and also by depolarization caused by rain, particularly at frequencies above 10 GHz.

Demand Assigned Multiple Access (DAMA) A transponder assignment mode that allows multiple users to share a common link with each user only required to put up a request to the control station or agency for the same as and when it requires to use the link.

Descending Node This is the point where the satellite orbit cuts the equatorial plane when the satellite passes from the northern hemisphere to the southern hemisphere.

De-Spun Antenna An antenna system placed on a platform that is spun in a direction opposite to the direction of spin of the satellite body. This ensures a constant pointing direction for the satellite antenna system.

Direct Sequence CDMA In this form of CDMA, the information signal, which has a relatively lower bit rate, is multiplied by a pseudorandom bit sequence with a much higher bit rate, with the result that the carrier frequency spectrum is spread over a much larger bandwidth. Interference between multiple channels is avoided as each transmitter uses a unique pseudorandom code sequence.

Earth Coverage Surface area of the Earth that is possible to cover by a satellite.

Eccentricity Referring to an elliptical orbit, this is the ratio of the distance between the centre of the Earth and the centre of the ellipse to the semi-major axis of the ellipse. It is zero for a circular orbit and between 0 and 1 for an elliptical orbit.

Eclipse An eclipse is said to occur when sunlight fails to reach the satellite's solar panel due to an obstruction from a celestial body. The major and most frequent source of an eclipse is due to the satellite coming in the shadow of Earth, known as the solar eclipse. Another type of eclipse known as the lunar eclipse occurs when the Moon's shadow passes across the satellite.

Effective Isotropic Radiated Power (EIRP) This is given by the product of the transmitter power and the antenna gain. An antenna with a power gain of 40 dB and a transmitter power of 1000 W would mean an EIRP of 10 MW; that is, 10 MW of transmitter power when fed to an isotropic radiator would be as effective in the desired direction as 1000 W of power fed to a directional antenna with a power gain of 40 dB in the desired direction.

Elevation Angle – Earth Station The elevation angle of an Earth station is the angle produced by the line of intersection of the local horizontal plane and a plane passing through the satellite, Earth station and centre of the Earth with the line joining the Earth station and the satellite.

Email This is the short name for electronic mail and is the method of transmitting messages electronically from one computer user to one or more recipients over a communication network

Equatorial Orbit An orbit in which the orbital plane coincides with the equatorial plane.

Equinox An equinox is said to occur when the angle of inclination of the equatorial plane with respect to the direction of the Sun as defined by the line joining the centre of the Earth and the Sun is zero. Such a situation occurs twice a year; one around the 21 March called the spring equinox and the other around the 21 September called the autumn equinox.

First Cosmic Velocity This is the injection velocity at which the apogee and perigee distances are equal, with the result that the satellite orbit is circular.

Fixed Satellite Services (FSS) This refers to the two-way communication between two Earth stations at fixed locations via a satellite.

Footprint See Earth coverage.

Free-Space Loss This is the loss of signal strength due to the distance from the transmitter. While free space is a theoretical concept of space devoid of all matter, in the present context it implies remoteness from all material objects or forms of matter that could influence propagation of electromagnetic waves.

Frequency Division Multiple Access (FDMA) A multiple access technique in which different Earth stations are able to access the total available bandwidth in satellite transponder/s by virtue of their different carrier frequencies, thus avoiding interference among multiple signals.

Frequency Hopping CDMA In the case of a frequency hopping spread spectrum system, the carrier is sequentially hopped into a series of frequency slots spread over the entire bandwidth of the satellite transponder. The transmitter hops its frequency over a given bandwidth several times per second, transmitting on one frequency for a certain period of time, and then hopping to another frequency and transmitting again.

Galileo Constellation This is a global satellite navigation system being built by the European Union and the European Space Agency. It is designed to provide European countries with a fully autonomous and reliable satellite-based positioning, navigation and timing capability independent of the US GPS and Russian GLONASS.

Geostationary Orbit (GEO) A satellite orbit with an orbit height at 35 786 km above the surface of the Earth. This height makes the orbital velocity equal to the speed of rotation of Earth, thus making the satellite look stationary from a given point on the surface of the Earth.

File Transfer Protocol (FTP) This is a standard based on IP used for transferring files between computers on the Internet.

Globalstar Constellation The US Globalstar satellite constellation comprises 48 satellites and four in-orbit spares and is designed to provide global voice, data, fax and messaging services.

GLONASS Constellation The GLONASS (global navigation satellite system) is a Russian satellite-based navigation system. The constellation comprises 21 active satellites providing continuous global services such as GPS.

GPS Constellation GPS (global positioning system) comprises 24 satellites and ground support facilities to provide three-dimensional position, velocity and timing information to users around the world on a 24/7 basis.

Ground Track This is an imaginary line formed by the locus of the lowest point on the surface of the Earth. The lowest point is the point formed by the projection of the line joining the satellite with the centre of the Earth on the surface of the Earth.

Guard Time Different bursts are separated from each other by a short time period, known as the guard time, which ensures that bursts from different stations accessing the satellite transponder do not overlap.

Heat Pipe This consists of a hermetically sealed tube filled with a liquid with a relatively low boiling point. Its inner surface has a wicking profile and depends upon repeated cycles of vaporization and condensation to transfer heat.

Horn Antenna A type of microwave antenna constructed from a section of a rectangular or circular waveguide.

Hybrid Topology Hybrid topology is the result of integration of two or more basic network topologies. The resultant hybrid topology combines the good and bad features of all constituent topologies.

Hypertext Transfer Protocol (HTTP) This is an application layer protocol built on the top of TCP. HTTP defines how messages are formatted and transmitted. The protocol provides a standard for communication of web browsers and servers.

Inclination Inclination is the angle that the orbital plane makes with the equatorial plane.

Inclined Orbit This is an orbit with an angle of inclination between 0° and 180°.

Injection Velocity This is the horizontal velocity with which a satellite is injected into space by the launch vehicle with the intention of imparting a specific trajectory to the satellite.

Internet Protocol (IP) This is the primary network protocol used on the internet and is generally used with TCP.

Iridium This is a type of low Earth orbit satellite constellation with 66 active satellites and 14 in-orbit spares designed to provide voice communication, data, fax and paging services.

Ion Propulsion In the case of ion propulsion, the thrust is produced by accelerating charged plasma of an ionized elemental gas such as xenon in a highly intense electrical field.

Intermodulation Interference This is caused due to generation of intermodulation products within the satellite transponder as a result of amplification of multiple carriers in the power amplifier, which is invariably a TWT amplifier.

Johnson Noise Another name for thermal or white noise.

Kepler's First Law The orbit of an artificial satellite around Earth is elliptical with the centre of the Earth lying at one of its foci.

Kepler's Second Law The line joining the satellite and the centre of the Earth sweeps out equal areas in the plane of the orbit in equal times.

Kepler's Third Law The square of the time period of any satellite is proportional to the cube of the semi-major axis of its elliptical orbit.

Lens Antenna An antenna made from dielectric material and depending upon refraction phenomenon for operation.

Link Budget This is a way of analysing and predicting the performance of a microwave communication link for given values of vital link parameters that contribute to either signal gain or signal loss. It is the algebraic sum of all gains and losses expressed in decibels when travelling from the transmitter to the receiver.

Low Earth Orbit (LEO) A satellite orbit with an orbit height of around 150–500 km above the surface of Earth. These orbits have lower orbital periods, shorter propagation delays and lower propagation losses.

Medium Earth Orbit (MEO) A satellite orbit with an orbit height around 10 000–20 000 km above the surface of the Earth.

Mesh Topology In mesh topology, each network node is connected to every other node. This is true mesh topology.

Multichannel per Carrier FDMA This is a type of FDMA where the Earth station frequency multiplexes several channels into one carrier base band assembly, which then frequency modulates an RF carrier and transmits it to the FDMA satellite transponder.

Multiple Access This means access to a given facility or resource by multiple users. In the context of satellite communication, the facility is the transponder and the multiple users are various terrestrial terminals under the footprint of the satellite.

Mobile Satellite Services (MSS) This refers to the reception by receivers that are in motion, like ships, cars, lorries and so on.

Network Protocol A network protocol defines the standard used for communication between different devices connected to a network such as a LAN, intranet or Internet. Network protocols include mechanisms for identification of network devices to make connections as well as formatting rules that specify how data is packaged into messages to be sent and received by different devices.

Molniya Orbit A highly inclined and eccentric orbit used by Russia and other countries of the erstwhile Soviet Union for providing communication services.

Noise Figure This is defined as the ratio of its signal-to-noise power at the input to the signal-to-noise power at the output.

Noise Temperature This is just another way of expressing noise performance of a device in terms of its equivalent noise temperature. It is the temperature of a resistance that would generate the same noise power at the output of an ideal (i.e. noiseless) device as that produced at its output by an actual device when terminated at its input by a noiseless resistance; that is, a resistance at absolute zero temperature.

Open Systems Interconnect (OSI) Reference Model The OSI model is at the core of open systems interconnect standard developed in 1984 by International Organization for Standardization (ISO). It comprises seven layers that define the different stages that data must go through when transported from one device to another in a network.

Orbcomm Satellite Constellation This is a satellite constellation of 32 satellites. A fully deployed constellation is capable of providing near real-time wireless data communication services worldwide.

Orbit A satellite trajectory that is periodically repeated.

Packet Switched Network In packet switched network technology there is no dedicated communication path or circuit identified prior to the start of communication for the entire duration of communication between the source and destination.

Payload This is that part of the satellite that carries the desired instrumentation required for performing its intended function and is therefore the most important subsystem of any satellite. The nature of the payload on any satellite depends upon its mission. The basic payload in the case of a communication satellite is the transponder, a radiometer in the case of a weather forecasting satellite, high-resolution cameras, multi-spectral scanners and so on, in the case of a remote sensing satellite and equipment like spectrographs, telescopes, plasma detectors, magnetometers and so on in the case of scientific satellites.

Perigee A point on a satellite orbit closest to the centre of the Earth. The perigee distance is the distance of the perigee point from the centre of the Earth.

Phased Array Antenna An antenna array in which the radiated beam axis can be electronically steered by having a certain phase difference between the signals fed to adjacent elements.

Polarization This is the direction of the electric field vector with respect to the ground in the radiated electromagnetic wave while transmitting and orientation of the electromagnetic wave again in terms of the direction of the electric field vector that the antenna responds to best while receiving.

Polarization Loss Polarization loss results if the received electromagnetic wave is of a polarization different from the one the antenna is designed for.

Polarization Rotation When an electromagnetic wave passes through a region of high electron content, such as the ionosphere, the plane of polarization of the wave is rotated due to interaction of the electromagnetic wave and the Earth's magnetic field. The angle through which the plane of polarization rotates is directly proportional to the total electron content of the ionosphere and inversely proportional to the square of the operating frequency. It also depends upon the state of the ionosphere, time of the day, solar activity, direction of the incident wave and so on. The directions of polarization rotation are opposite for transmit and receive signals.

Polar Orbit An orbit with an angle of inclination equal to 90°.

Power Supply Subsystem This is used to collect the solar energy, transform it to electrical power with the help of arrays of solar cells and distribute the electrical power to other components and subsystems of the satellite. In addition, a satellite also has batteries, which provide standby electrical power during eclipse periods, other emergency situations and also during the launch phase of the satellite when the solar arrays are not yet functional.

Pre-Assigned Multiple Access A transponder assignment mode in which the transponder is assigned either permanently for the satellite's full lifetime or at least for long durations. The pre-assignment may be that of a certain frequency band, time slot or a code.

Prograde Orbit Also called a direct orbit, an orbit where the satellite travels in the same direction as the direction of rotation of Earth. This orbit has an angle of inclination between 0° and 90°.

Project Iridium Project Iridium is a global communication system conceived by Motorola that makes use of satellites in low Earth orbits. A total of 66 satellites are arranged in a distributed architecture with each satellite carrying 1/66 of the total system capacity.

Propulsion Subsystem This is the satellite subsystem used to provide the thrusts required to impart the necessary velocity changes to execute all the manoeuvres during the lifetime of the satellite. This would include major manoeuvres required to move the satellite from its transfer orbit to the geostationary orbit in the case of geostationary satellites and also the smaller manoeuvres needed throughout the lifespan of the satellite, such as those required for station-keeping.

Quality of Service Quality of service is a collection of networking technologies and techniques to guarantee delivery of a certain performance level.

Random Multiple Access A transponder assignment mode in which access to the link or the transponder is by contention. A user transmits the messages without knowing the status of messages from other users. Due to the random nature of transmissions, data from multiple users may collide. In case a collision occurs, it is detected and the data are retransmitted. Retransmission is carried out with random time delays and sometimes may have to be done several times.

Reference Burst The reference burst is used to provide timing references to various stations accessing the TDMA transponder. It does not carry any traffic information.

Reflector Antenna This comprises a reflector and a feed antenna and is capable of offering a very high gain. Reflector antennas are made in a variety of shapes, sizes and configurations depending upon the type of reflector and feed antenna used.

Reliability The reliability of a network is a measure of the reliability of different components of the network and their interconnections. It is generally measured as mean time between failures (MTBF) or mean time to repair (MTTR).

Repeater This is a device used to connect two segments of the same network to extend its coverage. The primary function of a repeater is to regenerate data signals.

Retrograde Orbit An orbit where the satellite travels in a direction opposite to the direction of rotation of Earth. This orbit has an angle of inclination between 90° and 180°.

Right Ascension of the Ascending Node The right ascension of the ascending node indicates the orientation of the line of nodes, which is the line joining the ascending and descending nodes, with respect to the direction of the vernal equinox. It is expressed as an angle Ω measured from the vernal equinox toward the line of nodes in the direction of rotation of Earth. The angle could be anywhere from 0° to 360°.

Ring Topology In ring topology, all workstations are connected to one another in a closed loop.

Router A router is the network device used to connect multiple networks, either local a LAN or a LAN with an internet service provider (ISP) network. A router forward data packets along the networks using headers and routing tables to determine the best paths for forwarding them.

Satellite Constellation A satellite constellation is a group of satellites with their operation synchronized to provide coordinated ground coverage on a round the clock basis. This implies that the footprints of different satellites in the constellation sufficiently overlap to provide uninterrupted global or near global coverage on a 24/7 basis.

Scalability This is the capability of a network to accommodate increased data rate requirements and number of users. It also shows how well a network can adapt itself to new applications and the replacement of old components by new components with enhanced features.

Scintillation This is simply the rapid fluctuation of the signal amplitude, phase, polarization or angle-of-arrival. In the ionosphere, scintillation occurs due to small scale refractive index variations caused by local ion concentration. The scintillation effect is inversely proportional to the square of the operating frequency and is predominant at lower microwave frequencies, typically below 4 GHz.

Second Cosmic Velocity This is the injection velocity at which the apogee distance becomes infinite and the orbit takes the shape of a parabola. It equals $\sqrt{2}$ times the first cosmic velocity.

Security The objective of network security is to monitor and prevent unauthorized access, eavesdropping, misuse, modification or denial of use of a network and its resources to authorized users.

Signalling Channel The signalling channel is used to carry out system management and control functions.

Simple Mail Transfer Protocol (SMTP) This is a TCP/IP protocol used to transfer email messages between servers.

Single Channel per Carrier FDMA This is a type of FDMA in which each signal channel modulates a separate RF carrier, which is then transmitted to the FDMA transponder. The modulation technique used here could either be frequency modulation (FM) in the case of analogue transmission or phase shift keying (PSK) for digital transmission.

Sirius XM Radio This is a satellite radio constellation formed by the merger of XM Satellite Radio with Sirius Satellite Radio and comprises the Radiosat series of satellites (Radiosat-1 through Radiosat 4). It provides two satellite radio services, Sirius Satellite Radio and XM Satellite Radio in the USA.

SkyBridge Satellite Constellation This is a constellation of 80 low Earth satellites divided into two symmetrical Walker sub-constellations of 40 satellites each. The SkyBridge satellite constellation is designed to provide the communications infrastructure for a full range of broadband services, which includes interactive multimedia communication, high-speed data communications and Internet access.

Slant Range The line-of-sight distance between the satellite and the Earth station.

Solar Panel This is simply a series and parallel connection of a large number of solar cells to get the desired output voltage and power delivery capability.

Solstices Solstices are said to occur when the angle of inclination of the equatorial plane with respect to the direction of the sun as defined by the line joining the centre of the Earth and the Sun is at its maximum; that is, 23.4°. These are like equinoxes and also occur twice during the year, one around the 21 June called the summer solstice and the other around the 21 December called the winter solstice.

Space Division Multiple Access (SDMA) SDMA is a technique that primarily allows frequency reuse where adjacent Earth stations within the footprint of the satellite can use the same carrier transmission frequency and still avoid co-channel interference by using either orthogonal antenna beam polarization or narrow antenna beam patterns.

Spread Spectrum Communications A technique in which the carrier frequency spectrum is spread over a much larger bandwidth as compared to the information rate. This not only makes the system immune to interception by an enemy but also gives it an anti-jamming capability. CDMA and its different variants employ the spread spectrum technique.

Specific Impulse A parameter of the propulsion system, it is the ratio of thrust force to the mass expelled to produce the desired thrust. It is measured in seconds. A specific impulse indicates how much mass is to be ejected to produce a given orbit velocity increment.

Spin Stabilization A technique for stabilizing the attitude of a satellite in which the satellite body is spun around an axis perpendicular to the orbital plane. Like a spinning top, the spinning satellite body offers inertial stiffness, thus preventing the satellite from drifting from its desired orientation.

Star Topology In this topology, all work stations are connected to a central device with a point-to-point connection, unlike bus topology in which different devices are wired to a common cable that acts as a shared medium.

Station-Keeping This is the process of maintenance of the satellite's attitude against different factors that cause temporal drift.

Structural Subsystem This is the satellite subsystem that provides the framework for mounting other subsystems of the satellite and also an interface between the satellite and the launch vehicle.

Sun-Synchronous Orbit A sun-synchronous orbit, also known as a helio-synchronous orbit, is one that lies in a plane that maintains a fixed angle with respect to the Earth–Sun direction.

Switch Switches are multiport bridges that are used to link multiple computers in a network.

Teledesic Satellite Constellation This is a worldwide satellite network comprising 924 low Earth orbit satellites. The constellation offers a wide range of services, including multimedia conferencing, video conferencing, voice communication, video telephony and distance learning, providing seamless coverage to 100% of the population on Earth on a round the clock basis.

TDMA Frame In a TDMA network, each of the multiple Earth stations accessing a given satellite transponder transmits one or more data bursts. The satellite thus receives at its input a set of bursts from a large number of Earth stations. This set of bursts from various Earth stations is called the TDMA frame.

TDMA Frame Efficiency This is defined as the percentage of total frame length allocated for transmission of traffic data.

Thermal Control Subsystem This is the satellite subsystem that is used to maintain the satellite platform within its operating temperature limits for the type of equipment on board the satellite. It also ensures a reasonable temperature distribution throughout the satellite structure, which is essential to retain dimensional stability and maintain the alignment of certain critical equipment.

Thermal Noise This is generated in any resistor or resistive component of any impedance due to the random motion of molecules, atoms and electrons. It is called thermal noise as the temperature of a body is the statistical RMS value of the velocity of motion of these particles. It is also called white noise as, due to the randomness of the motion of particles, the noise power is evenly spread over the entire frequency spectrum.

Third Cosmic Velocity This is the injection velocity at which the satellite succeeds in escaping from the Solar System. It is related to the motion of Earth around the Sun. For injection velocities beyond the third cosmic velocity, there is a region of hyperbolic flights outside the Solar System.

Three-Axis Stabilization Also known as body stabilization, a technique for stabilizing the attitude of a satellite in which stabilization is achieved by controlling the movement of the satellite along the three axes; that is, yaw, pitch and roll, with respect to a reference.

Throughput This is defined as the average rate at which data is transferred through the network in a given time and is measured in bits per second (bps) and also sometimes in data packets per second or data packets transferred in a given time slot.

Time Division Multiple Access (TDMA) A multiple access technique in which different Earth stations in the satellite's footprint make use of a transponder by using a single carrier on a time division basis.

Time Hopping CDMA In this form of CDMA, the data signal is transmitted in rapid bursts at time intervals determined by a pseudorandom code sequence assigned to the user. Time hopping CDMA uses a wideband spectrum for short periods of time instead of parts of the spectrum all the time.

Topology This shows the way different components of the network are connected and the logical way data passes through the network from one component or device to the next irrespective of the physical structure of the network.

Transmission Control Protocol (TCP) This is one of the main protocols in the internet protocol suite and is used along with the IP to send packetized data between computers over the Internet.

Tree Topology This combines the features of star and bus topologies. In fact tree topology is an expanded star topology in which multiple star networks are interconnected by a common bus.

Tracking, Telemetry and Command (TT&C) Subsystem This is the satellite subsystem that monitors and controls the satellite from the lift-off stage to the end of its operational life in space. The tracking part of the subsystem determines the position of the spacecraft and follows its travel using angle, range and velocity information. The telemetry part gathers information on the health of various subsystems of the satellite encodes this information and then transmits the same. The command element receives and executes remote control commands to effect changes to the platform functions, configuration, position and velocity.

Trajectory A path traced by a moving body.

True Anomaly of a Satellite This parameter is used to indicate the position of the satellite in its orbit. This is done by defining an angle, ϑ, called the true anomaly of the satellite, formed by the line joining the perigee and the centre of the Earth with the line joining the satellite and the centre of the Earth.

Unique Word The function of a unique word is to establish the existence of burst and to enable determination of a timing marker, which can be used to establish the position of each bit in the remainder of burst.

User Datagram Protocol (UDP) This is one of the oldest protocols in use and is an alternative to TCP. Together with IP it is referred to as UDP/IP.

Voice over Internet Protocol (VoIP) This allows telephone calls to be made over a broadband internet network by converting analogue voice signals into digital data packets and using IP for the two-way transmission.

White Noise Another name for thermal noise or Johnson noise.

World Wide Web (WWW) This is a system of Internet servers that support documents formatted in HTML. It supports links to audio, video, graphics and other documents. Web browsers such as Firefox and Internet Explorer are applications that allow easy access to the World Wide Web.

Bibliography

1 Bander, R. (1998) *Launching and Operating Satellites: Legal Issues*, Kluwer Law International for Martinus Nijhoff Publishers.

2 Beyda, W.J. (1999) *Data Communications-From Basics to Broadband*, Prentice-Hall, New Jersey.

3 Calcutt, D. and Tetley, L. (1994), *Satellite Communications: Principles and Application*, Edward Arnold, a member of the Hodder Headline Group, London.

4 Capderou, M. and Lyle, S. (trans.) (2005) *Satellites: Orbits and Missions*, Springer-Verlag, France.

5 Chartant, R.M. (2004) *Satellite Communications for the Nonspecialist*, SPIE, Washington.

6 Elbert, B.R. (1999) *Introduction to Satellite Communication*, Altech House, Boston, MA.

7 Elbert, R.B. (2001) *Satellite Communication Ground Segment and Earth Station Handbook*, Artech House, Boston, MA.

8 Elbert, B.R. (1997), *The Satellite Communication Applications Handbook*, Artech House, Boston, MA.

9 Gatland, K. (1990) *Illustrated Encyclopaedia of Space Technology*, Crown, New York.

10 Gedney, R.T., Schertler, R. and Gargione, F. (2000) *Advanced Communications Technology*

11 Glibson, J.D. (2002) *The Communications Handbook (The Electrical Engineering Handbook Series)*, CRC Press, Boca Raton, FL.

12 Glisic, S.G. and Leppanen, P.A. (1997), *Handbook for Design, Installation and Service Engineers*, Kluwer Academic Publishers.

13 Harte, L. (2004) *Introduction to CDMA: Network, Services, Technologies, and Operation*, Althos.

14 Inglis, A.F. (1997) *Satellite Technology: An Introduction*, Butterworth-Heinemann, MA.

15 Kadish, J.E. (2000), *Satellite Communications Fundamentals*, Artech House, Boston, MA.

16 Levitan, B. and Harte, L. (2003) *Introduction to Satellite Systems: Technology Basics*, Market.

17 Logsdon, T. (1998), *Orbital Mechanics: Theory and Applications*, John Wiley & Sons, Inc., New York.

18 Lewis, E.G. (1992) *Communication Services via Satellite: A Handbook for Design, Installation and Service Engineers*, Butterworth-Heinemann.

19 Luther, A.C. and Inglis, A.F. (1997) *Satellite Technology: Introduction*, Focal Press, Boston, MA.

20 Lutz, E., Werner, M. and Jahn, A. (2000), *Satellites for Personal and Broadband Communications*, Springer, New York.

21 Maral, G. and Bousquet, M. (2002) *Satellite Communication Systems: Systems, Techniques and Technology*, John Wiley & Sons, Ltd, Chichester.

22 Montenbruck, O. and Gill, E. (2000) *Satellite Orbits: Models, Methods, Applications* (2000) Spinger-Verlag, Berlin, Heidelberg, New York.

23 Pattan, B. (1993) *Satellite Systems: Principles and Technologies*, Van Nostrand Reinhold, New York.

24 Perez, R. (1998) *Wireless Communications Design Handbook: Space Interference*, Academic Press, London.

25 Richharia, M. (2001) *Mobile Satellite Communications: Principles and Trends*, Addison-Wesley.

26 Richharia, M. (1999) *Satellite Communication Systems*, Macmillan Press Ltd.

27 Sarafin, T.P. (1995) *Spacecraft Structures and Mechanisms: From Concept to Launch*, Microcosm, Inc., USA and Kluwer Academic Publishers, the Netherlands.

28 Sherrif, R.E. and Hu, Y.F. *Mobile Satellite Communication Networks*, John Wiley & Sons, Ltd, Chichester.

29 Soop, E.M. (1994) *Handbook of Geostationary Orbits*, Kluwer Academic Publishers, Dordrecht, the Netherlands.

30 Tajmar, M. (2002) *Advanced Space Propulsion*, Springer, New York.

31 Verger, F., Sourbes-Verger, I., Ghirardi, R., Pasco, X., Lyle, S. and Reilly, P. (2003) *The Cambridge Encyclopaedia of Space*, Cambridge University Press.

5

Military Satellites

Military systems of today rely heavily on the use of satellites both during war as well as peacetime. Military satellites provide a wide range of services including communication services, intelligence gathering, weather forecasting, early warning, providing navigation information and timing data and so on. Military satellites have been launched in large numbers by many developed countries of the world, but more so by the USA and Russia. In this chapter, we shall deliberate on the various facets of military satellites related to their development and application potential. The chapter begins with an overview of military satellites, followed by a description of various types of military satellites. Salient features and current status of major international military satellite systems in each of these categories are also described in the chapter.

5.1 Military Applications of Satellites

Military satellites are considered 'Force Multipliers' as they form the backbone of most of the modern military operations. They facilitate rapid collection, transmission and dissemination of information, which is a major requisite in modern-day military systems. Space-based systems offer features like global coverage, high readiness, non-intrusive forward presence, rapid responsiveness and inherent flexibility. These features enable them to provide real-time or near real-time support for military operations in peacetime, crisis and throughout the entire spectrum of the conflict. They are also very useful during the planning phase of military operations as they provide information on enemy order of the battle, precise geographical references and threat locations.

The application sphere of military satellites extends from providing communication services to gathering intelligence imagery data, from weather forecasting to early warning applications and from providing navigation information to providing timing data. They have become an integral component of military planning of various developed countries, more so of the USA and Russia. As a matter of fact, the USA has the greatest number of military satellites in space, more than the rest of the world put together. The USA used the services of military satellites extensively during its military campaign in Iraq in 2003, against Afghanistan in 2001 and Yugoslavia in 1999. In the following paragraphs different applications of military satellites are outlined.

Handbook of Defence Electronics and Optronics: Fundamentals, Technologies and Systems, First Edition. Anil K. Maini.
© 2018 John Wiley & Sons Ltd. Published 2018 by John Wiley & Sons Ltd.

5.1.1 Application Areas of Military Satellites

1) *Military communication satellites*: These satellites link communication centres to the front line operators.
2) *Reconnaissance satellites*: Reconnaissance satellites, also known as spy satellites, provide intelligence information on the military activities of foreign countries. There are basically four types of reconnaissance satellites.
   a) Image intelligence or IMINT satellites
   b) Signal intelligence or ferret or SIGINT satellites
   c) Early warning satellites
   d) Nuclear explosion detection satellites.
3) *Military weather forecasting satellites*: These provide weather information, which is very useful in planning military operations.
4) *Military navigation satellites*: Navigation systems pinpoint the exact location of soldiers, military aircraft, military vehicles and so on. They are also used to guide a new generation of missiles to their targets.
5) *Space weapons*: These are weapons that travel through space to strike their intended target.

5.2 Military Communication Satellites

Satellite communication has been a vital part of the military systems of developed countries, the USA and Russia in particular. These satellites provide reliable, continuous, interoperable and robust communication services between the various military units and between these units and command centres. They help streamline military command and control and ensure information superiority in the battlefield. Military satellites, in general, provide the following services.

1) *Reliable Networks*
   a) Secured network of voice and broadband data services for command and control
   b) Secured telephony backbone services for remote locations and Wide Area Networking for data applications
2) *Field services*
   a) Voice, data, broadband and video services between military forces in the deployment areas and headquarters
3) *Terrestrial backup*
   a) Backup communication for disaster areas where the existing infrastructure is damaged
   b) Backup technical coordination links for critical locations
4) *Air Traffic Control*
   a) Secure, reliable communication among control towers as well as relaying information between pilots and towers
5) *Video conferencing and telemedicine network*
   a) Secure broadband communication between field medical crews and major hospitals
   b) Full support of file transfers (X-ray, medical files) and video conferencing equipment for virtual meetings
6) *Border control and custom network*

   a) Secure global communication services for surveillance operation inside and outside of the country
   b) Full support of captured surveillance video images.

Military communication systems serve a large number of users, ranging from those who have medium to high rate data needs using large stationary ground terminals to those requiring low to medium data rate services using small, mobile terminals and to those users who require extremely secure communication services. Each of these user groups has different requirements and is characterized by their own satellite and earth terminal designs. Depending upon the intended user group, military communication satellite systems can be further subcategorized as follows.

1) Wideband satellite systems
2) Tactical satellite systems
3) Protected satellite systems.

Wideband satellite systems provide point-to-point or networked moderate to high-data rate communication services at distances varying from in-theatre to intercontinental distances. Typical data rates for these systems are greater than 64 kbps. Users of wideband segment primarily have fixed and mobile transportable land-based terminals with a few terminals on large ships and aircraft.

Tactical satellite systems are used for communication with small mobile land, airborne and ship-borne tactical terminals. Such systems offer low to moderate data rate services at distances ranging from in-theatre to transoceanic. Tactical satellites employ high power transmitters as they communicate with small terminals.

Protected satellite systems provide communication services to mobile users on ships, aircraft and land vehicles. These systems require an extremely protected link against physical, nuclear and electronic threats. They generally offer low to moderate data rate services.

5.3   Military Satellite Communication Systems

Since the development of first military communication satellite system in the late 1960s, satellite technology has made unprecedented progress. Military satellites of today are far more advanced in terms of transmission capability, robustness, and anti-jamming capability and so on as compared to their predecessors. In the past, only USA and Russia had these systems but now many other countries including Israel, France, UK and so on have developed their own military satellite communication systems. In this section, we shall talk about the evolution process of military communication satellites.

The first military communication satellite systems were developed by USA in the 1960s. The systems developed initially were experimental in nature. They demonstrated the feasibility of employing satellites for military communications. They also provided the basic experience required for the development of sophisticated systems meeting all the stringent military requirements; be it the anti-jamming features or the reliability and maintainability aspects and so on. The experimental systems included the SCORE, Courier (Figure 5.1), Advent, Lincoln Experimental Satellites (LES) and West Ford satellites.

The first operational military system was developed by USA in the late 1960s. It was named the Initial Defense Communications Satellite Program or IDCSP. A total of 28 satellites were launched under the programme in a period of 3 years from 1966 to 1968. Each satellite had a single repeater with a capacity of around 10 voice circuits or 1 Mbps data communication rate. The system was used during the Vietnam War in 1967 to transmit data from Vietnam to Hawaii through one satellite and on to Washington DC through another. The complete system was declared operational in 1968 and its name changed to Initial Defense Satellite Communication System (IDSCS). IDSCS was a wideband system used for strategic communication applications between fixed and transportable ground stations and large ship-borne equipment, all having large antennas.

Figure 5.1 Courier satellite. (*Source:* Courtesy of the US Army.)

In the 1970s and 1980s, only USA and Russia had military communication satellites. But, today many other developed countries of the world like UK, France, Italy, Israel, China and so on have such systems. In the following paragraphs, we shall briefly discuss military satellites developed by various nations. Salient features, operational parameters and development status of major military communication satellite systems are presented in a latter section before concluding discussion on this category of military satellites.

5.3.1 American Systems

MILSATCOM architecture was proposed in the USA in 1976 to guide the development of military satellite communication systems in the country. Three types of military systems were proposed to be developed under this architecture; namely the wideband systems, mobile and tactical systems (or narrowband) and protected (or nuclear capable) systems. The wideband systems developed were the *Defense Satellite Communication Systems* (DSCS)-II and -III and *Global Broadcast Service* (GBS) *payload* on the *UHF Follow-On* (UFO) satellite.

Systems developed under the category of the mobile and tactical segment include the *Fleet Satellite Communication System*, the *Leasat programme* and the *UHF Follow-On Programme* (UFO). Satellites developed under the category of protected systems include the *MILSTAR system*, *Air Force Satellite Communication* (AFSATCOM) and the *Extremely High Frequency (EHF) payloads*. Figure 5.2 shows the satellites in the three types of systems.

5.3.1.1 Wideband Systems

The IDCSP satellites mentioned before represented phase I of the DSCS space segment programme. Phase II of the programme, named DSCS-II, began with the launch of six satellites launched in pairs, with the first pair launched in 1971. These satellites suffered some major technical problems and hence failed to operate one or two years after launch. Certain modifications were made in the next launches in order to remove these problems. By the year 1989, a total of 16 satellites had been launched. The DSCS-II constellation comprised of at least four

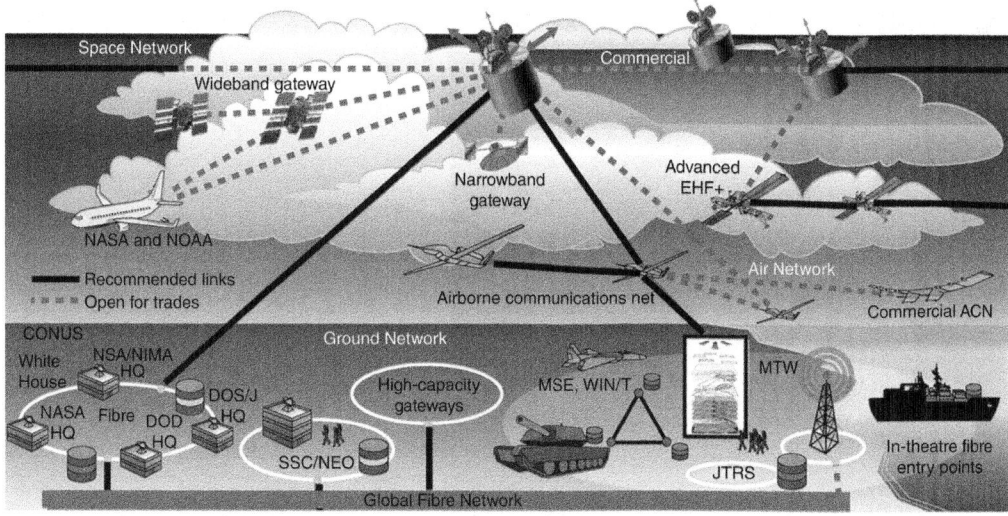

Figure 5.2 MILSATCOM architecture.

Figure 5.3 DSCS-III satellite (Reproduced by permission of Lockheed Martin. (*Source:* Courtesy of the US Airforce.)

active and two spare satellites. DSCS-II satellites offered increased capability than DSCS-I satellites and also had a longer lifetime.

The DSCS programme was initially developed for providing long-distance communication services between major military locations. However, by the 1990s, DSCS satellites served a large number of small, transportable and ship-borne terminals. DSCS-III (Figure 5.3) satellites were developed to operate in this diverse environment. They had increased communication capacity, particularly for mobile terminal users and improved survivability. The first DSCS-III

Figure 5.4 Wideband gapfiller satellite. (*Source:* Courtesy of the US Airforce.)

satellite was launched in 1982. A total of 12 satellites have been launched to date. The DSCS-III satellite system has a constellation of five operational satellites providing the required coverage.

The GBS is another part of the MILSATCOM's wideband architecture. It provides high-data rate intelligence, imagery, map, video and data communication services to tactical forces using small portable terminals. GBS was planned to be developed in three phases. Phase I employed Ku band transponders on a commercial satellite and a limited number of commercial receive terminals. Phase II employs four GBS transponders operating in Ka band on the UFO-8, -9 and -10 satellites. Phase III will provide global coverage GBS system on Advanced Wideband satellites.

The Wideband Gapfiller satellite programme (Figure 5.4) will supplement the military X band communications capability currently provided by the DSCS satellite system and the military Ka band capability of the GBS. In addition, the programme will include a high-capacity two-way Ka band capability to support mobile and tactical personnel. This programme will be succeeded by the Advanced Wideband System, which is in the planning stages.

5.3.1.2 Mobile and Tactical Systems

Developmental testing of tactical communication satellite systems began in the late 1960s with the launch of the *Lincoln Experiment Satellites* (LES) 5 and 6, and the Tactical Communications Satellite named TACSAT-I. All the three satellites operated on the UHF and SHF frequency bands. These satellites tested the feasibility of supporting small, mobile antenna users. Fleet Satellite Communication (FLTSATCOM) system was USA's first operational military system for tactical users of military satellites. The system was an outgrowth of these experimental satellites. A total of five FLTSATCOM satellites were launched in a span of 3 years between 1978 and 1981.

The FLTSATCOM system was followed by the *Leasat satellite system* (Figure 5.5). First operational Leasat satellite was launched in 1984. Leasat satellites operated in the UHF-band. Five satellites were launched in the Leasat system in a span of 6 years. These satellites primarily served the US Navy, Air Force, ground forces and mobile users. Leasat satellites have been

Figure 5.5 Leasat satellite. (*Source:* Courtesy of NASA.)

Figure 5.6 Block IV UFO (UFO-11) satellite. (*Source:* Courtesy of the US Navy.)

replaced by the UFO satellites. Four blocks of UFO satellites were launched in a span of 10 years from 1993 to 2003, namely the Block-I, -II, -III and -IV satellites. Block I and III satellites contained three satellites each, Block-II had four satellites and one Block-IV (Figure 5.6) satellite has been launched. UFO satellites support the global communications network of US Navy and a variety of other US military fixed and mobile terminals. They are compatible with ground- and sea-based terminals already in service.

UFO satellites have been replaced by Advanced Narrowband systems to provide global narrowband communication services to tactical users. The system has been operational since 2013.

5.3.1.3 Protected Satellite Systems

Protected satellite systems serve the nuclear capable forces. These satellites provide global coverage and have maximum survival capability. Satellites developed under the category of protected systems include the *MILSTAR system, Air Force Satellite Communication*

Figure 5.7 Advanced Extremely High Frequency. (AEHF) System. (*Source:* Courtesy of the US Airforce.)

(AFSATCOM) *programme* and the *Extremely High Frequency* (EHF) payloads. The MILSTAR system was designed to provide increased robustness and flexibility to the users. The MILSTAR programme includes two Block-I and four Block-II satellites. The Block-I satellites were launched in 1994 and 1995. The first Block-II satellite was lost during launch. The second one was launched in 2001 and the third and the fourth satellites were launched in 2002.

Protected satellite systems of the future include the *Advanced Extremely High Frequency System* and the *Advanced Polar System*. The Advanced Extremely High Frequency System (Figure 5.7), also referred to as the MILSTAR-3 system, is fully operational. It has 12 times the total throughput as compared to MILSTAR-II system, in some scenarios. Single-user data rates will increase to 8 Mbps. The system provides a large increase in the number of spot beams, which will improve user accessibility. The Advanced Polar satellite system will have two satellites in highly inclined Molniya orbits to provide communication services to the Polar Regions.

5.3.2 Russian Systems

Military communication satellites developed by Russia include the Parus, Potok (Geizer), Raduga (Gran), Raduga-1 (Globus), Strela-1, Strela-1M, Strela-2, Strela-2M and Strela-3 series. The Parus satellite system was the first military communication satellite system of Russia and is currently operational. It was developed to provide location information for the Parus navigation system. Parus communication satellites also provide store-and-dump communication services and relay data for ocean surveillance satellites. A total of 96 Parus satellites have been launched from 1974 to 2005. The last satellite of the series, Parus 96 was launched in January 2005.

The first Raduga satellite was launched way back in 1976. A total of 34 Raduga satellites have been launched, with the last satellite being launched in 1999. Raduga-1 satellites are improved version of Raduga satellites. Seven Raduga-1 satellites have been launched with the first and the last launch occurring in 1989 and 2004, respectively.

THE Potok series, code named Geizer, were military relay satellites designed to handle communications between the ground stations and the electro-optical reconnaissance satellite, Yantar. The first Potok satellite was launched in 1982. Ten Potok satellites have been launched to date with the last satellite, Potok-10, launched in 2000.

The Strela series of satellites are Russian tactical communication satellites. The Strela communication satellite system comprised of a constellation of medium orbit store-dump satellites that provided survivable communications for Soviet military and intelligence forces. Twenty-one experimental satellites were launched in the Strela-1 series in a span of a year between 1964 and 1965. Strela-1 satellites were followed by Strela-1M satellites. Around 370 Strela-1M satellites were launched between 1970 and 1992.

Five satellites were launched in the Strela-2 series, with the first launch taking place in 1965 and the last launch occurring in 1968. They were followed by Strela-2M satellite series. The Strela-2M series comprised of 52 satellites. The first Strela-2M satellite was launched in 1970 and the last satellite of the series was launched in 1994. All these satellites represent first generation of strategic store-dump military communication satellites of Russia.

Strela-3 was the second generation of Russian strategic store-dump military communication satellites. The operational constellation comprised of 12 spacecraft in two orbital planes, spaced 90° apart. The first satellites in the series were deployed in 1985 and the system was accepted into military service in 1990. In total, 136 satellites were launched in the Strela-3 series. The last Strela-3 satellite was launched in 2004.

5.3.3 Satellites Launched by Other Countries

Many other countries, including the UK, Italy, Israel, China and France, have launched their own military communication satellites. UK operates the Skynet series of satellites. It launched 10 satellites from 1969 to 2001. Italy has launched communication satellites named SICRAL-1 in 2000 and SICRAL-1b in 2005. Israel has launched two satellites; Amos-1 and -2 in GEO orbit in 1996 and 2003, respectively. France operates the Telecom-1 and -2 series of military communication satellites. The Telecom-1 series comprises of three satellites namely Telecom-1A, -1B and -1C, launched between 1984 and 1988. Telecom-2 satellites are advanced version of Telecom 1 satellites and comprise of four satellites Telecom-2A, -2B, -2C and -2D, launched during the period 1991–1996. It also plans to launch Syracuse-3A, -3B and -3C satellites in the near future.

China has launched several series of military communication satellites including the DFH-2 series, DFH-2A series, DFH-3 series, FH-1, FH-2 series, Spacenet 1, 2, 3, 3R series and ZX-7 series.

5.3.4 Frequency Spectrum Utilized by Space Systems

As mentioned in the chapter on satellite communication applications, the bands of interest for satellite communications lie above 100 MHz including the VHF, UHF, L, S, C, X, Ku, Ka and Q bands. Out of these bands the main bands of interest for military satellite systems are the X band, K band, Ka band and Q band. It must be emphasized here that the military communication needs are fundamentally distinct from those of commercial communications. Military spectrum requirements are based on the need for high-volume communications with

Table 5.1 Frequency bands used by commercial and military satellite systems.

Segment	Band	Bandwidth used	User	Satellites
UHF	200–400 MHz	160 KHz	Military	FLTSAT, LEASAT
	L (1.5–1.6 GHz)	47 MHz	Commercial	MARISAT, INMARSAT
	C (6/4 GHz)	200 MHz	Commercial	INTELSAT, DOMSATs, Anik E
	X (8/7 GHz)	500 MHz	Military	DSCS, Skynet and Nato
SHF	Ku (14/12 GHz)	500 MHz	Commercial	INTELSAT, DOMSATs, Anik E
	Ka (30/20 GHz)	2500 MHz	Commercial	JCS
	Ka (30/20 GHz)	1000 MHz	Military	DSCS IV
EHF	Q (44/20 GHz)	3500 MHz	Military	MILSTAR
	V (64/59 GHz)	5000 MHz	Military	Crosslinks

continuous uninterrupted service during war time. Table 5.1 lists the various bands used by both commercial and military satellite systems.

Use of high frequencies (K, Ka and Q band) helps military satellites achieve a high degree of survivability during both electronic warfare and physical attack. It also offers advantages like reliable communications in nuclear environment, minimal susceptibility to enemy jamming and eavesdropping, and the ability to achieve smaller secure beams with modest-sized antennas. The military satellites of USA operate in three main operational frequency segments namely the UHF, SHF and the EHF segments. The frequency bands of interest in the UHF segment are the 200–400 MHz band and the X band (8/7 GHz). The Ka band (30/20 GHz) and Q band (44/20 GHz) in the SHF segment and the EHF segment, respectively, are also used extensively for these applications. Mobile and the tactical satellite systems operate in the UHF-band (225–400 MHz). Wideband satellite systems operate in the X band (8/7 GHz) and the Ka band (30/20 GHz). Protected satellite systems operate in the EHF spectrum (44/20 GHz).

Russian military communication satellites mainly include the Raduga and the Strela series. The Raduga satellites operate in the C band (6.2 GHz up-link and 3.875 GHz down-link).

5.3.5 Dual-Use Military Satellite Communication Systems

Communication satellites intended for military applications are quite different from their civilian counterparts. They have better protection against jamming, better flexibility to rapidly extend services to new regions of the globe and to reallocate system capability as needed. Moreover, they employ better encryption techniques, enhanced TTC&M security, hardening against radiation and so on. They use special frequencies for transmitting the signals. Because of these unique design features, they cost as much as three times as compared to their equivalent civilian counterpart satellites. Due to the high costs involved, commercial satellite systems have been used for non-strategic and non-tactical military applications.

Since the mid-1990s, many commercial civilian communication satellites are being used for military services of non-tactical nature. These satellites are used for providing radio and television services to the armed forces, telephone or other services that allow the overseas forces to talk to their relatives and many other such services that do not require any special security protection. Keeping in mind their possible military usage, commercial satellite systems have adapted to this situation in terms of capacity availability, flexibility of geographical coverage and various types of security and encryption requirements. Digital Video Broadcast (DVB) services have also been developed to meet the military requirements.

5.4 Major International Military Communication Satellites

In the following paragraphs salient features and current status of major international military communication satellites are presented.

5.4.1 Defence Satellite Communication Systems (DSCS) Series

DSCS satellites are US military communication satellites providing high security data and voice communication services to the US Department of Defense (DOD). To date, three series of DSCS satellites have been launched. These include DSCS-1, -2 and -3. The DSCS-1 or the IDCSP (Initial Defence Communication Satellite Programme), comprising of 27 satellites provided the Pentagon with its first geosynchronous military communication satellite system. DSCS-2, comprising of 16 satellites, provided secure voice and data communications for the US military. DSCS-3 are geostationary communication satellites that provide a robust anti-jam, nuclear hardened capability that supports US DOD worldwide requirements, White House and diplomatic communications. It provides uninterrupted high priority secure voice and high-data rate communication services such as the exchange of wartime information between defence officials and battlefield commanders. DSCS-3 series comprises of 14 satellites, DSCS-3 1 to -3 14.

5.4.1.1 DSCS-3 Series

DSCS-3 4 (A2, USA-44), -3 5 (B14, USA-78), -3 6 (B12, USA-82), -3 7 (B9, USA-93)

Development Agency: Lockheed–Martin Missiles and Space, USA
Launch:
DSCS-3 4: 4 September 1989
DSCS-3 5: 16 February 1992
DSCS-3 6: 2 July 1992
DSCS-3 7: 19 July 1993
DSCS-3 4 was launched on Titan-34D, DSCS-3 5, -3 6 and -3 7 were launched on Atlas-2. All these satellites were launched from the Cape Canaveral launch centre, USA
Orbit: GEO
Weight: 1235 kg each
Payload: Six independent Super High Frequency (SHF) transponders, one special purpose single channel transponder operating in both the SHF and UHF bands, three receive antennas and five transmit antennas
Stabilization: three-axis stabilization
Operational life: design life of 10 years.

DSCS-3 8 (B10, USA-97), -3 9 (B7, USA-113), -3 10 (B13, USA-134), -3 11 (B8, USA-148), -3 12 (B111, USA-153), -3 13 (A3, USA-167), -3 14 (B6, USA-170)

Development Agency: Lockheed–Martin Missiles and Space, USA
Launch:
DSCS-3 8: 28 November 1993
DSCS-3 9: 31 July 1995

DSCS-3 10: 25 October 1997
DSCS-3 11: 21 January 2000
DSCS-3 12: 20 October 2000
DSCS-3 13: 11 March 2003
DSCS-3 14: 29 August 2003
All these satellites were launched from the Cape Canaveral launch centre, USA. DSCS-3 8 was launched on Atlas-2, DSCS-3 13, -3 14 on Delta 4M and DSCS-3 9, -3 10, -3 11 and -3 12 on Atlas-2A
Orbit: GEO
Weight: 1235 kg each
Payload: Six independent Super High Frequency (SHF) transponders and one special purpose single channel transponder operating in both the SHF and UHF bands
Stabilization: three-axis stabilization
Operational life: design life of 10 years.

5.4.2 Geizer (Potok Series)

Potok satellites (code named *Geizer*) are Russian military communication satellites. Potok was one element of the second generation global command and control system (GKKRS). Ten Potok satellites have been launched. Potok handled communications between ground stations and the Yantar-4KS1 and Yantar-4KS1M electro-optical reconnaissance satellites.

5.4.2.1 Potok-10
Development Agency: Applied Mechanics NPO Lavochkin of Russia
Launch: 4 July 2000 from Baikonur cosmodrome in Kazakhstan
Launch Vehicle: Proton-K
Orbit: GEO 80°E
Weight: 2400 kg
Stabilization: three-axis stabilization
Payload: Slav-2 and Sintez transponders (C band)
Operational life: design life of 5 years.

5.4.3 Globus Series

The Globus series of satellites represent second generation of Russian military communication satellites, replacing older Raduga ("Rainbow") satellites. They are also referred to as the Raduga 1 series of satellites. Eight satellites have been launched in the series including Raduga-1 1, -1 2, -1 3, -1 4, -1 5, -1 6, -1 7 and -1 8.

5.4.3.1 Raduga-1 4, -1 5, -1 6, -1 7, 1 8
Development Agency: Applied Mechanics NPO PM of Russia
Launch:
Raduga-1 4: 28 February 1999
Raduga-1 5: 28 August 2000
Raduga-1 6: 6 October 2001
Raduga-1 7: 27 March 2004
Raduga-1 8: 28 February 2009
All the satellites were launched on Proton-K from the Baikonour cosmodrome in Kazakhstan
Orbit:

Raduga-1 4: GEO 35°E
Raduga-1 5: GEO 500°E
Raduga-1 6: GEO
Raduga-1 7: GEO 85°E
Raduga- 1 8: GEO
Weight:
Raduga-1 5, 1 8: 2400 kg
Raduga-1 6, -1 7: 2000 kg each
Payload: 'Tor' C band transponders working at 20, 42 and 44 GHz
Stabilization: three-axis stabilization
Operational life: design life of 3 years.

5.4.4   Leasat Series (Syncom-4)

The Leasat or Syncom-4 (Figure 5.8) series of satellites are US military communication satellites providing worldwide communication services to mobile air, surface, subsurface and fixed Earth stations of the Navy, Marine Corps, Air Force and Army. Five Leasat satellites, namely Leasat-1, -2, -3, -4 and -5, have been launched to date. Leasat-1, -2 were launched in 1984 followed by Leasat-3, -4 in 1985 and Leasat-5 in 1990.

5.4.4.1   Leasat-5 (Syncom-4 5)

Development Agency: Hughes Space Systems, USA (now Boeing Systems)
Launch: 9 January 1990 from the Cape Canaveral launch centre, USA
Launch Vehicle: *Space Shuttle Columbia*
Orbit: GEO 1770°W
Weight: 7711 kg

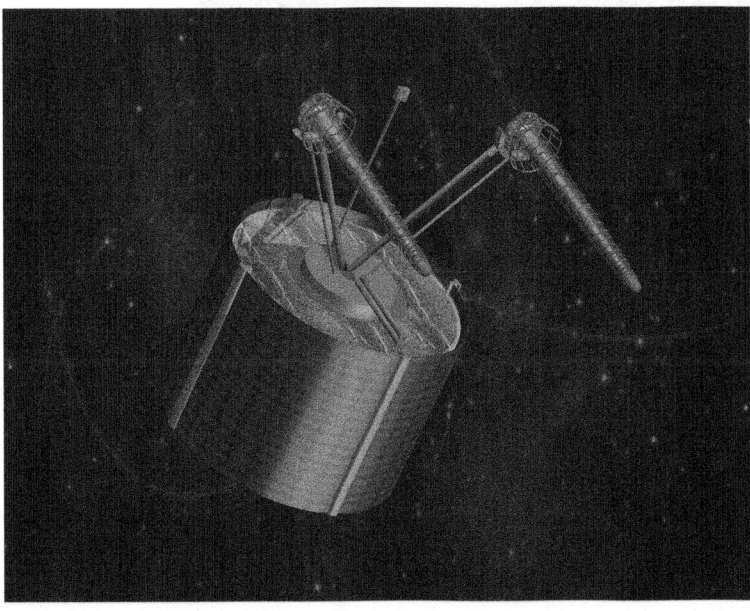

Figure 5.8 Leasat satellite (*Source:* Courtesy of the US Navy.)

Payload: five antennas (2 UHF helices, 1 Transmit and 1 Receive), two X band horns (one beacon, one receive), 12 UHF repeaters
Stabilization: spin stabilization
Operational life: design life of 7 years.

5.4.5 MILSTAR Series

MILSTAR (Military Strategic and Tactical Relay) is a tactical and strategic multiservice satellite system designed to provide survivable communication services for US forces worldwide. They provide voice, data, imagery and video communication services to users on foot, ships, submarines and aircraft. Two series of MILSTAR namely the satellites MILSTAR-1 and MILSTAR-2 series have been launched. The third series more popularly known as the AEHF (Advanced Extreme High Frequency) system represents the next-generation of highly secure military satellites. The first satellite of the AEHF system was launched in August 2010.

5.4.5.1 MILSTAR-1 Series
The MILSTAR-1 series (Figure 5.9) comprises two satellites: MILSTAR-1 1 (MILSTAR-1) and -1 2 (MILSTAR-2).

Development Agency: Lockheed–Martin Missiles and Space, USA and TRW, USA
Launch:
MILSTAR-1 1: 7 February 1994
MILSTAR-1 2: 6 November 1995
Both these satellites were launched from the Cape Canaveral launch centre, USA on *Titan-4A*
Orbit:
MILSTAR-1 1: GEO 1200°W
MILSTAR-1 2: GEO 40°E

Figure 5.9 MILSTAR-1. (*Source:* Courtesy of the US Airforce.)

Weight: 5150 kg each
Payload: LDR and cross-link payloads
Stabilization: three-axis stabilization
Operational life: design life of 10 years.

5.4.5.2 MILSTAR-2 Series

The MILSTAR-2 series (Figure 5.10) comprises three satellites, MILSTAR-2 2 (MILSTAR-4), -2 3(MILSTAR 5) and -2 4 (MILSTAR-6). MILSTAR-2 1 (MILSTAR-3), launched on 30 April 1999 was a launch failure. The MILSTAR-2 satellites have extended the communications capabilities of MILSTAR satellites to higher data rates.

Development Agency: Lockheed–Martin Missiles and Space, USA, TRW, USA and Boeing
    Space Systems, USA
Launch:
MILSTAR-2 F2: 27 February 2001
MILSTAR-2 F3: 15 January 2002
MILSTAR-2 F4: 8 April 2003
All the three satellites were launched from the Cape Canaveral launch centre, USA on Titan-4B
Orbit: GEO
Weight:
MILSTAR-2 F2: 4670 kg
MILSTAR-2 F3, -2 F4: 4500 kg
Payload: LDR, MDR and cross-link payloads
Stabilization: three-axis stabilization
Operational life: design life of 12 years.

5.5   Reconnaissance Satellites

Reconnaissance satellites, also known as spy satellites, provide intelligence information on the military activities of foreign countries. They can also detect missile launches or nuclear explosions in space. These satellites can catch and record radio and radar transmissions while passing over any country. Reconnaissance satellites can be further subcategorized into the following four types, depending upon their applications:

1) Image intelligence (IMINT) or photo-surveillance satellites
2) Signal intelligence (SIGINT) or ferret satellites
3) Early warning satellites
4) Nuclear explosion detection satellites.

   IMINT and SIGINT satellites are collectively referred to as surveillance satellites.

5.5.1   Image Intelligence (IMINT) Satellites

Image intelligence satellites provide detailed high-resolution images and maps of geographical areas, military installations and activities, troop positions and other places of military interest. These satellites constitute the largest category of military satellites. They are generally placed in low, near-polar orbits at altitudes of 500–3000 km as they take high-resolution close-up images.

   The resolution of images provided by these satellites is of the order of a few centimetres. Due to large atmospheric drag at these altitudes, image intelligence satellites generally have small lifetimes of the order of a few weeks. These satellites were widely used by the USA during operation Desert Storm in 1992. They provided warning of the Iraqi invasion of Kuwait nearly a week before it occurred, including both the timing and the magnitude of the assault. It should be mentioned here that some high-resolution non-military Earth observation satellites have also been used for military applications. These include the ORBIMAGE-4 (Orbital Imaging Corporation) and the QuickBird series of satellites.

IMINT satellites can be classified as close-look IMINT satellites and area survey IMINT satellites, depending upon their mode of operation. Close-look IMINT satellites provide high-resolution photographs that are returned to Earth via a re-entry capsule, whereas area survey IMINT satellites provide lower resolution photographs that are transmitted to Earth via radio. Recently launched IMINT satellites have the capability to take both close-look images as well as area images. IMINT satellites can also be classified into the following three types depending upon their wavelength band of operation.

1) PHOTOINT or optical imaging satellites
2) Electro-optical imaging satellites
3) Radar imaging satellites.

5.5.1.1 PHOTOINT or Optical Imaging Satellites

These satellites have visible light sensors that detect missile launches and take images of enemy weapons on the Earth's surface. These satellites can either be film-based or television-based. Film-based systems employ a film for recording the images and were the first type of systems to be used by reconnaissance satellites. They are no longer in use now. The system comprised two parts: the camera and the recovery capsule. In this case, after the pictures were taken, the film would spool-up in the return capsule. The capsule was released from the orbiting satellite once it had taken all the pictures. The capsule was then recovered in the Earth's atmosphere by an aircraft. The whole process of film retrieval took around 1–3 days. The image was then processed and analysed. Due to this time lag, they were used for strategic planning rather than in tactical combat situations. Moreover, these images could not be taken in cloudy conditions or in darkness and are susceptible to camouflaging. Figure 5.10 shows the operation of film-based PHOTOINT satellites.

Another type of PHOTOINT satellite system is the television-based system that takes pictures in the conventional manner. After the images are taken, the film is scanned for electronic retransmission back to Earth. Due to the complexities involved, the system was phased out rather quickly. Some of the famous PHOTOINT satellites include the USA's KH-1 (KeyHole), KH-2, KH-3, KH-4, KH-4A, KH-4B, KH-5, KH-6, KH-7, KH-8 and KH-9 satellites and Russian Araks, Orlets, Yantar and Zenit series.

5.5.1.2 Electro-Optical Imaging Satellites

Electro-optical imaging satellites provide full-spectrum photographic images in the visible and the IR bands. They use CCD cameras to take images. The CCD camera assigns different digital number values to represent varying light levels in the image. Digital enhancement techniques are used to further sharpen the images and remove the background noise. The digital information is then transmitted to the ground station via electronic communication links and the image is then 'reassembled' by the ground station computer. Electro-optical imaging satellites are able to image heat sources during the night but not objects having normal temperatures. Moreover, they do not work in cloudy conditions and are only slightly less susceptible to camouflaging as compared to the PHOTOINT satellites. The USA's KH-11 and KH-12 satellites and Russian Yantar-4KS1 and 4KS2 satellites are examples of electro-optical imaging satellites.

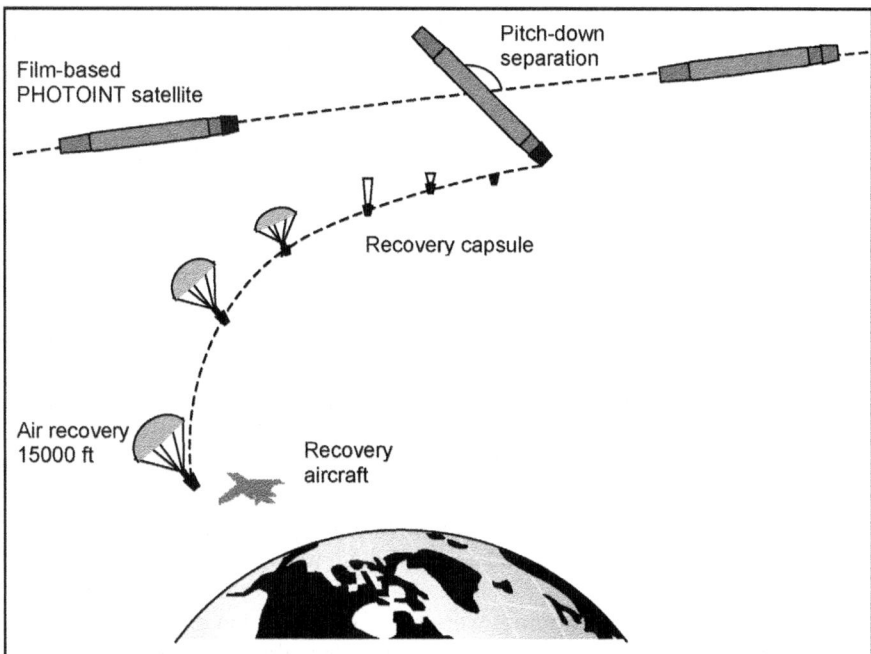

Figure 5.10 Operation of the film-based PHOTOINT satellites.

5.5.1.3 Radar Imaging Satellites

Both the PHOTOINT and the electro-optical imaging satellites were unable to take images under cloudy conditions. Radar imaging satellites overcome this problem. However, their resolution is poor when compared with the PHOTOINT and the electro-optical imaging satellites. Moreover, they suffer from the problem of 'backscatter noise' and are susceptible to active jamming. These satellites mostly employ synthetic aperture radar (SAR) to take images in the microwave band. Here, microwave pulses are transmitted toward the Earth's surface by SAR. These pulses penetrate the cloud cover and hit the various objects on the Earth's surface.

Taking into consideration the time taken by the reflected pulses to reach the satellite and the signal strength of the return beam, images are created. Different digital numbers are assigned to various light levels and then this information is transmitted electronically to Earth in the same manner as that for the electro-optical satellites. Other radar-based technologies employed by these satellites include the Doppler radar technology and the GMTI (Ground moving target indication) radar. Doppler radar technology is used to spot the movement of ships and aircraft and GMTI radar is useful for detecting ground movement of vehicles. Some of the radar imaging satellites includes the USA's Lacrosse, Quill and Indigo satellites and the Russian Almaz series of satellites. Development of IMINT satellites is discussed in the following section.

5.5.1.4 Development of IMINT Satellites

The first IMINT satellites were launched by the USA followed by the erstwhile Soviet Union. IMINT satellites launched initially belonged to the PHOTOINT category. The first PHOTOINT satellite systems were the *Discoverer* and the *Satellite and missile observation system* (SAMOS) of the USA Discoverer satellites circled the Earth in polar orbits. They used photographic films that were returned to Earth through a re-entry capsule. The first Discoverer satellite was launched in 1959. There were 38 public launches of the satellites under the programme.

The Discoverer-14 satellite, launched on 18 August 1960, was the first satellite to successfully return film from orbit. This satellite marked the beginning of the age of satellite reconnaissance. The Discoverer programme officially ended in 1962 with the launch of Discoverer-38 satellite. However, the programme continued under the secret code name CORONA until the year 1972, carrying out a total of 148 launches. CORONA's major accomplishment was to provide photographs of missile launch complexes of the Soviet Union. It also identified the Plesetsk missile test range of Soviet Union and provided information on the types of missiles being developed, tested and deployed by Soviet Union. The SAMOS programme launched heavier payloads to collect photographic and electromagnetic reconnaissance data, which was electronically transmitted back to Earth.

The National Reconnaissance Office (NRO) was formed in 1961 to design, build, operate and manage US reconnaissance satellites. Even today, it manages and operates all the reconnaissance satellites launched by the US. The CORONA programme lasted for 13 years and comprised four satellite generations named KH-1, KH-2, KH-3 and KH-4. The KH-4 family of satellites was further classified as KH-4, KH-4A and KH-4B. The KH (KeyHole) designation is used to refer to all photographic US reconnaissance satellites. The KH-1 satellites are sometimes referred to as the USA's first 'spy' satellites. The satellites launched initially had a resolution of the order of 10 m and a lifetime of around a week, which was later improved to 3 m and 19 days, respectively, in the KH-4B series.

The SAMOS and the CORONA programmes were the first generation of IMINT satellites that returned high-resolution images to Earth using re-entry capsules. Other first-generation satellites included the Argon and the LANYARD series of satellites. Argon was the code name given to the KH-5 satellites, designed for large-scale mapmaking. LANYARD satellites, or the KH-6 satellites, were used for gathering important intelligence information. Twelve satellites were launched in the KH-5 series and three satellites were launched in the KH-6 series. The KH-6 series was followed by KH-7, KH-8 and KH-9 series. All the satellites from KH-1 to KH-9 were film-based 'close-look PHOTOINT' satellites that returned high-resolution images to Earth using small re-entry capsules and were part of the KeyHole (KH) series of satellites.

They orbited in low Earth orbits at an altitude of around 200 km. Around 150 satellites were launched in the KH-1 to KH-9 series during the period 1960–1972. The use of PHOTOINT satellites employing return capsules was discontinued in the early 1980s. Satellites that took wide area images were advanced versions of IMINT satellites and transmitted images back to Earth via an electronic telemetry link. These satellites were referred to as the 'electro-optical' imaging satellites. The first electro-optical imaging satellite series was KH-11 series, code named Crystal/ Kennan. The first satellite under this series was launched in December 1976. Sixteen satellites have been launched under the series in a span of 36 years from 1976 to 2013. The KH-11 satellites orbited in higher orbits compared to their predecessors. They had the capability to take visible, near-IR and thermal-IR images.

The Russian IMINT satellites include the Zenit series, Yantar series, Orlets-1 and -2 and Araks series. The Zenit series comprised Zenit-2, -2M, -4, -4M, -4MK, -4MKM, -4MKT, -4MT, -6 and -8 series of satellites. The Zenit-2 series was the first to be launched, with 21 Zenit-2 satellites launched in a span of 30 years from 1961 to 1990. These satellites were film-based low resolution photo-intelligence satellites. Satellites in other Zenit series were high-resolution film-based satellites. The Yantar series comprised the Yantar-1K, -2K, -4K1, -4K2, -4K2M, -4KS1 and -4KS1M series of satellites. Yantar-1K, -2K, -4K1, -4K2 and -4K2M series of satellites comprised film-based photo-intelligence satellites whereas Yantar-4KS1 and -4KS1M were electro-optical imaging satellites. Orlets-1 and -2 were also film-based reconnaissance satellites. Eight satellites in the Orlets-1 series and two in the Orlets-2 series have been launched. Araks are the most recent reconnaissance satellite series with a resolution of 2–10 m. Two satellites have been launched in the series.

The development of radar-based reconnaissance satellites started in the 1970s. Quill was the first radar-based reconnaissance satellite. It was launched by the USA in 1964. It was followed by the Indigo satellite launched in 1976. The most important radar-based intelligence satellite project was the American project Lacrosse, whose first satellite was launched in 1988. This is an active radar imaging satellite system using synthetic aperture radar for observing tactical and strategic military targets. It also uses GMTI radar.

The Lacrosse constellation comprises two operational satellites orbiting in low Earth orbits at an altitude of around 650 km. Five Lacrosse satellites have been launched, with the last one launched in April 2005. The USA launched the NROL 21 satellite in 2006, and the Topaz 1 and 2 satellites in 2010 and 2012, respectively. The erstwhile Soviet Union launched its first radar imaging satellite series known as the Almaz series in the late 1980s. Three satellites were launched in the series in a span of five years from 1986 to 1991. It also launched the Kondor-1 satellite in 2013.

Countries like the UK and Japan have launched their own IMINT satellites. Japan has launched four electro-optical imaging satellites named IGS (Intelligence gathering satellite)-Optical-1, -2, -3V and -5V in the years 2003, 2006, 2007 and 2009, respectively. It has also launched four radar reconnaissance satellites IGS-Radar-1, -2, -3 and -4 in the years 2003, 2007, 2011 and 2013, respectively. The UK has also launched its first reconnaissance satellite, TopSat-1 (Topographic satellite), in 2005. TopSat-1 is a photo-imaging satellite. Israel has Ofeq-3, -4, -5, -6, -7 and -9 optical imaging satellites. These satellites orbit in unusual retrograde orbits.

Helios is a European optical reconnaissance satellite system funded by France, Italy and Spain. It comprises the Helios-1 and -2 series each having two satellites, namely Helios-1A (1995), -1B (1999), -2A (2004) and -2B (2009), respectively. China has launched several optical reconnaissance satellites since the 1970s. It has launched the FSW-0 (Fanhui Shi Weixing), FSW-1, FSW-2, FSW-3 and FSW-4 series of satellites. In addition, it has also launched high-resolution military imaging satellites named ZY-2A, ZY-2B, ZY-2C, Yaogan 2, 4, 5, 7, 8, 9, 11, 12, 14, 15, 16 and 17 satellites.

Israel has launched its radar-based reconnaissance satellite named TECHSAR. Other technology demonstrator satellites named Ofeq-1 and -2 were launched in 1988 and 1990, respectively. Ofeq-3, -5, -7 and 9 satellites are spy satellites of Israel. Germany has launched radar reconnaissance satellites SAR-Lupe -1, -2, -3, -4 and -5 with resolution less than 1 m. China has launched radar imaging satellites Yaogan-1, -3, -6, -10 and -13.

5.5.2 SIGINT Satellites

Signal intelligence or SIGINT satellites detect transmissions from broadcast communication and non-communication systems such as radar, radio and other electronic systems. These satellites intercept and decrypt government, military and diplomatic communications transmitted by radio, intercept ESM signals, receive telemetry signals during ballistic missile tests and relay radio messages from CIA agents in foreign countries. These satellites are essentially super sophisticated radio receivers that can capture radio and microwave transmissions emitted from any country and send them to sophisticated ground stations equipped with supercomputers for analysis. SIGINT is considered to be the most sensitive and important form of intelligence.

These satellites provided one of the first warnings of the possibility of an Iraqi invasion of Kuwait. SIGINT satellites, however, are not capable of intercepting landline communications. SIGINT satellites need to intercept radio communications over a very large frequency range, typically from 100 MHz to 25 GHz. It is difficult to cover this wide frequency range in one satellite, hence different types of SIGINT satellites operating in different parts of the radio frequency spectrum are operated simultaneously. The USA employs Rhyolite, Chalet, Vortex and Aquacade satellites, all operating in different parts of the radio frequency spectrum. Intercepted radio data are transmitted to Earth on a 24 GHz down-link using a narrow-beam antenna.

The main missions carried out by these satellites are outlined as under.

1) Interception and decryption of governmental, military and diplomatic communications transmitted by radio,
2) Interception of ESM (electronic support measure) signals that characterize the operating modes of the higher command organizations, installations of air-defence, missile forces and also the combat readiness of foreign armed forces,
3) Reception of telemetry signals during ballistic missile tests,
4) Relay of radio messages from CIA agents in foreign countries.

SIGINT satellites can be further categorized as communication intelligence (COMINT) or electronic intelligence (ELINT) satellites depending upon their intended function. COMINT or communication intelligence satellites perform covert interception of foreign communications in order to determine the content of these messages. As most of these messages are encrypted, they use various computer-processing techniques to decrypt the messages. The information collected is used to obtain sensitive data concerning individuals, government, trade and international organizations. COMINT satellites of today collect economic intelligence information and information about scientific and technical developments, narcotics trafficking, money laundering, terrorism and organized crime.

ELINT or electronic intelligence satellites are used for the analysis of non-communication electronic transmissions. This includes telemetry from missile tests (TELINT) or radar transmitters (RADINT). The most common ELINT satellites are designed to receive radio and radar emanations of ships at sea, mobile air-defence radar, fixed strategic early warning radar and other vital military components for the purpose of identification, location and signal analysis.

5.5.2.1 Development of SIGINT Satellites

The USA and Russia have the largest number of SIGINT satellites. However, some other countries like France and China have also developed their own SIGINT satellite systems.

5.5.2.1.1 USA Satellites

The first SIGINT satellites were launched by the USA in the early 1960s. These satellites orbited in LEO orbits. The limited and intermittent operation of these satellites suggested that for continuous monitoring and interception of communication channels, these satellites need to be placed at higher altitudes. In addition, satellites orbiting at higher attitudes are able to carry out both COMINT and ELINT operations.

The USA developed SIGINT satellites called 'Jumpseat' in the 1970s to be placed in the Molniya orbit. The basic task of these satellites was to intercept radio communications transmitted by communications satellites of the erstwhile Soviet Union orbiting in Molniya orbit. From 1971 to 1987, seven Jumpseat satellites were launched. Another series of satellites launched in the Molniya orbit was the Trumpet series. Three satellites were launched in the series in a span of 3 years from 1994 to 1997. The USA launched another series of satellites named the *Spook Bird* series, beginning in 1968, for radio interception from satellites of erstwhile Soviet Union orbiting in the geosynchronous orbit. Spook Bird satellites were launched in quasi-stationary orbits having an inclination of 3–10°, apogee distances of 39 000–42 000 km and perigee distances of 30 000–33 000 km. Spook Bird satellites move in a complex elliptical

Figure 5.11 Vortex satellite.

trajectory, enabling them to view broad regions. After two experimental launches production models of these satellites, named Rhyolite, were launched.

The Rhyolite constellation consisted of four operational satellites intercepting signals in the lower frequency UHF and VHF bands. They carried out a wide variety of missions in intercepting microwave communication transmissions and missile telemetry data from erstwhile Soviet Union and China. Four Rhyolite satellites were launched between 1970 and 1978. These satellites were later renamed Aquacade.

Another SIGINT satellite series developed by the USA during the 1970s was named Chalet. The first satellite of the series was launched in 1978 to intercept conversations carried on UHF radio links. The name of Chalet satellites was changed to Vortex in 1981. Six satellites were launched between 1978 and 1989. Vortex satellites (Figure 5.11) were a modernized versions of Chalet satellites with better onboard equipment for the purpose of expanding the range of interceptable radio frequencies in the direction of the centimetric band. Mercury satellites (Advanced Vortex satellites), the successors to the Chalet/Vortex satellites, are used to pinpoint radar locations. These satellites are in the GEO orbits as opposed to quasi-stationary orbits of Chalet/Vortex satellites. Three satellites were launched in this series in 1994, 1996 and 1998.

Magnum/Orion satellites were deployed at the end of the 1980s to replace the Rhyolite series of satellites as they reached the end of their operating lifetimes. Targets for these satellites include telemetry, VHF radio, cellular mobile phones, paging signals and mobile data links. Two Magnum satellites were launched, one in 1985 and the other in 1989. The Magnum series of satellites was replaced by the Mentor satellites. Six Mentor satellites were launched in a span of 17 years between 1995 and 2012. The USA had 6–8 operational SIGINT satellites during the 1980s and l990s. The frequency of launches of SIGINT satellites dropped in the beginning of the l990s.

The first space-based ELINT system of the USA was named GRAB (Galactic radiation and background). A total of five GRAB satellites were launched between 1960 and 1962. A primary ELINT programme named 'White Cloud' is a satellite constellation that is the US Navy's principal means of over-the-horizon reconnaissance and target designation for its weapons systems.

5.5.2.1.2 Russian Satellites

The first SIGINT satellite launched by the erstwhile Soviet Union was an ELINT satellite named Cosmos 189, launched in 1967. Until now, more than 200 SIGINT satellites have been launched under the Tselina satellite system (Figure 5.12). It basically comprised the low-sensitivity Tselina-O satellites and the high-sensitivity Tselina-D satellites. Tselina-O and -D satellites represent the first generation of Russian ELINT satellites. Tselina-2, Tselina-OK and Tselina-R represent the second generation of Tselina satellite system. The Tselina satellites detected and located the source of radio transmissions as well as determined the type, characteristics and performance modes of their targets.

5.5.2.1.3 Other Countries

France has launched several SIGINT satellites. The first SIGINT satellite of France was the Cerise satellite launched in 1995. It was a technology demonstrator satellite. It was followed by the Clementine satellite launched in 1999 and the Essaim series of satellites. Essaim is a system of four microsatellites that analyse the electromagnetic environment of the Earth's surface. All four Essaim satellites were launched in 2004. China has also launched SIGINT satellites, named the JSSW (Ji Shu Shiyan Weixing) series of satellites, comprising six satellites launched between 1973 and 1976.

5.5.3 Early Warning Satellites

Early warning satellites constitute a significant part of military systems. They provide timely information on the launch of missiles, military aircraft and nuclear explosions by the enemy to military commanders on the ground. This information enables them to ensure treaty compliance as well as provide an early warning of missile attack for appropriate action.

Figure 5.12 Tselina satellite. (*Source:* Courtesy Brian McMullin, Defense Intelligence Agency.)

Space-based infrared satellite systems are also being developed, which could track ballistic missiles throughout their trajectory and provide the earliest possible trajectory estimates to the command centre. In other words, these satellites would provide the earliest information of the start of a major missile attack and will be used to track long term patterns of space programmes of foreign countries.

Early warning (EW) satellites constitute an important part of the missile defence programme of USA, which aims to intercept and destroy missiles by shooting them down before they hit the target. EW satellites detect the launch of the missile, track the initial trajectory of the missile and relay this information to a missile defence command centre on the ground. The USA and Russia have developed extensive early warning satellite systems. In the following paragraphs, these systems will be discussed briefly.

5.5.3.1   Major Early Warning Satellite Programmes

MIDAS was the first early warning satellite system developed by the USA. It employed 24 satellites in low Earth orbits for detecting the launch of intercontinental ballistic missiles (ICBM) by Russia. However, the MIDAS programme was not very successful. The attention then shifted to launching early warning satellites in GEO orbits, as only four GEO satellites would be required for global coverage. The first geostationary early warning satellite system was the Defence support programme (DSP) of USA. DSP satellites detected the launch of intercontinental and submarine launched ballistic missiles, using IR and optical sensors. They also provided information on nuclear explosions. Over 19 DSP satellites (Figure 5.13) have been launched during 1970 to 1984. During the Persian Gulf War, DSP satellites provided effective warning of the launch of Scud missiles by Iraq.

The Space-Based InfraRed System (SBIRS) is intended to be the next-generation missile warning and tracking system. It will replace the DSP satellite system. The system (Figure 5.14) comprises a constellation of 24 satellites orbiting in three types of orbits, namely the GEO, HEO and LEO orbits. The constellation will have four satellites in the GEO orbit, two satellites

Figure 5.13 DSP satellite. (*Source:* Courtesy of the US Airforce.)

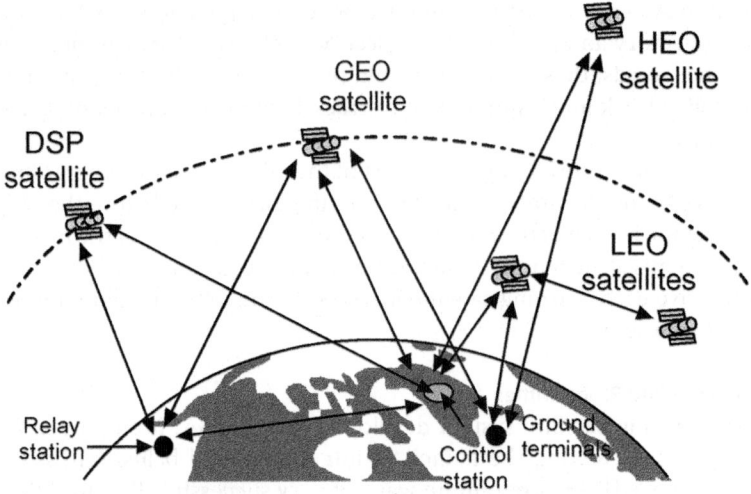

Figure 5.14 SBIRS architecture.

Figure 5.15 SBIRS-high component. (*Source:* Courtesy of the US Airforce.)

in the HEO orbit and 18 satellites in the LEO orbit. The GEO and HEO satellites constitute the SBIRS-high component (Figure 5.15) and the LEO satellites form the SBIRS-low component. Two SBIRS-GEO satellites have been launched in 2011 and 2013. Two satellites, SBIRS HEO-1 and -2, have been launched in the HEO orbit in 2006 and 2008, respectively.

The SBIRS-low component has been renamed the space tracking and surveillance system (STSS). SBIRS system is a part of the National Missile Defence (NMD) programme of the USA. SBIRS-high satellites are three-axis stabilized satellites and their sensors monitor the ground continuously, thereby providing much more accurate data. They will replace the DSP satellites.

STSS will track missiles as they fly above the horizon, offering much more accurate information on their trajectories. Such information is necessary for an effective anti-ballistic missile defence.

The first early warning satellite launched by the erstwhile Soviet Union was a test satellite launched in 1972. The first operational early warning satellite was launched 5 years later in 1977. These satellites, named Prognoz, orbited in Molniya orbits. This orbit enabled the satellite sensors to view the missiles against the cold background of space rather than the warm background of Earth. However, nine satellites were required to make the constellation fully operational. The Soviet Union government was unable to maintain the system and in the 1990s only half of the constellation was working. In the mid-1980s Soviet Union launched geostationary early warning satellites named Oko satellites, but they were not very successful. France has launched its early warning satellite programme named SPIRALE (*Systeme Preparatoire Infrarouge pour alerte*, or Preparatory System for IR Early Warning), comprising two satellites, SPIRALE-A and -B, to detect ballistic missiles in their boost phase.

5.5.4 Nuclear Explosion Detection Satellites

Vela satellites were developed by the USA to detect nuclear explosions on Earth and in space in order to monitor worldwide compliance with the 1963 nuclear test ban treaty. A total of 12 Vela satellites were launched during the period 1963 to 1970. In the 1970s, the nuclear explosion detection mission was taken over by the DSP system, and in the late 1980s, by the GPS system. The programme is now referred to as the *Integrated Operational Nuclear Detection System* (IONDS).

Two experimental satellites, namely the *Array of Low-Energy X-ray Imaging Sensors* (ALEXIS) satellite and the *Fast On-orbit Recording of Transient Events* (FORT'E) satellite, were launched by the USA in 1993 and 1997, respectively. The ALEXIS satellite sensors provide near real-time information on transient, ultra-soft X-rays. In addition, they also offer unique astrophysical monitoring capabilities. The FORT'E satellite features an electromagnetic pulse sensor. The sensor provides wideband radio frequency signal detection. The FORT'E satellite integrates with related technology to help discriminate between natural (e.g. lightening) and manmade signals.

5.6 Major International Reconnaissance Satellites

In the following paragraphs salient features and current status of major international military communication satellites are presented.

5.6.1 Defence Support Programme (DSP) Series

DSP satellites are a key part of North America's early warning system. These satellites detect missile launches, space launches and nuclear detonations. The first launch of a DSP satellite took place in the early 1970s and since that time, these satellites have provided an uninterrupted early warning capability to the United States. To date DSP Block-1 or Phase-I (DSP-1, -2, -3 and -4 satellites), Block-2 or Phase II (DSP-5, -6 and -7 satellites), Block-3 or Phase II-MOS/PIM (DSP-8, -9, -10 and 11 satellites), Block-4 or Phase II Upgrade (DSP-12 and -13 satellites) and Block-5 or Phase-III (Figure 5.16) (DSP-14, -15, -16, -17, -18, -19, - 20, -21, -22 and -23) series of satellites have been launched. The DSP-23 Phase-III satellite is scheduled for launch in the near future.

Figure 5.16 DSP Block-5. (*Source:* Courtesy of the US Airforce.)

5.6.1.1 DSP-18, -19, -20, -21, -22, -23

Development Agency: TRW Space and Electronics, USA

Launch:

DSP-18: 23 February 1997

DSP-19: 9 April 1999 (Failure)

DSP-20: 8 May 2000

DSP-21: 6 August 2001

DSP-22: 14 February 2004

DSP-23: 11 November 2007

All these satellites, except for DSP-23, were launched on Titan-4B. DSP-23 was launched on Delta-4H. All the satellites were launched from the Cape Canaveral launch centre, USA.

Orbit: GEO

Weight: 2386 kg each

Payload: Telescope with 6000 element IR array, nuclear explosion detectors, particle detection monitors.

Stabilization: spin stabilization

Operational life:

DSP-18: design life of 5 years

DSP-20, -21,-22, -23: design life of 7–9 years.

5.6.2 Oko Series

Russia's ballistic missile early warning network comprises two series of satellites, namely the Oko satellites and Prognoz satellites. The constellation was designed with nine operational Oko and seven operational Prognoz satellites. Up until December 2005, 86 Oko satellites were launched.

5.6.2.1   Oko-81, -82, -83, -84, -85, -86

Development Agency: Lavochkin, Russia
Launch:
Oko-81: 9 April 1997
Oko-82: 14 May 1997
Oko-83: 7 May 1998
Oko-84: 27 December 1999
Oko-85: 1 April 2002
Oko-86: 24 December 2002
All these satellites were launched on Molniya-M from the Plesetsk launch site in northern Russia
Orbit: Molniya, apogee of 39 700 km, perigee of 600 km, inclination of 630
Weight: 2400 kg each
Payload: infrared sensors, telescopes
Stabilization: three-axis stabilization
Operational life: design life of 3–5 years.

5.6.3   Geosat Follow-On Series

The Geosat Follow-On Series (GFO) satellite is used by the US Navy to provide continuous worldwide oceanographic data for ships at sea and for navy facilities on shore, directly supporting naval operations such as ship routing, anti-submarine warfare and amphibious operations.

5.6.3.1   GFO-1

Development Agency: Ball Aerospace, USA
Launch: 10 February 1998 from Vandenberg Air Force Base in California, USA
Launch Vehicle: Taurus-2210
Orbit: LEO, sun-synchronous 775 × 878 km, 1080
Weight: 410 kg
Payload: precise radar altimeter (resolution <5 cm)
Stabilization: spin stabilization
Operational life: design life of 8 years.

5.6.4   Okean-O 1

OKEON-O is a Russian-Ukrainian remote sensing satellite that enables monitoring of ocean salinity, waves and ice conditions.

Development Agency: Ukrainian Yuzhnoye Company
Launch: 17 July 1999 from the Baikonour Cosmodrome in Kazakhstan
Launch Vehicle: Zenit-2
Orbit: LEO, sun-synchronous 664 × 662 km, and inclination 98.1°
Weight: 6150 kg
Stabilization: three-axis stabilization
Payload: electro-optical and radar sensors
Operational life: design life of 3 years.

5.6.5   Advanced Orion Series

These satellites are geostationary signal intelligence satellites operated by the US Air Force. They replaced the Magnum/Orion series of satellites. The mission objective of Orion or Mentor satellites is to intercept communication transmissions especially line-of-sight microwave links and missile telemetry interception from the Soviet Union and China. Three satellites, namely Mentor-1, -2 and -3, have been launched.

Launch:
Mentor-1: 14 May 1995
Mentor-2: 8 May 1998
Mentor-3: 9 September 2003
Mentor-1 was launched on Titan-4A and Mentor-2 and -3 were launched from Titan-4B. All three satellites were launched from the Cape Canaveral launch centre, USA
Orbit: GEO
Stabilization: three-axis stabilization
Payload: large dish antenna.

5.6.6 Arkon-1 Series

The Arkon-1 series is a new area reconnaissance satellite of Russia having resolution in the order of 2–10 m. It has an unusually high (1510–2747km) orbit for a photo-reconnaissance satellite. To date, four Arkon-1 satellites have been launched in 1983, 1989, 1997 and 2002.

Development Agency: NPO Lavochkin, Russia
Launch:
Araks-1 (Kosmos-2344): 6 June 1997
Araks-2 (Kosmos-2392): 25 July 2002
Both the satellites were launched on Proton-K from the Baikonour Cosmodrome in Kazakhstan
Orbit: LEO, sun-synchronous 1500 × 1836 km, 64.4°
Weight
Araks-1: 6000 kg
Araks-2: 2600 kg
Stabilization: three-axis stabilization
Operational life: design life of 2 years.

5.6.7 Helios Series

The Helios programme is Europe's military optical reconnaissance satellite system. Two series of Helios satellites, namely Helios-1 and -2, have been launched.

5.6.7.1 Helios-1 Series
Two satellites have been launched in this series, namely Helios-1A and -1B.

Development Agency: Matra Marconi Space, Europe
Both the satellites were launched on Ariane-40 from Kourou in French Guiana, France.
Orbit: LEO, sun-synchronous 672 × 676 km, 98.1°
Weight: 2500 kg each
Payload: Panchromatic, high-resolution and wide-angle optical instruments
Stabilization: three-axis stabilization
Operational life: design life of 5 years

5.6.7.2 Helios-2 Series
One satellite (Helios-2A) has been launched in this series. Second satellite of the series named Helios-2B is planned be launched in 2008.
Two satellites (Helios-2A and 2B) have been launched in this series.

Helios-2A
Development Agency: EADS Astrium, Europe
Launch: 18 December 2004 from Kourou in French Guiana, France
Launch vehicle: Ariane-5G
Orbit: LEO, sun-synchronous, mean altitude of 680 km, 98.1°
Weight: 4200 kg
Payload: imagers in visible and infrared bands
Stabilization: three-axis stabilization
Operational life: design life of 5 years.

5.6.8 KH Series

The KH (KeyHole) designation is used to refer to all photographic American reconnaissance satellites. They are operated by National Reconnaissance Office (NRO). Thirteen KH series satellites have been launched. KH-1 satellites are sometimes referred to as USA's first 'Spy' satellites. The satellites launched initially had a resolution of the order of 10 m and lifetime of around 1 week, which was later improved to 3 and 19 days, respectively, in the KH-4B series. The KH-1 to KH-4 series of satellites was codenamed CORONA.

The SAMOS and the CORONA programmes were the first generation of the intelligence imagery satellites that returned high-resolution images to Earth using re-entry capsules. Other first-generation satellites included the ARGON and the LANYARD series of satellites. ARGON was the code name given to the KH-5 satellites, designed for large-scale mapmaking. LANYARD satellites or the KH-6 satellites were used for gathering important intelligence information. Twelve KH-5 and three KH-6 satellites were launched. The KH-6 series was followed by KH-7, KH-8 and KH-9 series. All the satellites from KH-1 to KH-9 were film-based 'Close-look PHOTOINT' satellites that returned high-resolution images to Earth using small re-entry capsules and were part of KeyHole series of satellites. They orbited in low Earth orbits at an altitude of around 200 km. Around 150 satellites were launched in KH-1 to KH-9 programmes during the period of 1960–1972.

The use of PHOTOINT satellites employing return capsules was discontinued in the early 1980s. Satellites that took wide area images were advanced version of IMINT satellites and transmitted images back to earth via an electronic telemetry link. These satellites were the 'electro-optical' satellites. The first electro-optical satellite series was KH-11 code named Crystal/Kennan, first launched in December 1976. Nine satellites were launched under the series in a span of 12 years from 1976 to 1988. KH-11 satellites orbited in higher orbits as compared to their predecessors.

They had the capability to take visible, near-IR and thermal-IR images. The KH-11 series was followed by Advanced KeyHole or KH-12 series of satellites. Five satellites have been launched under this series, from 1992 to 2005. KH-12 satellites provided real-time images in the visible, near-IR and thermal-IR bands. The KH-13 series is a potential successor to the KH-12 reconnaissance satellites. Not much information is available on KH-13 satellites.

5.6.8.1 KH-12 Series

Development Agency: Lockheed–Martin Missiles and Space, USA
Launch:
KH-12 1 (USA 86): 28 November 1992
KH-12 2 (USA 116): 5 December 1995
KH-12 3 (USA 129): 20 December 1996
KH-12 4 (USA 161): 5 October 2001
All the satellites were launched from Vandenberg Air Force Base in California, USA. KH-12 1, 12

2 and 12 3 were launched on Titan-4A, KH-12 4, 12 5 were launched on Titan-4B.

Orbit: Near-polar, sun-synchronous orbit

KH-12 1: 198 × 207 km, 620

KH-12 2: 256 × 911 km, 97.70

KH-12 3: 153 × 949 km, 97.90

Weight: 19 600 kg each

Payload: optical sensors operating in the visible, near-IR and thermal-IR bands and electronic cameras

Stabilization: three-axis stabilization

Operational life: design life of 10–12 years.

5.6.8.2   KH-13 Series (EIS Series, 8X Series)

Development Agency: Lockheed–Martin Missiles and Space, USA

Launch: 22 May 1999 from Vandenberg Air Force Base in California, USA

Launch vehicle: Titan-4B.

5.6.9   Lacrosse Series

Lacrosse (Onyx) satellites, operated by National Reconnaissance Office of USA are radar reconnaissance satellites. Up to December 2005, five Lacrosse satellites were launched, namely Lacrosse-1, -2, -3, -4 and -5.

5.6.9.1   Lacrosse 4 (Onyx 4)

Development Agency: Lockheed–Martin Astronautics, USA

Launch:

Lacrosse-4: 17 August 2000

Lacrosse-5: 30 April 2005

Lacrosse-4 was launched from Vandenberg Air Force Base in California, USA and Lacrosse-5 was launched from the Cape Canaveral launch centre. Both the satellites were launched on Titan-4B.

Orbit: LEO, sun-synchronous

Lacrosse-4: Mean altitude 690 km, 68°

Lacrosse-5: Mean altitude 714 km, 57°

Payload: Synthetic Aperture Radar (SAR)

Weight: 14 500 kg each

Operational life: design life of 9 years.

5.6.10   Neman Series (Yantar-4KS1M)

Neman satellites are Russian electro-optical reconnaissance satellites. To date, nine satellites have been launched in this series. These include Neman-1 to Neman-9.

5.6.10.1   Neman-9 (Cosmos 2370)

Development Agency: Photon Company of Russia

Launch: 3 May 2000 from the Baikonour Cosmodrome in Kazakhstan

Launch Vehicle: Soyuz-U

Orbit: LEO, sun-synchronous 200 × 270 km, 64.90

Stabilization: three-axis stabilization

Operational life: design life of 3–5 years.

5.6.11 Tselina Series

Tselina satellites are Russian ELINT satellites, comprising the Tselina-O, -D, -R and -2 series. The Tselina-2 series is the latest of the Tselina series of satellites.

5.6.11.1 Tselina-2

In total, 22 Tselina satellites have been launched, Tselina-2 1 to Tselina-2 22.

Development Agency: Yuzhone Company of Russia
Launch:
Tselina-2 21: 3 February 2000
Tselina-2 22: 10 June 2004
Both the satellites were launched from the Baikonour Cosmodrome in Kazakhstan on Zenit-2
Orbit: LEO, sun-synchronous, Mean altitude 850 km, 710
Weight: 3200 kg each
Stabilization: three-axis stabilization
Operational life: design life of 3 years

5.6.12 US-PM Series

US-PM satellites are passive ocean surveillance satellites. To date, 12 satellites, namely US-PM1 to US-PM12, have been launched in this series.

5.6.12.1 Cosmos-2367 (Kosmos-2367, US-PM10), Cosmos-2383 (Kosmos-2383, US-PM11), Cosmos-2408 (Kosmos-2408, US-PM12)

Development Agency: Arsenal design Bureau of Russia
Launch:
US-PM10: 24 December 1999
US-PM11: 21 December 2001
US-PM12: 28 May 2004
All three satellites were launched on Tsyklon-2 from the Baikonour Cosmodrome in Kazakhstan
Orbit: LEO, sun-synchronous, mean altitude 420 km, 650
Weight: 3300 kg each
Payload: Radio-Technical Reconnaissance system and systems for electronic camouflage and self- protection
Stabilization: three-axis stabilization
Operational life: design life of 5 years.

5.7 Military Weather Forecasting Satellites

Weather forecasting satellites provide high-quality weather information to the operational commanders in the battlefield. This helps in effective deployment of weapon systems, protection of Department of Defense (DOD) resources and for exploits deep in enemy territory. The weather forecasting satellites provided useful information to the American forces during the Persian Gulf War.

The Defense Meteorological Satellite Programme (DMSP), originally known as the Defense System Applications Programme (DSAP), is the USA's military weather satellite programme to monitor the meteorological, oceanographic and solar-geophysical environment of Earth in order to support DOD operations. It provides visible and IR cloud cover imagery and other

Figure 5.17 DMSP-5D3 satellite. (*Source:* Courtesy of the US Airforce.)

meteorological, oceanographic, land surface and space environmental data. The first DMSP satellite was launched in 1966. Since then 12 series of DMSP satellites, namely DMSP-1A, -2A, -3A, -3B, -4A, -4B, -5A, -5B, -5C, -5D1, -5D2 and -5D3 (Figure 5.17), have been launched. All satellites launched have had tactical (direct readout) as well as strategic (stored data) capacity. The satellites orbit in near-polar sun-synchronous orbits. The DMSP constellation comprises a constellation of two active satellites. In December 1972, DMSP data were declassified and made available to the civil/scientific community.

5.8 Military Navigation Satellites

Satellite navigation systems have proved to be a valuable aid for military forces. Military forces around the world use these systems for diverse applications including navigation, targeting, rescue, disaster relief, guidance and facility management both during wartime as well as peacetime. The main satellite navigation systems operational today are the GPS system of the USA and the GLONASS system of Russia. The GPS and GLONASS receivers are used by soldiers and also have been incorporated on aircraft, ground vehicles, ships and spacecraft.

In addition, the Galileo navigation satellite system of Europe, the Beidou navigation satellite system of China and the IRNSS system of India are recent satellite-based navigation systems. The Galileo navigation system will comprise 30 satellites in three planes in a medium earth orbit. Four Galileo IOV satellites have so far been launched. The Beidou satellite navigation system comprises two separate satellite constellations: a limited test system and a truly global navigation system also referred to as Compass. Compass will comprise 35 satellites and will be fully operational by 2020. The basic constellation requires three satellites. BD-1, BD-2, BD-2I and BD-2M series of navigation satellites have been launched by China to date. Four satellites have been launched in the BD-1 series, six satellites in the BD-2 series, two satellites in the BD-2I series and five in the BD-2M series. IRNSS is being developed by India. It will comprise a constellation of seven satellites and a ground control segment. The first satellite of the constellation IRNSS-1A was launched in July 2013 and the complete constellation will be operational by 2014–2015.

5.8.1 Military Applications of Navigation Satellites

Satellite navigation systems have proved to be a valuable aid for military forces. Military forces around the world use these systems for diverse applications including navigation, targeting, rescue, disaster relief, guidance and facility management, both during wartime as well as peacetime. GPS and GLONASS receivers are used by soldiers and also have been incorporated on aircraft, ground vehicles, ships and spacecraft. Some of the main applications are briefly described in the following paragraphs.

5.8.1.1 Navigation
Navigation systems are invaluable for soldiers to navigate their way in unfamiliar enemy territory. They are replacing the conventional magnetic compass used by soldiers for navigation. They can also be used by Special Forces and crack teams to reach and destroy vital enemy installations. As an example, GPS receivers were used extensively by the US soldiers during Operation Desert Storm and Operation Iraqi Freedom. The soldiers using this system were able to move to different places in the desert terrain even during sandstorms or at night. More than 9000 such receivers were used during the mission.

5.8.1.2 Tracking
The services of navigation satellites are also utilized to track potential targets before they are declared hostile to be engaged by various weapon platforms. The tracking data is fed as input to modern weapon systems such as missiles and smart bombs and so on.

5.8.1.3 Bomb and Missile Guidance
The GPS and GLONASS systems are used to guide bomb and missiles to targets and position artillery for precise fire even in adverse weather conditions. Cruise missiles commonly used by the USA use multichannel GPS receivers to determine accurately their location constantly while in flight. The multiple launched rocket system (MLRS) vehicle uses GPS-based inertial guidance to position itself and aim the launch box at a target in a very short time. The GPS system was also used extensively by the US military during the Balkans bombing campaign in 1999, the Afghanistan campaign in 2001–2002 and in Iraq in 2003.

5.8.1.4 Rescue Operations
Satellite navigation systems prove invaluable to the military for determining the location of causality during operations and in navigating rescue teams to the site.

5.8.1.5 Map Updating
These systems augment the collection of precise data necessary for quick and accurate map updating.

5.8.2 Principle of Satellite Navigation

Satellite navigation systems developed in the initial phase were based on the 'Doppler effect'. Later 'trilateration'-based systems came into the picture. In this section, the various development stages of both these systems are briefly discussed, with more emphasis on the 'trilateration'-based systems, as contemporary satellite navigation systems use this technique.

5.8.2.1 Doppler Effect-Based Satellite Navigation Systems

The first satellite navigation system was the Transit system developed by the US Navy and John Hopkins University of the USA back in the early 1960s. The first satellite in the system, Transit I, was launched on 13 April 1960. It was also the first satellite to be launched for navigation applications. The system was available for military use in 1964 and to civilians three years later in 1967. The system employed six satellites (three active satellites and three in-orbit spares) in circular polar LEOs at altitudes of approximately 1000 km. The last Transit satellite was launched in 1988. The main limitations of the system were that it provided only two-dimensional services and was available to users for only brief time periods due to low satellite altitudes. Moreover, high-speed receivers were not able to use the system. The system was terminated in 1996. The Transit system was followed by the Nova navigation system, which was an improved system with better accuracy. Russians launched their first navigation satellite, Kosmos-158, in 1967, 7 years after the first American navigation satellite launch. The satellite formed a part of the Tsyklon system. It provided services similar to that provided by the Transit system. It was operational until 1978. The system was superseded by the Parus and the Tsikada systems. Parus is a military system comprising satellites in six orbital planes spaced at 30° longitude intervals, thus having an angular coverage of 180°. Ninety-eight satellites have been launched in the Parus system with the last satellite Parus 98 launched on 21 July 2009. The system is operational and is mainly used for data relay and store-dump communication applications. Tsikada is a civilian system covering the rest of the 180°. It comprised satellites in four orbital planes at 45° intervals. Twenty Tsikada satellites have been launched with the last satellite launched on 21 January 1995.

All the systems discussed here are based on the Doppler effect. The satellite transmits microwave signals containing information on its path and timing. The pattern of the Doppler shift of this signal transmitted by the satellite is measured as it passes over the receiver. The Doppler pattern coupled with the information on the satellite orbit and timing establishes the location of the receiver station precisely. One satellite signal is sufficient for determining the receiver locations. However, the systems mainly transmit two frequencies as it improves the accuracy of the system. As an example, the Transit system transmitted signals at the 150 MHz and 400 MHz frequencies. The positioning accuracy was around 500 m for single frequency users and 25 m for dual frequency users.

5.8.2.2 Trilateration-Based Satellite Navigation Systems

Doppler-based navigation systems have given way to systems based on the principle of 'trilateration', as they offer global coverage and have better accuracy as compared to the Doppler-based systems. In this case, the user receiver's position is determined by calculating its distance from three (or four) satellites whose orbital and the timing parameters are known. The receiver is at the intersection of the invisible spheres, with the radius of each sphere equal to the distance between a particular satellite and the receiver, with the centre being the position of that satellite (Figure 5.18). Two such systems are in operation today, namely the Global Positioning System (GPS) of the USA and the Global Navigation Satellite System (GLONASS) of Russia (Figure 5.19).

Another trilateration-based navigation system is the European system named Galileo. It is currently in the development phase. The first test satellite of the constellation was launched on 28 December 2005, while the second test satellite was launched on 26 April 2008. The third test satellite will be launched in the near future. These satellites will characterize the critical technologies of the system.

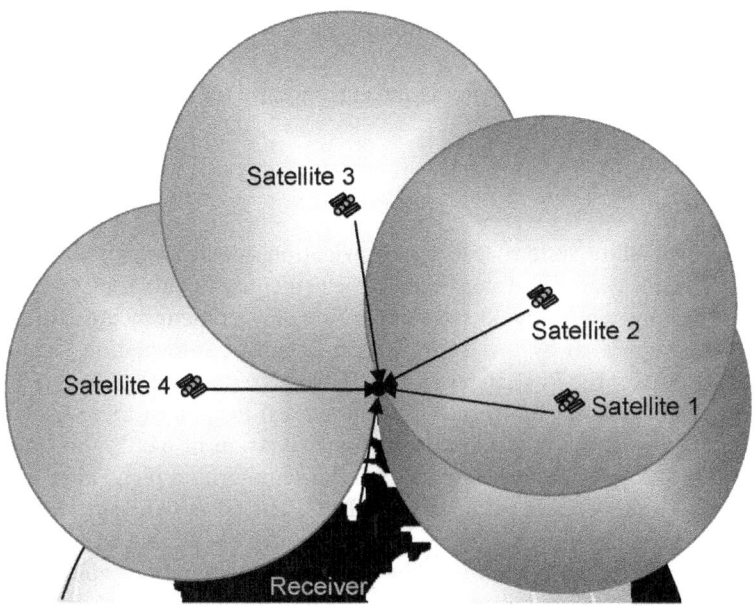

Figure 5.18 Principle of operation of trilateration-based satellite navigation systems.

Figure 5.19 GLONASS-series navigation satellite.

5.9 Major International Navigation Satellites

In the following paragraphs are presented salient features and current status of major international military navigation satellites.

5.9.1 GLONASS Satellite System

GLONASS (Global Navigation Satellite System) is Russian navigation satellite system. The system is a counterpart to the US GPS system and shares the same principles in data transmission and positioning methods. It is managed for the Russian federation Government by the Russian space forces and the system is operated by the Coordination Scientific Information Centre (KNITs) of the ministry of Defence of the Russian federation. The space segment of GLONASS is formed by 24 satellites (21 operational and three on-orbit spares) located in three orbital planes in a circular orbit of altitude 19 100 km and with an inclination of 64.8°. The three orbital planes are separated by 120° and the satellites within the same orbital plane are separated by 45°. The first GLONASS satellites were launched in 1982. The GLONASS system was declared fully operational on 24 September 1993. Since then, two series of GLONASS satellites namely GLONASS and GLONASS-M have been launched. The third-generation satellites of GLONASS system named GLONASS-K are scheduled for launch in the near future.

5.9.1.1 GLONASS Series (Uragan Series)
A total of 87 satellites was launched in this series up until December 2005.

Uragan-72 (Kosmos-2362), -73 (Kosmos-2363), -74 (Kosmos-2364), -75 (Kosmos-2374), -76 (Kosmos-2375), -77 (Kosmos-2376), -78 (Kosmos-2382), -79 (Kosmos-2381), -80 (Kosmos-2394), -81 (Kosmos-2395), -82 (Kosmos-2396), -83 (Kosmos-2402), -84 (Kosmos-2403), -85 (Kosmos-2411), -86 (Kosmos-2412), -87 (Kosmos-2417)
Development Agency: Russian company NPO PM
Launch:
Uragan-72, -73, -74: 30 December 1998
Uragan -75, -76, -77: 13 October 2000
Uragan -78, -79: 1 December 2001
Uragan -80, -81, -82: 25 December 2002
Uragan -83, -84: 10 December 2003
Uragan -85, -86: 26 December 2004
Uragan -87: 25 December 2005
All the satellites were launched on Proton-K from the Baikonour Cosmodrome in Kazakhstan
Orbit: in three planes, MEO Circular Mean altitude of 19 100 km, 64.80
Weight: 1300 kg
Payload: L band navigation payload
Stabilization: three-axis stabilization
Operational life: design life of 3 years.

5.9.1.2 GLONASS-M (Uragan-M) Series
The GLONASS-M series satellites (Figure 5.20) are second generation Russian Navigation satellites.

Uragan-M#1 (Kosmos-2382), -M#2 (Kosmos-2404), -M#3 (Kosmos-2413), -M#4 (Kosmos-2418), M#5 (Kosmos-2419), -M#6 (Kosmos-2424), -M#7 (Kosmos-2425), -M#8 (Kosmos-2426), M#9 (Kosmos-2431), M#10 (Kosmos-2432), M#11 (Kosmos-2433), M#12

Figure 5.20 GLONASM-M series navigation satellite. (*Source:* Vitaly V. Kuzmin, https://commons.wikimedia.org/wiki/File:Glonass-M_spacecraft_-_ParkPatriot2015part13-491.jpg. CC BY-SA 4.0.)

(Kosmos-2434), M#13 (Kosmos-2435), M#14 (Kosmos-2436), M#15 (Kosmos-2442), M#16 (Kosmos-2443), M#17 (Kosmos-2444), M#18 (Kosmos-2447), M#19 (Kosmos-2448), M#20 (Kosmos-2449), M#21 (Kosmos-2456), M#22 (Kosmos-2457), M#23 (Kosmos-2458), M#24 (Kosmos-2459), M#25 (Kosmos-2460), M#26 (Kosmos-2461)

Development Agency: Russian company NPO PM

Launch:

Uragan-M#1: 1 December 2001

Uragan-M#2: 10 December 2003

Uragan-M#3: 26 December 2004

Uragan-M#4, -M#5: 25 December 2005

Uragan-M#6, -M#7, -M#8: 25 December 2006

Uragan-M#9, -M#10, -M#11: 26 October 2007

Uragan-M#12, -M#13, -M#14: 25 December 2007

Uragan-M#15, -M#16, -M#17: 25 September 2008

Uragan-M#18, -M#19, -M#20: 25 December 2008

Uragan-M#21, -M#22, -M#23: 14 December 2009

Uragan-M#24, -M#25, -M#26: 1 March 2010

Uragan-M#1 to −M#11 were launched on Proton-K and Uragan-M#12 to −M#26 were launched on Proton-M. All the satellites were launched from the Baikonour Cosmodrome in Kazakhstan.

Orbit: in three planes, MEO Circular Mean altitude of 19 100 km, 64.8°

Weight: 1480 kg

Payload: L band navigation payload

Stabilization: three-axis stabilization

Operational life: design life of 5–7 years.

5.9.2 Global Positioning System (GPS)

GPS, also known as the Navigation Satellite Timing and Ranging (NAVSTAR), is the US satellite-based navigation system developed by the US DOD. Originally envisioned as primarily a military system, GPS is now a dual-use system, used for both military as well as civilian applications. The GPS navigation system employs a constellation of 24 satellites and ground support facilities to provide three-dimensional position, velocity and timing information to all the users worldwide 24 h a day. The GPS receivers calculate their location on the basis of ranging, timing and position information transmitted by GPS satellites (GPS satellites transmit information at two frequencies, 1575.42 MHz: L1 and 1227.6 MHz: L2).

The first GPS satellite was launched on 22 February 1978. It marked the beginning of first-generation GPS satellites, referred to as Block I satellites. Eleven satellites were launched in this block and were mainly used for experimental purposes. These satellites were out of service by the year 1995. The second generation of GPS satellites comprised of Block II and Block-IIA satellites. Block-IIA satellites were advanced versions of Block II satellites. A total of 28 Block II and Block-IIA satellites (nine satellites in the Block II series and 19 satellites in the Block-IIA series) were launched over the span of 8 years, from 1989 to 1997. The GPS system was declared fully functional on 17 July 1995 ensuring the availability of at least 24 operational, non-experimental GPS satellites.

Currently, third-generation GPS satellites, referred to as Block-IIR satellites, are being launched. The first satellite in this series was launched in 1997. Twenty-one satellites are planned to be launched in this block. Up until December 2009, 13 Block-IIR satellites had been launched. One of the potential advantages of Block-IIR satellites over previous satellites is that they have reprogrammable satellite processors enabling upgradation of satellites while in orbit. These satellites can calculate their own positions using inter-satellite ranging techniques. Moreover, they have more stable and accurate clocks on board them compared to the Block-II and -IIA satellites. Block-IIR satellites have three rubidium atomic clocks (having accuracy of one second in 300 000 years), whereas Block-II and -IIA satellites have two caesium atomic clocks (with an accuracy of 1 s in 160 000 years) and two rubidium atomic clocks (with an accuracy of 1 s in 300 000 years). Eight of the planned Block-IIR satellites have been improved further and are named Block-IIR-M satellites. These satellites will carry a new military code on both the frequencies (L1 and L2) and a new civilian code on the L2 frequency. The dual codes will provide increased resistance to jamming and the new civilian code will provide better accuracy to civilian users by increasing capability to compensate for atmospheric delays. Eight Block-IIR-M satellites have been launched Up until June 2010. Block-IIR satellites will be followed by Block-IIF satellites. Twelve Block-IIF satellites were planned to be launched by 2011. However, only Block-IIF-1 and Block-IIF-2 could be launched by the end of 2011. The remaining satellites, Block-IIF-3 to Block-IIF-12 were launched during 2012–2016. The launch of Block-IIF-12 took place on 5 February, 2016. These satellites have a third carrier signal, L5, at 1176.45 MHz. They also have larger design life, fast processors with more memory and a new civilian code. The GPS-III phase of satellites is in the planning stage. These satellites will employ spot beams, enabling the system to have better position accuracy (less than a metre). They will be positioned in three orbital planes having non-recurring orbits.

5.9.2.1 Navstar-IIR (GPS-IIR)

Navstar-IIR (Navigation System using Timing and Ranging) is the third evolutionary stage of the second generation of the Navstar GPS satellites (Figure 5.21). Block-IIR satellites provided dramatic improvements over previous blocks as they could determine their own position by

Figure 5.21 Navstar-IIR navigation satellite. (*Source:* Courtesy of the US Government.)

performing inter-satellite ranging with other Navstar-IIR vehicles. They also had reprogrammable satellite processors enabling problem fixes and upgrades in flight, increased satellite autonomy and radiation hardness.

Development Agency: General Electric Aerospace, USA (now Lockheed Missiles and Space)
Launch:
Navstar-IIR 1: 16 January 1997
Navstar-IIR 3: 7 October 1999
Navstar-IIR 4: 11 May 2000
Navstar-IIR 5: 16 July 2000
Navstar-IIR 6: 10 November 2000
Navstar-IIR 7: 30 January 2001
Navstar-IIR 8: 29 January 2003
Navstar-IIR 9: 31 March 2003
Navstar-IIR 10: 21 December 2003
Navstar-IIR 11: 20 March 2004
Navstar-IIR 12: 23 June 2004
Navstar-IIR 13: 6 November 2004
All the satellites were launched on Delta-7925 from the Cape Canaveral launch centre, USA
Orbit: 20 200 × 20 200 km, 55.0°
Weight: 2032 kg each
Payload: Antennas to transmit L band frequencies: L1 = 1575.42 MHz and L2 = 1227.6 MHz with precise atomic clocks
Stabilization: three-axis stabilization
Operational life: design life of 10 years.

5.9.2.2 GPS-IIR-M (Navstar-IIR-M)

Block-IIR-M satellites are improved GPS-IIR satellites and will carry a new military code on both the frequencies (L1 and L2) and a new civilian code on the L2 frequency. Eight GPS-IIR-M satellites were launched up until June 2010.

Development Agency: General Electric Astrospace, USA (now Lockheed Missiles and Space)
Launch:
Navstar-IIR-M 1: 26 September 2005
Navstar-IIR-M 2: 25 September 2006
Navstar-IIR-M 3: 17 November 2006
Navstar-IIR-M 4: 17 October 2007
Navstar-IIR-M 5: 20 December 2007
Navstar-IIR-M 6: 15 March 2008
Navstar-IIR-M 7: 24 March 2009
Navstar-IIR-M 8: 17 August 2009
All the satellites have been launched from the Cape Canaveral launch centre, USA on Delta-7925.
Orbit: 20 200 × 20 200 km, 55.0°
Weight: 2032 kg each
Payload: Antennas to transmit L band frequencies: L1 = 1575.42 MHz and L2 = 1227.6 MHz and precise atomic clocks
Stabilization: three-axis stabilization
Operational life: design life of 10 years.

5.10 The Future of Satellite Navigation Systems

Satellite-based navigation systems are being further modernized so as to provide more accurate and reliable services. The modernization process includes development of newer satellite navigation systems, launch of new more powerful satellites, use of new codes, enhancement of ground system and so on. In fact satellite-based systems will be integrated with other navigation systems so as to increase their application potential.

The GPS system is being modernized so as to provide more accurate, reliable and integrated services to the users. The first efforts in modernization began with the discontinuation of selective availability feature, so as to improve accuracy of the civilian receivers. In continuation of this step, Block IIRM satellites carry a new civilian code on the L2 frequency. This will help in further improving the accuracy by compensating for atmospheric delays and will ensure more navigation security. Moreover, these satellites carry a new military code (M-code) on both the L1 and L2 frequencies. This will provide increased resistance to jamming. The satellites will also have more accurate clock systems. Block-IIF satellites have a third carrier signal, L5, at 1176.45 MHz. They will also have larger design life, fast processors with more memory and a new civil signal. GPS-III phase of satellites are in the planning stage. These satellites will employ spot beams. Use of spot beams results in increased signal power, enabling the system to be more reliable and accurate, with the system accuracy approaching a metre. The first GPS IIIA launch is scheduled for 2018. GPS IIIA series will be followed by GPS IIIF series satellites scheduled for launch during 2025–2034.

5.11 Space-Based Weapons

Space-based weapons are categorized as weapons that travel through space to strike their intended targets. The intended target may be located on the ground, in the air or in space. Space weapons include anti-satellite weapons that can target the space systems of the adversary from a ground-based, aerial or space-borne weapon system and also space-based weapon systems that attack targets on the ground or intercept missiles travelling through space. Space-based weapons have been the subject of intense discussion and debate among scientists, technologists, defence strategists and policy makers for more than 50 years. It began during pre-Cold War days when it was triggered by the possibility of bombardment of satellites carrying nuclear weapons.

The second time was during the period that followed the end of the cold war and this time it involved the possibility of spaced based defence against nuclear missiles. This period witnessed the Strategic Defence Initiative (SDI) programme of the United States. Today it is again an area of focused research and development activity for developed and some developing countries to offer defence against ballistic missiles, safeguard space assets and project force. In the following sections, we describe the different types of space weapons in terms of the technologies involved, international status, capabilities, limitations and deployment issues. Some prominent systems that are briefly discussed in terms of their features and facilities are also discussed in detail towards the end.

5.11.1 Classification of Space Weapons

Space weapons may be classified on the basis of physical location of the weapon and intended target as follows. Each of the three previously mentioned categories includes both kinetic as well as directed-energy weapons.

1) Space-to-Space weapons
2) Earth-to-Space weapons
3) Space-to-Earth weapons

5.11.1.1 Space-to-Space Weapons

The idea of using space platforms for military purposes has its origin in the Cold War era and was the brain child of the USA and the erstwhile Soviet Union. The Almaz programme of the then Soviet Union and the MOL programme of the USA exemplify the idea of use of manned space platforms for carrying out military missions. The Almaz programme of the Soviet Union comprised a series of military space stations called Orbital piloted stations (OPS). These space stations were launched under the cover of the Salyut programme as the Soviet authorities didn't want to disclose the existence of the top secret Almaz programme. As a consequence, the Almaz orbital piloted stations OPS-1, OPS-2 and OPS-3 were named Salyut-2, Salyut-3 and Salyut-5, respectively. Figure 5.22 shows the Almaz manned space station. OPS-1 (Salyut-2) was launched on 3 April 1973 from Baikonur, but days after the launch an accident left the spacecraft disabled and depressurized. OPS-2 (Salyut-3) was launched on 25 June 1974. OPS-2 was also deorbited in January 1975. OPS-3 (Salyut-5) was launched on 22 June 1976.

The space station was visited by two crews during 1976--1977. OPS-3 finally burned up in the Earth's atmosphere on 8 August 1977. The next Almaz space station OPS-4 that promised a number of upgrades never became a reality. This space station was to be the first space station to be launched with synthetic aperture radar (SAR) and a manned reusable return vehicle and there was a plan to replace the Shchit-1 defence gun with Shchit-2 space-to-space cannon. The space station has remained grounded with the result that OPS-3 remained the last manned space station under the Almaz programme.

Each of the Almaz space stations was equipped with a reconnaissance payload that comprised a colossal telescope called Agat-1, an optical sight that permitted the crew to come to a standstill over a facility and infrared and topographic cameras. The telescope was approximately 1 m in diameter and had a focal length of 6.4 m. The reconnaissance payload was used to take images of military installations such as airfields, missile complexes with a resolution better than 50 cm.

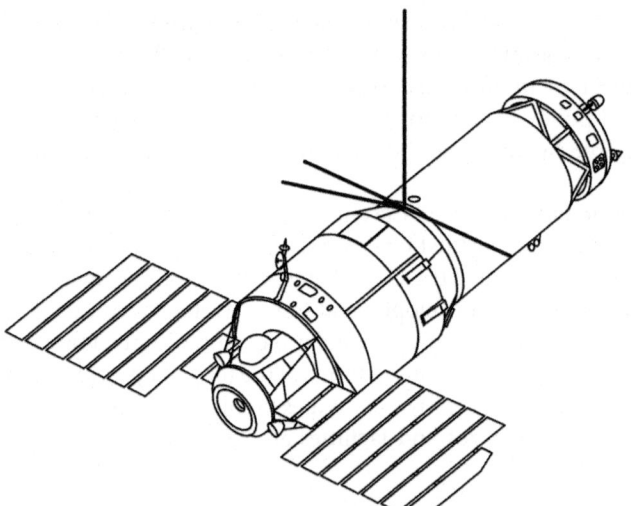

Figure 5.22 Almaz manned space station. (*Source:* Courtesy of NASA.)

Figure 5.23 Manned orbital laboratory. (*Source:* Courtesy of NASA.)

The data from the reconnaissance payload could also be relayed to the ground via a radio link. It appears that the camera films were developed on board. These were then scanned and transmitted to ground via the link. In addition to the reconnaissance payload described before, Almaz space stations were also reported to have been equipped with a 23 mm Nudelman Rikhter (NR-23) rapid-fire self-lubricating cannon capable of firing 950 rounds per minute. However, the entire station had to be reoriented toward the threat in order to aim the gun. It is reported that OPS-2 carried out a successful test firing on a target satellite.

The manned orbital laboratory (MOL) was proposed by the United States Air Force and was initially intended to test the military worthiness of humans in orbit. Figure 5.23 shows the MOL. The programme was planned as a successor to the cancelled X-20 Dyna-Soar project. It was thought having a man in loop would facilitate in-orbit repair, target selection and ability to shoot through cloud cover. The space station was configured around a modified Gemini-B spacecraft that could be attached to a laboratory vehicle. The space station was planned to be launched on board the Titan IIIC rocket. The space station was equipped with optical telescope and gyro stabilized cameras to be operated by astronauts to gather photo intelligence on Soviet military assets. The programme was launched in December 1963. One mock-up mission was launched on 3 November 1966. The proposed missions under the MOL programme included MOL-1 (1 December 1970), MOL-2 (1 June 1971), MOL-3 (1 February 1972), MOL-4 (1 November 1972), MOL-5 (1 August 1973), MOL-6 (1 May 1974) and MOL-7 (1 February 1975). MOL-1 and MOL-2 were proposed as unmanned missions while MOL-3, MOL-4, MOL-5, MOL-6 and MOL-7 were proposed as manned missions. The mission was cancelled in June 1969 due to budget constraints and the escalating war in Vietnam. Another reason for premature closure of the programme was the feeling that the features and facilities of unmanned spy satellites that followed thereafter met or exceeded the capabilities of manned MOL missions.

5.11.1.2 Earth-to-Space Weapons

Earth-to-space weapons are anti-satellite weapons that are designed to incapacitate or destroy satellites intended for strategic military applications. These satellites are mainly in low Earth orbits. Countries like the United States, Russia and China are believed to have developed and successfully field tested either kinetic energy or directed-energy weapon systems or both for anti-satellite applications. These weapon systems are both land-based as well as mounted on aerial platforms. These countries in the past have used these weapon systems to destroy their own satellites that have malfunctioned while in orbit and were rendered useless. Some of these experiments are briefly mentioned in the following paragraphs.

One such test was conducted by the United States on 13 September 1985 when an anti-satellite missile ASM-135 was used to destroy US satellite P78–1. P78–1, also known as Solwind, was launched on 24 February 1979 and was of the type of orbiting solar observatory (OSO) with a solar oriented sail. The payload comprised of a gamma ray spectrometer, a high latitude particle spectrometer, a white light spectrograph, an ultraviolet spectrometer, an aerosol monitor and an X-ray monitor. The satellite was the backbone of coronal research for more than 6 years. The satellite was brought down on 13 September 1985 using ASM-135 missile. ASM-135 (Figure 5.24) is an air-launched anti-satellite multi-stage missile that was first produced in 1984 and had a kinetic energy kill warhead. On 13 September 1985, ASM-135 was fired from an F-15A aircraft about 200 miles west of Vandenberg Air Force base and destroyed the Solwind satellite flying at an altitude of 345 miles.

Another test of same type was carried out on 21 February 2008 when the US spy satellite USA-193 was brought down using the RIM-161 standard missile 3 (RIM-161 SM-3). USA-193, also called NRO-21, was a US spy satellite launched on 14 December 2006 aboard the Delta-II rocket. The satellite malfunctioned shortly after deployment and was brought down intentionally on 21 February 2008. The satellite was shot down using an RIM-161 SM-3 missile (Figure 5.25) fired from a US warship near Hawaii. The exact purpose of the satellite was kept as a closely guarded secret but it is believed that the satellite carried high-resolution radar to generate images for the National Reconnaissance Office. RIM-161 is basically a ship-borne anti-ballistic missile that evolved from the well proven SM-2 Block-IV design. Like the Block-IV missile, it uses the same booster and a dual thrust rocket motor for the first and second stages. It also uses the same steering control section and guidance mechanism. The missile is equipped with a hit-to-kill kinetic energy warhead.

Figure 5.24 ASM-135 anti-satellite missile (Attribution: Lorax).

Figure 5.25 RIM-161 SM-3 missile. (*Source:* Courtesy of the US Navy.)

Russia too has been experimenting with the use of land-based and aerially delivered anti-satellite weapons of both kinetic energy and directed-energy types. The erstwhile Soviet Union tested ground-based lasers from the 1970s onwards for anti-satellite applications. A number of US spy satellites were reportedly blinded temporarily during the 1970s and 1980s. The Terra-3 programme is an example. The Terra-3 complex was a laser testing centre built in the 1970s on the Sary Shagan anti-ballistic missile testing range in Kazakhstan. The complex was equipped with high power/energy carbon dioxide and ruby lasers for anti-ballistic and anti-satellite applications. However, the laser energy from these sources was not sufficient for any anti-ballistic applications. Initial use therefore was limited to

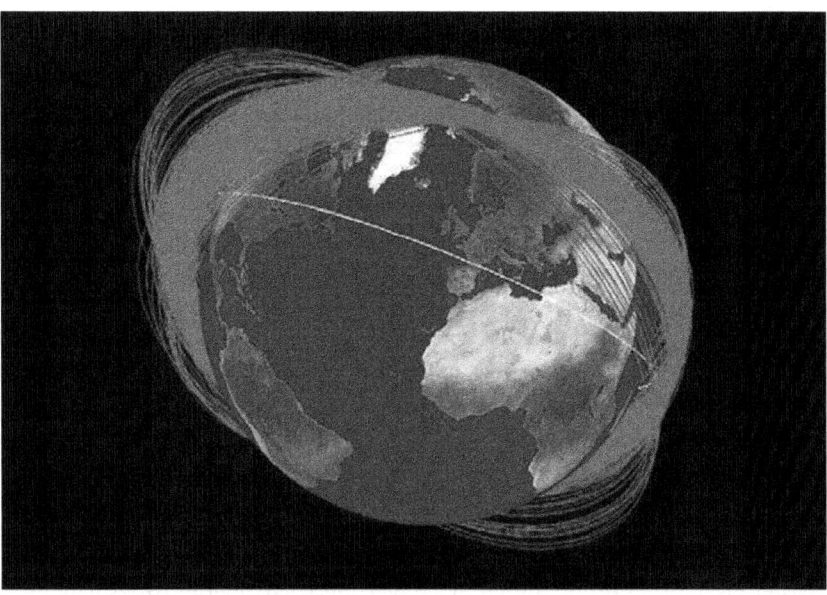

Figure 5.26 Orbital planes of the space debris of FY-1C weather satellite. (*Source:* Courtesy of NASA.)

anti-satellite applications primarily to blind sensors. One such experiment was executed on 10 October 1984 when a low-energy laser beam was directed at US space shuttle *Challenger* (OV-99) causing some of the onboard equipment to malfunction and also causing discomfort to crew members. The Soviet Union also researched directed-energy weapons under the Fon project from 1976 onwards. They also started development of air-launched anti-satellite weapons in early 1980s. Modified MiG-31 Foxhounds were used as the launch platform.

China has also successfully tested the anti-satellite missile, named SC-19, with a kinetic kill warhead. SC-19 has been reported to be based on a modified DF-21 ballistic missile or its commercial derivative KT-2. The ASAT missile is guided by an infrared imaging seeker. The test demonstrated use of a ground platform launched kinetic kill anti-satellite missile to destroy a near Earth orbit satellite. The satellite was a defunct Chinese weather satellite FY-1C in the Feng Yun series and the test was carried out on 11 January 2007 when the satellite in its polar orbit at 865 km was destroyed by a kinetic kill vehicle travelling with a speed of 8 km/s in the opposite direction. The missile was launched from a Transporter-erector-launcher (TEL) vehicle. Figure 5.26 shows the orbital planes of the space debris of the satellite one month after its disintegration by the Chinese ASAT missile.

The *Strategic Defensive Initiative* (SDI) programme of the United States, nicknamed 'Star Wars', proposed by the then US President Ronald Reagan on 23 March 1983 and with the objective of developing a defensive system to offer protection against enemy intercontinental ballistic missiles (ICBM) has also given a major boost to the ASAT programmes of the United States and Russia. While the ASAT projects were adapted for anti-ballistic missile applications, the reverse was also true. It may be noted that interception of a satellite with a static orbit is a much easier proposition than intercepting a warhead on a ballistic trajectory. This is mainly due to the low-level of uncertainty encountered in the case of satellite orbits and also due to the availability of relatively much longer tracking and manoeuvring times in an anti-satellite intercept.

5.11.1.3 Space-to-Earth Weapons

In the category of space-to-earth weapons, concepts of orbital weaponry and orbital bombardment have been designed by both the United States and the Soviet Union during the cold war era. The fractional orbital bombardment weapon system deployed by the Soviet Union during 1968–1983 is one such system. In this system, a nuclear warhead could be placed in a low Earth orbit and then later at the time of strike deorbited to hit any location on the surface of the Earth. Presently, there are no known operative orbital weapons. This has been largely due to the coming into existence of several international treaties prohibiting deployment of weapons of mass destruction in space. Fractional orbital bombardment system was also phased out in 1983. However, other weapons like kinetic bombardment weapons do exist as they don't violate these treaties.

The *Space-Based Laser* (SBL) programme of the United States is a technology demonstration programme with the objective of establishing the capability of shooting down a ballistic missile in its boost phase with a space-based high power laser. SBL aims to providing global boost phase intercept of ballistic missiles. Under the programme, it is proposed to put an experimental high power laser system into space and follow it up with the experiment of shooting down a missile. The outcome of this experiment, known as the Integrated Flight Experiment (IFX), is likely to determine the efficacy of SBL to protect the United States and its allies from ballistic missile threat as a part of layered defence.

Another space-based laser programme aimed at converting solar energy to laser light in space is the collaborative effort of the Japan aerospace exploration agency (JAXA) and Osaka University. This space generated laser light could then be transmitted to the Earth to generate electricity or to power a massive 'death ray'. It is estimated to put this novel laser system into space by 2030. Figure 5.27 shows the concept.

5.11.2 Strategic Defence Initiative

The Strategic Defence Initiative (SDI) was the brain child of the then US President Ronald Reagan. The programme was unveiled by him on 23 March 1983 through which he proposed to use ground-based and space-based systems to offer protection to the United States

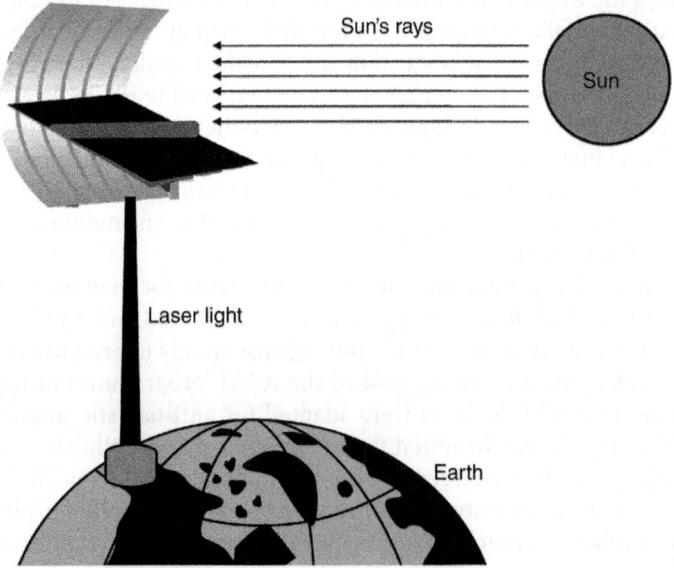

Figure 5.27 Collaborative Space based laser concept of JAXA and Osaka University.

and its allies from strategic nuclear warhead equipped ballistic missiles. The programme was nicknamed the Star Wars programme after the popular 1977 film directed by George Lucas. The SDI programme as envisaged by the US President was studied in detail by the Strategic Defence Initiative Organization (SDIO) set up in 1984 within the United States Department of Defense. Defence strategists and scientists described the programme as highly ambitious and felt that its implementation was not feasible with the then existing technology. Subsequently in 1993, the programme was renamed as the Ballistic Missile Defence Organization (BMDO) by the then US President Bill Clinton. The programme was modified with the emphasis shifting from national missile defence to theatre missile defence and its scope reduced from global coverage to regional coverage. Though the programme was never fully realized as envisaged, the research work carried out and the technologies developed under the programme have led to development of some of the contemporary anti-ballistic missile systems. The SDI programme witnessed the initiation and development of many technologies and products, some successful and some not-so-successful and some unsuccessful, which included ground-based programmes, directed-energy weapon (DEW) programmes, space programmes, sensor programmes and countermeasure programmes. In the following paragraphs, the major technologies and systems initiated under the SDI programme are briefly discussed.

5.11.2.1 Ground-Based Programmes

Prominent ground-based programmes included the Extended-Range Interceptor (ERINT), Homing Overlay Experiment (HOE) and Exoatmospheric Re-entry Vehicle Interception System (ERIS). Each one of these is briefly described in the following paragraphs.

5.11.2.1.1 *Extended-Range Interceptor (ERINT)*

The ERINT programme was an extension of the Flexible lightweight agile guided experiment (FLAGE) involving the development of a small, agile radar homing hit-to-kill vehicle. FLAGE was tested successfully by targeting a MGM-52 Lance missile in flight. The test was conducted at the White Missile Range in 1987. ERINT was the follow-on to the FLAGE experiment. ERINT used a new solid propellant rocket motor, which allowed the missile to fly faster and higher than FLAGE. ERINT also had an upgraded design including addition of aerodynamic manoeuvring fins and attitude control motors, which increased the range. With the new guidance technology, the missile was designed to be used primarily against manoeuvring tactical missiles and secondly against air breathing aircraft and cruise missiles. The first flight test of ERINT (Figure 5.28) was conducted at the White Sands Missile Range in June 1992 followed by another successful test in August 1992. These two preliminary tests did not attempt to hit target missiles. Preliminary testing with three direct hits simulating theatre missile defence was concluded in November 1993. This was followed by another test in June 1994, when it was used to destroy a drone to establish the accuracy of its guidance system. ERINT was subsequently selected as the new missile for the Patriot advanced capability-3 system (PAC-3) mainly because of its increased range, accuracy and lethality, all in a smaller package.

5.11.2.1.2 *Homing Overlay Experiment (HOE)*

The HOE by the US Army was the first to demonstrate the concept of exoatmospheric hit-to-kill to intercept and destroy ballistic missiles. The US army started a technology demonstration programme in the mid-1970s to validate the emerging technologies designed to have nonnuclear hit-to-kill intercepts of Soviet ballistic missiles in space. Planning began in 1976 and the contract for development of interceptor was awarded to Lockheed–Martin in August 1978. The interceptor of the HOE programme consisted of Minuteman-I launch stages

Figure 5.28 Flight test of ERINT. (*Source:* Courtesy of the US Army.)

carrying the homing and kill vehicle. The Kinetic Kill Vehicle (KKV) was equipped with an infrared seeker, guidance electronics and a propulsion system. The infrared seeker allowed the interceptor to guide itself into the path of an incoming ballistic missile warhead and collide with it. Four flight tests were carried out in February 1983, May 1983, December 1983 and June 1984. Each of the four tests involved launching a target from Vandenberg Air Force Base in California and an HOE interceptor from the Kwajalein Missile Range in the Republic of the Marshall Islands in the Pacific. Figure 5.29 shows the HOE test vehicle. The first three tests did not achieve a successful intercept with the targeted vehicle. In the fourth test in June 1984, the kinetic kill vehicle interceptor did find the Minuteman intercontinental ballistic missile re-entry vehicle in space and guided itself to an intercept and finally destroyed the target through collision. Both target and interceptor had sensors, which along with ground-based radars and airborne optical sensors produced data to show that the target was destroyed by the collision of the interceptor and not by an explosive charge after a near miss. The data also produced evidence that the interceptor guided itself to the target with the help of its infrared homing sensor. Using an explosive charge to destroy the target in the event of a near miss was a part of the deception programme, which was reportedly discontinued before the third flight test. In the first two tests in place though, it could not alter the result as the interceptor missed the target by large distances.

The intercept vehicle had a fixed fragment net intended to increase the lethal radius of the interceptor. It consisted of 36 aluminium ribs with stainless steel fragments that increased the interceptor size to achieve greater probability of target hit. The structure of the ribs was kept folded in flight and was deployed shortly before intercept. Once deployed, this umbrella-like web had a spread of about 4 m diameter.

Figure 5.29 Homing Overlay Experiment (HOE) test vehicle. (*Source:* Cliff, https://commons.wikimedia.org/wiki/File:ABM_test_vehicle_(Homing_Overlay_Experiment).jpg. CC BY 2.0.)

5.11.2.1.3   *Exoatmospheric Re-Entry Vehicle Interception System (ERIS)*

Development of ERIS began in 1985. The ERIS programme was an extension of the HOE programme and was built on the technologies tested during the HOE programme. ERIS was made up of the second and third stages of Minuteman ICBM and had a kill vehicle equipped with a long wave infrared scanning seeker. The sensor and guidance technology of the ERIS KKV (kinetic kill vehicle) was based on the experience learned from the HOE tests. The ERIS KKV with its inflatable octagonal kill enhancer was significantly smaller and lighter than the HOE KKV.

The first test of the ERIS KKV was conducted on 28 January 1991. The intercept vehicle successfully detected and intercepted a mock ICBM warhead launched from Vandenberg Air Force Base. It was the first time that an SDI experiment attempted an interception in a countermeasures environment by discriminating against decoys. The target re-entry vehicle deployed two balloon decoys on either side. The KKV was pre-programmed to hit the centre target that was the warhead.

The second and final test was conducted on 13 May 1992, when the intercept vehicle was targeted against a Minuteman-I ICBM. Though the test was a partial failure and the kill vehicle did not achieve a direct intercept, nevertheless the test met the primary targeted objectives of collection of radiometric data on the target and decoys, acquisition and resolution of threat and demonstration of target handover. Two of the originally planned four tests were cancelled. Due to the change in the global situation after the end of the cold war, the SDI programme was reoriented in the early 1990s and the ERIS programme was not developed into an operational system. The experiences of the ERIS programme were used to advantage in the successful development and deployment of the next-generation of exoatmospheric kill vehicles.

5.11.2.2   Directed-Energy Weapon Programmes

The prominent directed-energy weapon programmes included a nuclear explosion powered X-ray laser cluster aimed at targeting multiple warheads simultaneously, a chemical laser for use as anti-ballistic missile and anti-satellite weapon, a particle beam accelerator and a hyper-velocity rail gun. Each of these programmes is briefly described in the following paragraphs.

5.11.2.2.1 Nuclear Explosion Powered X-Ray Laser

The programme involved development of a nuclear explosion powered cluster of X-ray lasers that would be deployed using a series of submarine launched missiles or satellites. This curtain of nuclear energy powered X-ray lasers was intended to be used to shoot down many incoming warheads simultaneously. The first test, known as the Cabra event, was performed in March 1983 and was a failure. The failure of the first test was one of the primary reasons for opposition to the programme from critics who argued that X-ray lasers would not offer any significant advantage as an option for ballistic missile defence. However the programme offered many spin-off benefits. The knowledge gained from the programme led to the development of X-ray lasers for biological imaging, 3D holograms of living organisms and advanced materials research.

5.11.2.2.2 Chemical Lasers

Under this programme, SDIO (Strategic Defence Initiative Organisation) funded the development of a Deuterium Fluoride (DF) laser system called Mid-Infrared Advanced Chemical Laser (MIRACL). The MIRACL system (Figure 5.30) was first tested in 1985 in a simulated setup at the White Sands Missile Range. The test setup simulated the conditions the booster was likely to be in during the boost phase of its launch. The laser was subsequently tested on drones simulating cruise missiles with some success. The laser was also tested on an US Air Force satellite to demonstrate its capability as anti-satellite weapon, though with mixed results. The technologies developed during the MIRACL programme were subsequently used to develop the Tactical High-Energy Laser (THEL) system, which is in use against artillery shells. Airborne Laser (ABL) and Advanced Tactical Laser (ATL) are the other key chemical laser systems that have been successfully developed and tested after the closure of SDI. Both ABL and ATL are Chemical Oxy-Iodine Laser (COIL) systems configured on aerial platforms.

Figure 5.30 MIRACL system. (*Source:* Courtesy of the US Army.)

5.11.2.2.3 Particle Beam Accelerator

This is a programme aimed at establishing the operation of particle beam accelerators in space called BEAR (Beam Experiment Aboard Rocket) using a sounding rocket to carry a neutral-particle beam accelerator into space. The experiment conducted in July 1989 successfully established that a particle beam would propagate in space as predicted. A spin-off of the technology was its use for management of nuclear waste by reducing the half-life of nuclear waste using transmutation technology driven by an accelerator.

5.11.2.2.4 Hypervelocity Rail Gun

The SDI hypervelocity rail gun experiment was named the Compact High-Energy Capacitor Module Advanced Technology Experiment (CHECMATE). A hyper velocity rail gun is similar to a particle accelerator in the sense that it converts electrical potential energy into kinetic energy that is imparted to the projectile. It differs from conventional mass accelerators as here no gases are used. It differs from conventional electromagnetic accelerators in the sense that in the case of rail gun, the magnetic field trails behind the projectile at all times. A conductive pellet, which constitutes the projectile in this case, is attracted down the rails by the magnetic forces produced as a result of a gigantic current impulse of the order of hundreds of thousands of amperes flowing through the rail thereby generating muzzle velocities greater than 35 km per second.

Hypervelocity rail guns were considered as an attractive alternative to the space-based defence system because of their projected capability to quickly shoot at multiple targets. There are however many technological challenges. Early prototypes were essentially single use weapons due to rapid erosion of rail surfaces as a result of very high values of current and voltage. Another challenge is the survivability of projectile, which experiences acceleration force of greater than 100 000 g. Any onboard guidance system would also need to withstand same level of acceleration force.

5.11.2.3 Space-Based Programmes

Space-based programmes under the SDI saw the development of space-based interceptors. One such activity was a nonnuclear system of satellite-based miniature missiles called Brilliant Pebbles. These mini missiles used high velocity kinetic energy warheads. The system was designed to operate in conjunction with the Brilliant Eyes sensor system to detect and destroy the target missiles. The Brilliant Pebbles system was designed and developed by Lawrence Livermore National Laboratory during the period 1988–1994.

5.11.2.4 Sensor Programmes

Prominent activities under the SDI's sensor programme included the Boost Surveillance and Tracking System (BSTS), Space Surveillance and Tracking System (SSTS) and Brilliant Eyes. BSTS was designed to assist detection of missiles during the boost phase. SSTS was originally designed to track ballistic missiles during the midcourse phase. The Brilliant Eyes system was a derivative of SSTS and was designed to operate in conjunction with the Brilliant Pebbles system. Yet another programme that was used to test several sensor related technologies was the Delta 183 programme. The programme was so named as per the designation of the launch vehicle. The Delta 183 programme was initially conceived as a collaborative effort between the erstwhile Soviet Union and the United States. The Soviet Union subsequently withdrew from the programme and the United States proceeded without Soviet participation. The programme was reconfigured to carry several sensor payloads, which included an ensemble of imagers and photo sensors covering visible and ultraviolet bands, long wave infrared imager, laser detection and ranging device and a UV intensified CCD video camera. Figure 5.31 shows

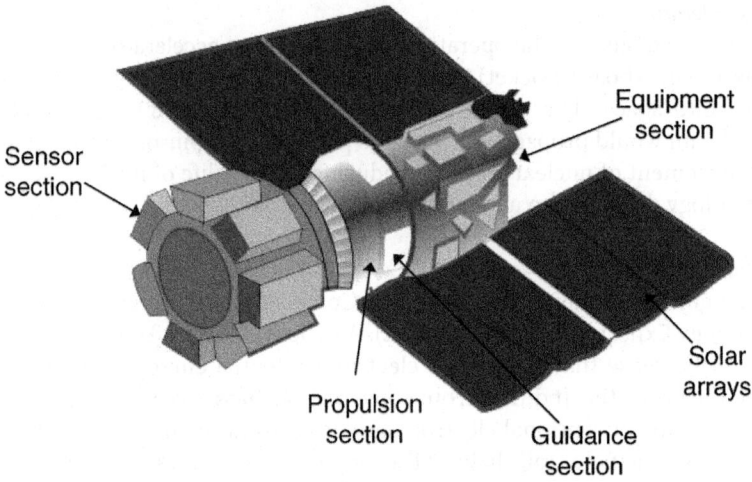

Figure 5.31 Delta Star spacecraft.

the exploded view of the Delta Star spacecraft. The long wave infrared imager was adapted from the guidance and control section of a Maverick missile. Different sensor payloads on board Delta Star were used to observe several missile launches. A great deal of data was generated on the performance of sensors. In some of these experiments, sensor performance was evaluated in the presence of countermeasures. The countermeasures scenario was created by the release of liquid propellant during launch of the missile.

5.11.3 Directed-Energy Laser Weapons

Kinetic energy weapons transport mass to target in order to cause the destructive effect. Kinetic energy weapons, unguided or guided, have their respective advantages and disadvantages. However they have a common drawback, which is inherent in the mode of their travel from source to target and the mechanism of transfer of energy to the target. Both types transfer the energy to the target through a physical object such as a projectile, which must travel a certain distance through the medium from source to target. One would like the time taken by the projectile to travel from the launch source to the target to be as short as possible. However, practical considerations put a limit on the maximum possible projectile velocity and hence the minimum achievable travel time. Efforts are on to increase the projectile velocity by developing a device called a rail gun that employs plasma driven by a magnetic field to accelerate the projectile to velocities exceeding 40 km/s. Use of high-energy laser weapons overcomes all the limitations of conventional kinetic energy weapons besides offering many new advantages. Belonging to the category of directed-energy weapons, these high-energy laser weapons once deployed on a mass scale will render obsolete many weapon systems hitherto considered unbeatable. Directed-energy laser weapons are described in detail in Chapter 12.

Illustrated Glossary

Boost Surveillance and Tracking System (BSTS) BSTS was designed to assist detection of missiles during the boost phase.

Brilliant Eyes System The Brilliant Eyes system was a derivative of SSTS and was designed to operate in conjunction with the Brilliant Pebbles system.

COMINT (COMmunication INTelligence) Satellites These satellites perform covert interception of foreign communications in order to determine the content of these messages. As most of these messages are encrypted, they use various computer-processing techniques to decrypt the messages.

DSCS Satellites DSCS stands for defence satellite communication systems. Launched by the USA, satellites in this series are intended for providing wideband military communication services.

DMSP Satellites DMSP stands for defence meteorological satellite programme. It is an American military weather forecasting satellite programme.

Early Warning Satellites These provide timely information on the launch of missiles, military aircraft and nuclear explosions to military commanders on the ground.

Earth-to-Space Weapons These are anti-satellite weapons that are designed to incapacitate or destroy satellites intended for strategic military applications. These satellites are mainly in low Earth orbits.

Electro-Optical Satellites These provide full-spectrum photographic images in the visible and the IR bands.

ELINT (Electronic Intelligence) Satellites These are used for the analysis of non-communication electronic transmissions. This includes telemetry from missile tests (TELINT) or radar transmitters (RADINT).

ERINT (Extended-Range Interceptor) This programme was an extension of the Flexible lightweight agile guided experiment (FLAGE) involving the development of small, agile radar homing hit-to-kill vehicle. ERINT was the follow-on to the FLAGE experiment. ERINT used a new solid propellant rocket motor, which allowed the missile to fly faster and higher than FLAGE. ERINT also had an upgraded design including addition of aerodynamic manoeuvring fins and attitude control motors, which increased the range.

Exoatmospheric Re-entry Vehicle Interception System (ERIS) The ERIS programme was an extension of the HOE programme and was built on the technologies tested during the HOE programme. ERIS was made up of the second and third stages of Minuteman ICBM and had a kill vehicle equipped with a long wave infrared scanning seeker.

Global Positioning System (GPS) GPS, also known as the Navigation Satellite Timing and Ranging (NAVSTAR), is a US satellite-based navigation system developed by the US Department of Defense (DOD). Originally envisioned primarily as a military system, GPS is now a dual-use system, used for both military as well as civilian applications. GPS navigation system employs a constellation of 24 satellites and ground support facilities to provide three-dimensional position, velocity and timing information to all the users worldwide 24 h a day.

GLONASS Satellite Navigation System GLONASS (Global Navigation Satellite System) is the Russian navigation satellite system. The system is a counterpart to the US GPS system and shares the same principles in data transmission and positioning methods. It is managed for the Russian federation Government by the Russian space forces and the system is operated by the Coordination Scientific Information Centre (KNITs) of the ministry of Defence of the Russian federation. The space segment of GLONASS is formed by 24 satellites (21 operational and three on-orbit spares) located in three orbital planes in a circular orbit of altitude 19 100 km and with an inclination of 64.8°. The three orbital planes are separated by 120°, and the satellites within the same orbital plane are separated by 45°.

Homing Overlay Experiment (HOE) The HOE of the SDI programme was used to demonstrate the concept of exoatmospheric hit-to-kill to intercept and destroy ballistic missiles.

Hyper Velocity Rail Gun These were considered as an attractive alternative to the space-based defence system because of their projected capability to quickly shoot at multiple targets. A hyper velocity rail gun is similar to a particle accelerator in the sense that it converts electrical potential energy into kinetic energy that is imparted to the projectile. It differs from conventional mass accelerators as here no gases are used. It differs from conventional electromagnetic accelerators in the sense that in the case of rail gun, the magnetic field trails behind the projectile at all times.

IMINT (Image Intelligence) Satellites IMINT satellites provide detailed high-resolution images and maps of geographical areas, military installations and activities, troop positions and other places of military interest.

Mid-Infrared Advanced Chemical Laser (MIRACL) The Mid-Infrared Advanced Chemical Laser (MIRACL) is a Deuterium Fluoride laser-based high power laser system. The MIRACL system was first tested in 1985 in a simulated setup at the White Sands Missile Range. The technologies developed during the MIRACL programme were subsequently used to develop the Tactical High-Energy Laser (THEL) system, which is in use against artillery shells. Airborne Laser (ABL) and Advanced Tactical Laser (ATL) are the other key chemical laser systems that have been successfully developed and tested after the closure of SDI.

Military Communication Satellites These satellites link communication centres to the front line operators.

Military Navigation Satellites Navigation systems pinpoint the exact location of soldiers, military aircraft, military vehicles and so on. They are also used to guide a new generation of missiles to their targets.

Military Weather Forecasting Satellites These provide weather information, which is very useful in planning military operations.

MILSATCOM Architecture MILSATCOM architecture was proposed in USA in 1976 to guide the development of military satellite communication systems in the country. Three types of military systems were proposed to be developed under this architecture; namely wideband, mobile and tactical (or narrowband) and protected (or nuclear capable) systems.

MILSTAR Satellites MILSTAR satellites are American military communication satellites belonging to the category of protected satellite systems.

Protected Satellite Systems These provide communication services to mobile users on ships, aircraft and land vehicles.

PHOTOINT or Optical Imaging Satellites These satellites have visible light sensors that detect missile launches and take images of enemy weapons on the ground.

Reconnaissance Satellites Also known as spy satellites, these provide intelligence information on the military activities of foreign countries.

SIGINT (Signal Intelligence) Satellites These satellites detect transmissions from broadcast communication systems such as radar, radio and other electronic systems. They can also intercept and track mobile phone conversations, radio signals and microwave transmissions.

Space-Based Laser (SBL) The SBL programme of the United States is a technology demonstration programme with the objective of establishing the capability of shooting down a ballistic missile in its boost phase with a space-based high power laser.

Space Surveillance and Tracking System (SSTS) SSTS was originally designed to track ballistic missiles during the midcourse phase.

Space Weapons These are categorized as weapons that travel through space to strike their intended targets. The intended target may be located on the ground, in the air or in space. Space weapons include anti-satellite weapons that can target the space systems of the adversary from a ground-based, aerial or space-borne weapon system and also space-based weapon systems that attack targets on the ground or intercept missiles travelling through space.

Strategic Defence Initiative (SDI) Programme SDI was the brain child of Ronald Reagan, then US President. The programme was unveiled by him on 23 March 1983 through which he proposed to use ground-based and space-based systems to offer protection to the United States and its allies from strategic nuclear warhead equipped ballistic missiles.

Tactical Satellite Systems These are used for communication with small mobile land-based airborne and ship-borne tactical terminals.

Trilateration Principle In trilateration-based satellite navigation systems, the user receiver's position is determined by calculating its distance from three (or four) satellites whose orbital and the timing parameters are known. The receiver is at the intersection of the invisible spheres, with the radius of each sphere equal to the distance between a particular satellite and the receiver, with the centre being the position of that satellite.

Vela Satellites These are US satellites of the 1960s intended for detection of a nuclear explosion.

Wideband Satellite Systems These systems provide point-to-point or networked moderate to high-data rate communication services at distances varying from in-theatre to intercontinental distances.

Bibliography

1 Burkett, D.L. (1989) *The U.S. Anti-Satellite Programme: A Case Study in Decision Making*, National Defence University, National War.

2 Gatland, K. (1990) *Illustrated Encyclopaedia of Space Technology*, Crown, New York.

3 Long, F.A., Hafner D. and Boutwell, J. (1986) *Weapons in Space*, W.W. Norton & Company, New York.

4 Vacca, J. (1999) *Satellite Encription*, Academic Press, California.

5 Verger, V., Sourbes-Verger, I., Ghirardi, R., Pasco, X., Lyle, S. and Reilly, P. (2003) *The Cambridge Encyclopaedia of Space*, Cambridge University Press.

6

Electronic Warfare

Electronic warfare systems' classifications, involved technologies and systems are comprehensively described first in this chapter. Major topics covered under electronic warfare systems include different categories of electronic warfare systems; electronic support measures (ESM) such as signal intelligence, radiation intelligence and telemetry intelligence; passive and active electronic countermeasures (ECM) such as chaff, decoys and various types of jammers and electronic counter-countermeasures (ECCM). Stealth technologies are also discussed in the chapter. Salient features of major international electronic warfare systems and their deployment scenarios is another highlight of this chapter. The next major topic discussed here relates to Electro-Optic Countermeasures (EOCM). EOCM systems play an important role in the present day warfare due to widespread use of lasers and other electro-optic systems. Both passive as well as active electro-optic countermeasures will be discussed in this chapter with particular emphasis on laser warning and countermeasures and missile approach warning sensor and infrared (IR) countermeasures. Active protection systems are briefly discussed towards the end of the chapter.

6.1 Introduction to Electronic Warfare

As stated above, electronic warfare refers to weapons or warfare disciplines in which one side uses equipment and techniques to deny to the opposition use of their systems of defence and attack on land, in the air and at sea without compromising the full use of their own systems meant for the same purpose. The methods, for instance, used by an attack aircraft to temporarily disable or confuse the enemy's radars to avoid detection, which would enable it to penetrate deep into enemy territory without making its intentions known, would be classified under the heading of electronic warfare. Similarly, the equipment used on land, in the air and at sea to provide information on the capabilities of enemy's systems such as their communication systems, radars, weapons and so on would also be categorized as EW equipment. Knowledge of the type of sensor or weapon, its location and information, such as a fire-control radar being locked onto the aircraft or a missile has been launched towards the target, allows air crew to avoid radar cover or to take evasive action.

Another example of EW is use of techniques to make your command/communication systems, radars, missiles and so on immune to the tactics employed by the enemy's forces to confuse, disable or deceive your systems. For instance, if your radar is so designed that it continues to do its job in spite of passive and active techniques employed by a target aircraft, use of such techniques would also fall in the category of electronic warfare. Electronic warfare

Handbook of Defence Electronics and Optronics: Fundamentals, Technologies and Systems, First Edition. Anil K. Maini.
© 2018 John Wiley & Sons Ltd. Published 2018 by John Wiley & Sons Ltd.

therefore spans from electronic reconnaissance used to gather information on enemy's command and communication setup, radars and other sensors to use of advanced designs for these sensors to make them completely immune to these reconnaissance methods and information gathering systems and further to the use of passive and active techniques to avoid identification and detection by enemy's sensors.

6.2 Types of Electronic Warfare Systems

In order to meet the primary objective of electronic warfare techniques and systems to deny to the enemy the effective use of electromagnetic spectrum and to protect the friendly systems using electromagnetic spectrum against an EW attack, EW techniques and equipment can be grouped into the following three broad categories.

1) Electronic Support Measures (ESM) or Electronic Warfare Support (ES)
2) Electronic Countermeasures (ECM) or Electronic Attack (EA)
3) Electronic Counter-Countermeasures (ECCM) or Electronic Protection (EP).

6.2.1 Electronic Support Measures (ESM) or Electronic Warfare Support (ES)

ESM or ES refer to that division of electronic warfare (EW), which involves actions either tasked by or under the direct control of operational commander taken to search for, intercept, locate and identify sources of unintentional or intentional radiated electromagnetic energy for the purpose of immediate threat recognition. EW support equipment and technologies that operate in the radio frequency (RF) region and also those equipment and technologies that operate in optical region of electromagnetic spectrum come under this category. While Radar Warning Receivers (RWRs) and electronic reconnaissance systems such as Electronic Intelligence (ELINT) and Communications Intelligence (COMINT) are examples of ES equipment operating in the RF domain, Night Vision Devices (NVDs, Electro-Optical Imaging Systems, Laser-Warning Receivers (LWRs), Missile Approach Warning Sensors (MAWS) and InfraRed Search and Track (IRST) sensors belong to the type of ES equipment operating in optical region of electromagnetic spectrum. Though electronic support measures provide a source of information for immediate action involving electronic countermeasures and counter-countermeasures, evasive actions, targeting and so on, and mainly comprise of various types of intercept receivers including RWRs and LWRs, the systems of electronic reconnaissance such as Communications Intelligence (COMINT), Electronic Intelligence (ELINT), also considered to belong to the category of electronic support measures, are mainly used for analysing and identifying the intercepted transmission in terms of, for example, operating frequency, bandwidth, modulation, polarization and so on.

6.2.2 Electronic Countermeasures (ECM) or Electronic Attack (EA)

Having detected, identified and located threatening sensors such as radars or weapons such as missiles, the *ECM* or *EA* systems do the job of attacking personnel, equipment and facilities with the intent of degrading, neutralizing or completely destroying adversary's combat potential. Degradation or neutralization of performance, for example, could lead to reduction in the detection range of the radar or deceiving the guided missile or laser-guided munitions away from its intended path. Electronic countermeasures use active techniques to achieve this. It could be some kind of jamming technique like noise jamming for radars or a high-energy laser used to saturate the front-end sensor of the receiving channel of a laser target designator. Weapons could be made ineffective by using some kind of deception technique or deploying decoys. Fundamentally, effective deployment of electronic countermeasures buys time for the user to evade enemy's defences and succeed in its mission. Different types of radio frequency (RF) jammers used against radars, high-energy electro-optic countermeasures (EOCM) class lasers used against electro-optical equipment, directed infrared countermeasures (DIRCM) equipment used against IR-guided missiles all are examples of active electronic countermeasures. Use of smoke and aerosol screens against approaching laser-guided munitions represents passive electronic countermeasures.

6.2.3 Electronic Counter-Countermeasures (ECCM) or Electronic Protection (EP)

While the basic purpose of using electronic countermeasures is do deny the radar detection of target or deceive it to follow a wrong target or an electronically created non-existent target, the purpose of ECCM techniques is to defeat this and make systems such as radars, missiles and others immune to countermeasures. This is achieved by using techniques to nullify his jamming and/or decoy systems. ECM and ECCM are intimately connected. It is an extremely sensitive area in that any disclosure of ECCM or EP measures designed into a system is likely to inform the enemy of its vulnerability to ECM or EA. Similarly, information on the counter-measures adopted by a side enables the other side to employ effective counter-countermeasures against it. It is an unending evolutionary process. Every countermeasure action is to lead to a counter countermeasure action that would make it immune to the countermeasure adopted by the adversary. Similarly, for every counter-countermeasures action, there has to be a countermeasure that would defeat the counter countermeasure. Use of frequency agility in which the transmissions are made to hop over a large frequency band in a random fashion is an effective and common ECCM technique against jamming systems. In this case, the jammer is forced to spread its power over the entire band causing loss of jamming signal strength on desired on a particular frequency thereby degrading its efficacy. Use of stealth technologies such as designing the platform structure to avoid sharp corners and flat surfaces and use of radar and laser absorbent materials aimed at minimizing radar and IR signatures of military platforms is another important ECCM feature.

6.3 Electronic Support Measures

ESMs refer to the branch of electronic warfare that includes techniques and equipment used for searching, intercepting and identifying the type and location of radiated electromagnetic energy for the purpose of immediate threat recognition. Electronic reconnaissance used for gathering information on the technical characteristics, deployment mode and operational capabilities of both the potential allies as well as potential enemies are vital components of ESMs. The other equipment that is included in the category of ESM, as outlined in the previous section, are mainly the different types of warning receivers such as radar warning receivers, laser-warning receivers, missile approach warning receivers and so on. ESM equipment operating in RF region of electromagnetic spectrum, which includes electronic reconnaissance equipment such as ELINT and COMINT equipment, and Radar Warning Receivers are discussed in this section. Laser-Warning Sensors and Missile Approach Warning Sensors (MAWS) are discussed in the subsequent sections.

6.3.1 Electronic Reconnaissance

Success in electronic warfare to a great extent depends upon comprehensive knowledge and understanding of various aspects of the electronic environment that is likely to prevail in the battlefield. Electronic reconnaissance is defined as the detection, identification, evaluation and location of foreign electromagnetic radiation emanating from sources other than radioactive phenomenon or nuclear detonations. Electronic reconnaissance is further classified as follows.

1) Radiation Intelligence (RINT)
2) Telemetry Intelligence (TELINT)
3) Signal Intelligence (SIGINT).

RINT is basically the intelligence derived from collection and analysis of non-information bearing electromagnetic radiation emitted unintentionally by foreign equipment, devices and so on other than those generated by detonation of nuclear weapons. *TELINT* is the intelligence derived from the interception and analysis of foreign telemetry. *SIGINT* is further subdivided into *Electronic Intelligence* (ELINT) and *Communications Intelligence* (COMINT). While ELINT is the intelligence information derived from non-communications type electromagnetic radiation emanating from foreign sources other than radioactive sources or nuclear detonations; COMINT is the intelligence information derived from communications type electromagnetic radiation emanating from foreign sources by those other than the intended recipients.

6.3.2 Signal Intelligence (SIGINT)

The SIGINT system is usually an integrated setup of high performance antennas, receivers, signal processing equipment, wide band recorders and so on. SIGINT equipment should not be confused with other forms of equipment such as radar, laser-warning sensors and so on used as electronic support measures because the two types widely differ in purpose. While the ESM equipment serves the main purpose of providing information that would lead to an immediate action in the form of initiating electronic countermeasures or counter-counter-measures or even in the form of tactical deployment of forces; SIGINT equipment basically provides both communications and non-communications information for a long term strategic use including generation of electronic data bases to support tactical operations.

The SIGINT system can be installed on a variety of ground-based, airborne or sea platforms. The location and type of platform and the nature of SIGINT equipment depends upon the location and technical characteristics such as emission frequency of the target emitters and also the geographical and political constraints faced by the side gathering information. If the target emitter was radiating around the HF range, the information could be collected by the SIGINT station from a distance far more than the radio horizon. If the target emitter used HF sky wave propagation, the signal from the emitter could be intercepted from a distance of thousands of kilometres. For targets emitting in higher frequency, interception and location is limited to line-of-sight distances. Airborne and space-borne platforms extend the radio horizon still further. Many developed countries have satellite-based SIGINT collection capability. Satellites can overfly the intended enemy territory during peace time as well as war time to gather intelligence. Functionally, the SIGINT equipment on the satellite will be similar to that carried in an aircraft. Since the satellite orbits can be predicted to a very good accuracy, SIGINT can be denied to the interceptor by switching off the transmitter while the satellite is above the horizon. Satellites with SIGINT payload were discussed earlier in Section 5.5 in Chapter 5 on *Military Satellites*. SIGINT, as outlined in an earlier paragraph, has further two sub-divisions; namely ELINT and COMINT. Each of them is discussed briefly in the following paragraphs.

6.3.2.1 Electronic Intelligence (ELINT)

ELINT relates to collection and analysis of information of intelligence value extracted out of electromagnetic radiations emanating from non-communications and non-atomic detonation foreign sources. Some common examples of electromagnetic radiations originating from non-communications and non-atomic detonation sources include radiations from missiles and missile guidance devices, radiations from developmental laboratories and field testing stations working on electronic devices, radars, navigational aids, guided missiles, guided missile launchers, air-to-air or air-to-ground identification signals and so on. The raw data collected by ELINT equipment is comprehensively processed to identify the characteristics of the foreign source of radiation. For example, by intercepting and analysing electromagnetic radiation emanating from radar, its characteristics such as operating wavelength or frequency, modulation, pulse repetition rate or pulse repetition interval, pulse duration or pulse width, pulse shape, antenna pattern characteristics, pulse-to-pulse signal amplitude variation and so on and direction of arrival can be estimated and the radar can be identified. Pulse repetition rate measurements enable determination of maximum unambiguous range of the radar. Pulse-to-pulse signal amplitude variation provides information regarding the antenna scan pattern, antenna rotation period (ARP), sidelobe level and so on. The direction of arrival measurement facilitates determination of location of the emitter by triangulation. However, the measurement of emitter parameters becomes increasingly difficult with the increase in the emitter density and as the radar signals become more complex. Emitter may be identified by comparing measured and derived parameters with those stored in the threat library, which contains the known parameters of the various operating modes of different radars.

The processed information is then passed on to the intelligence analyst who can combine this with other knowledge to assess the overall competence and possible intentions of the adversary. In the radar example, information on radar characteristics may be compared with known information on various radars to ascertain its range, location, use and so on required to evaluate its capability as radar and its vulnerability to countermeasures.

It is pertinent to mention here that the Armed Forces including Army, Air Force and Navy are primarily interested in information on location and capability of all enemy radars on a current basis commonly referred to as *Radar Order of Battle* (ROB). Information on location, concentration and capability of radars deployed by adversary helps the friendly forces, the Air Force for instance, to either skirt the area or use suitable countermeasures to avoid detection by the adversary. Another objective of ELINT is interception and analysis of new and unusual electronic signals. These signals emanating from R&D laboratories, field test stations and so on are equally important to be analysed to know the kind of sources that would be encountered in battlefield in the future. This may not be immediate priority for the Armed Forces fighting a tactical battle but certainly is one for strategic planning.

The ELINT equipment mainly comprises *collection equipment* housed on a fixed station or an airborne or maritime platform and *analysis equipment* used on ground to analyse the collected information. Major subsystems of ELINT collection system are the antenna, receiver, recorder, direction finder and analyser. The antenna subsystem intercepts the signal. Desirable characteristics of ELINT antenna subsystem include continuous and fixed broad area coverage, broad electronic spectrum coverage, high gain and an inherent capability for giving directional information. Some of the requirements are contradictory. One may have to be achieved at the cost of another. The decision often rests on the frequency range of interest and also on the specific ELINT target under consideration. As an example, broad area coverage may be obtained by either using a fixed broad beam antenna or a scanning narrow beam antenna. While the former compromises antenna gain, the latter limits the coverage area at any given time. The intercepted signal is passed on to the receiver. The function of the receiver is to transform the intercepted information into a measurable and recordable form. There are two basic types of receivers in use: the crystal video receiver and superheterodyne receiver. The crystal video receiver is characterized by poor sensitivity and frequency resolution, but has a relatively much shorter search time. The superheterodyne receiver, on the other hand, has inherently high sensitivity and good frequency resolution but suffers from a prohibitively long search time. The signal is next stored on a recorder for the purpose of future analysis and record. Recording also facilitates evaluation of short duration transmissions where there is a likelihood of transmission ending before it can be evaluated in real time. Direction-finding equipment is used to know the direction of arrival of intercepted transmission. Analysers in the ELINT collection system provide a preliminary observation of the type of modulation, repetition rate, duration and general shape of intercepted signal pulses. The ELINT analysis system's type and complexity depends upon the intended purpose, whether it is order of battle analysis or it is identification of new and unusual signals.

6.3.2.2 Communications Intelligence (COMINT)

COMINT, as outlined in an earlier paragraph, is a subdivision of Signal Intelligence (SIGINT) and is defined as technical information and intelligence derived from foreign communications including voice communications, text messages and online interactions by other than the intended recipients.

SIGINT equipment has both tactical and strategic roles. The fundamental difference between tactical SIGINT and strategic SIGINT lies in the purpose served by the two. While tactical intelligence enables the Armed Forces to formulate plans to accomplish their military objectives within a certain operational area, strategic intelligence is used for formulating national policy and strategic level military plans. Another difference lies in the importance or significance of a certain piece of information when seen in the light of a defined mission objective. For example, if a country A were conducting military training exercises in friendly country B and discovered that a company from another country, C, was carrying out mineral exploration in the host country, the information may not be of primary intelligence concern to the commander of the unit conducting training. On the other hand, the same piece of information would be of strategic importance to the policy makers of the country A indicating to them about the growing influence of country C in country B. The focus or the primary objective also differs in the two cases. While in the case of tactical SIGINT the focus could be to triangulate adversary radio transmissions to direct artillery fire on their reconnaissance assets, at the strategic level the primary focus could be to break enemy codes to know their intentions. The nature of collection sources also differ in the two cases. Sources that provide information on adversary mobilization procedures, weapons of mass destruction (WMD) capabilities or critical national level assets are of strategic nature. At the tactical level, a potential source would be one that can provide information on possible safe routes for an infantry battalion to infiltrate through the adversary's forward positions.

SIGINT equipment may be housed on ground-based, airborne, naval, submarine or space platforms. SIGINT equipment on any of these platforms could be intended for either tactical or strategic roles. Space-based COMINT equipment may be placed in either low earth orbit (LEO) or in geosynchronous orbit (GEO). COMINT payloads on LEO satellites are required to monitor low power transmissions that have short-range and/or antenna beam patterns that confine most of the signal to the horizontal plane. However, LEO satellites will have the intended target source of transmission in sight only for short time, which limits the spectral bandwidth that can be monitored using COMINT asset. LEO COMINT satellites therefore need to be frequently re-tasked to monitor different small segments of the radio spectrum of interest. Geosynchronous COMINT satellites remain stationary or nearly stationary with respect to a ground station. Geosynchronous satellite-based COMINT payloads therefore can monitor a large part of the Earth's surface and the entire radio spectrum almost in real time and downlink the collected information to a fixed station in friendly territory for analysis. However, due to the COMINT payload in this case being at a great distance of about 36 000 km from the source of transmission, there is need for large receiving antennas to receive low power signals.

6.3.3 Representative SIGINT Equipment

SIGINT systems are available from a large number of manufacturers for mounting on land-based, naval, airborne and space-based platforms designed to perform ELINT, COMINT, precision geolocation and direction-finding (DF) functions to support warfighter missions. Some representative SIGINT payloads designed for various platforms include *Tactical SIGINT Payload (TSP)* and *S-3000 Signals Intelligence (SIGINT) and Information Operations (IO) Systems* from BAE Systems, the *ASIP family of SIGINT systems* from Northrop–Grumman, *Common SIGINT System 4000* (CSS-4000) from Northrop–Grumman, *TOP SCAN Airborne ELINT/ESM System* from Rafael and *CERES Space-borne Signal Intelligence (SIGINT) System* from Thales. Each of these SIGINT systems is briefly described in the following paragraphs.

The TSP from BAE Systems (Figure 6.1) is an advanced sensor system that, due to its software-defined architecture, is capable of processing a host of military signals for a wide range of mission profiles. Its modular and scalable design provided by ruggedized commercially off-the-shelf (COTS) hardware makes it easily integrable on a variety of existing and future manned and unmanned aerial platforms. It is currently operating on the US Army's MQ-1C Gray Eagle (Figure 6.2), which is a technologically advanced derivative of the combat-proven Predator and next-generation tactical Unmanned Aircraft System (UAS) solution to meet challenging service requirements for persistent reconnaissance, surveillance and target acquisition and attack operations. TSP is the next-generation airborne signals intelligence (SIGINT) system that integrates a unique software-defined open architecture while also using COMINT collection, direction-finding and geolocation. TSP operates remotely over both Line-of-Sight (LOS) and Beyond Line-of-Sight (BLOS) networks enabling combat commanders receive a 360-degree aerial view of their operational area in near real time. Its capability to simultaneously collect multiple signal protocols and re-configurability on the fly allows it to support single and multiple operators from anywhere in the world.

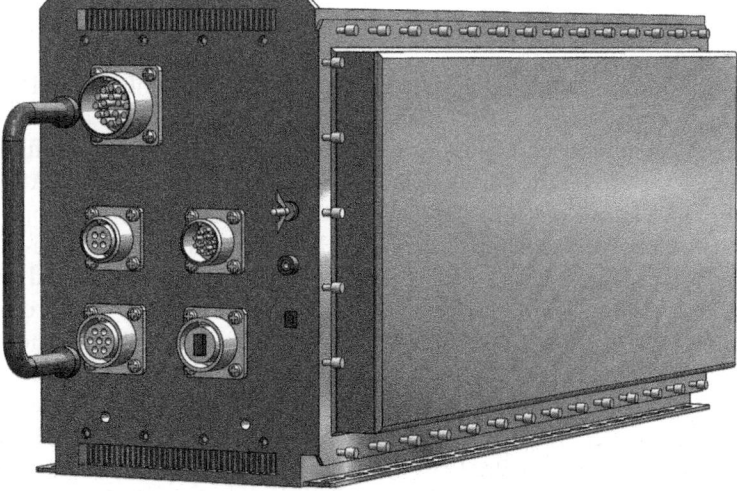

Figure 6.1 Tactical SIGINT Payload (TSP).

Figure 6.2 MC-1Q Gray Eagle UAS. (*Source:* Courtesy of the US Army.)

The *S-3000 family of SIGINT and IO systems* from BAE Systems, with its wide operational frequency range from high frequency (HF) to Ku band, open commercial off-the-shelf (COTS) architecture enabling third party integration, software-defined that allows re-configurability on-the-fly, multiple antenna types and locations and advanced transmit and receive capabilities can be seamlessly integrated on to a wide variety of platforms such as unmanned aerial vehicles, manned aircraft, land vehicles and surface vessels for a range of mission profiles. The system with its powerful computational resources provides operators with advanced signal location and exploitation capabilities to counter the present day battlefield threats of modulated, encrypted and multiplexed signals. This is achieved by extracting actionable intelligence from large volumes of data, enabling secure operation in the battlefield. It is available in direct air, conduction-cooled and liquid-cooled versions and its rugged construction allows operation in extreme environments.

Northrop–Grumman's *Airborne Signals Intelligence Payload* (*ASIP*) is a family of SIGINT systems designed with open service oriented architecture and scalable hardware and software components to meet requirements of a wide variety of platforms and mission profiles. The core of the ASIP SIGINT systems is the baseline ASIP sensor. Different variants of ASIP sensor have been developed for various platforms and applications. It was initially developed for US Air Force U2 programme and subsequently for Global Hawk platform. U2 is a state-of-the-art high-altitude intelligence, surveillance and reconnaissance platform developed by Lockheed–Martin and operated by the United States Air Force. Global Hawk of Northrop–Grumman originated from High-Altitude Endurance Unmanned Aerial Vehicle Advanced Concept Technology Demonstration (HAE UAV ACTD) programme initiated by the Defence Advanced Research Projects Agency (DARPA) and the Defence Airborne Reconnaissance Office (DARO). Scaled variants namely ASIP 1C and ASIP 2C of ASIP SIGINT payload have been developed for the US Air Force MQ-1 Predator and MQ-9 Reaper unmanned aircraft systems. ASIP is also the chosen SIGINT payload for the US Army's Guardrail Modernization programme.

The *Common SIGINT System 4000 (CSS-4000)* is one of the several products from Northrop–Grumman's airborne SIGINT product line. It is the next-generation of lightweight SIGINT payloads providing rapid interception, precision geolocation and communications signal processing functions to support a variety of mission profiles for fighter aircraft. CSS-4000 SIGINT payload provides simultaneous support for both overwatch and standoff missions requiring steep to shallow depression direction-finding (DF). High accuracy imaging intelligence (IMINT) cross-cueing performance, flexible signal processing architecture through onboard programmable FPGAs, support for active SIGINT enabling CyberISR and electronic attack missions and simultaneous support for fast scan, staring and copy functions are the other highlights. Major specifications include the 0.5–6500 MHz frequency range, eight configurable channels for transmit or receive with 40 MHz of instantaneous bandwidth per channel, 85 dB of dynamic range and 1320 Giga-Flops (GFLOPS) of peak processing capacity.

TOP SCAN is a SIGINT/DF payload developed by Rafael Advanced Defence Systems Ltd for airborne platforms. The system is configured around an advanced direction-finding and localization system using dual-axis interferometer for direction-finding and localization sensors. The system provides collection, detection, identification and localization of surface radar emitters with high accuracy in a dense emitter environment. These radar emitters that are the integral constituent of command, control and guidance of missiles and other air-defence systems are the potential threats. The TOP SCAN SIGINT/DF payload is capable of detecting, identifying and locating these threats even in the presence of multipath and interference. The system features wide spatial coverage, wide frequency coverage, modular and flexible configuration and high probability of intercept.

CERES (CapacitÉ de Renseignement Électromagnétique Spatiale) is a signals intelligence (SIGINT) system comprising a constellation of three satellites (Figure 6.3) as the space segment and ground segment including *ground control segment* to command and control the spacecraft and *user ground segment* to perform mission planning and data processing functions. AIRBUS

Figure 6.3 Constellation of CERES satellites. (*Source:* Courtesy of Airbus.)

Defence and Space, Thales Airborne Systems and Thales Alenia Space are the prime contractors. Airbus Defence and Space have responsibility for the space segment, comprising the three satellites that will integrate the SIGINT payloads. Thales Alenia Space will supply the bus/platform under subcontract to AIRBUS Defence and Space. The ground control segment is to be provided by the French space agency CNES (Centre National d'Etudes Spatiales).

CERES SIGINT system is designed to provide to the French Armed Forces space-based signals intelligence capabilities, which include detection, location and characterization of electromagnetic signals emanating from enemy's communications signals and radars on ground. The CERES constellation will be located in low Earth orbit at an inclination between 70° and 80°, to detect and locate electromagnetic signals simultaneously at different time intervals. The satellites will be positioned in close proximity to each other. CERES satellites carry signals intelligence (SIGINT) payloads to detect high-frequency radio waves and other electronic signals and the information intercepted by the payloads will be down-linked for processing by ground segment. The system also provides space-based early warning and ballistic missile detection capabilities. The system is scheduled to enter service by 2020.

6.3.4 ESM/Radar Warning Receivers (RWR)

While electronic reconnaissance and surveillance functions involve a long term observation and characterization of a potential opposition, a warning system in general is meant for the purpose of detecting, identifying and locating a threat approaching the system to be protected. All warning scenarios have a protected entity, which could be an aircraft, a ship and a ground-based armoured vehicle, an immediate danger such as illumination by a radar emission or a laser beam or an approaching missile and an environment containing a variety of benign objects that must be distinguished from the potential threats. Based on the nature of threat, common warning receivers or systems include the *Radar Warning Receivers* (RWRs), *Laser-Warning Receivers* (LWRs) and *Missile Approach Warning Receivers* (MWRs).

A RWR is a passive electronic warfare system that intercepts RF signals from radar emitters of the adversary to identify type and location of radar emitter and generates appropriate cues for the aircraft crew enabling activation of suitable countermeasures through its interface with other systems. RWRs and other ESM receivers are capable of detecting emissions, mainly radar emissions, over a wide band and are therefore different from conventional radar receivers that are designed to receive the well-defined signal waveforms used by the particular radar system. The RWRs not only intercept emissions over a wide band, they do it without losing any of the characteristics of each emission that is likely to be used for identification process. The frequency range of an ESM/RWR is typically from 0.5 to 20 GHz corresponding to C–J frequency bands. Efforts are on to extend the frequency range to 40 GHz in near future.

The parameters on which RWR and ESM receivers can derive information from the intercepted signal include the RF, the pulse width, polarization, pulse repetition rate (PRR), direction and time of arrival, frequency agility and so on. If the emission can be intercepted for a period more than fraction of a second, it can reveal a lot of information on the scanning characteristics of the emitter. As an example, if the emitter were fire-control radar; one could determine the time instant of lock-on, which is vital for initiating countermeasures or taking other evasive action. Information on the previously mentioned parameters also allows various emitters to be classified into categories such as search, fire-control, navigation radars and so on.

There is a large variety of receivers available for the purpose of ESM and radar warning. The performance of different types of receivers is compared on the basis of characteristics like noise figure, receiver sensitivity, measurement accuracy, bandwidth, dynamic range, probability of intercept, false alarm rate (FAR), throughout and so on.

Figure 6.4 Block-schematic arrangement of a crystal video receiver.

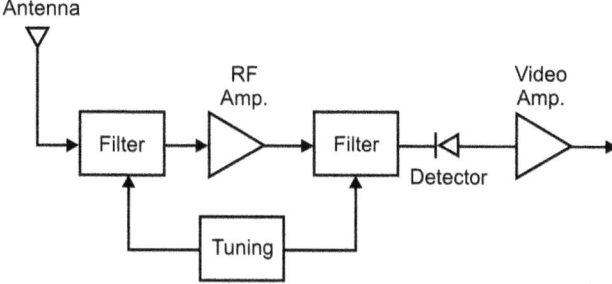

Figure 6.5 Block-schematic arrangement of a superheterodyne receiver.

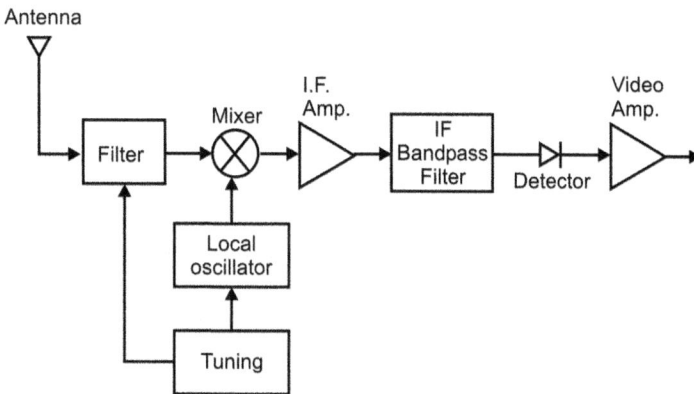

A *crystal video receiver* is the simplest and most commonly used radar warning receiver. In a crystal video receiver, the impinging microwave signal from hostile radar falls upon a wide band receiving antenna from which it is fed into a bank of simple receivers comprising a cascade arrangement of filter, detector and amplifier. The receivers are each tuned to consecutive slices of the covered band which allows simultaneous reception and discrimination of radars operating in various parts of the band. In its basic form, it is inherently extremely wide band and with a suitable antenna pattern, it has the potential of providing nearly a 100% probability of intercept for received signals exceeding threshold. Figure 6.4 shows the block-schematic arrangement. The diagram is self-explanatory. The low-cost, high probability of intercept and possibility of finding informative on direction-of-arrival (DOA) make this type an attractive option for radar warning applications.

The superheterodyne configuration is the other receiver type used for the purpose. A superheterodyne receiver contains a combination of amplification with frequency mixing. It is by far the most popular architecture for a microwave receiver. The basic block-schematic of a superheterodyne receiver is shown in Figure 6.5. The tunable RF filter helps to eliminate spurious outputs caused by intermodulation products generated in the mixer. The diagram is otherwise self-explanatory. Superheterodyne receiver configuration is widely used with various direction-finding techniques particularly where angle-of-arrival (AoA) is derived from phase information. This is due to the reason that frequency and phase information is preserved in a superheterodyne receiver during conversion of intermediate frequency (IF) and can therefore be measured accurately. The high selectivity of receiver leads to only a small portion of electromagnetic spectrum being sampled by the receiver with the result that overall probability of intercept is low. However, the relatively narrow bandwidth gives the best combination of sensitivity and dynamic range. High sensitivity allows target emitters at far-off ranges to be intercepted through their side lobe and back lobe radiations.

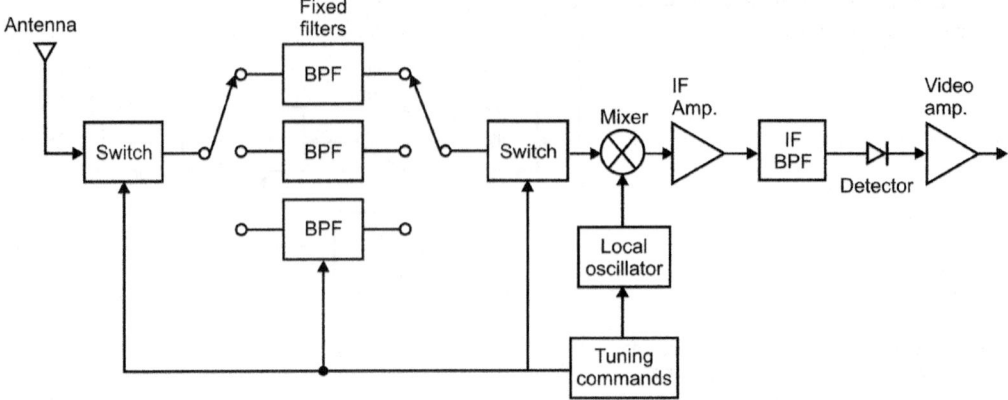

Figure 6.6 Block-schematic arrangement of a superheterodyne receiver with fixed frequency filters.

A wider total bandwidth can be achieved in scanning superheterodyne receivers where the receiver respectively scans in frequency across the desired bandwidth. Such a receiver can be made very effective against long pulse and continuous-wave (CW) signals by integrating the output of the video detector. If prior information on the emission frequency of the potential target is available from other intelligence sources, the scanning superheterodyne receiver can be programmed for optimum search of a portion of the spectrum corresponding to specific emitters rather than sweeping the entire band.

Another approach to increasing the overall bandwidth of a superheterodyne type receiver is to use a set of fixed frequency filters at the input before the mixer. The outputs of the filters can be switched to the mixer followed by intermediate frequency (IF) amplifier/filter and an appropriate detector. Figure 6.6 shows the block-schematic arrangement of such a receiver. Yet another approach is to simultaneously convert the outputs of filters to a common intermediate frequency (IF). This effectively produces a set of parallel superheterodyne receivers. This receiver possesses the sensitivity, the dynamic range and the frequency resolution of a simple superheterodyne receiver and at the same time has capability to process signals across a very wide radio frequency (RF) spectrum, typically 2–20 GHz. The probability of intercept is nearly 100%.

The *channelized receiver* described above is usually very complex and expensive. The same result could be achieved by sweeping a single receiver across the same band in a time period equal to or less than the shortest pulse to be intercepted. In this type of receiver called *micro-scan receiver* or *compressive receiver*, the local oscillator is swept in frequency linearly across a bandwidth equal to sum of RF and IF bandwidths of the receiver. The mixer output is fed to a dispersive delay line (DDL) with linear characteristics. That is, the delay introduced is proportional to signal frequency. These delay lines can be easily realized with surface acoustic wave (SAW) devices. The DDL produces at its output a compressed pulse at a time measured from the local oscillator frequency sweep. Multiple input RF signals appear as a series of narrow pulses at the output (Figure 6.7). These receivers have excellent sensitivity and probability of intercept approaching 100%. They are particularly useful for intercepting pulse-to-pulse frequency agile emitters.

The *acousto-optic receiver*, also known as a Bragg-cell receiver, is another type of receiver used for the purpose. The operation of an acousto-optic receiver is based on the deflection of a laser beam as it interacts with acoustic wave propagating in a piezoelectric material such as lithium niobate. This device is also known as a Bragg cell. The acoustic wave is excited on the piezoelectric substrate by the RF signal with acoustic frequency depending upon RF frequency.

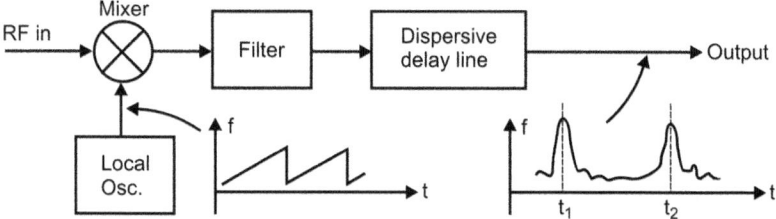

Figure 6.7 Block-schematic arrangement of a microscan/compressive receiver.

The acoustic wave causes periodic variation in refractive index of the material as it propagates. It is this refractive index variation that causes the laser beam to deflect. The deflected laser light is then made to fall on an array of photodiodes.

A *Bragg-cell receiver* consists of a laser source with the laser beam passing through the piezo-electric crystal that is acoustically excited by an RF signal. The laser beam is scattered through an angle that is proportional to the ratio of the wavelength of laser to the wavelength of source in the crystal. The scattered light is collected on the diode array that is read out electronically.

Simultaneous signals on different frequencies will each produce its own deflected beam at an angle proportional to its frequency. It is effectively a channelized receiver and the total bandwidth of an ESM receiver based on this technology can be extended by tuning the front-end before it is down converted to Bragg-cell operating frequency.

6.3.5 Representative ESM/Radar Warning Receivers

Most manufacturers of Defence equipment in general and electronic warfare equipment in particular offer ESM and Radar Warning Receivers for various platforms and mission profiles. Some of the better known systems include the *BOW RWR and ESM System* from SAAB, the *AN/ALR-67* and its variants *AN/ALR-67(V)*, *AN/ALR-67(V)2* and *AN/ALR(V)3* from Northrop–Grumman and Raytheon, *CATS-150 Compact Airborne Threat Surveyor* from Thales, the *APR-39D(V)2 RWR and EW Management System, AN/ALQ-218 RWR/ESM/ELINT Sensor System* and the *LR-100 RWR/ESM/ELINT Receiver System*, all from Northrop–Grumman. Each one of them is briefly described in the following paragraphs.

The *BOW family* of RWRs with ESM functions from SAAB, Sweden, is designed for fighter aircraft applications. It provides to the crew of fighter aircraft enhanced situational awareness through detection and evaluation of potential threats in dense signal environments with 100% of probability of intercept, high selectivity and sensitivity as well as the possibility to geo-locate RF emitters. Its highly modular and scalable architecture makes an easy adaptation to changing requirements. In the basic receiver configuration of BOW family of RWRs, a wideband receiver provides good threat warning capabilities and when complemented with a narrow band receiver, it offers better selectivity in signal dense environments and increased measurement performance for more demanding applications. The high end configurations of the family include digital receiver and interferometer antenna arrays, which provide long detection range and capability of fingerprinting of complex signals and high performance emitter location. All receiver subsystems cover basic frequency ranges of the E–J band in the basic system and C/D and K/L bands as an option. The azimuth coverage is 360° and the elevation coverage is ±45° for the basic system and ±90° as an option. Direction-finding (DF) accuracy is 7° RMS for the basic system and 1° RMS as an option. The BOW radar warning system has been implemented in different configurations on the Tornado, Gripen and SAAB Airborne Early Warning & Control (AEW&C) aircraft. Figure 6.8 shows the photograph of BOW radar warning receiver and ESM system.

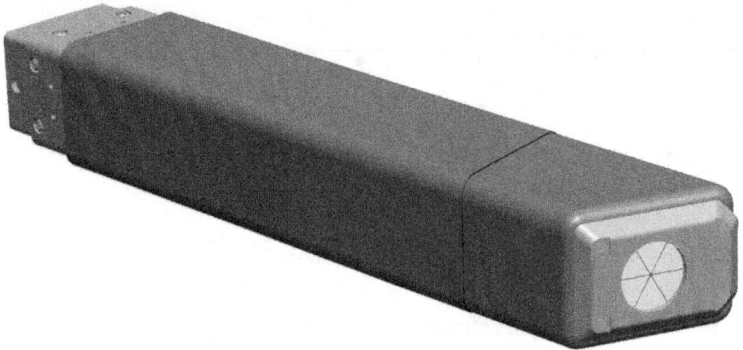

Figure 6.8 BOW radar warning receiver and ESM system.

AN/ALR-67 is a family of airborne RWR and countermeasures control systems designed to provide to the aircraft's crew warning on any hostile radar activity by detecting, identifying and locating radars and radar-guided weapon systems of the adversary. The system does so in the C–J frequency bands extending from 0.5 to 20 GHz. As a countermeasures and control system, it interfaces with a range of onboard systems, which includes jammers, fire-control radars, data links, anti-radiation missiles and missile approach warning sensors. The AN/ALR-67 family of RWRs has had three variants namely AN/ALR-67(V), AN/ALR-67(V)2 and AN/ALR-67(V)3. While Northrop–Grumman has been the main contractor for production of AN/ALR-67(V) and AN/ALR-67(V)2; Raytheon has been contracted for AN/ALR-67(V)3. AN/ALR-67 (V) is in use in large numbers with the US Navy and Marine Corps and the Air Forces of Australia, Canada, Finland, Kuwait, Malaysia, Spain and Switzerland. AN/ALR-67(V)2 is an improvement over the AN/ALR-67(V) in terms of detection range when in the presence of a wingman's radar signals. It makes use of inertial guidance system (INS) stabilization for accurate display in high 'g' and high roll manoeuvres. Major subsystems comprising AN/ALR-67(V)2 RWR include four small spiral high-band antennas to provide 360° azimuth RF coverage, four wideband, high-band quadrant receivers, a low-band array plus receiver to provide 360° azimuth low-band coverage, a narrowband superheterodyne receiver for signal analysis functions, twin CPU, an azimuth display unit (ADU) and a control unit. AN/ALR-67(V)2 with its multiple receiver architecture and state-of-the-art processing is the standard threat warning system for US Navy strike fighter aircraft and all international F/A-18 Hornet, F-14 Tomcat and AV-8B Harrier aircraft.

The *AN/ALR-67(V)3*, called an *Advanced Special Receiver* (ASP), is an upgrade to AN/ALR-67(V)2 systems intended to supersede the latter system. It incorporates many new features such as extended capabilities in detection and processing for Air-Defence radar threats, threat localization and potential lethality thereby enhancing the survivability of aircraft equipped with the system. It provides visual and aural alerts to air crew upon detection of ground-based, ship-based or airborne radar emitters thereby enhancing pilot situational awareness by providing accurate identification, lethality and azimuth displays of hostile and friendly emitters. The system adopts a channelized receiver architecture that allows successful detection of radar emitters in high signal density and detection of weak distant signals despite interference from strong nearby transmitters. The channelized radar warning receiver is packaged into a lightweight 100-pound system. The AN/ALR-67(V)3 is fully integrated with all avionics equipment on the F/A-18 A/B/C/D Hornets, as well as that on the F/A-18E/F Super Hornets. This fourth generation radar warning receiver is a major component of the Super Hornet's Integrated Defensive Electronic Countermeasures set.

Compact Airborne Threat Surveyor (CATS) is the state-of-the-art family of radar warning receivers from Thales group designed for helicopters. The system provides real-time battlefield situation awareness to avoid potential threats or hand over targets to weapon system in a sensor-to-shooter concept. The system operates in the E–K band covering a frequency range of 2–40 GHz with optional C (0.5–1 GHz) and D (1–2 GHz) bands, provides 360° azimuth coverage and ±45° elevation coverage, DF accuracy of better than 10° and a library of greater than 5000 modes. The modular architecture of CATS not only allows easy integration on different types of helicopters; its modular concept and open architecture allows it to be tailored to customers need and mission requirements. Its main features include instantaneous radar warning for pulse, pulse Doppler and CW multiple radar emitters associated with air and surface threats with 100% of probability of intercept, automatic instantaneous threat identification, fully user programmable library, extensive onboard data recording, Self-Protection System (SPS) processing (radar, laser, missile warning, chaff and flare dispenser, jammer) to provide enhanced EW situational awareness and optimization of countermeasures employment and compatibility with off-board transmission to other platforms and multi-platform cooperation.

The *APR-39D(V)2 Radar Warning Receiver (RWR)/Electronic Warfare Management System (EWMS)* belongs to the APR-39 family of RWRs from Northrop–Grumman. It is designed to protect a variety of fixed-wing, rotary-wing and tilt-wing aircraft from radar threats including pulsed, pulse Doppler, CW, scanning type radar emitters and emitters with frequency agility and low probability of intercept. It does so by enhancing air crew situational awareness through interactive management of all onboard sensors and countermeasures providing them sufficient time to execute manoeuvres and deploy countermeasures thereby maximizing survivability. The APR-39D(V)2 RWR provides 360° coverage to automatically detect and identify types of threats, their bearing and lethality. It is capable of handling new and complex radar emitters including millimetre wave radar emitters. APR-39D (V)2 provides full mission data recording and post-mission playback capability. Mission data recording includes all sensor information as well as other mission data such as GPS, time and maintenance activity. Post-mission playback capability allows for timely review and analysis. The APR-39D(V)2 is currently in development with the US Navy and Army. More than 7000 systems of APR-39 family of RWRs have been installed to date on both domestic and international AH-1W/Z, UH-1N/Y, MV-22B, KC-130T, UH-60, OH-58D, CH-53, CH-46, AH-64A/D and CH-47 aircraft.

The *AN/ALQ-218 RWR/ESM/ELINT sensor system* from Northrop–Grumman is designed for airborne situational awareness and signal intelligence gathering functions. It protects the warfighter by detecting, identifying, locating and analysing sources of RF emission thereby enabling air crews and/or commanders make the right decisions at the right time to ensure maximum protection of friendly forces and assets. ALQ-218 provides the initial verification for the correlation between the planned *Electronic Order of Battle* (EOB) and the actual EOB prevailing in the battlefield. ALQ-218 system features two independent receiver groups namely the primary receiver group comprising four channelized and four cued receivers operating in tandem and the auxiliary receiver group. The primary group of receivers provides immediate signal acquisition, accurate parameter measurement, timely updates and precision direction-finding/geolocation. The auxiliary receiver group is used to offload the primary receiver from long dwell-time measurements. It also provides an extended frequency capability and aids in the recognition of intra-pulse modulation plus the updating of range estimates for geolocation. The ALQ-218 employs a combination of short, medium and long baseline interferometers to provide geolocation of emitters for cueing jammers and other onboard sensors. Its look-through software allows periodic surveillance of the threat environment while jamming is in progress, thereby allowing the ALQ-218 receiver to operate in close proximity to onboard high power jammers. ALQ-218 is integrated onboard the EA-6B Prowler that flies the electronic warfare mission for the United States Navy, Air Force and Marine Corps. It gives this tactical jamming aircraft the ability to precisely identify and pinpoint the location of enemy radar sites for a more effective use of HARM missiles. The ALQ-218 receiver is used on EA-18 Advanced Electronic Attack aircraft, a derivative of the F/A-18F Super Hornet, and also installed on the P-8A Poseidon aircraft designed by Boeing for long-range anti-submarine warfare; anti-surface warfare and intelligence, surveillance and reconnaissance missions. The ALQ-218 follows a spiral development model that allows for installation on any air, sea or land-based platform. The system is also being considered for positioning on unmanned aerial vehicles and subsurface platforms.

Northrop–Grumman's *LR-100* is a combat-proven RWR/ESM/ELINT receiver system providing RWR, ESM and ELINT functionality in a single system, thereby eliminating need and cost of integrating a unique system for each function. It is ideally suited for installation on virtually any airborne, sea-based or land-based platform. The standard LR-100 configuration employs dual adaptive bandwidth superheterodyne receivers operating in the 2–18 GHz frequency band. The LR-100 RWR/ESM/ELINT system utilizes phase interferometry and a patented passive ranging technology to provide situational awareness with geolocation of emitters that is used for cueing other onboard sensors. LR-100 is easily upgradable due to its open architecture, VME backplane and Ada software. It can accommodate various antenna configurations to meet diverse platform and mission requirements. LR-100 systems have been installed on the *Global Hawk UAV, SH-2G SuperSea Sprite helicopter* by Kaman Aerospace, the *S-70B Naval Seahawk helicopter* manufactured by Sikorsky Aircraft, anti-submarine and maritime surveillance aircraft *P-3 Orion* by Lockheed–Martin, the *Hunter reconnaissance UAV* by Northrop–Grumman, the *ASCIET Aerostat*, the *Predator UAV* by General Atomics, the *C-28 aircraft* and *Special Operations aircraft, submarines* and *patrol boats*.

6.3.6 Laser-Warning Receivers

Laser-Warning Receivers (LWRs), like RWRs, also have to encounter two related but inherently different scenarios. One of them is to use these receivers for general monitoring of the battlefield to look for various types of laser systems operational in the battlefield, their characteristics, location and so on for subsequent use to formulate battle strategies. In this type of situation, the information provided by the LWRs is not used immediately. In the other scenario which is tactical in nature, the purpose is to protect the platform on which the LWR is mounted. The platform could be an aircraft, a helicopter, a ship, a satellite or even a ground vehicle. An LWR in such a situation provides information on the impending threat, which could be from a laser designator illuminating the platform to be protected from laser-guided munitions attack. This allows the platform to take an evasive action or initiate a countermeasure, an electro-optic countermeasure in this case.

LWRs have assumed great significance in the modern warfare environment due to rapid proliferation of different types of lasers used in a variety of military applications. Larger the number of lasers in use in the battlefield, greater is the need to equip oneself with ways and means that can offer protection against them. The lasers that are a usual sight in the modern warfare include Nd:Glass and Nd:YAG lasers used as rangefinders and designators, semiconductor lasers used in beam rider applications and communications, CO_2 lasers also in laser radar and beam rider applications and so on. Eye-safe lasers in the 1–3 µm band are fast replacing Nd:YAG lasers. Solid-state vibronic lasers like alexandrite and titanium-sapphire lasers that offer tunable output wavelengths in the visible and near infrared reduce the countermeasure vulnerability of fixed wavelength YAG lasers. Then there are lasers emitting the 3–5 µm band as countermeasures for heat-seeking missiles, 8–12 µm lasers for FLIR countermeasures. Weapon class lasers is another category and the lasers belonging to this group mainly include high-energy solid-state lasers, high power CO_2 lasers, chemical and Excimer lasers and free electron lasers. Such a wide variety of lasers covering a wavelength range from UV to IR put very stringent requirements on LWRs. An LWR, to be effective, must respond to this wide band, a situation similar to the one encountered in the case of RWRs. The other important considerations are the high probability of detection, low False Alarm Rate (FAR) and a high sensitivity.

The laser parameters of importance, or in other words the laser parameters that the laser-warning receivers or sensors should be capable of determining to achieve the desired objective, depend upon the intended application. For LWRs that serve the sole purpose of warning or alerting the platform to be protected of the potential threat, what would usually be sufficient is to make coarse measurement of laser wavelength, pulse duration and pulse repetition frequency (PRF). The information is considered adequate for distinguishing between weapon grade lasers, designators/rangefinders, communication lasers and lasers for countermeasures. Weapon class lasers emit in specific wavelengths and in long duration pulses, rangefinders are Q-switched lasers operating at a low PRR and designators are similar to rangefinders but have relatively higher pulse energy and pulse repetition frequency. Countermeasure lasers are also similar to rangefinders but have a relatively much higher pulse energy. Communication lasers are either modulated CW lasers or high PRR pulsed ones.

On the other hand, for LWRs that are directly linked to countermeasure lasers, it would be required to measure some more characteristics in addition to those listed previously. Also, the parameters would be required to be measured with a much greater accuracy. The issue of threat localization, that is, determining the location of the threat laser, is far more complex and demanding in the case of LWRs than it is in the case of RWRs. Often, the receiver intercepts energy coming from a direction other than direction of the source due to scattering/reflection and the directional data provided by the LWR could be misleading. If the receiver intercepts the direct beam, threat localization is simpler. The level of accuracy required in finding this information also depends upon the scenario. In most of the cases, quadrant localization is adequate whereas in some airborne LWRs, it may be necessary to determine the same to an accuracy of a few degrees.

Another aspect of LWRs is providing warning on strategic target observables that have prominent EOI/IR signatures such as Intercontinental Ballistic Missiles (ICBMs), Submarine Launched Ballistic Missiles (SLBMs), cruise missiles, strategic aircraft and so on. ICBMs and SLBMs emit strongly in the infrared as a result of the intense IR radiation from their rocket exhaust plumes. This is due to tremendous amount of heat energy being released in the rocket exhaust due to large quantity of fuels being burned to achieve the desired thrust. The most often considered emission bands for ICBM/SLBM missile warning are the $2.7\,\mu m$ and $4.3\,\mu m$ bands. Bands considered for cruise missiles and strategic aircrafts are in the 8–$12\,\mu m$ atmospheric window. In these bands, the EO/IR observables result from the air vehicle itself rather than its exhaust plumes.

Figure 6.9 shows a generalized block-schematic arrangement of a simple laser-warning receiver. It consists of receiving optics at the front-end to collect the laser energy followed by optical filters to select the laser wavelengths of interest. These filters are very narrow band pass

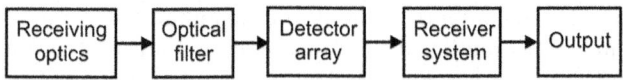

Figure 6.9 Generalized block-schematic arrangement of a laser warning receiver.

filters that pass only the desired wavelength. The radiation after passing through the filter is focused onto the detector or scanned across an array of detectors. The electrical signals are processed to extract the information on laser parameters.

6.3.7 Representative Laser-Warning Sensors

State-of-the-art laser-warning sensors detect, classify and identify potential battlefield laser threats illuminating the platform to be protected. The sensor system gives warning of the hostile laser threats and provides suitable interface to activate defensive or offensive countermeasures. Laser-warning sensor systems are available for a variety of ground-based, naval and airborne platforms and mission objectives from leading manufacturers of electro-optic and optronic systems for military applications. Some representative laser-warning sensor systems are briefly discussed in this section. These include the *Laser-Warning System* from Thales, the *High Angular Resolution Laser Irradiation Detector (Harild)* from Excelitas, the *Laser Irradiation Detector* (LIRD) from Fotona, the *LIAS Laser-Warning Receiver system* from ASELSAN, the *Naval Laser-Warning System* (NLWS) from SAAB, the *LWS-200CV* and *LMS-300 CV Laser-Warning System* from Avitronics, *Elbit's Laser-Warning System* (E-LAWS) from Elbit Systems of America and the *SHTORA-1 Defensive Aids Suite* from Electromachina JSC, Russia.

The *Laser-Warning System* (LWS) designed by Thales provides an early warning to the platform crew on laser range finder and target designator threats enabling them to take an evasive action thus enhancing platform survivability. The sensor system is capable of detecting threats from single, pulse, multiple pulse and continuous laser covering a wavelength band of 400–1600 nm with an extremely low false alarm rate of better than 1 in 24 h and typical response time of less than 100 ms. Detection range depends upon type of laser threat, laser power, beam divergence and atmospheric conditions and is typically 10 km. The sensor's spectral response of 400–1600 nm allows it to detect threats from a wide range of battlefield lasers including Nd:YAG, Nd:glass, frequency-doubled Nd:YAG, gallium arsenide, erbium-glass and Raman-shifted Nd:YAG lasers. The LWS is capable of discriminating between laser range finder and laser target designator threats and can be integrated into platform management system to provide laser-warning to Defensive Aid System (DAS). LWS and the grenade launchers are a key building block of Thales's Cerberus DAS system. Each sensor array is capable of providing azimuth coverage of 195° with resolution ±15° and elevation coverage of −12–+90°. Multiple sensor arrays can be used to provide azimuth coverage of 360°. The sensor system has been integrated onto a wide range of platforms ranging from small wheeled vehicles up to large tracked vehicles.

Excelitas' *High Angular Resolution Laser Irradiance Detector (HARLID)* is not a complete laser-warning receiver, but is a sensor configured around an array of silicon (Si) and indium gallium arsenide (InGaAs) detectors designed for use in laser-warning receiver systems to detect and provide precise angle-of-arrival (AoA) information from direct and indirect scattered light from laser range finders, laser target designators and active laser-based electro-optic systems. The HARLID sensor makes use of nine-element Si and InGaAs detector arrays organized in a sandwich configuration. In conjunction with light guides and a 6-bit digital Grey code mask, the angle-of-arrival of incident radiation is encoded into a binary digital pattern that varies with angle-of-arrival. Salient features and major specifications of HARLID include spectral response of 500–1650 nm provided by combination of silicon and indium gallium arsenide detectors, field-of-view of ±45° in both azimuth and elevation directions, angle-of-arrival accuracy of ±0.8° in both azimuth and elevation directions and wide dynamic range provided by low and high-sensitivity channels.

Figure 6.10 LIRD-4B detector. (*Source:* Courtesy of Fotona.)

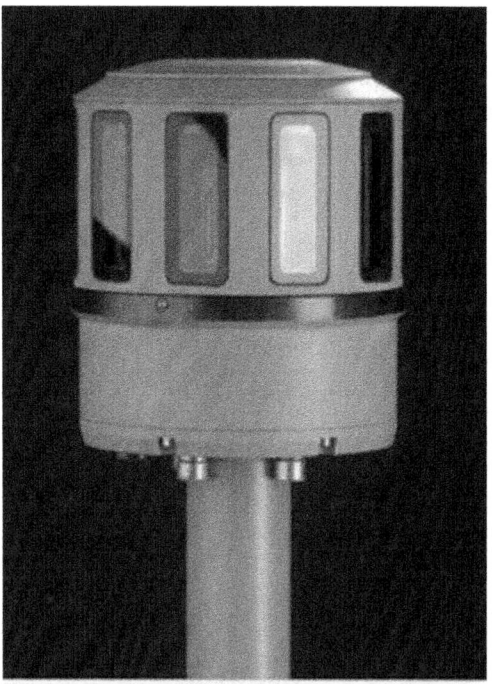

LIRD-4 detector system comprises several detection modules integrated in a hermetically sealed assembly and an indicator unit. The detector assembly detects and classifies the laser threat irradiating the platform on which the detector assembly is mounted along with its angle-of-arrival in horizontal sector. The type of threat, its angle-of-arrival and other threat features are displayed on the indicator unit. The system is capable of detecting laser threats from laser range finders, laser target designators and laser beam riders. The laser-warning provided by LIRD-4 enables the platform crew to take appropriate evasive action or deploy suitable countermeasures such as smoke screen. LIRD-4 is available in two variants, namely LIRD-4A and LIRD-4B. LIRD-4A has a unique feature of detecting scattered or indirect illumination as the laser sources may be aimed in the vicinity to avoid detection. LIRD-4B (Figure 6.10) has the beam rider detection capability in addition to the one for laser range finders and target designators. The LIRD-4 offers a spectral response of 400–1800 nm to include a wide range of laser sources such as frequency-doubled Nd:YAG, Nd:YAG, Nd:Glass, various semiconductor diode lasers, erbium-glass and Raman-shifted Nd:YAG lasers covering the visual and near infrared

Figure 6.11 LIAS laser warning sensor.

part of the spectrum. Fields-of-view in azimuth and elevation directions, respectively, are 360° provided by 20 sectors of 18° each and −20° to +60°. In the indirect illumination mode (LIRD-4A), the spectral response is 1000–1600 nm. The fields-of-view in this case are 360° again provided by 20 sectors of 18° each in azimuth direction and −7° to +13° in elevation. LIRD-4 can memorize up to the last 20 threat detection events. The indicator unit displays threat with resolution of 20 × 18°. The threats are characterized as LRF (Laser Rangefinder), Laser Designator (LD), Indirect Radiation (IN) and Beam Rider (BR). The detector provides interfaces to smoke grenade discharging systems and intercom for audio warning signal.

The *LIAS Laser-Warning Receiver System* is an advanced laser threat warning sensor system designed by ASELSAN to detect, classify and identify hostile laser threats illuminating friendly platforms and give warning to the platform crew enabling them to activate appropriate manoeuvre or countermeasure system available on the platform either directly or via a host computer. LIAS is designed to detect almost all types of military laser threats including LRF, Laser Target Designator (LTD) and Laser Beam Rider (LBR) threats operating on various optical bands.

LIAS is comprised of one processor unit and four sensor units installed on the body of the platform. Each of the four units provides azimuth coverage of 90° and elevation of ±40°. Figure 6.11 shows photograph of a LIAS sensor unit. Depending upon the size of the platform, the number of sensors to be used on the platform can be increased to eight to guarantee complete coverage of platform. The sensor covers a wavelength range of 0.5–1.1 μm (Band-I), 1.1–1.65 μm (Band-II), 0.8–1.1 μm (Band-III) and an optional 8–12 μm (Band-IV). The Direction-of-Arrival (DOA) accuracy offered by the sensor in different wavelength bands is ≤ ±1° RMS (Bands I and II), ≤ ± 10° RMS (Band-III) and ≤ ±22.5° RMS (Band-IV). The processor unit gathers the information from the sensor units, evaluates the signal parameters, classifies and identifies tracks and declares laser threats to a host computer. Available interfaces include RS-422, MIL-STD-1553B and fast Ethernet (100 Mbit). The open-architecture design of the system allows an easy integration on a helicopter or transport aircraft.

The *Naval Laser-Warning System* (NLWS) from SAAB is capable of detecting, classifying, identifying and analysing potential laser threats emanating from an adversary's LRFs and LTDs used for targeting, and active EOCM class lasers used for creating confusion in blue-water and littoral combat environments, thereby providing to the naval platform's command team the vital situational awareness about the presence of laser activity. This information on presence of potential laser threats is vital to efficient deployment of countermeasures. The NLWS consists of a number of sensors and a laser-warning controller. Different sensors interface with the

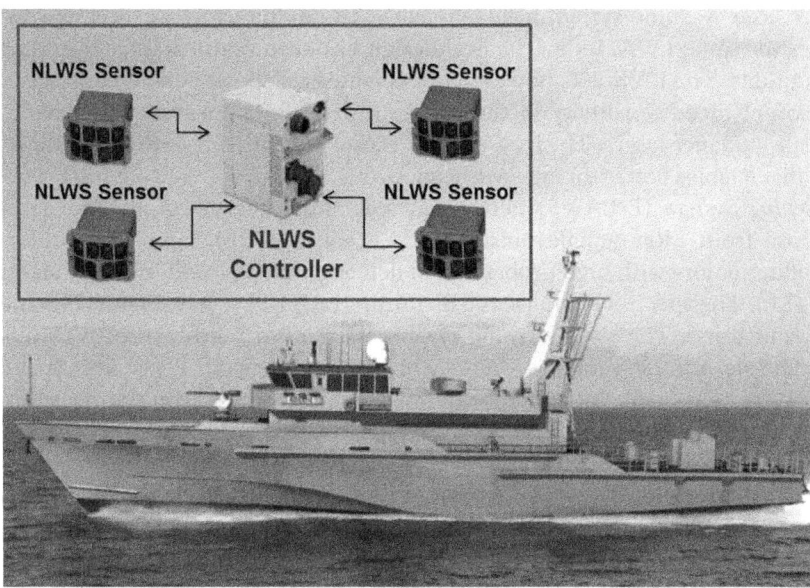

Figure 6.12 Naval Laser Warning System.

NLWS controller. Each sensor has a spatial coverage of 110° in azimuth with Angle-of-Arrival (AoA) accuracy of better than 7.5° RMS and 70° in elevation. The spectral response of the sensor ranges 0.5–1.7 μm to cover a variety of LRFs, LTDs and EOCM class lasers including laser dazzlers. The NLWS consists of a number of sensors and the number of sensors to be employed to guarantee complete azimuth coverage is selected in accordance with the size of the naval vessel to be protected. The NLWS through NLWS controller interfaces with the vessel's Command Managements System (CMS) and also with countermeasures such as Multi-Ammunition Soft-kill System (MASS) supplied by Rheinmetall Munitions available onboard the platform. MASS provides unique level of protection to different types of naval vessels against modern sensor-guided missiles. NLWS systems are in use by the German Navy and UAE Navy. Figure 6.12 shows NLWS mounted on a naval vessel.

The LWS-200 CV laser-warning system from Avitronics like other similar systems detects, classifies and identifies potential laser threats such as those from LRFs, LTDs and missile guidance systems, such as laser beam riders, with a probability of interception of greater than 99%, and alerts the platform crew to take evasive action or deploy countermeasures to protect the platform. Though originally developed for the South African Tank Technology Demonstrator (TTD) programme, the system is now being offered for installation on other platforms. The laser-warning systems comprises multiple laser-warning sensors and a LWC. A minimum of four sensors are required to provide azimuth coverage of 360°. More than four sensors can provide hemispherical coverage. Each laser sensor provides azimuth coverage of 110° and elevation coverage of greater than 70° with AoA detection accuracy of 15° RMS for all types of threats. The LWC performs the control, processing and threat display functions. Up to six LWS-200 laser sensors can be interfaced with the LWC. It also provides interfaces with audio warning, automatic/semi-automatic countermeasure dispensing and automatic/semi-automatic slewing of the turret towards the threat. Standard serial interfaces are also available for integration with onboard fire-control and battle management systems.

The LWS-300 CV laser-warning system has identical system architecture as the LWS-200 CV system and uses the same LWC. Its mechanical design has been optimized for the harsh environmental conditions. The LWS-300, however, offers increased spectral band coverage of 0.5–1.8 μm and also enhanced sensitivity in the 0.904 μm band for improved performance against typical missile guidance lasers. The LWS-300CV sensor also provides additional threat spectral band data that enables better threat classification.

Elbit's Laser-Warning System (E-LAWS) detects, locates and categorizes potential laser threats such as those from LRFs (single pulse), LTDs (2–50 Hz PRF), laser beam riders (2–70 kHz) and IR illuminators with high probability of detection (99.9%) and low false alarm rate (less than 1 in 18 h). The spectral response is 0.6–1.6 μm (LRF, LTD), 0.9–1.1 μm (BR) and 0.95–1.1 μm (IR illuminators). The laser-warning system provides azimuth coverage of 360° with four sensors. The elevation coverage is −20– + 90°. AoA accuracy is ±1° in azimuth and ±2.5° in elevation. The system can be integrated with full range of advanced countermeasures. The system's performance is immune to reflections from nearby objects, gun fire, lightning and fire.

The *Shtora-1 Defence Aids System* (DAS) was developed by the Zenit Research and Production Corporation to enhance battlefield survivability of Armoured Fighting Vehicles (AFVs) against attack from Anti-Tank Guided Weapons (ATGWs) such as TOW and HOT with a semi-automatic command-to-line-of-sight (SACLOS) guidance system, as well as missiles and laser-guided ATGWs such as Hellfire and Copperhead. Shtora-1 DAS consists of four laser sensors located on top of the turret, two electro-optical interference emitters operating in the 0.7–2.7 μm band and located on each side of the gun, a specialized computer/control panel and racks of dedicated 81 mm anti-laser smoke grenades that block radiation in the 0.4–14 μm band. The Shtora-1 DAS defeats IR-guided ATGMs by aligning the turret front to the incoming ATGM and using IR emitters to send false signals that confuse the ATGM guidance system and force the missile away from its intended path toward the target. The laser-guided weapons are defeated by deploying anti-laser smoke grenades. The smoke screen enshrouds the AFV and breaks or degrades the laser seeker lock-on to the target. The laser-warning sensors of Shtora-1 DAS operate in 0.65–1.6 μm band. The DAS comprises an array of coarse and fine resolution laser sensors. Each coarse sensor has azimuth coverage of 135° with AoA accuracy of 3.75°. The fine resolution sensors have azimuth coverage of 45°. The elevation coverage in both cases is −5– + 25°. The protected sector for the IR emitters is 20° (in azimuth) and 4° (in elevation). Shtora-1 DAS can be installed on vehicles as they are built or retrofitted. The first known application of the system is on the T-90 Main Battle Tank (MBT).

6.3.8 Missile Approach Warning Receivers

Missile approach warning is very important as according to a claim, more than 90% of all military aircraft losses are due to passive IR-guided missiles. Thus there is a continuing requirement for a missile approach warning system that is effective against IR-guided missiles and is suitable for mounting on a variety of aircraft. Detection range is often the key factor determining the choice of system as this would decide the time available to the platform to be protected to take an evasive action or initiate countermeasures. In other words, the detection range is a function of the required minimum warning time, the speed of the aircraft, speed of approaching missile and direction of attack.

Detecting a passively guided missile from a tactical aircraft platform effectively enough as to allow the air crew take an evasive action or initiate a countermeasure is a complex job. There are two commonly used approaches to building missile approach warning receivers (MWRs). The first uses an active radar system, usually a pulse Doppler or CW radar, which provides information on the range, velocity and position of the missile. The other types of MWRs are passive IR systems that depend for their operation on their ability to detect some emission characteristics of a guided missile. There are three primary emissions that the designer of MWR can exploit. The first is the radiation from the exhaust during the launch or boost phase of the missile's flight when the fuel is burned at the maximum rate and radiant intensity is also at its maximum. In this case, the receiver operates in the UV or 3–5 µm band. The second is the radiation produced from the combustion products of sustainer motor if missile uses one. The radiation intensity in this case is low and the emission is in the IR band. For this, receiver operates in the 8–14 µm band with an atmospheric window of low attenuation. These systems are capable of detecting missiles at larger ranges prior to their burnout. Active radar systems are overt in nature and suffer from limited detection range. On the other hand, passive IR systems offer covertness at the expense of limited information on range and velocity. One of the approaches to designing an MWR system is to make use of good points of both active and passive techniques. The overall system in that case would be a combination of approach warning radar and a passive missile approach warning receiver.

The major problems faced by the designers of MWRs are the same as the ones faced by the designers of RWRs or LWRs. These are the need to have a high probability of detection and a low false alarm rate. A low false alarm rat is very crucial as the countermeasures must be initiated very rapidly. If the countermeasures involved the use of consumables, a high false alarm rate could rapidly exhaust the supply or expose the presence of the aircraft to the enemy.

6.3.8.1 Types of MAW Systems

As outlined in the previous paragraphs, the missile approach warning systems are either active or passive. Three main technologies that have been exploited for design of Missile Approach Warning Systems (MAWS) include the use of pulse Doppler radar (Active Systems), IR sensors and UV sensors (passive systems). While active system such as pulse Doppler radar detects and tracks the missile by means of the radar signal returned from the skin of the missile; the passive systems like IR- and UV--based systems do so by detecting the exhaust plume signatures in the respective wavelength band. *Radar-based MAW systems* have the advantages of being capable of measuring range and speed of approaching missiles that enables determining the time-to-impact and optimizing the timing of deployment of countermeasures, immunity to weather conditions and its operation being independent of missile motor burning. The disadvantages of Pulse Doppler-based MAWS include the following. Being an active system, radiating electromagnetic waves might reveal the aircraft position and make it vulnerable to enemy attack. It has limited detection range for missiles with a small radar target cross-section, which also leads to a significantly reduced warning time. False alarm rate can be high due to emissions from other RF emitters. In the absence of measurement of approach direction with the desired accuracy, employment of Directed InfraRed counterMeasures (DIRCM) can be less effective.

Infrared-based MAW systems offer higher detection ranges at high altitudes where there is no ground clutter. In good weather conditions, as compared to UV sensor-based MAW systems, the atmospheric transmission of IR radiation tends to be better than that of solar-blind UV radiation. Also, it provides good AoA information for pointing a DIRCM. The disadvantages include the following. MWIR (Mid-Wave InfraRed) and LWIR (Long Wave InfraRed) sensors become almost blind in the presence of water content between the threat missile and the sensor. Due to this, IR-based MAW systems are not considered as all-weather systems. IR background clutter is a major problem while detecting surface-to-air missiles leading to accentuation of false alarm rate problem. Use of two-colour detectors solves the problem to some extent, but it makes the system far more complex. IR-based MAW systems don't provide actual range information and require use of large detector arrays to achieve azimuth coverage of 360°.

Ultraviolet-based MAW systems perform relatively better compared to IR-based MAW systems in high-clutter background environment and also in terms of false alarm rate due to its operation in the solar-blind UV spectral region. UV sensors are not affected by water vapour, which makes UV based MAW systems all-weather systems. They provide good AoA information efficient decision making for manoeuvring, deploying countermeasures and DIRCM. However, their operation depends upon high burning temperatures associated with solid fuel rocket motors. Like IR-based MAW systems, they too don't provide actual range information though time-to-impact information may be derived from the rapidly increasing amplitude of signal. Detection range could get limited due to advances in rocket motors leading to reduced IR/UV signatures.

6.3.9 Representative Missile Approach Warning Receivers

The MAWS senses approaching missiles and activates countermeasures such as chaff, decoys and flares to defeat an increasingly lethal array of air-to-air and surface-to-air missiles. In this section, we shall briefly discuss some representative missile approach warning systems belonging to the three categories of radar-, IR- and UV-based systems. MAWS briefly covered in this section include the *AN/ALQ-156 (V)* from BAE Systems, *MWS-20* from Thales and *EL/M-2160 (V1)* of Israel Aerospace Industries Elta Systems Ltd (all radar-based systems), *PAWS-2* from Raytheon and Elisra, *AN/AAR-56* from Lockheed–Martin, *MIRAS* from EADS (all IR-based systems), *MAW-300* from SAAB Avitronics, *AN/AAR-60* from EADS and *AN/AAR-47* from Orbital ATK (all UV-based systems).

The *ALQ-156(V) Missile Approach Warning System* sponsored by the US Army and produced by BAE Systems makes use of pulse Doppler radar to detect incoming potential missile threats by illuminating the threat missile with RF radiation and detecting reflection of RF signal from missile's skin to determine missile's range and to provide the optimum triggering of suitable countermeasures such as chaff, decoys or flare to protect the host platform. Its digital architecture allows it to be interfaced with other on board sensors such as laser-warning and radar warning sensors. The ALQ-156(V) measures both range and velocity of the incoming threat, which helps in eliminating false alarms from non-threat targets and improving overall effectiveness of the system. It recognizes missile threats by comparing closure rates and other ballistic parameters with those stored in a library. It can operate both as a standalone missile approach warning receiver and as part of a comprehensive multi-threat protection suite. When coupled with radar and laser-warning sensors, the ALQ-156(V) can automatically select the appropriate countermeasure for the detected threat. It measures range and velocity to eliminate false alarms on missiles fired at other targets in the battle area, which increases overall effectiveness by reducing responses to non-threat targets. Though it was originally designed to be installed on slow/low-flying fixed-wing aircraft and helicopters, the AN/ALQ-156 (V) has until now been integrated onto a large variety of airborne platforms including RC-12D/H/K, C-23B, C-130 Hercules, EH-1H, VH-3D, CH-47 Chinook, EH-60A, RU-21A/B/C/D, RV-1D and OV-1D Bronco.

AN/ALQ-156 (V) 1, AN/ALQ-156 (V) 2 and AN/ALQ-156 (V) 3 are the variants of AN/ALQ-156 (V) MAWS. AN/ALQ-156 (V) 1 comprises a transceiver, a control indicator and two identical blade antennas. Interfaced with M-130 dispenser with flares, AN/ALQ-156 (V) 1 is designed to protect aircraft against IR-guided homing missile threats. It is carried on board Boeing's CH-47D Chinook helicopter. AN/ALQ-156 (V)2 comprises a transceiver, a control indicator and four circular horn antennas. Like AN/ALQ-156 (V)1, when interfaced with M-130 dispenser with flares, AN/ALQ-156 (V)2 can be used to protect aircraft against IR-guided homing missile threats. Upon detection of an approaching missile, the radar sensor automatically initiates a signal that triggers the M-130 dispenser system, which then releases a flare to decoy the IR missile away from the aircraft. The system has been integrated on to Sikorsky's EH-60A helicopter, RC-12 special electronic mission aircraft and C-23B small military transport aircraft built by Short Brothers. AN/ALQ-156 (V)3 comprises a transceiver, a control indicator and four identical planar slot antennas. The transceiver has a built-in-test capability. Like the other two variants, it too is interfaced with M-130 dispenser with flares to provide protection to host platform against IR homing missile threats. It has been integrated on OV-1D light attack and observation aircraft manufactured by Grumman.

MWS-20 of Thales is an active Doppler radar-based missile approach warning system MAW capable of detecting and continuously tracking surface-to-air missiles, Man-Portable Air-Defence Systems (MANPADS) and other active and passive missiles during their entire flight phase providing precise and reliable information on missile range and velocity, direction-of-arrival and missile time-to-impact. Information on time-to-impact enables selection and triggering of appropriate countermeasures at the optimum time. MWS-20 has two variants; one for transport aircraft and the other for helicopters. The system comprises one single trans-mit-receive processing unit and four easy-to-install antennas, combining in itself excellent detection performance with an extremely low false alarm rate of less than 1 in 10 h in all flight and weather conditions. Easy installation on a variety of platforms and capability to interface with advanced self-protection systems are other has been integrated on a variety of airborne platforms including features. MWS-20 is in service on Cougar, EC725 and Puma helicopters, C-130 and Head of State (HOS) aircraft.

EL/M-2160 (V1) from Israel Aerospace Industries is an airborne missile approach warning system based on pulse Doppler radar. The system comprising a transceiver processing unit and six antennas gives warning on approaching missile threats in boost, sustain and post-burnout phases of missile flight in 360° coverage. The information on missile parameters such as range, velocity and DoA is used to determine time-to-impact, which in turn is used for timely activation of countermeasures such as chaff and flares for effective missile deception. Chaff and flares ejected by a dispenser unit divert the approaching missile from its intended track thereby protecting the platform from missile hit. EL/M-2160 (V1) has been integrated on a variety of airborne platforms including military aircraft, helicopters, special mission aircraft and commercial aircraft.

PAWS-2 from Raytheon and Elisra is an IR sensor-based missile warning system designed to provide protection to fighter aircraft, helicopters, transport and commercial aircraft against attack from surface-to-air missiles (SAM) attack and MANPADS. PAWS employs IR imagery and signal processing to detect and track approaching missile's hot plume as it appears within a protective sphere surrounding the aircraft. The system uses information on missile trajectory to discriminate between threatening and non-threatening missiles. On detecting the impending missile threat, the system makes a threat assessment by tracking it frame by frame and analysing target manoeuvrability, relative position, inertial data and angular velocity and intensity. The system automatically alerts the pilot with a warning signal and activates appropriate countermeasures such as IR flares. It also provides directional information to DIRCM. PAWS-2 also provides threat information to other EW systems as part of an EW suite, as well as to avionics subsystems. The system is already in use on Israel's F-16s. PAWS-2 has also been installed on Gripen E/F fighter aircraft.

The *AN/AAR-56 Missile Launch Detection (MLD) system* from Lockheed–Martin is one of the key defensive systems used by high-speed fixed-wing aircraft for long-range detection of both surface-launched and airborne missile threats. Mounted on the skin of the aircraft, the MLD system is a complex network of optical components and assemblies that includes six sensors, each with seven lenses, three computer-interfacing processing cards and six low-observable window frame assemblies. AN/AAR-56 MLD system is currently in production for the U.S. Air Force and its operational performance has been defined against all clutter environments. The AN/AAR-56 is mounted on the fifth generation F-22 Raptor of Lockheed–Martin. High-resolution and multi-spectral sensor variants with expanded algorithm base are under development at Lockheed–Martin. These sensors would incorporate situational awareness and defensive IR search and track against airborne targets enabling F-22 Raptor to use AN/AAR-56 as an IRST sensor against aerial targets.

Multi-colour Infrared Alert Sensor (MIRAS) from EADS and Thales is a passive IR sensor based missile warning system designed to protect airborne platforms from the complete range of MANPADS and air-to-air missiles. The system features long-range detection capability, high detection probability, short reaction time and highly reduced false alarm rate enabled by use of single chip multi-colour IR detection technology. Use of band subtraction algorithms enables rejection of background (air, land or sea) and solar reflection while revealing a threat's hot spot. The system comprises one to three Missile Sensor Units (MSU), each sensor unit providing hemispheric coverage. Different sensor units interface with missile warner signal processor. MIRAS can be easily installed and integrated on a variety of transport, tanker and fighter aircraft.

MAW-300 is a passive UV-based missile approach warning sensor. It is part of Integrated Defensive Aids Suite (IDAS) family of self-protection systems designed by SAAB to provide an EOB and self-defence for airborne platforms in diverse and dense threat environments. The IDAS family includes CIDAS (Compact Integrated Defensive Aids Suite), IDAS-1 and IDAS-3. IDAS provided multi-spectral radar, laser and missile warning with automatic countermeasures decoy dispensing. CIDAS is the lightweight variant of IDAS family designed for protection of aircraft against MANPADS and laser threats. IDAS-3 is a high end system configurable with full multi-spectral radar warning, laser-warning and missile approach warning. The modular architecture allows system configuration for any combination of three types of warning sensors. IDAS system is integrated with SAAB's BOP-L countermeasures dispensing system. A BOP-L dispenser is controlled by a fully integrated chaff and flare dispenser controller that resides in the Electronic Warfare Controller (EWC). Dispensing techniques are defined in the threat library of EWC. EWC selects suitable dispensing technique depending upon identified threat. The system has provision for manual, semi-automatic and automatic firing of the dispenser. IDAS has a vast list of customers in Europe, Asia, Africa and the Middle East who have either ordered or installed the system on helicopters, commercial transport aircraft as well as fighter aircraft. Some representative platforms where Integrated Defensive Aids Suite or its components have been installed include utility helicopter *Oryx* by the Atlas Aircraft Corporation, advanced light helicopter *Dhruv* by Hindustan Aeronautics Limited, the medium utility/transport helicopter Puma by Aerospatiale, medium utility helicopter *Cougar* from Aerospatiale/Eurocopter, the military helicopter *Super Lynx 300* by Augusta Westland, heavy lift helicopter *Chinook* from Boeing Aerospace, medium lift utility helicopter *Hawk* by Sikorsky, military transport aircraft C-130 by Lockheed–Martin, the fighter aircraft Su-30 by Russia's Sukhoi Aviation Corporation, the medium-sized, multirole military helicopter NH-90 from NH Industries, Russian *Mi-17 helicopter*, commuter airliner *Embraer-120* by Embraer, multirole fighter aircraft *Gripen* by SAAB, *Erieye* Airborne Early Warning and Control system installed on SAAB-2000 aircraft, turboprop airliner *Dash-8* from Bombardier Aerospace and multirole combat aircraft Tornado from Panavia aircraft GmbH.

The missile approach warning sensor type MAW-300 uses passive UV-based sensors operating in solar-blind UV spectrum. Each sensor provides conical coverage of 110°, which limits the unprotected hole below the platform and allows good overlap. Four sensors provide azimuth coverage of 360°. Full spherical coverage is achieved with six sensors. The system caters for

Figure 6.13 AN/AAR-60 MLDS. (*Source:* Hunini, https://commons.wikimedia.org/wiki/File:JASDF_
UH-60J(58-4598)_AN_AAR-60(V)_and_J_APR-7_at_Komaki_Air_Base_March_13,2016.JPG. CC BY-SA 4.0.)

adding up to eight sensors. MAW-300's multi-threat detection capability allows simultaneous tracking of multiple targets with near 100% probability of warning. MAW-300 comprises multiple sensors and missile approach warning controller. Each sensor has a dedicated MAW controller residing in Electronic Warfare Controller (EWC) of IDAS. The IDAS system has been tested against live missile firings under in-flight dynamic flight conditions.

The *AN/AAR-60 MILDS* (*Missile Launch Detection System*) from EADS (Figure 6.13) is a passive missile warning system configured around one to six UV imaging sensor heads including optics, filters and image processing and capable of detecting and tracking UV emissions from the propulsion systems of potential missile threats including shoulder launched MANPADS. AN/AAR-60 provides azimuth and elevation coverage of 360° and 95°, respectively, and is capable of engaging up to eight targets simultaneously. It can be integrated with a range of countermeasures. The imaging sensor heads of the MILDS system determine the angle of attack and priority. It gives the crew warning of the approaching missile threat enabling them to begin required evasive manoeuvres. The system also initiates proper countermeasures. High spatial resolution of MLDS coupled with its advanced temporal processing significantly increases the probability of detection and virtually eliminate false alarms. AN/AAR-60 MLDS and its variant AN/AAR-60 (V)2 are already in operational use on a variety of helicopters, transport aircraft and fighter aircraft. The airborne platforms using the system include NH-90 tactical support helicopter, Tiger helicopter, CH-47 and CH-53 helicopters, S-70A helicopters, Mi-8, Mi-17 and Mi-171 helicopters, C-130 military transport aircraft, AP-3C Orion aircraft, VIP/HOS A340 aircraft. It is optimized for mounting on pods, pylons and fuselage and is also in operational use with Air Forces operating F-16 aircraft.

AN/AAR-60 MLDS is the heart of the Airborne Missile Protection System (AMPS) developed by AIRBUS Defence and Space. In addition to AN/AAR-60 MLDS, it also features an Electronic Warfare Suite Controller (MCDU: Figure 6.14), an Inertial Measurement Unit (IMU) as well as a fully integrated Countermeasure Dispensing System (CMDS).

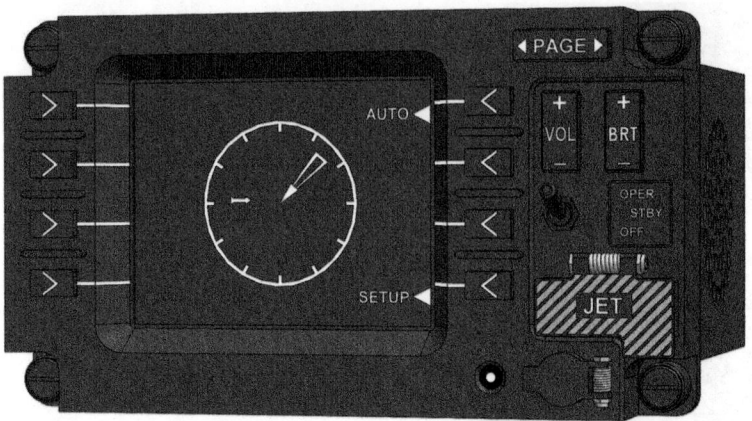

Figure 6.14 MCDU from AMPS.

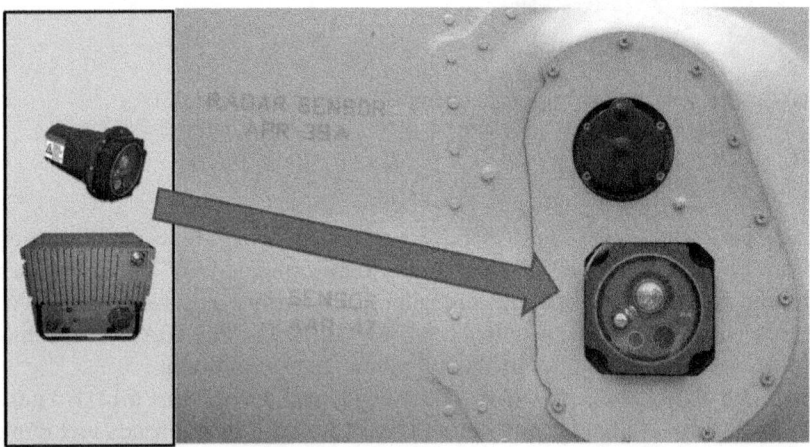

Figure 6.15 AN/AAR-47 Missile Warning System.

The *AN/AAR-47* (Figure 6.15) is an EWS providing a multi-threat warning in one integrated system against laser assisted threats, IR-guided missile threats and unguided munitions. Upon detection of a potential threat, the system provides audio and visual warning to the pilot. In the case of detection of an IR-guided missile threat, the system automatically initiates countermeasures by sending a command signal to the countermeasures dispensing set. The AN/AAR-47 is characterized by high probability of detection and timely warning and low false alarm rate. It includes sensor pre-processing for improved performance in high-clutter environments. The system interfaces with the radar warning receiver (AN/APR-39), countermeasures dispensing sets (AN/ALE-39, AN/ALE-40 or AN/ALE-47), multifunction display cockpits and the pilot's intercom system. Different versions of AN/AAR-47 include AN/AAR-47(V)1, AN/AAR-47(V)2, AN/AAR-47A (V)2, and AN/AAR-47B (V)2.

The system is in service on a wide variety of fixed- and rotary-wing aircraft. It is installed on AH-1W/Z, UH-1N/Y, CH-53D/E, HH/SH/MH-60, P-3, C- 130, CH-46, V-22 (US Navy and Marine Corps); C-5, C-141, C-17, MH-53J, HH-60G, C-130E/H/J, HC-130N/P, A-10, C-27J (US Air Force); and DHC-7, MH-47E, MH-60 (US Army). AN/AAR-47 is also installed on S-61/ SH-3, AH-6/MD-530, C-12, AW-139, B212/214/412, DHC-8, HB-350, CN-235, BT-67 and C208.

6.4 Electronic Countermeasures (ECM)

While different types of warning receivers, such as RWRs, LWRs and MAWs discussed in the earlier part of the chapter, can be used to detect and identify the enemy's sensors and weapons to enable the target take evasive action or initiate countermeasures, the stealth technologies discussed in the latter part of the chapter help the target avoid the cover of those sensors by being undetectable to them until it has gone very close to the sensor. ECMs are used to degrade the performance of enemy's sensors and weapons like reducing the detection range of a radar or forcing the missiles to deviate from their intended trajectory and move away from the target. ECMs basically buy time for the user to evade enemy's defences during the crucial stage of an engagement or battle. In the paragraphs to follow, we shall briefly describe both passive and active countermeasure techniques including different types of jamming, use of chaff and different types of decoys.

6.4.1 Jamming Techniques

Jamming refers to transmitting some kind of signal so as to either swamp the target return or other wanted signals or to induce range, velocity or angle errors into the target system. The former is known as *noise jamming* and the latter goes by the name of *deception jamming*. While a noise jammer is usually effective against a much wider range of threats, a deception jamming system is limited in its application to a very small group of threats. The deception jammer requires transmission of much lower jamming power, is smaller in size but is very complex in design. On the other hand, a noise jammer requires higher power transmission in the jamming signal and is larger in size and volume. The noise jammers are sometimes used by the enemy sensors to their own advantage when they use the jamming signal as a beacon and select the home-on-jam mode in a guided missile. The two types of jamming technique are described in the following paragraphs.

6.4.1.1 Noise Jamming

A *noise jammer* basically transmits a noise like signal with its main objective being to maximize the noise power in the receiver's bandwidth. The noise signal completely masks the desired signal, which could be a communication signal or an echo signal in the case of radar. The efficacy of noise jamming clearly depends upon the ratio of the desired signal strength to the jamming signal strength and also on the nature of the noise signal. The first condition is obvious and the second arises from the fact that the noise signal should be such that it is indistinguishable from the receiver's internally generated noise once it enters the receiver. It is because of this reason that the ideal noise waveform is white Gaussian noise, that is, it has a Gaussian amplitude distribution and a uniform spectral density.

Some of the methods used for generating the noise signal include *direct noise amplification* (DINA), where thermal noise is amplified over the bandwidth of interest using wideband amplifiers such as TWTs, *frequency modulation* of a carrier with wideband noise or a combination of wideband noise and a sawtooth or a similar waveform, *frequency modulation* of carrier by a narrowband noise, *amplitude modulation* by noise and so on.

The performance of a noise jammer can be explained by writing the equation for the jamming signal to the desired signal ratio at the receiver. In the case of a radar receiver, it is expressed by eqn. 6.1.

$$\frac{J}{S} = \frac{4\pi P_J G_J G_{RJ} R_T^{\,4}}{P_T G_{\mathrm{T}^2} \sigma R_J} \tag{6.1}$$

J = Received power due to the jamming signal (undesired signal)
S = Received power due to the target (desired signal)
P_J = Jammer power
G_J = Jammer antenna gain in the direction of the radar
R_J = Range from jammer to radar
G_{RJ} = Radar antenna gain in the direction of jammer
P_T = Radar peak power or equivalent power if radar integrates a number of pulses
G_T = Radar antenna gain in the direction of the target
R_T = Range from radar to target
λ = Radar signal wavelength
σ = Radar cross-section of the target

If the jammer signal bandwidth B_J is greater than that of the receiver B_R, the J/S equation gets modified to eqn. 6.2.

$$\frac{J}{S} = \frac{4\pi P_J G_J G_{RJ} R_T^{\,4}}{P_T G_{T^2} \sigma R_J^{\,2}} \times \frac{B_R}{B_J} \tag{6.2}$$

Ratio $\left(\dfrac{B_R}{B_J}\right)$ is taken to be equal to unity in the case of B_J being equal to or less than B_R.

6.4.1.1.1 Self-Protection Jammer

If the noise jammer is carried on the aircraft for self-protection, as is the case for a *self-protection jammer*, then the aircraft itself is also the radar target. In that case $R_J = R_T$ and $G_{RJ} = G_T$ and the *JIS expression* gets further modified to eqn. 6.3.

$$\frac{J}{S} = \frac{4\pi P_J G_J B_R R_T^{\,2}}{P_T G_T \sigma B_J} \tag{6.3}$$

The equation for J/S can be used to estimate the range at which the radar starts detecting the target or the jammer stops shielding it from the radar detection. This range, called the *burnthrough range*, can be determined for a self-protection jammer from eqn. 6.4.

$$R_B = \left[\frac{P_T G_T \sigma}{4\pi P_J G_J} \times \frac{B_J}{B_R} \times \frac{1}{S/J} \right]^{1/2} \tag{6.4}$$

The derivation of eqn. 6.4 assumes that the radar receiver noise is much smaller than the jamming noise and is therefore ignorable. Equation 6.4 also assumes that the jamming signal has the same polarization as that of the radar signal. In general, the jamming effectiveness falls off as the self-protection jammer approaches the radar because the target signal as received by the radar is varying as $(1/R)^4$ whereas the jamming signal varies as $(1/R)^2$ only.

6.4.1.1.2 Standoff and Escort Noise Jamming

In the case of *standoff* or *escort jamming*, the noise jammer is not carried by the aircraft to be protected. Instead, it is carried by another aircraft flying outside the range of local air defences and providing the noise jamming power to protect the attacking aircraft as it flies to and from the target in the case of standoff jamming. In the case of *escort jamming*, the jammer is onboard a specialized ECM aircraft that escorts the attack aircraft. The burnthrough range of a standoff jammer is much greater than that of a self-protection jammer.

6.4.1.1.3   Spot Jamming

A *spot jammer* is the one that concentrates its power into a narrow band centred on the frequency of the enemy's sensor approximately matching its bandwidth. Therefore, for given effective radiated jammer power, a spot jammer produces maximum noise power spectral density in the receiver.

6.4.1.1.4   Barrage Jamming

A *barrage jammer* is a broad band jammer. It spreads its radiated energy over a large bandwidth. The noise power spectral density at the receiver gets diluted in the case of barrage jamming.

6.4.1.1.5   Swept Spot Jamming

A *swept spot jammer* provides the advantage of high noise power spectral density of a spot jammer and large bandwidth of a barrage jammer. It does so by rapidly sweeping the spot noise signal across a wide band of frequencies. This jammer can be made very effective by appropriately choosing the sweep rate and the bandwidth.

6.4.1.1.6   Cross-Polarized Jamming

Cross-polarized jammer radiates a jamming signal whose polarization is at 90° to the polarization of the wave radiated by the radar to be jammed. It is particularly effective against radars that employ sidelobe blanking technique to improve the quality of their PPI display. The sidelobe blanking ensures that the radar does not respond to the signals entering via side lobes. It does so by using an auxiliary antenna channel whose effective gain is lower than that of the radar antenna main lobe but higher than the radar antenna side lobes. The radar receivers does not respond to any signal whose amplitude in the auxiliary antenna channel is greater than that of the main antenna channel (Figure 6.16). If the jamming signal has a cross-polarized component of significant strength; there is every possibility of cross-polarized main lobe pattern (shown dotted in Figure 6.16) falling below the effective gain of auxiliary antenna with the result that all signals in the main lobe could get suppressed while the side lobe signals would be accepted. One way to reduce the sensitivity of the radar receiver to cross-polarized jamming is to use another auxiliary channel with a cross-polarized antenna with its gain being larger than the cross-polarized gain of the main antenna and much less than its gain for the designed polarization.

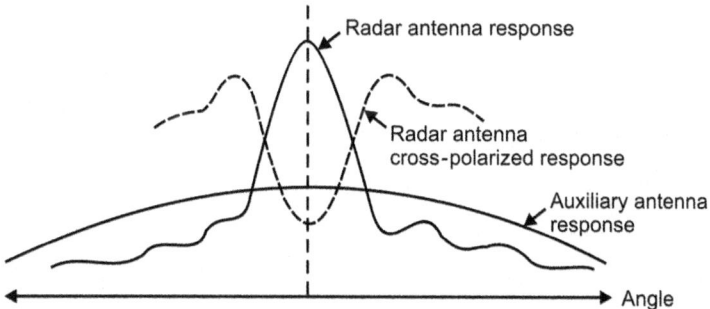

Figure 6.16 Cross-polarized jamming.

6.4.1.2 Deception Jamming

While a noise jammer attempts to swamp the radar echo signal by transmitting towards the radar a noise signal of sufficient power over the bandwidth of the radar receiver, a *deception jammer* transmits a corrupted signal towards the radar that it accepts as a genuine one. A deception jammer in most of the cases accepts the radar signal, carries out a suitable modification and then retransmits the same toward the radar. The modification usually involves changing an appropriate characteristic of the signal so as to induce deception in range, velocity or angle errors. In other words, the jammer creates a number of false targets for the radar to detect.

Figure 6.17 shows the block-schematic arrangement of a basic deception jammer. The jammer accepts the radar signal, which is then modified as per the deception programme. The modified signal is then amplified before it is retransmitted. A deception jammer is usually a self-protection device, one of the basic requirements for the aircraft. The required total gain of the system is given by eqn. 6.5.

$$\left[\frac{4\pi\sigma}{\lambda^2} \times \frac{J}{S} \right] \tag{6.5}$$

6.4.1.2.1 Deception in Range

Range deception is particularly effective against search and track radars. The jammer in this case creates a large number of false targets at different ranges and angles from the true target position and thus confuses and overloads the radar operator as well as radar processing circuits. So that false target pulses radiated by the jammer are indistinguishable from the real target echo pulses, one of the methods is to transmit replicas of radar pulses shortly after or before the arrival of the radar pulse. If the radar pulses are stable in terms of PRF, RF, their arrival time can be anticipated and false target pulses could possibly be transmitted either before or after arrival of the pulse. But in case the radar employs a varying PRF or pulse to pulse RF agility, the false target pulses can be sent only after the receipt of the real radar pulses. The range deception jammer first receives the radar pulse, amplifies it to make it stronger than the target skin echo and transmits it back towards the radar without introducing any change in the range information. This is done to ensure that the signal is received by the radar range gate along with the genuine signal. The stronger false signal overloads the processing circuitry and the operator tends to reduce the sensitivity of the receiver. The false signal this way succeeds in capturing the range gate. Having done that, the jammer progressively introduces a change in range all the time staying within the range gate, the phenomenon being known as the *Range Gate Pull-Off* (RGPO).

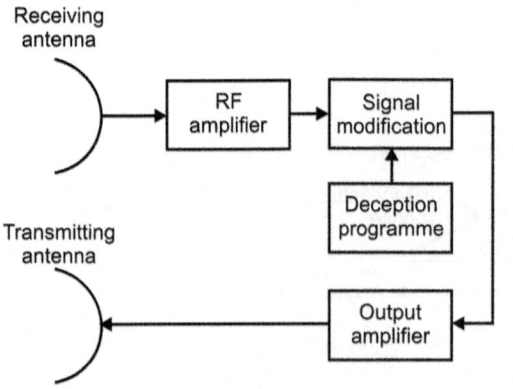

Receiving antenna

Transmitting antenna

Figure 6.17 Block-schematic arrangement of a deception jammer.

6.4.1.2.2 Deception in Velocity

A *velocity deception jammer* maximizes jammer energy in radar's Doppler filters. The deception jammer first receives the radar signal, amplifies it and then retransmits it without any change in frequency so that it is received by the Doppler filter processing the skin echo signal. The jammer signal is kept 5–10 dB stronger so that it captures the filter and the velocity gate. The jammer then progressively shifts the RF of its signal pulling the target away from the true target, a phenomenon similar to that of RGPO.

6.4.1.2.3 Angle Deception

Inducing range and velocity errors degrades the tracking performance of the radar. The tracking errors induced, however, are very small. If large tracking errors are required to be induced; angle deception would be needed. The techniques used for angle deception would depend upon the tracking methodology employed by the radar. For instance, technique used for tracking radar employing sequential lobing is entirely different from the deception technique to be employed against radar using monopulse tracking. Detailed description of these techniques would be beyond the scope of the present text. Modern aircraft employ jammers that have the capability to produce a variety of jamming signals. These jammers could either be pod mounted or configured inside the aircraft. The airborne noise and deception jamming system (Type *AN/ALQ-167(V)*) is an example.

6.4.2 Chaff

Chaff and flares are defensive mechanisms employed from military aircraft to avoid detection and/or attack by adversary air-defence systems. *Chaff* is basically an airborne cloud of light weight reflecting objects typically consisting of strips of aluminium foil, metal coated fibres or nylon that produce clutter echoes in a region of space for the purpose of confusing, screening or otherwise adversely affecting the opposition sensor system such as radar. Chaff elements are dipoles cut to a length that is approximately equal to half of the wavelength of the radar radiation. It is packaged in the form of cartridges that can be easily deployed. AN/ALE-40 and its improved version AN/ALE-47 Countermeasure Dispenser Systems from BAE Systems, BOL Advanced Countermeasure Dispenser from SAAB and BOZ-107 from SAAB Bofors Dynamics are some representative examples of chaff/flare dispenser systems. These and some more representative countermeasures systems including ECM jammers, chaff/flare dispensers and decoys are briefly discussed in a latter section. Each cartridge carries a range of dipoles of slightly varying lengths so as to make the chaff effective over a wide frequency band. These are self-protection systems providing protection against radar- and IR-guided weapons.

The important characteristics of chaff include its fall rate, bloom time, effective echoing area, polarization and bandwidth. *Fall rate* is the rate at which chaff falls to the ground and it depends upon factors like type of chaff (aluminium foil, aluminized fibre, silver coated nylon etc.), altitude from which it is dispersed and local meteorological conditions. Typical fall rate in still air is 0.5 m/s for aluminium foil, 0.6 m/s for silver coated nylon and 0.3 m/s for aluminized fibre. *Bloom time*, as the name suggests, is the time taken by the chaff to reach its maximum echoing area or a sufficient echoing area for a given application. It mainly depends on the method employed to disperse the chaff. It is an important feature and the requirements vary depending upon the application. For instance, chaff used for self-protection must bloom very rapidly, typically in a few milliseconds. The *Echoing Area* depends upon the size of the chaff and also the bandwidth it is supposed to cover. A narrow band chaff of a given size provides a much larger echoing area than a broad band chaff of the same size. *Polarization* is another important feature. As the chaff cloud is out of the turbulent region; majority of the dipoles acquire a horizontal attitude.

A portion of the chaff cloud that acquires a vertical orientation tends to fall more rapidly and thus offering a lower echoing area. The polarization performance of the chaff must be considered while deciding the quantity of chaff required for a given application. Chaff *Bandwidth* depends upon the variety of dipole lengths packed into a dispenser and subsequently forming the chaff cloud. Another very important feature of the chaff cloud is its spectrum. Ideally, the chaff cloud should produce the same Doppler spectrum as the one produced by the aircraft the chaff is supposed to protect. The *chaff spectrum* is usually narrow and depends upon the wind velocity because most of the time it spends in the air, it drifts with the wind. However, if the chaff is dispensed into a turbulent atmosphere, for the brief period, it would have a wide Doppler spectrum and this additional spectral width of the chaff spectrum can succeed in breaking the lock of tracking radar. There are other techniques that can be used to overcome the spectrum limitations of the chaff. A discussion on those techniques is beyond the scope of the text.

6.4.3 Flares

Flares are nothing but high-temperature heat sources ejected from an airborne platform such as aircraft to mislead or decoy heat-seeking missiles away from their intended trajectory towards the targeted aircraft. They do so by achieving burn temperatures greater than those generated by the exhausts of aircraft, thereby producing IR signatures similar to those characterizing the aircraft to be protected. Flares have been extensively used as IR countermeasures by aircraft with the objective of providing protection against IR-guided surface-to-air and air-to-air missiles.

There are two types of flares, namely *pyrotechnic* and *pyrophoric*. *Pyrotechnic flares* produce highly visible white light and smoke when ignited (Figure 6.18). They generate intense IR energy for time duration of 5–10 s, which is enough for confusing or distracting the attacking

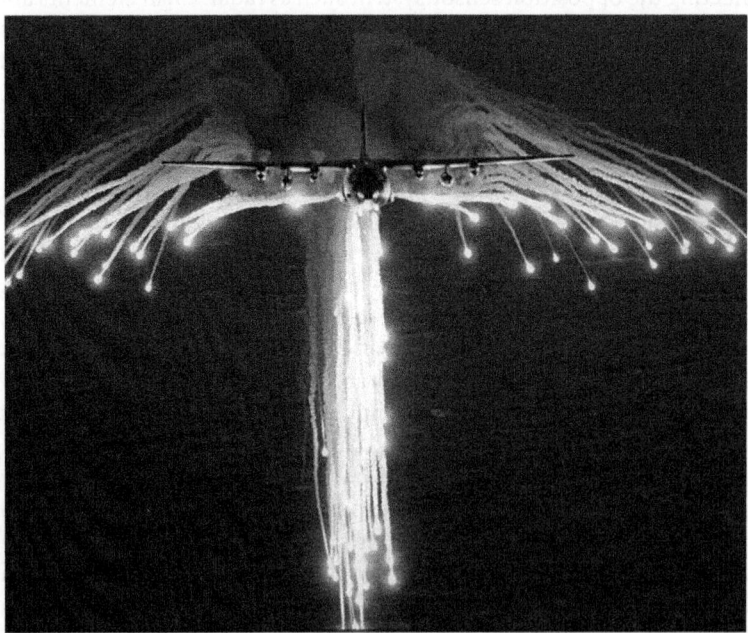

Figure 6.18 Ejection and ignition of a typical pyrotechnic infrared decoy flare. (*Source:* TidusTia, https://commons.wikimedia.org/wiki/File:C-130_Hercules_10.jpg.CC BY-SA 3.0.)

missile. Military aircraft generally carry pyrotechnic flares. Some flares, to be more effective, even contain a propulsion system for propelling the flare over a flight path similar to that of the aircraft to be protected but in a different direction. *Pyrophoric flares* are small pieces of foil and are much less visible. They oxidize very quickly producing heat thereby generating IR energy. They cool in the atmosphere as they fall to the ground as rusted metal debris.

Military self-protection flares vary in composition. The flare has a primary flare body to which are attached additional compounds that are more sensitive than the main flare body. The primary flare body comprises a moulded mixture of magnesium and polytetrafluoroethylene (Teflon). Additional compounds include the first fire mixture, the intermediate fire mixture and the dip coat. They help to ensure proper ignition. The flare material is wrapped with an aluminium filament reinforced tape and then packaged in a primarily aluminium casing. The aluminium case is closed with a felt spacer and a small plastic end cap. The top of the case has a pyrotechnic impulse cartridge that is activated electrically to push flare material out of the aircraft into the air stream. Intensity of the emitted IR radiation and its spectral distribution are the important parameters of IR flares. Parameters such as the mass flow rate, the energy of the chemical reaction or the emitting species have been exploited to achieve desired results.

Pyrophoric flares are made of thin foils or screens of nickel or iron, or steel or alloys of these materials with each other or with other metals. When a cartridge containing pyrophoric flares is fired, the pyrophoric foils are blown out in a scattered mass. Their blocking coating promptly evaporates leading them to undergo pyrophoric reaction as they flutter down. When discharged at high altitudes, the temperature dwell time is significantly longer than what it would be at sea level. Operational analysis has revealed that pyrophoric flares hold a lot of promise in decoying new generation of heat-seeking missiles as the spectral signature of pyrophoric materials such as alkyl aluminium compounds more closely resembles a jet aircraft's spectral signature. The dual spectral flare countermeasures already deployed by warfighters is another significant development.

The Alliant Kilgore Flares Company and Tracor have pioneered development and manufacture of various kinds of countermeasure products that offer protection against IR-guided missiles and other air-defence systems. While Alliant Kilgore develops and produces IR flares such as M-206, MJU-7A/B, MJU-10/B, MJU-32/B and MJU-38/B and a wide spectrum of pyrotechnic devices, Tracor is best known for its AN/ALE-47 countermeasures dispenser system, which is the most advanced system capable of deploying chaff, flares, active radio frequency decoys and other decoys from military aircraft.

6.4.4 Decoys

A *decoy* is another electronic countermeasure. There are passive as well as active decoy systems. A decoy is a dummy radar target much smaller in size than the aircraft it is intending to protect but whose reflectivity, or in other words the radar cross-section, is made larger than the actual target by using reflectors or other means. It confuses the radar operator as well as the radar processing circuits and thus increases the survivability of the attacking aircraft. Chaff discussed in the preceding paragraphs can also be considered as an expendable radar decoy but it has its own limitations, a narrow Doppler spectrum being one of the major ones. Decoy provides solution to some of these limitations. There are two commonly used modes of deployment of decoys. We have *towed decoys*, which are towed by the vehicle to be protected and *free falling decoys* deployed in a fashion similar to that of chaff.

The purpose of the towed decoy (which could be passive or active) is to provide an echo signal that is much stronger than that from the towing vehicle so as to force the missile tracker or radar to lock-on to the decoy rather than the actual target. It has an obvious advantage that the radar or the missile seeker cannot distinguish the decoy from the target on the basis of velocity discrimination. As long as the two, towing vehicle and decoy, remain unresolved by the sensor (radar, missile seeker etc.), the sensor will either track the signal centre of gravity of vehicle-decoy system or lock-on to the stronger source of signal return; that is, the decoy. In the case of aircraft towed decoys, the separation between the aircraft and the decoy is a few hundred metres. This countermeasure is particularly effective against semi-actively guided missiles. However, the efficacy of the system strongly depends upon the engagement geometry. The *Ariel towed radar decoy* from Selex Galileo, a Finmeccanica company, is an example of off-board electronic countermeasures to protect aircraft from radar directed weapons. AN/ALE-50 towed decoy from Raytheon and AN/ALE-55 fibreoptic towed decoy from BAE Systems are other examples.

Miniature active expendable decoys are also being increasingly used. This has been made possible by rapid advances in the field of Monolithic Microwave Integrated Circuits (MMlCs). Essentially, these decoys are miniaturized transponder jammers comparable in size to chaff dispensers or cartridges. Once deployed, they receive, modify and re-transmit an amplified version of the radar signal to simulate semblance of a real target that they are trying to protect. In fact, they are miniaturized version of deception jammers. When fired from chaff dispensers, they fall free. They can be fitted to a rocket and fired forward from the aircraft. These decoys have very stringent design requirements particularly due to their small size and mode of deployment. They should be capable of withstanding the launch forces, must have a long storage life once integrated with their storage batteries and so on.

6.4.5   Representative ECM Systems

In the preceding sections, we have discussed various categories of electronic countermeasure (ECM) systems including different types of jammers, chaff, IR flares and decoys. In this section, we shall briefly describe some representative systems belonging to each of these categories. The ECM systems discussed in this section include the *VHF/UHF Tactical Jamming System AN/ALQ-99* from Raytheon, *Self-Protection Jamming Pod AN/ALQ-184* from Raytheon, *Radar Jamming System AN/ALQ-187* from Raytheon, *Miniature Air-Launched Decoy Jammer (MALD)* from Raytheon, *Next-Generation Jammer (NGJ)* from Raytheon, *AN/ALE-40* and its improved version *AN/ALE-47 Countermeasure Dispenser Systems* from BAE Systems, *BOL Advanced Countermeasure Dispenser* from SAAB and *BOZ-107* from SAAB Bofors Dynamics, Chemring Kilgore's M-206, MJU-7A/B, MJU-10/B and MJU-32/B, MJU-49/B, MJU-50/B and MJU-51/B IR flares, *Ariel towed radar decoy* from Selex Galileo, *AN/ALE-50 towed decoy* from Raytheon and *AN/ALE-55 fibreoptic towed decoy* from BAE Systems.

The AN/ALQ-99 is an airborne jamming system capable of intercepting and then jamming the received RF signals. The system was designed and manufactured by EDO Corporation, acquired by ITT Corporation in 2007. The system comprises multiple receivers that monitor threats in four specific frequency bands and jamming transmitters. The receivers and antennas are mounted in a fin-tip pod and the jamming transmitters are carried in individual pods underneath the wings and the fuselage centre line. The RF signals intercepted and processed by the receivers are fed to a central mission computer to determine signal characteristics and also the direction-of-arrival of the signal. This information is used to operate jamming transmitters. Different versions of the AN/ALQ-99 jammer include AN/ALQ-99, AN/ALQ-99A that has enhanced frequency coverage with eight frequency bands instead of four in the case of AN/ALQ-99, AN/ALQ-99B (an upgraded version of AN/ALQ-99A with improved reliability, AN/ALQ-99C (an upgraded AN/ALQ-99B), AN/ALQ-99D introduced with EA-6B ICAP-II covering a wider frequency range, AN/ALQ-99E developed for the EF-111A Raven aircraft

Figure 6.19 AN/ALQ-184 ECM Pod. (*Source:* Courtesy of US Airforce.)

operated by the US Air Force offering much faster threat acquisition and identification and the AN/ALQ-99F that replaced the AN/ALQ-99D used by EA-18 Growler. The AN/ALQ-99 has been used in a number of wars, which includes the Vietnam War (1972–1973), Operation El Dorado Canyon (1986 American raid in Libya), the 1991 Gulf War, Operation Northern Watch (1992–2003), Operation Southern Watch (1997–2003), the 1999 Balkans War, the 2003 Second Gulf War, and the 2011 Operation Odyssey Dawn.

The *ALQ-184* (Figure 6.19) from Raytheon is a self-protect electronic countermeasures (ECM) pod designed to provide protection to the aircraft platform and the aircraft crew against radar-guided threats. It is a radar jammer that primarily comprises computer-controlled multi-beam receivers and travelling wave tube (TWT) amplifiers and operates in both receive and transmit modes to direct high power jamming signals against emitters. It can also tow AN/ALE-50 RF decoy. The AN/ALQ-184 is characterized by increased effective radiated power (ERP), reduced countermeasure response time, high probability of detection, frequency independent direction-finding on every received signal and improved reliability, maintainability and availability over previous self-protection pods used by the United States Air Force. The AN/ALQ-184 ECM pod is an improved version of AN/ALQ-119 noise and deception jamming pod modified with a Raytheon-supplied kit replacing most of the analogue hardware by plug-in circuit boards employing digital technology for the upper two frequency bands, and substituting electrically scannable antennas for the fixed antennas. AN/ALQ-184 ECM pod is used on F-16 Fighting Falcon and A-10 Thunderbolt tactical aircraft. It is also used on the F-4 Phantom II, F-111, F-15 Eagle, A-7 and C-130 Hercules aircraft. Different variants of AN/ALQ-84 are available in two-band (mid-band and high-band) and three band (low-band, mid-band and high-band) versions.

The *AN/ALQ-187 ECM pod* from Raytheon is an internally mounted version of the AN/ALQ-184(V). Advanced Countermeasure Electronic System (ACES) from Raytheon designed to detect, identify and counter contemporary threats in a high density environment is the next-generation of electronic warfare systems for the F-16 aircraft. It integrates Raytheon's ALQ-187(V)2 jammer, the ALR-93 radar warning receiver and the ALE-47 countermeasures dispenser system.

Figure 6.20 MALD. (*Source:* Courtesy of US Airforce.)

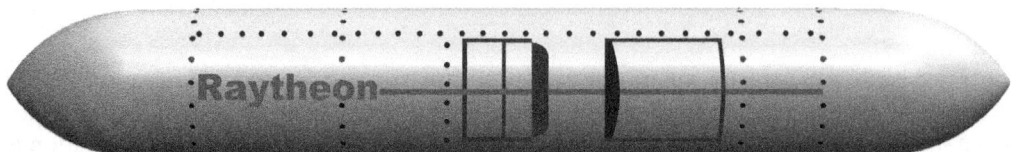

Figure 6.21 Next Generation Jammer Pod.

The *Miniature Air-Launched Decoy – Jammer* (*MALD-J*) from Raytheon is the jammer variant of the miniature air-launched decoy (MALD), which is an expendable air-launched flight vehicle with combat profiles and signatures such as the radar cross-section of fighter and bomber aircraft particularly of the United States and allied forces. Figure 6.20 shows a photograph of multiple MALDs mounted on the aircraft. After it is launched from its host aircraft, MALD flies a pre-programmed mission. The jamming variant of MALD can work both as a traditional decoy as well as a tactical jamming and spoofing asset, similar to the EA-18G Growler. Although it doesn't have the same capability as the EA-18G Growler aircraft in terms of number of threats that can be handled at one time or the operational ranges, it makes up for it by its capability to operate in close proximity to the target. A swarm of MALD-Js, even paired with standard MALDs, could completely overwhelm, confuse and blind a whole lot of air-defence installations along a certain route and during a certain period of time, which would allow attack assets to safely traverse the threat area on the way to their targets. MALD-J operates much closer to the victim radar during jamming operation and is able to loiter in the target area for an extended period of time, allowing it plenty of time to complete the mission. Though MALD-J can operate alone or in a swarm formation, it is designed to work with and leverage other electronic warfare platforms. MALD and MALD-J is used with F-16 C/D and B-52H aircraft.

The *Next-Generation Jammer* (*NGJ*) *electronic attack pod* (Figure 6.21) is being developed by Raytheon as a replacement for the ALQ-99 tactical jamming pod on the US Navy's EA-18G Growler aircraft. The system represents a marked improvement over ALQ-99 jamming pod in terms of its software-based digital architecture and use of high power active electronically scanned arrays based on gallium nitride technology. The pod is designed to provide enhanced airborne Electronic Attack (EA) capabilities to degrade and disrupt adversary's air-defence and ground communications systems. The development of the next-generation pod has transitioned into its engineering and manufacturing development phase. The next generation pod integrates the most advanced electronic attack technologies facilitated by use of a combination of high-powered, agile beam-jamming techniques and cutting edge solid-state electronics to meet the US Navy's current mission requirements. Its open systems architecture is also conducive to future upgradation.

The *AN/ALE-40 Counter Measure Dispenser System* (CMDS) manufactured by Tracor, now part of BAE Systems, is based on two types of expendable countermeasures namely *chaff* to protect the host aircraft from radar and radar-guided threats and *flares* to offer protection against IR heat-seeking missiles. The chaff payload of AN/ALE-40 comprises tiny strands of aluminium foil cut to specific lengths to match specific wavelengths of enemy's radars. The flare payload launches flares typically consisting of white hot magnesium that burns at a temperature to simulate IR signatures of the aircraft to be protected so as to decoy the approaching missile to take the flare for the actual aircraft.

The *AN/ALE-47 Countermeasure Dispenser System* (CMDS) manufactured by BAE Systems is an integrated, reprogrammable, computer-controlled system that uses information from integrated electronic warfare sensors such as RWRs and MAW receivers to determine the type of countermeasures to be dispensed by the crew to defeat IR- and radar-guided missiles. The aircrew can select the mode of operation of the dispenser for fully automatic, semi-automatic and manual operation. It is interchangeable with obsolete AN/ALE-40/39 and M-130 systems. The AN/ALE-47 is the standard aircraft countermeasures dispensing system for the United States Army, Air Force, Navy and Marine Corps. Other than the United States, AN/ALE-47 countermeasures dispenser system has its presence in a large number of countries including Argentina, Belgium, Botswana, Brazil, Chile, Egypt, Finland, Greece, Italy, Japan, Jordan, Kuwait, Malaysia, Netherlands, New Zealand, Norway, Poland, Portugal, Singapore, South Korea, Sweden, Taiwan, United Kingdom, Australia, Canada, Spain, Switzerland and Turkey. The AN/ALE-47 offers seamless integration onto many different types of platforms including fixed-wing fighter aircraft and rotary-wing helicopters. It is installed on the A-4, A-7, F-2, F-4, F-15, F-16, F/A-18, F-5, Hawk, P-3, E-8C, C-5, C-17, C-27A, C-130, C-130J, C-141, C-295 and C-235 (fixed-wing aircraft), UH-1N, SH-2G, MH-47, MH-60, HH-60, S-70B, EH-101, MV-22, Superlynx, AH-1W, VH-3D, CH-47, UH-60, SH-60J, BO-105, Bell 412, CV-22, Puma and AH-64 (rotary-wing helicopters). Figure 6.22 shows a photograph of Boeing's C-17 Globemaster-III large transport aircraft deploying IR flares from AN/ALE-47 CMDS installed on the aircraft.

The *BOL Countermeasure Dispenser System* from SAAB is a high-capacity dispenser of chaff or IR flares used by the crew to protect the host aircraft from radar and radar-guided threats and IR heat-seeking missiles including the advanced IR-guided missiles. One of the significant features of the countermeasures dispenser system is its high payload-to-volume ratio made possible by its elongated shape housing a long stack of chaff and IR flare payload packs (Figure 6.23) giving aircraft crew the sustained defensive capability needed for successful accomplishment of the mission. The other important features include a non-pyrotechnic electro-mechanical release mechanism forcing initial dispersion of the payload, which is further enhanced by BOL internal vortex generators (air scoops) and vortex fields behind the aircraft, multiple interface options including an MIL-1553 data bus and a RS-485 data link as well as 28 V discretes and easy installation alternatives for retrofit and also for new aircraft. The aircraft already equipped with pyrotechnic dispensers, BOL retrofit can significantly increase its total payload capacity thereby enhancing protection. It is currently operational on a number of

Figure 6.22 Infrared flares deployed from AN/ALE-47 CMDS installed on a C-17 Globemaster-III aircraft.

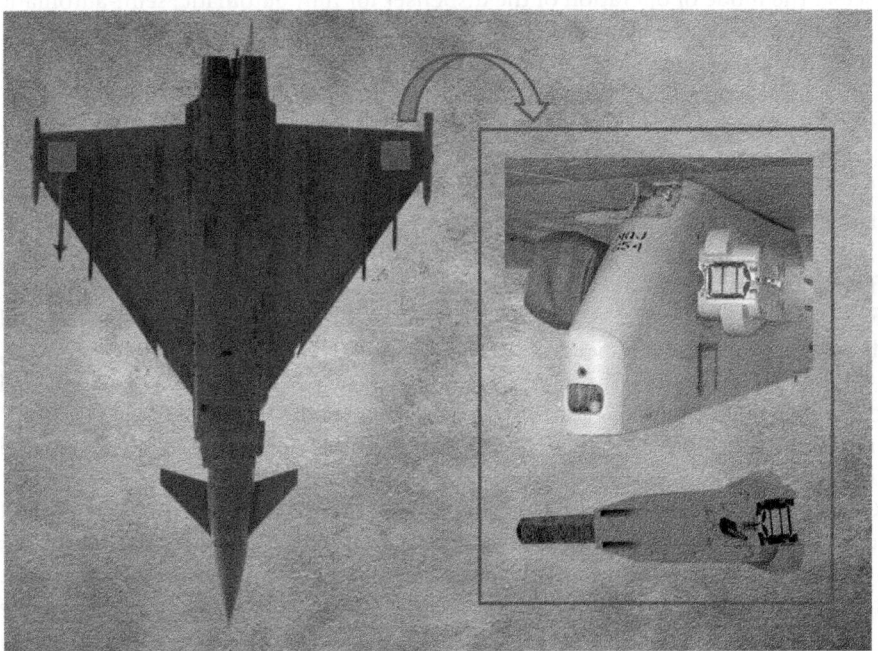

Figure 6.23 BOL Countermeasure Dispenser.

fighter aircraft including F-14, F-15, Gripen, Typhoon, SAAB-2000 AEW&C (Figure 6.24) and RAAF F-18. Interest for mounting on transport and bomber aircraft has also grown significantly. BOL has been successfully integrated with a range of missile launchers including the LAU-7 Sidewinder launcher, the LAU- 127/128/129 family of AMRAAM launchers, and the Conventional Rotary Launcher (CRL) and Multipurpose Rocket Launcher (MPRL).

Figure 6.24 BOL Countermeasures Dispenser integrated with SAAB-2000 AEW&C. (*Source:* MilborneOne, https://commons.wikimedia.org/wiki/File:Saab2000AEW-SE-045-246.jpg.CC BY-SA 3.0.)

Figure 6.25 BOZ-101 dispenser pod.

The *BOZ-107 Countermeasures Dispenser System* from SAAB Bofors Dynamics pod is a self-contained, microprocessor-controlled, reprogrammable dispenser capable of releasing either a chaff cloud as decided by the crew depending upon the threat scenario. BOZ-101, a model produced for the German Air Force Luftwaffe (Figure 6.25) and BOZ-102, a model produced for Italian Air Force Aeronautica Militare are the other versions of the system. BOZ-107 is specifically designed for the Royal Air Force's Tornado GR4 aircraft. It is one of the main ECM defence systems installed on the multirole Tornado GR4 aircraft, one of the versions of the Tornado-IDS (Tornado-Interdictor/Strike) aircraft. Different versions of Tornado-IDS include Tornado-GR1, Tornado-GR1A, Tornado-GR1B, Tornado-GR4, Tornado-GR4A and Tornado-IDS. The countermeasures dispenser is used to protect the aircraft from radar and IR-guided missiles. It is generally applied in conjunction with Sky Shadow ECM pod of GEC-Marconi. It uses either the Bofors BOZ-EC, which is CIDAS-100 Compact

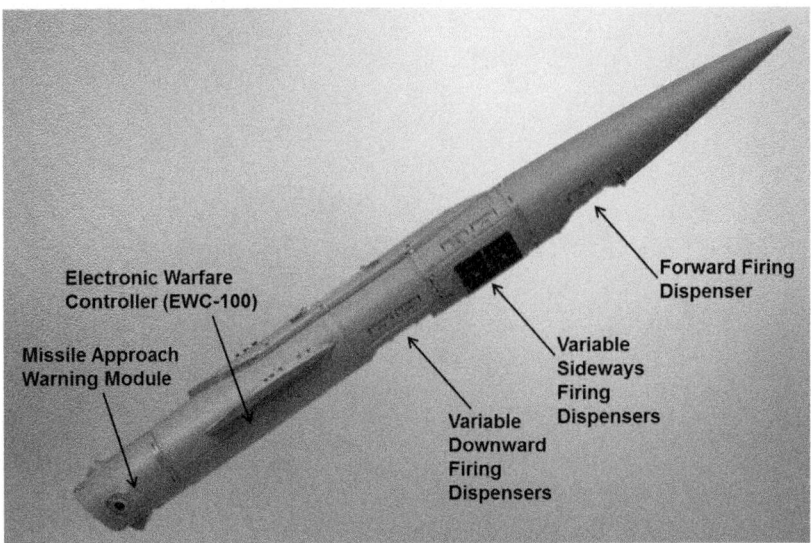

Figure 6.26 BOZ-EC Countermeasures pod.

Figure 6.27 IR decoy flares.

Integrated Defensive Aids Suite in BOZ pod (Figure 6.26), or the Bofors BOZ-100 IRCM (InfraRed CounterMeasures) as its sensor systems.

The Kilgore Flares Company, now Chemring Countermeasures USA, produces a wide range of IR decoy flares in different shapes, sizes and form factors (Figure 6.27) to provide protection to aircraft from heat-seeking missiles by decoying them away from the aircraft.

Some of these flares include the M-206, MJU-7A/B, MJU-10/B, MJU-32/B, MJU-49/B, MJU-50/B and MJU-51/B. These IR decoy flares are in use on some of the worlds most sophisticated and advanced airborne platforms. The M-206 decoy flare is a pyrotechnic flare measuring $26 \times 26 \times 205$ mm. M-206 IR Decoy Flare provides protection against heat-seeking missiles for helicopters and low-altitude aircraft such as the C-130, A-10, and RC-12, AH-64, CH-47 and UH-60. Dispensing systems include M130, AN/ALE-40, AN/ALE-45 or AN/ALE-47. MJU-7A/B IR Decoy Flare is also a pyrotechnic flare using Magnesium/Teflon/Viton (MTV) IR payload and measuring 26×52 mm in cross-section and 205 mm in length. It provides protection against heat-seeking missiles for aircraft such as the F-4, F-5, F-15, F-16 and F-104, as well as transport and cargo aircraft such as the C-130. Dispensing systems include the AN/ALE-40, AN/ALE-45 and AN/ALE-47. MJU-10/B is also MTV based IR decoy flare. It offers protection against heat-seeking missiles for the F-15 and heavy transport aircraft such as the C-5 and C-17. The MJU-10/B measures 52×65 mm in cross-section and 205 mm in length. Dispensing Systems include the AN/ALE-40, AN/ALE-45 and AN/ALE-47. The MJU-32/B IR Decoy Flare devices are magnesium-Teflon based cylindrical flares measuring 36 mm in diameter and 148 mm in length. It produces the same fire ball result as the Mk46 Mod 1c IR decoy flares but incorporate a safer igniter design that requires an external ignition source. It is used on the A-4, A-6, A-7, F-4, F-14, AV-8A and P-3C Orion (Aircraft) and CH-15, CH-46, CH-47, CH-47 and UH-1 (helicopters). The MJU-49/B decoy is a Special Material Decoy (SMD) Pyrophoric IR Countermeasure developed and manufactured for the US Navy. It is designed to protect helicopters and fighter aircraft from MANPAD and air-air missile threats. The MJU-49/B is manufactured in a round 36-mm standard from factor and can be employed using the ALE-39 series or any countermeasure dispensing system capable of employing a round 36-mm flare. The MJU-50/B IR countermeasure flare is a $26 \times 26 \times 205$ mm expendable containing pyrophoric special material. The MJU-51/B IR flare is a $26 \times 52 \times 205$ mm expendable also containing pyrophoric special material. Both MJU-50/B and MJU-51/B are compatible with the ALE-40, ALE-45 and ALE-47 dispenser systems.

Ariel from Selex Galileo (Figure 6.28), previously GEC-Marconi, is a compact and lightweight towed radar decoy capable of providing off-board countermeasures to protect the host airborne platform from attack by RF guided weapons including the state-of-the-art weapons employing error cancelling monopulse tracking techniques. Ariel employs angle deception techniques to defeat monopulse radars, semi-active missiles and home-on-jam weapon systems and is more effective than conical scan deception, cross-polar and cross-eye jamming. Also, it provides protection for a longer period of time than that provided by chaff or ejected countermeasures. Ariel is towed behind the aircraft and it communicates with the aircraft's onboard systems via a fibreoptic cable to transmit specific deception techniques from the threat library to defeat the

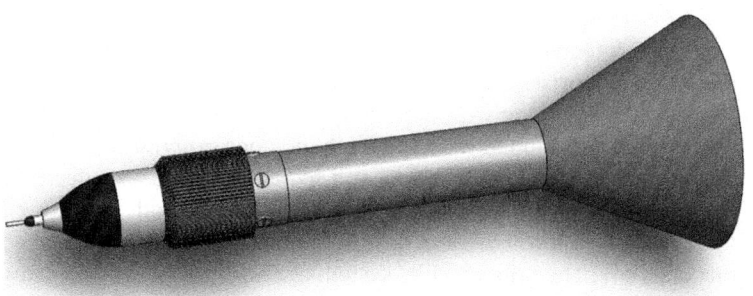

Figure 6.28 Ariel towed radar decoy.

Figure 6.29 AN/ALE-50 towed radar decoy. (*Source:* Courtesy of MKFI.)

incoming threats. Ariel can be installed and operated from all types of fixed-wing aircraft and is available in winched, unwinched, podded or internal configurations depending upon the requirements of platform. The complete system comprises a launcher and a launch controller, both installed on the aircraft, and one or more expendable towed decoys. Different variants of towed decoy are in use on Eurofighter, Typhoon, Tornado and Nimrod aircraft. Ariel employs fully user programmable countermeasures techniques enabling the decoy to be used either in a standalone mode or as part of an integrated self-protection system.

AN/ALE-50 Advanced Airborne Expendable Decoy (AAED), a joint development venture of the United States Air Force, Navy and Raytheon, is a towed expendable that offers protection to the host airborne platform against incoming radar-guided missiles (Figure 6.29). It does so by providing to the hostile missile threat a radar cross-section that is much larger than what is provided by the platform to be protected thereby becoming a preferential target for the attacking missile. This lures the missile away from the aircraft increasing its survivability. The AN/ALE-50 can be operated both as a standalone device and as an integrated system where it is controlled by the AN/ALE-47 countermeasures dispenser system. The towed decoy is amenable to easy installation on virtually any airborne platform and is currently operational on F-16, F/A-18E/F and B-1B aircraft. It is a battle-proven system that has demonstrated its efficacy in Kosovo, Afghanistan and Iraq wars.

The *AN/ALE-55 fibreoptic towed decoy* (Figure 6.30) is a RF countermeasures system of BAE Systems designed to provide self-protection to the host airborne platform against attack by radar-guided missiles. The system comprises an onboard signal conditioning assembly and a fibreoptic towed decoy. It works in conjunction with the aircraft's EW system to provide three layers of defensive jamming against radar-guided threats, which includes *suppression* preventing radar from tracking, *deflection* that breaks radar locks and *seduction* that uses a decoy to present itself as the target to the approaching missile. In the suppression mode, the towed decoy tries to jam the radar signal while it is still in the acquisition mode and has not established lock-on to the target aircraft or decoy. The onboard EW system analyses the radar signal and transmits appropriate jamming signal to the towed decoy through fibreoptic cable, which in

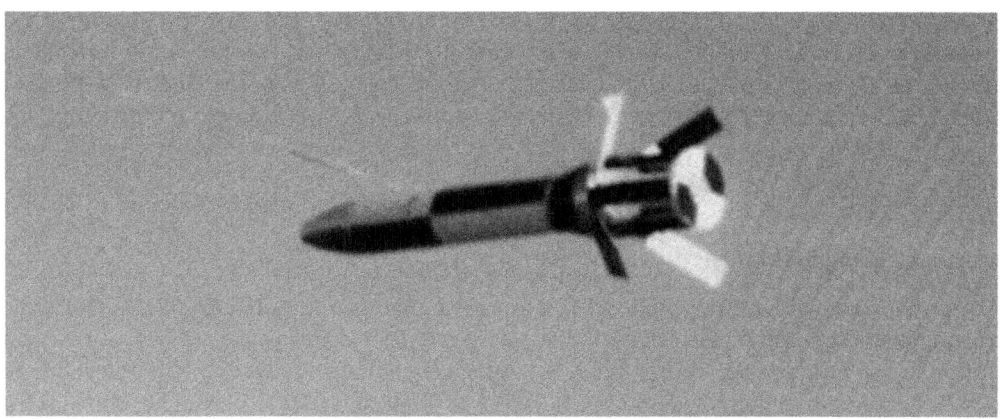

Figure 6.30 AN/ALE-55 fibreoptic towed radar decoy. (*Source:* Courtesy of the US Navy.)

turn transmits the high power jamming signal. In deflection mode, the jamming operation involves breaking radar lock and is carried out after the radar has established lock. In seduction mode, the towed decoy provides aircraft's radar signature to the attacking missile. AN/ALE-55 fibreoptic towed decoy can be installed on a variety of aircraft from fighter aircraft to bombers to transports. It is currently operational on the F/A-18E/F Super Hornet with the Royal Australian Air Force, Royal Saudi Air Force and the US Navy.

6.5 Electro-Optic Countermeasures

Electro-optic countermeasures are employed against systems operating in the optical spectrum from UV to IR. Laser designators and range finders, laser radar, electro-optic sensors such as laser seekers and IR seekers, respectively. used in laser-guided munitions and IR-guided weapons are some common battlefield electro-optic systems. The basic EW concepts with reference to the RF spectrum such as noise and deception jamming, use of warning receivers, need to minimize target signatures and use of stealth technologies, deployment of decoys, use of chaff and obscurants are equally valid when we talk about electro-optic countermeasures. What a radar warning receiver or sensor is to electronic countermeasures, a laser-warning receiver or sensor is to electro-optic countermeasures. Also, what chaff is to electronic countermeasures, IR flares are to electro-optic countermeasures. While electronic jammers are used to confuse or neutralize hostile radars and radar-guided threats, high-energy lasers are used to neutralize laser target designators. While chaff is used to obscure radar signals in the case of electronic countermeasures, smoke and aerosol screens may be used to obscure laser radiation.

LWRs have been discussed in the earlier part of the chapter. An effective countermeasure against IR homing missiles is the use of IR jammers. These jammers transmit spurious modulated IR energy towards the target so that what the missile receives is the target return as well as the spurious signal. This degrades a missile's tracking performance. The jammer radiation could also be used to saturate the seeker's front-end detector with a little larger power or even damage the front-end optics or detector or IR dome with a high-energy laser. An active IR jammer of this type should radiate energy in the spectral bandwidth of the seeker to provide adequate jamming to signal ratio at the seeker. The radiation also needs to be modulated with an optimum waveform.

Use of obscurants is a very effective electro-optic countermeasure. Smoke can be very effective at optical wavelengths. The obscurant can be scattering type or absorptive or both. It is less effective at IR wavelengths. It is not an attractive proposition for airborne deployment due to the problem for aircraft to deploy smoke in the right place, at the optimum time and in large quantities and also due to the problem of wind drift. IR decoy flares discussed in an earlier part of the chapter are also very common.

6.5.1 Need and Relevance

One of the relatively newer types of laser system that has found recognition and wide acceptance by the Armed Forces over the last couple of decades is a laser system that can offer effective countermeasures against the laser systems already in use. The relevance of EOCM equipment in general, and EOCM lasers of different types in particular, stems from the fact that whenever a particular defence technology has matured to a high level leading to widespread usage, it has triggered investment in research and development efforts in the corresponding countermeasures technologies and systems. It happened before in the case of radar and other related military systems operating in the RF spectrum of electromagnetic radiation; it has happened subsequently in the case of lasers and related devices operating in the optical spectrum.

There is a large-scale use of electro-optic devices such as sighting and observation devices and laser systems such as those used for range finding and target designation applications. Laser technology is quite mature today, which has led to induction of laser devices and systems for many new applications not thought of earlier. There has been a planned and concerted effort on the part of developed countries to deploy EOCM systems.

The need to develop and deploy EOCM class of laser systems that can offer effective countermeasures against similar systems deployed by the Armed Forces of the adversary is definitely more pertinent and relevant in the present day scenario. No military platform, be it land-based, aerial or ship-borne, today is free from the risk of being exposed to laser radiation. The activities of these platforms are under constant surveillance by various kinds of electro-optic devices and optronic sensors. Rendering these devices and sensors ineffective in the battlefield therefore makes a huge difference to the battlefield competence of a nation. Deployment of EOCM devices and systems designed to incapacitate or neutralize the more conventional laser devices and systems would act as a force multiplier as a platform incapacitated in the enemy camp is a platform added to your own. This goes a long way in enhancing the survivability quotient of the Armed Forces equipped with such a capability.

6.5.2 Passive and Active Countermeasures

Electro-optic countermeasures are broadly classified into passive and active countermeasures. *Passive countermeasures* for platform protection include use of armour, camouflage, fortification and other protection technologies such as a self-sealing fuel tank and so on. Armour is nothing but the protective covering used on the platform to prevent damage caused to it through the use of direct contact weapons or projectiles. Use of reactive armour is a common example. Camouflage is a form of visual deception. With reference to electro-optic countermeasures, it is a methodology that helps military platforms remain unnoticed by the adversary and is achieved by blending with the environment or by resembling something else. Fortifications are military constructions designed to provide protection to specific equipment, platforms and military bases. Self-sealing fuel tank technology prevents the fuel tanks of aircraft from leaking and getting ignited when attacked.

Active countermeasures further comprise soft- and hard-kill countermeasures. *Soft-kill countermeasures* change the electromagnetic, acoustic or other forms of signatures of the platform to be protected. This in turn adversely affects the tracking or sensing capability of the incoming threat. With reference to electro-optic countermeasures, the incoming threat could be a laser-guided bomb or projectile or an IR-guided missile. In the case of providing protection to land-based platforms from laser assisted threats, one common method of achieving this objective is through use of smoke or aerosol screen to block the radiation from the laser target designator to deprive the laser seeker in the guided munitions the required guidance information. In the case of active countermeasures, means are adopted to counterattack the incoming threat. This may be achieved by neutralizing one of the elements responsible for functioning of the guided threat or by physically destroying the incoming threat by launching a counter projectile in that direction. These are also known as active protection systems.

One example of neutralizing functionality of the threat is in the case of laser-guided munitions. A laser sensor on the platform to be protected senses the existence, PRF code and direction of arrival of laser radiation from the laser target designator and then either generates a smoke/aerosol screen by launching a salvo of grenades or sends another high power/energy laser beam in the same direction, thereby incapacitating the laser target designator. Another technique in serious contention is to use the high-energy laser radiation to illuminate a dummy target 100–200 m away from the platform to be protected thereby forcing the incoming laser-guided threat to land on the dummy target. This is a very attractive proposition for protection of critical and high-value military assets such as aircraft shelters, ammunition depots and so on during war. Use of IR flares discussed in an earlier section and used by target aircraft to protect it from the attack of IR-guided surface-to-air or air-to-air IR-guided missiles is yet another example.

The other type of hard-kill active protection system that relies on launching a counter projectile towards the incoming threat is generally used in the terminal phase of the incoming threat. This leads to physical destruction of the threat by means of either blast and/or fragment action. The consequences of blast and/or fragment action could be in the form of destabilization of kinetic energy penetrator, premature initiation of charge, destruction of airframe and so on. *ARENA* from Russia's Kolomna-based engineering design bureau (KBM) and *TROPHY* from Rafael, Israel are well-known active protection systems of this type designed for protection of land-based armoured fighting vehicles. Active protection systems are briefly discussed in a subsequent section.

6.5.3 Types of EOCM Equipment

There are two broad categories of EOCM equipment including the EOCM class of lasers such as anti-sensor lasers and laser dazzlers and the support devices such as laser threat detection systems. A laser dazzler from the viewpoint of countermeasures terminology may be viewed as a type of anti-sensor laser only where the target sensor is the human eye. Laser dazzlers are discussed in detail in Chapter 12 on *Directed-Energy Weapons*, Section 12.6. In the category of anti-sensor systems, we have systems capable of causing only a temporary disability of electro-optic devices and optronic sensors deployed by the adversary and also the systems that are capable of inflicting a permanent damage. In both cases, the target is the front-end optics and optronic sensors. These are sometimes referred to as soft-kill systems. These systems have by no means the capability of inflicting physical or structural damage to the platform carrying weapons. There are hard-kill EOCM systems that are capable of inflicting a physical damage to the front-end optics of any electro-optic system. These systems are usually vehicle-mounted

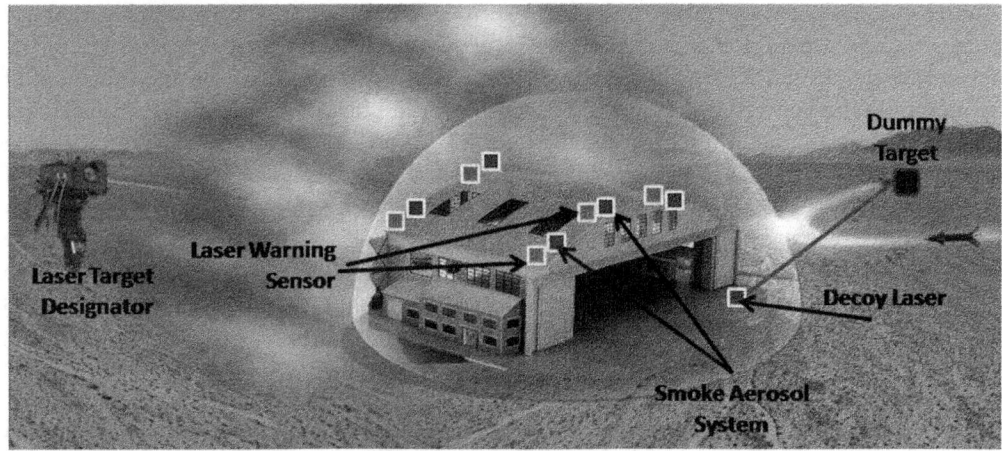

Figure 6.31 EOCM system as a decoy for protection of high value assets.

and are much larger in size and weight than their soft-kill counterparts. The pulse energy level in such lasers is of the order of several kilo-joules as compared to a few joules in the case of soft-kill systems.

A *laser-warning sensor system* is an indispensable component of any EOCM system whether it is designed for soft-kill or hard-kill. It provides information on the type and direction-of-arrival of laser threats emanating from the adversary camp. A laser-warning system provides valuable information that enables the platform crew to make optimum choice of the type and modus operandi of the countermeasure system to be used. That is why laser-warning sensor is an integral part of every EOCM system. When used in conjunction with a smoke/aerosol screening system, the information provided by the sensor is used to trigger and deploy an optimal aerosol/smoke screen in the direction of the threat and block the incoming radiation. This allows the crew of the platform to take an evasive action during those crucial 50–60 s for which the smoke screen remains effective. This type of defensive countermeasure action is particularly effective against laser-guided munitions attack. In the case of an active electro-optic countermeasures system equipped with high-energy lasers, the information on the incoming laser threat is fed to the servo control system, which either reorients the high-energy laser precisely in the direction of laser threat and neutralizes the source of radiation or illuminates a dummy target to divert the incoming threat from its intended course. Figure 6.31 shows a typical deployment scenario of an EOCM system used as a decoy system for protection of high-value assets such as aircraft shelters from laser assisted threats. Laser warning systems are available internationally covering a spectral response range from visible to the far infrared thus encompassing the whole gamut of battlefield solid-state lasers typically operating at 1064 nm and 1540 nm, mid-IR operating in the 3–5 μm band and far IR lasers operating in the 8–14 μm band. It may be mentioned here that mid- and far IR lasers are relevant to IR countermeasures to be discussed in the following paragraphs. Operational ranges of 8–10 km and AoA measurement resolution up to fraction of a degree are available in commercial systems. Laser-warning sensors have been discussed in detail in Section 6.3 on ESM.

6.5.4 Representative EOCM Equipment

In the category of integrated EOCM equipment, we have the whole range of systems, both application and platform specific. These are intended for soft-kill and hard-kill countermeasure

Figure 6.32 ROSY-L Rapid Obscuring System.

operations. It is not feasible to discuss each one of those systems here. An overview of some of the better known EOCM systems is presented here. ROSY (Rapid Obscuring System) Smoke/Obscurant Protection System (ROSY-L for land platforms and ROSY-N for naval platforms) and the MASS (Multi-Ammunition Soft-kill System) Naval Decoy Countermeasure system from Rheinmetall Defence, the Laser Countermeasure (LaCM) System jointly developed by the Office of Naval Research (ONR), Naval Research Laboratory (NRL) and Northrop–Grumman under EWISSP (Electronic Warfare Integrated System for Small Platforms) programme, the DHY-322 Laser Decoy System from CILAS, the AN/VLQ-7 STINGRAY Combat Defence System developed by Martin Marietta, Inc., the AN/VLQ-6 HARDHAT Missile Countermeasures System from Loral, the LARC Airborne Laser Ranging and Countermeasures System developed jointly by the United States and Britain, the COROLLA PRINCE Airborne Laser Weapon developed by the United States and the SHTORA-1 Defensive Aids Suite from Zenit Research and Production Corporation are some representative EOCM systems designed and built to protect the host platforms or the high value assets from laser-guided munitions attack. SHOTRA-1 was described in Section 6.3.6. Others are briefly described in the following paragraphs.

ROSY (Rapid Obscurant System) from Rheinmetall Defence has two variants, namely ROSY-L for providing protection land and ROSY-N for protection on sea. *ROSY-L (Rapid Obscurant System – Land)* (Figure 6.32) is the smoke/obscurant protection system designed to provide protection to crew and passengers of light military and civilian vehicles against surprise attacks, such as those experienced during reconnaissance patrols or while moving in convoy, from a wide range of guided weapon threats including TV-, electro-optic (EO), IR, imaging infrared (IIR) and semi-active command to line-of-sight (SACLOS) guided weapons. It is particularly suited to force protection high risk areas of operation. ROSY-L system comprises a basic system with a manual control device and one to four ROSY launchers per vehicle. It can be very easily installed and de-installed. Unlike the conventional smoke protection systems in use, Rosy-L is capable of instantaneously generating large area, multi-spectral dynamic smoke screens

Figure 6.33 MASS decoy system. (*Source:* Courtesy of MKFI.)

leading to interruption of the line-of-sight that can shield even moving vehicles. Also, its multi-mission capability assures 360° protection from multiple attackers such as stream and wave attacks. A variant of the ROSY-L system called ROSY MOD is designed for use on small weapon stations and light vehicles used by special operations forces. Its integration with the platform without the need for a surface-mounted launcher makes it undetectable. The ROSY-N (Rapid Obscurant System – Navy) is designed to protect naval and coastguard units from missiles and asymmetric attack in littoral zones and inland waters. It is particularly suited to installation on patrol vessels, speed boats, fast attack craft and landing craft.

The *MASS* (*Multi-Ammunition Soft-Kill System*) from Rheinmetall Defence is a compact and fully automatic ship-borne soft-kill launcher system originally designed to provide protection against anti-ship missiles and asymmetric threats employing RF, IR and/or EO seekers through deployment of programmable multi-spectral *Omni-Trap decoy rounds* with radar, IR, laser, EO and UV payloads (Figure 6.33). MASS can be used in standalone mode and also as an integrated setup with a command and control system. The latest version of the system includes the Omni-Trap decoy with millimetre waveband payload to further improve anti-ship missile defence capability and an off-board Corner Reflector (OCR) decoy developed in cooperation with Airborne Systems to enhance capability to defend against state-of-the-art RF seekers. CANTO anti-torpedo acoustic decoy has also been integrated with MASS system in cooperation with DCNS (*Direction des Constructions Navales Services*) France to counter torpedo threat. MASS is in use on a wide range of naval vessels. This includes Brandenburg class frigates, Braunschweig class corvettes, Frankenthal class mine hunters, Kulmbach class mine hunters and the future Baden-Wurttemberg class frigates from Germany; Halifax-class frigates from Canada; Tariq class frigates from Pakistan; Hamina class missile boats, Hämeenmaa class minelayers and Rauma-class missile boats from Finland; Skjold class corvettes from Norway; Khareef-class corvettes from Oman; Goteborg class corvettes, Stockholm class corvettes and Visby-class corvettes from Sweden; the Baynunah class corvette, Abu Dhabi class corvette and Falaj-2 class patrol vessel from the United Arab Emirates; the Lupo class frigate from Peru and LST-II-class landing ships from South Korea.

The *Laser Countermeasure (LaCM) system* was jointly developed by Office of Naval Research (ONR), Naval Research Laboratory (NRL) and Northrop–Grumman under the EWISSP (Electronic Warfare Integrated System for Small Platforms) programme to protect small surface platforms from attack by laser-guided munitions. Major subsystems of the LaCM system included a countermeasure laser to confuse the threat weapon, suitable optics to direct the countermeasure laser energy towards the approaching laser-guided threat and an erectable mast to provide an improved defensive geometry. Laser sensors on the platform to be defended provide warning of the impending attack on detecting and identifying the laser radiation illuminating the platform. On receiving the alert of an attack, the LaCM mast is kept in a suitably packed position so as to maintain the platform's relatively low profile. The mast is then deployed to be followed up by optics subsystem at the top of the mast directing the countermeasure energy to jam the threat sensor thereby preventing the threat weapon from hitting the targeted platform. Both omnidirectional and spot ground jamming beams can be used to counter the full range of laser-guided threats.

DHY-322 Laser Decoy System from CILAS, France mainly comprises of laser detection system and high-energy laser. A high-energy laser is mounted on a controlled turret. Laser detection system detects the origin of hostile laser radiation, deciphers its PRF code and then synchronizes the high-energy laser to operate at the decoded PRF code. The high-energy laser radiation is then used to illuminate a dummy spot where the incoming projectile cannot do any significant damage. DHY-322 countermeasure laser source in essence generates false laser spots on chosen field area chosen with higher energy and identical coding. This diverts approaching laser-guided munitions to a false laser spot on the chosen field area. DHY-322 is specifically designed to protect ships, strategic sites and other high-value assets.

Stingray Laser Detection and Countermeasures System from Lockheed–Martin is another EOCM system designed to protect the frontline forces by accurately locating and neutralizing optical and electro-optical fire-control systems of the adversary. The Stingray system captures optical and electro-optical systems using a cat's eye effect. The system with its kilowatt-class CO_2 laser and Nd:YAG and frequency-doubled Nd:YAG lasers are capable of blinding photoelectric sensors up to a distance of 8 km and anti-personnel operation over even greater distances. Sensor blinding adversely affects adversary platform's mobility. Subsequently, the adversary platform may be destroyed using anti-tank missiles.

AN/VLQ-6 HARDHAT is a vehicle-mounted Missile Countermeasures Device (MCD) developed by Loral, Inc. The system is used as a part of a comprehensive warning and threat response countermeasures system that detects and intercepts laser signals to provide warning of an imminent attack. The system then emits IR energy to disrupt the threat's missile/command unit tracking loop. The system is typically mounted high above the turret, which allows it to scan the frontal arc to detect and decoy away most of the widely used Anti-Tank Guided Missiles (ATGMs). *LARC* is an Airborne Laser Ranging and Countermeasures System developed jointly by the US and Britain. The system comprises four sensors that cover the lower hemispheric region beneath the aircraft to compute the angle-of-arrival of the laser threat. It employs high-energy countermeasure beams to neutralize the threat. *COROLLA PRINCE* developed by the United States is an airborne laser weapon based on Stingray technology. It features higher output power and larger operational ranges than Stingray and is designed to primarily blind ground-based optical and optronic tracking systems. The laser grenade is another significant development in the field of laser countermeasures. *Laser grenades* can generate high efficiency laser beams through High-Explosive (HE) oscillation coupled with thermal inert gases. They can be used to disrupt the normal operation of optical sighting devices, LRF and also the front-end detection systems of laser-guided munitions.

Figure 6.34 Concept of operation of the Active Protection System.

6.5.5 Active Protection Systems

There is yet another category of hard-kill countermeasure systems known as the *Active Protection Systems* (APS). APS's technology uses sensors and radar, computer processing, fire-control technology and interceptors to find, target and knock down or intercept approaching threats from Rocket Propelled Grenades (RPGs) and Anti-Tank Guided Missiles (ATGMs). Figure 6.34 illustrates the concept of APS. The illustration is self-explanatory.

Russian ARENA and Israel's TROPHY are common examples of APSs. *ARENA* is an APS designed to protect armoured fighting vehicles from anti-tank weapons including anti-tank guided missiles and missiles with top attack warheads. It uses a Doppler radar to detect incoming warheads and then fires a defensive rocket in the direction of incoming threat that detonates near the approaching warhead and destroys it before it hits the platform.

Trophy, from Rafael Israel, like ARENA, is also an APS designed to supplement the armour of both light and heavy armoured fighting platforms. It intercepts the incoming kinetic energy threats and destroys them by a shotgun-like blast. Figure 6.35 shows the photograph of the system. The sequence of operations includes *Threat Detection and Tracking* to detect, identify and verify the threat implemented by several sensors including flat-panel radars placed at strategic locations around the platform, *launching* and *intercept functions*. Once an incoming threat is detected, identified and verified, the counter projectile is launched automatically into a ballistic trajectory intercepting the incoming threat at a relatively safe distance.

6.6 Infrared Countermeasures

Infrared countermeasures (IRCMs) are employed to protect aircraft from surface-to-air and air-to-air heat-seeking missiles. The IRCM systems achieve this by confusing the missiles' IR guidance system and forcing them to deviate from their intended trajectory. The IR missile seeker head uses either a spinning reticle and stationary optics called spin scan or a stationary reticle and rotating optics called a conical scan. In both cases, a modulated signal is generated,

Figure 6.35 The Trophy Active Protection System. (*Source:* Zachi Evenor, https://commons.wikimedia.org/wiki/File:Trophy-APS--Merkava-4M-pic01-Zachi-Evenor_(cropped).jpg. CC BY 3.0.)

which allows the tracking logic to determine where the source of IR energy is with respect to the direction of flight of the guided missile. The command system of the missile steers it towards the source of IR energy, which is the intended target.

IR flares deployed by the aircraft are one common form of IR countermeasures. Flares create IR targets with a much stronger IR signature than the one from aircraft's engines and other parts. The flares mimic the target and force the missile to make incorrect steering decision, which leads to the missile breaking off a target lock-on. This further causes the missile to deviate from its intended trajectory. IR flares were discussed in Section 6.4.3. State-of-the-art missiles are designed so as not to be deceived by flares and to be able to discriminate between real target and flares. This has led to development of active IR countermeasures.

6.6.1 Active IRCM Systems

Active IRCM systems use a modulated source of IR radiation with a higher intensity than the emission from the target. If this modulated radiation were seen by missile seeker and if the modulation scheme matched the one used by the seeker of the target missile, the tracking logic of the missile confuses it as coming from the target aircraft and generates a false tracking command. As a result of this, the missile begins to deviate from the target forcing the IR seeker to go out of lock. Once the lock is broken, it is very hard for the missile to regain locking condition. The aircraft to be protected could then use IR flares to force the incoming IR guided missile to lock-on to the flare. The efficacy of IRCM system is determined by the ratio of jamming signal intensity J to the target signal intensity S and also on how close are the modulation frequencies to the actual missile frequencies. For spin-scan missiles, the required J/S is relatively far lower than what it is for newer missiles.

There are two broad categories of active IRCM systems, namely those employing non-coherent IR sources and those employing coherent IR sources. Standard IRCM systems use incoherent and non-directional sources of IR energy usually configured around IR emitting arc lamps. They suffer from the disadvantage that the probability of jamming signal penetrating

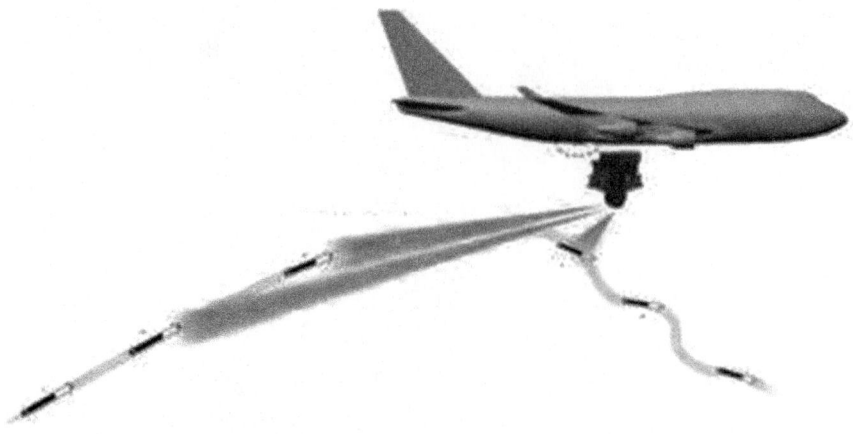

Figure 6.36 Concept of the DIRCM system.

the seeker head is low. Further, if the jamming signal were not effective against a particular seeker system, the IRCM acts as a strong source of IR energy thereby enhancing the ability of the missile to track the aircraft. This shortcoming is overcome in *Directional Infrared Countermeasures* (DIRCM).

A DIRCM *System* overcomes this potential drawback by integrating the IR source of jamming signal with a missile approach warning system. The IR source in this case is a tunable multiband laser source. The IR jammers are mounted on a movable turret and operated in conjunction with a missile approach warning system. The system operates as follows. When an IR-guided missile such as the MANPADS is fired, it radiates electromagnetic energy in different wavelength bands. Most DIRCM systems use UV sensors to detect the approach of the missile. Detection in the IR bands in addition to the UV band has also been exploited. The detection system provides a cue to the tracking system, which uses information from the detection system to track the incoming missile. The IR jammers direct the IR jamming energy at the seeker. The modulation scheme can be cycled to try to defeat a variety of seekers. The efficacy of DIRCM depends on how comprehensively the approaching threat can be analysed and how accurately it can be tracked. Laser safety is another issue of concern in a case where higher power needs to be employed in order to defeat advanced missiles. Figure 6.36 illustrates the concept of a DIRCM system.

6.6.2 Some Representative IRCM Systems

In the following paragraphs some representative DIRCM systems are discussed. These include BAE Systems' AN/ALQ-212 Advanced Threat Infrared Countermeasures (ATIRCM), *Multi-Spectral Infrared Countermeasures* (MUSIC) from Elbit Systems, Israel, and Northrop–Grumman's *GUARDIAN* and Common Infrared Countermeasures (CIRCM) programmes.

The *AN/ALQ-212 Advanced Threat Infrared Countermeasures* (*ATIRCM*) *system* developed by BAE systems in partnership with the US Army is designed to defend the host airborne platforms from the shoulder fired MANPADs providing protection against an array of highly lethal missile attacks in all IR threat bands. In conjunction with AN/AAR-57 Common Missile Warning System (CMWS), ATIRCM system is capable of handling multiple missile attacks with its one or more IR jamming heads. AN/AAR-57 CMWS detects the incoming missile and communicates its position relative to the host platform to be protected. ATIRCM then locates

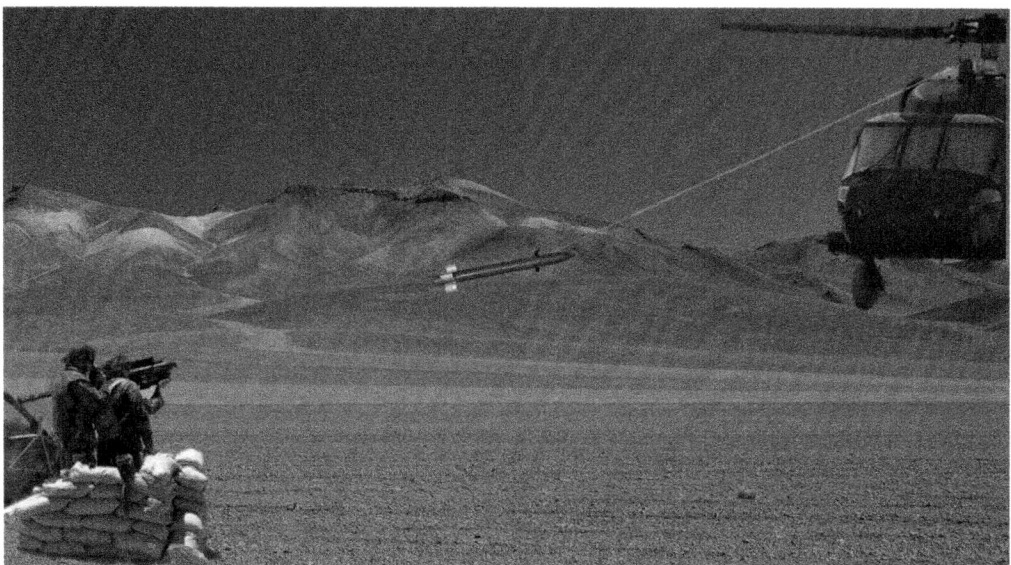

Figure 6.37 ATIRCM in use on helicopter.

and tracks the incoming threat and directs a high-energy IR laser beam to defeat the missile's IR seeker. The high-energy laser effectively blinds its guidance system thereby preventing it from homing in on the aircraft. ATIRCM is a part of directional IR countermeasures suite and is fielded on US Army CH-47 Chinook helicopters. Figure 6.37 illustrates the use of ATIRCM on a helicopter. ATIRCM is a combat-proven system that has established its efficacy during wars in Iraq and Afghanistan against state-of-the-art heat-seeking missile threats.

Elbit Systems' MUSIC family of DIRCM systems developed by Elbit Systems' wholly owned subsidiary Elbit Systems Electro-Optics ELOP, Ltd provides protection for airborne platforms such as fixed and rotary-wing aircraft and helicopters from IR-guided surface-to-air missiles such as MANPADS and other Short-Range Air-Defence systems (SHORADS) and air-to-air missiles. The MUSIC family of DIRCM systems comprise a high-energy fibre-laser emitter; a high rate thermal tracking camera-based infrared tracking system and a small, highly dynamic mirror turret to provide effective and reliable protection to all types of aircraft and under all operational conditions. It is used in conjunction with the PAWS family of combat-proven missile warning systems capable of detecting even the most advanced missile threats. On receiving a missile threat alert, PAWS tracks the threat and directs the DIRCM high-energy IR laser beam from a fibre-laser emitter through the mirror turret to illuminate and disrupt the threat's guidance, altering its flight course off the protected aircraft. The Elbit Systems' MUSIC family of DIRCM systems integrated with the PAWS IR-based missile warning system has undergone extensive testing with very successful results. Elbit Systems has developed several DIRCM variants, including MUSIC, C-MUSIC, J-MUSIC and MINI-MUSIC. C-MUSIC has been selected by the Israeli Government to protect passenger aircraft. It has been installed on Boeing 707, 737, 747, 757, 767 and 777 and AIRBUS 320 platforms. Unlike the MUSIC system, introduced into a military aircraft as an integrated system, C-MUSIC is self-contained in a pod, with missile warning, signal processing, laser emitter, beam conductor, beam director and cooling. C-MUSIC is based on MUSIC systems currently in use on helicopters and medium turbo prop fixed-wing aircraft and is designed to protect large jet aircraft against man-portable heat-seeking surface-to-air missiles. C-MUSIC's fibre-laser DIRCM technology creates a laser beam that is directed towards the

Figure 6.38 J-MUSIC DIRCM.

tracked missile, effectively defeating the incoming threat. J-MUSIC is also an advanced multi-spectral IR fibre-laser based DIRCM system based on C-MUSIC technology rapidly acquiring and tracking incoming missiles and deflecting them from the target using a high-energy IR laser beam. J-MUSIC (Figure 6.38) is a distributed system, which allows its Line Replaceable Units (LRUs) to be installed in various locations on board the aircraft, enabling an optimized installation solution for protection of the aircraft. It is installed on AIRBUS A400 military transport aircraft. MINI-MUSIC (Figure 6.39) is designed to protect small rotary and fixed-wing aircraft against heat-seeking MANPADS threats.

The MUSIC family of DIRCM systems has been extensively and successfully tested by various countries and are in full-scale production for a range of customers including Germany, Italy, Brazil Israel and others for installation on a variety of airborne platforms around the world. The systems are also fully operational with Israel's commercial aviation fleet, protecting commuter jet aircraft serving domestic and international destinations.

The GUARDIAN DIRCM system developed by Northrop–Grumman provides 360° protection to the host airborne platform against IR-guided MANPADS. GUARDIAN is the role fit version of the proven AN/AAQ-24 (V) DIRCM system (Figure 6.40) installed on hundreds of Department of Defense, VIP and other military aircraft around the world. It comprises two major subsystems, namely a UV sensor based missile warning system and a multiband laser pointer/tracker. The entire system is housed in a single pod that can be mounted underneath the aircraft's fuselage. When GUARDIAN detects a MANPADS launch, it tracks the incoming missile, and then directs a high-energy IR laser beam to jam the missile's guidance system forcing it to deviate from its intended trajectory and consequently missing the target aircraft. The entire process is completed in about 2–5 s and requires no action on the part of the aircraft crew. Figure 6.41 shows GUARDIAN DIRCM pod fitted underneath the aircraft fuselage.

Common Infrared Countermeasures (CIRCM) system from Northrop–Grumman is designed specifically to protect rotary-wing and medium fixed-wing aircraft from IR-guided missiles. CIRCM programme envisages use of fibre coupling for transport of laser energy to the jam

Figure 6.39 Mini-MUSIC DIRCM.

Figure 6.40 AN/AAQ-24 (V) DIRCM.

Figure 6.41 GUARDIAN DIRCM pod. (*Source:* Alan Radecki Akradecki, https://commons.wikimedia.org/wiki/File:Fedex-md11-N525FE-051109-08-16.jpg. CC BY-SA 3.0.)

Figure 6.42 CIRCM system. (*Source:* Courtesy of the US Army.)

head. The fibre coupled approach enables remote location of larger and more complex laser components that would be very difficult to mount on the jam head. CIRCM's jam head interface supports both direct coupled as well as fibre coupled architectures. Figure 6.42 shows the photograph of CIRCM system.

6.7 Electronic Counter-Countermeasures

The purpose of using support measures such as radar warning receivers and laser-warning receivers on a platform was to detect, identify and locate enemy's sensor system such as radar and laser emitters and subsequently follow it up with an appropriate action, which could be a manoeuvre or initiating a countermeasure. The purpose of using countermeasures was to make these sensor systems ineffective or degrade their performance during the brief period of engagement or attack. Even a momentary unlocking of a fire-control radar or degradation in tracking performance of the radar could give sufficient time to the crew of the attacking aircraft to accomplish their intended mission.

While the main job of the countermeasure is to minimize the efficacy of the enemy's sensor systems; the counter-countermeasures are a counter to the countermeasures. The purpose of the counter-countermeasures is to equip the systems like radar and other communication systems to defeat the objectives of countermeasures. The primary objective of ECCM techniques, for instance, when applied to a radar system is to design the radar in such a way that it accomplishes its intended task of fire control, tracking, guidance, search and so on depending upon the type of radar even in the presence of countermeasures employed by the attacking aircraft. Radar that is designed and built to avoid saturation of radar receiver front-end, maximize signal-to-jamming ratio, reject false targets created by deception jammers, discriminate against directional interference and so on is said to be equipped with ECCM features and have survivability in the battlefield.

Hughes' Air-Defence Radar (HADR) and is one of the many examples of radars with inbuilt ECCM features. This radar performs very effectively in severe clutter and ECM environments. The radar employs a range of techniques to reduce the adverse effects of ECM. These include use of coded pulses, frequency agility, multiple pulse widths and multiple PRFs. While a low sidelobe antenna, use of pencil beams in azimuth and elevation, high receiver dynamic range and sharp channel selectivity provide the primary radar with a low vulnerability to jamming, impact of residual jamming is countered by automatic threshold control, target detection correlation and sidelobe blanking. Special modes for use in an ECM environment are burn through and automatic frequency selection. *AN/TPS-78 Air-Defence and Surveillance Radar* from Northrop–Grumman (Figure 6.43) is another example of radar that is designed to offer superior performance in some of the harshest, most demanding cluttered environments enabled by its stacked-beam architecture. With its sophisticated filters, the radar can counter sea clutter caused by Anomalous Propagation (AP) phenomenon. The STIR 2.4 HP Radar Tracking and missile illumination system from Thales is yet another example of radar with extensive ECCM and anti-clutter features enabling operation under hostile ECM environment with strong jamming conditions.

As an another example, a laser designator used on an aerial platform to illuminate the ground targets for guiding weapons along the scattered radiation employ some form of pulse repetition frequency (PRF) coding in the transmitted laser pulses that is known to the seeker subsystem. This is done to defeat the opposition design of directing the weapon such as missile away from its intended course towards the real target to some other false target if it chooses to do so by illuminating a false target with a laser of the same type.

When it comes to building ECCM features into a radar system, one would need to take a note of all the subsystems comprising the radar system namely the antenna, the transmitter, the receiver and the signal processor. In the following paragraphs, we shall briefly discuss ECCM techniques as related to various subsystems of radar.

6.7.1 Antenna Related ECCM Features

The antenna parameters that can be exploited for building ECCM features into the radar system include gain, directional pattern, antenna scan rate and polarization. Some types of deception

Figure 6.43 AN/TPS-78 air defence and surveillance radar.

jammers work on anticipation of the scan rate. Their purpose can be defeated by using random scanning. Also a high gain antenna can be used to spotlight a target and burn through the jammer. Antennas with multiple scans allow rejection of beam containing the jammer signal and still maintain detection capability with the other beams. *Low side lobe level* is desirable in most of the cases except when the consequent increase in the main beam width worsens the problem of main beam jamming. Other techniques used to prevent jamming from entering through side lobes are the *Side Lobe Blanking* (SLB) and *Side Lobe Cancelling* (SLC). The purpose of an SLB system is to prevent the detection of strong targets and interference pulses through side lobes. The purpose of SLC is also similar in nature and is to suppress noise interference received through side lobes. This is achieved by estimating the direction of arrival of the jamming signal and jamming signal power with the help of an array of auxiliary antennas and then modifying the receiver directional pattern to place nulls in the jammer direction. The *polarization* characteristics of antenna can be effectively exploited to build ECCM capabilities either by keeping antenna cross-polarization pattern as low as possible to prevent jamming through cross-polarization pattern or by making use of cross-polarization pattern and copolarization pattern and then discriminating the useful signal from undesired signal from chaff or jammer on the basis of polarization.

6.7.2 Transmitter Related ECCM Features

When it comes to building ECCM capabilities into the transmitter portion of the radar, the parameters normally exploited include transmitted power, emission wavelength and the waveform. Increase in transmitted power coupled with burnthrough mode can be used to defeat noise jamming. Burnthrough mode, however, is not effective against countermeasures like chaff, decoys, repeaters and so on.

Frequency agility and *diversity* are very effective methods of ECCM. A frequency agile radar transmitter usually has an ability to change its emission frequency on pulse-to-pulse or batch-to-batch basis. Pulse-to-pulse frequency agility is, however, not compatible with Doppler processing. The same, however, is possible if the frequency agility is on a batch-to-batch basis. Frequency diversity refers to use of more than one complementary radar emissions at different frequencies either from single radar or from several radars. The objective of using frequency agile radar or resorting to frequency diversity is to force the jammer to spread its energy over the entire bandwidth covered by the frequency agile radar with the result that jammer output power spectral density reduces. This makes the ECM less effective.

Narrowing antenna's beamwidth is also a very effective counter countermeasure against noise jamming countermeasure as it restricts the sector that would be blanked by main beam jamming. In addition, it also provides a strobe in the direction of the jammer, which, along with two or three spatially separated radars, would have the same effect as that of using a larger antenna.

6.7.3 Radar Receiver and Signal Processing Related ECCM Features

The other important aspect is to make the radar receiver and signal processing chain immune to the ECM such as relatively high power noise jamming signals, signals from deception jammers, reflection from chaff and so on. Jamming signals that are able to penetrate the resistance offered by the antenna's ECCM abilities, if large enough, can saturate the radar processing chain, which can lead to near complete elimination of information from the targets. Solution for avoiding such a situation lies in the use of a wide dynamic range receiver. Two commonly used approaches for implementing such a receiver are the log and linear-log receivers.

A *log receiver* is the one whose video output is proportional to logarithm of the RF input signal envelope over a specified range. It allows the radar receiver to detect target returns that are larger than the noise jamming signal. The log receiver has certain disadvantages. One of them is that it is only effective against high level noise jamming signals and allows a low-level noise jammer to be more effective by amplifying low-level jamming signals more than high level target returns. The other disadvantage is that log characteristics widen the spectrum of received echoes which makes the life difficult for radars like MTI and pulse Doppler radars.

The shortcomings of log receivers are largely overcome in a linear-log receiver in which the output signal is nearly proportional to the RF signal envelope for high input signal amplitudes and directly proportional to the envelope amplitude for low signal amplitudes.

6.8 Stealth Technology

Stealth technology refers to all those techniques used to minimize target signatures. In the earlier paragraphs, we have discussed different types of warning receivers used by air crew to warn them of a possible detection by enemy radar (RWRs), or an electro-optical system (LWRs) or of a possible missile attack (MAWs). These systems assist the air crew take an evasive action and route their flight away from the cover of enemy sensors and weapons. Another way of looking at this problem is to look for ways and means to avoid detection, to be more practical, minimize the detection range of radar and other sensors. This will allow the air crew to penetrate deep into the enemy's territory and approach the enemy radar very close without being detected. All this can be achieved by minimizing the target signatures, which include all those characteristics of the target, an aircraft for instance, that can be used as the basis of detection by enemy sensors and weapons. For example, it would be radar cross-section of the on-the-target echoing area in the case of a radar sensor, target contrast in addition to the echoing area in the case of an electro-optical sensor or emission from the hot portions of the aircraft body or radiation from the combustion products in the case of passive IR-guided missiles.

The relevance of using stealth technology was amply demonstrated in the recent past during the Gulf War of 1991 when the extremely low radar and IR signatures of F-117 A and B-2 aircraft made them virtually undetectable and played havoc on difficult-to-attack Iraqi targets. Use of stealth technologies reduces the enemy's reaction time by reducing the detection range.

The radar cross-section (RCS) is one parameter that designer of the stealth technologies have exploited the most firstly because a reduction in RCS reduces the detection range and secondly because a reduced RCS leads to a reduced S/N ratio at the radar receiver input, which will degrade its tracking accuracy. It may also be mentioned here that the detection range of radar is proportional to only the fourth root of the RCS and a 100:1 reduction in RCS would only yield a 3.16-times reduction in detection range. However, the jamming power or the decoy size required for an effective electronic countermeasure would be directly proportional to the RCS. So, reduction in RCS gives an overall benefit.

6.8.1 Reduction of Radar Cross-Section

The basic purpose of reducing RCS is to reduce the amplitude of the reflected signal. This could be achieved by either shaping the vehicle in such a way as to minimize reflections or using some special materials or paints that absorb electromagnetic energy.

When an aircraft is designed with the primary objective of achieving an optimum aerodynamic performance, which has been the approach until quite recently, the resulting aircraft is more likely to have physical features that would act as strong reflectors of electromagnetic energy and thus make the aircraft highly vulnerable as it approaches the enemy sensors. These features include skin joints, panel edges, plane surfaces, re-entrant cavities of the structure and systems under dielectric panels. Reflections from internal and external surfaces, angles and sharp edges of air intakes can get reinforced by those from complex surfaces of engine's compressor, stator blades, flat plate radar antennas and bulkheads behind nose radomes.

There are two approaches to reshaping the aircraft. The first is to reduce the RCS to a minimum by eliminating at the design stage as many as possible of all those features that are strong reflectors of electromagnetic energy. The amount of energy reflected from such features depends upon their orientation to the incident wave and upon the relationship between the phases of individual reflected waves. Resonant conditions need to be avoided as far as possible by eliminating small components.

In another approach, the aircraft is so designed as to be made of such faces and wing leading and trailing edge angles that reflect radar energy away from the source. F/117 A (Figure 6.44) and B-2 fighter aircraft are good examples of use of stealth technologies.

6.8.2 Radar Absorbent Materials

The other approach is to build the aircraft structure as far as possible with materials that absorb radar energy. Though there is an increasing use of composite materials with good RF energy absorbing ability; use of metallic parts cannot be avoided completely. There are two basic types of material. The second type depends upon the interference phenomenon to effectively reduce the amplitude of the energy reflected from the material. The energy reflected from the air-material surface primary boundary and the other reflected from the material backing interfere destructively to minimize reflections. While absorbing materials are usually effective over a wide frequency range, materials relying on interference are effective over a limited frequency range.

Some radar absorbent paints have also been developed for the purpose. These paints can be used to cover most of the surface of the aircraft and are effective in reducing surface currents and creeping waves that would otherwise radiate as a component of the reflected signal.

Figure 6.44 F-117 Nighthawk Stealth Fighter. (*Source:* Courtesy of the US Airforce.)

The aircraft also need to minimize their visible and infrared signatures necessitated by increasing use of IR and other electro-optical sensors. The main sources of IR radiation are engine components, jet pipes, jet exhaust and aerodynamic heating in the case of high-speed aircraft. All these factors need to be looked into by property shielding the jet pipes, cooling the hot components, avoiding direct line-of-sight with hot turbine components and so on. Detection in the visible band depends for its operation on the contrast between the target and the background. The contrast can be minimized by using suitably coloured finishes. Another possible approach could be to equip the aircraft with lights capable of generating lighting of intensity and colour matching that of the background thus considerably reducing the visible signatures. Reduction of acoustic signatures is particularly important in the case of helicopters.

6.8.3   Examples of Stealthier Platforms

Use of stealth technologies with the objective of reducing the radar, visible, IR and acoustic signatures of the platforms is increasing with increase in their vulnerability to different types of sensors used by the adversary to detect, track and attack these platforms. Use of stealth technologies is no longer restricted only to airborne platforms; it is assuming significance in naval and land-based platforms also. Also, it is not confined only to reduction of RCS; the objective of incorporating stealth extends to reducing visible, IR, thermal and acoustic signatures as well. In the following paragraphs, we shall briefly discuss some representative airborne, land-based and sea-based platforms with stealth features.

Shenyang J-31, also called *Gyrfalcon* and *Falcon Hawk*, is a fifth generation multirole stealth aircraft designed and built by Shenyang Aircraft Corporation, an affiliate of Aviation Industry Corporation of China (AVIC), for the People's Liberation Army Air Force. It is characterized by wing span of 11.8 m, length of 16.9 m, height of 4.8 m, maximum take-off weight of 25 tonnes, maximum range of 2000 km when fitted with external fuel tanks and a top speed of 2200 km/h. Stealth features of the aircraft include forward-swept intake ramps with diverterless supersonic

Figure 6.45 F-22 Raptor. (*Source*: Rob Shenk, https://commons.wikimedia.org/wiki/File:Lockheed_Martin_F-22A_Raptor_JSOH.jpg. CC BY-SA 2.0.)

inlet bumps, trapezoid-shaped wings and two-piece canopy. As reported claimed by AVIC, the aircraft is stealthy against L band and Ku band radars and low-observable against a number of multi-spectrum sensors.

Chengdu J-20 is a twin-engine, fifth generation stealth fighter jet under development at Chengdu Aerospace Corporation for the People's Liberation Army Air Force and expected to enter service between 2018 and 2020. Salient features of the Chengdu J-20 aircraft include a wing span of 15 m, length of 23 m, height of 5 m, maximum take-off weight of 36 tonnes, maximum range of 3400 km and top speed of 2100 km/h. Stealth features of the J-20 are comparable to Lockheed–Martin's F-35 Lightning-II with far fewer reported technical setbacks and safety concerns. The J-20's weapons are carried internally. The central bay is believed to contain four Beyond-Visual-Range Air-to-Air Missiles (BVRAAMs) as well as two short-range Air-to-Air Missiles (AAMs).

The *F-22 Raptor* (Figure 6.45) is a twin-engine tactical fighter aircraft currently in use by the United States Air Force. It is developed by Lockheed–Martin. General characteristics of the aircraft include wing span of 13.56 m, length of 18.9 m, height of 5.05 m, and maximum take-off weight of 38 tonnes, maximum range of 2960 km, and combat radius of 852 km and top speed of 2410 km/h. The aircraft boasts of a number of stealth characteristics. It has a precisely engineered airframe with fixed-geometry serpentine inlets to reduce visibility and uses radar absorbent materials (RAM). It has reduced radio frequency emissions, IR and acoustic signatures. These features enable the Raptor to operate 24/7 over hostile areas while remaining virtually undetected by enemy air or ground forces.

The *F-35 Lightning-II* (Figure 6.46) is a single engine multirole stealth fighter aircraft under development at Lockheed–Martin. Its primary role is in aerial reconnaissance, ground attack and air-defence operations. General characteristics of aircraft include wing span of 10.7 m, length of 15.67 m, height of 4.33 m, maximum take-off weight of 31.8 tonnes, maximum range of 2220 km on internal fuel, combat radius of 1080 km on internal fuel and top speed of 1930 km/h. The F-35's stealth capabilities are unprecedented in tactical fighter aviation. An integrated airframe design, advanced materials and other features make the F-35 virtually undetectable to enemy radar. It uses radar absorbent materials (RAM) and is designed to have

Figure 6.46 F-35 Lightning-II. (*Source:* Courtesy of the US Airforce.)

Figure 6.47 PAK FA T-50 stealth fighter. (*Source:* Alex Beltyukov, https://commons.wikimedia.org/wiki/
File:Sukhoi_T-50_Beltyukov.jpg. CC BY-SA 3.0.)

a low RCS. It incorporates systems that help in reducing its IR and visual signatures. The acoustic signatures are, however, high being twice as loud during take-off and four times as loud during landing when compared to the McDonnell Douglas F-15 Eagle.

The *PAK FA* (*Prospective Airborne Complex of Frontline Aviation*) *T-50* (Figure 6.47) is a fifth generation twin-engine fighter jet being prototyped by Sukhoi for use by the Russian Air Force. It will be the first Russian aircraft to use stealth technologies and is intended to replace MiG-29 and Su-27 aircraft. General characteristics of the aircraft include a wing span of 13.95 m, length of 19.8 m, height of 4.74 m, maximum take-off weight of 35 tonnes, maximum range of 3500 km

Figure 6.48 T-14 Armata Main Battle Tank. (*Source:* Vitaly V. Kuzmin, https://commons.wikimedia.org/wiki/File:VDayRehearsal05052016-28.jpg.CC BY-SA 4.0.)

(subsonic) and 1500 km (supersonic) and a top speed of 2140 km/h. It has reduced RCS through use of plan form edge alignments. Its weapons are carried inside the air frame and it has recessed antennas. It uses RAMs in its rear portions and canopy. It will serve as the basis of the FGFA (Fifth Generation Fighter Aircraft) to be co-developed by Sukhoi and Hindustan Aeronautics Limited (HAL) for the Indian Air Force.

In the category of land-based vehicles, the Infantry Combat Vehicle (ICV) *CV-90* of Sweden and in service with Sweden, Denmark, Finland, Netherlands, Norway and Switzerland has been designed to minimize radar and IR signatures. The platform uses heat absorbing filters to avoid detection by thermal imaging, image intensifiers and IR cameras. It also has reduced acoustic signatures.

BOXER Multirole Armoured Vehicle (MRAV) manufactured by ARTEC GmbH industrial group is another land vehicle designed to have low thermal, acoustic and radar signatures. Yet another stealthier land combat vehicle is the *T-14 Armata Main Battle Tank* designed and manufactured by Uralvagonzavod (UVZ). Its design is based on *Armata Universal Combat Platform* and is practically invisible to IR and other spectra due to use of stealth technology. Reportedly, UVZ plans to deliver the tank in large numbers of about 500 per year to Russian Armed Forces beginning 2017. Figure 6.48 illustrates the concept of T-14 Armata Main Battle Tank.

The *PL-01 tank* (Figure 6.49) designed and manufactured by OBRUM of Poland with support from BAE Systems of UK is based on the Swedish CV-90 ICV and uses stealth technology. One of the primary objectives of the PL-01 tank design is to try and eliminate the IR, radar and visual signature of the traditional tank to a large degree, while also relying on guile to take identity masking a step further. With its thermal camouflage, it is harder to detect by thermal sensors. It is also proposed to cover the whole vehicle with radio absorbent material. The angular shapes of tank might also contribute towards reducing RCS.

Figure 6.49 PL-01 tank. (*Source:* Courtesy of Ministerstwo Obrony Narodowej.)

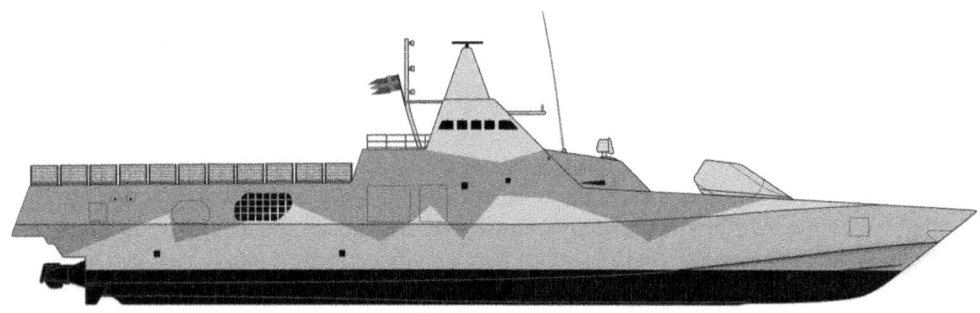

Figure 6.50 Visby-class Corvette. (*Source:* Marcusroos, https://commons.wikimedia.org/wiki/File:Visby_class_corvette.png.CC BY-SA 3.0.)

Like airborne and land-based platforms, serious efforts are on to design and develop stealthier sea-based platforms. The*Sea Shadow* (*IX-529*) experimental demonstrator, Visby-class corvette, Fayette-class, Shivalik-class Frigates and Zumwalt-class Destroyer are some representative examples.

Sea Shadow (Figure 6.50) designed by Lockheed–Martin for the United States Navy is an experimental stealth demonstrator developed under a combined programme by Defence Advanced Research Project Agency (DARPA), the United States Navy and Lockheed Missiles & Space Company. Its purpose was to explore a variety of new techniques for surface ships, one of which was radar and sonar signature control. The *Visby-class stealth corvette* built by Kockums for Swedish navy is designed to have low visibility, RCS and IR signatures. The La Fayette-class multipurpose stealth frigates were developed by DCNS, a French Naval Shipbuilding Company and European leader in naval defence, and built at the DCN Lorient

Figure 6.51 La Fayette-class frigate. (*Source:* Jean-Michel Roche, https://commons.wikimedia.org/wiki/File:FS_Courbet_6.jpg. CC BY-SA 3.0.)

Naval Dockyard. These stealth frigates are in use by the French Navy (*Marine Nationale*). Its variants are in use by the Royal Saudi Navy of Saudi Arabia (Al Riyadh class), Singapore by the Republic of Singapore Navy (Formidable class) and Taiwan by the Republic of China Navy (Kang Ding class). The stealth features of La Fayette-class frigates include reduced RCS facilitated by the sides of the vessel sloped at 10°, surfaces having been coated with radar absorbent paint and the reduced profiles of external features. Figure 6.51 shows an outline drawing of a Kang Ding (La Fayette-class) frigate. *Shivalik Class frigates* are multirole stealth craft built for the Indian Navy. The structural, thermal and acoustic stealth features of these frigates make them harder to detect by the adversary's sensors. The radar systems and engines are further modified to reduce noise levels and avoid detection. They will be succeeded by the Project 17-A Class frigates. *Zumwalt* (Figure 6.52) is a multi-mission stealth naval vessel designed to be a guided missile destroyer for the United States Navy. It has reduced RCS, IR and acoustic signatures. Reportedly, the RCS is comparable to that of a fishing boat, which makes it almost 50 times more difficult to detect by radar than an ordinary destroyer. The acoustic signatures are comparable to those of Los Angeles class submarines.

6.9 Current and Future Trends in Electronic Wafare

On one hand, there have been further developments in the existing EW concepts and systems such as development of ultra-wideband radar, multi-static radar and advances in the field of decoys and so on; on the other hand, some new concepts have also emerged. The definition of EW has been gradually undergoing a change over the last one decade or so. The earlier

Figure 6.52 Zumwalt class destroyer. (*Source:* Courtesy of the US Navy.)

definition of EW did not include the use of Anti-Radiation Missiles (ARM) and Directed-Energy Weapons (DEWs) as a part of Electronic Warfare. DEWs make use of high-energy laser radiation, high power RF energy, primarily microwaves – or particle beams with the objective of placing very high-energy density on the target so as to damage it in such a way as to defeat its mission intentions. It could be blinding a seeker, dazzling the crew or cause direct destruction through thermal or mechanical shock. The ARM, with its own antenna, receiver and processor, has the objective of homing on to the victim radar and destroying it completely. It gets its initial cues from the ESM system and then continues to home on to the victim radar emission through side lobes or flash of energy in the main beam. The concept of EW has been recently expanded to not only include ARM and DEWs but also other weapons of physical destruction of equipment, facilities and personnel.

The other concept that has emerged fast and assumed significance is the *Information Warfare* (IW). The main objective of the doctrine of IW is attainment of total information superiority over the opposition. The main reasons for the rapid development of this concept are ever increasing dependence of nations and their armed forces on information technology and also the fact that all information processors and systems are theoretically vulnerable to corruption or falsification of information, disruption, destruction and manipulation. Information technology is viewed in military spheres as a major force multiplier due to its capability to transfer and manipulate huge quantities of data. A false or biased data subtly introduced into potential adversary's decision making process can, for instance, seriously mislead them on our intentions and thus influence their response.

There has also been increasing use of *Knowledge Based Systems* (KBS) and *Artificial Neural Networks*. An area in which a KBS is already being used is in the radar ESM particularly in de-interleaving several interleaved radar emissions received by the RWR. This technology can also be applied for identification and prioritization of threats and assisting in making an optimum choice and timing of countermeasure. Neural networks are particularly attractive in solving a wide range of signal processing problems in the fields of pattern recognition, image processing, radar signal extraction and so on. Artificial neural networks together with knowledge based system capabilities are expected to provide the best possible response to the EW environment in a given situation.

Illustrated Glossary

ACES ECM Jammer The Advanced Countermeasure Electronic System (ACES) by Raytheon is designed to detect, identify and counter contemporary threats in a high density environment. It is the next-generation of EWS for the F-16 aircraft. It integrates Raytheon's ALQ-187(V) 2 jammer, the ALR-93 radar warning receiver and the ALE-47 countermeasures dispenser system.

Acousto-Optic Receiver This is also known as a Bragg-cell receiver. The operation of an acousto-optic receiver is based on the deflection of a laser beam as it interacts with acoustic wave propagating in a piezoelectric material such as lithium niobate. This device is also known as a Bragg cell. The acoustic wave is excited on the piezoelectric substrate by the RF signal with acoustic frequency depending upon RF frequency. The acoustic wave causes periodic variation in refractive index of the material as it propagates. It is this refractive index variation that causes the laser beam to deflect. The deflected laser light is then made to fall on an array of photodiodes. An acousto-optic receiver or Bragg-cell receiver consists of a laser source with the laser beam passing through the piezoelectric crystal which is acoustically excited by an RF signal. The laser beam is scattered through an angle that is proportional to the ratio of the wavelength of laser to the wavelength of source in the crystal. The scattered light is collected on the diode array which is read out electronically.

Active Countermeasures Use of these means to counterattack the incoming threat. This may be achieved by neutralizing one of the elements responsible for functioning of the guided threat or by physically destroying the incoming threat by launching a counter projectile in that direction. They further comprise soft- and hard-kill countermeasures. Soft-kill countermeasures change the electromagnetic, acoustic or other forms of signatures of the platform to be protected. This in turn adversely affects the tracking or sensing capability of the incoming threat. Hard-kill countermeasures cause physical destruction of the threat.

Active Protection Systems These use sensors and radar, computer processing, fire-control technology and interceptors to find, target and knock down or intercept approaching threats from Rocket Propelled Grenades (RPGs) and Anti-Tank Guided Missiles (ATGMs). ARENA and TROPHY are prominent examples of active protection systems.

Airborne Signals Intelligence Payload (ASIP) ASIP is a family of SIGINT systems from Northrop–Grumman and designed with open service oriented architecture and scalable hardware and software components to meet requirements of a wide variety of platforms and mission profiles.

ALQ-156 (V) MAWS The ALQ-156(V) Missile Approach Warning System (MAWS) sponsored by the US Army and produced by BAE Systems makes use of pulse Doppler radar to detect incoming potential missile threats by illuminating the threat missile with RF radiation and detecting reflection of RF signal from missile's skin to determine missile's range and to provide the optimum triggering of suitable countermeasures such as chaff, decoys or flare to protect the host platform. AN/ALQ-156 (V) 1, AN/ALQ-156 (V) 2 and AN/ALQ-156 (V) 3 are the variants of AN/ALQ-156 (V) MAWS.

AN/AAR-47 These is an EWS providing multi-threat warning in one integrated system against laser assisted threats, IR-guided missile threats and unguided munitions.

AN/AAR-56 MAWS The AN/AAR-56 Missile Launch Detection System (MLDS) from Lockheed–Martin is one of the key defensive systems used by high-speed fixed-wing aircraft for long-range detection of both surface-launched and airborne missile threats.

AN/AAR-60 MILDS MAWS The AN/AAR-60 MILDS by EADS is a passive missile warning system configured around between one and six UV imaging sensor heads including optics, filters and image processing, and is capable of detecting and tracking UV emissions from the propulsion systems of potential missile threats including shoulder launched MANPADS. Upon detection of a potential threat, the system provides audio and visual warning to the pilot.

AN/ALE-40 Countermeasure Dispenser System The AN/ALE-40 Counter Measure Dispenser System (CMDS) manufactured by Tracor, now part of BAE Systems, is based on two types of expendable countermeasures, namely chaff to protect the host aircraft from radar and radar-guided threats and flares to offer protection against IR heat-seeking missiles.

AN/ALE-47 Countermeasure Dispenser System This CMDS, manufactured by BAE Systems, is an integrated, reprogrammable, computer-controlled system that uses information from integrated electronic warfare sensors such as radar warning receivers and missile approach warning receivers to determine the type of countermeasures to be dispensed by the crew to defeat IR- and radar-guided missiles.

AN/ALE-50 Expendable Decoy The AN/ALE-50 Advanced Airborne Expendable Decoy (AAED) is a towed expendable that offers protection to the host airborne platform against radar-guided missiles. It is a joint development venture of the United States Air Force, Navy and Raytheon.

AN/ALE-55 Fibreoptic Decoy The AN/ALE-55 fibreoptic towed decoy is a RF countermeasures system by BAE Systems designed to provide self-protection to the host airborne platform against attack by radar-guided missiles.

AN/ALQ-99 ECM Jammer This is an airborne jamming system capable of intercepting and then jamming the received radio frequency (RF) signals. The system comprises multiple receivers that monitor threats in four specific frequency bands and jamming transmitters.

ALQ-184 ECM Jammer The ALQ-184 by Raytheon is a self-protect electronic countermeasures (ECM) pod designed to provide protection to the aircraft platform and the aircraft crew against radar-guided threats.

AN/ALQ-187 ECM Jammer The AN/ALQ-187 ECM pod from Raytheon is an internally mounted version of the AN/ALQ-184(V).

AN/ALQ-212 ATIRCM The AN/ALQ-212 Advanced Threat Infrared Countermeasures (ATIRCM) system is developed by BAE systems in partnership with the US Army and is designed to defend the host airborne platforms from the shoulder fired MANPADs providing protection against an array of highly lethal missile attacks in all IR threat bands.

AN/ALQ-218 Radar Warning Receiver AN/ALQ-218 RWR/ESM/ELINT sensor system from Northrop–Grumman is designed for airborne situational awareness and signal intelligence gathering functions. It protects the warfighter by detecting, identifying, locating and analysing sources of RF emission thereby enabling air crews and/or commanders make the right decisions at the right time to ensure maximum protection of friendly forces and assets.

AN/ALR-67(V)3 The AN/ALR-67(V)3, called the Advanced Special Receiver (ASP), supersedes the AN/ALR-67(V)2 system and incorporates many new features such as extended capabilities in detection and processing for Air-Defence radar threats, threat localization and potential lethality, thereby enhancing the survivability of aircraft equipped with the system. It provides visual and aural alerts to air crew upon detection of ground-based, ship-based or airborne radar emitters, thereby enhancing pilot situational awareness by providing accurate identification, lethality and azimuth displays of hostile and friendly emitters.

AN/ALR-67 Radar Warning Receiver AN/ALR-67 is a family of airborne radar warning receiver and countermeasures control systems designed to provide warning to the aircraft's crew on any hostile radar activity by detecting, identifying and locating radars and radar-guided weapon systems of the adversary.

Angle Deception Jammer This induces errors in angle tracking circuits of radar receiver by creating false targets in each bearing. An angle deception jammer is capable of creating much larger tracking errors than are created by range and velocity deception jammers.

AN/VLQ-6 HARDHAT This is a vehicle-mounted Missile Countermeasures Device (MCD) developed by Loral, Inc. The system is used as a part of a comprehensive warning and threat response countermeasures system that detects and intercepts laser signals to provide warning of an imminent attack. The system then emits IR energy to disrupt the threat's missile/command unit tracking loop. The system comprises four sensors that cover the lower hemispheric region beneath the aircraft to compute the angle-of-arrival of the laser threat. It employs high-energy countermeasure beams to neutralize the threat.

APR-39D (V)2 Radar Warning Receiver APR-39D (V)2 Radar Warning Receiver (RWR)/ Electronic Warfare Management System (EWMS) belongs to the APR-39 family of RWRs from Northrop–Grumman. It is designed to protect a variety of fixed-wing, rotary-wing and tilt-wing aircraft from radar threats including pulsed, pulse Doppler, continuous-wave, scanning type radar emitters and emitters with frequency agility and low probability of intercept.

ARENA Active Protection System This is an active protection system (APS) designed to protect armoured fighting vehicles from anti-tank weapons including anti-tank guided missiles and missiles with top attack warheads.

Ariel Towed Radar Decoy This is a compact and lightweight towed radar decoy from Selex Galileo, previously GEC-Marconi, and capable of providing off-board countermeasures to protect the host airborne platform from attack by RF guided weapons including the state-of-the-art weapons employing error cancelling monopulse tracking techniques.

Barrage Jammer This is a broad band jammer. It spreads its radiated energy over a large bandwidth. The noise power spectral density at the receiver gets diluted in the case of barrage jamming.

BOL Countermeasure Dispenser System This system from SAAB is a high-capacity dispenser of chaff or IR flares used by the crew to protect the host aircraft from radar and radar-guided threats and IR heat-seeking missiles including the advanced IR-guided missiles.

BOW Radar Warning Receiver The BOW family of RWRs from SAAB, Sweden, is designed to provide to the crew of fighter aircraft enhanced situational awareness through detection and evaluation of potential threats in dense signal environments with high probability of intercept, high selectivity and sensitivity as well as the possibility to geo-locate RF emitters.

BOXER Multirole Armoured Vehicle The BOXER MRAV manufactured by the ARTEC GmbH industrial group is a land vehicle designed to have low thermal, acoustic and radar signatures.

BOZ-107 Countermeasures Dispenser System This system from the SAAB Bofors Dynamics pod is a self-contained, microprocessor-controlled, reprogrammable dispenser capable of releasing either a chaff cloud as decided by the crew depending upon the threat scenario.

Bragg-Cell Receiver See *Acousto-Optic Receiver.*

CATS Radar Warning Receiver Compact Airborne Threat Surveyor (CATS) is the state-of-the-art family of radar warning receivers from Thales group designed for helicopters. The system provides real-time battlefield situation awareness to avoid potential threats or hand over targets to weapon system in a sensor-to-shooter concept.

CERES   This is a signal intelligence (SIGINT) system comprising a constellation of three satellites as the space segment and ground segment including ground control segment to command and control the spacecraft and user ground segment to perform mission planning and data processing functions.

Chaff   This is an airborne cloud of light weight reflecting objects typically consisting of strips of aluminium foil, metal coated fibres or nylon that produce clutter echoes in a region of space for the purpose of confusing, screening or otherwise adversely affecting the opposition sensor system such as radar. The important characteristics of chaff include its fall rate, bloom time, effective echoing area, polarization and bandwidth.

Channelized Receiver (RWR)   A channelized receiver increases the overall bandwidth of a superheterodyne type receiver. It does so by either using a set of fixed frequency filters whose outputs can be switched to the mixer followed by intermediate frequency (IF) amplifier/filter and an appropriate detector or by simultaneously converting the outputs of filters to a common intermediate frequency (IF) thereby effectively producing a set of parallel superheterodyne receivers. The channelized receiver possesses the sensitivity, the dynamic range and the frequency resolution of a simple superheterodyne receiver and at the same time has capability to process signals across a very wide radio frequency (RF) spectrum, typically 2–20 GHz. The probability of intercept is nearly 100%.

Chengdu J-20   This is a twin-engine, fifth generation stealth fighter jet under development at the Chengdu Aerospace Corporation for the People's Liberation Army Air Force and is expected to enter service between 2018 and 2020.

Common Infrared Countermeasures (CIRCM) System   This system from Northrop–Grumman is designed specifically to protect rotary-wing and medium fixed-wing aircraft from IR-guided missiles.

Common SIGINT System 4000 (CSS-4000)   The CSS-4000 is one of the several products from Northrop–Grumman's airborne SIGINT product line. It is the next-generation of lightweight SIGINT payloads providing rapid interception, precision geolocation and communications signal processing functions to support a variety of mission profiles for fighter aircraft.

Communications Intelligence (COMINT)   COMINT is the intelligence information derived from communications type electromagnetic radiation emanating from foreign sources by those other than the intended recipients.

Compressive Receiver   See *Micro-Scan Receiver*.

COROLLA PRINCE   COROLLA PRINCE, developed in the United States, is an airborne laser weapon based on Stingray technology.

Cross-Polarized Jammer   This radiates a jamming signal whose polarization is at 90° to the polarization of the wave radiated by the radar to be jammed. It is particularly effective against radars that employ sidelobe blanking technique to improve the quality of their PPI display.

Crystal Video Receiver   This is the simplest and most commonly used radar warning receiver. In a crystal video receiver, the impinging microwave signal from hostile radar falls upon a wide band receiving antenna from which it is fed into a bank of simple receivers comprising a cascade arrangement of filter, detector and amplifier. The receivers are each tuned to consecutive slices of the covered band that allows simultaneous reception and discrimination of radars operating in various parts of the band.

CV-90 ICV   The CV-90 is an Infantry Combat Vehicle (ICV) of Sweden and is in service with Sweden, Denmark, Finland, Netherlands, Norway and Switzerland. It has been designed to minimize radar and IR signatures.

Deception Jammer   This transmits a corrupted signal towards the radar, which it accepts as a genuine one. It is unlike a noise jammer that attempts to swamp the radar echo signal by transmitting towards the radar a noise signal of sufficient power over the bandwidth of the radar receiver.

Decoy This is a dummy radar target much smaller in size than the aircraft it is intending to protect but whose reflectivity, or in other words the RCS, is made larger than the actual target by using reflectors or other means. It confuses the radar operator as well as the radar processing circuits and thus increases the survivability of the attacking aircraft.

DHY-322 Laser Decoy System The DHY-322 from CILAS, France is specifically designed to protect ships, strategic sites and other high-value assets. This system mainly comprises a laser detection system and high-energy laser.

Directional Infrared Countermeasures System (DIRCM) This uses a high-energy IR source of jamming signal, which is a tunable multiband laser source, with a missile approach warning system. The warning system provides information on the approaching missile threat in terms of its angle-of-arrival, time-to-impact and so on, and the IR laser energy is directed towards the threat to create confusion in its tracking circuits forcing it to deviate from its intended trajectory.

E-LAWS Laser-Warning System E-LAWS, developed by Elbit Systems detects, locates and categorizes potential laser threats such as those from laser range finders (single pulse), laser target designators (2–50 Hz PRF), laser beam riders (2–70 kHz) and IR illuminators with high probability of detection (99.9%) and low false alarm rate (less than 1 in 18 h).

Electronic Attack (EA) See *Electronic Countermeasures.*

Electronic Countermeasures (ECMs) ECMs refer to the division of electronic warfare that refers to systems and techniques to deny a sensor, such as radar or laser detection of target, or deceive it to follow a wrong or an electronically created non-existent target.

Electronic Counter-Countermeasures (ECCM) The purpose of ECCM techniques is to defeat countermeasures employed by the adversary and make the systems like radars, missiles and other similar sensor systems immune to countermeasures. This is achieved by using techniques to nullify his jamming and/or decoy systems.

Electronic Intelligence (ELINT) ELINT is the intelligence information derived from non-communications type electromagnetic radiation emanating from foreign sources other than radioactive sources or nuclear detonations.

Electronic Protection (EP) See *Electronic Counter-Countermeasures.*

Electronic Reconnaissance This is defined as the detection, identification, evaluation and location of foreign electromagnetic radiation emanating from sources other than radioactive phenomenon or nuclear detonations. Electronic reconnaissance is further classified as Radiation Intelligence (RINT), Telemetry Intelligence (TELINT) and Signal Intelligence (SIGINT).

Electronic Support Measures (ESM) ESMs refer to the division of electronic warfare (EW) that involves actions either tasked by or under the direct control of operational commander taken to search for, intercept, locate and identify sources of unintentional or intentional radiated electromagnetic energy for the purpose of immediate threat recognition.

Electronic Warfare Support See *Electronic Support Measures.*

Electro-Optic CounterMeasures (EOCM) EOCM is that part of EW that deals with systems employed against enemy systems operating in the optical spectrum from UV to IR. Laser designators and range finders, laser radar, electro-optic sensors such as laser seekers and IR seekers, respectively, used in laser-guided munitions and IR-guided weapons are some common battlefield electro-optic systems.

EL/M-2160 (V1) MAWS EL/M-2160 (V1) from Israel Aerospace Industries is an airborne missile approach warning system based on pulse Doppler radar. The system comprises a transceiver processing unit and six antennas to give warning on approaching missile threats in boost, sustain and post-burnout phases of missile flight in 360° coverage. The information on missile parameters such as range, velocity and direction-of-arrival is used to determine time-to-impact, which in turn is used for timely activation of countermeasures such as chaff and flares for effective missile deception.

Escort Jammer This is a type of noise jammer. In the case of *escort jamming*, the jammer is on board a specialized ECM aircraft that escorts the attack aircraft.

F-22 Raptor Fighter Aircraft This is a twin-engine tactical fighter aircraft developed by Lockheed–Martin and currently in use by the US Air Force.

F-35 Lightning-II Fighter Aircraft The F-35 Lightning-II is a single engine multirole stealth fighter aircraft under development at Lockheed–Martin. Its primary role is in aerial reconnaissance, ground attack and air-defence operations.

Flares There are high-temperature heat sources ejected from an airborne platform such as aircraft to mislead or decoy heat-seeking missiles away from their intended trajectory towards the targeted aircraft. They do so by achieving burn temperatures greater than those generated by exhaust of aircraft thereby producing IR signatures similar to those characterizing the aircraft to be protected.

GUARDIAN DIRCM System This is developed by Northrop–Grumman and it is a Common Infrared Countermeasures (CIRCM) system that provides 360°-protection to the host airborne platform against IR guided MANPADS.

HARLID Laser Sensor The High Angular Resolution Laser Irradiance Detector (HARLID) sensor from Excelitas is not a complete laser-warning receiver; but is a sensor configured around an array of silicon (Si) and indium gallium arsenide (InGaAs) detectors designed for use in laser-warning receiver systems to detect and provide precise angle-of-arrival (AoA) information from direct and indirect scattered light from laser range finders, laser target designators and active laser-based electro-optic systems.

Infrared Countermeasures (IRCM) IRCMs are employed to protect aircraft from surface-to-air and air-to-air heat-seeking missiles. The IRCM systems achieve this by confusing the missiles' IR guidance system and forcing them to deviate from their intended trajectory.

La Fayette-Class Frigate These multipurpose stealth frigates were developed by DCNS, a French Naval Shipbuilding Company and European leader in naval defence, and built at the DCN Lorient Naval Dockyard. These stealth frigates are in use by French Navy (Marine Nationale).

Laser Countermeasure (LaCM) The LaCM system was jointly developed by Office of Naval Research (ONR), Naval Research Laboratory (NRL) and Northrop–Grumman under EWISSP (Electronic Warfare Integrated System for Small Platforms) programme to protect small surface platforms from attack by laser-guided munitions.

Laser-Warning Receiver (LWR) An LWR is an electro-optic support measures component of electronic warfare system. It intercepts laser radiation from laser emitters, such as the adversary's LRFs and LTDs, to identify the type and location of these emitters and generate appropriate cues for the aircraft crew enabling activation of suitable defensive and/or offensive countermeasures through its interface with other systems.

Laser-Warning System (LWS) The LWS is designed and developed by Thales Group. It provides an early warning to the platform crew on LRF and target designator threats enabling them to take an evasive action thus enhancing platform survivability.

LIAS Laser-Warning Receiver This system is an advanced laser threat warning sensor system designed by ASELSAN to detect, classify and identify hostile laser threats illuminating friendly platforms and give warning to the platform crew enabling them to activate the appropriate manoeuvre or countermeasure system available on the platform, either directly or via a host computer.

LIRD-4 Laser Detector The system detects and classifies the laser threat irradiating the platform along with its angle-of-arrival in horizontal sector. The type of threat, its angle-of-arrival and other threat features are displayed on the indicator unit. The system is capable of detecting laser threats from LRFs, LTDs and laser beam riders.

LR-100 Radar Warning Receiver Northrop–Grumman's LR-100 is a combat-proven RWR/ESM/ELINT receiver system providing RWR, ESM and ELINT functionality in a single system, thereby eliminating need and cost of integrating a unique system for each function.

LWS-200 CV Laser-Warning System This system by Avitronics detects, classifies and identifies potential laser threats such as those from LRFs, LTDs and missile guidance systems such as laser beam riders with a high probability of interception and alerts the platform crew to take evasive action or deploy countermeasures to protect the platform.

LWS-300 CV Laser-Warning System This system is similar to LWS-200 CV in system architecture. Its mechanical design has been optimized for the harsh environmental conditions and it offers increased spectral band coverage of 0.5–1.8 μm and also enhanced sensitivity in the 0.904 μm band for improved performance against typical missile guidance lasers.

M-206 Decoy Flare A pyrotechnic flare. Measuring $26 \times 26 \times 205$ mm, M-206 IR Decoy Flare provides protection against heat-seeking missiles for helicopters and low-altitude aircraft

MALD-J ECM Jammer (MALD-J) The MALD-J from Raytheon is the jammer variant of the miniature air-launched decoy (MALD), which is an expendable air-launched flight vehicle with combat profiles and signatures such as the RCS of fighter and bomber aircraft, particularly of the US and allied forces.

MASS (Multi-Ammunition Soft-Kill System) MASS from Rheinmetall Defence is a compact and fully automatic ship-borne soft-kill launcher system originally designed to provide protection against anti-ship missiles and asymmetric threats employing RF, IR and/or EO seekers through deployment of programmable multi-spectral Omni-Trap decoy rounds with radar, IR, laser, EO and UV payloads.

MAWS-300 MAWS This is a passive UV-based missile approach warning sensor. It is part of Integrated Defensive Aids Suite (IDAS) family of self-protection systems designed by SAAB to provide Electronic Order of Battle (EOB) and self-defence for airborne platforms in diverse and dense threat environments. The missile approach warning sensor type MAW-300 uses passive UV-based sensors operating in a solar-blind UV spectrum. Each sensor provides conical coverage of 110°, which limits the unprotected hole below the platform and allows good overlap. Four sensors provide azimuth coverage of 360°.

Micro Scan Receiver Also known as a compressive receiver, this achieves the objectives of a channelized receiver by sweeping a single receiver across the same band in a time period equal to or less than the shortest pulse to be intercepted. In this type of receiver, the local oscillator is swept in frequency linearly across a bandwidth equal to sum of RF and IF bandwidths of the receiver. The mixer output is fed to a Dispersive Delay Line (DDL) with linear characteristics. The DDL produces at its output a compressed pulse at a time measured from the local oscillator frequency sweep. Multiple input RF signals appear as a series of narrow pulses at the output. These receivers are particularly useful for intercepting pulse-to-pulse frequency agile emitters.

Missile Approach Warning System (MAWS) A MAWS is a sensor system used to detect missile threats and cue the aircraft crew to take a defensive manoeuvre or deploy suitable countermeasures such as chaff, flares or IR countermeasures. There are three main categories of MAWS including radar-based, IR-based and UV-based sensors. While active system such as pulse Doppler radar detects and tracks the missile by means of the radar signal returned from the skin of the missile; the passive systems like IR- and UV-based systems do so by detecting the exhaust plume signatures in the respective wavelength band.

MJU-7A/B Decoy Flare This is a pyrotechnic flare that uses a Magnesium/Teflon/Viton (MTV) IR payload and measures 26×52 mm in cross-section and 205 mm in length.

MJU-10/B Decoy Flare This is also an MTV-based IR decoy flare. It offers protection against heat-seeking missiles for the F-15 and heavy transport aircraft such as the C-5 and C-17. The MJU-10/B measures 52×65 mm in cross-section and 205 mm in length.

MJU-32/B Decoy Flare The MJU-32/B IR Decoy Flare devices are magnesium-Teflon based cylindrical flares measuring 36 mm in diameter and 148 mm in length.

MJU-49/B Decoy Flare This is manufactured in a round 36-mm standard form factor and can be employed using the ALE-39 series or any countermeasure dispensing system capable of employing a round 36-mm flare.

MJU-50/B Decoy Flare The MJU-50/B IR countermeasure flare is a $26 \times 26 \times 205$ mm expendable containing special pyrophoric material.

MJU-51/B Decoy Flare This is an expendable pyrophoric flare measuring $26 \times 52 \times 205$ mm.

Multi-Colour Infrared Alert Sensor (MIRAS) MIRAS from EADS and Thales is a passive IR sensor-based MWS designed to protect airborne platforms from the complete range of MANPADS and air-to-air missiles.

MUSIC Family, DIRCM Systems The MUSIC family by DIRCM systems has been developed by Elbit Systems' wholly owned subsidiary Elbit Systems Electro-Optics ELOP, Ltd. The MUSIC family provides protection for airborne platforms such as fixed and rotary-wing aircraft and helicopters from IR-guided surface-to-air missiles such as MANPADS and other Short-Range Air-Defence systems (SHORADS) and air-to-air missiles. Different variants include MUSIC, MUSIC-C, MUSIC-J and Mini-MUSIC DIRCM systems.

MWS-20 MAWS The MWS-20 by Thales is an active Doppler radar-based missile approach warning system capable of detecting and continuously tracking surface-to-air missiles, MANPADS and other active and passive missiles during their entire flight phase providing precise and reliable information on missile range and velocity, D0A and missile time-to-impact. Information on time-to-impact enables selection and triggering of appropriate countermeasures at the optimum time.

Naval Laser-Warning System (NLWS) NLWS from SAAB is capable of detecting, classifying, identifying and analysing potential laser threats emanating from an adversary's LRFs and LTDs used for targeting and active EOCM class lasers used for creating confusion in blue-water and littoral combat environments thereby providing to the naval platform's command team the vital situational awareness about the presence of laser activity.

Next-Generation Jammer (NGJ) The NGJ electronic attack pod has been developed by Raytheon as a replacement for the ALQ-99 tactical jamming pod on the US Navy's EA-18G Growler aircraft. The system represents a marked improvement over ALQ-99 jamming pod in terms of its software-based digital architecture and use of high power active electronically scanned arrays based on gallium nitride technology. The pod is designed to provide enhanced airborne Electronic Attack (EA) capabilities to degrade and disrupt adversary's air-defence and ground communications systems.

Noise Jammer This transmits a noise-like signal maximizing the noise power in the receiver's bandwidth to completely mask the desired signal, which could be a communication signal or an echo signal in the case of radar.

PAK FA T-50 Fighter Aircraft The PAK FA (Prospective Airborne Complex of Frontline Aviation) T-50 is a fifth generation twin-engine fighter jet being prototyped by Sukhoi for use by Russian Air Force.

Passive Countermeasures Passive countermeasures employed by a platform to be protected include use of armour, camouflage, fortification and other protection technologies such as a self-sealing fuel tank and so on.

PAWS-2 MAWS PAWS-2 from Raytheon and Elisra is an IR sensor-based missile warning system designed to provide protection to fighter aircraft, helicopters, transport and commercial aircraft against attack from surface-to-air missiles (SAM) attack and MANPADS.

PL-01 tank The PL-01 tank designed and manufactured by OBRUM of Poland with support from BAE Systems of UK is based on Swedish CV-90 ICV and uses stealth technology.

Pyrophoric Flares These are small pieces of foil and are much less visible. They oxidize very quickly producing heat thereby generating IR energy. They cool in the atmosphere as they fall to the ground as rusted metal debris.

Pyrotechnic Flares These produce highly visible white light and smoke when ignited. They generate intense IR energy for time duration of 5–10 s, which is enough for confusing or distracting the attacking missile.

Radar Order of Battle (ROB) ROB is the information on location and capability of all enemy radars on a current basis.

Radar Warning Receiver (RWR) A RWR is a passive electronic warfare system that intercepts RF signals from radar emitters of the adversary to identify type and location of radar emitter and generates appropriate cues for the aircraft crew enabling activation of suitable countermeasures through its interface with other systems.

Radiation Intelligence (RINT) RINT is basically the intelligence derived from collection and analysis of non-information bearing electromagnetic radiation emitted unintentionally by foreign equipment, devices and so on other than those generated by detonation of nuclear weapons.

Range Deception Jammer This creates a large number of false targets at different ranges and angles from the true target position and thus confuses and overloads the radar operator as well as radar processing circuits. Range deception is particularly effective against search and track radars.

ROSY-L (Rapid Obscurant System-Land) ROSY-L from Rheinmetall Defence is the smoke/obscurant protection system designed to provide protection to crew and passengers of light military and civilian vehicles against surprise attacks, such as those experienced during reconnaissance patrols or while moving in convoy, from a wide range of guided weapon threats including TV-, EO-, IR-, IIR- and SACLOS-guided weapons.

ROSY-N (Rapid Obscurant System – Navy) ROSY-N (is designed to protect naval and coast guard units from missiles and asymmetric attack in littoral zones and inland waters. It is particularly suited to installation on patrol vessels, speed boats, fast attack craft and landing craft.

S-3000 family SIGINT The S-3000 family are SIGINT and IO systems from BAE Systems. With a wide operational frequency range from high frequency (HF) to Ku band, open commercial off-the-shelf (COTS) architecture enabling third party integration, software-defined that allows re-configurability on-the-fly, multiple antenna types and locations, and advanced transmit and receive capabilities can be seamlessly integrated on to a wide variety of platforms such as unmanned aerial vehicles, manned aircraft, land vehicles and surface vessels for a range of mission profiles.

Sea Shadow This is an experimental stealth demonstrator developed under a combined programme by Defence Advanced Research Project Agency (DARPA), the United States Navy and Lockheed Missiles & Space Company.

Shenyang J-31 This is also called the Gyrfalcon and Falcon Hawk and is a fifth generation multirole stealth aircraft designed and built by Shenyang Aircraft Corporation, an affiliate of Aviation Industry Corporation of China (AVIC), for People's Liberation Army Air Force.

Shivalik-Class Frigate This is a multirole stealth craft built for the Indian Navy. The structural, thermal, and acoustic stealth features of Shivalik-class frigate make them harder to be detected by the adversary sensors.

SHTORA-1 Defensive Aids Suite (DAS) The Shtora-1 DAS developed by the Zenit Research and Production Corporation is designed to enhance battlefield survivability of Armoured

Fighting Vehicles (AFVs) against attack from Anti-Tank Guided Weapons (ATGWs) such as TOW and HOT with a SACLOS guidance system as well as missiles and laser-guided ATGWs such as Hellfire and Copperhead.

Side Lobe Blanking (SLB) The purpose of SLB is to prevent the detection of strong targets and interference pulses through side lobes.

Side Lobe Cancelling (SLC) The purpose of SLC is similar in nature to sidelobe blanking and is to suppress noise interference received through side lobes. This is achieved by estimating the direction of arrival of the jamming signal and jamming signal power with the help of an array of auxiliary antennas and then modifying the receiver directional pattern to place nulls in the jammer direction.

Signal Intelligence (SIGINT) The SIGINT system is usually an integrated setup of high performance antennas, receivers, signal processing equipment, wideband recorders and so on to provide both communications and non-communications information for a long term strategic use including generation of electronic data bases to support tactical operations. SIGINT is further subdivided into Electronic Intelligence (ELINT) and Communications Intelligence (COMINT).

Spot Jammer This concentrates its power into a narrow band centred on the frequency of the enemy's sensor approximately matching its bandwidth.

Standoff Jammer This is a noise jammer not carried by the aircraft to be protected. Instead, it is carried by another aircraft flying outside the range of local air defences and providing the noise jamming power to protect the attacking aircraft as it flies to and from the target. The burnthrough range of a standoff jammer is much greater than that of a self-protection jammer.

Stealth Technology This refers to all those techniques used to minimize target signatures.

Stingray Laser Detection and Countermeasures System This system from Lockheed–Martin is an EOCM system designed to protect the frontline forces by accurately locating and neutralizing the adversary's optical and electro-optical fire-control systems.

Superheterodyne Receiver This contains a combination of amplification with frequency mixing. Irrespective of the received radio frequency, the receiver converts the received signal to a fixed lower frequency called intermediate frequency, which can be conveniently processed.

Swept Spot Jammer This is characterized by the high noise power spectral density of a spot jammer and large bandwidth of a barrage jammer

T-14 Armata Main Battle Tank This is designed and manufactured by Uralvagonzavod (UVZ). Its design is based on Armata Universal Combat Platform and is practically invisible to IR and other spectra due to use of stealth technology.

Tactical SIGINT Payload (TSP) The TSP from BAE Systems is an advanced sensor system that, due to its software-defined architecture, is capable of processing a host of military signals for a wide range of mission profiles.

Telemetry Intelligence (TELINT) TELINT is the intelligence derived from the interception and analysis of foreign telemetry.

TOP SCAN This is a SIGINT/DF payload developed by Rafael Advanced Defence Systems Ltd for airborne platforms. The system is configured around an advanced direction-finding and localization system using dual-axis interferometer for direction-finding and localization sensors.

TROPHY Active Protection System TROPHY, from Rafael, Israel, is an active protection system designed to supplement the armour of both light and heavy armoured fighting platforms. It intercepts the incoming kinetic energy threats and destroys them by a shotgun like blast.

Velocity Deception Jammer This maximizes jammer energy in a radar's Doppler filters. The deception jammer first receives the radar signal, amplifies it and then retransmits it without any change in frequency so that it is received by the Doppler filter processing the skin echo signal.

Visby-Class Corvette This is built by Kockums for the Swedish Navy and is designed to have low visibility, RCS and IR signatures.

Zumwalt-Class Destroyer This is a multi-mission stealth naval vessel designed to be a guided missile destroyer for the US Navy.

Bibliography

1 Vakin, S.A., Shustov, L.N. and Dunwell R.H. (2001), *Fundamentals of Electronic Warfare*, Artech House Publishers.

2 Adamy, D.L. (2001) *EW-101: A First Course in Electronic Warfare*, Artech House Publishers.

3 Adamy, D.L. (2004) *EW-102: A Second Course in Electronic Warfare*, Artech House Publishers.

4 Adamy, D.L. (2009) *EW-103: Communications Electronic Warfare*, Artech House Publishers.

5 Adamy, D.L. (2015) *EW-104: Electronic Warfare Against a New Generation of Threats*, Artech House Publishers.

6 Schleher, D.C. (1999), *Electronic Warfare in Information Age*, Artech House Publishers.

7 Adamy, D.L. (2006), *Introduction to Electronic Warfare Modelling and Simulation*, SciTech Publishing.

8 Graham, A. (2010), *Communications, Radar and Electronic Warfare*, John Wiley & Sons, Ltd.

9 Hannen, P.J. (2013), *Radar and Electronic Warfare Principles for the Non-Specialist*, SciTech Publishing.

10 Boyd, J.A., Harris, D.B., King, D.D. Jr. and Welch, H.W. (1978) *Electronic Countermeasures*, Peninsula Publishers.

11 Pollock, D.H. (1993), *The Infrared & Electro-optical Systems Handbook, Volume 7: Countermeasure Systems*, Infrared Information Analysis Centre and SPIE Optical Engineering Press.

7

Laser Fundamentals

Since the early 1980s, lasers have penetrated almost every conceivable area of application from fundamental science and technology to industry, from medicine and healthcare to entertainment and from tactical battlefield to the strategic domain. The applications have grown at a very fast rate not only in already existing domains, but laser devices have also found their presence in many new areas. While the expansion of non-military applications of lasers is primarily driven by availability of a large number of wavelengths followed by ever increasing power levels and diminishing price tags at which these wavelengths could be generated; military applications of lasers and related electro-optic devices have grown mainly because of the technological maturity of lasers that were born in the late 1960s and early 1970s. Technological advances in optics, optoelectronics and electronics leading to more rugged, reliable, compact and efficient laser devices are largely responsible for making them indispensable in modern warfare. Laser systems meant for existing military applications continue to improve in terms of performance specifications and system engineering; a lot of work is being done particularly in technologically advanced countries to exploit the potential of lasers and optoelectronic devices in newer areas. The present chapter gives an overview of the operational basics of lasers and the properties and characteristic parameters that define a laser system as well as the major types that are less or more commonly used in the military domain.

7.1 Operational Basics

In the following paragraphs, the operational principle of a laser and associated topics such as two-, three- and four-level lasers, different parts that make up a laser system and modes of a laser are discussed.

7.1.1 Principles of Laser Operation

The basic principle of operation of a laser device is evident from the expanded form of the acronym 'LASER', which says that it produces a light output due to stimulated emission of radiation. In the case of ordinary light, such as that from the Sun or an electric bulb, different photons are emitted spontaneously due to various atoms or molecules releasing their excess energy on their own. In the case of stimulated emission, an atom or a molecule holding excess energy is stimulated by another photon emitted earlier to release that energy in the form of a photon. As we shall see in the following paragraphs, *population inversion* is an essential condition for the stimulated emission process to take place. To understand the process of population inversion subsequently

Handbook of Defence Electronics and Optronics: Fundamentals, Technologies and Systems, First Edition. Anil K. Maini.
© 2018 John Wiley & Sons Ltd. Published 2018 by John Wiley & Sons Ltd.

leading to stimulated emission and laser action, a little brief on quantum mechanics and optically allowed transitions would be well worth its place here.

7.1.1.1 Rules of Quantum Mechanics

All particles, big or small, according to the basic rules of quantum mechanics, have discrete energy levels or states. Various discrete energy levels correspond to different periodic motions of its constituent nuclei and electrons. While the lowest allowed energy level is also referred to as the *ground state*, all other relatively higher energy levels are called *excited states*. The discrete energy levels that exist in any form of matter are not necessarily only those corresponding to periodic motion of electrons. There are many types of energy levels other than the simple-to-describe electronic levels. The nuclei of different atoms constituting the matter themselves have their own energy levels. Molecules have energy levels depending upon vibrations of different atoms within the molecule and molecules also have energy levels corresponding to the rotation of the molecules. When we study different types of lasers, we shall see that all kinds of energy levels, electronic, vibrational and rotational, are instrumental in producing laser action in some of the very common types of lasers.

Transitions between electronic energy levels of relevance to laser action correspond to the wavelength range from ultraviolet to near infrared. Lasing action in neodymium lasers (1064 nm) and argon-ion lasers (488 nm) are some examples. Transitions between vibrational energy levels of atoms correspond to infrared wavelengths. The carbon dioxide laser (10 600 nm) and hydrogen fluoride laser (2700 nm) are two examples. Transitions between rotational energy levels correspond to a wavelength range from 100 μm to 10 mm.

In a dense medium such as a solid, a liquid or a high-pressure gas, atoms and molecules are constantly colliding with each other thus causing atoms and molecules to jump from one energy level to another. What is of interest to a laser scientist however is an *optically allowed transition*. An optically allowed transition between two energy levels is the one that involves either absorption or emission of a photon satisfying the resonance condition of $\Delta E = h\upsilon$, where ΔE is the difference in energy of the two involved energy levels, h is Planck's constant $(= 6.6260755 \times 10^{-34}$ Joules^{-s} or $4.1356692 \times 10^{-15}$ eV$^{-s})$ and υ is the frequency of the photon emitted or absorbed.

7.1.1.2 Absorption, Spontaneous Emission and Stimulated Emission

An electron or an atom or a molecule makes a transition from a lower level to a higher level only if suitable conditions exist. These include:

1) The particle that has to make the transition should be in the lower energy level.
2) The incident photon should have energy $(= h\upsilon)$ equal to the transition energy, which is the difference in the energies of the two involved energy levels; that is, $\Delta E = h\upsilon$.

If these conditions are satisfied, the particle may make an absorption transition from the lower level to the higher level (Figure 7.1a). The probability of occurrence of such a transition is proportional to both the population of the lower level and also the related Einstein coefficient.

There are two types of emission processes, namely *spontaneous emission* and *stimulated emission*. The emission process involves transition from a higher excited energy level to a lower energy level. Spontaneous emission is the phenomenon in which an atom or molecule undergoes a transition from an excited higher energy level to a lower level all by itself without any outside intervention or stimulation and in the process emits a resonance photon [Figure 7.1(b)]. The rate of the spontaneous emission process is proportional to the related Einstein coefficient. In the case of stimulated emission [Figure 7.1(c)], first there exists a photon called a stimulating photon with an energy equal to the resonance energy, $h\upsilon$. This photon perturbs another excited

(a)

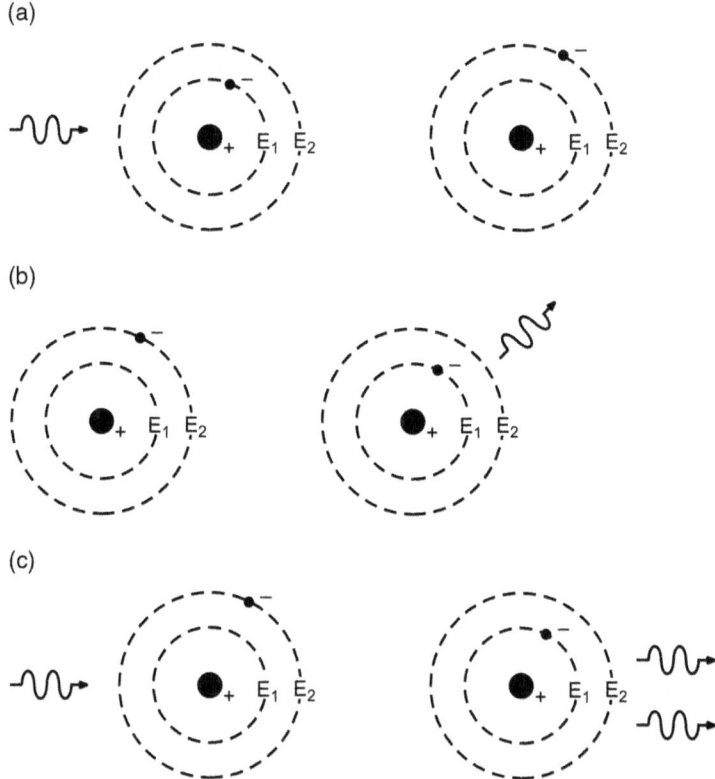

(b)

(c)

Figure 7.1 Absorption and emission processes: (a) absorption, (b) spontaneous emission and (c) stimulated emission.

species (atom or molecule) and causes it to drop to the lower energy level, in the process emitting a photon of the same frequency, phase and polarization as that of the stimulating photon. The rate of stimulated emission process is proportional to the population of the higher excited energy level and the related Einstein coefficient. Please note that, in the case of spontaneous emission, the rate of emission process does not depend on the population of the energy state from where the transition has to take place, as is the case in absorption and stimulated emission processes. According to rules of quantum mechanics, absorption and stimulated emission are analogous processes and can be treated similarly.

We have seen that absorption, spontaneous emission and stimulated emission are all optically allowed transitions. Stimulated emission is the basis for photon multiplication and the fundamental mechanism underlying all laser action. In order to arrive at the necessary and favourable conditions for stimulated emission and set the criteria for laser action therefore, it is important to analyse the rates at which these processes are likely to occur. The credit for defining relative rates of these processes goes to Einstein, who gave us the well-known 'A' and 'B' constants known as Einstein's coefficients. The A coefficient relates to the spontaneous emission probability and the B coefficient relates to the probability of stimulated emission and absorption. Remember that absorption and stimulated emission processes are analogous phenomena. The rates of absorption and stimulated emission processes also depend on populations of lower and upper energy levels, respectively.

Consider, for the purpose of illustration, a two-level system with a lower energy level '1' and an upper excited energy level '2' with populations of N_1 and N_2, respectively, as shown in Figure 7.2(a). The Einstein's coefficients for the three processes are B_{12} (absorption), A_{21} (spontaneous

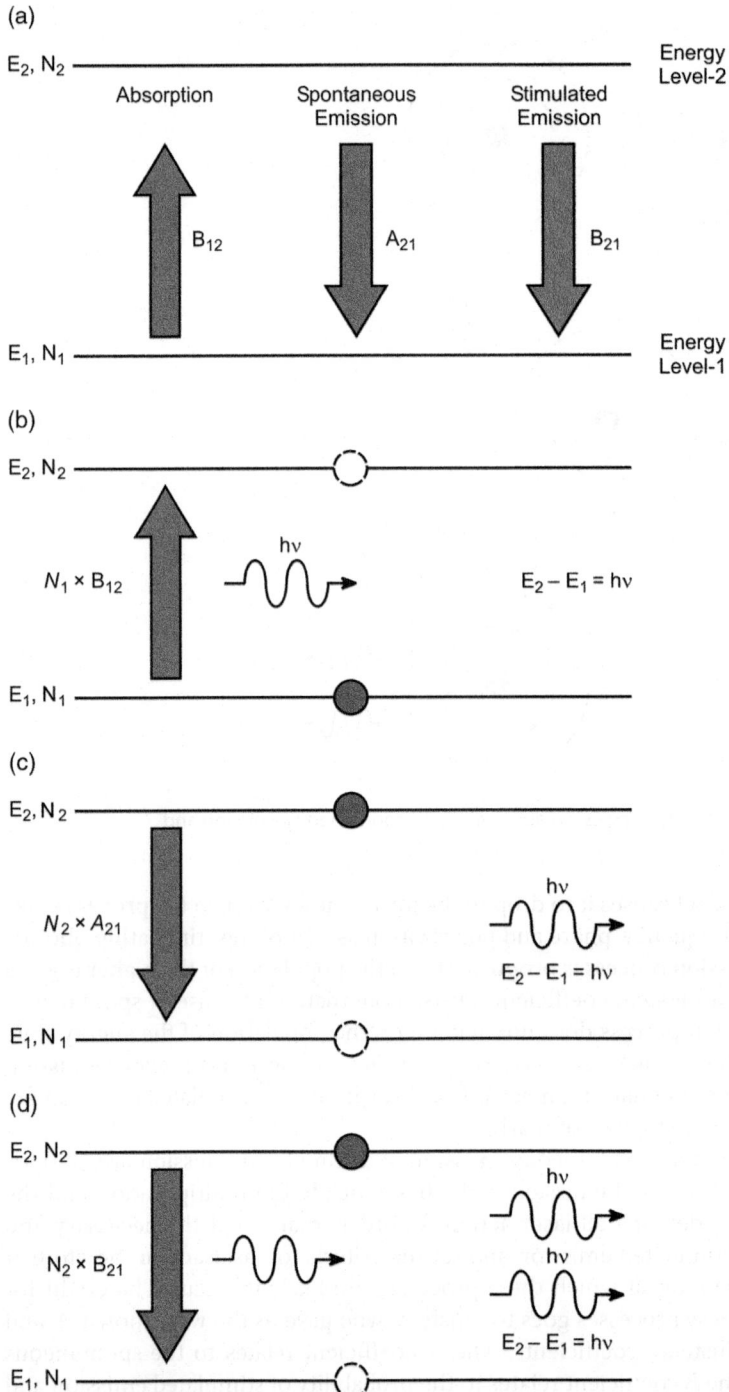

Figure 7.2 Absorption, spontaneous emission and stimulated emission processes in a two-level system: (a) two level system, (b) absorption, (c) spontaneous emission and (d) stimulated emission.

emission) and B_{21} (stimulated emission). The subscripts to the Einstein coefficients here represent the direction of transition. For instance, B_{12} is the Einstein coefficient for transition from level 1 to level 2. Also, since absorption and stimulated emission processes are analogous according to laws of quantum mechanics, $B_{12} = B_{21}$. According to Boltzmann statistical thermodynamics, under normal conditions of thermal equilibrium, atoms and molecules tend to be at their lowest possible energy level with the result that population decreases as the energy level increases. If E_1 and E_2 are the energy levels associated with level 1 and level 2, then the populations of these two levels can be expressed by eqn. 7.1.

$$\frac{N_2}{N_1} = \exp\left[-\left(E_2 - E_1\right)/kT\right] \tag{7.1}$$

Where,

k = Boltzmann Constant = 1.38×10^{-23} Joules/Kelvin or 8.6×10^{-23} eV/Kelvin
T = absolute temperature in Kelvin

Under normal conditions, N_1 is greater than N_2. When a resonance photon ($E = h\upsilon$) passes through the species of this two-level system, it may interact with a particle in level 1 and get absorbed, in the process raising it to level 2. The probability of occurrence of this is given by $(B_{12} \times N_1)$ (Figure 7.2b). Alternatively, it may interact with a particle already in level 2 leading to emission of a photon with same frequency, phase and polarization. The probability of the occurrence of this process known as stimulated emission is given by $(B_{21} \times N_2)$ (Figure 7.2b). Yet another possibility is that a particle in the excited level 2 may drop to level 1 on its own without any outside intervention emitting a photon in the process. The probability of this spontaneous emission is A_{21} (Figure 7.2c). The spontaneously emitted photons have the same frequency but have random phase, propagation direction and polarization.

If we analyse the competition between the three processes, it is clear that if $(N_2 > N_1)$, which is not the case under the normal conditions of thermal equilibrium, there is a possibility of an overall photon amplification due to enhanced stimulated emission. This condition of $(N_2 > N_1)$ is known as *Population Inversion* as under normal conditions $(N_1 > N_2)$.

7.1.1.3 Population Inversion

It can be analytically verified that under conditions of thermodynamic equilibrium, practically all atoms or molecules are in the lower level. When the population of the lower level is much larger than the population of the excited upper level, the probability of each spontaneously emitted photon hitting an atom or molecule in the lower level and getting absorbed is also much higher than the same stimulating another excited atom or molecule in the upper level. The same concept underlies the expressions for the probability of absorption, spontaneous emission and stimulated emission outlined earlier and reproduced here once again to reiterate what we have just described:

Probability of Absorption $\propto (B_{12} \times N_1)$
Probability of Spontaneous Emission $\propto (A_{12})$
Probability of Stimulated Emission $\propto (B_{21} \times N_2)$

If we want the stimulated emission to dominate over absorption and spontaneous emission, we must have more number of excited species in the upper level than the population of lower level. Such a situation is known as *population inversion*. It is so called as, under normal circumstances, the population of lower level is much more than the population of upper level. So, population inversion is an essential condition for laser action. Spontaneous emission depletes the excited upper level population (N_2 in the present case) at a rate proportional to A_{21}, producing undesired photons with random phase, direction of propagation and polarization. Due to this loss and other losses associated with the laser cavity discussed in Section 7.7.2.3, each laser has a certain minimum value of $N_2 - N_1$, which can produce a laser output. This condition of population inversion is known as the *inversion threshold* of the laser. *Lasing threshold* is another analogous term.

There are two possible ways to produce population inversion. One is to populate the upper level by exciting extra atoms or molecules to the upper level. The other is to depopulate the lower laser level involved in the laser action. In fact, for a sustained laser action, it is important to both populate the upper level and depopulate the lower level. Two commonly used pumping or excitation mechanisms are *optical pumping* and *electrical pumping*. Both electrons and photons have been successfully used to create population inversion in different laser media. While optical pumping is ideally suited to solid-state lasers such as ruby, Nd:YAG and Nd:glass lasers, electrical discharge is the common mode of excitation in gas lasers such as helium-neon and carbon dioxide lasers.

The excitation input, optical or electrical, usually raises the atoms or molecules to a level higher than the upper laser level from where it rapidly drops to the upper laser level. In some cases, the excitation input excites atoms other than the active species. The excited atoms then transfer their energy to the active species to cause population inversion. A helium-neon laser is a typical example of this kind where the excitation input gives its energy to helium atoms, which subsequently transfer the energy to neon atoms to raise them to the upper laser level.

The other important concept essential for laser action is the existence of a *metastable state* as the upper laser level. For stimulated emission, the excited state needs to have a relatively longer lifetime of the order of few microseconds to a millisecond or so. The excited species need to stay in the excited upper laser level for a longer time so as to allow interaction between photons and excited species, which is necessary for efficient stimulated emission. If the upper laser level had a lifetime of a few nanoseconds, most of the excited species would drop to the lower level as spontaneous emission. The crux is that for efficient laser action, buildup of population of the upper laser level should be faster than its decay. A longer upper laser level lifetime helps achieve this situation.

7.1.2 Two-, Three- and Four-Level Lasers

Another important feature that has a bearing on laser action is the energy level structure of the laser medium. As we shall see in the following paragraphs, energy level structure, particularly the energy levels involved in the population inversion process and the laser action, significantly affect the performance of the laser.

7.1.2.1 Two-Level Laser System

In a *two-level laser system*, there are only two levels involved in the total process. That is, the atoms or molecules in the lower level, which is also the lower level of the laser transition, are excited to the upper level by the pumping or excitation mechanism. The upper level is also the upper laser level. Once the population inversion is achieved and its extent is above the inversion threshold, the laser action can take place. Figure 7.3 shows the arrangement of energy levels in a two-level system. A two-level system is, however, a theoretical concept only as far as lasers are concerned. No laser ever has been made to work as a two-level system.

7.1.2.2 Three-Level Laser System

In a *three-level laser system*, the lower level of laser transition is the ground state (the lowermost energy level). The atoms or molecules are excited to an upper level higher than the upper level of the laser transition (Figure 7.4). The upper level to which atoms or molecules are excited from the ground state has relatively much shorter lifetime as compared to the lifetime of the upper laser level, which is a metastable level. As a result, the excited species rapidly drop to the metastable

Figure 7.3 Two-level laser system.

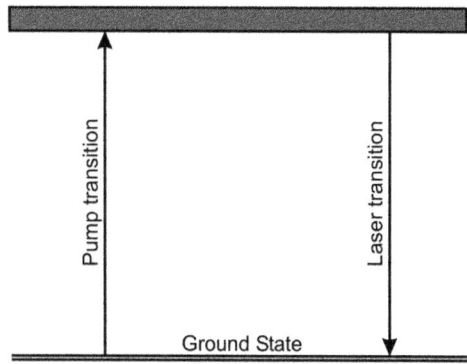

Figure 7.4 Three-level laser system.

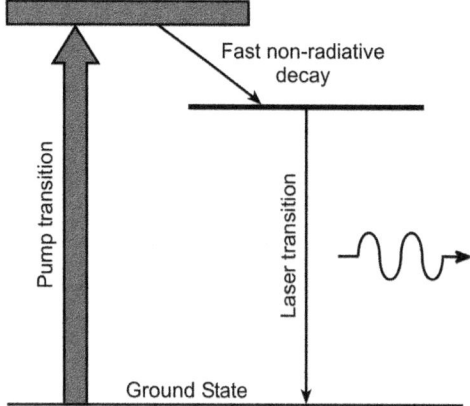

level. A relatively much longer lifetime for the metastable level ensures a population inversion between the metastable level and the ground state provided that at least more than half of the atoms or molecules in the ground state have been excited to the uppermost short-lived energy level. The laser action occurs between the metastable level and the ground state.

The ruby laser is a classic example of a three-level laser. Figure 7.5 shows the energy level structure for this laser. One of the major shortcomings of this laser and also all three-level lasers is due to the lower laser level being in the ground state. Because under thermodynamic equilibrium conditions, almost all atoms or molecules are in the ground state, it requires at least more than half of this number to be excited out of that state to achieve laser action. This implies that a much larger pumping input would be required to exceed population inversion threshold. This makes it very difficult to sustain population inversion on a continuous basis in three-level lasers. That is why the ruby laser cannot be operated in continuous-wave (CW) mode.

It would be ideal if the lower laser level were not in the ground state so it had fewer atoms or molecules in the thermodynamic equilibrium condition. This would solve the problem encountered in three-level laser systems. Such a desirable situation is possible in four-level laser systems in which the lower laser level is above the ground state as shown in Figure 7.6.

7.1.2.3 Four-Level Laser System

In a *four-level laser system*, the atoms or molecules are excited out of the ground state to an upper highly excited short-lived energy level. Remember that the lower laser level here is not the ground state. In this case, the number of atoms or molecules required to be excited to the upper level

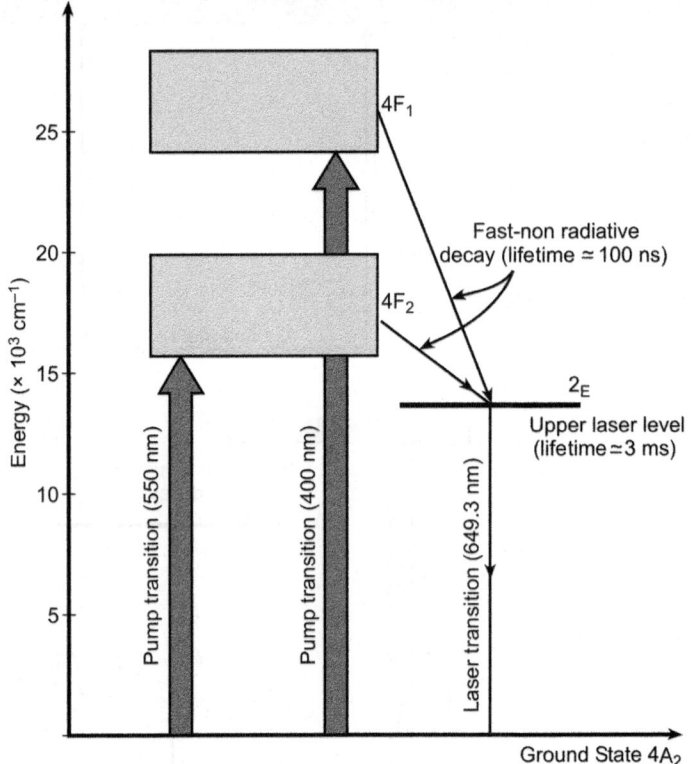

Figure 7.5 Energy level diagram of a ruby laser.

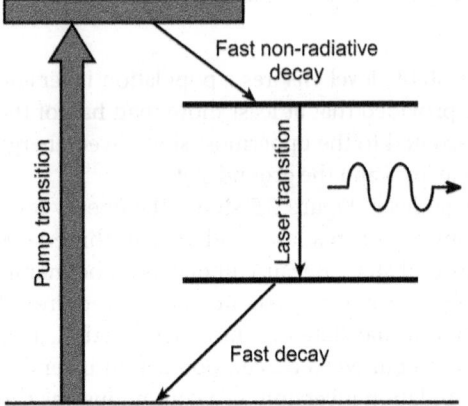

Figure 7.6 Four-level laser system.

would depend upon the population of the lower laser level, which is much smaller than the population of the ground state. Also if the upper level to which the atoms or molecules are initially excited and the lower laser level have a shorter lifetime and the upper laser level (metastable level) a longer lifetime, one can visualize that it would be much easier to achieve and sustain population inversion. This comes from two major events in such a four-level laser. The first is rapid population of the upper laser level, which comes from extremely rapid dropping of the excited species from the upper excited level where they find themselves with an excitation input to the upper laser level accompanied by the longer lifetime of the upper laser level. The second event is the depopulation of the lower laser level due to the shorter lifetime there. And once it is simpler to

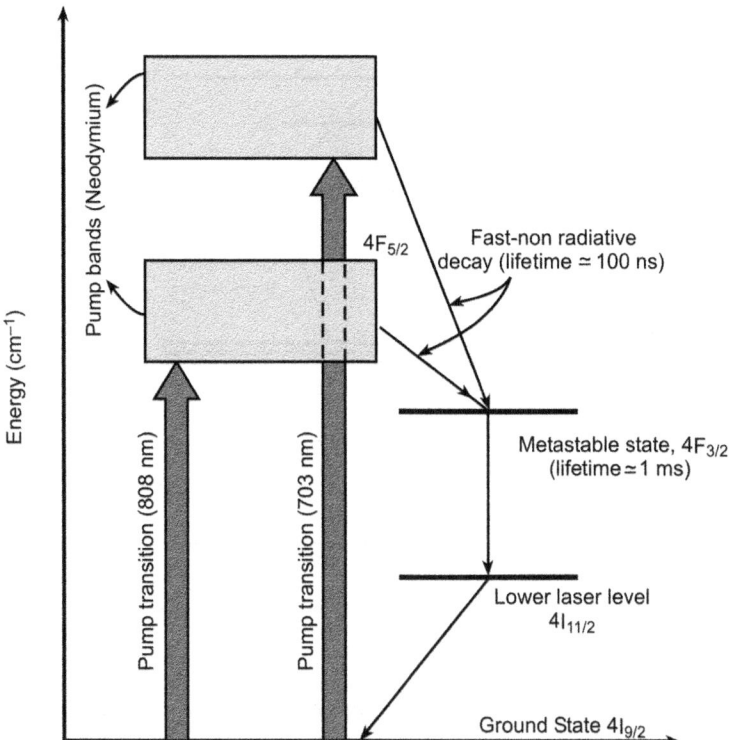

Figure 7.7 Energy level diagram of an Nd-YAG laser.

sustain population inversion, it becomes easier to operate the laser in CW mode. This is one of the major reasons why a four-level laser like an Nd:YAG laser or a helium-neon laser can be operated in CW mode while a three-level laser like a ruby laser can be operated only as a pulsed laser.

Nd:YAG, helium-neon and carbon dioxide lasers are some of the very popular lasers with a four-level energy level structure. Figure 7.7 shows the energy level structure of Nd:YAG laser. The pumping or excitation input raises the atoms or molecules to the uppermost energy level, which in fact is not a single level but instead a band of energy levels. It is a desirable feature as would be clearer a little while later when we discuss pumping mechanisms. The excited species rapidly fall to the upper laser level (metastable level). This decay time is about 100 ns. The metastable level has a metastable lifetime of about 1 ms. The lower laser level has a decay time of 30 ns. If we compare the four-level energy level structure of Nd:YAG laser with that of Nd:YLF (neodymium-doped yttrium lithium fluoride), which is another solid-state laser with a four-level structure, we find that there is a striking difference in the lifetime of the metastable level. Nd:YLF has a higher metastable lifetime (typically few ms) as compared to 1 ms of Nd-doped YAG. This gives the former a higher storage capacity of the excited species in the metastable level. In other words, this would mean that an Nd:YLF rod could be pumped harder to extract more laser energy than an Nd:YAG rod of the same size.

7.1.2.4 Energy Level Structures of Practical Lasers
In the case of real lasers, the active media do not have the simple three- or four-level energy level structures in the previous pages. The energy level structures in the case of practical lasers are far more complex. For instance, the short-lived uppermost energy level to which the atoms

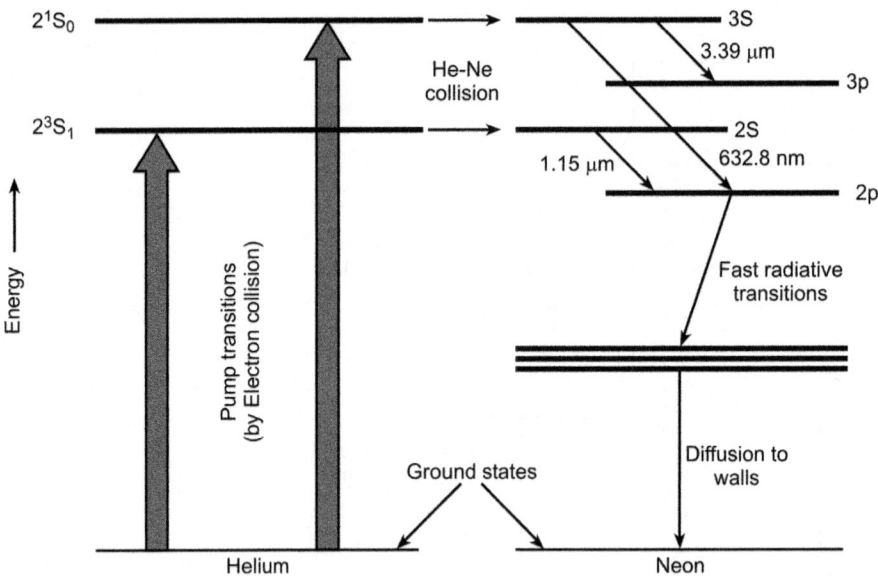

Figure 7.8 Energy level diagram of a He-Ne laser.

or molecules are excited out of the ground state and from where they drop rapidly to the metastable level is not a single energy level. It is in fact a band of energy levels. It is a desirable feature as it makes pumping more efficient and a larger part of the pumping input is converted into a useful output to produce population inversion. Also the energy levels involved in producing laser output are not necessarily single levels in all lasers. There could be multiple levels in the metastable state, multiple levels in the lower energy state of the laser transition or multiple levels in both states. This gives the capability to produce stimulated emission at more than one wavelength to such lasers. Helium-neon and carbon dioxide lasers are typical examples showing this phenomenon. Figure 7.8 shows the energy level structure of a helium-neon laser.

Also, it is not always the active species alone that constitute the laser medium or laser material. Atoms or molecules of other elements are sometimes added with specific objectives. In some cases, such as in a helium-neon laser, the active species producing laser transition are neon atoms. Free electrons in the discharge plasma produced as a result of electrical pumping input excite the helium atoms first as that can be done very efficiently. The excited helium atoms, when colliding with neon atoms, transfer their energy to them. As another example, in a carbon dioxide laser, the laser gas mixture mainly consists of carbon dioxide, nitrogen and helium. While nitrogen participates during the excitation process and plays the same role as that played by helium in a helium-neon laser, helium in a carbon dioxide laser helps in depopulating the lower laser level.

7.1.3 Constituent Parts of a Typical Laser

A typical laser comprises of three main parts; namely the active medium, the pumping source and the resonant structure. The active medium is usually in the form of rod or slab comprising of a host material doped with a lasing species in a solid-state laser, a gas mixture comprising of lasing species and other gases enabling efficient laser action in a gas laser and a semiconductor material in a semiconductor laser. The pumping mechanism required for producing population inversion is usually optical in the case of solid-state lasers and electrical in the case of gas

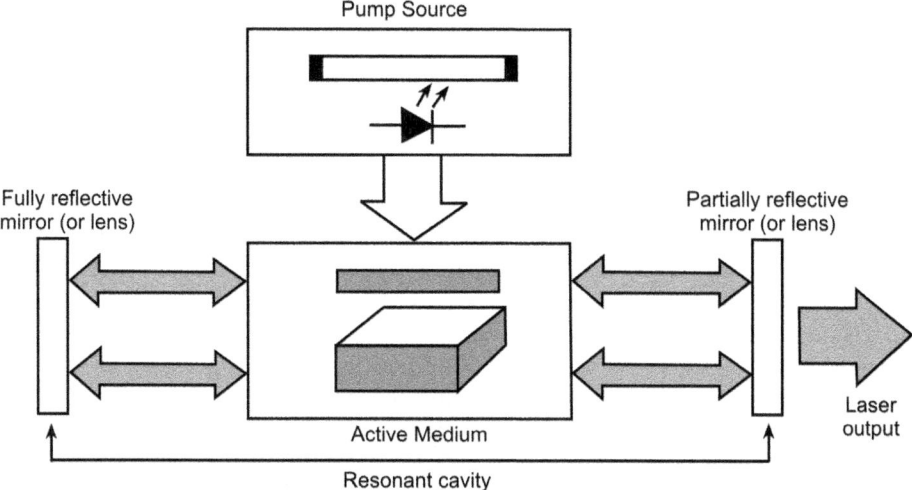

Figure 7.9 Constituent parts of a typical laser.

and semiconductor material. The resonant structure is constituted by the active medium bound by a pair of mirrors (or lenses) on its two ends. The output mirror (or lens) is partially reflective to allow usable laser output and the other mirror is fully reflecting. Figure 7.9 shows constituent parts of a typical laser.

7.1.3.1 Gain of Laser Medium

Gain of the laser or active medium is the extent to which this medium can produce stimulated emission. The gain of the medium is defined more appropriately as a *gain coefficient*, which is the gain expressed as a percentage per unit length of the active medium. When we say that the gain of a certain laser medium is 10% per centimetre, it implies that 100 photons having the same transition energy, as that of excited laser medium become 110 photons after travelling 1 cm of the medium length. The amplification or the photon multiplication offered by the medium is expressed as a function of the gain of the medium and the length of the medium by eqn. 7.2.

$$G_A = e^{\alpha x} \tag{7.2}$$

Where,

G_A = Amplifier Gain or Amplification Factor
α = Gain Coefficient
x = Gain Length

This expression for gain can be rewritten as given in eqn. 7.3.

$$G_A = (e^{\alpha})^x = (1+\alpha)^x \ldots\ldots\ldots\ldots \text{ for } \alpha \ll 1 \tag{7.3}$$

Therefore, to a reasonably good approximation, we can write

$$\text{Amplification factor} = (1 + \text{Gain coefficient})^{\text{length of medium}}$$

This implies that, when the medium with a gain coefficient of 100% is excited and population inversion is created, a single spontaneously emitted photon will become two photons after this spontaneously emitted photon travels 1 cm of the length of the medium. The two photons cause further stimulated emission as they travel through the medium. This amplification continues and the number of photons emitted by stimulation process keeps building up just as the principal amount builds up with compound interest. One can use this relationship to compute the amplification. It would be interesting to note how photons multiply themselves as a function of length. For instance, though 10 photons become 11 photons after travelling 1 cm for a gain coefficient of 10% per centimetre, the number reaches about 26 after 10 cm and 1173 after travelling gain length of 50 cm. But all this happens as long as there are enough excited species in the metastable state to ensure that stimulated emission predominates over absorption and spontaneous emission. On the other hand, it is also true that for a given pump input, there is a certain quantum of excited species in the upper laser level. As the stimulated emission initially triggered by one spontaneously emitted photon picks up, the upper laser level is successively depleted of the desired excited species and the population inversion is adversely affected. This leads to reduction in the growth of stimulated emission and eventually saturation sets in, what we call *gain saturation*.

Another aspect that we need to look into is to find whether the typical gain coefficient values the majority of active media used in lasers have is really good enough for building practical systems or not. Let us do a small calculation. If a 5 mW CW helium-neon laser were to operate for just 1 s, it would mean an equivalent energy of 5 mJ. Each photon of He-Ne laser output at 632.8 nm would have energy of approximately 3×10^{-19} J, which further implies that the previously mentioned laser output would necessitate generation of about 1.7×1016 photons. Now with the kind of gain coefficient the helium-neon laser plasma has, one can compute the gain length required for the purpose. For any useful laser output, therefore, the solution lies in having a very large effective gain length if not a physically large gain length.

If we enclose the laser medium within a closed path bounded by two mirrors, as shown in Figure 7.10, we can effectively increase the interaction length of the active medium by making the photons emitted by stimulated emission precess back and forth. One of the mirrors in the arrangement is fully reflecting and the other has a small amount of transmission. This little transmission, which also constitutes the useful laser output, adds to the loss component. This is true because the fraction of the stimulated emission of photons taken as the useful laser output is no longer available for interaction with the excited species in the upper laser level. Quite obviously, the maximum power that can be coupled out of the system must not exceed the total amount of losses within the closed path. For instance, if the gain of the full length of the active medium is 5% and the other losses such as those due absorption in the active medium, spontaneous emission, losses in the fully reflecting mirror (which will not have an ideal reflectance of 100%) and so on are 3%, the other mirror can have at the most a transmission of 2%.

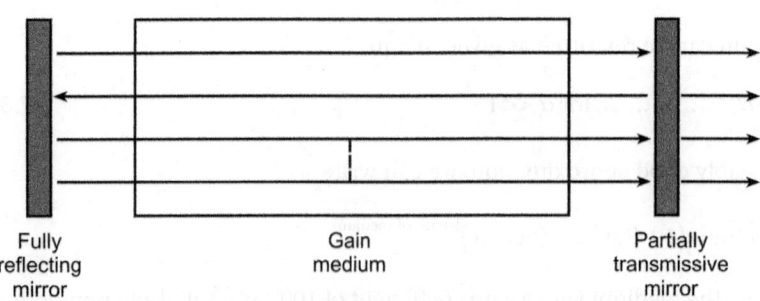

Fully   Gain   Partially
reflecting   medium   transmissive
mirror   mirror

Figure 7.10 Lasing medium bounded by mirrors.

In a closed system like this, the power inside the system is going to be much larger than the power available as useful output. For instance, for 1% transmission and assuming other losses to be negligible, if the output power is 1 mW, the power inside the system would be 100 mW.

7.1.3.2 Laser Resonators

The active laser medium within the closed path bound by two mirrors, as shown in Figure 7.10, constitutes a basic laser resonator provided it meets certain conditions. Also, resonator structures of most practical laser sources would be more complex than the simplistic arrangement of Figure 7.10. It is clear that, if we want the photons emitted as a result of stimulated emission process to continue to add to the strength of those responsible for their emission, it is necessary for the stimulating and stimulated photons to be in phase. The addition of mirrors should not disturb this condition. If this happens, then all those photons stimulated by this photon would also satisfy this condition. This can be possible if we satisfy the condition given in eqn. 7.4.

$$\text{Round trip length} = 2L = n\lambda \tag{7.4}$$

Where,

L = Length of the resonator
λ = Wavelength
n = Integer

This expression can be rewritten as

$$f = \frac{nc}{2L} \tag{7.5}$$

Where,

c = Velocity of electromagnetic wave
f = Frequency

Based on the type of end mirrors and the inter-element separation, which largely dictate the extent of interaction between the emitted photons and the laser medium and also the immunity of the laser resonator to misalignment of end components, the resonators can be broadly classified as stable and unstable resonators. A *stable resonator* is the one in which the photons can bounce back and forth between the end components indefinitely without getting lost out by the sides of the components. In such a resonator, due to the focusing nature of one or both components, the light flux remains within the cavity. A *plane-parallel resonator* (Figure 7.11) in which both end components are plane mirrors and are placed precisely at right angle to the laser axis would be a stable resonator. In practice, however, this is not true. A slight misalignment of even one of the mirrors would ultimately lead to light flux escaping the laser cavity after several

Figure 7.11 Plane-parallel resonator.

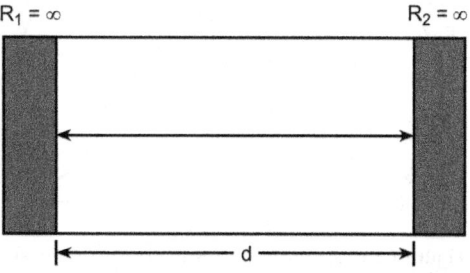

$R_1 = \infty$ $R_2 = \infty$

d

reflections from the two mirrors. Nevertheless, such a resonator encompasses a large volume of the active medium. It is not used in practice, as it is highly prone to misalignment.

This problem can be overcome by using one plane and one curved mirror, as is the case in *hemispherical* and *hemifocal* resonators respectively shown in Figure 7.12(a) and (b) or both curved mirrors, as is the case in *concentric* and *confocal* resonators, respectively, shown in Figure 7.13(a) and (b).

Though the problem of sensitivity of the plane-parallel resonator to misalignment of cavity mirrors is largely overcome in different stable resonator configurations discussed in the previous paragraphs and shown in Figures 7.12 and 7.13, not all of them have emitted photons interacting with a large volume of the excited species, which is also equally desirable. It is also true that In the case of low gain media with consequent very low transmission output mirrors, the photons travel back and forth a large number of times within the cavity before their energy appears at the output. This makes the resonator alignment more critical. That is why a plane-parallel resonator may never be the choice in a low gain laser medium.

On the other hand, in a high gain medium, a certain amount of light flux leakage can be tolerated. This fact is made use of in an *unstable resonator* configuration, which otherwise achieves interaction of the emitted photons with a very large volume of the excited species. Figure 7.14 shows one possible type of unstable resonator. You will notice that photons escape from the sides of the mirror after one or two passes within the cavity. This light leakage, which also constitutes the useful laser output, is more than compensated for by high gain medium and also by large interaction volume. Also, since the photons have to make relatively fewer passes within the cavity as compared to a low gain stable resonator configuration before drifting out, the alignment becomes that much less critical.

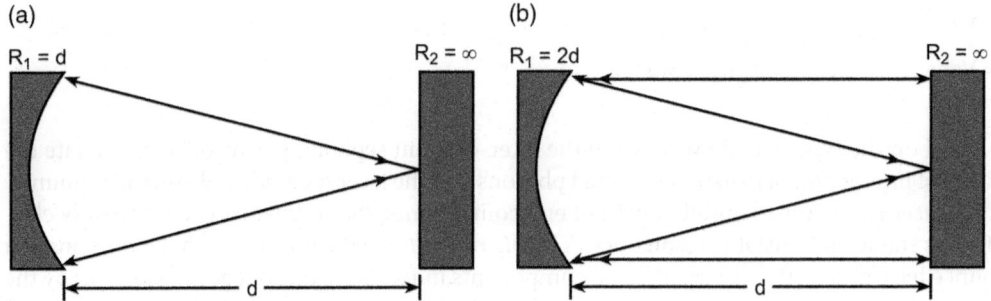

Figure 7.12 (a) Hemispherical resonator. (b) Hemifocal resonator.

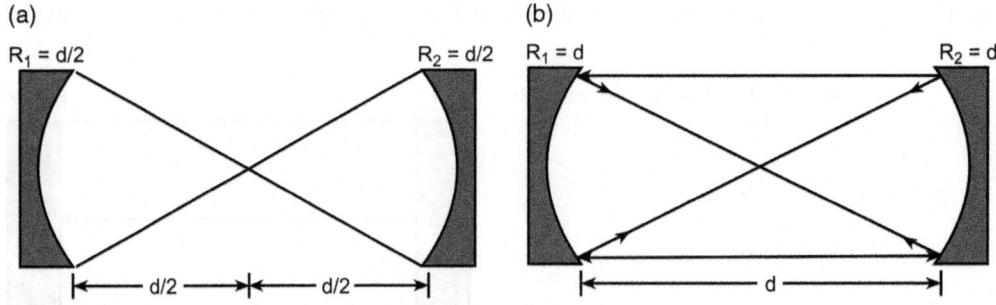

Figure 7.13 (a) Concentric resonator. (b) Confocal resonator.

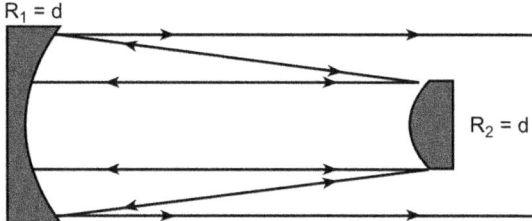

$R_1 = d$

$R_2 = d$

Figure 7.14 Unstable resonator.

7.1.3.3 Pumping Mechanism

Pumping mechanism is used to create population inversion of the lasing species. Commonly employed pumping mechanisms include:

1) Optical pumping
2) Electrical pumping
3) Other mechanisms, such as pumping by chemical reactions, electron beams and so on.

One aspect that is common to all pumping mechanisms is that the pumping energy/power must be greater than the laser output energy/power. When applied to optical pumping, it is obvious that the optical pump wavelength must be smaller than the laser output wavelength. Another aspect that is common to all schemes is that pumping efficiency largely affects the overall laser efficiency. For instance, if the energy difference of the pump transition is much greater than that of the laser transition, the laser efficiency is bound to be relatively poorer. An argon-ion laser is a typical example. Yet another aspect that is common to all pumping mechanisms is that the topmost pump level is not a single energy level but rather a band of closely spaced energy levels with allowed transitions to a single and in some cases more than one metastable level. When applied to optical pumping, this allows use of optical sources, such as flash lamps, with broad band outputs.

7.1.3.3.1 Optical Pumping

Optical pumping is employed for those lasers that have a transparent active medium. Solid-state and liquid dye lasers are typical examples. The most commonly used pump sources are the flash lamp in the case of pulsed and the arc lamp in the case of CW solid-state lasers.

Flash lamps are pulsed sources of light and are widely used for pumping of pulsed solid-state lasers. These are available in a wide range of arc length (from a few centimetres to as large as more than a metre, though arc length of 5–10 cm is common), bore diameter (typically in the range of 3–20 mm), wall thickness (typically 1–2 mm) and shape (linear, helical). Figures 7.15 and 7.16 show constructional features of typical linear (Figure 7.15) and helical (Figure 7.16) flash lamps.

Flash lamps for pumping solid-state lasers are filled with usually a noble gas such as xenon or krypton at a pressure of 300–400 torr. Two electrodes are sealed into the envelope that is usually made up of quartz. An electrical discharge created between the electrodes leads to a very high value of pulsed current, which further produces an intense flash. The electrical energy to be discharged through the lamp is stored in an energy storage capacitor/capacitor bank.

Figure 7.15 Linear flash lamps.

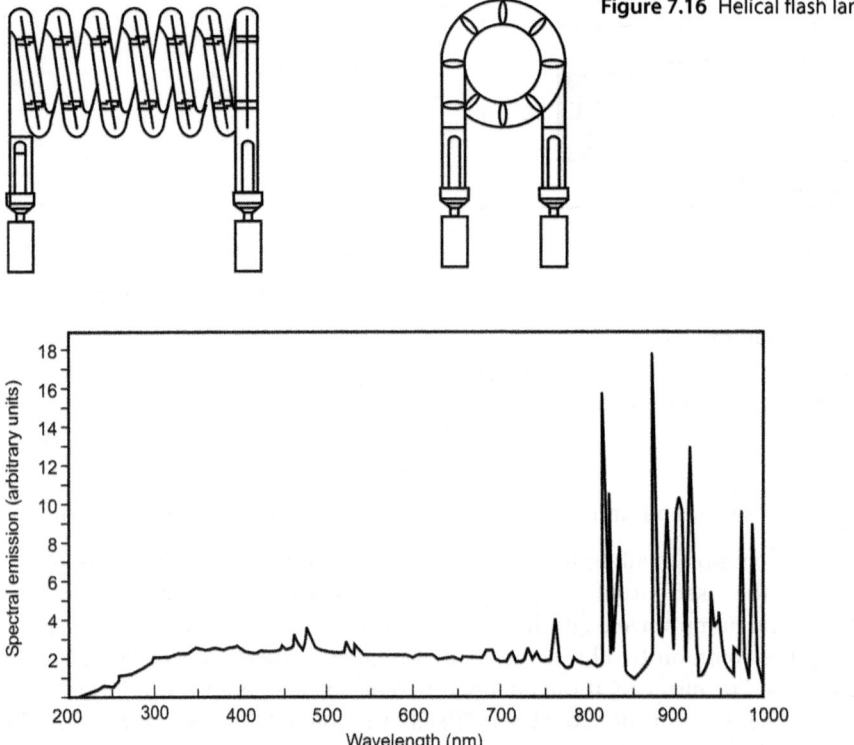

Figure 7.16 Helical flash lamp.

Figure 7.17 Emission spectrum of a xenon-filled flash lamp.

Xenon filled lamps produce higher radiative output for a given electrical input as compared to krypton-filled lamps. Krypton, however, offers a better spectral match, more so with Nd:YAG. That is, emission spectrum of krypton flash lamp is better matched to the absorption spectrum of Nd:YAG. Emission spectra in the case of xenon and krypton-filled lamps are given in Figures 7.17 and 7.18, respectively. The absorption spectrum of Nd:YAG is given in Figure 7.19.

The electrical characteristics of flash lamps are discussed in Chapter 7 on solid-state laser electronics under the heading of flash lamp power supplies. Major electrical parameters include the flash lamp impedance parameter, maximum average power and maximum peak current, minimum trigger voltage and explosion energy. Impedance characteristics of flash lamp are extremely important as they determine the energy transfer efficiency from energy storage capacitor, where it is stored, to the flash lamp.

Arc lamps are used for CW pumping of solid-state lasers. Like flash lamps, arc lamps too are gas discharge devices. Arc lamps suitable for solid-state laser pumping are linear lamps (Figure 7.20), which are very much like linear flash lamps except for electrode design. Arc lamps, as is evident from Figure 7.20, use pointed cathodes rather than rounded ones used in the case of flash lamps. Arc lamps are filled with xenon or krypton at a pressure of 1–3 atmospheres. Krypton-filled linear arc lamps are more common because of their relatively better spectral match to Nd:YAG absorption band. Bore diameter of 4–7 mm and arc length in the range of 50–150 mm are common.

However, efficiency with which pump output is usefully transferred to excite the lasing species is definitely lower in the case of broadband optical pumping provided by flash lamps and arc lamps. Optical pumping at a single wavelength in a laser having absorption level corresponding

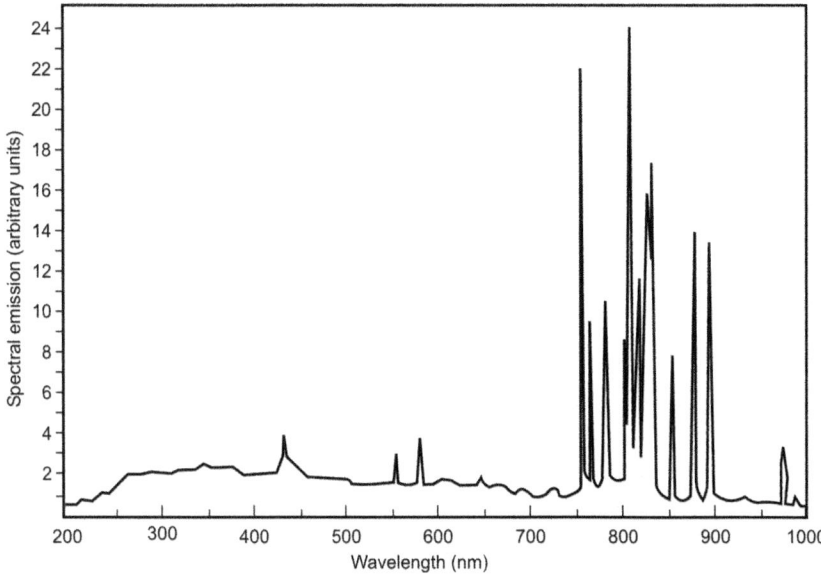

Figure 7.18 Emission spectrum of a krypton-filled flash lamp.

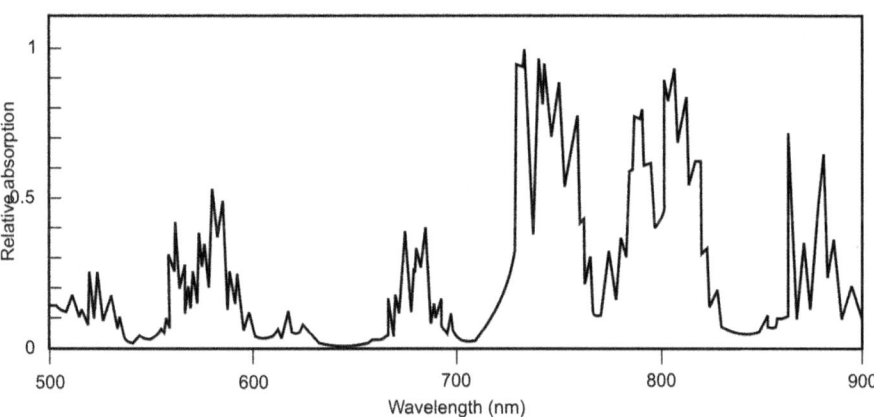

Figure 7.19 Absorption spectrum of an Nd:YAG.

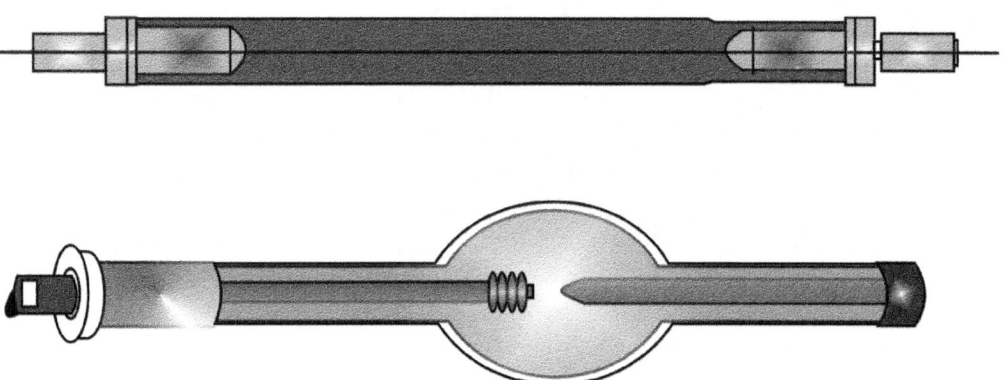

Figure 7.20 Construction of linear arc lamp.

to that wavelength in the pump band achieves a relatively higher pumping efficiency, which leads to higher overall laser efficiency. Optical pumping of solid-state lasers by semiconductor lasers in what are better known as diode-pumped solid-state lasers achieves an efficiency that is 25–30 times what is currently achievable in the case of flash lamp pumped solid-state lasers.

Laser diode arrays for solid-state laser pumping are available in various package configurations. The basic element in these arrays, also called stacks, is the laser diode bar [Figure 7.21(a)]. Each bar has multiple emitters. Laser bars are available in both conduction-cooled as well as liquid-cooled varieties. State-of-the-art bars offer up to 100 W of CW power. Stacks of these

(a)

(b)

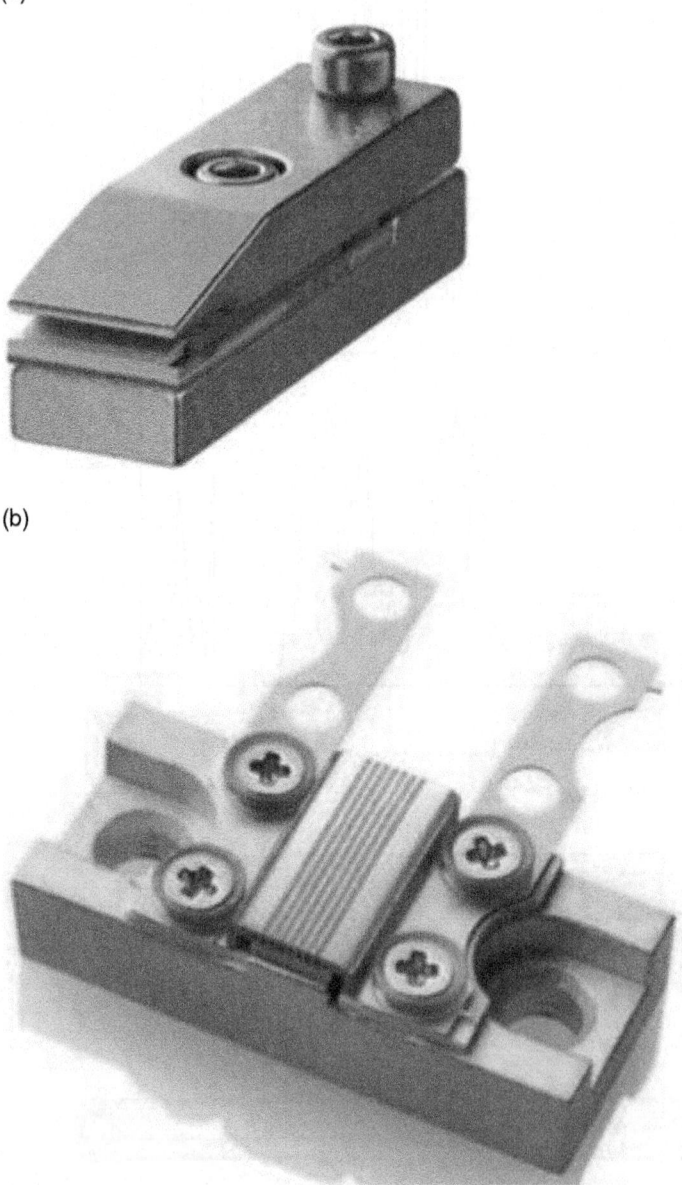

Figure 7.21 (a) Laser diode bar and (b) laser diode stack.

bars are also available for higher pump power requirement [Figure 7.21(b)]. However, maximum pump power available from diode laser arrays is still relatively much lower than what is possible with flash lamps.

7.1.3.3.2 Electrical Pumping

Pumping by electrical discharge is common in gas lasers. The excited electrons in the gas discharge plasma transfer their energy to the lasing species either directly or indirectly through the atoms or molecules of another element. A helium-neon laser is a typical example of indirect transfer of pump energy. The electrons first transfer the energy to helium atoms and then the excited helium atoms transfer the energy to neon atoms. A high voltage initially ionizes the gas and once the discharge is struck, it can be sustained by a relatively much lower voltage and current. In a typical He-Ne laser, initiating voltage is of the order of 8–10 kV while the sustaining voltage is around 1.5–2 kV.

Diode lasers are also electrically pumped, but not really the same way, as are the gas lasers. In the case of diode lasers, the electrical current in the forward-biased diode frees electrons to create electron-hole pairs. The electrons and holes recombine to emit photons. While doing so, electrons drop back to the lower state.

7.1.3.3.3 Other Methods of Pumping

Some of the other methods of pumping or creating population inversion, which are specific to certain types of lasers, include excitation by *combustion reaction* as in gas dynamic CO_2 lasers, *chemical reaction* as in chemical lasers such as hydrogen fluoride (HF) laser, deuterium fluoride (DF) laser and chemical oxygen iodine laser (COIL) and *electron acceleration* as in free electron lasers. As an example, in the case of a gas dynamic laser, combustion reaction produces a high-temperature high-pressure mixture of CO_2 and other gases required in a CO_2 laser. This gas mixture is then rapidly expanded through a set of nozzles to a very low pressure, low temperature condition. Though the temperature and pressure drop rapidly, a large number of molecules still remain in the excited state thus creating population inversion.

7.1.4 Longitudinal and Transverse Modes

The expression for frequency of laser oscillations in a laser resonator as given by eqn. 7.5 indicates that there could be a large number of frequencies for different values of integer n satisfying this resonance condition. Most laser transitions have gain for a wide range of wavelengths. Remember that we are not referring to lasers that can possibly emit at more than one wavelength such as a helium-neon laser. Here, we are referring to the gain bandwidth of one particular transition. Gas lasers such as He-Ne and CO_2 lasers have Doppler broadened gain curves with He-Ne laser having a bandwidth of about 1400 MHz for 632.8 nm transition and CO_2 laser with a bandwidth of about 60 MHz at 10 600 nm.

Therefore it is possible to have more than one resonant frequency, each one of them called a *longitudinal mode*, to be simultaneously present unless special measures are taken to prevent this from happening. As is clear from the expression for frequency, the inter-mode spacing is given by $c/2L$. As a typical case, for a He-Ne laser with a cavity length of 30 cm, inter-mode spacing would be 500 MHz, which may allow three longitudinal modes to be simultaneously present as shown in Figure 7.22(a). Interestingly, one could reduce the cavity length to a point where the inter-mode spacing exceeds the gain bandwidth of the laser transition to allow only a single longitudinal mode to prevail in the cavity. For instance, a 10 cm cavity length leading to an inter-mode spacing of 1500 MHz would allow only a single longitudinal mode [Figure 7.22(b)]. However, we will appreciate that there are other important criteria that decide the cavity length.

(a)

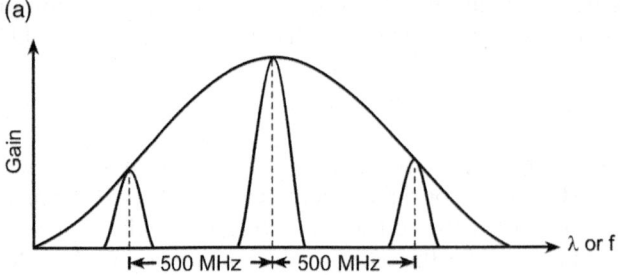

(b)

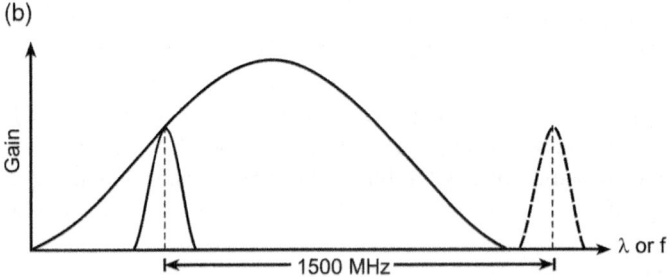

Figure 7.22 Longitudinal modes. (a) Three longitudinal modes present simultaneously and (b) a single longitudinal mode.

Another laser parameter that we are interested in and that is also largely influenced by the design of the laser resonator is the *transverse mode* structure of the laser output. The transverse modes basically tell us about the irradiance distribution of the laser output in the plane perpendicular to the direction of propagation or in other words along the orthogonal axes perpendicular to the laser axis. To illustrate it further, if z-axis is the laser axis, then intensity distribution along x-axis and y-axis would describe the transverse mode structure.

TEM_{mn} describes the transverse mode structure, where m and n are integers indicating the order of the mode. In fact, integer m and n are the number of intensity minima or nodes in the spatial intensity pattern along the two orthogonal axes. Conventionally, first integer represents the electric field component and the second subscript indicates the magnetic field component. Those who are well versed with electromagnetic theory should not find this difficult at all to grasp. Remember that transverse modes must satisfy the boundary conditions like having zero amplitude on the boundaries. The simplest mode, also known as the fundamental or the lowest order mode, is designated as the TEM_{00}-mode. The two subscripts here indicate that there are no minima along the two orthogonal axes between the boundaries. The intensity pattern in both the orthogonal directions has a single maximum with the intensity falling on both sides following a well-known mathematical distribution called Gaussian distribution. The Gaussian distribution (Figure 7.23) is given by eqn. 7.6.

$$I(r) = I_0 \exp\left[-2r^2/w^2\right] \tag{7.6}$$

Where,

$I(r)$ = Intensity at a distance of r from the centre of the beam
w = Beam radius at $(1/e^2)$ of peak intensity point, which is about 13.5% of the peak intensity

Also,

$$I_0 = \frac{2P}{\pi w^2} \tag{7.7}$$

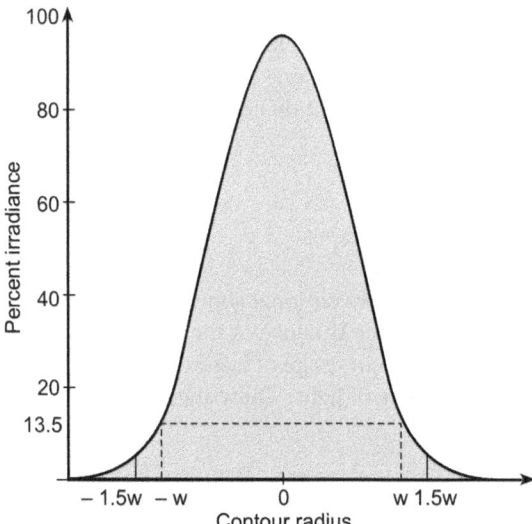

Figure 7.23 Gaussian distribution.

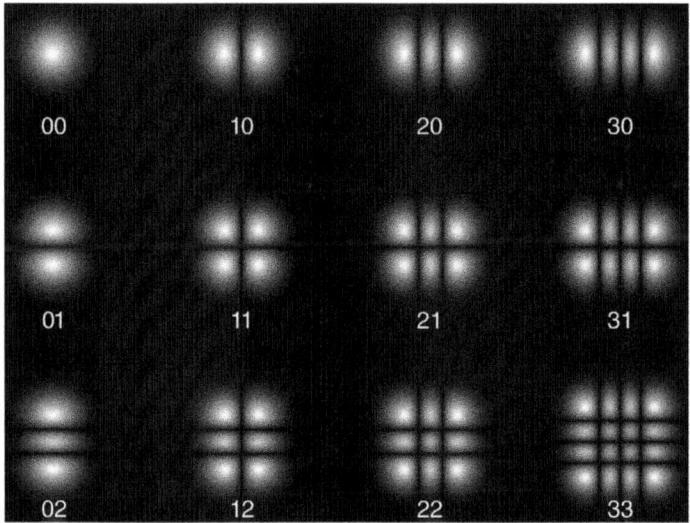

Figure 7.24 Spatial intensity distribution for various transverse modes.

Where,

P = Total power in the beam

Before we discuss the definite advantages that the operation at lowest order or fundamental mode TEM_{00} offers, we shall have quick look at higher-order modes and also as to how different transverse mode appear with respect to their intensity distributions. Figure 7.24 shows the spatial intensity distribution of the laser spot for various transverse mode structures of the laser resonator.

Going back to the fundamental mode, we can appreciate that this mode has the least power spreading. To add to this, this mode has the least divergence; it has the minimum diffraction loss and therefore can be focused to the smallest possible spot. The transverse mode structure is also critically dependent upon laser medium gain and type of laser resonator. There are established resonator design techniques to ensure operation at the fundamental mode. Often, lasers optimized to produce maximum power output operate at one more than higher-order modes. Also, lasers with low gain and stable resonator configuration can be made to conveniently operate at fundamental mode.

7.2 Laser Characteristics

Laser radiation can be distinguished from the light from conventional sources on the basis of its special characteristics and the effects it is able to produce because of these characteristics. It is these characteristics that have led to explosive growth in usage of laser devices in the last more than 45 years after the invention of this magic source of light. These include:

1) Monochromaticity
2) Coherence, temporal and spatial
3) Directionality

We shall discuss each one of these in some detail in the following paragraphs.

7.2.1 Monochromaticity

Monochromaticity refers to single frequency or wavelength property of the radiation. Laser radiation is monochromatic and this property has its origin in the stimulated emission process by which a laser emits light. The stimulated photon has the same frequency, phase and polarization as those of the stimulating photon. As we shall see in the following paragraphs, monochromaticity is one of the essential requirements for the laser radiation to be coherent. Though every monochromatic radiation is not necessarily coherent, a coherent radiation is necessarily monochromatic.

7.2.1.1 Line Broadening Mechanisms

Laser radiation is not perfectly monochromatic. There could be various factors responsible for the spread in the frequency or wavelength, called its line width. The uncertainty principle causes slight variations in the wavelength of different photons emitted during the stimulated emission process. This implies that the frequency spread in the case of pulsed lasers would depend upon the pulse width of radiation. Shorter pulses would have a larger spread. This also means that CW laser outputs are likely to be more monochromatic than the pulsed laser outputs.

Interaction of lasing species with other atoms and molecules is another cause of line broadening. In the case of gas lasers, dependence of line width on gas pressure is an example of this phenomenon. Line width is observed to increase with gas pressure. Increased pressure decreases time interval between successive collisions, which in turn affects energy transfer. Line broadening due to increased pressure is known as *pressure broadening* or collisional broadening.

Yet another factor contributing to increase in line width is the inherent random motion of atoms and molecules. This phenomenon is called Doppler broadening. The emitted wavelength corresponding to a randomly moving atom or molecule is such that its Doppler shifted component matches with the nominal transition wavelength. Random motion of atoms and molecules causes frequency spread. As an example, gas lasers have a Doppler broadened gain curve. The frequency spread in the case of a multimode He-Ne laser at 632.8 nm could be as much as 1400 MHz. The same figure for a CO_2 laser at 10 600 nm is 60 MHz.

Another reason for frequency spread could be the simultaneous oscillation of more than one longitudinal mode within the laser resonator. The spread in the case of a semiconductor laser is much more, of the order of 6×10^6 MHz.

Fortunately, we have techniques, some of which will be discussed in the latter part of the book, which can be used to stabilize the frequency at one point on the broad gain bandwidth curve. We have techniques by which we can ensure that the laser oscillates at a single longitudinal mode. There are techniques for narrowing the line width. As an example, an intra-cavity etalon in an Nd:YAG laser could be employed to reduce its line width to less than $0.2\,\text{cm}^{-1}$. As another example, frequency stabilized helium-neon and carbon dioxide lasers have line width that is several orders of magnitude narrower than that of unstabilized counterparts.

There are standard techniques and equipment available for precise measurement of wavelength and line width of laser radiation. Some of the common ones include optical spectrum analysis based on diffraction gratings, conversion of frequency variation into intensity variations with the help of an unbalanced interferometer or a high finesse reference cavity, self-heterodyne technique, heterodyne technique using two independent lasers and frequency combs. These are briefly described in the latter part of the chapter. It may also be mentioned here that standard beam diagnostic equipment that make use of the above mentioned or other techniques are available for the purpose of laser beam analysis in terms of its wavelength, line width and modal profile. These include wave meters, optical spectrum analysers, beam profilers and so on. Some of the commercially available diagnostic equipment are briefly mentioned in the latter part of the chapter.

7.2.2 Coherence

Coherence is the most important property that distinguishes the laser radiation from the ordinary light. Light is said to be coherent when different photons (or the waves associated with those photons) have the same phase and this phase relationship is preserved as a function of time (Figure 7.25). That is, this phase relationship is preserved as the radiation wave front travels with time. There are two types of coherence called *temporal coherence* and *spatial coherence*.

7.2.2.1 Temporal Coherence

Temporal coherence is the preservation of phase relationship with time and that is what we have been referring to as coherence until now. *Spatial coherence*, which is preservation of phase across the width of the beam, is discussed in the next section.

The necessary conditions for temporal coherence are that all photons should be emitted with same phase and that all photons should have same wavelength. If the starting phase is the same

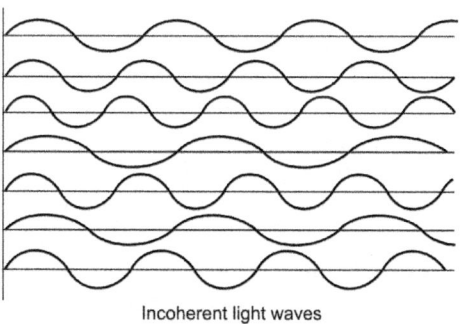
Incoherent light waves

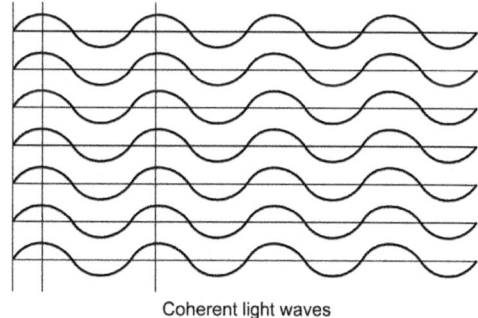
Coherent light waves

Figure 7.25 Coherence.

and all have same wavelength, the phase will be preserved irrespective of time. Though, ideally speaking, both these conditions are ensured by stimulated emission process, but in reality, things are never perfect. Reasons for wavelength spread were explained earlier under monochromaticity. If all photons originate with same phase, temporal coherence would then depend upon the wavelength or frequency spread. The formation of fringes in the Michelson interferometer illustrates the phenomenon of temporal coherence. Temporal coherence is measured as *coherence length* or *coherence time*. The two are interrelated as given in eqn. 7.8.

$$\text{Coherence length} = c \times \tau_c \tag{7.8}$$

Where,

τ_c = coherence time
c = speed of light

Coherence length can be computed from the known value of wavelength spread from eqn. 7.9.

$$\text{Coherence length} = \frac{\lambda^2}{2\Delta\lambda} \tag{7.9}$$

The same relationship can be written in terms of frequency spread by substituting $\lambda = c/f$ and $\Delta\lambda = (c/f^2) \times \Delta f$.

$$\text{Coherence length} = \frac{c}{2\Delta f} \tag{7.10}$$

A simple calculation tells you that the coherence length for ordinary light may be in the order of a fraction of a micron. On the other hand, it could be tens of kilometres for an actively frequency stabilized CO_2 laser. For instance, a multimode He-Ne laser emitting at 632.8 nm with a frequency spread of 1400 MHz would have a coherence length of about 10 cm. The same laser if frequency stabilized to say 1 MHz or so would have a coherence length of 150 m. A CO_2 laser actively stabilized to within 10 kHz, which is possible with some of the techniques, will have a coherence length of 15 km. And looking at ordinary light that emits all wavelengths from 400–900 nm, expected value of coherence length is 0.36 μm assuming an average wavelength of 600 nm.

7.2.2.2 Spatial Coherence

The other type of coherence is the *spatial coherence*. It tells about the correlation in phase of different photons transverse to the direction of travel. It is the area in the plane perpendicular to the direction of travel over which the radiation preserves the coherence. Spatial coherence depends on the transverse mode discrimination property of the laser resonator. Laser radiation operating in the lowest order mode TEM_{00} will certainly be more spatially coherent than a multimode laser radiation. When a laser is operating in a single transverse mode, the radiation will be spatially coherent across the diameter of the beam over reasonable propagation distances. Young's double slit experiment and formation of fringes thereof is the best illustration of the phenomenon of spatial coherence.

Temporal and spatial coherence are independent of one another. While monochromaticity leads to temporal coherence, it is the transverse mode discrimination characteristics of the resonator that decide the spatial coherence.

7.2.3 Directionality

The *directionality* of laser radiation has its origin in the coherence of the stimulated emission process. All photons emitted as a result of stimulated emission process have the same frequency, phase, direction and polarization. These photons when emitted carry no information regarding the location of the excited atom or molecule responsible for its emission. It appears as if all photons were emitted from a tiny volume with dimensions that are of the order of a wavelength. If a photon is emitted off-axis, spatial coherence makes it appear as if it were emitted from the axis. Similarly, a photon that is emitted away from the beam waist on the same axis, temporal coherence makes it appear as if it were emitted from the beam waist.

7.3 Laser Parameters

In many applications including industrial, medical, military and scientific applications, measurement of laser power or energy is often not adequate and it becomes necessary to measure many other beam parameters related to shape and intensity profile. In addition, one is also interested to know how these parameters change as the laser beam propagates through the atmosphere. For example, in the case of a laser target designator, in addition to laser energy per pulse and pulse width, beam divergence is another important parameter that determines the operational range of the system. On the other hand, in the case of a laser intended for spectroscopy applications, the line width and wavelength stability are the parameters that demand special attention. In the following paragraphs are discussed important parameters, which both designers as well as users of laser systems are interested in:

1) Wavelength
2) Pulse energy (pulsed lasers)
3) CW power (CW lasers), peak power (pulsed lasers) and average power (pulsed lasers)
4) Pulse width (pulsed lasers)
5) Pulse repetition frequency (pulsed lasers)
6) Duty cycle (pulsed lasers)
7) Rise and fall times
8) Irradiance
9) Radiance
10) Beam divergence
11) Spot size
12) M^{2-}value
13) Wall plug efficiency

7.3.1 Wavelength

Wavelength is of course the first and the foremost parameter with which the laser is identified. It is in a way laser specific. But there are lasers, which can possibly emit at more than one wavelength. While Nd:YAG is always associated with 1064 nm, a He-Ne laser can emit at 632.8, 543, 1150 and 3390 nm. There are some lasers that can be tuned across a band. The vibronic class of solid-state lasers, dye lasers and free electron lasers belong to this category. Table 7.1 lists wavelengths of some of the well-known lasers.

Table 7.1 Wavelengths of common lasers.

Laser type	Wavelength (nm)	Laser type	Wavelength (nm)
Nd-YAG	1064	helium-cadmium	441.6
Nd-YLF (polarized)	1053	helium-cadmium	325
Nd-YLF (unpolarized)	1047	carbon dioxide	9600
Nd-YVO4	1064	carbon dioxide	10 600
Nd-phosphate glass	1054	copper vapour	511
Nd-silicate glass	1062	copper vapour	578
Nd-fused silica glass	1080	gold vapour	628
Ruby (R1-line)	694.3	argon-ion	488
Ruby (R2-line)	692.9	argon-ion	514.5
Alexandrite	701–826	krypton-ion	647
Titanium-sapphire	660–986	nitrogen	337.1
Cr-GSGG	742–842	argon fluoride	193
Erbium-YAG	2940	krypton fluoride	249
Erbium-YLF	1730	xenon chloride	308
Erbium-glass	1540	xenon fluoride	350
Helium-neon	632.8	diode laser (gas)	904
Helium-neon	1150	diode laser (gaalas)	720–900
Helium-neon	1523	diode laser (ingaas)	1060
Helium-neon	3390	diode laser (ingaasp)	1300–1550

7.3.2 Pulse Energy

Pulse energy is defined with respect to pulsed lasers. It is in fact the area under the power versus time curve representing the laser pulse. If the laser pulse is considered as a rectangular one with amplitude equal to the peak power, the pulse energy then is the product of peak power and the pulse width. Also, if the laser pulse, which has a Gaussian profile, is approximated as an isosceles triangle as shown in Figure 7.26 and the pulse width is measured as the full width at the points of half of the peak power, the area under the curve turns out to be product of peak power and the pulse width.

7.3.3 Continuous-Wave, Peak and Average Power

CW power is the laser power that is constant over time. It is defined for lasers that produce a CW output. The typical laser output power level may vary from a fraction of a milliwatt in a He-Ne laser or a semiconductor diode laser used in laser pointers to hundreds of kilowatts or even several megawatts in high power lasers. Other power parameters are the peak power and average power defined with reference to pulsed lasers.

Peak power is the highest instantaneous optical power in the laser pulse (Figure 7.27). Though peak power occurs only for a time duration, which is a small fraction of the pulse width, it is considered to be present during the entire pulse width. This greatly simplifies average power and energy calculations. Peak power is of particular relevance to laser systems like laser range finders and laser target designators, as it is one of the important parameters that determine the

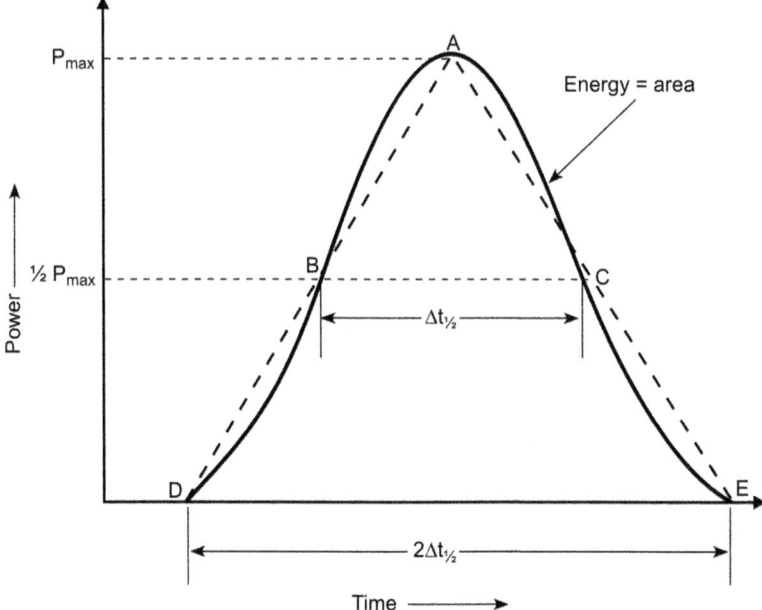

Figure 7.26 Computation of the pulse energy of a Gaussian pulse.

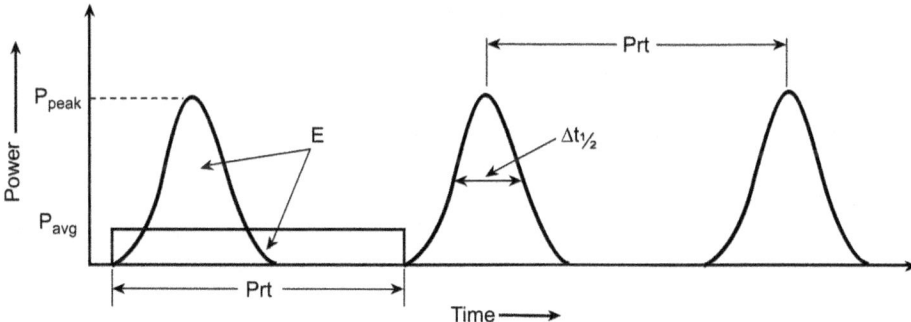

Figure 7.27 Laser pulse parameters.

maximum operational range of these systems. The peak power specification in the case of military laser range finders and target designators configured around Q-switched solid-state lasers is in the range of 1–10 MW.

Average power is the product of peak power and the duty cycle (Figure 7.27). It can also be written as product of pulse energy and the repetition rate. A laser target designator producing 100 mJ of pulse energy in a 20 ns pulse at a repetition rate of 20 PPS will have peak power and average power specifications of 5 MW and 2 W, respectively.

7.3.4 Pulse Width

Pulse width or *pulse duration* in the case of a pulsed laser is usually measured as full width at half maximum (FWHM) as shown earlier in Figure 7.26. Pulse width is intimately related to bandwidth. The narrower the pulse width, the higher the required bandwidth. In fact,

Heisenberg's uncertainty principle puts a limit on the minimum possible laser pulse width for a given value of available bandwidth. That is,

$$\text{Minimum pulse width} = \frac{0.441}{\text{Bandwidth}} \tag{7.11}$$

$$\text{Minimum bandwidth} = 0.441/\text{Pulse width} \tag{7.12}$$

Pulse width could be a few femtoseconds to hundreds of femtoseconds in the case of some mode-locked lasers, a few nanoseconds to several tens of nanoseconds in the case of Q-switched solid-state lasers and few tens of microseconds to hundreds of microseconds in the case of free-running lasers.

7.3.5 Pulse Repetition Frequency (PRF)

PRF of a pulsed laser is number of laser pulses produced per second. It is equal to the reciprocal of time interval between two successive laser pulses (Figure 7.27). PRF is of particular significance in laser systems like laser target designators used for guided weapon delivery applications.

7.3.6 Duty Cycle

The duty cycle is the ratio of pulse width to the time interval between two successive pulses (Figure 7.27). Duty cycle in the case of a laser target designator producing 20 ns wide laser pulses at a repetition rate of 20 PPS is 0.0000004. Peak power, average power, pulse width, pulse energy, repetition rate and duty cycle are interrelated. The following mathematical expressions sum up their inter-relationship.

$$\text{Pulse energy} = \text{Peak power} \times \text{Pulse width} \tag{7.13}$$

$$\text{Average power} = \text{Peak power} \times \text{Duty cycle} = \text{Pulse energy} \times \text{Repetition rate} \tag{7.14}$$

$$\text{Peak power} = \text{Pulse energy}/\text{Pulse width} \tag{7.15}$$

$$\text{Duty cycle} = \text{Pulse width} \times \text{Repetition rate} \tag{7.16}$$

7.3.7 Rise and Fall Times

Rise and *fall times* refer to the time duration between 10 and 90% of the peak amplitude of the pulse respectively during rising and falling portions of the laser pulse. Pulse rise time becomes particularly important while designing optoelectronic front-end circuits for converting a laser radiation pulse into an equivalent electrical signal. The bandwidth of the current-to-voltage converter needs to be commensurate with the rise time specification. For example, a 10 ns laser pulse with a rise time of 2 ns not only needs an optical sensor with a rise time specification of 2 ns or better; the transimpedance amplifier used to transform photocurrent into an equivalent voltage pulse also needs to have a bandwidth of 175 MHz (=350/2) or better to faithfully reproduce the pulse.

7.3.8 Irradiance

Irradiance also referred to as power density is defined as the power per unit area of the laser radiation falling on the target. It is expressed in W/m^2. This parameter is particularly important when the laser radiation is used to illuminate a receiving system. The laser power actually entering the receiver system depends upon the receiving aperture and the power density available at that plane. A typical example where irradiance parameter assumes importance is the use of a high-energy laser system in a countermeasure mode to saturate or damage distant electro-optic sensor systems such as laser range finders and laser target designators. In such a situation, the laser source needs to produce a certain minimum irradiance at the target plane to be able to produce the desired effect.

7.3.9 Radiance

Radiance, also referred to as *brightness*, is usually defined with respect to the laser source. It is the power emitted per unit area per unit solid angle. It is expressed as watts/m^{2-}steradian. For small values of angle, planar angle θ is related to solid angle (Ω) by [$\Omega = (\pi/4)\theta^2$]. Quite obviously, a small exit beam diameter and lower divergence mean higher radiance or brightness.

7.3.10 Beam Divergence

Beam divergence is an indicator of the spread in the laser beam spot as it travels away from the source. It is a function of the wavelength λ and size of output optics. If d is diameter of output optics, then

$$\theta = \left[(1.27\lambda)/d\right] \tag{7.17}$$

Where,
θ is the divergence in radians

Thus for a given λ, larger d leads to smaller divergence. This of course is the minimum value of divergence the laser can have assuming that it is transmitting the fundamental transverse mode, TEM_{00}. In the presence of higher-order transverse modes, the beam divergence increases more rapidly. The divergence parameter determines the spot size in the far field. If θ is full angle divergence in radians and d the spot diameter at a distance R in metres (Figure 7.28), then $\theta = (d/R)$, assuming that the value of ϑ is very small. If θ is measured in milliradians and R is in kilometres, then d is in metres. In fact, the exact expression for the spot diameter d is given by eqn. 7.18.

$$d = D + R\theta \tag{7.18}$$

This approximates to $d = R\theta$ for large values of R.

Beam divergence is one of the important parameter that decides the maximum operational range of laser systems like laser range finders and laser target designators. Divergence decides the laser spot diameter at the target and therefore the power density for a given peak power. It is the power density and not the absolute value of power that would determine the power

Figure 7.28 Definition of beam divergence.

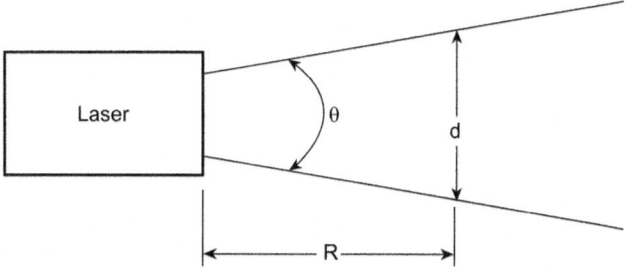

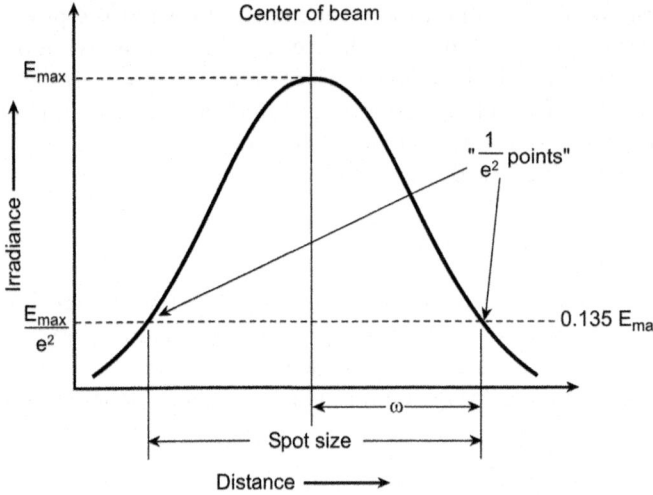

Center of beam

E_{max}

"$\dfrac{1}{e^2}$ points"

Irradiance

$\dfrac{E_{max}}{e^2}$

0.135 E_{max}

ω

Spot size

Distance

Figure 7.29 Spot size.

entering the receiver channel of these systems. Also, when the divergent laser beam is made to fall on a positive lens, the diameter of the focused spot is the product of the focal length of the lens and the divergence of the incident laser radiation. For a given focal length, the focused spot size is therefore a function of the divergence of the incident laser beam.

7.3.11 Spot Size

Spot size or the beam diameter is defined as the distance across the centre of the beam for which the irradiance equals $0.135 \, (= 1/e^2)$ times the maximum value at the centre (Figure 7.29). This implies that, if the laser beam were made to fall on a circular aperture of diameter equal to the laser beam diameter as defined previously with the centre of the beam coinciding with centre of aperture, not all laser power is transmitted through the aperture. In fact, fractional transmission through the aperture can be computed from eqn. 7.19.

$$T = 1 - \exp\left[-2(r/w)^2 \right] \tag{7.19}$$

Where,

r is radius of aperture
w is spot radius

7.3.12 M^2-Value

M^2-value is a measure of beam quality. When the laser beam propagates through space, the divergence in the case of an unfocused pure Gaussian beam is given by $(4\lambda/\pi D)$, where D is the diameter of the beam waist. In the case of real beams, the divergence is higher due to various factors such as presence of additional modes and the equation for divergence is usually written as $[(M^2 \times 4\lambda)/(\pi D)]$ with the value of M^2 being greater than one. M^2 is therefore defined as the ratio of the divergence of the real beam to that of a theoretical diffraction-limited beam of the same waist size with a Gaussian beam profile (TEM_{00}-mode). It is also referred to as the beam propagation ratio as per the ISO-11146 standard. Also, the closer a real beam is to the

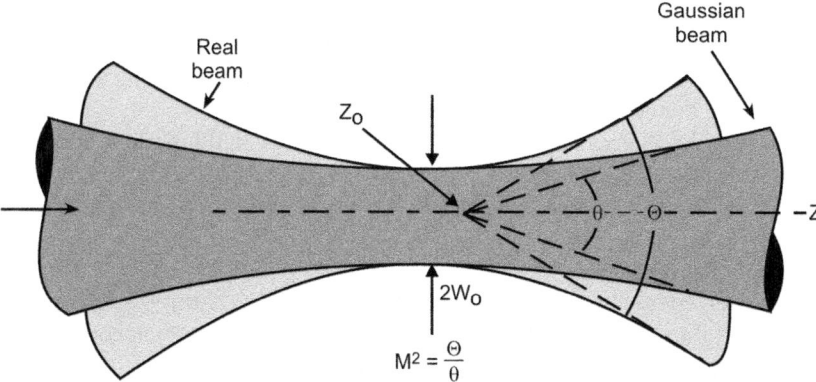

Figure 7.30 Propagation of real and perfect Gaussian beams.

diffraction-limited beam, the more tightly it can be focused, the greater the depth of field will be and the smaller the size of beam handling optics. The focal spot size in the case of real beam shall be M^2 times the focal spot size of the pure Gaussian beam. Also, the angular size of the real beam in the far field will be M^2 times that of a perfect Gaussian beam (Figure 7.30).

7.3.13 Wall Plug Efficiency

Wall plug efficiency is the overall efficiency of the laser system. It is the ratio of laser power produced (CW power or average power as applicable) to the power drawn from source of input. As an example, a military laser designator producing 20 ns, 100 mJ pulses at 20 PPS and drawing 12.5 A at 24 VDC would have a wall plug efficiency of $(100 \times 10^{-3} \times 20)/(24 \times 12.5) = 2/300 = 0.67\%$.

7.4 Measurement of Laser Parameters

In the previous sections, important parameters of laser radiation were discussed. In this section the experimental techniques and diagnostic instrumentation available for measurement of various parameters are discussed. The parameters for which measurement solutions are discussed in the following paragraphs include the following.

1) Power (CW lasers)
2) Average power and pulse energy (pulsed lasers)
3) Repetition rate (repetitively pulsed lasers)
4) Spot size
5) Divergence
6) M^2 value

7.4.1 Measurement of Power, Energy and Repetition Rate

A wide range of laser power/energy meters is commercially available from host of manufacturers covering wavelength range from short ultraviolet to the far infrared, CW power levels from nanowatts to kilowatts, pulse energies from nanojoules to hundreds of joules and repetition rates up to several kilohertz. Also, meters are available for making these measurements on both collimated as well as widely diverging beams such as those from diode lasers.

These measurement systems are invariably modular in nature comprising of a sensor head that is connected to a meter/display unit. Almost all manufacturers offer meters that are compatible to a range of sensor heads, which enables the user to have a wide range of measurement capability by having a single meter and multiple sensor heads.

Measurement of power, pulse energy and repetition rate are interrelated functions and most of the state-of-the-art laser power/energy meters are capable of measuring these parameters. It is because of this that measurement of these parameters is covered under a common heading.

7.4.1.1 Choosing the Right Sensor

Choosing the right power/energy measurement solution is usually a three-step process involving selection of the right sensor head, understanding the required meter capabilities and ensuring compatibility of the chosen sensor head with the selected meter.

Selection criteria for the sensor head are based on the nature and range of expected values of the laser parameters to be measured. There are three broad categories of sensor heads commonly used for the purpose. These include *thermal* or more precisely *thermopile* sensors, *photodiode* sensors and *pyroelectric* sensors. Each of the three types has its own characteristic features suiting a particular measurement requirement.

In the case of thermopile sensors, the incident laser radiation is absorbed and converted into heat. The heat ultimately flows to a heat sink maintained at a near constant temperature by either water or air cooling. Differential temperature between the absorber temperature and that of heat sink represents the amplitude of absorbed laser energy. A thermocouple junction converts this temperature difference into an equivalent electrical signal.

These sensors are characterized by a broad spectral response from ultraviolet to far infrared, wide dynamic range from tens of microwatts to several kilowatts and a uniform spatial response. Also, their response is not affected by beam size or position or uniformity. The disadvantage with these sensors is their sluggish response. Response time ranges from 1–50 s depending upon sensor size increasing with increase in sensor size. Therefore, these sensors are best suited for measurement of CW power, average power in repetitively pulsed lasers and energy of long laser pulses.

In the case of photodiode sensors, photons in the incident laser radiation generate charge carriers, which can be sensed either as a current or a voltage. These are characterized by a limited spectral response (typically from 200 to 1800 nm), high sensitivity (typically few nanowatts), low noise, fast response time (typically few nanoseconds) and relatively lower spatial uniformity. Lower spatial uniformity particularly affects measurements on non-uniform beams and beams that wander over the detector active area between successive measurements. These sensors particularly suit low power measurements in CW lasers. These sensors saturate above a power density of about $1 \, \text{mW/cm}^2$, which necessitates use of optical attenuators in case measurement of higher power levels is desired.

Pyroelectric sensors are also a type of thermal sensor like thermopile sensor with the difference these respond to rate of change of temperature rather than absolute value of temperature difference. These sensors essentially act like capacitors and therefore integrate pulses to produce a signal with a peak proportional to laser energy. They are ideally suited to measurement of parameters of pulsed lasers. Like thermopile sensors, these are also characterized by a broad spectral response.

7.4.1.2 Choosing the Right Meter

As outlined earlier, state-of-the-art meters offer both power and energy measurement options and are compatible with all three categories of sensor heads. These metres when coupled to different sensors offer measurement of CW power, average power, pulse energy and repetition rate. One such meter is the FieldMaxII-Top from Coherent. It is capable of power measurement in the range of $10\,\mu W$–$30\,kW$ when used with a thermopile sensor head and $1\,nW$–$300\,mW$ when used with a photodiode sensor head. Energy measurement in the range of $1\,nJ$–$300\,J$ is possible with a pyroelectric sensor head. Maximum repetition rate in the case of pulse energy measurement is $300\,Hz$.

7.4.2 Measurement of Spot Size

A simple technique for measuring spot size is by measuring fractional transmission through a known aperture placed in the path of the beam. Refer to Figure 7.31. The aperture is placed in the path of the laser beam and adjusted in such a way that the power metre records maximum power. Aperture size should be such that the recorded power is in the range of 50–90% of the power recorded without aperture. Beam radius can then be computed from eqn. 7.20.

$$(r/w) = \left[\{\ln(1-T)\}/2\right]^{1/2} \tag{7.20}$$

Where,

r = Aperture radius
T = Fractional transmission through aperture = Power with aperture/Power without aperture
w = Beam radius.

7.4.3 Measurement of Divergence

Beam divergence can be computed by measuring the beam diameter at two points at known distances as shown in Figure 7.32. Full angle beam divergence in this case is given by eqn. 7.21.

$$\theta \approx \left[(d_2 - d_1)/(R_2 - R_1)\right] \tag{7.21}$$

Figure 7.31 Measurement of spot size.

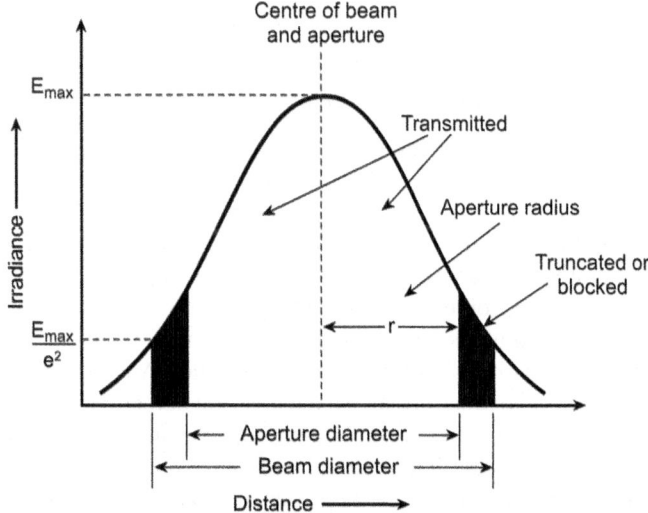

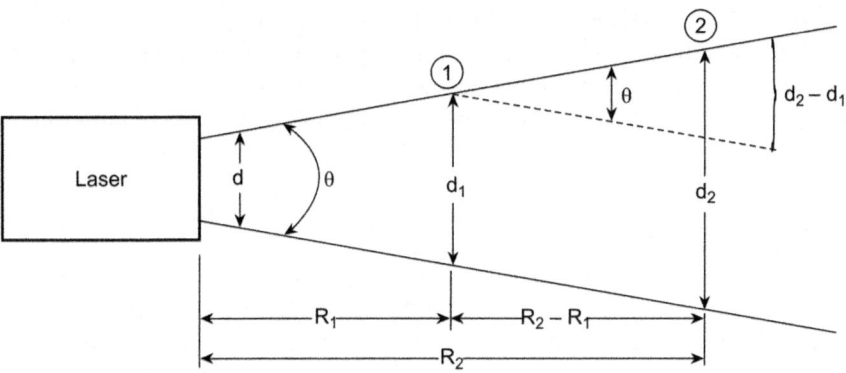

Figure 7.32 Measurement of beam divergence.

The expressions for computing beam divergence given above are valid for the region in the *far field* defined by eqn. 7.22.

$$\text{Far field} \geq \left[\left(100 \times D^2 \right) / \lambda \right] \tag{7.22}$$

Where,

D is the diameter of output optics
λ is the wavelength of the laser beam.

There is another region called the *near field* defined by eqn. 7.23.

$$\text{Near field} \leq \left[D^2 / \lambda \right] \tag{7.23}$$

Where,

D is the diameter of output optics
λ is the wavelength of the laser beam.

It may be mentioned here that near field divergence may be different from far field divergence. Description of reasons for this difference is beyond the scope of the book.

7.4.4 Measurement of M^2 Value

The M^2 value in the case of real beams is measured by focusing the beam with the help of a lens with a known focal length. A beam profiler arrangement configured around either moving knife-edges or special cameras is used to make measurement of beam waist. The M^2 value cannot be determined from a single beam profile measurement. It is computed from the data generated from a series of measurements. Figure 7.33 shows one such setup that uses a camera-based beam profile measurement. The setup uses a fixed position lens and a moving detector (camera in this case) to make multiple measurements in the beam waist region. This design is used in Spiricon's M^2–200 Beam Propagation Analyser.

An alternative setup uses a fixed position detector and a scanning lens assembly (Figure 7.34). The Mode Master PC M^2 Beam Propagation Analyser from Coherent is built around this design philosophy. It uses a dual knife-edge beam profiler integrated with a diffraction-limited precision scanning lens. The focusing lens creates an internal beam waist and the two

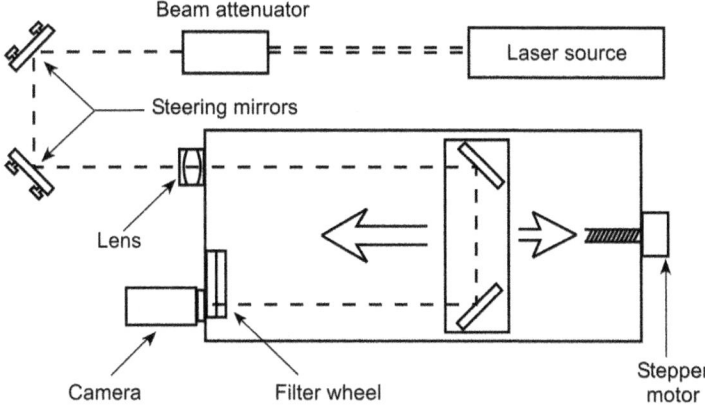

Figure 7.33 M^2-value measurement setup using a fixed position lens and moving detector.

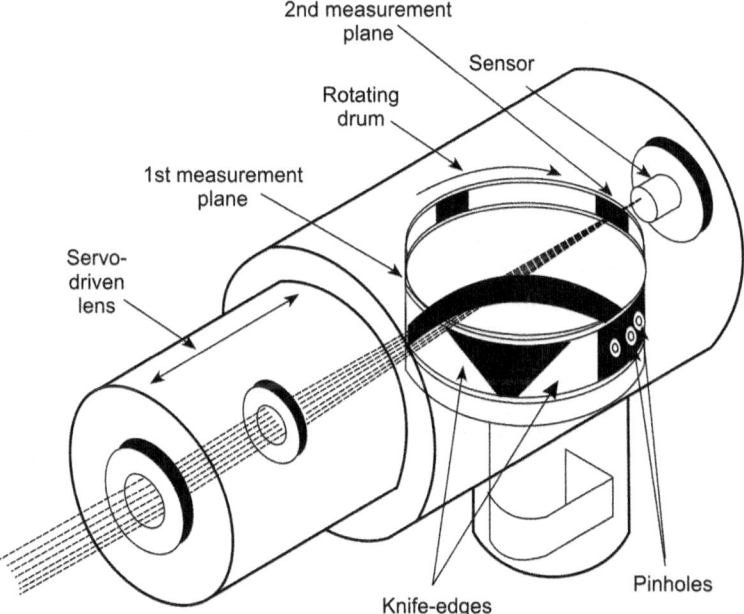

Figure 7.34 M^2-value measurement setup using a scanning lens and fixed position detector.

orthogonal knife-edges mounted on a rotating drum measure the beam diameter and the corresponding beam axis location at multiple planes along the beam waist. The M^2 value is then computed from the multiple beam width measurements along with their location on the beam axis and the known characteristics of the focusing lens.

7.5 Laser Beam Diagnostic Equipment

In the previous pages the basic techniques available to us for the measurement of important laser parameters were briefly described. While on one hand, a wide range of laser power and energy meters are commercially available to meet different requirements of measurement of laser power, energy and other related parameters, on the other hand, laser beam diagnostic equipment allow precise measurement of parameters related to shape and profile. The list of these test and measurement instruments includes wavelength meters, laser spectrum analysers, beam profilers and beam propagation analysers. The functions performed by each one of them along with other salient features available in modern diagnostic equipment are briefly described in the following paragraphs.

7.5.1 Wavelength Meter

A wavelength meter is used for measurement of wavelength of the laser radiation. One such piece of equipment is the WaveMaster™ Laser Wavelength Meter from Coherent. It measures the wavelength of both CW and pulsed lasers of any PRF in the wavelength range of 380–1095 nm. Wavelength reading can be displayed in nanometres, wave numbers or GHz. Measurements of wavelength in both air and vacuum are possible. Wavelength is measured with a resolution of 0.001 nm and accuracy of 0.005 nm. The equipment comes with a built-in RS-232 and an optional GPIB interface.

7.5.2 Laser Spectrum Analyser

A laser spectrum analyser is used to carry out modal analysis of lasers operating in more than one longitudinal or transverse mode of the laser cavity. The spectrum analyser typically uses a scanning Fabry–Perot interferometer cavity. The resonant frequency of the interferometer is scanned by varying the spacing between the cavity mirrors with the help of a piezoelectric spacer segment placed between the two mirrors. Laser light transmitted through the cavity to a detector has only those line spectra that match with the resonant spectral frequency of the cavity. An oscilloscope synchronized to the cavity scan rate then displays the detected spectrum of different laser lines.

One such piece of equipment is SAPlus Laser Spectrum Analyser from Artisan Scientific. The equipment provides modal analysis in the wavelength range of 450–1800 nm. It offers a free spectral range choice of 2 or 8 GHz. It is capable of measuring the linewidth, longitudinal mode structure and frequency stability of narrowband lasers.

Figure 7.35 shows a schematic diagram of the spectrum analyser. As shown in the figure, it utilizes a scanning confocal Fabry–Perot interferometer cavity comprising two concave mirrors with the spacer separating the two mirrors with a piezoelectric section. The piezoelectric

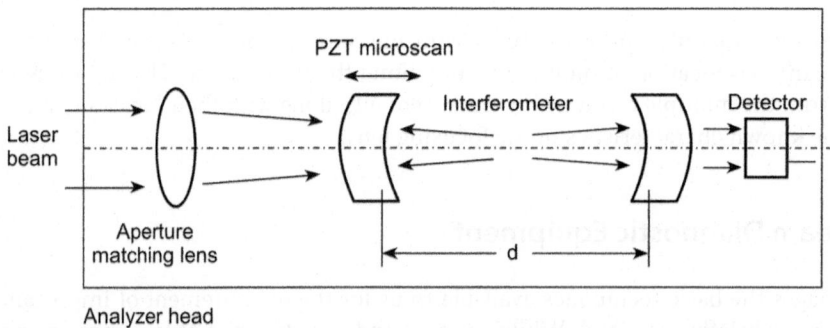

Figure 7.35 Schematic arrangement of a spectrum analyser.

section, when driven electrically from the controller, can be used to vary the mirror spacing and thus the resonance frequency of the cavity. Free spectral range in this case is given by $c/4d$ where d is the mean mirror separation. The purpose of the lens assembly is to match the laser beam to the input aperture of the confocal cavity. The laser beam can be directed straight into the spectrum analyser or through a beam splitter that allows only a fraction of the laser beam (about 10% in this case) to the spectrum analyser.

7.5.3   Laser Beam Profiler

The laser beam profiler is used for measurement of spatial intensity profile and other related parameters of the laser beam. As the laser beam propagates, the width and spatial intensity profile of the beam changes. Spatial intensity distribution of laser beam is a very important parameter when it comes to studying its behaviour in any application.

There are two types of profilers in use. One of the types uses special cameras as detectors and offers an excellent solution to fast and detailed analyses of CW and pulsed lasers. The other type uses moving knife-edges, which is particularly attractive for carrying out measurements on small and focused beams and in the case of requirement of large dynamic range. The Beam View Analyser (Figure 7.36) and Beam Master Profiler (Figure 7.37) from M/s Coherent are examples of camera-based and moving knife-edge based laser beam profilers.

Figure 7.36 The Beam View Analyser from Coherent.

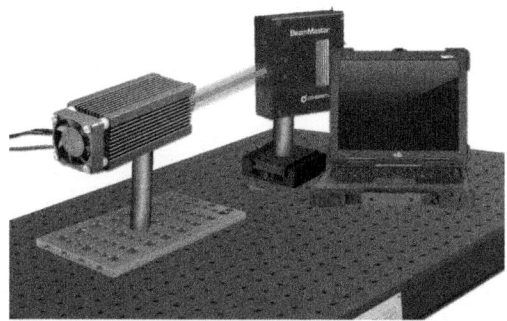

Figure 7.37 The Beam Master Profiler from Coherent.

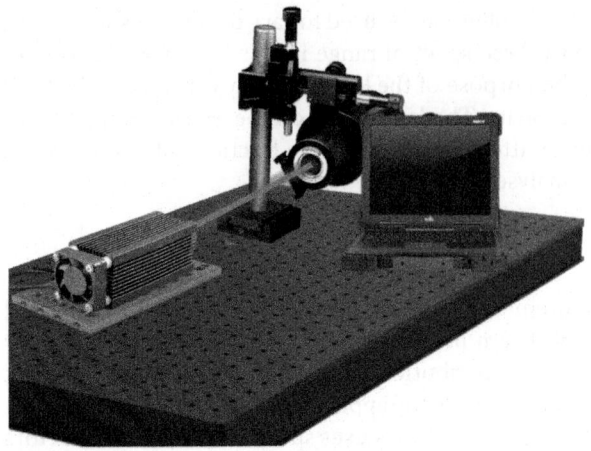

Figure 7.38 ModeMaster PC beam propagation analyser from M/s Coherent.

7.5.4 Beam Propagation Analyser

A laser beam propagation analyser is used for measurement of beam quality (M^2 value) and other beam propagation analysis functions for CW lasers. State-of-the-art beam propagation analysers offer measurement functions, which include beam quality (M^2 value), beam diameter, second moment diameter, divergence, divergence asymmetry, astigmatism, astigmatism asymmetry, pointing stability, waist diameter and location, waist asymmetry, beam profiles, power density and Rayleigh range. The ModeMaster PC^{TM} (Figure 7.38) is one such beam propagation analyser from M/s Coherent. It comprises of a dual knife-edge beam profiler integrated with a servo-controlled diffraction-limited scanning lens. The lens focuses the beam to create an internal beam waist and the two orthogonal knife-edges mounted on a rotating drum measure beam diameter and corresponding beam axis location at 256 planes along the beam waist. The software provided with the system derives different parameters listed before from the data generated during the course of large number of measurements.

7.6 Types of Lasers

Lasers have been classified on the basis of various parameters such as nature of active medium, pumping mechanism, nature of laser output in terms of power/energy level, nature of laser output in terms of wavelength and so on. Based on type of active medium, lasers are classified as:

1) Solid-state lasers
2) Semiconductor lasers and
3) Gas lasers

Based on pumping mechanism, lasers are classified as:

1) Optically pumped lasers
2) Gas dynamic lasers
3) Electrically pumped lasers

Based on nature of laser output in terms of power/energy level, lasers are classified as:

1) Pulsed lasers
2) CW lasers

Based on operational mode, lasers are classified as:

1) Q-switched lasers
2) Mode-locked lasers
3) Cavity-dumped lasers

Based on output wavelength, lasers are classified as:

1) Ultraviolet (UV) lasers
2) Visible lasers
3) Infrared (IR) lasers

The most important way of classification of lasers seems to be on the basis of nature of lasing medium. On this basis, solid-state lasers, semiconductor lasers and gas lasers are the three major categories. In addition, there are a large number of other varieties of lasers that do not fit into any of the previously mentioned broad categories. These include dye lasers, Excimer lasers, metal vapour lasers, free electron lasers, X-ray lasers and so on.

7.7   Solid-State Lasers

As mentioned previously, most lasers can be grouped with any of the three broad categories of lasers; namely *solid-state lasers*, *gas lasers* and *semiconductor diode lasers.* The basis of this classification is though the type of laser medium, quite a few of other aspects related to the operation of the lasers belonging to these categories are to some extent dictated by the type of the medium used. For instance, solid-state lasers are almost invariably pumped optically; the gas lasers are excited by an electrical discharge while semiconductor lasers are pumped by an electrical current flowing through a forward-biased diode junction. In the following paragraphs in this section, we shall discuss the nature of the active medium, the operational modes and the common types of solid-state lasers.

7.7.1   Active Medium

The active medium In the case of solid-state lasers is the lasing species embedded into a crystalline or a glass host material. The characteristics of the host material are no less important than those of the lasing species. The host material should have such optical, mechanical and thermal properties as to favour homogenous propagation of light through the crystal and thus a good beam quality (optical properties), high average power operation, capability to withstand severe operating conditions of practical laser systems (mechanical and thermal properties). For instance, the host material should be reasonably transparent to the pump radiation. It should not absorb radiation at either the pumping wavelength or laser wavelength. It should have high thermal conductivity so as to allow CW operation or operation at high repetition rates In the case of pulsed operation leading to high average power operation.

Relatively poorer thermal conductivity in the case of a glass host disallows the operation of an Nd:glass laser in CW mode, not even at high repetition rates. On the other hand, an Nd:YAG laser, where the host material is *Yttrium Aluminium Garnet* (YAG), can be made to operate in CW because of the far superior thermal conductivity of the host. Yet another point is the interaction of light-emitting species with the host material. This influences the energy level structure of the active species thus modifying the energy levels involved in the laser action. As an illustration, neodymium in the YAG host in the case of an Nd:YAG laser emits at 1064 nm while the same neodymium when doped in phosphate based glass acting as host emits at 1054 nm. Going a step further, neodymium in silicate glass emits at 1062 nm and in fused silica glass at 1080 nm.

Yttrium Lithium Fluoride (YLF) is another important host material popular with neodymium-based solid-state lasers. It has fewer heat related problems compared to YAG. Also, Nd:YLF can store more energy than its Nd:YAG counterpart and therefore is capable of generating higher energy Q-switched laser pulses. Due to its birefringent nature, it can produce laser output at 1047 nm and 1053 nm, each with its own polarization orientation. Birefringence is the property of polarization-dependent refractive index.

Among synthetic garnets, other than *yttrium aluminium garnet* ($Y_3Al_5O_{12}$ and abbreviated to YAG), some of the popular host materials include *Gadolinium Gallium Garnet* ($Gd_3Ga_3O_{12}$ and abbreviated to GGG) and *Gadolinium Scandium Gallium Garnet* ($Gd_3Sc_2Ga_3O_{12}$ and abbreviated to GSGG). These host materials have good thermal properties, are hard and optically isotropic. Good thermal properties permit operation at high average power levels. *Yttrium-doped Vanadate* (YVO_4) is yet another important host material. Neodymium-doped YVO_4 has a very large stimulated emission cross-section, which makes it a high gain laser material, and a strong broad band absorption around 808 nm, which makes it a highly suitable material for laser diode pumping.

This interaction of active species with the host material sometimes assumes interesting proportions in the case of a special class of solid-state lasers called vibronic (vibrational-electronic) lasers where the electronic energy levels of the light-emitting species interact with vibrational levels of the host and get broadened. The lasing levels instead of being single levels get transformed to upper and lower lasing bands. This feature makes this class of lasers tunable over a range of wavelengths. Important solid-state lasers in this category include titanium doped *sapphire* (sapphire is the host) and alexandrite, which is chromium-doped *chrysoberyl* (chrysoberyl is the host).

Neodymium, chromium, titanium and erbium are common lasing species for solid-state lasers. Neodymium and chromium are the most widely exploited lasing species. Chromium is used in ruby (chromium-doped aluminium oxide), alexandrite (chromium-doped chrysoberyl, $BeAl_2O_4$), and chromium-doped GSGG. Titanium is used in titanium-sapphire (titanium-doped Al_2O_3) lasers. Neodymium is used in Nd:YAG and Nd:glass lasers. Erbium is another lasing species, which in YAG and glass hosts give the active medium for a rapidly emerging class of solid-state lasers called eye-safe lasers. The word *eye-safe* comes from the fact that these lasers produce output at 1540 nm, which poses a relatively lower eye hazard compared to neodymium lasers. Neodymium lasers produce output around 1064 nm, which poses serious eye hazards.

7.7.2 Operational Modes

Operational modes here mean different resonator designs leading to different laser output formats. On this basis, different operational modes or more appropriately different operational methodologies, lasers could produce at the output any of the following output formats.

1) CW output
2) Free-running output
3) Q-switched output
4) Cavity-dumped output
5) Mode-locked output

7.7.2.1 Continuous-Wave (CW) Output

In this operational mode, which is the simplest mode of operating a laser, the laser produces a continuous output as long as the laser is pumped. Pumping needs to be continuous in this case. The laser resonator in this case in its simplest form comprises of just the active medium and the resonator mirrors and does not contain any switches. In a CW laser, the steady state gain equals the loss and the excess pump power is converted into laser output. Output of such a laser will fluctuate due to the many ways in which different longitudinal and transverse modes couple to each other via the active medium unless means are used to control longitudinal and transverse modes. A laser producing a single longitudinal mode and a single transverse mode may have output fluctuation of less than 1%.

7.7.2.2 Free-Running Output

This is a quasi-CW mode of operation in which the laser operates in the CW mode for a time period that is of the order of a few hundreds of microseconds to a few milliseconds. This time period generally equals or is slightly longer than the storage time of the active medium, which in turn is of the same order as the pump input pulse. Refer to Figure 7.39, which shows the timing sequence of production of free-running pulsed output with respect to the application of pump input and gain and loss response of the laser resonator.

As is clear from the timing diagram, in the case of free-running output too, the steady state gain equals the loss and the excess pump power appears as the laser output. As the pump input

Figure 7.39 Timing diagram of free-running output.

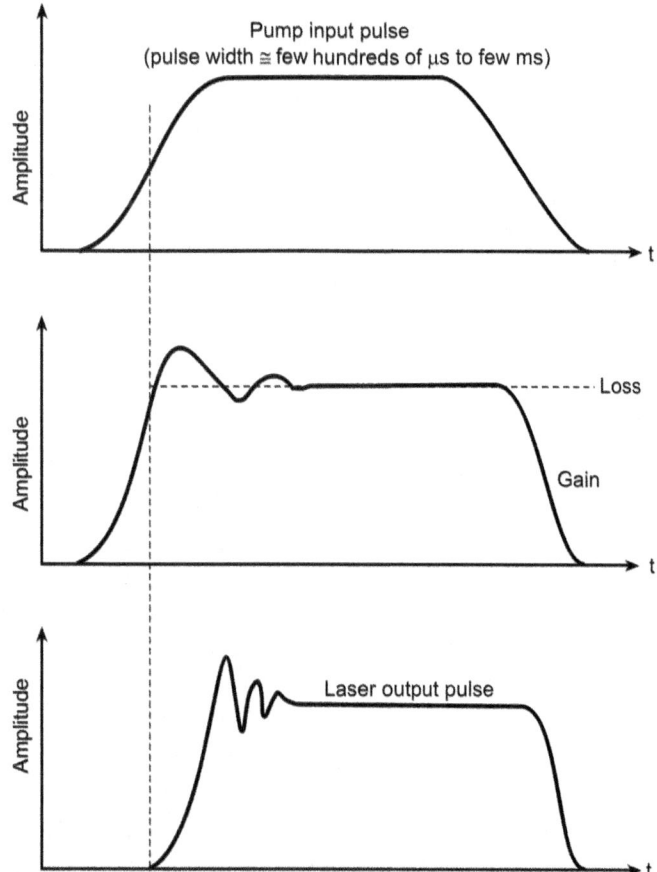

exceeds a certain threshold where the gain equals the loss. Laser radiation begins to appear at the output. Both gain and laser output respond with overshoots, which ultimately damp out after a few cycles and the output settles down to the CW like output as shown. Overshoots depend upon the rise time of the pump input. A faster rise time leads to a larger overshoot.

7.7.2.3 Q-Switched Output

Q-switching is a mechanism of producing short laser pulses with a pulse width in the order of a few nanoseconds. The term Q-switching here means switching of the quality factor of the resonator cavity, which is the ratio of energy stored per cycle to the energy lost per cycle, rapidly from low value to a high value. Refer to the timing diagram in Figure 7.40. Initially, the Q-factor of the resonator cavity is kept at a very low value by introducing some kind of optical switch such as a Pockels cell and a polarizer. When the laser is pumped, it is prohibited from lasing like it does In the case of a CW output. The obvious consequence of this is that the active medium builds up a much higher population inversion density or gain than would have been possible in the case of a CW laser. When the inversion density reaches almost its peak value, the Q-factor of the resonator cavity is rapidly switched to the high value leading to a steep fall in the loss value. This manifests itself in the production of a short laser pulse at the output as shown. The output laser pulse begins to appear at a time instant where the loss becomes less than the gain and reaches the peak value at the time instant when the diminishing gain equals the loss. After

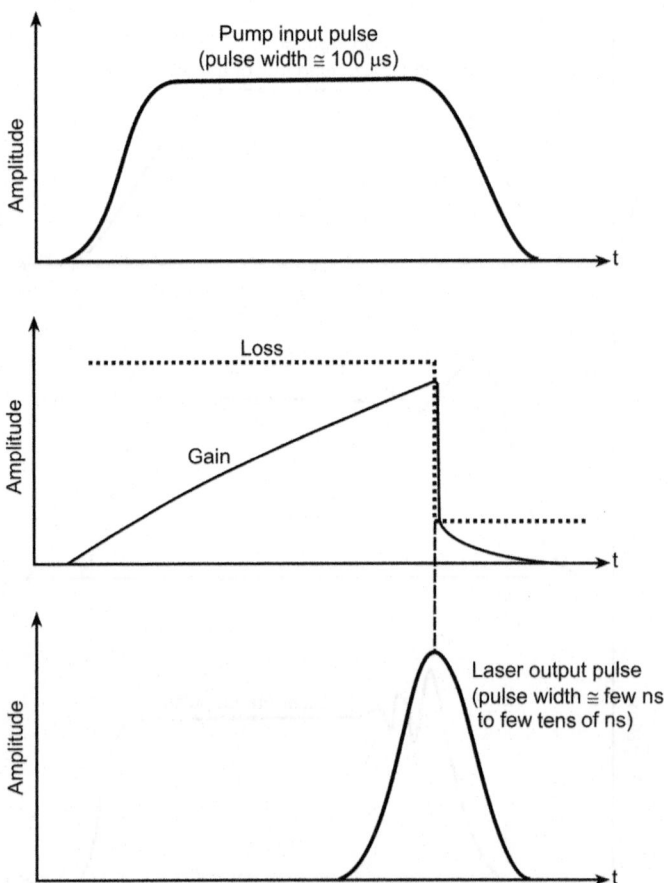

Figure 7.40 Timing diagram of Q-switched output.

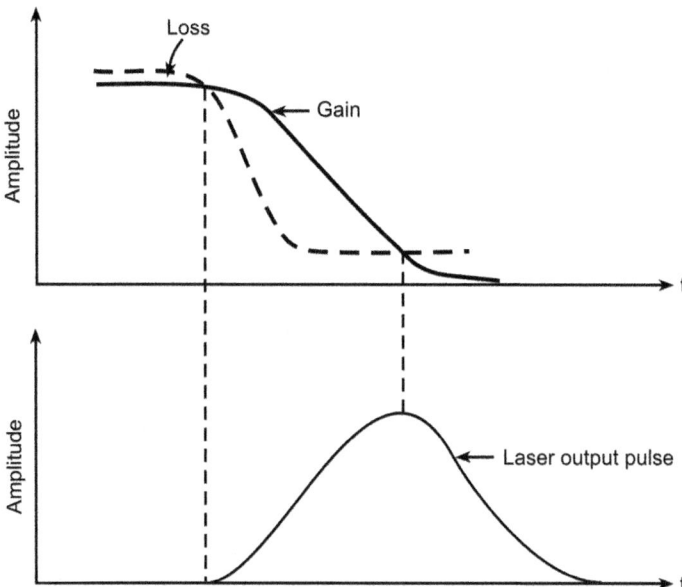

Figure 7.41 Q-switched output on an expanded time scale.

that the gain becomes less than the loss, which manifests itself in the form of a falling laser output. These time instants can be observed more clearly if we expand the time scale around the point where the Q-factor is switched from a low value to a high value (Figure 7.41).

The pulse duration of a Q-switched pulse output depends upon resonator round trip time, reflectivity of the output coupler and the extent to which the active medium is pumped above the threshold. To be a little more specific, the pulse width depends primarily on the resonator cavity's round trip time and the resonator gain to loss ratio. Without going into any intensive mathematics, it would suffice to say here that in an optimized Q-switched laser oscillator, the output laser pulse width is in the order of a few times the cavity round trip time and may be typically in the range of 10–50 ns. To a good approximation, Q-switched pulse width can be computed from eqn. 7.24.

$$t_{pq} = \frac{t}{(1-R)} \tag{7.24}$$

Where,

t_{pq} = Q-switched pulse width
t = Laser cavity round trip time
R = Reflectivity of output mirror.

Laser cavity round trip time in turn can be computed from eqn. 7.25.

$$t = \frac{2Ln}{c} \tag{7.25}$$

Where,

L = Length of laser cavity
n = Refractive index of laser medium.

7.7.2.4 Cavity-Dumped Output

The phenomenon of cavity dumping is a slight variation of the Q-switching process with the difference that, in the case of the former, both of the resonator cavity mirrors are 100% reflective. When the active medium is pumped, energy is initially stored in the population inversion as is done in the case of Q-switching. During this time, the loss is at a very low value. Since both of the resonator mirrors are 100% reflective, the amplified light remains trapped within the cavity. As peak irradiance is reached, the loss is again switched to a high value thus ejecting the intra-cavity circulating energy in the form of a pulse. One of the advantages of cavity dumping technique is that here the output pulse width depends upon the cavity length only and is independent of the gain characteristics of the active medium. To be more specific, the pulse width equals the round trip cavity transit time provided that the Q-switch employed for switching the quality factor of the cavity is also switched within the same time period. This technique allows generation of output pulses a few nanoseconds' duration and the lower limit of the pulse width achievable with this technique is 2–3 times smaller than what is achievable with Q-switching. Figure 7.42 shows the timing diagram of generation of a cavity-dumped output pulse.

Pulse width in the case of cavity-dumped output can be computed from eqn. 7.26.

$$t_{pc} = \frac{2Ln}{c} \tag{7.26}$$

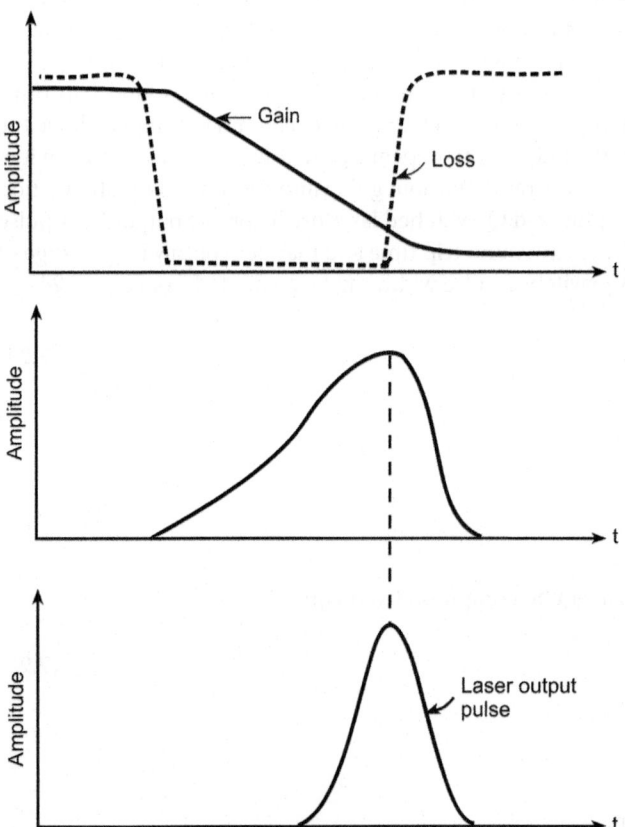

Figure 7.42 Timing diagram of a cavity dumped output.

Where,

t_{pc} = Q-switched pulse width
L = Length of laser cavity
n = Refractive index of laser medium.

7.7.2.5 Mode-Locked Output

Minimum pulse width achievable from the laser is intimately related to its frequency bandwidth. It was mentioned in the previous chapter that Heisenberg's uncertainty principle puts a limit on the minimum possible laser pulse width for a given value of available bandwidth. That is,

$$t_p = \frac{0.441}{B} \tag{7.27}$$

Where,

t_p = Minimum pulse width
B = Bandwidth.

But, it is not practicable to reach this limit with every pulse forming technique. For instance, In the case of Q-switching, the minimum achievable pulse width is of the order of 10 ns or so because of the required pulse buildup time. Cavity dumping overcomes this shortcoming to some extent and pulse widths of the order of 1–2 ns are achievable. But still, it is nowhere close what is theoretically achievable for a given solid-state laser. For instance, the frequency bandwidth of an Nd:YAG is about 150 GHz for a homogeneously broadened line. For a Gaussian pulse, the minimum achievable pulse width would be (0.441/150) ns ≈ 3 ps. Mode locking helps achieve pulse widths approaching the theoretical limit.

We know that in the absence of special measures a laser usually oscillates at several transverse modes and typically hundreds of longitudinal modes. The number of oscillating longitudinal modes depends upon the inter-mode spacing and gain bandwidth of the transition line. Each mode oscillates independent of the others and their phases are randomly distributed in the range of + π to − π radians. As a result of oscillations at large number of longitudinal modes and various modes with random amplitude and phase relationship, the laser output is a time averaged statistical mean value and is observed to fluctuate in amplitude with respect to time as shown in Figure 7.43. The process of mode locking forces different longitudinal modes to oscillate with a fixed phase relationship with respect to each other, which produces an ultrashort pulse with well-defined amplitude as a function of time. Figure 7.44 shows a mode-locked pulse generated from phase locking of different longitudinal modes. In the case of an ideal mode-locked laser pulse, the intensities of different longitudinal modes follow a Gaussian

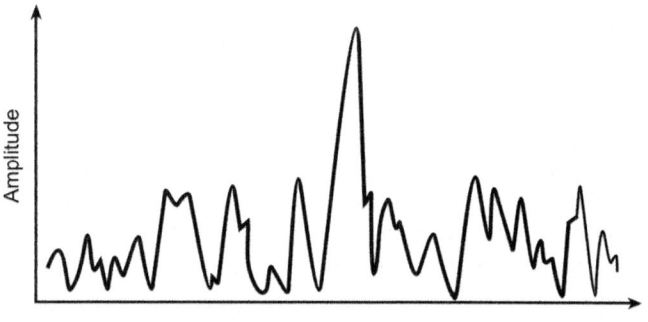

Figure 7.43 CW Laser output in the absence of mode locking.

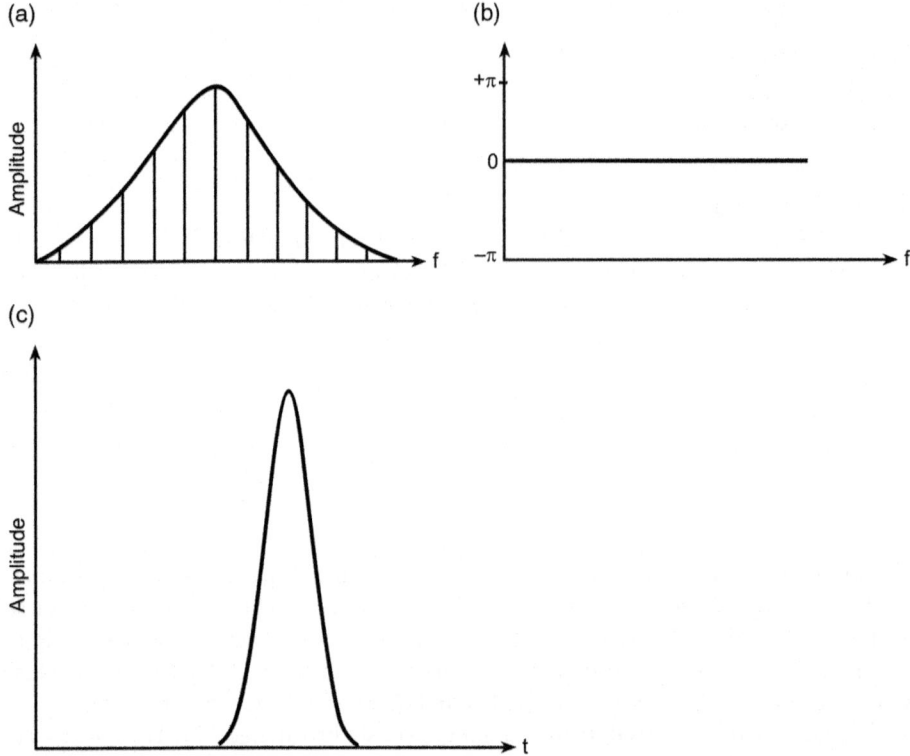

Figure 7.44 Mode-locked pulse. (a and b) For an ideal mode-locked laser pulse, the intensities of different longitudinal modes follow a Gaussian distribution and the spectral phases are identically zero. (c) The ideal mode-locked pulse in the time domain, which is a single Gaussian pulse.

distribution and the spectral phases are identically zero as shown in Figure 7.44(a) and (b), respectively. Figure 7.44(c) shows the ideal mode-locked pulse in the time domain, which is a single Gaussian pulse.

Phase locking is achieved by introducing into the resonator cavity a suitable nonlinear element such as a saturable absorber or an externally driven optical modulator. While the former is called the process of passive mode locking, the latter is called active mode locking. The phenomenon of mode locking can be considered to be a process that locks together in phase a cluster of photons. The mode locking element transmits this cluster every time it passes through it while bouncing back and forth between the cavity mirrors. The repetition rate of the mode-locked pulses is equal to the round trip transit time of the resonator cavity. That is,

$$PRF = \frac{1}{\Delta T} = \frac{c}{2L} \tag{7.28}$$

It is also possible to combine the processes of Q-switching and mode locking or the processes of cavity dumping and mode locking. Q-switching and mode locking can be achieved simultaneously by introducing Q-switching elements in to the cavity in addition to the mode locking element. The output in that case has the pulse envelope of a Q-switched pulse comprising of individual short pulses obtained from a mode locking process as shown in Figure 7.45.

Figure 7.45 Mode locked pulses with a Q-switch envelope.

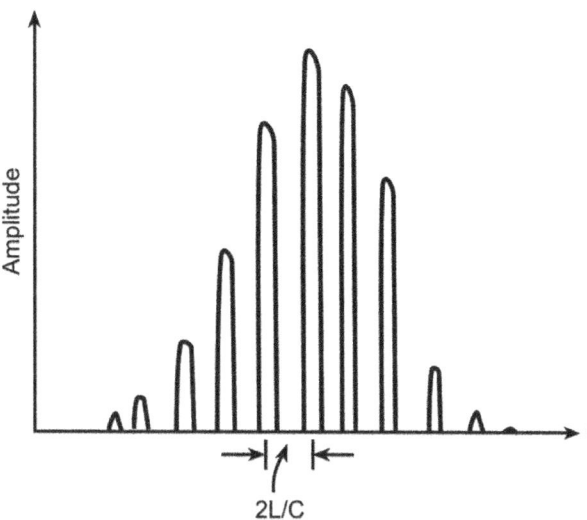

7.7.3 Ruby Lasers

The active medium here is sapphire; that is, aluminium oxide (Al_2O_3) doped with 0.01–0.5% of chromium. In fact, doping process replaces the small percentage of Al^{+3} by Cr^{+3} by adding small amounts of Cr_2O_3 to the melt of highly purified Al_2O_3. Figure 7.46 shows the energy level diagram of a ruby laser. As can be seen from the figure, it has two pump absorption bands centred around 400 nm (blue band) and 550 nm (green band), both with a life time of 100 ns. There are two closely spaced metastable levels with a lifetime of 3 ms from where they drop to the ground state. The ground state here acts as the lower laser level. The laser emits at 694.3 nm (R_1 line). Another possible emission line is at 692.9 nm (R_2 line) but the former is the predominant emission line. The three-level energy structure of ruby is responsible for its poor efficiency (0.1–1%).

Its properties rapidly degrade with increasing temperature with the result that ruby lasers are operated as pulsed lasers at relatively lower repetition rates. The fact that a ruby laser is a three-level laser is also a determinant factor responsible for its inability to operate as a CW Laser. Figure 7.47 shows the photograph of the SINON Q-switched ruby laser system from M/s Quantel-Derma GmbH. Major performance specifications include wavelength = 694 nm, pulse width in Q-switched mode = 20 ns, pulse width in long pulse mode = 4 ms, pulse repetition rate = 0.5–2.0 Hz and an adjustable beam diameter of 3–6 mm. The system has primary application in dermatology and intended for removal of pigmentation and tattoos. Some of the earlier versions of military laser range finders used a Q-switched ruby laser as the source. These have been completely replaced by neodymium- and erbium-doped laser sources.

7.7.4 Neodymium-Doped Lasers

Neodymium-doped solid-state lasers are the most widely used and exploited type of not only solid-state laser but also lasers in general as well. Different neodymium-doped lasers differ in the host structure that is doped with neodymium. Yttrium aluminium garnet (YAG), yttrium

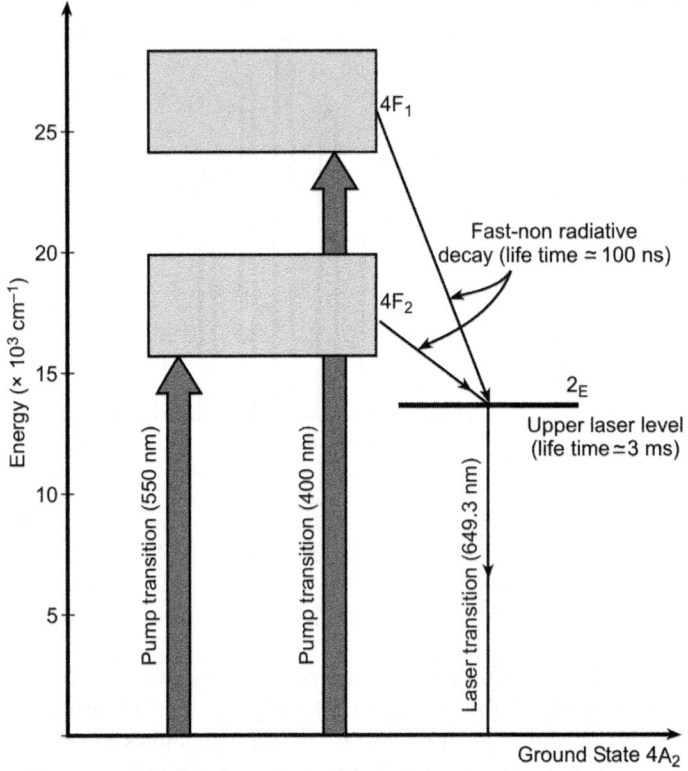

Figure 7.46 Energy level diagram of a ruby laser.

Figure 7.47 Q-switched output ruby laser system.

lithium fluoride ($YLiF_4$) and yttrium-doped vanadate (YVO_4) are the commonly used crystalline hosts while silicate, phosphate and fused silica are the popular glass hosts. Figure 7.48 shows the energy level diagram of a Nd:YAG laser. All neodymium-doped lasers, whether crystalline host based or glass host based have an energy level diagram similar in structure to the one shown in Figure 7.48. Interaction of neodymium with the host may lead to a slight change in the output wavelength, to the tune of 1% or so, from one Nd-doped laser to another. For instance, Nd:YAG, Nd:YLF, Nd:YVO$_4$, Nd:glass (silicate), Nd:glass (phosphate) and Nd:glass (fused silica) have wavelengths of 1064, 1047/1053, 1064, 1062, 1054 and 1080 nm, respectively.

Nd:YAG and Nd:YLF Lasers can be operated both as pulsed as well as CW Lasers. Nd:glass lasers due to poorer thermal conductivity can, however, be only operated as pulsed lasers; that too at very low repetition rates. Glass as a laser material, however, has the advantage of high-energy storage capability and the ease with which it can be grown in large sizes with the desired optical quality. Rod sizes of 5–6 cm diameter and 100 cm in length are not uncommon. These are used to build relatively high-energy solid-state lasers with pulse energy of the order of several kilojoules. Also glass lasers have a broad line width, which makes them better adaptable to mode locking phenomenon to generate much shorter pulse widths, in the order of a picosecond or so. The minimum pulse width that can be achieved with mode locking in this

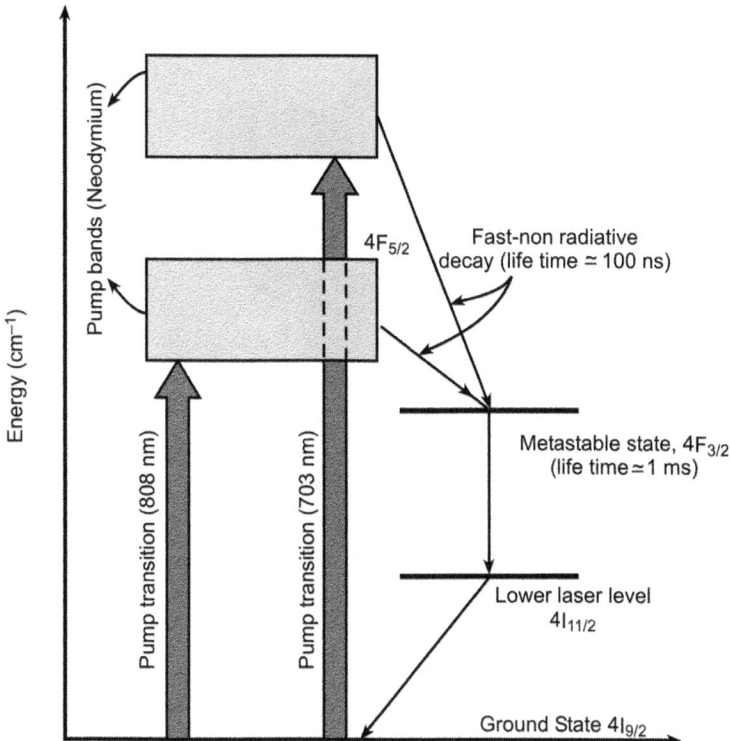

Figure 7.48 Energy level diagram of an Nd:YAG laser.

case is of the order of few tens of femtoseconds. The minimum achievable mode-locked pulse width in a given laser is given by eqn. 7.29.

$$\tau = \frac{0.44}{\Delta \nu} \tag{7.29}$$

Where,
$\Delta \nu$ = Line width.

7.7.4.1 Nd:YAG Lasers

Among neodymium-doped lasers, Nd:YAG is the most important and most widely used solid-state laser because of its high gain and good thermal and mechanical properties. In Nd:YAG, trivalent neodymium replaces trivalent yttrium. Some of its noteworthy properties include hardness, stability of structure from very low temperatures to up to the melting point, good optical quality and high thermal conductivity of the YAG host. In addition, the cubic structure of YAG host favours a narrow line width leading to high gain and low lasing threshold. Nd:YAG has absorption peaks around 540 nm, 590, 750 and 810 nm. Pump bands around 540 and 590 nm are relevant for flash lamp pumping as they emit a large amount of radiation in visible region. Wavelengths of 750 and 810 nm are relevant for pumping by CW arc lamps. 810 nm is relevant for diode-pumped Nd:YAG lasers. The output laser wavelength is 1064 nm.

Due to its good thermal and optical properties, Nd:YAG lasers can be used both in CW as well as high repetition rate Q-switched pulsed mode. Average and peak powers in excess of 1 kW and 100 MW, respectively, can be achieved in these lasers.

Figure 7.49 Laser target designator model No: AN/PEQ-17.

High repetition rate (up to 20 Hz) Q-switched (5–20 ns pulse width) Nd:YAG lasers find extensive use in a variety of battlefield applications such as range finding and target designation. Figure 7.49 shows the photograph of a portable light weight laser target designator model AN/PEQ-17 from M/s Elbit Systems, typically used for LGB (Laser-Guided Bomb) delivery applications. The laser produces a 1064 nm wavelength, 50–70 mJ/pulse Q-switched laser pulses having output beam divergence equal to 0.3 mrad. The system operates on pulse repetition frequencies of NATO Stanag band I/II and offers a designation range of 5 km.

7.7.4.2 Nd:YLF Lasers

The advantages of Nd:YLF lasers include fewer heat related problems due to smaller change in its refractive index with temperature and higher energy storage capability leading to its ability to generate higher energy Q-switched pulses. The fluorescence life time in fact is twice as long as it is in the case of Nd:YAG. In addition, Nd:YLF crystal is birefringent, which allows generation of two output wavelengths at 1047 nm and 1053 nm, each with its own polarization orientation. Figure 7.50 shows the simplified energy level diagram of an Nd:YLF laser. Depending upon the polarization, two lines each are possible around 1.05 and 1.3 μm. Both emissions (1.05 and 1.3 μm) originate from the same metastable upper laser level as shown in Figure 7.50. Nd:YLF is predominantly used to produce either of the two lines around 1.05 μm. An intra-cavity polarizer can be used to select either of the two emission lines at 1053 nm (ordinary) or 1047 nm (extraordinary).

Emission at 1053 nm in the case of an Nd:YLF laser matches very well with the peak gain of neodymium-doped phosphate and fluorophosphate glasses. This interesting coincidence makes it highly suitable as an oscillator for pumping Nd:glass based laser amplifier chains used to generate very high peak power Q-switched pulses needed for fusion research.

Figure 7.51 shows a photograph of an LDY300-series diode-pumped Q-switched frequency-doubled Nd:YLF laser that generates laser pulse energy in the range of 10–30 mJ at a repetition

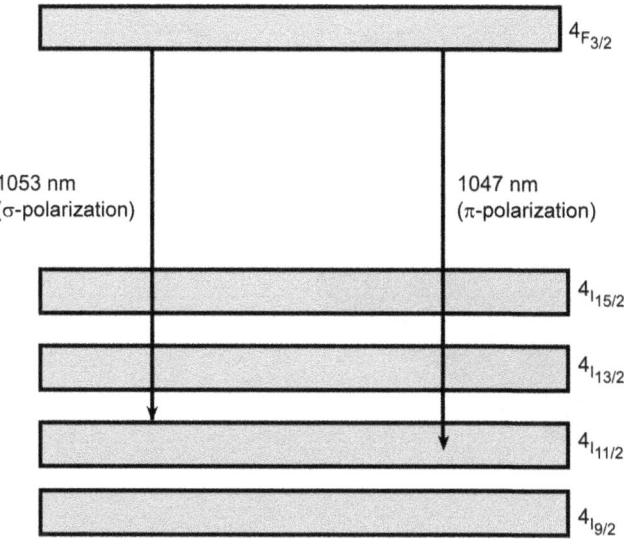

Figure 7.50 Energy level diagram of an Nd:YLF laser.

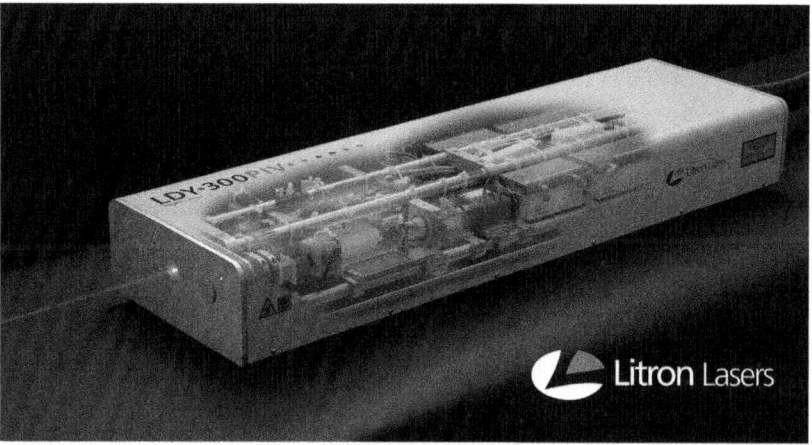

Figure 7.51 LDY300-series diode pumped Q-switched Nd:YLF laser. (Source: Courtesy Litron Lasers UK.)

rate varying from 0.2 to 20 kHz. Output wavelength is 527 nm and typical applications include particle imaging velocimetry (PIV), particle sizing and pumping of Ti-sapphire lasers.

7.7.4.3 Nd:YVO$_4$ Lasers

Neodymium-doped yttrium vanadate as a laser material is important because of several properties that make it particularly attractive for laser diode pumping. These include a large stimulated emission cross-section and a strong broadband absorption around 808 nm. Also, high gain of Nd:YVO$_4$ coupled with strong absorption of laser diode pump radiation around 808 nm obviate the need for a large crystal size, which had been a serious problem in the early stages of development of this material. Yet another significant characteristic of this laser material is its natural birefringence. The laser output is linearly polarized in the extraordinary direction and the absorption coefficient for the laser diode pump radiation polarized in the same direction is much larger (about four times) in Nd:YVO$_4$ than it is in the case of Nd:YAG.

Figure 7.52 Nd:YVO$_4$ laser module Type No PULSELAS-P-1064-150-HE. (Source: Courtesy M/s ALPHALAS, Germany.)

Some of the disadvantages of Nd:YVO$_4$ as a laser material include a shorter fluorescence life time and a slightly poorer thermal conductivity compared to Nd:YAG. In fact, a higher stimulated emission cross-section, which contributes in making it a low lasing threshold material, is partially offset by a shorter fluorescence life time. It may be mentioned here that the lasing threshold depends upon product of fluorescence lifetime and the stimulated emission cross-section. Nd:YVO$_4$ is particularly attractive for laser diode end-pumped CW Lasers. These lasers are quite often internally frequency-doubled to produce 532 nm (green) output. Figure 7.52 shows a photograph of a compact Nd:YVO$_4$ Q-switched laser module from M/s Alphalas, Germany. The module produces 1 ns wide 1.5 mJ pulses at a repetition rate from 0 to 100 Hz. Other features include TEM_{00} beam profile, >100:1 polarization ratio and beam diameter of 0.3 mm.

7.7.4.4   Nd:Cr:GSGG Lasers

Nd:Cr:GSGG as a laser material is primarily important for the reason that its absorption band has a better spectral match to the emission from flash lamps as compared to Nd:YAG with the result that Nd:Cr:GSGG lasers offered significantly higher electrical input to laser output efficiency as compared to Nd:YAG. In the case of Nd:Cr:GSGG, the pump energy is absorbed by the broad absorption bands of chromium, which is further transferred non-radiatively to neodymium atoms. This concept was tried in a YAG host initially but was unsuccessful due to an inefficient energy transfer process. Inefficiency was primarily due to a prohibitively large transfer time of about 6 ms, which had to be shorter than the fluorescence decay time of 0.23 ms. The problem was overcome with GSGG as the host material. The transfer time in this case was 17 µs.

It may be mentioned here that a higher efficiency of Nd:Cr:GSGG laser in no way means that it is a better laser system compared to Nd:YAG and other better known solid-state lasers. Nd:Cr:GSGG not only has a significantly poorer thermal conductivity compared to Nd:YAG, but also the additional absorption bands of chromium in blue and red regions that lead to higher pump efficiency produce significant quantum defect heating due to large difference in pump photon and laser photon energies. This phenomenon leads to a far more severe thermal lensing and thermally induced birefringence.

7.7.4.5 Nd:Glass Lasers

After the Nd:YAG laser, Nd:glass is the other important neodymium-doped solid-state laser. While Nd:YAG is mainly known for its high gain and good thermal and mechanical properties, Nd:glass is capable of being grown in large sizes with diffraction-limited optical quality. As a result, Nd:YAG is the laser of choice in a wide range of applications requiring either CW or high repetition rate pulsed operation. An Nd:glass laser on the other hand is the preferred laser for high-energy and high peak power applications requiring either single pulse or low repetition rate operation such as laser fusion research.

The host materials for Nd:glass lasers are silicate, phosphate and fused silica glasses with silicate and phosphate being the more common ones. Silicate and phosphate glasses are, respectively, based on SiO_2 (silicon dioxide) and P_2O_5 (phosphorus pentoxide) materials with the latter being the material of choice due to its large emission cross-section and a lower nonlinear coefficient.

Like other neodymium-doped crystalline lasers, the Nd:glass laser is also a four-level laser. Figure 7.53 shows a simplified energy level diagram of an Nd:glass laser. As shown in the diagram, laser transition is from lower lying level of $^4F_{3/2}$ to a lower lying level of $^4I_{11/2}$.

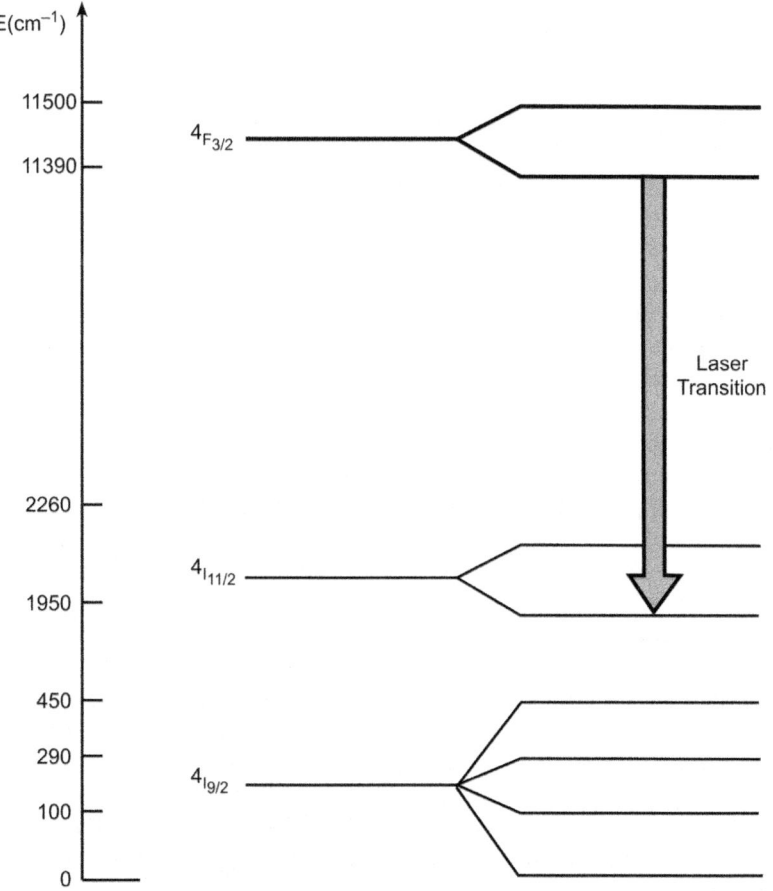

Figure 7.53 Simplified energy level diagram of an Nd:Glass laser.

The emitted wavelength is typically 1062 nm in the case of silicate glasses and 1054 nm in the case of phosphate glasses.

Also, fluorescent line width in the case of Nd:Glass lasers is insensitive to temperature variations, which makes it practically feasible to operate these lasers over a wide temperature range, as wide as $-100-+100\,°C$, with very little change in its performance characteristics.

7.7.5 Erbium-Doped Lasers

Erbium-doped lasers have some potential medical and military applications because of the two wavelengths they are capable of generating when doped in YAG and Glass hosts. These wavelengths are 2940 nm (erbium:YAG) and 1540 nm (erbium:glass) and their importance arises from water absorbent characteristics of these wavelengths. While 2940 nm holds promise for medical applications in the field of plastic surgery due to its extremely large absorption by water in tissue, 1540 nm is attractive as an eye-safe alternative to neodymium-doped YAG (or glass) based military laser rangefinders and laser target designators. Neodymium lasers with emissions around 1064 nm pose a serious eye hazard. So, when it comes to using laser range finders and laser target designators for training exercises and war games, eye-safe lasers are definitely a much better option.

7.7.5.1 Erbium-Doped YAG Laser

As mentioned in the previous section, the Er:YAG laser produces an output wavelength of 2940 nm. Figure 7.54 shows energy level diagram of Er:YAG. The lasing action occurs between an upper laser level with a fluorescence life time of 0.1 ms and a lower laser level with a relatively much larger lifetime of 2 ms. The longer lifetime of the lower level is in fact a big disadvantage as accumulation of population at the lower level inhibits laser action and also disallows the Q-switching operation. Some of the other erbium-doped crystalline hosts that have been exploited for laser action include Er:YLF, Er:YAlO$_3$ and Er:Cr:YSGG. These materials produce wavelengths in the range of 2710–2920 nm.

7.7.5.2 Erbium-Doped Glass Laser

Erbium-doped glass produces an output at 1540 nm. Figure 7.55 shows the energy level diagram of erbium:glass. Three-level behaviour of erbium leads to a low laser efficiency. The problem is further worsened by the weak absorption of pump radiation by erbium ions. In order to overcome these shortcomings, ytterbium (Yb^{+3}) and chromium (Cr^{+3}) ions are added. Ytterbium acts as a sensitizing agent. It helps in absorbing pump radiation in the wavelength region (0.9–1 µm) where erbium is more or less transparent. Chromium ions also do a similar job. They help in matching the emission spectrum of flash lamps with the absorption spectrum of ytterbium, erbium and glass.

Q-switched Er:glass lasers find their main application in eye-safe handheld laser range finders. These are lasers with low repetition rates, typically in the range of 5–20 pulses per minute though in some cases it may be as high as 2 Hz. A large number of manufacturers offer handheld Er:glass laser range finders. The LH-40 Eye-Safe Laser Finder from M/s Eloptro South Africa, LRB-21K and LRB-25000 Eye-Safe Laser Range Finders, both from M/s Newcon Optik are some examples. Most of the devices in this category have similar performance specifications in terms of operational range, range accuracy and pulse repetition rate. Figure 7.56 shows a photograph of the LRB-25000. It has a maximum operational range of 25 km, range accuracy of ±5 m and PRF of 0.15 Hz. High repetition rate military lasers of the eye-safe variety currently employ Nd:YAG lasers whose output is wavelength shifted using an optical parametric oscillator (OPO). The laser range finder Type LDM-38 from M/s Carl Zeiss Optronics and the laser range finder Type G-TOR from M/s SAAB Sweden (Figure 7.57) are examples. Both these laser range finders are configured around OPO-shifted Nd:YAG lasers emitting at 1570 nm. G-TOR offers pulse repetition rate as high as 25 Hz.

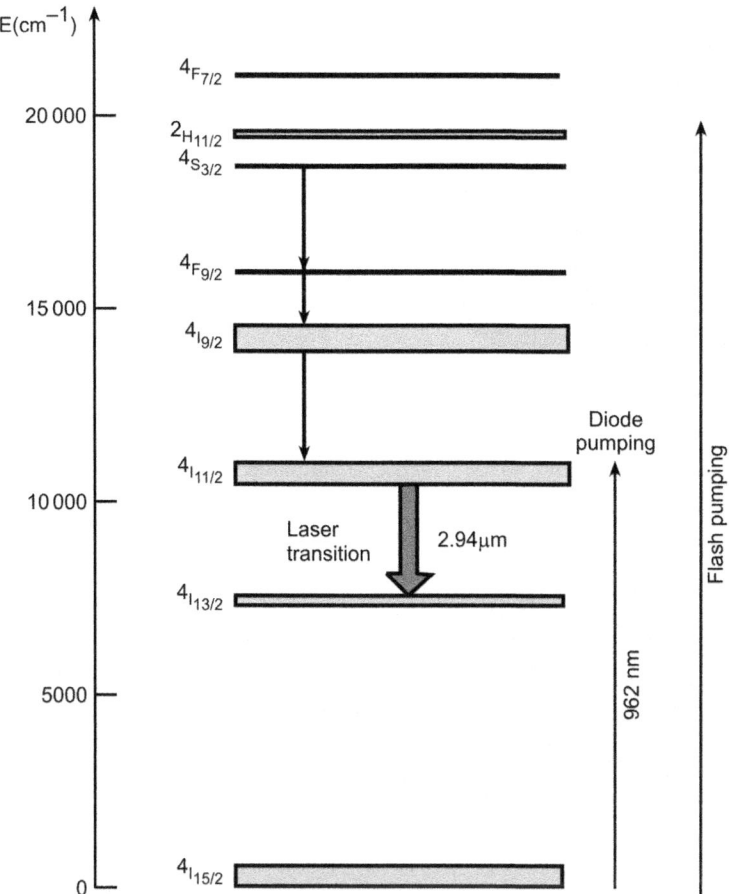

Figure 7.54 Energy level diagram of Er:YAG.

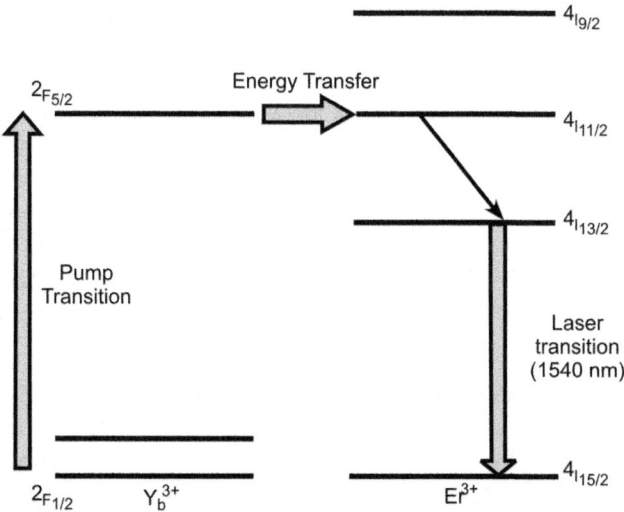

Figure 7.55 Energy level diagram of erbium:glass (phosphate).

Figure 7.56 Er:glass laser range finder, Type LRB-25000.

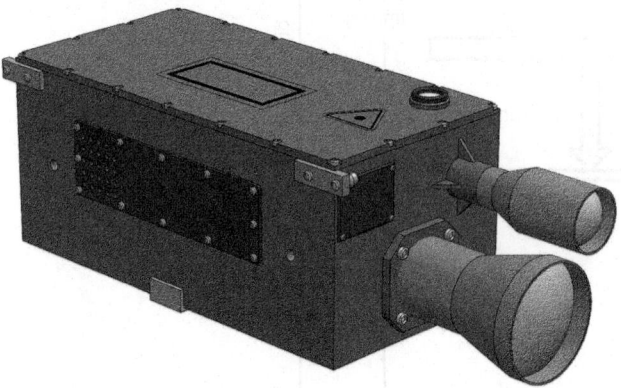

Figure 7.57 OPO shifted Nd:YAG Laser range finder, Type G-TOR.

7.7.6 Vibronic Lasers

What distinguishes a vibronic laser from the more commonly used and better known neodymium-doped and ruby lasers is in their energy level structure with a lower lasing level in the form of a band rather than a single discrete energy level. This energy band results from the interaction of the electronic energy levels of the active species with the vibrational levels of the crystalline lattice giving rise to vibrational-electronic sub-levels. The laser transition occurs between the lowest level of the upper band to anywhere in the lower band. This characteristic feature of vibronic lasers makes this class of lasers tunable.

These lasers have assumed significance due to not only the tunability aspect associated with them but also due to their capability to generate wavelengths not available from other solid-state laser media. Alexandrite (chromium-doped $BeAl_2O_4$), Cr:GSGG (chromium-doped $Gd_3Sc_2Go_3O_{12}$) and titanium-sapphire (titanium-doped Al_2O_3) with respective tunability ranges of (700–850) nm, (740–850) nm and (660–1180) nm are more popular varieties of vibronic lasers. The alexandrite laser with its intense red output coupled with tunability makes it a good candidate for developing a *Laser Dazzler*, a device employed for causing temporary dazzling of human eyes.

7.7.6.1 Alexandrite Laser

Alexandrite is the common name for chromium-doped chrysoberyl. Chromium concentration is in the range of 0.1–0.4 atomic percent. Alexandrite has optical and mechanical properties that are similar to that of ruby. It has many of the physical and chemical properties of a good laser host material including high thermal conductivity, hardness, chemical stability and a high thermal fracture limit. Thermal conductivity and thermal fracture limit of alexandrite is about twice and five times that of Nd:YAG, respectively, which enables alexandrite to be pumped at high average powers without thermal fracture.

Figure 7.58 shows an energy level diagram of alexandrite. Vibronic lasing action occurs between the upper level ($^4T_2$) and any of the excited vibronic states near the ground level ($^4A_2$). This characteristic makes it a tunable laser. The output wavelength is tunable in the range of 700–850 nm. As shown in the energy level diagram, there is another energy level about 800 cm^{-1} below the upper level. This energy level (^2E) is in thermal equilibrium with the upper level. ($^4T_2$) and (^2E) levels, respectively, have lifetimes of 6.6 μs and 1.54 ms. The effective fluorescent life time of the upper level here is according to the combined influence of ($^4T_2$) and (^2E) states and is approximately 260 μs.

Another interesting property of Alexandrite is the way its performance is influenced by increase in temperature. The stimulated emission cross-section and hence the gain of alexandrite increases with increase in temperature. The gain peak shifts to longer wavelength due to population of terminals levels particularly those closer to ground level, which decreases the population inversion for lower emission wavelengths. The net result is that laser performance is positively affected with increase in temperature for larger wavelengths in the tuning range. The other effect of increase in temperature is reduced fluorescent lifetime. It decreases from about 260 μs at room temperature to half that at 100 °C.

The ability of an alexandrite laser to sustain high gain at elevated temperatures coupled with tunability makes it different from conventional solid-state lasers.

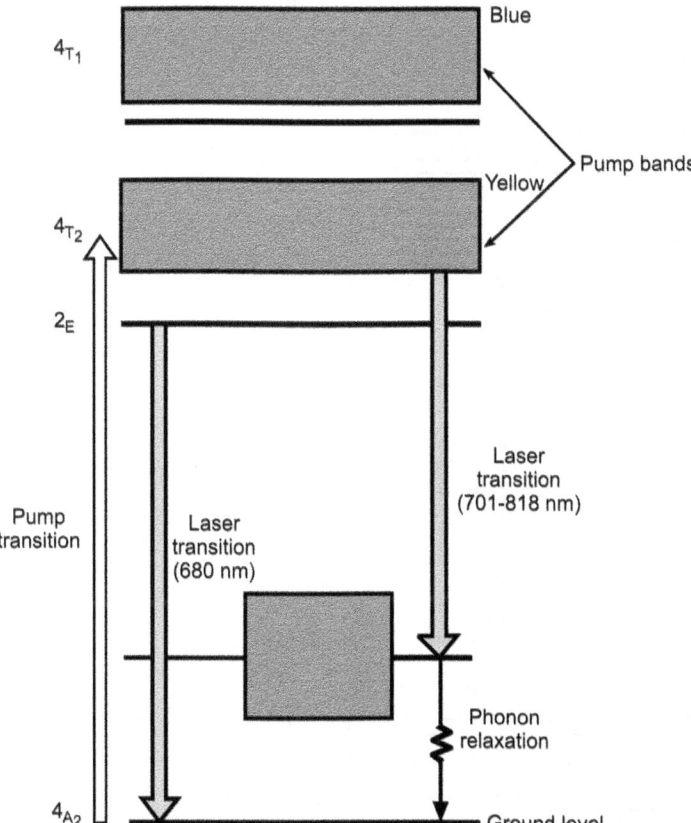

Figure 7.58 Energy level diagram of an Alexandrite laser.

7.7.6.2 Titanium-Sapphire Laser

The titanium-sapphire is the most widely used tunable solid-state laser because of its wide tunability and good material characteristics. Output wavelength is tunable from 660 to 1180 nm with the peak of the gain curve located around 800 nm. The host material, that is, sapphire, has very high thermal conductivity, mechanical rigidity and exceptionally high chemical inertness. Concentration of titanium ions in the laser crystal is about 0.1%.

Titanium-sapphire lasers cannot be efficiently pumped by flash lamps and the reason for this is their too short a fluorescence life time. It may be mentioned here that the population inversion required in a laser to exceed the lasing threshold is inversely proportional to the product of fluorescence life time and stimulated emission cross-section. An extremely small fluorescence life time of 3.2 μs in titanium-sapphire makes this product very small. The consequence is that a very high pump flux would be needed to achieve the threshold of lasing. Also, the absorption band has a peak around 500 nm, which makes it again very difficult to use laser diode pumping as 500 nm is too short a wavelength for diode lasers.

In view of these reasons, titanium-sapphire lasers are usually pumped by frequency-doubled neodymium lasers (Nd:YAG, Nd:YLF), argon-ion lasers and copper vapour lasers. These are usually pumped by argon-ion lasers to obtain CW output and by frequency-doubled Nd:YAG or Nd:YLF lasers and copper vapour lasers to obtain pulsed output. The neodymium lasers in turn may be diode-pumped. It may be mentioned here that strong emission of argon-ion lasers at 488 nm and 514.5 nm, frequency-doubled YAG lasers at 532 nm, frequency-doubled YLF lasers at 527 nm and copper vapour lasers at 510 nm match very well with the absorption band of titanium-sapphire. Figure 7.59 shows a simplified energy level diagram of a titanium-sapphire laser. Argon-ion laser pumped Ti-sapphire lasers are commercially

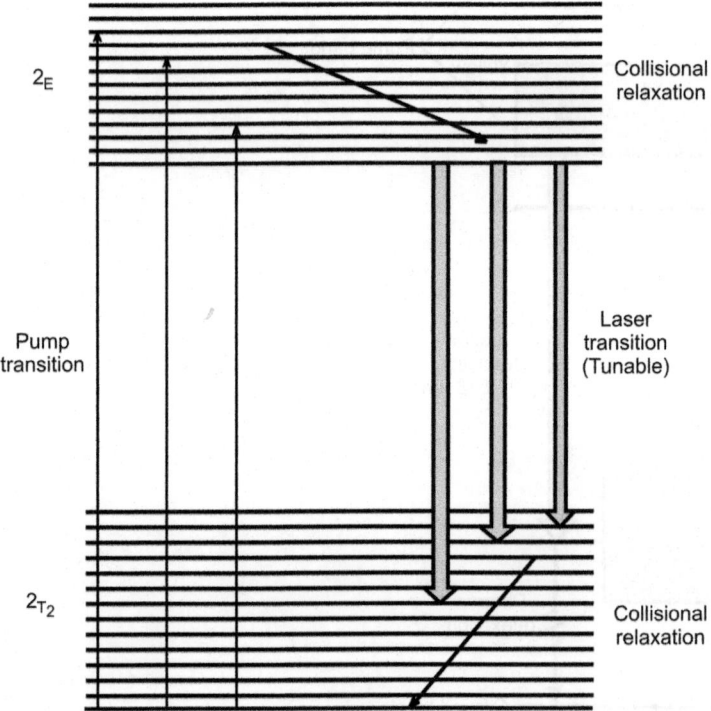

Figure 7.59 Simplified energy level diagram of a Ti-sapphire laser.

available producing few a watts of CW output, as well as frequency-doubled Nd:YAG or Nd:YLF laser pumped Ti-sapphire lasers producing several watts of CW power output or pulsed output of several millijoules at 1 kHz or in excess of 100 mJ at 20 PPS.

7.8 Fibre Lasers

The fibre laser is a type of solid-state laser where the gain medium is rare earth ion doped fibre rather than being in the form of a rod or slab. It is an alternative to bulk solid-state lasers in a wide range of industrial and military applications requiring high power levels with high beam quality in a compact and rugged package configuration. Inherent to the fibre-laser design and operational regimes are excellent performance characteristics, which include high level of immunity to mis-alignment, high beam quality, compactness and long term stability. In addition to these, fibre lasers exhibit outstanding thermo-optical properties arising out of large surface area-to-volume ratio. Advantages and limitations of fibre lasers are discussed in detail in the following paragraphs.

The gain medium in the case of fibre lasers is a glass fibre doped with rare earth element ions such as neodymium (Nd^{3+}), erbium (Er^{3+}), ytterbium (Yb^{3+}), thulium (Tm^{3+}), holium (Ho^{3+}) or praseodymium (Pr^{3+}). Fibre coupled semiconductor laser diodes or fibre lasers are used as the pump source and resonator comprises of active medium bounded by dielectric mirrors or fibre Bragg gratings. In the following paragraphs are described fibre lasers in terms of their principle of operation, comparison with bulk solid-state lasers, operational regimes and typical applications.

7.8.1 Basic Fibre Laser

In its simplest form, a fibre laser comprises of rare earth element ions doped fibre as the active medium, fibre coupled semiconductor diode laser or another fibre-laser as the pump source and dielectric mirrors or fibre Bragg gratings to form the resonant cavity. Figure 7.60 shows a simplified arrangement of different components of the basic fibre-laser. Both pump and laser radiation is guided through the waveguide structure constituted by the core and the cladding of the single clad fibre. Dielectric mirrors along with single clad fibre constitute the resonant cavity. In practical fibre lasers, in most cases, fibre Bragg gratings are used instead.

A fibre Bragg grating is type of distributed Bragg reflector constructed in a short segment of fibre as shown in Figure 7.61(a). The grating is created by a periodic variation of the refractive index of the core as illustrated in Figure 7.61(a). Figure 7.61(b) shows input, transmitted and reflected power spectra. The rare earth ion doped fibre core guides the light and this necessi-tates that the pump radiation is spatially coherent.

As the power available from single-mode semiconductor diode lasers is usually limited to few watts only, such a configuration cannot be used to build relatively higher output power lasers. This limitation is overcome by using a double clad fibre design. In this case, an active doped

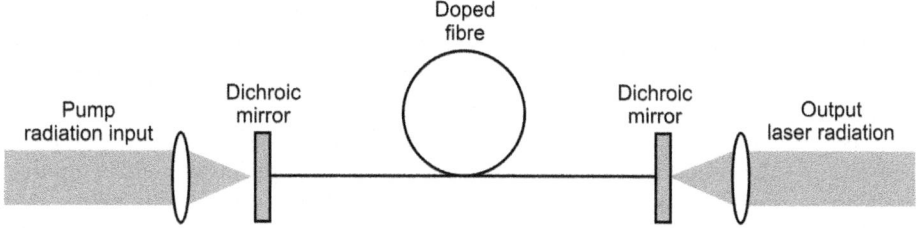

Figure 7.60 A basic fibre laser.

(a)

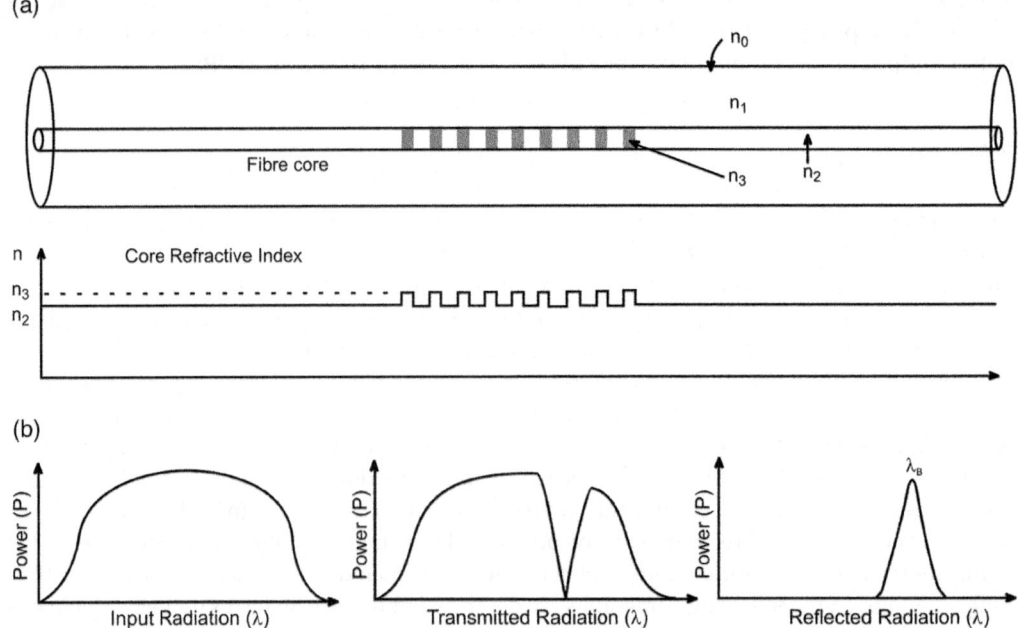

(b)

Figure 7.61 Principle of the fibre Bragg grating. (a) construction and refractive index variation in the grating (b) input, transmitted and reflected power spectra.

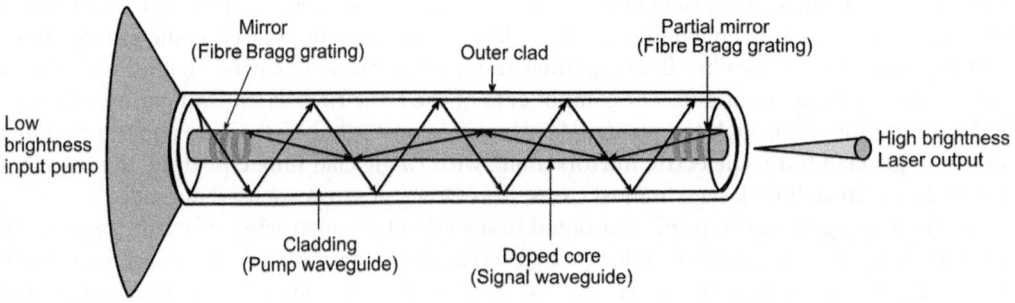

Figure 7.62 Basic double clad fibre laser.

core is surrounded by a second waveguide structure constituting the inner cladding also known as the pump core. Figure 7.62 shows the arrangement.

Double clad fibre-laser design allows the use of multimode semiconductor diode lasers as the pump source. Pump radiation in this case is launched in the inner cladding. The pump radiation is gradually absorbed over the entire length of the fibre, which with laser action is converted into single-mode high brightness laser radiation. Unlike bulk solid-state lasers where the intensity is limited to Rayleigh length by diffraction, in the case of the double clad fibre-laser the intensity is maintained over the entire fibre length due to confinement of both pump and laser radiation. Consequently, the gain of the active medium, which is defined by product of light intensity in the gain medium and the interaction length, is significantly higher than what it would be in the case of a bulk solid-state laser. This property gives fibre lasers an inherent high single pass gain and low pump threshold values.

(a)

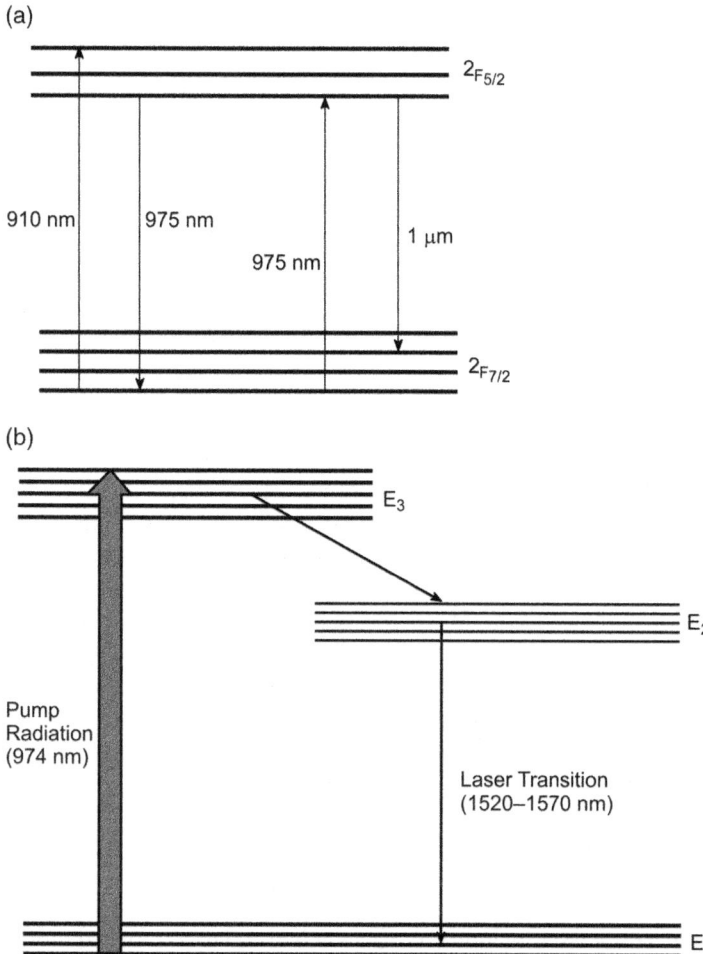

(b)

Figure 7.63 Energy level diagrams: (a) ytterbium and (b) erbium.

Wavelengths emitted by common fibre lasers are in the regions of 1.0–$1.1\,\mu m$ from ytterbium-doped lasers, 1.52–1.57 from erbium-doped lasers and 1.9–$2.1\,\mu m$ from thulium and holmium doped lasers. Figure 7.63(a) and (b) shows simplified energy level diagrams of ytterbium and erbium, respectively. The diagrams are self-explanatory. Ytterbium, because of a low quantum defect (less than 10%), is the dopant material of choice, particularly in high average power fibre lasers.

One drawback of the double clad fibre-laser concept is reduced absorption of pump radiation by the active ion core due to existence of intensity distributions in the inner cladding that has no overlap with the active medium. As a result of this, optical-to-optical efficiency reduces. This problem is overcome by breaking the cylindrical symmetry of inner cladding, usually done by using D-shaped or rectangular pump core geometries. Periodic bending when used in symmetrical fibres also enhances absorption of the pump radiation.

7.8.1.1 Fibre Lasers versus Bulk Solid-State Lasers

As compared to bulk solid-state lasers, fibre lasers exhibit relatively much *higher wall plug efficiency*. Wall plug efficiency approaching 50% is achievable in the case of fibre lasers against a figure of 20% to 30% for solid-state lasers. However, fibre lasers need to use pump diodes producing a higher beam quality than the beam quality required in the case of bulk solid-state lasers.

Fibre lasers produce a laser with a *higher beam quality* at higher average output power levels compared to bulk solid-state lasers. The fibre structure and the associated waveguide effect coupled with large surface area-to-volume ratio give fibre lasers excellent thermo-optical properties. As a result, fibre lasers are able to produce high output power levels with high beam quality. Though both fibre and bulk solid-state lasers can generate multi-kilowatt power levels, it is far more difficult in the case of latter to achieve the beam quality of the former for a given output power level.

Fibre lasers inherently offer a relatively *higher gain bandwidth*. The glass host broadens the optical transition in rare earth ion dopant. This makes fibre lasers continuously tunable in the near infrared spectral region from 1–2 μm by using a suitable dopant from ytterbium, erbium and thulium. Higher gain bandwidth also allows fibre lasers to generate ultra-short pulses by using passive mode locking, though sometimes it is not feasible to fully exploit their large gain bandwidth to achieve the shortest possible pulse width due to excessive nonlinearity.

Fibre lasers have *broad absorption bands* with good absorption, which makes the pumping process less critical in terms of pump wavelength. This further implies that the pump diode lasers could possibly be used without temperature stabilization. All fibre lasers, due to absence of any free-space optics are immune to any misalignment and are more robust and exhibit excellent long term stability. While fibre lasers offer far superior performance when it comes to high average power lasers with high beam quality, it is the bulk solid-state lasers that have an edge in high pulse energy, high peak power applications.

Fibre lasers suffer from excessive Kerr nonlinearity arising from the long length and small mode area of the fibre. Nonlinearity often limits the shortest achievable pulse width and single frequency operation of the laser. Nonlinear effects are much less severe in the case of bulk solid-state lasers.

7.8.2 Operational Regimes

Common operational regimes of a fibre laser include CW, Q-switched and mode-locked modes of operation. The *CW* mode of operation of a fibre laser with output power levels extending from milliwatts to kilowatts is the most sought after variety in a wide range of industrial and military applications. The *Q-switched* mode of operation can be used to generate pulse energies in the range of several millijoules to several tens of millijoules and pulse widths in the range of several tens of nanoseconds to hundreds of nanoseconds. The *mode-locked regime* is used to generate ultra-short pulses with pulse widths of the order of tens to hundreds of femtoseconds. Both active as well as passive mode locking is possible. Another operational regime that is attractive in the case of fibre lasers is the *up-conversion mode* of operation in which lasing action is achieved in difficult-to-operate lasing transitions.

Single frequency low gain operation is also possible in fibre lasers by using short length resonator with narrow bandwidth fibre Bragg gratings as reflectors on the two ends. These single frequency fibre lasers can be used to typically generate CW power levels of few milliwatts to few tens of milliwatts. A power level approaching a watt has also been achieved. Single frequency operation is also possible using distributed feedback (DFB) laser concept. In a DFB laser, a single fibre Bragg grating with a phase shift in the middle is used.

In the distributed Bragg reflector (DBR) based single frequency laser design, as shown in Figure 7.64, a short cavity length in the order of few centimetres is used between two narrow bandwidth fibre Bragg gratings with a spectral line width of a fraction of a nanometre. Short cavity length ensures a longer separation between adjacent longitudinal modes forcing only a

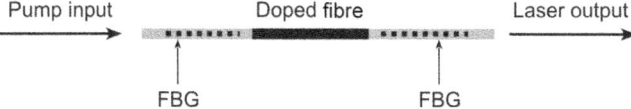

Figure 7.64 DBR based single frequency fibre laser design.

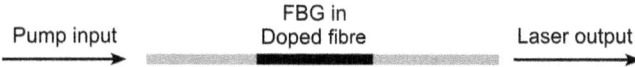

Figure 7.65 DFB laser based single frequency fibre laser design.

single longitudinal mode to propagate. Shorter spectral line width of fibre Bragg grating reflectors ensures that only a single longitudinal mode experiences gain above lasing threshold.

In the case of DFB laser design, as shown in Figure 7.65, a longer fibre Bragg grating with a $\pi/2$ phase jump is used to construct the resonator. DFB operational mode due to its precise phase control over the long FBG combined with $\pi/2$ phase jump gives a relatively much lower spectral line width and more single frequency operation.

7.8.3 Photonic Crystal Fibre Lasers

Photonic crystal fibre lasers use a special type of fibre called photonic crystal fibre (PCF) that is different from the conventional step index fibre. In comparison to the conventional step index fibre that derives its wave guiding properties from a spatially varying glass composition, the basic photonic crystal fibre is an optical fibre with a structured pattern of air holes also called voids that run parallel to the axis all along its length (Figure 7.66). Unlike conventional fibres, in the case of PCF, both core and cladding are made from the same material. Most PCFs are made from pure fused silica, which is highly compatible with PCF fabrication techniques though PCFs have also been made of other materials such as heavy metal soft glasses and polymers. The PCF also derives its wave guiding properties from the presence of air holes. One of the methods of fabrication of PCFs known as the stacked tube technique involves using a preform with larger holes made by stacking capillary and/or solid tubes and inserting them into

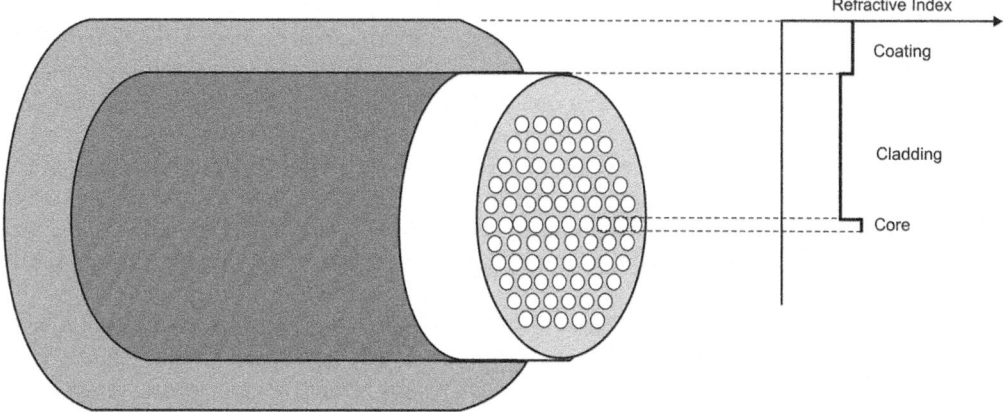

Figure 7.66 Construction of photonic crystal fibre.

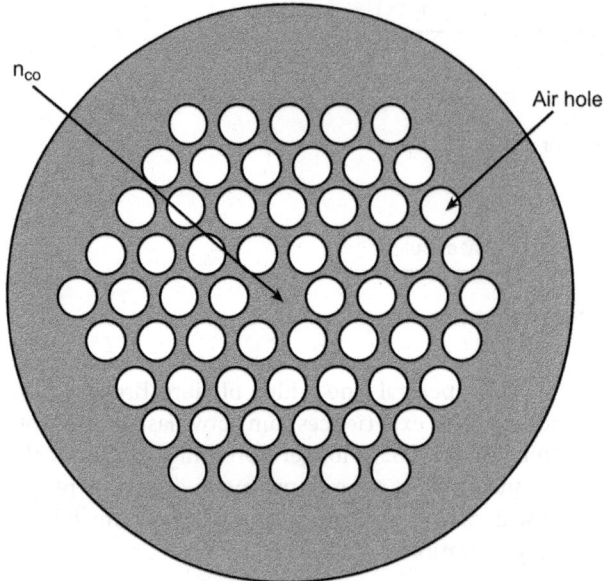

Figure 7.67 Basic photonic crystal fibre.

n_{co}

Air hole

a larger tube. The preform is first drawn to a diameter of approximately 1 mm, which is subsequently further drawn into a fibre with a final diameter in the order of 100 μm or so.

The most common type of PCF has a hexagonal array of air holes with one or more missing holes in the centre forming the core (Figure 7.67). The region with missing hole has a higher refractive index analogous to the core of a conventional step index fibre.

There are two distinct mechanisms of guiding light in the case of PCF. These include *photonic band gap guiding* and *index guiding*. In the case of the photonic band gap guiding mechanism, the wavelengths falling within the band gap of the photonic crystal cannot propagate out and are therefore confined to the core. Index guiding mechanism is similar to conventional step index fibre operation.

A typical index-guided PCF has a solid core and a structured pattern of air holes surrounding it. Presence of air holes reduces refractive index. In comparison to conventional step index fibre, an index-guided PCF structure offers a much finer and accurate control over magnitude of refractive index. Guiding mechanism is due to total internal reflection.

Conventional high power fibre lasers use double clad step index fibres with a polymer outer cladding and a core doped with rare earth ions such as ytterbium or erbium. This configuration has an upper limit to achievable pump and output power levels while maintaining a single-mode output. This is due to the fact that operation at single transverse mode necessitates use of small core size of the order of few microns resulting in higher power densities at high pump and output power levels. High power density further leads to detrimental nonlinear effects though these effects can be minimized to an extent by increasing the core size while still maintaining a single-mode output. The techniques used include controlling index and initial excitation profiles and introducing micro bending losses to preferentially inhibit propagation of higher-order modes.

Dual clad fibre-laser architecture based on PCF technology provides a good alternative to scale the output power to higher levels. In the double clad PCFs (Figure 7.68), the pump cladding is surrounded by an air cladding region. Due to a large refractive index contrast, the pump cladding has a very high NA, typically in the range of 0.6–0.7. A large NA significantly lowers the pump source requirements with respect to beam quality and brightness and permits

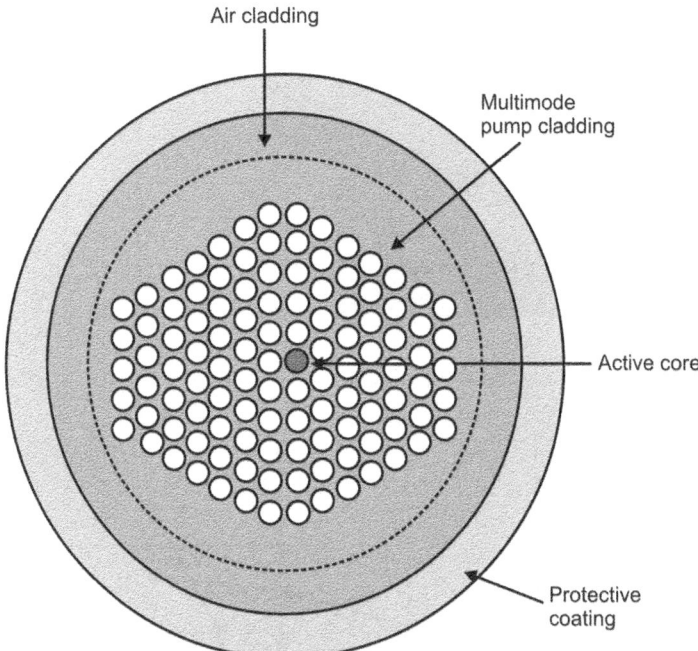

Figure 7.68 Double clad PCF structure.

efficient pumping with relatively lower cost large emitting area pump diodes. Within the inner cladding is another micro structured rare earth doped core. The single-mode core can be expanded to a large mode area to facilitate operation at higher power levels with single-mode output. With such PCF designs, problems due to nonlinearity are substantially reduced and there is a good overlap between the pump mode and laser mode.

7.8.4   Applications

High power fibre lasers with average power levels in the range of tens of watts to multi-kilo-watt levels are finding use in a wide range of industrial and military applications. Figure 7.69 shows the photograph of a CW fibre-laser from M/s SPI lasers, UK. The PRISM series lasers produce output at 1080 nm and are capable of generating CW output power up to 1500 W. The lasers are intended for a wide range of industrial and medical/biomedical applications. Up until the 1990s, telecommunications industry was the largest consumer of fibre lasers. Fibre-laser, in particular the high power fibre-laser growth has immensely benefited from technological developments during the days of telecommunications industry expansion. Many of the components including the single emitter fibre coupled pump diodes used in high power fibre lasers today were developed during the 1980s and 1990s. Salient features of fibre lasers and advantages they offer with respect to bulk solid-state lasers have already been discussed in an earlier section. Power scalability while retaining the beam quality in the case of fibre lasers has been the key factor in their increasing popularity and acceptance over carbon dioxide and bulk solid-state lasers used earlier in majority of industrial and military applications. While the highest achieved power level in the case of carbon dioxide and neodymium-doped YAG lasers became saturated after the year 2000, fibre lasers continue to evolve with single-mode fibre-laser achieving 10 kW of power level and broad band multimode fibre

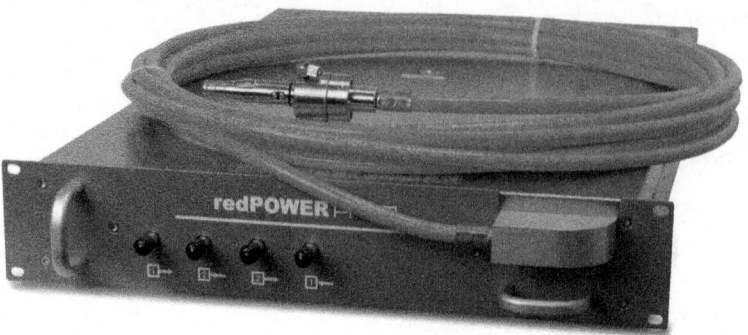

Figure 7.69 CW fibre laser type R4. (Source: Courtesy SPI Lasers, UK.)

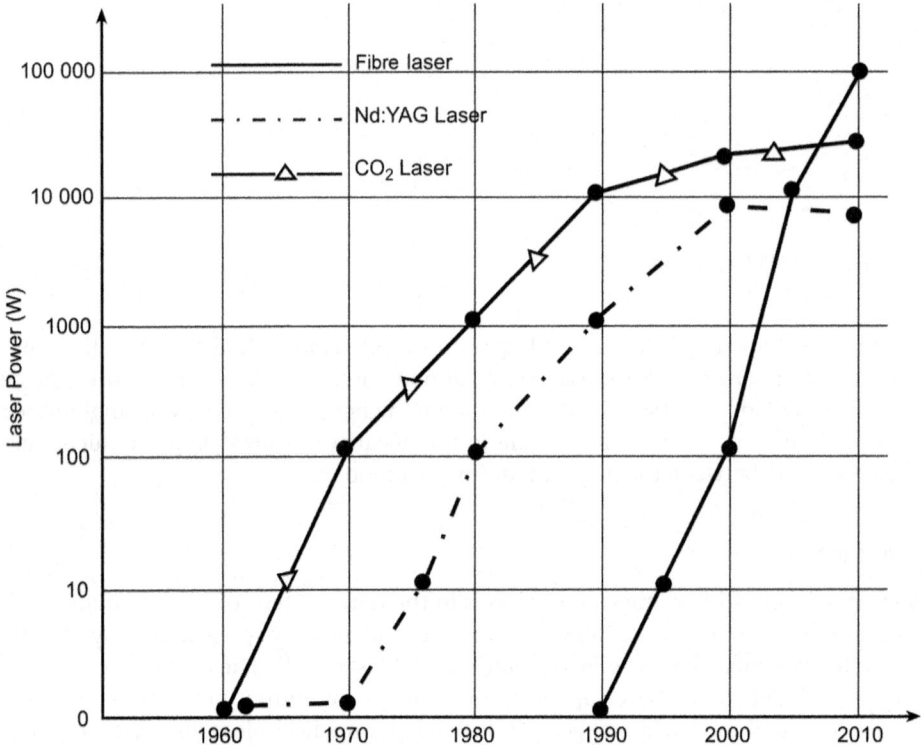

Figure 7.70 Power levels achieved in the Nd:YAG, CO_2 and fibre lasers.

lasers scaled up to a power level of 50 kW. Figure 7.70 shows the power levels achieved in Nd:YAG, CO_2 and fibre lasers over the last six decades. The power level is continuing to grow for fibre lasers.

High power pulsed and CW fibre lasers are fast replacing CO_2 and Nd:YAG lasers for a wide range of industrial applications in automotive, aerospace and medical industries where they are mostly used for cutting, welding and cladding operations. Single-mode fibre lasers with moderate power levels in the range of tens of watts to hundreds of watts are ideally suited to micro welding applications and for cutting thin materials. Fibre lasers with average power levels of few hundreds of watts and peak power levels of few kilowatts are very attractive for host of industrial applications including cutting, welding, marking, engraving, sintering and soldering. Single-mode and also low-order multimode fibre lasers with power level of 1 kW or more have found wide acceptance in a range of material processing applications due to their more reliable and consistent performance, low operating and maintenance costs and faster rate of materials processing.

One of the major applications of high power fibre lasers is in directed-energy weapon applications in defence. High power at high beam quality, compactness, ruggedness, reliability and fibre delivery are the key features that give them an edge over chemical lasers and bulk solid-state lasers when it comes to military applications. Fibre lasers have already been used in a number of directed-energy weapon technology demonstrators for a range of applications including counter explosive devices, counter RAM (Rocket Artillery Mortar) and counter UAV (Unmanned Aerial Vehicle) applications. In future, directed-energy laser weapons for ballistic missile defence and anti-satellite applications cannot be ruled out.

7.9 Gas Lasers

The family of gas lasers, unlike lasers belonging to the other two major families of solid-state and semiconductor diode lasers, has widely varying characteristics including the wavelength range, power levels and to some extent even the pumping mechanism. The available output powers from gas lasers vary from a fraction of a milliwatt in a small helium-neon laser used for optical alignment to megawatt level in a gigantic high power chemical laser used as a weapon. The wavelength range also spans almost the entire optical spectrum from UV to far IR with thousands of laser wavelengths discovered throughout the region. Another significant feature that is common to all gas lasers is that they have Doppler broadened gain versus frequency curve. A practical implication of this is that it allows more than one longitudinal mode to simultaneously oscillate in the laser cavity unless special measures are taken for mode selection. As an illustration, a helium-neon laser emitting at 632.8 nm has a Doppler broadened gain curve with FWHM (Full Width at Half Maximum) of 1400 MHz. One longitudinal mode in this case typically has 1 MHz bandwidth with the result that the gain curve may accommodate three longitudinal modes assuming a 500 MHz inter-mode spacing, which is the case for a 30 cm long laser cavity. Doppler broadening occurs in gas lasers due to random motion of atoms. Such a broadening mechanism is, however, absent in solid-state and semiconductor diode lasers as the atoms are fixed in these lasers

7.9.1 The Active Media

The active medium in the case of a gas laser is almost invariably a mixture of more than one gases with the gases other than the actual lasing species performing certain subtle functions like acting as an intermediate step during transfer of energy from pump source to lasing species (helium in helium-neon lasers and nitrogen in carbon dioxide lasers), assisting in heat transfer (helium in carbon dioxide lasers) and depopulating the lower lasing level (helium in helium-neon laser). The gas mixture fills a tube at a pressure that again depends upon a number of parameters. Low pressure in the order of a small fraction of atmospheric pressure suitable for longer periods of discharge is mostly the case in CW lasers. In the case of pulsed lasers where discharge stability is required for a shorter period, the laser gas mixture could be filled at a pressure close to atmospheric pressure and sometimes in excess of 1 atm. The optimum gas pressure for lasers of a given type also depends upon the laser design.

The active media of different gas lasers may not be in the same form. For example, in the case of argon-ion and krypton-ion lasers, it is the ionized atoms of rare gases argon and krypton, respectively. In the case of metal vapour lasers such as copper vapour and gold vapour lasers, the active medium is the hot metal vapour. And in yet another case of helium-cadmium laser, the metal vapour is ionized as well.

Inter-level transitions in the case of most gas lasers are electronic except for carbon dioxide lasers, hydrogen fluoride (HF) lasers, deuterium fluoride (DF) lasers and carbon monoxide lasers that involve vibrational transitions. Some far infrared lasers producing wavelengths greater than 30 μm have vibrational or rotational transitions.

7.9.2 Pumping Mechanism

Most of the gas lasers are excited by electrical discharge as shown in the generalized arrangement of a gas laser in Figure 7.71. The active medium is usually excited either by passing an electric discharge current along the length of the tube known as longitudinal excitation, as shown in Figure 7.71, or by an electric discharge perpendicular to the length of the laser tube known as transverse excitation, as shown in Figure 7.72. The former is used in relatively low power CW lasers while the latter is employed in high power pulsed or CW lasers.

In the case of CW lasers, a high DC voltage is initially required to ionize the gas. Once the ionization takes place, the DC voltage is brought to a much lower value needed to sustain the plasma. In the case of a pulsed laser, a hefty capacitor is charged to the required DC voltage and then made to discharge through the laser medium. Some gas lasers like those generating far IR wavelengths are optically pumped. Another gas laser of shorter wavelength such as a carbon dioxide laser is used as the pump laser.

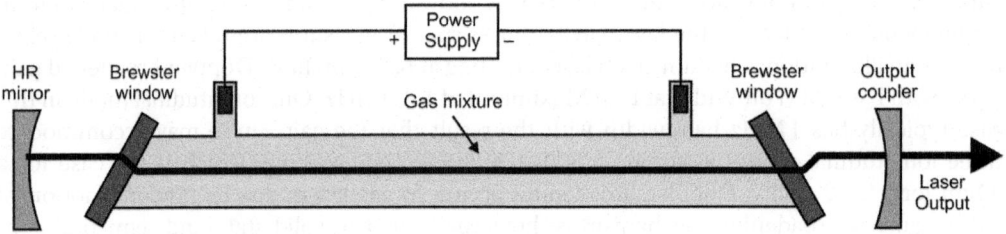

Figure 7.71 Generic gas laser: longitudinal excitation.

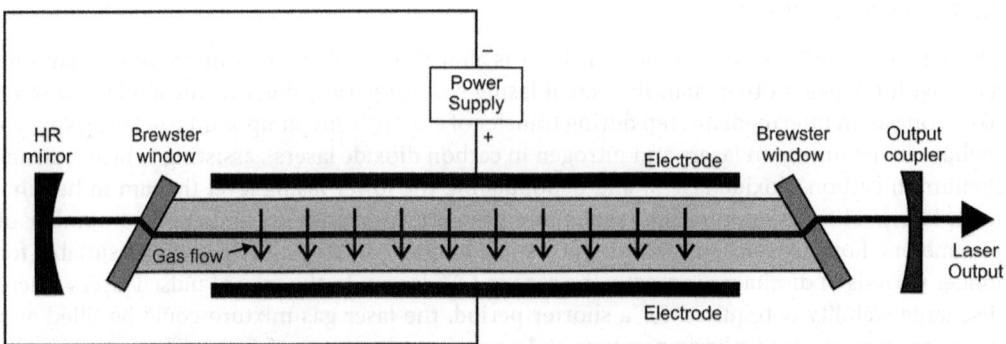

Figure 7.72 Generic gas laser: transverse excitation.

7.9.3 Helium-Neon Lasers

A helium-neon (He-Ne) laser is one of the most commonly used types of gas lasers. Though we are more familiar with the red beam from a He-Ne laser at 632.8 nm, lasing action is also possible at infrared wavelengths of 1.153 μm, 3.391 μm and visible wavelength of 543.5 nm. In fact, first successful operation of a CW laser was achieved in a He-Ne laser at 3.391 μm by Javan Benett and Herriott at Bell Labs following the first ever lasing action demonstrated by Maiman in a ruby laser. Incidentally, He-Ne lasers are usually identified by a red output; the lasing transition in He-Ne that that has the highest gain is the one at 3.391 μm. Figure 7.73 shows the energy level structure of a He-Ne laser. The active medium is the laser gas mixture of helium and neon, which is predominantly helium with only 10–20% of neon. The gas mixture is excited by an electrical discharge. Initially, a high voltage in the order of 10 kV ionizes the gas mixture. The voltage is then brought down to about 2 kV or so, which is sufficient to sustain the discharge by supplying a few milli-amperes of current required by it. The free electrons in the plasma excite helium atoms to higher energy levels. The excited helium atoms then transfer their energy to neon atoms, which have energy levels (4s and 5s levels of neon) coincident with excited helium levels (2s levels of helium) as shown in the figure. The transfer of energy is through a collisional process. 4s and 5s levels of neon are metastable levels having a relatively much larger life time, which facilitates laser action. The metastable life time here is in the order of 1 ms and the laser action takes place between metastable levels (4s or 5s of neon) and the lower lasing levels (3p or 4p of neon).

As outlined earlier, a He-Ne laser can possibly produce laser output at either of the visible (543.5 nm or 632.8 nm) or IR (1.153 μm or 3.391 μm) wavelengths depending upon the operating conditions and choice of optics. A value of 632.8 nm has become the standard He-Ne laser wavelength and is also the most widely used one.

A He-Ne laser has a low gain. Therefore, cavity losses need to be very low to make the laser action possible. Efficiency is also very low in the range of 0.01–0.1% due to the involved lasing levels being far above the ground state as is clear from the energy level diagram of Figure 7.73.

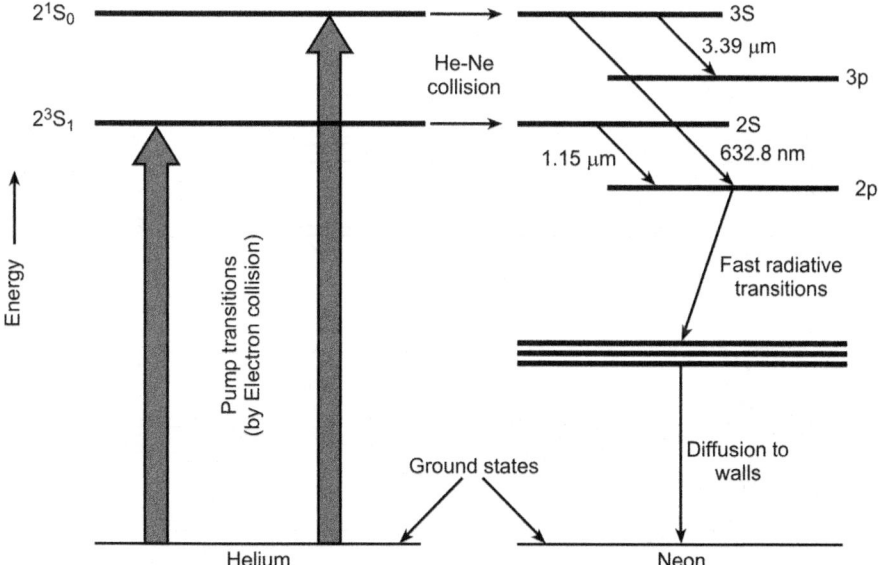

Figure 7.73 Energy level diagram of a helium-neon laser.

A He-Ne laser has a Doppler broadened gain curve, which is about 1400 MHz at 632.8 nm. Doppler broadening is proportional to frequency. Superimposed on the Doppler broadened gain curve is the cavity resonance function with inter-longitudinal mode spacing given by $\Delta D = (c/2L)$, L being the length of the cavity. A short cavity, let us say 15 cm long, would support only a single longitudinal mode as in that case the inter-mode spacing would be 1000 MHz A longer cavity opens up the possibility of more than one longitudinal mode sustaining simultaneously in the cavity. The uncertainty in frequency of single longitudinal mode would be dictated by the Doppler broadened line width due to temperature changes causing cavity length variation. The same phenomenon causes power fluctuation in the output beam. However, the single longitudinal mode could be forced to occur at the centre of the gain curve and stay there irrespective of temperature and other factors causing cavity length variation by employing what is known as active means of frequency stabilization. These are discussed at length in Chapter 8.

7.9.3.1 Constructional Features

Figure 7.74(a) and (b) shows the constructional features of a typical helium-neon laser. In one of the geometries, as shown in Figure 7.74(a), mirrors are directly bonded to the two ends of the tube by a high-temperature process called hard sealing. This minimizes helium leak, which is so vital for the long life of the laser. One of the disadvantages of direct bonding of laser mirrors is that it exposes the mirror coatings to the discharge, which could adversely affect the life of the laser. This could, however, be overcome by using mirrors with high damage threshold and an improved laser geometry. In another type, as shown in Figure 7.74(b), one of the ends of the tube is sealed with a Brewster window, which practically eliminates any reflection losses for the plane polarized light. In plane polarized light, the plane of polarization is parallel to the plane of incidence. The other mirror component, output coupler or the high reflectivity mirror in this case, is placed outside the tube. It may or may not be part of the overall housing. This proposition turns out to be more expensive but is definitely a better option for relatively high power helium-neon lasers producing output power in the range of 10–60 mW. Also, it makes possible selection of different wavelengths. Lower power lasers invariably employ the first type of construction.

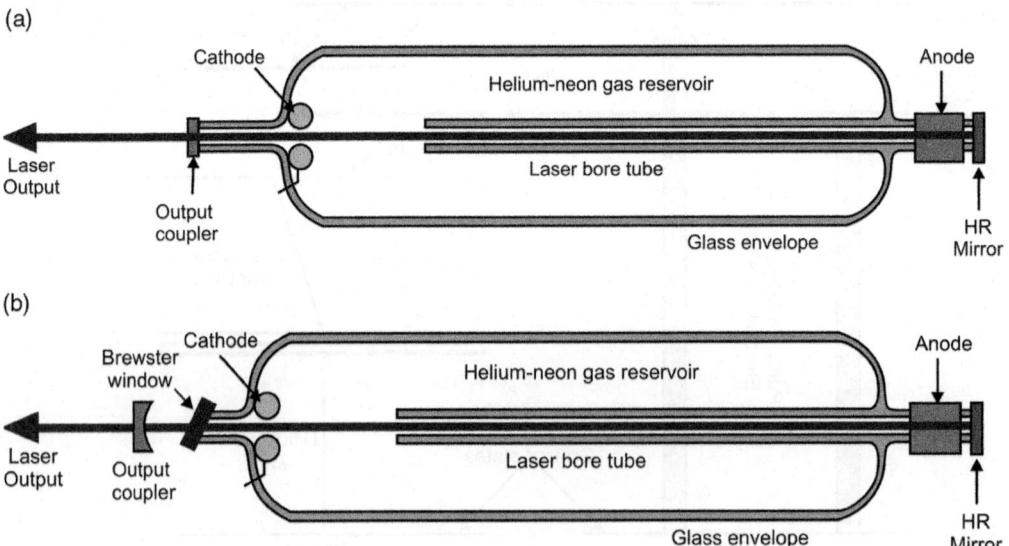

Figure 7.74 Helium-neon laser construction: (a) both mirrors sealed and (b) Brewster window sealed.

Coming back to He-Ne laser tube construction, the mirrors need to have low losses to compensate for the low gain of the He-Ne laser. One of the mirrors is 100% reflective. The output mirror transmits a few percent of irradiance inside the cavity and reflects most of it back into the laser cavity. The output mirror has a transmission of typically 2%. Another notable feature of laser tube construction is that the electric discharge applied to the ends of the tube is concentrated in a narrow bore of a few millimetres in diameter. This raises laser excitation efficiency and improves beam quality. Rest of the tube volume acts as a gas reservoir and it contains extra helium and neon. Gas pressure within the tube is about 2 to 3 torr, which is 0.3–0.4% of the atmospheric pressure.

It may be mentioned here that the power output of a helium-neon laser depends upon several parameters, of which the tube length, gas pressure and the discharge bore diameter are the predominant ones. In fact, product of gas pressure (in torr) and bore diameter (in mm) is the figure of merit and its optimum value is 3.5–4. That is, the output power for a given tube length is the highest when the figure of-merit equals approximately 3.5–4. As an illustration, He-Ne laser tubes with a bore diameter in the range of 1.0–1.5 mm have a gas fill pressure in the order of 3.0–4.0 torr.

7.9.3.2 Commercial He-Ne Lasers

He-Ne lasers are commercially available as both sealed-off tubes and packaged types. The package types are further available as lasers with a separate power supply as well as self-contained lasers for OEM applications. Figure 7.75 shows photographs of the different types of He-Ne laser including self-contained lasers suited to OEM applications [Figure 7.75(a)] and

(a)

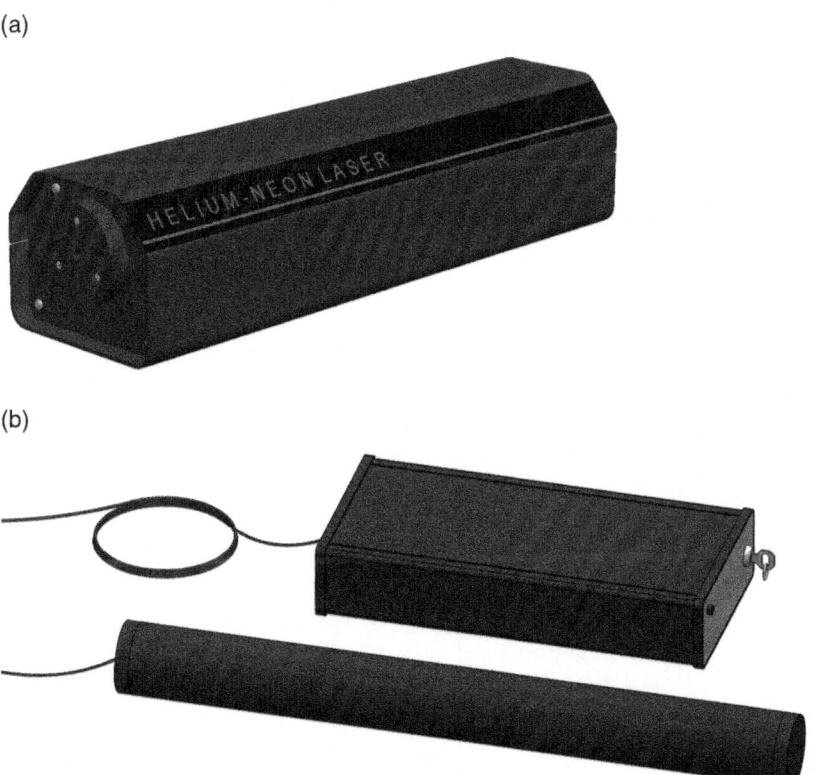

(b)

Figure 7.75 Helium-neon lasers. (a) Self-contained laser and (b) one with its own power supply.

those available with separate power supplies [Figure 7.75(b)]. He-Ne lasers are increasingly being replaced by semiconductor diode lasers, particularly in those applications where the high beam quality of a helium-neon laser is not a necessity.

7.9.4 Carbon Dioxide Lasers

Among the family of gas lasers, the carbon dioxide (CO_2) laser is the most widely used and diversely exploited type of gas laser. The laser finds a myriad of applications in industry, medical diagnosis and treatment, science and technology and warfare.

The laser medium in this case is a gas mixture of CO_2, helium (He) and nitrogen (N_2). CO_2 is the lasing species and laser transitions correspond to the energy levels associated with molecular vibration and rotation modes of CO_2 molecule. Nitrogen participates in the process of creation of population inversion by acting as an intermediate step in the same way as helium does in the case of He-Ne laser. Helium in the case of a CO_2 laser helps in depopulating the lower laser level.

Unlike the gas lasers such as the He-Ne laser, the energy levels responsible for laser action in a CO_2 laser do not correspond to excitation and de-excitation of electrons. Instead they correspond to vibrational/rotational levels of CO_2 molecule. The CO_2 molecule has three types of vibrations; namely asymmetric stretching, symmetric stretching and bending written in descending order of energy level. The free electrons in the gas discharge plasma transfer their energy very efficiently to N_2 molecules, which raise themselves to an appropriate level coincident with CO_2 energy level corresponding to asymmetric stretching. Figure 7.76 shows the energy level diagram in the case of a CO_2 laser. The N_2 molecules transfer the energy to CO_2

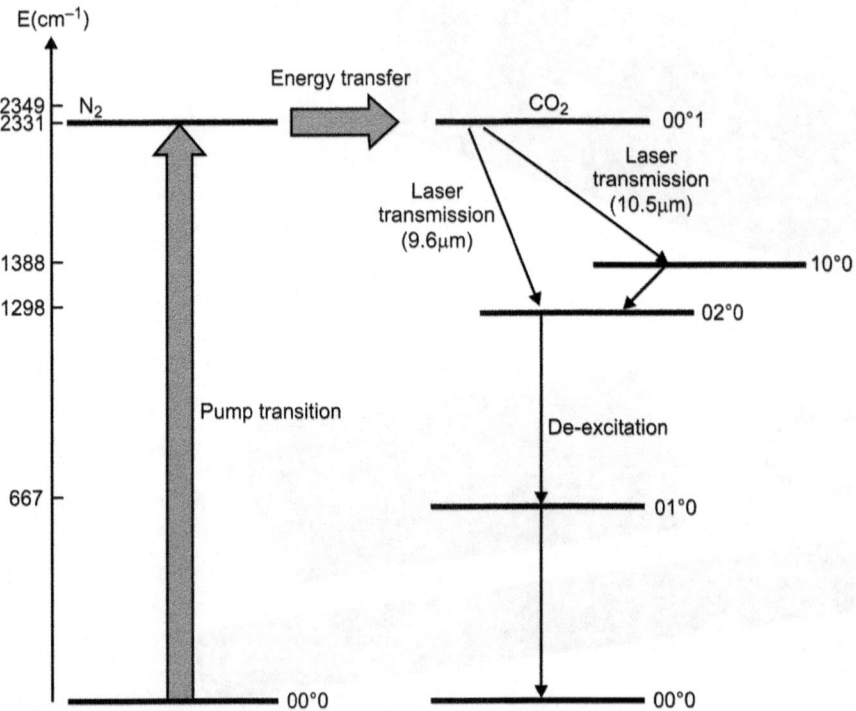

Figure 7.76 Energy level diagram of a CO_2 laser.

molecules. This is the upper lasing level. Laser transitions correspond to the CO_2 molecules dropping from higher energy asymmetric stretching mode to the lower energy symmetric stretching or bending modes. Transition to a lower level corresponding to symmetric stretching produces a 10.6 μm output while its transition to another lower level corresponding to symmetric bending produces a 9.6 μm output.

In fact, around both 9.6 μm and 10.6 μm outputs, there are a large number of closely spaced lines. This is the outcome of rotational motion of the molecule. The rotational transition energy is, however, much smaller than either the thermal or the vibrational transition energy. The molecule can either speed up or slow down a little in its rotational motion while moving between the vibrational levels. When it speeds up, it takes energy from the vibrational transition and when it slows down, it gives energy to the vibrational transition. In the two cases, the effective transition energy either increases (in the case of slowing down of rotational motion) or decreases (in the case of speeding up of rotational motion) thereby decreasing or increasing the emitted wavelength, respectively, around the nominal value.

A CO_2 laser too has a Doppler broadened gain curve that is 60 MHz wide at 10.6 μm output. The frequency uncertainty in the case of an unstabilized laser could be as much as close to 60 MHz. The laser could, however, be frequency stabilized either on the centre of gain curve or anywhere on the gain curve using some of the established frequency stabilization techniques to an accuracy of better than ±100 kHz. Some techniques allow the laser to be stabilized down to ±10 kHz, thus making it an ideal tool for carrying out Doppler free spectroscopic studies. Frequency stabilization techniques are discussed at length in later part of the book in Chapter 8.

7.9.4.1 Construction

CO_2 lasers employ either a sealed tube or flowing gas construction. The gas mixture is excited by an electric discharge. Both DC as well as RF excitation has been successfully and widely used in the case of CO_2 lasers. As the electric discharge breaks down CO_2 molecules to form carbon monoxide and oxygen, a catalyst is added to regenerate CO_2. Figure 7.77 shows the photograph of a sealed-off CO_2 laser of 48-series from M/s Synrad Inc. The laser employs robust sealed-off construction. Three different models in the series produce CW output power levels of 10, 25 and 50 W with output beam divergence of 4 mrad and mode quality factor M^2 that is better than 1.2.

Figure 7.77 Sealed-off CO_2 laser.

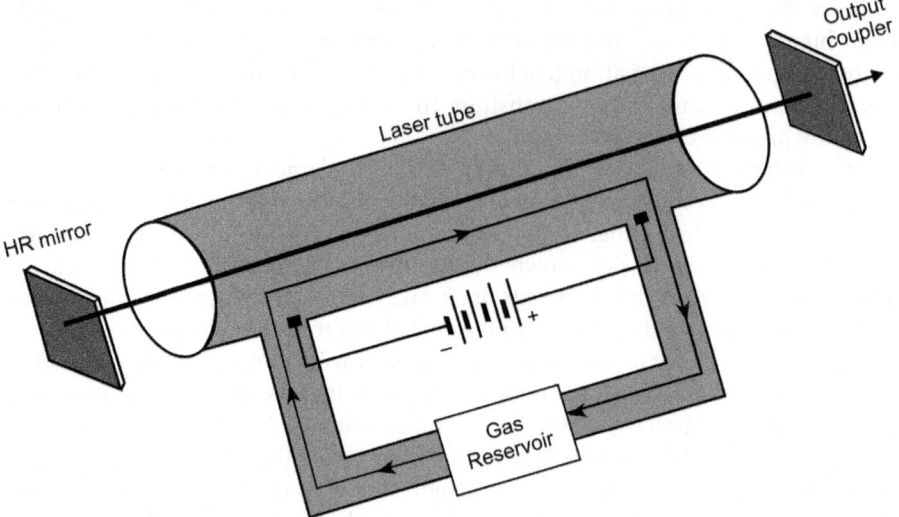

Figure 7.78 Longitudinal gas flow CO_2 laser.

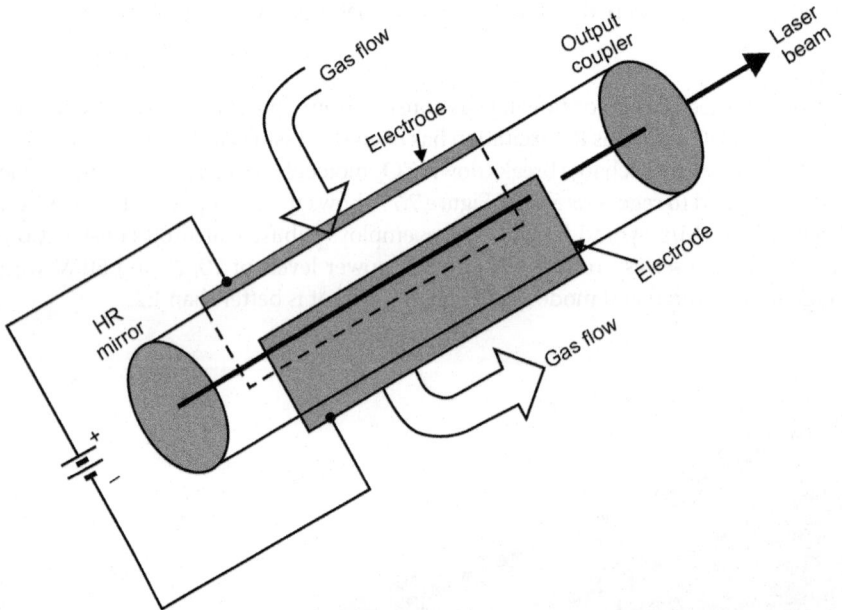

Figure 7.79 Transverse gas flow CO_2 laser.

Flowing gas type of CO_2 lasers may be of longitudinal flow or transversal flow type. In the case of former, the gas flows along the axis of the cavity while in the case of latter, it flows perpendicular to the cavity axis. In longitudinal flow lasers (Figure 7.78), the gas pressure is low, the output power is also relatively lower and lasers operate as CW lasers. Transverse flow lasers (Figure 7.79) are normally used to get higher power outputs. Gas pressures in these lasers can be higher. In the case of longitudinal flow lasers, the gas is usually recycled with some fresh gas added. In the case of transverse flow lasers, the gas flow through the laser cavity is much faster, which removes waste heat and also the contaminants. Pressure broadening of the line width also allows for mode-locked operation thus making possible generation of pulses as short as 1 ns. For pressures approaching 10 atm, the line broadening is large enough to allow near-continuous tuning across most of the wavelength range. Transverse flow is usually used in relatively higher output power lasers. While in the case of sealed-off and longitudinal flow CO_2 lasers, as a rule of thumb, power in excess of 50 W per metre of gain length is achievable; the same figure in the case of transverse flow lasers could be as high as 10 kW per metre.

7.9.4.2 TEA CO_2 Laser

Another type of CO_2 laser is one that is transversely excited and the gas pressure is about 1 atm. The CO_2 lasers described in the earlier paragraphs have low gas pressures and usually produce CW output. CW lasers cannot have high-pressure gas mixture as it is not practical to have a stable continuous discharge at pressures above about 1/10 atm. Pulsed electric discharge is, however, possible at higher pressures and it works very well if the discharge is transverse to the laser axis. That is what is done in a Transversely Excited Atmospheric (TEA) pressure CO_2 laser (Figure 7.80). A transversely excited atmospheric pressure (TEA) CO_2 laser is a high power pulsed CO_2 laser. The gas pressure is around 1 atm and the discharge is transverse to the laser axis. TEA CO_2 lasers invariably use a form of pre-ionization to uniformly ionize the space between the electrodes. The primary attraction of a TEA CO_2 laser is in its ability to generate short intense pulses and extraction of high power per unit volume of laser gas mixture. Pulse durations of few tens of nanoseconds to few microseconds and pulse energies from a few millijoules to hundreds of joules at repetition rates up to a few hundred hertz are achievable.

A waveguide CO_2 laser is another type of CO_2 laser that renders compactness. RF excited sealed-off waveguide CO_2 lasers producing several watts to several tens of watts of CW output hold lot of promise. Figure 7.81 shows the picture of one such sealed-off RF excited waveguide CO_2 laser, Model No. v30 from M/s Synrad Inc. The laser produces 30 W of CW output with a beam divergence and beam quality factor M^2 of better than 7 mrad and 1.2, respectively.

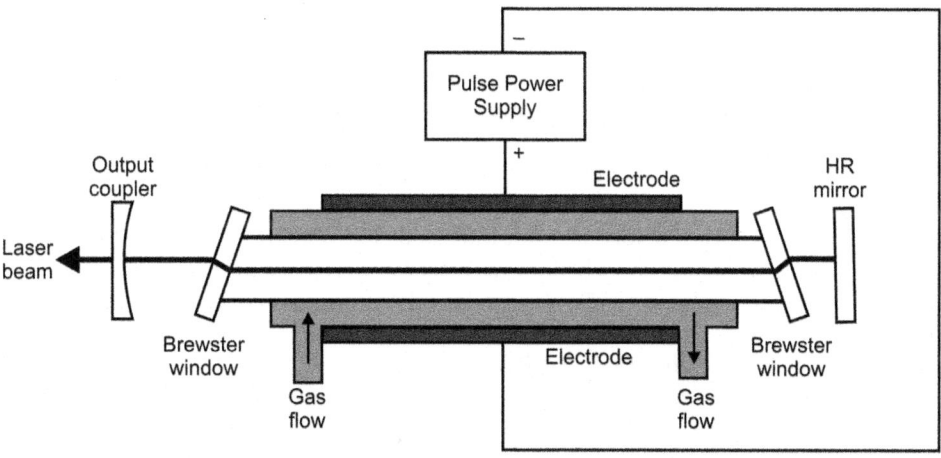

Figure 7.80 TEA CO_2 laser.

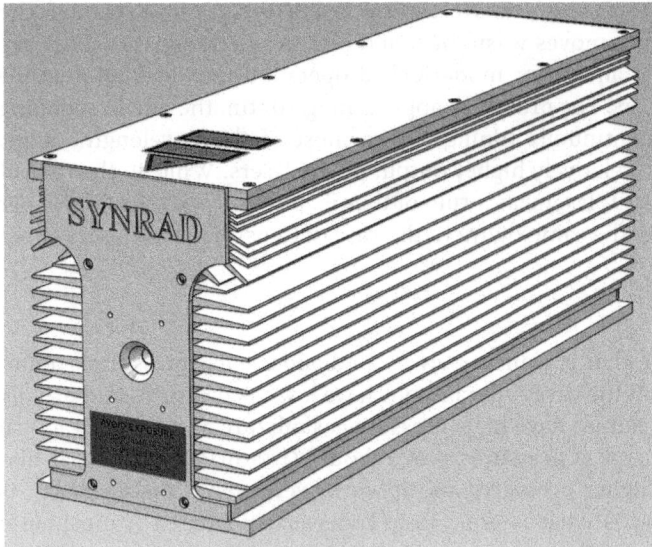

Figure 7.81 RF excited waveguide carbon dioxide laser.

7.9.5 Metal Vapour Lasers

The active medium in the case of metal vapour lasers is either ionized metal atom as is the case in helium-cadmium family of lasers or neutral atoms as in the case of copper vapour and gold vapour lasers. The helium-cadmium lasers produce emissions at 325 nm and 353.6 nm (ultraviolet) and 441.6 nm (blue). The neutral metal vapour lasers emit in the visible spectrum. Copper vapour lasers emit at 511 nm (green) and 578 nm (yellow). Gold vapour laser emits at 628 nm (red). Pumping takes place through electric discharge.

7.9.5.1 Helium-Cadmium Laser

In the case of a helium-cadmium laser, electrons in the discharge plasma excite helium atoms to higher energy levels. Excited helium atoms transfer their energy to cadmium atoms, ionizing them and raising them to upper laser levels in the process. It may be mentioned here that it is much easier to ionize cadmium than it is in the case of helium due to the former having only two electrons in its incompletely filled outer shell and the latter having a filled outer shell. Figure 7.82 shows an energy level diagram of a helium-cadmium laser. As shown in the diagram, cadmium atoms are trapped in two metastable states. One of them produces the two UV lines with the two lower laser levels and the other metastable state produces the blue line with one of the lower laser levels.

Cadmium at room temperature is in solid form and it is heated to about 250 °C to vaporize it to a pressure in the order of a few milli-torr required for laser operation to take place. Helium pressure is in the order of a few torr, about a thousand times higher than that of cadmium. Helium-cadmium tubes usually have a helium gas reservoir to replace helium gas that may leak out. Also, cadmium vapours may condense on relatively cooler parts of the tube during laser operation. The tube is so designed as to have certain portions that can collect surplus cadmium so that it does not deposit on the optical surfaces.

Like a He-Ne laser, helium-cadmium lasers too have one of two optical designs, one with resonator mirrors sealed to the ends of the discharge tube and the other with a tube sealed with Brewster windows and the resonator mirrors placed outside the tube. Again, the emitted wavelength depends upon the chosen optics.

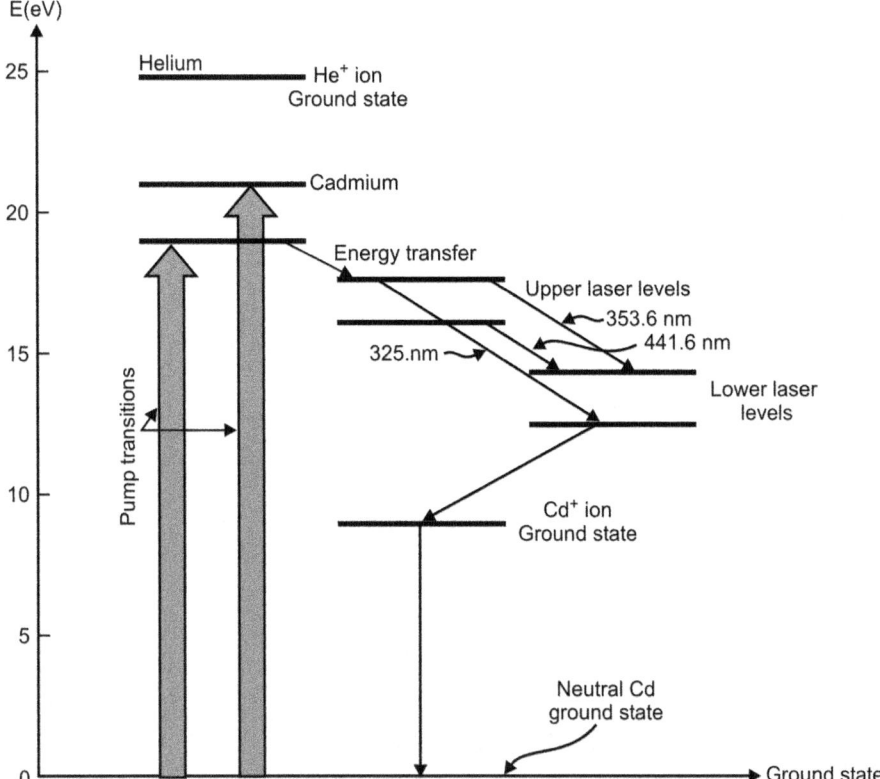

Figure 7.82 Energy level diagram of a helium-cadmium laser.

7.9.5.2 Copper Vapour and Gold Vapour Lasers

Important members of the family of metal vapour lasers where the lasing species is the neutral atom vapour are the copper vapour and gold vapour lasers. As mentioned earlier, copper vapour lasers have two emissions lines at 511 nm and 578 nm and the gold vapour laser emits at 628 nm. Copper vapour and gold vapour lasers are pulsed lasers with pulse repetition rate typically in the range of several kilohertz and achievable average power in the order of tens of watts.

Figure 7.83 shows the energy level diagram of a copper vapour laser. Gold vapour laser would also have a similar energy level structure. The energy level structure of copper vapour laser can be used to explain why such a laser cannot be used to produce a CW output. The description is equally valid for a gold vapour laser. Let us see as to what happens when an electric discharge passes through the metal vapour. Electrons in the discharge plasma collide with the metal vapour atoms, copper vapour atoms with reference to the energy level diagram of Figure 7.83, and raise them to an excited state. The excited state has a life time in the order of 10 ns or so if the metal vapour pressure is low. This small life time is not long enough to produce laser action. The effective life time can be increased to about 10 ms, which is long enough to produce stimulated emission and laser action, by increasing the vapour pressure to about 100 milli-torr. To get this vapour pressure, metallic copper or gold must be heated to a temperature in excess of 1500 °C in the laser tube.

The problem begins with the long life time of the lower laser level, which is in the order of tens to hundreds of microseconds. This leads to accumulation of lasing species at the lower

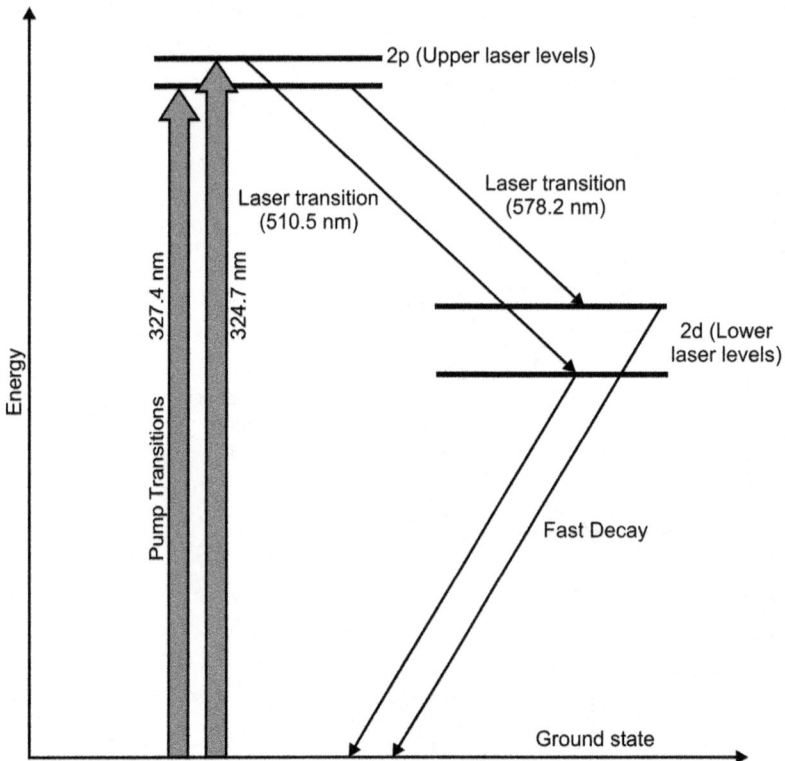

Figure 7.83 Energy level diagram of a copper vapour laser.

level inhibiting the laser action on a continuous basis. Laser action can be resumed after the lower laser level gets depopulated to the required level. This phenomenon allows these lasers to be operated only on pulsed basis, though the pulse repetition frequency can be sufficiently high. Addition of a rare gas such as helium, neon or argon helps in depopulating the lower laser level.

Some of the significant features of metal vapour lasers include a high gain of the laser medium (10–30% per cm), high laser efficiency (typically several tenths of a%) and high average power (approaching 100 W).

7.9.6 Rare Gas Ion Lasers

Argon-ion and krypton-ion lasers are the popular candidates belonging to the family of rare gas ion lasers. The laser emission in the case of argon-ion and krypton-ion lasers occurs from the transitions made by ionized argon and krypton atoms. These lasers emit a range of wavelengths from UV through the entire visible spectrum. Shorter wavelengths (less than 400 nm) come from doubly ionized argon (Ar^{+2}) and krypton (Kr^{+2}) atoms. Longer wavelengths come from singly ionized argon (Ar^{+}) and krypton (Kr^{+}) atoms. These lasers can generate output powers from a few milliwatts to few tens of watts, much higher than their other CW gas laser counterparts like He-Ne and helium-cadmium lasers.

Though both argon-ion and krypton-ion lasers emit a range of wavelengths from ultraviolet through visible to near infrared, argon-ion laser has strong emission at 488 nm (blue) and 514.5 nm (green) and krypton-ion laser emits strongly at 647.1 nm (red). The two gases can be

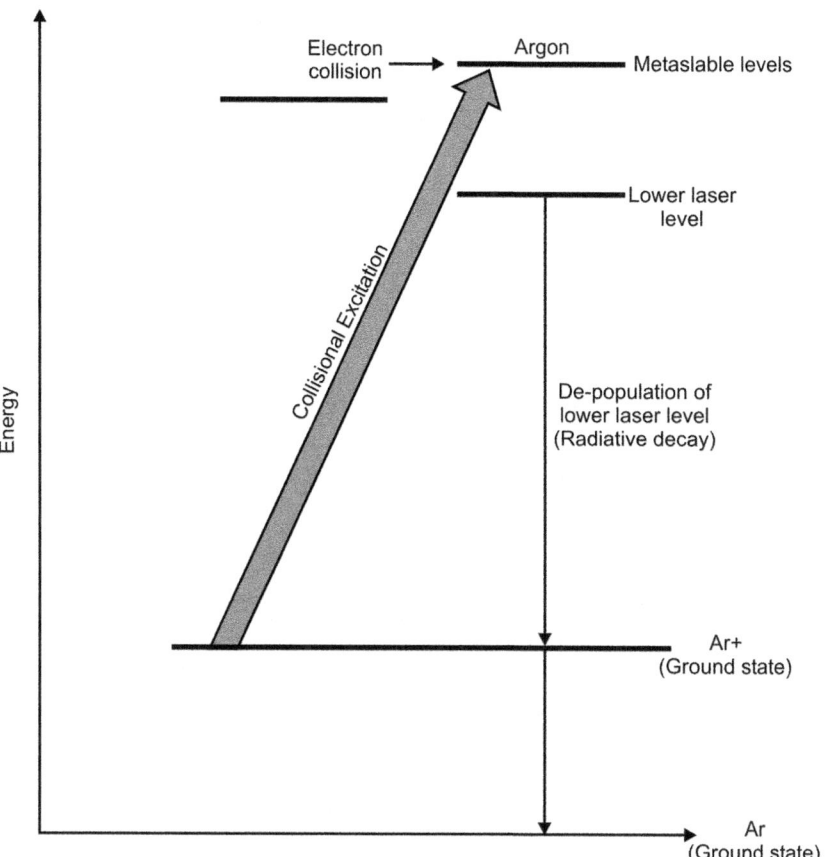

Figure 7.84 Energy level diagram of an argon-ion laser.

put into a single tube and excited simultaneously to get laser lines throughout the visible spectrum. Such mixed gas lasers are used in laser light shows.

Laser action in rare gas ion lasers can be explained with the help of energy level diagram of argon-ion laser as shown in Figure 7.84. A similar energy level structure exists for krypton-ion laser too. A high initial voltage ionizes the gas. Electrons in the electric current passing through the ionized gas transfer energy directly to argon atoms ionizing them and raising the resulting ions to a cluster of high-energy levels. Ions drop from populated metastable upper laser levels to a cluster of lower laser levels. The emission wavelength depends upon the pair of levels involved in the transition. In fact, if permitted by optics, several different wavelengths can be made to oscillate simultaneously. The ions drop from the lower laser level with very short life time to ion ground state by emitting extreme UV wavelength at 74 nm.

As regards construction of these lasers, it is not much different from that of helium-neon lasers. These lasers usually have Brewster window sealed plasma tubes with external cavity mirrors. The gas pressure is about 0.7–0.8 torr. Wavelength selection optics can be inserted between the Brewster window and the rear mirror, if so desired. Like He-Ne lasers, the discharge is confined to a narrow bore diameter in the centre of the tube. A series of metal discs with central holes may also define the bore in these lasers. A gas reservoir is also a part of the structure to replace the depleted gas. Wall plug efficiency of these lasers is in the range of 0.01–0.001%. Because of lower efficiency, high output power rare ion lasers need to be cooled, either by forced air cooling (in the case of medium power lasers) or liquid cooling (in the case of higher output power lasers. One of the most common applications of argon-ion lasers has been in laser light shows. They have been largely replaced by diode-pumped solid-state lasers for these applications as diode-pumped solid-state lasers are inherently smaller, more efficient and reliable, and increasingly more versatile and less expensive than rare ion lasers. Argon-ion lasers, due to their broad line width of about 5 GHz, adapt themselves well to mode locking. Mode-locked laser pulses from these lasers are suitable for pumping titanium-sapphire and tunable dye lasers.

7.9.7 Excimer Lasers

Excimer lasers are pulsed lasers capable of providing pulse energies in the order of Joules. These are the most powerful practical lasers emitting in the UV region. Except for the fluorine Excimer laser, all others have an active medium consisting of molecules made up of one rare gas atom such as argon, xenon or krypton and one halogen such as fluorine, chlorine or bromine. So, the active medium is a rare gas halide. Some of the more popular and practical Excimer lasers include F_2 (157 nm), ArF (193 nm), KrCl (222 nm), KrF (249 nm), XeCl (308 nm) and XeF (350 nm).

The process of creation of population inversion and subsequent laser emission in the case of Excimer lasers is unique. The two atoms constituting the molecule are bound to each other only when the molecule is in the excited state, which is the upper laser level. The molecule falls apart when it drops to the ground state, which is the lower laser level. Figure 7.85 shows the energy level diagram of a rare gas halide molecule depicting its behaviour both in the excited state where it is a molecule as well as in the ground state where the molecule breaks apart as individual atoms. Figure 7.85 shows the energy level as a function of the spacing between the rare gas atoms and the halogen atoms. Absence of any dip in the

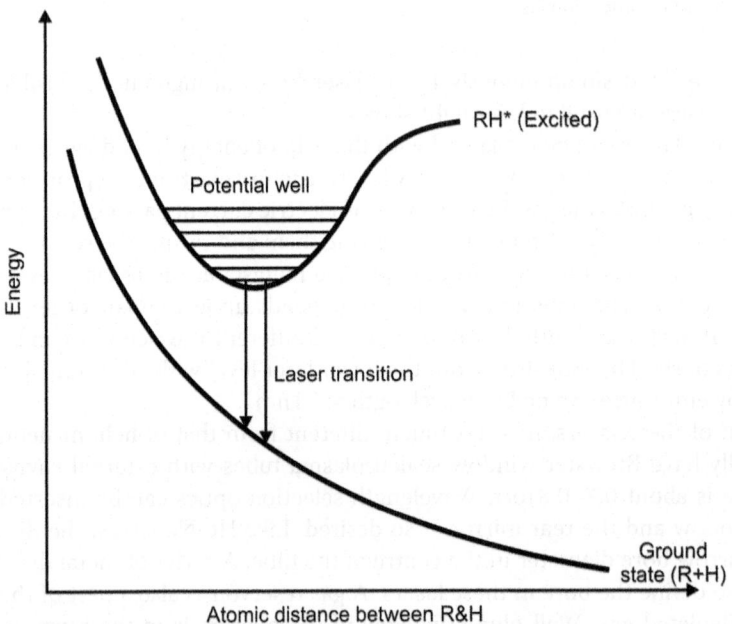

Figure 7.85 Energy levels of a rare gas halide molecule.

energy level indicates that the atoms are not bound to each other and that their energy decreases as the inter-atomic separation increases. The region in an excited state with a dip in energy level indicates that the rare gas and halogen atoms are bound to each other to form molecules. These excited molecules can occupy any of the several vibrational energy levels as shown by the horizontal lines. These molecules fall apart as they fall to the ground state as there is no bonding energy available in the relatively lower energy ground state to hold them together.

The gas mixture comprises of the derived rare gas, the halogen and the buffer gas such as helium or neon. In fact, more than 90% of the gas mixture is the buffer gas. The mixture has a small percentage of rare gas. Present in an even smaller fraction is the source of halogen atoms. As pure halogens can pose serious handling problems due to their high reactivity, it may be supplied in the form of molecules containing desired halogen atoms. It is more so for fluorine, which is considered as the most dangerous to handle of all the halogens and therefore may be supplied as nitrogen trifluoride (NF_3). The gas mixture is excited by passing a short intense electrical pulse. Electrons in the plasma transfer energy to the gas mixture. The halogen molecules break up to form electronically excited molecules and create population inversion. The upper state life time is about 10 ns. The molecules drop to the ground state and break up emitting UV radiation in the process.

These are the best available pulsed lasers and find application in semiconductor manufacturing industry, research and medicine. Pulse energies from several micro-joules to a few joules are available at repetition rates of tens to a few hundred hertz. Repetition rates of a few kilohertz are achievable but then the pulse energies available at those repetition rates are relatively lower. Krypton-fluoride (KrF) and xenon-chloride (XeCl) are the most energetic of all Excimer lasers and the average power in these lasers can be in excess of 200 W.

7.9.8 Chemical Lasers

A chemical laser derives the energy required to produce population inversion and consequent laser emission from a chemical reaction. Importance of chemical lasers lies in their ability to generate CW output power level reaching several megawatts. This coupled with the fact that the range of wavelengths generated by these lasers is well absorbed by metals, which makes them suitable for cutting and drilling operations in industry and more so as high power laser sources for directed-energy weapon applications.

There are two categories of chemical lasers. One category has hydrogen fluoride (HF) and deuterium fluoride (DF) and the other has the two iodine lasers including chemical oxygen iodine laser (COIL) and all gas-phase iodine laser (AGIL). Both HF/DF and COIL systems have demonstrated their capability to be operated at megawatt-class CW output power. Lasing action has been demonstrated in AGIL and efforts are on to scale the output to higher power levels. Given their importance as potent high power laser sources for directed-energy weapon applications, these are discussed in detail in Chapter 12 on *Directed-Energy Weapons*.

7.9.9 Carbon Dioxide Gas Dynamic Lasers

The gas dynamic laser is an unconventional carbon dioxide laser that derives its energy from the combustion of a suitable fuel-oxidizer mixture, which means that it does not require any electrical energy for its operation. This concept was experimented with in late 1970s and early 1980s for development of high power CW laser with potential application in directed-energy weapons. This laser is further discussed in Chapter 12.

7.9.10   Dye Lasers

This is a popular laser device because of its inherent ability to produce a coherent output that is tunable over a wide range of wavelength from near UV through visible to the near IR region. Of course, a single dye would not account for wavelength tunability from UV to IR. One would need to switch dyes. The vibronic class of solid-state lasers is the only close competitor when one is looking for a tunable laser. Dye lasers are particularly important in scientific research and some medical applications where a large wavelength tuning range and their inherent ability to produce ultra-short pulses are used to advantage. While tunable output from dye lasers was initially considered for use in spectroscopic applications, applications of dye lasers have expanded over the years to include medical applications for both diagnostic and therapeutic purposes. The tunability property of dye lasers results from the energy level diagram of the active medium. The active medium is an organic dye characterized by upper and lower laser levels that are bands of a large number of sub-levels rather than single energy levels. Their ability to produce ultra-short pulses results from the wide gain bandwidth profile characteristic of dye lasers. A titanium-sapphire laser in the vibronic class of solid-state lasers also has a wide gain bandwidth profile and therefore is capable of producing short pulses. Dye lasers, however, have an advantage that they cover wavelength range from UV to near IR unlike titanium-sapphire lasers that produce output in the near IR only.

Minimum pulse length producible from the laser is a function of gain bandwidth profile, a phenomenon that was described earlier in the chapter. To briefly mention, pulse length in nanoseconds is given by (0.441/bandwidth), where bandwidth is in MHz.

7.9.10.1   Active Medium

The active medium in a dye laser is an organic dye. Organic dyes come from a family of complex molecules characterized by electronic-vibrational levels. The complex sets of electronic and vibrational levels associated with these molecules gives them the bright colour and hence the name dyes. The vibrational levels in fact create sub-levels of the excited electronic levels thus leading to a band of upper and lower laser levels in the same way as we find them in vibronic class of solid-state lasers. The similarity between the two is clearly visible in the energy level diagram of an organic dye as shown in Figure 7.86. The dye is usually dissolved in a liquid solvent. The solvent acts as the host material in the same way as the crystals and glasses do in the case of solid-state lasers.

In the case of low power CW lasers and pulsed lasers with modest average power, the laser dye solution is contained in a sealed glass container called cuvette. In the case of higher power levels, the dye solution is usually made to flow rapidly through the cell to prevent it from getting heated and subsequently degraded. The dye solution may also be made to flow in the form of an unconfined jet with the help of a specially designed nozzle.

Some examples of laser dyes include Rhodamine-6G, Fluorescein and Coumarin. Rhodamin-6G is a commonly used laser dye pumped by a frequency-doubled Nd:YAG laser beam as it has a very high photo-stability, high quantum yield, low cost and more importantly close proximity to the absorption maximum (approximately 530 nm) of the dye. The tunable range is from 555 to 585 nm with a maximum at 566 nm.

7.9.10.2   Pump Mechanisms

Dye lasers are optically pumped. They are either flash lamp pumped or pumped by pulsed lasers. In the case of flash lamp pumped dye lasers, there are two types. In one case, a linear flash lamp pumps a parallel linear dye cell in the same way it pumps the laser rod in a solid-state laser. In the other case, a coaxial flash lamp is used and the dye flows through the centre of the

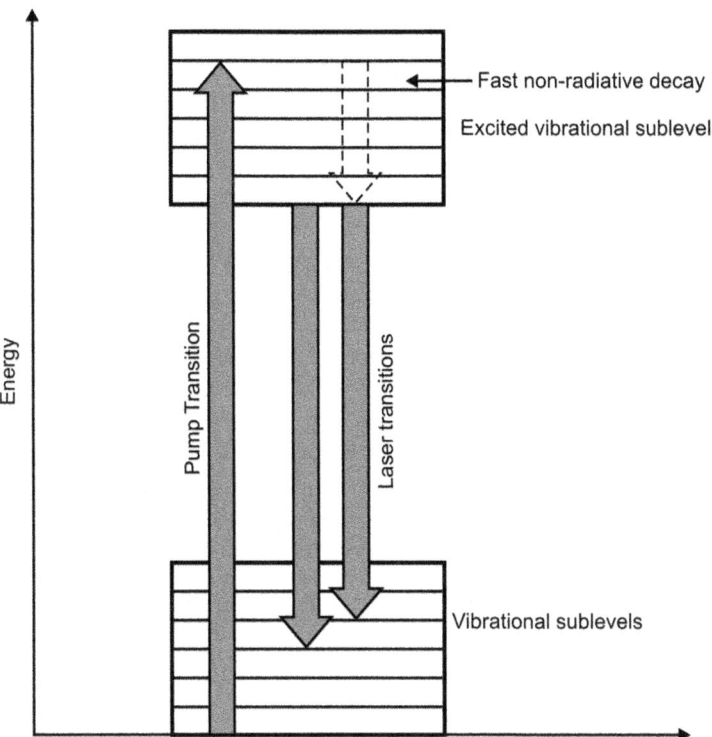

Figure 7.86 Energy level diagram of a dye laser.

flash lamp. Flash lamp pumped dye lasers produce pulsed laser output with pulse repetition rate in the range of few pulses per minute (in the case of high-energy pulses) to few hundreds of pulses per second (in the case of low-energy pulses).

Pulsed lasers are very usefully employed for the purpose of pumping dye lasers. Most lasers suitable for dye laser pumping emit in short wavelength end of the visible spectrum and in ultraviolet as dye lasers mostly emit in visible wavelengths. Some of these lasers include frequency-doubled neodymium lasers (532 nm), frequency-tripled neodymium lasers (355 nm), frequency-quadrupled neodymium lasers (266 nm), the copper vapour laser (510 nm and 578 nm), krypton-fluoride Excimer laser (249 nm), xenon-chloride Excimer laser (308 nm) and xenon fluoride Excimer laser (351 nm). CW lasers are also used for pumping dye lasers to get CW outputs. Argon, krypton and neodymium lasers are used for the purpose.

7.9.10.3 Wavelength Selection

Wavelength selection is done with the help of a tuning element such as a diffraction grating, a prism or some other wavelength selective element such as a tunable etalon in the laser resonator. Diffraction gratings and prisms disperse light at different angles depending upon the wavelength, which allows either these elements or some other optical components to be used for the purpose of wavelength tuning. Etalon, on the other hand, with its two reflective surfaces functions like a miniature optical cavity and is used to limit oscillations to a narrow range of wavelengths. Wavelength selection is made by altering the distance between the two reflective surfaces.

7.9.11   Free Electron Lasers

A free electron laser is a unique laser, where the process of light amplification is a bit unconventional regarding what we have seen in respect of lasers discussed earlier in this chapter. Unlike conventional lasers that rely on bound atomic or molecular states, free electron lasers use a relativistic electron beam as the active medium. In the case of a free electron laser, lasing medium is a beam of free electrons completely unattached to any atoms.

It may be mentioned here that a relativistic particle is a particle moving at a speed close to the speed of light such that effects of special relativity are important for its behaviour. Mass less particles such as photons always move with speed of light and are therefore always relativistic. Particles with some mass are considered relativistic when their kinetic energy is comparable to or greater than the energy mc^2 corresponding to their rest mass. This implies that their speed is close to that of light. Such particles are generated in particle accelerators. In the context of a free electron laser, a beam of electrons is accelerated to relativistic speeds and then made to pass through a periodic or more precisely alternating transverse magnetic field. The transverse magnetic field is produced with the help of an array of magnets with alternating poles placed along the beam path. The array of magnets is sometimes called the *Wiggler* and the magnetic field produced by it the *Wiggler field*. It is so-called as it forces the electron beam to assume a sinusoidal path. As the beam travels through the magnetic field, it releases some of its energy as light before it exits from the other end of the field. The emitted wavelength is given by eqn. 7.30.

$$\lambda = \frac{p}{\left[2\left(1 - v^2/c^2\right)\right]} \qquad (7.30)$$

Where,

p = Period of Wiggler field
v = Velocity of electrons

The same relationship may also be written in the form of eqn. 7.31.

$$\lambda = \frac{(0.131p)}{(0.511 + E)^2} \qquad (7.31)$$

Where,
E = Electron energy in MeV

The free electron laser is tunable over a wide range of wavelengths in the region from X-rays to microwaves. The tunability is achieved by varying E. The accelerated beam of electrons having desired energy can be obtained from well-established technology of charged-particle accelerators. Figure 7.87 shows the simplistic arrangement of various components in a free electron laser.

Free electron lasers though thought of as potential sources for DEW applications, these lasers have their problems when it comes to generating higher powers at shorter wavelengths. Smaller free electron lasers with power output up to 100 W are particularly important for research and medical applications.

7.9.12   X-Ray Lasers

How complex it can be to generate a coherent beam of X-rays is evident from a simple calculation, which tells that you would need an energy level difference in the order of 100 eV or so

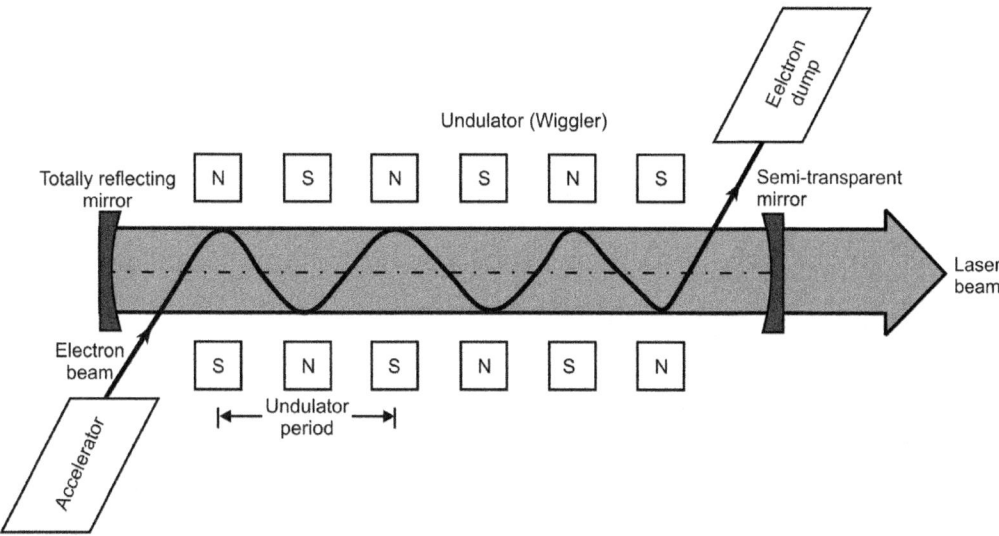

Figure 7.87 Free electron laser.

between the two laser levels in order to generate a 10 nm laser output. Creating a population inversion in an X-ray laser is a gigantic task. That is why these lasers are either excited by nuclear explosions or by massive Q-switched Nd:Glass lasers meant for fusion research. The star war programme initiated the development of X-ray lasers. However, due to technological complexities and infrastructure requirements, X-ray lasers are not considered an attractive proposition.

7.10 Semiconductor Lasers

Most semiconductor lasers are diode lasers that are pumped by an electrical current. These are also known as injection diode lasers. However, there are also optically pumped semiconductor lasers that are pumped by optical radiation either from a direct coupled single emitter or fibre coupled laser diode array and quantum cascade lasers that utilize intra-band rather than inter-band transitions of the conventional semiconductor lasers for laser action. Most of the discussion in this section is centred on semiconductor diode lasers, which typically emit in visible to near IR bands of the electromagnetic spectrum. Optically pumped semiconductor lasers, quantum cascade lasers, lead salt lasers and antimonide lasers are also covered briefly. Topics covered in this section include operational fundamentals, different types of semiconductor diode lasers, characteristic parameters and handling precautions.

7.10.1 Operational Basics

The active medium in a semiconductor laser is a semiconductor material. The optical gain in this case is usually achieved by a process of stimulated emission at an inter-band transition triggered by prevailing conditions of high carrier density in the conduction band. In the case of diode lasers, high carrier density in conduction band is caused by injection current. The emission of radiation is due to recombination of holes and electrons in a forward-biased PN junction diode. An *injection diode laser* powered by an injected electrical current is the most practical form of diode laser to distinguish it from an optically pumped diode laser.

(a)

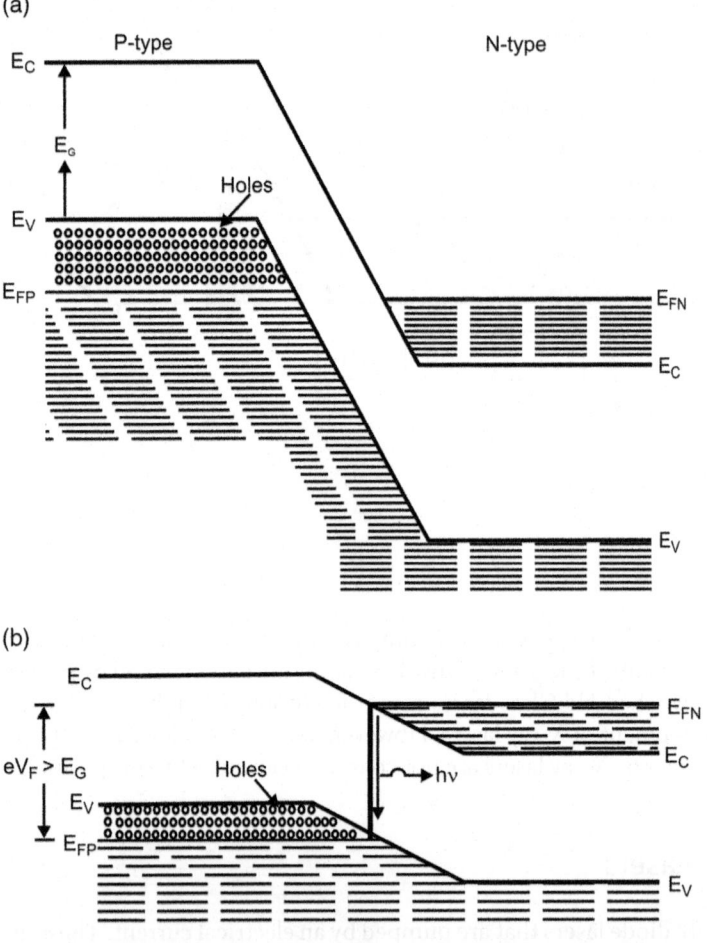

Figure 7.88 (a) Zero applied forward bias. (b) Applied forward bias greater than E_G.

Figure 7.88 illustrates the operational basics of a semiconductor diode laser action. Figure 7.88(a) and (b), respectively, depicts conditions of no forward bias and applied forward bias greater than the band gap energy. As shown in the diagrams, energy levels up to Fermi level are occupied by electrons. Also Fermi level is continuous across the junction under zero applied forward bias conditions. When a forward bias with the corresponding energy greater than the band gap energy E_g is applied across the junction, the barrier potential reduces to zero. Fermi level on the N-side (E_{FN}) and P-side (E_{FP}) are separated by eV_F. There are more electrons in the conduction band near E_C than there are electrons in the valence band near E_V. This leads to population inversion near E_C and E_V energy levels. This region of population inversion along the junction is called inversion layer or active region. Due to presence of electrons and holes in the active region, there is a likelihood of their recombination leading to spontaneous emission of a photon where energy of emitted photon equals the difference in energy levels of electron and hole states involved in the recombination. In the process, the electron may re-occupy the energy state of the hole. The injected electrons and holes constitute the injection current and those involved in the recombination process constitute the spontaneously emitted photon output.

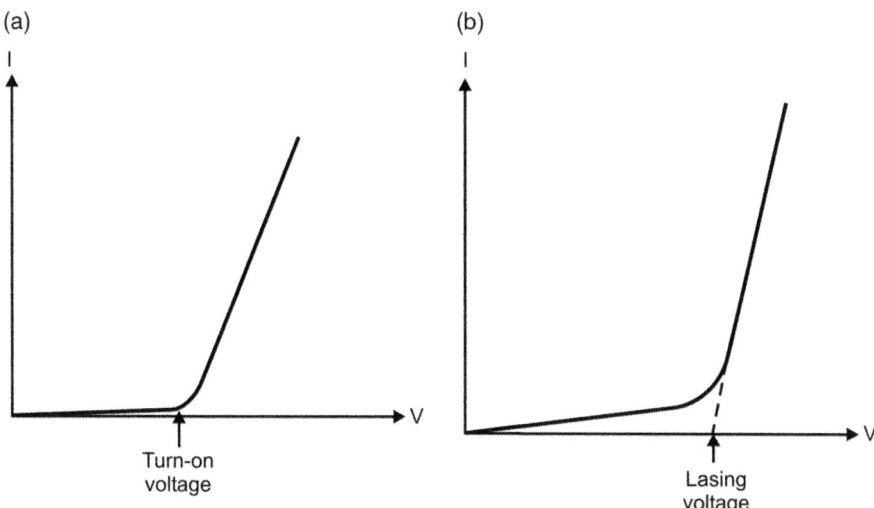

Figure 7.89 I-V characteristics of the: (a) LED and (b) semiconductor diode laser.

If the injection current exceeds a certain minimum value, called the lasing threshold, the number of electrons and holes available for recombination becomes sufficiently large so as to create a possibility where a spontaneously emitted photon having energy equal to involved recombination energy stimulates an electron-hole pair to recombine to emit a photon of the same frequency, phase and polarization as that of the stimulating photon. This phenomenon is called *stimulated emission*. Surrounding the recombination region also called gain region by a suitable optical cavity facilitates the process of stimulated emission. The cavity in the case of diode laser is made by cleaving the two ends of the crystal to form perfectly smooth, parallel edges forming a Fabry–Perot resonator. Since the semiconductors have a high refractive index, the smooth surfaces offered by cleaved ends reflect about 30% of light back into the material to get sustained laser action in a high gain semiconductor laser material. The stimulated emission produces light amplification as the photons travel back and forth between the two end faces of the cavity. And when the gain due to stimulated emission exceeds the losses due to absorption or imperfect reflections and so on, sustained lasing action is produced.

Light-emitting diodes (LEDs) too operate the same way with a major difference in the forward-biased current. While the current in the case of an LED is in the order of a few milli-amperes, the same in the case of laser diodes emitting few milliwatts of laser power is in the order of 80–100 mA. At low levels of drive current, spontaneous emission predominates. When the drive current is more than the lasing threshold, the light output is predominantly due to stimulated emission. Figure 7.89(a) and (b) shows the I-V characteristics of a typical LED and a semiconductor diode laser. As shown in the figure, the LED turns ON for a forward bias voltage greater than or equal to the turn-on voltage and the current flowing through the device is nearly zero. In the case of a laser diode, though the device is turned on and a small magnitude of current does start flowing; the lasing action starts only after the forward voltage exceeds the lasing threshold voltage.

This phenomenon can also be explained with the help of output optical power versus drive current characteristics of the laser diode shown in Figure 7.90. As shown in the figure, stimulated emission becomes predominant thereby leading to sustained lasing action only after the drive current exceeds a certain lasing threshold current value. For drive currents less than the threshold value, spontaneous emission predominates and what we get is an LED-like emission.

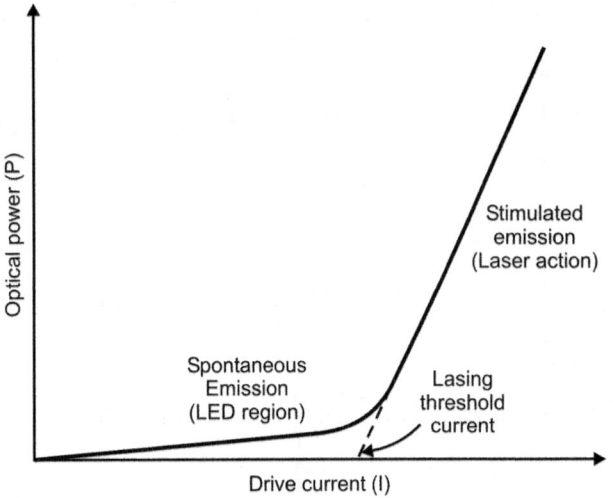

Figure 7.90 Optical power versus drive current characteristics of a semiconductor diode laser.

7.10.1.1 Semiconductor Laser Materials

Only direct band gap semiconductor materials are suitable for building diode lasers and LEDs though some compound semiconductors with indirect band gap can be used to make LEDs. In the case of direct band gap materials such as gallium arsenide (GaAs), gallium nitride (GaN), gallium aluminium arsenide (GaAlAs) and so on, a transition from conduction band to valence band releases all its energy in the form of light. In the case of indirect band gap materials like silicon, some of it goes into vibrations of the crystal lattice, which makes it difficult for this class of semiconductor materials to emit light by recombination of charge carriers. Compound semiconductors are used for making laser diodes. For laser applications, the most important of these are those comprising of equal amounts of elements from III-A (gallium, aluminium and indium) and V-A (arsenic, antimony, phosphorus and nitrogen) groups of the Periodic Table. There are compounds with three elements called the ternary compounds and compounds with four elements called the quaternary compounds.

Commonly used semiconductor material compositions for semiconductor diode laser fabrication include gallium arsenide (common laser wavelength: 905 nm @ room temperature), gallium nitride (common laser wavelength: 405 nm), gallium aluminium arsenide (common laser wavelengths: 785 and 808 nm), aluminium gallium arsenide (common laser wavelength: 1064 nm), indium gallium arsenide (common laser wavelength: 980 nm), indium gallium nitride (common laser wavelengths: 405 and 445 nm), gallium antimonide arsenide (common laser wavelengths: 1877, 2004, 2330, 2680, 3030 and 3330 nm), Indium gallium arsenide phosphide (common laser wavelengths: 1310, 1480, 1512, 1550 and 1625 nm) and aluminium gallium indium phosphide (common laser wavelengths: 635, 657, 670 and 760 nm).

7.10.2 Types of Semiconductor Lasers

Depending upon the structure of various semiconductor materials used to fabricate semiconductor lasers, these can be categorized as under.

1) Homojunction and heterojunction lasers
2) Quantum well diode lasers
3) Distributed feedback (DFB) lasers

4) Vertical cavity surface emitting lasers (VCSEL)
5) Vertical external cavity surface emitting lasers (VECSEL)
6) External cavity semiconductor diode lasers
7) Optically pumped semiconductor lasers
8) Quantum cascade lasers
9) Lead salt lasers

Each one of these is briefly described in the following paragraphs.

7.10.2.1 Homojunction and Heterojunction Lasers

The boundaries between the active layer and the adjacent layers play an important role in deciding some of the vital parameters such as efficiency of the lasers. In the case of *homojunction lasers*, all layers are of the same semiconductor material. One such example is GaAs/GaAs laser. However, this simple laser diode structure is highly inefficient and therefore could be used only to demonstrate pulsed operation. In the case of *heterojunction lasers*, the active layer and either one or both of the adjacent layers are of different material. If only one of the adjacent layers is of a different material, it is called a simple heterojunction and if both are different, it is called a double heterojunction. Some of the popular semiconductor laser types and the corresponding emission wavelength band include AlGaInP/GaAs (heterojunction) at 620–680 nm, $Ga_{0.5}In_{0.5}P$/GaAs (heterojunction) at 670–680 nm, GaAlAs/GaAs (heterojunction) at 750–870 nm, GaAs/GaAs (homojunction) at 904 nm and InGaAsP/InP (heterojunction) at 1100–1650 nm.

In heterostructure diode lasers, a layer of low-band gap material is sandwiched between layers of a high-band gap material as shown in Figure 7.91. One such commonly used pair of materials is gallium arsenide (GaAs) and aluminium gallium arsenide ($Al_xGa_{1-x}As$). In these devices, the active region (the region where free electrons and holes exist simultaneously) is confined to the thin middle layer. As a result of this, relatively much larger number of electron-hole pairs can contribute to the amplification process. In addition to this, light is reflected from heterojunction, which helps in confining the emitted photons to the active region. These factors are responsible for high efficiency of heterostructure diode lasers.

Figure 7.91 Heterojunction diode laser structure.

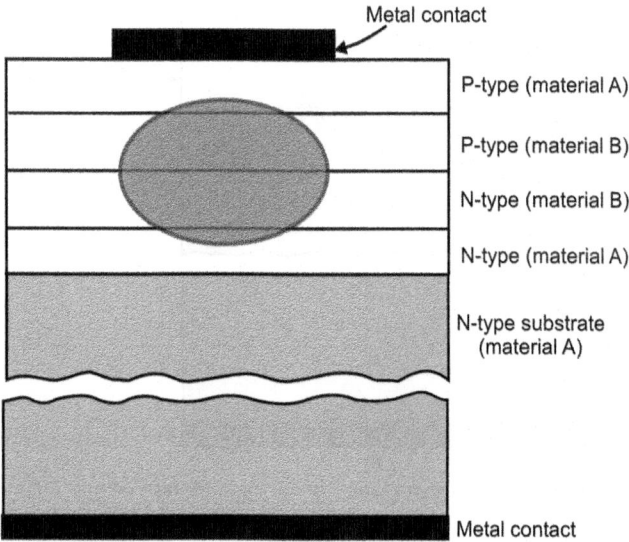

7.10.2.2 Quantum Well Diode Lasers

For semiconductor layers that are only a few nanometres thick, the assumption that the material is a continuum is no longer valid and the quantum mechanical properties of atoms and electrons become important. When a thin layer of a semiconductor material with a relatively smaller band gap is sandwiched between two thick layers with larger band gaps, a structure known as a *quantum well* is formed and the electrons passing through the semiconductor are captured in the thin layer. Though these electrons have sufficient energy to be free in the small band gap quantum well layer, it is not enough to allow them to enter the large band gap thicker layers. The thicker layers thus help in confining the captured electrons to the quantum well. If the quantum well is placed in the semiconductor junction, the concentration of electrons as they recombine with holes to emit photons leads to increase in efficiency and decrease in lasing threshold. The fact that the quantum well layer and the outer thicker layers are made from semiconductor materials of different refractive indices further helps the cause by confining the emitted photons to that narrow region. Figure 7.92 shows the basic structure of a simple quantum well semiconductor diode laser.

One of the drawbacks of the simple quantum well structure of Figure 7.92 is that the thin layer is too small to effectively confine the light. A modification of this basic quantum well structure is shown in Figure 7.93. In this, two additional layers are added outside three existing layers. The additional layers made of material-A in Figure 7.93 have a refractive index that is smaller than the refractive indices of centre layers made of materials B and C. Such an arrangement succeeds in confining the light with greater efficacy than is feasible with simple quantum well structure. Semiconductor diode lasers having this modified structure are called *Separate Confinement Heterostructure* (SCH) lasers. Most practical diode lasers use SCH quantum well structure.

Lasers with more than one quantum well layer are known as multiple quantum well lasers. These are nothing but a stacked arrangement of quantum wells and confinement layers. Multiple quantum well structures can generate relatively higher power outputs.

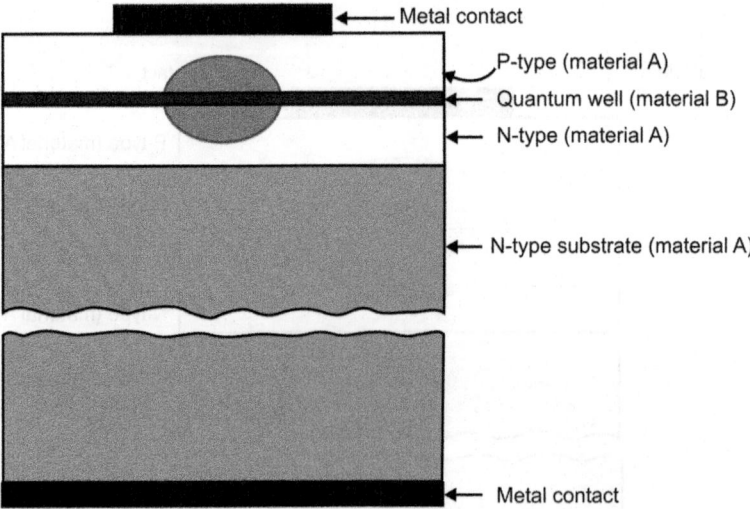

Figure 7.92 Quantum well semiconductor diode laser: simple quantum well.

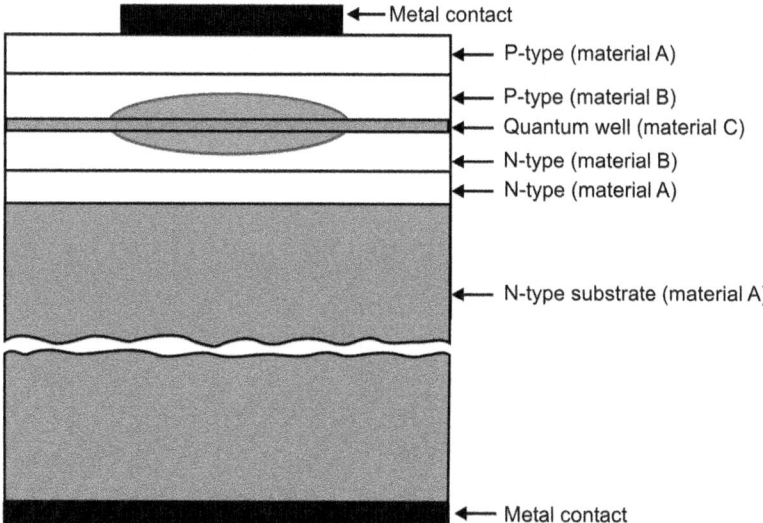

Figure 7.93 A simple quantum well semiconductor diode laser and (b) SCH structure.

Extension of the quantum well concept has led to development of what are known as *quantum wire* and *quantum dot* structures. While a quantum well structure can confine electrons in only one dimension, that is, in the plane of quantum well layer, quantum wire and quantum dot structures provide confinement respectively in two and three dimensions. This leads to further improvement in laser efficiency. A quantum dot laser uses quantum dot structures as the active medium. As a result of tight confinement of charge carriers, quantum dot semiconductor diode lasers offer improved performance as compared to lasers built on bulk or quantum well active media in terms of modulation bandwidth, lasing threshold, relative intensity noise, temperature sensitivity and so on. Also, the quantum dot active region may be engineered by varying the dot size and composition to operate on different wavelengths earlier not possible with semiconductor laser technology.

Yet another type is the *quantum cascade laser* in which laser transition occurs between different energy levels of quantum well and not the band gap. This allows operation at longer wavelengths. The wavelength can be tuned by changing the thickness of quantum layer. Quantum cascade lasers are described in Section 7.10.2.8.

7.10.2.3 Distributed Feedback (DFB) Lasers

Diode lasers have a gain bandwidth curve that is broad enough to accommodate several longitudinal modes. In a conventional diode laser, therefore, due to changes in temperature and other operating conditions, the dominant longitudinal mode may hop from one value to the next adjacent value leading to instability in output power and wavelength. There are applications where this instability is highly undesirable. One such example is fibreoptic communication where instability in output wavelength causes chromatic dispersion in the fibre thus limiting its utility.

A Distributed Feedback Laser (DFB) offers a mechanism of ensuring a very narrow output wavelength band. In this, a diffraction grating is etched very close to the active layer (Figure 7.94). The grating provides a wavelength selective feedback to the gain region. The feedback causes interference and only the narrow wavelength region for which the interference is constructive

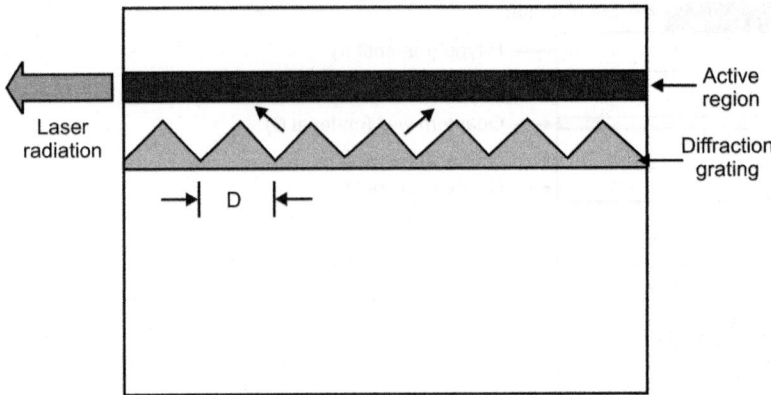

Figure 7.94 Distributed Feedback (DFB) laser.

is allowed to sustain. The DFB lasers offer single longitudinal and transverse mode operation. For a diffraction grating with pitch D, wavelength λ is given by eqn. 7.32.

$$\lambda = \frac{2nD}{m} \tag{7.32}$$

Where,

n is the refractive index of the medium
m is a positive integer

Typical values of m are 1 or 2 indicating order of distributed feedback coupling. A slight shift in output wavelength is possible due to temperature dependence of n.

In a DFB laser, since the grating provides the feedback required for lasing, end faces are not required to do that. That is why one of the end faces is anti-reflection coated. DFB lasers because of narrow line width and wavelength tunability find extensive use in applications such as optical communications, spectroscopy, gas sensing and metrology. In fact, DFB lasers today are the workhorses of fibreoptic communication systems. Figure 7.95 shows photographs of some DFB laser types from M/s Eagleyard Photonics, Germany. These lasers are available in the wavelength range of 760–1083 nm and output power levels in the range of 10–150 mW. The FWHM spectral width, temperature coefficient of wavelength ($d\lambda/dT$) and current coefficient of wavelength ($d\lambda/dI$) of this series of DFB laser diodes are 2.0 MHz, 0.06 nm/K and 0.003 nm/A, respectively. These are ideally suited to applications such as spectroscopy, metrology and gas sensing.

7.10.2.4 Vertical Cavity Surface Emitting Laser (VCSEL)

In the diode laser structures discussed so far, the optical cavity is perpendicular to the direction of current flow. In the case of vertical cavity surface emitting laser (VCSEL), the optical cavity is along the direction of flow of injection current as shown in Figure 7.96. The laser beam in this case emerges from the surface of the wafer rather from its edges. Having mirrors on both sides of the active medium forms the resonant structure. Mirrors are usually of the distributed Bragg reflector type formed by a multi-layer structure of alternating low and high refractive index semiconductor materials. Mirrors in this case need to be highly reflective as the overall gain is low due to short length of the gain medium. Compared to edge emitting lasers, these lasers produce relatively lower output power levels. However, VCSEL require a very small chip

Figure 7.95 DFB lasers from Eagleyard Photonics, Germany.

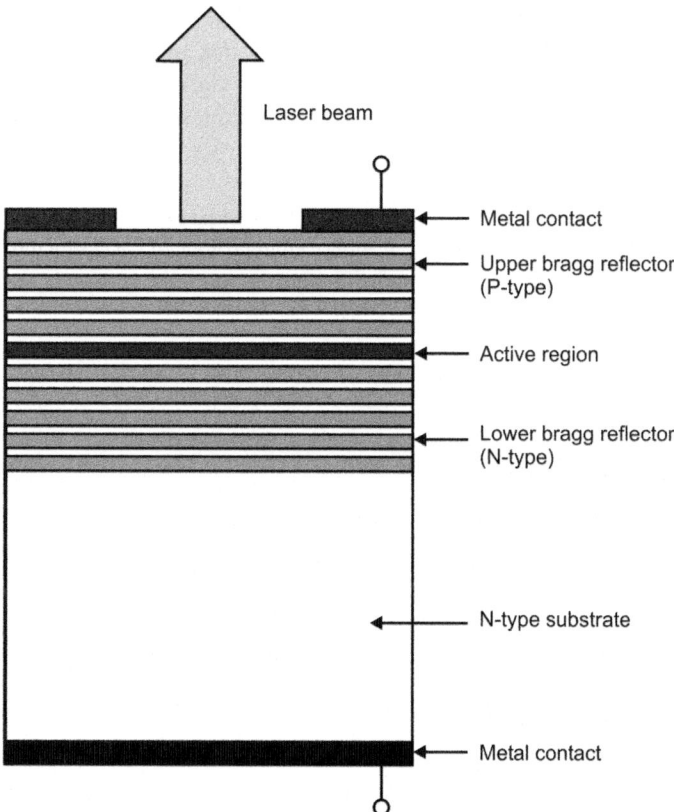

Figure 7.96 Vertical cavity surface emitting laser. (VCSEL) structure.

area, typically a few tens of μm² with the result that a large number of such lasers can be tightly packaged in an array structure on a single chip. These lasers are characterized by a very small threshold current, which could be as low as 1 μA.

7.10.2.5 Vertical External Cavity Surface Emitting Lasers (VECSEL)

In the case of VCSEL, the two mirrors are either grown epitaxially as a part of the diode structure or grown separately and then bonded to the semiconductor chip having the active region. Vertical external cavity surface emitting laser (VECSEL) is a variant of VCSEL where the resonator is completed with a mirror placed external to the diode structure thus introducing a free-space region in the resonant cavity as shown in Figure 7.97. Compared with other types of semiconductor diode lasers, VECSELs are capable of generating relatively much higher optical powers with high beam quality. It may be mentioned here that when mode area is large, external resonator is necessary for achieving high beam quality. That is why VCSELs are not suitable for generating high power levels with high beam quality, though they have a large mode area.

In the case of VECSELs, the resonant cavity may contain some additional optical elements such as an optical filter to facilitate wavelength tuning or a saturable absorber for passive mode locking to generate picoseconds and femtosecond pulses at a repetition rate of few GHz. Insertion of a nonlinear crystal in the cavity allows frequency doubling, which, for example, could lead to fabrication of devices with red, green and blue outputs. It may be mentioned here that a VECSEL is a low gain device and the intra-cavity power is much higher than the output power. This facilitates intra-cavity frequency doubling.

A typical VECSEL has a gain structure with a Bragg reflector and a gain medium with multiple quantum wells. Following are the typical material combinations used to generate different wavelength regions. One possible combination is a GaAs wafer, InGaAs quantum wells in the gain medium and a Bragg reflector grown from AlAs and GaAs. This structure is suitable for generating wavelengths in the region of 960–1030 nm. Emission in the 1.5 μm

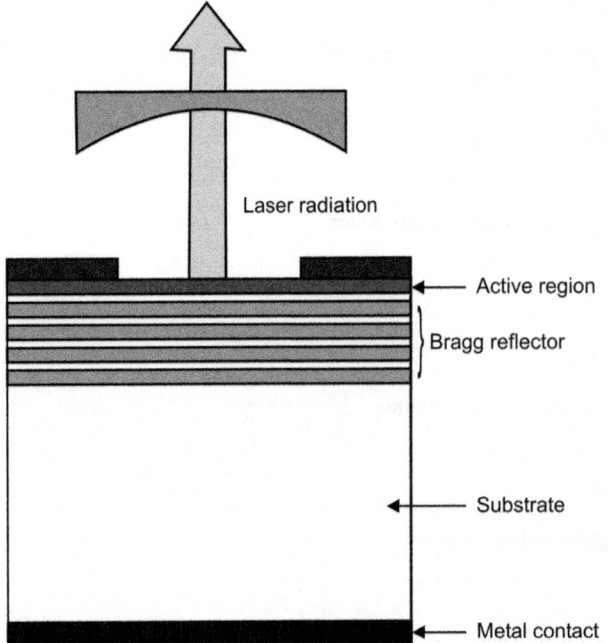

Figure 7.97 Vertical external cavity surface emitting laser (VECSEL) structure.

Laser radiation

Active region

Bragg reflector

Substrate

Metal contact

region is possible with indium phosphide (InP) as the wafer material, indium gallium arsenide phosphide (InGaAsP) quantum wells and indium phosphide – indium gallium aluminium arsenide (InP/InGaAlAs) combination for a Bragg mirror. For longer wavelengths such as 2.0 μm, gallium indium antimonide (GaAlSb) quantum wells could be used on gallium antimonide (GaSb) wafers.

7.10.2.6 External Cavity Semiconductor Diode Lasers

The basic operational concept of an external cavity diode laser (ECDL) is similar to the one explained in the case of VECSEL. Both VECSEL and ECDL, as we shall discover in the following paragraphs, have some common characteristic features. It may be mentioned here that VECSELs are external cavity semiconductor lasers, which are usually not diode lasers.

External cavity diode laser, as suggested by the name itself, is a semiconductor diode laser whose resonator is completed with one or more optical components outside the diode laser chip. In the simplest form, the diverging output from one of the end faces that is anti-reflection coated is collimated by an external lens. The collimated beam is made to fall on a partially reflecting mirror that provides optical feedback and also acts as output coupler. An external cavity laser allows for a longer resonator, which in turn leads to lower phase noise and narrow emission line width resulting from increased damping time of intra-cavity light. An external resonator also opens up possibility of introducing suitable intra-cavity optical elements for wavelength selectivity and tuning, mode locking and so on.

A diffraction grating is usually employed for wavelength selectivity and tuning. Two common configurations are the *Littrow configuration* and *Littman-Metcalf configuration*. Figure 7.98 shows the basic schematic arrangement of the Littrow configuration. A diffraction grating is used as the end mirror with first order diffracted beam providing optical feedback. Wavelength tuning is achieved by changing grating orientation. This configuration has a drawback that the direction of output beam changes with wavelength tuning. This shortcoming is overcome in the Littman–Metcalf configuration shown in Figure 7.99. In this configuration, grating orientation is fixed and an additional mirror is used to provide first order feedback. The mirror is rotated for wavelength tuning. The direction of output beam in this case is fixed. The beam undergoes wavelength dependent diffraction twice per resonator round trip resulting in higher wavelength selectivity and narrower line width. External cavity diode lasers find extensive use in applications such as spectroscopy where a tunable narrow line width source is important.

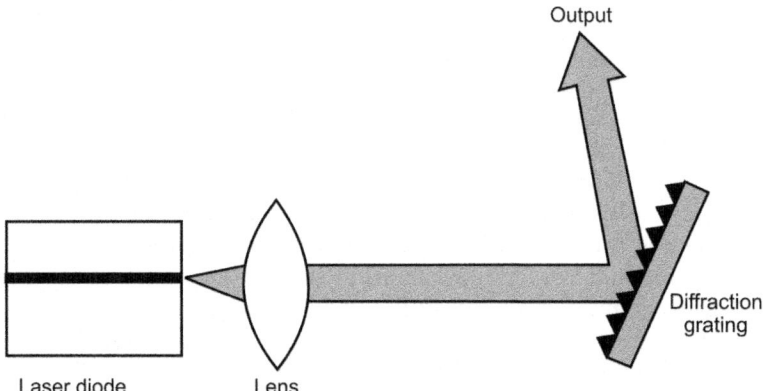

Figure 7.98 External cavity diode laser: Littrow configuration.

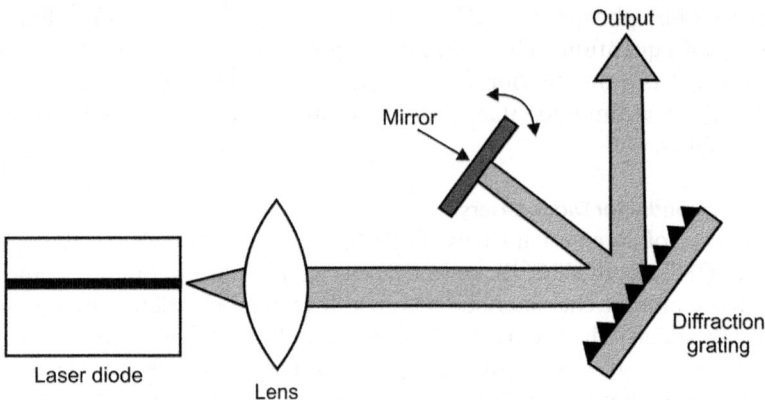

Figure 7.99 External cavity diode laser: Littman–Metcalf configuration.

7.10.2.7 Optically Pumped Semiconductor Lasers

Some of the shortcomings of the edge emitting conventional semiconductor lasers such as high divergence, asymmetry and astigmatism are largely overcome in vertical cavity surface emitting lasers (VCSELs). Features of VCSELs are further improved by introducing an external cavity to widen the horizon of its capabilities in vertical external cavity semiconductor lasers (VECSELs). However, the conventional electrically pumped VCSELs and VECSELs cannot match the output power capability of edge emitting devices as it is not practical to flood a large area with charge carriers through electrical pumping without using extended electrodes. Extended electrodes introduce too much of loss and therefore are not desirable. Solution lies in employing optical pumping by another diode laser as shown in the simplified schematic arrangement of Figure 7.100.

An optically pumped semiconductor laser (OPSL) is nothing but a vertical external cavity semiconductor laser that is pumped optically. The active region in this case comprises of alternate layers of a binary semiconductor material and tertiary semiconductor material quantum wells. The emission wavelength depends upon the stoichiometry and physical dimensions of quantum wells.

As is evident from Figure 7.100, the pump radiation enters the chip at an angle. The off-axis pump geometry facilitates use of multiple pumps arranged azimuthally for generation of higher output power. Off-axis pump geometry also ensures that the intra-cavity beam is not obstructed by pump beam relay optics. The non-collinear pumping fills the mode volume and introduces very little ellipticity.

While OPSL offers several advantages over their diode-pumped solid-state laser counter-parts in terms of far less stringent requirement of pump source wavelength band and absence of thermal lensing, the biggest advantage is in its ability to be readily customized over a wide range of emission wavelength and its superior longitudinal mode characteristics. For example, intra-cavity frequency-doubled and tripled outputs in optically pumped InGaAs gain chip cover almost full visible band from 355–577 nm.

OPSLs, because of their previously mentioned characteristic features, find extensive use in medical therapeutics, life sciences, forensics, light shows and scientific research. OPSL technology facilitates the user to think many new applications or improve many of the existing applications previously denied or limited by fixed wavelength feature of conventional electrically pumped semiconductor lasers. OPSL allows the use of best wavelength to fit the intended application rather than having to work around the closest fit wavelength.

Figure 7.101 shows a photograph of OPSLs of Genesis MX-series from M/s Coherent Inc. Genesis MX-series are high power CW optically pumped semiconductor lasers. Different

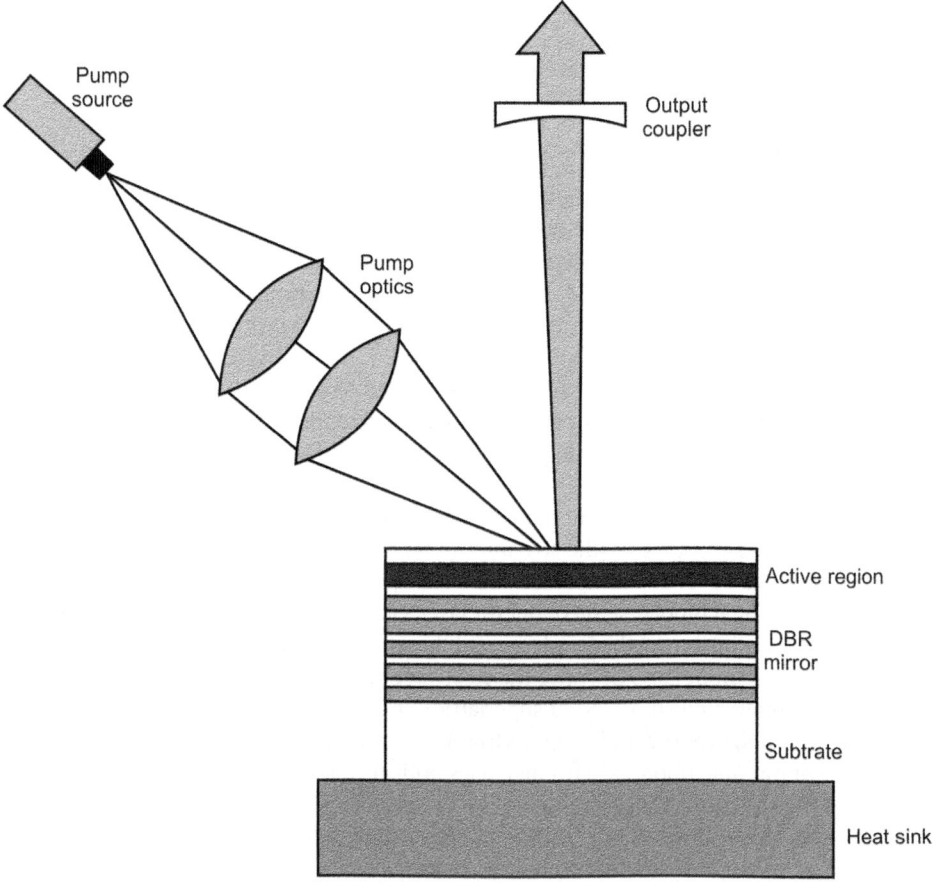

Figure 7.100 Optically pumped semiconductor laser schematic arrangement.

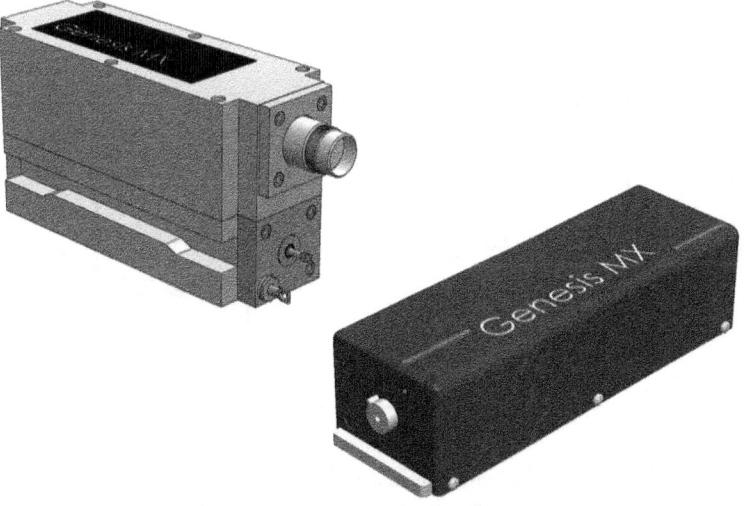

Figure 7.101 Genesis MX-series OPSLs.

variants of this series of OPSLs offer output powers in the range of 500 mW–20 W and emission wavelength band of 460–1154 nm. Genesis MX-series lasers are intended for use in medical, life sciences, Defence and scientific research applications.

7.10.2.8 Quantum Cascade Lasers

Quantum cascade lasers (QCLs) are compact high power wavelength agile semiconductor lasers that emit in mid- to far IR wavelength band. The upper limit of emission wavelength can extend even to the terahertz region. The semiconductor lasers described in earlier sections are all inter-band devices where emission of laser radiation takes place due to recombination of electrons in the conduction band and holes in the valence band across the band gap of the semiconductor material. The earliest double heterostructure devices had an operating wavelength that was exclusively dependent on the band gap energy as shown in Figure 7.102(a). In the case of quantum well structures, as shown in Figure 7.102(b), carriers are confined to energy levels within these wells opening up possibility of lower transition energies thereby extending the operating wavelength. Semiconductor diode lasers are bipolar devices, quantum cascade lasers on the other hand are unipolar devices and laser emission in this case occurs across the inter-subband, also called intra-band transitions of electrons in the conduction band as shown in Figure 7.102(c).

The gain region comprises of a periodic structure of an active region and an injector region as shown in Figure 7.103. The overall downward trend in energy levels is due an applied electric field. In the case of a conventional bipolar semiconductor laser, electrons and holes are annihilated after they undergo recombination across the band gap and therefore can play no further role in photon generation. In the case of unipolar QCL, after the electron has undergone an intra-band transition and generated a photon in a given period of superlattice, it can tunnel into the next period of the structure causing emission of another photon. As the electron traverses through the QCL structure, it causes emission of multiple photons. This increases optical gain leading to generation of higher output power. The intra-band transition energy in the case of QCL depends upon the design parameters such as layer thickness of quantum wells rather than the material system properties. As a result, emission wavelength can be tailored to be anywhere in a wide spectral band from mid-infrared to far infrared for the same semiconductor material.

In order to make a light-emitting device, the gain region is confined in an optical waveguide, which makes it possible to direct the photon emission into a collimated beam and also allows

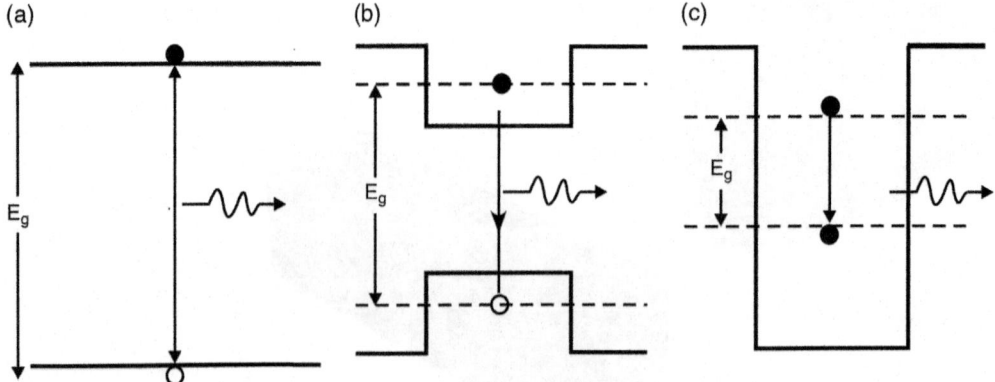

Figure 7.102 Inter-band and intra-band laser transitions in semiconductor lasers: (a) inter-band transition in a double heterostructure device, (b) inter-band transition in a quantum well structure and (c) intra-band transition in quantum cascade laser.

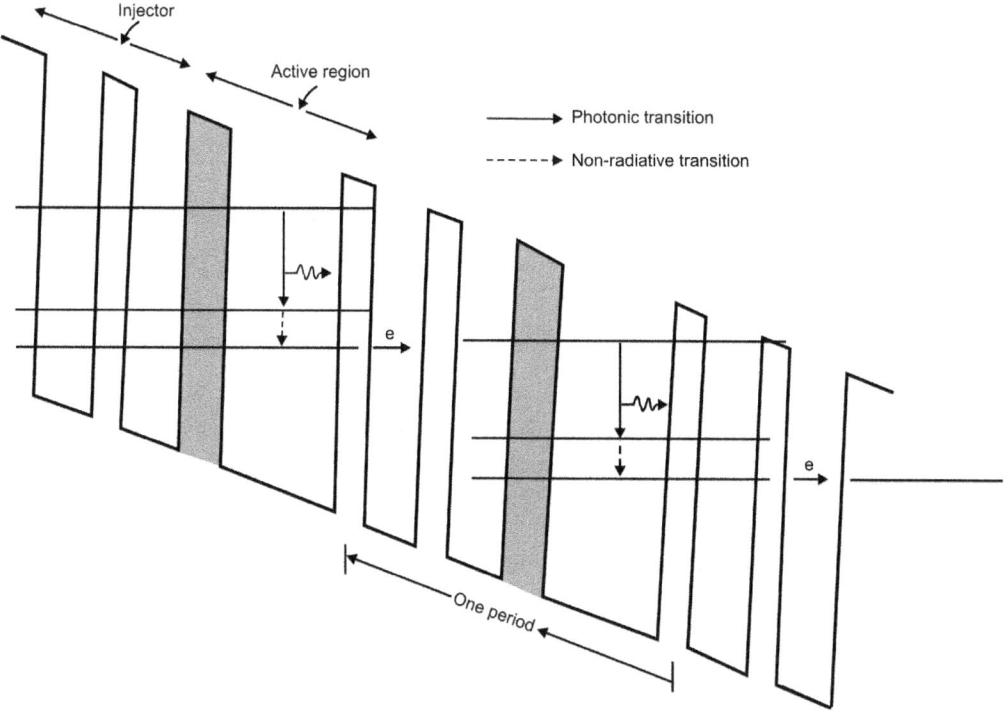

Figure 7.103 Gain region of a quantum cascade laser.

building of a laser resonator. Two types of optical waveguides, namely ridge waveguide and buried heterostructure waveguide are in common use. Detailed description of the two types of waveguides is, however, beyond the scope of the text.

Several material systems are in use for fabrication of QCLs. One of the earlier QCLs was fabricated by using InGaAs/InAlAs material system in an InP substrate. The quantum well depth of 0.52 eV of InP devices produces emission in the mid-IR spectral band. InP based QCLs have achieved high levels of performance with the development of high power CW devices operating above room temperature. Another common material system used for fabrication of QCLs is GaAs/AlGaAs. This material system offers variable quantum well depth depending on the aluminium fraction in the barriers. These devices have been successfully used to generate output in terahertz region. Yet another material system in use for QCLs is InGaAs/AlAsSb. With a quantum well depth of 1.6 eV, QCL devices with emissions at 3.0 μm have been fabricated. The emission wavelength gets further reduced to about 2.5 μm in InAs/AlSb QCLs with a quantum well depth of 2.1 eV. An important point worth mentioning here is that intra-band optical transitions are independent of relative momentum of conduction band and valence band minima with the result that QCLs based on indirect band gap materials have appeared feasible. Si/SiGe is one such material system that has shown promise for fabricating QCL devices. The alternating layers of two different semiconductor materials forming the quantum heterostructure are grown on to a substrate using various methods such as molecular beam epitaxy (MBE), metal organic vapour phase epitaxy (MOVPE) and metal organic chemical vapour deposition (MOCVD).

Different types of resonators are in use for building QCL devices. These include Fabry–Perot lasers, distributed feedback lasers and external cavity lasers. The Fabry–Perot cavity QCL is

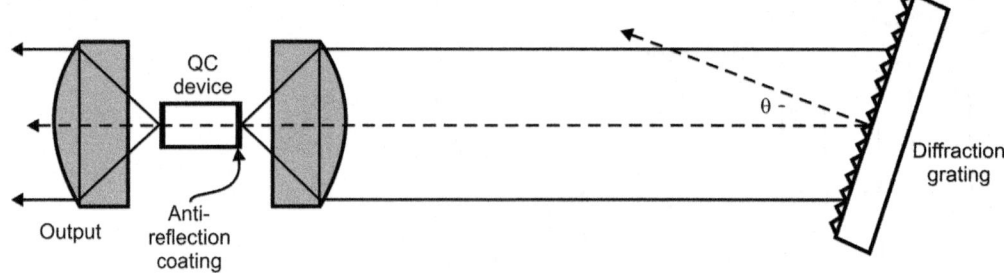

Figure 7.104 External cavity quantum cascade laser.

the simplest of all. The ends of the crystalline semiconductor device are cleaved on either end of the waveguide to form two parallel mirrors. The residual reflectivity offered by the cleaved facets creates the resonant cavity. Fabry–Perot QCLs are capable of generating high powers, though the output tends to be multimode at higher operating currents. The emission wavelength can be tuned by changing the operating temperature of the device. The DFB QCL has a distributed Bragg reflector built on the top of the waveguide to prevent it from emitting at other than the desired wavelength, otherwise it is similar to Fabry-Perot configuration. This allows the DFB QCL to operate at a single wavelength even at higher operating currents. The laser wavelength can be tuned by changing device temperature. Another mode of operation of DFB QCL is the chirped mode in which the laser is pulsed and the wavelength of laser is rapidly chirped during the course of the pulse. This facilitates rapid scanning of a given wavelength band. In the case of external cavity QCL, one or both of the cleaved facets are given anti-reflection coating. As a result, cleaved facets no longer provide cavity action. The optical cavity is created by mirrors placed external to the device. A frequency selective optical element such as a diffraction grating may be used in the external cavity to achieve both single wavelength operation and wavelength tuning (Figure 7.104).

High output power in the mid- to far IR spectral region extending to terahertz, tunability and room temperature operation make quantum cascade lasers highly suitable for a wide range of applications in remote sensing of environmental gases and atmospheric pollutants, vehicular cruise control and collision avoidance in poor visibility conditions, medical diagnostics, industrial process control and homeland security.

7.10.2.9 Lead Salt Lasers

Lead salt semiconductor lasers are semiconductor diode lasers generating wavelength tunable pulsed and CW output in the mid-IR spectral band, which is particularly useful in high-resolution absorption spectroscopy used for detection and identification of trace gases with high sensitivity and selectivity. In general, tunable CW mid-IR laser sources are classified as class A and class B sources. The former category includes sources that generate tunable mid-IR laser radiation directly from the gain media, which could be a gas discharge, a semiconductor material, rare earth and transition metal doped bulk solid-state material or optical fibre. The class B category includes those laser sources where mid-infrared output is generated by optical parametric frequency conversion of a near infrared radiation. Most of the mid-infrared laser sources using solid-state, gas and semiconductor lasing media have been covered earlier in the chapter and previous chapters on solid-state lasers and gas lasers. In the paragraphs to follow are briefly covered operational basics and characteristic features of lead salt lasers.

The lead salt semiconductor laser is a P-N junction diode laser that consists of a single crystal of lead telluride (PbTe), lead selenide (PbSe), lead sulfide (PbS) or their alloys with themselves or with strontium selenide (SnSe), strontium telluride (SnTe), cadmium sulfide (CdS) and other materials. The structure of laser cavity is similar to the conventional semiconductor diode laser employing a Fabry–Perot resonator cavity comprising of two parallel end faces. The injection current populates the near empty conduction band and laser radiation is emitted by the process of stimulated emission across the band gap. The band gap energy in the case of lead salt semiconductor lasers is very small, in the range of 0.25–0.30 eV. These lasers require cryogenic cooling for population inversion and laser action. The band gap energy depends on semiconductor composition and temperature. The band gap energy and hence the operating wavelength may be tailored by either varying the stoichiometry between lead and other elements or by using different alloys of lead. Also, wavelength tuning over $100\,cm^{-1}$ by changing temperature or $10\,cm^{-1}$ by changing injection current are possible. While temperature tuning is basically due to change in band gap energy, injection current tuning is due to change in refractive index of the active region. A temperature tuning rate of $2.0–5.0\,cm^{-1}/K$ and injection current tuning rate of $0.02–0.07\,cm^{-1}$ are achievable. The temperature tuning mechanism is relatively sluggish as it involves changing the temperature of the entire laser package. On the other hand, the current tuning process is very rapid.

Lead salt lasers emit in 3–30 μm band and are characterized by large beam divergence and astigmatism, which puts particularly stringent requirements on the first optical element of the collection optics. Typical CW power level from lead salt lasers is in the range of 0.1–0.5 mW. CW operation has been reported for temperatures above 200 K and pulsed operation to above 80 °C.

7.10.3 Characteristic Parameters

Important characteristic parameters of diode lasers relate either to their I-V characteristics or the output beam characteristics. These include threshold current, slope efficiency and linearity of laser operation in the case of former and beam divergence, line width and beam polarization in the case of latter. Most of the characteristic parameters are sensitive to changes in temperature. Different parameters and their sensitivity to temperature changes are briefly described in the following paragraphs.

7.10.3.1 Threshold Current

Threshold current is the minimum forward-biased injection current needed to achieve sustained laser action. When the injection current is below the threshold value, most of the input electrical energy is dissipated as heat and conversion to light output is highly inefficient. Higher threshold current therefore means that more electrical energy must be dissipated as heat energy. Current density is defined as ratio of threshold current to the active area and measured in Amperes per cm^2. It is also referred to as the figure of-merit of the diode laser, as it predominantly decides the laser's lifetime. Higher threshold current density shortens the lifetime of the laser diode. I-V characteristics of the laser diode indicating threshold current were shown in Figure 7.89(b) and the light power versus drive current characteristics of a semiconductor diode laser indicating lasing threshold current are shown in Figure 7.90.

Threshold drive current is a strong function of temperature and increases rapidly with temperature. Change in threshold current $I_s(\Delta T)$ due to incremental change in temperature ΔT can be computed from eqn. 7.30.

$$I_S(\Delta T) = I_S(T) \times \left[e^{(\Delta T/T_0)} - 1 \right] \tag{7.33}$$

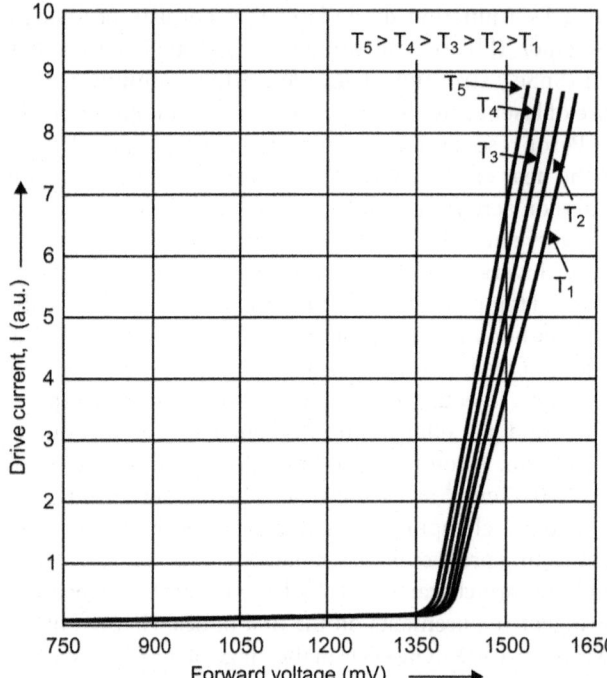

Figure 7.105 Temperature dependence of I-V characteristics.

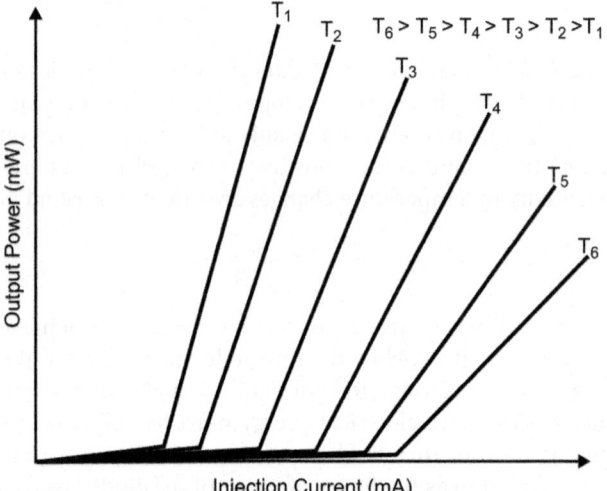

Figure 7.106 Temperature dependence of light power versus drive current characteristics.

Where,

$I_s(T)$ is the threshold current at absolute temperature T

T_0 is a substrate specific characteristic temperature

As an example, T_0 is in the range of 120—230 K for GaAlAs diode laser and in the range of 60–80 K for InGaAsP lasers. The lower the value of T_0, the more sensitive the laser is to changes in temperature. A shift in threshold current occurs primarily due to temperature dependency of carrier concentration in active layer and probability of non-radiative recombination processes. Figures 7.105 and 7.106 show the effect of change in temperature on I-V

Figure 7.107 Beam divergence in two orthogonal planes in edge emitting semiconductor lasers.

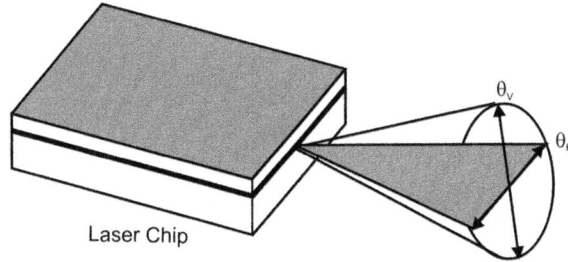

Laser Chip

characteristics and light power versus drive current characteristics of semiconductor lasers, respectively. Increase in temperature also adversely affects the lifetime of the laser as increase in temperature leads to increase in threshold current density. Lifetime approximately doubles for every $10°$ decrease in chip temperature. This is taken care of by mounting the chip on a heat sink.

7.10.3.2 Slope Efficiency

Slope efficiency is determined by the slope of the characteristic I-V curve above the threshold current and is measured in mW/mA (or W/A). In the case of a pigtailed device, slope is reduced by a factor depending upon the coupling efficiency of laser power into the fibre.

Slope efficiency is strongly dependent upon temperature and decreases within increase in temperature as shown in Figure 7.106 earlier while describing the temperature dependency of threshold current.

7.10.3.3 Beam Divergence

Laser diodes characteristically produce a highly diverging laser beam with the exception of surface emitting diode lasers. Higher divergence is primarily influenced by diffraction of light waves as they are coupled out of the chip. Also, due to rectangular shaped active light-emitting area with strongly differing edge lengths, the divergence in the two orthogonal planes is different as shown in Figure 7.107. Divergence in the plane parallel to the plane of the active layer is relatively much smaller than that in the plane perpendicular to the active layer. As a result, the laser beam appears as an elliptical spot at some distance from the laser. If required, it is possible to circularize the elliptical beam with the help of a cylindrical lens, which refracts the light in parallel direction. VCSEL produce a relatively more symmetrical beam and the beam divergence is also relatively lower due to their large emitting area.

7.10.3.4 Line Width

Line width is another important specification. In the case of the gain-guided lasers described in Section 10.4, the envelope of gain profile typically has a 3 dB width of 2–3 nm, which corresponds to a frequency range of about 1000 GHz at 800 nm. Such a gain bandwidth profile can support a large number of longitudinal modes. In the case of an index-guided laser (described in Section 7.10.4), one spectral line is dominant with the result that the line width is much narrower, typically 10^{-2} nm. This corresponds to a frequency spread of few GHz at 800 nm. In the case of DFB lasers, which are also index guided, the line width is still smaller, typically in the order of 10^{-4} nm, which corresponds to a frequency spread of few MHz at 800 nm.

The spectral profile of semiconductor lasers is also strongly affected by temperature changes. Both the gain profile as well as individual lines shift to longer wavelengths with increase in temperature. Figure 7.108 shows the effect of change in temperature on gain profile. The effect

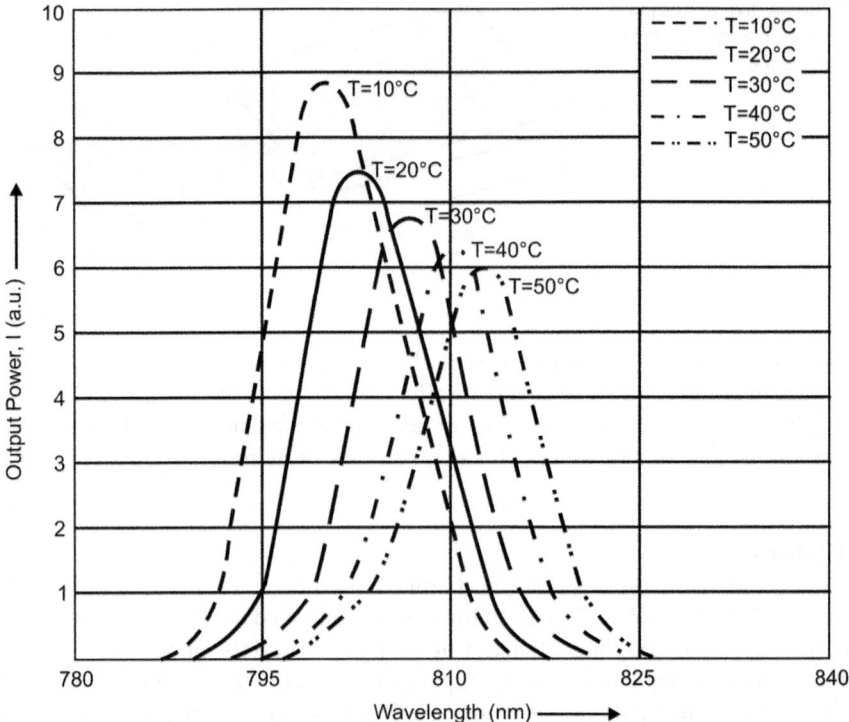

Figure 7.108 Effect of change in temperature on gain profile.

on individual lines is due to increase in resonator length accompanied by increase in refractive index. Typical values of temperature coefficient are as follows.

$$\text{Temperature coefficient}\,(\text{GaAlAs Lasers}) = 0.24\,\text{nm/K}\,(\text{envelope variation})$$
$$= 0.12\,\text{nm/K}\,(\text{individual line variation})$$

$$\text{Temperature coefficient}\,(\text{InGaAsP Lasers}) = 0.3\,\text{nm/K}\,(\text{envelope variation})$$
$$= 0.08\,\text{nm/K}\,(\text{individual line variation})$$

7.10.3.5 Beam Polarization

Beam polarization is yet another important parameter. Diode lasers emit almost linearly polarized light if driven above threshold, which is influenced by polarization dependency of the reflection factor of the emission area of the crystal.

7.10.4 Gain-Guided and Index-Guided Diode Lasers

The narrow strip of the active layer that is responsible for generating laser output defines whether the laser is gain guided or index guided. In the case of *gain-guided lasers*, an insulating layer on the top of the laser chip confines the flow of injection current only to a narrow stripe in the active layer with the result that only this region has sufficient gain to produce sustained laser action. There is no gain at the sides, so those regions do not emit light, even though no physical boundary separates the stripe from the rest of the active layer.

In the case of *index-guided lasers,* the stripe from where laser emission takes place is defined by semiconductor composition and associated change in refractive index. The current flows only through the central 'mesa' and the stripe of the active layer buried below it. The layers that border the sides of the active layer have a different composition and different refractive index, trapping the light in the active stripe as the adjacent layers confine light in the active layer in a double heterostructure laser. Index-guided lasers produce better beam quality and are used in most laser applications.

7.10.5 Handling Semiconductor Diode Lasers

Diode lasers exhibit high reliability and a long lifetime exceeding 100 000 h provided certain precautions are taken while handling them and also in the design of the driver circuits used to power them. Diode lasers are particularly sensitive to electrostatic discharge, short duration electric transients such as current spikes, injection current exceeding the prescribed limit and reverse voltage exceeding the breakdown limit. Damage to the laser diode often manifests itself in the form of reduced output power, shift in the value of threshold current and its inability to be able to be focused to a sharp spot. The electrostatic discharge caused by human touch is the most common cause of premature failure of a diode laser.

In order to protect the diode lasers from these failure modes, the driver circuit should be carefully designed and should have all the features recommended by the manufacturer of the laser. The driver should be a constant current source with inbuilt features like soft start, protection against transients, interlock control for the connection cable to the laser and safe adjustable limit for injection current. In case the laser is to be operated in the pulsed mode, the injection current should be pulsed between two values above the lasing threshold rather than between cut-off and lasing mode. Leading manufacturers of diode lasers offer a wide variety of suitable current sources for low, medium and high power laser diodes. They also offer temperature controllers to stabilize the output wavelength. Drivers and temperature controllers for diode lasers are discussed in Chapter 8.

Illustrated Glossary

Absorption In an absorption transition, an electron or an atom or a molecule makes a transition from a lower level to a higher level only if the incident photon has energy ($= h\nu$) equal to the transition energy, which is the difference in the energies of the two involved energy levels; that is., $\Delta E = h\nu$.

Alexandrite Laser This is type of vibronic laser. Alexandrite is the common name for chromium-doped chrysoberyl. The output wavelength is tunable in the range of 70 850 nm.

Carbon Dioxide Laser Among the family of gas lasers, carbon dioxide (CO_2) laser is the most widely used and diversely exploited type of gas laser. The laser medium in this case is a gas mixture of carbon dioxide (CO_2), helium (He) and nitrogen (N_2). CO_2 is the lasing species and laser transitions correspond to the energy levels associated with molecular vibration and rotation modes of CO_2 molecule. Transition to a lower level corresponding to symmetric stretching produces 10.6 μm output while its transition to another lower level corresponding to symmetric bending produces 9.6 μm output. Nitrogen participates in the process of creation of population inversion by acting as an intermediate step in the same way as helium does in the case of a He-Ne laser. Helium in the case of a CO_2 laser helps in depopulating the lower laser level. Carbon dioxide laser has a Doppler broadened gain curve that is 60 MHz wide at 10.6 μm output. Carbon dioxide lasers employ either a sealed tube or flowing gas construction. The gas mixture is excited by an electric discharge. Both DC as well as RF excitation has been successfully and widely used in the case of CO_2 lasers.

Cavity Dumping This phenomenon is a slight variation of the Q-switching process with the difference that in the case of the former, both of the resonator cavity mirrors are 100% reflective. When the active medium is pumped, energy is initially stored in the population inversion as is done in the case of Q-switching. Initially the loss is kept at a very low value. Since both of the resonator mirrors are 100% reflective, the amplified light remains trapped within the cavity. As peak irradiance is reached, the loss is switched to a high value thus ejecting the intra-cavity circulating energy in the form of a pulse.

Chemical Laser This derives the energy required to produce population inversion and consequent laser emission from a chemical reaction. Importance of chemical lasers lies in their ability to generate continuous-wave output power level reaching several megawatts. There are two categories of chemical lasers. One category has hydrogen fluoride (HF) and deuterium fluoride (DF) and the other has the two iodine lasers including a chemical oxygen iodine laser (COIL) and all gas-phase iodine laser (AGIL).

Coherence Light is said to be coherent when different photons (or the waves associated with those photons) have the same phase and this phase relationship is preserved with respect to time.

Concentric Resonator A resonator in which both end components are spherical mirrors (radius of curvature of each component = $d/2$ where d is resonator length).

Confocal Resonator A resonator in which both end components are spherical mirrors (radius of curvature of each component = d where d is resonator length).

Continuous-Wave (CW) Output In CW output, the laser produces a continuous output as long as the laser is pumped. Pumping needs to be continuous in this case. The laser resonator in this case in its simplest form comprises of just the active medium and the resonator mirrors and does not contain any switches.

Distributed Feedback (DFB) Laser DFB lasers offer single longitudinal and transverse mode operation. Here, a diffraction grating is etched very close to the active layer. The grating provides a wavelength selective feedback to the gain region. The feedback causes interference and only the narrow wavelength region for which the interference is constructive is allowed to sustain.

Divergence (Laser Beam) Beam divergence is an indicator of the spread in the laser beam spot as it travels away from the source. It is a function of the wavelength λ and size of output optics.

Doppler Broadening This is an increase in line width due to inherent random motion of atoms and molecules.

Duty Cycle This is the ratio of pulse width to the time interval between two successive pulses.

Dye Laser This uses organic dye as the active medium. A dye laser's importance arises out of its inherent ability to produce a coherent output that is tunable over a wide range of wavelength from near ultraviolet through visible to near infrared region. The tunability properties of dye lasers result from the energy level diagram of the active medium. The active medium is an organic dye characterized by upper and lower laser levels that are bands of a large number of sub-levels rather than single energy levels. Dye lasers are optically pumped. They are either flash lamp pumped or pumped by pulsed lasers. Also, their ability to produce ultra-short pulses results from the wide gain bandwidth profile characteristic of dye lasers.

Electrical Pumping This employs pumping by electrical discharge and is common in gas lasers. The excited electrons in the gas discharge plasma transfer their energy to the lasing species either directly or indirectly through the atoms or molecules of another element.

Erbium:Glass Laser An erbium-doped glass laser produces output at 1540 nm, which is an attractive wavelength as an eye-safe alternative to neodymium-doped YAG (or glass) based military Laser Rangefinders and Laser Target Designators.

Erbium:YAG Laser An erbium-doped YAG laser produces output at 2940 nm. This value holds promise for medical applications in the field of plastic surgery due to its extremely large absorption by water in tissue.

Excimer Laser Excimer lasers are pulsed lasers capable of providing pulse energies in the order of Joules. These are the most powerful practical lasers emitting in the ultraviolet region. The active medium is a rare gas halide. Some of the more popular and practical Excimer lasers include F_2 (157 nm), ArF (193 nm), KrCl (222 nm), KrF (249 nm), XeCl (308 nm) and XeF (350 nm).

External Cavity Diode Laser This is a semiconductor diode laser whose resonator is completed with one or more optical components outside the diode laser chip

Fall Time This refers to the time duration between 10 and 90% of the peak amplitude of the pulse, respectively, during the falling portion of the laser pulse.

Fibre Laser This is a type of solid-state laser where the gain medium is a glass fibre doped with rare earth element ions such as neodymium (Nd^{3+}), erbium (Er^{3+}), ytterbium (Yb^{3+}), thulium (Tm^{3+}), holium (Ho^{3+}) or praseodymium (Pr^{3+}). Fibre coupled semiconductor laser diodes or fibre lasers are used as the pump source and resonator comprises of active medium bounded by dielectric mirrors or fibre Bragg gratings. All fibre lasers, due to absence of any free-space optics are immune to any misalignment and are more robust and exhibit excellent long term stability. While fibre lasers offer far superior performance when it comes to high average power lasers with high beam quality, it is the bulk solid-state lasers that have an edge in high pulse energy, high peak power applications. Wavelengths emitted by common fibre lasers are in the regions of 1.0–1.1 µm from ytterbium-doped lasers, 1.52–1.57 from erbium-doped lasers and 1.9–2.1 µm from thulium and holmium doped lasers.

Four-Level Laser In a four-level laser system, the atoms or molecules are excited out of the ground state to an upper highly excited short-lived energy level. Remember that the lower laser level here is not the ground state. In this case, the number of atoms or molecules required to be excited to the upper level would depend upon the population of the lower laser level, which is much smaller than the population of the ground state.

Free Electron Laser In the case of a free electron laser, the lasing medium is a beam of free electrons completely unattached to any atoms. Unlike conventional lasers that rely on bound atomic or molecular states, free electron lasers use a relativistic electron beam as the active medium. The accelerated beam of electrons having desired energy can be obtained from well-established technology of charged-particle accelerators. The free electron laser is tunable over a wide range of wavelengths in the region from X-rays to microwaves. The tunability is achieved by varying electron energy, E.

Free-Running Output This is a quasi-CW mode of operation in which the laser operates in the CW mode for a time period that is in the order of a few hundreds of microseconds to a few milliseconds. This time period generally equals or is slightly longer than the storage time of the active medium, which in turn is of the same order as the pump input pulse.

Gain Coefficient (Lasing Medium) Gain of the active medium is the extent to which this medium can produce stimulated emission. Gain coefficient is the gain expressed as a percentage per unit length of the active medium. Gain coefficient of 10% per centimetre implies that 100 photons with the same transition energy as that of excited laser medium become 110 photons after travelling 1 cm of the medium length.

Gas Dynamic Carbon Dioxide Laser The gas dynamic laser is an unconventional carbon dioxide laser that derives its energy from the combustion of a suitable fuel-oxidizer mixture, which means that it does not require any electrical energy for its operation. A gas dynamic laser achieves population inversion by rapid expansion of high-temperature high-pressure laser gas mixture produced during combustion to a near vacuum in an adiabatic process through an integrated supersonic nozzle bank. Though the expansion reduces the gas temperature, a large number of excited molecules are still in the upper laser level. Population inversion is created if the reduction in pressure and temperature downstream of the nozzle bank takes place in a time that is much shorter than the vibrational relaxation time of the upper laser level corresponding to asymmetric stretching mode of carbon dioxide coupled with nitrogen.

Gain Guided Laser In the case of gain guided lasers, an insulating layer on the top of the laser chip confines the flow of injection current only to a narrow stripe in the active layer with the result that only this region has sufficient gain to produce sustained laser action. There is no gain at the sides, so those regions do not emit light, even though no physical boundary separates the stripe from the rest of the active layer.

Helium-Neon Laser The He-Ne laser is one of the most commonly used types of gas lasers. The active medium is the laser gas mixture of helium and neon, which is predominantly helium with only 10–20% of neon. The gas mixture is excited by an electrical discharge. A helium-neon laser can possibly produce laser output at either the visible (543.5 nm or 632.8 nm) or IR (1.153 μm or 3.391 μm) wavelengths, depending upon the operating conditions and choice of optics. The wavelength of 632.8 nm has become the standard He-Ne laser wavelength and is also the most widely used. A He-Ne laser has a Doppler broadened gain curve that is 1400 MHz wide at 632.8 nm output.

Hemifocal Resonator A resonator in which one of the end components is a plane mirror (radius of curvature = infinity) and the other is a spherical mirror (radius of curvature = $2d$ where d is resonator length).

Hemispherical Resonator A resonator in which one of the end components is a plane mirror (radius of curvature = infinity) and the other is a spherical mirror (radius of curvature = d where d is length of resonator).

Heterojunction Lasers (Semiconductor Lasers) In the case of heterojunction lasers, the active layer and either one or both of the adjacent layers are of different material. If only one of the adjacent layers is of a different material it is called a simple heterojunction and if both are different it is called a double heterojunction.

Homojunction Lasers (Semiconductor Lasers) In the case of homojunction semiconductor lasers, all layers are of the same semiconductor material. One such example is a GaAs/GaAs laser.

Index-Guided Laser In the case of index-guided lasers, the stripe from where laser emission takes place is defined by semiconductor composition and associated change in refractive index. The current flows only through the central 'mesa' and the stripe of the active layer buried below it. The layers that border the sides of the active layer have a different composition and different refractive index, trapping the light in the active stripe as the adjacent layers confine light in the active layer in a double heterostructure laser. Index-guided lasers produce better beam quality and are used in most laser applications.

Irradiance Irradiance, also referred to as power density, is defined as the power per unit area of the laser radiation falling on the target. It is expressed as W/m^2. This parameter is particularly important when the laser radiation is used to illuminate a receiving system.

Laser Spectrum Analyser This is used to carry out modal analysis of lasers operating in more than one longitudinal or transverse mode of the laser cavity.

Lead Salt Semiconductor Laser This is a P-N junction diode laser that consists of a single crystal of lead telluride (PbTe), lead selenide (PbSe), lead sulfide (PbS) or their alloys with themselves or with strontium selenide (SnSe), strontium telluride (SnTe), cadmium sulfide (CdS) and other materials. The structure of laser cavity is similar to the conventional semi-conductor diode laser employing a Fabry–Perot resonator cavity comprising two parallel end faces. The injection current populates the near empty conduction band and laser radiation is emitted by the process of stimulated emission across the band gap.

Longitudinal Mode Longitudinal modes are different resonant frequencies that oscillate simultaneously in a laser resonator when the inter-mode spacing as dictated by the cavity length is significantly smaller than the line width.

Metal Vapour Laser This is type of gas laser where the active medium is either ionized metal atom as is the case in helium-cadmium family of lasers or neutral atoms as in the case of copper vapour and gold vapour lasers. The helium-cadmium lasers produce emissions at 325 nm and 353.6 nm (UV) and 441.6 nm (blue). The neutral metal vapour lasers emit in the visible spectrum. Copper vapour lasers emit at 511 nm (green) and 578 nm (yellow). Gold vapour laser emits at 628 nm (red). Pumping takes place through electric discharge.

Mode Locking The phenomenon of mode locking can be considered to be a process that locks together in phase a cluster of photons. The mode locking element transmits this cluster every time it passes through it while bouncing back and forth between the cavity mirrors. The repetition rate of the mode-locked pulses is equal to the round trip transit time of the resonator cavity. Mode locking technique can be used to generate ultra-short laser pulses with pulse width s approaching the theoretical limit given by $0.441/B$ where B is line width.

M^{2-}value This is a measure of beam quality. M^2, therefore, is defined as the ratio of the divergence of the real beam to that of a theoretical diffraction-limited beam of the same waist size with a Gaussian beam profile (TEM_{00}-mode). It is also referred to as the beam propagation ratio as per ISO-11146 standard.

Monochromaticity This refers to single frequency or wavelength property of the radiation. Laser radiation is monochromatic and this property has its origin in the stimulated emission process by which a laser emits light. The stimulated photon has the same frequency, phase and polarization as those of the stimulating photon.

Nd:Cr:GSGG Laser This is a type of neodymium-doped solid-state laser. Nd:Cr:GSGG as the laser material is primarily important for the reason that its absorption band has a better spectral match to the emission from flash lamps as compared to Nd:YAG with the result that Nd:Cr:GSGG lasers offered significantly higher electrical input to laser output efficiency as compared to Nd:YAG.

Nd:Glass Laser Nd:glass is the other important neodymium-doped solid-state laser after the Nd:YAG laser. The host materials for Nd:glass lasers are silicate, phosphate and fused silica glasses with silicate and phosphate being more the common ones. While Nd:YAG is mainly known for its high gain and good thermal and mechanical properties, Nd:Glass is capable of being grown in large sizes with diffraction-limited optical quality. The Nd:Glass laser is the preferred option for high-energy and high peak power applications requiring either single pulse or low repetition rate operation such as laser fusion research.

Nd:YAG laser Nd:YAG is the most important and most widely used of all neodymium-doped lasers because of its high gain and good thermal and mechanical properties. The active medium is neodymium-doped yttrium aluminium garnet where trivalent neodymium replaces trivalent yttrium.

Due to its good thermal and optical properties, Nd: YAG lasers can be used both in CW as well as high repetition rate Q-switched pulsed mode. Average and peak powers in excess of 1 kW and 100 MW, respectively, can be achieved in these lasers.

Nd:YLF laser This is a type of neodymium-doped solid-state laser. The active medium is neodymium-doped yttrium lithium fluoride. The advantages of Nd:YLF lasers include fewer heat related problems due to smaller change in its refractive index with temperature and higher energy storage capability leading to its ability to generate higher energy Q-switched pulses. In addition, the Nd:YLF crystal is birefringent, which allows generation of two output wavelengths at 1047 nm and 1053 nm, each with its own polarization orientation.

Nd:YVO4 Laser This is a type of neodymium-doped solid-state laser. Neodymium-doped yttrium vanadate as a laser material is important because of several properties that make it particularly attractive for laser diode pumping. These include a large stimulated emission cross-section and a strong broadband absorption around 808 nm.

Optical Pumping Optical pumping employs optical radiation. Optical pumping is employed for those lasers that have a transparent active medium. Solid-state and liquid dye lasers are typical examples. The most commonly used pump sources are the flash lamp in the case of pulsed and the arc lamp in the case of continuous-wave solid-state lasers. Semiconductor diode lasers are also used for pumping solid-state lasers.

Photonic Crystal Fibre-Laser These use a special type of fibre called a photonic crystal fibre (PCF) that is different from the conventional step index fibre. In comparison to the conventional step index fibre that derives its wave guiding properties from a spatially varying glass composition; the basic photonic crystal fibre is an optical fibre with a structured pattern of air holes, also called voids, that run parallel to the axis all along its length. The PCF derives its wave guiding properties from the presence of air holes.

Plane-Parallel Resonator A plane-parallel resonator in which both end components are plane mirrors (radius of curvature of both components is infinity) and are placed precisely at right angle to the laser axis. It is a type of stable resonator.

Population Inversion Population inversion is a condition in which the number of excited species in an upper energy level is higher than the population of a lower energy level. If we want the stimulated emission to dominate over absorption and spontaneous emission, establishing the condition of population inversion is an essential requirement.

Pressure Broadening This, also called collisional broadening, is line broadening due to increased pressure.

Pulse Energy (Pulsed Lasers) This is defined with respect to pulsed lasers. It is in fact the area under the power versus time curve representing the laser pulse.

Pulse Repetition Frequency (PRF) PRF of a pulsed laser is the number of laser pulses produced per second.

Q-Switching This is a mechanism of producing short laser pulses with pulse width in the order of a few nanoseconds. The quality factor (Q-factor) of the resonator cavity is switched from low value to a high value. Initially, when the laser is pumped, a low value of Q-factor of the resonator cavity prohibits laser action. When the inversion density reaches almost its peak value, the Q-factor of the resonator cavity is rapidly switched to the high value leading to a steep fall in the loss value. This manifests itself in the production of a short laser pulse at the output.

Quantum cascade laser (QCLs) QCLs are compact high power wavelength agile semiconductor lasers that emit in the mid- to far IR wavelength band. The upper limit of emission wavelength can extend even to the terahertz region. Conventional semiconductor lasers are all inter-band devices where emission of laser radiation takes place due to recombination of electrons in the conduction band and holes in the valence band across the band gap of the semiconductor material. Quantum cascade lasers on the other hand are unipolar devices and laser emission in this case occurs across inter-subband also called intra-band transitions of electrons in the conduction band.

Quantum Dot Laser (Semiconductor Lasers) This type of laser uses quantum dot structures as the active medium. While a quantum well structure can confine electrons in only one dimension; that is, in the plane of quantum well layer, quantum dot structures provide confinement in three dimensions. This leads to further improvement in laser efficiency over quantum well lasers.

Quantum Wire Laser (Semiconductor Lasers) This type of laser uses quantum wire structures as the active medium. While a quantum well structure can confine electrons in only one dimension, that is, in the plane of quantum well layer, quantum wire structures provide confinement in two dimensions. This leads to further improvement in laser efficiency over quantum well lasers.

Quantum Well Laser (Semiconductor Laser) A quantum well laser uses quantum well structures as the active medium. When a thin layer of a semiconductor material with a relatively smaller band gap is sandwiched between two thick layers with larger band gaps, a structure known as a quantum well is formed and the electrons passing through the semiconductor are captured in the thin layer. Though these electrons have sufficient energy to be free in the small band gap quantum well layer, it is not enough to allow them to enter the large band gap thicker layers. The thicker layers thus help in confining the captured electrons to the quantum well. If the quantum well is placed in the semiconductor junction, the concentration of electrons as they recombine with holes to emit photons leads to increase in efficiency and decrease in lasing threshold.

Radiance Radiance, also referred to as brightness, is usually defined with respect to the laser source. It is the power emitted per unit area per unit solid angle. It is expressed in W/M^2 steradian.

Rare Gas Ion Lasers These belong to the category of gas lasers. Argon-ion and krypton-ion lasers are the popular candidates belonging to the family of rare gas ion lasers. The laser emission in the case of argon-ion and krypton-ion lasers occurs from the transitions made by ionized argon and krypton atoms. Though both argon-ion and krypton-ion lasers emit a range of wavelengths from ultraviolet through visible to near infrared, argon-ion laser has strong emission at 488 nm (blue) and 514.5 nm (green) and krypton-ion laser emits strongly at 647.1 nm (red). The two gases can be put into a single tube and excited simultaneously to get laser lines throughout the visible spectrum.

Resonator (Laser) The active laser medium within the closed path bounded by two mirrors (or lenses) constitutes the basic laser resonator provided it satisfies the condition of $2L = n\lambda$ where L = length of resonator and n = integer. Also, resonator structures of most practical laser sources would be more complex than the simplistic arrangement.

Rise Time Rise time refers to the time duration between 10 and 90% of the peak amplitude of the pulse, respectively, during the rising portion of the laser pulse.

Ruby Laser This is a type of solid-state laser where the active medium is sapphire; that is, aluminium oxide (Al_2O_3) doped with 0.01–0.5% of chromium. The laser emits at 694.3 nm (R_1-line). Another possible emission line is at 692.9 nm (R_2-line) but the former is the predominant emission line. The three-level energy structure of ruby is responsible for its poor efficiency (0.1–1%).

Semiconductor Laser Semiconductor lasers use a semiconductor material as the active medium. The optical gain in this case is usually achieved by a process of stimulated emission at an inter-band transition triggered by prevailing conditions of high carrier density in the conduction band. In the case of diode lasers, high carrier density in conduction band is caused by injection current. The emission of radiation is due to recombination of holes and electrons in a forward-biased PN junction diode. An injection diode laser powered by an injected electrical current is the most practical form of diode laser distinguished from an optically pumped diode laser. Some of the popular semiconductor laser types and the corresponding emission wavelength band include AlGaInP/GaAs (heterojunction) at 620–680 nm, $Ga_{0.5}In_{0.5}P$/GaAs (heterojunction) at 670–680 nm, GaAlAs/GaAs (heterojunction) at 750–870 nm, GaAs/GaAs (homojunction) at 904 nm and InGaAsP/InP (heterojunction) @1100–1650 nm.

Slope Efficiency Slope efficiency is determined by the slope of the characteristic I-V curve above the threshold current and is measured in mW/mA (or W/A). Slope efficiency is strongly dependent upon temperature and decreases within increase in temperature.

Spatial Coherence Spatial coherence is preservation of phase across the width of the beam. It describes the correlation in phase of different photons transverse to the direction of travel.

Spontaneous Emission Spontaneous emission is the phenomenon in which an atom or molecule undergoes a transition from an excited higher energy level to a lower level all by itself without any outside intervention or stimulation and in the process emits a resonance photon. The rate of spontaneous emission process is proportional to the related Einstein coefficient.

Spot Size Spot size or the beam diameter is defined as the distance across the centre of the beam for which the irradiance equals 0.135 ($1/e^2$) times the maximum value at the centre.

Stable Resonator A stable resonator is the one in which the photons can bounce back and forth between the end components indefinitely without getting lost out by the sides of the components. In such a resonator, due to the focusing nature of one or both components, the light flux remains within the cavity. Plane parallel, hemispherical, hemifocal, concentric and confocal resonators are examples.

Stimulated Emission In the case of stimulated emission, there first exists a photon called a stimulating photon with an energy equal to the resonance energy ($h\nu$). This photon perturbs another excited species (atom or molecule) and causes it to drop to the lower energy level, in the process emitting a photon of the same frequency, phase and polarization as that of the stimulating photon. The rate of stimulated emission process is proportional to the population of the higher excited energy level and the related Einstein coefficient.

TEA CO_2 Laser A transversely excited atmospheric pressure (TEA) CO_2 laser is a high power pulsed CO_2 laser. The gas pressure is around one atmosphere and the discharge is transverse to the laser axis. The primary attraction of a TEA CO_2 laser is in its ability to generate short intense pulses and extraction of high power per unit volume of laser gas mixture. Pulse durations of few tens of nanoseconds to few microseconds and pulse energies from few millijoules to hundreds of joules at repetition rates up to few hundreds of hertz are achievable.

Temporal Coherence This is preservation of phase relationship with time. Formation of fringes in Michelson interferometer illustrates the phenomenon of temporal coherence. Temporal coherence is measured as coherence length or coherence time.

Three-Level Laser In a three-level laser system, the lower level of laser transition is the ground state (the lowermost energy level). The atoms or molecules are excited to an upper level higher than the upper level of the laser transition. This upper level has a relatively much shorter lifetime compared to the lifetime of the upper laser level, which is a metastable level. As a result, the excited species rapidly drops to the metastable level. A relatively much longer lifetime for the metastable level ensures a population inversion between the metastable level and the ground state provided that at least more than half of the atoms or molecules in the ground state have been excited to the uppermost short-lived energy level. The laser action occurs between the metastable level and the ground state.

Threshold Current Threshold current is the minimum forward-biased injection current needed to achieve sustained laser action.

Titanium-Sapphire Laser This is type of vibronic laser. Titanium-sapphire is the most widely used tunable solid-state laser because of its wide tunability and good material characteristics. Output wavelength is tunable from 660 nm to 1180 nm with the peak of the gain curve located around 800 nm. The host material, that is, sapphire, has very high thermal conductivity, mechanical rigidity and exceptionally high chemical inertness.

Transverse Mode The transverse mode tells us about the irradiance distribution of the laser output in the plane perpendicular to the direction of propagation or in other words along the orthogonal axes perpendicular to the laser axis.

Two-Level Laser In a two-level laser system, there are only two levels involved in the total process. That is, the atoms or molecules in the lower level, which is also the lower level of the laser transition, are excited to the upper level by the pumping or excitation mechanism. The upper level is also the upper laser level. Once the population inversion is achieved and its extent is above the inversion threshold, the laser action can take place. A two-level system is, however, a theoretical concept only as far as lasers are concerned.

Unstable Resonator This achieves interaction of the emitted photons with a very large volume of the excited species. Photons escape from the sides of the mirror after one or two passes within the cavity. This light leakage, which also constitutes the useful laser output, is more than compensated for by high gain medium and also by large interaction volume. Also, since the photons have to make relatively fewer passes within the cavity as compared to a low gain stable resonator configuration before drifting out, the alignment becomes that much less critical.

Vertical Cavity Surface Emitting Laser (VCSEL) With a VCSEL, the optical cavity is along the direction of flow of injection current.

Vertical External Cavity Surface Emitting Laser (VECSEL) A VECSEL is a variant of VCSEL where the resonator is completed with a mirror placed external to the diode structure thus introducing a free-space region in the resonant cavity. In the case of VCSEL, the two mirrors are either grown epitaxially as a part of the diode structure or grown separately and then bonded to the semiconductor chip having the active region.

Vibronic Laser This is a type of solid-state laser. What distinguishes a vibronic laser from the more commonly used and better known neodymium-doped and ruby lasers is in their energy level structure with a lower lasing level in the form of a band rather than a single discrete energy level. The laser transition occurs between the lowest level of the upper band to anywhere in the lower band. This characteristic feature of vibronic lasers makes this class of lasers tunable. These lasers have assumed significance due to not only the tunability aspect associated with them but also due to their capability to generate wavelengths not available from other solid-state laser media.

Wall Plug Efficiency This is the overall efficiency of the laser system. It is the ratio of laser power produced (CW power or average power as applicable) to the power drawn from source of input.

Bibliography

1 Silfvast, T.W. (2008), *Laser Fundamentals (Second Edition)*, Cambridge University Press.
2 Csele, M. (2004), *Fundamentals of Light Sources and Lasers*, Wiley-Interscience.
3 Thyagarajan, K. and Ghatak, A. (2010), *Lasers: Fundamentals and Applications (Second Edition)*, Springer.
4 Hecht, J. (1999), *The Laser Guidebook (Second Edition)*, McGraw Hill Education.
5 Svelto, O. (2010), *Principles of Lasers*, Plenum Press.
6 Hecht, J. (2008), *Understanding Lasers: An Entry Level Guide (Third Edition)*, IEEE Press.
7 Shimoda, K. (1984), *Introduction to Laser Physics*, Springer-Verlag.
8 O'Shea, D.C., Callen, W.R. and Rhodes, W.T. (1977), *Introduction to Lasers and Their Applications*, Addison-Wesley Publishing Co.

9 Arechhi, F.T. and Schulz-Dubois, E.O. (1976), *Laser Handbook*, Amsterdam North-Holland Publishing Co.

10 Schawlow, A.L. (1969), *Lasers and Light: Readings from Scientific American*, W.H. Freeman and Company.

11 Das, P. (1991), *Lasers and Optical Engineering*, Springer-Verlag.

12 Thyagarajan, K. and Ghatak, A.K. (1981), *Laser Theory and Applications*, Plenum Press.

13 Siegman, A.E. (1986), *Lasers*, University Science Books.

14 Webb, C.E. and Jones, J.D.C. (2003), *Handbook of Laser Technology and Applications: Volume I*, Institute of Physics Publishing.

15 Koechner, W. (2006), *Solid State Laser Engineering*, Springer.

16 Sennaroglu, A. (2006), *Solid State Lasers and Applications*, CRC Press.

17 Koechner, W. and Bass M. (2003), *Solid State Lasers: A Graduate Text*, Springer-Verlag.

18 Hardwell, T.O. (2008), *Solid State Lasers: Properties and Applications*, Nova Publishers.

19 Thompson, B.J. (2007), *Solid State Lasers and Applications*, Taylor and Francis Group, LLC.

20 Cheo, P.K. (1989), *Handbook of Solid State Lasers*, New York: M. Dekker.

21 Scheps, R. (2002), *Introduction to Laser Diode-pumped Solid State Lasers*, SPIE Press.

22 Webb, C.E. and Jones, J.D.C. (2003), *Handbook of Laser Technology and Applications: Volume I, II and III*, CRC Press.

23 Endo, M. and Walter, R.F. (2006), *Gas Lasers*, CRC Press.

24 Duley, W.W. (1976), *CO_2 Lasers Effects and Applications*, Academic Press.

25 Junji, O. (2017), *Semiconductor Lasers*, Springer.

26 Choi, H.K. (2004), *Long Wavelength Infrared Semiconductor Lasers*, John Wiley & Sons, Ltd.

27 Thompson, G.H.B. (1980), *Physic of Semiconductor Laser Devices*, John Wiley & Sons, Ltd.

28 Kapon, E. (2011), *Semiconductor Lasers-II: Materials and Structures*, Academic Press.

29 Kapon, E. (1999), *Semiconductor Lasers-I: Fundamentals*, Academic Press.

30 Chow, W.W. and Koch, S.W. (1999), *Semiconductor Laser Fundamentals: Physics of the Gain Materials*, Springer.

31 Nakamura, S., Fasol, G. and Pearton, S.J. (2000), *The Blue Laser Diode – The Complete Story*, Springer.

32 Injeyan, H. and Goodno, G.D. (2011), *High Power Laser Handbook*, McGraw-Hill.

33 Hertsens, T. (2000), *An Overview of Laser Diode Characteristics: Application Note # 5*, ILX Lightwave Photonic Test and Measurement.

8

Laser Electronics

The advances in the technologies of electronics devices and circuitry over the last two decades has been one of the key factors leading to explosive growth and maturity of laser systems designed for military usage in diverse application scenarios. The electronics that goes along with a military laser system is much more than a power supply and involves complex technologies. There are two main factors that have played a major role in enhancing the application potential of lasers. The first of course is the development and subsequent arrival on the commercial scene of a large variety of lasers covering a very large portion of the electromagnetic spectrum and producing a wide range of output power levels at an affordable price tag. The second main reason for the widespread use of lasers is the development that has taken place in the field of electronics that goes along with a laser or a laser-based system. The role of an electronics engineer working in the field of lasers is far more challenging today than it was in the early stages of development of lasers. Laser electronics today involves a large number of complex technologies, some of them specific to lasers. The chapter begins with a brief description of basic building blocks of electronics relevant to different laser sources and laser systems configured around these laser sources to be followed by comprehensive coverage of operational basics and design of electronics packages for the three common laser sources namely the solid-state, gas and semiconductor diode lasers and the military systems configured around them.

8.1 Basic Building Blocks of Laser Electronics

In this section are discussed some basic building blocks of electronics commonly used to design electronics packages of most of the laser sources and systems configured around them. The intention is to familiarize the readers with the operational basics of these building blocks. This will make it easier for them to understand the specific laser electronics packages discussed in the latter part of the chapter. This section will particularly benefit laser and optoelectronics students and professionals who do not have a comprehensive knowledge of electronics.

8.1.1 Linear Power Supplies

The power supply does the job of providing required DC voltages from available AC mains in the case of mains operated systems and DC input in the case of portable systems. Power supply constitutes the heart of the laser source that different laser systems are configured around. It could be the power supply used for charging energy storage capacitor in a flash-pumped solid-state laser or a laser diode driver in the case of a diode-pumped solid-state laser or a

Handbook of Defence Electronics and Optronics: Fundamentals, Technologies and Systems, First Edition. Anil K. Maini.
© 2018 John Wiley & Sons Ltd. Published 2018 by John Wiley & Sons Ltd.

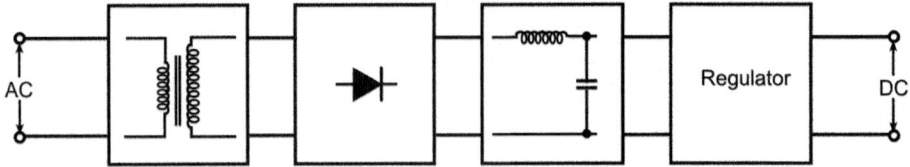

Figure 8.1 Constituents of a linear power supply.

semiconductor diode laser. In the case of a gas laser, one would generally require a high voltage power supply to initiate the plasma discharge. Power supplies are often classified into linear power supplies or switched mode power supplies, depending upon the nature of regulation circuit. In the following paragraphs the basic constituents of a linear power supply are discussed, which include transformers, rectifiers, filters and voltage regulators. Switched mode power supplies are discussed in the next section.

8.1.1.1   Constituents of a Linear Power Supply

A linear power supply essentially comprises a *mains transformer*, a *rectifier circuit*, a *filter circuit* and a *regulation circuit* (Figure 8.1). The transformer provides voltage transformation and produces across its secondary winding/s AC voltage/s required for producing desired DC voltages. It also provides electrical isolation between the input power supply, that is, AC mains and the DC output. Step-down transformers required for generating common DC voltages and load current ratings are commercially available. Step-up transformers for generating higher output voltages could be custom designed.

The rectifier circuit changes the AC voltage appearing across a transformer secondary to DC or, more precisely, a unidirectional output. Commonly used rectifier circuits include the half-wave rectifier, conventional full-wave rectifier requiring a tapped secondary and the full-wave bridge rectifier.

The rectifier voltage will always have some AC content known as a power supply ripple. The filter circuit smoothens the ripple of the rectifier voltage. The regulator circuit is a type of feedback circuit that ensures the output DC voltage does not change from its nominal value due to any change in line voltage or load current.

In a linearly regulated power supply, the active device, usually a bipolar junction transistor is operated anywhere between cut-off and saturation. Commonly used regulator circuit configurations include emitter follower regulator, series-pass transistor regulator and shunt regulator. The emitter follower regulator is in fact, now available in an IC package in both fixed output voltage as well as variable output voltage varieties. These are popularly known as *three-terminal regulators* and are discussed in the following sections.

All power supplies have inbuilt protection circuits. Common protection features include current limit, short circuit protection, thermal shut down and crowbar protection. These are also discussed in the following sections.

8.1.1.2   Rectifier Circuits

Figure 8.2 shows the three rectifier circuits for positive output voltages and Figure 8.3 shows the same for negative output voltages along with input and output waveforms. The rectifier circuits shown in Figures 8.2(a) and 8.3(a) are half-wave rectifier circuits. Circuits shown in Figures 8.2(b) and 8.3(b) are conventional full-wave rectifier circuits. Figures 8.2(c) and 8.3(c) are full-wave bridge rectifier configurations. The circuits are self-explanatory. Some common terms used with reference to rectifier circuits are defined in the following paragraphs.

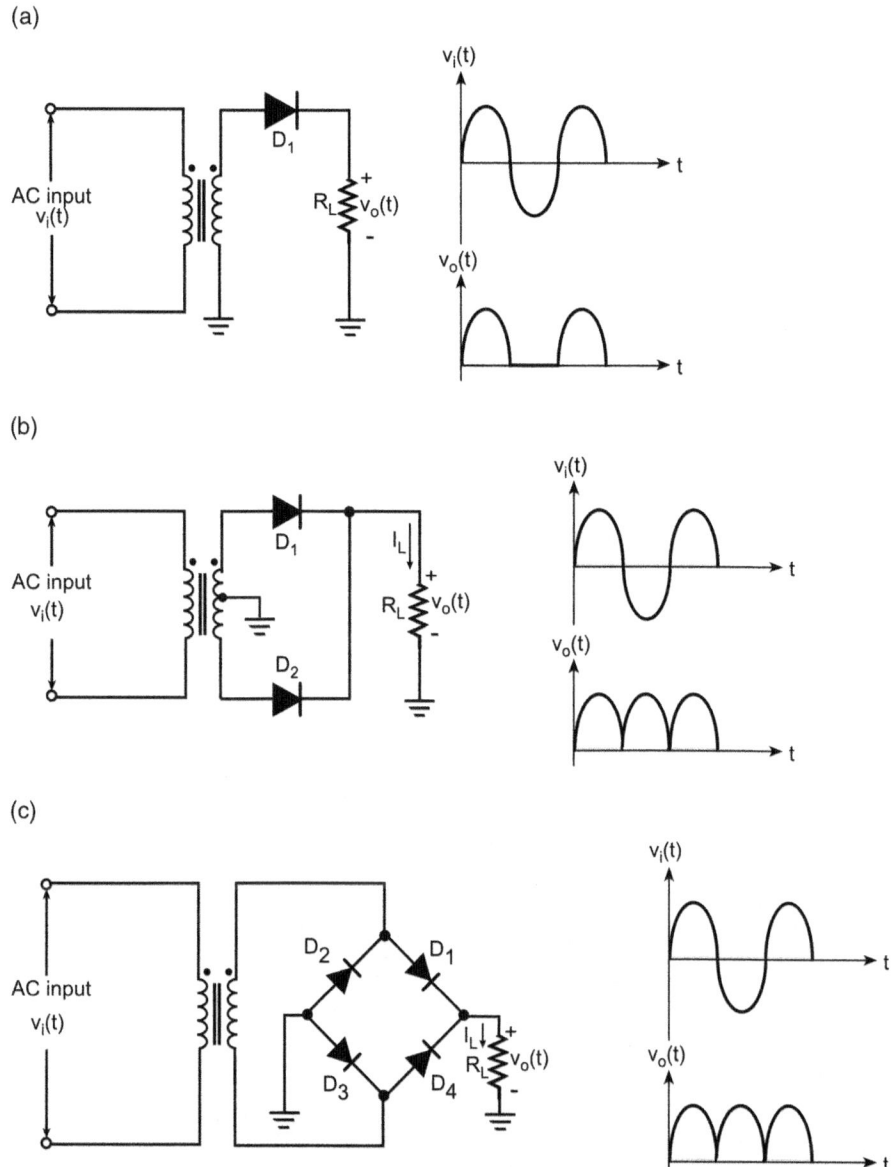

Figure 8.2 Rectifier circuits for positive output voltages: (a) Half wave rectifier, (b) conventional full wave rectifier, (c) full wave bridge rectifier.

Ratio of rectification is the ratio of DC power delivered to the load to the AC power rating of the transformer secondary. It is the lowest (=0.287) in the case of a half-wave rectifier, 0.693 in the case of conventional full-wave rectifier and highest (=0.812) for a bridge rectifier. *Ripple frequency* in the case of a half-wave rectifier is f, while in the case of full-wave rectifiers (conventional and bridge) it is 2f, where f is the frequency of AC mains. *Ripple factor* is ratio of RMS amplitude of AC component to the DC component. It is 1.21 for half-wave and 0.482 for full-wave rectifiers.

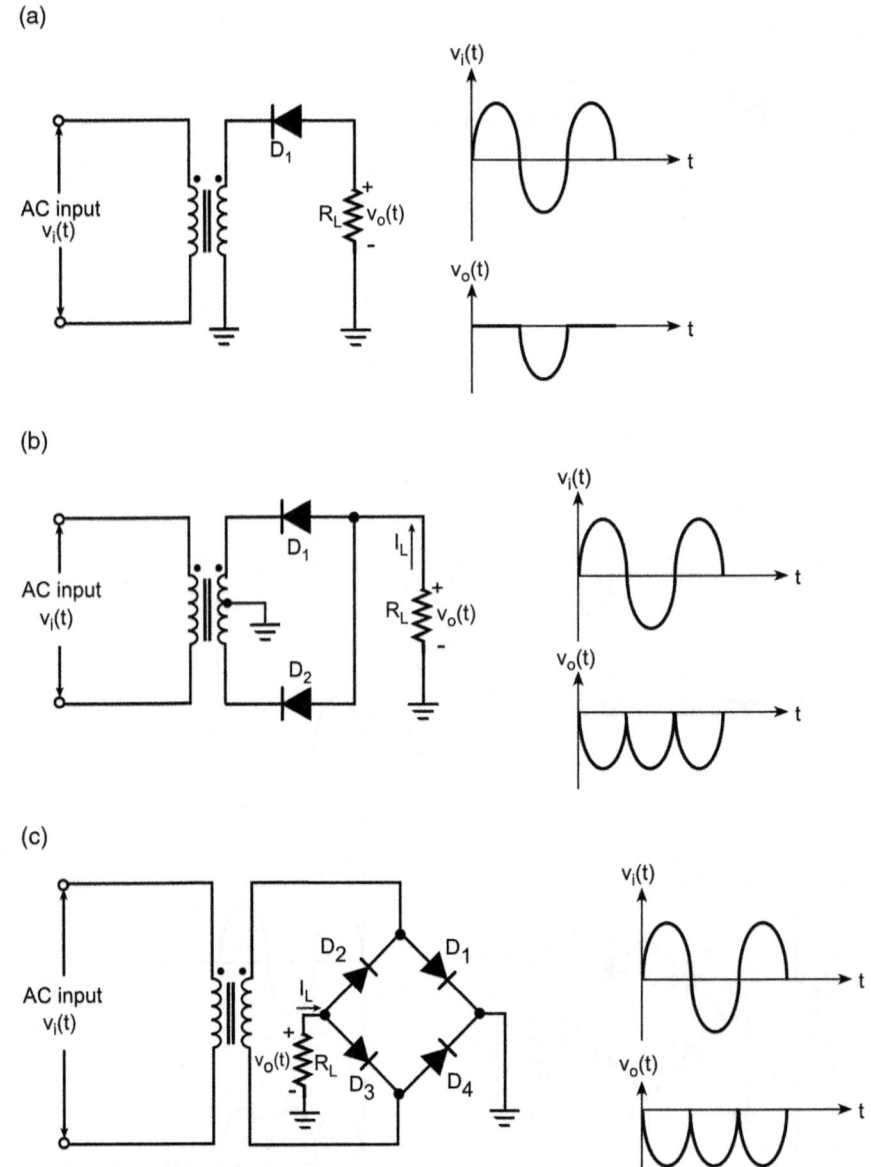

Figure 8.3 Rectifier circuits for negative output voltages: (a) Half wave rectifier, (b) conventional full wave rectifier, (c) full wave bridge rectifier.

Peak inverse voltage that appears across the rectifier diodes in the three cases is V_m in a half-wave rectifier, $2V_m$ in a conventional full-wave rectifier and V_m in a bridge rectifier. V_m is peak value of the AC voltage appearing across the secondary.

8.1.1.3 Filters

The filter in a power supply helps in reducing the ripple content, which in the rectifier wave-form is so large that the waveform can hardly be called a DC. Inductors, capacitors and an inductor-capacitor combination are used for the purpose of filtering. The fact that an inductor

(a)

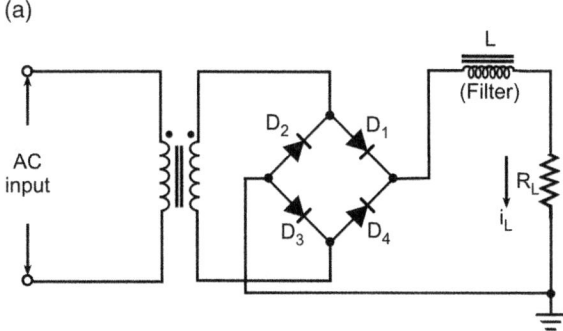

(b)

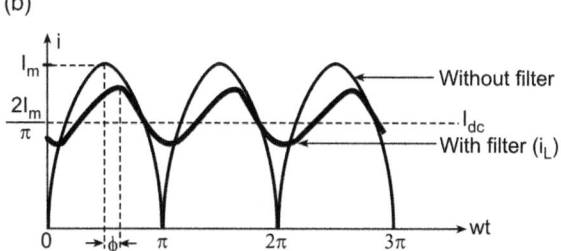

Figure 8.4 Inductor (choke) filter: (a) basic circuit and (b) output current waveform with and without filter.

offers high reactance to AC components is the basis of filter action provided by inductors. In the case of an inductor filter, also referred to as the choke filter, the ripple factor r is given by eqn. 8.1.

$$r = \frac{R_L}{3\sqrt{2}(\omega L)} \tag{8.1}$$

Equation 8.1 equals $(R_L/1330L)$ for a power line frequency of 50 Hz. Here, L is in henries and R is in ohms. Figure 8.4 shows an inductor filter used at the output of a full-wave bridge rectifier along with input and output waveforms.

As is clear from expression 8.1, the ripple factor is directly proportional to load resistance, R_L. That is, ripple content increases with increase in load resistance. In other words, choke filter is not effective for light loads (or high values of load resistance) and is preferably used for relatively higher load currents.

The filtering action of a capacitor connected across the output of the rectifier comes from the fact that it offers a low reactance to AC components. Figure 8.5(a) shows the capacitor filter connected across the output of a bridge rectifier. The AC components are bypassed to ground through the capacitor and only the DC is allowed to go through to the load. The capacitor charges to the peak value of the voltage waveform during the first cycle and as the voltage in the rectified waveform is on the decrease, the capacitor voltage is not able to follow the change as it can discharge only at a rate determined by the (CR_L) time constant. In the case of light loads (or high values of load resistance), the capacitor would discharge only a little before the voltage in the rectified waveform exceeds the capacitor voltage, thus charging it again to the peak value [Figure 8.5(b)]. The ripple content is inversely proportional to C and R_L.

(a)

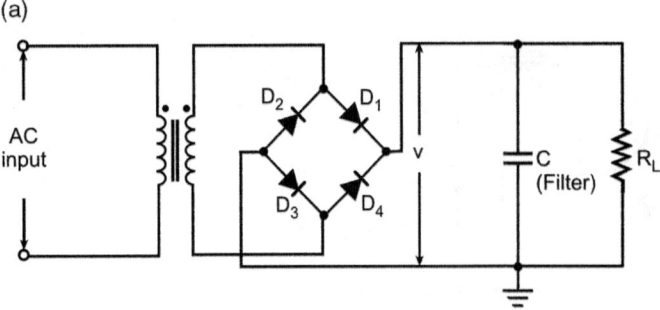

(b)

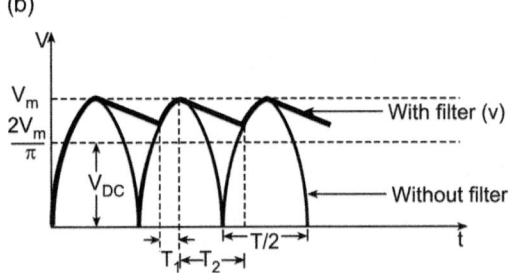

Figure 8.5 Capacitor filter: (a) basic circuit and (b) output voltage waveform with and without filter.

Ripple can be reduced by increasing C for a given of R_L. For heavy loads when R_L is small, even a large capacitance value may not be able to provide ripple within acceptable limits. Ripple factor r is given by eqn. 8.2.

$$r = \frac{1}{4\sqrt{3}\left(fCR_L\right)} \tag{8.2}$$

Equation 8.2 equals $(2886/CR_L)$ for a power line frequency of 50 Hz. Here, C is in microfarads and R_L is in ohms.

We have seen that while a choke filter is effective only for heavy load currents; a capacitor filter provides adequate filtering only for light loads. The performance of inductor and capacitor filters deteriorates fast as the load resistance is increased in the case of the former or decreased in the case of the latter. Apparently, an appropriate combination of L and C could give us a filter that would provide adequate filtering over a wide range of R_L values. For all practical purposes, the ripple factor in a choke input LC filter is independent of R_L and is given by $(1.2/LC)$ for power line frequency of 50 Hz. In this expression, L is in henries and C is in microfarads. Figure 8.6 shows an LC filter connected across the output of a full-wave rectifier.

The inductance value in the choke input LC filter should be such that the current i through the inductance never falls to zero, which amounts to saying that the negative peak of the ripple current should never exceed the DC value of the current. The minimum inductance that achieves this is known as critical inductance. The chosen value of inductance should be greater than equal to the critical inductance. Critical inductance is given by $L_C = (R_L/755)$ for a power line frequency of 50 Hz. Here, R_L is in ohms and L_C is in henries.

8.1.1.4 Linear Regulators

As outlined earlier, the regulator circuit in the power supply ensures that the load voltage (in the case of voltage regulated power supplies) or the load current (in the case of current regulated power supplies) is constant irrespective of variations in line voltage or load resistance. In the

Figure 8.6 LC (Choke input) filter.

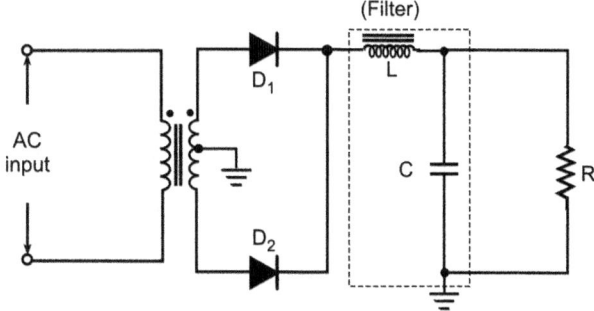

Figure 8.7 Emitter follower regulator for positive output voltages.

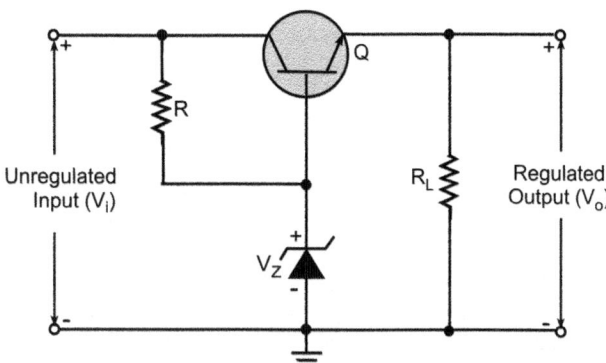

present section different types of voltage regulator circuits are discussed. Constant current sources are discussed in a later section. Three basic types of linear voltage regulator configurations include the emitter follower regulator, series-pass regulator and shunt regulator. Each one of these is briefly described in the following paragraphs.

8.1.1.4.1 Emitter Follower Regulator

Figure 8.7 shows the basic positive output emitter follower regulator. The emitter voltage, which is also the output voltage, remains constant as long as the base voltage is held constant. A zener diode connected between the base terminal and ground ensures this. The regulated output voltage in this case is given by eqn. 8.3.

$$V_O = (V_Z - 0.6) \tag{8.3}$$

Where,
V_Z is the voltage of the Zener diode in volts.

Due to high inherent current gain of the series-pass transistor, a low power zener diode can be used to regulate high value of load current. The base current in this case needs only to be $(1/h_{FE})$ times the load current. Figure 8.8 shows the emitter follower regulator circuit for negative output voltages. If the load current is so large that it is beyond the capability of a zener diode to provide the requisite base current, a Darlington combination can be used instead of a single transistor series-pass element (Figure 8.9 and Figure 8.10).

8.1.1.4.2 Series-Pass Regulator

Figure 8.11 shows the basic constituents of a series-pass type linear regulator. The series-pass element, bipolar transistor Q in the circuit of Figure 8.11, works like a variable resistance with

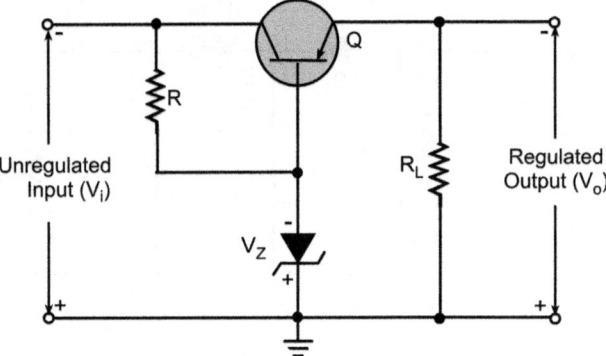

Figure 8.8 Emitter follower regulator for negative output voltages.

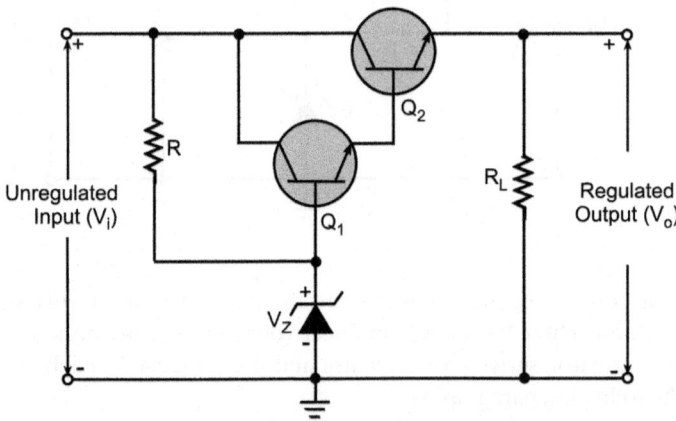

Figure 8.9 Emitter follower regulator with Darlington element (positive output voltage).

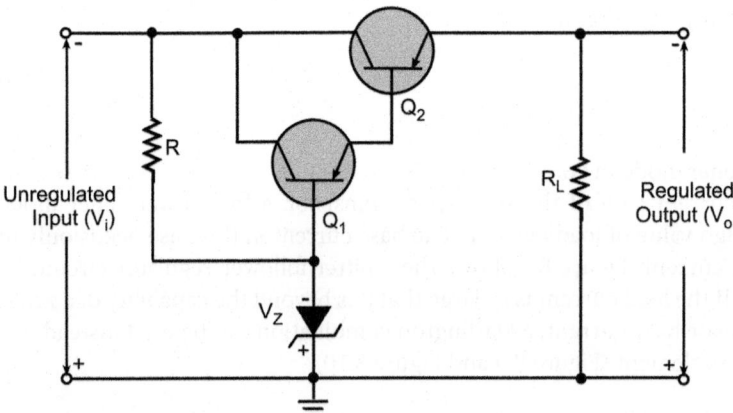

Figure 8.10 Emitter follower regulator with Darlington element (negative output voltage).

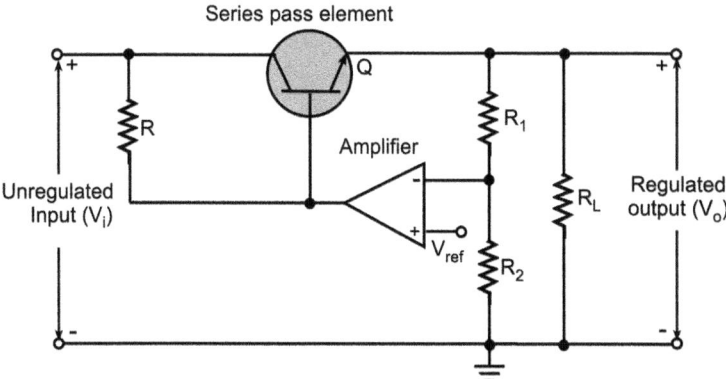

Figure 8.11 Series pass linear regulator.

the conduction of the transistor depending upon the base current. The regulator circuit functions as follows.

A small fraction of the output voltage is compared with a known reference DC voltage and their difference is amplified in a high gain DC amplifier. The amplified error signal is then fed back to the base of the series-pass transistor to alter its conduction so as to maintain essentially a constant output voltage. The regulated output voltage in this case is given by eqn. 8.4.

$$V_O = V_{REF} \times \left[(R_1 + R_2)/R_2 \right] \tag{8.4}$$

As the output voltage tends to decrease due to decrease in input voltage or increase in load current, the error voltage produced as a result of this causes the base current to increase. The increased base current increases transistor conduction thus reducing its collector-emitter voltage drop, which compensates for the reduction in the output voltage.

Similarly, when the output voltage tends to increase due to increase in input voltage or decrease in load current, the error voltage produced as a consequence is of the opposite sense. It tends to decrease transistor conduction thus increasing its collector-emitter voltage drop thus maintaining a constant output voltage. The regulation provided by this circuit depends upon the stability of the reference voltage and the gain of the DC amplifier. A typical series-pass regulator circuit using a bipolar junction transistor as the error amplifier is shown in Figure 8.12.

The power dissipated in the series-pass transistor is the product of its collector-emitter voltage and the load current. As the load current increases within a certain range, the collector-emitter voltage decreases due to the feedback action keeping the output voltage as constant. The series-pass transistor is so chosen as to safely dissipate the power under normal load conditions. If there is an overload condition for some reason or other, the transistor is likely to get damaged if such a condition is allowed to persist for long. In the worst-case, if there were a short circuit at the output, the whole of the unregulated input would appear across the series-pass transistor increasing the power dissipation to a prohibitively large magnitude, eventually destroying the transistor. Even a series-connected fuse does not help in such a case, as the thermal time constant of transistor is much smaller than that of the fuse. Thus, it is always desirable to build overload protection or short circuit protection in the linearly regulated power supply design. One such configuration is shown in Figure 8.13.

Under normal operating conditions, transistor Q_3 is in saturation. Thus, it offers very little resistance to the load current path. In the event of an overload or a short circuit, diode D_1 conducts thus reducing the base drive to transistor Q_3. Transistor Q_3 offers an increased resistance

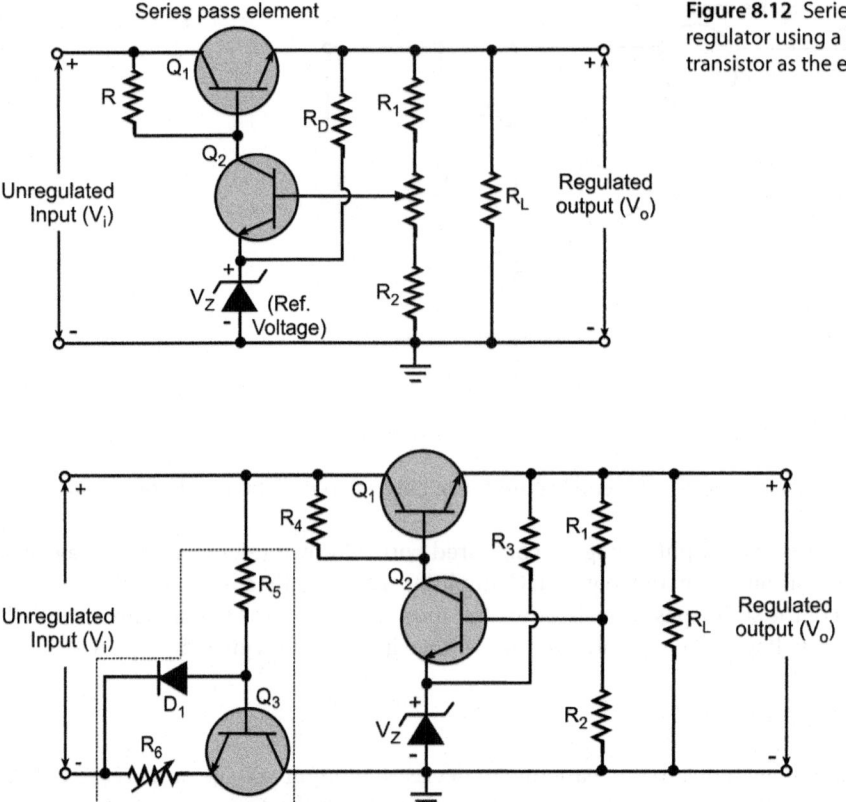

Figure 8.12 Series pass linear regulator using a bipolar junction transistor as the error amplifier.

Figure 8.13 Series pass linear regulator with overload protection.

to the flow of load current. In the event of a short circuit, the whole of input voltage would appear across Q_3. Transistor Q_3 should be so chosen as to safely dissipate power given by the product of worst-case unregulated input voltage and the limiting value of current. Diode D_1 and transistor Q_3 should preferably be mounted on the same heat sink so that the base-emitter junction of Q_3 and a diode's P-N junction are equally affected by temperature rise and the short circuit limiting current is as per the preset value. There can be other possible circuit configurations that can provide the desired protection. Discussion on all of them is beyond the scope of the present book. Other types of protection feature that are usually built into power supplies include a crowbar and thermal shutdown. Crowbar protection is a type of over voltage protection and thermal shut down disconnects the input to the regulator circuit in the event of temperature of the active device/s exceeding a certain upper limit. It may be mentioned here that the control and protection functions are usually provided by an integrated circuit (IC) in a modern power supply. A wide range of control ICs is available for both linear and switched mode power supplies.

8.1.1.4.3 *Shunt Regulator*
In a series type linear regulator, the pass element is connected in series with the load and any decrease or increase in the output voltage is accompanied by a decrease or increase in the collector-emitter voltage of the series-pass element. In the case of a shunt type linear regulator (Figure 8.14), regulation is provided by a change in the current through the shunt transistor to

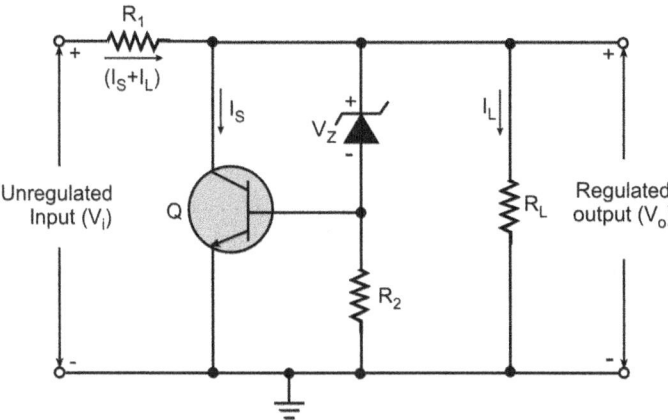

Figure 8.14 Shunt regulator.

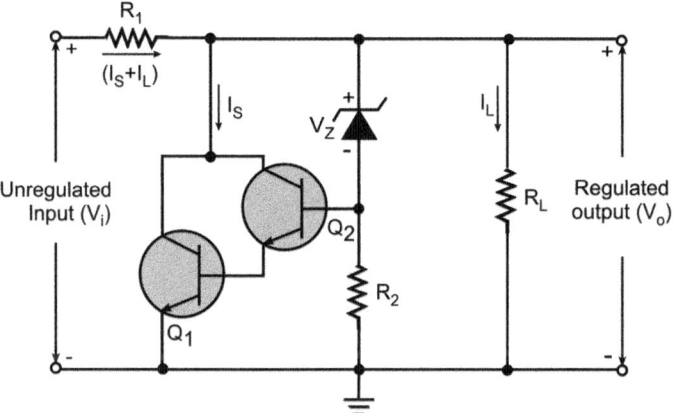

Figure 8.15 Shunt regulator with Darlington arrangement.

maintain a constant output voltage. The regulated output voltage in a shunt regulated linear power supply is the unregulated input voltage minus drop across a resistance R_1. Now, the current through R_1 is the sum of load current I_L and current through shunt transistor I_s. As the output voltage tends to decrease, the base current through the transistor reduces with the result that its collector current I_s reduces too. This reduces drop across R_1 and the output voltage is restored to its nominal value. Similarly, any tendency of the output voltage to increase is accompanied by an increase in current through the shunt transistor consequently increasing voltage drop across R_1, which in turn maintains a constant output voltage. A Darlington combination in place of shunt transistor enhances the current capability (Figure 8.15).

A shunt regulator is not as efficient as a series regulator for the simple reason that the current through the series resistor in the case of a shunt regulator is the sum of load current and shunt transistor current and it dissipates more power than the series-pass regulator with same unregulated input and regulated output specifications. In a shunt regulator, the shunt transistor also dissipates power in addition to the power dissipated in the series resistor. The only advantage with a shunt regulator is its simplicity and that it is inherently protected against overload condition.

8.1.1.4.4 Linear IC Voltage Regulators

In the preceding paragraphs are discussed series and shunt regulator circuits designed with discrete components. Present day regulating circuits are almost exclusively configured around one or more integrated circuits. These are known as IC voltage regulators. IC voltage regulators are available to meet a wide range of requirements. Both fixed output voltage (positive and negative) and adjustable output voltage (positive and negative) IC regulators are commercially available in a wide range of voltage, current and regulation specifications. These have built-in protection features such as current limit, thermal shut down and so on.

IC 723 is one such general-purpose adjustable output voltage regulator that is capable of being operated in positive or negative power supplies as series, shunt and switching regulator. The internal schematic arrangement of IC 723 resembles the typical circuit for a series-pass linear regulator and comprises of a temperature compensated reference, an error amplifier, a series-pass transistor and a current limiter with access to remote shutdown. Figure 8.16 shows the internal schematic arrangement of regulator IC 723. Figures 8.17 and 8.18 show the basic circuits for building low positive output voltage (2–7 V) and high positive output voltage

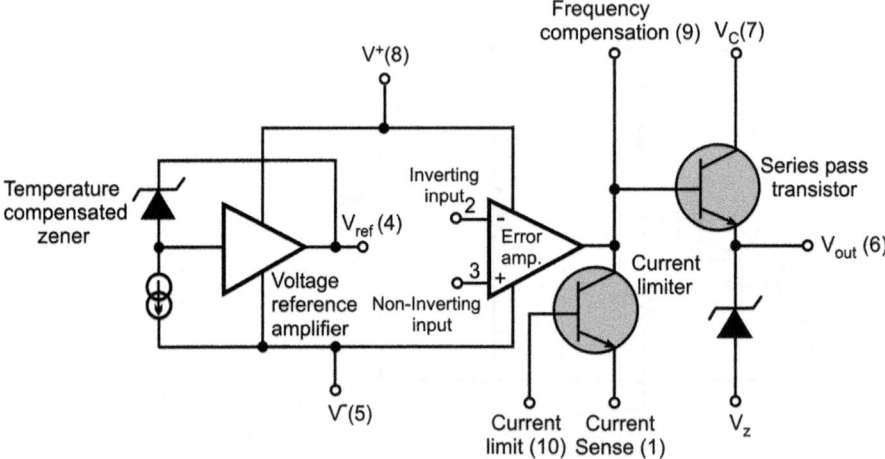

Figure 8.16 Internal schematic arrangement of regulator IC 723.

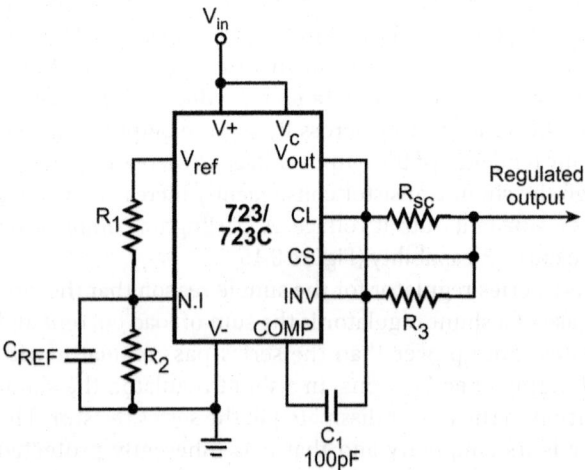

Figure 8.17 Low positive output voltage regulator using IC 723.

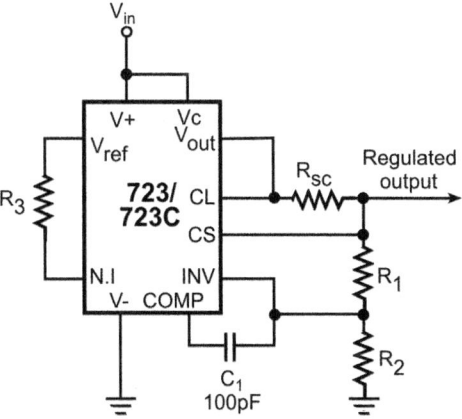

Figure 8.18 High positive output voltage regulator using IC 723.

(7–37 V) regulator circuits. In the case of the circuit arrangement of Figure 8.17, the regulated output voltage is given by eqn. 8.5.

$$V_O = V_{REF} \times \left[R_1 R_2 / \left(R_1 + R_2 \right) \right] \tag{8.5}$$

In the case of the circuit arrangement of Figure 8.18, the output voltage is given by eqn. 8.6.

$$V_O = V_{REF} \times \left[\left(R_1 + R_2 \right) / R_2 \right] \tag{8.6}$$

In both cases, the recommended value of R_3 is $[R_1 R_2 / (R_1 + R_2)]$ and $R_{SC} = 0.6 / I_{SC}$, where I_{SC} is short circuit limiting value.

A more popular and more extensively used type of linear IC voltage regulator is the *three-terminal regulator*. In their basic operational mode, three-terminal regulators require virtually no external components. These are available in both fixed output voltage (positive and negative) as well as adjustable output voltage (positive and negative) types with current ratings of 100 mA, 500 mA, 1.5 A and 3.0 A. Popular fixed positive output voltage three-terminal regulators include LM/MC 78XX-series and LM 140-XX/340-XX series. LM117/217/317 is a common adjustable positive output voltage regulator. Popular fixed negative output voltage three-terminal regulators include LM/MC 79XX-series and LM 120-XX/320-XX series. LM137/237/337 is a common adjustable negative output voltage regulator. A two-digit number in place of 'XX' indicates the regulated output voltage.

Figure 8.19 shows the basic application circuits using LM/MC 78XX-series and LM/MC 79XX-series three-terminal regulators. C_1 and C_2 are decoupling capacitors. C_1 is generally used when the regulator is located far from the power supply filter. Typically, a 0.22 µF ceramic disc capacitor is used for C_1. C_2 is typically a 0.1 µF ceramic disc capacitor. LM 140XX/340XX series and LM 120XX/320XX series regulators are also used in the same manner. In the case of fixed output voltage three-terminal regulators, if the common terminal instead of being grounded were applied a DC voltage, the regulated output voltage in that case would be greater than the expected value by a quantum equal to the voltage applied to the common terminal. For details on the salient features and performance specifications of three-terminal regulators and their application circuits, one can refer to data sheets and application notes provided by manufacturers of these devices.

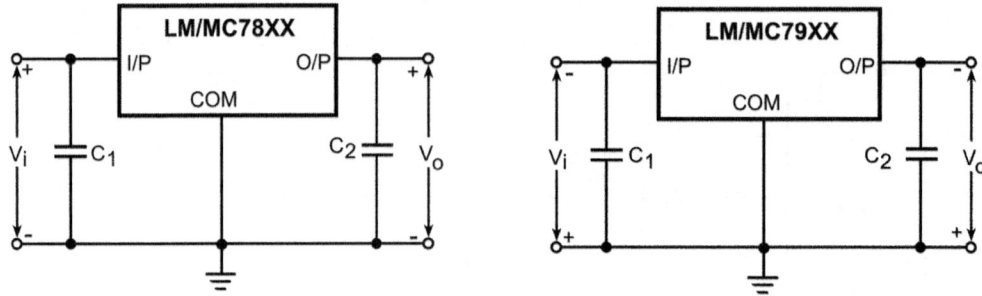

Figure 8.19 Basic application circuits using three-terminal regulators.

8.1.2   Switched Mode Power Supplies

As outlined earlier, based on the regulation concept, power supplies are classified as either linear or switched mode power supplies. Conventional AC/DC power supplies comprising a transformer, rectifier, filter and regulator (series or shunt type) constitute the linear power supply. The active device in the regulator circuit of a linear power supply is always operated in the active or linear region of its output characteristics. Any change in the output voltage due to change in the input voltage or load current results in change in the voltage drop across the regulator transistor (in the case of a series regulator) or a change in the current through the regulator transistor (in the case of a shunt regulator) so as to maintain a constant output voltage across the load.

DC/DC converters and DC/AC inverters belong to the category of switched mode power supplies (SMPS). Also, there are switching supplies operating from AC mains. These are called offline switching supplies. An offline switching supply can be distinguished from a conventional AC/DC supply as in the case of the former the AC mains is rectified and filtered without using an input transformer, and the DC voltage so obtained is then used as an input to a switching type DC/DC converter.

In a switching power supply, the active device that provides regulation is always operated in the switched mode; that is, it is operated either in cut-off or in saturation region of the output characteristics. The input DC is chopped at a high frequency (typically 10 kHz to 100 kHz) using an active device (bipolar junction transistor, power MOSFET, IGBT or SCR) and the converter transformer. The transformed chopped waveform is rectified and filtered. A sample of the output voltage is used as feedback signal for the drive circuit of the switching device to achieve regulation.

8.1.2.1   Linear versus Switched Mode Power Supplies

Linear power supplies are well-known for their extremely good line and load regulation, low output voltage ripple and almost negligible RFI/EMI. Switching power supplies, on the other hand, have a much higher efficiency (typically 80–90% against 50% in the case of linear supplies) and reduced size/weight for a given power delivering capability. Quite often, compactness and efficiency are two major selection criteria. Improved efficiency and reduced size/weight are particularly significant when designing a power supply for a portable system particularly when there is a requirement for a number of different regulated output voltages. Also, unlike linear supplies, efficiency in switching supplies does not suffer as the unregulated input to regulated output differential becomes large. In portable systems operating from battery packs and requiring higher DC voltages for their operation, the switching supply is the only option. We cannot use a linear regulator to change a given unregulated input voltage to a higher regulated output voltage.

8.1.2.2 Different Types of Switched Mode Power Supplies

Switched mode power supplies are designed in a variety of circuit configurations depending upon the intended application. Almost all switching supplies belong to one of the following three broad categories.

1) Flyback converters
2) Forward converters
3) Push-pull converters

There are variations in the circuit configuration within each of these categories of switched mode power supplies. For instance, in the category of flyback converters, we have self-oscillating flyback converters and the externally driven flyback converters. Again, in the category of externally driven type flyback converters, there are isolation and non-isolation type configurations. Also, there are DC/DC and offline flyback converters.

Similarly, there are different circuit configurations in the other two categories of switching supplies also. Although these configurations differ to an extent, the basic operational principle and the design criteria for different types belonging to one category remain more or less the same.

8.1.2.2.1 *Flyback Converters*

The self-oscillating type flyback DC/DC converter is the most basic converter based on the flyback principle (Figure 8.20). A switching transistor, a converter transformer, a fast recovery rectifier and an output filter capacitor make up a complete DC/DC converter. It is a constant output power converter.

During the conduction time of the switching transistor, the current through the transformer primary starts ramping up linearly with a slope equal to V_{in}/L_p. The voltages induced in the secondary and the feedback windings make the fast recovery rectifier reverse biased and hold the conducting transistor 'on'. When the primary current reaches a peak value I_p, where the core begins to saturate, the current tends to rise very sharply. This sharp rise in current cannot be supported by the fixed base drive provided by the feedback winding. As a result, the switching transistor begins to come out of saturation.

Figure 8.20 Self-oscillating fly back converter.

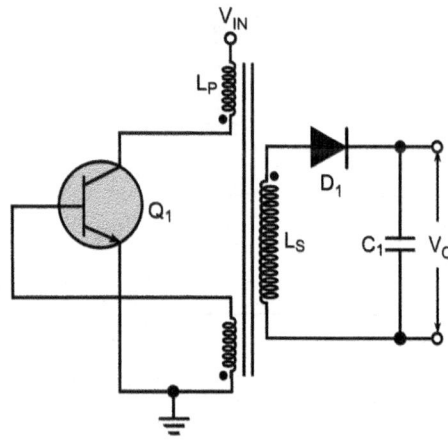

This is a regenerative process with the transistor getting switched off. The magnetic field produced by the current flowing in the primary winding collapses, thus reversing the polarities of the induced voltages. The fast recovery rectifier is forward biased and the stored energy is transferred to the capacitor and the load through the secondary winding. Thus, energy is stored during the ON-time and transferred during the OFF-time of the switching transistor.

The output capacitor supplies the load current during the ON-time of the transistor when no energy is being transferred from the primary side. It is a constant output power converter. The converter can deliver an output power given by eqn. 8.7.

$$P_O = \frac{1}{2}\eta L_P I_P^2 f \tag{8.7}$$

Where,

L_P is the primary inductance
I_P is the peak value of primary current
f is the switching frequency
η is the conversion efficiency.

The output voltage reduces as the load increases and vice versa. Utmost care should be taken to ensure that the load is not accidentally taken off the converter. In that case, the output voltage would rise without limit till any of the converter components gets damaged. It is suitable for low output power applications due to its inherent nature of operation and may be used with advantage up to an output power of 150 W. It is characterized by high output voltage ripple.

A variation of this circuit is the externally driven flyback converter. Figure 8.21 shows the circuit diagram of the basic externally driven flyback type DC/DC converter. The basic principle remains the same. Energy is stored during ON-time of the active device and transferred during OFF-time of the active device. The feedback loop consisting of a comparator and the resistance divider provides the voltage sense as well as some degree of regulation.

Extension of the converter circuit of Figure 8.21 is the externally driven flyback converter with pulse width modulation (PWM) control to achieve regulation, shown in Figure 8.22. Pulse width modulation is the most widely used control technique in conjunction with flyback converters. As

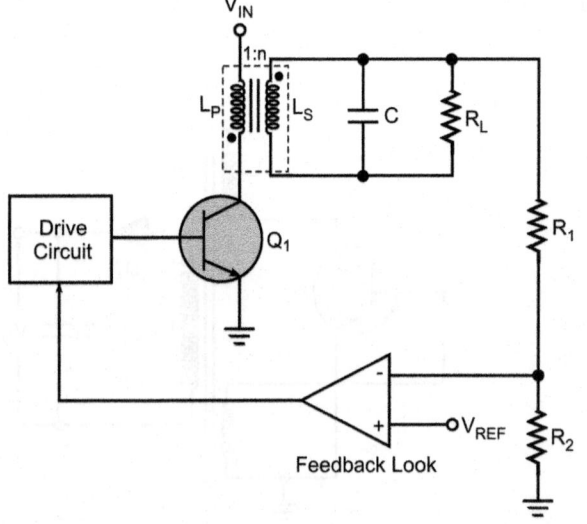

Figure 8.21 Basic externally driven fly back converter.

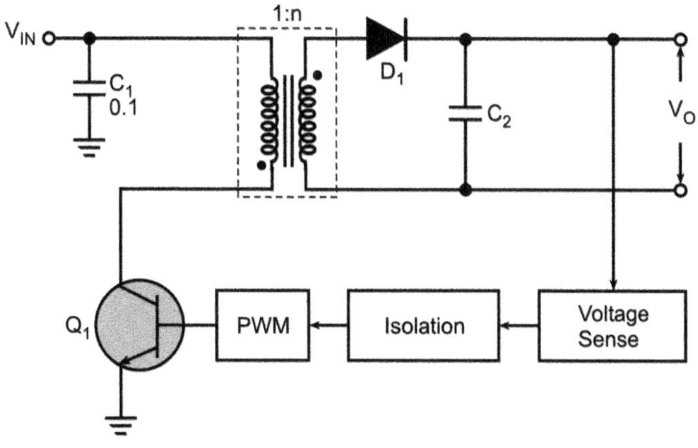

Figure 8.22 Externally driven fly back converter with PWM control.

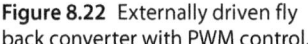

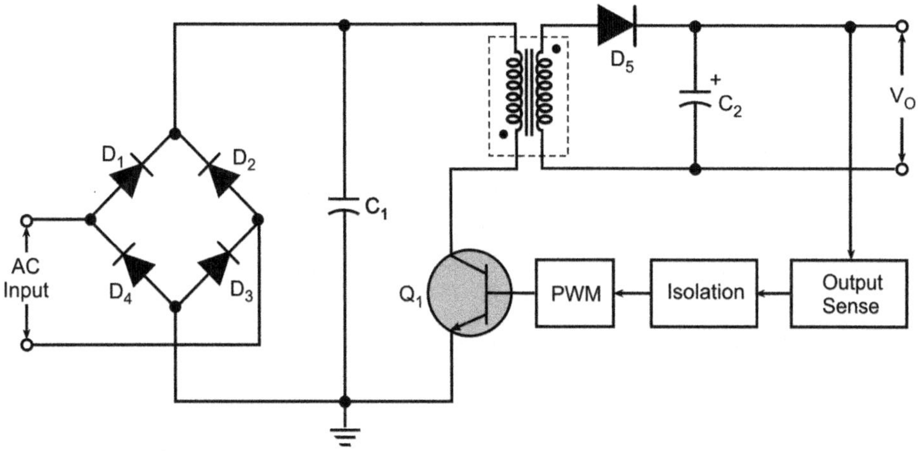

Figure 8.23 Offline externally driven fly back converter.

the load current increases, the output voltage tends to fall. The PWM control senses the change and increases the ON-time of the active device so as to increase the power delivering capability (increased ON-time means increased stored energy) and restores the output voltage. Similarly, an increase in the output voltage causes a reduction in the ON-time of the active device.

There are other control circuitries, which provide regulation by changing the OFF-time rather than the ON-time. A reduced OFF-time increases the drive frequency and hence the power delivering capability and vice versa. A number of integrated circuits have been developed to provide drive and control functions for DC/DC converters. Some of these ICs provide PWM control while others offer constant ON-time and variable frequency operations. These ICs have built-in features like over voltage protection, current limit and so on. Such ICs (TL497, TL494, TL594 and SG3524 to name a few) have considerably simplified the drive and control circuit design.

Most switching supplies used in consumer and industrial systems are offline. Figure 8.23 shows an offline externally driven flyback type DC/DC converter. It is called offline because the input voltage to the transistor switch is developed right from the AC line without first going through 50/60 Hz transformer. Bridge rectifier and the filter capacitor accomplish this in the circuit of Figure 8.23. The feedback loop in an offline supply must have isolation so that the DC output is isolated from the AC line. A small transformer or an opto-isolator usually

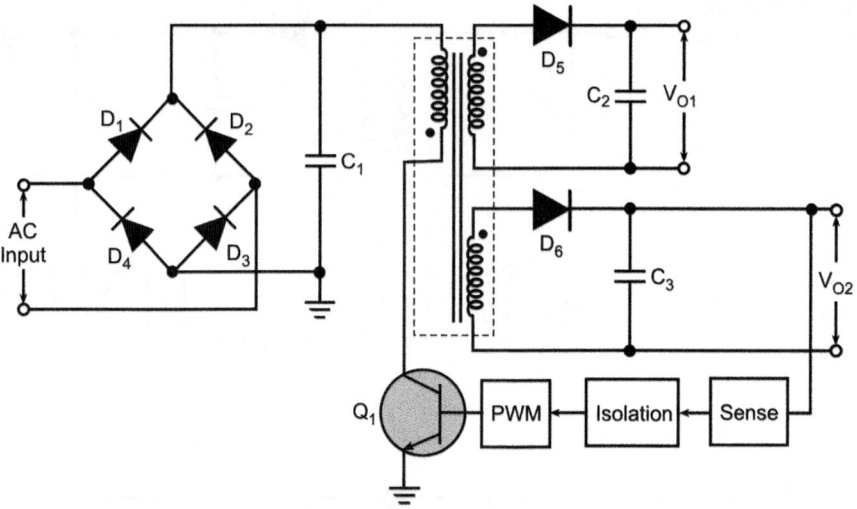

Figure 8.24 Offline fly back DC/DC converter with multiple outputs.

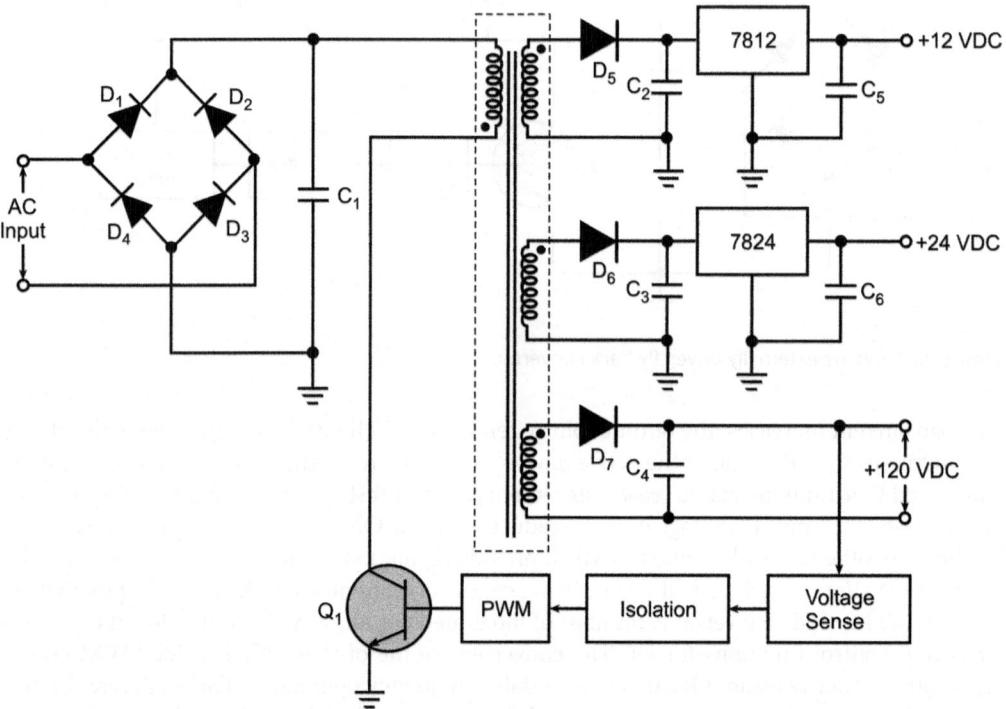

Figure 8.25 Multi-output switching supply with post regulation.

accomplishes this. Most switching supplies are required to produce more than one regulated DC voltage. Figure 8.24 shows an offline multiple output flyback DC/DC converter.

In case more stringent regulation is required in respect of one or more outputs, linear post-regulator can be used as shown in Figure 8.25. Three-terminal IC regulators have been used here for the purpose.

8.1.2.2.2 Forward Converter

The forward converter is another popular switching supply configuration. Figure 8.26 shows the basic circuit diagram of an offline forward converter. There are some fundamental differences between a flyback converter and a forward converter. In the case of circuit diagram shown in Figure 8.26, when the transistor Q_1 is switched on, the polarities of the transformer windings as indicated by the position of dots are such that diode D_5 is forward biased and diodes D_6 and D_7 are reverse biased. Most of the energy in a forward converter is stored in the output inductor rather than the transformer primary used to store energy in a flyback converter. When the transistor switch is turned off, the magnetic field collapses. Diode D_5 is reverse biased and diodes D_6 and D_7 are forward biased. As the current through an inductor cannot change instantaneously, the output current continues to flow through the output and the forward-biased diode D_6 provides the current path. Unlike a flyback converter, current flows from the energy storage element during both halves of the switching cycle. Thus, for the same output power, a forward converter has much less output ripple than a flyback converter. Controlling the duty cycle of the transistor switch provides output regulation.

In the absence of the third winding and diode D_7, a good fraction of energy stored in the transformer primary is lost. This effect is more severe at higher switching frequencies. The third winding and the forward-biased diode D_7 return the energy, which would otherwise be lost and reset the transformer core after each operating cycle. This not only increases converter efficiency but also makes the converter transformer core immune to saturation problems.

8.1.2.2.3 Push-Pull Converter

Push-pull converter is the most widely used switching supply belonging to the family of forward converters. There are several different circuit configurations within the push-pull converter sub-family. These circuits differ only in the mode in which the transformer primary is driven. These include the conventional two-transistor, one-transformer push-pull converter (both self-oscillating and extremely driven type) two-transistor, two-transformer push-pull converter, half-bridge converter and full-bridge converter.

Figure 8.27 shows the conventional self-oscillating type of push-pull converter. Its operation can be explained by considering it equivalent to two alternately operating self-oscillating flyback converters. When transistor Q_1 is in saturation, energy is stored in the upper half of the

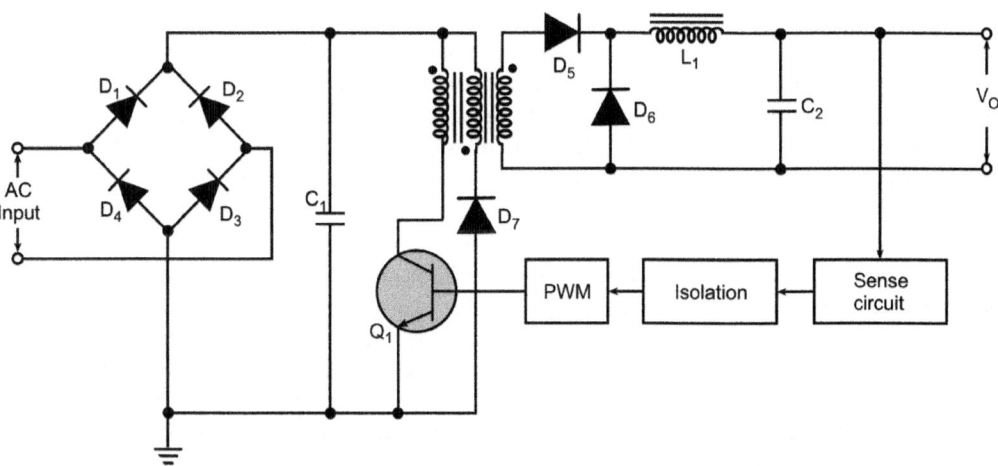

Figure 8.26 Forward converter.

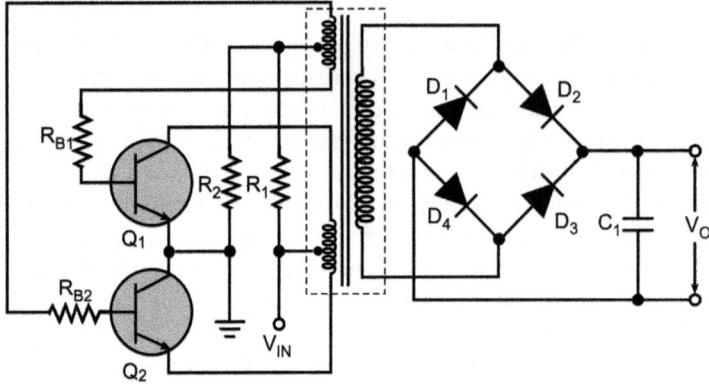

Figure 8.27 Self-oscillating push-pull converter.

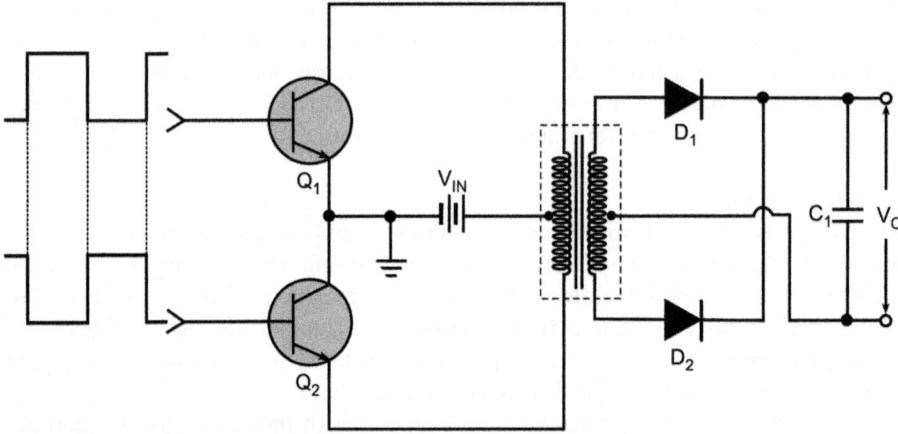

Figure 8.28 Externally driven push-pull converter.

primary winding. When the linearly rising current reaches a value where the transformer core begins to saturate, the current tends to rise sharply, which is not supported by a more or less fixed base bias. The transistor starts to come out of saturation. This is a regenerative process and ends up in switching off transistor Q_1 and switching on transistor Q_2. Thus transistors Q_1 and Q_2 switch on and off alternately. When Q_1 is on, energy is being stored in the upper half of the primary and the energy stored in the immediately preceding half cycle in the lower half of the primary winding (when transistor Q_2 was on) is getting transferred. Thus, energy is stored and transferred at the same time. The voltage across secondary is a symmetrical square waveform, which is then rectified and filtered to get the DC output.

As the primary is centre-tapped, and only half of the primary winding is active at one time, the main transformer is not utilized as good as it is in the case of other forms of push-pull converter, like half-bridge and full-bridge converters. Also, in a push-pull converter, switching transistors operate at collector stress voltages of at least twice the DC input voltage. As a result, a push-pull converter is not a highly recommended choice for offline operation. A push-pull converter that has wider applications than its self-oscillating counterpart is the externally driven push-pull converter. Figure 8.28 shows the block-schematic arrangement of externally driven push-pull converter. This has been possible due to availability of a variety of SMPS drive and control ICs.

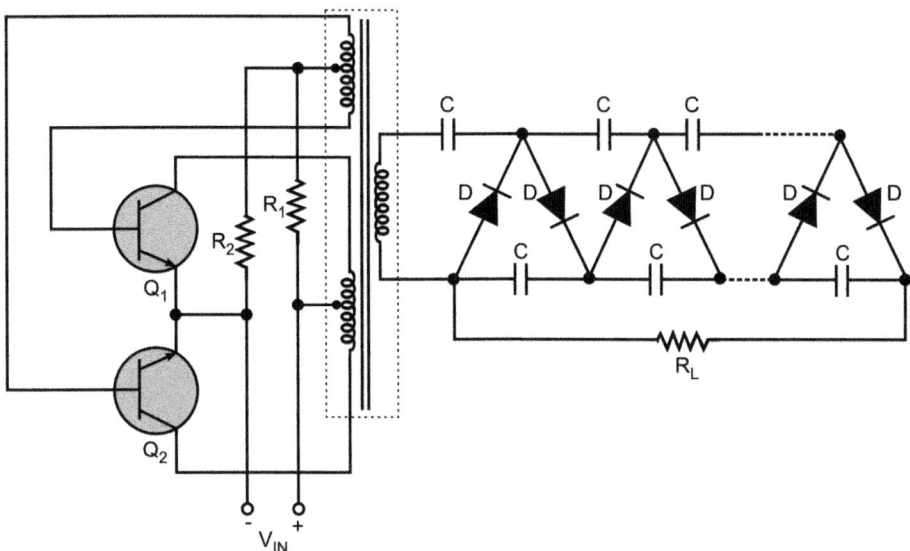

Figure 8.29 Push-pull converter with voltage multiplier chain.

Self-oscillating push-pull converters are frequently used along with a voltage multiplier chain to design a high voltage low current power supply (Figure 8.29). This configuration is particularly useful for designing helium-neon laser power supplies. The basic push-pull converter converts the low DC input voltage to a stepped up square waveform, which is then multiplied using a chain of diodes and capacitors. Voltage multiplier circuits are discussed in detail in Section 8.4 on gas laser electronics.

In the self-oscillating two-transistor, one-transformer push-pull converters, the transformer provides both power transformation as well as power switching. This circuit has some disadvantages. Firstly, as the power switching is done at output power levels, the converter efficiency lowers quite a bit in the case of a high-power converter. Secondly, the peak collector current depends upon the available base voltage, transistor gain and input characteristics and is dependent on load. As there is a wide variation in the input characteristics from device to device, the circuit performance depends upon the chosen device. Also, the transformer core must be the expensive square loop material with a large maximum flux density rating.

These problems are overcome in the two-transformer, two-transistor push-pull converter (Figure 8.30). Power switching is done at base power level and the output transformer performs power transformation only. Capacitors C_1 and C_2 are the speed-up capacitors (also known as commutating capacitors) used to achieve a faster turn off of the respective transistors.

The half-bridge converter, as shown in Figure 8.31, is recommended for high power applications. Transistors Q_1 and Q_2 operate alternately. The half-bridge converter has the advantage that it allows the use of transistors with lower breakdown voltages.

The full-bridge converter as shown in Figure 8.32 has the advantage that the highest voltage any transistor is subjected to is only V_{in} against $2V_{in}$ as in the case of push-pull converter. Due to reduce voltage and stress on the transistors, full-bridge converter offers a great reliability.

8.1.2.2.4 *Switching Regulators*
Commonly used switching regulator configurations include step-down or buck regulator, step-up or boost regulator and inverting regulator. Figure 8.33 shows the basic buck regulator. It resembles the conventional forward converter discussed in a latter section except for the fact

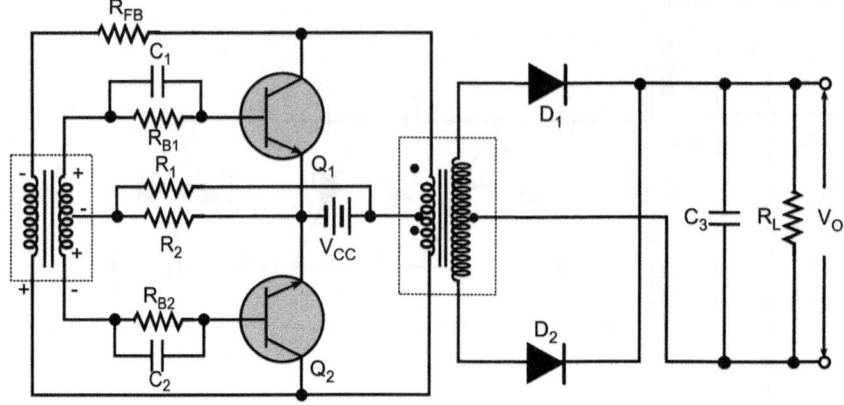

Figure 8.30 Two-transformer, two-transistor push-pull converter.

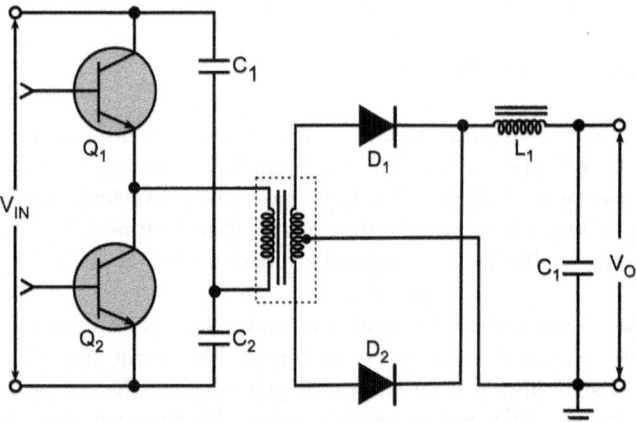

Figure 8.31 Half-bridge converter.

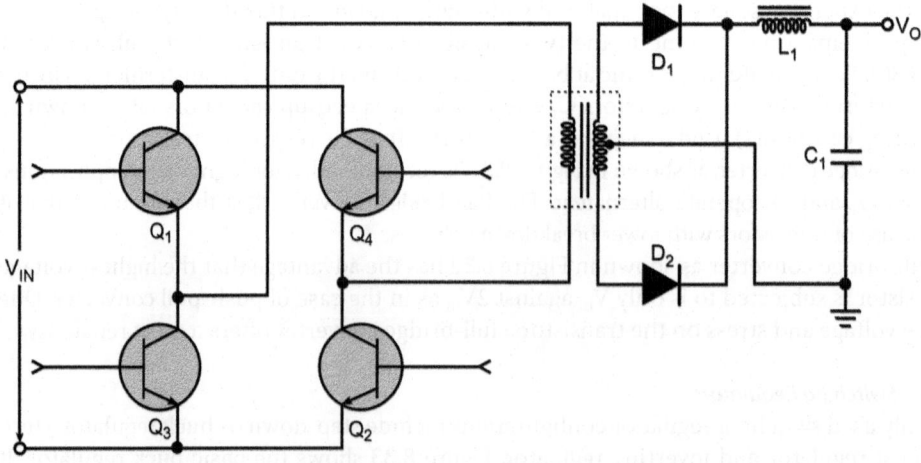

Figure 8.32 Full-bridge converter.

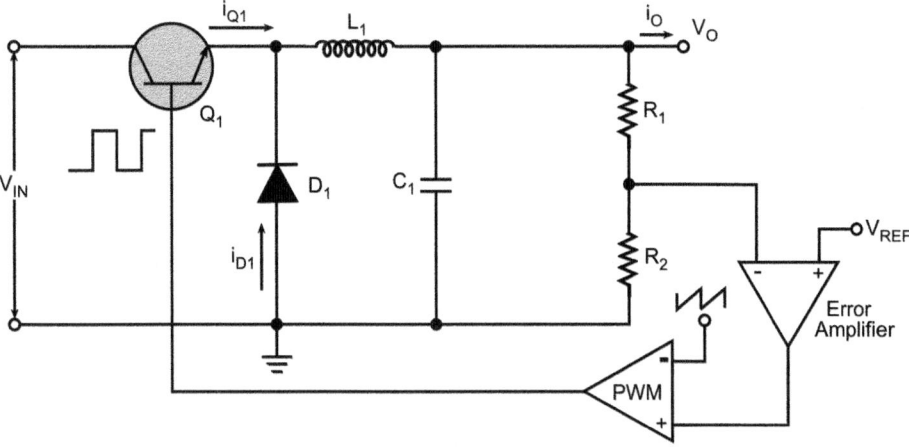

Figure 8.33 Buck regulator.

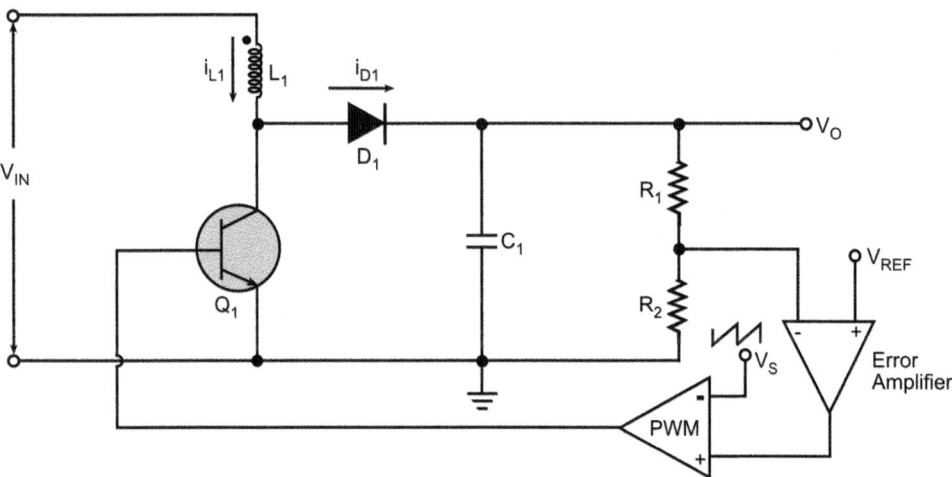

Figure 8.34 Boost regulator.

that it does not use a transformer and there is no input-output isolation. Output voltage is always less than the input voltage and is given by eqn. 8.8.

$$V_O = DV_{IN} \tag{8.8}$$

Where,
D is the duty cycle $(= T_{on}/T)$ of the drive waveform to the transistor switch.

Regulation is achieved by pulse width modulation of the transistor switch. It is a very popular circuit configuration for fabrication of high efficiency three-terminal switching regulators.

The step-up switching regulator, also called the boost regulator (Figure 8.34), is based on the flyback principle. It resembles the basic flyback converter except that it is of non-isolating type. The energy storage and transfer element in this case is an inductor rather than a transformer. The energy is stored in the magnetic field of the inductor during the conduction time of the switching transistor. Energy stored equals $\frac{1}{2}(L_p I_p^2)$. As the diode is reverse biased during conduction, the energy cannot get transferred while it is being stored. When the switching transistor is driven to

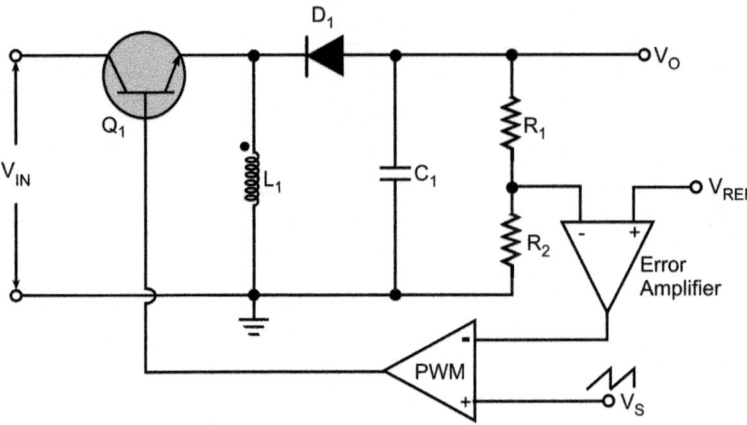

Figure 8.35 Inverting regulator.

cut-off, the diode gets forward biased and the stored energy is delivered to the load along with the energy from DC input voltage. The voltage across the load equals the DC input voltage plus the voltage due to the energy stored in the inductor. The output voltage in this case is given by eqn. 8.9.

$$V_O = \frac{V_{IN}}{(1-D)} = V_{IN}\frac{T}{T_{off}} \tag{8.9}$$

Where D is the duty cycle and T is the total time period equal to $T_{on} + T_{off}$.

The power output capability of this circuit is given by eqn. 8.10.

$$P_O = \frac{1}{2}L_P I_P^2 f \tag{8.10}$$

Where f is the switching frequency.

As the DC input voltage is not electrically isolated from the output, the output voltage in this case cannot be exclusively determined from the power rating of the circuit.

An inverting regulator (Figure 8.35) is another circuit configuration based on the flyback converter principle. For a positive input, it produces a negative output. Energy is stored in inductor L during the conduction time of the transistor. The diode is reverse biased during this time period. The stored energy is transferred during the OFF-time. The circuit delivers a constant output power to the load.

The output voltage is given by eqn. 8.11.

$$V_O = -\sqrt{P_O R_L} = -V_{IN}\frac{T_{on}}{T_{off}} \tag{8.11}$$

Regulation is achieved by controlling the duty cycle of the drive waveform. In the inverting regulator configuration, it is possible to have an output voltage that is either less than or greater than the input. It is also sometimes referred to as a buck-boost regulator.

8.1.2.2.5 Three-Terminal Switching Regulators

The basic step-down (buck) regulator in Figure 8.36 has been widely exploited in the form of a three-terminal switching regulator. Figure 8.36 shows the typical circuit configuration found inside such a regulator. Except for the switching transistor and the output inductor, all other component blocks are integrated on the chip.

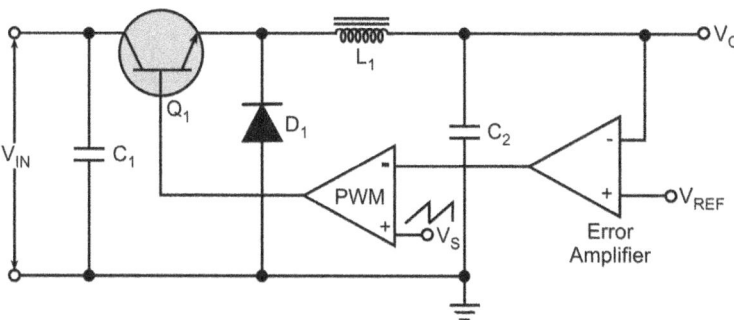

Figure 8.36 Three-terminal switching regulator.

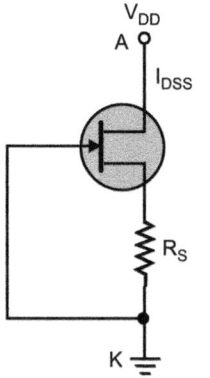

Figure 8.37 Current regulating diode.

The output voltage is compared with a reference voltage and the difference is amplified to drive a pulse width modulator, which in turn operates the switch. The three-terminal regulator can be used to construct a step-down switching supply that works very well for a wide input voltage range (typically 4–1). Output power levels of 300 W are conveniently achievable.

8.1.3 Constant Current Sources

A constant current source delivers a predefined constant current to a load despite variations in the supply voltage and the load resistance. In this section, we shall discuss some common constant current sources configured around active devices. It may be mentioned here that a constant current source is an important building block of diode laser electronics and finds extensive use in diode-pumped solid-state lasers and semiconductor diode lasers as drive circuits.

8.1.3.1 Junction FET-Based Constant Current Source

A junction FET with constant gate-source voltage acts like a constant source if the drain-source voltage were greater than a certain voltage called pinch-off voltage. This is evident from the drain current versus drain voltage characteristics of the junction FET. In the saturation region, the drain current is almost independent of drain-source voltage. For gate-source voltage of zero, the drain current is maximum equal to I_{DSS}. In the case of JFET used as a constant current source, if gate terminal were shorted to the source terminal, the constant current obtained would be (I_{DSS}).

Current regulating diodes are nothing but junction FETs with gate terminal internally shorted to source terminal. Figure 8.37 shows the internal schematic and circuit symbol of a current regulating diode. Inclusion of source resistor R_S as shown in Figure 8.37 allows the magnitude of constant current to be tuned to the desired value.

8.1.3.2 Transistor-Based Constant Current Source

Figure 8.38 shows a simple constant current source configured around a bipolar transistor. The base voltage of the bipolar transistor (equal to the breakdown voltage of the zener diode, V_Z) and hence its emitter voltage ($= V_Z - V_{BE}$) remain constant irrespective of the supply voltage variations as long as current through the zener diode is more than a certain

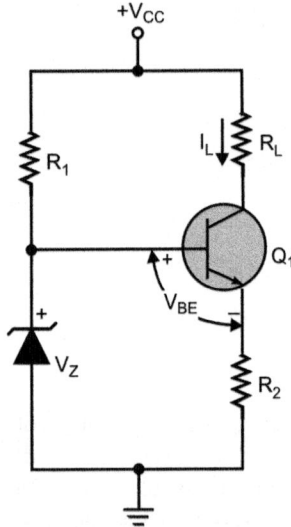

Figure 8.38 Constant current source configured around a bipolar transistor.

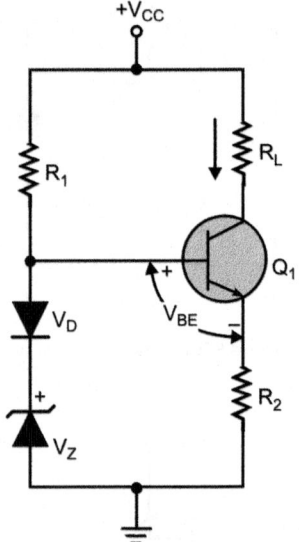

Figure 8.39 Constant current source with temperature compensation.

minimum value. Resistor R_1 supplies both the zener diode current as well as the base current for the transistor. Current through resistor R_2 is given by $[(V_Z - V_{BE})/R_2]$. Since V_{BE} (typically 0.6–0.7 V for a silicon transistor) is also constant for a given temperature, emitter current is constant. If the h_{FE} of the transistor is sufficiently large, emitter current is nearly equal to the collector current or the load current. Therefore, the load current in the circuit of Figure 8.38 is constant irrespective of the variations in the load resistance value and the supply voltage. The circuit behaves as constant current source. Varying R_2 changes the value of the constant current.

In the circuit of Figure 8.38, the load current is likely to vary with changes in the operating temperature due to temperature dependence of V_{BE}. This can be taken care of by connecting a diode (made of the same semiconductor material as the transistor) in series with the zener diode as shown in Figure 8.39. In this case, the emitter current equals $[(V_Z + V_D - V_{BE})/R_2]$, which further equals (V_Z/R_2) for $V_D = V_{BE}$. Another possible circuit configuration makes use of an LED in the place of the zener diode (Figure 8.40). This configuration has the advantage that the voltage across the LED tracks variation in V_{BE} due to change in temperature thus eliminating the need for the series-connected diode. The emitter current and hence the load current in this case can be computed from $[(V_{LED} - V_{BE})/R_2]$.

8.1.3.3 Opamp-Controlled Constant Current Source

Yet another constant current source that holds promise is the one configured around an operational amplifier. Figure 8.41 shows one such circuit. The non-inverting terminal of the opamp has a constant reference voltage applied. The opamp output drives the transistor to conduction and due to virtual earth phenomenon, the voltage at the inverting input of the opamp and hence across the sense resistor R_{sense} equals the reference voltage. The constant current through the load resistance then equals (V_Z/R_{sense}) and is independent of V_{BE} of the transistor.

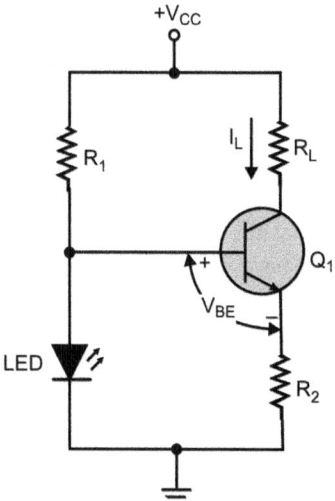

Figure 8.40 Constant current source with an LED as the reference voltage source.

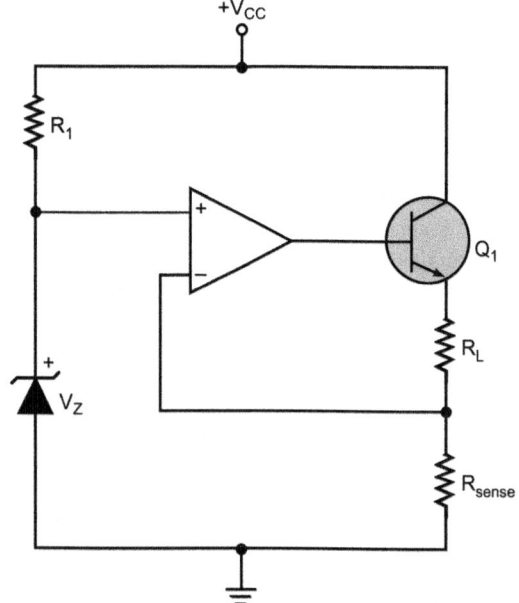

Figure 8.41 Opamp-based constant current source.

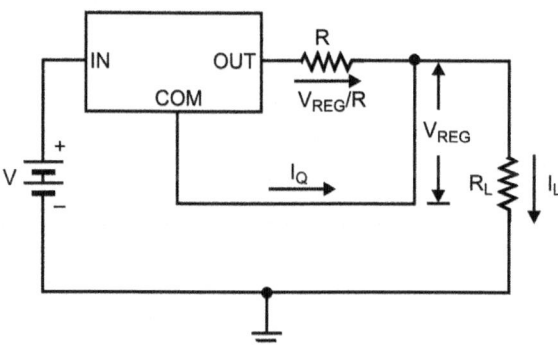

Figure 8.42 Constant current source with a three-terminal regulator.

8.1.3.4 Constant Current Source Using Three-Terminal Regulators

Fixed and adjustable output three-terminal IC voltage regulators can be conveniently used to construct constant current sources. Figure 8.42 shows the basic circuit. The load current in this case is given by $[(V_{REG}/R) + I_Q]$. V_{REG} is the regulated output voltage and I_Q is the quiescent current of the three-terminal regulator.

8.1.3.5 Current Mirror Configurations

A current mirror circuit mirrors or copies the current flowing through an active device by controlling the current flowing through another active device. Conceptually, an ideal current mirror is nothing but an ideal current amplifier with a unity gain. There are various current mirror circuits exploited for designing constant current sources. In the following paragraphs, the basic current mirror circuits and the variations to the basic current mirror including Widlar and Wilson current sources are briefly described.

8.1.3.5.1 Basic Current Mirror

Figure 8.43 shows the basic current mirror circuit. If the two transistors were perfectly matched, they could be assumed to have same values of base-emitter voltage drops and DC current gains. This is possible to achieve when the two transistors are fabricated on the same IC chip at the same time. Transistor Q_1 with its collector terminal shorted to its base terminal is wired like a diode. A semiconductor diode has not been used instead as that would have made the matching of characteristics of the diode and the base-emitter junction diode of the transistor Q_2 difficult to achieve. With the help of simple mathematics, it can be proved that current I flowing through Q_2 equals the current I_N flowing through R_1. If the two transistors have identical values of DC current gain β and base current I_B, they would have identical collector and emitter currents as shown in Figure 8.43, which gives $I_{IN} = I + (2I/\beta) = [(\beta + 2)/\beta]I \cong I$.

Thus, the current I_{IN} set by V_{CC} and R_1 is mirrored as the collector current of Q_2. It may be mentioned here that more than one matched transistor can be connected to the base of Q_1. The current through the collector of each of the transistors equals the current flowing through the collector of left-half transistor. The higher the value of β, closer the mirrored current approximates the set value. Figure 8.44 shows another current mirror circuit in which the JFET current regulating diode sets the constant current. The set value of current in this case equals I_{DSS}, which makes $I = I_{DSS}$. It may be mentioned here that in the circuit arrangement of Figure 8.43, current through R_1 would vary with variation in the supply voltage with the result that the current I though a mirror image of current I_{IN} will not be constant. The current mirror configuration of Figure 8.44 where transistor Q_1 is driven by a constant current source overcomes this problem.

8.1.3.5.2 Widlar Current Source

A constant current source should ideally have infinite incremental output impedance. The Widlar current source, as shown in Figure 8.45, is an improvement over the basic current mirror circuits of Figures 8.43 and 8.44 and offers a much higher incremental output impedance. This is achieved by connecting a resistor (R_2 in this case) in series with the emitter terminal of Q_2. This resistor introduces a current feedback around Q_2, thus increasing its output impedance by a factor of $1 + G$. G is the loop gain given by $1 + g_{m2}R_2$ where g_{m2} is the transconductance of Q_2. Again, in order to have a constant load current, the circuit must be driven by a constant current rather than R_1.

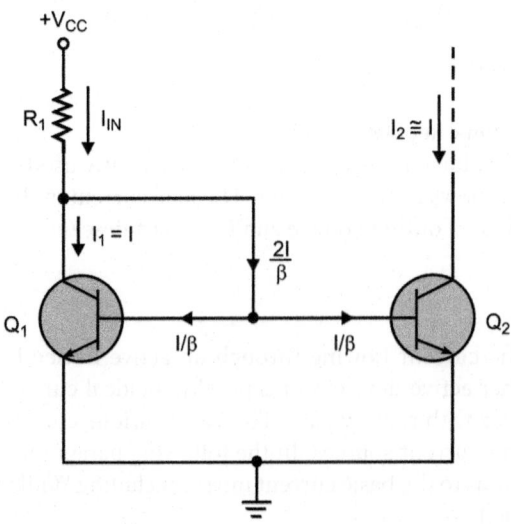

Figure 8.43 Current mirror implemented with a bipolar transistor.

Figure 8.44 Current mirror with JFET providing a constant current.

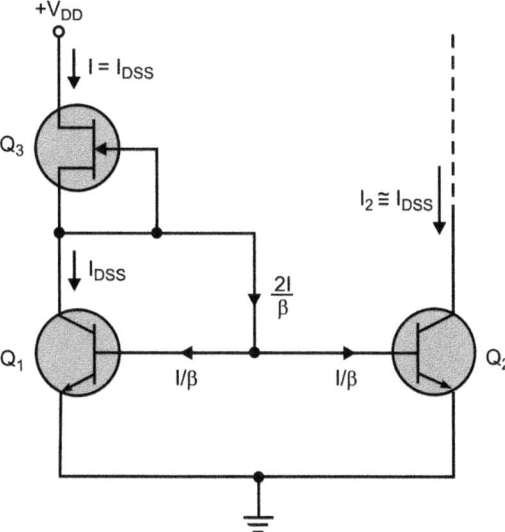

Figure 8.45 Widlar current source.

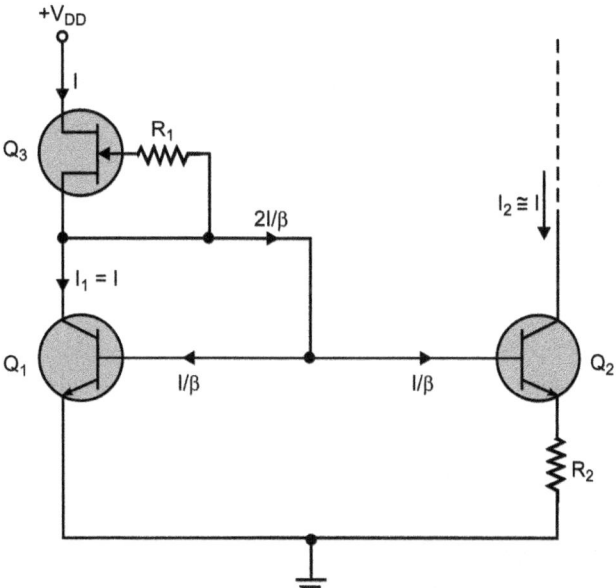

8.1.3.5.3 *Wilson Current Source*

Wilson current source as shown in Figure 8.46 has the advantage of virtually eliminating the base current mismatch of the conventional current mirror configuration. The other advantage is its higher output impedance.

8.1.4 Transimpedance Amplifier

There are many situations where the signal of interest is current rather than voltage. As an example, the electrical signal produced by a photo sensor as a result of light falling on it is a current. In order to process the signal to extract the intended information, more often than

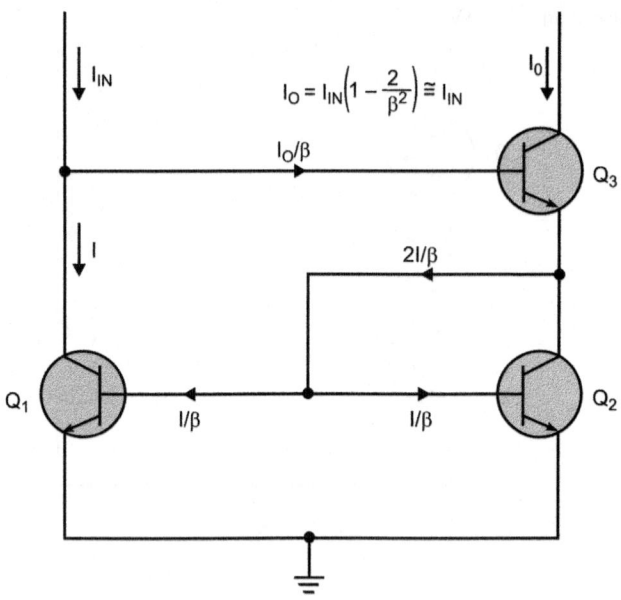

Figure 8.46 Wilson current source.

$$I_O = I_{IN}\left(1 - \frac{2}{\beta^2}\right) \cong I_{IN}$$

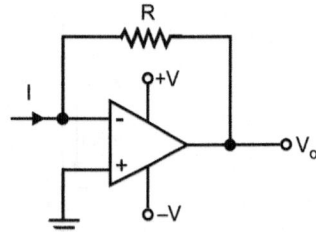

Figure 8.47 Basic transimpedance amplifier circuit.

not, the current needs to be converted into an equivalent voltage. This function is performed with a current-to-voltage converter circuit, which is also known as transimpedance amplifier for obvious reasons.

The current-to-voltage converter or the transimpedance amplifier constitutes an important building block of laser electronics. All laser receiver circuits, light meters, laser range finders, laser-warning sensors and laser communication receivers, to mention a few, use some kind of photosensor. The desired information is in the form of a continuous or pulsed electric current, which needs to be converted into an equivalent voltage signal for further processing. One common application of a transimpedance amplifier in military laser systems is its use as the front circuit in the receiving channel of a laser range finder. Here, the photo current pulse generated in the avalanche photo diode as a result of back scattered laser radiation pulse falling on it is transformed into an equivalent voltage pulse to provide the stop pulse. A PIN photo diode is also used to receive a small fraction of the transmitted laser pulse energy, which is converted into a voltage pulse in another transimpedance amplifier circuit to provide the start pulse.

Figure 8.47 shows the basic opamp-based current-to-voltage converter fed at its input by a current source I. The analysis of the circuit is very simple. Since the non-inverting of the opamp is tied to ground and the circuit has negative feedback, due to the virtual earth phenomenon, the inverting input also behaves as if it were grounded. In fact, high open loop gain of the opamp coupled with negative feedback forces the two inputs to remain at the same potential for a finite output. With this background, the magnitude of current flowing through the feedback resistor R would be (V_O/R). Since no current flows into the opamp due to its infinite input impedance (under ideal conditions), applying Kirchhoff's current law at the inverting terminal node, we get $[I + (V_O/R) = 0]$. This gives $V_O = -IR$.

Figure 8.48 shows the basic transimpedance amplifier circuit fed at its input from a photodiode. The photodiode here is used in photovoltaic mode, that is, the bias voltage is zero. The photo diode could as well be used in the photoconductive mode with a reverse bias applied to the photo diode.

8.1.5 Peak Detector Circuits

A peak detector circuit produces at its output a voltage equal to the peak value of the signal applied at its input. The circuit could be designed to detect either the positive or the negative peak. Like current-to-voltage converter, peak detector is also an important building block of laser receiver electronics in systems such as laser-warning sensors, laser power and energy meters and so on. It is particularly important while designing front-end detection circuits for Q-switched laser pulses. The peak detector circuit in that case acts as a pulse stretcher. The peak of the stretched pulse equals the peak of the input pulse. The stretched pulse allows the designer to use relatively lower speed processing circuitry subsequent to the front-end.

Figure 8.49 shows a positive peak detector circuit. The circuit functions as follows. In the case of a sinusoidal input or any other bidirectional input, the diode is forward biased during

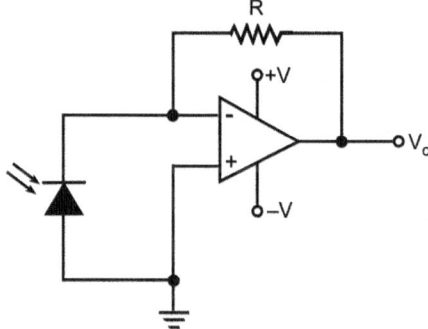

Figure 8.48 Photodiode current-to-voltage converter.

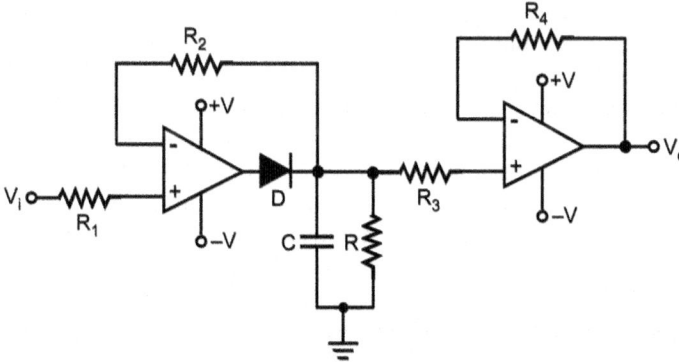

Figure 8.49 Positive peak detector circuit.

the positive half cycles. The capacitor rapidly charges to the positive peak from the output of the opamp through relatively much smaller on-resistance of the forward-biased diode. In the case of a positive pulse input, the same happens during the duration of the pulse. As the input begins to decrease beyond the peak value, the diode gets reverse biased thus isolating the capacitor from the output of the opamp. The capacitor can discharge only through the parallel combination of the two resistors. One of these is the resistor connected across the capacitor and the other is the series combination of the feedback resistor and the input resistor of the opamp at the inverting input. This effective discharge path resistance is relatively much larger than the on-resistance of the forward-biased diode. It may be mentioned here that the resistor across the capacitor is there to allow a discharge path for the circuit to respond to decreasing amplitude of the signal peaks. The unity gain buffer at the output is to prevent capacitor discharge due to loading effects of the following circuit.

The parallel R-C circuit time constant is typically 100 times the time period corresponding to minimum frequency of operation. The R-C time constant also controls the response time to changing amplitude. Higher R-C time constant gives lower ripple but a sluggish response while lower R-C time constant gives faster response at the cost of increased ripple. The chosen value is a trade-off between two conflicting requirements.

Slew rate is the main opamp parameter to be considered while choosing one for this application. The desired slew rate should be such that the slew rate limited frequency, which is a function of the peak-to-peak output swing and the slew rate, is at least equal to the highest frequency of operation. Peak-to-peak voltage swing at the output of the opamp is equal to $[V_{PEAK} - (-V_{SAT})] = (V_{PEAK} + V_{SAT})$. In the case of pulse input, the slew rate should at least equal (V_O/T) volts per microsecond. V_p is the peak amplitude of the pulse and T is its width.

8.1.6 Timer Circuits

In this section, we shall discuss IC-based timer circuits configured around digital integrated circuits and linear integrated circuit timers.

8.1.6.1 Digital IC-Based Timer Circuits

Some of the commonly used digital integrated circuits that can be used as timer circuits in the monostable multivibrator configuration include the 74121 (single monostable multivibrator), 74221 (dual monostable multivibrator), 74122 (single retriggerable monostable multivibrator), 74123 (dual retriggerable monostable multivibrator), all belonging to the TTL family and 4098B (dual retriggerable monostable multivibrator) belonging to the CMOS family. Figure 8.50 shows the use of the IC 74121 as a monostable multivibrator along with trigger input. The IC provides features for triggering on either low-to-high or high-to-low edges of the trigger pulses. Figure 8.50(a) shows one of the possible application circuits for high-to-low edge triggering and Figure 8.50(b) shows one of the possible application circuits for low-to-high edge triggering. Output pulse width depends on external R and C. The output pulse width can be computed from $T = 0.7RC$. Recommended ranges of values for R and C, respectively, are 4–40 kΩ and 10 pF–1000 μF. IC provides complementary outputs. That is, we have a stable LOW or HIGH state and corresponding HIGH or LOW-state available on (Q) and ($\bar{Q}$) outputs.

Figure 8.51 shows the use of the 74123, a retriggerable monostable multivibrator. Like the 74121, this IC too provides features for triggering on either low-to-high or high-to-low edges of the trigger pulses. Output pulse width depends on external R and C. It can be computed from $T = 0.28RC[1 + (0.7/RC)]$, where R and C are, respectively, in kilo-ohms and pico-farads and T is in nanoseconds. This formula is valid for $C > 1000$ pF. The recommended range of values for R is 5–50 kΩ. Figure 8.51 parts (a) and (b), respectively, gives application circuits for

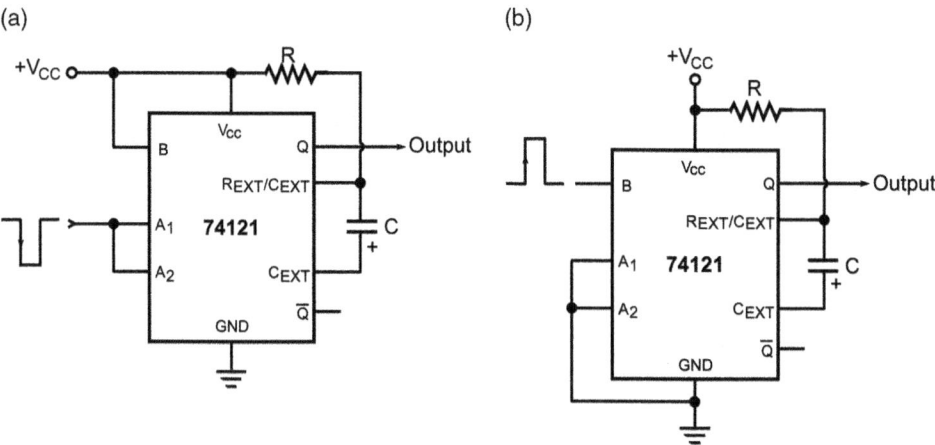

Figure 8.50 74121-based monostable multivibrator circuits: (a) HIGH-to-LOW triggering and (b) LOW-to-HIGH triggering.

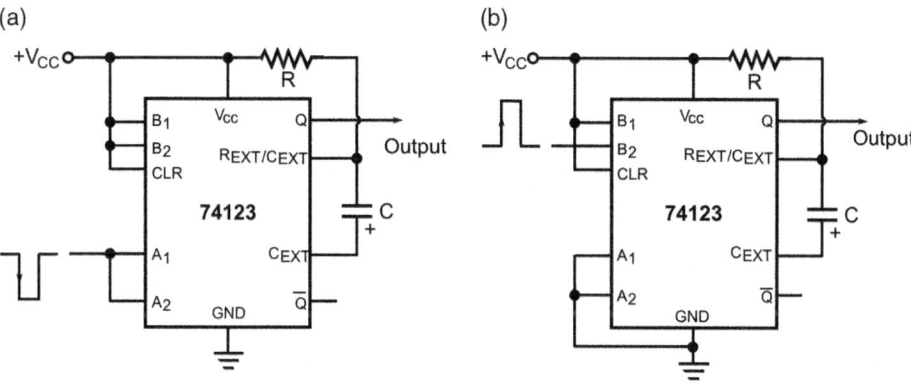

Figure 8.51 74123-based monostable circuits: (a) HIGH-to-LOW triggering and (b) LOW-to-HIGH triggering.

high-to-low and low-to-high triggering. It may be mentioned that, there can be other triggering circuit options for both low-to-high and high-to-low edge of monostable multivibrator.

8.1.6.2 IC Timer-Based Multivibrators

The IC timer 555 is one of the most commonly used general-purpose linear integrated circuits. The simplicity with which monostable and astable multivibrator circuits can be configured around this IC is one of the main reasons for its wide use. Figure 8.52 shows the internal schematic of timer IC 555. It comprises of two opamp comparators, a flip-flop, a discharge transistor, three identical resistors and an output stage. The resistors set the reference voltage levels at the non-inverting input of the lower comparator and inverting input of the upper comparator at $+V_{CC}/3$ and $+2V_{CC}/3$, respectively. Outputs of two comparators feed SET and RESET input of the flip-flop and thus decide the logic status of its output and subsequently the final output. The flip-flops' complementary outputs feed the output stage and the base of the discharge transistor. This ensures that the output is HIGH, when the discharge transistor is off and LOW when the discharge transistor is on. Different terminals of the timer 555 are designated as Ground (Pin 1), Trigger (Pin 2), Output (Pin 3), RESET (Pin 4), Control (Pin 5),

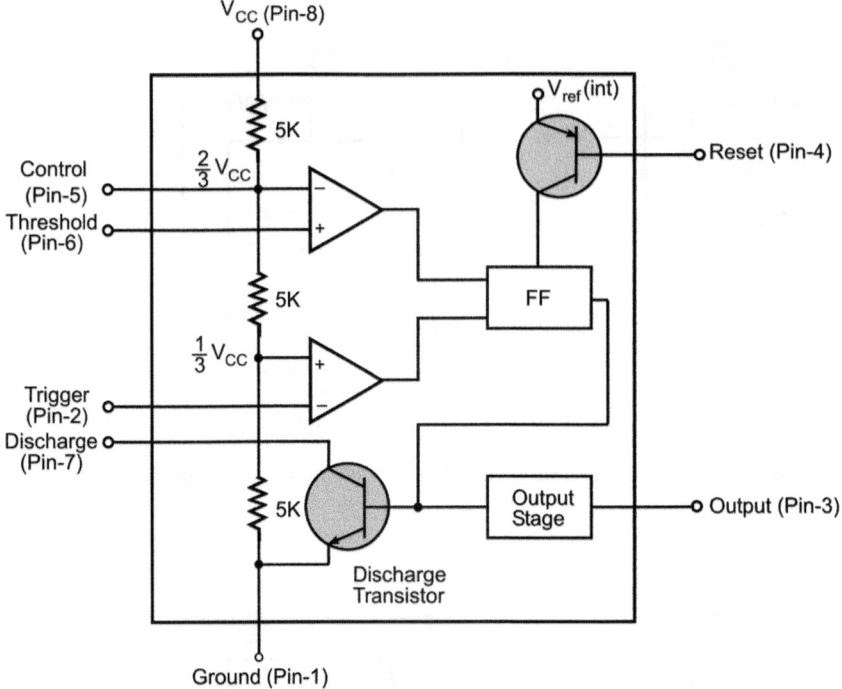

Figure 8.52 Internal schematic arrangement of IC timer 555.

Threshold (Pin 6), Discharge (Pin 7) and + V_{CC} (Pin-8). With this background, we shall now describe the astable and monostable circuits configured around timer 555.

8.1.6.2.1 Astable Multivibrator Using Timer IC 555

Figure 8.53(a) and (b) shows the basic 555 timer-based astable multivibrator circuit along with relevant waveforms. Initially, capacitor C is fully discharged, which forces output to go to the HIGH state. An open discharge transistor allows the capacitor C to charge from + V_{CC} through R_1 and R_2. When the voltage across C exceeds + $2V_{CC}/3$, the output goes to the LOW-state and the discharge transistor is switched ON at the same time. Capacitor C begins to discharge through R_2 and the discharge transistor inside the IC. When the voltage across C falls below + $V_{CC}/3$, output goes back to HIGH state. The charge and discharge cycles repeat and the circuit behaves like a free-running multivibrator. Terminal 4 of the IC is the RESET. Usually, it is connected to + V_{CC}. If the voltage at this terminal is driven below 0.4 V, output is forced to LOW-state overriding command pulses at terminal-2 of the IC. HIGH-state and LOW-state time periods are governed by the charge (+ $V_{CC}/3$ to + $2V_{CC}/3$) and discharge (+ $2V_{CC}/3$ to + $V_{CC}/3$) timings.

$$\text{HIGH} - \text{state time period}, T_{HIGH} = 0.69C(R_1 + R_2)$$

$$\text{LOW} - \text{state time period}, T_{LOW} = 0.69R_2C$$

Remember that when the astable multivibrator is powered, first cycle HIGH-state time duration is about 30% longer as the capacitor is initially discharged and it charges from 0 (rather than + $V_{CC}/3$) to (+ $2V_{CC}/3$).

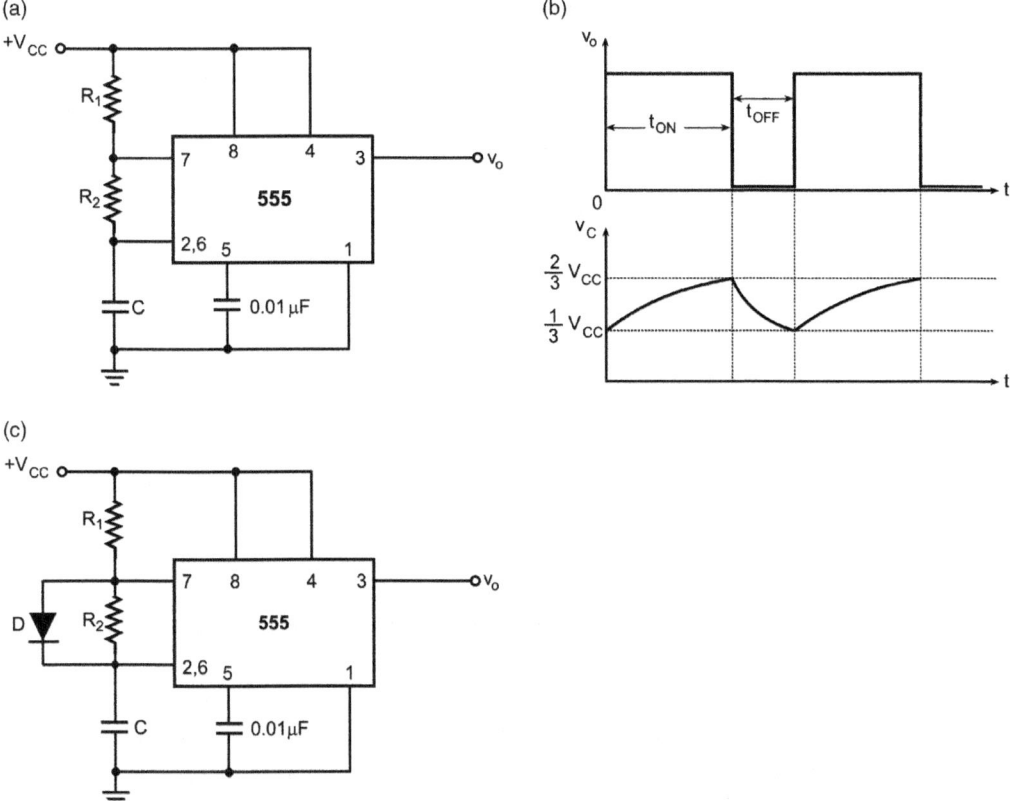

Figure 8.53 Basic 555 timer-based astable multivibrator: (a) circuit, (b) relevant waveforms, (c) modified 555 timer based astable multivibrator.

In the case of the astable multivibrator circuit in Figure 8.53(a), the HIGH state time period is always greater than the LOW-state time period. Figure 8.53(c) shows modified circuit where HIGH-state and LOW-state time periods can be chosen independently. For the astable multivibrator circuit of Figure 8.53(c), the two time periods are given by the following equations.

$$\text{HIGH} - \text{state time period}, T_{HIGH} = 0.69\,R_1 C$$

$$\text{LOW} - \text{state time period}, T_{LOW} = 0.69\,R_2 C$$

For $R_1 = R_2 = R, T = 1.38RC$ and $f = (1/1.38RC)$.

8.1.6.2.2 Monostable Multivibrator Using Timer IC 555

Figure 8.54(a) shows the basic monostable multivibrator circuit configured around timer 555 along with relevant waveforms as shown in Figure 8.54(b). Trigger pulse is applied to pin-2 of the IC, which should initially be kept at $+V_{CC}$. A HIGH at pin 2 forces the output to LOW-state. A HIGH-to-LOW trigger pulse at pin 2 sets the output in the HIGH state and simultaneously allows the capacitor C to charge from $+V_{CC}$ through R. Remember that Low level of the trigger pulse needs to go at least below $+V_{CC}/3$. When the capacitor voltage exceeds $+2V_{CC}/3$, the output goes back to the LOW-state. We shall need to apply another trigger pulse to pin2 to

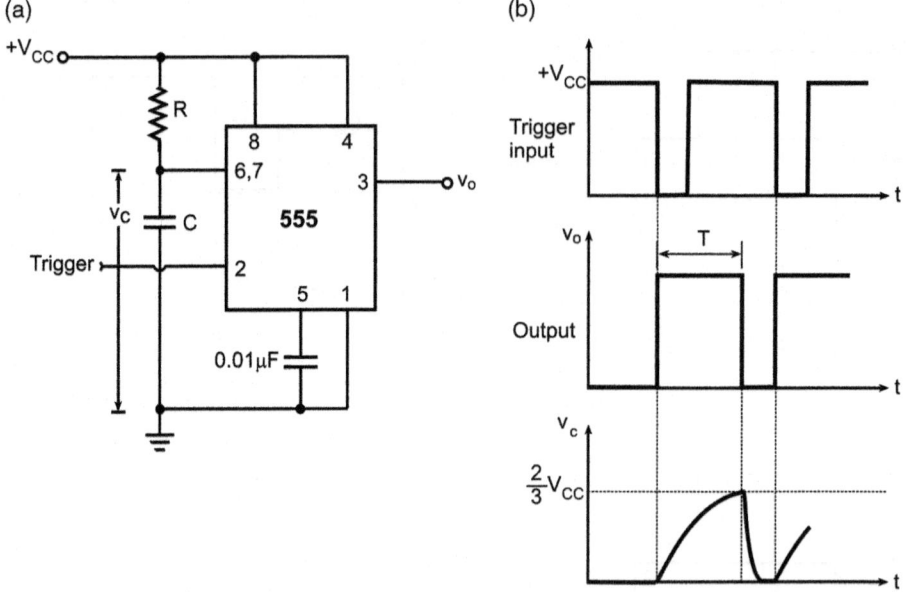

Figure 8.54 Basic 555 timer-based monostable multivibrator: (a) circuit and (b) relevant waveforms.

make the output go to HIGH state again. Every time, the timer is appropriately triggered, the output goes to HIGH state and stays there for a time period taken by capacitor to charge from 0 to $+2V_{CC}/3$. This time period, which equals the monoshot output pulse width, is given by $T = 1.1RC$.

It is often desirable to trigger a monostable multivibrator either on the trailing (high-to-low) or leading edges (low-to-high) of the trigger waveform. In order to achieve that, we shall need an external circuit between the trigger waveform input and pin 2 of timer 555. The external circuit ensures that pin 2 of the IC gets the required trigger pulse corresponding to the desired edge of the trigger waveform. Figure 8.55(a) shows the monoshot configuration that can be triggered on the trailing edges of the trigger waveform. Figure 8.55(b) shows relevant waveforms. R_1-C_1 constitutes a differentiator circuit. One of the terminals of resistor R_1 is tied to $+V_{CC}$ with the result that the amplitudes of differentiated pulses are $+V_{CC}$ to $+2V_{CC}$ and $+V_{CC}$ to Ground corresponding to leading and trailing edges of the trigger waveform, respectively. Diode D clamps the positive going differentiated pulses to about $+0.7$ V. The net result is that the trigger terminal of timer 555 gets the required trigger pulses corresponding to HIGH-to-LOW edges of the trigger waveform.

Figure 8.56(a) shows the monoshot configuration that can be triggered on the leading edges of the trigger waveform. Figure 8.56(b) shows relevant waveforms. R_1-C_1 combination constitutes the differentiator producing positive and negative pulses corresponding to LOW-to-HIGH and HIGH-to-LOW transitions of the trigger waveform. Negative pulses are clamped by the diode and the positive pulses are applied to the base of a transistor switch. Collector-terminal of the transistor feeds the required trigger pulses to pin 2 of the IC.

For the circuits shown in Figures 8.55 and 8.56 to function properly, values of R and C for the differentiator should be chosen carefully. Firstly, differentiator time constant should be much smaller than the HIGH time of the trigger waveform for proper differentiation. Secondly, differentiated pulse width should be less than the expected HIGH time of the monoshot output.

Figure 8.55 555-timer monostable configuration triggering on HIGH-to-LOW edges: (a) monoshot configuration and (b) relevant waveforms.

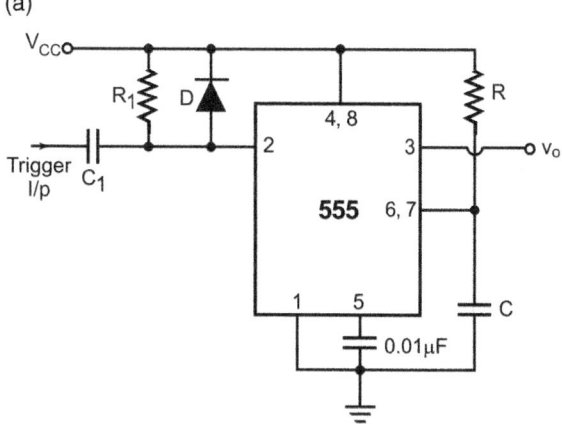

(a)

(b)

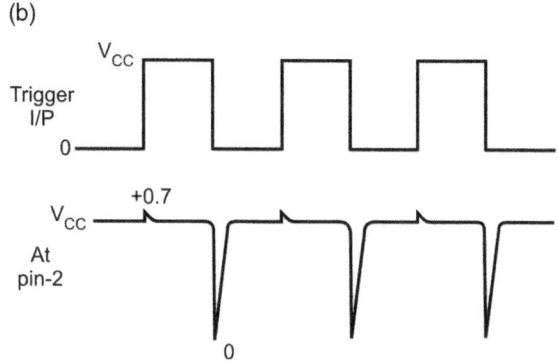

Figure 8.56 555-timer monostable configuration triggering on LOW-to-HIGH edges: (a) monoshot configuration and (b) relevant waveforms.

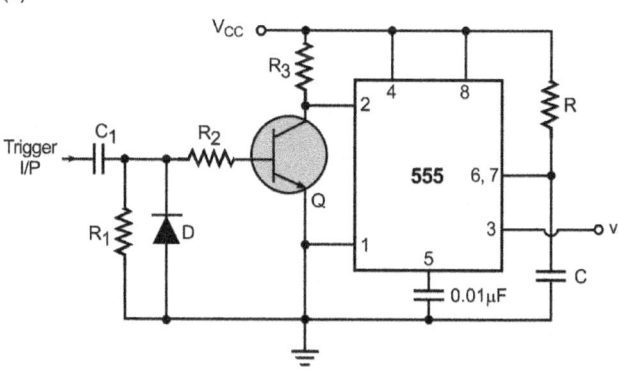

(a)

(b)

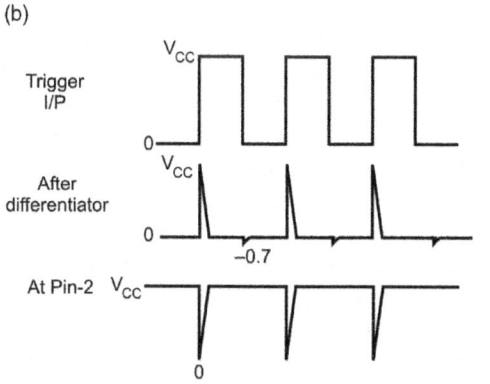

8.1.7 High Voltage Trigger Circuits

A high voltage trigger circuit is yet another important building block of laser electronics. It is an essential component of the electronics package that goes along with any flash lamp pumped solid-state laser where it performs the function of initiating an electrical discharge in the gas filled inside the tube to create a low resistance path between the two electrodes. High voltage trigger pulses are also used to initiate plasma discharge in gas lasers such as helium-neon and carbon dioxide lasers. The amplitude of the high voltage pulses is typically in the range of 5–20 kV and the pulse shape is not critical. High voltage pulses are also required to operate the Q-switch in the case of electro-optically Q-switched solid-state lasers. The amplitude of the high voltage pulse required for the Q-switch is typically in the order of 2–3 kV and the pulse has a well-defined shape in terms of its rise time and pulse width.

Figure 8.57(a) shows typical high voltage trigger pulse generation circuit used for flash lamp triggering. A bipolar junction transistor, an enhancement MOSFET or an insulated gate bipolar transistor (IGBT) can also be used as the electronic switch in place of SCR. The circuit functions as follows. When the SCR is in the off-state, capacitor C charges through resistor R and the transformer primary winding. When the capacitor is fully charged to the applied DC voltage (V), the SCR is triggered to the on-state by application of a suitable trigger pulse at its gate terminal. The capacitor rapidly discharges through the primary winding of the transformer generating a pulse across the primary winding. This pulse is stepped up to produce the desired amplitude of high voltage pulse across the secondary winding. The polarity of the pulse depends upon the winding polarity. In the case of the circuit shown in Figure 8.57(a), the polarity of pulse at the dotted end of the primary winding is negative, which means that the polarity of pulse on the dotted end of the secondary winding will also be negative. As a result, we get a positive polarity output pulse. Figure 8.57(b) shows the relevant waveforms.

The value of resistance R should be such that the current V/R is less than the holding current value of the SCR. Otherwise, the SCR once turned on will continue to conduct even after the capacitor has fully discharged through it and the primary winding. Figure 8.58 parts (a) and (b) shows the same circuit implemented with an N-channel enhancement MOSFET and relevant waveforms. Please note that in the case of circuits of Figure 8.58(a), value of resistance R has no lower limit it had in the case of SCR circuit. Also, in the case of MOSFET based trigger circuit, MOSFET remains ON during the duration of the input pulse and the capacitor starts charging only after the MOSFET goes back to the OFF-state. The IGBT based circuit would be similar to MOSFET based circuit.

8.1.8 Analogue-to-Digital Converter Circuits

An A/D converter is another very important building block of laser electronics. It forms an essential interface when it comes to analysing analogue data with digital circuits. It is an indispensable part of a digital communication system where the analogue signal to be transmitted is digitized at the sending end with the help of an A/D converter. It is invariably used in all optronic sensor systems where the optronic front-end converts the received optical radiation into an electrical signal that is analogue in nature. A/D converter is used to transform the analogue signal into an equivalent digital signal that is processed digitally to extract the desired information. Be it a laser range finder receiver or a laser-warning sensor or the laser seeker in a laser-guided bomb, an A/D converter is invariably used to convert the voltage signal available at the output of corresponding transimpedance amplifiers into digital bit patterns for further processing.

(a)

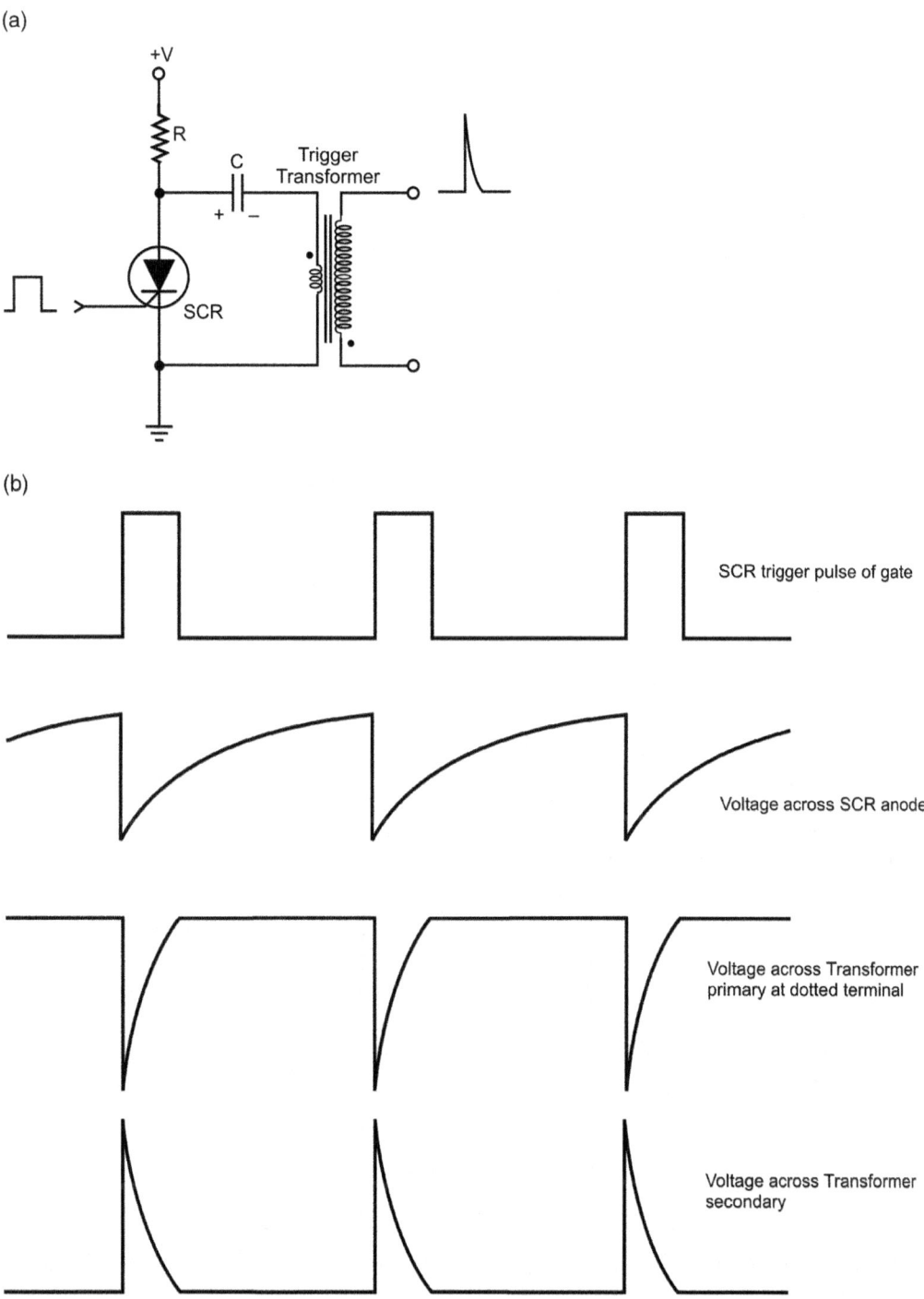

(b)

SCR trigger pulse of gate

Voltage across SCR anode

Voltage across Transformer primary at dotted terminal

Voltage across Transformer secondary

Figure 8.57 High voltage trigger pulse generator: (a) typical high voltage trigger pulse generation circuit and (b) relevant waveforms.

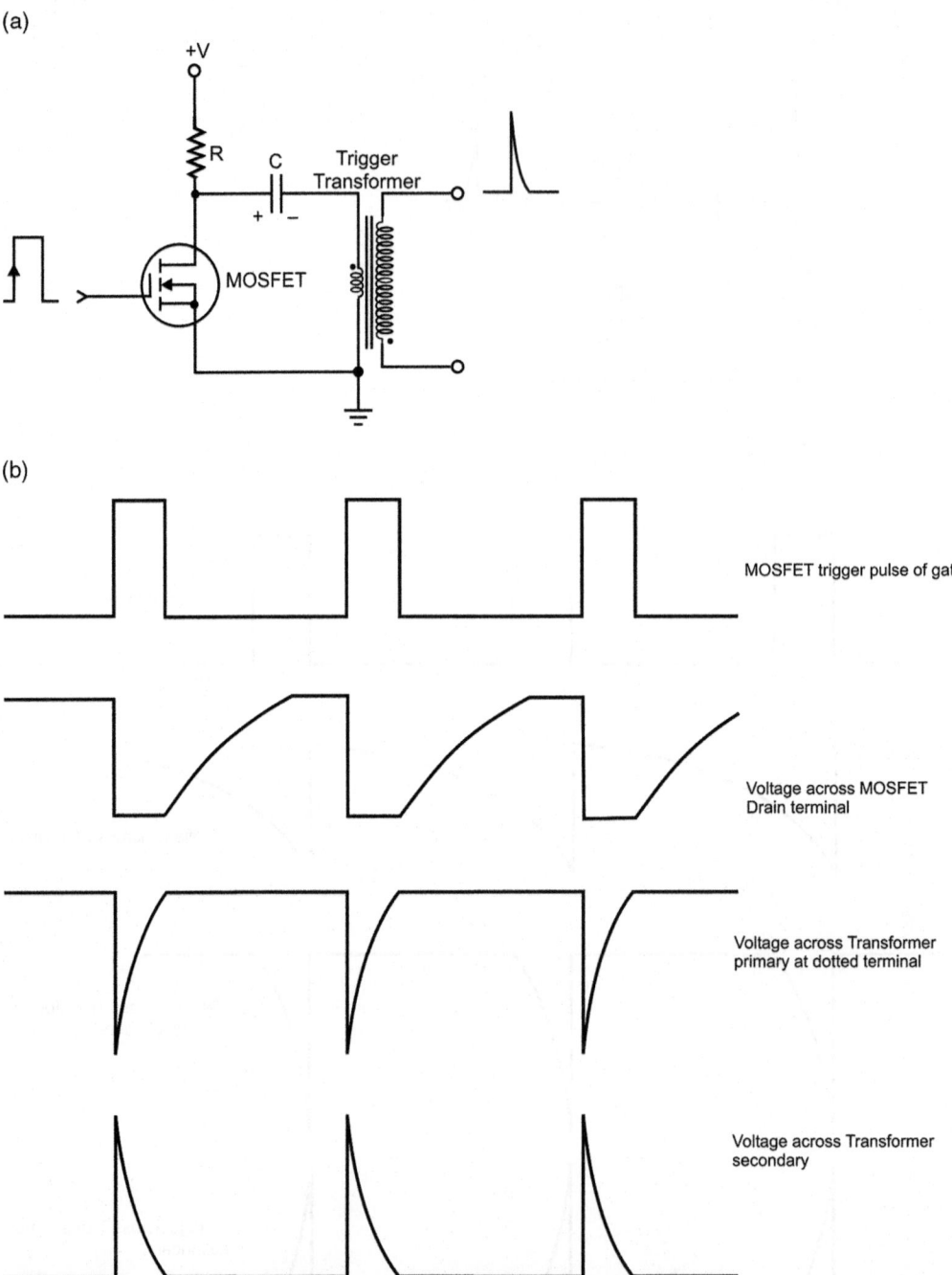

Figure 8.58 MOSFET-based high voltage trigger circuit: (a) N-channel enhancement MOSFET and (b) relevant waveforms.

An A/D converter takes at its input an analogue voltage and after a certain amount of time, produces a digital output code representing the analogue input. The A/D conversion process is generally more complex than the D/A conversion process. There are various techniques developed for the purpose of A/D conversion and these techniques have different advantages and disadvantages with respect to one another, which have been utilized in the fabrication of different categories of A/D converter ICs. A D/A converter circuit, as we shall see in the following paragraphs, forms a part of some of the A/D converter types.

We begin with a brief interpretation of the terminology and the major specifications that are relevant to the understanding of A/D converters. The idea is to enable the designers make a judicious choice of an A/D converter suitable for their application. A brief comparative study of different types of A/D converters and the suitability of each one of them for a given application requirement is also discussed.

8.1.8.1 A/D Converter Specifications

The major performance specifications of an A/D converter include the following.

1) Resolution
2) Accuracy
3) Gain and offset errors
4) Gain and offset drifts
5) Sampling frequency and the aliasing phenomenon
6) Quantization error
7) Nonlinearity
8) Differential nonlinearity
9) Conversion time
10) Aperture and acquisition times
11) Code width

Each one of these is briefly described in the following paragraphs.

Resolution of an A/D converter is the quantum of input analogue voltage change required to increment its digital output between one code change and the next code change. An n-bit A/D converter can resolve 1 part in 2^n. It may be expressed as a percentage of full-scale or in bits. Resolution of an 8-bit A/D converter, for example, can be expressed as 1 part in 256 or as 0.4% of FS or simply as 8-bit resolution.

Accuracy specification describes the maximum sum of all the errors, both from analogue sources (mainly the comparator and the ladder resistors) as well as the digital sources (quantization error) of the A/D converter. These errors mainly include the gain error, the offset error and the quantization error. The accuracy describes the actual analogue input and full-scale weighted equivalent of the output code corresponding to actual analogue input.

Gain error is the difference between the actual full-scale transition voltage and the ideal full-scale transition voltage. It is expressed either as percentage of full-scale range (percentage of FSR) or in LSBs. *Offset error* is the error at analogue zero for an A/D converter operating in the bipolar mode. It is measured in percent of FSR or in LSBs.

Gain drift is the change in the full-scale transition voltage measured over the entire operating temperature range. It is expressed in FS per degree Celsius or PPM of FS per degree Celsius or LSBs. *Offset drift* is the change with temperature of the analogue zero for an A/D converter operating in the bipolar mode. It is generally expressed in PPM of FS per degree Celsius or LSBs.

If the rate at which the analogue signal to be digitized is sampled is at least twice the highest frequency in the analogue signal, which is what is embodied in the Shannon–Nyquist sampling theorem, then the analogue signal can be faithfully reproduced from its quantized values by using a suitable interpolation algorithm. The accuracy of the reproduced signal is, however, limited by quantization error. In case the sampling rate is inadequate, that is, it is less than the Nyquist rate, then in that case the reproduced signal is not a faithful reproduction of original signal and these spurious signals called aliases are produced. The frequency of aliased signal is difference between the signal frequency and the sampling frequency. This problem is called *aliasing* and in order to avoid aliasing, the analogue input signal is low pass filtered to remove all frequency components above half of the sampling rate. This filter, called the *anti-aliasing filter*, is used in all practical A/D converters.

Quantization Error is inherent to digitizing process. It can be reduced by increasing the number of digitized levels. An A/D converter with an *n*-bit output can only identify 2^n output codes while there exist an infinite number of analogue input values adjacent to the LSB of the A/D converter that are assigned the same output code. Expressed as a percentage, the quantization error in an 8-bit converter is 0.4% or 1 part in 256.

Nonlinearity (INL) specification (also referred to as the Integral Nonlinearity, INL by some manufacturers) of an A/D converter describes its departure from a linear transfer curve. Nonlinearity error does not include gain, offset and quantization errors. It is expressed as a percentage of FS or in LSBs.

Differential nonlinearity (DNL) indicates the worst-case difference between the actual analogue voltage change and the ideal 1 LSB voltage change, DNL specification is as important as the INL specification as an A/D converter with a good INL specification may have a poor-quality transfer curve if DNL specification is poor. DNL is also expressed as a percentage of FS or in LSBs. DNL in fact explains the smoothness of the transfer characteristics and is thus of great importance to the user. Figure 8.59(a) shows the transfer curve for three-bit A/D converter with 7 V full-scale range, (1/4) LSB INL and 1LSB DNL. Figure 8.59(b) shows the same for 7 V FS range, 1LSB NL and 1/4 LSB DNL. Although the former has much better INL specification, the latter with better DNL specification, has a much better and smoother curve and may thus be preferred. Too high a value of DNL may even grossly degrade the converter resolution. In a 4-bit converter with ±2LSB DNL, the 16-step transfer curve may be reduced to

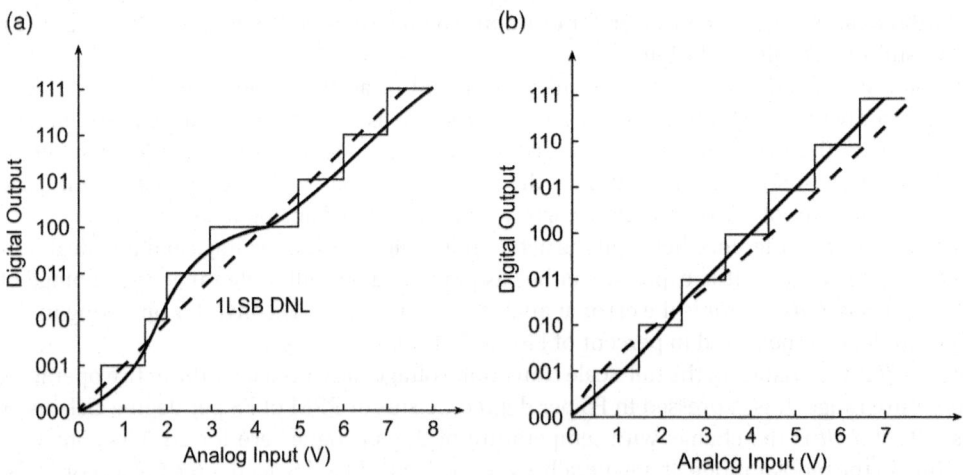

Figure 8.59 Transfer curve of a 3-bit A/D converter: (a) ¼ LSB INL and 1 LSB DNL and (b) 1 LSB INL and ¼ LSB DNL.

a six-step curve. DNL specification should in no case be ignored unless the INL specification is tight enough to guarantee the desirable DNL.

Conversion time is the time that elapses from the time instant of the start of conversion signal until the conversion complete signal occurs. It ranges from a few nanoseconds for flash type A/D converters to a few microseconds for successive approximation type A/D converters and may be as large as tens of milliseconds for dual slope integrating A/D converters.

When a rapidly changing signal is digitized, the input signal amplitude would have changed even before the conversion is complete with the result that output of the A/D converter does not represent the signal amplitude at the start. A *sample and hold* circuit with a buffer amplifier is used at the input of the A/D converter to overcome this problem. *Aperture and acquisition times* are the parameters of the sample and hold circuit. The signal to be digitized is sampled with an electronic switch that can be rapidly turned on and off. The sampled amplitude is then stored on the hold capacitor. The A/D converter digitizes the stored voltage and after the conversion is complete, a new sample is taken and held for the next conversion. *Acquisition time* is the time required by the electronic switch to close and the hold capacitor to charge while *aperture time* is the time that the switch takes to completely open after the occurrence of hold signal. Ideally, both the times should be zero. The maximum sampling frequency is thus determined by the aperture and acquisition times in addition to the conversion time.

The quantum of input voltage change that occurs between the output-code transitions expressed in LSBs of full-scale is the *code width*. *Code width uncertainty* is the dynamic variation or *jitter* in the code width due to noise.

8.1.8.2   A/D Converter Terminology

Some of the more commonly used terms while interpreting the specifications and salient features of A/D converters are briefly described in the following paragraphs.

In the *unipolar mode of operation*, the analogue input to the A/D converter varies from 0 to full-scale voltage of one polarity only. In the *bipolar mode of operation*, an A/D converter is configured to convert both positive as well as negative analogue input voltages.

Coding defines the nature of A/D converter output data format. Commonly used formats include straight binary, offset binary, complementary binary, 2s complement, low byte and high byte. In A/D converters with a resolution greater than 8-bits, some products are offered in *high byte* or *low byte* formats to simplify their interface with 8-bit microprocessor systems. The low byte output contains the least significant bit and some or all of the lower 8 bits of the A/D converter output. In the high byte, output contains the MSB and some or all of the upper 8 bits. In the byte oriented data output format, data bit sets shorter than 8 bits are placed starting with the right side of the data output transfer register. This could apply to upper or lower byte. For example, a 12-bit ADC will have four extra bits that could be *right justified*. In the *left justified data*, data bit sets shorter than 8-bits are placed starting with the left side of data output transfer register. This could apply to the lower or the upper byte. For example, a 12-bit ADC will have four extra bits that could be left justified.

Command register is an internal register of the ADC that can be programmed by the user to select various modes of operation such as unipolar or bipolar mode selection, range selection, data output format selection and so on. *Status register* indicates the current status of the analogue-to-digital conversion with a 'Busy' or 'Conversion Complete' signal. Digital input/output pins that activate/monitor and control ADC operation are called *control lines*. Some examples are Chip Select, Write, Start Convert, Conversion Complete and so on.

8.1.8.3 Types of A/D Converters

A/D converters are often classified according to the conversion process or the conversion technique used to digitize the signal. Based on various conversion methodologies, common types of A/D converter include the following:

1) Flash or simultaneous or direct conversion A/D converter
2) Half-flash A/D converter
3) Counter-type A/D converter
4) Tracking A/D converter
5) Successive approximation type A/D converter
6) Single slope, dual slope and multi-slope A/D converters
7) Sigma-Delta A/D converter.

Each of these types of A/D converter is described in the following paragraphs.

8.1.8.3.1 Simultaneous or Flash A/D Converters

The simultaneous method of A/D conversion is based on using a number of comparators. The number of comparators needed for n-bit A/D conversion is $(2^n - 1)$. As an example, Figure 8.60 shows the arrangement of a 3-bit simultaneous type A/D converter. The construction of a simultaneous A/D converter is quite straightforward and relatively easy to understand. However, as the number of bits in the desired digital signal increases, the number of comparators required performing A/D conversion increases very rapidly and it may not be feasible to use this approach once the number of bits exceeds six or so. The greatest advantage of this technique lies in its capability to execute extremely fast analogue-to-digital conversion.

8.1.8.3.2 Half-Flash A/D Converter

A *half-flash A/D converter*, also known as a *pipeline A/D converter*, is a variant of flash type converter that overcomes to a large extent the primary disadvantage of requirement of a prohibitively large number of comparators in high-resolution full flash converters without significantly degrading its high-speed conversion performance. When compared to a full flash converter of certain resolution, while the number of comparators and associated resistors reduce drastically in a half-flash converter, conversion time increases approximately by a factor of 2. For an n-bit flash converter, number of comparators required is 2^n [$(2^n - 1)$ for encoding of amplitude and one comparator for polarity], the same for an equivalent half-flash converter would be $2 \times (2^{n/2})$. In the case of an 8-bit converter, the number is 32 (for half-flash) against 256 (for full flash).

Figure 8.61 shows the architecture of 8-bit half-flash A/D converter. A half-flash converter uses two full flash converters with each full flash converter having a resolution equal to half the number of bits of the half-flash converter. That is, an 8-bit half-flash converter uses two 4-bit flash converters. In addition, it uses a 4-bit D/A converter and an 8-bit latch. The timing and control circuitry is omitted for the sake of simplicity. The circuit functions as follows.

The most significant 4-bit A/D converter converts the input analogue signal into a corresponding 4-bit digital code, which is stored in the most significant 4 bits of the output latch. This 4-bit digital code, however, represents the low-resolution sample of the input. Simultaneously, it is converted back into an equivalent analogue signal with a 4-bit D/A converter. The approximate value of the analogue signal so produced is then subtracted from the sampled value and the difference is converted into digital code using least significant 4-bit A/D converter. The least significant A/D converter is referenced to $1/16$ ($=1/2^4$) of the reference voltage used by the most significant A/D converter. The new 4-bit digital output is stored in least significant 4 bits of the output latch. The latch now contains the 8-bit digital equivalent of the analogue input. It may

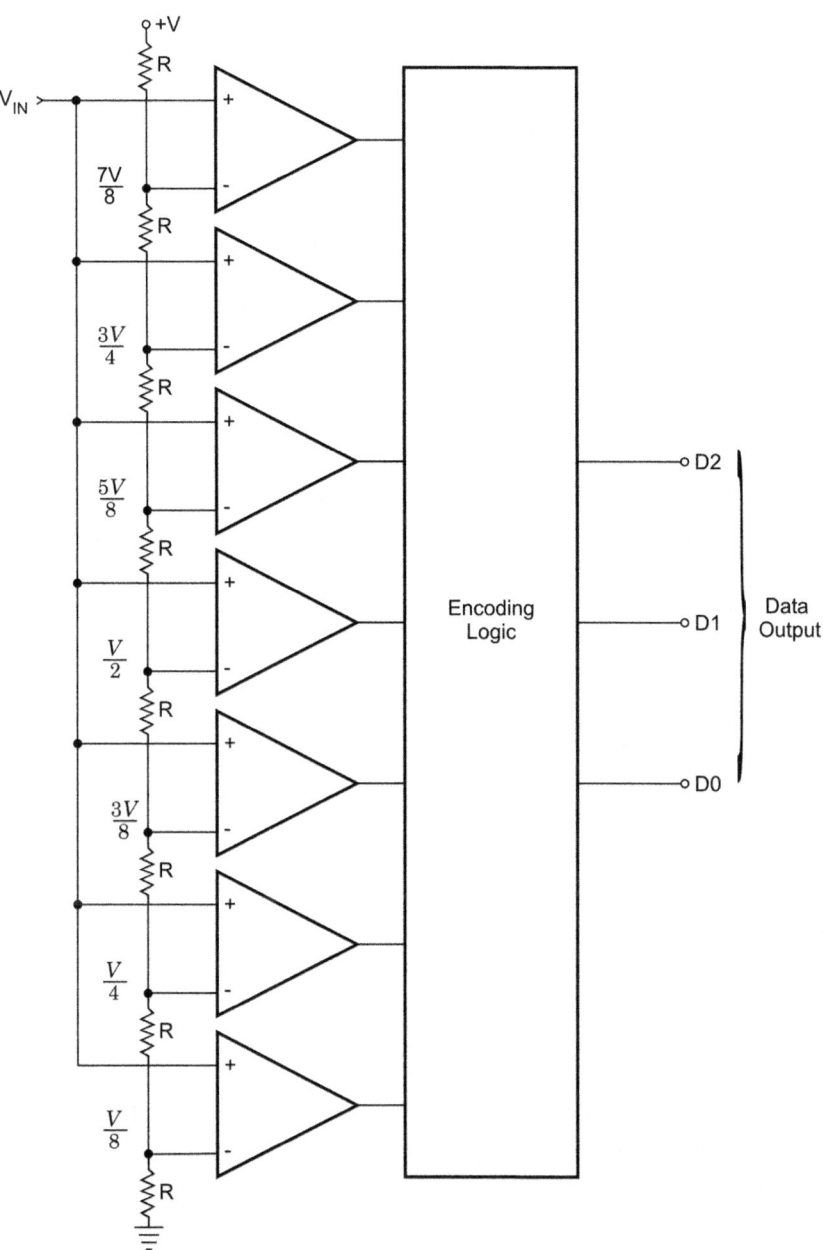

Figure 8.60 A 3-bit flash type A/D converter.

also be mentioned here that an 8-bit half-flash converter can be either used as a 4-bit full flash converter or 8-bit half-flash converter. Some half-flash converters use a single full flash converter and reuse it for both conversions. This is achieved by using additional sample and hold circuitry.

8.1.8.3.3 Counter-Type A/D Converter
It is possible to construct higher resolution A/D converters with a single comparator by using a variable reference voltage. One such A/D converter is the *counter-type A/D converter*

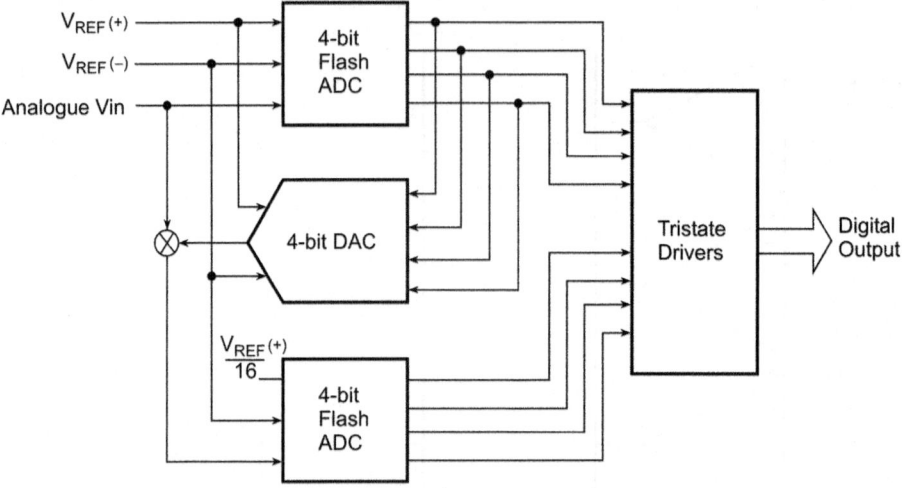

Figure 8.61 Architecture of an 8-bit half-flash converter.

Figure 8.62 Block-schematic of a generalized counter-type A/D converter.

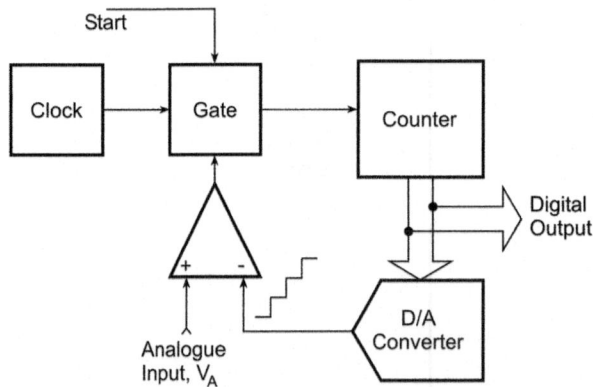

represented by the block-schematic in Figure 8.62. The circuit functions as follows. To begin with, the counter is reset to all 0s. When a convert signal appears on the start-line, the input gate is enabled and the clock pulses are allowed to the counter's clock input. The counter advances through its normal binary count sequence. A counter output feeds a D/A converter and the staircase waveform generated at the output of D/A converter forms one of the inputs of the comparator. The other input to the comparator is the analogue input signal. Whenever the analogue output of D/A converter exceeds the analogue input voltage the comparator changes state. The gate is disabled and the counter stops. The counter output at that instant of time is then the required digital output corresponding to analogue input signal.

The counter-type A/D converter provides a very good method for digitizing to a high-resolution. This method is much simpler than the simultaneous method for higher resolution A/D converters. The drawback with this converter is that the required conversion time is longer. Since the counter always begins from all 0s position and counts through its normal binary sequence, it may require as many as 2^n counts before conversion is complete. The average conversion time can be taken to be $2^n/2 = 2^{n-1}$ counts. This makes the counter-type of A/D converter unsuitable for digitizing rapidly changing analogue signals.

8.1.8.3.4 Tracking Type A/D Converter

In the counter-type A/D converter described previously, the counter is reset to zero at the start of each new conversion. The D/A converter output staircase waveform always begins at zero and increases in steps until it reaches a point where analogue output of D/A converter exceeds the analogue input to be digitized. As a result, the counter-type A/D converter of the type discussed before is slow. A *tracking type A/D converter*, also called a *delta encoded A/D converter*, is a modified form of counter-type converter that overcomes to some extent the shortcoming of the latter. In the modified arrangement, the counter, which is primarily an UP-counter, is replaced by an UP/DOWN counter. It counts in an upwards sequence whenever D/A converter output analogue voltage is less than the analogue input voltage to be digitized and it counts in the downwards sequence whenever D/A converter output analogue voltage is greater than analogue input voltage. In this type of converter, whenever a new conversion is to begin, the counter is not reset to zero; in fact, it begins counting either up or down from its last value depending upon the comparator output. The D/A converter output staircase waveform contains both positive going and negative going staircase signals that track the input analogue signal.

8.1.8.3.5 Successive Approximation Type A/D Converter

A successive approximation type A/D converter aims to approximate the analogue signal to be digitized by trying only one bit at a time. The process of A/D conversion by this technique can be illustrated with the help of an example. Let us take a 4-bit successive approximation type A/D converter. Initially, the counter is reset to all 0s. The conversion process begins with MSB being set by the start pulse. That is, the flip-flop representing the MSB is set. The counter output is converted into an equivalent analogue signal and then compared with the analogue signal to be digitized. A decision is then taken whether the MSB is to be left-in (i.e. flip-flop representing MSB remains set) or it is to be taken out (i.e. the flip-flop is reset) when the first clock pulse sets the second MSB. Once the second MSB is set, again a comparison is made and a decision taken whether the second MSB is to remain set or not when the subsequent clock pulse sets the third MSB. The process continues until we go down to LSB. In general, the number of clock cycles required for each conversion will be (n) for n-bit A/D converter of this type. Figure 8.63 shows the block-schematic representation of a successive approximation type of A/D converter. Since only one flip-flop (in the counter) is operated upon at one time, a ring counter which is nothing but a circulating register (serial shift register with outputs Q and $\bar{Q}$ of the last flip-flop connected to J and K inputs respectively of the first flip-flop) is

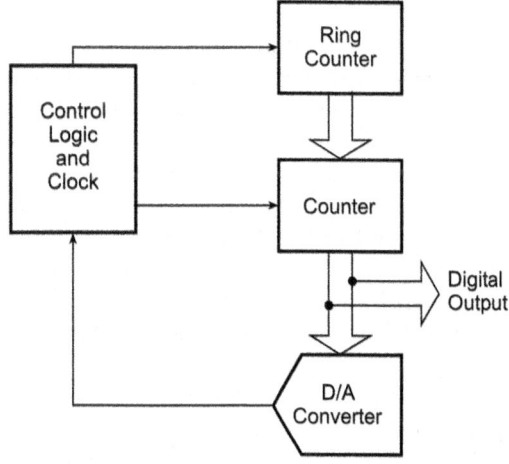

Figure 8.63 Architecture of a successive approximation A/D converter.

used to do the job. This type of A/D converter is much faster than the counter-type A/D converter. In an n-bit converter, the counter-type A/D converter on an average would require (2^{n-1}) clock cycles for each conversion whereas a successive approximation type converter requires only (n) clock cycles.

8.1.8.3.6 Single, Dual and Multi-Slope A/D Converters

In a *single slope A/D converter*, one of the inputs to the comparator is a ramp of fixed slope while the other input is the analogue input to be digitized. The counter and the ramp generator are initially reset to 0s. The counter starts counting with the first clock cycle input. The ramp is also synchronized to start with the first clock input. The counter stops when the ramp amplitude equals the analogue input. In this case, the counter count is directly proportional to the analogue signal. It is a low-cost, reasonably high accuracy converter but it suffers from the disadvantage of loss of accuracy due to changes in the characteristics of the ramp generator. This shortcoming is overcome in dual slope integrating type A/D converter. There are multi-slope converter architectures too aimed at further enhancing the performance of integrating A/D converters. For example, the *triple-slope architecture* is used to increase the conversion speed at the cost of added complexity. Bias currents, offset voltages and gain errors associated with operational amplifiers used as integrator and comparator do introduce some errors. These can be cancelled by using additional charge/discharge cycles and then using the results to correct the initial measurement. One such A/D converter is *quad-slope converter* that uses two charge/discharge cycles as compared to one charge/discharge cycle in the case of dual slope converter. Quad-slope A/D converters have a much higher accuracy than dual slope counterparts.

8.1.8.3.7 Sigma-Delta A/D Converter

The *sigma-delta A/D converter* employs a different concept from what has been discussed so far in the case of various types of A/D converters. While the A/D converters covered so far rely on sampling of analogue signal at the Nyquist frequency and encode the absolute value of the sample, in the case of sigma-delta converter, as explained in the following paragraphs, the analogue signal is over sampled by a large factor (i.e. sampling frequency is much larger than the Nyquist value) and also it is not the absolute value of the sample but the difference between the analogue values of two successive samples that is encoded by the converter.

In the case of A/D converters discussed prior to this and sampled at Nyquist rate f_s, the RMS value of the quantization noise is uniformly distributed over the Nyquist band of DC to $f_s/2$ as shown in Figure 8.64(a). The signal-to-noise ratio for a full-scale sine wave input in this case is given by $S/N = (6.02n + 1.76)$ dB, n being the number of bits. The only way to increase the signal-to-noise ratio is by increasing number of bits. On the other hand, a sigma-delta converter attempts to enhance signal-to-noise ratio by over sampling the analogue signal, which has the effect of spreading the noise spectrum over a much larger bandwidth, and then filtering out the desired band. If the analogue signal were sampled at a rate of Kf_s, the quantization noise gets spread over DC to $Kf_s/2$ as shown in Figure 8.64(b). K is called the oversampling ratio. An enhanced S/N ratio means a higher resolution, which is achieved by other types of A/D converters by way of increasing number of bits.

It may be mentioned here that if we simply use over sampling to improve resolution, it would be required to oversample by a factor of 2^{2N} to achieve an N-bit increase in resolution. Sigma-delta converter does not require to be oversampled by such a large factor because it not only limits the signal pass band but also shapes the quantization noise in such a way that most of it falls outside this pass band as shown in Figure 8.64(c).

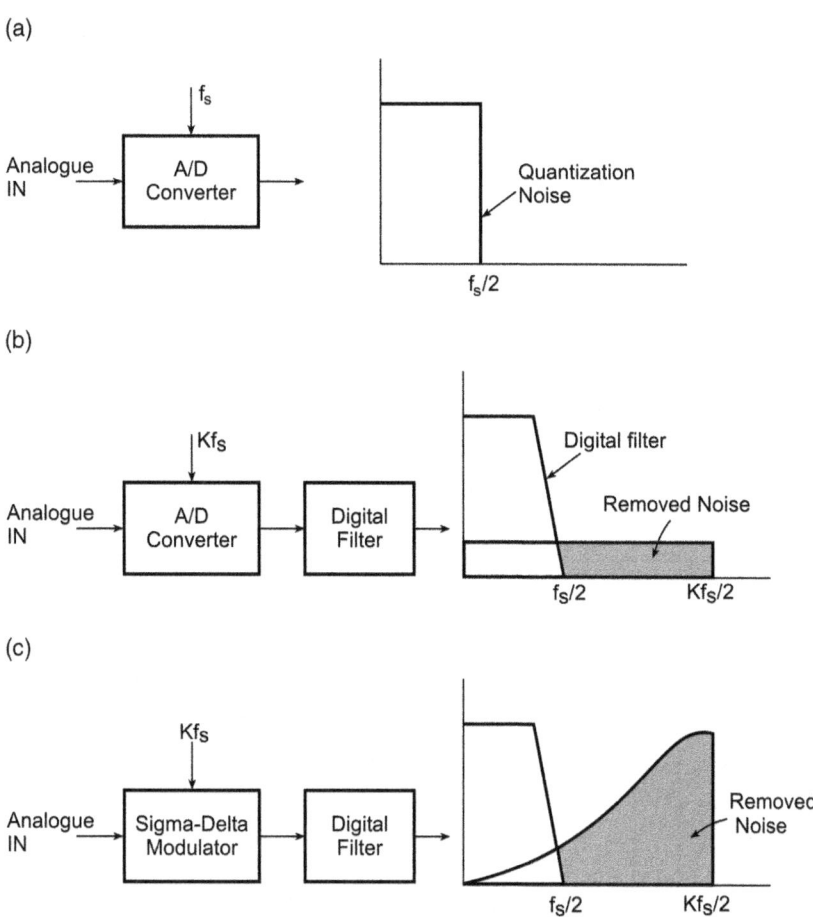

Figure 8.64 Sigma-delta A/D converter: (a) conventional A/D converter sampled at Nyquist rate, (b) effect of oversampling and (c) advantage with sigma-delta converter.

8.1.9 Microcontrollers

Microcontrollers and Field Programmable Gate Arrays (FPGA) are the key components of embedded systems used to perform dedicated processing functions. Microcontrollers are embedded inside surprisingly large number of product categories including automobiles, entertainment and consumer products, test and measurement equipment, military systems to name some prominent ones. Any device or system that measures, stores, controls, calculates or displays information is sure to have an embedded microcontroller as a part of the device or system. A brief description of these two important categories of electronic devices is presented in this and the following sections. A detailed description including different architectures, interfacing with peripheral devices and so on, is beyond the scope of the text.

8.1.9.1 Comparison with a Microprocessor

A microcontroller may be considered a specialized computer-on-a-chip or a single-chip computer. The word 'micro' suggests that the device is small and the word 'controller' suggests that the device may be used to control one or more functions of objects, processes or events. It is also called an embedded controller as microcontrollers are often embedded in the device or system that they control.

A microcontroller contains a simplified processor, some memory (RAM and ROM), I/O ports and peripheral devices such as counters/timers, analogue-to-digital converters and so on, all integrated on a single chip. It is this feature of the processor and peripheral components available on a single chip that distinguishes it from a microprocessor based system. A microprocessor is nothing but a processing unit with some general-purpose registers. A microprocessor based system also has RAM, ROM, I/O ports and other peripheral devices to make it a complete functional unit but all these components are external to the microprocessor chip. This difference is illustrated in the block schematics of Figure 8.65(a) for microprocessor and (b) for microcontroller. While a microprocessor based system is a general-purpose system that may be programmed to do any of the large number of functions it is capable of doing,

(a)

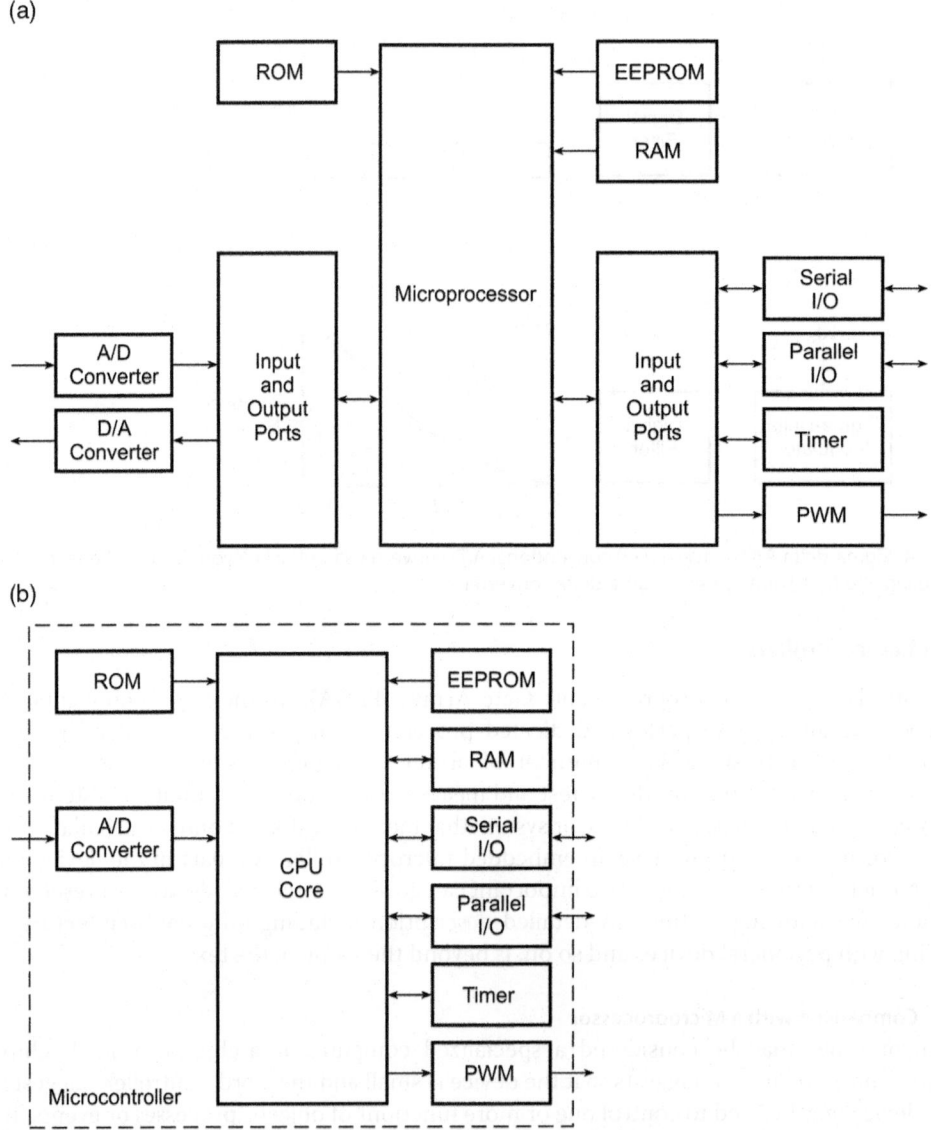

(b)

Figure 8.65 Comparison of a microcontroller and microprocessor: (a) microprocessor and (b) microcontroller.

microcontrollers are dedicated to one task and run one specific programme. This program is stored in the ROM and generally does not change.

8.1.9.2 Microcontroller Hardware

Figure 8.65(b) shows the block-schematic arrangement of various components of a microcontroller. As outlined earlier, a microcontroller is an integrated chip with an on-chip CPU, memory, I/O ports and some peripheral devices to make a complete functional unit. A typical controller as depicted in Figure 8.65(b) has the following components:

1) Central processing unit (CPU)
2) Random access memory (RAM)
3) Read only memory (ROM)
4) Special function registers
5) Peripheral components including serial and/or parallel ports, timers and counters, analogue-to-digital (A/D) converters and digital-to-analogue (D/A) converters.

The *CPU* processes the programme. It executes the instructions stored in the programme memory pointed to by the programme counter in synchronization with the clock signal. The processor complexity could vary from simple 8-bit processors to sophisticated 32-bit or even 64-bit processors. Some common microcontrollers using 8-bit processors include 68HC11 (Freescale Semiconductor – earlier part of Motorola), the 80C51 family of microcontrollers (Intel and Dallas Semiconductor), Zilog-eZ8 and Zilog-eZ80 (Zilog) and XC800 (Infineon). Examples of microcontrollers using 16-bit processors include the 8096 family (Intel), 68HC12 and 68HC16 (Freescale Semiconductor), F2MC family (Fujitsu) and XC166 family (Infineon). Examples of microcontrollers using 32-bit processors include 683XX, MPC 860 (PowerQUICC), MPC 8240/8250 (PowerQUICC-II), MPC 8540/8555/8560 (PowerQUICC-III), all from the Freescale Semiconductor, TRICORE family (Infineon) and FR/FR-V family (Fujitsu).

Random Access Memory (RAM) is used to hold intermediate results and other temporary data during the execution of the programme. Typically, microcontrollers have few hundreds of bytes of RAM. As an example, microcontroller type numbers 8XC51/80C31, 8XC52/80C32 and 68HC12, respectively, have 128 bytes, 256 bytes and 1024 bytes of RAM.

Read Only Memory (ROM) holds the programme instructions and the constant data. Microcontrollers use one or more of the following memory types for the purpose. These include ROM (Mask programmed ROM), PROM (one-time programmable ROM, which is not field programmable), EPROM (field programmable and usually UV erasable), EEPROM (filed programmable, electrically erasable, byte erasable) and flash (similar to EEPROM technology). Microcontroller type numbers 8XC51, 8XC51FA, 8XC52 have 4K, 8K and 16K of ROM. As another example, a 68HC12 16-bit microcontroller has 32K of flash EEPROM, 768 bytes of EEPROM and 2K of erase protected boot block.

Special function registers control various functions of a microcontroller. There are two categories of these registers. The first type includes those registers that are wired into the CPU and do not necessarily form part of addressable memory. These registers are used to control program flow and arithmetic functions. Examples include the status register, program counter and stack pointer. These registers are, however, taken care of by compilers of high level languages and therefore programmers of high level languages such as C, Pascal and so on do not need to worry about them. The other category of registers is the one that is required by peripheral components. Contents of these registers could, for instance, set a timer or enable serial communication. As an example, special function registers available on 80C51 family of microcontrollers (80C51, 87C51, 80C31) include the program counter, stack pointer, RAM address register, program address register and PC incrementer.

Peripheral components such as analogue-to-digital converters, I/O ports, timers and counters, are available on majority of the microcontrollers. These components perform functions as suggested by their respective names. In addition to these, microcontrollers intended for some specific or relatively more complex functions come with many more on-chip peripherals. Some of the common ones include pulse width modulator, serial communication interface (SCI), serial peripheral interface (SPI), inter-integrated circuit (I^2C) two-wire communication interface, RS 232 (UART) port, IrDA (Infrared port), USB port, controller area network (CAN) and local interconnect network (LIN).

8.1.10 Field Programmable Gate Arrays

A *Field Programmable Gate Array* (FPGA) uses an array of logic blocks, which can be configured by the user. The word field *programmable* here signifies that the device is programmable outside the factory where it is manufactured. Internal architecture of an FPGA device has three main parts namely array of logic blocks, programmable interconnects and I/O blocks. Figure 8.66

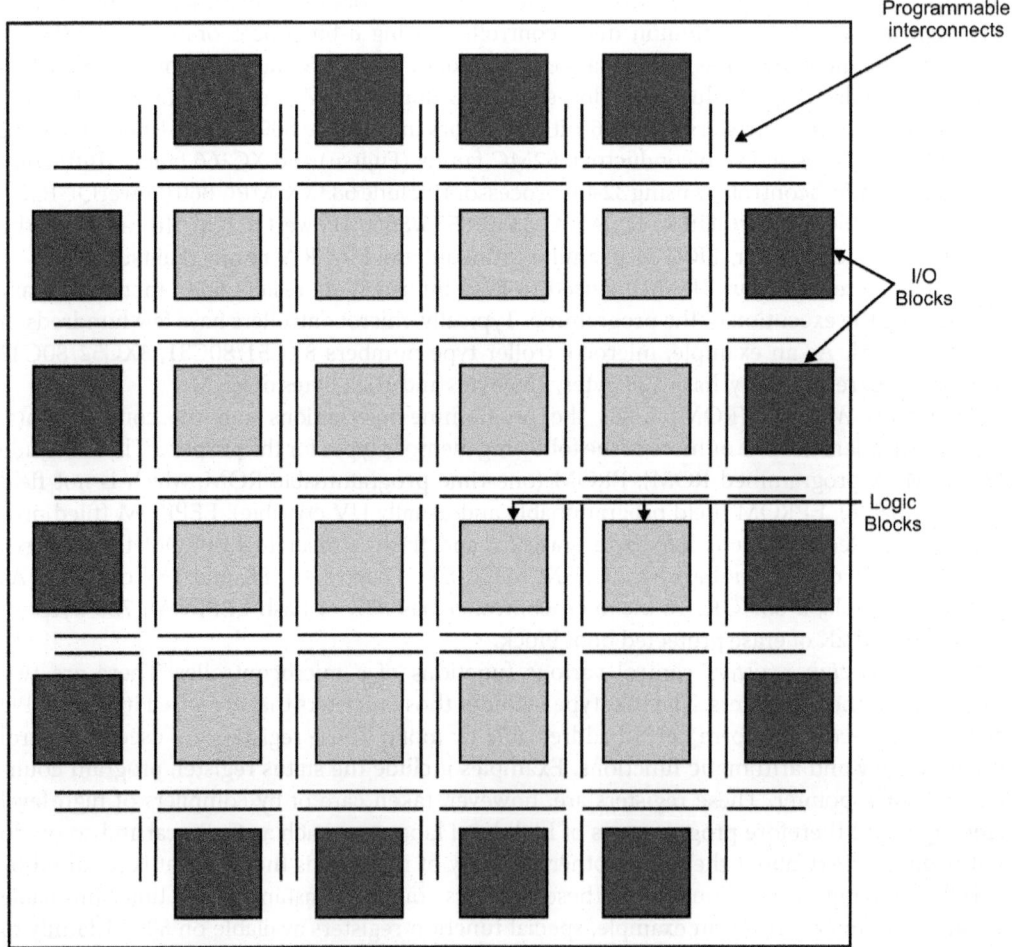

Figure 8.66 Simplified architecture of an FPGA (a) and (b).

shows the architecture of a typical FPGA. Each of the I/O blocks provides individually selectable input, output or bidirectional access to one of the general-purpose I/O pins on the FPGA package. Logic blocks in an FPGA are no more complex than a couple of logic gates or a look-up table feeding a flip-flop. Programmable interconnects connect logic blocks to logic blocks and also I/O blocks to logic blocks.

FPGAs offer a much higher logic density and much larger performance features compared to CPLDs. Some of the contemporary FPGA devices offer logic complexity equivalent to that of eight million system gates. Also, these devices offer features like built-in hardwired processors, large memory, clock management systems and support for many of the contemporary device-to-device signalling technologies. FPGAs find extensive use in a variety of applications, which include data processing and storage, digital signal processing, instrumentation and telecommunications.

8.1.10.1 Internal Architecture

An FPGA, as outlined previously, consists of an array of uncommitted configurable logic blocks, programmable interconnects and I/O blocks. The basic architecture of an FPGA is shown in Figure 8.66. The basic difference between a Complex Programmable Logic Device (CPLD) and an FPGA lies in their internal architecture. CPLD architecture is dominated by relatively smaller number of programmable sum-of-products logic arrays feeding small number of clocked flip flops, which makes the architecture less flexible but with more predictable timing characteristics. On the other hand, FPGA architecture is dominated by programmable interconnects and the configurable logic blocks are relatively simpler. This feature makes these devices far more flexible in terms of range of designs that can be implemented with these devices. Contemporary FPGAs have on-chip presence of higher level embedded functions, and embedded memories. Some of them even come with an on-chip microprocessor and related peripherals to constitute what is called complete 'System on a programmable chip'. Virtex-II Pro and Virtex-4 FPGA devices from Xilinx are examples. These devices have one or more PowerPC processors embedded within FPGA's logic fabric.

Figure 8.67 shows a typical logic block of an FPGA. It consists of a four-input look-up table (LUT) whose output feeds a clocked flip-flop. The output can either be a registered output or the unregistered LUT output. Selection of output takes place in the multiplexer. LUT is nothing but a small one-bit wide memory array with its address lines representing the inputs to the logic block and one-bit output acting as the LUT output. A LUT with n inputs can realize any logic function of n inputs by programming the truth table of the desired logic function directly into the memory. Logic blocks can have more than one LUTs and flip flops also to give them capability to realize more complex logic functions.

FPGAs today offer a complete system solution on a single chip, though very complex systems might be implemented with more than one FPGA devices. Some of the major application areas of FPGA devices include Digital signal processing, Data storage and processing, software-defined radio, ASIC prototyping, speech recognition, computer vision, cryptography, medical imaging, defence systems, bioinformatics, computer hardware emulation and reconfigurable computing.

Figure 8.67 Logic block diagram of FPGA.

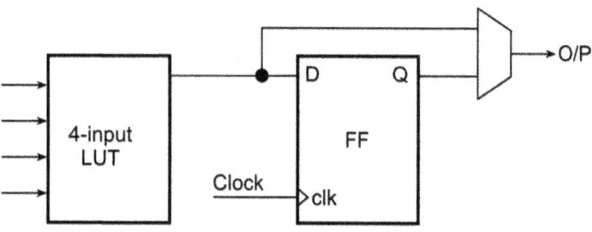

8.1.10.2 Programmable Interconnect Technologies

Programmable interconnect technologies that have evolved over the years for programming programmable logic devices (PLDs) include fuse, EPROM or EEPROM floating gate transistors, static RAM and anti-fuse.

Fuse is an electrical device that has a low initial resistance and is designed to permanently break an electrically conducting path when current through it exceeds a specified limit. It uses bipolar technology, is non-volatile and one-time programmable.

Floating gate transistor switch interconnect technology is based on the principle of placing the floating gate transistor between two wires in such a way as to facilitate a wire-AND function. This concept is used in EPROM and EEPROM devices and that is why floating gate transistor is sometimes referred to as EPROM or EEPROM transistor. This technology is commonly used in SPLDs and CPLDs. Floating gate transistor-based switch matrix, however, requires a large number interconnects and therefore transistors. For example, a CPLD with 128 macro-cells with four inputs and one output each would require as many as 65 536 interconnects for 100% routability. Large number of interconnects also adds to the propagation delay. Use of multiplexers can reduce this number significantly and can also address the problem of increased propagation delay. MUX-based interconnect matrix is being used in CPLDs. CPLD type XPLA3 from Xilinx is an example.

Static RAM (SRAM) is basically a semiconductor memory and the word 'static' implies that it is a non-volatile memory. That is, memory retains its contents as long as power is on. SRAM with m address lines and n data lines is referred to as $2^m \times n$ memory and is capable of storing 2^m n-bit words. Figure 8.68 shows the basic SRAM cell comprising of six MOSFET switches with four of them connected as cross-coupled inverters. Basic SRAM cell can store one bit of information. The reading operation is carried out by pre-charging both the bit lines (BL and $\overline{BL}$) to logic '1' and then asserting the WL-line. The writing operation is done by giving desired logic status to BL-line and its complement to the $\overline{BL}$ line and then asserting the WL-line.

Anti-fuse is an electrical device with a high initial resistance and designed to permanently create an electrically conducting path typically when voltage across it exceeds a certain level.

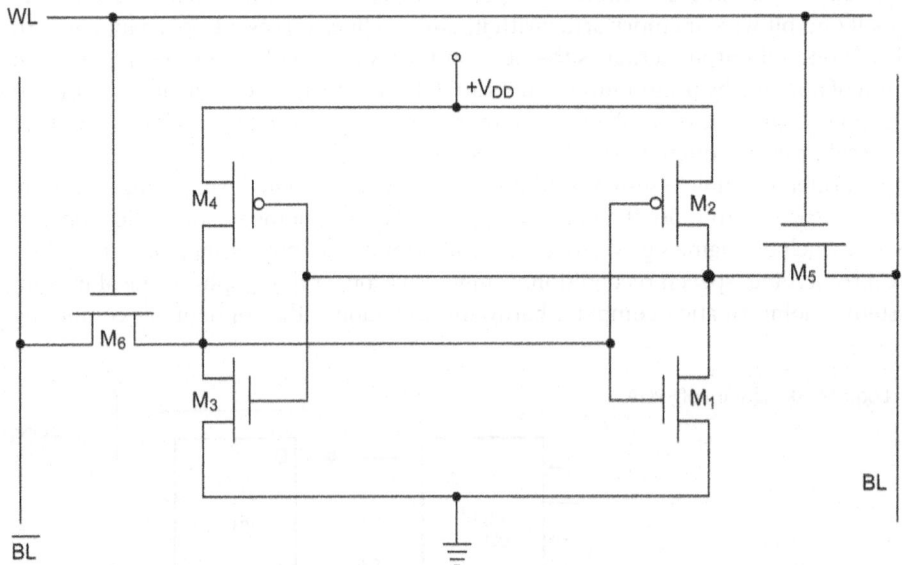

Figure 8.68 Basic SRAM cell.

Anti-fuse uses CMOS technology, which is one of the main reasons for its wide use in PLDs, FPGAs in particular. A typical anti-fuse consists of an insulating layer sandwiched between two conducting layers. In the unprogrammed state, insulating layer isolates top and bottom conducting layers. When programmed, the insulating layer gets transformed to a low resistance link. Typically, it uses metal for conductors and amorphous silicon as insulator. Application of high voltage across amorphous silicon permanently transforms it into polycrystalline silicon-metal alloy having a low resistance. There are other anti-fuse structures too. Anti-fuses are widely used as programmable interconnects in PLDs. Anti-fuse PLDs are one-time programmable in contrast to SRAM controlled interconnect based PLDs, which are reprogrammable. It may be mentioned here that reprogrammable feature helps the designers fix logic bugs or add new functions. Anti-fuse PLDs have advantages of non-volatility and usually higher speeds. Anti-fuses may also be used in PROMs. In that case, each bit contains both a fuse and an anti-fuse. The device is programmed by triggering one of the two.

8.2 Laser Electronics and Related Technologies

Electronics plays an important role in the demonstrated capabilities, operational efficiency and achievable system compactness of military lasers and optoelectronics systems. Majority of military laser systems employ one of the three common types of lasers namely solid-state, semiconductor and gas lasers. *Laser Range Finders* and *Target Designators*, for instance, are mainly configured around solid-state lasers and in some cases semiconductor diode lasers and gas lasers. *Precision-guided munitions* use solid-state laser-based laser target designators. *Electro-optic countermeasure* systems are based on high-energy solid-state laser sources. *Inertial navigation sensors* use a He-Ne ring laser cavity, *laser proximity sensors* and *laser aiming devices* employ semiconductor diode lasers and *directed-energy laser weapons* are projected to be using high power bulk solid-state or fibre lasers. On the optoelectronics front, laser sensors, laser seekers, LADAR sensors and LIDAR sensors constituting the bulk of battlefield optoelectronics mainly use silicon, indium gallium arsenide, indium antimonide, mercury cadmium telluride and image intensifier photo sensors.

To summarize, if we were to look at the spectrum of electronics that goes along with military lasers and optoelectronics systems, we would need to consider the electronics of solid-state, semiconductor diode and gas lasers. When we set out to discuss the role of electronics in lasers and laser-based systems, we usually talk about only those subsystems that are essential from the operational viewpoint of the laser in question. There is much more than the power supply circuit that is used to operate the laser. In the following paragraphs, we shall take a closer look at the areas of electronics encountered in the case of three most commonly used laser sources namely solid-state, gas and semiconductor lasers. Prominent military laser systems configured around these laser sources are also briefly discussed.

8.2.1 Solid-State Laser Electronics

When we discuss electronics package of a flash lamp pumped pulsed solid-state laser (Nd:YAG, Nd:Glass etc.), it is the power supply unit needed to charge the energy storage capacitor to store the requisite quantum of energy that is considered representative of the electronics package. It is simply because of the reason that It is not only the most complex of the electronics circuit modules used in the case of flash lamp pumped solid-state lasers; its electrical conversion efficiency plays a key role in deciding the wall plug efficiency and the size/weight of the laser. Capacitor charging power supplies intended for flash lamp pumped solid-state laser applications are

Figure 8.69 Capacitor charging power supply.

commercially available both as bench top models and modular units for OEM applications. Figure 8.69 shows a photograph of one such capacitor charging power supply suitable for flash lamp pumped solid-state lasers.

These capacitor charging power supplies are available for a range of input voltage (AC or DC), DC output voltage and charging rate specifications. DC output voltage from a few hundreds of volts to several kilovolts and charging rate in the range of tens to thousands of J/s are common. In addition, these power supplies offer many control and protection features relevant to flash lamp pumped laser power supplies. Some of these features include end of charge status indication, peak output voltage hold, output voltage monitor, over voltage and over temperature protection and so on.

We also talk about, though with less enthusiasm, the other circuit modules such as the simmer power supply, which is invariably used in high repetition rate flash lamp pumped solid-state lasers, or the Q-switch driver used in the Q-switched lasers or even the pulse forming network (PFN), which ensures a critically damped current pulse through the flash lamp when the energy storage capacitor is made to discharge through it. But the electronics does not end here. When the same laser is transformed into a laser rangefinder, there is a lot more, such as a low noise front-end with the requisite bandwidth to process the received laser pulse, high-speed counting circuitry, range gating circuitry, control logic and so on.

Nd:YAG, frequency shifted Nd:YAG and erbium:glass, respectively, operating at 1064, 1540 and 1550 nm are the most commonly used laser sources for laser range finding, target designation and laser-guided munitions delivery applications. While Nd:YAG is almost invariably used for target designation and laser-guided munitions delivery, lasers operating at 1540/1550 nm are preferred for range finding applications considering the eye safety of operating personnel. It may be mentioned here that as per ANSI standards for eye safety, 1540 nm is relatively much safer than 1064 nm, which is hazardous to the human eye. Semiconductor diode lasers are also used for range finding applications both for commercial and military domains, their use is mainly restricted to relatively shorter ranges with maximum measurable range generally not exceeding 5 km. On the other hand, portable and handheld Nd-YAG laser-based range finders are available for maximum range measuring capability in excess of 25 km. Figure 8.70 shows a photograph of one such handheld laser range finder.

In addition to conventional target range finding, there are many related applications where the basic range finding concept is put to use in different military laser systems. Some of the prominent ones include proximity sensors, obstacle avoidance sensors and bathymetry for sea bed mapping. These devices also make use of either solid-state or semiconductor diode lasers. To summarize the whole range of electronics used in military laser systems configured around solid-state laser sources, one has to work with almost all categories of power supplies including different topologies of switched mode power supplies and protection circuits on one hand and

Figure 8.70 Handheld eye safe laser range finder.

a wide range of electronics building blocks including low noise optoelectronic front-end circuits, large bandwidth amplifiers, signal conditioning circuits, counting and timing circuits on the other.

8.2.2 Gas Laser Electronics

He-Ne and carbon dioxide lasers are the most commonly used gas laser sources. While carbon dioxide lasers are commonly used in some types of laser range finders, He-Ne lasers mainly find application in inertial navigation sensor. What most of us know about the He-Ne laser electronics is a high voltage power supply that initiates and subsequently sustains the plasma. The plasma current stabilization may not be important if it is to be used for the purpose of alignment. But when it comes to using the same laser in an inertial grade rotation rate sensor such as a *Ring Laser Gyroscope* (Figure 8.71), the current would need to be stabilized to a level better than 100 ppm. In addition, you would also need to stabilize its frequency to better than + 1 MHz on its Doppler broadened gain curve, which is about 1400 MHz wide for 632.8 nm output wavelength. To further add to the design complexity of plasma current initiation and control circuit, the difference between the plasma currents in the two counter-propagating laser beams needs to be stabilized to an order of magnitude better current stability than the absolute current stability of each arm.

Frequency stabilization of gas lasers is a complete field in itself. There are scientists who have worked in this area for decades to discover new methods to stabilize the laser frequency using active means or improve upon those already existing. Different frequency stabilization

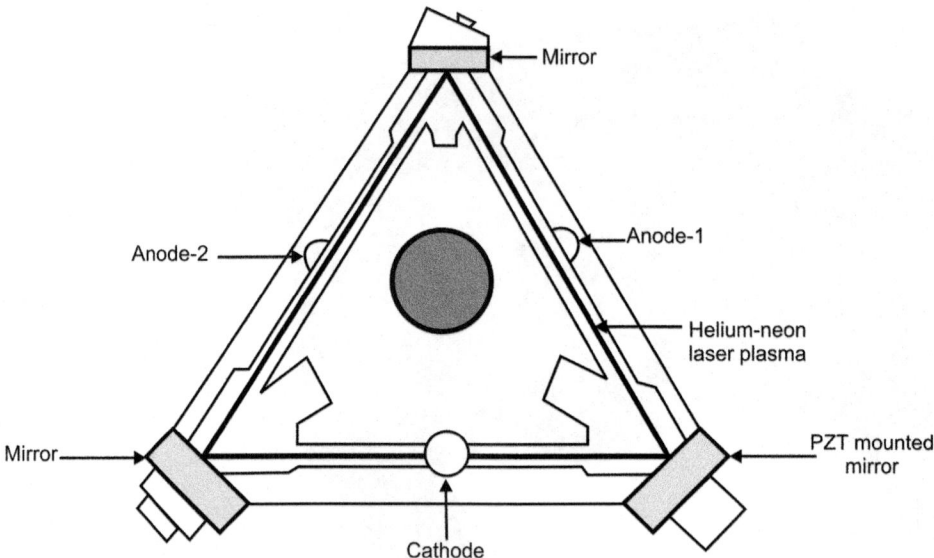

Figure 8.71 Typical He-Ne ring laser gyro cavity.

techniques include Dither stabilization, Optogalvanic stabilization and Stark cell stabilization. It is possible to actively stabilize the frequency of a carbon dioxide lasers to better than few kHz using Stark cell stabilization by stabilizing the frequency to the centre of Lamb dip on the gain versus frequency curve.

In the category of carbon dioxide lasers, we have both DC excited as well as RF excited lasers. While the design of power supply for a DC excited carbon dioxide laser is similar in concept to the one used in the case of He-Ne lasers except of current and voltage levels, in the case of RF excited lasers a typical excitation source comprises of an RF source operating in the frequency range of 50–150 MHz and an impedance matching network. The RF source may further be split up into an RF oscillator and a cascade arrangement of RF amplifiers depending upon the output power delivering capability. RF power is fed to the laser cavity through impedance matching network.

8.2.3 Semiconductor Diode Laser Electronics

The design of drive and control circuits needed to power semiconductor diode lasers should consider certain handling and protection issues if they are to have the prescribed life and reliability performance. It is more so for semiconductor diode lasers used in military applications. Diode lasers are particularly sensitive to electrostatic discharge, short duration electric transients such as current spikes, injection current exceeding the prescribed limit and reverse voltage exceeding the breakdown limit.

In order to protect the diode lasers from these failure modes, the driver circuit should be carefully designed and should have all the features recommended by the diode laser manufacturer. The driver should be a constant current source with inbuilt features like soft start, protection against transients, interlock control for the connection cable to the laser and safe adjustable limit for injection current. In case the laser is to be operated in the pulsed mode, the injection current should be pulsed between two values above the lasing threshold rather than between cut-off and lasing mode. A laser diode when used in a laser printer or a laser pointer

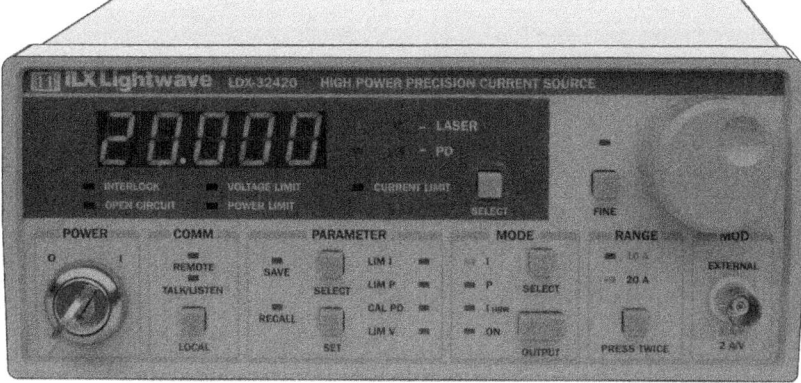

Figure 8.72 Commercial laser diode driver.

or even a compact disc player may need a conventional constant current source without too stringent a requirement on the current stabilization to do the job. Drive current and diode temperature stabilization to a high degree become extremely important when the intended application demands a stable output wavelength. One such application area is in laser-based Raman sensor used for detection and identification of chemical warfare agents and explosive materials. The concepts of drive current and diode temperature stabilization have been put to use very effectively in tuning the diode laser output wavelength, which is a requirement in laser systems designed for detection and identification of chemical warfare agents. It may be mentioned here that it is the ability to stabilize the parameter that allows you to vary it.

Leading manufacturers of semiconductor diode lasers offer a wide range of current sources for low, medium and high power laser diodes to suit different requirements. Both general-purpose bench top models and modular units for OEM applications are commercially available from a fairly large number of manufacturers. Figure 8.72 shows a photograph of a bench top precision laser diode driver. Most of the commercial laser diode drivers offer operation in both constant current and constant power modes and have inbuilt protection features including adjustable current and voltage limits, intermittent contact protection and so on.

They also offer temperature controllers to stabilize the output wavelength. Temperature controllers for diode laser temperature control are also available in a wide range of performance parameters to suit different requirement specifications. Both bench top units and OEM modules are commercially available. Figure 8.73 shows a photograph of a typical bench top thermoelectric laser diode temperature controller. Most of these temperature controllers are capable of operating in constant temperature, constant power or constant current modes with temperature stability of better than 0.003 °C. Integrated laser diode controllers offering both current drive as well as thermoelectric temperature control in a single instrument are also commercially available from a number of manufacturers.

8.2.4 Laser Sensors

In a large number of optoelectronic systems, in particular battlefield optoelectronic systems, the primary function is detection of laser radiation with or without its important parameters depending upon the intended application. Some such systems include laser-warning sensors, laser position sensors and laser seekers. A laser-warning sensor system is an essential constituent of both passive and active Electro-Optic Countermeasures (EOCM) systems. Different battlefield

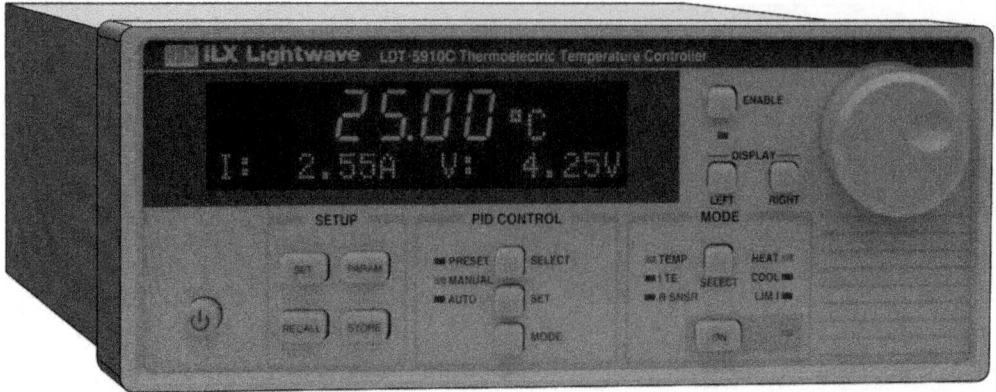

Figure 8.73 Temperature controller for semiconductor diode lasers.

applications demand laser-warning sensor systems with different functional features, thus they have different design complexities. In the simplest form, a laser-warning sensor may be used to detect the existence of a laser threat without giving any information on the direction of arrival of the threat. In another case, it may also have the direction sensing feature with the resolution of direction sensing also varying with the intended application from a coarse sensing capability in the range of ±5° to ±45° to high-resolution sensing with direction-of-arrival sensing capability in the range of ±0.1° to ±1.0°. In addition, the operational wavelength band may also vary from one application to another. More complex laser-warning systems indicate the type of laser threat, the wavelength and the direction of arrival of the threat. A typical high end laser-warning sensor is capable of characterizing threats from laser range finders, laser target designators and laser beam riders and these are available for mounting on helicopters, main battle tanks and light armoured vehicles.

While one of the most common military applications of a laser-warning sensor is as a subsystem in an integrated EOCS for armoured fighting platforms, there are many other application scenarios where laser sensors are deployed. Emerging trends indicate the use of an array of laser sensors interfaced with a high-energy laser to protect critical and strategic assets such as aircraft shelters, ammunition depots, strategic buildings, naval vessels and so on from laser-guided munitions attack. An array of laser sensors in this case detect laser threat and decode its parameters. These parameters are used to control the operation of a high-energy laser source. The laser source in turn illuminates a dummy target to misguide the incoming laser-guided munitions toward the dummy target.

Laser-guided munitions use a laser sensor called a laser seeker, which is also a kind of position sensor. A laser seeker is the heart of the guidance system of a laser-guided weapon such as laser-guided bomb or missile. Figure 8.74 shows the laser-guided bomb integrated with a laser seeker head. A typical laser seeker employs a quadrant sensor for determination of direction of arrival of laser radiation scattered off the intended target when illuminated by a laser target designator. The laser target designator and the laser seeker work in harmony. The two operate on the same Pulse Repetition Frequency (PRF) code, which allows the bomb to home on to the source of laser scatter.

A LADAR sensor (Figure 8.75) is one of the most contemporary forms of laser seekers usually used for ultra-high precision hitting of strategic targets. These are mainly used in conjunction with other guidance systems on strategic payloads for intended target discrimination from advanced decoys and aim point selection. It is also well suited for combat identification, navigation of autonomous vehicles and topography. LADAR is also suitable in finding targets hidden

Figure 8.74 Typical laser guided bomb integrated with laser seeker head. (*Source:* Courtesy of the US Airforce.)

Figure 8.75 LADAR sensor.

by camouflage nets and foliage. LADAR seeker can detect and identify specific features of the target with very high definition up to a resolution of few cm from a distance of few km. LADAR sensors are usually employed on loitering systems that look at the target from different angles, verify a target's identity and select the best attack position for desired results. The sensor in essence generates a 3D image of the intended target. The 3D image is compared with various 3D templates stored in weapon's memory before the mission and it facilitates identification of target and selection of aim point.

8.2.5   Test and Evaluation of Lasers

Measurement of laser power, energy, pulse width and so on is yet another area that is electronic or more precisely optoelectronic in nature. We have today all kinds of lasers producing CW or pulsed or Q-switched pulsed laser outputs. While we would be mainly interested in the output power in CW lasers such as gas lasers, it is the energy and the pulse width that are of interest in the case of pulsed lasers. In the case of Q-switched pulsed lasers such as solid-state lasers, we would like to measure energy per pulse, average power and also the peak power. Equipment capable of measuring one or more of these parameters is commercially available. Figure 8.76 shows one such commercial meter capable of measuring laser power and pulse energy over a

Figure 8.76 Laser power/energy meter.

wide range when used in conjunction with suitable sensor heads. While commercially available test and measuring equipment can be used for carrying measurement of important laser parameters and therefore be useful to the engineers responsible for maintaining military laser equipment, there are cases where dedicated test systems are needed to perform health checks. Sometimes, these quick health checks, also called serviceability checks, have to be performed on the systems integrated on the platform. Test systems for serviceability checks of precision-guided munitions, performance evaluation of laser range finders and target designators and checking efficacy of EOCM systems are some examples.

The point that we are trying to drive home is that, while talking or writing about laser electronics, it would be far from being justified to confine the discussion to just power supplies for different types of lasers because then we would probably be covering no more than 30% of the electronics that concerns contemporary lasers and laser-based systems. In the present section, we have presented different areas of usage of laser and optoelectronics systems in defence and we have briefly touched upon the role of electronics in each of those areas. Details are presented in the following sections of the chapter.

8.3 Solid-State Laser Electronics

Solid-state lasers are at the core of a vast majority of military laser systems intended for tactical applications. Pulsed solid-state lasers operating in the Q-switched mode and emitting at 1064 nm and 1540 nm are the most commonly used laser types. In continuation of Part I, focus in this article is on solid-state laser electronics confining the discussion mainly to requirements, design criticalities and circuit options.

Laser Range Finders and *Laser Target Designators* are the most abundantly used laser systems for tactical battlefield scenarios. Laser range finders are used both as standalone devices as well as an integral part of an overall Electro-Optic Fire-Control (EOFC) system of armoured fighting platforms. In the standalone mode, the device is used for determining the range of adversary's targets by troops and platforms on observation and reconnaissance missions. In an EOFC system, the range data on the target as produced by the range finder is used by a computer to control the gun position to facilitate precise target hit. The laser source used in all these types of range finders is either a Q-switched neodymium-doped yttrium aluminium garnet (Nd:YAG) emitting at 1064 nm or an Nd:YAG laser whose output wavelength has been shifted to an eye-safe wavelength of 1540 nm using an optical parametric oscillator (OPO). Erbium-doped glass (Er:glass) lasers are also used to generate eye-safe wavelength but their use is restricted to range finders operating at relatively low repetition rates.

Laser target designators are almost invariably configured around high repetition rate Q-switched Nd:YAG laser sources. These are used in laser-guided munitions delivery applications. In a typical laser-guided munitions delivery operation, the laser target designator, which could be land-based or airborne, irradiates the intended target with laser pulses of a pre-decided pulse repetition frequency (PRF) code, which is also known to the laser seeker head of the guided weapon. The laser seeker head in the laser-guided weapon senses the laser radiation scattered from the target, deciphers the PRF code and then commands the weapon to home on to source of scatter.

When it comes to electronics that goes along with laser range finders and target designators configured around Q-switched solid-state lasers, we need to discuss the electronics package required for operating the laser source at desired specifications and also the electronics required to make range measurement. Further pulsed solid-state lasers are either flash lamp pumped or pumped by laser diodes. Earlier lasers of 1980s and 1990s were almost exclusively flash-pumped. They have been largely replaced by laser diode-pumped versions over the last 10–15 years due to much higher optical-to-optical and wall plug efficiency figures. In the following paragraphs are described the electronics packages for flash lamp pumped and laser diode-pumped Q-switched solid-state lasers in general and Nd:YAG laser in particular. A brief outline on the functions of different modules constituting the overall electronics package with particular reference to the importance of each module is presented first, which is followed by detailed description of each of the important modules along with preferred schematic options and design guidelines.

8.3.1 Electronics for Flash-Pumped Solid-State Lasers

In the case of flash-pumped Nd:YAG and Nd:glass lasers, the gain medium is optically pumped by a flash lamp such as a xenon or krypton flash lamp whose output optical spectrum matches with the absorption spectrum of the gain medium. There are two possible operational modes in which requisite quantum of energy can be delivered to the flash lamp. In one of the modes employed in earlier lasers and called the non-simmer mode of operation, the electrical energy stored in a capacitor is discharged through the flash lamp by application of a high voltage trigger pulse to ionize the gas fill and create a low resistance path. In the other mode used in present-day flash-pumped lasers and called the simmer mode of operation, a low resistance path is maintained through the flash lamp. The lamp is kept isolated from the energy storage capacitor by an electronic switch, which is triggered to the on-state by a TTL/CMOS pulse forcing the stored energy to discharge through the flash lamp.

Figure 8.77 shows the detailed block-schematic arrangement of an electronics package of a flash lamp pumped Q-switched solid-state laser. The heart of the system is the *main power supply*, which is invariably a switched mode one used to charge an energy storage capacitor to a voltage so as to store the required quantum of energy per pulse to be delivered to the flash lamp. The main power supply is also called the *capacitor charging power supply*. The capacitor must charge to the desired voltage in a certain time, which is at the most equal to the reciprocal of the repetition rate of the laser. In practice, it should be slightly less, allowing for some minimum time for flash lamp quenching. The average power that this supply is expected to deliver at its output is the product of the energy per pulse and the repetition rate. The power supply accounts for more than 90% of the total electrical input to the system. The efficiency of this

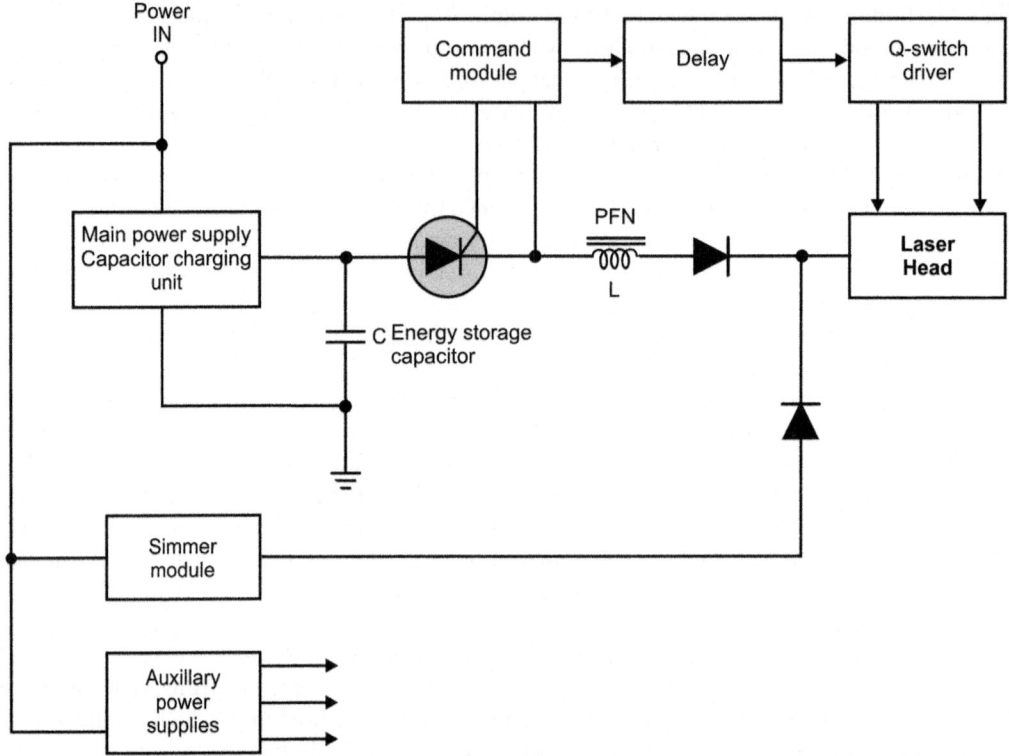

Figure 8.77 Block-schematic of Q-switched flash lamp pumped solid state laser electronics.

supply is therefore the prime determinant factor for the overall electrical efficiency of the laser. The conversion efficiency also directly affects the size and weight of the overall system, a parameter particularly important in the military applications of Q-switched, flash lamp pumped solid-state lasers.

The *simmer module* maintains a relatively low-amplitude keep-alive current through the flash lamp at all times irrespective of whether the lamp is flashing or not. The current varies typically from a few tens of milli-amperes to several hundreds of milli-amperes depending upon the characteristics of the flash lamp. This mode of operation called the *simmer mode* has many advantages. From the operational viewpoint, it allows one to use a low-voltage (TTL, CMOS etc.) trigger pulse to transfer the energy stored in the capacitor to the flash lamp. Second, it significantly enhances the flash lamp life, offers tremendous improvement on the pulse-to-pulse jitter, and overcomes most of the electromagnetic interference problems present in non-simmer mode of operation. It may be mentioned here that in the non-simmer mode of operation of the flash lamp, the triggering of the flash lamp is done by applying high voltage trigger pulses with an amplitude in the order of 10–15 kV. These pulses appear at a rate equal to the repetition rate of the laser and are the major source of electromagnetic interference. Operation in simmer mode overcomes this shortcoming. There is a *pseudo simmer mode* of operation also, which is a slight variation from the traditional simmer mode of operation. In this mode, the simmer current flows for a short time, starting a little ahead of the energy discharge operation. It has all the advantages of simmer mode operation and in addition saves on the power but at the expense of added circuit complexity.

Q-switch driver is another important module for solid-state lasers and generates the drive signal for an electro-optic Q-switch. The driver needs to generate a high-voltage pulse with the desired amplitude (typically, 2.5–3.5 kV), pulse width that could be in the range of 200–500 ns and a rise time that should not be more than a few tens of nanoseconds. The *pulse forming network (PFN)* ensures that the discharge current pulse through the flash lamp has the desired pulse width and is critically damped, thus giving the most optimum energy transfer.

The *command module* generates flash lamp firing command pulses and also delayed trigger pulses for the Q-switch driver module. It may be mentioned that the Q-switch drive pulse is applied after a certain known time delay from the time instant of application of flash lamp trigger command pulse to allow for the population inversion to build up to its peak value. The flash lamp command pulses in the case of simmer mode operation are low-voltage pulses (TTL, CMOS etc.) and high voltage trigger pulses in the case of non-simmer mode of operation. The delayed trigger pulses for the Q-switch driver are always low-voltage pulses and the Q-switch driver produces the desired high voltage pulses for the electro-optic Q-switch. In addition, there is an *auxiliary module* that generates the regulated low-voltage DC power supplies from the input source of power for the operation of different circuit modules.

In the non-simmer mode of operation, the electronics is similar to simmer mode, except that there is no electronic switch and simmer power supply and the command module feeds a high voltage trigger generation circuit.

8.3.2 Capacitor Charging Power Supply

The capacitor charging power supply is the most important of all the modules for reasons outlined previously. The power supply output needs to charge a high-value capacitor, typically 20–50 µF in the case of designators and range finders and as high as thousands of microfarads in high power pulsed lasers producing laser pulse energies of several kilo-joules meant for EOCM and laser weapon applications. The switched mode concept is invariably used for the design of capacitor charging power supply. An externally driven fly back converter is the preferred topology. Its design, however, is not as straightforward as it would be in the case of a resistive load. The reason is as follows. In the case of capacitive load, energy storage and energy transfer mechanisms are relatively more complex. Each time an energy packet is stored in the primary of the transformer and subsequently transferred to the capacitor, the time needed transfer the packet of energy depends upon the quantum of voltage it would impart to the capacitor. As a result, for the same energy quantum, the time required for transfer continuously reduces as the voltage builds up across the capacitor from zero to the final value due to diminishing voltage quantum. Therefore, it is not advisable to use a fixed frequency or fixed off-time switching supply. The drive waveform needs to be a variable frequency one with the off-time periods governed by the charge status of the energy storage capacitor. It can be mathematically proved that the size of voltage packet varies from V when the capacitor is fully discharged to $\left[\sqrt{N} - \sqrt{(N-1)} \right] \times V$ where N is the number of packets required to charge the capacitor to the final value. If the off-time of the drive waveform could be varied or decreased to be more precise in accordance with the voltage buildup across the capacitor, the power supply would operate at the highest possible conversion efficiency. Figure 8.78 shows the preferred block-schematic arrangement of a capacitor charging power supply configured around an externally driven fly back converter.

The circuit is divided into two major parts namely the drive circuit and the feedback circuit. The *drive circuit* comprises of a cascaded arrangement of a voltage controlled oscillator

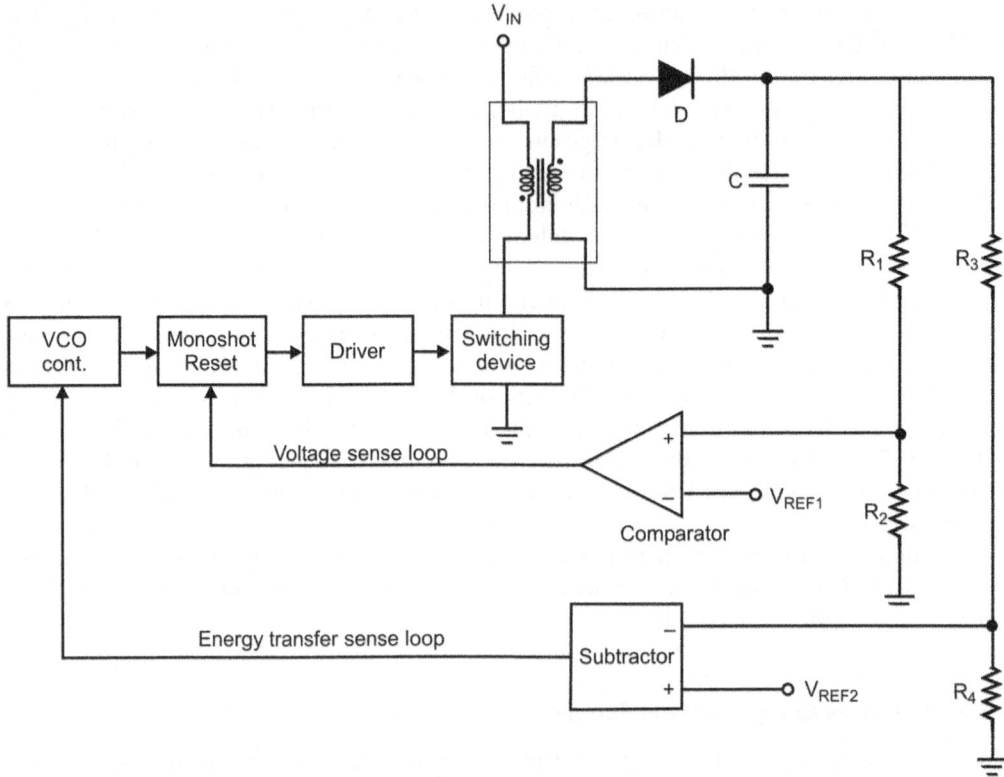

Figure 8.78 Block-schematic of a capacitor charging power supply.

(VCO), a monoshot circuit and a drive circuit. Output of VCO feeds the trigger input of the monoshot. The pulse width of the monoshot is chosen to be equal to the desired on-time of the switching device. The frequency of monoshot output and hence the off-time of the wave-form is governed by the frequency of the VCO output, which in turn depends upon the volt-age applied to its control input. The drive circuit provides the required drive current and/or voltage depending upon the type of switching device used. The feedback circuit VCO output frequency is configured around a comparator and a subtractor. The comparator circuit is used to reset the monoshot circuit and therefore withdraw drive current from switching device as and when the output voltage reaches the desired output voltage. The subtractor output controls the frequency of the VCO and therefore the off-time of the drive waveform to the switching device. The subtractor output in turn depends upon the voltage across the energy storage capacitor decreasing with increase in voltage. Decrease in control voltage to VCO increases the output frequency thereby reducing the off-time at the output of the mon-oshot. What is important here is that pattern of reduction of off-time is linearly related to reduction of control voltage, which in turn is linearly related increase in voltage across the capacitor. Thus off-time reduces in accordance with increase in capacitor voltage. This design yields the most optimum results.

Compact capacitor charging power supplies covering a range of input-output voltage and power output specifications are available from different manufacturers of laser electronics. One such unit intended for OEM applications is shown in Figure 8.79.

Figure 8.79 Capacitor charging power supply module for OEM applications.

8.3.3   Simmer Power Supply

The simmer power supply maintains a steady state partial ionization of the flash lamp during the time the lamp is not flashing by maintaining a low keep-alive current though it. Simmer power supply must be designed with due consideration to I-V characteristics of the flash lamp expressed by $v \propto (i)^{-0.3}$ where v and I, respectively, are voltage across and current through the flash lamp. The simmer power supply is a high voltage DC power supply producing an output voltage in the range of 800–1500 V depending upon the characteristics of the flash lamp and the required magnitude of the simmer current. A high voltage trigger pulse, typically 10–15 kV, creates pre-ionization before the simmer power supply can take over and deliver the keep-alive current through the flash lamp. The output of the simmer power supply is applied to the flash lamp through a series resistor called ballast resistor. The magnitude of the simmer current therefore depends upon the difference between the simmer supply output voltage, voltage across the flash lamp in the simmer mode and the value of the ballast resistance. The value of the ballast resistance should be slightly higher than the negative impedance offered by the flash lamp in the simmer regime.

Simmer power supply is generally designed around an externally driven fly back converter topology with output power delivery capability equal to product of required open circuit output voltage and magnitude desired simmer current. Open circuit voltage is further equal to sum of voltage across the lamp in simmering condition and voltage drop across the ballast resistor. Figure 8.80 shows a simmer power supply interface with the flash lamp. A TTL/CMOS pulse applied to the gate of SCR switches it on thereby producing high voltage trigger pulse across the secondary of trigger transformer. High voltage trigger produces required pre-ionization forcing simmer current to flow through it. This circuit, however, has a drawback that if simmer current extinguished due to some reason, there is no inbuilt mechanism to restore it. This shortcoming is overcome in the circuit schematic of Figure 8.81. In this case, in the event of simmer current getting extinguished, an astable multivibrator controlled by a comparator restores normal operation. The astable multivibrator operates typically at 20–30 Hz. Simmer power supply modules, like many other laser electronics subsystems such as capacitor charging power supplies, flash lamp trigger circuits and so on, are also commercially available for OEM manufacturers. One such module is shown in Figure 8.82.

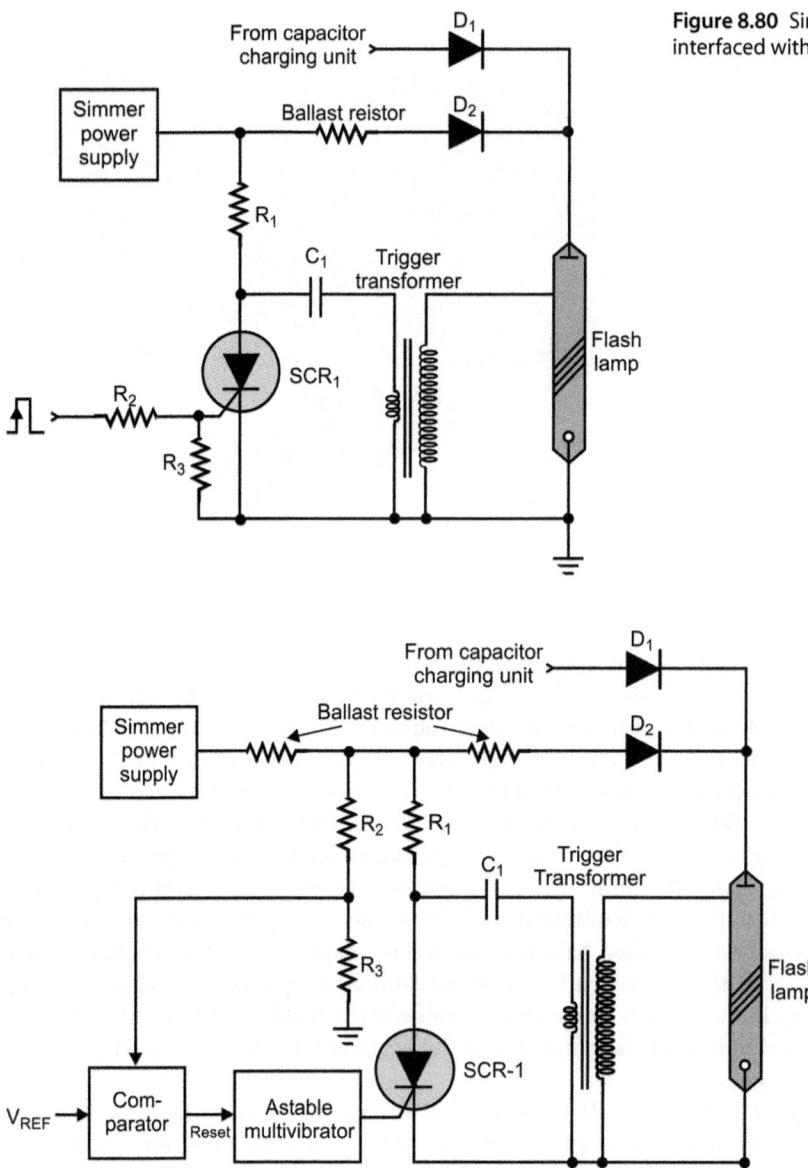

Figure 8.80 Simmer power supply interfaced with flash lamp.

Figure 8.81 Modified simmer interface circuit.

For pulse repetition frequencies that are not very high, the simmer mode of operation leads to significant power loss during the long time intervals between the two flashes. That is why this mode of operation is usually preferred for PRFs in excess of 50 Hz. For relatively lower PRFs, the pseudo simmer mode of operation is employed. It has all the advantages of the simmer mode with some added circuit complexity. In this, the lamp remains in the non-conducting state for most of the time between two consecutive flashes and the partial state of ionization (or simmer mode) is activated about 100–200 μs before every flash trigger pulse. Figure 8.83 shows the typical schematic arrangement. A simmer initiation trigger

Figure 8.82 Simmer power supply module.

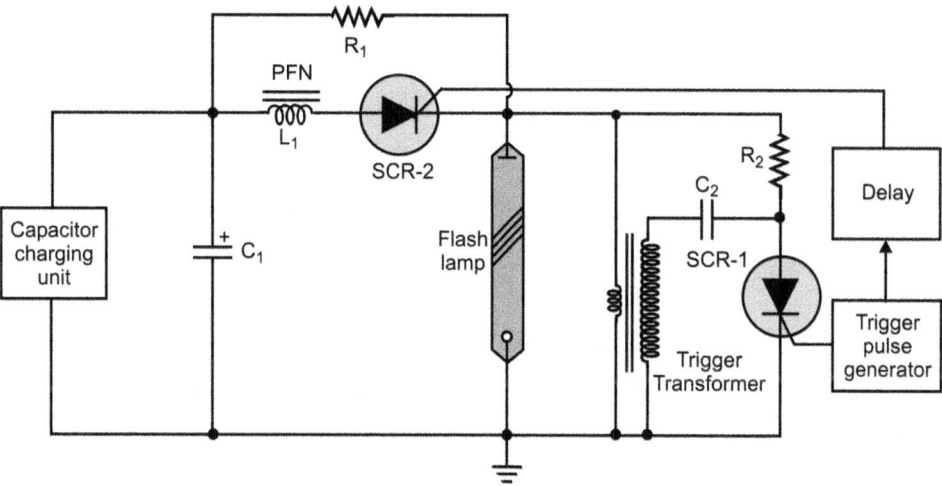

Figure 8.83 Pseudo simmer mode of operation.

pulse activates the partial discharge state. A delayed trigger pulse fires the SCR allowing the energy stored in the capacitor to discharge through the lamp producing a flash.

8.3.4 Pulse Forming Networks

The pulse forming network is so designed as to produce a critically damped current pulse through the flash lamp when the energy storage capacitor is discharged through it. This the most efficient way of energy transfer, which also minimizes reverse voltage appearing across the capacitor. Figure 8.84 shows single-stage pulse forming network. In the case of critical damping, the flash lamp resistance, R is given by $R = 2\sqrt{(L/C)}$ and it corresponds to damping factor α being equal to 0.8. There are no current reversals and the rise time is faster than it is in the case of the over-damped condition [expressed by $R > 2\sqrt{(L/C)}$] and slightly slower than it is in the case of the under-damped condition [expressed by $R < 2\sqrt{(L/C)}$]. Also, current pulse rise time is more or less equal to the current decay time, which results in better efficiency and optimal peak intensities. A slightly under-damped current pulse may be used sometimes to

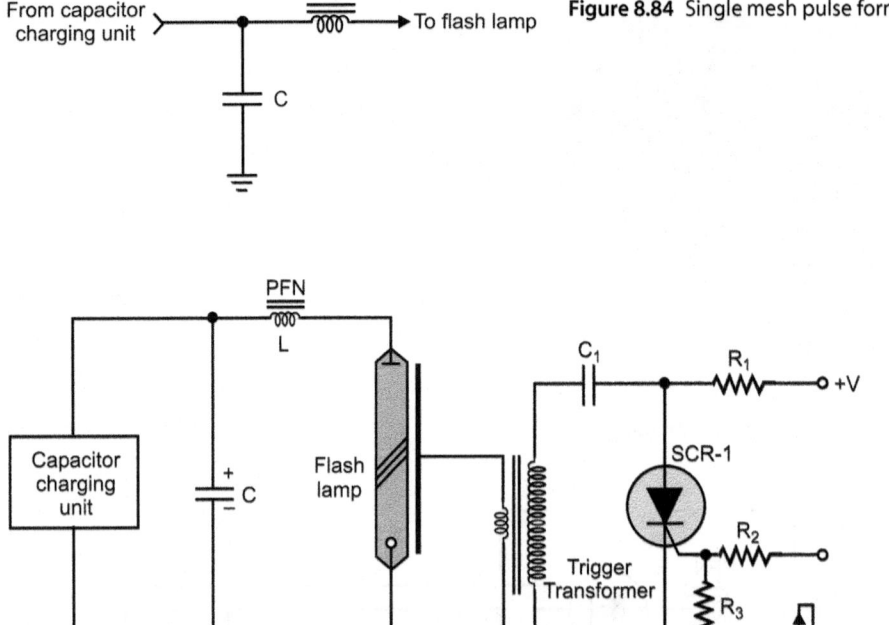

From capacitor charging unit → To flash lamp C

Figure 8.84 Single mesh pulse forming network.

Figure 8.85 External triggering circuit.

achieve shorter rise time and consequent higher value of peak current provided care is taken to keep current reversal well within the prescribed limits of the flash lamp and energy storage capacitor. The value of energy storage capacitor depends upon stored energy E_0, desired pulse width t_p and flash lamp impedance parameter K_0 and is given by $C = \left(0.09 E_0 t_p^2 K_0^{-4}\right)^{1/3}$. Also, $t_p = \sqrt{LC}$ and $L = t_p^2/9C$. Multi-mesh pulse forming networks are also there but they are less common as far as their application in flash-pumped solid-state lasers is concerned.

8.3.5 Flash Lamp Trigger Circuits

A flash lamp trigger circuit is required only in the non-simmer mode of operation. Common modes of flash lamp triggering include over voltage triggering, external triggering, series triggering and parallel triggering. Over voltage and parallel triggering schemes are less popular. External and series triggering circuits are more common. External and series triggering circuits are shown in Figures 8.85 and 8.86, respectively.

The main advantage of the *external triggering* is that it does not interfere with the main energy discharge circuit. The disadvantage is that a high voltage trigger point is exposed and therefore needs to be properly isolated from the environment lest it causes problems at high altitude or in humid conditions. External triggering is recommended for low repetition rate, low-energy systems where the flash lamp is air cooled. In the case of *series triggering circuit*, the series trigger transformer is so designed that the transformer core saturates and the saturated secondary winding inductance serves the purpose of the PFN inductor also. Series triggering offers the advantages of reliable and reproducible triggering.

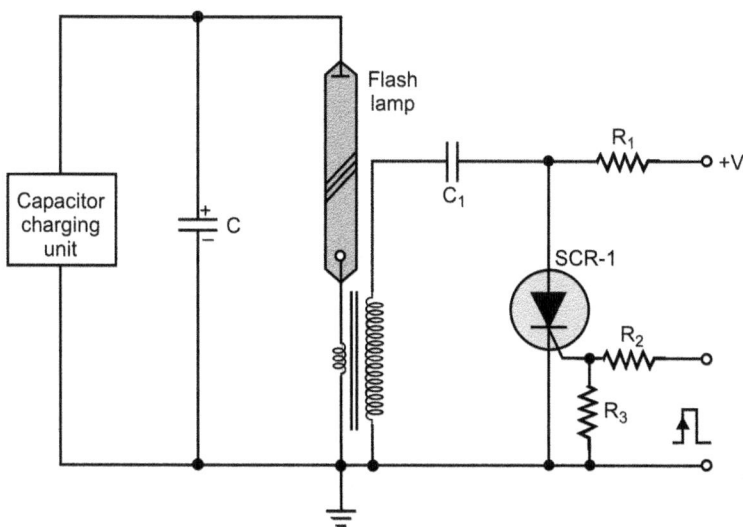

Figure 8.86 Series triggering circuit.

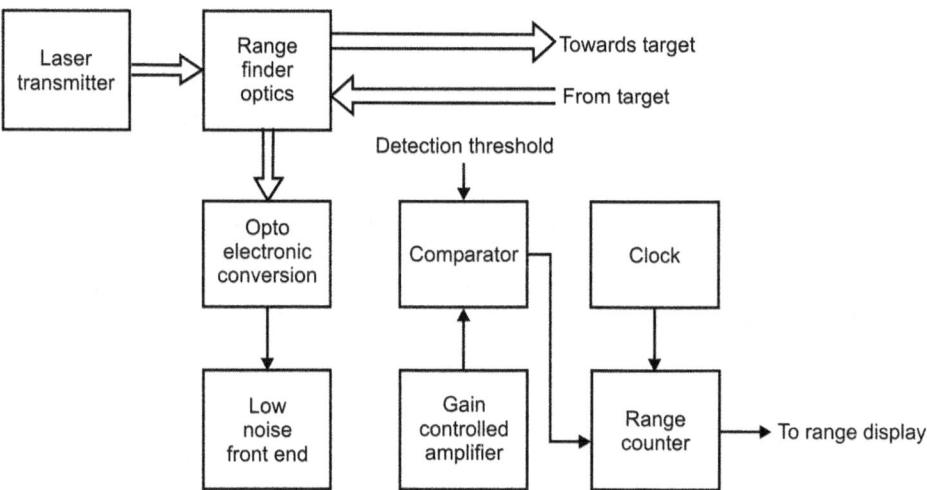

Figure 8.87 Receiver electronics for a laser range finder.

8.3.6 Receiver Electronics for a Laser Range Finder

Figure 8.87 shows a block-schematic arrangement of different building blocks of receiver electronics for a typical laser range finder. An optoelectronics front-end circuit is one of the most critical building blocks of the receiver section. The front-end circuit is supposed to transform the received laser pulse, whose pulse width could be anywhere in the range of 10–20 ns, to an equivalent electrical signal. The peak power of the received laser pulse could be as low as a few tens of nanowatts when ranging a far-off target and as high as a few tens of milliwatts when the target is close by. This implies that the amplifier portion of the front-end needs to have a dynamic range as high as 80–100 dB. This is usually achieved partly in the avalanche photodiode (APD) by controlling the responsivity of the device through its reverse-bias variation and partly in the gain-controlled amplifier stage. Laser pulse width

Figure 8.88 Range receiver electronics.

decides the bandwidth of the front-end and is given by the bandwidth $= 350/t_R$ where t_R is rise time of the laser pulse in nanoseconds. Range counter clock frequency determines the range measurement accuracy and is given by $\pm(c/2 f_{CLK})$ where $c = 3 \times 10^8$ m/s. A large number of manufacturers, such as Analogue Modules Inc., offer different modules of laser range finder electronics including photo detector amplifiers, fast pulse peak stretchers, range counters and so on (Figure 8.88).

In the preceding paragraphs, the major electronics subsystems of flash lamp pumped pulsed solid-state laser source based military systems, such as laser range finders and target designators, were discussed. Present day laser range finders and target designators are largely configured around diode-pumped pulsed solid-state lasers. In that case, the transmitter electronics is nothing but the drive and control circuitry required for laser diodes used to optically pump the gain medium. Laser diode drive and control electronics mainly includes current source to drive the laser diodes and temperature controller to operate the laser diodes at the desired temperature. Different building blocks of laser diode electronics are discussed in Section 8.5.

8.4 Gas Laser Electronics

Carbon dioxide and helium-neon lasers are the two commonly used gas lasers when it comes to tactical military applications. High power lasers such as carbon dioxide gas dynamic lasers, hydrogen fluoride/deuterium fluoride lasers and chemical oxy-iodine lasers with potential of generating megawatt levels of CW power for directed-energy weapon applications are also broadly classified as gas lasers, though a discussion on these lasers is not relevant as far as role of electronics is concerned. These lasers are pumped by gas dynamics or by chemical reactions. Focus in this section is on gas laser electronics with particular reference to role of electronics in gas lasers having potential for tactical military applications. Most of the discussion is centred on helium-neon and carbon dioxide lasers. The electronics for metal vapour lasers, ion lasers and Excimer lasers are also briefly covered.

8.4.1 Gas Laser Electrical Discharge

Carbon dioxide laser-based laser range finders and laser radar and helium-neon laser cavity based ring laser gyroscope inertial navigation sensors are the common military applications exploiting use of gas lasers. In addition, gas lasers such as gas dynamic carbon dioxide laser, chemical oxy-iodine laser, all gas-phase iodine laser, hydrogen fluoride and deuterium fluoride lasers, though not electrically or optically pumped, are the gas lasers largely exploited to build high power directed-energy weapon technology demonstrators. We shall, however, confine our discussion to electrically excited gas lasers in particular those with military applications.

In both carbon dioxide and helium-neon lasers, the power supply used to initiate and subsequently sustain electrical discharge through the gas mixture contained in a sealed envelope constitutes the primary and essential component of electronics. The active medium is usually excited either by passing an electric discharge current along the length of the tube known as longitudinal excitation both types of gas lasers (Figure 8.89) or by an electric discharge perpendicular to the length of the laser tube known as transverse excitation common in carbon dioxide lasers only (Figure 8.90). Frequency stabilization electronics used in the case of actively stabilized Doppler broadened gas lasers such as helium-neon and carbon dioxide lasers is another area that relates to gas laser electronics.

8.4.1.1 Negative Resistance Characteristics

Gas discharge characteristics when excited electrically are the key to design of power supplies for gas lasers. Typical gas discharge characteristics as applicable to carbon dioxide and helium-neon lasers, exhibit negative resistance in their current-voltage relationship.

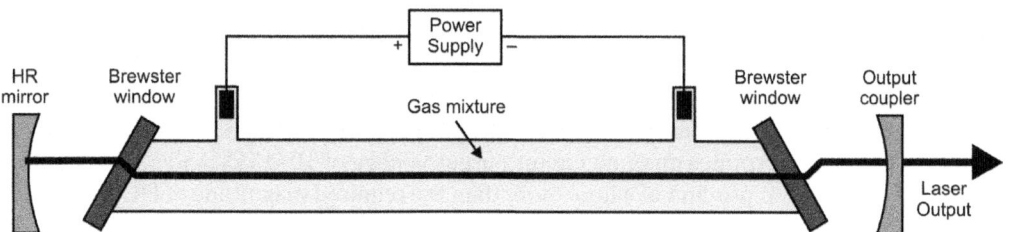

Figure 8.89 Longitudinally excited gas laser.

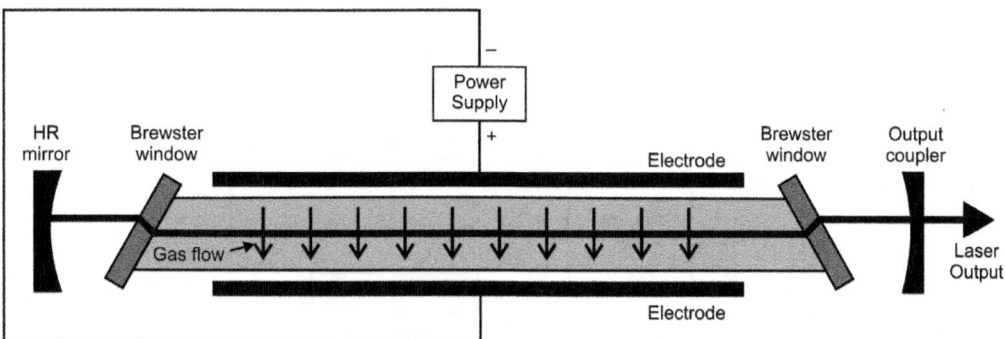

Figure 8.90 Transversely excited gas laser.

The current is zero initially, which may increase to a few nano amperes after the applied voltage exceeds a certain threshold causing some ionization of the gas mixture. The current increases slowly until we reach the breakdown point. At the breakdown point, large number of molecules in the gas mixture is ionized and conductivity increases significantly. Increase in discharge current causes further reduction in discharge resistance with the result that the voltage required to sustain the discharge actually decreases with increase in current. This gives rise to what is called the *negative resistance region* in gas discharge current-voltage characteristics.

8.4.2 Helium-Neon Power Supplies

Figure 8.91 shows generalized block-schematic representation of a helium-neon power supply. The generalized form of a gas laser power supply essentially comprises of a *high voltage generation circuit* that provides the starting voltage either in the form of a high voltage trigger pulse or a DC voltage level to initiate the gas discharge and a *power supply* with current limiting feature to deliver the steady state current to sustain the discharge. Amplitude of a high voltage trigger pulse needs to be greater than the breakdown voltage of the gas mixture in question and the current limiting feature is provided by a resistance called *ballast resistance*, which limits the discharge current to desired value.

There are three commonly used circuit topologies for designing helium-neon laser power supplies. In one of the topologies, an AC-DC or DC-DC power supply producing an output voltage slightly higher than the voltage required sustaining the discharge plasma is cascaded to a voltage multiplier chain of diodes and capacitors to produce the starting voltage. The voltage at the output of multiplier chain drops to almost the voltage level present at the output of DC power supply due to inherently poor regulation of the multiplier chain.

In the second commonly used circuit topology, the DC power supply is connected directly to the plasma tube through the Ballast resistance. A high voltage trigger pulse of 10–15 kV is applied to the tube to initiate the discharge. Once the gas mixture is ionized, the DC power supply takes over to sustain the discharge.

In the third topology, constant power output fly back converter is used. The DC-to-DC converter is designed to produce an open circuit output voltage of 10–15 kV and a power output delivery rating equal to product of a little more than the required magnitude of DC voltage and desired discharge current. As a consequence, the converter output voltage rapidly drops to the desired value after the discharge is initiated.

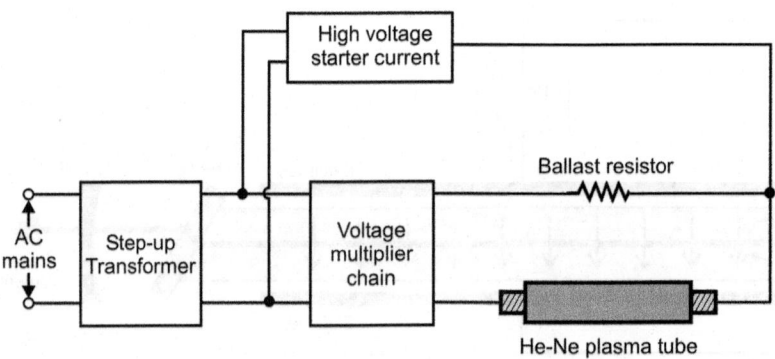

Figure 8.91 Generalized block-schematic representation of helium-neon laser power supply.

8.4.2.1 Diode-Capacitor Multiplier Chain

There are two common voltage multiplier configurations, one for the odd multiplication factors (Figure 8.92) and the other for even multiplication factors (Figure 8.93). The output voltage in the case of the multiplier configuration of Figure 8.92 is given by eqn. 8.12.

$$V_O = \frac{nV_m}{1+[n(n^2-1)/12\,fCR_L]} \tag{8.12}$$

In the case of a multiplier chain with an even multiplication factor, the output voltage is given by eqn. 8.13.

$$V_O = \frac{nV_m}{1+[n(n^2/2+1)/6\,fCR_L]} \tag{8.13}$$

In both cases, $C_1 = C_2 = C_3 = C_4 = C_5 = C_6 = C_7 = C_8 = C$, n is the multiplication factor, f is frequency of operation equal to power line frequency for AC-DC power supply and the switching frequency in the case of a switched mode power supply. V_P is the peak amplitude of

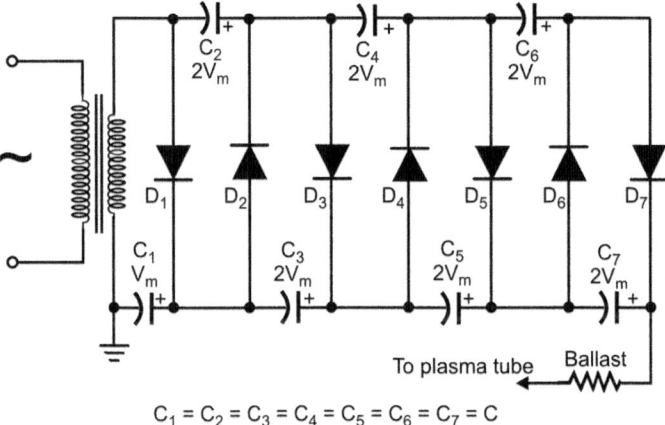

$$C_1 = C_2 = C_3 = C_4 = C_5 = C_6 = C_7 = C$$

Figure 8.92 High voltage multiplier chain for odd multiplication factors.

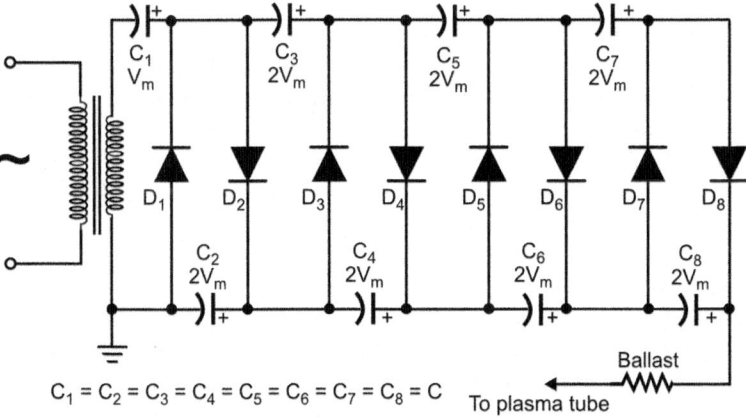

$$C_1 = C_2 = C_3 = C_4 = C_5 = C_6 = C_7 = C_8 = C$$

Figure 8.93 High voltage multiplier chain with even multiplication factors.

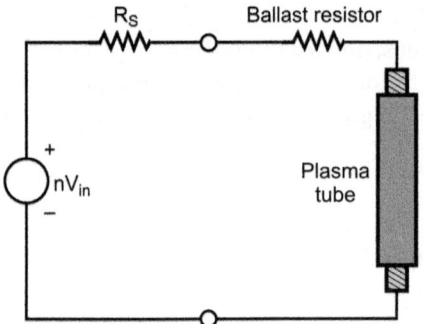

Figure 8.94 Equivalent circuit of a cascade multiplier chain.

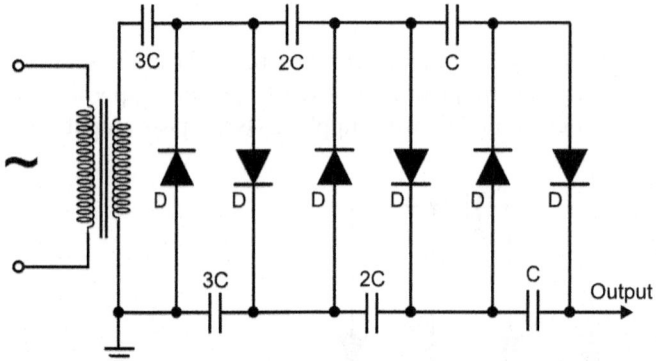

Figure 8.95 Voltage multiplier chain with unequal capacitors.

the voltage applied to the input of multiplier chain. Also in both cases, multiplied output voltage is equal to nV_P provided the bracketed term in the denominator is zero or negligible. This is possible for large value of capacitance C.

In fact, the multiplier chain as a whole offers a series resistance which drops a voltage across it when current is drawn from it. The multiplier chain may be represented by an equivalent circuit of the type shown in Figure 8.94. This equivalent series source resistance to a very good approximation increases as the cube of the multiplication factor. This implies that if the multiplication factor is doubled, then for a given operating frequency, the capacitance will have to be increased by a factor of eight if the multiplier circuit were to maintain the same output voltage on load.

Another voltage multiplier circuit that offers a superior performance is the one that uses unequal capacitors. Figure 8.95 shows the circuit configuration. The configuration shown here is that of a multiplier with an even multiplication factor. The circuit works at an efficiency that is better than that of the circuit using equal value capacitances. The circuits shown in Figures 8.92, 8.93 and 8.95 are half-wave circuits. A full-wave voltage multiplier that gives a far better ripple performance is shown in Figure 8.96. The circuit shown provides a multiplication factor of 4.

8.4.2.2 Helium-Neon Power Supply Circuits

The most commonly used helium-neon laser power supply circuit is a push-pull type high-frequency switched mode DC-to-AC power supply cascaded to a diode-capacitor type voltage multiplier chain as shown in Figure 8.97. The push-pull circuit generates a high-frequency square waveform across a secondary transformer, which is then multiplied to generate the starting high voltage for the discharge tube. The push-pull circuit here is configured around a

Figure 8.96 Full wave voltage multiplier.

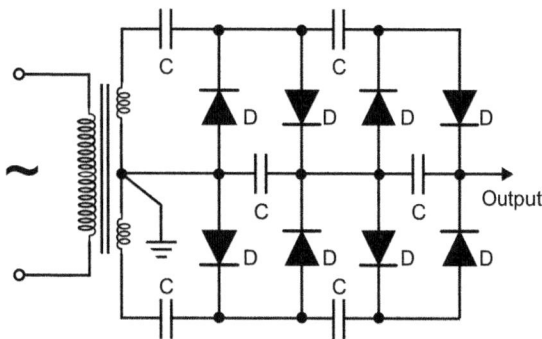

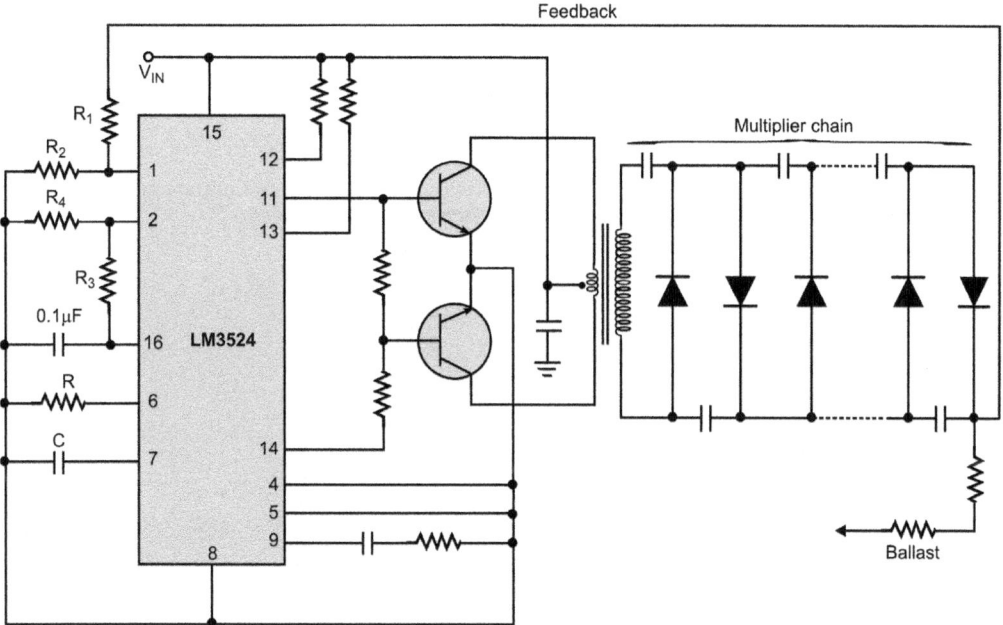

Figure 8.97 Push pull converter with cascaded multiplier chain.

switching control IC type LM 3,524. It may be mentioned here that a large number of control ICs intended for switching power supply design are commercially available. Some of the other popular type numbers include TL 497, TL 594, and TDA 8130. Their data sheets contain typical application circuits for switching supply design. As the discharge is truck, the voltage falls to a lower value to supply the sustaining discharge current through the ballast resistance. The rest of the voltage is dropped across the multiplier chain components. In some cases, the sustaining voltage may be tapped from an earlier point in the chain for better overall performance. In that case, the output of the multiplier chain feeds the discharge tube through a relatively high resistance of the order of tens of mega ohms and the sustaining voltage feeds the tube through the ballast resistance.

The other common circuit topology is one where the power supply generates the voltage required to sustain the discharge and a high voltage trigger circuit generates the starting voltage to initiate the discharge. Figure 8.98 shows a typical circuit. The power supply is configured

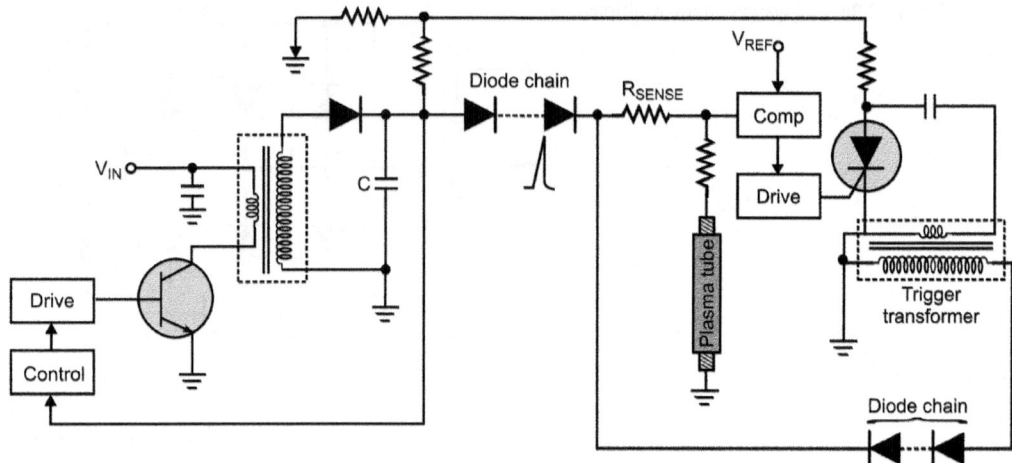

Figure 8.98 Fly back converter with high voltage trigger circuit.

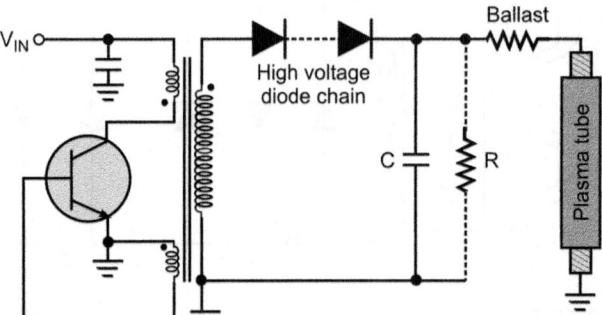

Figure 8.99 Constant power output fly back converter.

around an externally driven fly back type DC-to-DC converter. The ballast resistance is split into two. Initially when there is no discharge, the voltage appearing after R_{SENSE} is equal to open circuit voltage. A fraction of this voltage is compared with V_{REF} and the comparator drives the high voltage trigger circuit configured around SCR, R-C components and trigger transformer. As soon as discharge is struck, voltage appearing after R_{SENSE} drops and the drive pulses are withdrawn from SCR gate. The power supply supplies the discharge current through a series connection of resistors constituting ballast resistance.

In the third circuit topology as shown in the circuit implementation of Figure 8.99, the converter circuit is designed to deliver an output power equal to the product of the required sustaining voltage at the power supply output and the discharge current. Since it is a constant output power converter, the output voltage would increase without limit in the ideal case in the event of zero current drawn from the power supply. When the output voltage exceeds the required initiating voltage, the plasma current drawn from the power supply forces the output voltage to fall to the sustaining voltage governed by the output power capability of the converter.

Compact helium-neon laser power supplies are commercially available both as AC operated bench top versions as well as DC operated supplies for OEM market (Figure 8.100). Helium-neon lasers with an integrated power supply are also available (Figure 8.101).

Figure 8.100 Commercial helium-neon laser power supplies.

Figure 8.101 Helium neon laser with inbuilt power supply.

8.4.3 Helium-Neon Power Supplies for a Ring Laser Gyro

There are applications where the power supply needs to provide ultra-stable discharge current. The basic arrangement of the power supply where the output voltage would adjust itself so as to maintain a constant current though the plasma tube is shown in Figure 8.102. In this case, it is the plasma current and not a fraction of the output voltage that is compared with a reference to generate an error voltage. The error voltage in turn controls one of the parameters of the drive waveform such as pulse width or frequency to maintain a constant discharge current. The constant value of discharge current can be varied by varying the reference voltage. Also, He-Ne lasers have a tendency to degrade in output power with the passage of time. This can be checked by using an adjustable reference that decides the constant plasma current delivered by the power supply. This reference can be set for a little higher current in case the laser output power has deteriorated. However, current cannot be increased and should not be increased indiscriminately as there is well-defined plasma current range over which the plasma operates most optimally. Such a control should be used only for fine adjustment.

One application where an ultra-stable plasma discharge current is an essential operational requirement is in a ring laser gyroscope sensor. A ring laser gyroscope comprises a helium-neon ring laser cavity with two counter-propagating laser beams. A ring laser gyroscope, which is the heart of the inertial navigation systems of modern aircraft, missiles and so on, makes use

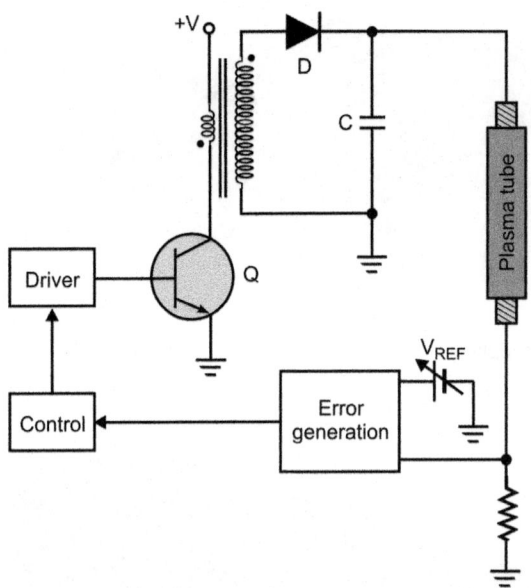

Figure 8.102 Current regulated He-Ne power supply.

of a ring He-Ne laser in a closed cavity. Without going into details of its operational principle, it would suffice to mention here that stability of discharge current in each arm of the ring laser cavity is very crucial to proper functioning of the device. It is also important to have a stable value of difference between the currents in two arms. A power supply configuration that achieves this is shown in Figure 8.103. As shown in the figure, there are two current sense loops controlled by the same reference voltage. Another possible approach could be to use a precision reference for one arm and then use the actual controlled current flowing through that arm as the reference for controlling current in the other arm. The preferred circuit topology for the power supply is an externally driven PWM controlled fly back converter generating a regulated output voltage equal to the sustaining voltage at the required plasma current.

8.4.4 Ballast Resistance

As outlined earlier, ballast resistance is a fixed resistance connected from the output of the power supply in series with the plasma tube to ensure that the power supply sees an overall positive resistance while feeding the plasma tube whose I-V characteristics exhibit a negative resistance. Incorrect value of ballast resistance leads to a highly unstable plasma that has a tendency to extinguish. Choice of ballast resistance is therefore governed by the plasma tube for which the power supply is being designed. The resistance typically varies between 50 and 100 kΩ. The commercial He-Ne laser with associated power supplies take care of this aspect. Plasma tube manufactures specify the required value of the ballast resistance for the tubes manufactured by them. So, if you are designing a power supply for a given plasma tube, it is important to consult the manufacturer's catalogue for this information.

Another important aspect of the ballast resistance is to connect it as close to the plasma tube electrode as possible. Placement of the ballast resistance close to the electrode ensures that the stray capacitance is minimized. Stray parasitic element leads to noisy plasma. If the power supply is to be in a separate package, the ballast resistance in that case finds its place in the plasma tube housing. Also, the resistance should preferably be of the carbon film or carbon composition type and, in case the required wattage specification forces you to use a wire wound

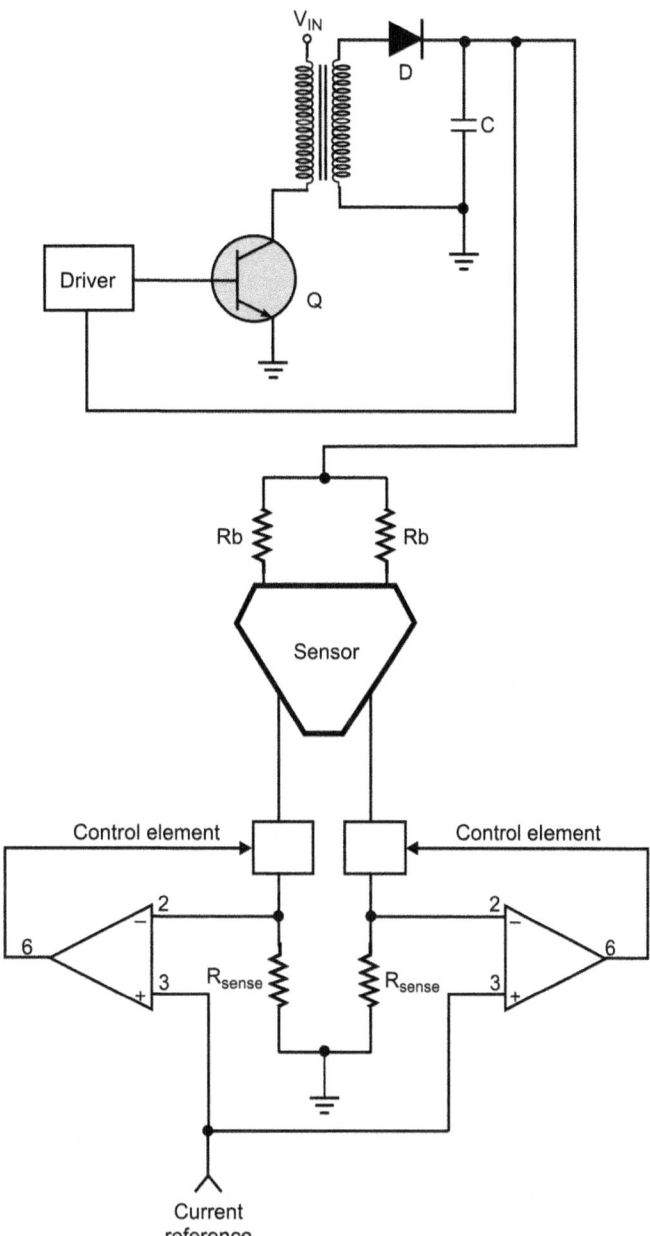

Figure 8.103 Helium-neon power supply for a ring laser gyro cavity.

type, it should preferably be of the non-inductive variety and the rated wattage of the resistor should ideally be 10 times the actual power dissipated in the tube.

8.4.5 Carbon Dioxide Laser Power Supplies

Carbon dioxide lasers are operated in both CW and pulsed output modes. CW lasers are either DC excited or RF excited. Pulsed output lasers are mainly TEA (Transversely Excited Atmospheric pressure) lasers. Power supply circuits for DC excited lasers are similar to the ones described in earlier paragraphs in the case of helium-neon lasers. Different power supply circuits discussed in the case of helium-neon lasers are equally valid for carbon dioxide lasers.

The basic power supply arrangement employed in the case of pulsed CO_2 laser such as a TEA laser is to charge an energy storage capacitor to a high voltage from a high voltage DC power supply and then discharge the same through the gas mixture. The discharge process is controlled by a HV switch. Commonly used HV switches include spark gaps, thyratron tubes and solid-state switching devices. It may be mentioned here that the CO_2 laser's performance is not as badly affected by change in discharge impedance during pulsed discharge as is that for some of the other types of gas lasers such as ion lasers and metal vapour lasers. However, discharge in the case of the TEA CO_2 laser does have a tendency to break down into bright line arcs. One way of solving this problem is to have a cathode in the form of large number of pins (Figure 8.104) each with its ballast resistor in series with it. The ballast resistors limit the current through their respective pins and therefore inhibit any arc formation.

Present day TEA CO_2 lasers use Marx bank capacitor systems. In the Marx bank system, the arrangement of individual capacitors is such that they are in parallel during the charging process and in series during the discharge operation. This means if each capacitor in the bank were charged to a voltage V, then the total voltage available across the tube for discharge operation would be nV, n being the number of capacitors. Figure 8.105 shows the Marx bank power supply

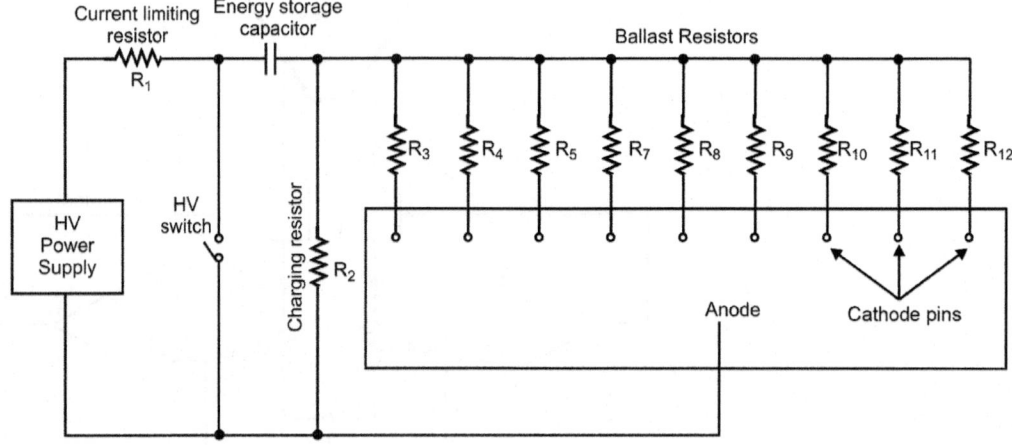

Figure 8.104 TEA CO_2 laser with a cathode that has a large number of pins.

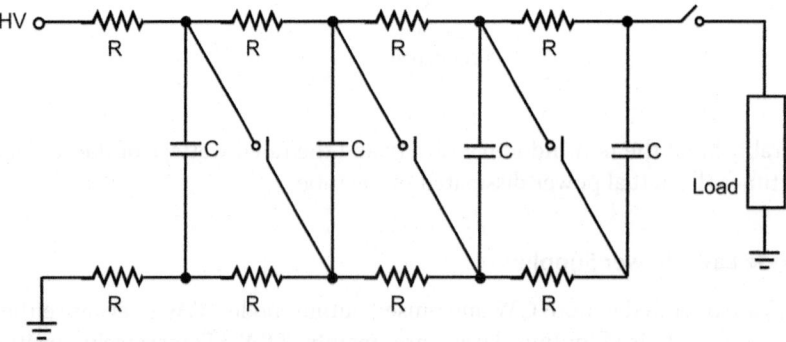

Figure 8.105 Marx bank power supply arrangement.

Figure 8.106 Power supply for an RF excited carbon dioxide laser.

arrangement. In the arrangement shown in Figure 8.105, switches are open during the charging operation. They are triggered simultaneously to the closed position to initiate discharge. Though the Marx bank shown here uses four stages, Marx bank systems with higher number of stages are also used. Marx banks are particularly suitable for producing fast rising pulses.

The power supply of an RF excited carbon dioxide laser comprises of an RF source and an impedance matching network (Figure 8.106). The RF source may further be split up into an RF oscillator and a cascade arrangement of RF amplifiers depending upon the output power delivering capability. Frequency of operation is in the range of 50–150 MHz. RF power is fed to the laser cavity through impedance matching network. The laser cavity's mechanical structure combined with RF discharge can be considered an electrical load comprising complex impedance. Power transfer efficiency from the RF source to the discharge is a key operational feature of the laser that is made more complicated by the fact that the discharge impedance is power dependent.

Carbon dioxide laser power supplies too are commercially available both as bench top laboratory systems as well as compact versions for the OEM market (Figure 8.107).

8.4.6 Metal Vapour Laser Power Supplies

Metal vapour lasers are categorized in two broad classes, namely the helium-cadmium family of lasers where the ionized cadmium atoms constitute the active species and the second class where neutral atoms constitute the lasing species. Copper vapour and gold vapour lasers are the predominant candidates of the second category of metal vapour lasers. While the former type operates as CW lasers, the latter are essentially pulsed lasers.

The basic power supply arrangement for CW metal vapour lasers such as helium-cadmium lasers is similar to the one discussed in the case of helium-neon lasers. In the case of pulsed metal vapour lasers like copper vapour and gold vapour lasers, the basic power supply would be similar to the one discussed in the case of TEA CO_2 lasers. Figure 8.108 shows the basic power supply for a pulsed metal vapour laser. The energy storage capacitor here is charged through the inductor and then discharged through the laser tube by triggering the thyratron switch to the closed state.

Figure 8.107 Carbon dioxide laser power supplies.

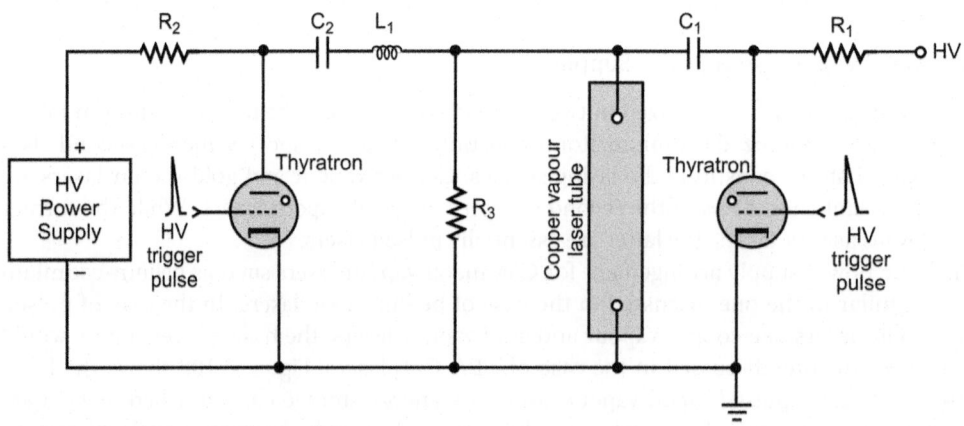

Figure 8.108 Basic power supply for a pulsed metal vapour laser.

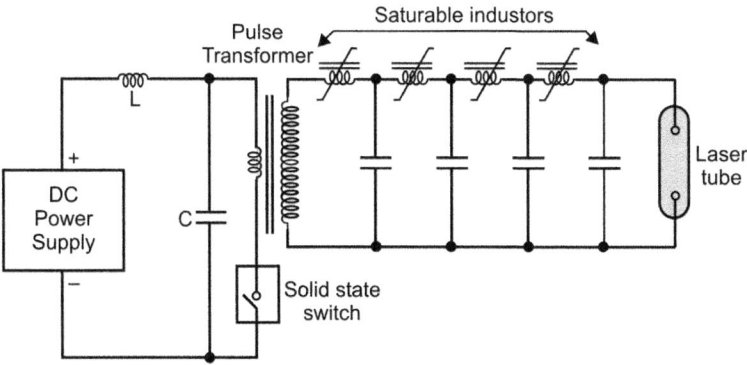

Figure 8.109 Preferred power supply configuration for pulsed metal vapour lasers.

Figure 8.109 shows a preferred power supply configuration for pulsed metal vapour lasers. In this case, the energy storage capacitor is charged to a relatively much lower voltage than is required across the discharge tube. The capacitor is made to discharge through the primary of the step-up pulse transformer by triggering the switch connected in series with the primary of the pulse transformer to the closed state. The discharge circuit is a multi-mesh pulse forming network comprising saturable inductors. The pulse forming network acts like a magnetic pulse-compression circuit. It shortens the discharge pulse duration from a few microseconds to about 100–150 ns. This configuration leads to improved conversion efficiency as the reduced pulse duration is much better matched to the requirements of the discharge. Also, use of pulse transformer allows capacitor to be charged to relatively lower voltage. This further allows use of solid-state switching devices. Due to reduced quantum of energy being stored in the capacitor, the life time of the switch also increases.

8.4.7 Ion Laser Power Supplies

Gas lasers such as helium-neon, carbon dioxide, Excimer and metal vapour lasers operate at high voltage and relatively lower current. Noble gas ion lasers on the contrary operate at relatively lower voltage of the order of 200 V and much higher current values in the range of 10–100 A. The excitation mechanism in ion lasers is electron impact in the gas in a DC discharge with high current density in the range of several hundreds of A/cm^2. In some cases, such as an argon-ion laser, high value of current density is maintained by using a magnetic field that confines the discharge. Operation at high current density necessitates that there is a suitable arrangement to transfer heat out of the tube and also that the tube is fabricated from such materials to withstand hostile operating conditions. Another requirement is that of maintaining uniform gas pressure in the tube, which is achieved by using a pressure sensor and a gas fill regulator. In addition, the power supply circuit needs to have the feedback control loops to be able to operate the laser either in the constant output power or in constant discharge current mode.

It may be mentioned here that it is the current regulator rather than voltage regulation that is implemented while designing the power supply for argon/krypton-ion lasers. It is because of the reason that the incremental resistance of the discharge is very low of the order of few ohms. This implies that a small change in voltage across the discharge will result in a relatively large change in the discharge current. Thus, discharge current regulation definitely makes more sense.

Figure 8.110 shows the generalized block-schematic arrangement of a typical ion laser power supply along with auxiliary circuits. Depending upon the output power level, discharge current

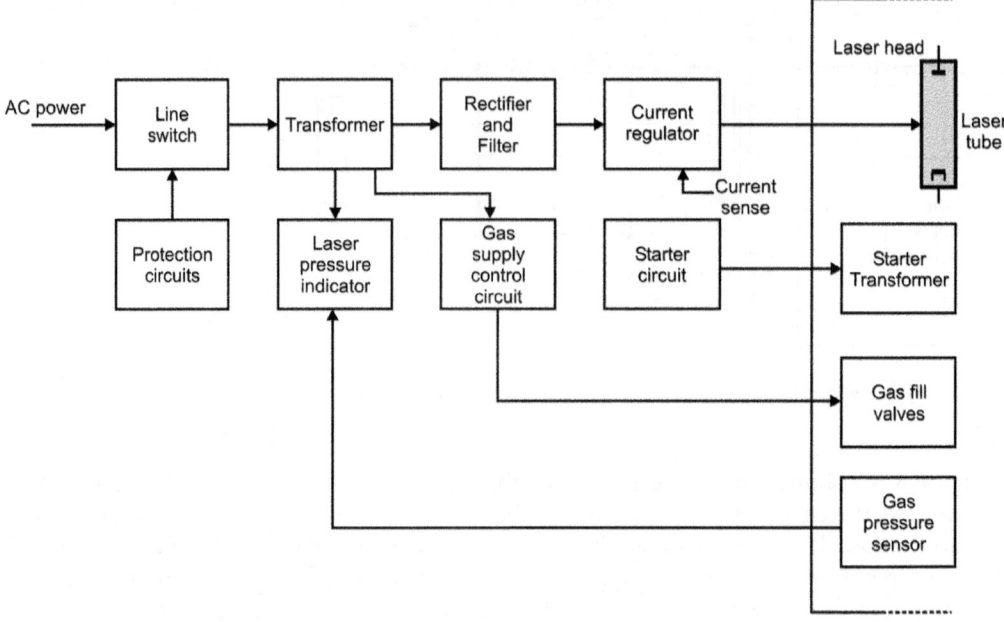

Figure 8.110 Generalized block-schematic of an ion laser power supply.

in the case of ion lasers could be of the order of several tens of Amperes. The block-schematic shown is that of a conventional current regulated AC/DC power supply. While a DC voltage of the order of 200 V is needed to sustain a steady discharge in an ion laser; it needs a higher voltage of the order of few kilovolts to initiate the discharge. This is provided by the starter circuit with the high voltage trigger transformer located inside the laser head. The line switch is nothing but a kind of circuit breaker operated by protection circuits such as over current limit, over temperature cut-off and so on. In the case of modern ion lasers the linearly current regulated conventional AC/DC supply of Figure 8.110 may be replaced by a switched mode power supply based on buck-boost topology. The inverted polarity available from the buck-boost topology allows easy connection to the cathode terminal. The anode terminal is generally used for applying the starting voltage. The ion laser power supplies also have auxiliary circuits for control of gas fill valves and monitoring of gas pressure.

8.4.8 Excimer Laser Power Supplies

Excimer lasers are pumped either by pulsed electric discharge or by high-energy electron beams. Those excited by high-energy electron beams are capable of producing very high pulse energy in the order of tens of kilo-joules. Such giant lasers find applications in areas such as laser assisted thermonuclear fusion and defence research. Detailed discussion on this form of excitation is beyond the scope of the present text. On the other hand, electrically excited Excimer lasers are relatively much smaller and far less expensive. These find use in a large number of industrial applications such as semiconductor fabrication, material processing, photochemistry and so on.

Excimer lasers are excited by short duration high peak current discharge pulses with pulse rise time of the order of tens of nanoseconds and peak value of discharge current pulse in the range of few kilo amperes. Excitation of the active medium is usually a two-stage process. Energy is stored in a capacitor in a relatively slow process in one part of the circuit and then

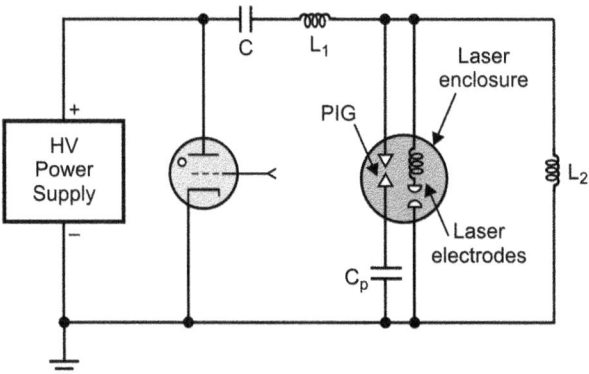

Figure 8.111 Basic power supply for an Excimer laser.

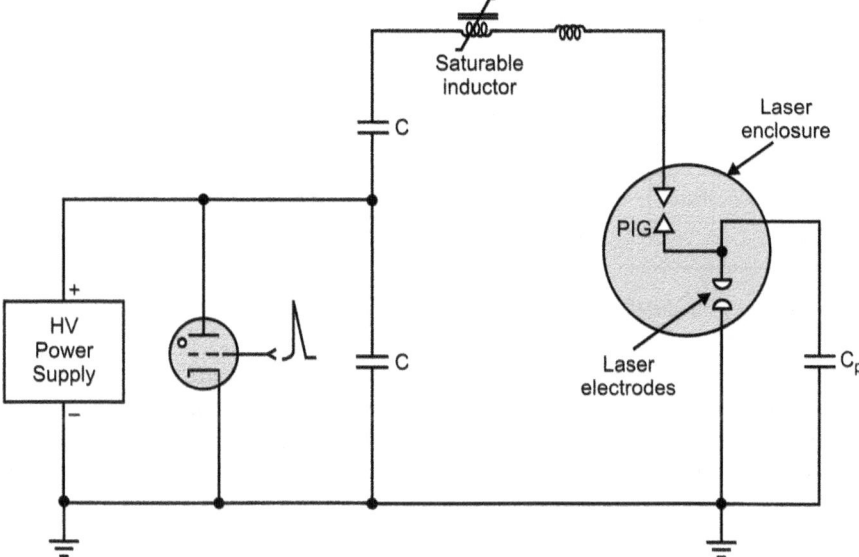

Figure 8.112 Modified power supply circuit for an Excimer laser.

transferred to a peaking capacitor that enables a much faster discharge in another part of the circuit. Figure 8.111 shows one such power supply circuit.

In the circuit of Figure 8.111, C is the charging capacitor. Stored energy is transferred to the peaking capacitor C_p across the pre-ionization spark gap. The basic circuit is adequate for Excimer lasers with moderate values of pulse energy and repetition rate for the following reasons. When the discharge is initiated, voltage across the tube rises very rapidly. As it reaches the breakdown voltage, current begins to flow. Increasing current leads to rapid fall in voltage due to negative resistance characteristics of discharge. As a result, electrons in the discharge plasma cannot get accelerated to velocities required to excite the lasing species. This forces the designer to store and switch more than the required energy. This leads to reduction in the efficiency and life time of the thyratron switch.

Commercial Excimer laser power supplies employ a modified circuit where the charging part of the circuit is decoupled from the fast discharge part of the circuit. One such circuit where the decoupling between charge and discharge parts of the circuit is provided by a saturable inductor is shown in Figure 8.112.

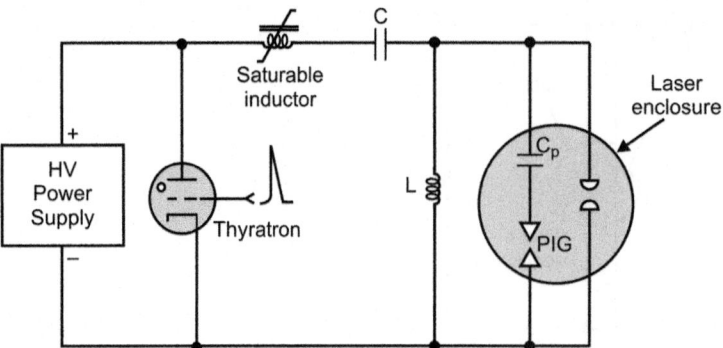

Figure 8.113 Modified power supply circuit for an Excimer laser with a peaking capacitor inside the laser enclosure.

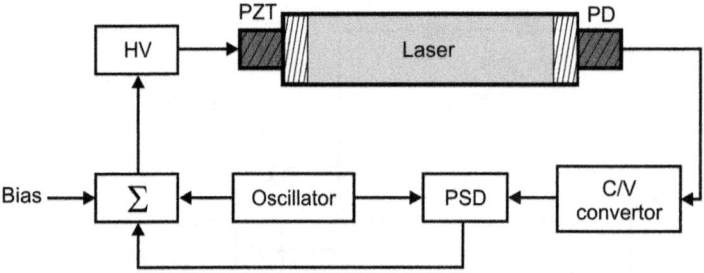

Figure 8.114 Dither frequency stabilization.

Another circuit using a peaking capacitor that is inside the laser enclosure close to the main electrodes is shown in Figure 8.113. When the peaking capacitor is located internal to the laser housing, it not only minimizes the number of feedthroughs into the enclosure; it also allows the designer to have a discharge circuit with low inductance producing faster rise time discharge pulses.

8.4.9 Frequency Stabilization of Carbon Dioxide and Helium-Neon Lasers

Carbon dioxide and helium-neon lasers have Doppler broadened gain versus frequency curves, which is 60 MHz in the case of carbon dioxide laser and 1400 MHz for a helium-neon laser emitting at 632.8 nm. With active stabilization techniques such as dither stabilization, optogalvanic stabilization and Stark cell stabilization, these lasers can be frequency stabilized to the order of ±1 MHz.

Figure 8.114 shows the schematic arrangement of a *dither stabilization* scheme. The grating allows selection of various lines within the laser tuning curve. The output of the low frequency oscillator when applied to the PZT moves the grating along the axis and hence varies the output of detector. The other input to the frequency synchronous detector is the error signal, which corresponds to the deviation in the laser output frequency from the line centre. In fact, the magnitude and phase of the error signal determine the location of operational point with respect to the centre of the Doppler broadened gain curve. While the amplitude decides how far or close to the line centre the operational point is, the phase decides which side of the line centre it is located. Though dither stabilization is simple to implement, the disadvantage of this technique is the presence of frequency modulation in the output as a result of laser dithering.

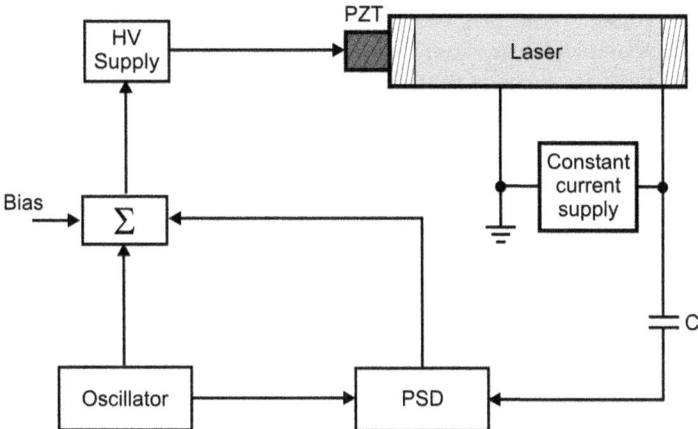

Figure 8.115 Optogalvanic stabilization schematic.

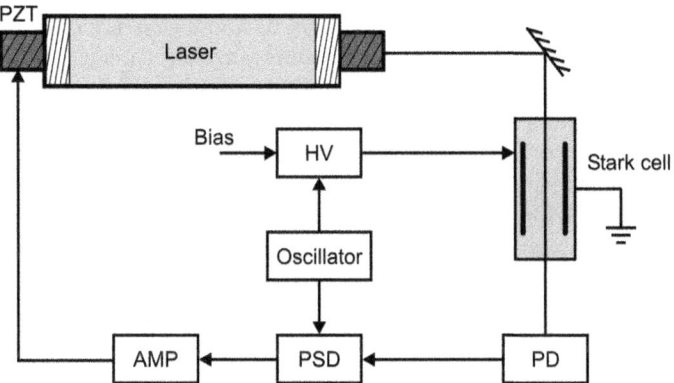

Figure 8.116 Stark cell stabilization schematic

Optogalvanic stabilization technique is particularly suitable for carbon dioxide lasers. It makes use of variation in discharge impedance due to alteration in internal radiation field intensity. Figure 8.115 shows a typical opto-galvanic frequency stabilization setup. An AC signal is applied to the cavity length transducer resulting in frequency modulation of the laser and therefore modulation of the plasma-tube impedance. The impedance variation, which is proportional to the slope of the output power versus frequency curve of the laser, is measured by exciting the plasma-tube by a high-speed current regulated power supply and measuring the resulting variation in voltage drop across the plasma-tube. This voltage is applied to the synchronous detector. The other input to the synchronous detector is the reference signal used to modulate the laser cavity length with the help of PZT. The error signal produced at the output of synchronous detector is used to stabilize the laser frequency.

The *Stark cell stabilization* technique allows dither free frequency stabilization. It also allows stabilizing frequency anywhere on the Doppler broadened gain versus frequency curve of the lasers. Figure 8.116 shows a block-schematic arrangement of Stark cell stabilization.

A part of the output laser beam is made to pass through the stark cell. An appropriate value of DC voltage is applied to one of the stark plates to make the stark-shifted molecular transition of the gas contained in the cell precisely coincide with the carbon dioxide laser output line of interest. A slowly varying linear voltage ramp may be applied instead. Also superimposed on this is a relatively high-frequency sinusoidal or square waveform. The frequency is typically 5 kHz. The beam at the output of the Stark cell is thus modulated in both amplitude and frequency by the applied high-frequency signal around the line of interest. The laser beam is phase sensitively detected with high-frequency signal applied to the Stark cell as the reference. The output of phase sensitive detector is the frequency discriminant error signal. The error signal is amplified and then fed back to the PZT through a suitable interface. One of the end elements of the cavity is mounted on the PZT. The error signal adjusts the length of the cavity in the feedback mode to lock the laser frequency to the centre of the Doppler profile of the line of interest.

8.5 Semiconductor Diode Laser Electronics

For tactical military applications of lasers, semiconductor diode lasers are close second to solid-state lasers. Laser aiming modules fitted on small arms to enhance night fighting capability of infantry soldiers, short-to-medium range eye-safe laser range finders for observation and surveillance, laser proximity sensors and short-range laser dazzlers are some prominent examples. While semiconductor diode lasers or simply laser diodes are extensively used in a range of laser devices intended for tactical military applications such as short and medium-range laser range finders, proximity sensors, short-range laser dazzlers and laser aiming devices, it needs to be emphasized here that they are also used as the optical pumping source for all Nd:YAG laser-based military laser systems. Laser diode electronics primarily comprises of a constant current source that can provide to the forward-biased laser diode the desired magnitude of current and also has inbuilt features to provide protection to the device against all those parameters it is adversely sensitive to. The other important circuit block is the temperature controller that can maintain the laser diode junction at the desired value of constant temperature irrespective of ambient temperature.

The need for a precise constant current source and constant temperature operation arises from the dependence of laser diode wavelength on drive current and operating temperature. Not all laser diode applications including military applications have stringent requirements for an ultra-stable wavelength necessitating a high degree of current and temperature stabilization, though for efficient and reliable operation, stability of drive current and operating temperature are always desirable. Another important aspect of laser diode drive and control electronics is the need for these drive circuits to have features that protect laser diodes against damage due to excessive drive current, electrostatic discharge and transients. Though laser diodes exhibit excellent reliability under ideal operating conditions, they are highly susceptible to damage due to excessive drive current, electrostatic discharge (ESD) and transients.

8.5.1 Constant Current Mode

There can be two types of circuit topologies the laser diodes can be operated in. These are constant current and constant output power modes. Operating temperature is of course an important factor in both modes. In view of dependence of laser diodes' performance on operating temperature and drive current, the preferred mode of their operation is the constant drive current with precise control of operating temperature. Constant current operation without temperature control is generally not desirable. Even at constant drive current, output power would increase with decrease in operating temperature. With significant decrease in temperature, output power could easily go past the absolute maximum value. The circuit topology in the case of constant current mode of operation is usually configured around a current sensing element that continuously senses the drive current and produces a proportional voltage. This is then compared with a reference voltage representing the desired value of drive current to generate

Figure 8.117 Constant current mode of operation.

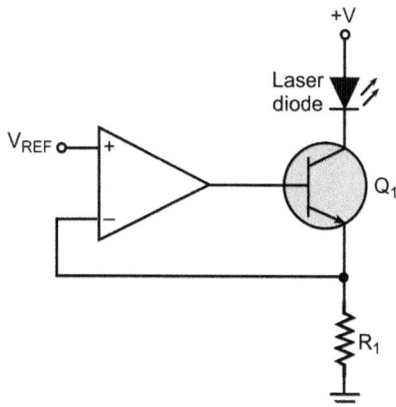

the error signal. The error signal after suitable conditioning is then fed back to restore the drive current to the desired value. Figure 8.117 shows the basic drive circuit to drive a laser diode in the constant current mode. In the circuit of Figure 8.117, the constant drive current equals V_{REF}/R_1. Practical circuits with inbuilt protection features are far more complex. Several constant current drive circuit topologies are discussed in the following sections.

8.5.2 Constant Power Mode

There are applications where constant output power is more relevant. In such a situation, direct dependence of output power on drive current can be exploited to maintain a constant output power. Even when a laser diode is driven by a constant current, the heat dissipated at the laser diode junction leads to rise in temperature and hence fall in output power. Increase in temperature could be because of absence of any active temperature control mechanism or even inadequacy of heat sink in the presence of active temperature control. This reduction in output power can be compensated by increasing the drive current. The current source can be designed in such a way that it adjusts the drive current in a feedback mode to maintain a constant output power instead of maintaining a constant drive current. Figure 8.118 shows the basic circuit topology. The basic circuit topology of a constant output power drive circuit is similar to that

Figure 8.118 Constant power mode of operation.

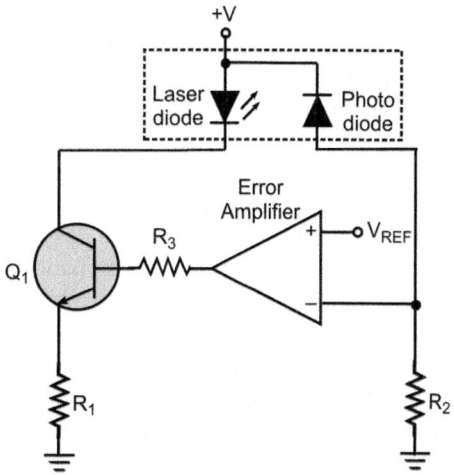

Figure 8.119 Laser diode with integral photodiode.

of a constant current drive circuit except for the nature of sense signal, which in the present case is a photo current proportional to the output power.

A constant output power drive circuit needs to be operated with an absolute current limit to prevent any thermal runaway problem caused by uncontrolled increase in drive current. Laser diode modules with an integral photodiode (Figure 8.119) facilitate constant output power operation, though noise intrinsic to the integral photodiode manifests itself in the form of noisy and unstable output in the case of constant output power mode of operation. The device usually has three terminals including either an anode or cathode of the laser diode, either an anode or cathode of the photodiode and a common terminal obtained by connecting the remaining terminals of the laser diode and photodiode. In some cases, all four terminals, two of the laser diode and two of the photodiode, are brought out.

8.5.3 Laser Diode Drive Circuit – Constant Current Mode

Figure 8.120 shows the basic constant current source circuit for grounded laser diode configuration. Laser diode current is sensed differentially by measuring voltage across R_{SENSE} wired in series with laser diode. Laser diode current in this case can be computed from $I_O = V_{IN} \times [R_4/(R_3 \times R_{SENSE})]$, $R_1 = R_3$ and $R_2 = R_4$. C_1 is the compensation capacitor connected for stable operation. V_{IN} may be derived from a band gap reference. In case the control voltage V_{IN} is of negative polarity, it is connected to R_1 and R_3 is grounded instead. A digitally controlled current source may use a voltage output digital-to-analogue converter to generate V_{IN}.

A similar circuit for a floating load is shown in Figure 8.121. Laser diode current in this case can be computed from $I_O = V_{IN} \times [R_2/(R_1 \times R_{SENSE})]$. The voltage controlled constant current drive circuits shown in Figures 8.121 and 8.122 provide constant current drive to the laser diode, provided the DC supply voltage, the voltage at the emitter terminal and the resistance in the emitter lead were all constant. It would be a reasonably good assumption if the source voltage were derived from a precision band gap reference; the resistors used had stability specifications equal to or better than the desired level of current stability. But when it comes to achieving a higher level of current stability, say ±10 ppm or better, a feedback loop that in situ samples the diode current and applies a correction in the case of drift in drive current becomes essential. A negative feedback loop of this kind would reduce the drift or error in the drive current by a factor that equals the loop gain.

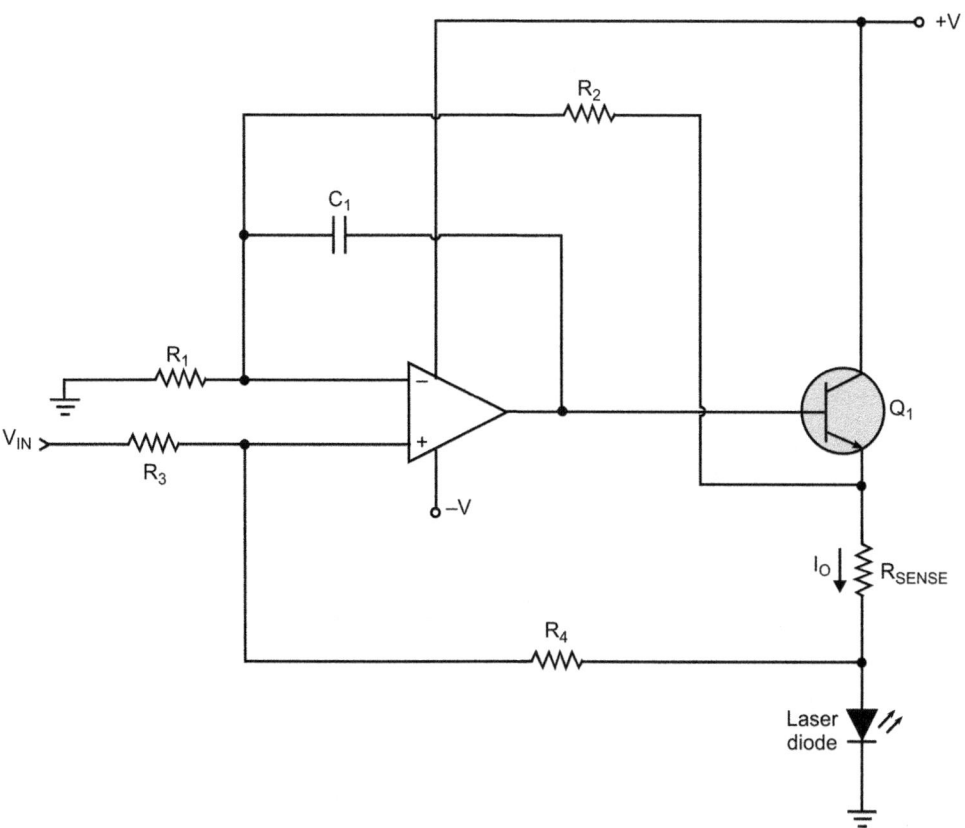

Figure 8.120 Constant current laser diode drive circuit for a grounded load.

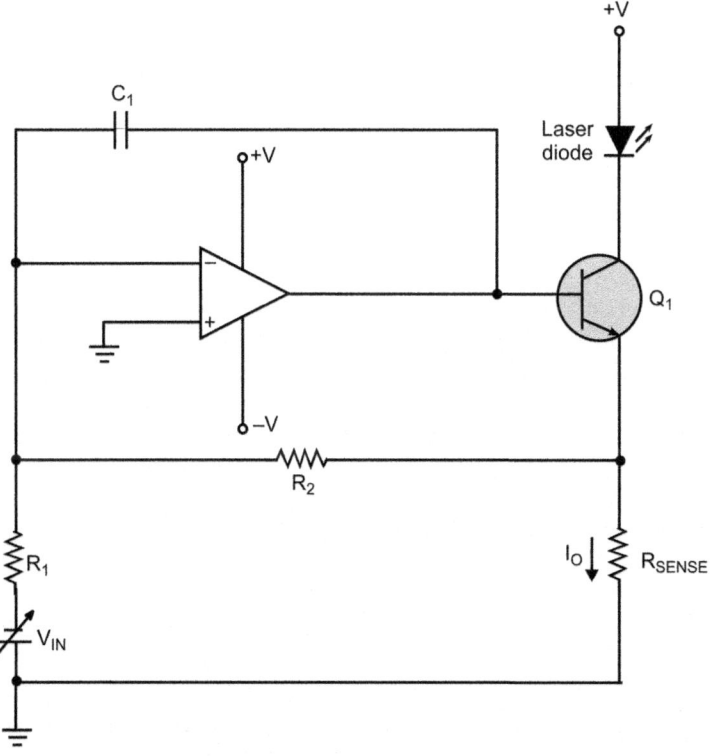

Figure 8.121 Constant current laser diode drive circuit for a floating load.

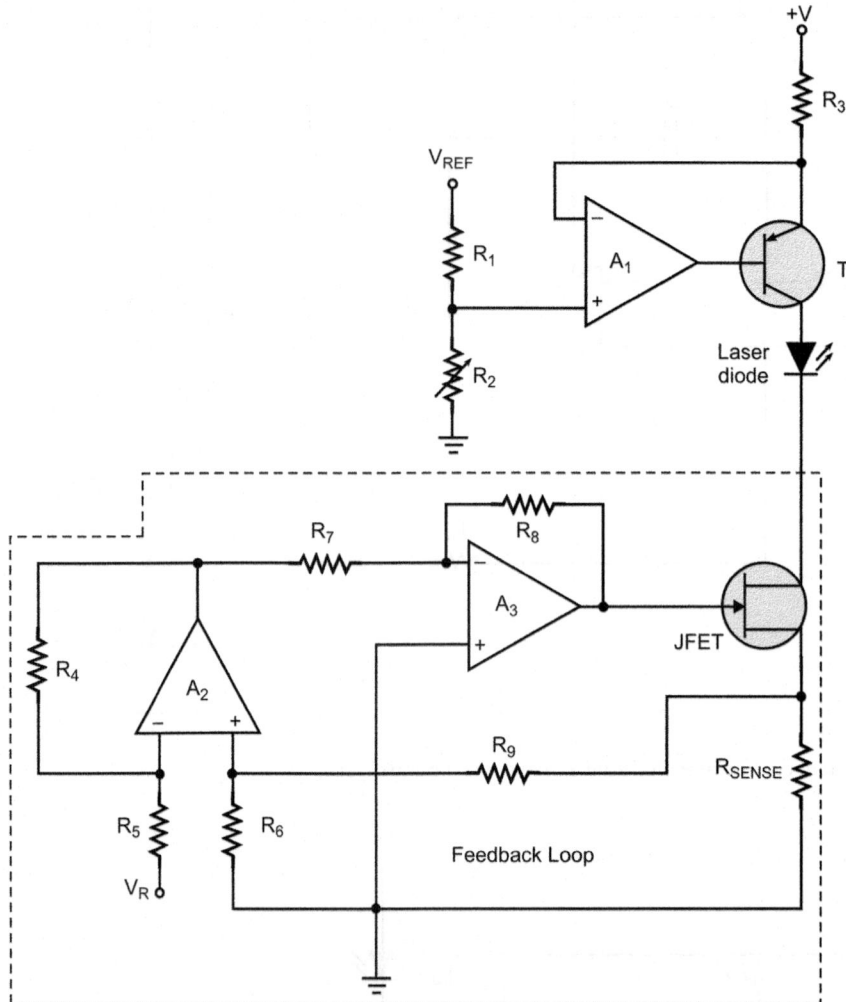

Figure 8.122 Laser diode driver with an in situ current monitor.

One such circuit is shown in Figure 8.122. The basic circuit topology is similar to the one used in driver circuits described in the previous pages. The only change is inclusion of a junction FET (JFET) connected in series with the sense resistor and wired as a voltage variable resistor. A small variation in drive current is compensated by an appropriate variation in the drain-source resistance of the JFET. Initially, at the nominal value of drive current, circuit parameters are so adjusted as to ensure that JFET with the feedback loop closed gets a negative gate voltage to keep it nearly in the middle of its VVR characteristics. This is done to exploit fully the voltage dependent resistance range. Precision constant current laser diode drivers are commercially available today from host of manufacturers for OEM applications. A representative photograph is shown in Figure 8.123.

Figure 8.124 shows a constant current laser diode drive circuit with modulation capability and inbuilt protection features. The protection features are further discussed in Section 8.5.7. Here, we shall briefly describe protection features provided by the drive circuit. The Schmitt comparator at the input provides a delay of the order of few tens of milliseconds after the

Figure 8.123 Precision constant current laser diode driver for OEM applications.

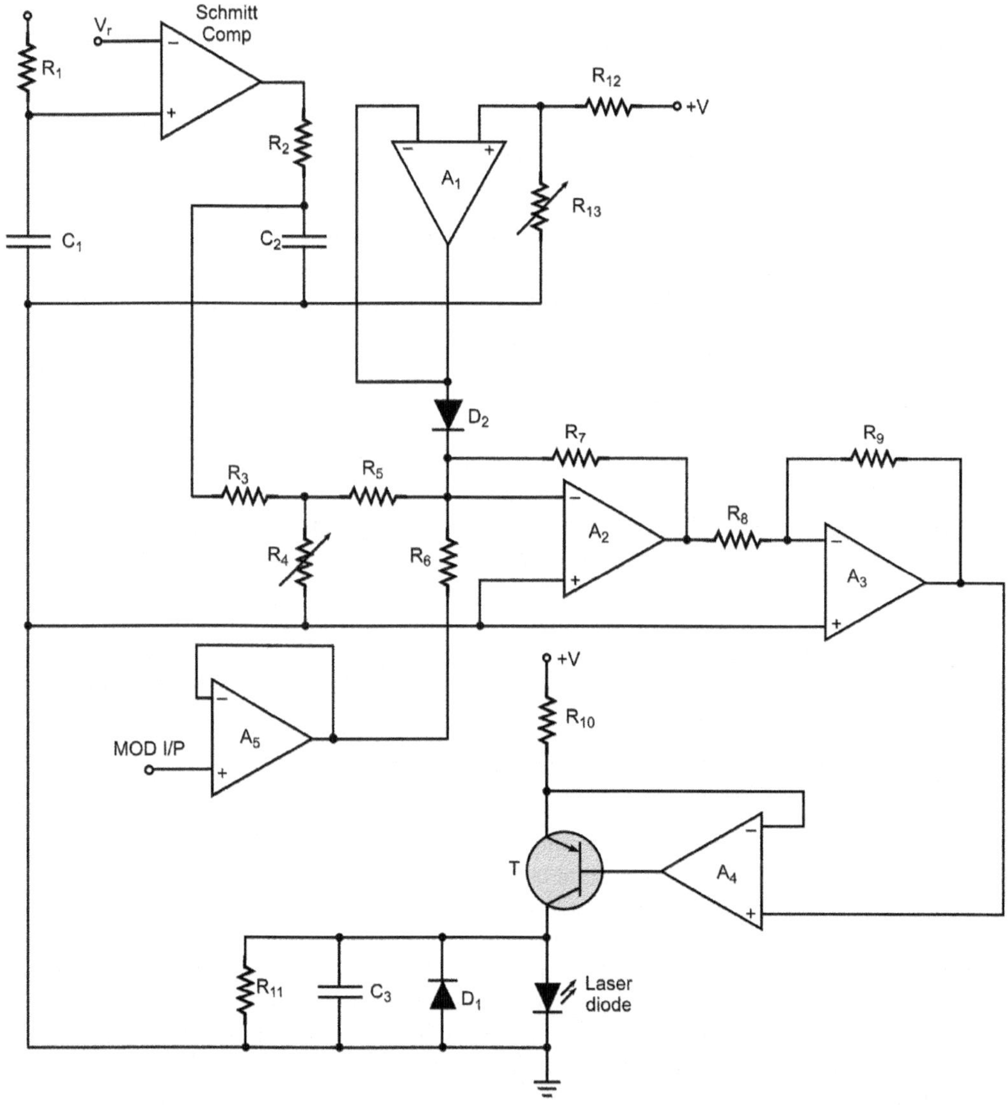

Figure 8.124 Constant current drive circuit with protection features.

switch-on to offer protection against switch-on transients. The time delay is decided by the $R_1 C_1$ time constant. R_2, C_2 provide the desired slow or soft start and decay during switch on and switch off. $R_2 C_2$ time constant is of the order of $1-2$ s. R_{11} and C_3 provide additional protection against transients; D_1 protects the laser diode against reverse voltages.

The constant value of drive current is decided by the voltage present at the non-inverting input of Opamp A_4, which equals the voltage present at the output of Opamp A_2 provided that $R_8 = R_9$. This is further equal to sum of voltage levels due to modulation input and reference voltage V_R after potential divider constituted by R_3 and R_4 and also assuming that $R_5 = R_6 = R_7$. This is true only if voltage V after potential divider arrangement R_{13} and R_{14} and appearing at output of Opamp A_1 did not interfere. This is true as long as diode D_2 is reverse biased, which it is normal operation. If the drive current exceeds a certain preset limit governed by voltage at the output of Opamp A_1, diode D_2 gets forward biased and clamps the voltage at the non-inverting input of Opamp A_4 to the voltage at the output of Opamp of A_1 thereby providing over current limit. The circuit may be modified to include a JFET and associated feedback loop components as shown earlier in Figure 8.122 to improve current stability. Figure 8.125 shows the typical circuit. The circuit is self-explanatory.

8.5.4 Laser Diode Drive Circuit – Constant Output Power Mode

A laser diode can be driven to maintain a constant output power by varying the drive current in accordance with the output power. The feedback loop circuit is such that it increases or decreases the drive current in response to decrease or increase in output power. Figure 8.126 shows the constant output power drive circuit. The integral photodiode produces current proportional to optical output power from the laser diode. This photocurrent is converted into a proportional voltage using a transimpedance amplifier configured around opampA4. The voltage representative of laser power is then summed up with a reference voltage representing the desired power level in an inverting summer configuration A_5. The output of the summer, which is null when the actual power level equals the desired power level, feeds an integrator A_6 that provides the correction signal even for infinitesimally small deviations from the desired power level. The integrator output summed up with a bias voltage feeds control element, which is a junction FET in this case. The drive current and hence the output power is governed by voltage present at non-inverting input of opamp A_3. Evidently, reduction in laser diode output power causes reduction in photodiode current, which finally leads to gate terminal of JFET becoming more negative. This further causes a reduction in voltage at A_3 non-inverting input increasing the drive current to restore the output power. Similarly, an increase in laser diode power reduces drive current to restore the power at nominal value.

8.5.5 Pulsed Mode Operation of Laser Diodes

There are two possible modes of operation of laser diodes to produce a pulsed output. In one of the operational modes of relevance to their use in military devices, such as laser range finders employing time-of-flight principle and laser proximity sensors, the laser diode is driven by current pulses that are a few tens of to a few hundred nanoseconds wide. In the other operational mode, called quasi-CW mode, laser diode is driven by current pulses that are typically hundreds of microseconds to few milliseconds wide. This operational mode is invariably used in the case of laser diode arrays pumping solid-state lasers including those for military range finders, target designators, EOCM and so on. Quasi-CW operation of laser diodes for optical pumping of solid-state lasers is at relatively low repetition rates of typically few Hz to few tens of Hz. This allows them to be operated at relatively high peaks powers, which is made possible due to low duty cycle of operation of quasi-CW devices, which keeps the average power low.

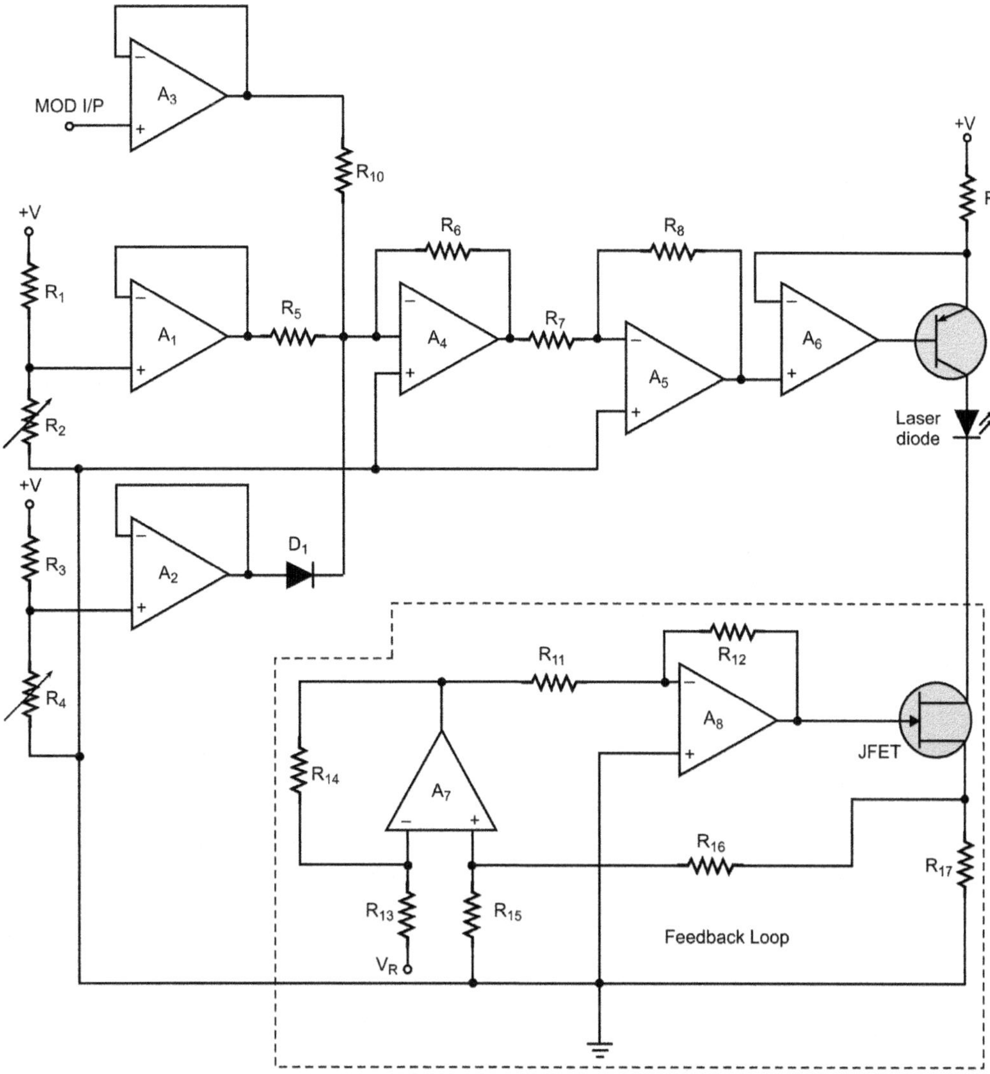

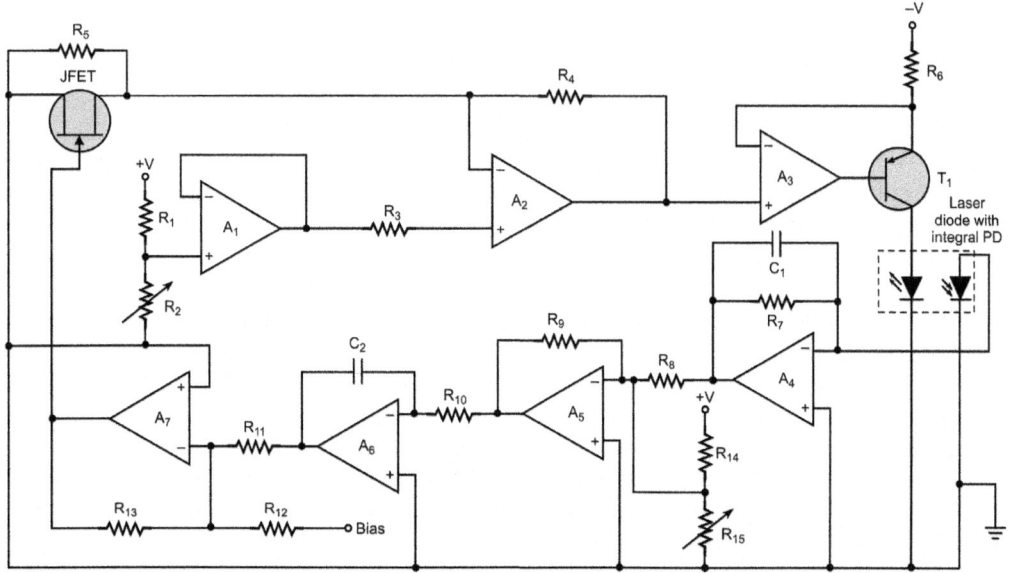

Figure 8.125 Laser diode driver with feedback control and protection features.

Figure 8.126 Laser diode drive circuit for constant output power.

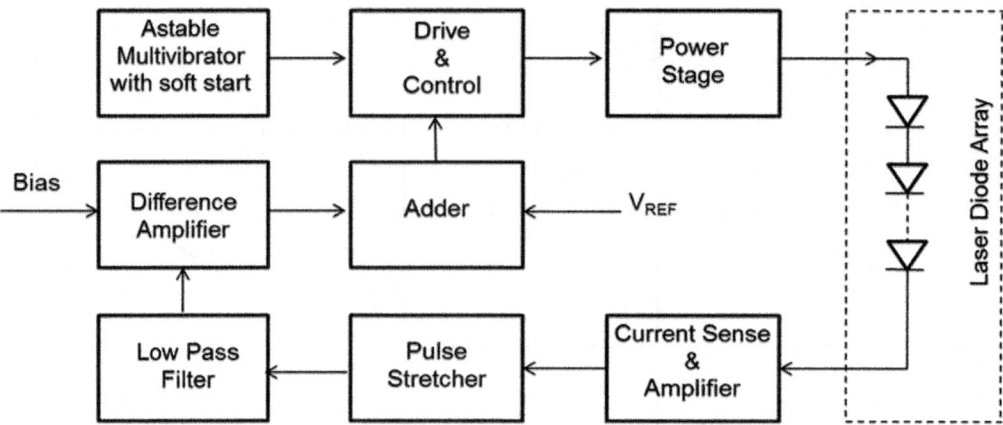

Figure 8.127 Laser diode driver for quasi-CW operation.

An important consideration while designing laser diode drive circuits for pulsed operation, conventional or quasi-CW operation, the laser diode/s must not be switched between cut-off and the nominal maximum value. This is highly detrimental to the life of laser diodes. They must always be operated between lower values slightly greater than the lasing threshold and the nominal value. Laser diode driver circuits for conventional pulsed and quasi-CW operation are configured around same basic building blocks as those described earlier in the case of CW operation. The laser diode drive circuit in Figure 8.124 can be used for operation in conventional pulsed mode by applying the desired pulsed waveform at the modulation input. The lower value of drive current is governed by an R_3-R_4 potential divider.

In the case of quasi-CW operation of laser diode arrays used for optical pumping of solid-state lasers, peak current pulse amplitude typically varies from few tens to few hundreds of amperes. The current drive stage in this case is usually a cascade arrangement of a drive and control stage and a power stage. Drive and control stage is a preamplifier stage with some kind of control feature that allows to vary the amplifier gain and hence the drive current to the power stage. Figure 8.127 shows a block-schematic arrangement of a typical quasi-CW laser diode drive circuit. The circuit operates as follows. The astable multivibrator produces a train of pulses having pulse width and repetition rate at which the laser diode or laser diode array needs to be driven. A soft start inbuilt into the circuit provides protection against damage due to instant power on. Power stage contains the active device/s and associated circuitry capable of delivering required current pulse amplitude.

The drive and control stage could be configured either around discrete bipolar transistors/ MOSFETs or an opamp with JFET as one of the gain determining elements. The power stage could be configured around one or parallel connection of more than one bipolar transistors or MOSFETs. The sensing element is usually a resistor connected in series with the diode or diode array. The peak pulse voltage across the sense resistor representing the peak amplitude of current pulse is amplified and then subsequently stretched. The stretched pulse train after filtering is fed to a differential amplifier stage where it is compared with a standard reference to produce an error voltage. The error voltage here represents deviation of peak amplitude of current pulse from the desired value. The error voltage is added to a fixed bias and fed to the control element in the drive stage to control the drive current to the power stage. Again, a large number of manufacturers internationally offer compact laser diode drivers for the OEM market catering to the requirements of CW, conventional pulsed and quasi-CW operational modes. Figure 8.128 shows a photograph of one such laser diode driver for quasi-CW operation. One cannot miss

Figure 8.128 Quasi-CW laser diode driver.

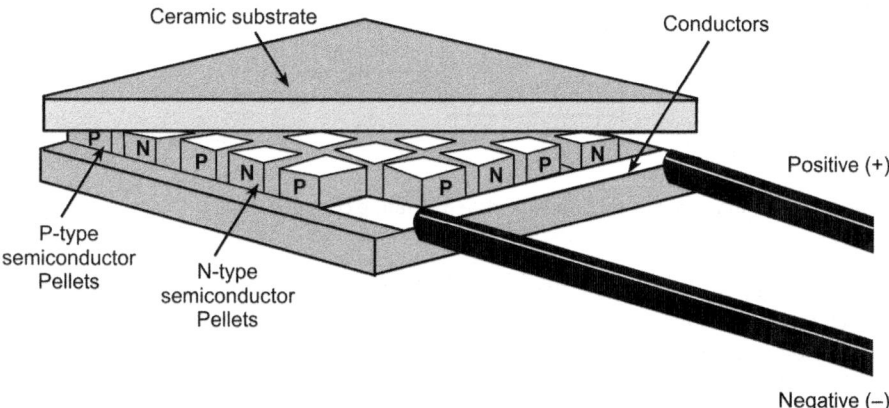

Figure 8.129 TE module construction.

the bank of capacitors connected across the DC supply input to provide high peak current drive capability required in such an operation.

8.5.6 Laser Diode Temperature Control Circuits

Laser diode junction gets heated due to power dissipated at the junction and one of the methods to remove heat away from the junction is by using a heat sink. This passive technique of heat removal using a heat sink becomes impractical for moderate to large power diode lasers. Also, it cannot be used to operate the laser diode at a temperature lower than the ambient temperature nor can it be used to stabilize the temperature. Use of an active temperature stabilization mechanism therefore becomes essential while working with moderate to high power laser diodes and also in low power laser diodes where application demands precise temperature control. A thermoelectric cooling device based on the Peltier effect is the heart of such a system. A practical TE module is usually a two-dimensional array of P-N couples connected electrically in series and sandwiched between two thermally conducting and electrically insulating faces (Figure 8.129). Both single and multiple stage TE cooler modules (Figure 8.130) are commercially available. Multiple stage modules offer higher cooling/heating capacity and maximum differential temperature specifications.

Figure 8.130 Single stage and multi-stage TE modules.

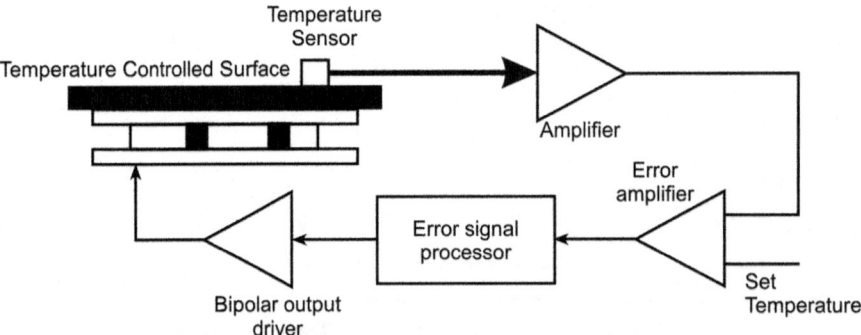

Figure 8.131 Schematic arrangement of 1 thermoelectric temperature control circuit.

A thermoelectric (TE) cooler based active temperature stabilization circuit uses TE module as the control element, a temperature sensor and a properly designed feedback loop. The TE cooler is a reversible solid-state heat pump whose operation is based on the Peltier effect. According to the Peltier effect, when electric current is passed through a junction of dissimilar metals, heat is created or absorbed at the junction depending upon the direction of flow of current. The heat transfer according to this effect takes place in the direction of flow of charge carriers.

A TE module literally pumps heat from one place to another heating one face and cooling the other in the process. Whether a particular face becomes hot or cold depends upon the direction of current flow through the TE module. Thus, by controlling the magnitude and direction of the current drive to the TE module, it can be used as a control element in a temperature stabilization circuit. The TE module pumps the heat dissipated in the laser diode on to a heat sink mounted on the supposedly hot face of the TE module. The heat is transferred from heat sink to the ambient. TE modules are characterized by maximal performance parameters namely maximum temperature difference ΔT_{MAX} at zero heat load, maximum heating capacity Q_{MAX} for zero temperature difference, device current I_{MAX} measured at ΔT_{MAX}, terminal voltage V_{MAX} corresponding to I_{MAX} with no heat load and coefficient of performance COP, which is a ratio of pumped heat load to electrical power supplied to the device.

Figure 8.131 shows a basic block-schematic of the TE module based temperature control circuit. Key building blocks of the drive and control circuit include temperature sensor, error amplifier, error signal processor and a bipolar output drive circuit. A bipolar output driver provides the required power to drive the TE module. The output stage in most cases is designed to allow flow of drive current in either direction through the TE module. This enables both cooling and heating of the device to maintain its temperature at the specified value regardless of ambient temperature being higher or lower.

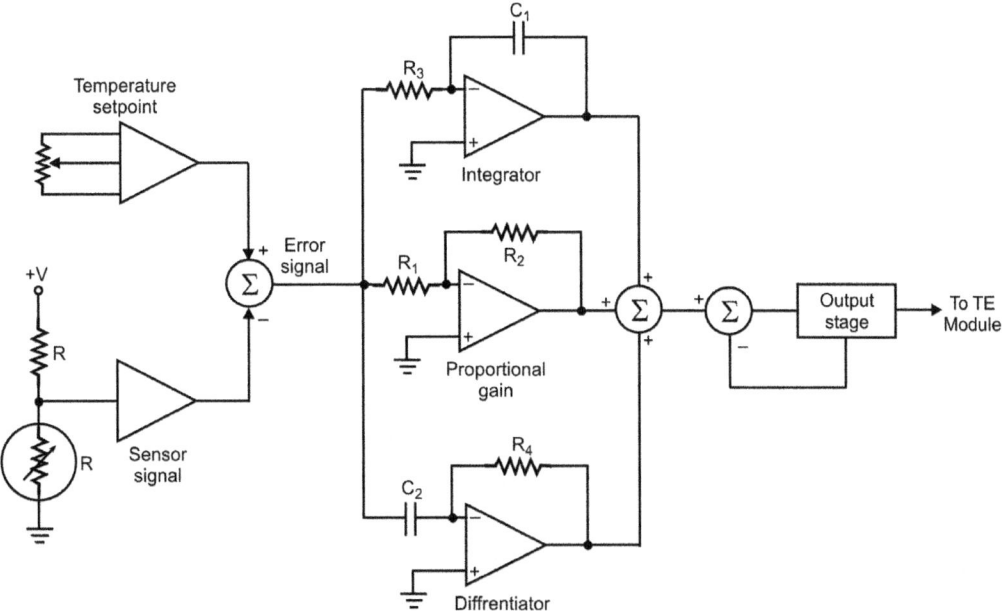

Figure 8.132 PID controller and other building blocks.

A negative temperature coefficient (NTC) thermistor is the most commonly used temperature sensor due to its high sensitivity. RTD and semiconductor sensors have better linearity but suffer from poor sensitivity. Semiconductor sensors are also available in integrated circuit form producing a current linearly related to absolute temperature. AD 590 is one such sensor. A balanced bridge circuit configuration that generates a differential output is generally used. An error amplifier is configured around an opamp. An error signal processor could be anything from simple on-off controller to a proportional controller or proportional-integral (PI) or even fully digital proportional-integral-differential (PID) controller. An on-off controller is never used in practice. In *proportional control* the drive signal to the TE module is proportional to the difference between the actual and desired temperatures. In a proportional controller there is always a residual error even after the controller has settled to the final state. This error is proportional to the difference between the desired temperature and the actual temperature and is inversely proportional to the gain of the control loop.

The problem of residual error of a proportional controller can be overcome by the addition of an integrator in the control loop. The result is a *proportional-integral controller.* One disadvantage of PI control is that it would be slow to respond to large residual errors. A *proportional-integral-differential controller* (PID) overcomes the problem of slow response to large residual errors encountered in PI controllers. Addition of the derivative term improves the loop's transient response. This type of controller is mainly used in applications where large thermal loads must be controlled rapidly and accurately. Figure 8.132 shows a typical PID controller interfaced with other building blocks including temperature sensor, reference generating circuit and error amplifier.

The output stage provides necessary drive power to the TE module. In most cases, electronic systems are designed to operate from a single positive DC voltage supply. Also, in most applications, TE modules need to be driven in the bipolar mode to cater for both heating and cooling operations. A commonly used circuit topology to provide bipolar drive to TE modules while

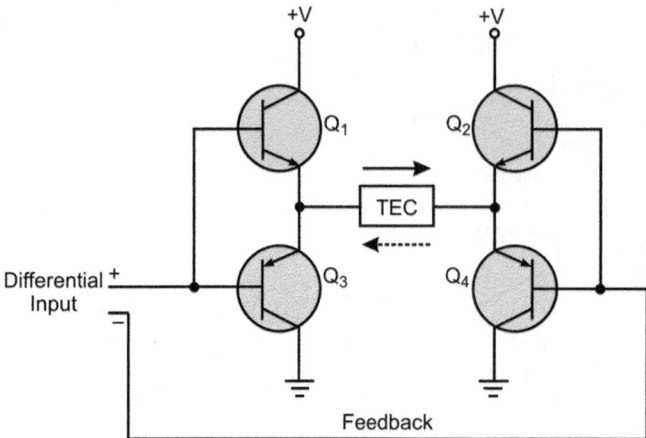

Figure 8.133 Output stage configured around a half-bridge circuit.

Figure 8.134 Integrated laser diode drive and temperature controller.

operating from a single DC voltage is the half-bridge circuit topology (Figure 8.133). The circuit is usually driven at the input by a driver amplifier stage with a differential output. Transistors Q_1–Q_4 and Q_2–Q_3 conduct alternately to provide bidirectional operation of TE module. A cascade connection of two half-bridge circuits may be used to enhance voltage and current drive capability of the output stage.

Thermoelectric drive and control modules for laser diode temperature control and stabilization are commercially available both as general-purpose bench top equipment as well as for OEM applications. Even integrated laser diode drive and temperature control modules are also available for OEM applications. Figure 8.134 shows a representative photograph.

8.5.7 Laser Diode Protection

Laser diodes exhibit excellent reliability under ideal operating conditions with life time often exceeding 100 000 hours. However, they are highly susceptible to damage due to excessive drive current, electrostatic discharge (ESD) and transients. Laser diode damage manifests itself in the form of one or more of the following effects, which includes reduced output power, shift in threshold current, increase in beam divergence and failure to laser action thus producing an LED-like output only. Laser diode protection strategy needs to be multi-pronged one with desired focus on laser diode instrumentation, system setup, power line conditioning and laser diode handling. Laser diode instrumentation includes current source used to drive the laser diode and temperature controller and the relevant damage mechanisms are over current, overheating, current spikes and power line surges. System setup includes cables, proper grounding and shielding and improper setup leads to radiated electrical transients. Improper power line conditioning also leads to severe fast transients. Electrostatic discharge caused by improper handling during storage, transport and mounting is considered as the single leading cause of premature laser diode failure. It may be noted that typical laser diodes have rise times in the range of tens of pico seconds and therefore are highly sensitive to fast events. Each of the four levels of laser diode protection strategy is briefly described in the following paragraphs.

8.5.7.1 Protection for Laser Diode Drive and Control Circuit

Laser diode instrumentation includes the *current source* used to drive the laser diode and the *temperature controller* used to maintain a constant operating temperature regardless of the ambient conditions. The current source should be so designed as to provide an independent adjustable limit of the drive current, guard against diode junction becoming reverse biased and suppress both power line and current transients. Temperature control circuit should not allow the current driving thermoelectric module used to control the temperature to exceed its maximum rated value. Also, the temperature control feedback loop should guard against any thermal oscillations. It may be noted that many important laser diode operating parameters such as operating wavelength, threshold current, electrical input to laser output efficiency and life time are strongly dependent on operating temperature and temperature stability.

Protection features desirable for laser diode current source include slow start, current limit, over voltage protection and power line transient suppression. Commercial laser diode drivers maintain a short circuit across the output leads to which the laser diode is to be connected by using a shorting relay or an FET switch. The shorting is maintained until the output is turned on and also when power to the instrument is off. Shorted output provides protection against ESD damage. Figure 8.135 shows the basic topology of a laser diode current source with protection features.

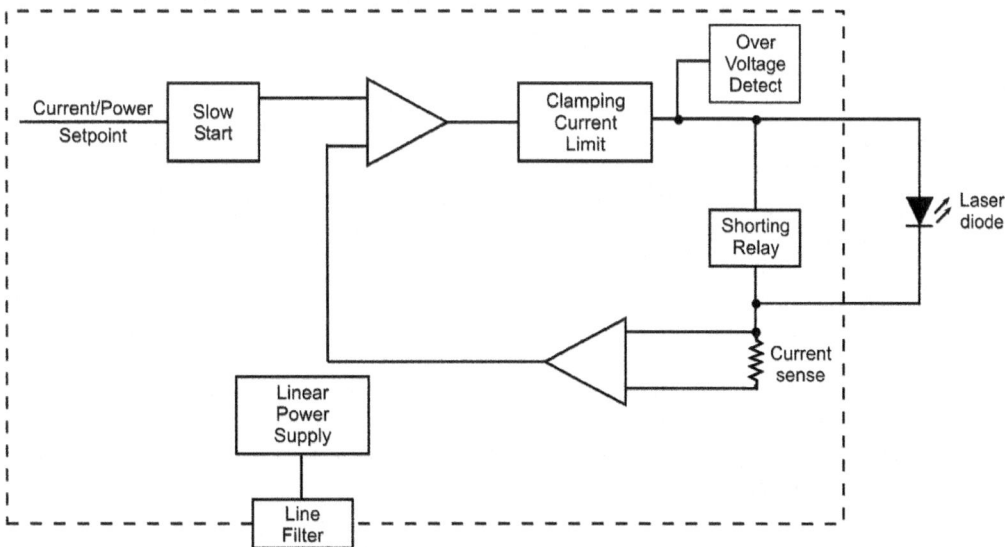

Figure 8.135 Laser diode current source with protection features.

Slow start protects the laser diode from turn-on transients by forcing the drive current through the laser diode to rise slowly to the final operating value. The start-up time is typically longer than 100 ms. Start-up time ensures that the control circuits are fully active and circuit transients have died out before the drive current reaches its operating value.

The current source should have an inbuilt *adjustable current limit* that is effective irrespective of whether it is operated in DC, pulsed or constant power output modes. The current limit should preferably be independent so when set it overrides any other condition that can lead to the drive current exceeding the maximum limit of the device. Excessive drive current leads to what is called catastrophic facet damage (CFD). Damage to diode at the output facet usually occurs when the optical energy density exceeds 10 000 W/mm^2. In a typical laser diode, the area of emitting surface for a given output laser power is such that the power density at the output facet is of the order of 500–700 W/mm^2. Although the material of active region is transparent in the pumped part of the active region, due to non-radiative combination, it becomes absorbing at the facet surface. Temperature rises faster than normal and this temperature increase produces further increase in absorption at the surface, which leads to further rise in temperature. This thermal runaway kind of occurrence causes damage to the facet and/or facet coating.

Over voltage protection is another important feature to be built into the current source designed to drive a laser diode. A constant current source can maintain a constant current through a load only up to a certain voltage across the load known as compliance voltage. Also, if the load impedance were to suddenly drop the current control loop would drive the output current to short circuit limit. The laser diode in this case will be over driven though for a brief period of time until the feedback control circuit adjusts itself to reduce the drive voltage. This short-term overdrive is generally sufficient to damage the device. The laser diode current source should be so designed as to guard against this condition.

Power line transients originate from a variety of sources, which include poor power conditioning, electronic equipment using switched mode power supplies and computers. The laser diode driver and temperature controller should be so designed as to efficiently suppress transients that make their way to the power input stage. An appropriate combination of filters could do the job. However, it may not be possible to prevent transients that are radiatively coupled into the laser or drive cables.

Protection needs of the *temperature controller* are also very stringent. The laser diode lifetime, as outlined earlier, is inversely proportional to the operating temperature. The lifetime has been observed to improve by an order of magnitude for every 30 °C reduction in the laser diode case temperature as shown in Figure 8.136. So, there is a great motivation to operate laser diodes at as low a temperature as possible. The output wavelength is also a strong function of junction temperature which, in turn, depends on the environmental temperature, power dissipated at the junction and relevant thermal resistances. Wavelength shifts by 0.4 nm for each unit change in temperature. Heat removal could make use of passive techniques where, at best, the case temperature can equal the ambient temperature or active techniques where there is no such limitation.

The TE module (also called the Peltier cooler) based temperature controller is used when the application demands a high degree of temperature stability. The current limit for the TE module should always be set below its maximum rating. Thermal oscillations should be avoided by appropriately setting the gain of the temperature controller's feedback loop. The chosen TE module should have a high current rating and the current limit for the TE module's drive circuit should be set just above the operating value.

To sum up, a laser diode drive circuit should be a current source and not a voltage source. The current source should have inbuilt protection features such as shorting output, independently adjustable current limit, over voltage protection, slow start and immunity to transients. Current

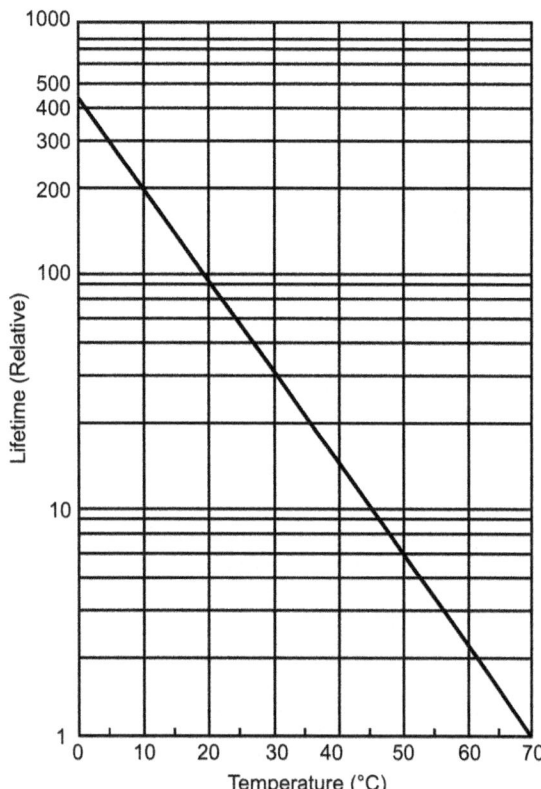

Figure 8.136 Laser diode lifetime versus operating temperature.

limit should be set with due consideration to recommendations of laser diode manufacturer. In the absence of any specific guideline on the desirable current limit it should be set to just above the expected maximum operating current. The laser diode mount should have adequate heat sinking capability and must not allow the laser diode to go to thermal runaway. The laser diode should be operated at the lowest temperature possible depending upon type of laser diode and intended application.

8.5.7.2 Interconnection Cables and Grounding

Use of proper cabling and grounding methodology is crucial to deciding susceptibility of laser diode instrumentation (laser diode driver, temperature controller) to radiated noise. Main sources of radiated noise include transformers, fluorescent lights, switched mode power supplies, high-speed data communication channels, motorized equipment, lightening and so on. Identification of all possible coupling mechanisms of the radiated transient noise, which in turn decides the radiated susceptibility, is usually a difficult proposition. Primary concerns though remain improper mounting, cabling and grounding methodologies. It may be noted here that proper laser diode mounting, choice of right cable type and correct cable routing and grounding practice are as important as having all protection features discussed in the preceding paragraphs built into the laser diode driver and temperature controller.

First and foremost, the precaution to be observed in the case of *cabling* is to make sure that the laser diode driver is turned off prior to connecting the laser diode. A loose cable connection could be very dangerous. Intermittent or a bouncing cable connection is seen by the laser diode driver as intermittent high impedance load with the result that it tries to drive the laser diode harder.

Both twisted pair and shielded cables can be used depending upon the application requirement. Twisted pair cable is preferred in the case of long cable runs. Twisted pair cable, on account of reduction in loop area and, consequently, the inductive term in the coupling equation, helps in reducing the low frequency noise that would have inductively coupled into the system. It may be noted here that twisted pair cable offers practically no resistance to high-frequency noise as the coupling mechanism in this case is capacitive. Use of a shielded cable helps in this case. Use of a shielded cable alone is not enough though. Improper shield termination often limits its effectiveness. The shield on the cable provides a low resistance path for the high-frequency noise to ground. As a result of this, it not only restricts radiation of energy from the internal conductor, more importantly, it prevents high-frequency noise radiated by other sources of noise from getting coupled into the system. Thus, it significantly improves both radiated emission as well as radiated susceptibility performance.

Also, a cable shield should never be used as a current carrying conductor. Both ends of the shield should preferably be terminated with a low inductance ground. In case it causes ground loop formation leading to possible inductive coupling of low frequency noise one of the ends of the shield, preferable on the diode mount side, should be connected to ground through a capacitor of few hundred pico-farads. The capacitor serves the purpose of breaking the loop for low frequency noise signals while maintaining a low-impedance path for high-frequency noise signals.

The *laser diode mount* also plays an important role in overall noise performance. If the mount were not properly grounded and/or the radiation environment were extremely harsh; the mount itself might act as an antenna and couple radiated noise signals from external sources into the system. Also, laser diode mount should be selected with due consideration to the type of laser diode package.

Grounding in the context of laser diode instrumentation serves two purposes. The first purpose is to reference the laser diode with respect to ground and the second is to provide a return path for the device current. Ground loops should be avoided. Ideally, it should be a single point ground connection. If multiple ground nodes were unavoidable, ground loop current should be minimized. With reference to second requirement, one should avoid using laser return path as a return path for other circuits. The final implementation of grounding methodology is a compromise between two conflicting requirements of avoiding ground loops on one hand and providing low inductance, high-frequency grounds to suppress fast transients. Multiple low inductance ground connections are preferred when confronted mainly with high-frequency noise. If low frequency noise were a bigger worry a single point ground connection should be preferred.

To sum up, with reference to mounting, cabling and grounding requirements, as far as possible the laser diode return path should be kept separate from other current paths; ground loops should be avoided; appropriate shielded cables that are properly fastened and terminated should be used to minimize radiated emission and susceptibility and the laser diode mount should not be such as to couple the radiated noise into the system.

8.5.7.3 Transient Suppression

Laser diodes are highly prone to damage due to all types of electromagnetic interference including fast transients. These transients are caused by equipment, such as soldering irons, motors, compressors and so on, requiring large power line surge currents and systems such as computers using switched mode power supplies. Though use of shielded cables and proper grounding helps take care of radiated transients, often fast transients beat the best defences. Immunity to fast transients can be further improved by avoiding sharing of power line used for laser diode driver with other equipment. Isolation transformers or/and surge suppressors should be used

Figure 8.137 Laser power output versus drive current for different values of reverse biased ESD stress.

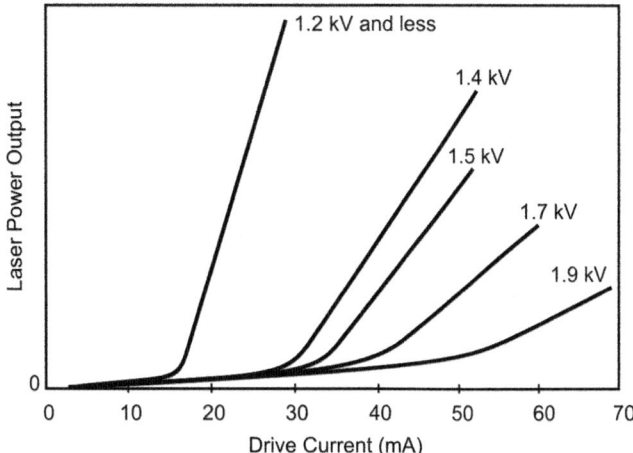

on the power line. In case you are forced to share the power line with other equipment, make sure that you use a separate surge protector for the laser diode driver.

8.5.7.4 Electrostatic Discharge

To ensure efficient and reliable operation of laser diode throughout its life time, proper handling of laser diodes during storage, transport and mounting is no less important than the design requirements of laser diode instrumentation discussed in the preceding paragraphs. Electrostatic discharge is considered the single most predominant cause of premature device failure. AlGaAs and InGaAsP laser diodes may be highly prone to damage by ESD voltages as low as 1200 V. Figure 8.137 shows a laser power output versus drive current characteristics of a typical InGaAsP laser diode for different values of reverse biased ESD stress. As is evident from the family of curves shown in Figure 8.137, there is an onset of failure at 1400 V. More often than not, ESD causes latent damage to the laser diode without any immediate visible symptoms. Consequent performance degradation may appear long after initial occurrence of ESD caused damage. Laser diodes can be protected from ESD damage by taking certain precautions, highlighted in the following paragraph.

Caution should be exercised while handling laser diodes. Use of anti-static gloves, wrist straps and clothing, grounded soldering stations and other equipment are strongly recommended. Also recommended is the use of anti-static floor coverings, ionized air blowers and dissipative work surfaces. Charge generating materials must be kept at least 30 cm away from the laser diode to prevent charge accumulation due to inductive coupling. When not in use and during transportation and storage, laser diodes should be fully enclosed in a conductive material. Also, when not in use, laser diode pins should be kept shorted by inserting them into conductive foam.

Illustrated Glossary

Analogue-to-Digital Converter This is used to transform the analogue signal into an equivalent digital signal that is processed digitally to extract the desired information. It forms an essential interface when it comes to analysing analogue data with digital circuits. It is an indispensable part of a digital communication system where the analogue signal to be transmitted is digitized at the sending end with the help of an A/D converter. It is invariably used in all optronic sensor systems where the optronic front-end converts the received optical radiation into an electrical signal that is analogue in nature.

Ballast Resistance (Gas Lasers) This is a fixed resistance connected from the output of the power supply in series with the plasma tube to ensure that the power supply sees an overall positive resistance while feeding the plasma tube whose I-V characteristics exhibit a negative resistance. An incorrect value of ballast resistance leads to a highly unstable plasma that has a tendency to extinguish. Choice of ballast resistance therefore is governed by the plasma tube for which the power supply is being designed.

Boost Regulator This, also called the step-up regulator, is based on the flyback principle. It resembles the basic flyback converter except that it is of non-isolating type. The energy storage and transfer element in this case is an inductor rather than a transformer. The output voltage is given by $V_O = V_{IN}/(1 - D)$ where D is the duty cycle of switching waveform.

Buck Regulator This resembles the conventional forward converter except for the fact that it does not use a transformer and there is no input-output isolation. Output voltage is always less than the input voltage and is given by $V_O = DV_{IN}$.

Capacitor Charging Power Supply (Solid-State Laser) This is used to charge an energy storage capacitor to a voltage so as to store the required quantum of energy per pulse to be delivered to the flash lamp in a flash-pumped solid-state laser.

Constant Current Source This delivers a predefined constant current to a load despite variations in the supply voltage and the load resistance.

Carbon Dioxide Laser Power Supplies These are operated in both CW and pulsed output modes. CW lasers are either DC excited or RF excited. Pulsed output lasers are mainly TEA (transversely excited atmospheric pressure) lasers. Power supply circuits for DC excited lasers are similar to the ones described previously for the case of helium-neon lasers. The basic power supply arrangement employed in the case of pulsed CO_2 lasers such as TEA lasers is to charge an energy storage capacitor to a high voltage from a high voltage DC power supply and then discharge the same through the gas mixture. The discharge process is controlled by a HV switch. Commonly used HV switches include spark gaps, thyratron tubes and solid-state switching devices. Power supply of an RF excited carbon dioxide laser comprises of an RF source and an impedance matching network. The RF source may further be split up into an RF oscillator and a cascade arrangement of RF amplifiers depending upon the output power delivering capability. Frequency of operation is in the range of 50–150 MHz.

Constant Current Mode (Semiconductor Diode Lasers) In the constant current mode of operation, a semiconductor diode laser is driven by a constant current source that maintains a constant drive current irrespective of power supply voltage and load impedance variations.

Constant Power Mode (Semiconductor Diode Lasers) In the constant power mode of operation, a semiconductor diode laser is driven by a variable current that maintains a constant output power. The output power is sensed and the drive current is increased or decreased corresponding to decrease or increase in output power.

Counter-Type A/D Converter A counter circuit feeds a D/A converter whose output feeds one of the inputs of a comparator with analogue signal to be digitized feeding the other input. The comparator output feeds the counter. The counter status at the time of comparator changing state is the digital output. The counter-type A/D converter provides a very good method for digitizing to a high-resolution. This method is much simpler than the simultaneous method for higher resolution A/D converters. The drawback with this converter is that the required conversion time is longer. Since the counter always begins from an all 0s position and counts through its normal binary sequence, it may require as many as 2^n counts before conversion is complete. The average conversion time can be taken to be $2^n/2 = 2^{n-1}$ counts. This makes counter-type of A/D converter unsuitable for digitizing rapidly changing analogue signals.

Current Mirror This circuit mirrors or copies the current flowing through an active device by controlling the current flowing through another active device. Conceptually, an ideal current mirror is nothing but an ideal current amplifier with a unity gain.

Current Regulating Diode This is a junction FET with a gate terminal internally shorted to the source terminal. The regulated current is given by drain-to-source saturation current (I_{DSS}).

Dither Stabilization (Frequency Stabilization of Gas Lasers) In the dither stabilization technique, laser cavity length is modulated by applying a low frequency sinusoidal signal to a PZT-mounted grating. The frequency modulation of laser produced due to cavity length modulation leads to amplitude modulation of laser output. This signal is phase sensitively detected with respect to the reference signal used to modulate laser cavity length and the resultant error signal is used in a feedback loop to force the laser operate at the centre of a Doppler broadened gain versus frequency curve.

Dual slope A/D converter A modified form of the single slope A/D converter to overcome shortcomings of the single slope A/D converter.

Excimer Laser Power Supplies Excimer lasers are excited by short duration high peak current discharge pulses with pulse rise time of the order of tens of nanoseconds and peak value of discharge current pulse in the range of few kilo amperes. Excitation of the active medium is usually a two-stage process. Energy is stored in a capacitor in a relatively slow process in one part of the circuit and then transferred to a peaking capacitor that enables a much faster discharge in another part of the circuit.

External Triggering (Flash Lamp) External triggering of a flash lamp is a technique in which the high voltage trigger pulse is applied to a trigger wire wrapped around the outside envelope of the flash lamp. In the case of liquid-cooled solid-state lasers, the trigger pulse may be applied to the metal laser cavity. The main advantage of the external trigger is that it does not interfere with the main energy discharge circuit. The disadvantage is that a high voltage trigger point is exposed and therefore needs to be properly isolated from the environment lest it causes problems at high altitude or in humid conditions. External triggering is recommended for low repetition rate, low-energy systems where the flash lamp is air cooled.

Field Programmable Gate Array (FPGA) An FPGA uses an array of logic blocks, which can be configured by the user. The word *programmable* here signifies that the device is programmable outside the factory where it is manufactured. Internal architecture of an FPGA device has three main parts, namely an array of logic blocks, programmable interconnects and I/O blocks.

Flash or Simultaneous A/D Converter This method of A/D conversion is based on using a number of comparators. The number of comparators needed for n-bit A/D conversion is $(2^n - 1)$. As the number of bits in the desired digital signal increases, the number of comparators required to perform A/D conversion increases very rapidly and it may not be feasible to use this approach once the number of bits exceeds 6 or so. The greatest advantage of this technique lies in its capability to execute extremely fast analogue-to-digital conversion.

Half-Flash A/D Converter This, also known as a pipeline A/D converter, is a variant of the flash type converter that overcomes to a large extent the primary disadvantage of requirement of a prohibitively large number of comparators in high-resolution full flash converters without significantly degrading its high-speed conversion performance. When compared to a full flash converter of a certain resolution, while the number of comparators and associated resistors reduces drastically in a half-flash converter, conversion time increases approximately by a factor of 2. For an n-bit flash converter, number of comparators required is 2^n [$(2^n - 1)$ for encoding of amplitude and one comparator for polarity], the same for an equivalent half-flash converter would be $2 \times (2^{n/2})$.

Helium-Neon Laser Power Supplies There are three commonly used circuit topologies for designing helium-neon laser power supplies. In one of the topologies, an AC-DC or DC-DC power supply producing an output voltage slightly higher than the voltage required sustaining the discharge plasma is cascaded to a voltage multiplier chain of diodes and capacitors to produce the starting voltage. The voltage at the output of multiplier chain drops to almost the voltage level present at the output of DC power supply due to inherently poor regulation of the multiplier chain. In the second commonly used circuit topology, the DC power supply is connected directly to the plasma tube through ballast resistance. A high voltage trigger pulse of 10–15 kV is applied to the tube to initiate the discharge. Once gas mixture is ionized, the DC power supply takes over to sustain the discharge. In the third topology, a constant power output fly back converter is used. The DC-to-DC converter is designed to produce an open circuit output voltage of 10–15 kV and a power output delivery rating equal to product of a little more than the required magnitude of DC voltage and desired discharge current. As a consequence, the converter output voltage rapidly drops to the desired value after the discharge is initiated.

Ion Laser Power Supplies Noble gas ion lasers operate at relatively lower voltage of the order of 200 V and much higher current values in the range of 10–100 A. The excitation mechanism in ion lasers is electron impact in the gas in a DC discharge with high current density in the range of several hundreds of A/cm^2. While a DC voltage of the order of 200 V is needed to sustain a steady discharge in an ion laser, it needs a higher voltage in the order of few kilovolts to initiate the discharge. This is provided by a starter circuit with the high voltage trigger transformer located inside the laser head.

Inverting Regulator This is based on the flyback converter principle. For a positive input, it produces a negative output. Energy is stored in an inductor during the conduction time of the transistor. The diode is reverse biased during this time period. The stored energy is transferred during the OFF-time. The circuit delivers a constant output power to the load. The output voltage is given by $V_O = -V_{IN} \times (T_{ON}/T_{OFF})$.

Linearly Regulated Power Supply This supply is one in which the active device, usually a bipolar junction transistor, is operated anywhere between cut-off and saturation. Commonly used regulator circuit configurations include emitter follower regulator, series-pass transistor regulator and shunt regulator. The regulator circuit in the power supply ensures that the load voltage (in the case of voltage regulated power supplies) or the load current (in the case of current regulated power supplies) is constant irrespective of variations in line voltage or load resistance.

Longitudinal Excitation (Gas Lasers) In the case of longitudinal excitation of gas lasers, the active medium is excited by passing an electric discharge current along the length of the tube.

Marx Bank Capacitor System In this system, the arrangement of individual capacitors is such that they are in parallel during the charging process and in series during the discharge operation. This means if each capacitor in the bank were charged to a voltage V, then the total voltage available across the tube for discharge operation would be nV, n being the number of capacitors. The Marx bank capacitor system is commonly used in TEA CO_2 laser power supplies.

Metal Vapour Laser Power Supplies The basic power supply arrangement for CW metal vapour lasers, such as helium-cadmium lasers, is similar to the power supplies used for helium-neon lasers. In the case of pulsed metal vapour lasers like copper vapour and gold vapour lasers, the basic power supply would be similar to the one used in TEA CO_2 lasers.

Microcontroller This may be considered a specialized computer-on-a-chip or a single-chip computer. The word 'micro' suggests that the device is small and the word 'controller' suggests that the device may be used to control one or more functions of objects, processes or events. It is also called an embedded controller as microcontrollers are often embedded in the device or system that they control. Microcontrollers and also Field Programmable Gate Arrays (FPGA) are the key components of embedded systems used to perform dedicated processing functions. Any device or system that measures, stores, controls, calculates or displays information is sure to have an embedded microcontroller as a part of the device or system. A microcontroller contains a simplified processor, some memory (RAM and ROM), I/O ports and peripheral devices such as counters/timers, analogue-to-digital converters and so on, all integrated on a single chip.

Optogalvanic Stabilization (Frequency Stabilization of Gas Lasers) The optogalvanic stabilization technique makes use of variation in discharge impedance caused by internal radiation field intensity that is in turn produced by cavity length modulation. Variation in discharge impedance is measured by driving the plasma tube by a constant current source. The voltage signal thus produced is phase sensitively detected with respect to reference signal and the error signal is used to stabilize frequency at line centre of gin versus frequency curve.

Over Voltage Triggering (Flash Lamp) This is a technique in which the DC voltage across the energy storage capacitor is sufficient to break down the gas inside the flash lamp and initiate the main discharge. The DC voltage appears across the flash lamp once the series-connected high voltage switch is turned on. The switch may be an SCR or a MOSFET or a triggered spark gap. Though the trigger circuit is simple, it is not common in solid-state lasers.

Parallel Triggering (Flash Lamp) This is a technique in which the secondary of the high voltage trigger transformer is connected in parallel. The circuit has all the advantages of series triggering and in addition uses a small trigger transformer. In this case, however, the secondary of the transformer needs to be isolated from the energy storage capacitor by using a capacitor or a diode. This method is rarely used due to prohibitively high cost of protection components.

Peak detector This produces at its output a voltage equal to the peak value of the signal applied at its input. The circuit could be designed to detect either the positive or the negative peak.

Peltier effect According to the Peltier effect, when an electric current is maintained in a circuit of material consisting of two dissimilar conductors, there is the cooling of one junction and the heating of the other. The effect is particularly pronounced in circuits comprising dissimilar semiconductors.

Programmable Interconnect Technologies These are techniques used for programming programmable logic devices. The interconnect techniques used for programming programmable logic devices (PLDs) include fuse, EPROM or EEPROM floating gate transistors, static RAM and anti-fuse.

Pseudo Simmer Mode of Operation (Solid-State Lasers) This is a slight variation from the traditional simmer mode of operation. In this mode, the simmer current flows only for a short time, starting a little ahead of the energy discharge operation. It has all the advantages of simmer mode operation and in addition saves on the power but at the expense of added circuit complexity.

Pulse Forming Network (Solid-State Laser) This is the interface between the capacitor charging power supply and the flash lamp. It is so designed as to produce a critically damped current pulse through the flash lamp when the energy storage capacitor is discharged through it. This the most efficient way of energy transfer, which also minimizes reverse voltage appearing across the capacitor.

Quad-Slope A/D Converter A modified form of a single slope A/D converter to overcome shortcomings and enhance performance.

Series Regulator In a series type linear regulator, the pass element is connected in series with the load and any decrease or increase in the output voltage is accompanied by a decrease or increase in the collector-emitter voltage of the series-pass element.

Series Triggering (Flash Lamp) This is a technique in which the secondary winding of the trigger transformer is in series with the energy storage capacitor. The secondary winding carries the flash lamp discharge current. The series trigger transformer is so designed that the transformer core saturates and the saturated secondary winding inductance serves the purpose of the PFN inductor also. Series triggering offers the advantages of reliable and reproducible triggering. Triggering is reliable even for low-energy storage capacitor voltages.

Shunt Regulator In the case of a shunt type linear regulator, regulation is provided by a change in the current through the shunt transistor to maintain a constant output voltage. The regulated output voltage in a shunt regulated linear power supply is the unregulated input voltage minus drop across a series resistance.

Sigma-Delta A/D Converter This employs a different concept from what is followed in the case of various other types of A/D converters. While other known A/D converters rely on sampling of analogue signal at the Nyquist frequency and encode the absolute value of the sample, in the case of sigma-delta converter, the analogue signal is over sampled by a large factor (i.e. sampling frequency is much larger than the Nyquist value) and also it is not the absolute value of the sample but the difference between the analogue values of two successive samples that is encoded by the converter. This leads to a higher S/N ratio and increased resolution for a given number of bits.

Simmer Power Supply (Solid-State Laser) This is a DC supply used to maintain a relatively low-amplitude keep-alive current through the flash lamp at all times irrespective of whether the lamp is flashing or not. The current varies typically from a few tens of milli-amperes to several hundreds of milli-amperes depending upon the characteristics of the flash lamp.

Single Slope A/D Converter Here, one of the inputs to the comparator is a ramp of fixed slope while the other input is the analogue input to be digitized. The counter and the ramp generator are initially reset to 0s. The counter starts counting with the first clock cycle input. The ramp is also synchronized to start with the first clock input. The counter stops when the ramp amplitude equals the analogue input. In this case, the counter count is directly proportional to the analogue signal. It is a low-cost A/D converter that has reasonable accuracy.

Stark Cell Stabilization (Frequency Stabilization of Gas Lasers) This technique is used for the frequency stabilization of carbon dioxide lasers. It allows dither free frequency stabilization. It also allows stabilizing frequency anywhere on the Doppler broadened gain versus frequency curve of the lasers. In this case, it is an external cell that is dithered rather than the laser cavity. Stark-shifted molecular transition of the gas contained in the cell precisely coincides with the carbon dioxide laser output line of interest. The detector signal at the output of Stark cell is phase sensitively detected with respect to reference signal used to drive Stark cell. The error signal is used to lock laser frequency to any point on Doppler broadened gain versus frequency curve of the laser.

Successive Approximation A/D Converter This aims to approximate the analogue signal to be digitized by trying only 1 bit at a time. This type of A/D converter is much faster than the counter-type A/D converter. In an n-bit converter, the counter-type A/D converter on an average would require 2^{n-1} clock cycles for each conversion whereas successive approximation type converter requires only n clock cycles.

Switched Mode Power Supply In a switched mode power supply, the active device that provides regulation is always operated in the switched mode,; that is, it is operated either in cut-off or in saturation region of the output characteristics. The input DC is chopped at a high frequency (typically 10–100 kHz) using an active device (bipolar junction transistor, power MOSFET, IGBT or SCR) and the converter transformer. The transformed chopped waveform is rectified and filtered. A sample of the output voltage is used as feedback signal for the drive circuit of the switching device to achieve regulation.

Thermoelectric (TE) Cooler A TE is a reversible solid-state heat pump whose operation is based on the Peltier effect. A TE module literally pumps heat from one place to another heating one face and cooling the other in the process. Whether a particular face becomes hot or cold depends upon the direction of current flow through the TE module. Thus, by controlling the magnitude and direction of the current drive to the TE module, it can be used as a control element in a temperature stabilization circuit.

Tracking Type A/D Converter This, also called a delta encoded A/D converter, is a modified form of counter-type converter that overcomes to some extent the shortcoming of the latter. In the tracking type A/D converter, the counter is an UP/DOWN counter and not an UP-counter as was the case in counter-type A/D converter. It counts in an upwards sequence whenever D/A converter output analogue voltage is less than the analogue input voltage to be digitized and it counts in the downwards sequence whenever D/A converter output analogue voltage is greater than analogue input voltage.

Transimpedance Amplifier This is a current-to-voltage conversion circuit. A typical application of Transimpedance amplifier is transforming the photo current generated in a photodiode as a result of light falling on it into an equivalent voltage.

Transverse Excitation (Gas Lasers) In transverse excitation of gas lasers, the active medium is excited by passing an electric discharge current perpendicular to the length of the laser tube.

Widlar Current Source A type of current mirror configuration.

Wilson Current Source A type of current mirror configuration.

Bibliography

1 Jung, W. (2004), *Opamp Applications Handbook*, Newnes.
2 Clayton, G. and Winder, S. (2003), *Operational Amplifiers*, Newnes.
3 Carter, B. and Mancini, R. (2009), *Opamps for Everyone*, Elsevier Inc.
4 Wilson, J. and Hawkes, J. (1997), *Optoelectronics: An Introduction*, Prentice Hall.
5 Ganguly, A.K. (2007), *Optoelectronic Devices and Circuits: Theory and Applications*, Alpha Science International Ltd.
6 Maini, A.K. and Agrawal, V. (2009), *Electronic Devices and Circuits*, Wiley India Pvt, Ltd.
7 Billings, K.H. and Morey, T. (2010), *Switched Mode Power Supply Handbook*, McGraw-Hill.
8 Bird, B.M., King, K.G. and Pedder, D.A.G. (1992), *An Introduction to Power Electronics*, John Wiley & Sons, Ltd.
9 Brown, M. (2001), *Power Supply Cookbook*, Newnes.
10 Carr, J.J. (1996), *DC Power Supplies: A Technician's Guide*, TAB Books.
11 Chetty, P.R.K. (1986), *Switched Mode Power Supply Design Handbook*, TAB Professional and Reference Books.
12 Davis, B. (1996), *Understanding DC Power Supplies and Oscillators*, Prentice-Hall.
13 Ferencz, I.O. (1987), *Power Supplies, Part A and Part B*, Elsevier.
14 Gottlieb, I.M. (1993), *Power Supplies, Switching Regulators, Inverters and Converters*, TAB Books.
15 Mazda, F.F. (1997), *Power Electronics Handbook*, Newnes.
16 Pressman, A.I., Billings, K.H. and Morey, T. (2009), *Switching Power Supply Design*, McGraw-Hill Education.

17 Yariv, A. (1991), *Optical Electronics*, Saunders College Publishing.

18 Verdeyen, J.T. (1995), *Laser Electronics*, Pearson.

19 Gagnon, W., Albrecht, G., Trenholme, J. and Newton, M. (2008), *Pulsed Power for Solid State Lasers*, Lawrence Livermore National Laboratory LLNL-Book-400175.

20 Injeyan, H. and Goodno, G.D. (2011), *High Power Laser Handbook*, H. Injeyan and G.D. Goodno, (eds), McGraw-Hill.

21 Webb, C.E. and Jones, J.D.C. (2003), *Handbook of Laser Technology and Applications: Volume I, II and III*, CRC Press.

22 Koechner, W. (2006), *Solid State Laser Engineering*, Springer.

23 Csele, M. (2004), *Fundamentals of Light Sources and Lasers*, John Wiley & Sons, Ltd.

24 Sun, H. (2012), *Laser Diode Beam Basics, Manipulations and Characterizations*, Springer.

25 Numai, T. (2004), *Fundamentals of Semiconductor Lasers*, Springer Series in Optical Sciences, Springer.

26 Epperlein, P.W. (2013), *Semiconductor Laser Engineering, Reliability and Diagnostics: A Practical Approach to High Power and Single Mode Devices*, John Wiley & Sons, Ltd.

9

Photo Sensors and Related Devices

Optoelectronics is mainly related to the study of electronic devices that emit and detect light. These devices are collectively referred to as optoelectronic devices. Commonly used optoelectronic devices include photo emitters, photo sensors and displays. While photo emitters are electrical-to-optical transducers that are used to convert the electrical energy into output light, photo sensors are optical-to-electrical transducers used for converting the incident light energy into an electrical signal. Light-emitting diodes and displays constitute the important devices in the category of photo emitters. Photoconductors, photo diodes, photo transistors, photo multiplier tubes and image intensifiers are some of the commonly used photo sensors. Photo sensors have a significant role to play in the design and development of lasers and laser-based systems. They constitute the heart of a variety of systems ranging from the simple gadgets like light meters to the most complex of military systems like precision-guided munitions, laser range finders, target trackers and so on, remote sensing systems, from fibreoptic and laser-based communication applications to spectrophotometry and photometry applications and so on. This chapter discusses in detail the fundamentals and application circuits of different types of photo emitter, photo sensors and related optoelectronic devices with established military applications. The chapter begins with a broad classification of optoelectronic devices including photo emitters and photo sensors, followed by the definition of commonly used radiometric and photometric terms. Different types of photo sensors and photo emitters are discussed next in terms of principle of operation, characteristic parameters and application circuits. The chapter concludes with a description of night vision technologies and related military devices. Both image enhancement type and thermal imaging type night vision devices are discussed in detail with reference to involved technologies, representative military devices and applications.

9.1 Classification of Photo Sensors

Optoelectronic devices are broadly classified as photo emitters, photo sensors and optocouplers. Photo emitters are further grouped as light-emitting diodes (LED), lasers and displays. Photo sensors are classified into two major categories, namely (1) photo electric sensors and (2) thermal sensors. *Photo electric sensors* can be further classified in to two types; the devices that depend on the external photo effect for their operation and the devices that make use of some kind of internal photo effect. The two types of devices together include the following.

1) Photo emissive sensors
2) Photoconductors
3) Junction type photo sensors

Handbook of Defence Electronics and Optronics: Fundamentals, Technologies and Systems, First Edition. Anil K. Maini.
© 2018 John Wiley & Sons Ltd. Published 2018 by John Wiley & Sons Ltd.

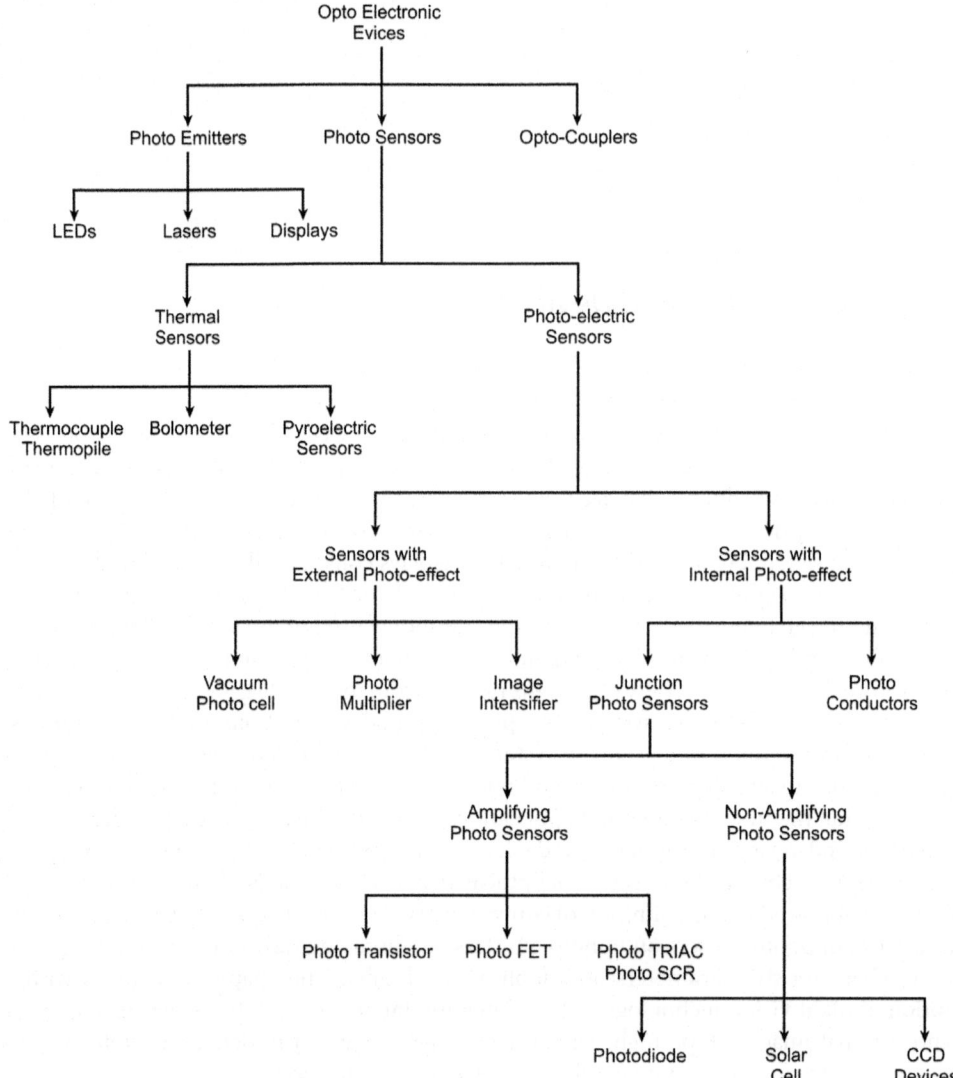

Figure 9.1 Classification of photo sensors.

Of the three types mentioned, *photo emissive sensors* belong to the first category based on the external photo effect. The other two device categories use an internal photo effect. Common photo emissive sensors include non-imaging sensors such as vacuum photo cells and photo multiplier tubes and imaging sensors such as image intensifier tubes. Photoconductors are bulk semiconductor devices whose resistance decreases with increase in incident light intensity. They are also known by name of *photo resistors, light dependent resistors* and *photo cells. Junction type photo sensors,* further fall into the amplifying or non-amplifying type. The amplifying type of junction photo sensors include photo transistors, photo thyristors and photo FETs. Non-amplifying types of junction photo sensors include photo diodes, solar cells and charge coupled devices. A charge coupled device is also a type of imaging sensor. In the category of thermal sensors, we have (1) thermocouple (or thermopile) type sensors, (2) bolometric sensors and (3) pyroelectric sensors.

Figure 9.1 shows a classification of optoelectronic devices in general and photo sensors in particular. While thermal sensors absorb incident radiation and operate on the resulting temperature rise, photo electric sensors, on the other hand, are based on quantum effect. Thermal sensors are relatively sluggish in their response to the incident radiation than the photo electric sensors. Thermal sensors, however, offer a much wider operational wavelength band than the photo electric sensors.

9.2 Radiometry and Photometry

Radiometry is the study of properties and characteristics of electromagnetic radiation and the sources and receivers of electromagnetic radiation. Radiometry covers a wide frequency spectrum; however, for the present chapter we will limit our discussion to infrared (IR) and ultraviolet (UV) frequencies.

Photometry is the science that deals with visible light and its perception by human vision. The most important difference between radiometry and photometry is that in radiometry, the measurements are made with objective electronic instruments whereas in the case of photometry, measurements are done with reference to the response of human eye. In this section, we define the commonly used radiometric and photometric quantities.

9.2.1 Radiometric and Photometric Flux

Flux is defined as a flow phenomenon or a field condition occurring in space. It is a measure of the total power emitted from a source or incident on a particular surface. The symbol for radiometric flux is ϕ_R and for photometric flux it is ϕ_P. Radiometric flux or luminous flux is measured in watts (W) while the photometric flux is measured in lumens (lm). Lumen is defined as the amount of photometric flux generated by 1/683 W of radiometric flux at 555 nm where the photo opic vision sensitivity of eye is the maximum.

Efficacy of a radiation source is defined as the ratio of photometric flux to the total radiometric flux from the source. It is given by eqn. 9.1.

$$K = \frac{\phi_P}{\phi_R} \tag{9.1}$$

Where,

K = Efficacy (lm/W)
ϕ_P = Photometric flux (lm)
ϕ_R = Radiometric flux (W).

9.2.2 Radiometric and Photometric Intensity

Intensity function describes the flux distribution in space. Radiometric Intensity I_R is defined as the radiometric flux density per steradian. It is given by eqn. 9.2 and is expressed in watts per steradian (W/Sr).

$$I_R = \frac{\phi_R}{\Omega} \tag{9.2}$$

Where,

ϕ_R = Radiometric flux (W)
Ω = Solid angle (Sr).

Photometric or luminous Intensity I_P is defined as the ratio of luminous flux density per steradian. It is given by eqn. 9.3.

$$I_P = \frac{\phi_P}{\Omega} \tag{9.3}$$

Where,

ϕ_P = Photometric flux (lm)
Ω = Solid angle (Sr).

The unit of photometric flux density is the Candela (Cd) and is equal to luminous flux density of one lumen per steradian (lm/Sr).

9.2.3 Irradiance and Illuminance

Irradiance, also called *Radiant Incidence* (E_R) defines the radiometric flux distribution on a surface. It is expressed as

$$E_R = \frac{\phi_R}{A} \tag{9.4}$$

Where,

ϕ_R = Radiometric flux (W)
A = Area of flux distribution (m^2).

Illuminance (E_P) defines the photometric flux distribution on a surface and is expressed as

$$E_P = \frac{\phi_P}{A} \tag{9.5}$$

Where,

ϕ_P = Luminous flux (Lu)
A = Area of flux distribution (m^2).

Two very commonly used units to define Illuminance are lux and foot-candle. Lux is defined as the illumination of one lumen of luminous flux evenly distributed over an area of 1 m^2. Foot-candle is an old English unit and is defined as an illumination of one lumen of luminous flux evenly distributed over an area of 1 square foot. A measurement of 1 foot-candle is equal to 10.764 lux.

9.2.4 Radiance and Luminance

Radiance, also called radiant sterance, is defined as the ratio of luminous flux per unit solid angle per unit area. Its units are W/Sr/m^2. Luminance is defined as the ratio of luminous flux per unit solid angle per unit area. It is expressed as lm/Sr/m^2.

9.3 Characteristic Parameters

Major characteristic parameters used to characterize the performance of photo sensors include the following.

1) Responsivity
2) Noise-Equivalent Power (NEP)

3) Sensitivity usually measured as detectivity and D-star
4) Quantum efficiency
5) Response time
6) Noise

9.3.1 Responsivity

Responsivity is defined as the ratio of electrical output to radiant light input determined in the linear region of the response. It is measured in amperes per watt (A/W) or V/W if the photo senor produces a voltage output rather than a current output. Responsivity is a function of wavelength of incident radiation and band gap energy. Spectral response is a related parameter. It is a curve that shows variation of responsivity as a function of wavelength. Most photo electric sensors have a narrow spectral response whereas most thermal sensors have wide spectral response. As an example, the spectral responses of silicon, germanium, indium gallium arsenide photo diodes are in the range of 200–1100, 500–1900 and 700–1700 nm, respectively, whereas that of thermistors is from 0.5–10 μm.

Silicon photo diodes exhibit a response from the UV through the visible and into the near IR part of the spectrum. With the band gap energy of silicon being 1.12 eV at room temperature, its spectral response peaks in the near IR region between 800 and 950 nm. Peak responsivity figures for silicon PIN photo diodes are in the range of 0.4–0.6 A/W whereas for avalanche photo diodes, it is in the range of 40–80 A/W. Thermal sensors have poorer responsivity as compared to photo electric sensors. As an example, the responsivity figure for pyroelectric sensors is in the range of 0.5–5 μA/W.

The shape of the spectral response curve of silicon photo sensors, particularly in the blue and UV part of the spectrum, can be altered by choosing an appropriate manufacturing process. Figure 9.2 shows typical spectral response curves of normal, blue enhanced, UV enhanced and low noise silicon photo diodes. The low noise spectral response corresponds to photo voltaic mode of operation of photo diodes where no external bias is applied to the photo diode. Since dark current is a function of bias magnitude, photo voltaic mode of operation

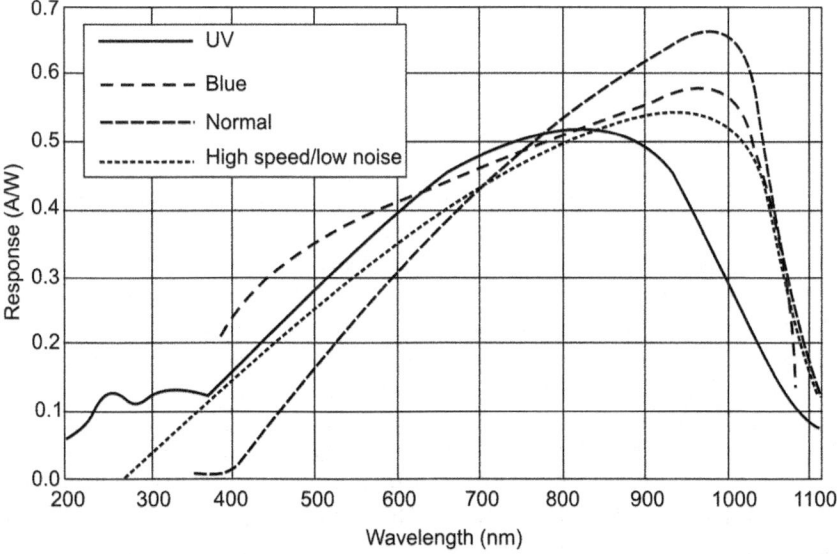

Figure 9.2 Spectral response of a silicon photodiode.

eliminates dark current as a source of noise. In this case, the NEP will be lower, thereby allowing greater sensitivity at lower wavelengths but at the cost of slightly lower responsivity at higher wavelengths.

Silicon becomes transparent to radiation of longer than 1100 nm in wavelength. On the contrary, wavelengths in the UV region are absorbed in the first 100 nm thickness of the silicon. Even the most careful surface preparation leaves some surface damage which reduces the collection efficiency for this wavelength. High absorption coefficient of silicon in blue and UV enhanced spectral region leads to generation of carriers within heavily doped p + (or n+) contact surface of P-N and PIN photo diodes. This causes rapid decrease in quantum efficiency in this region due to shorter life time and surface recombination. The spectral response in blue and UV enhanced region is enhanced in photo diodes by minimizing near-surface recombination. This is achieved by using techniques such as thin and highly graded p + (or n + or metal Schottky) contacts, lateral collection to minimize percentage of heavily doped surface area and using passivation layer. A passivation layer such as silicon nitride, silicon dioxide or titanium dioxide can reduce surface recombination.

Anti-reflection coatings may be used to enhance the responsivity to about 25% at the required wavelength at the cost of reduction in efficiency at other wavelengths that they reflect. The package window also plays an important role in shaping the spectral response. The standard glass window absorbs wavelengths shorter than 300 nm. For UV detection, a fused silica or UV transmitting glass window is used. Various filter windows are also available to tailor the spectral response to suit the application. Optical filters can also be added to change the spectral response. A common example is the use of a specific filter to modify the normal silicon response to approximate the spectral response of the human eye.

In the case of photo diodes, the responsivity is typically highest in a region with photon energies slightly greater than the band gap energy. It declines sharply for photon energies in the region of the band gap, where the absorption decreases. It can be calculated by using eqn. 9.6.

$$R_v = \frac{\eta \times e}{h \times \upsilon} \tag{9.6}$$

Where,

h = Planck's constant (=6.625 × 10^{-34} J-s)
υ = Frequency of incident radiation (Hz)
η = quantum efficiency
e = electron charge (=1.6 × 10^{-16} C).

Responsivity increases slightly with applied reverse bias due to improved charge collection efficiency in the photo diode. Responsivity also exhibits dependence on temperature variations due to variation in band gap energy. The band gap energy varies inversely as the change in temperature. Figure 9.3 shows change in responsivity with temperature. As is evident from the family of curves shown in the figure, responsivity is more or less independent of temperature from 500 to 900 nm. For wavelengths less than 500 nm, responsivity decreases or increases gradually with increase or decrease in temperature respectively. For wavelengths greater than 900 nm, it increases or decreases rapidly with increase or decrease in temperature, respectively. This can be explained from the nature of temperature dependence of band gap energy.

The term *responsivity* should not be confused with *sensitivity*; the latter is the lowest detectable light level that is typically determined by detection noise. It is also significantly influenced by the required detection bandwidth. Also, a photo senor should ideally be operated in a spectral region

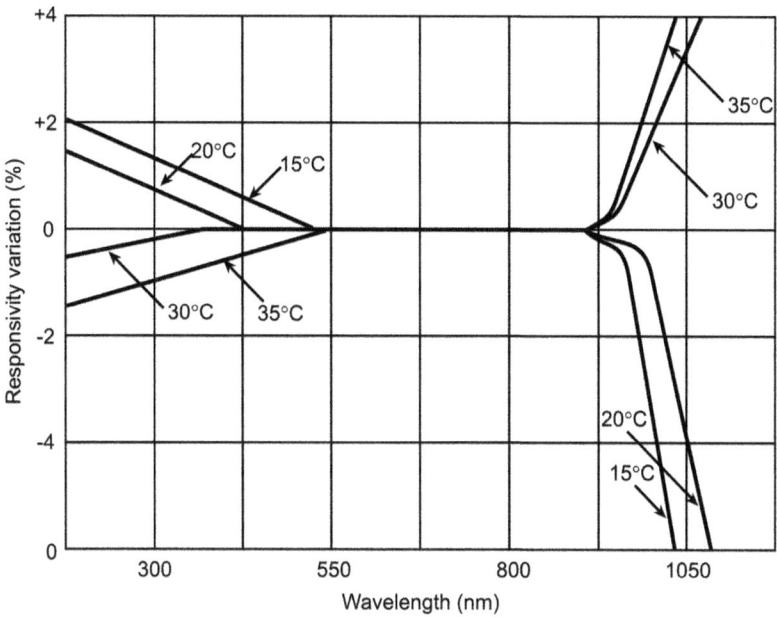

Figure 9.3 Temperature dependence of spectral responsivity.

where its responsivity is not far below the highest possible value, because this leads to the lowest possible detection noise and thus to a high signal-to-noise and high sensitivity.

9.3.2   Noise Equivalent Power (NEP)

NEP is the input power to a sensor that generates an output signal current equal to the total internal noise current of the device, which implies a signal-to-noise ratio of one. In other words, it is the minimum detectable radiation level of the sensor. Obviously, a low NEP is desirable. NEP depends on the wavelength, since that influences the responsivity of the sensor. The lowest NEP is achieved for those wavelengths where the responsivity is the highest.

$$NEP = \frac{I_N}{R_v} \tag{9.7}$$

Where,

I_N = Total noise current (A)
R_v = Responsivity in (A/W).

For a given value of responsivity, NEP depends upon magnitude of noise current. For zero and extremely low bias voltages, the Johnson noise current usually dominates over shot noise current. At larger reverse bias voltages, shot noise current is significantly higher than the Johnson noise current. As a consequence, NEP is observed to be higher for relatively large reverse bias voltages.

The noise power and thus the noise-equivalent power depend on the assumed detection bandwidth. If one were to use the full detection bandwidth of the device to compute NEP; then the NEP would not allow a fair comparison of sensors with different bandwidths. Therefore, it is a common practice to assume a bandwidth of 1 Hz, which is usually far below the detection bandwidth. NEP is usually specified in units of W/√Hz rather than watts. In that case, noise current is specified as amperes per square root of Hertz (A/√Hz). Effectively, computation of NEP is based on power spectral density (PSD) rather than on power. Power is nothing but the power spectral density computed over a bandwidth of 1 Hz. Since the photo diode light power to current conversion depends on wavelength, the NEP is quoted at a particular wavelength. Also, the NEP like responsivity is nonlinear over the wavelength range.

9.3.3 Detectivity and D-Star

Detectivity of a sensor is the reciprocal of its NEP. A sensor with higher value of detectivity is more sensitive than a sensor with lower detectivity value. Detectivity, like NEP depends upon noise bandwidth and sensor area. To eliminate these factors, a normalized figure of detectivity referred to as D-star is used. It is defined as the detectivity normalized to an area of $1 \, \text{cm}^2$ and a noise bandwidth of 1 Hz. The value of D-star (D^*) can be calculated using eqn. 9.8.

$$D^* = D\sqrt{A\Delta f} \tag{9.8}$$

Where,

$D^* = $ D-star ($\text{W}^{-1}\text{cmHz}^{1/2}$)
$D = $ Detectivity (W^{-1})
$A = $ Sensor area (cm^2)
$\Delta f = $ Bandwidth (Hz)

9.3.4 Quantum Efficiency

An ideal photo senor should produce one photo electron per incident photon of light. This is not true for practical sensors. The ratio of the number of photo electrons released to the number of photons of incident light absorbed is referred to as the *quantum efficiency* of the sensor. It is the percentage of input radiation power converted into photo current. In other words, it is the intrinsic efficiency of the sensor. The value of quantum efficiency η is computed as the ratio of actual responsivity to the ideal responsivity as given by eqn. 9.9.

$$\eta = \frac{R_{Actual}}{R_{Ideal}} \tag{9.9}$$

Where,

$R_{Actual} = $ Actual or observed responsivity
$R_{Ideal} = $ Ideal responsivity.

$$R_{Ideal} = \frac{\text{Electron charge}}{\text{Photon energy}} = \frac{e \times \lambda}{h \times c} \tag{9.10}$$

Where,

$h = $ Planck's constant $= 6.63 \times 10^{-34}$ Js
$c = 3 \times 10^{8}$ m/s.

Substituting the values of h, c and e in eqn. 9.10, $R_{Ideal} = 8.044 \times 10^{5} \lambda$

Substituting the value of R_{Ideal} in eqn. 9.9,

$$\eta = \frac{R_{Actual}}{8.044 \times 10^5 \lambda} = \frac{1.24 \times 10^3 R_{Actual}}{\lambda}$$

$$\eta = \frac{1240 \times R_{Actual}}{\lambda} \tag{9.11}$$

9.3.5 Response Time

This is expressed as a *rise/fall time* parameter in photo electric sensors and as *time constant* parameter in thermal sensors. Rise and fall times are the time durations required by the output to change from 10 to 90% and 90 to 10% of the final response, respectively. It determines the highest signal frequency to which a sensor can respond. The *time constant* is defined as the time required by the output to reach to 63% of the final response from zero initial value.

Bandwidth of photo electric sensors is related to its rise time by eqn. 9.12.

$$BW = \left[\frac{0.35}{t_r} \right] \tag{9.12}$$

Where,

BW = Bandwidth (MHz)
t_r = Rise time (μs).

Response time of a photo diode is governed by three parameters, namely:

1) Drift time, t_{DRIFT}, which is the charge collection time of the carriers in the depletion region of the photo diode.
2) Diffusion time, $t_{DIFFUSION}$, which is the charge collection time of the carriers in the undepleted region of the photo diode.
3) RC time, t_{RC}, is the RC time constant of the diode-external circuit combination.

RC time, $t_{RC} = 2.2RC$, where R is the sum of the photo diode's series resistance and the load resistance and C is the sum of the photo diode's junction and the stray capacitances. Also, the junction capacitance is directly proportional to the diffused area of the photo diode and inversely proportional to the applied reverse bias. As a consequence of this, photo diodes with smaller area and larger applied reverse bias have faster rise time. In addition, stray capacitance can be minimized by using short leads and a careful electronic circuit layout. The total rise time is given by eqn. 9.13.

$$t_R = \sqrt{\left[t_{DRIFT}^2 + t_{DIFF}^2 + t_{RC}^2 \right]} \tag{9.13}$$

In photovoltaic mode of operation where there is no applied reverse bias, drift time can be considered to be negligible and the rise time is dominated by the diffusion time for diffused areas less than 5 mm^2 as smaller diffused areas offer very small junction capacitance and by *RC* time constant for larger diffused areas for all wavelengths. When operated in photoconductive mode where there is an applied reverse bias, the dominant factor is the drift time in case the photo diode is fully depleted and by all the three parameters if not fully depleted.

9.3.6 Noise

Noise is the most critical factor in designing sensitive radiation detection systems. Noise in these systems is generated in photo sensors, radiation sources and post-detection circuitry. Photo senor noise mainly comprises of the following components.

1) Johnson noise
2) Shot noise
3) Generation-recombination noise
4) Flicker noise

Johnson noise also known as *Nyquist noise* or *thermal noise* is caused by the thermal motion of charged particles in a resistive element. The RMS value of the noise voltage depends on the resistance value, temperature and the system bandwidth and is given by eqn. 9.14.

$$V_{RMS} = \sqrt{4KRT\Delta f} \qquad (9.14)$$

Where,

V_{RMS} = RMS noise voltage (V)
R = Resistance value in Ω
K = Boltzmann constant (1.38×10^{-23} J/K)
T = Absolute temperature (K)
Δf = System Bandwidth (Hz).

Shot noise in a photo senor is caused by the discrete nature of the photo electrons generated. It is related to the statistical fluctuation of both dark current and the photo current. It depends on the average current through the photo senor and system bandwidth and is given by eqn. 9.15.

$$I_{SRMS} = \sqrt{2eI_{av}\Delta f} \qquad (9.15)$$

Where,

I_{SRMS} = RMS shot noise current (A)
I_{av} = Average current through the photo senor (A)
e = Charge of an electron (=1.60×10^{-19} C)
Δf = Detection bandwidth (Hz).

It is the dominant source of noise in the case of photo diodes operating in photoconductive mode.

Generation-recombination noise is caused by the fluctuation in current generation and the recombination rates in a photo senor. This type of noise is predominant in photoconductive sensors operating at IR wavelengths. The generation-recombination noise can be calculated using eqn. 9.16.

$$I_{GRMS} = 2eG\sqrt{\eta EA\Delta f} \qquad (9.16)$$

Where,

I_{GRMS} = RMS generation-recombination noise current (A)
e = Charge of an electron (1.60×10^{-19} C)
Δf = Detection bandwidth (Hz)
E = Radiant intensity (W/cm^2)

A = Sensor receiving area (cm^2)
G = Photoconductive gain
η = Quantum efficiency.

Flicker noise or 1/f noise occurs in all conductors where the conducting medium is not a metal and exists in all semiconductor devices that require bias current for their operation. Its amplitude is inversely proportional to the frequency. Flicker noise is usually predominant at frequencies below 100 Hz.

The total equivalent noise (I_{NEQ}) is calculated by eqn. 9.17.

$$I_{NEQ} = \sqrt{\left(I_{JRMS}^{2} + I_{SRMS}^{2} + I_{GRMS}^{2} + I_{FRMS}^{2}\right)} \tag{9.17}$$

9.4   Photoconductors

Photoconductors also referred to as photoresistors, light dependent resistors (LDRs) and photocells, are semiconductor photo sensors whose resistance decreases with increasing incident light intensity. They are bulk semiconductor devices with no P-N junction with a structure as shown in Figure 9.4(a). When light is incident on the photoconductor, electrons jump from the valence band to the conduction band. Hence, the resistance of the semiconductor material decreases. The resistance change in a photoconductor is in the order of six decades, ranging from few tens of mega ohms under dark conditions to few tens or hundreds of ohms under bright light conditions. Other features include wide dynamic response, spectral coverage from UV to far IR and low cost. However, they are sluggish devices with a response time in the order of hundreds of milliseconds.

The resistance-illuminance relation in photoconductors is described by eqn. 9.18.

$$R_a = R_b \times \left(\frac{E_a}{E_b}\right)^{-\alpha} \tag{9.18}$$

(a)   (b)

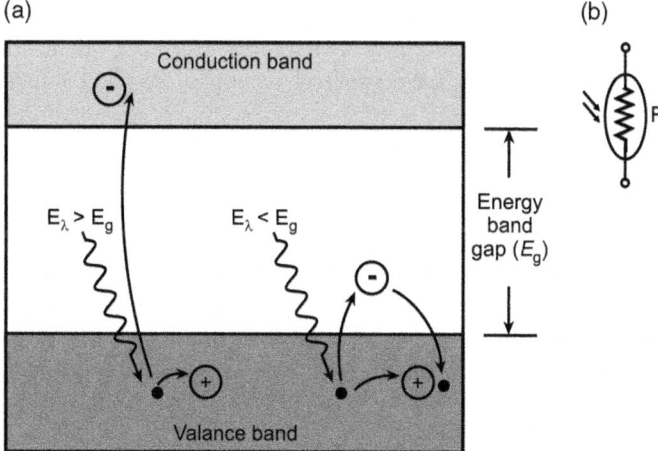

Figure 9.4 Photo conductor: (a) cross-section and (b) circuit symbol.

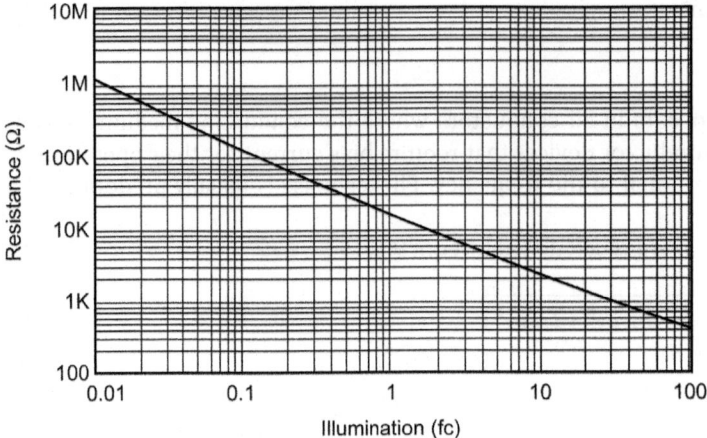

Figure 9.5 Typical resistance-illumination curve of a photo conductor.

Where,

R_a and R_b are the resistances at illumination levels of E_a and E_b, respectively.
E_a and E_b are the illumination levels in lux or foot candles.
α is the characteristic slope of the resistance-illumination curve.

The value of α is in the range of 0.55–0.9. Figure 9.4(b) shows the circuit symbol of a photo conductor. Photo conductors are made in a various shapes and sizes. Figure 9.5 shows the typical resistance-illumination curve of a photo conductor. Commonly used materials in photo conductors are cadmium sulfide (CdS), lead sulfide (PbS), lead selenide (PbSe), mercury cadmium telluride (HgCdTe) and germanium copper (Ge:Cu). The spectral response of some of these photo conductor materials is shown in Figure 9.6. Inexpensive CdS photo conductors are used in many consumer items like camera light meters, clock radios, security alarms, street lights and so on. On the other hand, Ge:Cu cells are used for IR astronomy and IR spectroscopy applications. Figure 9.7 shows photographs of some representative photo conductors.

Photo conductors are further classified into intrinsic or extrinsic, depending upon whether an external dopant has been added or not to the semiconductor material. Intrinsic photo conductors operate at shorter wavelengths as the electrons have to jump from the valence to the conduction band. Extrinsic photoconductors have spectral response covering longer wavelengths.

9.4.1 Application Circuits

Photo conductors are usually used for detection of IR radiation. When a bias is applied to the photo conductor in the absence of radiation, a current is generated that can be referred to as the dark current. When light is incident on the photoconductor, its resistance decreases and the current flowing through it increases. The photo signal is the increase in the current caused by radiation. Generally, this photo signal is much smaller (in the order of few parts in thousand) than the dark current. Extracting this small signal from the dark current is the primary task of the front-end circuit.

Figure 9.8 parts (a) and (b) show some simple circuits using photoconductors. However, using photoconductors in these configurations reduces the responsivity of the conductor as the

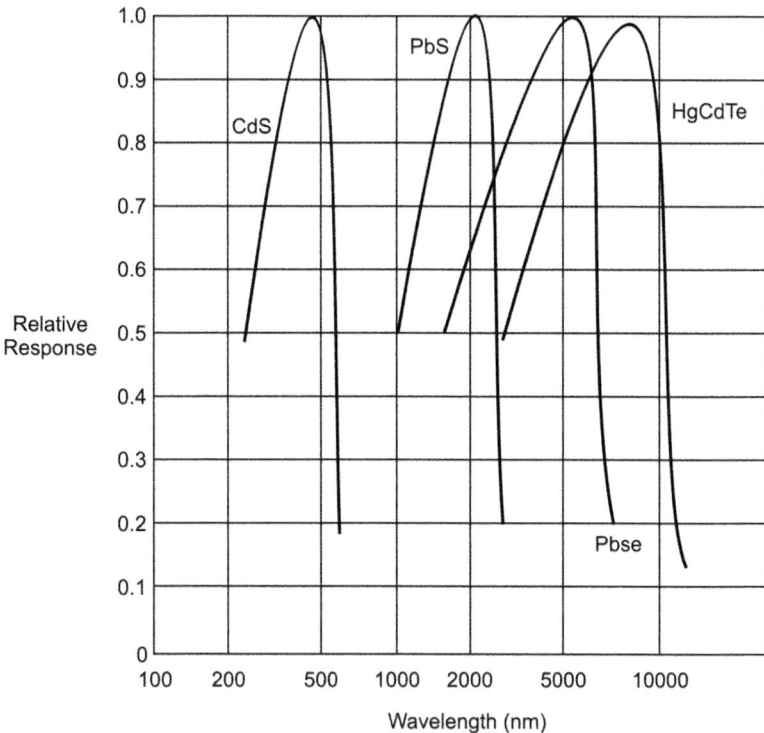

Figure 9.6 Spectral response of commonly used photo conductor materials.

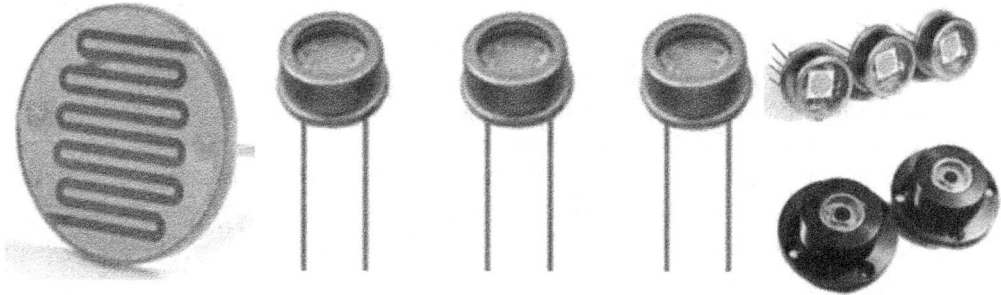

Figure 9.7 Some representative types of photoconductors.

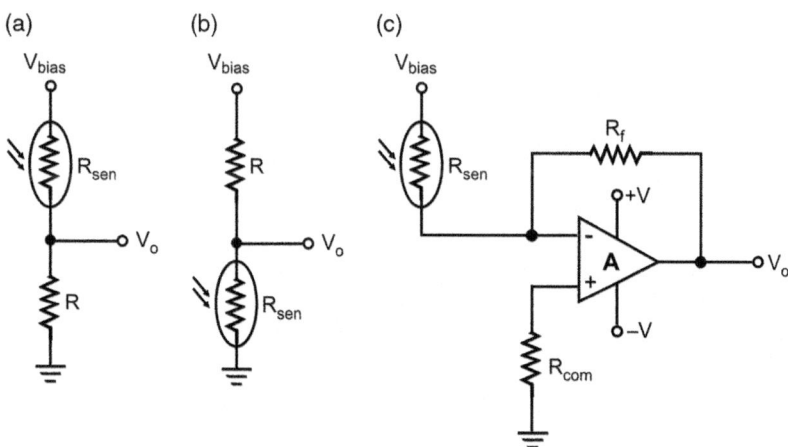

Figure 9.8 Application circuits: (a) and (b) the simplest application circuits using photoconductors, (c) application circuit of a photo conductor using an opamp in the trans-impedance mode.

relative change in the circuit resistance is smaller because of the load resistance I. The choice of R and R_{sen} also affects the output voltage from the circuit. For Figure 9.8(a), higher the value of R, the higher the output voltage but the relative responsivity is poorer. Similarly, in the case of circuit of Figure 9.8(b), higher the value of R, lower is the output voltage but the relative responsivity is better.

To overcome these problems, photo conductors are used in conjunction with amplifiers to obtain both better responsivity and high output voltage. There are two possible circuit configurations namely voltage mode amplifiers and current mode or transimpedance amplifiers. The basic transimpedance amplifier is shown in Figure 9.8(c). The non-inverting input of the opamp is connected to ground through resistance R_{com} to minimize the DC offset voltage.

The output voltage V_o is given by eqn. 9.19.

$$V_o = -\left(R_f / R_{sen}\right) \times V_{bias} \tag{9.19}$$

The gain of the transimpedance amplifier should be so set that the amplifier does not saturate at the maximum expected radiation intensity. Also, if the bias voltage of the photoconductor is more than the maximum rated input voltage of the opamp, then a zener diode should be connected between the inverting input of opamp and the ground terminal. Theoretically, the signal voltage can be obtained by subtracting the output voltage in dark conditions from the voltage signal in eqn. 9.19 and is given by eqn. 9.20.

$$V_o = -\left[\left(R_f / R_{sen}\right) - \left(R_f / R_{dark}\right)\right] \times V_{bias} \tag{9.20}$$

Where,

R_{dark} is the resistance value of the photoconductor in the absence of radiation.

However, it is not a practically feasible solution as the dark resistance of the photo conductor is a strong function of temperature and even a slight increase in temperature decreases the value of dark resistance by a large amount and vice-a-versa. So, the sensor temperature has to be controlled to the order of 0.01 °C or better, which is often not feasible. The most common method used to extract the signal is to modulate the incident radiation at a specific frequency, either by placing a mechanical chopper in front of the sensor or by electrically modulating the radiation source. The signal generated due to radiation is now an AC signal while the dark current is a DC signal. The AC signal can be separated from the DC background signal using an AC coupled amplifier. A voltage mode amplifier using AC coupling is shown in Figure 9.9.

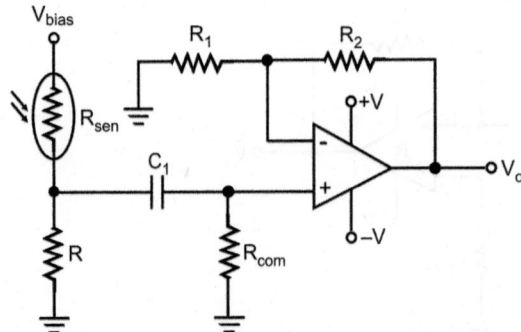

Figure 9.9 Application circuit of photo conductor using a voltage mode amplifier with AC coupling.

9.5 Photo Diodes

Photo diodes are junction type semiconductor light sensors that generate current or voltage when the P-N junction in the semiconductor is illuminated by light of sufficient energy. The spectral response of the photo diode is a function of the band gap energy of the material used in its construction. The upper cut-off wavelength of a photo diode is given by eqn. 9.21.

$$\lambda_c = \frac{1240}{E_g}$$ (9.21)

Where,

λ_c = Cut-off wavelength (nm)
E_g = Band gap energy (eV).

Photo diodes are mostly constructed using silicon, germanium, indium gallium arsenide (InGaAs), lead Sulfide (PbS) and mercury cadmium telluride (HgCdTe). Figure 9.10 shows the spectral characteristics of these photo diodes.

9.5.1 Types of Photo Diodes

Depending upon their construction there are several types of photo diodes. These include PN photo diodes, PIN photo diodes, Schottky type photo diodes and avalanche photo diodes (APDs).

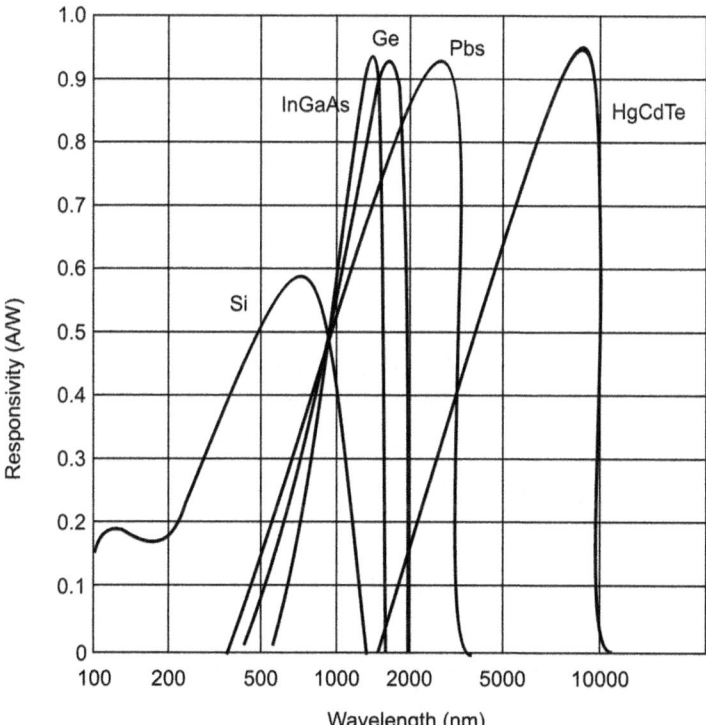

Figure 9.10 Spectral characteristics of photo diodes.

9.5.1.1 PN Photo Diodes

A PN photo diode comprises of a PN junction as shown in Figure 9.11(a). When light with sufficient energy strikes the photo diode, the electrons are pulled up into the conduction band, leaving behind holes in the valence band. These electron-hole pairs occur throughout the p-layer, depletion layer and N-layer materials. When the photo diode is reverse biased, the photo induced electrons will move down the potential hill from the P-side to the N-side. Similarly, the photo induced holes will add to the current flow by moving across the junction to the P-side. Shorter wavelengths are absorbed at the surface while the longer wavelengths penetrate deep into the diode. Figure 9.11(b) shows the mechanism of conversion of incident light photons into electric current in a PN photo diode. PN photo diodes are used for precision photometry applications like medical instrumentation, analytical instruments, semiconductor tools and industrial measurement systems.

9.5.1.2 PIN Photo Diodes

In PIN photo diodes, an extra high resistance intrinsic layer is added between the P and the N layers as shown in Figure 9.12. This has the effect of reducing the transit or diffusion time of the photo induced electron-hole pairs which in turn results in improved response time. PIN photo diodes

(a)

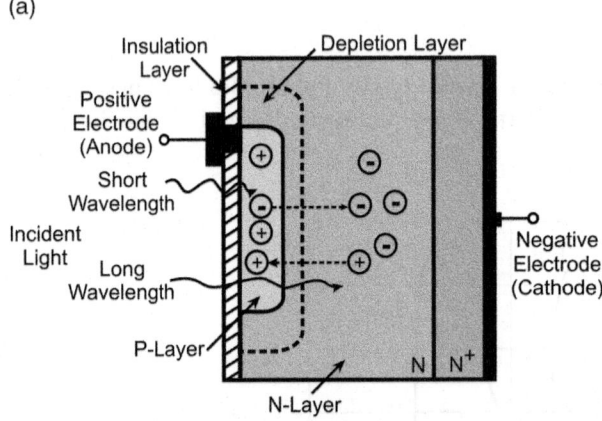

(b)

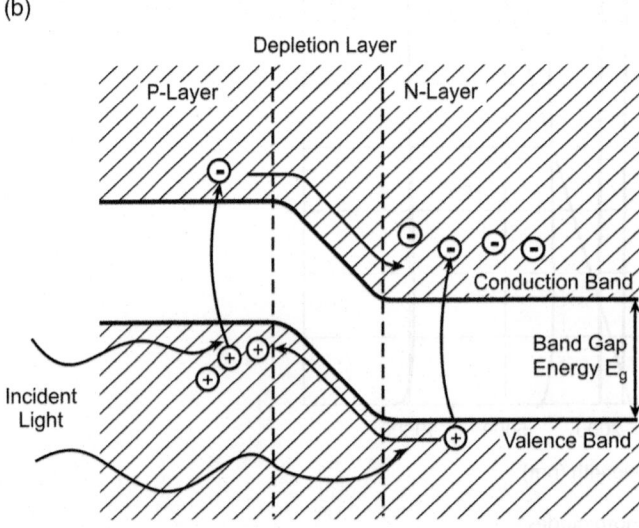

Figure 9.11 (a) Cross-section of a P-N photo diode. (b) Generation of current in a PN photo diode.

feature low-capacitance, thereby offering high bandwidth making them suitable for high-speed photometry as well as optical communication applications. Single element PIN photo diodes, quadrant photo diodes and one- and two-dimensional arrays of photo diodes find extensive application in a variety of sensor systems for military applications. Prominent among these are laser-warning sensor suites on armoured fighting vehicles and airborne platforms, laser seekers in laser-guided munitions, laser receivers in laser communication systems and focal plane arrays in LADAR sensors. Photo diodes and photo diode arrays integrated with peripheral components such as low noise preamplifier circuits are also available in a single package. Figure 9.13 shows photographs of single, quad and linear and two-dimensional arrays of PIN photo diodes available in various package configurations. The assortment of device packages in the figure shows some of the more common ones. Photo diodes are available in many more packages including customized ones.

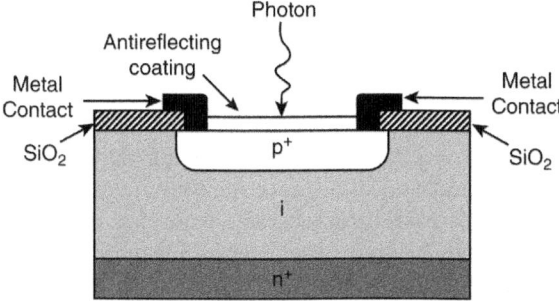

Figure 9.12 PIN photo diode.

Figure 9.13 PIN photo diodes in different configurations and packages.

9.5.1.3   Schottky Photo Diodes

In Schottky type photo diodes, a thin gold coating is sputtered on to the N-material to form a Schottky effect P-N junction. Schottky photo diodes have an enhanced UV response.

9.5.1.4   Avalanche Photo Diodes (APDs)

APDs are high-speed; high-sensitivity photo diodes utilizing an internal gain mechanism that functions by applying a relatively higher reverse bias voltage than that is applied in the case of PIN photo diodes.

Figure 9.14 shows the cross-section of an avalanche photo diode. Avalanche photo diodes are so constructed as to provide a very uniform junction that exhibits the avalanche effect at reverse bias voltages between 30 and 200 V. The electron-hole pairs that are generated by incident photons are accelerated by the high electric field to force the new electrons to move from the valence band to the conduction band. In this way, the multiplication in the order of 50–100 is achieved. Avalanche photo diodes have fast response times similar to that of PIN photo diodes. Also, responsivity figures for silicon PIN photo diodes are in the range of 0.4–0.6 A/W whereas for APDs they are between 40 and 80 A/W, around 100 times more than that of PIN photo diodes. Moreover, they offer excellent signal-to-noise ratio in the order of offered by photo multiplier tubes. Hence, they are used in a variety of applications requiring high sensitivity such as long-distance optical communication and optical distance measurement. Military laser range finders based on time of flight principle have their receivers' front-end invariably employing a silicon or indium gallium arsenide avalanche photo diode depending upon whether it is an Nd:YAG laser range finder or an eye-safe laser range finder. Avalanche photo diodes just like PIN photo diodes are also available as both single detectors as well as linear or two-dimensional arrays in similar package configurations.

9.5.2   Equivalent Circuit

A photo diode is electrically represented by a current source in parallel with an ideal diode. Figure 9.15 shows the equivalent circuit of a photo diode. The current source represents the photo current generated by incident radiation and the diode represents the PN junction. In addition, the current source is also shunted by a *junction capacitance* (C_j) and a *shunt resistance* (R_{SH}) across it. The parallel arrangement of four elements is then connected in series with a *series resistance* (R_S).

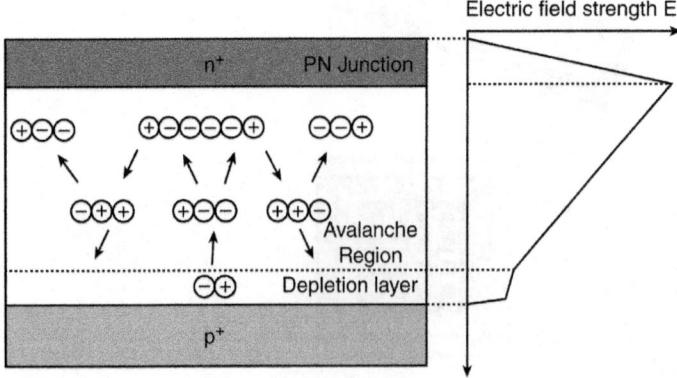

Figure 9.14 Cross-section of an avalanche photo diode.

Figure 9.15 Equivalent circuit of a photo diode.

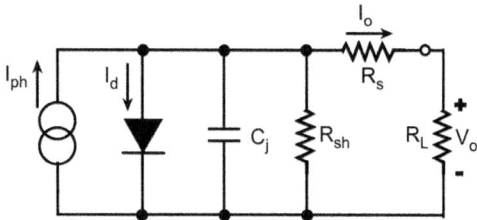

Shunt resistance is the slope of the current-voltage characteristic curve of the photo diode at the origin, where the applied bias voltage is zero. The value of R_{SH} in the case of an ideal photo diode is infinity. However, in the case of practical devices, values range from tens to thousands of mega ohms. A higher shunt resistance is always desirable. Shunt resistance is used to determine the noise current in the photo diode in the photovoltaic mode of operation where the applied bias voltage is zero.

Junction capacitance is formed by the boundaries of the depletion region acting as the plates of a parallel plate capacitor and the depletion region acting as the dielectric medium. Consequently, it is directly proportional to the diffused area and inversely proportional to the width of the depletion region. Also, higher resistivity substrates offer lower junction capacitance. The value of junction capacitance depends upon the applied reverse bias according to eqn. 9.22.

$$C_J = \frac{\varepsilon_0 \varepsilon A}{\sqrt{\left[2\varepsilon \mu_0 \rho \left(V_A + V_{bi} \right) \right]}} \tag{9.22}$$

Where,

ε = Dielectric constant of semiconductor material (=11.9 for silicon)
ε_0 = Permittivity of free space = 8.85×10^{-12} F/m
μ_0 = Mobility of electrons (=1400 cm^2/V^{-s} at 300 K)
ρ = Resistivity of silicon
V_A = Applied reverse bias
V_{bi} = Built-in voltage of silicon.

Figure 9.16 shows the variation of junction capacitance as a function of applied reverse bias for a silicon PIN photo diode type number BPX 65.

Series resistance of a photo diode arises from the resistance of the contacts and the resistance of the undepleted region. It is given by eqn. 9.23.

$$R_S = \left(\frac{\rho W_S W_D}{A} \right) + R_C \tag{9.23}$$

Where,

W_S = Substrate thickness
W_D = Depletion region width
A = Diffused area
R_C = Contact resistance
P = Resistivity of substrate material.

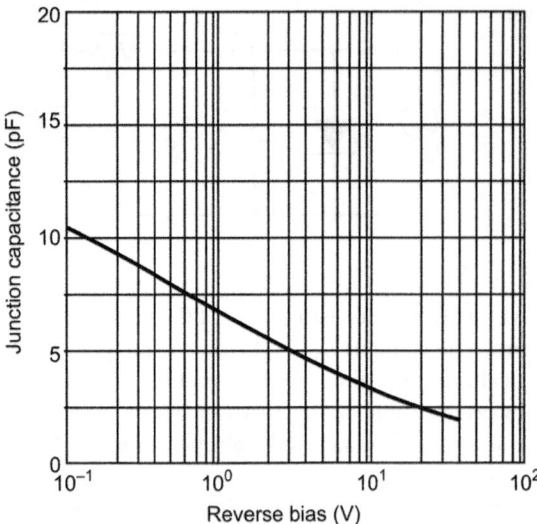

Figure 9.16 Variation of junction capacitance of a photo diode as a function of reverse bias.

Series resistance in the case of an ideal photo diode is zero. In the case of practical devices, values range from 10 to 1000 Ω. It is used to determine the linearity of the photo diode response in photovoltaic mode of operation.

9.5.3 V-I Characteristics

Figure 9.17 parts (a) and (b) show the circuit symbol and V-I characteristics of a photo diode, respectively. As we can see from the V-I characteristics of a photo diode, the curve in the dark state is similar to that of a conventional rectifier diode. However, when light strikes, curve shifts downwards with increasing intensity of light. If the photo diode terminals are shorted, a photocurrent proportional to the light intensity will flow in a direction from anode to cathode. If the circuit is open, then an open circuit voltage will be generated with the positive polarity at the anode. It may be mentioned here that the short circuit current is linearly proportional to light intensity while the open circuit voltage has a logarithmic relationship with light intensity.

Photo diodes can be operated in two modes namely the *photovoltaic* mode and *photo conductive* mode. In the photovoltaic mode of operation, no bias voltage is applied and due to the incident light, a forward voltage is produced across the photo diode. In photo conductive operational mode, a reverse bias voltage is applied across the photo diode. This widens the depletion region, resulting in higher speed of response. As a rule of thumb, all applications requiring a bandwidth less than 10 kHz can use photo diodes in photovoltaic mode. For all other applications, photo diodes are operated in photo conductive mode. Moreover, the linearity of a photo diode is also improved when it is operated in the photo conductive mode. However, there is an increase in the noise current of the photo diode when it is operated in the photo conductive mode. This is due to the reverse saturation current, referred to as the dark current, flowing through the photo diode. The value of dark current is typically in the range of 1–10 nA at a specified reverse bias voltage. When the photo diode is operated in the photovoltaic mode, the value of dark current is zero.

9.5.4 Application Circuits

As discussed previously, photo diodes can be operated in two modes, namely photovoltaic and the photo conductive. In the photovoltaic mode, the photo diode is operated with zero external bias voltage and is generally used for low speed applications or for detecting low light

(a)　　　　　　　(b)

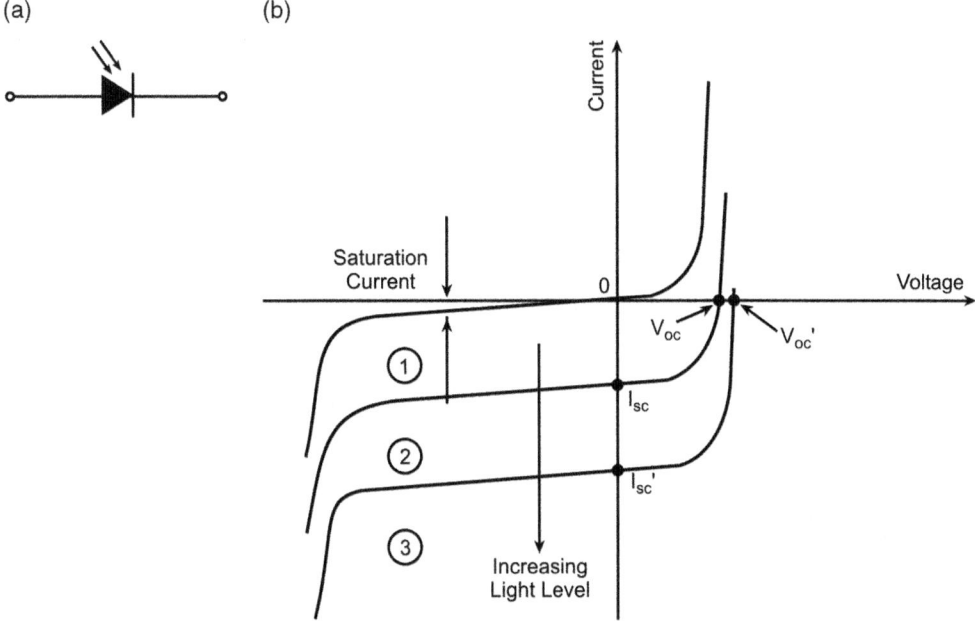

Figure 9.17 (a) Circuit symbol of a photo diode. (b) V-I characteristics of a photo diode.

(a)　　　　　　　(b)

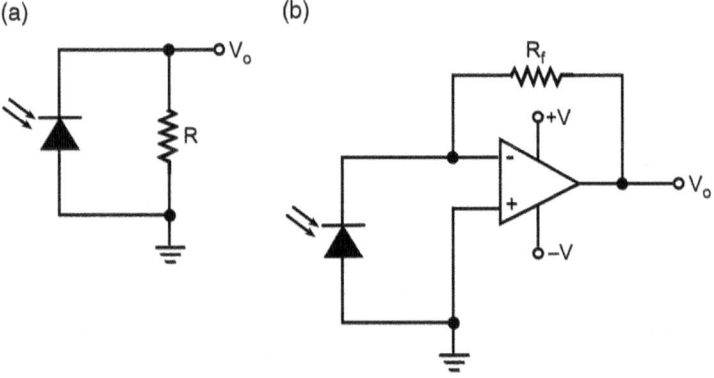

Figure 9.18 Application circuits of photodiodes in photovoltaic mode: (a) resistive load and (b) opamp based current-to-voltage converter.

levels. Figure 9.18(a) and (b) shows two commonly used application circuits employing photo diodes in the photovoltaic mode. The output voltages for these circuits are given by $I_{det} \times R$ and $I_{det} \times R_f$, respectively, where I_{det} is the current through the photo diode. The circuit in Figure 9.18(b) offers better linearity than the circuit in Figure 9.18(a) as the equivalent input resistance across the photo diode in this case is R_f/A, where A is the open loop gain of the operational amplifier. It is obvious that value of R_f/A is much smaller compared to that of R in the case of Figure 9.18(a). Figure 9.19 shows the load line analysis of a photo diode operating in photovoltaic mode. The load line corresponding to the smaller load is closer to the current axis and the one corresponding to a larger load is close to the voltage axis. As is evident from the figure, for a better linear response the equivalent resistance across the photo diode should be as small as possible.

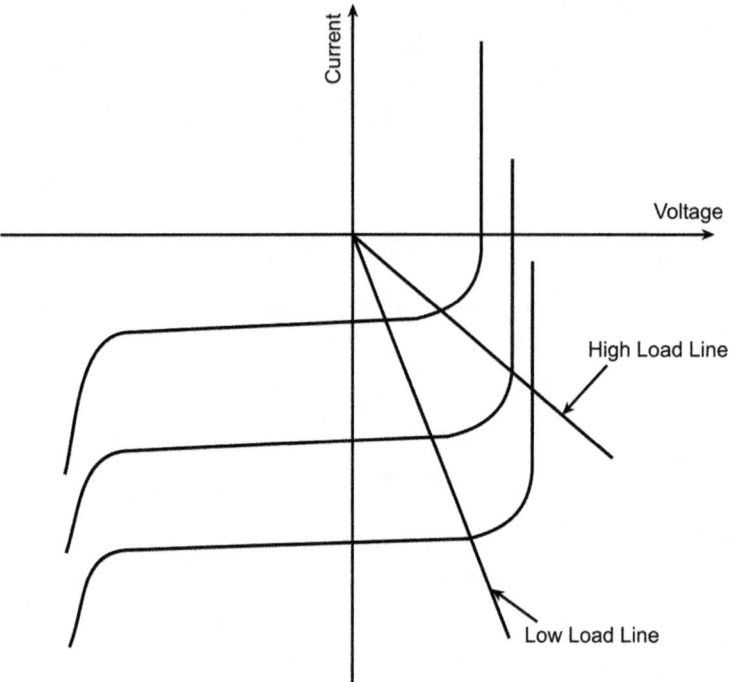

Figure 9.19 Load line analysis of a photo diode in photovoltaic mode.

Figure 9.20(a)–(d) shows four possible circuits of using photo diodes in the photoconductive mode. In Figure 9.20(b), the operational amplifier is used as a voltage amplifier whereas in Figure 9.20(c) and (d), the operational amplifier is used in the transimpedance mode. For circuit in Figure 9.20(b), the output voltage and the effective resistance across the photo diode are $I_{det} \times R$ and R, respectively. I_{det} is the current flowing through the photo diode. The output voltage and effective resistance across the photo diode in Figure 9.20(c) and (d) are $I_{det} \times R_f$ and R_f/A respectively, where I_{det} is the photo diode current and A is the open loop gain of the operational amplifier. The response of the photo diode for different loads operating in photoconductive mode is shown in Figure 9.21. As we can see circuits with lower resistance load line offers better linearity.

Avalanche photo diodes (APDs) are also connected in a similar manner to that discussed before except that a much higher reverse bias voltage is required for its operation. Also, the power consumption of APDs during operation is much higher than that of PIN photo diodes and is given by the product of input signal, sensitivity and reverse bias voltage. Hence a protective resistor is added to the bias circuit as shown in Figure 9.22 or a current limiting circuit is used.

As the gain of APDs changes with temperature, if they are operated over a wide temperature range some temperature offset circuit has to be added that changes the reverse bias voltage with temperature so as to compensate for the change in gain with temperature. As an alternative, a temperature controller can be added to keep the temperature of APD constant. For detecting low signal levels, shot noise from the background light should be limited by using optical filters, by source modulation and by restricting the field-of-view.

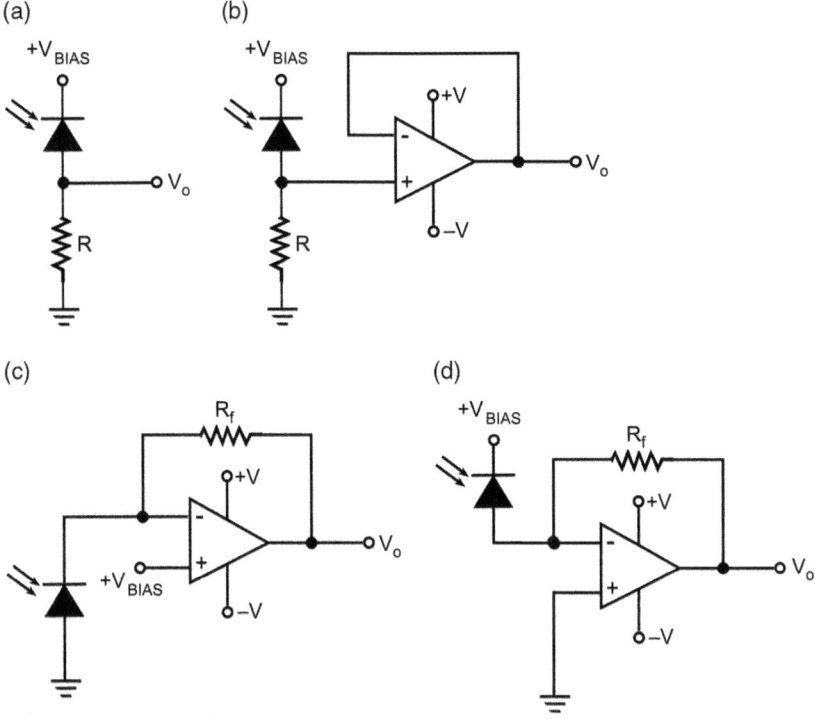

Figure 9.20 Application circuits of photo diodes operating in the photo conductive mode: (a) resistive load, (b) resistive load with opamp wired as a voltage follower, (c) trans-impedance amplifier with positive output voltage and (d) trans-impedance amplifier with negative output voltage.

Figure 9.21 Load line analysis of photo diodes operating in photo conductive mode.

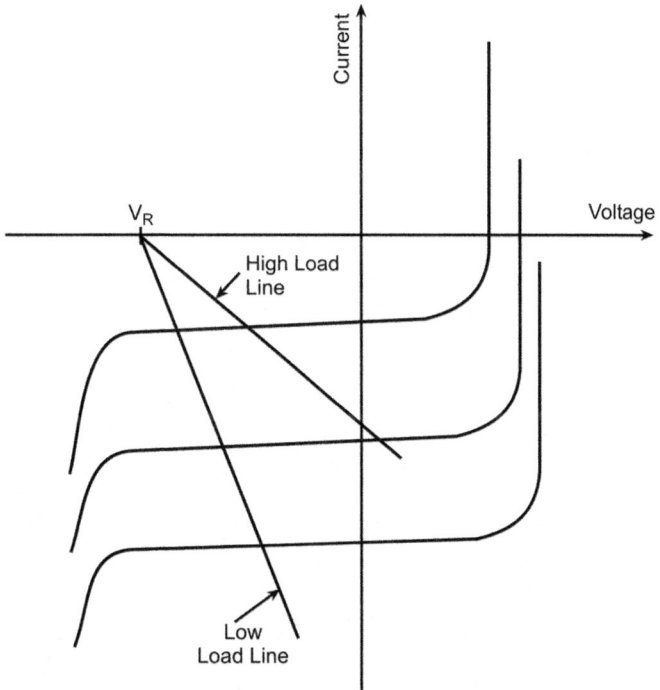

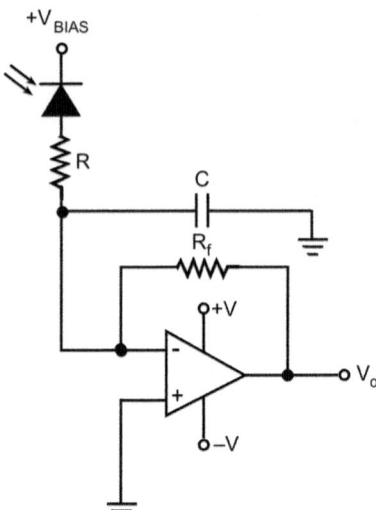

Figure 9.22 Application circuit using an APD.

9.6 Solar Cells

Solar cells are devices whose operation is very similar to that of a photo diode in the photovoltaic mode. The operational principle of the basic solar cell is based on the photovoltaic effect. As mentioned above, due to the photovoltaic effect, there is a generation of an open circuit voltage across a P-N junction when it is exposed to light, which is the solar radiation in the case of a solar cell. This open circuit voltage leads to the flow of electric current through a load resistance connected across it as shown in Figure 9.23.

As is evident from the figure, the impinging photon energy leads to generation of electron-hole pairs. The electron-hole pairs either recombine and vanish or start drifting in the opposite directions with electrons moving towards the N-region and holes moving towards the P-region. This accumulation of positive and negative charge carriers constitutes the open circuit voltage. This voltage can cause a current to flow through an external load or when the junction is shorted, the result is a short circuit current whose magnitude is proportional to input light intensity. The voltage output and the current delivering capability of an individual solar cell are very small for it to be any use as an electrical power input to any system. As an example, a typical solar cell would produce 500 mV output with a load current capability of about 150 mA. The series-parallel arrangement of solar cells is done to get the desired output voltage with required power delivery capability. The series combination is used to enhance the output voltage while the parallel combination is used to enhance the current rating. Figure 9.24 shows some representative solar cell panels.

Figure 9.25 shows the current-voltage and power-voltage characteristics of a solar cell. As is evident from the figure, the solar cell generates its maximum power at a certain voltage. The power-voltage curve has a point of maximum power called the Maximum Power Point (MPP). The cell voltage and the corresponding current are less than the open circuit voltage (V_{OC}) and the short circuit current (I_{SC}), respectively, at the maximum power point.

Solar cell efficiency is the ratio of maximum electrical solar cell power to the radiant light power on the solar cell area. The efficiency figure for some crystalline solar cells is in excess of 20%. The most commonly used semiconductor material for making solar cells is silicon. Both crystalline and amorphous forms of silicon are used for the purpose. Another promising material for making solar cells is gallium arsenide (GaAs). Gallium arsenide solar cells, when perfected, will be lightweight and more efficient.

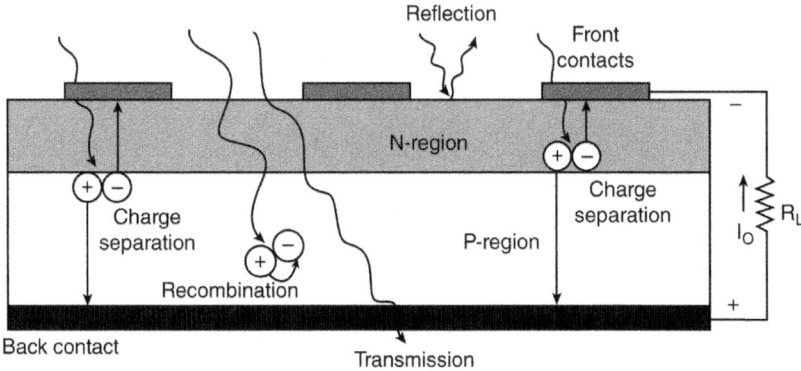

Figure 9.23 Principle of operation of a solar cell.

Figure 9.24 Different package styles of solar cells.

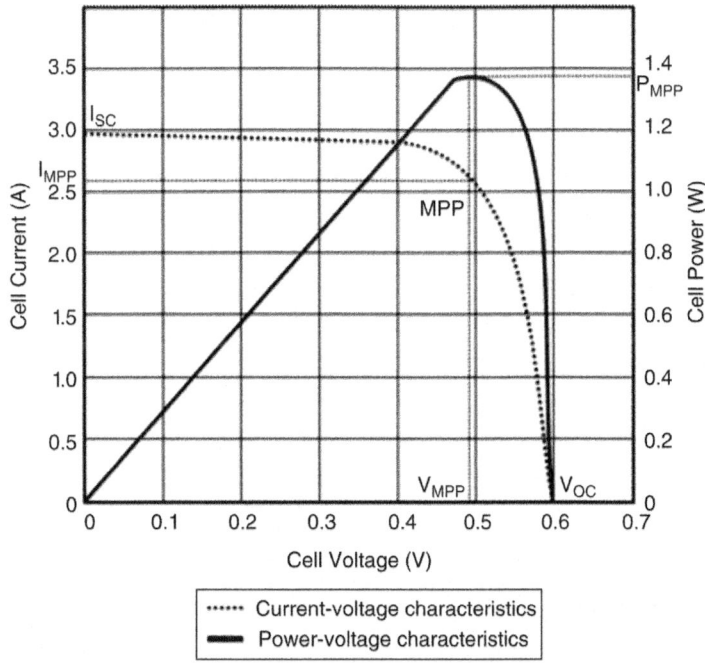

Figure 9.25 Current-voltage and power-voltage characteristics of a solar cell.

9.7 Photo Transistors

Figure 9.26 shows the construction of a photo transistor. Photo transistors are usually connected in a common emitter configuration with base open and the radiation is concentrated on the region near the collector-base junction. Figure 9.27(a) shows a circuit symbol for the photo transistor and Figure 9.27(b) shows the typical V-I characteristics of a photo transistor. When there is no radiation incident on the photo transistor, the collector current is due to the thermally generated carriers and is given by eqn. 9.24.

$$I_C = (\beta + 1) I_{CO} \qquad (9.24)$$

Where,
I_{CO} is the reverse saturation current.

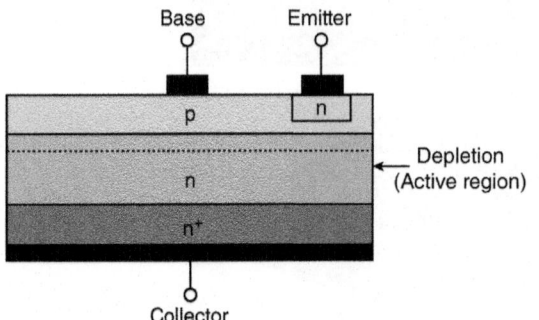

Figure 9.26 Cross-section of a photo transistor.

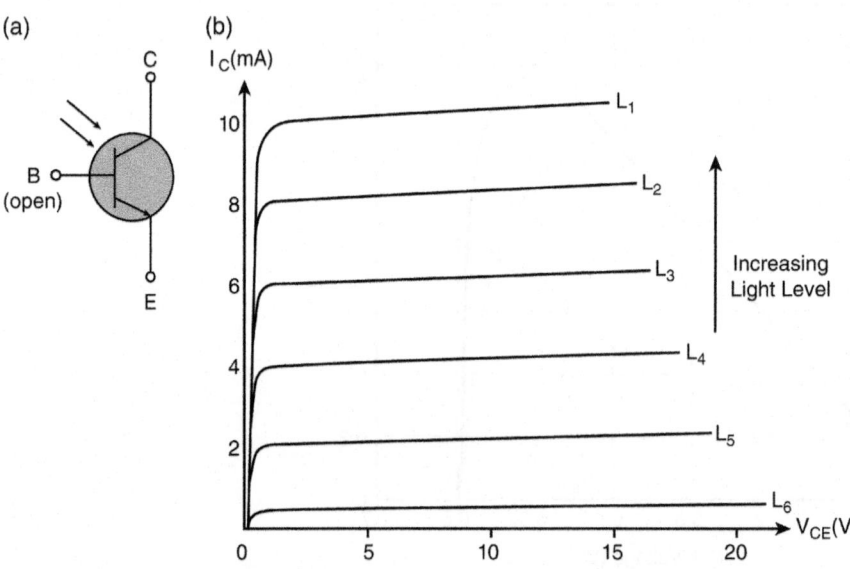

Figure 9.27 (a) Circuit symbol of a photo transistor. (b) V-I characteristics of a photo transistor.

In photo transistors, this current is referred to as the dark current. When light is incident on the photo transistor, photocurrent is generated and the magnitude of the collector current increases. The expression for the collector current is given by eqn. 9.25.

$$I_C = (\beta + 1)(I_{CO} + I_\lambda) \tag{9.25}$$

Where,

I_λ is the current generated due to incident light photons.

Photo transistors are available in package styles similar to those of conventional bipolar transistors.

Some of the representative package styles are shown in Figure 9.28.

9.7.1 Application Circuits

Photo transistors can be used in two configurations, namely the common emitter configuration [Figure 9.29(a)] and the common collector configuration [Figure 9.29(b)]. In the common emitter configuration, the output is high and goes low when light is incident on the photo transistor

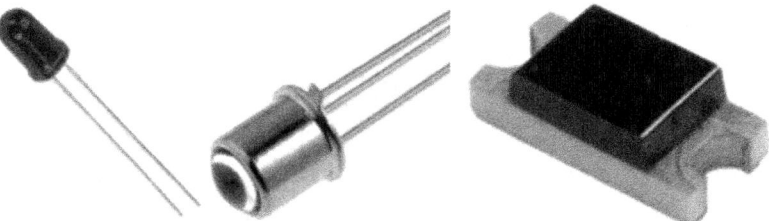

Figure 9.28 Photo transistor package configurations.

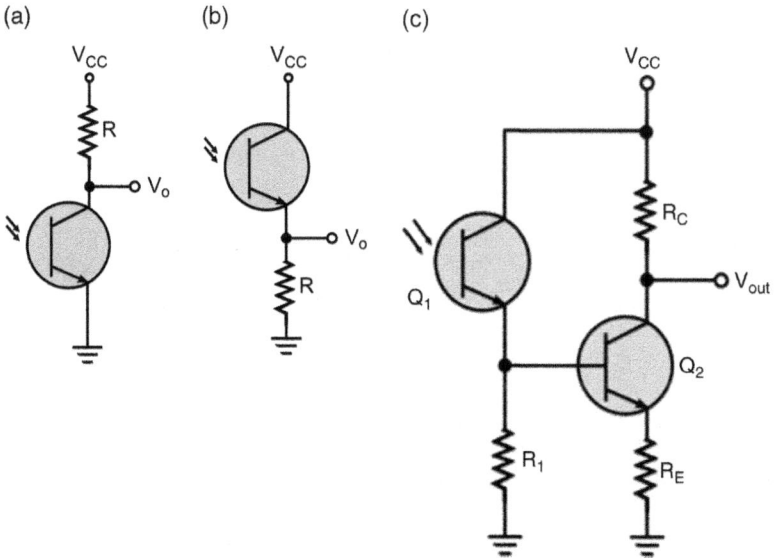

Figure 9.29 Application circuits of photo transistors: (a) common emitter configuration, (b) common collector configuration and (c) a photo transistor wired in a common collector configuration feeding a common emitter transistor amplifier.

whereas in the common collector configuration, the output goes from low to high when light is incident on the photo transistor. The transistor in both the configurations can act in two modes, namely the active mode and the switched mode. In the active mode, the transistor operates in the active region of its characteristics and the output is proportional to input light intensity. In the switched mode, the photo transistor is switched between cut-off and saturation and output is in the HIGH and LOW states, respectively. The modes are controlled by the value of the resistor R. The output of the photo transistor can be amplified using an opamp or a transistor-based amplifier circuit. Figure 9.29(c) shows a photo transistor wired in a common collector configuration feeding a common emitter transistor amplifier.

9.8 Photo FET, Photo SCR and Photo TRIAC

In this section, we shall discuss the three other important photo sensors, namely *photo FETs*, *photo SCRs* and *photo TRIACs*. While photo SCRs and photo TRIACs are latching type of photo sensors, photo FETs are non-latching photo sensors like photo diodes and photo transistors.

9.8.1 Photo FETs

Photo FETs are light sensitive FET devices wherein the diode formed by the reverse-biased gate-channel junction acts as a photo diode. Incident light generates additional photo carriers resulting in increased conductivity level. Gate current flows if the gate is connected to an external resistor. Figure 9.30(a) shows the circuit symbol of a photo FET. When no light is incident on the photo FET, the gate impedance is very high. When light is incident on the photo FET, the value of gate impedance decreases. Figure 9.30(b) shows the typical application circuit using a photo FET. When no light is incident, the gate voltage is approximately equal to the voltage $-V_{GG}$. When light is incident, the negative gate voltage decreases resulting in increase in the value of drain current I_D and the value of the output voltage V_o decreases.

9.8.2 Photo SCR

Photo SCRs, generally referred to as Light Activated SCRs (LASCRs), are essentially the same as conventional SCRs except that they are triggered by light incident on the gate junction area. They comprise of a window and lens to focus more light on the gate junction

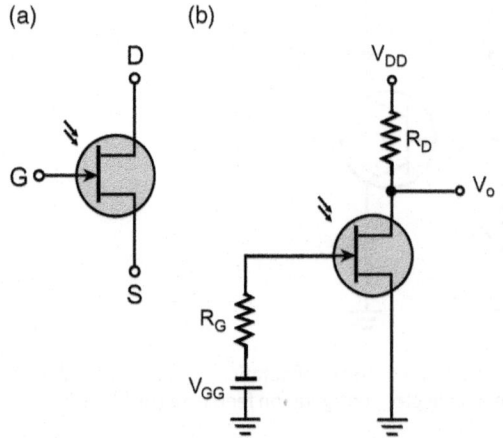

(a)   (b)

Figure 9.30 (a) Circuit symbol of a photo FET. (b) A simple application circuit using a photo FET.

Figure 9.31 (a) Circuit symbol of a photo SCR. (b) Simple application circuit using a photo SCR.

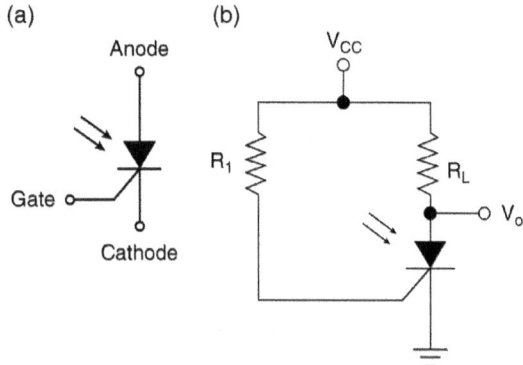

area, more specifically on the middle junction J_2 of the SCR. They conduct current in one direction when activated by a sufficient amount of light and continue to conduct till the current falls below a specified value. In other words, photo SCRs act as a latch that can be triggered on by the light incident on the gate junction but it does not turn off when the light source is removed. They can be turned off by reducing the current below its threshold value. Photo SCRs can handle large amount of current as compared to a photo diode or a photo transistor. They have high value of rate of change of voltage with time; that is, a high dv/dt rating, which is important for triggering the SCR on application of light input. Photo SCRs are most sensitive to light when their gate terminal is open. They are generally used in the receiving channel of optocouplers.

Figure 9.31(a) shows the circuit symbol of a photo SCR and Figure 9.31(b) shows a simple application circuit built around it. When no light is incident on the photo SCR, it is OFF and no current flows through the load resistor R_L. When light intensity above a certain specified threshold value impinges on the photo SCR, it is triggered to the ON state. It stays in the ON state even after the light pulse is removed.

9.8.3 Photo TRIAC

Photo TRIACs, also referred to as light activated TRIACs are bidirectional thyristors that are designed to conduct current in both directions when the incident light radiation exceeds a specified threshold value. Photo TRIACs are generally used as solid-state AC switches and as photo sensors in optocouplers to provide isolation from the driving source to the load. Figure 9.32 shows the circuit symbol for a photo TRIAC. The operation of a photo TRIAC is similar to standard TRIAC, except that the trigger current is generated indirectly in the case of a photo TRIAC by the light incident on it whereas in the case of a standard TRIAC it is supplied directly. One of the most important parameters to describe the performance of a photo TRIAC is the output dv/dt rating. Other important parameters are the breakdown voltage and the power rating of the device.

There are two different types of photo TRIACs available namely the non-zero-crossing photo TRIACs and zero-crossing photo TRIACs. The non-zero-crossing photo TRIACs are used for applications that require fine control involving small time constants. Zero-crossing photo TRIACs are used in applications where the control time constant is fairly large.

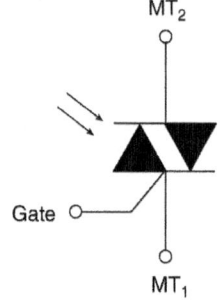

Figure 9.32 Circuit symbol of a photo TRIAC.

9.9 Image Sensors

Image sensors are used to store optical images in the form of electrical signals. An image sensor constitutes the heart of a digital camera. Two most common and widely used types of image sensors are the charge coupled device (CCD) sensor and the CMOS sensor. Both types of sensors use a two-dimensional array of thousands to millions of discrete pixels. The amount of light falling on each of the pixels generates free electrons with the number of electrons and hence the quantum of charge depending upon the intensity of impinging photons. The two types of sensors differ in the mode in which this charge is converted into voltage and subsequently read out of the chip for further processing. CCD and CMOS sensors are described in the following paragraphs in terms of their operational basics and salient features. A brief comparison between the two is also presented.

9.9.1 Charge Coupled Device (CCD)

CCD is basically an array of thousands to millions of light sensitive elements called pixels etched onto a silicon surface. Each of the pixels is a buried channel MOS capacitor. CCDs are typically fabricated on a p-type substrate and a buried channel is implemented by forming a thin n-type region on its surface. A silicon dioxide layer is grown on the top of the n-type region and an electrode, also called a gate, on top of the insulating silicon dioxide completes the MOS capacitor. The electrode could be metal, but is more likely to be a heavily doped polycrystalline silicon conducting layer (Figure 9.33).

The sensor is not actually flat, but has tiny cavities, like wells, that trap the incoming light and allow it to be measured. Each of these wells or cavities is a pixel. The size of the CCD is specified in mega-pixels. Megapixel value of the CCD can be computed by multiplying number of pixels in a row by number of pixels in a column. For example, 1000 pixels in a row and 1000 pixels in a column make a 1.0 megapixel CCD chip. When the light reflected off the target to be imaged is incident on this array of pixels, the impinging photons generate free electrons in the region underneath the pixels. In order to make sure that these free electrons don't

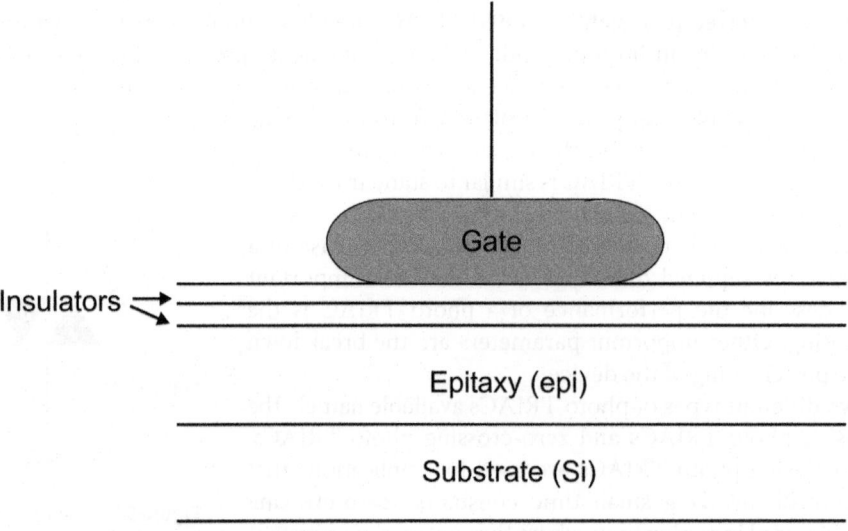

Figure 9.33 MOS capacitor.

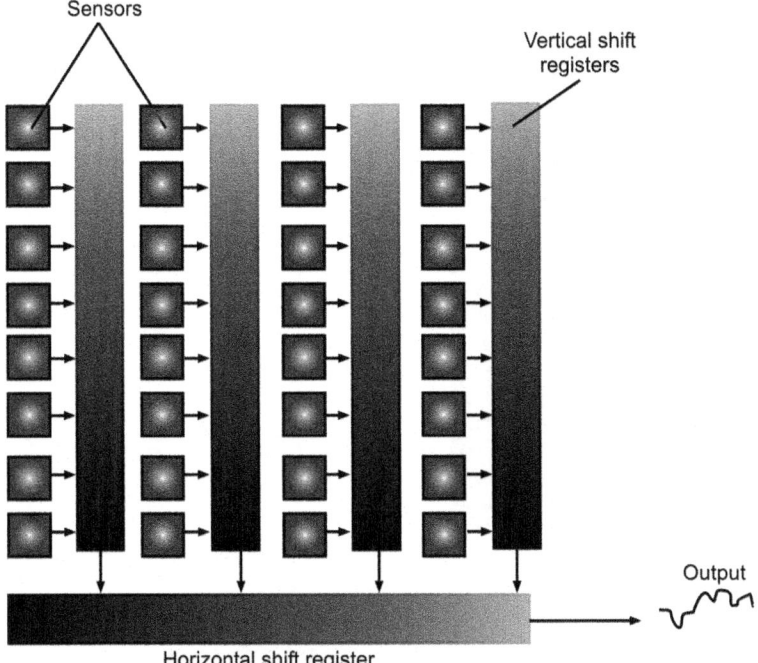

Sensors

Vertical shift
registers

Output

Horizontal shift register

Figure 9.34 CCD array.

combine with holes and disappear as heat energy; the electrons underneath each pixel are held there by applying a positive bias voltage to the pixels. If the sensor array were exposed to light for the same time; the number of electrons and hence the quantum of charge held under a certain pixel would vary directly as the luminous intensity that particular pixel is exposed to. This charge pattern represents the light pattern falling on the device. The charge is read out by suitable electronics and then converted into a digital bit pattern that can be understood and stored in a computer. This digital comes bit pattern then represents the image.

The charge held in the bins corresponding to different pixels is read out, converted into equivalent analogue voltages and then digitized with the help analogue-to-digital converter. Charge on a CCD is shifted in two directions namely parallel (or horizontal) and serial (or vertical). While parallel shift occurs from the right to the left, the serial shift is performed from top to bottom and directs the charge packets to the measurement electronics. One way to make the readout process faster is to split up into two or four different sections. Each section follows the process of parallel and serial shift. Figure 9.34 shows a typical CCD sensor with a single point read out. Figure 9.35 shows the photograph of a typical packaged CCD chip.

The CCD sensor described in the previous paragraphs can only determine number of photons collected by each pixel and therefore it carries no information about the wavelength or colour of those photons. As a result of that, the CCD sensor is capable of recording the image only in monochrome. In order to record image in full colour, a filter array is bonded to the sensor substrate. One such common colour filter array is the Bayer filter. Bayer's colour filter array (CFA) comprises of an arrangement of red, green and blue filters to capture colour information. It has an alternating red/green and blue/green arrangement and is sometimes called an RGBG filter. Bayer's CFA is made up of alternating rows of red/green and blue/green filters. Figure 9.36 shows a CCD array with a Bayer's filter bonded to its surface. A particular colour

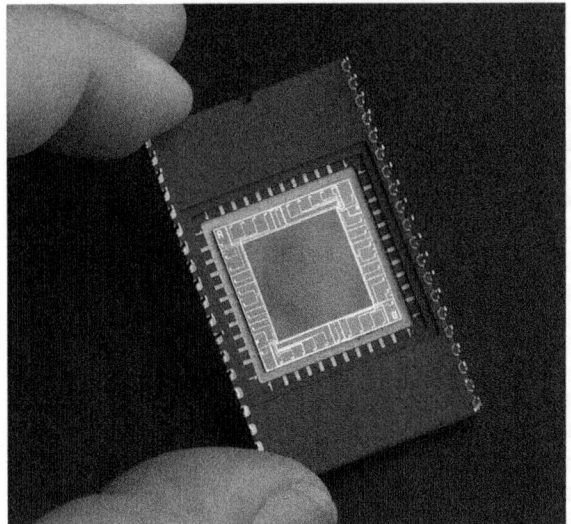

Figure 9.35 Packaged CCD chip. (*Source:* NASA).

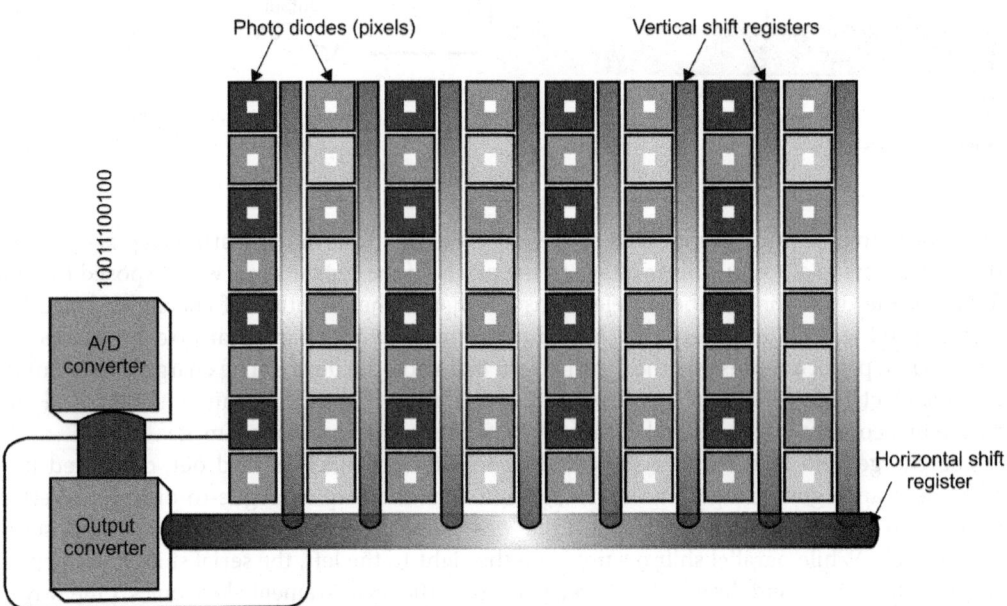

Figure 9.36 CCD array with Bayer's filter.

filter allows photons of only that colour to pass through to the pixel. This is illustrated in Figure 9.37. The number of photons collected by each pixel in this case corresponds to the colour allowed by the filter above it. In a Bayer pattern filter, there are twice as many green squares as there are red or blue. This is because the human eye is much more sensitive to green light than either red or blue, and has a much greater resolving power in that range.

There is space between adjacent light sensitive pixels and this space is used to house on-chip electronics. Micro lenses placed above the filter help in directing light into one or other of the adjacent pixels. As outlined earlier, a single pixel allows photons of one colour only. Full colour image is worked out by using a complex method called *demosaicing*. Simply stated, the camera

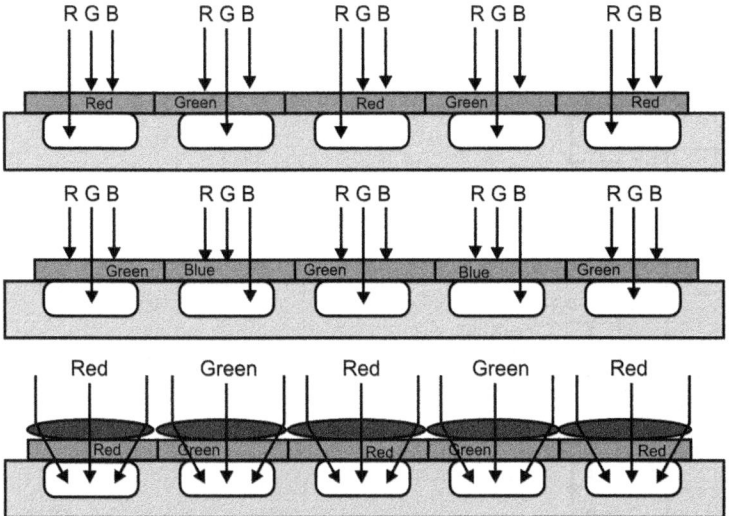

Figure 9.37 Colour discrimination by Bayer's filter.

treats each 2×2 set of pixels as a single unit, thereby providing one red, one blue and two green pixels. The actual colour is estimated based on the photon levels in each of these four pixels.

9.9.1.1 Electron Multiplication CCD

Electron-multiplying CCD (EMCCD) technology helps in overcoming the shortcoming of traditional CCD technology in offering high sensitivity at high-speed. Traditional CCD sensors backed by readout noise figures of typically less than 10 electrons offer high sensitivity. High sensitivity, however, comes at the cost of reduced speed of readout, which is typically less than 1 MHz. The speed constraint arises from the bandwidth limitation of the CCD charge amplifier. Increasing bandwidth, which is essential for high-speed operation, increases noise too. That is, high-speed amplifiers are relatively noisier than their slow speed counterparts. Electron-multiplying CCD overcomes this limitation by building a unique electron-multiplying structure into the chip. As a result of this, an EMCCD as an image sensor is capable of detecting single photon events without an image intensifier. EMCCD sensors achieve high sensitivity, high-speed operation by amplifying charge signal before the charge amplifier.

Most electron-multiplying CCDs utilize a frame-transfer CCD structure shown in Figure 9.38 In the case of frame-transfer CCD format, sensor area captures the image and storage area stores the image prior to read out. The storage area is normally identical in size to the sensor area and is covered with an opaque mask, normally made of aluminium. When the sensor area is exposed to light, an image is captured, which is automatically shifted downwards behind the masked region of the chip, and subsequently read out. During the readout process, another image is being captured by the sensor area. The aluminium mask therefore acts like an electronic shutter. As shown in Figure 9.38, there is a multiplication register between the normal serial readout register and the charge amplifier. To read out the sensor the charge is shifted out through the readout register and through the multiplication register where amplification occurs prior to readout by the charge amplifier. The multiplication register has several hundred stages or cells that use higher than normal clock voltages to achieve charge amplification. The amplification occurs in the multiplication register through a process known as *clock-induced charge* or *spurious charge* that occurs naturally in CCDs.

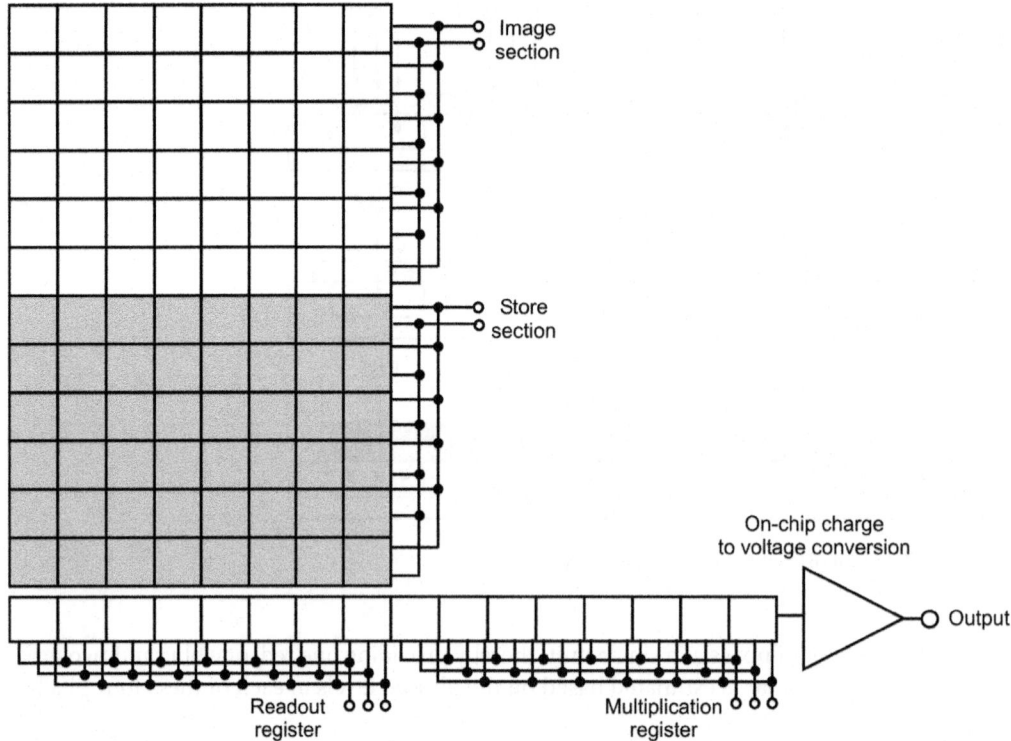

Figure 9.38 Electron multiplying CCD.

The multiplication register contains many hundreds of cells as outlined before and the amplification process occurs in each cell by harnessing clock-induced charge process. When clocking the charge through a register there is a very small but finite probability (typically less than 2%) that the charges being clocked can create additional charges by a process known as *impact ionization*. Impact ionization is the process by which a charge having sufficient energy creates another electron-hole pair. Hence a free electron charge in the conduction band can create another charge leading to amplification. The electron multiplication factor M may be computed from $M = (1 + p)^N$ where N is number of cells and p is the probability value. If the multiplication register for instance had 512 cells or stages and the probability of secondary electron generation were 1.3%, the multiplication factor would be around 744. The electron multiplication prior to the output amplifier ensures that the readout noise introduced by the amplifier has negligible effect.

The major advantages of EMCCD sensors include high sensitivity in low-light conditions, high-speed imaging capability, good daytime imaging performance and reduced likelihood of sensor damage while viewing in bright conditions. By elevating photon-generated charge above the readout noise of the device, even at high frame rates, the EMCCD has the capability of meeting the needs of ultra-low-light imaging applications without the use of external image intensifiers. The disadvantage is its relatively higher power consumption due to need for active cooling of the CCD. The extreme low-light capability of these EMCCDs lend themselves well to a range of applications including border and coastal surveillance, surveillance of ports and airports, protection of sensitive sites and critical assets and low-light scientific imaging such as in astronomy.

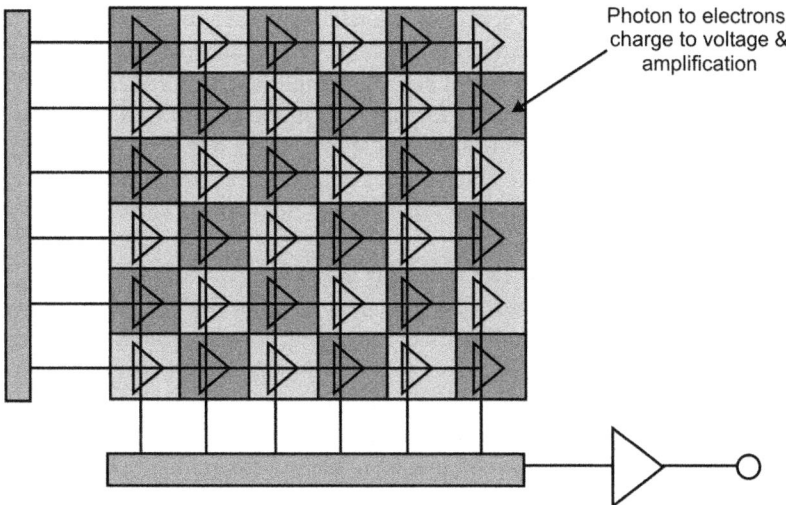

Photon to electrons,
charge to voltage &
amplification

Figure 9.39 CMOS sensor.

9.9.2 CMOS Sensors

Both CCD and CMOS sensors depend for their operation on photo electric effect to create electrical signal from light photons. CCD (charge coupled device) and CMOS (complementary metal oxide semiconductor) image sensors are two different technologies for capturing images digitally. Each has unique strengths and weaknesses providing advantages in different applications.

In a CCD sensor, quantum of charge held by different pixels is transferred through one or a very limited number of output nodes. Each pixel's charge is converted to a proportional voltage and after buffering sent off-chip as an analogue signal. In a CCD sensor, all of the pixel can be devoted to light capture and the output's uniformity, which is a key factor in image quality, is high. On the other hand, in a CMOS sensor, each pixel has its own charge-to-voltage convertor, amplifier and a pixel select switch (Figure 9.39). This is called *active pixel sensor architecture* in contrast to *passive pixel sensor architecture* used in a CCD sensor. Also, the sensor often also includes on-chip amplifiers, noise-correction and analogue-to-digital conversion circuits and other circuits critical to pixel sensors' operation. The chip in this case outputs digital bits. Inclusion of these functions reduces the area available for light capture. Also, with each pixel doing its own conversion, uniformity and consequently image quality is lower. While read out mechanism of a CCD sensor is serial, it is very much parallel in the case of a CMOS sensor allowing a high total bandwidth for high-speed.

9.9.3 CCD Sensors versus CMOS Sensors

In the following paragraphs, we shall discuss some of the major differences between CCD and CMOS sensors. Some of the key differences are as follows.

1) Fabrication of CMOS sensors involves standard CMOS technology well-established for fabrication of integrated circuits. This also allows easy integration of required electronics on the same chip thereby resulting in devices that are compact and cost effective. On the other hand, CCD sensor fabrication involves dedicated and costly manufacturing processes.

2) In comparison with CCD sensors, CMOS sensors have relatively poor sensitivity and uniformity. The former is due to poor fill factor in the case of CMOS sensors and a large fill factor in CCD devices. The latter is due to use of separate amplifiers for different pixels in the case of CMOS sensors and a single amplifier for all pixels in the case of CCD sensors.

3) CMOS sensors have a higher speed than CCD devices due to use of active pixels and having analogue-to-digital converters on the same chip thus leading to lesser propagation delays. Low end CMOS sensors have low power requirements, but high-speed CMOS cameras typically require more power than CCD devices.

4) Dynamic range of a CCD sensor is higher than that of a CMOS sensor.

5) As far as the noise performance is concerned, CMOS sensors score over CCD sensors when it comes to temporal noise due to lower bandwidth of amplifiers at each pixel. But in terms of fixed pattern noise performance, CCD sensors are better due to single point charge-to-voltage conversion.

6) CMOS sensors allow on-chip incorporation of a range of functions such as automatic gain control, auto-exposure control, image compression, colour encoding, motion tracking and so on.

7) Due to absence of shift registers in CMOS sensors, they are immune to smearing around over-exposed pixels as it is caused by spilling of charge into the shift register.

In view of these described differences, it is evident that CCD sensors are used in cameras that offer excellent photo sensitivity and focus on high-quality, high-resolution images. CMOS sensors on the other hand are generally characterized by lower image quality, resolution and photo sensitivity. CMOS cameras are usually less expensive and due to lower power consumption have great battery life. CMOS sensors are, however, fast improving to a level where they can achieve near parity with CCD devices in some applications.

9.10 Photo Emissive Sensors

The photo sensors discussed so far are based on internal photo effect where the photo electrons generated by the incident radiation remain within the semiconductor material. The other category of photo sensors is those that rely on an external photo effect where the photo generated electrons travel beyond the physical boundaries of the material. These sensors are also referred to as photo emissive sensors. In this section, we discuss some of the commonly used photo emissive photo sensors including vacuum photo diodes, photo multiplier tubes and image intensifier tubes.

9.10.1 Vacuum Photo Diodes

A vacuum photo diode is the oldest photo senor. It comprises of an anode and a cathode placed in a vacuum envelope. Cathode when irradiated, releases electrons that are attracted by the positively charged anode, thus producing a photocurrent proportional to the light intensity.

9.10.2 Photo Multiplier Tubes (PMTs)

PMTs are extremely sensitive photo sensors operating in the UV, visible and near IR spectrum. PMTs have internal gain in the order of 10^8 and can even detect single photon of light. They are constructed from a glass vacuum tube which houses a photocathode, several dynodes and an anode. When the incident photons strike the photocathode, electrons are produced as a result

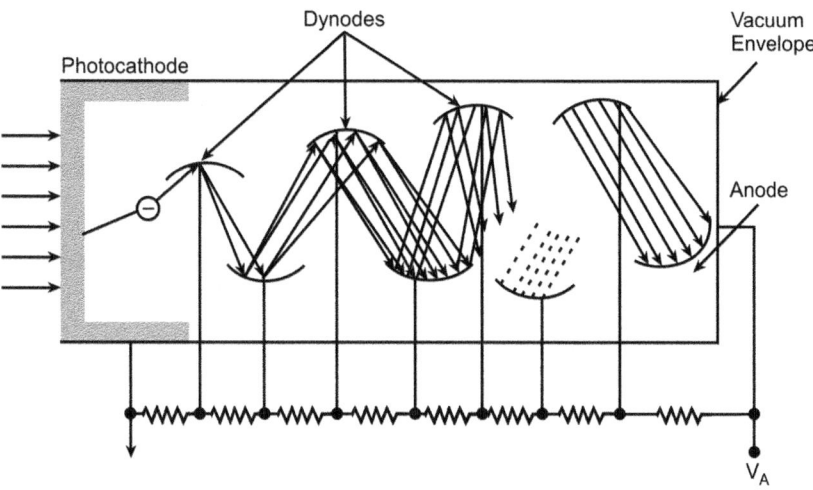

Figure 9.40 Cross-section of a photo multiplier tube.

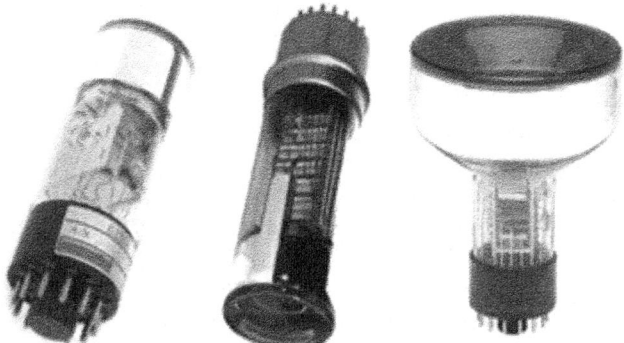

Figure 9.41 Photo multiplier tubes.

of the photo electric effect. These electrons accelerate towards the anode and in the process electron multiplication taken place due to secondary emission process from the dynodes. PMTs require a few kilovolts of biasing voltages for proper operation. Figure 9.40 shows the cross-section of a photo multiplier tube. As we can see from the figure, the dynodes are given progressively increasing positive voltages with the dynode nearest to the cathode having the lowest voltage and the dynode nearest to the anode with the maximum voltage.

Salient features of photo multiplier tubes include low noise, high-frequency response and large active area. By virtue of these features, PMTs are used in nuclear and particle physics, astronomy, medical imaging and motion picture film scanning. Avalanche photo diodes have replaced PMTs in some applications, but PMTs are still used in many cases. Figure 9.41 shows photographs of some representative PMTs.

9.10.3 Image Intensifiers

Image intensifiers are devices that amplify visible and near IR light from an image so that a dimly lit scene can be viewed by a camera or by human eye. Contemporary image intensifiers comprise of an objective lens, vacuum tube with photocathode at one end, tilted micro-channel

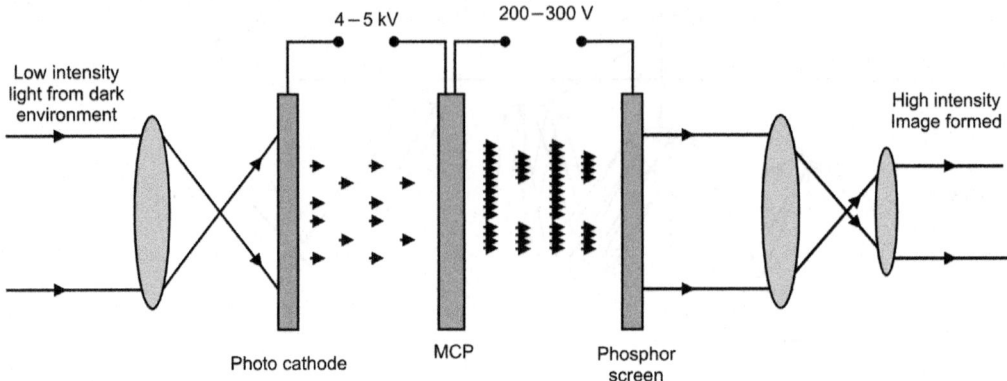

Figure 9.42 Image intensifier tube.

plate (MCP) and a phosphor screen (Figure 9.42). Objective lens focuses the image on to the photocathode. When the photons strike the photocathode, electrons are released due to the photo electric effect. These photo electrons are accelerated through around 4–5 kV into a tilted micro-channel plate where secondary electron multiplication takes place. The electrons all move together due to the potential difference across the tube and for each photo electron hundreds or even thousands of electrons are created. All these electrons hit the phosphor screen at the other end, releasing one photon for every electron. Thus, the screen converts the high-energy electrons into photons, which corresponds to the input image radiation but with the incident flux being amplified many times.

Image intensifiers are classified into the following categories; namely generation 0, 1, 2, 3 and 4. Generation 0 and 1 devices did not have the MCP and the stream of electrons generated by the photocathode was accelerated towards the phosphor screen by the applied potential. Generation 1 devices were a tremendous improvement upon generation 0 devices and had three times the photo sensitivity of generation 0 devices. Generation 2 devices introduced the concept of micro-channel plates. Generation 3 devices are the same as generation 2 devices except that the photocathode material in these devices is gallium arsenide (GaAs) whereas it was S-25 in the case of generation 2 devices. Also, generation 3 devices had a better MCP. Generation 3 ultra and generation 4 tubes are also available that offer slight improvement over generation 3 devices. Image intensifiers are used in night vision devices used for military applications. Night vision technologies based on image enhancement and thermal imaging are discussed in detail in a latter section. These are discussed with reference to operational fundamentals, advantages and limitations, related military devices and application areas.

9.11 Thermal Sensors

Thermal sensors absorb radiation, which produces a temperature change that in turn causes a change in the physical or the electrical property of the sensor. In other words, thermal sensors respond to change in their bulk temperature caused by the incident radiation. Thermocouple, thermopile, bolometer and pyroelectric sensors belong to the category of thermal sensors. Thermal sensors lack the sensitivity of photo electric sensors and are generally slow in response, but have a wide spectral response. Most of these sensors are passive devices, requiring no bias. In this section, we discuss different types of thermal sensors and their application circuits.

9.11.1 Thermocouple and Thermopile

Thermocouple sensors are based on the Seebeck effect; that is, the temperature change at the junction of two dissimilar metals generates an EMF proportional to the temperature change. The commonly used thermocouple materials are bismuth-antimony, iron-constantan and copper-constantan. Their temperature coefficients are $100\,\mu V/°C$, $54\,\mu V/°C$ and $39\,\mu V/°C$, respectively. To compensate for the changes in the ambient temperature, thermocouples generally have two junctions namely the measuring junction and the reference junction (Figure 9.43).

The responsivity of a single thermocouple is very low and therefore to increase the responsivity, several junctions are connected in series to form a thermopile (Figure 9.44). Thermopiles are series combination of around 20–200 thermocouples. The spectral response of thermocouples and thermopiles extends into the far IR band up to $40\,\mu m$. They are suitable for making measurements over a large temperature range up to 1800 K. However, thermocouples are less suitable for applications where smaller temperature differences need to be measured with great accuracy such as 0–100 °C measurement with 0.1 °C accuracy. For such applications, thermistors and RTDs are more suitable.

The responsivity of thermopiles is in the order of $10–100\,V/W$ and the typical signal output varies from few tens of microvolt to few millivolts. Hence, they need low noise and very low offset operational amplifier for providing the gain. The gain required varies from as small as 10 to as large as 10 000 or more. Generally, for gain less than 1000, a single-stage amplifier is used. For gain values more than 1000, two stages are used. Figure 9.45 shows the application circuit where two amplifier stages are used. As we can see from the figure, thermopiles require no bias voltage.

The thermopile signal would be positive or negative depending upon whether the temperature of the object filling the thermopile's field-of-view is greater than or less than that of the thermopile. Also, the output of the circuit varies with change in the ambient temperature. It is therefore necessary to compensate for the ambient temperature variations. Many thermopile modules have an inbuilt thermistor to compensate for the ambient temperature variations.

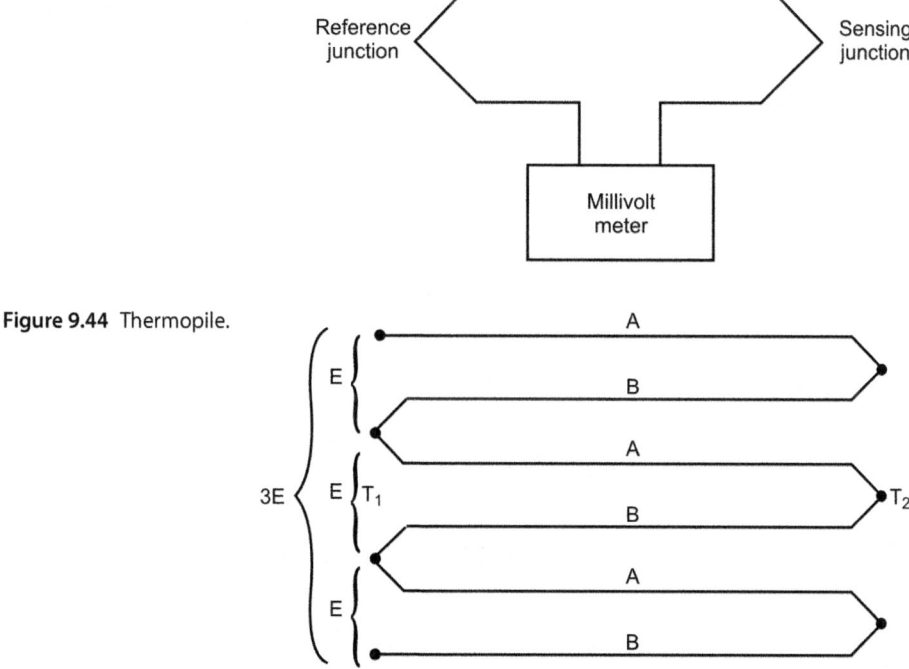

Figure 9.43 Thermocouple.

Figure 9.44 Thermopile.

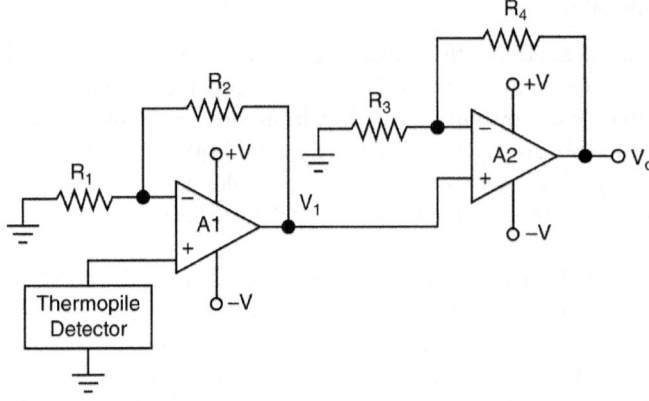

Figure 9.45 Application circuit using thermopiles.

9.11.2 Bolometer

The bolometer is the most popular type of thermal sensor. The sensing element in a bolometer is a resistor with a high-temperature coefficient. A bolometer is different from photoconductor as, in a photoconductor, a direct photon-electron interaction causes a change in the conductivity of the material whereas in a bolometer the increased temperature and the temperature coefficient of the element causes the resistance change. Bolometers can be further categorized as metal, thermistor and low temperature germanium bolometers.

A *metal bolometer* uses metals such as bismuth, nickel or platinum with temperature coefficient in the range of 0.3–0.5%/C. Thermistor bolometers are the most popular and they find applications in burglar alarms, smoke sensors and other similar devices. The sensor in this case is a thermistor, an element made of manganese, cobalt and nickel oxide. They have high-temperature coefficients up to 5%/°C. The temperature coefficient varies with temperature as $1/T^2$. They are classified as negative temperature coefficient (NTC) and positive temperature coefficient thermistors (PTC) depending upon whether their temperature coefficient of resistance is negative or positive. Figure 9.46(a) shows the circuit symbol of a thermistor. The spectral response of thermistors extends from 0.5 to 10 μm. More sensitive thermistors typically have an NEP and response time in the order of 10^{-10} W and 100 ms. Less sensitive thermistors

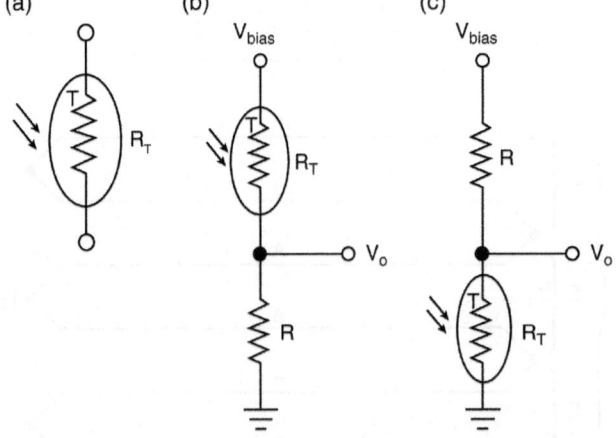

Figure 9.46 (a) Circuit symbol of a thermistor, (b) and (c) application circuits of a thermistor.

have the figures of 10^{-8} W and 5 ms, respectively. Figure 9.46 parts (b) and (c) shows the simplest possible configurations in which a thermistor can be used for measurement of light intensity. The figures are self-explanatory. The output of the circuits in Figure 9.46 parts (b) and (c) can be fed to an operational amplifier or to a comparator for linear light control or light on-off control respectively. A low temperature germanium bolometer is a sensitive laboratory type bolometer that uses germanium as the sensor. It has the highest responsivity when operated at few degrees above the absolute zero temperature.

9.11.3 Pyroelectric Sensor

Pyroelectric sensors are characterized by spontaneous electric polarization, which is altered by temperature changes as light illuminates these sensors. Pyroelectric sensors are low-cost, high-sensitivity devices that are stable against temperature variations and electromagnetic interference. Pyroelectric sensors only respond to modulating light radiation and there will be no output for a CW incident radiation. Figure 9.47(a) shows the circuit symbol of a pyroelectric sensor.

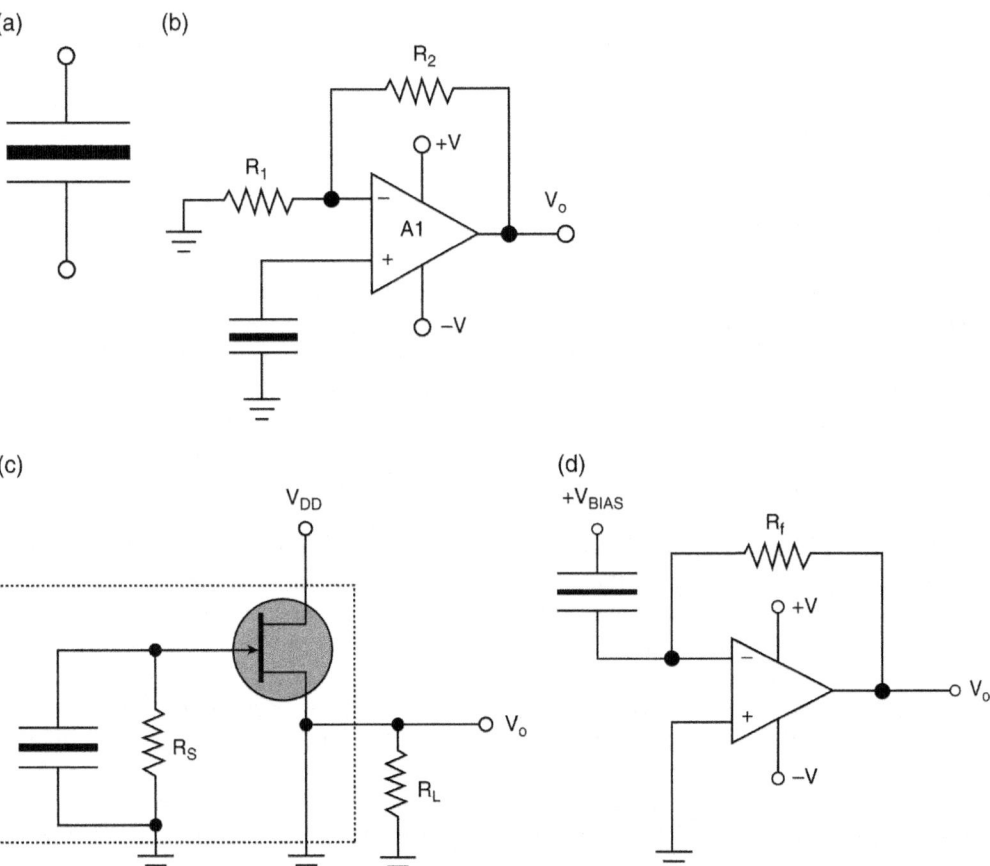

Figure 9.47 (a) Circuit symbol of a pyroelectric sensor, (b) and (c) voltage mode pyroelectric sensor application circuits, and (d) current mode pyroelectric sensor application circuit.

Pyroelectric sensors operate in two modes namely the voltage mode and the current mode. In the voltage mode, the voltage generated across the entire pyroelectric crystal is detected. In the current mode of operation, current flowing on and off the electrode on the exposed face of the crystal is detected. Voltage mode is more commonly used than the current mode.

The circuit for voltage mode is shown in Figure 9.47(b). The operational amplifier chosen should have very high input impedance in the order of 1012–1014 Ω. But the circuit is sensitive to ambient temperature variations. Ambient temperature variations can be compensated for by employing AC coupling between the amplifier stages or by adding a compensation crystal in opposition, either in series or parallel. One crystal is exposed to radiation and the other is shielded from radiation. As the ambient temperature changes, the surface charge generated on one crystal is cancelled by the equal and opposite charge generated on the other crystal. The incident radiation, however, generates charge only on one crystal and is not cancelled.

Voltage mode pyroelectric sensors are generally integrated with a FET. A shunt resistor (R_S) in the range of 1010–1011 Ω is added to provide thermal stabilization. External connections include a power supply and load resistor R_L as shown in Figure 9.47(c). The output voltage appears across R_L.

The circuit for current mode operation is shown in Figure 9.47(d). The modulation frequency can be much higher in the case of current mode operation than it is in the case of voltage mode operation. Hence, it is much easier to separate the signal from the ambient temperature drift.

9.12 Light-Emitting Diodes (LEDs)

An LED is a semiconductor P-N junction diode designed to emit light when forward biased. It is one of the most popular optoelectronic sources. LEDs consume very little power and are inexpensive.

When a P-N junction is forward biased, the electrons in the N-type material and the holes in the P-type material travel towards the junction. Some of these holes and electrons recombine with each other and in the process radiate energy. The energy will be released either in the form of photons of light or in the form of heat. In silicon and germanium diodes, most of the energy is released as heat and the emitted light is insignificant. However, in some materials like gallium phosphide (GaP), gallium arsenide (GaAs) and gallium arsenide phosphide (GaAsP) substantial photons of light are emitted. Hence, these materials are used in the construction of light-emitting diodes.

In the absence of an externally applied voltage, the n-type material contains electrons while the p-type material contains holes that can act as current carriers. When the diode is forward biased, the energy levels shift and hence there is a significant increase in the concentration of electrons in the conduction band on the n-side and that of holes in valence band on the p-side. These electrons and holes combine near the junction to release energy in the form of photons (Figure 9.48). It may be mentioned here that the process of light emission in a LED is that of spontaneous emission; that is, the photons emitted are not in phase and travel in different directions.

The energy of the photon resulting from this recombination is equal to the band gap energy of the semiconductor material and is expressed by the empirical formula given in eqn. 9.26.

$$\lambda = \frac{1240}{\Delta E}$$

(9.26)

Figure 9.48 P-N junction of a LED.

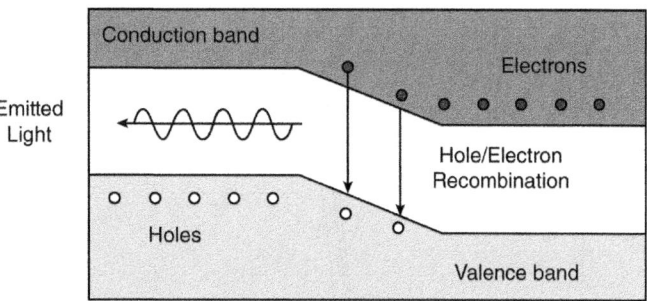

Table 9.1 Band gap energy and the typical wavelengths of commonly used LED materials.

Material	Band gap energy (eV)	Wavelength (nm)
GaAs	1.43	910
GaP	2.24	560
$GaAs_{60}P_{40}$	1.91	650
AlSb	1.60	775
InSb	0.18	6900

Where,

λ = Wavelength (nm)
ΔE = Band gap energy (eV).

Some of the commonly used semiconductor materials used for fabricating LEDs are gallium arsenide (GaAs), gallium phosphide (GaP), gallium arsenic phosphide (GaAsP), aluminium antimonide (AlSb) and indium antimonide (InSb). Table 9.1 enlists the band gap and the typical wavelengths emitted by these materials.

9.12.1 Characteristics

The characteristics of interest in a LED are the V-I characteristics, spectral distribution curve, light output versus input current and the directional characteristics.

1) *V-I characteristics*: Figure 9.49(a) shows the V-I characteristics of LEDs of different colours. As the LED is operated in the forward-biased mode, the V-I characteristics in the forward-biased region are shown. V-I characteristics of LEDs are similar to that of conventional P-N junction diodes except that the cut-in voltage in the case of LEDs is in the range of 1.3–3 V as compared to 0.7 V for silicon diodes and 0.3 V for germanium diodes.
2) *Spectral distribution curve* shows the variation of light intensity with wavelength. Figure 9.49(b) shows the typical spectral curves for yellow, green and red LEDs.
3) *Light output versus input current*: Figure 9.49(c) shows a typical light output versus input current curve depicting the dependence of emitted light on forward current flowing through the LED.
4) *Directional characteristics* refer to the variation in the light output with change in the viewing angle [Figure 9.49(d)].

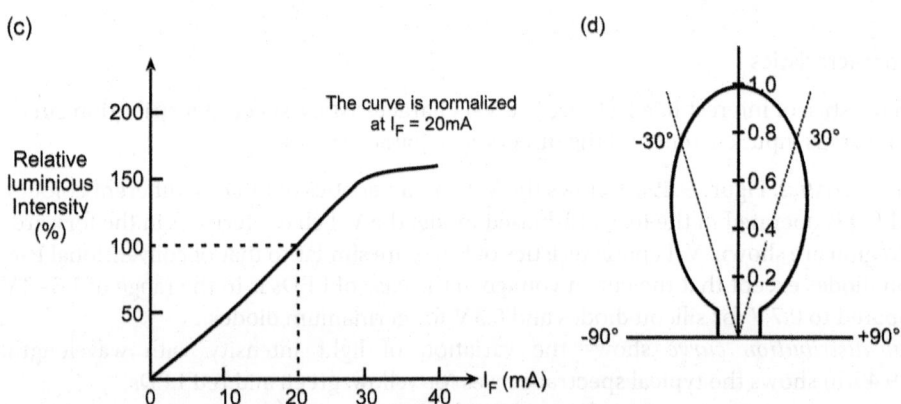

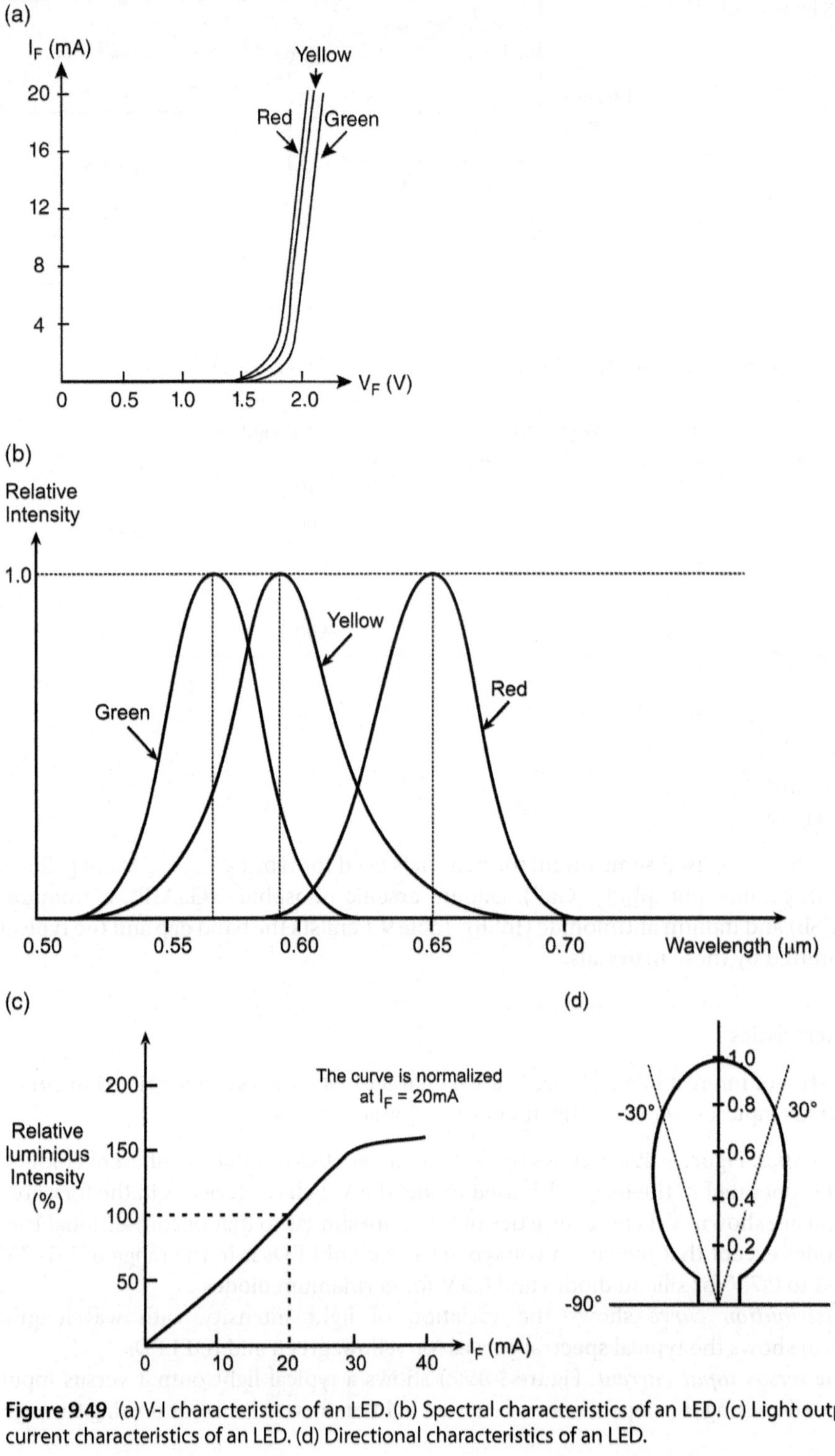

Figure 9.49 (a) V-I characteristics of an LED. (b) Spectral characteristics of an LED. (c) Light output versus input current characteristics of an LED. (d) Directional characteristics of an LED.

9.12.2 Parameters

The parameters of interest in the case of LEDs are forward voltage (V_F), candle power (CP), radiant power output (P_o), peak spectral emission (λ_P) and spectral bandwidth.

1) *Forward voltage* (V_F): is the DC voltage across the LED when it is ON. The typical values of V_F for LEDs are in the range of 1.3–3.0 V. As we can see from Figure 9.49(a), V_F is near 1.5 V for yellow, green and red LEDs.
2) *Candle Power* (CP): is a measure of the luminous intensity or the brightness of the light emitted by the LED. It is the most important parameter of a LED. It is a nonlinear function of LED current and the value of CP increases with increase in the current flowing through the LED.
3) *Radiant Power Output* (P_o): is the light power output of the LED.
4) *Peak spectral emission* (λ_P): λ_P is the wavelength where the intensity of light emitted by the LED is maximum.
5) *Spectral bandwidth*: gives an indication as to how concentrated the brightest colour is around its nominal wavelength.

9.12.3 Drive Circuits

LEDs are operated in the forward-biased mode. As the current through the LED changes very rapidly with change in the forward voltage above the threshold voltage, LEDs are current driven devices. Figure 9.50(a) shows a simple circuit for driving an LED. The resistor R is the current limiting resistor used to limit the current flowing through the LED. In this case the V-I characteristics of the LED is used to determine the voltage that needs to be applied to the LED to generate the desired forward current. A silicon diode can be placed inversely parallel to the LED for reverse polarity voltage protection. The current that will flow through the LED is given by eqn. 9.27.

$$I = (V_{cc} - V_f)/R \tag{9.27}$$

Where,

V_{CC} is the supply voltage
V_F is the forward diode voltage
R is the current limiting resistor.

However, any change in the forward voltage of the LED due to temperature changes or variation from device to device causes a change in the LED current. Moreover, there is power dissipation across the series resistor R, resulting in reduced efficiency. A better drive circuit configuration is the one that employs a constant current source as shown in Figure 9.50(b). The current flowing through the LED is determined by the reference voltage V_{REF} and the resistor R.

LEDs can also be used to display the logic output states. Figure 9.50 parts (c) and (d) show typical logic circuits that can be used to drive LEDs. Figure 9.50(c) uses a transistor-based switch while Figure 9.50(d) employs a logic gate/buffer. In both the circuits, the LED glows when the voltage V_{in} is in the logic HIGH state.

When the light emitted by one LED is not sufficient, several LEDs can be connected in series to enhance the light level to the desired value. LEDs can be connected in series as shown in Figure 9.51(a). In a series connection, the current flowing through each LED is the same. The value of the supply voltage V_{CC} should be sufficiently large to drive the desired number of LEDs. Also, it should be checked that the series current flowing in the circuit through each LED is in the operating range.

(a)

(b)

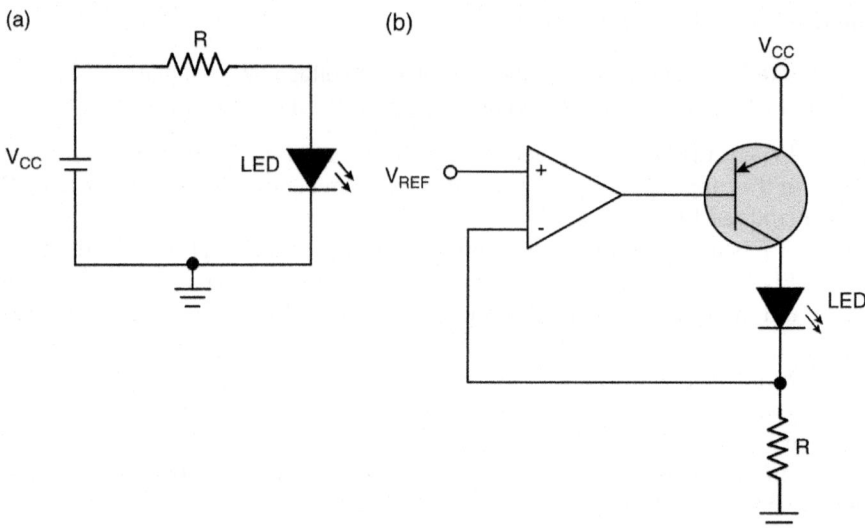

(c)

(d)

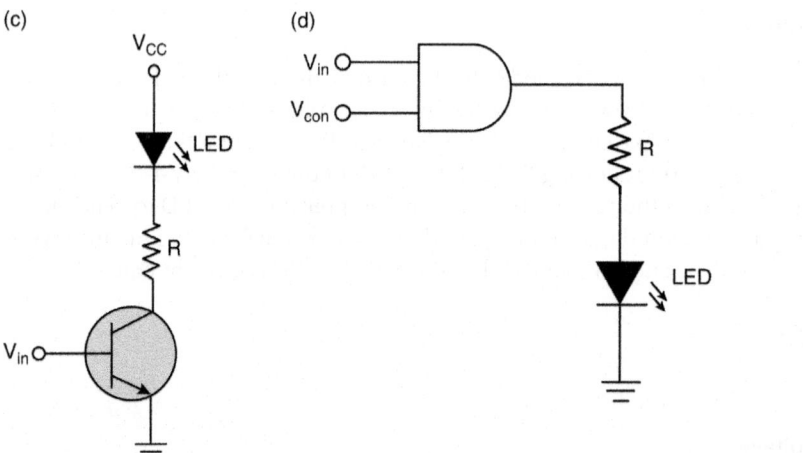

Figure 9.50 (a) Simple LED circuit. (b) Constant current source based LED driver circuit. (c), (d) Logic circuits for driving LEDs.

The value of the resistor R to be connected is given by eqn. 9.28.

$$R = \left[V_{cc} - \left(V_{f1} + V_{f2} + V_{f3} \ldots\ldots + V_{fn} \right) \right] / I \qquad (9.28)$$

Where,

V_{f1}, V_{f2}......V_{fn} are the forward voltages across LED_1, LED_2..... LED_n, respectively
I is the current flowing through the LEDs.

LEDs can also be connected in parallel to enhance the output light level. However, in the case of parallel connection, one needs to be more careful. Figure 9.51(b) shows the parallel connection of LEDs. The resistors R_1, R_2....R_n are used to protect the diodes. if these resistors were not used, then the LED with the lowest forward voltage will draw excess current and is likely to get damaged. Then the LED with the next smallest forward voltage will get damaged and this

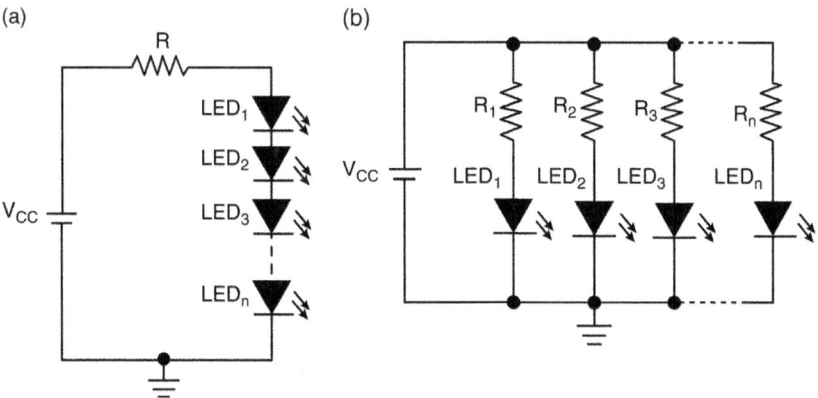

Figure 9.51 (a) Connecting LEDs in series. (b) Connecting LEDs in parallel.

process continues till all the LEDs get damaged. The values of these resistors determine the current flowing through individual LEDs.

9.13 Displays

Displays are output devices that are used for visual presentation of information. Displays form an interface between the machine and the human. In this section, we shall discuss different types of displays and the characteristic parameters used to define the quality of displays.

9.13.1 Characteristics

Three factors are critical for a good visual display, namely legibility, brightness and contrast.

1) *Legibility* is the property of a display by virtue of which the characters are easy to read with speed and accuracy. The factors which contribute to the legibility of the display are its style, size, character sharpness and shape.
2) *Brightness* refers to the perception of luminance by the visual world.
3) *Contrast* of a display depends on the background luminance and the source luminance. The readability of the display depends upon the contrast parameter. It is defined in different ways for passive and active displays. In the case of passive displays such as LCD, contrast is defined by eqn. 9.29.

$$C = (L_O - L_B)/L_O \tag{9.29}$$

Where,

L_O is the object or source luminance (Cd/m^2)
L_B is the background luminance (Cd/m^2).

Contrast can have values between 0 and 1, zero being the case when the object and the background luminance are the same and one when the background has zero luminance.

For active displays such as LED displays, the contrast parameter is defined in terms of contrast ratio, which is defined by eqn. 9.30.

$$CR = L_O/L_B \tag{9.30}$$

Where,

L_O is the object or source luminance (Cd/m^2)
L_B is the background luminance (Cd/m^2).

The contrast ratio can have values between one and infinity. For a contrast ratio equal to 1, the object and the background have the same luminance and the displayed characters are not visible at all. The background luminance is zero and the display has best visibility at contrast value equal to infinity.

9.13.2 Types of Displays

Displays can be categorized into different types depending upon the manner in which they display information. These include bar graph displays, segmented displays, dot-matrix displays and large displays.

Bar graph displays are composed of several bar elements as shown in Figure 9.52. Bar graph displays are replacing analogue instruments as indicators due to their simplicity and cost-effectiveness. *Segmented displays* are available in two configurations namely the seven-segment displays as shown in Figure 9.53(a) and 16-segment displays as shown in Figure 9.53(b).

A *seven-segment display* comprises of seven bars and one or two decimal points and is an industry standard for numeric displays. These displays are used for displaying numerals and limited alphabet. Sixteen-segment displays can present the entire upper-case alphabet and numerals. Segmented displays can present only limited information. *Dot-matrix displays* are the simplest displays that represent the set of lower-case and upper-case alphabets and numerals at reasonable cost and complexity. The most commonly used dot-matrix display is the 5×7 display, which comprises 35 display elements set in a pattern of five rows and seven

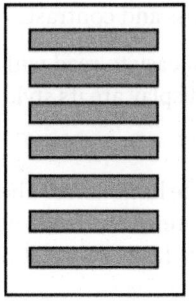

Figure 9.52 Bar graph display.

(a)　　　　　(b)

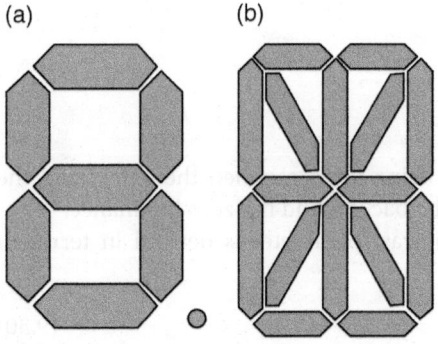

Figure 9.53 (a) Seven-segment displays and (b) 16-segment displays.

Figure 9.54 A 5×7 dot-matrix display.

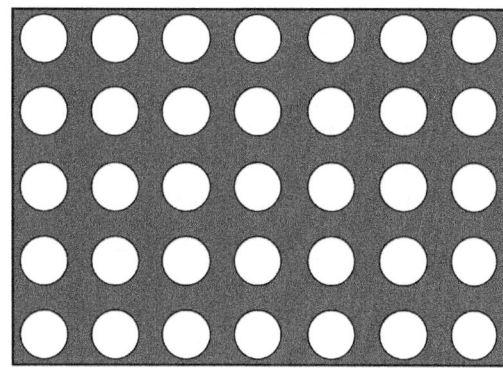

columns (Figure 9.54). Each element is addressed selecting the proper row and column. Bar, segmented and 5×7 dot-matrix displays can be constructed using light-emitting diodes (LED) or liquid crystal displays (LCD). Large-scale displays include cathode ray tube (CRT) displays, plasma displays, LCD TFT displays and so on.

9.13.3 Liquid Crystal Displays

Liquid crystals are materials that exhibit properties of both solids and liquids; that is, they are an intermediate phase of matter. They can be classified into three different groups, namely nematic, smectic and cholestric. Nematic liquid crystals are generally used in LCD fabrication with the twisted nematic material being the most common.

9.13.3.1 Construction

Figure 9.55(a) shows the construction of a twisted nematic LCD display. As we can see from the figure, it comprises of a cell of liquid crystal fluid, conductive electrodes, a set of polarizers and a glass casing. Polarizers are components that polarize light in one plane. The polarizer attached to the front glass is referred to as the front polarizer while the one attached to the rear glass is the rear polarizer. On the inner surface of the glass casing, transparent electrodes are placed in the shape of the desired image. The electrode attached to the front glass is referred to as the segment electrode while the one attached to the rear glass is the backplane or the common electrode. The patterns of the backplane and segment electrodes form the numbers, letters, symbols and so on. The liquid crystal is sandwiched between the two electrodes. The basic principle of operation of LCD is to control the transmission of light by changing the polarization of the light passing through the liquid crystal with the help of an externally applied voltage. As LCDs do not emit their own light, backlighting is used to enhance the legibility of the display in dark conditions. A variety of methods exist for backlighting LCD panels such as use of incandescent lamps, LEDs and electro luminescent lamps.

LCDs have the capability to produce both positive as well as negative images. A positive image is defined as a dark image on a light background. In a positive image display, the front and the rear polarizers are perpendicular to each other. Light entering the display is guided by the orientation of the liquid crystal molecules that are twisted by 90° from the front glass plate to the rear glass plate. This twist allows the incoming light to pass through the second polarizer as shown in Figure 9.55(b). When voltage is applied, the liquid crystal molecules straighten out and stop redirecting light. As a result light travels straight through and is filtered out by the second polarizer. Therefore, no light can pass through making this region darker compared to the

(a)

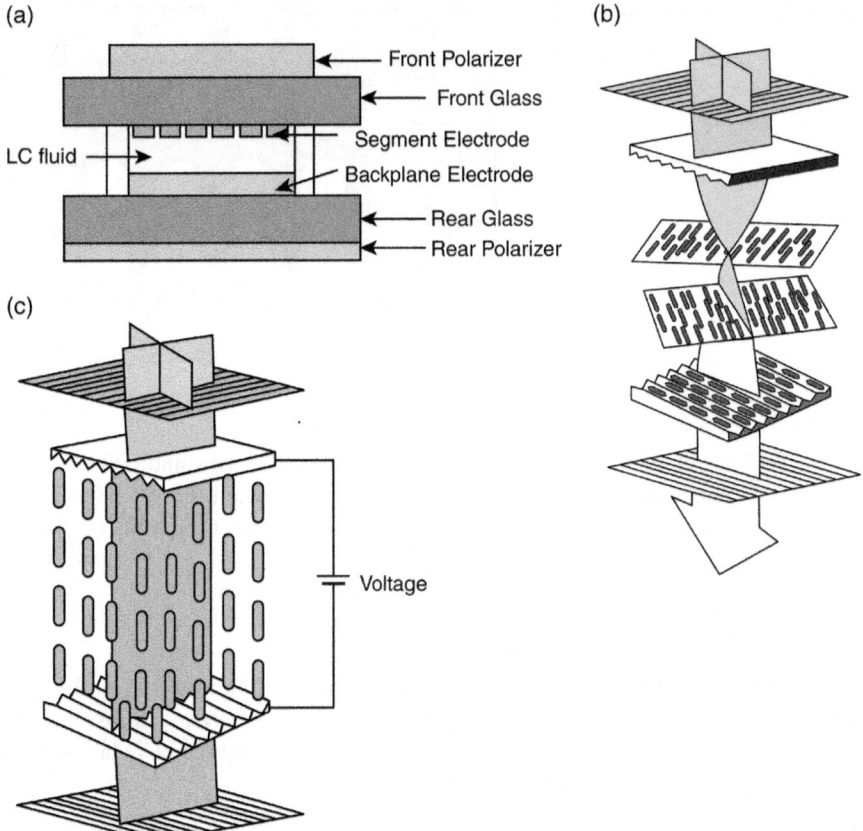

(b)

(c)

Voltage

Figure 9.55 (a) Cross-section of a twisted nematic LCD Display, and (b) and (c) Twisted nematic LCD operation.

rest of the screen as shown in Figure 9.55(c). Hence, in order to display characters or graphics, voltage is applied to the desired regions making them dark and visible to the eye.

A negative image is a light image on a dark background. In negative image displays, the front and the rear polarizer are aligned to each other.

9.13.3.2 Driving LCD

The LCD driver waveforms are designed to create a zero DC potential across all the pixels, as a DC voltage deteriorates the LC fluid such that it cannot be energized. LCDs are driven with symmetrical waveforms with less than 50 mV DC component. Figure 9.56 shows the brightness versus RMS drive voltage curve for LCDs. V_{ON} is the RMS voltage applied across the liquid crystal that creates an ON pixel that is typically at the 90% contrast level. V_{OFF} or V_{TH} is the RMS voltage across the liquid crystal when the contrast voltage reaches the 10% level. Another important specification is the discrimination ratio, which is defined as the ratio of V_{ON} to V_{OFF}. The discrimination ratio defines the contrast levels the LCD panel will achieve.

LCDs can be classified as direct drive and multiplex drive displays, depending upon the technique used to drive them. Direct drive displays also known as static drive displays have an independent driver for each pixel. The drive voltage in this case is a square waveform with two voltage levels, namely ground and V_{CC} as shown in Figure 9.57(a). In the figure, segment 0 is the ON segment whereas the segment 1 is the OFF segment. Direct drive displays offer the best contrast ratios over a wide operating temperature range. However, as the display size increases

Figure 9.56 Brightness versus drive voltage curve for LCD displays.

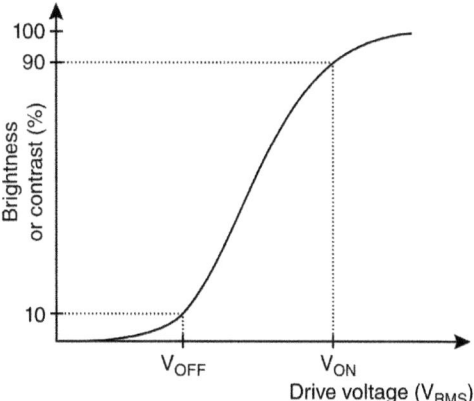

Figure 9.57 (a) Direct drive waveform. (b) Multiplexed drive waveform.

(a)

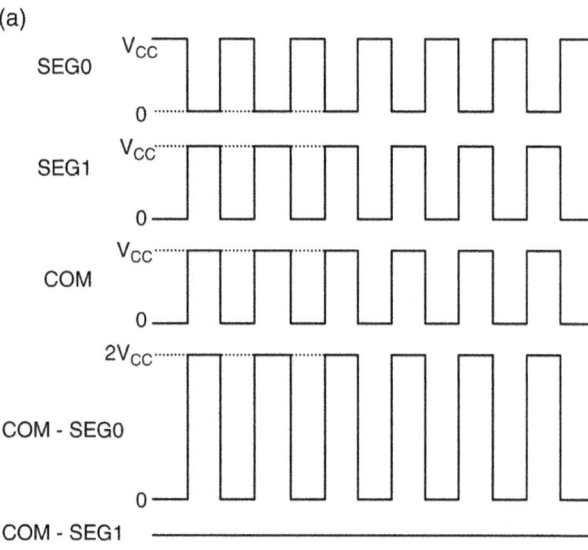

(b)

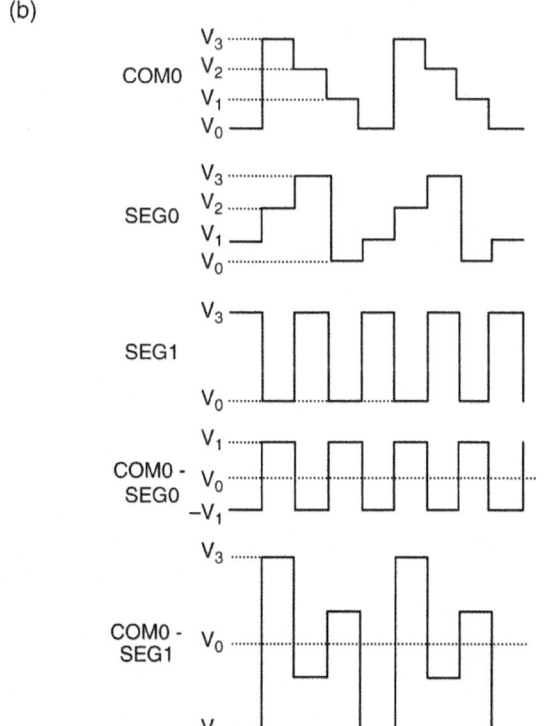

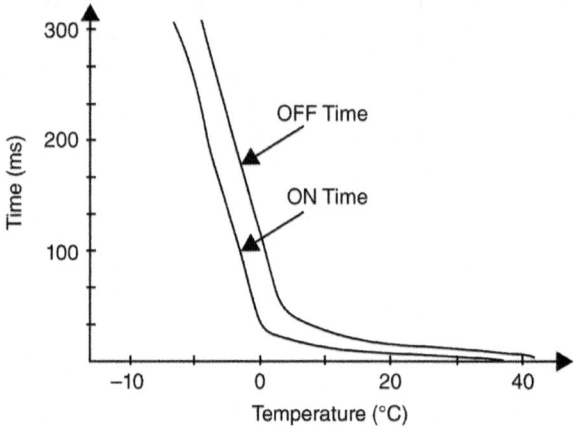

Figure 9.58 Variation of response time of an LCD display with temperature.

the drive circuitry becomes very complex. Hence, multiplex drive circuits are used for larger size displays. These displays reduce the total number of interconnections between the LCD and driver. They have more than one backplane and the driver produces an amplitude-varying, time synchronized waveforms for both the segments and backplanes. Figure 9.57(b) shows a typical multiplexed segment and the backplane drive waveforms. The segment 0 is inactive whereas segment 1 is active. Segment 0 is inactive as the voltage across the LCD never crosses its activation threshold voltage.

9.13.3.3   Response Time

The LCD response time is defined by the ON and OFF response times. ON time refers to the time required by an OFF pixel to become visible after the application of proper drive voltage. The OFF time is defined as the time required by the ON pixel to turn OFF after the application of proper drive voltage. The response time of LCDs varies widely with temperature and increases rapidly at low operating temperatures. Hence, LCDs can only operate at low temperatures when used along with temperature controllers. At high temperatures, the liquid crystal molecules begin to assume random orientations, resulting in the pixels on the positive image display becoming completely dark, while the pixels on the negative image becoming completely transparent. Figure 9.58 shows the typical variation of the ON and the OFF times of a LCD display with temperature.

9.13.3.4   Types

LCDs are non-emissive devices; that is, they do not generate light on their own. Depending upon the mode of transmission of light in a LCD, they are classified as reflective, transmissive and trans-reflective LCD displays.

Reflective LCD displays have a reflector attached to the rear polarizer, which reflects incoming light evenly back into the display. Figure 9.59 shows the principle of operation of reflective LCD displays. These displays rely on the ambient light to operate and do not work in dark conditions. They produce only positive images. The front and the rear polarizers are perpendicular to each other. These types of displays are commonly used in calculators and digital wrist watches.

In *transmissive LCD displays*, the back light is used as the light source. Most of these displays operate in the negative mode; that is, the text will be displayed in a light colour and the background is a darker colour. Figure 9.60 shows the basic construction of a transmissive display. Negative transmissive displays have front and rear polarizers in parallel with each other whereas positive transmissive displays have the front and the rear polarizers perpendicular to

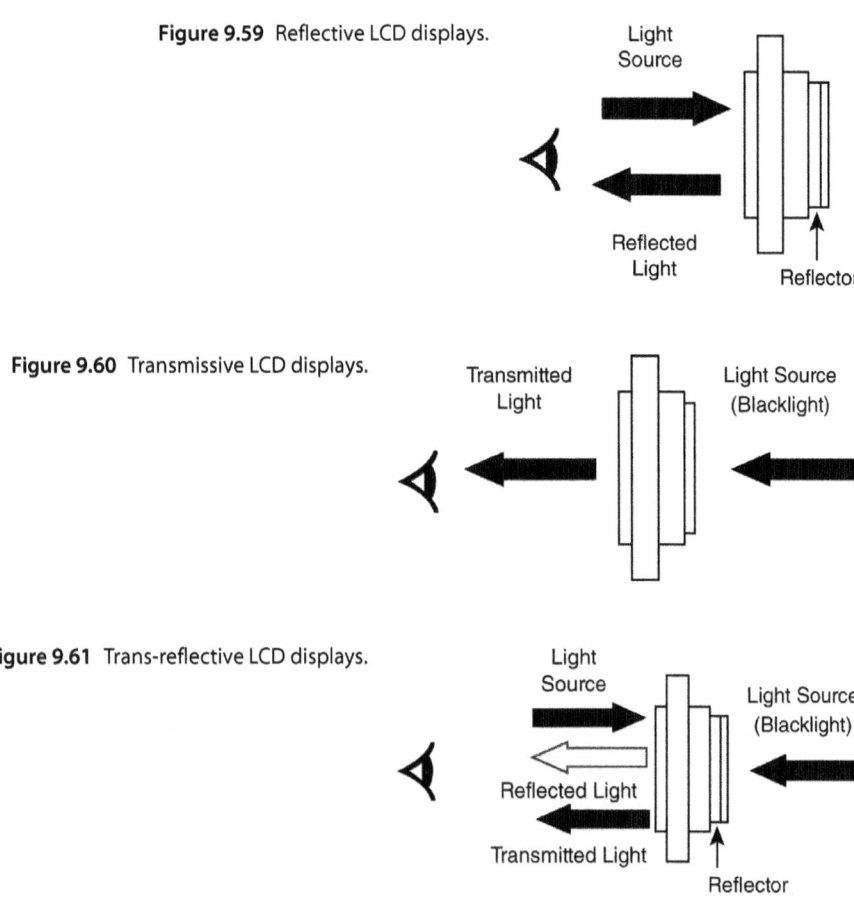

Figure 9.59 Reflective LCD displays.

Figure 9.60 Transmissive LCD displays.

Figure 9.61 Trans-reflective LCD displays.

each other. Transmissive displays are good for very low light level conditions. They offer very poor contrast when used in direct sunlight because sunlight swamps out the backlighting. Hence, these displays cannot be used in natural sunlight and provide good picture quality indoors. They are generally used in medical devices, electronics test and measuring equipment and in laptops.

Trans-reflective displays are a combination of reflective and transmissive displays (Figure 9.61). A white or silver translucent material is applied to the rear of the display, which reflects some of the ambient light back to the observer. It also allows the backlight to pass through. They are good for displays operating in varying light conditions. However, they offer poorer contrast ratios than reflective displays.

LCD displays can also be classified as passive LCD displays and active LCD displays depending upon the nature of the activation circuit. Passive displays use components that do not supply their own energy to turn 'ON' or turn 'OFF' the desired pixels. They are made up of a set of multiplexed transparent electrodes arranged in a row/column pattern. To address a pixel, the column containing the pixel is sent a charge and the corresponding row is connected to ground. Passive displays can have either direct drive or multiplexed drive circuitry. However, for larger displays it is not possible and economical to have separate connections for each segment. Also, as the number of multiplexed lines increase, the contrast ratio decreases due to the cross-talk phenomenon wherein a voltage applied to the desired pixel causes the liquid crystal molecules in the adjacent pixels to partially untwist.

These inherent problems of passive displays are removed in active displays. Active displays use an active device like a transistor or a diode into each pixel, which acts like a switch that precisely controls the voltage each pixel receives. Active displays are further classified as Thin Film Transistor (TFT) displays and Thin Film Diode (TFD) displays depending upon whether the active device used is a transistor or a diode. In both these devices a common electrode is placed above the liquid crystal matrix. Below the liquid crystal is a conductive grid connected to each pixel through a TFT or a TFD. Gate of each TFT is connected to the row electrode, the drain to the column electrode and the source to the liquid crystal. The display is activated by applying the display voltage to each row electrode line by line. One of the major advantages of active displays is that nearly all effects of cross-talk are eliminated.

9.13.3.5 Advantages and Disadvantages

As LCD displays are not active sources of light, they offer considerable advantages such as very low power consumption, low operating voltages and good flexibility. However, their response time is too slow for many applications. They offer limited viewing angles and are temperature sensitive.

9.13.4 Cathode Ray Tube Displays

Cathode ray tube (CRT) displays are used in a wide range of systems ranging from consumer electronic systems like television and computer monitors to measuring instruments like oscilloscopes to military systems like radar and so on. CRT display is a specialized vacuum tube in which the images are produced when the electron beam strikes the fluorescent screen. CRT displays can be monochrome displays as well as coloured displays.

Monochrome CRT displays comprise of a single electron gun, a fluorescent screen and an internal or external mechanism to accelerate and deflect the electron beam. Figure 9.62 shows the cross-sectional view of a CRT display. The electron gun produces a narrow beam of electrons that are accelerated by the anodes. There are two sets of deflecting coils, namely the horizontal coil and the vertical coil. These coils produce an extremely low frequency electro-magnetic field in the horizontal and vertical directions to adjust the direction of the electron beam. CRT tubes also have a mechanism to vary the intensity of the electron beam. In order

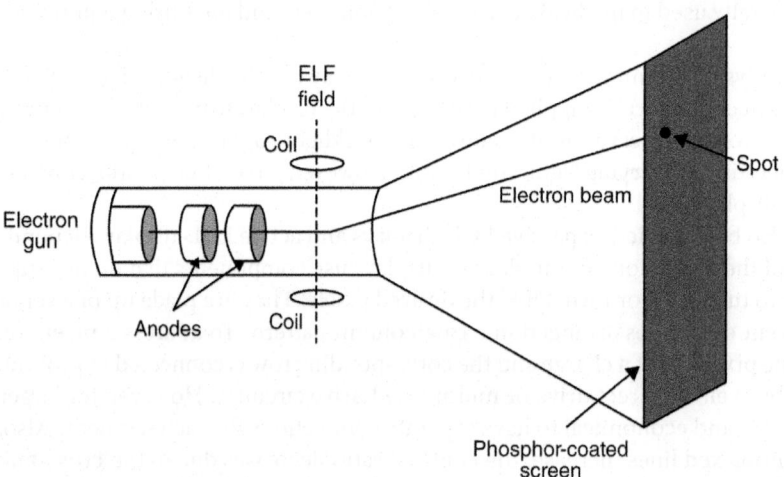

Figure 9.62 Cross-section of a monochrome CRT display.

to produce moving pictures in natural colours on the screen, complex signals are applied to the deflecting coils and to the circuitry responsible for controlling the intensity of the electron beam. This results in movement of the spot from right to left and from top to bottom in a sequence of horizontal lines referred to as a raster. The speed of the spot movement is so fast, that the person viewing the screen sees a constant image on the entire screen.

Coloured CRT displays comprise three electron guns; one each for the three primary colours, namely red, blue and green. The CRT produces three overlapping images; one in red (R), one in green (G) and one in blue (B). This is referred to as the RGB colour model.

9.13.4.1 Advantages and Disadvantages

CRT displays offer very high-resolution and, as these displays emit their own light, they therefore have very high values of peak luminance. Moreover, these displays offer wide viewing angles in the order of 180°. Also, CRT display technology is more mature compared to alternative display technologies and they are cheaper compared to other displays.

Inspite of the significant advantages offered by CRT displays as mentioned before, alternate display technologies are slowly replacing the CRT displays due to the drawbacks of the CRT displays. CRT displays are bulky and consume significant power. Moreover, they require high voltages to operate and they cause fatigue and strain to the human eye.

9.13.5 Emerging Display Technologies

This section gives introduction to the emerging display technologies including organic light-emitting diodes (OLEDs), digital light processing technology (DLP), plasma displays, field emission displays (FEDs) and electronic ink displays. All these display technologies are explained in brief in this section. Detailed description of these technologies is beyond the scope of the book.

9.13.5.1 Organic Light-Emitting Diodes (OLEDs)

OLEDs are composed of a light-emitting organic material sandwiched between two conducting plates, one of n-type material and the other of p-type. When an electric potential is applied between these plates, holes are ejected from the p-type plate and electrons are ejected from the n-type plate. Due to the recombination of these holes and electrons, energy is released in the form of light photons. The wavelength of light emitted depends upon the band gap energy of the semiconductor material used. In order to produce visible light, band gap energy of the semiconductor material is in the order of $1.5–3.5\,eV$.

Depending upon their basic structure, OLEDs can be classified into three types, namely small molecule OLEDs (SMOLEDs), polymer LEDs (PLEDs) and dendrimer OLEDs. OLEDs can be driven using passive as well as active matrix driver circuits.

As OLEDs are emissive devices, they offer significant advantages as compared to LCD displays like faster switching speeds, higher refresh rates, lower operating voltages and larger viewing angles.

9.13.5.2 Digital Light Processing Technology (DLP)

DLP technology makes use of an optical semiconductor device referred to as digital micro mirror device (DMD), which is basically a precise light switch that can digitally modulate light through a large number of microscopic mirrors arranged in a rectangular array. These mirrors are mounted on tiny hinges and can be tilted away or towards the light source with the help of the DMD chip and thus projecting a light or a dark pixel on the screen. Use of DLP systems is currently limited to large projection systems.

9.13.5.3 Plasma Display Panels (PDP)

Plasma displays are composed of millions of cells sandwiched between two panels of glass. Two electrodes, namely address electrodes and display electrodes, are also placed between the two glass plates covering the entire screen. The address electrodes are printed on the rear glass plate and the transparent display electrodes are located above the cells along the front glass plate. These electrodes are perpendicular to each other forming a grid network.

Each cell is filled with a xenon and neon gas mixture. The electrodes intersecting at a specific cell are charged to excite the gas mixture in each cell. When the gas mixture is excited, plasma is created releasing UV light, which then excites the phosphor electrons located on the sides of the cells. These electrons in turn release visible light and return to their lower energy state. Each pixel is composed of three cells containing red, green and blue phosphors, respectively.

Plasma displays offer several advantages such as each pixel generating its own light offering large viewing angles, generating superior image quality and image quality is not being affected by the area of the display. However, these displays are fragile in nature and are susceptible to burnout from static images.

9.13.5.4 Field Emission Displays (FEDs)

FEDs function much like the CRT displays with the main difference being that these displays use millions of small electron guns to emit electrons at the screen instead of just one as in the case of CRT displays. The extraction of electrons in FEDs is based on the 'tunnelling effect'. FED displays produce the same quality of image as produced by the CRT display without being bulky as the CRT display. In fact these displays can be as thin as LCD displays and as large as plasma displays.

9.13.5.5 Electronic Ink Displays

Also referred to as electronic paper, these are active matrix displays making use of pigments that resembles the ink used in print.

9.14 Night Vision Technologies

Night vision technologies and the associated night vision devices enable the users see in low light conditions. Contemporary devices allow viewing even in near total darkness. The ability to see in low light conditions is governed by two basic requirements namely sufficient spectral range and sufficient intensity range. Low values of spectral range and intensity range in the case of human eyes therefore become the limiting factors for their ability to see with an acceptable level of contrast in low light conditions. Use of technology to enhance both these parameters makes night vision possible. Night vision devices are extensively used by military for location of enemy targets, surveillance and navigation thereby playing a crucial role in enhancing night fighting capability of armed forces. They are also used by law enforcement and security agencies for surveillance. Night vision enabled cameras are being used even by big business houses to monitor surroundings. In the following paragraphs various approaches to night vision are discussed in detail, along with their capabilities, limitations and application potential. A comparison between the two major categories of night vision devices, namely image enhancement devices and thermal imaging devices, is also presented.

9.14.1 Basic Approaches to Night Vision

The two basic and widely different approaches to night vision include *image intensification* (or *enhancement*) and *thermal imaging*. Both techniques are briefly described in the following paragraphs.

9.14.1.1 Image Intensification (or Enhancement)

Image intensification or *enhancement* works on the principle of collecting small quanta of light reflected off the target scene to be viewed in visible and near IR bands of electromagnetic spectrum in low light conditions. The collected photons are amplified through the processes of photon-electron conversion, electron multiplication and electron-photon conversion. These processes take shape in what is called an image intensifier tube briefly described earlier in Section 9.9.3. The other important constituent parts of an image intensifier tube based night vision device include the objective lens used for collection of photons, eye piece for viewing intensified image and a power supply that generates required DC voltages for electron acceleration.

Active illumination is often used in conjunction with image intensifier tube in what is known as *active night vision technology* to enhance image resolution in very low-level light conditions. Illumination is generally provided by IR diodes emitting in the 700–1000 nm spectral band. Active night vision technology has the disadvantage that it can be detected by night vision goggles and therefore prone to giving away the location of the user particularly undesirable in tactical military operations.

A variation of the conventional night vision device is the *digital night vision* device. While in a conventional night vision device available light is collected through the objective lens and focused on an intensifier, most digital night vision devices process and convert the optical image into an electric signal through a highly sensitive charge coupled device (CCD) image sensor. This electrical signal is then transferred onto a micro-display, which is a type of LCD flat-panel display screen. The micro-display usually takes the form of an eyepiece which you look into to view the image rather than on an LCD screen that we find on most digital cameras. Digital night vision offers several advantages over conventional night vision. Digital night vision devices have relatively lower costs, are free from image distortions of photo cathode and blemishes of phosphorescent screen, are immune to damage by bright light exposure and offer image recording facility.

9.14.1.2 Thermal Imaging

Thermal imaging night vision technology works on the principle of detecting temperature difference between the objects in the foreground and those in the background. All objects above absolute zero emit IR energy. The magnitude of IR energy emitted by a hot body is proportional to fourth power of its absolute temperature (Stefan–Boltzmann law) and the peak emission occurs at a wavelength that is inversely proportional to its absolute temperature (Wien's displacement law). Therefore, the hotter the body, the higher the magnitude of IR energy emitted by the object and the lower the wavelength of peak emission. A thermal imaging device is essentially a heat sensor capable of detecting tiny differences in temperature of different points on the surface of the object to be viewed. The information on the temperature difference available in the form of IR energy is collected by the thermal imaging device and converted into an electronic image. The ability to detect tiny temperature differences that always exist not only between the desired object and the surroundings but also between different points of the object itself coupled with emission being in the IR region allows a thermal imaging device to 'see' in near total darkness.

Thermal imaging night vision devices are extensively used not only by military and law enforcement agencies for the purpose of target detection and acquisition, surveillance and monitoring, search and rescue operations, fire fighting and so on; but also used in fields as varied as medicine, archaeology, process monitoring, the sautomotive industry, meteorology and astronomy.

9.14.2 Image Intensifier Tubes

An *image intensifier tube* amplifies low light level images to levels that can be seen with the human eye or detected by digital image sensors. The principle of operation of an image intensifier tube was briefly described earlier in Section 9.3.3. Operational basics are summarized here again to have continuity while explaining some related aspects.

9.14.2.1 Operational Basics

Image intensifier tubes collect the existing ambient light originating from natural sources, such as starlight or moonlight, or from artificial sources such as streetlights or IR illuminators through the objective lens of the night vision device. Low-level light consisting of photons enters the night vision device through input window and strikes the photo cathode. The inside of the input window is generally coated with a thin layer of light sensitive material. This light sensitive layer acts as the photo cathode. The photo cathode is protected from any damage from oxidation by operating the image intensifier tube under vacuum in the order of 10^{-9} to 10^{-10} torr. Photo electrons released by the photo cathode are accelerated and focused by a high magnitude electric field towards the micro-channel plate (MCP). The MCP has millions of small channels and the electrons entering these channels are both accelerated by another high magnitude electric field within the MCP and multiplied by secondary emission resulting from electrons bouncing off the inner walls of these channels. For each electron entering the MCP, approximately 1000 electrons are generated and subsequently accelerated from the output of the MCP by a third electrical field towards the phosphor screen. The phosphor screen, which is a thin light-emitting layer deposited on the inside of the output window of the image intensifier tube, converts impinging electrons back to photons. For every photon entering the input window of the intensifier tube, tens of thousands of photons come out of the output window after emission from phosphor screen. This photon multiplication takes place due to electron acceleration in the region between the input window and the photo cathode, electron acceleration and secondary emission within the MCP channels and electron acceleration in the region between the MCP and the phosphor screen. This multi-stage process produces an intensified or amplified image of the object that is much brighter than the original image.

9.14.2.2 Construction

The constituent parts of an image intensifier tube include the input window, photo cathode, the micro-channel plate, the phosphor screen, the output window and the power supply. Figure 9.63 gives an overview of constructional features of an image intensifier tube.

The input window material is selected according to required sensitivity at shorter wavelengths. Common materials used for the input window are synthetic silica (transmitting wavelength of 160 nm or longer), fibreoptic plate (with transmission wavelength of 350 nm or longer), magnesium fluoride (with transmission wavelength of 115 nm or longer) and borosilicate glass (with transmission wavelength of 300 nm or longer). Figure 9.64 shows transmission characteristics.

The photo cathode that converts photons into photo electrons, the micro-channel plate that multiplies the photo generated electrons, the phosphor screen that reconverts the electrons back into photons and the power supply that produces the electric field responsible for acceleration of electrons in different regions are all arranged in close proximity in an evacuated ceramic case. The efficiency with which the photo cathode converts photons into electrons, also known as photo cathode radiant sensitivity or quantum efficiency, depends upon wavelength. A number of photo cathode materials are in use. Of these, semiconductor crystals gallium arsenide (GaAs) and gallium arsenide phosphide (GaAsP) offer extremely high sensitivity. Photo electrons are accelerated by an electric field produced by a high voltage applied between the photo cathode and MCP input surface.

The MCP is a thin glass disc about 0.5 mm thick consisting of an array of millions of tilted glass channels each of about 5–6 μm diameter bundled in parallel (Figure 9.65). A single-stage

Figure 9.63 Construction of an image intensifier tube.

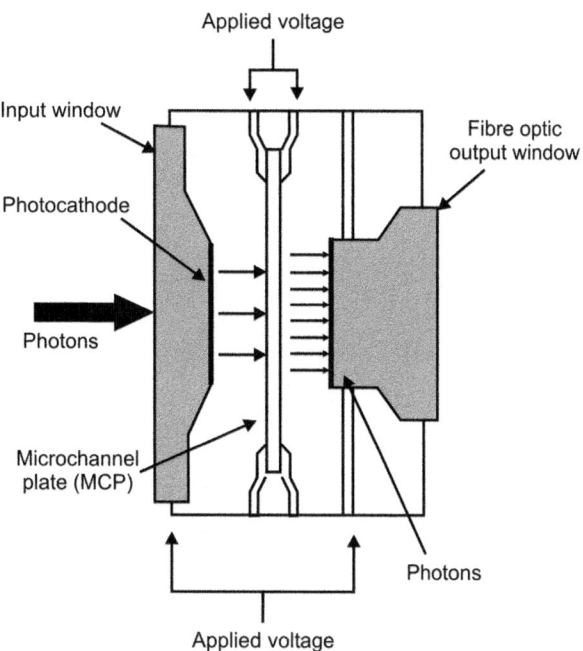

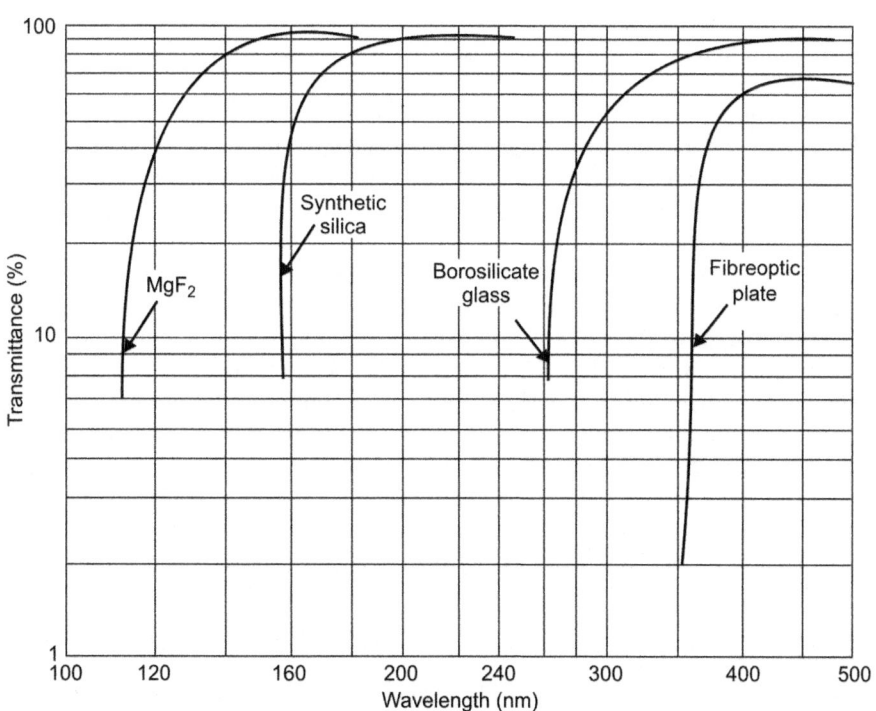

Figure 9.64 Transmission characteristics for different input window materials.

Figure 9.65 Micro-channel plate (MCP).

MCP used earlier in second generation intensifier tubes provides electron multiplication of about 10^3. Two- and three-stage MCPs produce gain of 10^5 and greater than 10^6. First-generation tubes didn't use an MCP. Number of stages to be used in the MCP depends upon the required value of gain. The strip current that flows through the MCP decides the dynamic range or linearity of the image intensifier tube. A low resistance MCP causing a large strip current to flow through MCP is desirable for achieving high linearity.

The *phosphor screen* reconverts the impinging electrons back to photons. Commonly used phosphor types include P24, P43, P46 and P47. Phosphor screens are characterized by peak emission wavelength, decay time, power efficiency and emission colour. Phosphor screen decay time is one of the most important parameters to be considered while selecting a suitable phosphor type. The chosen phosphor type is such that its decay time matches the readout method and its spectral emission matches the readout sensitivity. When used with a linear image sensor or high-speed CCD, a short decay time is recommended for the phosphor screen to avoid appearance of an after image in the next frame. On the other hand, a short decay time that minimizes flicker is recommended for night time viewing and surveillance applications.

The output window material is selected to match the readout method. Different output window types include borosilicate glass, fibreoptic plate (FOP) and twisted fibreoptics. Borosilicate glass window is used for relay lens read out. In this case, the relay lens is focused on the phosphor screen. Fibreoptic output plate is a standard output window and is ideal for direct coupling to a CCD with FOP input window. A fibreoptic plate consists of some millions to hundreds of millions of glass fibres bundled in parallel. FOP can transmit an optical image from one surface to another without causing any distortion. The diameter of the glass fibre matches with the channel diameter of the MCP. Twisted fibreoptics as an output window is used for night time viewing applications. Using twisted fibreoptics reduces eyepiece length thereby making night vision device more compact.

Image intensifier tube based night vision devices are made in a variety of package configurations according to the requirements of specific application scenarios. These mainly include night vision monoculars, binoculars, weapon sights and goggles. Figure 9.66 shows some representative photographs.

9.14.2.3 Operational Modes

There are two common modes of operation of an image intensifier tube. These are the *gated mode* and the *photon counting mode*. In the gated mode of operation, the intensified image can be gated to open or close the optical shutter by varying the potential difference between the

Figure 9.66 Different package configurations of night vision devices.

photo cathode and the inside surface of the MCP thereby either allowing or disallowing the formation of intensified image. In the *gate-on mode*, the potential of photo cathode is lower than that of MCP. As a result of this, the photo electrons are attracted towards the MCP to be subsequently multiplied and hit the phosphor screen producing an intensified image. In the gate-off mode, the potential of inside surface of MCP is less than that of photo cathode with the result that photo electrons revert back to the photo cathode. Therefore, there is no intensified image seen on the phosphor screen. In practice, MCP potential is fixed and the intensifier tube is turned on by applying a negative polarity pulse of about 200 V to the photo cathode. Gated operation is very effective in analysing high-speed optical phenomena.

As outlined earlier, image intensifier tubes using three-stage MCP have much higher sensitivity than the sensitivity of image intensifier tubes employing a single-stage MCP. This is particularly important when it comes to operating at extremely low light levels. When the light level is as low as 10^{-4} lux, a three-stage MCP helps in producing an image of an acceptable quality. However, when the light level falls below 10^{-5} lux, the incident photons are separated in time and space. It is no longer possible to capture an image with a gradation. At extremely low light levels when only a few light spots appear on the phosphor screen per second, a good quality image can be obtained by detecting each spot and its position and integrating them into an image storage unit. The brightness distribution in this case called the *photon counting mode* is given by difference in the number of photons at each position.

9.14.3 Different Generations of Image Intensifiers

Night vision technology has undergone substantial changes during its existence for more than 40 years. These changes have led to gigantic improvement in performance standards of night vision devices. Each substantial change in technology is associated with a generation with the result that we have seen several generations of night vision devices based on image intensifier tube technology. Beginning with generation 0, we are currently in the fourth generation.

Different generations of image intensifier based night vision devices were outlined earlier in Section 9.9.3. Salient feature of different generations are described in the following paragraphs.

9.14.3.1 Generation 0 Night Vision Devices

The earliest night vision devices existed during World War II and the 1950s. These devices, known as *Generation 0 devices*, were based on image conversion rather than image intensification. The night vision device primarily comprised of a photo cathode that converted incident photons into electrons. The electrons were accelerated towards an anode by applying a positive potential to the anode. The device also had an IR source of radiation called an IR illuminator mounted on the device. The IR illuminator irradiated the target scene with IR radiation. The IR radiation reflecting off the target back to the night vision device is collected by its objective lens and focused on to the photo cathode.

The M1 and M3 IR night sighting devices called *sniper scope* or *snooper scope* were used by the US Army during the Second World War and also the Korean War to assist snipers. These night vision devices as outlined previously used an IR source to illuminate the target. These devices employed an S-1 photo cathode primarily made up of silver, caesium and oxygen and an anode. Electrostatic inversion and electron acceleration were used to achieve gain.

Generation 0 night vision devices had limitations. (1) The use of night vision devices was highly prone to be detected by the adversary in possession of IR viewing devices. (2) Acceleration of electrons towards the anode produced image distortion and also reduced the life of the tube.

9.14.3.2 Generation 1 Night Vision Devices

Genearion 1 night vision devices were an adaptation of generation 0 technology in the sense that both used a photo cathode and an anode, the former for photon-to-electron conversion and the latter to accelerate photo electrons towards it. A major deviation in generation 1 night vision devices from generation 0 devices is in the absence of IR source used in the case of latter devices to provide scene illumination. Generation 1 devices depend upon ambient light provided by the Moon and stars. Generation 1 tubes were connected in series to enhance sensitivity. These devices suffer from the shortcomings of generation 0 devices, namely short tube life and image distortion. In addition, they don't perform as well in cloudy and moonless nights. Generation 1 devices were introduced during the Vietnam War in the 1960s. AN/PVS-2 Starlight Scope and PNV/57 Tanker Goggles are examples of generation 1 night vision devices.

9.14.3.3 Generation 2 Night Vision Devices

Generation 2 night vision devices were the first to use micro-channel plates (MCPs) for electron multiplication leading to a significant increase in device sensitivity. These devices were introduced during the 1970s. Since the MCP actually increased the number of available photo electrons for a given number of light photons incident on the photo cathode rather than merely accelerating them, the resultant images were relatively brighter and had significantly less distortion. The consequent light amplification was in the order of 20 000 times, which resulted in much improved performance even in very low-light level ambient conditions of cloudy and moonless nights. Introduction of MCPs into the intensifier tube also obviated the need to connect the tubes in series as was done earlier in the case of generation 1 devices. This significantly reduced the size of the night vision device and made handheld devices and helmet-mounted devices a reality. AN/PVS-4 and AN/PVS-5 are examples of generation 2 devices. Subsequently, generation 2 devices equipped with better optics, SUPERGEN tubes, led to the arrival of *generation 2+ night vision devices* on the scene.

Figure 9.67 AN/PVS-4 night sight.

AN/PVS-4 (Figure 9.67) was the first second generation passive night sight designed and manufactured by the Optic Electronic Corporation of Dallas and was extensively used by the US Army during the Gulf War and Iraq War. It has now been replaced by third-generation weapon night sights. Equipped with MX9644 image intensifier tube, the night sight had image resolution of 32 lp/mm and star light detection and recognition ranges of 600 m and 400 m, respectively. The AN/PVS-5 is a dual tube night vision goggle used for aviation and ground support. Manufactured by ITT Industries and Litton Industries and equipped with image intensifier tube MX 9,916, the night vision goggles had an image resolution of more than 20 lp/mm and star and moonlight detection ranges of 50 and 150 m, respectively.

9.14.3.4   Generation 3 Night Vision Devices

Compared to generation 2 devices, generation 3 night vision devices had two distinctive changes. These included use of gallium arsenide photo cathode and an ion barrier coating on the MCP. Due to its higher quantum efficiency or radiant sensitivity and its spectral response extending to the near IR region, a gallium arsenide photo cathode enables target detection at longer ranges and in darker conditions. The light amplification factor increases from 20 000 times in the case of generation 2 devices to a maximum of 50 000 times in the case of generation 3 devices. The ion barrier film improves the tube life from 2000 h in the case of generation 2 devices to 10 000 h in the case of generation 3 devices, even though it is at the cost of a slight reduction in radiance sensitivity due to fewer photo electrons being able to reach the MCP.

AN/PVS-7, ATN NVG7 and AN/PVS-14 are some representative night vision devices employing the third-generation image intensifier tube. Designed by ITT Industries and Litton Industries and manufactured by ITT Industries, Litton Industries, Northrop–Grumman and L3 Communications, the AN/PVS-7 is a set of single tube passive/active night vision goggles. Active night vision enabled by a built-in LED allows operation in low-light level situations. The device is waterproof and charged with dry nitrogen enabling operation in extreme temperature variations. Important technical specifications of the device are as follows. Resolution is greater than 64 lp/mm, field-of-view is 40° and starlight detection and recognition ranges are 325 and 225 m, respectively. The device was extensively used during Gulf War, Operation Enduring Freedom in Afghanistan and Armed Conflict in Iraq.

Figure 9.68 ATN NVG7-3 night vision goggles.

The ATN Night Vision Goggles NVG7–3 (Figure 9.68) is similar to AN/PVS-7. Available in head or helmet-mounted versions for hands-free use, this device too is equipped with an IR light source for close-up illumination in total darkness. It is available with a wide range of image intensifier options. Detection and recognition ranges are 180 and 100 m, respectively. Resolution and field-of-view are 40 lp/mm and 40°, respectively.

AN/PVS-14 is monocular auto-gated passive night vision device configured around a third-generation image intensifier tube type MX 11769. Designed by ITT Industries and manufactured by Litton Industries (now L3 Warrior Systems) and ITT Industries, it was introduced in service during the year 2000 and is extensively used by the US Armed Forces and its NATO allies. It has a resolution greater than 64 lp/mm and a field-of-view of 40°. Starlight detection and recognition ranges are 350 and 300 m, respectively. Auto-gating feature ensures optimum performance of the image intensifier tube by electronically adjusting the duty cycle of the photo cathode gating voltage.

Generation 3+ offers improved performance specifications over generation 3 devices. Two important features associated with generation 3+ night vision devices are an automatic gated power supply system and a thinned ion barrier layer. Absence of ion barrier layer or its thinning improves luminous sensitivity even though it is at the cost of slight reduction in life of the tube. Mean time to failure (MTTF) is typically 15 000 h in generation 3+ devices compared to 20 000 h in the case of generation 3 devices. Operationally, it is insignificant as an image intensifier tube seldom reaches 15 000 operational hours before it needs replacement.

Another common term encountered while discussing night vision technology is the Omnibus or OMNI. Omnibus or OMNI refers to the multi-year/multi-product contracts of the US Army for procurement of night vision devices from Exelis (formerly ITT Night Vision). Under these contracts, the company delivers generation 3 devices with increasingly higher performance. The current contract is for OMNI-VIII.

9.14.3.5 Generation 4 Night Vision Devices

As of now, there is nothing like generation 4 night vision technology. There are only four generations of devices from generation 0–3. Generation 4 night vision devices were initially conceived to use filmless and gated technology. The proposal was to remove the ion barrier film from the MCP introduced in generation 3 devices. Removal of film was aimed at reducing background noise and enhance signal-to-noise ratio. It also allowed more electrons reach the MCP so that the images were significantly less distorted and brighter. Introduction of automatic gated power supply for the photo cathode enabled the device adapt instantaneously to light level fluctuations from low light levels to high light levels or vice versa. Removal of the ion barrier was also intended to reduce the halo effect seen around bright spots or light sources. While device performance improved, the absence of an ion barrier film led to increased tube failure rates. As a consequence of this, the idea of film removal was abandoned in favour of use of a thinned film giving birth to what was known as generation 3+ described in the previous section.

9.14.4   Intensified CCD

Intensified CCD (ICCD) successfully exploits the optical amplification provided by an image intensifier to overcome limitations of the basic CCD sensor outlined in Section 9.9. Two important features of an intensified CCD are high optical gain and gateable operation. Both attributes are the characteristic features of the image intensifier tube. Though image intensifiers were initially developed for the military and law enforcement agencies for a range of deployment scenarios in surveillance, targeting and navigation, the development of ICCD technology and devices has extended its usage to many a scientific application in spectroscopy, scientific and industrial imaging and medical diagnostics. In fact, the development of image intensifier tubes is increasingly being driven by scientific applications.

9.14.4.1   Construction

An intensified CCD primarily comprises of an image intensifier tube whose light output is coupled to a CCD sensor. As outlined earlier, an image intensifier tube is an evacuated tube that comprises a photo cathode, micro-channel plate (MCP) and a phosphor screen. The photo cathode captures the incident image. Photo electrons generated by the photo cathode are driven towards the MCP by an electric field. The MCP channels accelerate electrons thereby resulting in electron multiplication through the process of secondary emission. This leads to a cloud of electrons exiting the MCP. Gains in excess of 10 000 can readily be achieved. The degree of electron multiplication depends on the electric field intensity and the number of stages of MCP. The output of the image intensifier is coupled to the CCD typically by a fibreoptic coupler. Fibre coupled systems are physically compact with low optical distortion levels. The high efficiency fibreoptic coupling also allows image intensifier tubes operate at lower gains, which in turn results in better dynamic range performance. A lens-coupled ICCD uses a lens between the output of the image intensifier and the CCD. Lens coupling offers the flexibility of using the ICCD sensor in a non-intensified mode by allowing the image intensifier to be removed. Disadvantages of lens-coupled ICCDs are the larger physical size, lower coupling efficiencies and increased scatter. Power supply is another important constituent part of the ICCD sensor. The power supply section generates DC voltage (typically 600 to 900 V) for MCP to achieve desired gain, DC voltage (typically 4–8 kV) for the phosphor screen and voltage pulses (typically 200 V) for gated operation of the photocathode. The gating pulse width and rise/fall time depend upon desired gating parameters. Gating pulse width of less than a nanosecond and rise/fall time of a small fraction of a nanosecond is achievable.

9.14.4.2   ICCD Characteristic Features

Important characteristic features of an ICCD include its spectral response, spatial resolution, gating time and repetition rate, noise, sensitivity, dynamic range and frame rate. Each one of these is briefly described in the following paragraphs.

The *spectral response* of an ICCD camera is primarily determined by the input window and photocathode materials and the photo cathode size used in the Image Intensifier tube. The early intensifier tubes (generation 2) used bismuth or multi-alkali photo cathodes, which were the materials used earlier in the photo cathodes of photo multiplier tubes. Present day generation 3 image intensifier tubes employ semiconductor photo cathodes made from semiconductor materials such as gallium arsenide, gallium arsenide phosphide. The photo cathodes of generation 2 intensifiers were made by coating the material on a quartz window, which extended its spectral response down to 160 nm. The quartwindow can be substituted for a magnesium fluoride window to provide response further down to approximately 120 nm. The mix and thickness of the photocathode can be adjusted to optimize the wavelength response in different regions. The spectral response of gallium arsenide, gallium arsenide phosphide and indium gallium arsenide, respectively, is in the range of 350–950 nm, 280–820 nm and 370–1100 nm.

The *noise* and hence *sensitivity* of the ICCD is also governed by the Image Intensifier. A noise component called the dark current component, also called Effective Background Illumination (EBI), exists that originates from thermally generated charge in the photocathode. The dark current is generally not an issue when using short gate times.

A significant advantage of an ICCD over EMCCD and CCD is their *optical shuttering properties*. An image intensifier inherently includes shutter functionality enabled by application of a control voltage between the photo cathode and the MCP. If the control voltage between the photocathode and the MCP is reversed, photo electrons don't reach the MCP and hence there is no electron multiplication, with the result that no light is emitted from the image intensifier. In this case no light falls onto the CCD, which means that the shutter is closed. The process of reversing the control voltage at the photocathode is called gating and therefore ICCDs are also called gateable CCD cameras. The Image Intensifier can be operated as a very fast optical switch, capturing an optical signal typically in a nanosecond. Gating on and off operation were discussed earlier in the section on *operational modes*. The minimum gate time depends on a number of factors but principally on the structure of the photocathode and the electronic gating circuitry. Gating operation allows the ICCD capture instantaneous images of high-speed optical phenomenon while excluding extraneous signals. The intensifier can be repetitively gated at rates of up to 50 kHz for standard operation or up to 500 kHz for specially requested cameras. Although the CCD section of the camera cannot be read out at this rate, there are advantages in operating the optical gating independently. A repetitive signal can be sampled and the output of the intensifier summed on the CCD to integrate up a larger signal that otherwise may not be visible.

Spatial resolution is usually defined as a number of line pairs that a camera can resolve per millimetre. Among several methods available for measuring the resolution of an optical system, Modulation Transfer Function (MTF) is in common use. The MTF is a quantitative measure of the ability of an optical system to transfer various levels of detail from object to image. The MCP and the phosphor in the image intensifier tube both contribute towards degrading the Modulation Transfer Function (MTF) compared to the CCD. The modulation transfer function is a measure of the transfer of modulation (or contrast) from the subject to the image. In other words, it measures how faithfully the optical system reproduces (or transfers) detail from the object to the image. When a black and white striped pattern with associated sine wave changes in brightness is focused on the photo cathode, the image contrast drops gradually as the stripe pattern density is increased. The relationship between the contrast and the stripe density in line pairs per mm (lp/mm) is the MTF. Recent developments in finer phosphor deposition, reducing gaps and reducing the bore of the micro channel plates has resulted in much better performance but typically resolution is limited to less than 60 lp/mm.

The *dynamic range* of the ICCD is governed by the CCD section and varies inversely with the gain of the ICCD. A higher dynamic range CCD used in the ICCD will result in a higher dynamic range ICCD camera. As the gain increases, the smaller signals that can be accommodated are compensated for by the lower read noise to keep the dynamic range constant. When the read noise drops below a single photon level, the dynamic range of the ICCD starts dropping as the gain increases further. The frame rates of an ICCD are governed by the CCD specifications, especially the number of pixels and pixel readout rate.

9.14.5 Thermal Imaging

CCD and CMOS sensors described in the previous paragraphs have a spectral response that is sensitive to visible region of electromagnetic spectrum with the trailing part extending slightly to near IR region. Thermal imaging sensors on the other hand use focal plane arrays comprising IR sensing elements that respond to mid (3–5 µm) and long IR (8–14 µm) regions. The IR energy radiated by the object to be imaged is incident on the thermal imaging sensor, which uses a series of complex mathematical algorithms to construct the image in visible for the viewer. As the thermal imaging sensor doesn't need visible light for operation, it is able to see in total darkness. Thermal imaging sensors are far more expensive than their visible spectrum counterparts. In view of their military applications, high end devices are often export restricted. In the following paragraphs, we shall discuss operational basics, types and applications of thermal imaging sensors.

9.14.5.1 Operational Basics

The concept of thermal imaging is based on black body radiation law according to which all objects above absolute zero emit IR radiation. As outlined earlier, the magnitude of emitted IR is proportional to fourth power of its absolute temperature as stated by the Stefan–Boltzmann law. According to the Stefan–Boltzmann law, $J = \sigma T^4$ where J is emitted energy per unit surface area of the black body across all wavelengths per unit time; σ is a constant of proportionality called the Stefan–Boltzmann constant and T is the absolute temperature of the black body. Also, peak emission occurs at a wavelength that is inversely proportional to its absolute temperature as stated by Wien's displacement law. According to Wien's displacement law, $\lambda_{MAX} = b/T$ where λ_{MAX} is the wavelength of peak emission and b is constant of proportionality called Wien's displacement constant.

A thermal imaging sensor functions as follows. It makes use of thermal radiation emitted by the target or scene of interest to generate its image. Essentially, it comprises a front-end optical system, a two-dimensional array of IR detector elements and image-processing circuitry to produce an output in the desired format. The front-end optical system focuses IR radiation emitted by all objects in view on a two-dimensional array of IR detector elements that create a detailed temperature pattern of it called a thermogram. The thermogram is generated from several thousand points in the field-of-view of the detector array. The thermal imager measures very small relative temperature differences and converts otherwise invisible heat patterns into clear, visible images that are seen through either a viewfinder or monitor. Most thermal imaging sensors scan at a rate of 30 times a second. They can sense temperatures in the range of $-20 - +2000\,°C$ and can sense temperature senses as small as $0.1\,°C$. In the next step, the temperature pattern is translated into electronic impulses. The signal processing unit converts these electronic impulses into data for the display. Figure 9.69 illustrates the concept of thermal imaging.

9.14.5.2 Types of Thermal Imaging Sensors

There are two distinctive detector technologies namely *direct detection* (or photon counting) and *thermal detection*. In the case of direct detection, the detector element translates the photons directly into electrons. The charge accumulated, the current flow, or the change in

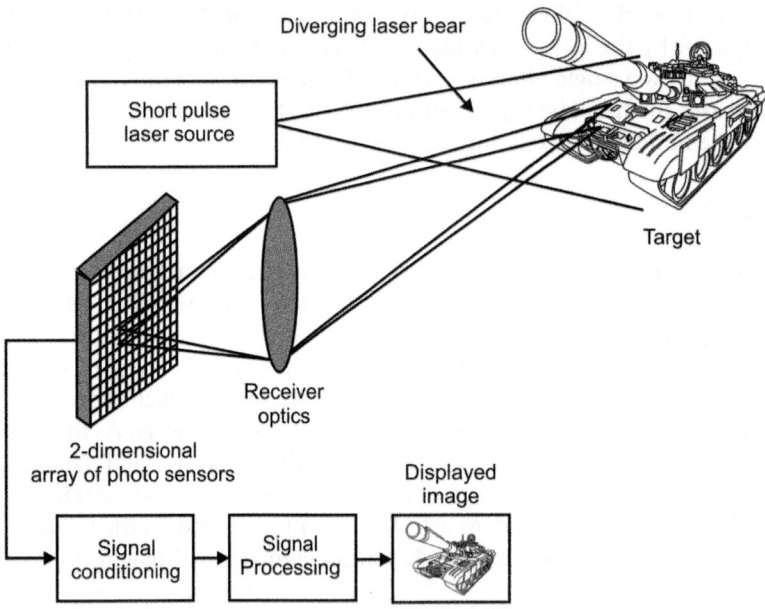

Figure 9.69 Thermal imaging.

conductivity is proportional to the radiance of objects in the scene. Detectors in this category include lead selenide (PbSe), mercury cadmium telluride (HgCdTe), indium antimonide (InSb), platinum-silicide (PtSi) and so on. All thermal imaging sensors based on direct detection technology except those working in short-wave IR (SWIR) use detectors cooled to cryogenic temperatures close to –200 °C. Newer photon type IR sensors operating at elevated temperatures are now available. This has allowed solid-state thermal electric coolers and sterling coolers to be used. Cooled thermal imagers are highly susceptible to damage from rugged use, have a long cooling time of typically few minutes, a limited MTBF of a few thousand hours, high cost, large size and weight and high electrical power consumption leading to short battery life. The biggest advantage of these detectors is excellent spatial resolution and sensitivity that results from detector cooling.

Thermal detection, on the other hand, uses uncooled detectors. They make use of secondary effects such as the relation between conductivity, capacitance, and expansion and detector temperature. Detectors in this category include bolometers, thermocouples, thermopiles, pyroelectric detectors and so on. These sensors operate at room temperature and are light weight. This feature for example allows microbolometer thermal imagers to be mounted on helmets.

9.14.5.3 Different Generations of Thermal Imaging Sensors

There have been different generations of thermal imaging sensors. Each successive generation has incorporated not only a major change in the type of detector but also a major change in optical systems used to image the target onto the detector.

Four distinct generations of thermal imagers have been designed, based on IR detector technologies developed during the last 35 years, and classified according to the number of elements contained in each group. The first generation thermal imagers contain single element detectors or detectors with only a few elements (1 × 3). A two-dimensional mechanical scanner was usually used in order to generate a two-dimensional image. Sensitivity of thermal imaging sensors of the first generation was limited by background radiation. This problem was overcome in second generation thermal imaging sensors by using modified front-end optics that reduced unwanted flux. However, it resulted in a fixed f-number for all fields-of-view. Second generation thermal imagers are vector detectors, usually containing 64 or more elements. The two-dimensional scanner was somehow simplified in the vertical direction to include only the interlace motion. Third-generation thermal imagers contain dual band two-dimensional arrays with several columns of elements and a dual/variable f-number optical system. These thermal imagers still scan in one direction and perform time delay integration of the signal in the scanning direction in order to improve the signal-to-noise ratio. Fourth-generation thermal imagers contain two-dimensional array detectors (160×120, 320×240, 680×480) called focal plane arrays that do not require any scanning mechanism for acquiring the two-dimensional picture.

9.14.6 Image Enhancement versus Thermal Imaging

Both image enhancement night vision devices and thermal imager have similar application areas. While choosing one or the other for a given application, the following considerations are important.

1) The first important consideration is the cost. A good night vision device including weapon mountable variants would cost a few hundred US dollars. On the other hand, price of a thermal imager may be anywhere from few thousands of US dollars with military qualified devices costing as much as tens of thousands of US dollars. Therefore, an image intensifier night vision device scores over a thermal imager when available budget is an important issue.
2) A night vision device needs light to operate even though a small amount of light may be adequate for achieving desired results. Thermal imagers on the other hand can work in total darkness. Therefore, lighting conditions must be considered before investing in one device or the other.
3) The type of environment plays a big role. Thermal imager is the only choice in the case of heavy fog or dense foliage as long wavelength band ($8-14\,\mu m$) penetrates much better through smoke, smog, dust and water vapour. Their ability to see through sand storms, which are a common occurrence in desert operations, makes them an ideal choice for such operations. Extreme cold makes night vision the better choice.
4) Thermal imaging is great for detection, but not ideal for recognition. On the other hand, a night vision device gives a better recognition once it is detected, but if that person wears camouflage or the animal stands stationary at a distance it can be difficult to find.

9.14.7 Applications of Night Vision Equipment

Night vision equipment, which includes image enhancement night vision devices and thermal imagers, finds extensive usage in a wide range of applications in military, law enforcement, surveillance, security, navigation, hunting, wild life observation and so on. These devices are used not only by military, police forces and law enforcement agencies for navigation, surveillance and targeting; they are also used by private detectives, fire fighters, hunters and nature enthusiasts. Both image enhancement and thermal imaging night vision equipment have preferred areas of application to which one is better suited than the other. These are briefly described in the following paragraphs.

9.14.7.1 Applications of Image Enhancement Night Vision Devices

Image Intensification night vision needs at least a small amount of light to operate. This ambient light in some cases may be provided by the Moon and stars. In the case of total darkness, the night vision device will be able to see no better than the naked eye. This shortcoming is overcome in many night vision devices with the use of an IR illuminator, which works like a flashlight

Figure 9.70 Night vision rifle scope and typical image.

Figure 9.71 AN/PVS-14 night vision scope. (*Source:* Courtesy of The US Army).

for night vision while staying invisible to the naked eye. Use of an IR illuminator though has a drawback that it makes your device highly prone to detection by the enemy's night vision device. Some of the application scenarios of image enhancement night vision equipment are briefly described in the following paragraphs.

1) One of the uses of night vision devices is by *hunters*. The device technology has improved to an extent to provide quality image and have become extremely rugged and tough enough to withstand recoil of a night vision rifle scope. Figure 9.70 shows photograph of a first-generation night vision rifle scope (ATN Aries MK 390 Palladium) and a typical image as seen from a rifle scope while hunting.
2) Night vision technology is used extensively by the military and police due to the instant tactical advantage they get through the ability to see in the dark and precisely detect and even identify the target even at a distance. The AN/PVS-14 (Figure 9.71) is one such versatile night vision scope extensively used by Armed Forces and special operations units.

It is a monocular night vision device designed for use by the individual soldier in a variety of ground-based night operations. While the dark-adapted unaided eye provides situational awareness and vision of close-range objects, the night vision device provides long-range vision of potential threats and targets thus making it a force multiplier in night time warfare. Salient features of AN/PVS-14 include a built-in IR illuminator, fully adjustable headmount and mil-spec multicoated optics, automatic brightness control, main gain control that allows the user to increase or decrease image tube brightness for best possible image contrast in high and low-light conditions and flexibility to use hands-free, weapons-mounted or attached to a camera or camcorder for night time photography. MIL-STD-1913 weapon rails allow use of the AN/PVS-14 with most types of assault rifles.

3) Night vision cameras are extensively used for surveillance and security applications especially for round the clock surveillance indoors and in controlled environments. Night vision security cameras provide enhanced security by improving video images in low-light conditions. With the help of IR illumination, they can be effectively used in near total dark conditions too. Perhaps the greatest feature of night vision cameras is that they are true 24-h cameras, as they are the only cameras with the ability to record any event in a protected area around the clock when connected to a digital video recorder. The true 24-h camera produces images in full colour in normal light conditions, such as those during the daytime. The camera automatically switches to monochrome imagesfor viewing in darkness when the intelligent sensor in the camera detects inadequacy of light. Night vision cameras are available in two basic styles, namely dome and bullet style cameras. The resolution typically varies from 400 to 700 TV lines. Security cameras with long-range night vision find extensive usage in monitoring large parking lots, huge and dark warehouses, apartment complexes and other similar assets on a 24 × 7 basis. Traffic monitoring and monitoring of suspicious activity in public places, security installations and critical national assets are the other important application areas of security cameras with night vision capability.

9.14.7.2 Applications of Thermal Imaging Night Vision Devices

Thermal imaging devices are used on land-based Armoured Fighting Vehicles, naval vessels and aerial platforms such as aircraft, helicopters and missiles for surveillance, target acquisition and tracking applications. Some of the common non-military applications of these systems include surveillance of living things, search and rescue operations during fire fighting, detection of gas leaks, monitoring of volcanoes, detection of heat in faulty electrical joints and so on.

In military applications, thermal imaging offers several distinct advantages. The first advantage is its immunity to detection by the adversary as it is a passive sensor that does not emit any radiation for generation of image. Secondly, it is extremely hard to camouflage the target from the sensor as it senses heat radiation. Thirdly, thermal imagers can see through smoke, fog, haze and other atmospheric obscurants better than sensors operating in visible spectrum. One of the limitations of thermal imaging sensors is that it is hard for it to discriminate friend from foe. Friendly forces can use heat beacons to overcome this shortcoming. Some of the application scenarios of thermal imaging night vision equipment are briefly described in the following paragraphs.

1) A thermal imager can be a great asset in *firefighting operations*. Their ability to see through smoke and debris allows fire fighters to find people who have passed out and also those fighting for survival and too afraid to come out. A thermal imager can also tell a fire fighter if a door is hot and possibly contains a fierce blaze on the other side. Thermal imagers designed for firefighting missions are designed to operate in very rough and tough environmental

Figure 9.72 FLIR K-50 thermal imaging camera for firefighting. (Courtesy FLIR Commercial Systems B.V, Netherlands.)

(a)   (b)

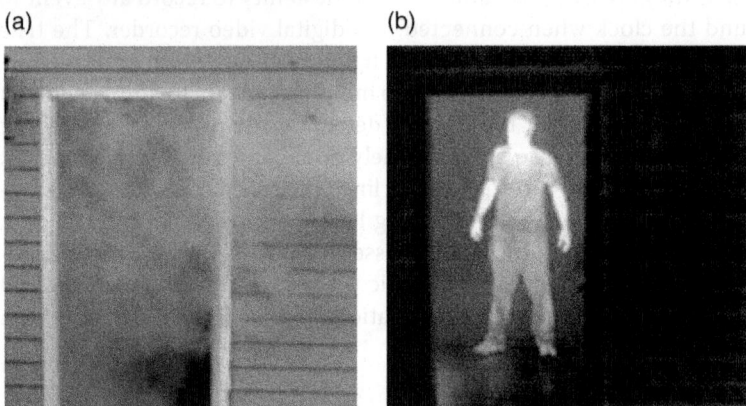

Figure 9.73 (a) Image as seen by the naked eye. (b) Thermal image.

conditions. Figure 9.72 shows a photograph of one such thermal imaging camera in the FLIR K-series (Model K-50). This camera is characterized by uncooled microbolometer focal plane array with 320×280 pixel resolution, thermal sensitivity of 30 mK and operating temperature range of $-20 - +85\,°C$ with capability to operate at elevated temperature of $+260\,°C$ for 5 min. The device is resistant to water and corrosion, heat and flame and immune to shock and vibration. Figure 9.73 parts (a) and (b) show two images of a fire fighting operation, one seen with unaided eyes [Figure 9.73(a)] and the other as seen through the thermal imager [Figure 9.73(b)]. The ability of a thermal imager to see through smoke is evident from the thermal image of Figure 9.73(b).

2) Thermal imagers are also very popular with farmers hunting hogs to protect their farms. The reason for this popularity is that hogs rarely go out during the day time and have the benefit of plant or tree cover. Thermal imagers allow the farmers see past their crops to find the hiding animals. Hunters also love thermal imagers for finding hidden deer. The FLIR Scout PS-series handheld thermal imaging cameras (PS-24 and PS-32) allows things such as traversing through rough terrain in pitch dark conditions, to spot nocturnal animals or livestock, locate missing people in adverse conditions and many more. Figure 9.74 shows a photograph of the Scout PS-24 thermal imaging camera.

Figure 9.74 Scout PS-24 thermal imaging camera. (Courtesy FLIR Commercial Systems B.V, Netherlands.)

Thermal imagers are extensively used by law enforcement agencies and the Armed Forces for a variety of applications in navigation, detection and targeting. Thermal imagers allow them to detect the potential threats without exposing their location to the adversary. The state-of-the-art thermal imaging rifle scopes have become rugged enough to withstand the abuse of recoil making them extremely popular with Armed Forces. While thermal imagers are incredibly effective when it comes to detecting human beings or animals; discrimination between a friend and a foe may be a challenge. This may be an issue in life and death situations. Thermal imaging cameras are one of the most effective tools for surveillance because they work equally well in the day and night. A regular CCTV camera is limited by its need for light, and night vision doesn't function during the day. The chance to see through smoke and fog also gives thermal a leg up on other surveillance techniques. Figure 9.75 shows a photograph of a thermal imaging sensor designed for a border and coastal surveillance array. The sensor offers continuous zoom from 25° and 2° and a long-range detection capability. The sensor is capable of detecting a man size target from greater than 15 km.

State-of-the-art sensors intended for surveillance, target detection and tracking applications employ multi-sensor configurations often combining a visible spectrum sensor with an IR sensor. Final image in this case is the result of fusion of imaging data from the two sensors. Figure 9.76 shows a photograph of one such sensor that combines a thermal imaging camera with a low-light CCD camera. The FLIR camera here is a 640×480 pixel uncooled vanadium oxide micro bolometer focal plane array operating in the 7.5–13.5 μm band. FLIR offers a thermal sensitivity of better than 50 mK.

Figure 9.75 HRC-E FLIR. (Courtesy FLIR Commercial Systems B.V, Netherlands.)

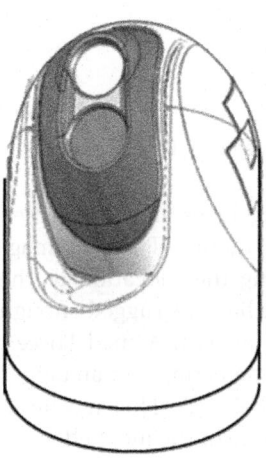

Figure 9.76 Multi-sensor FLIR, type M-series. (Courtesy FLIR Commercial Systems B.V. Netherlands.)

Illustrated Glossary

Avalanche Photo Diodes (APDs) APDs are high-speed, high-sensitivity photo diodes utilizing an internal gain mechanism that functions by applying a relatively higher reverse bias voltage than that applied in the case of PIN photo diodes.

Bayer's Colour Filter Array (CFA) Bayer's CFA comprises an arrangement of red, green and blue filters to capture colour information. It has an alternating red/green and blue/green arrangement and is sometimes called an RGBG filter. Bayer's CFA is made up of alternating rows of red/green and blue/green filters. A particular colour filter allows photons of only that colour to pass through to the pixel.

Bolometer The sensing element in a bolometer is a resistor with a high-temperature coefficient. A bolometer is different from photoconductor, as in a photoconductor a direct photon-electron interaction causes a change in the conductivity of the material whereas in a bolometer the increased temperature and the temperature coefficient of the element causes the resistance change.

Charge Coupled Device (CCD) A CCD is basically an array of thousands to millions of light sensitive elements called pixels etched onto a silicon surface. Each of the pixels is a MOS capacitor. When the sensor array were exposed to light for the same time; the number of electrons and hence the quantum of charge held under a certain pixel would vary directly as the luminous intensity that particular pixel is exposed to. This charge pattern represents the light pattern falling on the device. The charge is read out by suitable electronics and then converted into a digital bit pattern, which represents the image. In a CCD sensor, quantum of charge held by different pixels is transferred through one or a very limited number of output nodes. Each pixel's charge is converted to a proportional voltage and after buffering sent off-chip as an analogue signal.

CMOS Sensor A CMOS sensor records the intensities of light as variable charges similar to a CCD sensor. Unlike CCD sensors, CMOS image sensors do not use charge coupling, which transfers charges to a second bank of photo sites before sending them out for charge-to-voltage conversion, amplification and analogue-to-digital conversion. In a CMOS sensor, each pixel has its own charge-to-voltage convertor, amplifier and a pixel select switch. This is called active pixel sensor architecture in contrast to passive pixel sensor architecture used in a CCD sensor. Also, the sensor often includes on-chip amplifiers, noise-correction, analogue-to-digital conversion circuits and other circuits critical to pixel sensor operation. The chip in this case outputs digital bits.

Detectivity Detectivity of a sensor is the reciprocal of its NEP. A sensor with higher value of detectivity is more sensitive than a sensor with lower detectivity value. Detectivity, like NEP depends upon noise bandwidth and sensor area.

Digital Night Vision Device While in a conventional night vision device available light is collected through the objective lens and focused on an intensifier, most digital night vision devices process and convert the optical image into an electric signal through a highly sensitive charge coupled device (CCD) image sensor. This electrical signal is then transferred onto a micro-display, which is a type of LCD flat-panel display screen. The micro-display usually takes the form of an eyepiece which you look into to view the image rather than on an LCD screen that we find on most digital cameras.

Direct Detection (Thermal Imaging) In the case of direct detection, the detector element translates the photons directly into electrons. The charge accumulated, the current flow, or the change in conductivity is proportional to the radiance of objects in the scene. Detectors in this category include lead selenide (PbSe), mercury cadmium telluride (HgCdTe), indium antimonide (InSb), platinum-silicide (PtSi) and so on. All thermal imaging sensors based on direct detection technology, except those working in short-wave IR (SWIR), use detectors cooled to cryogenic temperatures close to $-200\,°C$. Newer photon type IR sensors operating at elevated temperatures are now available.

Electron-Multiplying CCD (EMCCD) EMCCD uses an electron-multiplying structure called a multiplication register between the readout register and the output charge amplifier. Electron multiplication occurs through a mechanism called impact ionization. EMCCD technology helps in overcoming the shortcoming of traditional CCD technology in offering high sensitivity at high-speed. As a result of this, EMCCD as an image sensor is capable of detecting single photon events without an image intensifier.

Flicker Noise Flicker noise or $1/f$ noise occurs in all conductors where the conducting medium is not a metal and exists in all semiconductor devices that require bias current for their operation. Its amplitude is inversely proportional to the frequency. Flicker noise is usually predominant at frequencies below 100 Hz.

Gated Mode (Image Intensifier Tube) In the gated mode of operation, the intensified image can be gated to open or close the optical shutter by varying the potential difference between the photo cathode and the inside surface of the MCP thereby either allowing or disallowing the formation of intensified image.

Generation 0 Night Vision Devices These are based on image conversion rather than image intensification. The night vision device primarily comprises of a photo cathode that converts incident photons into electrons. The electrons are accelerated towards an anode by applying a positive potential to the anode. The device also had an IR source of radiation called an IR illuminator mounted on the device. The IR illuminator irradiated the target scene with IR radiation. The IR radiation reflecting off the target back to the night vision device is collected by its objective lens and focused on to the photo cathode.

Generation 1 Night Vision Device These are an adaptation of generation 0 technology in the sense that both used a photo cathode and an anode, the former for photon-to-electron conversion and the latter to accelerate photo electrons towards it. A major deviation in generation 1 night vision devices from generation 0 devices is in the absence of IR source used in the case of latter devices to provide scene illumination. Generation 1 devices depend upon ambient light provided by the Moon and stars. Generation 1 tubes were connected in series to enhance sensitivity.

Generation 2 Night Vision Devices These use micro-channel plate (MCP) for electron multiplication leading to a significant increase in device sensitivity. Since the MCP actually increases the number of available photo electrons for a given number of light photons incident on the photo cathode rather than merely accelerating them; the resultant images are relatively brighter and have significantly less distortion.

Generation 3 Night Vision Devices Compared to generation 2 devices, these have two distinct changes. These include use of gallium arsenide photo cathode and an ion barrier coating on the MCP. Due to its higher quantum efficiency or radiant sensitivity and its spectral response extending to near IR region, a gallium arsenide photo cathode enables target detection at longer ranges and in darker conditions. The light amplification factor increases from 20 000 times in the case of generation 2 devices to a maximum of 50 000 times in the case of generation 3 devices. The ion barrier film improves the tube life from 2000 h in the case of generation 2 devices to 10 000 h in the case of generation 3 devices, even though it has the cost of a slight reduction in radiant sensitivity due to fewer photo electrons being able to reach the MCP.

Generation 4 Night Vision Devices These have all the features of generation 3 devices with the modification that it proposes to use a thinned ion barrier film or remove it altogether. Removal of film is aimed at reducing background noise and enhance signal-to-noise ratio. It also allows more electrons reach the MCP so the images are significantly less distorted and brighter.

Generation-Recombination Noise This is caused by the fluctuation in current generation and the recombination rates in a photo sensor. This type of noise is predominant in photo-conductive sensors operating at IR wavelengths. The generation-recombination noise can be calculated using $I_{GRMS} = 2eG\sqrt{\eta EA\Delta f}$.

Illuminance Illuminance, E_P, defines the photo metric flux distribution on a surface. Two very commonly used units to define Illuminance are lux and foot-candle. Lux is defined as the illumination of one lumen of luminous flux evenly distributed over an area of one square metre. Foot-candle is an old English unit and is defined as an illumination of one lumen of luminous flux evenly distributed over an area of 1 square foot: 1 foot-candle is equal to 10.764 lux.

Image Intensification Image intensification or enhancement works on the principle of collecting small quanta of light reflected off the target scene to be viewed in the visible and near IR bands of electromagnetic spectrum in low light conditions. The collected photons are amplified through the processes of photon-electron conversion, electron multiplication and electron-photon conversion. These processes take shape in what is called an image intensifier tube

Image Intensifier Tube This collects the existing ambient light originating from natural sources, such as starlight or moonlight, or from artificial sources such as streetlights or IR illuminators through the objective lens of the night vision device. Low-level light consisting of photons enters the night vision device through input window and strikes the photo cathode. Photo electrons released by the photo cathode are accelerated and focused by a high magnitude electric field towards the micro-channel plate (MCP). For each electron entering the MCP, approximately 1000 electrons are generated and subsequently accelerated from the output of the MCP by a third electrical field towards the phosphor screen. For every photon entering the input window of the intensifier tube, tens of thousands of photons come out of the output window after emission from phosphor screen. This multi-stage process produces an intensified or amplified image of the object that is much brighter than the original image.

Intensified CCD An ICCD successfully exploits the optical amplification provided by an image intensifier to overcome limitations of the basic CCD sensor. Two important features of an intensified CCD are high optical gain and gateable operation.

Irradiance Irradiance, also called *radiant incidence*, defines the radiometric flux distribution on a surface. It is measured in W/m^2.

Johnson Noise Johnson noise, also known as *Nyquist noise* or *thermal noise*, is caused by the thermal motion of charged particles in a resistive element. The RMS value of the noise voltage depends on the resistance value, temperature and the system bandwidth and is given by $V_{RMS} = \sqrt{4KRT\Delta f}$.

Light Dependent Resistors See *Photo Conductors*.

Light-Emitting Diode (LED) A, LED is a semiconductor P-N junction diode designed to emit light when forward biased.

Luminance This is defined as the ratio of luminous flux per unit solid angle per unit area. It is expressed as $lm/Sr/m^2$.

Micro-Channel Plate (MCP) The MCP offers a mechanism of electron multiplication. It has millions of small channels and the electrons entering these channels are both accelerated by a high magnitude electric field within the MCP and multiplied by secondary emission resulting from electrons bouncing off the inner walls of these channels. For each electron entering the MCP, approximately 1000 electrons come out of it.

Modulation Transfer Function (MTF) MTF is a measure of the transfer of modulation (or contrast) from the subject to the image. In other words, it measures how faithfully the optical system reproduces (or transfers) detail from the object to the image. When a black and white striped pattern with associated sine wave changes in brightness is focused on the photo cathode, the image contrast drops gradually as the stripe pattern density is increased. The relationship between the contrast and the stripe density in line pairs per mm (lp/mm) is the MTF.

Noise-Equivalent Power (NEP) NEP is the input power to a sensor which generates an output signal current equal to the total internal noise current of the device, which implies a signal-to-noise ratio of one. In other words, it is the minimum detectable radiation level of the sensor.

Organic LED (OLED) OLEDs are composed of a light-emitting organic material sandwiched between two conducting plates, one of n-type material and the other of p-type material. When an electric potential is applied between these plates, holes are ejected from the p-type plate and electrons are ejected from the n-type plate. Due to the recombination of these holes and electrons, energy is released in the form of light photons.

Photo Conductors These, also referred to as photo resistors, light dependent resistors (LDRs) and photocells, are semiconductor photosensors whose resistance decreases with increasing incident light intensity. They are bulk semiconductor devices with no P-N junction. When light is incident on the photoconductor, electrons jump from the valence band to the conduction band. Hence, the resistance of the semiconductor material decreases. The resistance change in a photoconductor is in the order of six decades, ranging from a few tens of mega ohms under dark conditions to few tens or hundreds of ohms under bright light conditions.

Photo Diode These are junction type semiconductor light sensors that generate current or voltage when the P-N junction in the semiconductor is illuminated by light of sufficient energy. The spectral response is a function of the band gap energy of the material used in its construction. The cut-off wavelength is given by $\lambda_c = 1240/E_g$. Photo diodes are mostly constructed using silicon, germanium, indium gallium arsenide (InGaAs), lead Sulfide (PbS) and mercury cadmium telluride (HgCdTe).

Photo FETs These are light sensitive FET devices where the diode formed by the reverse-biased gate-channel junction acts as a photo diode.

Photometric Flux Flux is defined as a flow phenomenon or a field condition occurring in space. It is a measure of the total power emitted from a source or incident on a particular surface. Photometric flux is measured in lumens (lm). A lumen is defined as the amount of photo metric flux generated by $1/683$ W of radiometric flux at 555 nm where the photopic vision sensitivity of eye is the maximum.

Photometric Intensity Photometric or luminous Intensity is defined as the ratio of luminous flux density per steradian. The unit of photo metric flux density is the Candela (Cd) and is equal to luminous flux density of one lumen per steradian (lm/Sr).

Photometry This is the science that deals with visible light and its perception by human vision. In the case of photometry, measurements are done with reference to the response of human eye.

Photo Multiplier Tube (PMT) PMTs are extremely sensitive photosensors operating in the UV, visible and near IR spectrum. PMTs have internal gain in the order of 10^8 and can even detect single photon of light. They are constructed from a glass vacuum tube which houses a photocathode, several dynodes and an anode. When the incident photons strike the photocathode, electrons are produced as a result of the photo electric effect. These electrons accelerate towards the anode and in the process electron multiplication takes place due to secondary emission process from the dynodes.

Photon Counting Mode (Image Intensifier Tube) At extremely low light levels when only few light spots appear on the phosphor screen per second, a good quality image can be obtained by detecting each spot and its position and integrating them into an image storage unit. The brightness distribution in this case called the *photon counting mode* is given by the difference in the number of photons at each position.

Photo SCRs Photo SCRs, generally referred to as Light Activated SCRs (LASCRs) are essentially the same as conventional SCRs except that they are triggered by a light incident on the gate junction area. They comprise a window and lens to focus more light on the gate junction area, more specifically on the middle junction J_2 of the SCR. They conduct current in one direction when activated by light of sufficient amount and continue to conduct till the current falls below a specified value.

Photo Sensor Noise This mainly comprises Johnson noise, shot noise, generation-recombination noise and flicker noise.

Photo Transistors These are light sensitive transistors. Photo transistors can be used in two configurations, namely the common emitter configuration and the common collector configuration

Photo TRIACS These are bidirectional thyristors designed to conduct current in both directions when the incident light radiation exceeds a specified threshold value. Photo TRIACs are generally used as solid-state AC switches and as photosensors in optocouplers to provide isolation from the driving source to the load.

PIN Photo Diodes In PIN photo diodes, an extra high resistance intrinsic layer is added between the P and the N layers. This has the effect of reducing the transit or diffusion time of the photo induced electron-hole pairs which in turn results in improved response time. PIN photo diodes feature low-capacitance, thereby offering high bandwidth making them suitable for high-speed photometry as well as optical communication applications.

Plasma Display Panel Displays Plasma displays are composed of millions of cells sandwiched between two panels of glass. Each cell is filled with a xenon and neon gas mixture. The electrodes intersecting at a specific cell are charged to excite the gas mixture in each cell. When the gas mixture is excited, plasma is created releasing UV light that then excites the phosphor electrons located on the sides of the cells. These electrons in turn release visible light and return to their lower energy state. Each pixel is composed of three cells containing red, green and blue phosphors, respectively.

Pyroelectric Sensors These are characterized by spontaneous electric polarization, which is altered by temperature changes as light illuminates these sensors. Pyroelectric sensors are low-cost, high-sensitivity devices that are stable against temperature variations and electromagnetic interference. Pyroelectric sensors only respond to modulating light radiation and there will be no output for a CW incident radiation.

Quantum Efficiency This is defined as ratio of the number of photo electrons released to the number of photons of incident light absorbed. It is the percentage of input radiation power converted into photo current. In other words, it is the intrinsic efficiency of the sensor. The value of quantum efficiency, η, is computed as the ratio of actual responsivity to the ideal responsivity

Radiance Radiance, also called radiant sterance, is defined as the ratio of luminous flux per unit solid angle per unit area. Its units are $W/Sr/m^2$.

Radiometric Flux This is defined as a flow phenomenon or a field condition occurring in space. It is a measure of the total power emitted from a source or incident on a particular surface. Radiometric flux or luminous flux is measured in watts (W).

Radiometric Intensity This is defined as the radiometric flux density per steradian. It is expressed in watts per steradian (W/Sr).

Radiometry This is the study of properties and characteristics of electromagnetic radiation and the sources and receivers of electromagnetic radiation. In radiometry, the measurements are made with objective electronic instruments.

Response Time This is expressed as rise/fall time parameter in photo electric sensors and as time constant parameter in thermal sensors. Rise and fall times are the time durations required by the output to change from 10 to 90% and 90 to 10% of the final response, respectively. It determines the highest signal frequency to which a sensor can respond. Time constant is defined as the time required by the output to reach to 63% of the final response from zero initial value.

Responsivity This is defined as the ratio of electrical output to radiant light input determined in the linear region of the response. It is measured in amperes per watt (A/W) or V/W if the photo sensor produces a voltage output rather than a current output. Responsivity is a function of wavelength of incident radiation and band gap energy. Spectral response is a related parameter.

Schottky photo diodes In Schottky type photo diodes, a thin gold coating is sputtered on to the N-material to form a Schottky effect P-N junction. Schottky photo diodes have an enhanced UV response.

Shot Noise In a photo sensor, shot noise is caused by the discrete nature of the photo electrons generated. It is related to the statistical fluctuation of both dark current and the photo current. It depends on the average current through the photo sensor and system bandwidth and is given by $I_{SRMS} = \sqrt{2eI_{av}\Delta f}$.

Solar Cells These are devices whose operation is very similar to that of a photo diode in the photovoltaic mode. The operational principle of the basic solar cell is based on the photovoltaic effect. Due to the photovoltaic effect, there is a generation of an open circuit voltage across a P-N junction when it is exposed to light, which is the solar radiation in the case of a solar cell. This open circuit voltage leads to the flow of electric current through a load resistance connected across it.

Thermal Detection (Thermal Imaging) This uses secondary effects such as the relation between conductivity, capacitance, expansion and detector temperature. They use uncooled detectors including bolometers, thermocouples, thermopiles, pyroelectric detectors and so on.

Thermal Imaging Thermal imaging night vision technology works on the principle of detecting temperature difference between the objects in the foreground and those in the background. All objects above absolute zero emit IR energy. A thermal imaging device is essentially a heat sensor capable of detecting tiny differences in temperature of different points on the surface of the object to be viewed. The information on the temperature difference available in the form of IR energy is collected by the thermal imaging device and converted into an electronic image. The ability to detect tiny temperature differences that always exist not only between the desired object and the surroundings but also between different points of the object itself coupled with emission being in the IR region allows a thermal imaging device to 'see' in near total darkness.

Thermal Sensors These absorb radiation, which produces a temperature change that in turn causes a change in the physical or the electrical property of the sensor. In other words, thermal sensors respond to change in their bulk temperature caused by the incident radiation. Thermocouple, thermopile, bolometer and pyroelectric sensors belong to the category of thermal sensors. Thermal sensors lack the sensitivity of photo electric sensors and are generally slow in response, but have a wide spectral response.

Thermocouple Sensors These are based on the Seebeck effect; that is, the temperature change at the junction of two dissimilar metals generates an EMF proportional to the temperature change. The commonly used thermocouple materials are bismuth-antimony, iron-constantan and copper-constantan.

Thermopile Sensors A thermopile sensor is a series connection of a number of thermocouples. The responsivity of a single thermocouple is very low and therefore to increase the responsivity, several junctions are connected in series to form a thermopile.

Vacuum Photo Diode This comprises an anode and a cathode placed in a vacuum envelope. A cathode when irradiated, releases electrons that are attracted by the positively charged anode, thus producing a photocurrent proportional to the light intensity.

Bibliography

1 Uiga, E. (1995), *Optoelectronics*, Prentice Hall.
2 Kasap, S.O. (2012), *Optoelectronics & Photonics: Principles and Practices*, Prentice Hall.
3 Rosencher, E., Vinter, B. and Piva, P.G. (2002), *Optoelectronics*, Cambridge University Press.
4 Desmarais, L. (1997), *Applied Electro Optics*, Prentice Hall.

5 Waynant, R. and Ediger, M. (2000), *Electro-Optics Handbook (Second Edition)*, McGraw-Hill Professional.

6 Wilson, J. and Hawkes, J. (1998), *Optoelectronics: An Introduction (Third Edition)*, Prentice Hall.

7 Saleh, B.E.A. and Teich, M.C. (2007), *Fundamentals of Photonics (Second Edition)*, Wiley-Interscience.

8 Hradaynath, R. (2002), *An Introduction to Night Vision Technology*, Defence Research & Development Organization.

9 Hradaynath, R. (2001), *Selected Papers on Night Vision Technology*, SPIE Press Book.

10 Vollmer, M. and Mollmann, K-P. (2010), *IR Thermal Imaging: Fundamentals, Research and Applications*, Wiley-VCH.

11 Williams, T. (2009), *Thermal Imaging Cameras: Characteristics and Performance*, CRC Press.

12 Holst, G.C. and Terrence, S. (2011), *CMOS/CCD Sensors and Camera Systems*, SPIE Press.

10

Military Laser Systems

That the laser as a device had gigantic application potential was evident from the interest it generated among scientists and technologists around the world immediately after first demonstration of laser action in the form of a ruby laser by Theodore Maiman in 1960. Since the early 1980s, lasers have penetrated almost every conceivable area of application from fundamental science and technology to industry; from medicine and healthcare to entertainment and from tactical battlefield to the strategic domain. The applications have grown at a very fast rate and not just in already existing domains; laser devices have found their presence in many new areas as well. The focus in the present chapter is on a wide range of military applications of lasers and laser-based systems. Major application areas discussed in this chapter include tactical military laser systems such as laser aiming devices, laser range finders and laser target designators, and laser-based sensor systems including proximity sensors, bathymetry sensors, navigation sensors such as ring laser and fibreoptic gyroscopes, sniper location sensors, LIDAR sensors for detection of chemical and biological warfare agents, LADAR sensors and spectroscopic sensors for detection of explosive agents. Most of the discussion is centred on the requirements of the application, various technological options, current state-of-the-art and future trends.

10.1   Military Applications of Lasers

While the expansion of non-military applications of lasers is primarily driven by availability of a large number of wavelengths followed by ever increasing power levels and diminishing price tags at which these wavelengths could be generated, military applications of lasers and related electro-optic devices have grown mainly because of technological maturity of the lasers that were born in the late 1960s and early 1970s. Technological advances in optics, optoelectronics and electronics leading to more rugged, reliable, compact and efficient laser devices are largely responsible for making them indispensable in modern warfare. On one hand, laser systems meant for existing military applications continue to improve in terms of performance specifications and system engineering, a lot of work is being done particularly in technologically advanced countries to exploit the potential of lasers in newer areas.

Most common of all tactical battlefield applications of lasers have been as *laser range finders* for surveillance and fire-control applications and *laser target designators* for munitions guidance. One cannot imagine a modern battlefield tank whose fire-control system does not utilize the services of a laser rangefinder. It is true for other forms of armoured fighting vehicles too. Short-range semiconductor diode laser-based range finders are finding use on squad weapons like assault rifles, light machine guns and so on (Figure 10.1).

Handbook of Defence Electronics and Optronics: Fundamentals, Technologies and Systems, First Edition. Anil K. Maini.
© 2018 John Wiley & Sons Ltd. Published 2018 by John Wiley & Sons Ltd.

Figure 10.1 Laser range finder on an M320 grenade launcher. (*Source:* Courtesy of The US Army.)

The laser target designator is an essential component of laser-guided munitions delivery system though there are other forms of electro-optic guidance like the one employed in the case of infrared-guided missiles. Laser-guided munitions including bombs, projectiles and missiles constitute an important class of precision strike weapons and the laser target designator plays a key role in the overall delivery system. While a laser target designator is the key element of a laser-guided munitions delivery system, optical gyroscope based inertial sensors such as ring laser gyroscopes (RLG) and fibreoptic gyroscopes find extensive use in the navigation systems of commercial airliners, ships, spacecraft, military aircraft and ballistic missiles.

There are other military applications that exploit the principles of laser range finding and target designation. *Laser-based proximity sensors, gap measuring devices* and *obstacle avoidance systems* are some examples employing a laser range finding principle. *Laser tracking* is another example where a laser target designator come range finder mounted on a two-axis gimbal platform can be used to determine 3D coordinates of a remote target and used to track the target.

Yet another application of basic laser range finding principle is in *laser bathymetry*. Laser bathymetry is swath surveying technique that is complementary to conventional multi-beam acoustic systems. Laser bathymetry is particularly attractive for surveying shallow coastal waters where the acoustic technique is not very effective due to limited swath width. With the arrival of airborne laser bathymetry systems that survey at aircraft speed and are effective in shallow and shoal infested waters, laser bathymetry has become an ideal complement to acoustic swath mapping of coastal waters. Also, airborne laser bathymetry offers a swath width that is independent of water depth.

Laser pointing is another common application. Small low-cost, low power semiconductor diode laser modules with provision of precise X-Y movements are finding a widespread use on squad weapons for target aiming and pointing particularly during night time operations. This increases weapon effectiveness by improving the single shot hit probability and reducing collateral damage. Figure 10.2 shows a photograph of one such integrated module combining brilliant a white light and a bright red laser aiming device in single package. The module can be integrated with the weapon through Picatinny or universal rails.

Figure 10.2 Laser aiming aid type SureFire X400.

Lasers and optoelectronics devices and systems are finding widespread use in *low intensity conflict (LIC) operations*. Many of the established military technologies and systems are being adapted for use in LIC scenario. A low intensity conflict is the most common form of warfare today and is likely to be so for the foreseeable future. Low intensity conflict poses an alarming threat to national security and is an area of concern for the whole of international community today. Its scope extends from requiring emergency preparedness and response to domestic intelligence activities to riot and mob control, from combating illegal drug trafficking to protection of critical infrastructure, from handling counter-insurgency and anti-terrorist operations to detection of chemical and biological agents, from detection and identification of explosive agents to detection of concealed weapons. Laser and optoelectronics technologies play an important role in handling low intensity conflict situations. The key advantage of using laser technology in such applications mainly stems from its near-zero collateral damage, speed of light delivery and potential for building nonlethal weapons. Some of the well-established laser devices in low intensity conflict (LIC) applications include *laser dazzlers* for close combat operations, mob/riot control and protection of critical infrastructures from aerial threats; *LIDAR sensors* for detection of chemical, biological and explosive agents; femtosecond lasers for imaging of concealed weapons and lasers for sniper and gun fire location identification. Uses of laser vibrometry and electron speckle interferometry techniques for detection of buried mines are the other applications.

Over the last decade or so, there has been the emergence of some new potential areas of usage of the laser systems. The initial uses of lasers in defence as outlined in earlier paragraphs were in the form of devices such as laser pointers, laser range finders and target designators for enhancing the accuracy of the conventional weapon systems. Due to the unique nature of laser light, namely its coherence, monochromaticity, high degree of collimation and high intensity, it was considered a potent weapon system for directed-energy applications. For these applications, the lethal laser energy is delivered to the target to cause some form of physical destruction or neutralize electro-optic sensors used on the target platforms. Directed-energy applications of lasers including nonlethal and lethal laser weapons are discussed in detail in the concluding chapter of the book.

10.2   Laser Aiming Devices

A very common application of a laser device when attached to a firearm is as an aiming device. Laser aiming modules are usually configured around low power semiconductor diode lasers emitting in visible or infrared (IR). The CW power level is in the range of 5–10 mW. Most laser aiming modules use red laser diodes emitting at either 635 or 650 nm. Green lasers based on diode-pumped solid-state laser technology emitting at 532 nm are also presently in use. One of the limitations of using visible lasers for aiming and targeting is that it is visible to the naked eye and thus inhibits a covert operation. Laser aiming modules configured around IR diodes to produce an aim point on the target invisible to the naked human eye are now available. These aim points are detectable with night vision devices usually fitted to the firearm.

The aiming module is aligned to emit a laser beam parallel to the barrel. Due to the extremely low value of divergence of the emitted laser beam, it makes very a small spot even at long distances up to hundreds of metres. As an illustration, full angle divergence of 0.5 mrad would produce a spot diameter of 50 mm at a 100 m distance to target. The user places the spot on the desired target and the barrel of the gun is aligned, not necessarily allowing for bullet drop, windage, distance between the direction of the beam and barrel axis and the target mobility during travel time of bullet. Dual wavelength laser aiming devices emitting at a visible wavelength, usually the red and a near IR wavelength, are also commercially available. These modules are equipped with a mode select switch that allows the user to select either of the wavelengths at a time or both wavelengths simultaneously. Some devices are also equipped with a mechanism to provide adjustment for windage and elevation. One example of such an aiming device is the Laser Aiming Module, LAM-10 M from M/s NewconOptik. Figure 10.3 shows a photograph of an LAM-10M laser aiming module. The module offers a visibility range of greater than 1000 m, both IR (830–850 nm) and visible (650 nm) wavelengths of operation, beam divergence of 0.5 mrad and windage/elevation adjustment of ± 20 mrad. It offers two modes of operation at high output power for maximum distance and low power to maximize battery life. It may be mentioned here that most devices intended for a similar role have comparable technical specifications.

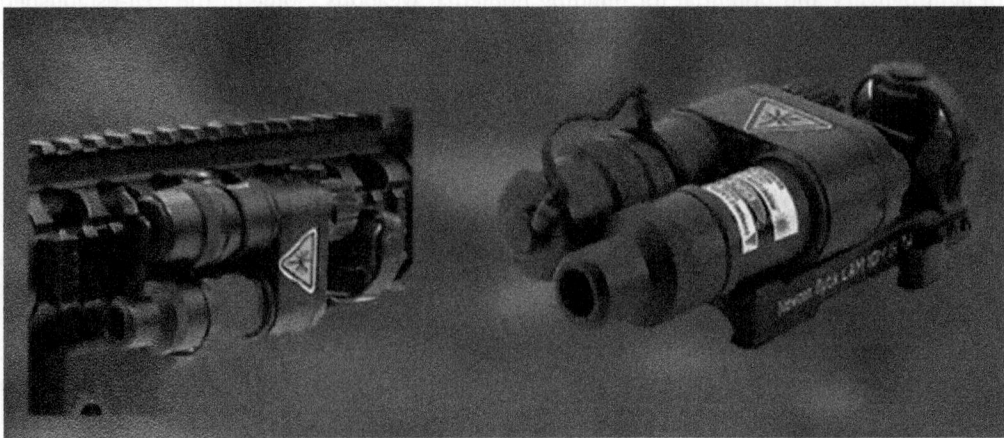

Figure 10.3 Dual wavelength laser aiming module, Model LAM-10 M.

10.3   Laser Range Finders (LRF)

An *LRF* is used in a military application to determine distance to the intended target. It is used both as a standalone device as well as an integral part of a fire-control system of main battle tanks and other armoured fighting vehicles. Laser range finders are available in a wide range of performance specifications in terms of operational range, range accuracy, size and weight to suit different application and platform requirements. Also, laser range finders are configured around different types of lasers including Nd:YAG, Nd:glass, Er:glass and semiconductor diode lasers. These are discussed in the latter part of this section. Laser range finders exploit one of the following four techniques for determining distance to the target. These include time-of-flight, triangulation, phase shift and FM-CW techniques. Each of the four techniques is briefly described in the following paragraphs.

10.3.1   Time-of-Flight LRF

In the *time-of-flight technique,* a narrow pulse width laser beam is transmitted towards the intended target. The target range is measured from the time taken by the laser pulse to travel to the target and back as shown in Figure 10.4. Target range or distance to target *d* is given by eqn. 10.1.

$$d = (c \times \Delta t)/2 \tag{10.1}$$

Where,

d = Target range (m)
c = Speed of light = 3×10^{8} m/s
Δt = Time interval between transmitted and received laser pulses.

   Range accuracy in this case depends on the receiver processing speed and rise and fall time of the laser pulses. Range here is measured as the time interval between the rising or falling edge of the transmitted pulse and the corresponding edge of the received pulse. Uncertainty due to finite values of rise or fall time causes range inaccuracy. Received pulse is processed in a high-speed counter and clock speed determines the processing speed. The time interval and hence range to target is measured in terms of number of clock pulses counted by the time interval counter that is started by the start pulse corresponding to the leading or trailing edge of transmitted laser pulse and stopped by a stop pulse corresponding to the leading or trailing edge of the relevant received pulse. In terms of number of clock cycles, distance to target is given by eqn. 10.2.

$$d = (c \times N)/(2 \times f_{CLK}) \tag{10.2}$$

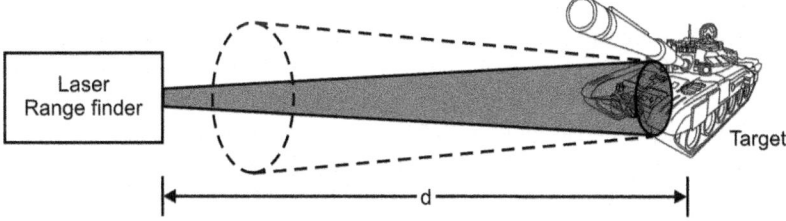

Figure 10.4 Time-of-flight principle of measuring target range.

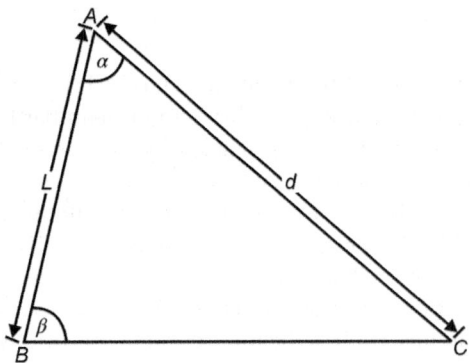

Figure 10.5 Triangulation method of range finding.

Where N is the number of clock pulses counted between start and stop signals and f_{CLK} is the clock frequency. Range inaccuracy in this case is given by $\pm c/2 f_{CLK}$.

The worst-case range inaccuracy equals one clock period. *Range resolution* is determined by the laser pulse width, which implies that the laser range finder based on the time-of-flight principle cannot discriminate between two targets separated in radial range by a distance corresponding to the distance travelled by light in a time period equal to pulse width.

As an illustration, a time-of-flight laser range finder that measures 2000 clock pulses of a 50 MHz clock signal between laser start and stop pulses measures a target range of 6000 m. The range accuracy is ± 3 m, which is of course independent of target range.

10.3.2 Triangulation Technique

In the *triangulation technique*, range is determined by using simple laws of trigonometry. The principle of operation can be best explained with the help of Figure 10.5. In Figure 10.5, a laser transmitter is located at point A, laser receiver is located at point B and point C indicates the target location. The distance to target d from laser transmitter location can be computed from known values of length l and angles α and β using eqn. 10.3.

$$d = l \times \frac{\sin \beta}{\sin(\alpha + \beta)} \tag{10.3}$$

As an illustration, in a triangulation laser range finder, if the transmitter and receiver were separated by 10 m and the angles made by target-to-transmitter and target-to-receiver lines of sight with the line joining transmitter and receiver were 85° and 90°, respectively, target range would be 115 m. Equation 10.3 is valid in flat or Euclidean geometry. The computed results become inaccurate if distances were appreciable as compared to the curvature of the Earth. In that case, more complicated expressions derived using spherical trigonometry should be used.

This technique and associated mathematical formulation turns out to be exceedingly difficult in the real world because of the trouble involved in measuring the baseline and the angles α and β accurately and repeatedly. Another approach again based on triangulation method that overcomes these limitations is described as follows. Refer to the geometry shown in Figure 10.6. In this figure, a laser source is used as the transmitter and a CCD sensor as the receiver not explicitly shown in the figure. The target point is illuminated by the laser beam and the laser spot is imaged on the image plane of the CCD sensor. In the figure shown, u, u_1 and u_2 represent the image positions on the image plane corresponding to target positions x, x_1 and x_2, respectively. The laser and the CCD sensor are mounted in such a way that the beam from the

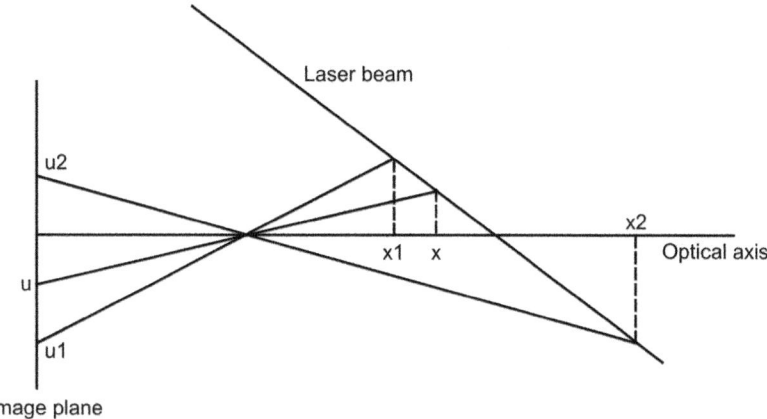

Figure 10.6 An alternative approach to the triangulation method.

laser passes through the optical axis of the camera as shown in Figure 10.6. It may be mentioned here that the placement of the laser with respect to the camera is completely irrelevant so long as their relationship is fixed. The CCD sensor, however, needs to be located close to the laser. Also, laser beam should cross the optical axis in the middle of the ranges of interest. As is evident from Figure 10.6 and eqn. 10.4, there is a linear relationship between the distance to an object and the position of its projection on the image plane when illuminated by a laser spot.

Target range is measured by first measuring points u_1 and u_2 on the image plane corresponding to known target positions x_1 and x_2. The unknown range in that case is given by eqn. 10.4.

$$x = \frac{N}{(ud - k)} \tag{10.4}$$

Where

$$d = x_2 - x_1$$

$$k = (u_2 x_2 - u_1 x_1)$$

$$N = (u_2 - u_1)x_1 x_2.$$

And u is the position of the image of x on the image plane relative to the optical axis, and N, k, and d are defined as shown. In order to measure an unknown range, spot u is located on the image plane and range is computed directly from eqn. 10.4. One may like to pre-compute different values of d, k and N for possible values of x in the range of interest and then store them in a look-up table. The table could then be addressed by the measured value of u. This would make processing very fast. If our application needed real speed we could pre-compute a set of possible values for x in the range of interest, store those values in a look-up table and address the table by the value of u. The resolution would then be limited by the amount of memory available for storing the look-up table.

10.3.3 Phase Shift Technique

In the *phase shift technique* of range finding, the laser beam with sinusoidal power modulation is transmitted towards the target and the diffused or specular reflection from the target is received. The phase of received laser beam is measured and compared with that of the

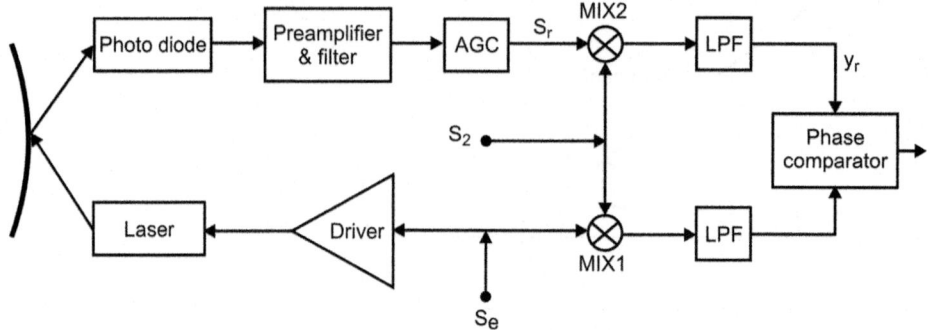

Figure 10.7 Block-schematic arrangement of a phase shift laser range finder.

transmitted laser beam. Figure 10.7 shows the block-schematic arrangement of the phase shift type of laser range finder. To improve the accuracy of this setup, the phase shift is not directly measured at the working high frequency but at an intermediate frequency using a heterodyne technique that preserves the phase shift versus distance. The signals of the two mixers outputs are filtered by a pass band circuit tuned on intermediate frequency. The phase shift is 2π times the product of the time-of-flight and modulation frequency. This allows us to compute the time-of-flight and hence the distance to target from known values of phase shift and modulation frequency. The phase shift $\Delta\phi$ is given by eqn. 10.5.

$$\Delta\phi = \omega\tau = \left(2\pi f\right)\left(\frac{2d}{c}\right) = \left(\frac{4\pi f d}{c}\right) \tag{10.5}$$

Where

f = Modulation frequency
d = Range
c = Velocity of propagation of light in free space
τ = Time of flight from source to target and back to source.

From eqn. 10.5, range d is given by eqn. 10.6.

$$d = \left(\frac{c\Delta\phi}{4\pi f}\right) \tag{10.6}$$

Higher modulation frequencies can result in a higher spatial resolution. The phase shift method appears similar to time-of-flight as the phase shift is proportional to the time-of-flight. However, time-of-flight method conventionally refers to the technique where the time delay is measured more directly. In the case of phase shift method, range measurement is ambiguous as phase shift varies periodically with increasing distance. Maximum unambiguous range is limited to a value corresponding to phase shift of 2π radians and is given by $c/2f$. This ambiguity can be removed by measuring phase shift at two different frequencies.

10.3.4 FM-CW Range Finding Technique

The FM-CW laser range finding technique is similar to the one followed in the case of its radar counterpart; that is, FM-CW radar. In this, the frequency of a narrow line width laser is modulated with a ramp or sinusoidal signal, collimated and then transmitted towards to the target.

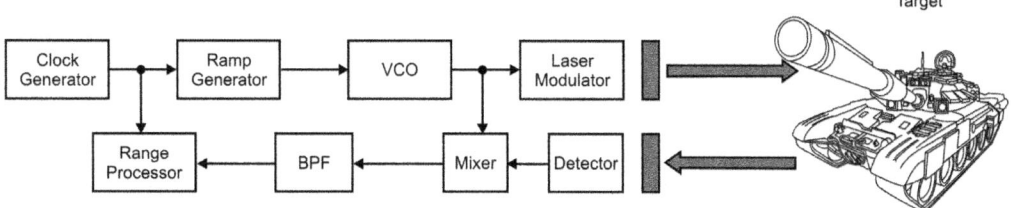

Figure 10.8 Block-schematic of FM-CW laser range finder.

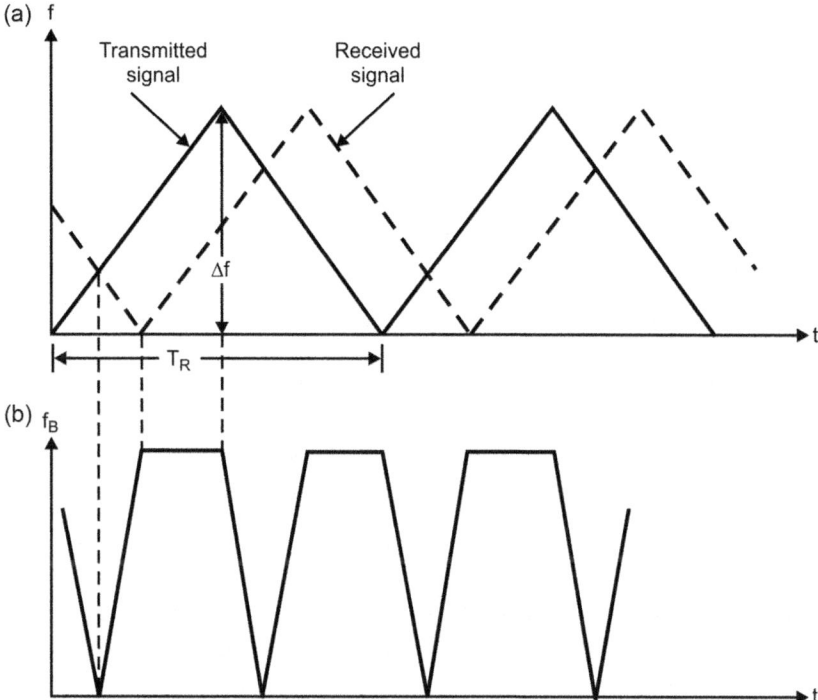

Figure 10.9 FM-CW radar waveforms: (a) VCO frequency versus time and (b) beat frequency versus time.

The received signal corresponding to the reflected laser beam, specular or diffused, is mixed with the reference signal representing the transmitted laser beam. The beat frequency produced as a result of homodyne detection is used to derive the range information. Figure 10.8 shows a basic block-schematic arrangement of an FM-CW laser range finder. The arrangement is self-explanatory. Figure 10.9(a) shows the frequency variation in transmitted and received laser signals as a function of time when the transmitted laser is modulated by a ramp signal. Received signal is time delayed from the transmitted signal as shown in the figure. The time delay and the corresponding beat frequency represent distance to target. Figure 10.9(b) shows beat frequency as a function of time. The beat frequency is measured in the portion where it is constant with respect to time and not in the transient region. The distance to target d is given by eqn. 10.7.

$$d = \frac{\left(f_B \times c \times T_R\right)}{\left(4 \times \Delta f\right)} \tag{10.7}$$

Where,

f_B is the beat frequency
T_R is the ramp waveform time period
Δf is modulation frequency bandwidth.

The minimum measurable range in this technique corresponds to half the period of beat frequency signal. Maximum measurable range equals half the distance travelled by light in half the time period of ramp signal. Minimum and maximum ranges are, respectively, given by eqns. 10.8 and 10.9.

$$d_{MIN} = \frac{c}{\left(4 \times \Delta f\right)} \tag{10.8}$$

$$d_{MAX} = \frac{\left(c \times T_R\right)}{4} \tag{10.9}$$

The measurement range and sensitivity in this case depends on the line width or coherence length of the laser. With a line width of a few kHz, it is possible to achieve range of several hundred km with an accuracy of better than 1.0 m. FM-CW laser range finders implemented with narrow line width fibre-laser sources offer significant improvement in dynamic range, sensitivity, compactness, ruggedness and mounting flexibility.

10.3.5 Choice of Laser

Different types of laser sources have been exploited to build laser range finders for military applications. These include different types of solid-state lasers, semiconductor diode lasers, fibre lasers and CO_2 lasers.

10.3.5.1 Nd:YAG and Nd:Glass Range Finders

Laser range finders used on land-based platforms either in a standalone mode or as a part of integrated fire-control system are usually low repetition rate systems producing laser pulses in the range of 5–30 pulses per minute (PPM) and are configured around Q-switched Nd:YAG or Nd:glass lasers. These are capable of ranging up to a target range of 25 km with accuracy better than ±5 m and are available as compact, handheld devices similar in appearance to a pair of binoculars. Nd:YAG and Nd:glass laser ranger finders are based on the time-of-flight principle. Laser parameters that influence the maximum ranging capability, range resolution and range accuracy, respectively, are peak transmitted power dictated by laser pulse energy, clock frequency driving the range counter and laser pulse width. Model DL-20 from Belarusian Optical and Mechanical Association (BelOMA) and Model LH-30 manufactured by Bharat Electronics are the two representative laser range finders using a Q-switched Nd:YAG laser source. It may be mentioned here that Nd:YAG and Nd:glass laser range finders emitting at eye-hazardous wavelengths have gradually been replaced by Er:glass and OPO-shifted Nd:YAG laser range finders emitting at eye-safe wavelengths. Eye-safe range finders are discussed in subsequent paragraphs.

Another class of range finders that combine the functions of target designation and range finding for the purpose of munitions guidance or those that form the subsystem of a tracking system have relatively much higher repetition rates in the range of 10–50 PPS. The inter-pulse period in this case depends upon the speed of the target to be tracked. These are invariably Q-switched Nd:YAG lasers primarily due to significantly superior thermal conductivity specification of YAG compared to glass as a host material. These lasers are not only becoming smaller in size, these are now available with many additional features to give them inbuilt counter countermeasure capability against similar systems deployed by the adversary in an electro-optic countermeasure role. The LRF-3M from Crystal Techno Ltd is one such repetition rate Nd:YAG laser range finder capable of operating at repetition rates up to 20 Hz. High repetition rate laser range finders have also been gradually replaced by their eye-safe counterparts. Most long-range eye-safe systems are configured around either Er:glass or OPO-shifted Nd:YAG laser sources.

10.3.5.2 Eye-Safe Laser Range Finders

Nd:YAG and Nd:glass range finders emit at 1064 nm, which is a serious eye hazard. The personnel using these devices always run the risk of being exposed to this hazardous radiation. They need to protect their eyes while using them in case some scattered radiation falls on their eyes and causes serious damage. The eye protection for the friendly troops comes in the form of safety goggles, which attenuate the laser radiation to an absolutely safe level. This has led to the development of eye-safe laser range finders and target designators. The low repetition rate class of range finders discussed earlier have invariably been replaced by eye-safe versions. LRB-21K from NewconOptik, STIR-6243/6244 from Sintec Optronics, LH-40 from Bharat Electronics, CELT-2 and MELT from Thales Australia and HEIMDALL from SAAB are some representative examples of Er:glass laser range finders. Even the high repetition rate laser range finders are being gradually replaced by systems operating at eye-safe wavelengths.

One of the techniques used for the purpose is to use a different active medium that would emit an eye-safe wavelength. Er:glass is one such active medium and Er:glass lasers emit at 1540 nm, a wavelength safe for human eyes. The STIR-6241A (repetition rate @ 20 Hz) from Sintec Optronics, ELEM-DP 10K (repetition rate @ 10 Hz) from JENOPTIK, CELT-HR (repetition rate @ 10 Hz) from Thales Australia, TOR (repetition rate @ 12.5 Hz) and FREJ (repetition rate @ 8 Hz) from SAAB are some representative examples of high repetition rate Er:glass laser range finders.

High repetition rate operation at eye-safe wavelengths is not ideal with Er:glass lasers due to poor thermal conductivity characteristics of the glass host, though some Er:glass laser range finders are now commercially available up to a maximum repetition rate of around 20 Hz. Several nonlinear techniques have been successfully used in conjunction with Nd:YAG lasers to generate eye-safe wavelengths. Raman-shifted YAG and OPO based YAG have emerged as very strong contenders for building high repetition rate laser sources. These techniques allow relatively higher repetition rate operation as compared to Er:glass systems. ODIN (repetition rate @ 4 Hz) from SAAB and LRF-3 (repetition rate @ 20 Hz) from Crystal Techno Ltd are some high repetition rate laser range finders configured around KTP OPO-shifted Nd:YAG laser sources.

Another category of eye-safe laser range finders employs semiconductor diode lasers mainly operating at 905 nm and 1550 nm. These range finders are mainly short-to-medium range devices with maximum operational ranges up to a few kilometres. These are commercially available from NewconOptik, Nikon, Bushnell, American Technologies Network Corporation, Leica Camera AG and a host of other manufacturers. The Rangemaster and Geovid range finders from Leica, G-Force DX and Scout DX 1000 ARC from Bushnell, LRB 7X50 and LRF Mod 3/3CI from NewconOptik, ATN Ranger Eye 1500 from the ATN Corporation and PROSTAFF series from Nikon are some representative examples.

10.3.5.3 CO$_2$ LRFs

Another laser type exploited as a range finder is the RF excited waveguide CO$_2$ laser. Advances in waveguide geometry and RF components have resulted in development of compact laser sources emitting at 10.6 μm. Though range finders built using these lasers perform much better in adverse climatic conditions as compared to their counterparts operating at 1064 nm, these have been replaced by solid-state eye-safe laser range finders. Carbon dioxide laser range finders employing TEA CO_2 laser sources in the transmitter and IR detectors cooled to cryogenic temperatures in the receiver were developed during 1970s and early 1980s but have been discontinued.

10.3.6 Applications

LRFs find extensive use as standalone devices for the purpose of observation and situational awareness of an adversary's movement of personnel and military assets. Most armoured fighting platforms are equipped with a laser range finder. While the basic range finder can be used to find target range, when combined with a digital magnetic compass and inclinometer, it can be used to also determine target coordinates. The other important application of a laser range finder is in integrated fire-control systems of armoured fighting platforms. A laser range finder when interfaced with a fire-control computer significantly enhances target accuracy. Modern fire-control systems of all main battle tanks today are invariably laser range finder assisted. Laser range finders when interfaced with night vision, thermal and day time optical aids lead to many a useful and effective battlefield asset for observation, surveillance and situational awareness.

High repetition rate laser range finders in target designators are used for munitions guidance. They are also at the heart of laser trackers and 3D scanners. Other prominent systems with a laser range finder at their core include gap measuring devices, obstacle avoidance systems, laser proximity sensors, laser bathymetric sensors, LIDAR and 3D imaging seekers.

10.3.7 Some Representative LRF Systems

In the following paragraphs salient features of some representative laser range finder systems are discussed, including handheld Nd:YAG and Er:glass range finders, OPO-shifted range finders and semiconductor diode laser range finders. The systems covered in this section include the LH-30 from Bharat Electronics, DL-20 from Belarusian Optical and Mechanical Association and LRF-3M from Crystal Techno Ltd (all Nd:YAG laser range finders); the LRB-21K from NewconOptik, STIR-6243/6244 from Sintec Optronics, TecnaEloptroLH-40/41/41C and MELT from Thales Australia, HEIMDALL, FREG-series, TOR-series and ODIN from SAAB, LIORA from Elbit Systems and LRF-3 from Crystal Techno Ltd (all Er:glass/OPO-shifted Nd-YAG eye-safe laser range finders) and the G-Force DX and Scout DX 1000 ARC from Bushnell, LRB 7X50 and LRF Mod 3/3CI from NewconOptik, ATN Ranger Eye 1500 from the ATN Corporation and PROSTAFF series from Nikon (all eye-safe semiconductor diode laser range finders).

The *Model LH-30* handheld laser range finder (Figure 10.10) manufactured by Bharat Electronics (India) has maximum operational range of 20 km with range accuracy specification of ±5 m. Other features include beam divergence of 1.0 mrad, pulse repetition rate options of 10 pulses per minute and 30 PPM, pulse energy of 6–12 mJ, built-in magnification of 6×, RS-422A serial interface and remote triggering and bite readout.

The *DL-20* laser range finder module (Figure 10.11) manufactured by the Belarusian Optical and Mechanical Association (BelOMA) is also configured around a Nd:YAG laser source and has an operational range of 200–20 000 m with ±3 m range accuracy. Other features include 40 mJ pulse energy, 12 PPM pulse repetition rate and RS-422/RS-232 serial interface. The system operates from 12 VDC and weighs 3.4 kg.

The *LRF-3M* is a high repetition rate Nd:YAG laser range finder capable of operating at pulse repetition frequencies up to 20 Hz. The Q-switched Nd-YAG laser source used in the system

Figure 10.10 The LH-30 laser range finder.

Figure 10.11 DL-20 laser range finder module.

Figure 10.12 LRB-21K laser range finder.

produces pulse energy in the range of 100–120 mJ with 0.7 mrad beam divergence. The system has maximum operational range of 30 km. Range accuracy and range resolution specifications are, respectively, 1.0 m and 15 m.

LRB-21K (Figure 10.12) manufactured by NewconOptik is a handheld Er:glass laser range finder operating at a 1540 nm wavelength and laser beam divergence of 1 mrad. It has operational range of 50–21 000 m with range accuracy of ± 5 m and maximum pulse repetition frequency of 2 Hz. It has built-in three-axis compass and embedded GPS receiver to facilitate instant calculation of target coordinates enabling more precise target triangulation. It has added feature of gating range of 100–6000 m with 100 m accuracy. The measurement result can be viewed through the eye piece or transmitted to a computer for further processing through the RS-232 serial interface. The range finder operates from a 12 V nickel-cadmium battery pack and is qualified for –35– + 50 °C.

The *STIR-6243/6244* manufactured by Sintec Optronics is yet another handheld eye-safe laser range finder. Important features include maximum range of 20 km (STIR-6243) and 10 km (STIR-6244), range accuracy of ±1 m, pulse repetition rate of 10 PPM and RS-422 serial interface. Both models are capable of simultaneous measurement on multiple targets.

Models LH-40, LH-41 and LH-41C (Figure 10.13) are also handheld eye-safe laser range finders configured around an Er:glass laser. All three operate at 1540 nm, have maximum operational range of 20 km with range resolution of 5 m and have pulse repetition rates of 10 PPM and 20 PPM (for a limited period). Also, all three have multiple target range measurement capability measuring the first and last target in the case of LH-40 and LH-41 and first, second and last in the case of LH-41C. LH-40 and LH-41 are equipped with an RS-422 serial interface for communicating range data for further processing or remote display. LH-41C has a RS-232 serial interface. LH-41C also features an integrated digital magnetic compass for target bearing and elevation data for fire-control application. It can be connected to an optional external GPS via a cable to get instant target location data and can be combined with a night vision device.

MELT from Thales, Australia is an eye-safe laser range finder designed for military environment. It is configured around an Er:glass solid-state laser emitting at 1535 nm with transmitted beam divergence of 0.4–0.6 mrad. This range finder based on single shot time-of-flight principle is designed to be used as a handheld electro-optic surveillance device and is also easily integrable to OEM systems. The MELT is designed to be integrated within turrets, payloads, handheld sighting systems, light armoured vehicles, remote weapon stations, fixed-wing aircraft, UAVs and other state-of-the-art multisensory suites. Salient features of the range finder include maximum ranging distance of 20 km with range accuracy of ±5 m, range resolution of 5 m and

Figure 10.13 LH-41/41C laser range finder.

Figure 10.14 HEIMDALL laser range finder.

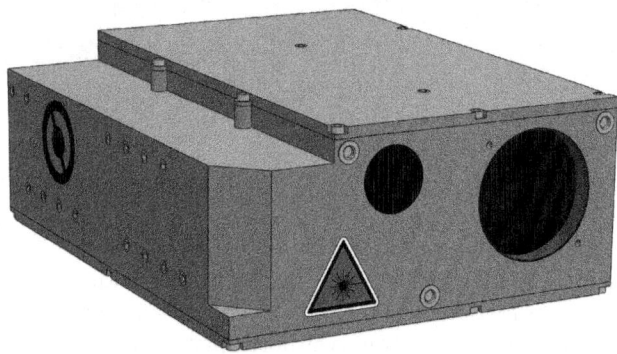

maximum pulse repetition frequency of 1 Hz. The system is also capable of simultaneous ranging up to three events (first, second and last) within the selected window. MELT features an extensive BIT facility and use monitoring system. It can also communicate range, status and diagnostic data to a host management system.

HEIMDALL from SAAB is yet another eye-safe laser range finder configured around a class-I Er:glass laser emitting at 1535 nm offering low hazard functionality (Figure 10.14). The HEIMDALL laser range finder is designed to range and track moving objects and its compact design is ideally suited to integration with advanced fire-control and other telemetric systems. Salient features include maximum operational range of 20 km with 2.5 m range resolution, maximum pulse repetition frequency of 1 Hz and an RS-422 serial interface for data communication.

HEIMDALL is designed for use in a variety of military and civilian applications. Its interface-ability to an IR or TV camera makes it ideally suited to area surveillance, both on the ground

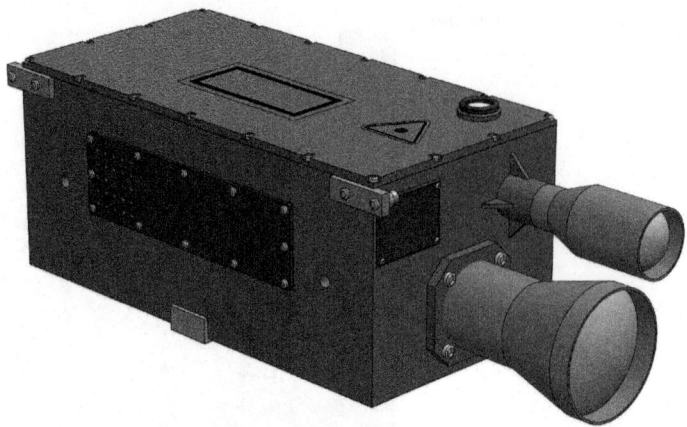

Figure 10.15 G-TOR laser range finder.

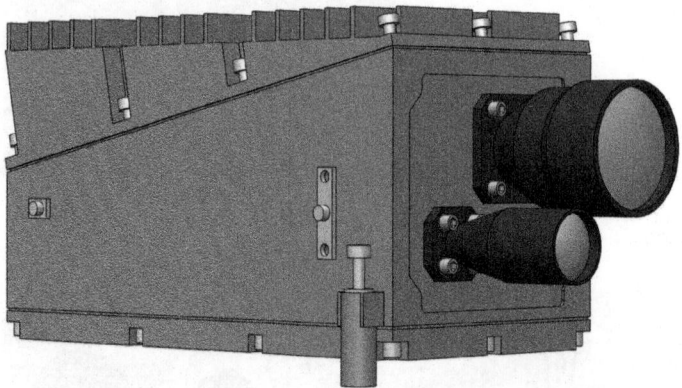

Figure 10.16 G-FREG laser range finder.

and in coastal environments. The HEIMDALL range finder can also be an attractive alternative to radars for detecting ground movement at airports particularly smaller airports. HEIMDALL may also be mounted on an airborne platform, helicopter or an aeroplane to fight fires by locating the centre of the fire allowing water to be dropped at the right moment, thereby avoiding the fire spreading.

The *FREG-series* (G-FREG and S-FREG), *TOR-series* (G-TOR and S-TOR) and *ODIN* from SAAB belong to the family of high repetition rate laser range finders configured around OPO-shifted Nd:YAG laser sources emitting at 1570 nm. While the TOR range finder is capable of repetition rate up to 25 Hz, ODIN and FREG can operate at repetition rates up to 4 and 8 Hz, respectively. The maximum operation range in each of the three cases is 20 km with range resolution of 2.5 m. All three types are equipped with an RS-422 serial interface bus for data communication to a remote station. Both G-TOR (Figure 10.15) and S-TOR are designed for easy integration with electro-optical systems to provide ranging support to radar-based anti-aircraft fire-control systems. The FREG-series range finders (Figure 10.16) are also modular in construction and are therefore easily integrable with electro-optical systems for various applications. These range finders are also available as complete range finders. ODIN range finders are designed for use on combat vehicles for surveillance and fire-control applications

Figure 10.17 LIORA laser range finder.

and also for ground ranging in anti-aircraft systems. Like FREG-series range finders, ODIN is also available both as a complete range finder for standalone use as well as a module for integration into an electro-optical system.

LIORA (Figure 10.17) from Elbit Systems Electro-Optics (Elop) Ltd is a multifunctional handheld device that incorporates a monocular day optical telescope, an eye-safe laser rangefinder, a digital compass and a GPS receiver. Designed and built as a compact unit for infantry units and special operations forces, it can also be integrated as part of a modular target acquisition or observation system. The range finder has all the features of state-of-the-art handheld eye-safe laser range finders in terms of maximum operational range, range accuracy and resolution. Other features include RS-232 or RS-485 serial communication interface and built-in digital compass and GPS for self and target location.

LRF-3 from Crystal Techno Ltd is an eye-safe laser range finder using an OPO-shifted Nd:YAG laser technology capable of operation at high pulse repetition frequencies up to 20 Hz. Maximum ranging distance is 30 km with range accuracy and range resolution specifications of 1 and 15 m, respectively. It can simultaneously up to three targets using range gating. Range data can be communicated to external systems through RS-422 or RS-232 serail communication interface.

G-Force DX and *Scout DX* (Figure 10.18), configured around semiconductor diode lasers, belong to a wide range of compact laser range finders from Bushnell. Both have similar performance specifications with a maximum ranging distance in the range of 900–1200 m and range accuracy of better than 50 cm. Equipped with selectable Bow and Rifle modes, each with their own specialized capabilities, these range finders perform like two units in one. The range finders are small enough to fit in a pocket and tough enough to withstand any weather or terrain.

NewconOptik manufactures a wide range of compact laser range finders with a maximum ranging distance from 1 to 25 km. These are available as monoculars, binoculars and modules for OEM applications. Some of the high end models are capable of measuring distance, speed, height, azimuth and elevation angles. LRB-7X50 (Figure 10.19) is a binocular range

Figure 10.18 Scout-DX ARC 1000 laser range finder.

Figure 10.19 LRB 7X50 laser range finder.

finder employing a class-I eye-safe semiconductor diode laser emitting at 905 nm. The range finder is capable of measuring range up to 1.5 km with range accuracy and resolution specifications of ±1 and 1 m, respectively. Weighing only 1.3 kg, this compact, lightweight and weatherproof laser range finder measures ranging distance in metres and yards, can store up to the last 10 measurements in memory and features target quality indicator and a variable reticle shape.

Laser range finder modules from Newcon Optik including *LRF MOD 3/3CI* (Figure 10.20) employ the same optics and electronics used by the monoculars and binoculars offered by the company. These modules are designed to typically get integrated with thermal imagers, day/night surveillance systems and a variety of aircraft optical systems to add range finding capabilities to bigger systems. All modules offer distance and speed measurement and support RS-232 interface. Modules with CI modification such as LRF MOD-3CI offer azimuth, elevation and height measurement in addition to distance and speed. LRF MOD-3 and LRF MOD-3CI have distance measurement range of 1–3000 m with range accuracy and resolution

Figure 10.20 LRF MOD 3/3CI.

Figure 10.21 ATN Ranger Eye 1500 laser range finder.

specifications of ±1 and 1 m, respectively. LRF MOD 3CI offers azimuth and elevation measurement ranges of 360° and ± 60° respectively with an accuracy of ±1°.

The *ATN Ranger Eye 1500* (Figure 10.21) from the ATN Corporation is an advanced range finding system with a maximum ranging distance of 1350 m with ±1 m accuracy. Configured around an eye-safe laser and a 7× monocular with multicoated optics, it features four different modes of operation. It can be used for a wide range of sport activities such as hunting, golf, archery, target shooting as well as for industrial, topography, safety and tactical applications.

Nikon's *PROSTAFF-5 laser range finder* offers a distance measurement range of 10–550 m. Salient features include a distance target priority mode particularly useful in wooded areas, 6× monocular with multi-layer coated optics for bright and clear images, large ocular (18 mm) for easy viewing, wider field-of-view (7.5°), dioptre adjustment function, compact, lightweight and ergonomic design and a waterproof package. Typical applications include survey measurements and hunting.

10.4 Laser Target Designators

A laser target designator is one of the most widely used pieces of electro-optical equipment in the tactical battlefield. It is an essential and a critical component in the laser-guided munitions delivery setup. In military parlance, laser designation of targets is done for guidance of laser-guided munitions including bombs, missiles and projectiles. The Paveway series of bombs and Lockheed–Martin's Hellfire missiles both launched from aerial platforms and the cannon-launched Krasnopol and Copperhead projectiles are some examples of laser-guided munitions that require a precise laser designation of the intended target. In the following paragraphs different aspects of the role of laser target designator in the delivery of laser-guided munitions are discussed.

10.4.1 Deployment

Laser target designators, laser target acquisition devices and the laser-guided munitions constitute the three important components of a laser-guided munitions delivery setup. The *laser target designator* and *range finder* marks the target and provides accurate information on target range, azimuth and elevation that is used to locate the intended target. These can be either ground-based systems or mounted on an aircraft. *Laser acquisition devices* such as laser spot trackers make use of the laser designator energy reflected off the target to aid visual acquisition of the target. These can be aircraft-mounted or be part of the laser-guided munitions. The other type of laser acquisition device is the seeker and guidance kit combination. This type of device is part of the laser-guided munitions. The *laser-guided munitions* home in on the source of the reflected laser energy, which is the intended target. With reference to the laser spot tracker, it is pertinent to mention here that ground laser designator operators have the luxury of more time, which the operators of airborne laser designators don't have. In the absence of a laser acquisition device, the aircrews would be seriously handicapped in successfully executing laser-guided munitions delivery on the intended target in the short time that they have at their disposal, more so when the target is camouflaged cannot be visually distinguished from other objects. Notwithstanding the limited field-of-view of the laser spot tracker, it allows aircrews to successfully execute the mission in such a situation.

As outlined earlier, laser guided munitions home in on to the source of reflected laser energy. In some cases, the target is irradiated with laser energy before the launch or release of the laser-guided munitions and continues to be irradiated during the entire duration of flight. In another case, it is irradiated after the launch or release and only during the time of the flight. In yet another deployment mode, the laser-guided munitions requires target illumination only during the terminal phase of the flight.

Various factors that can affect laser-guided munitions delivery operation include laser beam scatter due to atmospheric effects, reflection off the target in different directions depending upon target shape and composition, environmental considerations influencing line-of-sight and visibility condition. Laser designator operators therefore carefully plan laser designator locations and also consider various operating techniques such as offset designation, delayed designation and redundant designators. Safety of laser designator operators, particularly the ground operators, is another essential consideration. The protocol on blinding laser weapons prohibits the use of laser designed to cause permanent blindness to unenhanced vision. For other types of lasers used for target designation, range finding, communications and so on, signatories to the protocol have an obligation to take all feasible measures to avoid incidence of permanent blindness to enhanced vision.

In the case of aerial delivery, the aircraft attack heading should also be planned in such a way as to eliminate the possibility of guided weapon homing in on to the laser designator rather than the target. In the case of cannon-launched laser-guided projectiles using ground or aerial designation, there needs to be extensive coordination between the observer and the artillery firing unit. Pulse repetition frequency (PRF) code compatibility is another important issue. Some weapon systems use a three-digit code while others use a four-digit code. For joint task force operations, therefore, the interoperability of different systems needs to be ensured by the joint task force operations officer allocating laser codes.

10.4.1.1 Important Considerations
In the following are the important laser target designator deployment considerations to ensure an effective laser-guided munitions delivery operation.

1) Laser-guided munitions are generally considered fair weather weapons. Laser target designators work best in clear-weather conditions. The target hit probability significantly improves in favourable atmospheric and environmental conditions. Smoke, clouds, haze, fog and so on significantly attenuate and scatter laser energy, which degrades delivery accuracy. All these parameters manifest themselves in the visibility condition and the munitions guidance range decreases with decrease in visibility conditions. The maximum guidance range of a laser-guided munitions is often specified for a certain visibility condition, which is generally 20 km.

2) Existence of line-of-sight between the target the laser target designator and also between the target and the laser spot tracker or/and lase guided munitions is an essential requirement. In the case of laser-guided munitions, the line-of-sight must exist prior to launch or after launch depending on guided munitions' capabilities. The direction of attack should be such to fulfil the following two requirements. First: it should allow the laser-guided munitions or the laser spot tracker receive sufficient laser energy reflected off the designated target. Second: it should eliminate the possibility of the guided munitions homing in on to the laser target designator rather than the target. Use of eye-safe wavelength for range finding is a step in this direction.

3) The target should be laser designated at the correct time and for desired time duration. Designation of the target earlier than is necessary increases the probability of it being detected by the enemy's sensors, thereby facilitating use of countermeasures. In the contemporary warfare, military platforms are well equipped to detect and counter laser target designators and guidance systems.

4) The PRF code of the laser target designator should be compatible with the one programmed in laser-guided munitions or the laser spot tracker.

5) The guided munitions should be released in the specific weapon's delivery envelope to ensure that it finds itself in the laser basket on reaching the point of maximum guidance range.

10.4.1.2 Laser Target Marking and Acquisition
A laser target designator provides accurate target marking for precise delivery of air-to-surface laser-guided bombs and missiles and surface-to-surface cannon-launched laser-guided projectiles. The precision with which a laser target designator marks the target depends upon

the target size and shape, laser beam divergence and designation range. Target size also has a bearing on the ability of the aircrews to visually acquire the target with or without the laser spot trackers. It is more so if the target was camouflaged. Even the targets of relatively larger size are also difficult to visually distinguish from natural objects of the same size and colour.

Laser target marking has a decisive advantage as compared to ballistically delivered marks such as smoke and other visual marks. Target marking by smoke is highly susceptible to wind effects. Visual marks also compromise the observer's position and alert them about the impending attack. Visual marking if not deployed properly may even obscure the target. Laser marking is free from these limitations.

Both operators of ground-based laser target designators and crews of attack aircraft have their set of advantages and disadvantages. While operators of laser target designators have the luxury of time compared to the aircrews of attack aircraft, they have a line-of-sight issue and have to encounter lot of difficulty in precisely marking the target in the presence of smoke, haze or vegetation. Laser designator operators also run the risk of getting exposed to the enemy fire. On the other hand, laser designation from aircraft has generally an unrestricted line-of-sight and is less threatened from enemy's fire. Also, attack aircraft are better equipped with target acquisition and designation systems. The target acquisition systems allow the picking up of camouflaged targets from a distance and identification of the intended target when there are multiple targets in view. The laser spot tracker acquires the laser energy used to mark the target and this visible mark helps the aircrews attack even the camouflaged targets. The visible mark could also be used to align the laser seeker so that it finds itself in the laser basket after release. Laser target designators are PRF code compatible with laser spot trackers and laser-guided munitions.

10.4.2 Laser Designation and Munitions Delivery Considerations

Laser designation and laser-guided munitions delivery operations are governed by laser designator characteristics, target characteristics, laser seeker characteristics and environmental considerations. Each of these factors is described in the following paragraphs.

10.4.2.1 Laser Designator Characteristics

Modern laser target designators use a Q-switched Nd:YAG laser source and operate at 1064 nm. All target designators have range finding channel either at the same wavelength of 1064 nm or an eye-safe wavelength in the range 1535–1570 nm. It may be mentioned here that the range finding channel of almost all modern laser target designators uses an eye-safe wavelength. The system is equipped with a suitable sighting system and a digital magnetic compass. It is mounted on a suitable platform that can provide angular motion. There is a trend towards also building eye-safe laser target designators in the coming years. That would, however, necessitate change of laser seeker heads with a spectral response to cover 1540 nm.

Laser target designators for handheld applications as well as for land-based and aerial platforms are available. Typical performance specifications of laser target designators include laser wavelength, pulse energy, pulse width, pulse repetition frequency and laser beam divergence. Current laser target designators as outlined before operate at 1064 nm. Laser pulse energy is in the range of 50–120 mJ, pulse width is in the range of 5–50 ns, pulse repetition frequency is in the range of 5–20 Hz and laser beam divergence is typically anywhere between 0.1 and 0.5 mrad. Major technical specifications of Ground Laser Target Designator-III (GLTD-III) from Northrop–Grumman include a pulse energy of 80 mJ, beam divergence of 0.3 mead, designation range in excess of 10 km, ranging distance of 200–19 995 m with ±1 m accuracy and pulse repetition frequency in the band I/II.

Laser pulse energy together with the *pulse width* determine the peak transmitted power of the transmitted laser pulse. A 100 mJ laser pulse that is 20 ns wide would produce a peak power of 5 MW. Peak value of the transmitted laser power by the laser target designator is one of the major factors deciding the power density available at the cross-section of the laser seeker of the laser-guided munitions. The other major factors are laser beam divergence, designator-to-target and target-to-munitions distances, atmospheric attenuation and target reflectivity. Laser-guided munitions delivery parameters are discussed in detail in Section 11.4.4 in Chapter 11 on *Precision-Guided Munitions*.

Laser designators and laser seekers in the laser-guided munitions use a pulse coding system in which only certain discrete values of pulse repetition frequency are used. Each pulse repetition frequency expressed in the form of time interval between two successive laser pulses is assigned a code. During a laser-guided munitions delivery mission, laser target designator and laser seeker are made to work on the same PRF code. The seeker responds only to laser pulses with compatible PRF code provided the pulse energy is above a certain threshold value. *Pulse repetition frequency* (PRF) determines the rate at which the information on angular error between the line-of-sight to the target and the seeker axis is updated. Also, higher pulse repetition frequency increases the probability of target acquisition by the guided munitions in the limited time available for the purpose. That is why higher pulse repetition frequencies are preferred for important targets and difficult operating conditions. Higher pulse repetition frequencies, however, cause faster battery drain.

The *laser spot size* on the designated target affects the precision of marking and is determined by the laser beam divergence and designator-to-target distance. A 0.5 mead beam divergence will produce a 50 cm spot at a distance of 1 km. Laser beam divergence is so selected as to produce a spot size at the maximum designation range equal to no more than half the target area. The riflescope type of optics is used for the purpose of aiming with the crosshairs helping in selection of precise aim point on the target.

10.4.2.2 Laser Scatter and Reflection

As outlined earlier, laser-guided munitions home in on to the laser energy scattered or reflected off the target. There can be various possible modes of laser scatter and reflection that play an important role in the success of munitions delivery operation. One of the possible undesirable occurrences is the false target indication and false seeker lock-on due to transmitted laser energy returning scattered by suspended matter in the atmosphere. This phenomenon occurs at very short distances from the laser exit port and presents safety concerns for the operating personnel. In addition, it may cause a false seeker lock-on. That is, the seeker locks on to this stray reflected energy rather than the energy reflected off the intended target. This situation can be avoided by masking the laser designator from the field-of-view of the seeker. The masking may be done by natural means such as vegetation, terrain or by means of temporary screens.

There can be various types of reflections including specular or mirrorlike reflection, scattered reflection, spillover reflection, vertical reflection and aa mixture of specular and scattered reflections. In the case of *specular reflection*, as illustrated in Figure 10.22, the laser beam is reflected at an angle equal to the angle of incidence. This also implies that, for normal incidence, the laser beam is reflected directly towards the laser position. In the case of specular reflection, the laser beam remains narrow, which necessitates the laser seeker to be in this narrow region of reflection. Bare metal and glass surfaces and flat shiny surfaces produce a mirror-like reflection.

Scattered reflection takes place from flat surfaces that are not shiny. In this case, laser energy is reflected in a large arc (Figure 10.23).

(a)

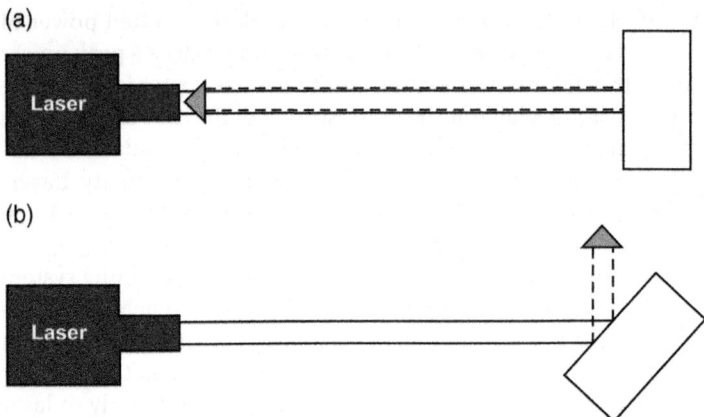

(b)

Figure 10.22 Specular reflection. (a) Normal incidence. (b) Oblique incidence.

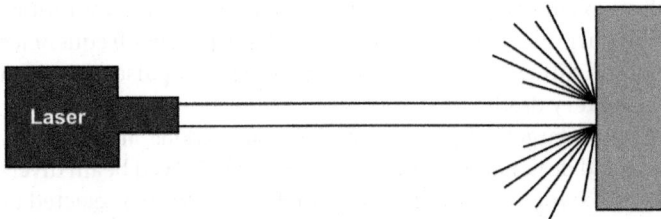

Figure 10.23 Scattered reflection.

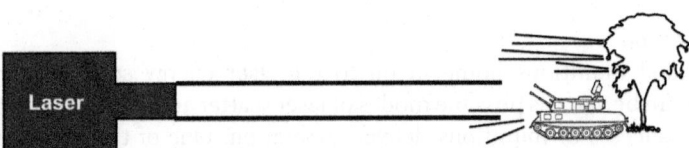

Figure 10.24 Spillover reflection.

When the laser spot is larger than the size of the target, there is spillover of the laser energy to nearby objects or terrain. In this case, there are scattered reflections from these nearby objects and terrain in addition to the scattered reflection from the intended target. This phenomenon, as illustrated in Figure 10.24, is called spillover reflection. Spillover reflection may also be caused by a jittery laser beam due to an unsteady target tracking by the laser designator. Another possibility is the underspill that causes reflection off the objects or terrain short of the intended target.

Reflection from a typical target surface is a combination of scattered reflection and specular (mirror-like) reflection. This implies that there is a strong reflected laser energy in the direction at which it would have been reflected had there been a mirror-like surface. This further implies that if the laser designator radiation were incident normal to the target surface, the strongest reflection would be along the designator-to-target line. The phenomenon is illustrated in Figure 10.25(a) and (b) for normal and oblique incidence. Knowledge about the angle of incidence of laser designator radiation on the target surface can therefore be used to predict the direction in which maximum reflected laser energy is available, thereby making it easier for the laser seeker to lock.

(a)

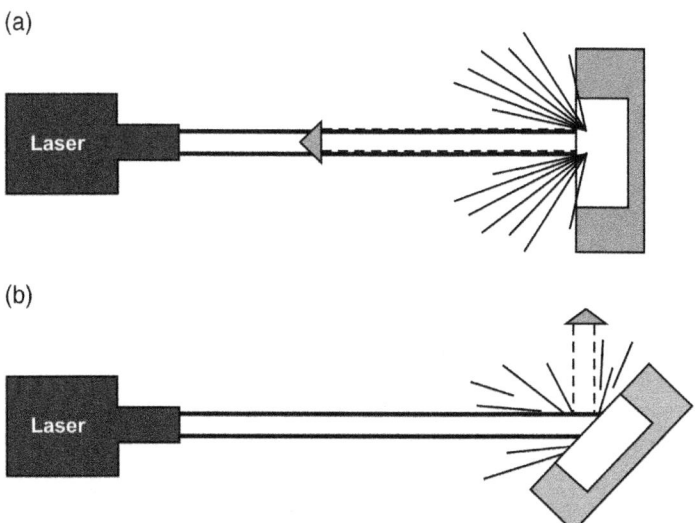

(b)

Figure 10.25 Target reflection. (a) Normal incidence. (b) Oblique incidence.

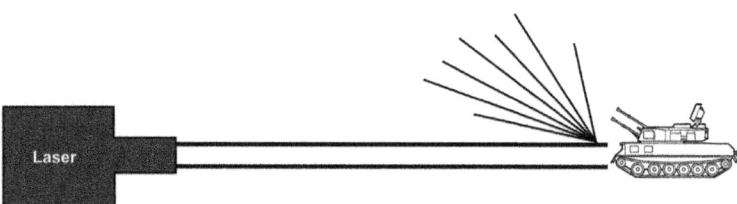

Figure 10.26 Reflection from the top third of target.

Detectable reflected laser energy is the strongest at predictable reflectance angles as shown in Figure 10.26. The height of laser spot on the target is an important factor and the optimum height is guided-munitions specific, in addition to being dependent on type of delivery and target characteristics. Generally, the top third of the target is irradiated.

Sometimes, the laser seeker gets blocked and doesn't see the reflected laser energy. This happens when after the guided munitions is released from the aircraft; the aircraft follows such a flight path as to irradiate the target on the side that is out of view of the seeker. This is known as the *podium effect*. If the direction from which the seeker would be heading towards the target was known, the laser designator can be so aligned as to produce the strongest reflection in that direction.

Boresight error can also lead to less or no reflected laser energy and consequently impairment of munitions delivery mission. It is essential that the laser beam and the sighting mechanism of the laser designator are matched to the same point in the entire operational range. The boresight error may be large enough for the laser beam to fall outside the target though the sighting mechanism may be indicative of aiming the target. This problem can be overcome by having a mechanism for checking the boresight error, if any, and carrying out the required correction in the field. Some designators do have this feature.

Target reflectivity is another important parameter determining the strength of laser energy reflected off the target and hence the probability that the laser seeker would find it exceeding its detection threshold. Target reflectivity depends upon the target material and different materials exhibit different reflectivities. As an example, unpolished aluminium reflects typically 55% while water reflects only 2%. Typical reflectivities of brick structure and vegetation are 55–90% and 30–70%, respectively.

10.4.2.3 False Lock-on Problems

As outlined earlier, there is a possibility of laser seeker falsely locking on to stray reflections rather than the laser energy reflected off the intended target. One such possibility is scattered reflection from suspended matter from the close-to-designator distances. Such a situation could possibly be prevented from occurring by masking the designator. Another false lock-on problem occurs in the case of laser spot tracker equipped aircraft. Generally, there is no safety hazard to the operators of ground laser designators in the entire optimal attack zone of the aircraft. But there is a possibility that the laser spot tracker shift to the laser designator rather than the intended target while the aircraft is operating in 120° attack zone. It is therefore advisable that the aircrews verify the target by means other than the laser spot tracker such as visual inspection, non-laser marking and so on. This would allow them to abort the munitions launch or turn off the laser designation prior to launch in case they found that the laser spot tracker was giving a false indication.

However, if the guided munitions have been released, turning off the laser designation wouldn't ensure the safety of the operating personnel of the friendly forces as in that case the weapon would continue to follow the ballistic trajectory in the direction of release.

10.4.2.4 Environmental Considerations

Line-of-sight, visibility, extreme temperatures and solar saturation are some of the environmental factors affecting a successful laser designation. For a successful laser-guided munitions delivery operation, there must exist an unrestricted *line-of-sight* between the laser designator and the intended target and also between the target and the laser-guided munitions or the laser spot tracker. Visibility degradation occurs due to fog, cloud cover, dust, smoke and other particulate matter in the atmosphere. Laser energy attenuation is a function of visibility condition and is discussed in detail in Section 11.4.4 in Chapter 11. The attenuation factor increases causing a decrease in the laser energy received by the laser seeker with decrease in visibility.

Presence of cloud cover has another implication too. The laser seeker is able to see the designated laser spot only after it comes out of the cloud cover with the result that it is left with very little time in which it has to acquire the intended target and lock-on to it. As outlined earlier, the atmospheric obscurents can also lead to false seeker lock-on as the seeker has to discriminate between laser energy reflected from the obscurents and that from the intended target. One of the remedies to address this problem is repositioning of the target designator system. Positioning it on a slightly higher ground reduces the attenuating effect obscurents such as smoke. Also, repositioning the designator to a non-obscured position helps.

Operating in dark conditions is also sometimes an issue. Though, laser energy doesn't attenuate in darkness, target identification and engagement becomes difficult in dark conditions. Laser target designators equipped with night vision capabilities don't have his limitation.

Another problem arises from non-reflecting or refracting targets such as tunnels. These targets cannot be directly illuminated. In such cases, it is advisable to designate a nearby reflecting target for the mission to be successful. Extreme temperatures not only adversely affect batteries, they can also adversely affect the designator laser beam causing it to refract away from the aim point as it travels through the atmosphere.

Solar saturation is the inability of the laser seeker to discriminate the desired laser spot on the target from the background in the presence of intense solar radiation. This generally happens against targets above the horizon after sunrise and before sunset and when the seeker dome is pitted, cracked or glazed.

10.4.2.5 Seeker Types and Characteristics

Laser seeker characteristics of direct relevance to successful munitions delivery operation include pulse repetition code (PRF) matching, field-of-view, sensitivity and the target acquisition time. A laser designator and laser-guided munitions' seeker or the laser spot tracker must operate on the same *PRF code*. This not only enables the seeker to discriminate against countermeasures deployed by the adversary, it also allows simultaneous attack on multiple targets. Laser seeker generally have a narrow *field-of-view*. Proportional field-of-view is typically $\pm 10°$. It is therefore important that the seeker is so oriented as to have the laser designated spot within its field-of-view. Seeker *sensitivity* is the minimum laser power that it must receive for desired level of performance. Sensitivity in turn determines the maximum guidance range as the received laser power falls with an increase in target-to-munitions distance for a given designator-to-target distance with other parameters remaining fixed. Target *acquisition time* available to the laser seeker or the laser spot tracker is limited. This is done to avoid detection by the adversary and also to conserve the battery energy. Limiting the laser designation time thereby limiting the target acquisition time may significantly degrade the laser-guided munitions accuracy. Laser-guided munitions delivery operations are also specific to type of munitions. Laser designation procedure could be different for laser spot trackers, laser-guided projectiles and laser-guided bombs for best results. Target type, whether it is a point or an area target, also influences the laser-guided munitions delivery operation. Point targets like tanks, guns, armoured vehicles, bunkers, communication sites and surface-to-air missile systems are best neutralized by precise laser-guided munitions attack. Area targets such as artillery positions, infantry formations, command posts, logistic sites, aircraft parking ramps and other large area targets are best neutralized with a large volume delivered through the intended target area. In this case, the laser designators may choose to designate the general area or specific targets within the area.

10.4.2.6 Offset, Delayed and Redundant Designation

In the case of *offset designation*, it is not the intended target but a nearby suitable reflecting object that is illuminated by the laser target designator. This provides immunity against detection by the targeted platform and countermeasures deployed by the adversary if it were equipped with laser-warning sensors. In the case of offset designation for a laser spot tracker equipped aircraft delivering unguided munitions, information on the bearing of the designated point and its distance from the intended target needs be passed on to the aircrew. The laser designator's aim point is moved from the offset position to the intended target in the case of laser-guided munitions employment on receiving directions from the aircrew.

Delayed designation refers to laser designation for a shortened time during the terminal phase of the munitions flight path. It is generally used in the case of low-level releases to prevent the guided munitions impacting short of the intended target. Additionally, delayed designation leaves the adversary with less time to detect the designation unit and use countermeasures even if it was equipped with laser-warning sensors. An early designation forces the weapon to guide on and look down towards the target prematurely thereby losing valuable energy and impacting short of the intended target. Laser designation activation at the right time and for the right duration is extremely important for delayed designation to be effective. Laser designation activation time and lasing duration are decided on the basis of known weapon impact time and laser-guided munitions' laser requirements. To achieve effective delayed designation, there must be a positive two-way communication between the laser designator operator and the aircrew. The Paveway-II series of laser-guided bombs employ delayed designation while executing low-level releases. The designation time is typically 8 s. Hellfire missiles while operating in lock-on after launch mode also employ delayed designation particularly when the target area has low cloud ceilings.

In the case of *redundant designation,* two or more laser designators are employed at different locations. The multiple designators are operated at the same PRF code and used to designate a single target for delivery of a single laser-guided munitions. Redundant designation has the advantage that the seeker may be able to acquire the intended target through the target reflected energy of the second designator if one were compromised. Also, two designators would prevent occurrence of any guidance failure due to temporary blockage in the case of moving targets. Redundant designation though useful for high priority targets has its own set of problems. If there was some backscatter produced from some suspended particulate matter in the atmosphere due to one of the designators, it may lead to the guided munitions impacting on the unintended point. If one of the laser designators were outside the ±10° safety zone or it were located significantly forward of the other, there would be an increased probability of the weapon guiding on to the designator rather than the laser spot on the target. In view of this, this technique is not employed routinely.

10.4.3 Laser Designation Procedures

Laser designation procedures may be classified as general guidelines and procedures specific to the nature of laser-guided munitions delivery mission such as laser designation for artillery support, close air support and so on. In the following paragraphs, some of the important factors common to various munitions delivery deployment scenarios followed up by specific procedures illustrated with practical cases are discussed.

10.4.3.1 General Procedures

Laser designator positioning, attack heading with due consideration of the operator's safety, laser designation timing, laser designator operator survivability, target concealment and seeker blockage related issues are the important factors that need to be carefully considered while executing a laser-guided munitions delivery mission.

Laser designation position is carefully chosen keeping in view a number of factors including requirement of uninterrupted line-of-sight, target location, terrain, weather condition, expected munitions trajectory, communications requirements and safety of friendly forces. The significance of these parameters has been discussed in the earlier sections of the chapter. Proper communications and coordination between designation units and delivery units is particularly important in the case of redundant designation mode.

Attack heading is another important factor. The designator-to-target line and the attack heading are generally pre-coordinated between the designation and guided munitions delivery units. Attack heading should be so chosen as to fulfil the following two major requirements. First: The attack heading should be such that the guided munitions acquire the laser energy reflected off the target. Second: It avoids false target indication. For this, the attack heading should lie outside the designator-to-target safety zone. The safety zone is defined as a cone with its apex at the target location extending ±10° on either side of the designator-to-target line. The vertical limit is also 20°, which makes this safety cone 20° wide in both horizontal and vertical directions. The minimum permitted altitude of the delivery aircraft is therefore governed by the slant range of the target to ensure that it is always outside the safety zone. The optimal attack zone is considered to be inside a ±60° cone around the designator-to-target line. Leaving the safety zone, the attack zone is left to be +10–60° on one side of the designator-to-target line and −10−−60° on the other side of it. While an increased offset angle improves the laser designator operator's safety, it degrades the ability of the laser spot tracker to acquire the laser spot and ±10−±45° is considered to be the best attack zone. The safety zone, optimal attack zone and best attack zone are illustrated in Figure 10.27.

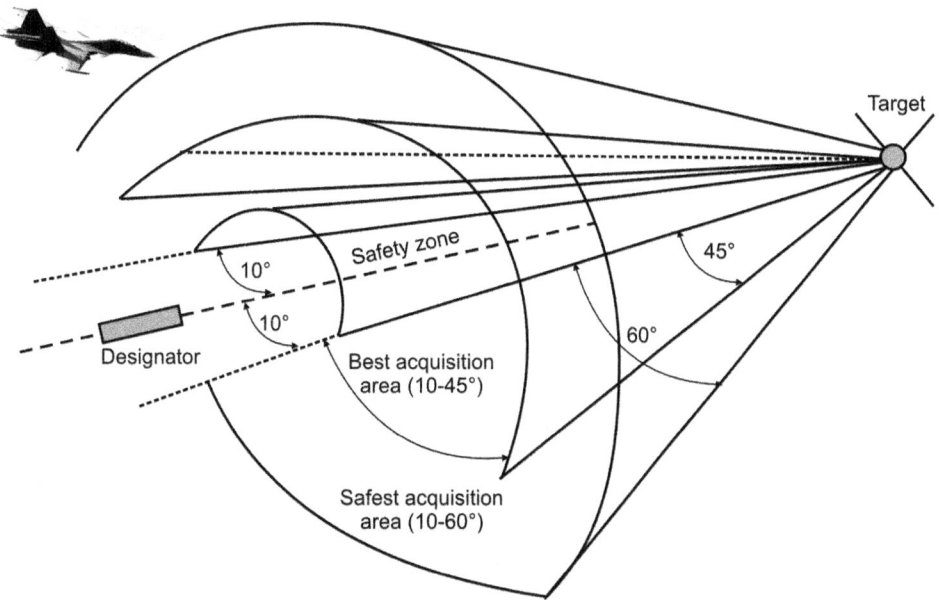

Figure 10.27 Attack heading.

Laser designation time is an important factor for successful operation of laser spot tracker or laser-guided munitions. Laser designation time depends upon a number of parameters including munitions' characteristics and requirements, laser designation time constraints due to overheating, susceptibility to detection by enemy sensors and employment of countermeasures. The operation therefore needs to be properly coordinated between the laser designation and munitions delivery units. Laser designation time should be kept to a minimum depending upon the requirements of the laser seeker. This enhances the survivability of the laser designator operator as a reduced designation time means reduced time for the adversary to detect, locate and neutralize the designator.

Sometimes the target might go out of the field-of-view of the laser seeker or the laser spot tracker due to *target concealment* or *movement*. In such situations, different options available to the laser designator operator are as follows. In case the concealment is partial, the operator may designate another point on the target so as to bring the seeker or the laser spot tracker within the field-of-view. In the case of total concealment of the target, the operator may designate a nearby object. If the laser seeker or the laser spot tracker were in locked condition just prior to the target getting out of the field-of-view, the laser designator operator should make sure that laser spot movement from the intended target to nearby object is smooth.

10.4.3.2 Laser Designation for Artillery

Laser designation by ground-based laser target designators is used to designate stationary or moving point and area targets for attack by laser-guided munitions and laser spot tracker equipped attack aircraft. Laser designation for artillery support is to guide the cannon-launched laser-guided projectiles such as the Copperhead and Krasnopol projectiles. There can be two possible deployment modes of laser designation for artillery. These include designation from a ground-based designator (Figure 10.28) and designation from an airborne designator. Targets are identified and prioritized in the engagement area. The fire support coordinator selects the planned aim points and this information is passed through chain of command to the artillery

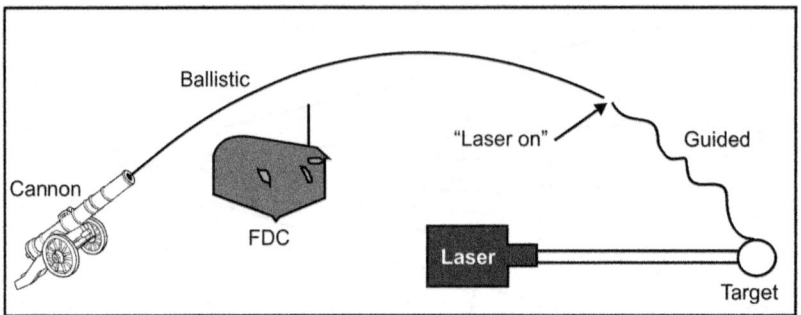

Figure 10.28 Typical delivery profile for a cannon-launched laser-guided projectile.

unit. Procedures for designation from an airborne platform are broadly the same as those for ground-based designators. In this case, the information on launch platform-to-target line is communicated to the aircrew thereby enabling them align themselves near this line to increase the probability of target engagement by the laser-guided projectile. Airborne designators have a greater mobility that makes it easy for the laser designator operator to acquire targets and maintain an uninterrupted line-of-sight. In the case of heavy enemy air defences, the attack aircraft should maintain a longer standoff distance and plan SEAD (Suppression of Enemy Air Defences) missions.

10.4.3.3 Laser Designation for Close Air Support

The first step in any laser designation procedure to be followed to assist a laser spot tracker equipped aircraft delivering non-laser-guided munitions or an aircraft intending to deliver laser-guided munitions is the *target acquisition*. A laser designator is used to mark the intended target. The probability of target acquisition on the first pass can be significantly improved by employing supplementary non-laser marking techniques such as use of smoke and illuminating flares and is therefore recommended if permitted by the tactical situation. While using these non-laser marking techniques, care must be taken to avoid target obscuration by a secondary mark.

In the case of laser designation for laser spot tracker equipped aircraft, the laser code information is passed on to the aircrew of the attack aircraft and the laser spot tracker looks for the coded laser spot on the intended target. In the case of laser-guided munitions, the PRF code is set prior to aircraft take-off and the it is the aircrew that must inform the code to the designation unit through the chain of command. Proper communications and coordination between the involved parties such as the laser designator operator, aircrew and forward air controller is essential to avoid confusion. Unless using offset laser designation, the laser designator operator must designate one target at a time and avoid moving the laser beam in search of a new aim point while the designation is on. Target acquisition is followed by munitions delivery. The munitions homes on to the source of reflected coded laser energy. Figure 10.29 shows one of the possible delivery profiles for laser-guided munitions with a ground-based laser target designator.

In the case of delivery of laser-guided munitions, the run-in heading is also an important factor. The forward air controller should carefully select the run-in heading to ensure safety of friendly ground forces and delivery on the first pass. Laser-guided munitions must be delivered within the delivery envelope to ensure that the target finds itself within the field-of-view of the seeker of the guided munitions during the guidance phase. The aircrew may ensure release within the delivery envelope by seeing the laser designated spot with the help of laser spot

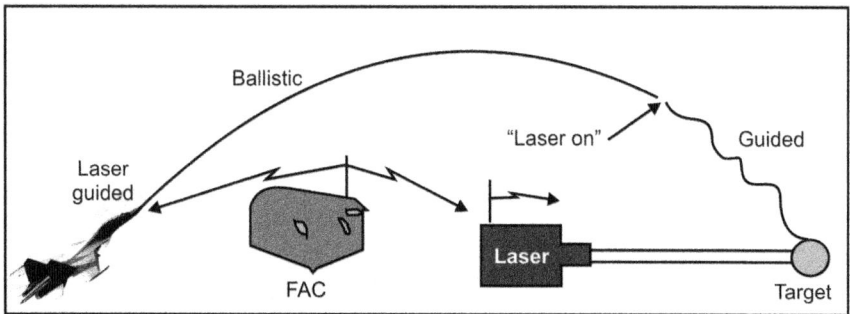

Figure 10.29 One of the possible delivery profiles of an aerially-delivered laser-guided munitions.

tracker or use cues from the onboard navigation sensors to generate bombing references. Laser-ON and laser-TERMINATE calls need to be properly coordinated. High-angle loft attacks should be avoided while releasing weapons behind friendly positions.

In case the attack aircraft is carrying both unguided bombs and laser-guided munitions, preference should be given to delivery of laser-guided munitions on the first pass. Delivery of unguided bombs as the first choice could degrade the target visibility. First pass delivery of laser-guided munitions has another advantage that it discloses the target location to all other aircrews in flight. However, the actual delivery strategy is decided by the tactical situation, aircraft capabilities, coordination by the forward air controller.

Another possibility is that of multiple aircraft attacking a single target. This provides redundancy to the mission and significantly increases the probability of target neutralization in the shortest possible time frame. The mission generally employs a single laser target designator. Multiple designators may be considered for redundancy again. The forward air controller should ensure that the multiple laser-guided munitions deliveries are appropriately spaced apart in time so that the hit accuracy of the munitions delivered in the follow-up attack is not adversely affected due to smoke, debris and so on of the first attack.

Target acquisition and consequently laser-guided munitions delivery during night time and under low light conditions is far more difficult than what it would be in daytime operations. This is due to serious degradation in the ability to use visual cues under the prevailing conditions. Not only is the target acquisition made more difficult; designation of intended target is not easy either, particularly when the target is moving. However, this problem is overcome to a great extent by the use of night vision devices such as night vision scopes, night vision goggles and FLIR (Forward Looking InfraRed).

10.4.4 Representative Laser Designation Systems

Laser designation systems in use for the purpose of laser-guided munitions delivery are available as tripod-mounted laser target designators for ground-based designation, laser target designators for designation from an airborne platform and targeting pods of which it is a part. Some representative laser designation systems in the three categories are briefly described in the following paragraphs.

10.4.4.1 Ground-Based Laser Target Designators

Some common ground-based laser target designators from reputed international manufacturers briefly described in the following paragraphs include the AN/PEQ-17 from M/s Elbit Systems, GLTD-III from Northrop–Grumman and DHY-307/DHY-307LW from CILAS.

Figure 10.30 Portable laser target designator Model AN/PEQ-17.

AN/PEQ-17 from M/s Elbit Systems (Figure 10.30) is a one man-portable robust, compact, low-cost and lightweight laser designation system. The unit is designed to provide designation to stationary or moving targets for all types of laser-guided munitions. It is configured around an Nd:YAG laser operating at 1064 nm. Major performance specifications include maximum ranging and designation distances of 10 and 5 km, respectively, adjustable laser pulse energy of 50–70 mJ, beam divergence of 0.3 mrad, direct view magnification of 8×, direct field-of-view of 5.5°, PRF codes in the NATO Stanag band I/II and an operating temperature range of −40−+70 °C. Other important features include built-in laser spot camera, remote firing facility, data link to thermal sight, remote data transmission, built-in test (BIT) and environmental qualification in accordance with MIL-STD-810.

The *GLTD-III* (*Ground Laser Target Designator-III*), manufactured by Northrop–Grumman Laser Systems (Figure 10.31) is an advanced version of its ground laser target designator GLTD-II. GLTD-II was used by Special Operations Forces, Joint Terminal Attack Controllers and Forward Air Controllers during Operation Enduring Freedom and Operation Iraqi Freedom and has proven its efficacy. GLTD-III was developed to make it more effective. GLTD-III is significantly smaller, lighter and more energy efficient than its predecessor made largely possible by replacement of the flash-pumped Nd:YAG laser of GLTD-II with the state-of-the-art athermal diode-pumped Nd:YAG laser that requires no active cooling. Use of athermal technology helps overcome major shortcomings of warm-up time and standby power consumption associated with conventional diode-pumped laser. Major performance specifications include an operating wavelength of 1064 nm, pulse energy of 80 mJ, beam divergence of 0.3 mrad, ranging distance of 200–19 995 m with ± 1 m accuracy, designation range in excess of 10 km, sighting optics field-of-view of 5° (Horizontal) and 4.4° (Vertical) and magnification of 10×, PRF codes in Stanag band I/II and user programmable PRF codes, RS-422 compatible data input and an output and operating temperature range of −32−+49 °C. The designator has provision for three mounting rails for interfacing night vision devices.

Figure 10.31 GLTD-III laser target designator.

Figure 10.32 DHY-307 laser target designator. (*Source:* David Monniaux, https://commons.wikimedia.org/wiki/File:DHY_307_laser_target_designator_P1220819.jpg.CC BY-SA 3.0.)

DHY-307 (Figure 10.32) manufactured by CILAS is designed to be operated by special operations forces either as a standalone system or mounted on an optronic turret for precise engagement of all types of laser-guided munitions including bombs, missiles and artillery shells. It is interfaceable with different types of thermal cameras and night vision devices. DHY-307 is configured around an Nd:YAG laser operating at 1064 nm. Major performance specifications include a laser pulse energy of greater than 80 mJ, beam divergence of 0.3 mrad, sighting optics magnification of 7× and field-of-view of 100 mrad, ranging distance of 300–20 000 m with ± 5 m accuracy, target marking or designation range in excess of 5 km, Stanag-3733, Russian and user specific PRF codes, MIL-STD-810 qualification for vibration, shock, humidity, rain, sand, dust and so on and operating temperature range of −40– + 50 °C. DHY-307LW is the evolution of combat-proven DHY-307 and is configured around an athermal diode-pumped Nd:YAG laser leading to a highly compact and energy efficient system.

10.4.4.2 Airborne Laser Target Designators

Laser target designators designed for airborne applications and briefly described in this section include the Rattler™ from Elbit Systems and DHX-26 from CILAS. Most laser designation systems intended for airborne applications are integrated as part of a targeting pod that has many more functional components. Targeting pods are discussed separately in the next section.

Rattler (Figure 10.33) from Elbit Systems is a lightweight, high performance laser target designator configured around an athermal diode-pumped Nd:YAG laser. Athermal technology reduces a designator's size, weight and power consumption. The laser specifications include pulse energy of 30 mJ, pulse width of 10–20 ns and beam divergence of better than 0.35 mrad. The operational concept of the Rattler designator involves target designation by the coded laser for aerial or cannon-fired delivery of laser-guided munitions, target marking by laser spot tracker and enslaving the weapon system to the laser spot.

DHX-26 from CILAS is low weight and volume, high performance laser target designator designed for precision long-range designation for guidance of all types of laser-guided munitions including bombs, missiles and projectiles conforming to Stanag 3733 PRF codes. DHX-26 belongs to the family of modular range finders and target designators. It is easily integrable on a targeting pod and other stabilized electro-optic payloads. The designator is configured around a flash-pumped Q-switched Nd:YAG laser operating at 1064 nm. An optical parametric oscillator (OPO) generates the eye-safe wavelength for range finding. The system is qualified for MIL-STD-810 for environmental specifications and MIL-STD-461 for EMI/EMC specifications.

10.4.4.3 Targeting Pods

Targeting pods are target designation tools used for identification of ground targets and guiding munitions to their intended targets. Targeting pods are equipped with other electro-optic systems such as a laser spot tracker to locate the laser pulses reflected from the designated target or/and laser range finder to determine target range. In some cases, same sensor performs the functions of laser spot tracking and range finding. These systems are known as Laser Ranger and Marked Target Seekers (LRMTS). Some targeting pods are equipped with a laser target designator to designate their own targets or for other friendly

Figure 10.33 Rattler laser target designator.

Figure 10.34 LITENING targeting pod. (*Source:* Courtesy of the US Government.)

units. Some common laser designator targeting pods include Raphael/Northrop–Grumman LITENING, Thales' ATLIS, CLDP, DAMOCLES and TALIOS and Lockheed–Martin's LANTIRN. Each of these is briefly described in the following paragraphs.

The LITENING targeting pod (Figure 10.34) is a self-contained, multi-sensor targeting and surveillance system. It is capable of performing both laser-guided munitions delivery as well as surveillance and reconnaissance functions by enabling aircrews to detect, acquire, auto-track and identify targets at extremely long ranges. The pod is currently flown by the US Marine Corps, USAF and international customers including Australia, Denmark, Finland, Israel, Italy, Netherlands, Portugal and Spain.

The targeting pod contains a high-resolution, FLIR sensor to give an IR image of the target with a wide field-of-view search capability and a narrow field-of-view acquisition and targeting capability, a CCD camera to produce target imagery in visible portion of electromagnetic spectrum, a navigation sensor on gimbal for automatic bore sighting capability, a laser target designator for precise delivery of laser-guided munitions and a range finder. Since 1995, Northrop–Grumman has teamed up with Raphael for further development and marketing of the LITENING targeting pod. Different variants of the LITENING pod currently operational include LITENING-II/ER/AT, LITENING-G4 and LITENING-SE. LITENING-II was an improvement on the basic pod. It incorporated a third-generation FLIR, laser marker and software upgrades. In LITENING-ER, 320×246 FLIR was replaced by a higher resolution 640×512 FLIR, thereby enhancing target detection range. LITENING-AT further extends target detection and recognition ranges, enhances target coordinate generation accuracy and provides multi-target cueing. LITENING-G4 included new sensors for significantly improved target recognition and identification features. LITENING-SE has provision for recording and data linking of generated data and imagery in addition to functions related to munitions delivery and navigation.

ATLIS stands for *Automatic Tracking and Laser Integration System* and is a laser/electro-optical targeting pod. ATLIS package includes a laser designator boresighted with a television camera. The ATLIS pod's inbuilt tracking function allows the pilot of the attack aircraft to keep the laser spot automatically focused on the target, thus enabling them deliver guided munitions and break away from the scene to avoid anti-aircraft fire or blast. ATLIS-II incorporates a CILAS ITAY-139 Nd:YAG laser designator/range finder and a dual-mode (visible and IR) television tracker. ATLIS is primarily a daylight and clear-weather system and is being gradually replaced by pods integrated with FLIR. ATLIS pod has been integrated with platforms such as the Jaguar, Mirage-F1, Mirage-2000N and F-16 Fighting Falcon.

CLDP stands for *Convertible Laser Designation Pod* and offers significant performance improvement over ATLIS. The IR imaging technology incorporated in CLDP provided the all-important night time operational capability to the aircrew. The CLDP family includes the CLDP-TC and timeCLDP-TCS models, the latter incorporating an improved day/night system. Two variants, namely CLDP-TC and CLDP-TCS are available. The latter has an improved day/night system.

Thales' *DAMOCLES* (Figure 10.35) is the third-generation multifunction pod incorporating a laser designator providing day/night smart munitions guidance capability and a full suite of sensors for navigation and air-to-air target identification roles. A mid-wave IR sensor operating in 3–5 μm band performs effectively in warm and/or humid conditions and provides day/night small target recognition at medium ranges as well as a long-distance reconnaissance capability. Display of FLIR imagery in pilot's heads-up display is another important feature of DAMOCLES. Its modular design allows for future upgrades, including integral image-processing software, additional display information and an optional bidirectional data link to exchange imagery data with ground forces. Damocles is currently in service on a range of platforms with France and a

Figure 10.35 Targeting pod DAMOCLES. (*Source:* David Monniaux, https://commons.wikimedia.org/wiki/File:NAVFLIR_DAMOCLES_P1220870.jpg. CC BY-SA 3.0.)

Figure 10.36 TALIOS targeting pod. (*Source:* Tiraden, https://commons.wikimedia.org/wiki/File:MBDA_Meteor_et_Thales_Talios.jpg.CC BY-SA 4.0.)

number of countries in the Middle East, North Africa and Asia. It is integrated with a wide range of aircraft, which includes Mirage- 2000, Mirage-F1, Rafale, Tornado, Typhoon and SU-30.

TALIOS is the new generation multifunction targeting pod designed around operational feedback from users and covering the entire critical decision chain from intelligence gathering to weapon delivery (Figure 10.36). Different functional units integrated with the pod provide IR imagery in the 3–5 μm band, TV imagery in the 0.7–0.9 μm band, laser range finding at an eyesafe wavelength of 1.5 μm, laser designation at 1.06 μm conforming to Stanag 3733 codes, laser spot tracking at 1.06 μm and laser marking at 0.8 μm. The pod is compatible with laser-guided munitions, INS/GPS guided missiles and imagery guided weapons. In the air-to-ground role, it can operate in an autonomous or cooperative mode using an integrated laser spot tracker and laser marker and offers long-range damage assessment capability. Real-time data-link transmission is another important feature. The pod provides day/night reconnaissance of small targets at medium ranges and day/night visual airborne target identification. With its open architecture, TALIOS is designed for easy integration on all existing and future fighter aircraft.

LANTIRN (Low-Altitude Navigation and Targeting Infrared for Night) from Lockheed–Martin comprises of a navigation pod and a targeting pod integrated and mounted beneath the aircraft. LANTIRN significantly increases the combat effectiveness of the host aircraft allowing them to fly at low altitudes, and in day and night conditions to attack ground targets with a variety of precision-guided and unguided weapons. The navigation pod includes a terrain following radar and a fixed IR sensor and the targeting pod has a high-resolution FLIR sensor to provide an IR image of the target, a laser designator and range finder for precise delivery of laser-guided munitions, a missile boresight correlator for automatic lock-on of imaging infrared (IIR) Maverick missiles and software for automatic target tracking. The pod is integrated with the F-15E Strike Eagle and F-16 Fighting Falcon. Figure 10.37 shows a photograph of a LANTIRN-ER targeting pod.

10.5 Laser Proximity Sensors

The laser proximity sensor is a type of laser range finder designed to accurately measure relatively smaller distances to target as compared to those encountered in conventional laser range finders used in tactical battlefield applications such as observation and surveillance and fire-control.

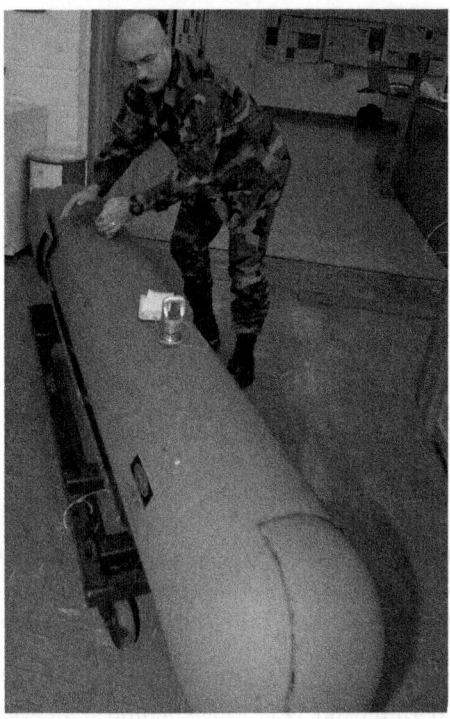

Figure 10.37 LANTIRN targeting pod. (*Source:* Courtesy of the US Airforce.)

While a conventional battlefield laser range finder measures distance to target and the range information is displayed to the observer when the device is used in observation and surveillance role or fed to a fire-control computer in a fire-control application; in the case of a proximity sensor, the processing circuitry generates a command signal when the distance to target is equal to the preset distance value within a certain specified tolerance. The command signal in turn could be used to perform a variety of control functions. In military parlance, the most common application of a proximity sensor is in a laser proximity fuse where the command signal is used to initiate detonation of warheads of large artillery shells, aviation bombs and guided missiles.

Laser-based methods of determining target range have been discussed in the preceding paragraphs while discussing laser range finders. Theoretically, all the previously described range finding techniques, namely time-of-flight, triangulation, phase shift and FM-CW techniques, can be used to build laser proximity sensors or laser proximity fuses. The optimum method for a given application, however, depends on the maximum range and range accuracy requirement of the intended application. The triangulation technique is much better suited to design of laser proximity sensors or fuses as it is capable of measuring shorter distances up to few metres with high accuracy. The range accuracy, however, falls off rapidly with increasing distance. Figure 10.38 shows the basic principle of operation of a laser proximity sensor. The sensor has a transmitting channel usually configured around a semiconductor diode laser or a passively Q-switched diode-pumped solid-state microchip laser along with associated transmit optics and a receiving channel comprising of receiving optics, PIN or APD sensor and range processing circuitry.

As shown in the figure, a transmitted laser beam reflected from the target located at three different distances produces beam images that are displaced across the active area of the photo sensor. The receiver is designed to produce the focused laser beam spot at the centre of the active area for a preset value of proximity distance. There will be a distance both longer and

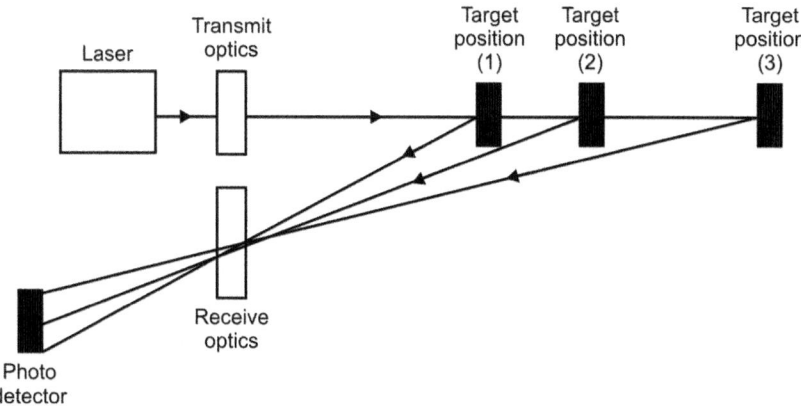

Figure 10.38 Laser proximity sensor principle of operation.

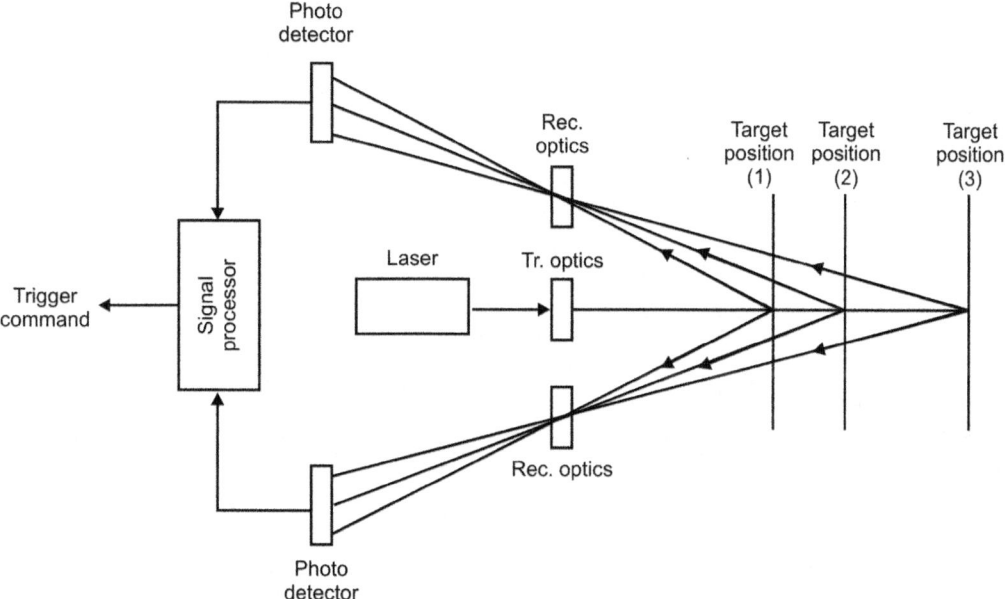

Figure 10.39 Laser triangulation proximity sensors in an axially symmetric configuration.

shorter than the desired distance where the focused beam spot falls just outside the active area. This forms the basis of operation of a laser proximity sensor in general and a laser proximity fuse in particular. The performance of the laser proximity sensor can be enhanced by using an axially symmetric arrangement of multiple aperture photo sensors as shown in Figure 10.39. The design offers better ballistics due to the centre of gravity being located on longitudinal axis of the ammunition round and a higher S/N ratio due to an averaging of multiple return signals. Figure 10.40(a) shows a photograph of a laser proximity fuse assembled in the M433 40 mm round and manufactured by the M/s Physical Optics Corporation. Figure 10.40(b) shows the basic block-schematic arrangement of the proximity sensor shown in Figure 10.40(a). Desired proximity distance can be selected by choosing an appropriate threshold. The block-schematic diagram is self-explanatory.

(a)

(b)

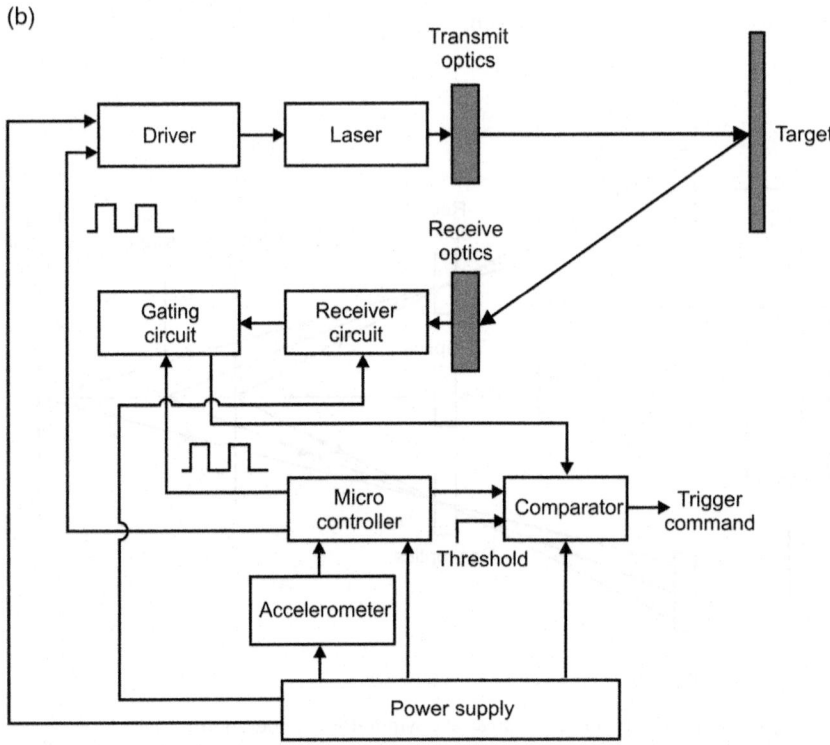

Figure 10.40 Laser proximity sensor type SOProF (Courtesy Physical Optics Corporation USA). (a) Photograph and (b) block-schematic arrangement.

10.6 Laser-Based Detection of Electro-Optic Targets

Yet another emerging application of laser technology is in detection and identification of battlefield optical and optoelectronic sighting systems, which includes optical scopes, night vision devices, thermal imagers, laser range finders, target designators and so on. The device operates on the principle of cat eye effect as shown in Figure 10.41. The target optical device is illuminated by a laser beam, the optical system returns fraction of it as backscattered energy. Backscattered energy is received by a sensitive receiver. With particular reference to homeland

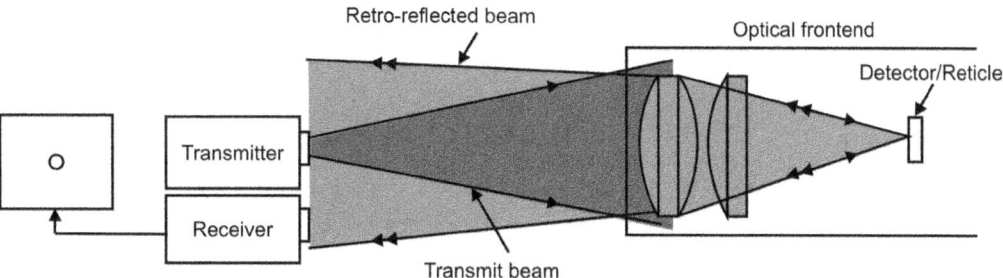

Figure 10.41 Cat eye effect.

Figure 10.42 Typical deployment scenario in an urban environment.

security, such a device could be very useful for detection of optical scopes employed by snipers. Another homeland security-related application could be surveillance of sensitive areas particularly in urban territories. Figure 10.42 shows typical deployment scenario of such a device in an urban environment when used for sanitization of the area.

One such international system is the Laser Sniper Detector that comes in two variants, namely SLD-400 and SLD-500 from CILAS, France. Figure 10.43 shows a photograph of the SLD-400. It can be fitted on a tripod or on a static vehicle. Its highly accurate threat detection and localization capability is compatible with the performance of the associated weapon or fire-control system. SLD-500 is an improvement over SLD-400. Figure 10.44 shows a photograph of another similar system called Optifinder-1200 from M/s Torrey Pines Inc.

The Optifinder-1200 is an active system designed to detect the retro-reflected signal caused by the front-end optics of the target system. It uses a combination of high-sensitivity detector and optical filter to offer maximum operational range of 1200 m. It also makes use of time-of-flight calculation of the transmitted signal and the retro-reflected signal to compute the target range. The chosen detector offers day/night operational capability. The system allows penetration through bad weather and vegetation by range gating. Three variants of the optifinder system are available with operational ranges of 350, 1200 and 2000 m.

Figure 10.43 Laser sniper detector type SLD-400.

Figure 10.44 Optifinder-1200 sniper detection system.

10.7   Laser Bathymetry Sensors

Hydrographic surveying and nautical charting are important for safe navigation in coastal waters and are therefore important not only for worldwide maritime commerce but also for military applications. Acoustic sensor based bathymetry and laser bathymetry are the two established techniques for the purpose. Laser bathymetry is a technique of measuring depths of water bodies usually of relatively shallow coastal waters using a scanning pulsed laser beam. Measurement is usually done from an airborne platform, which gives it the name Airborne Laser Bathymetry (ALB).

Laser bathymetry makes use of transmissive and reflective properties of water and the bottom of water body to measure the depth. Measurement of depth relies on differential timing of laser pulses reflected from surface of the water body and those reflected from the bottom. Laser bathymetry from an airborne platform offers distinct advantages over the conventional vessel mounted acoustic sensor technology as the latter provides a swath coverage that is approximately twice the water depth. As a result, it offers limited swath coverage in shallow waters of coastal zones as illustrated in Figure 10.45. Also, acoustic technology is vulnerable in shoal infected waters. Acoustic technology is more suitable in deep waters. Airborne laser bathymetry offers complimentary capabilities of high coverage rates in shallow waters, seamless data acquisition across land/sea interface and rapid deployment.

When the laser beam from the airborne platform hits the water column, a fraction of the incident laser energy is reflected off the surface and the remaining energy is mostly transmitted through the column. The laser beam transmitting through the water column suffers attenuation due to absorption, scattering and refraction. This limits the maximum measurable water depth. Maximum measurable depth by laser bathymetry depends on water clarity and is generally three times the Sechhi depth. Sechhi depth is an old method of quantifying water clarity and is equal to the depth at which a standard black and white disc is no longer visible to the naked eye.

Figure 10.46 shows the operational concept of a typical airborne laser bathymetry system. The laser transmitter sends out two collinear pulsed laser beams simultaneously, one at a near IR wavelength, usually 1064 nm, and the other at 532 nm. The two pulsed laser beams strike the

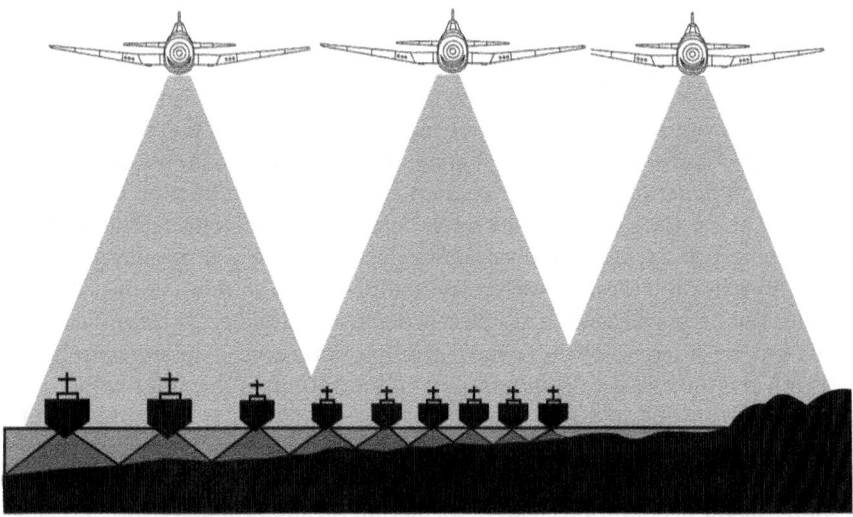

Figure 10.45 Swath coverage in airborne laser bathymetry and acoustic sensor technology.

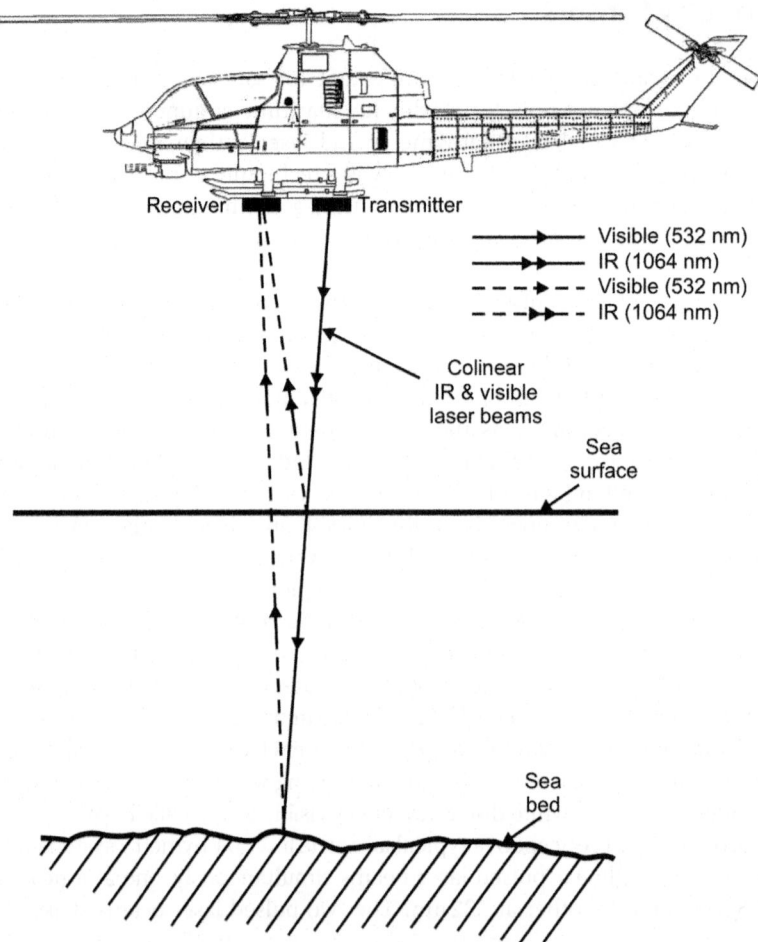

Figure 10.46 Operational concept of airborne laser bathymetry.

surface of water column at the same point and at the same time. The near IR beam is specularly reflected from the water surface. A part of it enters the receiver, which measures its intensity and time instant of arrival with reference to the time instant of transmitted pulse train. A small fraction that is transmitted is absorbed within few cm of the water column. In the case of green laser at 532 nm, most of it penetrates the water column and only a small fraction is specularly reflected. A fraction of green laser reaching the water column bed undergoes diffused reflection. A laser beam after diffused reflection travels upwards through the water column and only a small fraction of it reaches the receiver within its field-of-view. The receiver measures the intensity and time instant of arrival again with reference to transmitted pulse train. The water depth is computed from the difference in the round trip propagation time of the surface reflected laser beam at 1064 nm and the one at 532 nm reflected off the bottom of the water body.

The whole process of transmission, reception and signal processing is far more complex than it appears from the basic description presented here. For example, in a practical system, it is usually the specular reflection of the laser beam from the water surface when the conditions are windy, as these conditions lead to formation of small capillary waves that act like mirror facets. In calm conditions, the surface acts like one large mirror, thus directing the reflection away from the sensor. It may be mentioned here that the transmitted laser beam makes an angle of about 20° with respect to the vertical. In calm conditions, the system depends on Raman scattering and fluorescence signals from water volume just beneath the surface and uses an off-wavelength receiver to generate the received signal. Also, the green laser beam hitting the surface is refracted by about 15° depending on the surface wave structure at the time of impact. The sensor also needs to take into account the refraction effect while computing the water depth

Armed Forces around the world have equipped themselves with airborne laser bathymetry systems over the years for a range of applications. Some of these include acquisition of the CZMIL system by the US Navy in 2012 and US Army Corps of Engineers (USACE) in 2011, SHOALS 3000 by the Military Survey Department of UAE in 2010, SHOALS 3000TH by the US Navy in 2005, SHOALS 1000T by the US Army Corps in 2003, SHOALS 1000 by the Japan Coast Guard in 2003, Hawkeye by the Swedish Navy in 1995, SHOALS 200 by the US Army Corps of Engineers in 1994, ALARMS by DARPA in 1988 and FLASH by the Swedish Defence Research Institute in 1988.

The SHOALS (Scanning Hydrographic Operational Airborne Laser Survey) system from Optech Incorporated, Canada, is a commonly used airborne laser bathymetry system. Different variants of the system in SHOALS-200, SHOALS-1000 and SHOALS-3000 have found themselves on the inventory of Armed Forces over the years as outlined in the previous paragraph. The system meets International Hydrographic Organisation Order 2 requirements for accuracy. The system offers depth measurement accuracy of better than 20 cm and horizontal positioning accuracy of better than 1.5 m. With its special 'Shoreline Depths' processing mode, SHOALS can provide continuous topographic and bathymetric mapping through the shoreline from water onto land. The maximum depth measuring capability of the system is 40–50 m in clear ocean waters, 20–40 m in coastal waters and less than 20 m in more turbid inland waters. Other factors that limit system's depth measuring capability and accuracy include high surface waves, heavy fog and precipitation, sun glint, heavy bottom vegetation and fluid mud.

10.8   LADAR Sensors

Laser radar, also called *LADAR*, uses a laser beam instead of microwaves. That is, the transmitted electromagnetic energy lies in the optical spectrum in laser radars whereas in the case of microwave radar, it is in microwave region. The frequencies associated with laser radars are very high ranging from 30 to 300 THz and the corresponding wavelengths from 10 to 1.0 μm. Higher operating frequency means a higher operating bandwidth, greater time or range resolution and enhanced angular resolution. Another advantage of laser radars compared to microwave radars is their immunity to jamming. Higher frequencies associated with laser radars permit detection of smaller objects. This is made possible by the fact that laser radar output wavelengths are much smaller than the smallest sized practical objects. In other words, laser radar cross-section of a given object would be much larger than the microwave radar cross-section of the same. In fact, rain droplets and airborne aerosols too have significantly larger laser radar cross-sections to allow their range and velocity measurement, which is very important for many meteorological applications. A higher resolution of laser radar allows recognition and identification of certain unique target features such as target shape, size, velocity, spin, vibration and that forms the basis of their use for target imaging and tracking applications.

While on one hand there are numerous advantages that laser radars offer over microwave radars, the former are severely affected by adverse weather conditions. Also, the narrow beam width of laser radars is not conducive to surveillance applications. For surveillance

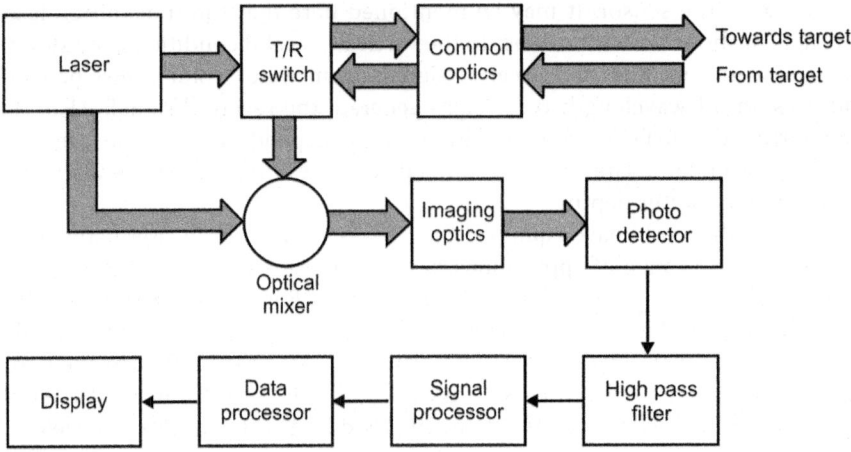

Figure 10.47 Block-schematic of coherent laser radar.

applications, laser radar needs to operate at very high repetition rates so that large volumes can be interrogated within the prescribed time. Alternatively, multiple simultaneous beams can be used.

The laser range finder discussed in previous paragraphs is also a type of laser radar. A conventional laser range finder uses incoherent or direct detection. The term laser radar is usually associated with systems that use coherent detection. Figure 10.47 shows a block-schematic arrangement of coherent laser radar. The laser beam is transmitted towards the target. A fraction of the transmitted power/energy reflected from the target is collected by the receiver. The block diagram shown in Figure 10.47 is that of a monostatic system in which transmitter and receiver share common optics made possible by the use of Transmit-to-Receive switch. In a bistatic arrangement, transmitting and receiving optics are separate. The received laser beam is coherently detected in an optical mixer. In the case of homodyne detection, a sample of transmitted laser power is used as local oscillator. In the case of heterodyne detection, another laser phase locked to transmit laser is used as local oscillator. Heterodyne detection is used when transmitter and receiver are not collocated. Output of optical mixer is imaged on to the photo sensor. The electrical signal generated by a photo sensor module is processed to extract the desired information about the target. The photo sensor module is a single sensor in the case of a scanning transmitted beam and a two-dimensional array of sensors in the case of a flash type laser imaging radar.

Figure 10.48 illustrates the concept of flash type 3D laser imaging. In this case, a diverging pulsed laser beam illuminates the entire scene of interest. The transmitted pulse time is referenced by an auxiliary photo sensor. The backscattered light is imaged on to a two-dimensional array of photo sensing elements called pixels. While a conventional camera would measure the light intensity of back scattered light pulse, in this case different sensor elements in the array measure time-of-flight. The time-of-flight is proportional to the distance between the point on the target from where the laser beam is reflected and the sensor element. The sensor array thus produces a 3D image of the target (Angle-Angle-Range). An alternative approach of generating target image is to use a scanning laser beam and a single sensor.

A common application of the LADAR concept is in LADAR seekers used mainly in conjunction with other guidance systems on strategic payloads for intended target discrimination from advanced decoys and aim point selection. It is also well suited to combat identification, navigation

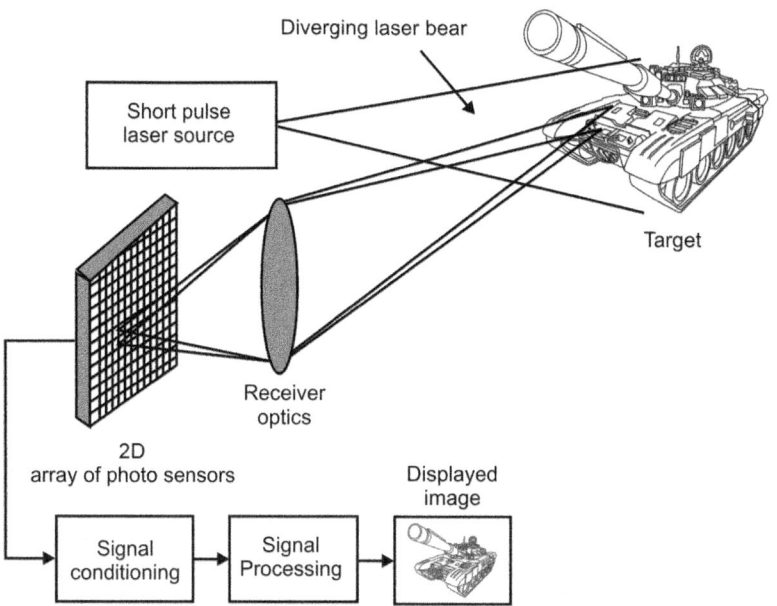

Figure 10.48 Flash type 3D laser imaging concept.

Figure 10.49 Multimode enhanced LADAR seeker.

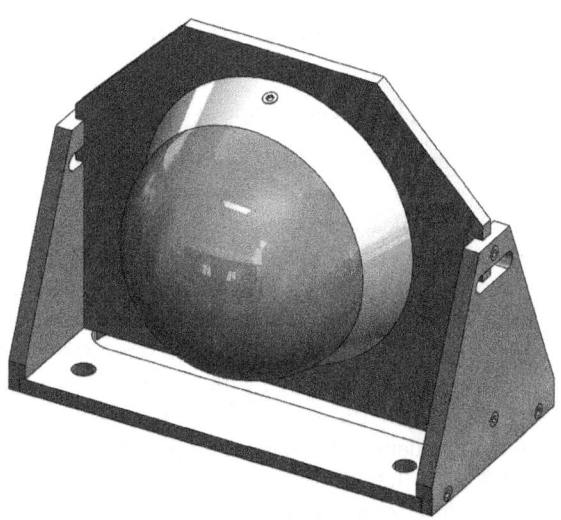

of autonomous vehicles and topography. LADAR is also suitable in finding targets hidden by camouflage nets and foliage. LADAR seeker can detect and identify specific features of the target with a very high definition up to a resolution of a few centimetres from a distance of few kilometres. An automatic target acquisition algorithm processes the images to identify and acquire targets based on 3D templates stored in the weapon's memory before the mission. LADAR sensors are usually employed on loitering systems that look at the target from different angles, verify target's identity and select the best attack position for desired results.

Figure 10.49 shows a photograph of an advanced multimode LADAR seeker from Lockheed–Martin. The seeker can operate as standalone semi-active laser (SAL) and LADAR modes as well as simultaneous SAL and LADAR modes for target identification, acquisition and tracking. The seeker is designed to conduct wide area search and identify actual or potential targets including those obscured by camouflage or foliage. Such LADAR seekers have been successfully tested on Loitering Attack Missile (LAM) missions under DARPA's NLOS-LS (Non-Line-of-Sight Launch System) and USAF's LOCAAS (Low-Cost Autonomous Attack System) programmes. LAM with its multimode LADAR seeker searches a large area and relays location of various targets back to the command centre where these targets are engaged by direct attack or by other assets. In the case of a priority target, LAM can be commanded to break off its search mission and attack the target.

10.9 Laser-Based Gyroscopic Sensors

A gyroscopic sensor constitutes an essential component of an inertial navigation sensor (INS). Inertial navigation sensors are extensively used on a range of aerospace vehicles, which includes commercial airliners, military aircraft and spacecraft. Inertial navigation sensors are extensively used for the mid-course correction of guided missiles. In the case of *inertial navigation guidance*, the vehicle uses onboard sensors to determine its motion and acceleration with the help of gyroscopes and accelerometers. The gyroscope is used to measure angular rotation and an accelerometer is used to measure the linear motion. The gyroscope and accelerometer are combined into a single unit along with a control mechanism and the unit is called Inertial Measurement Unit (IMU) or Inertial Navigation sensor (INS). In the following paragraphs are discussed two very important laser-based gyroscopic sensors namely the *Ring Laser Gyroscope* (RLG) and the *Fibreoptic Gyroscope* (FOG).

10.9.1 Ring Laser Gyroscope

A ring laser gyroscope (RLG) is an important optronic rate sensor and in essence the heart of any inertial navigation system. There are different types of gyroscopic sensors such as spinning wheel mechanical gyro, dynamically tuned gyro, ring laser gyro and fibreoptic gyro. A ring laser gyro offers superior performance as compared to conventional spinning type mechanical gyro as there are no moving parts and therefore no inherent drift terms due to absence of friction. Unlike a mechanical gyro, the ring laser gyro does not resist changes to its orientation. A combination of three orthogonally placed RLGs makes a rate sensor with three degrees of freedom.

Having briefly described different types of guidance techniques in the previous sections, keeping in view the scope of the present text, in the following paragraphs the operational principle, salient features and performance parameters of a ring laser gyroscope are discussed in a bit more detail.

A ring laser gyroscope is primarily a rate sensor, which can be used to measure rotation by integrating the rate information. It essentially consists of two counter-propagating laser beams over the same path. It operates on the principle of the Sagnac effect according to which the null points of the internal standing wave pattern produced by the counter-propagating laser modes shift in response to angular rotation. The shift in the nulls of the standing wave pattern manifests itself in the form of a moving interference pattern observed by combining the two laser modes.

The basic concept of a ring laser gyroscope can be explained as follows. The frequency of oscillation in a linear laser is such that that the laser cavity consists of an integral number of wavelengths. A linear laser cavity is constituted by a double pass of the distance between the two mirrors. Since it is imperative that the beam replicates itself for successive passes over the cavity length, there will always be nodes at the two mirrors and the two oppositely propagating laser beams comprise a standing wave. In the case of a ring laser, the two oppositely directed laser beams can be considered travelling waves and there is no such constraint of having a node at the mirror. The two beams in this case can be independent of each other and oscillate at different amplitude and frequency. The oscillation frequency of each is determined by the optical path length and not the geometrical path length. Thus, any mechanism, which is rotation in this case that produces an optical path difference, results in different frequencies of oscillation for the two laser beams. If the rotation were in a clockwise direction, the clockwise laser beam will see a larger optical path length than the path length encountered by the counter clockwise beam. For rotation in a counter clockwise direction, it would be the opposite. The frequency difference is proportional to the rotation rate of the cavity.

Thus, by measuring the frequency difference, rotation rate of the cavity and the platform it is strapped on to can be measured. According to the Sagnac equation, the frequency difference is given by eqn. 10.10.

$$\Delta v = \frac{4 A \Omega}{\lambda P} \tag{10.10}$$

Where

A = Encircled oriented surface area of the ring laser
P = Perimeter
λ = Wavelength
Ω = Rotation rate.

Figure 10.50(a) shows square ring laser geometry, which is one form of geometry in use. In the case of square ring laser geometry, as shown in Figure 10.50(a), four mirrors form a closed light path. A small part of the ring laser path houses the excited gain medium. The gain medium along with mirrors is responsible for producing the two counter-propagating laser beams. The difference frequency produced as result of rotation is obtained by interference of the two beams. The two beams are usually combined with the help of a corner prism as shown in the diagram. Figure 10.50(b) brings out the constructional features of the square type ring laser geometry. Another commonly used ring laser cavity structure is the triangular geometry as shown in Figure 10.51. This type of geometry is popular with ring laser gyroscopes from Honeywell.

As is evident from the Sagnac equation, the constancy of A and P is very important to achieve the kind of angular resolution the present day ring laser gyroscopes are capable of. It is due to this reason that ring laser cavity block is made from a material of exceptionally high thermal stability. ZERODUR is one such material. ZERODUR is a registered trade name of a near-zero thermal expansion transparent glass ceramic made by Schott, Germany. The material has a thermal expansion coefficient of $\pm 10^{-7}/K$ over a temperature range of 0–50 °C.

RLGs suffer from an effect called lock-in at very low rotation rates. When the rotation rate is very low, the frequencies of counter-propagating laser beams become almost identical, which leads to the two beams getting injection locked to a common frequency. As a result of this, it does not respond to rotation. This is illustrated in transfer characteristics of Figure 10.52. This problem is largely overcome by using forced dithering in which the ring laser cavity is rotated clockwise and counter clockwise about its axis using a mechanical spring driven at its resonance frequency. Typically, a peak dither rate of 1 Arc-sec/s is used and dither frequency is in the range of 400–500 Hz.

Important parameters that describe a ring laser gyroscope's performance include bias error, bias drift and instability, scale factor stability and angular random walk. Each of these is briefly described in the following paragraphs.

(a)

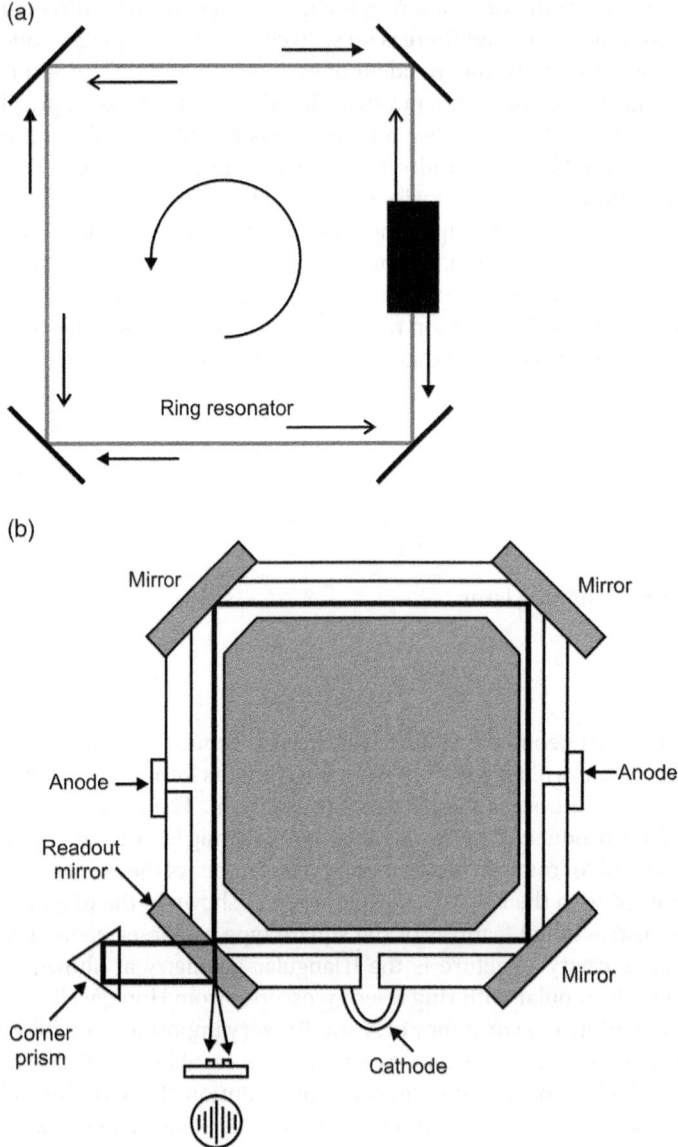

(b)

Figure 10.50 Square type ring laser geometry. (a) Four mirrors form a closed light path and (b) the constructional features.

Bias error of a rate gyroscope in general is the output produced by the gyroscope when it is not experiencing any rotation. Bias error is defined as the voltage or percentage of full-scale output representative of the rotational velocity in °/s as measured by the gyroscope in the absence of any rotation. The bias error is primarily caused by a number of factors including calibration errors, bias drift, bias variation with temperature and effect of shock or 'g' level.

Bias drift refers to variation of bias error over time assuming all other factors remain constant. This is a warm-up effect usually caused by self-heating of electrical and mechanical

Figure 10.51 Triangular geometry ring laser gyroscope.

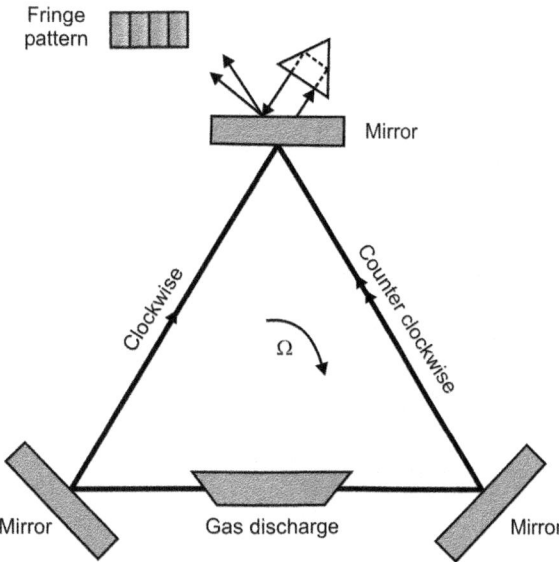

Fringe pattern

Mirror

Clockwise

Counter clockwise

Ω

Mirror   Gas discharge   Mirror

Figure 10.52 Lock-in effect in RLG.

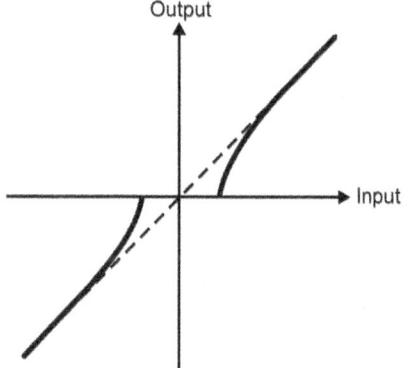

Output

Input

components of the gyroscope. This effect is more prevalent only during the warm-up time of first few seconds and is non-existent after that. It should not be confused with the more colloquial term drift rate.

Bias instability is the fundamental parameter that defines how good or bad the gyroscope is. It is defined as the minimum point on the Allan Variance curve measured in °/h. It represents the best bias stability achievable in a given gyroscope with bias averaging taken at the interval defined by Allan Variance minimum. As an example, bias stability error in the case of the Honeywell gyroscope type GG1320 is specified to be ≤ 0.002°/h. The Allan Variance technique is a very powerful mathematical method that allows a user to assess the real performance of any gyroscope in relation to the dynamics of his application, taking into account the effects of bias, noise, drift and long term sensor instability. Bias instability is always measured by averaging successive data samples. The challenge lies in selecting a suitable time over which to average sampled data. Averages taken over short time intervals will be dominated by noise while those over a longer period are dominated by longer term drift. The technique involves selecting a range of time intervals, typically from 0.01 to 500 s, over which to average data. The standard deviation thus obtained from one averaged time period to the next is calculated and plotted

Figure 10.53 MiniRLG2-based INS. (*Source: Courtesy of Teledyne Marine.*)

against the averaging interval in log-log form. The resulting graph has a characteristic 'bathtub' shape. By examination of this graph, it is possible to compute the key defining characteristics of the gyroscope, namely angular random walk, bias instability and rate random walk.

Scale factor and *scale factor stability* represent the transfer or output versus input characteristics of the gyroscope. It measures the signal produced in terms of voltage or bits for a given rotation rate. It is a measure of the slope of the best straight line through the points on a graph representing the gyroscope output against input rotation rate measured over the specified dynamic range of the gyroscope. As an example, for an analogue gyroscope, it is measured in volts per °/s. Scale factor stability figure for Honeywell gyroscope type GG 1320 is specified to be ≤ 10 PPM and maximum rate measuring capability of ±500°/s.

Angular random walk (ARW) is a measure of gyroscope noise and is measured in °/√hr or °/√s. It can be thought of as the standard deviation due to noise while integrating the output of a stationary gyroscope over a period of time. It is inversely proportional to the square root of integration time. As an illustration, if a gyroscope had ARW of 1°/√s; ideally speaking, the result of angular position measurement in the ideal case would be zero. But the longer the integration time, the larger the spread of the results away from zero. The spread will follow the law of being inversely proportional to the square root of integration time. In the case of Honeywell gyroscope, and this is also true for typical inertial grade gyroscopes, the random walk noise is specified to be ≤0.0018°/√h. As outlined in the beginning of the section, the INS would have three orthogonally placed RLGs to provide measurement capability of rotation in three degrees of freedom. Along with accelerometers, the RLGs constitute the INS. Figure 10.53 shows one such RLG based INS package. The system is a high grade inertial measurement unit that meets the accuracy requirements of attitude, orientation, position and navigation. The system is characterized by bias stability of 0.1°/h, scale factor stability of 75 PPM and angular random walk specification equal to 0.028°/√h.

The HG9900U from Honeywell Aerospace (Figure 10.54) is another navigation grade inertial measurement unit (IMU). The IMU mainly comprises of Honeywell GG1320AN digital laser gyros, Honeywell QA2000 accelerometers and associated electronics. The laser gyro in this INS package is characterized by one sigma error coefficients of < 0.003°/h (Bias), < 0.002°/√h (Random Walk) and <5 PPM (Scale Factor). Accelerometers are characterized by one sigma error coefficients of < 25 μg (Bias) and < 100 PPM (Scale Factor).

Figure 10.54 HG9900U IMU.

Figure 10.55 HG1700 IMU.

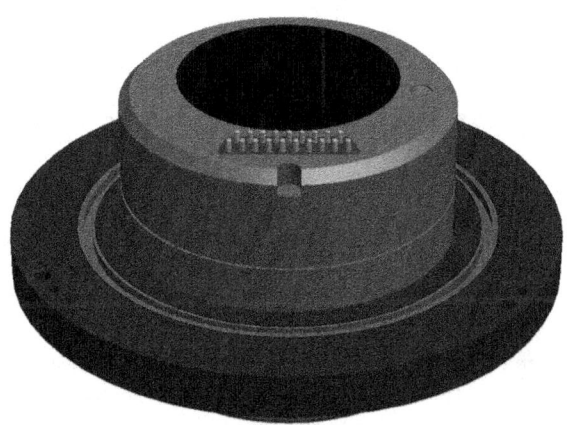

The HG1700 (Figure 10.55) from Honeywell Aerospace is another high performance tactical grade inertial measurement unit (IMU) designed for a wide range of guidance and control applications. The IMU package comprises of three ring laser gyroscopes and three quartz resonating beam accelerometers and associated electronics environmentally sealed in a rugged aluminium housing. The IMU is available in four different performance grades with one sigma gyro bias of 1, 2, 3 and 5°/h. Corresponding maximum random walk specifications are 0.125, 1, 2 and 3°/√h, respectively.

10.9.2 Fibreoptic Gyroscope

Unlike a classical mechanical gyroscope based on the principle of conservation of momentum and very much like the ring laser gyroscope described in the previous paragraphs, a fibreoptic gyroscope (FOG), has virtually no moving parts and no inertial resistance to movement. The operational principle of a FOG is also based on the Sagnac effect. It consists of a long coil of optical fibre typically few km in length and makes use of interference of light to detect mechanical rotation. The FOG provides extremely precise rotational rate information, due to

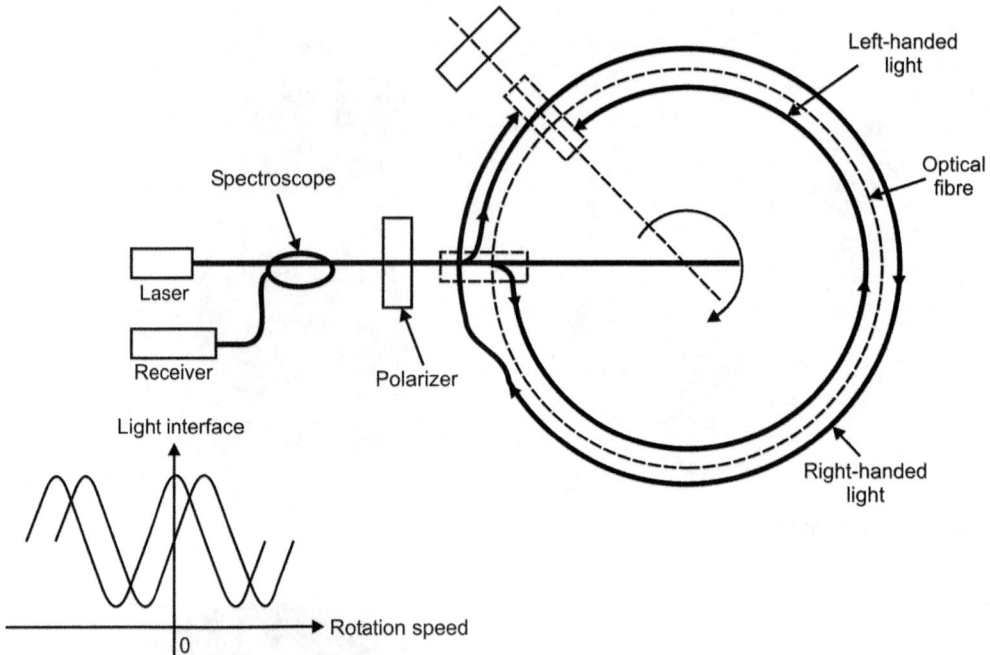

Figure 10.56 Principle of operation of a fibreoptic gyroscope.

its lack of cross-axis sensitivity to vibration, acceleration and shock. Furthermore, it does not require any starting calibration and consumes relatively little power. FOGs can be very compact as optical fibre lengths as long as few km can be wrapped around a small-diameter coil. Their development got a big boost and became a reality in the early 1970s due to technological advances in semiconductor diode lasers and low-loss single-mode optical fibre mainly triggered by the demands of the telecommunications industry.

Figure 10.56 shows a schematic representation of a FOG. Two split laser beams from a single laser are injected into the same fibre but in opposite directions. Due to the Sagnac effect, the beam travelling against the rotation experiences a slightly shorter path length delay than the other beam. The resulting differential phase shift is measured through interferometry, thus translating one component of the angular velocity into a shift of the interference pattern, which is measured photometrically. Beam splitting optics launches light from a laser diode into two waves propagating in the clockwise and counter clockwise directions through a coil consisting of many turns of optical fibre. The strength of the interference signal depends on the effective area of the closed optical path according to the Sagnac effect. This is not simply the geometric area of the loop but is enhanced by the number of turns in the coil. The FOG was first proposed by Vali and Shorthillin 1976. Development of both the passive interferometer type of FOG, or IFOG, and a newer concept, the passive ring resonator FOG, or RFOG, is proceeding in many companies and establishments worldwide.

A FOG provides extremely precise rotational rate information partly because of its lack of cross-axis sensitivity to vibration, acceleration and shock and also because it has no moving parts and doesn't rely on inertial resistance to movement. Hence, this is perhaps the most reliable alternative to the mechanical gyroscope. Because of their intrinsic reliability, fibreoptic gyroscopes are used for high performance space applications. As compared to a ring laser gyroscope, it typically offers a higher resolution but has relatively poorer drift and scale factor

Figure 10.57 Fibreoptic gyroscopes type EMP-1.2K.

Figure 10.58 A three-axis FOG based IMU.

performance specifications. Typically, its sensitivity figure of 0.1°/h still does not match any-where close to the demonstrated RLG sensitivity of better than 0.001°/h. Also, unlike ring laser gyroscopes where zero beat frequency always means zero angular velocity; in the case of a fibreoptic gyroscope, one needs to determine which indication corresponds to zero angular velocity. The reason for success of the FOG lies in its some desirable features like lightweight, small size, limited power consumption, projected long life time and low cost. This makes FOGs particularly attractive in applications such as automotive and robotics, which call for less demanding performance levels in terms of sensitivity and drift specifications and where small size and low cost are important considerations.

FOGs are used in a wide variety of applications. Some of the common ones include inertial navigation, navigation of remotely operated vehicles (ROV) and autonomous underwater vehicles (AUV), surveying, automotive industry and robotics, EO/FLIR/Radar stabilization, line-of-sight tracking and precision pointing, gunfire control systems, platform stabiliza-tion, gyroscope compassing and target acquisition systems, smart munitions and Earth observation, scientific and telecommunication satellites. Figures 10.57 and 10.58 show the photographs of two such FOGs. In Figure 10.57 is shown type EMP-1.2 K from Emcore Photonics Systems that provides precise navigation with 1 mile/h accuracy without GPS and a fast gyroscope compassing to 1 mrad. Major performance specifications of the device include short term and long term drift stability specifications of 0.005°/h and 0.01°/h, respectively, scale factor stability and linearity specifications of 50 PPM and 25 PPM, respectively, and a random walk specification that is 0.0015°/√h. In Figure 10.58, a three-axis fibreoptic gyroscope assembly from AL Cielo is shown. The sensors can be used in diverse navigation and control applications, such as for the stabilization of missiles, airborne sensor systems or land vehicle systems. State-of-the-art FOGs compare very closely with the specifications of ring laser gyroscopes.

10.10   LIDAR For Detection of Chemical and Biological Warfare Agents

The need and relevance of efficient and reliable technologies for standoff detection and identification of chemical, biological and explosive agents has become increasingly important in the present day scenario. Chemical and biological weapons are relatively inexpensive to produce and are capable of unleashing a bigger devastating effect as a terrorist weapon. The international community has already witnessed the devastating effects these weapons are capable of causing on more than one occasion. Millions have been affected by the destruction caused by these weapons of mass destruction in the past. The sarin attack by the AumShinrikyo cult in 1995 in a Tokyo subway, hydrogen cyanide, mustard gas attack by Iraq in the Anfal campaign against the Kurds, most notably in Halabja massacre in 1988 and international dispersal of Anthrax spores against USA in 2001, are some examples of chemical and biological weapons attack. Use of improvised explosive devices by terrorists and anti-national elements killing soldiers and civilians alike has become a routine affair. The incidents have brought to the fore the need for standoff detection and identification of chemical, biological and explosive agents. In the paragraphs to follow are discussed some laser-based methodologies/techniques that are either in use or being explored for the purpose

10.10.1   Detection of Chemical Warfare Agents – Differential Absorption LIDAR

Nerve agents such as organophosphate compounds and blister agents such as mustard compounds because of their acute toxicity, rapidity of action, indetectability by human senses and economic viability are the two potentially very harmful classes of chemical warfare agents. Nerve agents are organophosphoric esters, which initially stimulate and then paralyse certain nerve transmissions by interfering with the cholinesterase enzyme. Common chemical warfare agents belonging to this category include Tabun, Sarin, Soman, Vx, Dichlorvos and Malathion. Figure 10.59 shows the photograph of a victim of chemical warfare agent attack.

LIDAR-based detection is the most practically realizable technique for standoff detection of chemical and biological warfare agents. Commonly exploited physical phenomena forming the basis of chemical and biological agent detection include elastic backscattering, laser-induced fluorescence and differential absorption. They are superior to point detection systems such as IR and Raman spectroscopy because of their capability to range and discriminate the chemical and biological molecules in real time. *Ultraviolet laser-induced fluorescence* could be used for detection of biological molecule since most biological molecules fluoresces when excited by UV radiation.

Differential absorption LIDAR (DIAL) is the most frequently used technique used for detection of chemical warfare agents along with detection of toxic gases and pollutants. It uses two wavelengths; one corresponds to the wavelength of peak absorption of the targeted

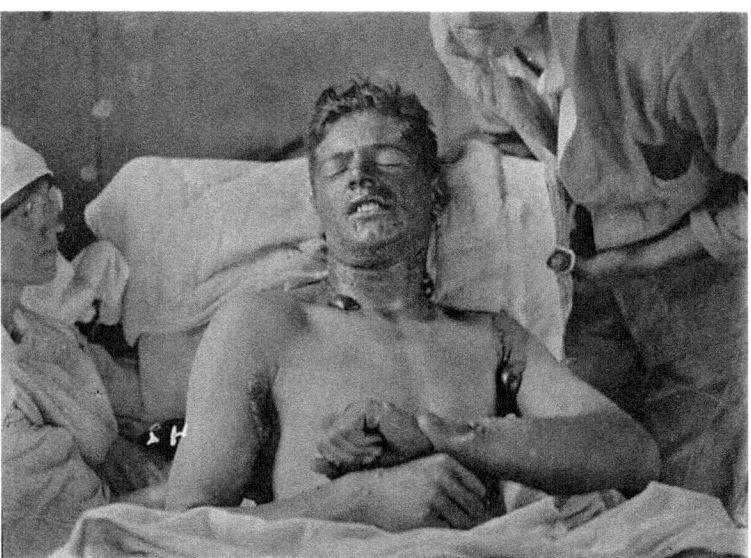

Figure 10.59 Effects of chemical weapon attack.

molecule and the other corresponds to the weak absorption of the targeted molecule. Ratio of the two received backscattered signals measures the concentration of the targeted chemical warfare agent. Figure 10.60 shows a block-schematic arrangement of the DIAL system. Two laser pulses, called ON and OFF wavelengths, are transmitted towards the targeted area of interest in the atmosphere that is suspected to be contaminated with a potentially harmful chemical warfare agent. As outline earlier, the ON wavelength corresponds to the peak absorption of the suspected harmful species and the OFF wavelength is slightly detuned from the ON wavelength to encounter significantly lower absorption from the same species. It may be mentioned here that the choice of ON and OFF wavelengths is unique for a given chemical species and is determined by using a tunable laser that scans the area of interest by transmitting pairs of these ON and OFF wavelengths. Obviously, an ON-wavelength called λ_{ON} encounters maximum absorption, with the result that the corresponding backscattered signal is relatively weak compared to the λ_{OFF} that encounters weak absorption and produces a relatively stronger backscattered signal. This forms the basis of knowing atmospheric constituents at that time. Ratio of the two backscattered signals gives concentration of the molecule of interest.

Many of the chemical warfare agents have significant absorption bands in the 3–4 µm and 9–11 µm bands. CO_2 lasers emitting in the 9–11 µm band have been commonly used for the detection of majority of chemical agents. Some DIAL systems do make use of both these bands and therefore employ both tunable mid-IR as well as CO_2 laser sources. The schematic arrangement of the DIAL system shown in Figure 10.60, however, only makes use of a CO_2 laser. The transmit laser beam after suitable collimation is directed towards the atmosphere in the desired direction with the help of a scanning gimbal mirror. The backscattered signal is received by the receiving telescope and is further focused on to the detector subsystem. An interference filter is used to block the undesired radiation and allow only the radiation of wavelength of interest to pass through it. The detected signal is then digitized and fed to the data processor to extract the desired information on the type and concentration of the chemical species.

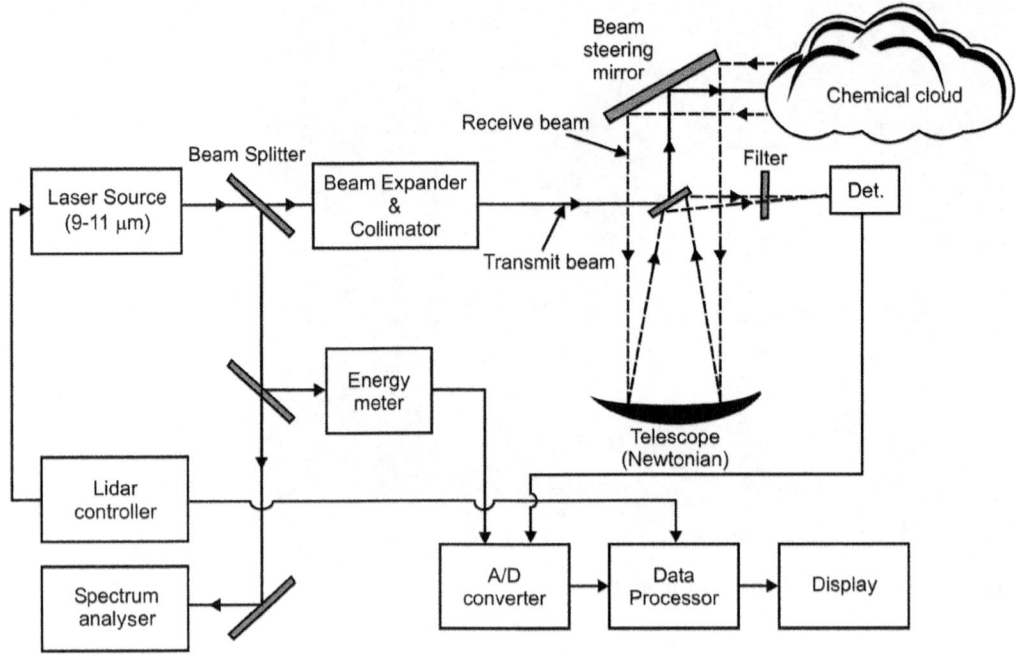

Figure 10.60 Block-schematic of differential absorption LIDAR.

10.10.2 Detection of Biological Warfare Agents – UV-LIF LIDAR

Standoff detection of biological warfare agents and their discrimination from background aerosols is extremely challenging as the distinction between innocuous ambient bacteria and virulent microbes amounts to subtle differences in their molecular make up. As these subtle differences involve a very small percentage of molecules; they produce a very small change in their optical signatures. This makes it extremely difficult to reliably detect and discriminate these potentially dangerous molecules. The processes of detection and discrimination are further complicated by the fact that there is a constant variation in the growth media and the contaminants associated with processing of biological warfare agents affect their optical signatures. Figure 10.61 shows the effects of a biological warfare agent attack.

The standoff detection of biological warfare agents is based on the concept of laser-induced fluorescence (LIF) effect. Laser-induced fluorescence is the emission from atoms or molecules after they have been excited to go to higher energy levels by excitation by another laser. The emission takes place at a wavelength that is higher than the wavelength of laser light exciting it. The biological warfare agent molecules mainly constitute aromatic amino acids and coenzymes. Aromatic amino acids such as tryptophan, tyrosine and phenylanine absorb laser radiation at 280–290 nm and fluoresce in 300–400 nm bands. It is therefore possible to detect biological warfare agents by using UV laser excitation at a suitable wavelength. Also, discrimination of biological agents can be achieved only from the LIF signal because the fluorescence cross-section for particles in the size range of 1–10 µm are sufficiently large enough to make single particle interrogation feasible.

Figure 10.62 shows a block-schematic arrangement of a typical monostatic UV-LIF LIDAR system. The principle components of this LIDAR system include a frequency-tripled or quadrupled ND-YAG laser respectively emitting at 355 nm (frequency-tripled output) and 266 nm (frequency-quadrupled output), a telescope used for both transmission and reception,

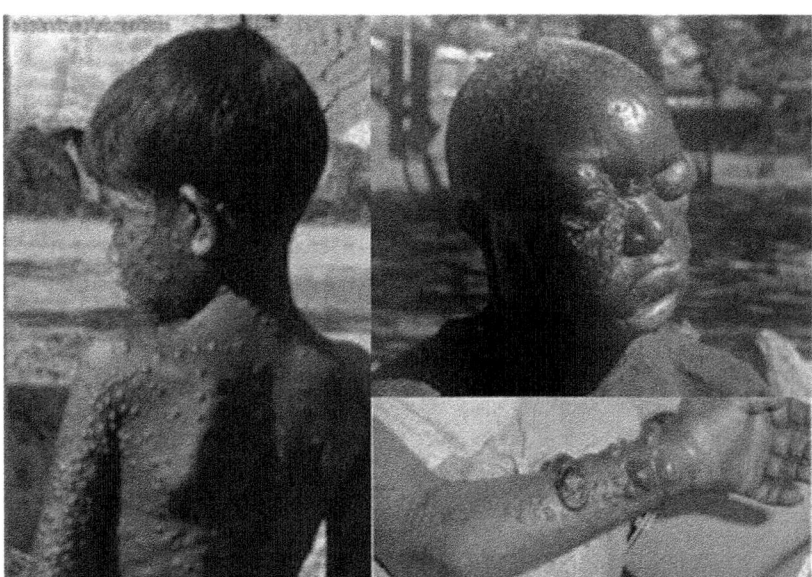

Figure 10.61 Effects of a biological warfare attack.

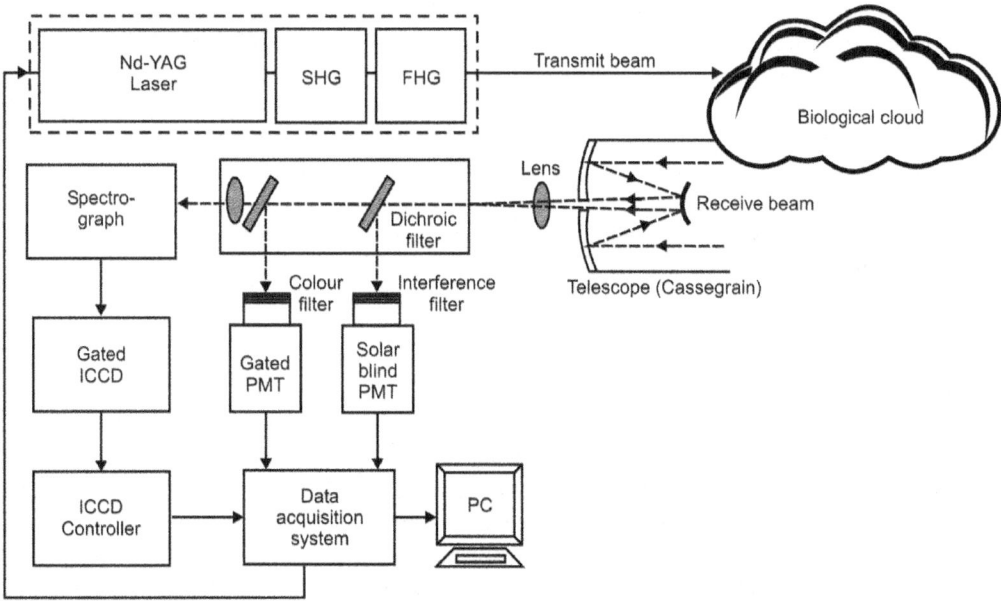

Figure 10.62 Block-schematic arrangement of a UV-LIF Lidar system for detection of biological agents.

a photomultiplier tube used for recording the backscattered signal and the bio-fluorescence signal, a spectrograph with a gated ICCD array for recording the dispersed fluorescence spectra and the LIDAR controller. The fourth harmonic at 266 nm is transmitted towards the biological cloud. The telescope receives the backscattered and the bio-fluorescence signal, which is then fed to the PMT channel. Backscattered signal is received by the gated PMT channel and is present whenever there is a cloud along the beam path. This channel is used to measure the distance to the cloud. A solar-blind PMT channel is used to receive fluorescence signal and is activated only when there is a suspicious cloud. A spectrograph with a gated ICCD is used to identify the nature of biomolecule that is responsible for the fluorescence. The bandwidth of the receiver channel is usually kept smaller so as to disallow unwanted background radiation from entering the channel.

10.11   Laser-Based Detection of Explosive Agents

Standoff detection of explosive agents using lasers, based on the trace detection method, is one of the most widely researched technologies internationally. For the technology to mature to an extent where it can be transformed into a product usable in the kind of environment and field conditions usually encountered in homeland security-related applications is a great technological challenge. One of the major problems comes from decrease in the intensity of a backscattered light signal due to wavelength dependent absorption and scattering losses and the intensity decreasing inversely with distance squared. The problem is compounded by the fact that the trace levels associated with common explosive agents are extremely low, in the range of fraction of ppb (parts per billion) to few ppm (parts per million) for common explosives. Second major problem pertains to unique identification of targeted explosive agent in a background of interferents. Many chemical agents have atomic compositions including sulfur, phosphorus, fluorine and chlorine in addition to nitrogen, oxygen, hydrogen and carbon present in organic molecules. Thus, the detection methodology needs to be highly sensitive and selective. Laser-based spectrometric methods have the potential of being fast, sensitive, and selective with ability to detect and identify a wide range of explosive agents and upgradable to handle new threats.

Atmospheric transmission at the wavelengths concerned is an important factor while assessing suitability of a given standoff detection methodology. Commonly used technologies are *laser-induced breakdown spectroscopy (LIBS), Raman spectroscopy* and its variants, *laser-induced fluorescence (LIF) spectroscopy* and *IR spectroscopy*. These are all trace detection methods, though there are bulk detection methods such as millimetre-wave imaging and terahertz spectroscopy.

10.11.1   Laser-Induced Breakdown Spectroscopy

Laser-induced breakdown spectroscopy focuses high-energy laser beam on the trace sample to break down a small part of the sample into plasma of excited ions and atoms. The plasma emits light that is characteristic of emissions from ionic, atomic and small molecular species. These light emissions are detected by a spectrometer to identify the elemental composition. Figure 10.63 shows a typical laser-induced breakdown spectroscopy setup. The diagram is self-explanatory. One of the challenges in use of LIBS is to assess its efficacy to detect and identify explosive species in real environment that is replete with many interfering substances. One way to get the desired selectivity is to use double pulse LIBS. In double pulse LIBS, first pulse is used to create so-called laser generated vacuum and the second pulse transmitted a few microseconds later generates the return signal. A double pulse LIBS is also observed to improve sensitivity in addition to enhancing selectivity. Selectivity can be further improved by adding temporal resolution to the LIBS emission analysis.

The 1064 nm wavelength has been widely employed for LIBS systems. Due to serious eye hazards posed by 1064 nm, scientists have also tried 266 nm. The 266 nm value also allows the designer to build Raman capability into the system. It may be mentioned here that 266 nm has 600 times higher maximum permissible exposure (MPE) limit compared to 1064 nm.

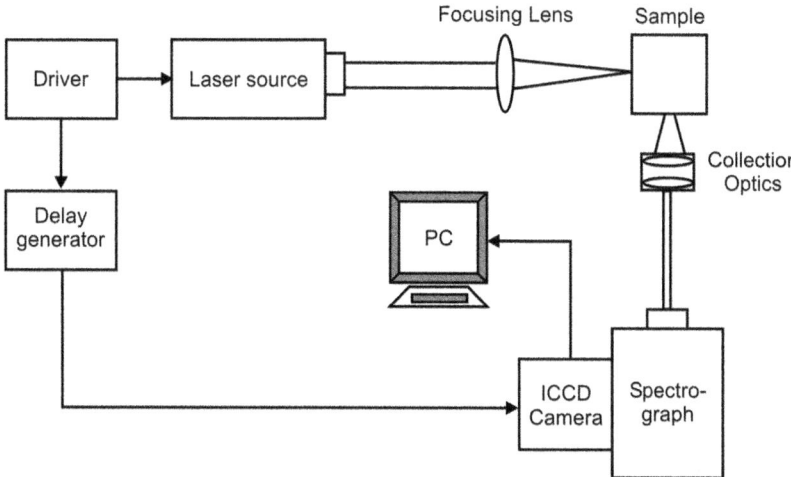

Figure 10.63 Laser induced breakdown spectroscopy setup.

10.11.2 Raman Spectroscopy

Raman spectroscopy offers another method for standoff detection of explosive agents. It has been extensively used for many years as a standard analytical tool for identification of chemical agents in the laboratory environment. The basis of detection in this case is the shift in the wavelength caused by inelastic Raman scattering by the target molecule. The inelastic scattering of impinging photons where some energy is lost to (or gained from) the target molecule returns scattered light with a higher (or lower) wavelength depending upon whether energy was lost to (or gained from) the target molecule. The difference is dictated by the energy of vibrational modes of the target molecule and therefore constitutes the fingerprint or the basis of identification. Complex mixtures are identified using algorithms for pattern recognition. Figure 10.64 shows a typical Raman spectroscopy setup. The diagram is self-explanatory.

A major drawback of Raman technique is its extremely poor sensitivity caused by the fact that Raman scattering occurs for one in about 10^7 photons impinging on the sample. Weak return signal intensity of Raman spectroscopy limits its use for trace detection as it also makes it sensitive to ambient light and fluorescence from the sample itself or other chemicals in the vicinity. The fluorescence masks the Raman signal. These problems are overcome by the use of resonant Raman spectroscopy. With a tunable laser, the wavelength can be chosen to match or nearly match a resonant absorption in the target molecule leading to intensity enhancement of the order of 10^6. The problem of fluorescence masking a Raman signal can be overcome by use of either IR or UV radiation. IR radiation does not have sufficient energy to cause fluorescence and UV radiation will cause fluorescence in the visible, which is well separated from Raman signal.

Laser-induced fluorescence (LIF) is another important tool for similar applications. Though a very valuable tool in combustion diagnostics and for studying decomposition of explosives, it has not been found very useful for detection of explosive agents.

Almost all the laser-based standoff detection methods have detection limits too high to make them suitable for detection of explosives in field conditions, which requires capability to detect traces in the vapour phase or in the form of particles. Another problem is insufficient selectivity to identify target explosive in the presence of interferents. In fact, none of the laser-based spectroscopic methodologies available today is ready for full functionality prototype

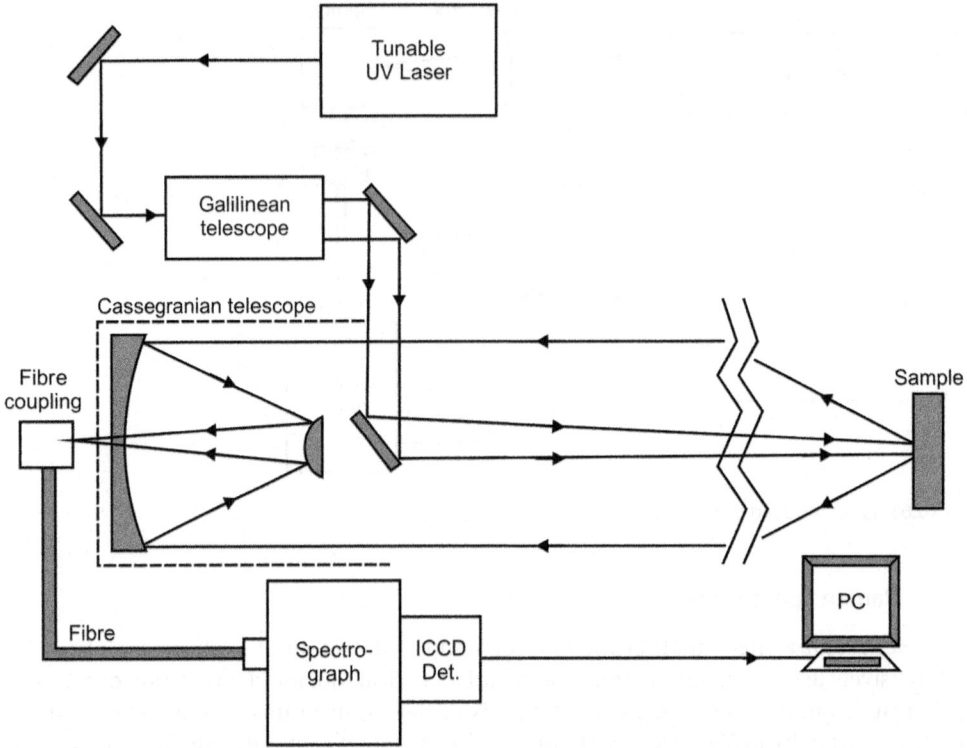

Figure 10.64 Raman spectroscopy setup.

manufacturing. There is lot of scope of further research to improve the performance of the more mature technologies in terms of both detection sensitivity and selectivity. The final solution would probably lie not in one type of sensor but integration of several technologies to derive the benefits of best of each.

Illustrated Glossary

Angular Random Walk (ARW) (RLG) An ARW is a measure of gyroscope noise and is measured in $°/\sqrt{h}$ or $°/\sqrt{s}$. It can be thought of as the standard deviation due to noise while integrating the output of a stationary gyroscope over a period of time.

Bias Drift (RLG) This refers to variation of bias error over time assuming all other factors remain constant. This is a warm-up effect usually caused by self-heating of electrical and mechanical components of the gyroscope.

Bias Error (RLG) Bias error of a rate gyroscope in general is the output produced by the gyroscope when it is not experiencing any rotation. Bias error is defined as the voltage or percentage of full-scale output representative of the rotational velocity in degrees as measured by the gyroscope in the absence of any rotation.

Bias Instability (RLG) Bias instability is the fundamental parameter that defines how good or bad the gyroscope is. It is defined as the minimum point on the Allan Variance curve measured in $°/h$. It represents the best bias stability achievable in a given gyroscope with bias averaging taken at the interval defined by Allan Variance minimum.

Boresight Error This occurs when the laser beam and the sighting mechanism of the laser designator are matched to the same point in the entire operational range. It can lead to less or no reflected laser energy and consequently impairment of munitions delivery mission. The boresight error may be large enough for the laser beam to fall outside the target though the sighting mechanism may be indicative of aiming the target.

Delayed Designation This refers to laser designation for a shortened time during the terminal phase of the munitions flight path. It is generally used in the case of low-level releases to prevent the guided munitions impacting short of the intended target.

Differential Absorption LIDAR (DIAL) DIAL is the most frequently used technique used for detection of chemical warfare agents along with detection of toxic gases and pollutants. It uses two wavelengths; one corresponds to the wavelength of peak absorption of the targeted molecule and the other corresponds to the weak absorption of the targeted molecule.

False Lock-On False lock-on of the laser seeker or the laser spot tracker occurs when it locks on to stray reflections rather than the laser energy reflected off the intended target. One such possibility is scattered reflection from suspended matter from the close-to-designator distances.

FM-CW Laser Range Finder The FM-CW laser range finding technique is similar to the one followed in the case of its radar counterpart; that is, FM-CW radar. In this the frequency of a narrow line width laser is modulated with a ramp or sinusoidal signal, collimated and then transmitted towards to the target. The received signal corresponding to the reflected laser beam, specular or diffused is mixed with the reference signal representing the transmitted laser beam. The beat frequency produced as a result of homodyne detection is used to derive the range information, which is given by the following equation.

$$d = \frac{(f_B \times c \times T_R)}{(4 \times \Delta f)}$$

Where,

f_B is the beat frequency

T_R is the ramp waveform time period

Δf is modulation frequency bandwidth.

Minimum and maximum ranges are, respectively, given by the following equations.

$$d_{MIN} = \frac{c}{(4 \times \Delta f)} \text{ and } d_{MAX} = \frac{(c \times T_R)}{4}$$

Laser Aiming Device These are usually configured around visible or IR semiconductor diode lasers aligned to emit a laser beam parallel to the barrel are used for precise target aiming particularly in night time operations. Most laser aiming modules use red laser diodes emitting at either 635 nm or 650 nm. Green lasers based diode-pumped solid-state laser technology emitting at 532 nm are also presently in use. Due to extremely low value of divergence of the emitted laser beam, particularly at night time, it makes very a small spot even at long distances up to hundreds of metres.

Laser Bathymetry This makes use of transmissive and reflective properties of water and the bottom of water body to measure the depth. Measurement of depth relies on differential timing of laser pulses reflected from surface of the water body and those reflected from the bottom. Laser bathymetry from an airborne platform offers distinct advantages over the conventional vessel mounted acoustic sensor technology as the latter provides a swath coverage that is approximately twice the water depth.

Laser-Induced Breakdown Spectroscopy (LIBS) *LIBS* focuses high-energy laser beam on the trace sample to break down a small part of the sample into plasma of excited ions and atoms. The plasma emits light that is characteristic of emissions from ionic, atomic and small molecular species. These light emissions are detected by a spectrometer to identify the elemental composition.

Laser-Induced Fluorescence This is the emission from atoms or molecules after they have been excited to go to higher energy levels by excitation by another laser. The emission takes place at a wavelength that is higher than the wavelength of laser light exciting it.

Laser Proximity Sensor This is a type of laser range finder designed to accurately measure relatively smaller distances to target as compared to those encountered in conventional laser range finders. In the case of a proximity sensor, the processing circuitry generates a command signal when the distance to target is equal to the preset distance value within a certain specified tolerance. The command signal in turn could be used to perform a variety of control functions. In military parlance, the most common application of a proximity sensor is in a laser proximity fuse where the command signal is used to initiate detonation of warheads of large artillery shells, aviation bombs and guided missiles.

Laser Range Finder (LRF) An LRF is used in a military application to determine distance to the intended target. It is used both as a standalone device as well as an integral part of a fire-control system of main battle tanks and other armoured fighting vehicles. Laser range finders exploit one of the following four techniques for determining distance to the target. These include time-of-flight, triangulation, phase shift and FM-CW techniques.

Laser Target Designator This is one of the most widely used pieces of electro-optical equipment in the tactical battlefield. It is an essential and a critical component in the laser-guided munitions delivery setup. Laser target designator provides accurate target marking for precise delivery of air-to-surface laser-guided bombs and missiles and surface-to-surface cannon-launched laser-guided projectiles. Laser target designators used for munitions delivery applications are configured around high repetition rate Q-switched Nd:YAG lasers. All target designators have a range finding channel either at the same wavelength of 1064 nm or an eye-safe wavelength in the range 1535–1570 nm. It may be mentioned here that range finding channel of almost all modern laser target designators uses eye-safe wavelength.

Offset Designation In the case of offset designation, it is not the intended target but a nearby suitable reflecting object that is illuminated by the laser target designator. This provides immunity against detection by the targeted platform and countermeasures deployed by the adversary if it were equipped with laser-warning sensors.

Phase Shift Laser Range Finder In the phase shift range finding technique, the laser beam with sinusoidal power modulation is transmitted towards the target and the diffused or specular reflection from the target is received. The phase of received laser beam is measured and compared with that of the transmitted laser beam. This allows computation of the time-of-flight and hence the distance to target from known values of phase shift and modulation frequency. The phase shift ($\Delta\phi$) is given by the following equation.

$$\Delta\phi = \omega\tau = (2\pi f)\left(\frac{2d}{c}\right) = \left(\frac{4\pi fd}{c}\right)$$

Where
f = Modulation frequency
d = Range
c = Velocity of propagation of light in free space
τ = Time-of-flight from source to target and back to source.

The distance d is given by

$$d = \left(\frac{c\Delta\phi}{4\pi f} \right)$$

Podium Effect This occurs when after the guided munitions is released from the aircraft; the aircraft follows such a flight path as to irradiate the target on the side that is out of view of the seeker. As a result, the laser seeker gets blocked and doesn't see the reflected laser energy.

Raman Spectroscopy This offers another method for standoff detection of explosive agents. It has been extensively used as a standard analytical tool for identification of chemical agents in the laboratory environment. The basis of detection in this case is the shift in the wavelength caused by inelastic Raman scattering by the target molecule. The inelastic scattering of impinging photons where some energy is lost to (or gained from) the target molecule returns scattered light with a higher (or lower) wavelength depending upon whether energy was lost to (or gained from) the target molecule. The difference is dictated by the energy of vibrational modes of the target molecule and therefore constitutes the fingerprint or the basis of identification.

Redundant Designation In the case of redundant designation, two or more laser designators are employed at different locations. The multiple designators are operated at the same PRF code and used to designate a single target for delivery of a single laser-guided munitions.

Ring Laser Gyroscope This is primarily a rate sensor, which can be used to measure rotation by integrating the rate information. It essentially consists of two counter-propagating laser beams over the same path. It operates on the principle of Sagnac effect according to which the null points of the internal standing wave pattern produced by the counter-propagating laser modes shift in response to angular rotation. The shift in the nulls of the standing wave pattern manifests itself in the form of a moving interference pattern observed by combining the two laser modes. According to the Sagnac equation, the frequency difference is given by the following equation.

$$\Delta v = \frac{4A\Omega}{\lambda P}$$

Where
A = Encircled oriented surface area of the ring laser
P = Perimeter
λ = Wavelength
Ω = Rotation rate.

Scale Factor (RLG) Scale factor and scale factor stability represent the transfer or output versus input characteristics of the gyroscope. It measures the signal produced in terms of voltage or bits for a given rotation rate.

Scattered Reflection This takes place from flat surfaces that are not shiny. In this case, laser energy is reflected in a large arc.

Solar Saturation This is the inability of the laser seeker to discriminate the desired laser spot on the target from the background in the presence of intense solar radiation. This generally happens against targets above the horizon after sunrise and before sunset and when the seeker dome is pitted, cracked or glazed.

Specular Reflection In the case of specular reflection, also called mirror-like reflection, the laser beam is reflected at an angle equal to the angle of incidence. This also implies that for normal incidence, the laser beam is reflected directly towards the laser position. Bare metal and glass surfaces and flat shiny surfaces produce a mirror-like reflection.

Spillover Reflection This occurs when the laser spot is larger than the size of the target. There is spillover of the laser energy to nearby objects or terrain. In this case, there are scattered reflections from these nearby objects and terrain in addition to the scattered reflection from the intended target. Spillover reflection may also be caused by a jittery laser beam due to an unsteady target tracking by the laser designator. Another possibility is the underspill that causes reflection off the objects or terrain short of the intended target.

Targeting Pod These are target designation tools used for identification of ground targets and guiding munitions to their intended targets. Targeting pods are equipped with other electro-optic systems such as laser spot tracker to locate the laser pulses reflected from designated target or/and laser range finder to determine target range.

Target Reflection Reflection from a typical target surface is a combination of scattered reflection and specular (mirror-like) reflection. This implies that there is a strong reflected laser energy in the direction at which it would have been reflected had there been a mirror-like surface. Knowledge of angle of incidence of laser designator radiation on the target surface can therefore be used to predict the direction in which maximum reflected laser energy is available, thereby making it easier for the laser seeker to lock.

Target Reflectivity This is an important parameter determining the strength of laser energy reflected off the target and hence the probability that the laser seeker would find it exceeding its detection threshold. Target reflectivity depends upon the target material and different materials exhibit different reflectivities. As an example, unpolished aluminium reflects typically 55% while water reflects only 2%. Typical reflectivities of brick structure and vegetation are 55–90% and 30–70%, respectively.

Time-of-Flight Laser Range Finder In the time-of-flight technique, a narrow pulse width laser beam is transmitted towards the intended target. The target range is measured from the time taken by the laser pulse to travel to the target and back. Range is given by the following equation.

$$d = (c \times \Delta t)/2$$

Where,
d = Target range (m)
c = Speed of light = 3×10^8 m/s
Δt = Time interval between transmitted and received laser pulses.
In terms of number of clock cycles, distance to target is given by the following equation.

$$d = (c \times N)/(2 \times f_{CLK})$$

N is the number of clock pulses counted between start and stop signals and f_{CLK} is the clock frequency. Range inaccuracy in this case is given by $\pm c/2 f_{CLK}$.

Triangular Laser Range Finder In the triangulation technique, range is determined by using simple laws of trigonometry. If the laser transmitter is located at point A, laser receiver is located at point B and point C indicates the target location, the distance to target d from laser transmitter location can be computed from known values of length l separating A and B, target-to-transmitter angle α and target-to-receiver angle β using the following equation.

$$d = l \times \frac{\sin \beta}{\sin(\alpha + \beta)}$$

This equation is valid in flat or Euclidean geometry. The computed results become inaccurate if distances were appreciable as compared to the curvature of the Earth. In that case, more complicated expressions derived using spherical trigonometry should be used.

Bibliography

1 Hecht, J. (2008), *Understanding Lasers: An Entry Level Guide (Third Edition)*, Wiley-IEEE Press.

2 Sennaroglu, A. (2006), *Solid State Lasers and Applications*, CRC Press.

3 McAulay, A.D. (2011), *Military Laser Technology for Defence*, Wiley-Interscience.

4 Waynant, R. and Ediger, M. (2000), *Electro-optics Handbook*, McGraw-Hill, Inc.

5 Armenise, M., Ciminelli, C., Dell'olio, F. and Passaro, V.M.N. (2011), *Advances in Gyroscopic Technologies*, Springer.

6 Yin, S., Ruffin, P.B. and Yu, F.T.S. (2008), *Fiberoptic Sensors*, CRC Press.

7 Accetta, J.S. (1993), *The Infrared and Electro-Optic Systems Handbook, Volume* 7, SPIE International Society for Optical Engineering.

8 Gething, M.J. (2008), *Jane's Electro-Optics Systems (2008–2009)*, Janes Information Group

9 Rouse, J. (2008), *Guided Weapons (Third Edition)*, Brasseys.

10 Baudelet, M. (2014), *Laser Spectroscopy for Sensing: Fundamentals, Techniques and Applications*, Woodhead Publishing Limited.

11 Weitkamp, C. (2005), *Lidar: Range Resolved Optical Remote Sensing of the Atmosphere*, Springer.

12 Demtroder, W. (2008), *Laser Spectroscopy, Volume 1: Basic Principles*, Springer.

13 Demtroder, W. (2008), *Laser Spectroscopy, Volume 2: Experimental Techniques*, Springer.

14 Cremers, D.A. and Radziemski, L.J. (2013), *Handbook of Laser Induced Breakdown Spectroscopy*, Wiley-Blackwell.

11

Precision-Guided Munitions

Precision has always been recognized by military strategists as one of the five most important attributes of a weapon with operational range, striking power, volume of fire and portability being the other four. The state-of-the-art precision-guided munitions (PGMs) combine all these attributes to make them a potent force multiplier in the contemporary battlefield. The emergence and subsequent maturity of precision-guided munitions, made possible mainly due to advances in electronics, optoelectronics and optics, is one of the most significant developments of twentieth-century warfare. After a brief introduction to precision-guided munitions, the present chapter focuses on guidance techniques employed in this class of weapons, different types of PGMs in terms of involved technologies, capabilities and limitations, deployment configurations, state-of-the-art and future trends. The PGMs discussed in the chapter include the laser-guided munitions, infrared (IR)-guided weapons, radar-guided weapons and GPS/INS-guided weapons.

11.1 Introduction

PGMs, also called smart munitions, belong to the group of advanced fire power munitions, which mainly includes projectiles fired from land or ship-based military platforms, surface-to-air and air-to-air missiles and aerially delivered bombs. These weapons called *smart weapons* employ one or more guidance techniques to hit the target more precisely with minimized collateral damage than would be possible with conventional unguided weapons also called dumb weapons. Launched from a variety of military platforms including land vehicles, aircraft, ships, submarines or even by individual soldiers on the ground, precision strike weapons exemplify the concept of a low-cost threat forcing a high-cost and complicated defence mechanism.

Though the concept of PGMs was first envisaged during the First World War, the scientific and technological capability of that time didn't allow it to become a practical reality. The PGM arrived on the battlefield in a rudimentary yet significant form during the Second World War. It was the success and the experience gained during the Korean and Vietnam conflicts and then the Gulf War that established the efficacy of precision strike weapons beyond any doubt and gave to us the generation of weapons that are incorporated into the arsenals of a large number of nations. This class of weapons has come a long way from the era of Second World War when the guided weapon was characterized by a CEP (Circular Error Probable) of about 1000 m to the state-of-the-art precision strike weapon with CEP of less than a metre. While the existing and better-established guidance technologies have been strengthened and improved over the years, new guidance mechanisms are being developed to further enhance the precision of

Handbook of Defence Electronics and Optronics: Fundamentals, Technologies and Systems, First Edition. Anil K. Maini.
© 2018 John Wiley & Sons Ltd. Published 2018 by John Wiley & Sons Ltd.

precision-guided weapons. More than one guidance technologies are being employed to make them all-weather weapons and eliminate the possibility of a mishit. Use of GPS/INS in conjunction with other guidance technologies such as radar or electro-optic guidance is an example. Joint Direct Attack Munitions (JDAM) is an important manifestation of use of multiple guidance technologies. The capability of a modern precision weapon has progressed from targeting a specific building to hitting a specific room and soon the next-generation technologies will turn the foot soldier with an advance chip in his boot into a precision strike weapon, able to navigate without GPS and fire-guided bullets at targets like would-be snipers before they have a chance to fire at him.

11.2   Types of Guided Weapons

Different types of guided weapons mainly include anti-radiation weapons, radar-guided weapons, laser-guided weapons, IR-guided weapons, wire-guided weapons, beam rider weapons and GPS/INS-guided weapons.

11.2.1   Anti-Radiation Weapons

Anti-radiation weapons (ARWs) are mainly designed to target ground-based radars. They do so by detecting radio emission from these radars and then homing onto the source of radio emission. ARWs can also be used to target jammers and also radios used for communication. Air-to-surface, surface-to-surface, surface-to-air and air-to-air variants of ARWs are in use. A common deployment of these weapons is in Anti-Radiation Missiles (ARMs) carried by specialist aircraft to target ground-based radars in SEAD (Suppression of Enemy Air-Defence) role. The AGM-88 anti-radiation missile (HARM) from the USA is an example of an air-to-surface anti-radiation weapon (Figure 11.1). Surface-to-surface ARMs such as the MM40 EXOCET (Figure 11.2) employ an active radar seeker whose receiver component is used to home onto the radar. These missiles are extremely hard to defeat with electronic countermeasures. Surface-to-air ARMs are used to target AEW (Airborne Early Warning) and AWACS (Airborne Warning and Control System) systems. The FT-2000 system in the People's Republic of China is an example. More recently, air-to-air ARM designs have also begun to appear on the scene. Russian Vympel R-27P is one such anti-radiation air-to-air missile. Such missiles, being passive in nature, do not trigger any radar warning receivers and therefore are relatively immune to countermeasures.

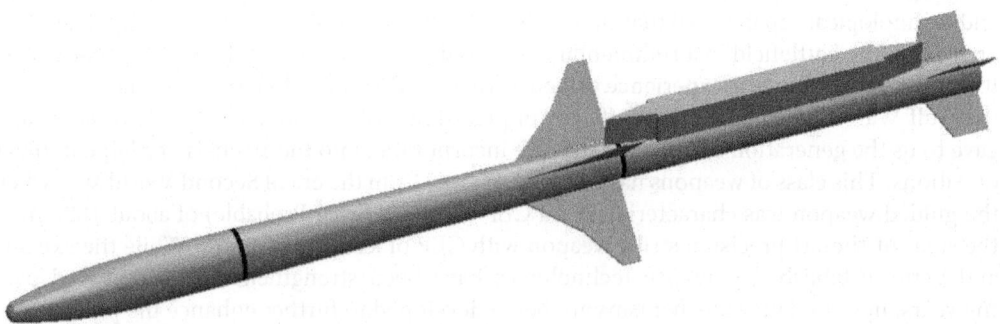

Figure 11.1 AGM-88 HARM ARM.

Figure 11.2 MM40 EXOCET ARM. (*Source:* Marcomogollon, https://commons.wikimedia.org/wiki/File:EXOCET_MM_40_BLOCK_3.JPG. CC BY-SA 4.0.)

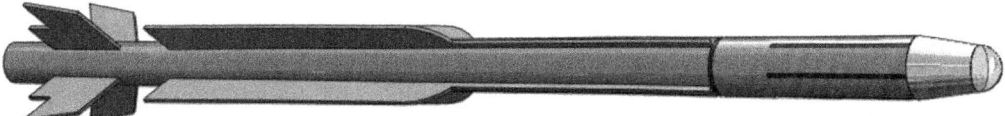

Figure 11.3 MBDA MICA.

11.2.2 Radar-Guided Weapons

Radar guidance, which includes semi-active radar guidance and active radar homing, is commonly used in long-range surface-to-air and air-to-air missiles with the former being the more widely used of the two. In the case of semi-active radar guidance, external radar irradiates the target and the missile seeker head makes use of the signal reflected off the target to home on to the target. Active radar homing missile has a radar transceiver onboard the missile, which finds and continuously tracks the target until it hits it. The MBDA MICA short and medium-range surface-to-air and air-to-air missiles (Figure 11.3) and MBDA EXOCET anti-ship missile in France, as well as the DRDO-Astra BVRAAM in India, are examples of active radar homing missiles.

11.2.3 Laser-Guided Weapons

Laser-guided weapons make use of a laser beam to guide the weapon (bomb, projectile or a missile) and hit the target precisely. Most laser-guided munitions, with the exception of laser beam riding, operate on the principle of semi-active laser homing similar to semi-active radar homing. A laser beam irradiates the intended target, a process called laser designation. The laser beam bounces off the target and gets scattered in all directions. The laser seeker in the munitions detects the direction of arrival of laser energy and guides it to the target. Targets employ laser absorbing paints, smoke screens and active protection systems as countermeasures. Laser-guided munitions are available as cannon-launched surface-to-surface projectiles and aerially delivered bombs and surface-to-surface, surface-to-air and air-to-surface missiles. Copperhead and Krasnopol (both cannon-launched projectiles), Paveway-II (aerially delivered bomb) and Hellfire (surface-to-surface and air-to-surface missile) are some examples. Figure 11.4 shows a photograph of a Hellfire missile depicting its inner parts.

Figure 11.4 Hellfire missile. (*Source:* Courtesy of the US Airforce.)

11.2.4 Infrared-Guided Weapons

IR-guided weapons make use of electromagnetic radiation emitted from the target predominantly in IR part of the spectrum to track the target and then home onto it. Such IR spectrum seeking missiles are also referred to as heat-seeking missiles for obvious reasons. The seeker head in this case is an IR sensor located on the tip or head of the missile. The IR seekers are designed to be sensitive to either the 3–5 μm band, in which case they are called single-colour seekers, or the 3–5 μm and 8–12 μm bands, referred to as two-colour seekers. IR missiles using two-colour seekers are far more immune to countermeasures such as flares. Another variant of IR-guided missile employs an Imaging Infrared (IIR) seeker head, which uses a IR/UV focal plane sensor array. Missiles employing this IIR seeker head are far more resistant to countermeasures and are less likely to be fooled into locking onto the Sun's radiation. The IRIS-T manufactured by Diehl BGT Defence as a part of Germany led multi-national programme (Figure 11.5) is an advanced air-to-air-guided missile employing an IIR seeker. It has three variants including IRIS-T (air-to-air-guided missile), IRIS-TSL (surface-launched medium-range guided missile) and IRIS-T (surface-launched short-range guided missile).

11.2.5 Wire-Guided Weapons

In the case of *wire guidance,* the missile is guided by electrical signals sent to it through a bundle of wires connected between the missile and the guidance mechanism located near the launch site. The wires reel out behind the weapon as it flies. Wire guidance is commonly used in anti-tank missiles where its suitability in limited line-of-sight availability is particularly advantageous. Also, the missile's limited range imposed by length of the wire is not an issue in anti-tank operations. The TOW (Tube-launched Optically tracked and Wire-guided) from the USA with an operational range of 3750 m and the MILAN of France (Figure 11.6) with an extended operational range of 3000 m (MILAN ER) are well-known service wire-guided missiles.

 The TOW family of missiles, including the TOW 2A, TOW 2B Aero and TOW Bunker Buster missiles, is the premier long-range, heavy assault-precision anti-armour, anti-fortification and

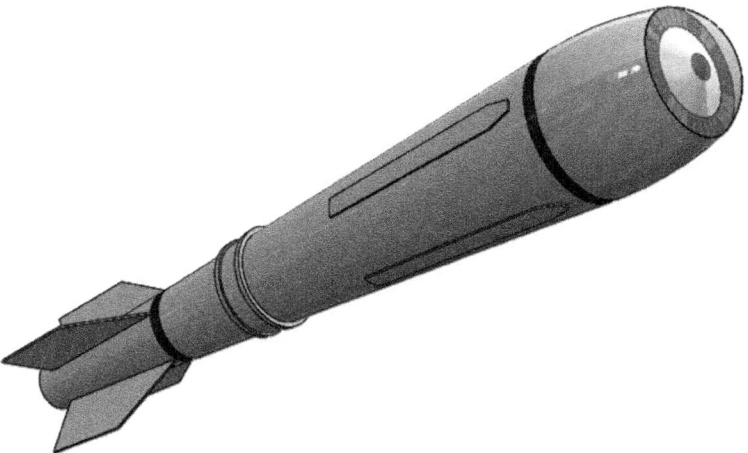

Figure 11.5 IRIS-T.

Figure 11.6 MILAN missile.

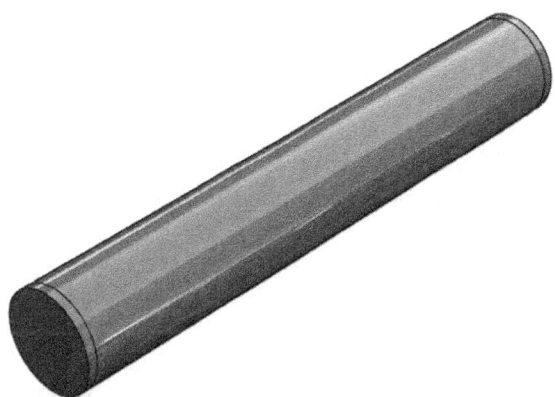

anti-amphibious landing weapon system. TOW missiles have been integrated on ground, vehicular and helicopter platforms and are in service in more than 40 international armed forces. MILAN is a portable medium-range anti-tank missile with later versions equipped with tandem heat warheads making it more effective against reactive armour. It is in use in about 30 countries worldwide.

11.2.6   Beam Riding Weapons

Beam riding weapons make use of a radar or laser beam for guidance. A narrow radar or laser beam is directed at the target usually a tank or an aircraft. The missile is launched in the direction of the target and sometimes after it is launched; it flies into the radar or laser beam. With the help of sensors and a computer on board the missile it keeps itself within the beam. The aiming station keeps the beam pointed at the target till it hits it. The inherent shortcoming of the radar beam riding guidance is that the beam spreads as it moves away from the aiming station. Laser beam riders don't have this limitation. Beam riding guidance is generally used for short-range air-defence and anti-tank-guided missile applications. *LAHAT* (a short-range anti-tank missile manufactured by Israel Aerospace Industries) and shown in Figure 11.7, *Starstreak*

Figure 11.7 LAHAT ATGM. (*Source:* Natan Flayer, https://commons.wikimedia.org/wiki/File:LAHAT.jpg. CC BY-SA 3.0.)

(a short-range air-defence system manufactured by Thales Air-Defence) and *RBS-70* (a short-range anti-tank-guided missile manufactured by SAAB Bofors Dynamics) are well-known laser beam riders.

LAHAT is an advanced laser homing attack missile that makes use of semi-active laser guidance. The target in this case can be designated either directly from the launching platform or by another land-based or aerial platform. It has an operational range of 8 km when fired from a ground platform, 13 km when fired from an aerial platform and has a hit accuracy of 0.7 m CEP. It is in use in Germany, Israel and India. Starstreak is a man-portable/vehicle-mounted high velocity missile with Mach 3.5 velocity designed to counter threats from conventional air threats and fast 'pop up' strikes by helicopter attacks. It has an operational range in excess of 7 km and is currently in the inventory of the Armed Forces in the UK, South Africa, Indonesia and Thailand. RBS-70 is not susceptible to any deception by countermeasures employed by the target aircraft in the form of chaff or flares. Major specifications of RBS-70 include an operational range of 0–6 km and a speed up to 1.6 Mach. The RBS-70 New Generation (RBS-70NG) includes an improved sighting system capable of night vision. The RBS-70 Mk-2 upgrade is called a Bolide missile. It is faster with a speed of 2 Mach against 1.6 Mach of standard RBS-70 and a range of 8 km as against 6 km in the case of standard RBS-70. Figure 11.8 shows a photograph of the RBS-70NG.

11.2.7 GPS/INS-Guided Weapons

GPS/INS-guided weapons make use of a multichannel GPS receiver to provide information on weapon's location and an inertial measurement unit (IMU) to monitor its attitude to adjust its flight path to precisely hit the intended target. GPS/INS guidance provides an effective and low cost means of precision targeting that is unaffected by weather and target concealment and is immune to countermeasures. GPS/INS-guided weapons are primarily used against fixed targets or relocatable targets whose location is likely to remain static for the duration of planning and attack. In another role, GPS/INS guidance is also used to adjust weapon's free fall

Figure 11.8 RBS-70 NG laser beam rider MANPADS. (*Source:* Courtesy of the US Navy.)

to hit a predefined point fed into the weapon prior to launch. Yet another application of GPS guidance is in midcourse correction of guided missiles and cruise missiles. Precision in the basic weapons is characterized by a CEP in the range of 1–10 m. However, CEP is considerably improved when GPS is used together with a semi-active laser (SAL) or an IIR seeker head.

11.3   Guidance Techniques

In the previous paragraphs in Section 11.2, we have described common types of guided weapons with a brief reference to the relevant operational principle. In the following paragraphs, we shall explain common guidance techniques that form the basis of functioning of guided weapons. As shall be evident from the description to follow, guided weapons may use one or more of these guidance mechanisms for improved performance. Different guidance techniques discussed in this section include the following.

1) Beam rider guidance
2) Command guidance
3) Homing guidance
4) Navigation guidance.

Homing guidance further comprises; (1) semi-active homing guidance, (2) active homing guidance, (3) passive homing guidance and (4) retransmission homing guidance. Navigation guidance further comprises; (1) inertial navigation guidance, (2) ranging navigation guidance, (3) celestial navigation guidance and (4) geophysical navigation.

11.3.1   Beam Rider Guidance

The concept of *beam riding guidance* of munitions as briefly explained in an earlier paragraph is based on a radar beam or a laser beam constantly pointed towards the target throughout the

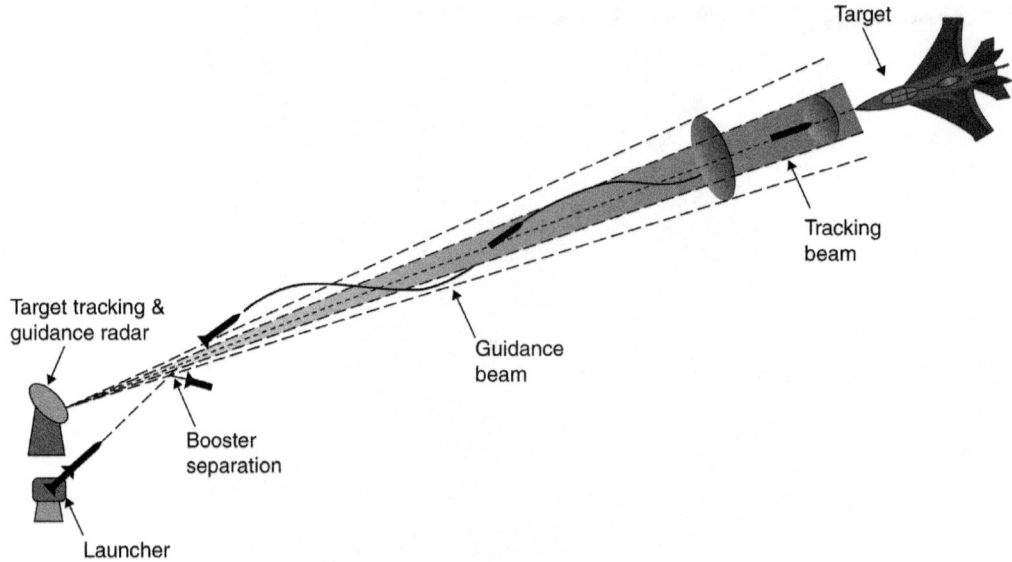

Figure 11.9 Beam rider guidance concept.

flight time of the munitions. After the missile is launched; it attempts to keep itself inside the beam, while the aiming station keeps the beam always pointed at the target. The missile's flight path control functions as follows. The missile's guidance sensors located at the rear of the missile receive information about the position of the missile within the beam. The missile interprets this information and generates its own correction signals. These correction signals are used to send command signals to the control surfaces of the weapon to keep the missile in the centre of the beam. The launch station keeps the beam pointed at the target throughout the engagement period and the missile rides the beam to the intended target. Both radar and laser beam rider guidance has been successfully employed for surface-to-surface, surface-to-air and air-to-ground weapons.

Figure 11.9 illustrates the concept of beam riding for a surface-to-air weapon using a radar beam. A laser beam could also have been used instead with similar performance. Figure 11.10 shows a laser beam rider missile launched from a helicopter against a tank target. As the beam moves farther away from the launcher and towards the target, it spreads out and it becomes difficult to keep the beam in the centre of the target. That is why the beam rider concept is effective only for short-to-medium operational ranges.

Laser beam riding guidance became more attractive particularly for short-range anti-air and anti-tank missiles in the 1980s and 1990s with the introduction of low-cost and highly portable laser designators. Laser beam riding also allows the designer to encode additional information in the beam using digital means. Laser beam rider missiles are inherently more accurate. Also, narrow laser beam guiding the weapon makes it more difficult to be detected and therefore immune to countermeasures.

11.3.2 Command Guidance

In the case of *command guidance*, the missile is commanded on an intercept course with the target. Conventionally, this is achieved by using two separate radars to continuously track the target and the missile. Tracking data from these radars is fed to a computer that computes

Figure 11.10 Laser beam rider concept.

the trajectories of the two vehicles. The computer in turn sends appropriate command signals over a radio link to the missile. A sensor on board the missile decodes the commands and operates the control surfaces of the missile to adjust its course so as to intercept the target in flight. Figure 11.11(a) shows the block-schematic representation of command guidance. The diagram is self-explanatory. Figure 11.11(b) shows the deployment scenario.

Wire-guided missiles are an example of command guidance. In the case of a wire-guided system, command signals are sent to the missile through a conventional wire or a fibreoptic cable that actually reels out from the rear of the missile up to the launch platform. The missile trajectory in this case is controlled with the help of command signals transmitted via a wired link rather than a radio link. Wire-guided missiles are commonly used for short-range anti-tank operations launched from either land-based platforms or helicopters. In many cases, even torpedoes fired from submarines also use wire guidance. TOW is a popular example of a wire-guided missile. Manufactured by the Hughes Aircraft Company, it is primarily used in anti-tank warfare and is a command to line-of-sight weapon. Current versions are capable of penetrating 30 inches of armour at a maximum range of greater than 3 km. It can be fired from a vehicular platform, a helicopter and even by infantrymen using a tripod stand. Figure 11.12 shows a TOW missile being launched from a land platform.

Command guidance can be classified as command line-of-sight (CLOS) and command off Line-of-Sight (COLOS) guidance. CLOS systems are further subdivided into four groups. The first is the Manual Command to Line-of-Sight (MCLOS) where target tracking, missile tracking and control functions are all performed manually. The second type is the Semi-Manual Command to Line-of-Sight (SMCLOS). In this case, target tracking is automatic but missile tracking and control functions are performed manually. The third category is called the Semi-Automatic Command to Line-of-Sight (SACLOS). Here, the target tracking is manual and the missile tracking and control functions are automatic. SACLOS is the most common form of guidance in use against ground targets such as bunkers and tanks. The Hellfire from Lockheed–Martin is a helicopter-launched fire-and-forget anti-armour air-to-ground weapon of the SACLOS category. The first three generations of the weapon use a laser seeker while the fourth

(a)

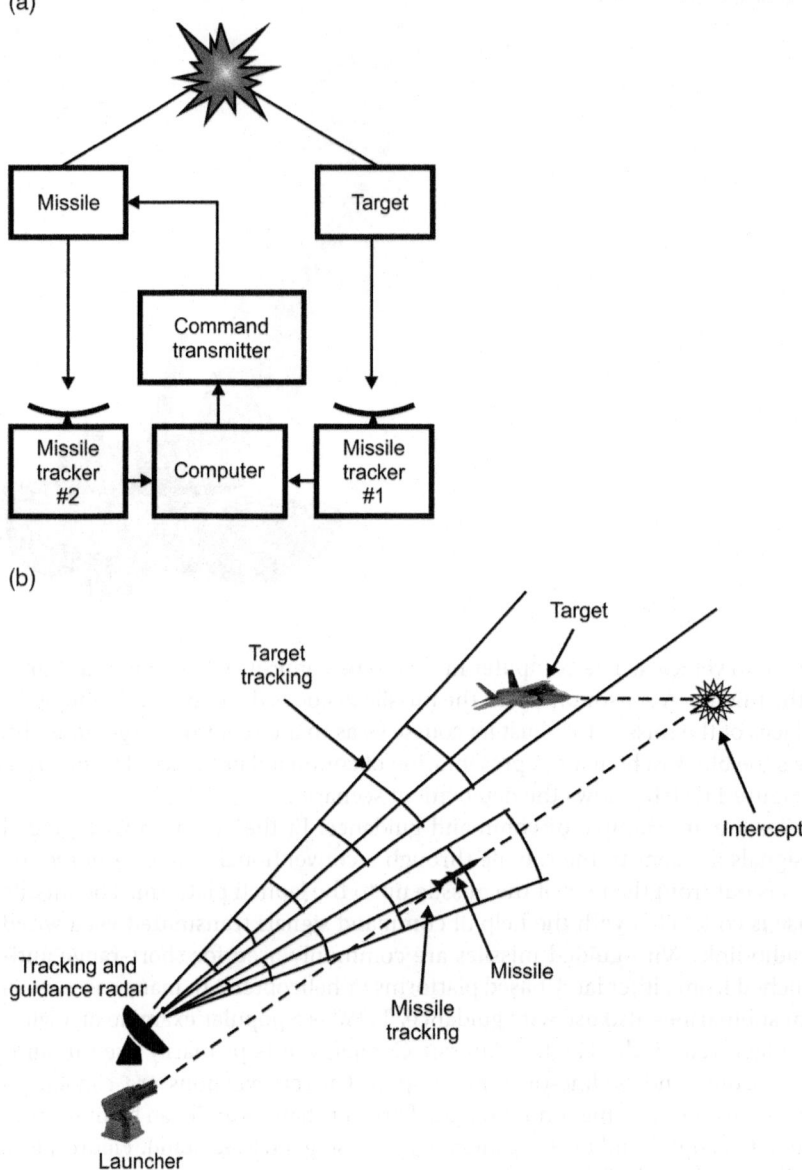

(b)

Figure 11.11 Command guidance. (a) block-schematic representation of command guidance and (b) the deployment scenario.

generation uses a radar seeker. The fourth subgroup is known as Automatic Command to Line-of-Sight (ACLOS). In this, all three functions are automatic.

The COLOS system, unlike CLOS system, does not depend on angular coordinates of the missile and the target. The guidance system ensures missile interception of the target by locating both missile and target in space for which a distance coordinate is needed. This can be possible only if both missile tracker and target tracker are active. It may also be mentioned here that, in the case of the COLOS system, missile and target tracker can be oriented in different directions.

Figure 11.12 TOW missile system. (*Source:* Courtesy of the US Army.)

11.3.3 Homing Guidance

Homing guidance is the most common form of guidance methodology used in surface-to-air and air-to-air-guided weapons. Homing guidance is further subdivided into four groups, namely semi-active homing, active homing, passive homing and track-via-missile homing, also known as retransmission homing. Each of the four types is briefly described in the following paragraphs.

In the case of *semi-active homing guidance*, the target is illuminated by an external source, which could be radar or laser. The electromagnetic energy reflected by the target is intercepted by the seeker head of the guided weapon. An onboard computer processes the intercepted signal and determines the target's relative trajectory. It sends appropriate command signals to the control surfaces of the weapon to make it intercept the target. Figure 11.13 illustrates the concept of semi-active homing guidance in the case of an air-to-air missile. Semi-active homing is similar to command guidance except for the fact that in the case of former, the command computer is on board the weapon. The type of seeker head, whether it is a radar or laser seeker, depends on the type of external source designating the target. Both radar as well as laser-guided semi-active homing weapons are in use. Sparrow air-to-air missile and laser-guided weapons of the Paveway family are examples of semi-active homing guidance. Laser-guided munitions and IR-guided missiles are described in detail in subsequent parts of the chapter.

In the case of *active homing guidance*, the source of target designation is also on board the weapon with the result that this methodology does not require an external source. These features put it in the category of fire-and-forget missiles as the launch platform does not need to continue to illuminate the target after the missile has been launched. Active homing guidance weapons are usually radar-guided. Also, in the case of active homing guidance, the transmitted and reflected waves are at the same angle with respect to the line-of-sight between the target and the missile. This is different from semi-active homing mechanism in which transmitted and reflected waves are at an angle. It is because of this reason that semi-active and active homing guidance systems are sometimes called bistatic and monostatic systems, respectively. The RBS-15 anti-ship missile from SAAB Bofors Dynamics, MBDA EXOCET anti-ship and MBDA MICA surface-to-air and air-to-air missiles from MBDA, AS-34 Kormoran anti-ship missile

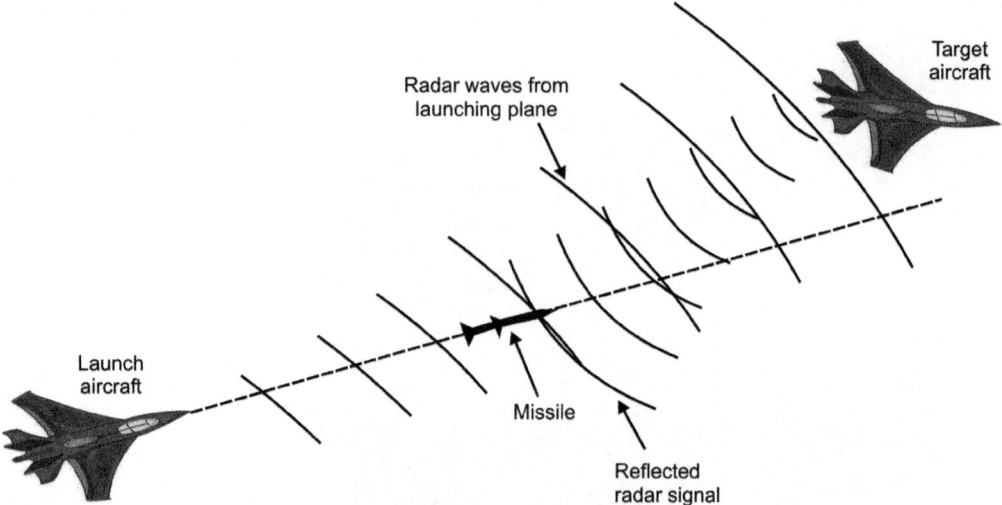

Figure 11.13 Semi-active homing guidance.

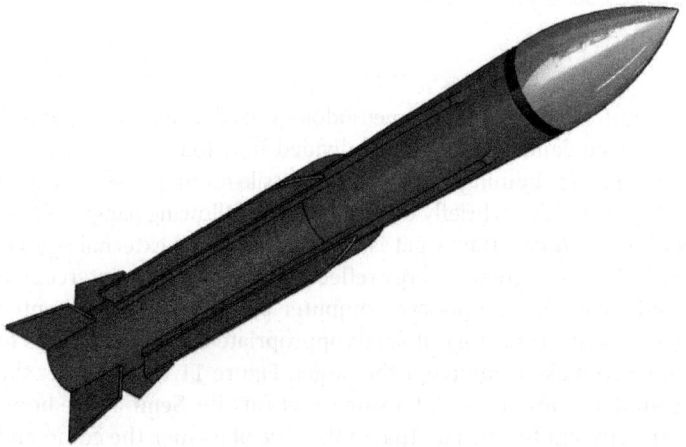

Figure 11.14 DRDO Astra BVRAAM air-to-air missile.

from EADS and Indian DRDO-Astra BVRAAM air-to-air missile (Figure 11.14) are some examples of missiles that use active radar homing in the terminal phase.

Passive homing guidance makes use of some form of energy emitted by the target. This energy is intercepted by the missile seeker, which is processed to extract guidance information to guide the missile to home on to the target. This energy could be in the form of heat energy generated by the target, which is made use of by the seeker in an IR-guided missile. IR-guided missiles constitute an important category of electro-optically guided precision strike weapons. These are discussed in detail in a subsequent part of the article. Anti-radiation missiles such as AGM-88 HARM air-to-ground missiles track the radio frequency energy emitted by ground-based radar stations to generate guidance signals. Passive torpedoes make use of sound waves generated by engines of the ships or sonars to attack their targets. There are missiles such as AGM-65 Maverick that are equipped with electro-optic sensors that rely on visual images to guide the weapon to the target.

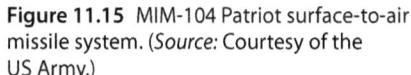
Figure 11.15 MIM-104 Patriot surface-to-air
missile system. (*Source:* Courtesy of the
US Army.)

In the case of *retransmission homing guidance*, the target is illuminated by external radar. The energy reflected by the target is intercepted by the missile sensor. In this case, the missile does not have an onboard computer to process the sensor signal and generate guidance command. Instead, the sensor signal is transmitted back to the launch platform for processing. The command signals generated at the launch platform are retransmitted back to the missile for use by the missile's control surfaces to guide the missile to home on to the target. The advantage with this guidance technique, also called 'Track-via-Missile', is that the expensive tracking and processing hardware is reusable and doesn't get destroyed along with the missile. The disadvantage is that it requires a high-speed communication link between the missile and the launch station. The MIM-104 Patriot surface-to-air missile system (Figure 11.15) by the Raytheon Company in the US is an example of Track-via-Missile homing guidance system.

11.3.4 Navigation Guidance

The term *guidance* not only refers to the determination of the desired path of travel also called the trajectory from the vehicle's current location to an intended target, it also refers to the desired changes in velocity, rotation and acceleration needed to be executed for following the desired path. The term *navigation* refers to the determination, at a given time, of the vehicle's present state vector defined by location and velocity and also its attitude. The term *control* refers to the manipulation of the forces, by way of steering controls, thrusters and so on needed to track guidance commands while maintaining vehicle stability. It is the combination of these three functions that leads what is known as *navigation guidance*.

Navigation guidance is further subdivided into inertial navigation, ranging navigation, celestial navigation and geophysical navigation. In the case of *inertial navigation guidance*, the vehicle uses on board sensors to determine its motion and acceleration with the help of inertial

Figure 11.16 AMRAAM missile.

measurement unit (IMU) or Inertial Navigation sensor (INS). Inertial navigation system basically works by telling the vehicle where it is at the time of launch and the vehicle's computer uses the signals from the inertial measurement unit to ensure that the vehicle travels along the programmed path. Inertial navigation systems are widely used on a range of aerospace vehicles, which includes commercial airliners, military aircraft and spacecraft. With reference to precision-guided munitions, navigation guidance is used for midcourse correction of guided missiles. Long-range all-weather subsonic cruise missile 'Tomahawk' and medium-range all-weather beyond visual range air-to-air missile 'AMRAAM' (Figure 11.16) are some examples that use inertial navigation for midcourse guidance.

While the inertial navigation guidance technique makes use of onboard sensors, *ranging navigation* depends on external signals for guidance, which are usually provided by radio beacons. Based on the direction and strength of the signals received by the aircraft, it navigates its way along the desired trajectory. Ranging navigation guidance has been largely rendered obsolete with the arrival of Global Positioning System (GPS) on the scene. GPS-based navigation has largely replaced radio beacons in both military and civilian applications. GPS is a key enabling technology for existing and future military precision navigation applications. The Joint Direct Attack Munitions (JDAM) series of guided bombs make use of integrated INS and GPS guidance techniques to determine where they are with respect to the locations of their targets. INS-GPS combination gives the precision-guided weapon a kind of all-weather capability and largely overcomes the vulnerability to adverse ground and weather conditions of weapons employing laser and imaging IR seekers. In fact, state-of-the-art precision strike weapons use a combination of guidance technologies including inertial navigation, global position sensing and laser/IR seeking to achieve higher performance levels.

Celestial navigation is one of the oldest navigation techniques that use the positions of the stars to determine location, especially latitude, on the surface of the Earth. This form of navigation guidance requires a good visibility of the stars, which makes it particularly useful at night or at very high altitudes. In celestial navigation, the missile compares the positions of the stars to an image stored in its memory to determine its flight path. The submarine launched ballistic missile (SLBM) 'Poseidon' by Lockheed–Martin carrying multiple independent re-entry technology and with an operational range in excess of 4500 km is an example of a ballistic missile using celestial navigation.

Geophysical navigation guidance depends for operation on the measurements made on the surface of Earth. It uses compasses and magnetometers to measure the Earth's magnetic field as well as gravitometers to measure the Earth's gravitational field. This technique has not found much application in missile guidance. Yet another guidance technique makes use of *terrain contour matching.* It uses a radar altimeter to measure height above the ground. By comparing the contours of the terrain against data stored aboard the missile, the missile's auto-pilot navigates its way to destined location. Terrain Contour Matching abbreviated as TERCOM is a navigation system used primarily by cruise missiles. A related technique to terrain matching is called *digital scene matching.* It is far more accurate than the terrain matching technique. This technique relies for guidance on comparing the image seen below the weapon to satellite or aerial images stored in the missile computer. If the scenes do not match, the computer sends commands to control surfaces to adjust the missile's course until the images match to a certain acceptable level. Digital scene matching is used on the Tomahawk cruise missile. In fact, Tomahawk's guidance system uses a combination of INS, GPS, TERCOM and Digital Scene Matching techniques.

11.4 Laser-Guided Munitions

Laser-guided munitions, due to high precision, efficacy and lethality, constitute an important weapon system in the arsenal of the armed forces worldwide. They have proven their lethality and efficacy beyond any doubt during several conflicts, which include their limited use by British forces in the Falklands War in 1982, during operation Desert Storm in the Gulf War in 1991 by the coalition forces against Iraq and during the Kosovo War in 1999. Laser-guided munitions were used in large numbers to great effect during the Gulf and Kosovo wars. They are of great tactical importance in the contemporary battlefield scenario.

11.4.1 Operational Basics

Laser-guided munitions use a type of optoelectronic position sensor subsystem known as the *seeker unit* that determines in real time the position of the weapon with respect to the target and feeds this maintains its orientation in the desired direction of the target so as to ultimately hit it precisely. Figure 11.17 shows the constructional features of laser-guided munitions.

In a laser-guided munitions delivery operation, the target is illuminated (called the target designation), by a pulsed solid-state laser producing high peak power pulses with a known pulse repetition frequency (PRF). Peak power, pulse width and PRF are typically in the range of 5–8 MW, 10–20 ns and 5–20 Hz, respectively. The laser seeker head in the weapon makes use of laser radiation scattered from the target to generate information on the angular error, which in turn is used to generate command signals needed to guide the weapon to source of scatter, which is the target (Figure 11.18).

Before the weapon locks on to the radiation scattered from the target, it makes sure that the radiation is the intended one. To achieve this, laser target designator and the laser seeker used

Figure 11.17 Constructional features of a laser-guided bomb.

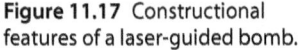

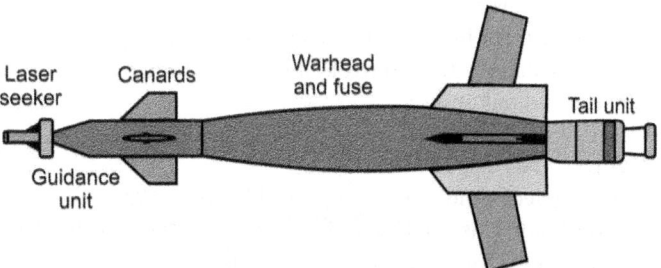

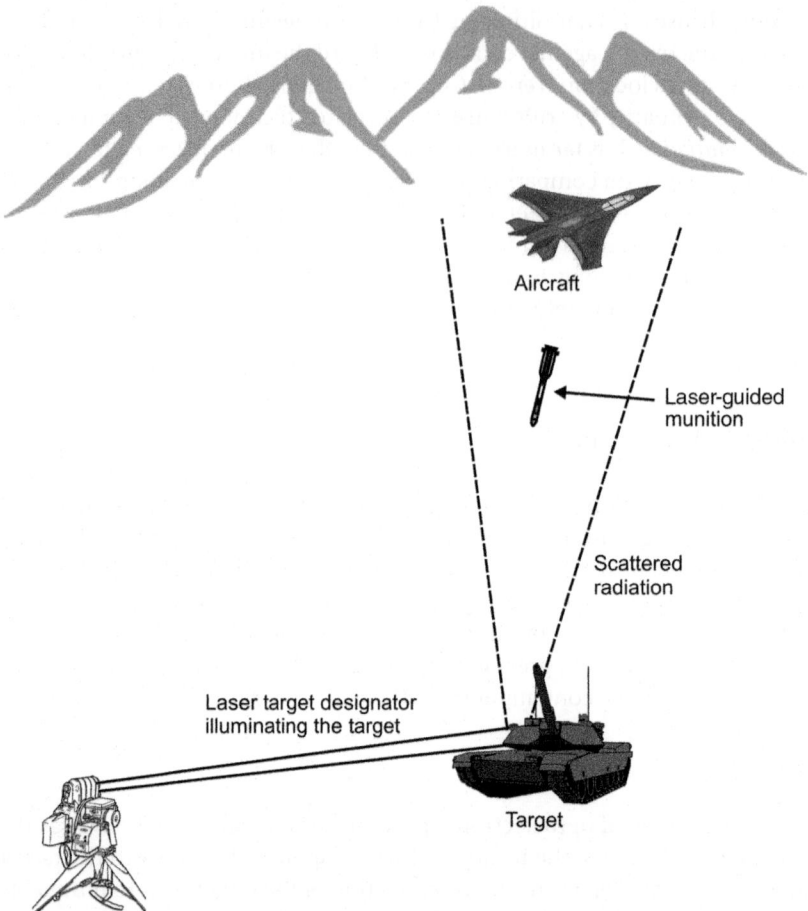

Figure 11.18 Laser-guided munitions delivery.

in the guided weapon delivery mission use the same PRF code and the PRF code compatibility check forms the basis of identification of the desired radiation. PRF code compatibility is therefore essential to weapon's functionality and mission success. The PRF code is generally chosen to an accuracy of $\pm 1 - \pm 2$ μs in the time interval between two successive laser pulses at a nominal value that is usually in the range of 50–200 ms. The optoelectronic position sensor in the front-end of the seeker unit determines orientation of the weapon with respect to the target. But before it does that, it deciphers the PRF of the received radiation and only if the PRF matches with the chosen PRF value within the specified tolerance, it is further processed to extract information on the angular position of the weapon with respect to the target. Information on angular error in azimuth and elevation is fed to the servo control subsystem in the specified format, which in turn controls the flight trajectory of the weapon with the help of the front canards and tail unit.

11.4.1.1 The Position Sensor
The optoelectronic sensor employed for the purpose of position sensing is usually a quadrant photo sensor. It is the heart of laser seeker. A two-dimensional array of photo sensors is also used in some cases. Figure 11.19 explains the principle of operation of a quadrant photo sen-

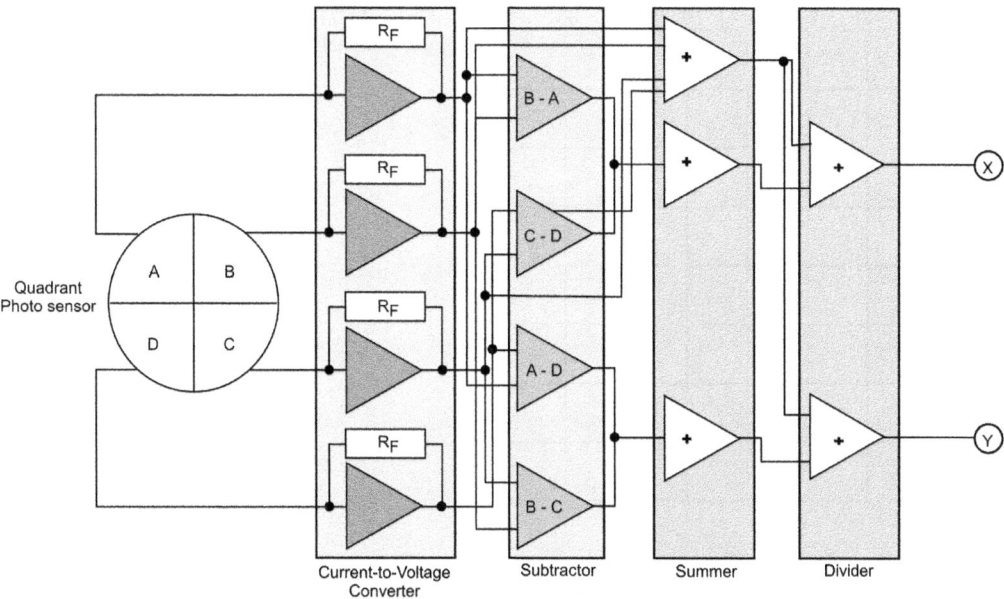

Figure 11.19 Principle of operation of a quadrant photo sensor.

Figure 11.20 Position of a laser beam spot: (a) laser beam aligned with the receiver's optical axis, (b) laser beam shifted in the positive X-direction, (c) laser beam shifted in the negative X-direction, (d) laser beam shifted in the positive Y-direction and (e) laser beam shifted in the negative Y-direction.

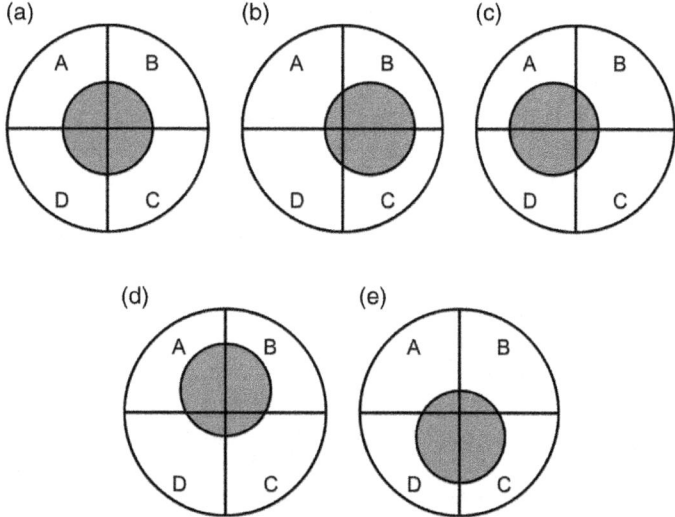

sor when used for position sensing application in laser-guided munitions. The quadrant photo sensor is placed before the focal plane of the front-end optics. The focal spot is symmetrical about the centre of the quadrant photo sensor when the perpendicular to focal plane of the detector is collinear with the axis of received laser radiation scattered from the intended target as shown in Figure 11.20(a). This is the case when the weapon is pointing precisely toward the target. If the laser radiation is impinging on the laser seeker cross-section at an angle, which will be the case when the weapon is not pointing toward the intended target, the centre of the focused laser spot will shift depending upon angular error in azimuth and elevation as shown in Figure 11.20(b)–(e). The beam positions in the X and Y directions are

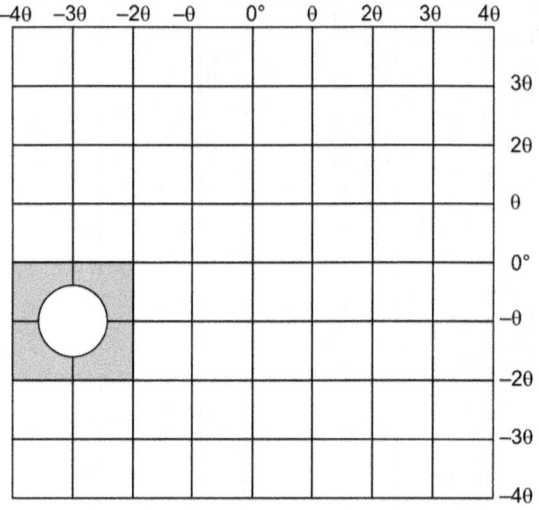

calculated using the following equations. X and Y represent angular errors in azimuth and elevation directions and are given by eqn. 11.1.

$$X = \frac{(B+C)-(A+D)}{(A+B+C+D)} \quad \text{and} \quad Y = \frac{(A+B)-(C+D)}{(A+B+C+D)} \tag{11.1}$$

Where A, B, C and D are the electrical voltages corresponding to laser power falling on the four quadrants. In the case of precise pointing towards the target, all four power levels are equal and therefore $X = 0$ and $Y = 0$. $(A+B+C+D)$ represents total power and division by total power ensures that the calculated position error is independent of laser intensity variations.

The output analogue signals proportional to the magnitude of laser power falling on four quadrants are digitized and then processed to compute X and Y. Error signals X and Y are then used to guide the weapon towards the desired position of null with the canards driven by a servo control system. Maximum value of proportional field-of-view offered by the laser seeker of this type in the X and Y directions is proportional to $\pm R$ where R is the radius of the focused laser spot. Larger spot size gives a larger field-of-view but a lower angular resolution. The radius of focused spot can at the most be equal to half of the radius of the quadrant active area.

In another position sensing technique that uses a two-dimensional array based sensor, each active element in the array is identified by a unique azimuth and elevation angle. The focused spot at any time covers more than one active element and the angular error in azimuth and elevation is computed from eqn. 11.2. Figure 11.21 explains the concept.

$$X = n \times \theta + \frac{[(B+C)-(A+D)]}{[(A+B+C+D)]} \quad Y = m \times \theta + \frac{[(A+B)-(C+D)]}{[(A+B+C+D)]} \tag{11.2}$$

Where θ is the proportional field-of-view of each miniature quadrant. n and m are constants depending upon the illuminating quadrant. A, B, C and D are the electrical voltages corresponding to laser power falling on the four quadrants.

11.4.2   Important Parameters

Major performance parameters of laser-guided munitions include *sensitivity, field-of-view, PRF code compatibility* and *response linearity.* In addition, immunity to false codes and response to desired code in the presence of false code are the other important parameters relevant to assessing performance efficacy of laser-guided munitions.

Sensitivity is the minimum value of laser power density impinging on the laser seeker cross-section in a plane orthogonal to the optical axis of the seeker head to which it can respond satisfactorily. It is a characteristic of seeker's front-end optics and photo sensor. Sensitivity of the seeker decides the maximum guidance range for known values of laser target designator parameters, target reflectivity, height of laser target designator (in the case of airborne laser designator) and laser seeker above sea surface and visibility condition.

Field-of-view determines the probability of the weapon finding itself within the laser basket at the maximum guidance range. Laser-guided munitions using a seeker head with a larger field-of-view would have higher probability of finding themselves in the laser basket and subsequently precisely hit the intended target.

PRF code compatibility is the primary requirement for the weapon to function. PRF code compatibility means the PRF code of the received laser radiation would be the same as the PRF code programmed in the guidance unit before start of mission. PRF code is usually expressed as the time interval between two successive laser pulses in milliseconds usually up to third decimal place. The two codes are considered compatible if the difference in time periods of the two codes is less than a certain specified value.

Response linearity predominantly decides the circular error probability (CEP). *Immunity to false PRF codes* and *capability to stay locked to the desired code in the presence of false PRF codes* enhances the probability of target hit. The former test is performed by irradiating the seeker head with a PRF code different from the programmed PRF code and later by irradiating the seeker head simultaneously with radiations of correct and false PRF codes.

11.4.3   Deployment Configurations

Different deployment configurations are used for laser-guided munitions delivery depending upon the nature of mission. The one illustrated in Figure 11.18 uses a ground-based laser target designator and aerially delivered bomb. There can be other possible scenarios. For example, in another case, the laser target designator is located on another aircraft, which is not the one that is carrying the laser-guided bomb. Such a deployment scenario is depicted in Figure 11.22.

In yet another deployment configuration, the target is designated from a ground-based designator and the guided munitions are also launched from a ground-based platform. This is true in the case of cannon-launched laser-guided projectiles. This is depicted in Figure 11.23. Yet another possible deployment configuration is one in which laser target designation and laser-guided munitions delivery is executed from the same airborne platform as shown in Figure 11.24.

While the delivery configuration shown in Figure 11.23 is applicable to cannon-launched laser-guided projectiles, the other three cases shown in Figures 11.18, 11.22 and 11.24 are different options used in the case of delivery of laser-guided bombs and missiles.

11.4.4   Laser-Guided Munitions Delivery Parameters

For a given sensitivity of the laser seeker head (minimum power density required to be present in the seeker plane orthogonal to its optical axis for it to function satisfactorily), the guidance range depends on the power density actually available at that plane. The available power density

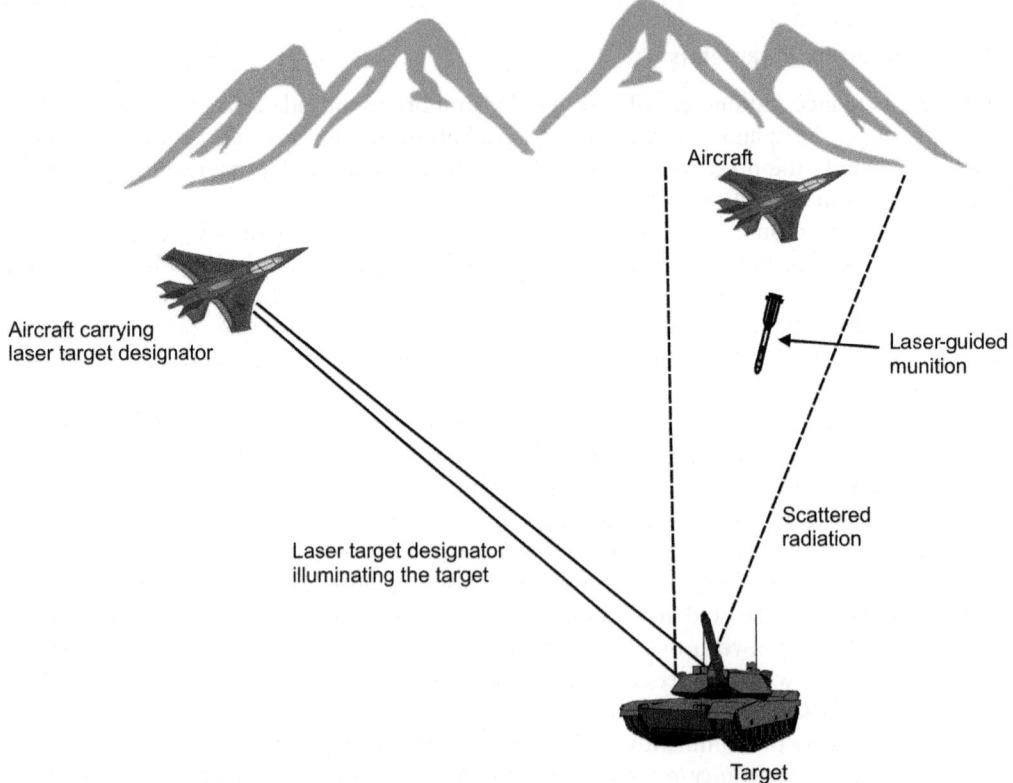

Figure 11.22 Laser-guided munitions delivery with a target designated from another aircraft.

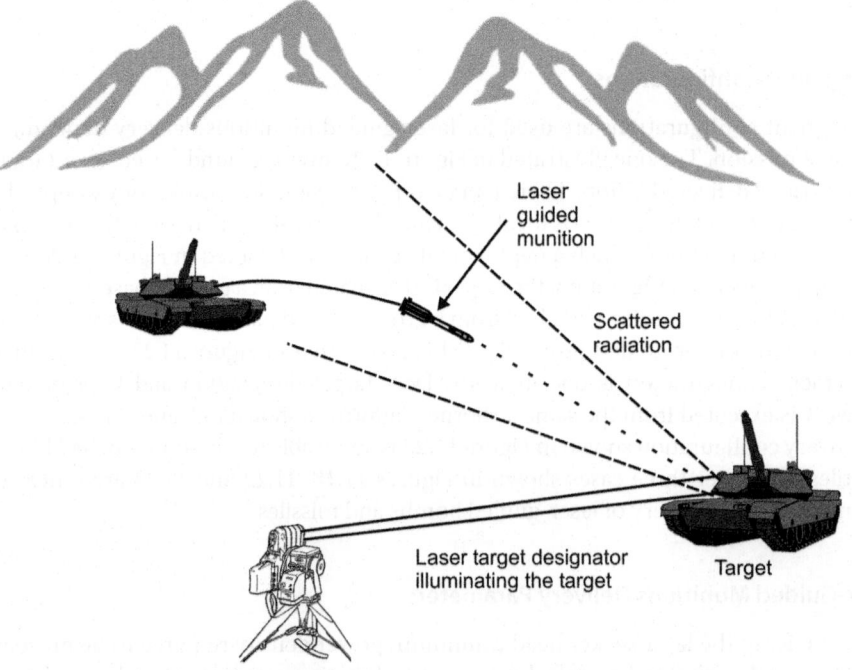

Figure 11.23 Laser target designator operation and laser-guided munitions delivery from ground-based platforms.

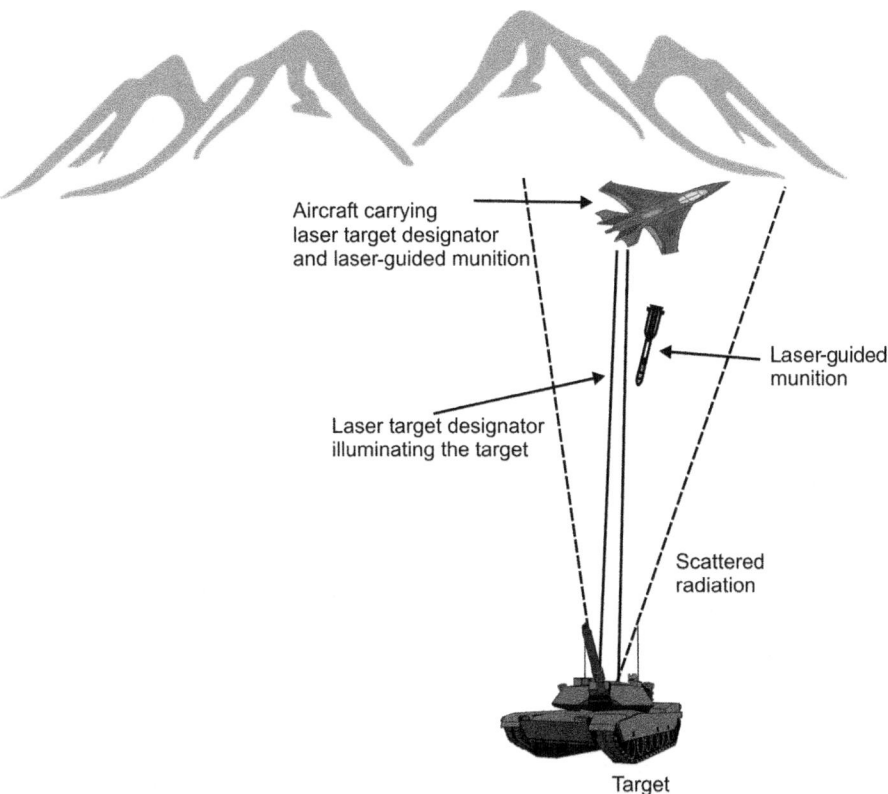

Aircraft carrying
laser target designator
and laser-guided munition

Laser-guided
munition

Laser target designator
illuminating the target

Scattered
radiation

Target

Figure 11.24 Laser target designator operation and laser-guided munitions delivery from the same airborne platform.

depends on a number of parameters, prominent among them being *peak transmitted power* from the laser target designator, transmitted beam diameter, transmitted beam width full angle, laser target designator-to-target distance, target-to-receiver distance, atmospheric attenuation coefficient, target reflectivity, angle between transmitter line-of-sight and normal to the target, angle between receiver line-of-sight and normal to the target, angle between receiver line-of-sight and normal to receiver aperture and target surface area.

The *transmitted beam diameter* and *full-width angle* decide the *laser beam spot area* at a given distance from the transmitter. The laser beam spot area is directly proportional to the laser beam full angle width and distance from transmitter. Smaller divergence (consequently, a lower laser beam full angle width) produces a relatively smaller laser beam spot area up to a longer distance and therefore is always desirable. *Power density at target location* is given by the laser power available at the target location divided by the laser beam spot area at that location. *Laser power reaching target location* depends upon transmitted laser power, atmospheric attenuation coefficient at operating wavelength and transmitter to target distance. Further, the *attenuation coefficient* depends on the visibility conditions and height of transmitter above sea level. *Target irradiance*, which is the laser power density on the target surface, is not the same as the power density at the target location. This is due to the angle made by the transmitter line-of-sight and normal to the target surface. *Target irradiance* is less than the power density at target location by a factor given by cosine of this angle. *Target brightness* is the laser power density per unit solid angle in the reflected beam. This depends on target reflectivity and mechanism of reflection, whether it is specular or diffused or a combination of the two. Target brightness together with the solid angle subtended by the receiver aperture and the attenuation coefficient from target-to-receiver decide the power density available at the plane normal to the receiver cross-section. The actual power density available is further reduced by a factor equal to cosine of angle between receiver line-of-sight and normal to receiver aperture. *Received power* is the product of power density multiplied by receiver aperture area. The value needs to be further multiplied by loss coefficient due to front-end optics.

In the following paragraphs, a detailed mathematical treatment of laser-guided munitions delivery is presented. The following parameters are considered for computing the magnitude of laser power entering the laser seeker head cross-section.

1) Peak transmitted power from the laser target designator, P_T (W)
2) Power received by receiving aperture, P_R (W)
3) Transmitted beam diameter, D_T (m)
4) Transmitted beam width full angle, θ_B (radians)
5) Laser target designator-to-target distance, R_T (m)
6) Target-to-receiver distance, R_R (m)
7) Sea level atmospheric attenuation coefficient, σ_W
8) Fractional decrease in atmospheric attenuation coefficient as a function of height of laser target designator above sea level, α_{HT}
9) Fractional decrease in atmospheric attenuation coefficient as a function of height of receiver above sea level, α_{HR}
10) Target reflectivity, ρ
11) Angle between transmitter line-of-sight and normal to the target, θ_T (radian)
12) Angle between receiver line-of-sight and normal to the target, θ_R (radian)
13) Angle between receiver line-of-sight and normal to receiver aperture, θ_A (radian)
14) Projected spot area in a plane orthogonal to transmitter line-of-sight, A_N (m^2)
15) Projected spot area in a plane orthogonal to receiver line-of-sight, A_M (m^2)
16) Laser beam area at a given distance, A_B (m^2)
17) Target surface area, A_T (m^2)
18) Area of receiving aperture, A_R (m^2).

Figure 11.22 shows a laser-guided munitions delivery operation in the case where a laser target designator and laser-guided munition are on different airborne platforms. The expression for received power for laser-guided munitions delivery for land-based laser target designator can be derived from the one used in the case of an airborne laser target designator.

As a first step, we shall compute laser beam spot area at the target location. Laser beam spot diameter and laser beam spot area at target location are, respectively, given by eqns. 11.3 and 11.4.

$$\text{Laser beam spot diameter at target location} = D_T + \theta_B R_T \tag{11.3}$$

$$\text{Laser beam spot area at target location} = \frac{\pi (D_T + \theta_B R_T)^2}{4} \tag{11.4}$$

Power density at the target location without considering the effect of atmospheric attenuation and considering the effect of atmospheric attenuation are, respectively, given by eqns 11.5 and 11.6.

Without considering the effect of atmospheric attenuation,

$$\text{Power density at target location} = \frac{4 P_T}{\pi (D_T + \theta_B R_T)^2} \tag{11.5}$$

Considering the effect of atmospheric attenuation,

$$\text{Power density at target location} = \left[\frac{4P_T}{\pi\left(D_T + \theta_B R_T\right)^2} \right] e^{-(\sigma_w \alpha_{HT} R_T)} \qquad (11.6)$$

Power density at the target surface, called target irradiance, is less than the power density at target location by a factor equal to cosine of angle between transmitter line-of-sight and normal to the target surface. This happens due to the fact that laser beam spot on the target is no longer circular as is the case of normal incidence. It is given by eqn. 11.7.

$$\text{Target irradiance} = \left[\frac{4P_T \cos\theta_T}{\pi\left(D_T + \theta_B R_T\right)^2} \right] e^{-(\sigma_w \alpha_{HT} R_T)} \qquad (11.7)$$

In the next step, we compute the laser power density reflected from the target. The laser energy reaching the target undergoes absorption, specular reflection and diffused reflection and the probability of occurrence of each of these phenomena depends upon the respective coefficient. The sum of magnitudes of all the three coefficients is 1. Assuming that the target is a perfectly diffuse reflector with a Lambertian radiation pattern, reflected power density is given by eqn. 11.8.

$$\text{Reflected power density} = \left[\frac{4\rho P_T \cos\theta_T}{\pi\left(D_T + \theta_B R_T\right)^2} \right] e^{-(\sigma_w \alpha_{HT} R_T)} \qquad (11.8)$$

Target brightness, which is the reflected power density per unit solid angle, is then computed from eqn. 11.9. It is given by reflected power density divided by π.

$$\text{Target brightness} = \left[\frac{4\rho P_T \cos\theta_T}{\pi^2\left(D_T + \theta_B R_T\right)^2} \right] e^{-(\sigma_w \alpha_{HT} R_T)} \qquad (11.9)$$

Power density available at the receiver location is computed by multiplying target brightness by the solid angle subtended by the receiving aperture and taking into consideration the effect of angle between receiver line-of-sight and normal to the target and the atmospheric attenuation effect. The magnitudes of subtended angle and the power collected by receiving aperture are given by eqns. 11.10 and 11.11, respectively.

$$\text{Solid angle subtended by receiving aperture on target} = \frac{A_R}{R_R^2} \qquad (11.10)$$

$$P_R = \left(\frac{A_R}{R_R^2}\right) \times \left[\frac{A_T \cos\theta_R \times 4\rho P_T \cos\theta_T}{\pi^2\left(D_T + \theta_B R_T\right)^2} \right] \times e^{-(\sigma_w \alpha_{HT} R_T)} \times e^{-(\sigma_w \alpha_{HR} R_R)}$$

$$\qquad (11.11)$$

$$= \left(\frac{A_R}{R_R^2}\right) \times \left[\frac{A_T \cos\theta_R \times 4\rho P_T \cos\theta_T}{\pi^2\left(D_T + \theta_B R_T\right)^2} \right] \times e^{-\sigma_w(\alpha_{HT} R_T + \alpha_{HR} R_R)}$$

In a case where the laser-guided munitions seeker head is not looking towards the target, effective receiving aperture would further get reduced by an additional cosine factor equal to $\cos \vartheta_A$. The modified expression for received power is given by eqn. 11.12.

$$P_R = \left[\frac{4\rho P_T A_R A_T \cos\theta_T \cos\theta_R \cos\theta_A e^{-\sigma_W(\alpha_{HT}R_T + \alpha_{HR}R_R)}}{\pi^2 \left(D_T + \theta_B R_T\right)^2 R_R{}^2} \right] \tag{11.12}$$

The expression for the received power is a generalized one with a designation from an airborne platform different from the laser-guided munitions delivery platform. In a case where the designation is from a land-based platform, which is generally the case, the parameter α_{HT} representing a fractional decrease in atmospheric attenuation with height of laser target designator will be absent. That is $\alpha_{HT} = 1$. In case laser target designator and laser-guided munitions are on the same airborne platform, $R_T = R_R = R$ and $\alpha_{HT} = \alpha_{HR} = \alpha_H$. The expression for received power is given by eqn. 11.13.

$$P_R = \left[\frac{4\rho P_T A_R A_T \cos\theta_T \cos\theta_R \cos\theta_A e^{-2\sigma_W \alpha_H R}}{\pi^2 \left(D_T + \theta_B R_T\right)^2 R^2} \right] \tag{11.13}$$

Atmospheric attenuation coefficient σ_W is a function of operating wavelength and visibility, V. It is approximated by eqn. 11.14.

$$\sigma_W = \frac{3.912}{V}\left(\frac{\lambda}{550}\right)^{-q} \tag{11.14}$$

Where,

σ_W is the atmospheric attenuation coefficient/km at sea level
λ is the wavelength in nm
V is the visibility in km
q is a coefficient.

The coefficient q depends on the particle size distribution.

$q = 1.6$ for V > 50 km and 1.3 for 6 km < V < 50 km
$q = 0.58V^{1/3}$ for V < 6 km.

For an operating wavelength of 1064 nm and visibility conditions of 10 km (taken as the worst case) and 20 km (taken as the best case), atmospheric attenuation coefficient at sea level can be computed from eqn. 11.14 to be equal to 0.165 and 0.08 km, respectively. Also, the atmospheric attenuation coefficient is different in the case of slant ranges and is a function of height above sea level as well as slant angle.

It has been observed and experimentally validated that the atmospheric attenuation coefficient decreases with increase in height and slant angle. According to a study, ratio of attenuation coefficient to its sea level value is measured for slant angles of 20°, 30°, 40° and 50°. The results are observed to follow the empirical relationships of eqns 11.15–11.18 for slant angles of 20°, 30°, 40° and 50°, respectively. These equations represent linear approximations of the experimental data recorded for different slant angles.

$$\alpha_H = -0.04514H + 0.9530 \text{ for slant angle} = 20° \tag{11.15}$$

$$\alpha_H = -0.05H + 0.9626 \text{ for slant angle} = 30° \tag{11.16}$$

$$\alpha_H = -0.0576H + 0.9608 \text{ for slant angle} = 40° \qquad (11.17)$$

$$\alpha_H = -0.0642H + 0.9663 \text{ for slant angle} = 50° \qquad (11.18)$$

Where H is the slant height in km.

11.4.4.1 Selection of Pulse Repetition Frequency

Pulse repetition frequency (PRF) code compatibility check forms the basis of identification of the desired radiation and is therefore essential to weapon's functionality and mission success. As outlined in the first chapter, the laser target designator and laser seeker used in the laser-guided munitions delivery mission use the same PRF code and this code is generally chosen to an accuracy of $\pm 1 - \pm 2$ μs in the time interval between two successive laser pulses in a nominal value usually in the range of 50–200 ms. Laser target designators designed for laser-guided munitions delivery operate on a set of known PRF codes, which are possible to be programmed in guided munitions used in conjunction with the laser target designator. This set of PRF codes is specific to the laser target designator–laser-guided munitions combination. As a result of this, test systems available internationally for performing serviceability check on laser-guided munitions are designed for specific laser-guided munitions. This is one of key limitations of these test systems. It is always desirable to build universality in the test system. The test system design should be so as to be able to generate any conceivable PRF code in the range of 50–200 ms in terms of time interval between two successive laser pulses with a resolution (in time interval between two laser pulses) of ± 1 μs.

11.4.4.2 Selection of Laser Wavelength and Pulse Width

State-of-the-art laser-guided munitions delivery systems almost invariably use a Q-switched Nd:YAG laser operating at the 1064 nm wavelength for target designation. Before, both Q-switched Nd:YAG and Nd:glass lasers were used with the former being a more commonly used option. Q-switched Nd:glass lasers operating at 1070 nm have more or less been completely replaced by Nd:YAG lasers. The test system should therefore be configured around a laser source that operates at 1064 nm. Since laser seeker unit used in guided munitions has an interference filter with a bandwidth of usually $\pm 5 - \pm 10$ nm as a part of the front-end optics; the operating wavelength of the chosen laser source for the test system should be accurate to better than ± 5 nm. In case a laser diode is used for the purpose; closed loop control of its temperature is necessary to take care of wavelength drift with temperature.

Pulse width of the laser target designator is usually in the range of 10–20 ns. Modern laser target designators employ a pulse width of around 10 ns. Earlier systems used around 20 ns. Any pulse width chosen to be in the range of 10–20 ns for the test system would serve the purpose.

11.4.4.3 Selection of Laser Beam Divergence and Spot Size

The laser source used for generating the test laser beam should be a collimated one and beam divergence should preferably be less than or equal to divergence specification of a typical laser target designator. Typical value of divergence specification of a laser target designator is 0.3 mrad or better. Also, laser spot should be such as to overspill the laser seeker head optical front-end cross-section. Usually, the laser spot diameter should be 2–3 times the receiver aperture diameter so as to allow angular movement of the laser beam required for testing the field-of-view and linearity of the laser seeker. A laser beam diameter of 60–90 mm serves the purpose. Since the test system-to-seeker under test distance is usually 1–2 m and the beam is a collimated one, the exit diameter of the laser beam is the same as the required spot diameter on the seeker cross-section.

11.4.5 Capabilities and Limitations

Laser-guided munitions with their diverse variants in surface and aerially launched laser beam riders, cannon-launched projectiles, aerially delivered bombs and helicopter launched missiles are today the preferred weapons of precision attack by the Armed Forces worldwide. This has happened due to constant upgradation of their capabilities over the years since Second World War.

One of the most important attributes of laser-guided munitions is their *precise weapon delivery*. In Second World War, it required 9000 bombs to hit a target of the size of an aircraft shelter; in Vietnam, the number was reduced to 300 and today only one laser-guided munition can achieve the objective. *Reduced collateral damage* is another big advantage of precision attack weapons in general and laser-guided munitions in particular. Since the days of the Second World War when a single air attack could kill tens of thousands of people without raising any moral outcry, attitudes toward both enemy and friendly or neutral have undergone a remarkable transformation. Pin-point accuracy offered by precision laser-guided munitions gives military planners the comfort and confidence of attacking the intended targets even in the midst of major cities. The precision of attack and near-zero collateral damage gives military planners and decision makers the freedom and flexibility to use military force closer to non-combatant-inhabited areas in enemy homeland or enemy-occupied territory than at any previous time in military history. Yet another advantage of laser-guided munitions is their *resistance to jamming* and *electronic countermeasures*. *Ability to operate at night, simplicity, reliability* and *maintainability* are the other positive attributes.

While laser-guided munitions are highly accurate and have demonstrated their efficacy in several conflicts in the past, the precision of attack and reliability of operation are guaranteed only under certain conditions. Factors that have a large bearing on the overall performance of a laser-guided munitions delivery mission include *accurate* and *uninterrupted target designation, atmospheric attenuation* due to prevailing environmental conditions and *mode of weapon release.*

Laser designation of the target should be such that the weapon's seeker head finds itself within the reflected radiation basket and also the received laser power is greater than its sensitivity. This is the first major challenge. Laser designator-to-target path length is sometimes an issue under adverse environmental conditions. Laser under these conditions is attenuated more than it would have under ideal or normal conditions. This may lead to laser power as received by the seeker falling below its threshold, which further causes failure of guidance system if the received power is inadequate. Improper designation may also lead to mishits. It was observed during the Gulf War when the laser radiation was reflected off sand surface rather than the intended target. Temporary blockage of laser radiation due to smoke, fog and dusty conditions and increased moisture content is another reason for guided munitions missing targets.

Correct weapon release is another challenge. The guidance system of earlier laser-guided munitions such as Paveway-II resulted in a rectilinear flight path that had a tendency to lag below the sight line. Currently, weapons are released on an unguided ballistic flight path. The weapon release should be such as to allow the weapon on a ballistic trajectory find itself within the laser basket at the time of start of terminal guidance. A relatively narrow field-of-view of laser seeker further complicates the problem.

Uninterrupted target designation is yet another challenge. The concept of terminal guidance reduces the overall designation time and improves reliability of uninterrupted designation. In the case of autonomous laser designation, that is, designation from weapon carrying aircraft, uninterrupted target designation is a big problem in the presence of smoke, fog, clouds and dusty conditions. In addition, leaves the aircraft vulnerable to enemy's attack by ground fire or air support.

Optimum weapon release height poses another challenge. Optimum altitude for weapon release is in the range of 6–9 km. This makes attack aircraft seriously vulnerable to surface-to-air missile (SAM) attack. During their 1981 raid on the Iraqi nuclear reactor at Osirak, The Israeli Air Force chose to use unguided Mark 84 bombs rather than laser-guided weapons during a raid on Iraqi nuclear reactor at Osirak in 1981 because they felt the need to designate the target would leave the attackers unacceptably vulnerable.

11.5   Major Laser-Guided Weapon Systems

Laser-guided munitions are broadly categorized as surface-launched projectiles, aerially delivered bombs and surface-to-surface, surface-to-air and air-to-surface missiles. A large number of laser-guided weapon systems in these categories are manufactured by international giants including Lockheed–Martin, SAAB Bofors Dynamics Israel Aerospace Industries, Matra, Raytheon and KBP Instrument Design Bureau. The more common and established weapons including Krasnopol/Krasnopol-M and Copperhead (cannon-launched laser-guided projectiles), Paveway-series and Griffin (aerially delivered bombs), LAHAT, RBS-70/RBS-70NG and Hellfire (laser-guided missiles) are briefly discussed in the following paragraphs.

11.5.1   Laser-Guided Projectiles

Krasnopol and Copperhead are cannon-launched fin-stabilized terminally laser-guided explosive projectiles designed to engage small hard point ground targets such as tanks, armoured vehicles, self-propelled artillery systems and other high-value targets such as bridges, defensive fortifications, C^4I (command, control, communications, computers and intelligence) centres and so on. The Krasnopol projectile is produced in two variants, namely Krasnopol and Krasnopol-M. The former is a 152 mm two-section projectile designed to operate with both towed and self-propelled guns and howitzers. However, it has a shortcoming that it is incompatible with the 2S19 auto-loader due to the projectile's length. Krasnopol-M is a 152/155 mm projectile. It is an improvement over Krasnopol and is fully compatible with the 2S19 auto-loader, which makes it usable with western-produced 155 mm howitzers. Other than that, both have the same attack profile (diving top attack as illustrated in Figure 11.25), targets engaged and the type of warhead used. The target ranges are similar; 20 km in the case of Krasnopol and 17 km for Krasnopol-M. Figure 11.26 shows a photograph of the Krasnopol projectile.

The Krasnopol projectile follows a ballistic trajectory approaching the target once it is fired. Subsequent to its firing, a forward observer illuminates the target with a laser designator at a maximum range of 7 km. The seeker in the front-end of the projectile locks on the target and the guidance and control system corrects its flight path in order to impact on the selected/ illuminated point. The Krasnopol projectile follows a top attack pattern as shown in Figure 11.25 to achieve an optimized probability of kill against the armoured target.

M-712 Copperhead is a 155-mm calibre terminally laser-guided projectile with a minimum and maximum range of 3 and 16 km, respectively. It has two operational modes: ballistic and glide. Ballistic mode is used when the cloud ceiling is high and visibility condition is good. Terminal guidance begins at 3 km from the target. Glide mode is used when cloud ceiling and/ or visibility is too low to allow use of ballistic mode. The attack profile in the case of Copperhead projectile is laser illuminated point attack. Figure 11.27 shows a Copperhead projectile in flight as it nears the target. The Copperhead projectile was successfully used during Operation Desert Storm in 1990–1991 and Operation Iraqi Freedom in 2003. It is in use by various Armed Forces including the Australian Army, US Army, Egyptian Army, Jordanian Armed Forces and the Taiwanese Army.

Figure 11.25 Diving top attack in the case of Krasnopol.

Figure 11.26 Krasnopol laser-guided projectile. (*Source:* Mike1979 Russia, https://commons.wikimedia.org/wiki/File:2K25_Krasnopol.jpg. CC BY-SA 3.0.)

11.5.2 Laser-Guided Bombs

Of all the variants of laser-guided munitions, the laser-guided bomb is the most widely exploited weapon if the number of user countries and number of laser-guided bombs used in warfare in the past are any indication. The Paveway family of laser-guided bombs has revolutionized tactical air-to-ground warfare by converting dumb bombs into smart precision-guided munitions. Paveway bombs are the preferred choice of Air Forces worldwide as they have proven their accuracy and efficacy in almost all major conflicts in the past. The family has evolved over the years and has seen continuous capability enhancement with newer versions. The Paveway family has seen four generations, namely Paveway-I, -II, -II Plus, -III and -IV.

Figure 11.27 Attack profile – Copperhead projectile. (*Source:* Courtesy of the US Army.)

Figure 11.28 Paveway-II.

Paveway-I used a gimballed seeker head, a computer control group (CCG) and a set of air foils. The seeker head operated in 'bang-bang' mode, which meant that the control surfaces were deflected either fully or not at all. This led to a sub-optimal flight trajectory. The bombs that could be fitted with Paveway-I LGB kit included M117, M118E1, MK-82, MK-83, MK-84, MK-20, CBU-74/B, CBU-75/B, CBU-79/B and CBU-80/B. More than 10 000 Paveway-I LGBs were used by US Air Force in South-East Asia with great success.

Paveway-II (Figure 11.28) has a nose-mounted seeker head and fins for guidance. Manufactured by defence contractors Raytheon and Lockheed–Martin, Paveway-II also uses the bang-bang guidance concept. That is, the fins deflect fully or don't deflect at all. *Paveway-III* is an improvement over Paveway-II and uses a more efficient proportional guidance technology. Produced by Raytheon, it was introduced to service in 1983.

Paveway-IV (Figure 11.29) is an advanced and highly accurate laser-guided weapon. It is the most recent member of the Paveway family. Manufactured by Raytheon Systems Ltd, UK, Paveway-IV entered Service in 2008. Paveway-IV will replace the Paveway-II and Enhanced Paveway-II weapon systems as well as the 1000 lb unguided general-purpose bomb. Paveway-IV employs a combination of semi-active laser guidance and INS/GPS guidance to combine the flexibility and accuracy of laser guidance and all-weather capability of INS/GPS to give significantly improved battlefield performance.

The *Griffin laser-guided bomb kit* is manufactured by Israel Aerospace Industries and is designed to retrofit the existing MK-82, MK-83 and MK-84 dumb gravity bombs. The kit

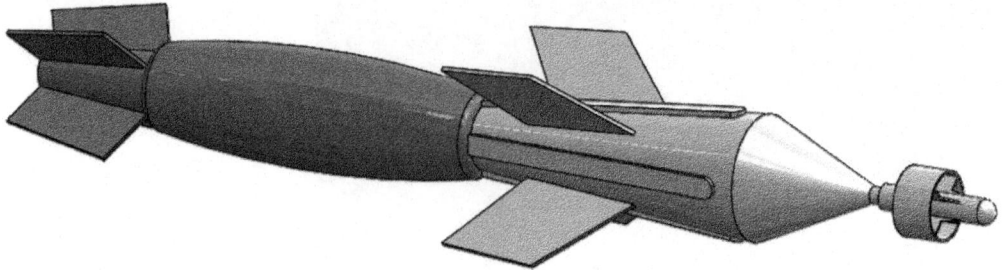

Figure 11.29 Paveway-IV.

employs a laser seeker head and a set of steerable tail planes for guidance. The CEP is estimated to be 5 m. It is in use by the Israeli Defence Forces, Indian Air Force and Colombian Air Force.

The *Sudarshan laser-guided bomb kit* developed by the Aeronautical Development Establishment of DRDO and manufactured by Bharat Electronics is another LGB kit. It was introduced in Indian Air Force in 2013. The CEP is estimated to be 10 m. In future, the Sudarshan LGB kit will incorporate GPS sensor to improve its performance.

11.5.3 Laser-Guided Missiles

Laser-guided missiles use both beam riding as well as semi-active laser guidance concepts. RBS-70/RBS-70NG and LAHAT are example of laser beam riding missiles. These were briefly described in the first section of the chapter. AGM-114 Hellfire-II is a combat-proven tactical surface-to-surface and air-to-surface missile system that uses semi-active laser homing. Figure 11.30 shows Hellfire-II fired from a land vehicle. The Hellfire family comprises the Longbow Hellfire and Hellfire-II missiles. The Hellfire-II missile has a maximum range of 7 km (direct fire) and 8 km (indirect fire). The missile can be launched from multiple air, sea and ground platforms either in autonomous mode or with remote designation. A variant designated

Figure 11.30 Hellfire-II fired from a land vehicle. (*Source:* Courtesy of the US Army.)

AGM-114L uses millimetre-wave radar guidance. Manufactured by Lockheed–Martin and introduced to service in 1984, its primary use is as air-to-surface to engage and defeat individual static or moving advanced armour, mechanized or vehicular targets, patrol craft, buildings and bunkers. AGM-114K, AGM-114M, AGM-114N and AGM-114R are laser-guided variants.

11.5.4 New Developments

While laser-guided bomb kits continue to improve in terms of hit accuracy, operational range, guidance technology and so on, in recent years there has been emphasis to improve guidance technology to improve a weapon's performance in adverse weather conditions. This has been made possible by combining laser guidance with Global Positioning System (GPS)/Inertial Navigation System (INS). Laser Joint Direct Attack Munition (LJDAM) is an example. Another major development has been the use of guidance technology in smaller ammunition. In the recent past, field trials have shown encouraging results in laser-guided bullets.

JDAM is a low-cost guidance kit used to convert existing unguided free-fall bombs into near-precision-guided weapons. The JDAM kit consists of a tail section that contains a GPS/INS and body strakes for additional stability and lift. JDAM is produced by Boeing. The Laser Joint Attack Direct Munition (Laser JDAM) expands the capabilities of the Joint Direct Attack Munition (JDAM) by combining a laser sensor kit with JDAM kit. Laser JDAM has the accuracy of a laser-guided weapon and all-weather capability and longer range of GPS/INS-guided weapons. LJDAM can precisely hit both stationary and mobile targets. Laser JDAM has been integrated with the GBU-38 and is operational on the US Air Force F-15E and F-16 and US Navy F/A-18 and A/V-8B platforms. It is planned to integrate LJDAM with the GBU-31 and GBU-32. Figure 11.31 shows the Boeing LJDAM on an F-16 fighter aircraft (lowermost weapon in the figure).

Figure 11.31 LJDAM integrated with an F-16C aircraft. (*Source*: Courtesy of the US Military.)

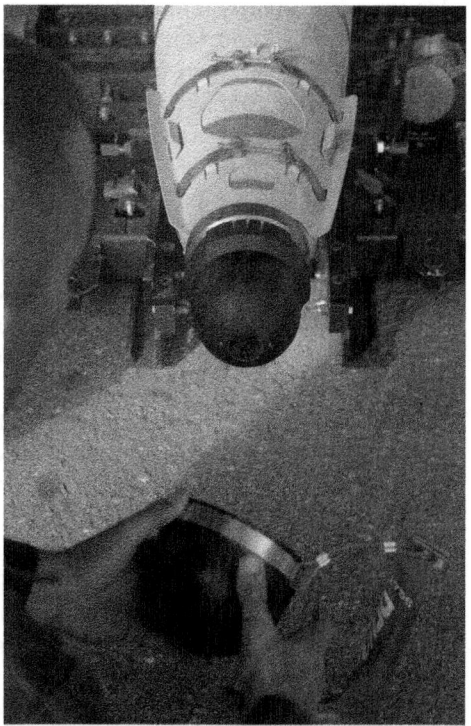

Figure 11.32 Prototype laser-guided bullet developed at the Sandia National Lab.

The *laser-guided bullet* development at Sandia National Laboratories is making headlines as it is expected to significantly increase the range of sharp shooting. Modern bullets gain their accuracy from a technique known as rifling, in which the rifle barrel has a series of spiralling grooves etched into it. The spiralled grooves give spin to the bullet thereby stabilizing its flight path. The laser-guided bullet developed at Sandia National Laboratories is fired from a smooth bore barrel and is stabilized by four steerable fins at its rear (Figure 11.32). The fin movement is controlled by a computer chip, which in turn is driven by signal from an optical sensor on bullet's nose. The intended target is illuminated by a laser beam and bullet uses steerable fins to adjust its mid-flight trajectory. The operation is similar to that of a laser-guided munition, which makes a laser-guided bullet nothing but a miniature laser-guided munition. According to one computer simulation, an unguided bullet fired at a target at 800 m would miss the target by about 9 m. The laser-guided bullet on the other hand would cut that inaccuracy to just 20 cm. Knowing the peculiarities of ballistics, the accuracy gets better for longer ranges.

11.6 Testing Laser-Guided Munitions

There are two broad categories of test systems developed by industry or discussed in literature for the purpose of testing electro-optically guided precision strike munitions. One type is primarily intended for carrying out comprehensive checks on the weapon in terms of all the important parameters. These test systems are usually bulky, operate on mains power and are non-portable. These systems are used when the weapon needs to be comprehensively characterized. The other type is the portable test systems used for performing what is called as serviceability check. Though comprehensive testing of the weapon in the simulated battlefield conditions may be done on a periodic basis, there is always a need for having portable test systems that can perform pre-flight functionality checks with the weapons strapped on to the launch platform. These functional checks referred to as serviceability checks, as mentioned before, perform Go/No-Go testing of the weapon by confining the evaluation process to one or two vital parameters of the weapon. The test system usually generates the target signatures as seen by the seeker unit. In the case of laser-guided munitions, it is the parameters of the laser beam scattered from the target reaching the seeker unit when designated by a laser target designator and in the case of IR-guided missiles, it is the IR spectrum emitted by different parts of the target aircraft falling on the seeker unit of the IR-guided missile.

Serviceability check when performed just before the mission gives that extra bit of confidence to the mission crew on the strength of the weaponry the platform is equipped with. Not only this, these tests may sometimes throw a surprise or two, for instance, by revealing that the laser-guided bomb is not locking to the chosen Pulse Repetition Frequency (PRF) code. As a consequence, the mission PRF code could be altered to another available one that is functional. Similarly, in the case of IR-guided missiles, the functionality of the weapon could be ascertained by irradiating the seeker head with IR radiation having the desired spectral profile. In the following paragraphs, an overview of testing requirements and capabilities/limitations of international test systems commercially available for the purpose is presented. This section covers laser-guided munitions. Testing requirements and available test systems for IR-guided weapons are covered in Section 11.9.

Testing laser-guided munitions primarily means testing of their guidance unit. The seeker head is the most important component of the guidance unit. Establishing functionality of the guided weapon therefore zeros down to establishing functionality of seeker head. Major parameters of interest include the following.

1) Sensitivity
2) Field-of-view
3) PRF code compatibility
4) Response linearity.

In addition to these parameters, immunity to false codes and response to desired code in the presence of false code are the other important parameters relevant to assessing performance efficacy of laser-guided munitions.

Sensitivity is the minimum value of laser power density impinging on the seeker cross-section in a plane orthogonal to the optical axis of the seeker head to which it can respond satisfactorily. It is a characteristic of seeker's front-end optics and photo sensor. Sensitivity of the seeker decides the maximum guidance range for known values of laser target designator parameters, target reflectivity, height of laser target designator (in the case of airborne laser designator) and laser seeker above sea surface and visibility condition.

Field-of-view determines the probability of the weapon finding itself with in the laser basket at the maximum guidance range. Laser-guided munitions using a seeker head with a larger field-of-view would have a higher probability of finding themselves in the laser basket and subsequently hit the intended target.

PRF code compatibility, as outlined earlier, is the primary requirement for the weapon to function. PRF code compatibility means the PRF code of the received laser radiation would be the same as the PRF code programmed in the guidance unit before start of mission. PRF code is usually expressed as time interval between two successive laser pulses in milliseconds up to third decimal place. The two codes are considered compatible if the difference in time periods of the two codes is less than equal to a specified limit, which is typically in the range of 2–10 μs.

Response linearity predominantly decides the circular error probability (CEP). It is an indicator of the delivery accuracy of a weapon system and is a parameter to assess the target damage. It is defined as the radius of a circle within which half of a missile's projectiles are expected to fall.

Immunity to false PRF codes and capability to stay locked to the desired code in the presence of false PRF codes enhances the probability of target hit. The former test is performed by irradiating the seeker head with a PRF code different from the programmed PRF code and the latter by irradiating the seeker head simultaneously with radiations of correct and false PRF codes.

In the present-day battlefield scenario of widespread use of countermeasures, precision strike munitions are designed with adequate counter-countermeasures capability. The designer's objective is to ensure that the weapon does not respond to laser radiation of the same wavelength but with a different PRF code, however close it might be to the desired PRF code. It may be mentioned here again that the chosen PRF code is accurate to the specified value, typically in the range of ±2– ± 10 μs, in the time interval between two successive pulses with the nominal value being in the range of 50–200 ms range. For example, if the desired accuracy were ±2 μs and the chosen PRF code had a time period of 52.475 ms, all radiation with the time interval between two successive pulses differing from 52.475 ms by more than ±2 μs will be rejected by the seeker unit. That is, radiation levels at 52.478 ms and 52.472 ms will also be rejected. The test system therefore needs to have a programmable PRF code generation feature to facilitate such a test.

Another important test is the mixed code test. In the mixed code test, the seeker unit is tested for its response to the correct PRF code in the presence of a false PRF code. In the realistic battlefield scenario, the seeker unit is likely to encounter a weak signal at the desired PRF code in the presence of a strong signal at the undesired PRF code. For example, if the seeker unit encountered two different laser radiations simultaneously, one at 52.475 ms (desired PRF code) with 10 μW/cm^2 power density and another at 52.485 ms (undesired PRF code) with 10 mW/cm^2 power density from the same or different directions, it should lock-on to and track the former.

All the parameters mentioned previously are tested for comprehensive evaluation of the laser-guided munition. In the case of serviceability check, one or two vital parameters that are essential for weapon to perform the intended function are tested. Even if the sensitivity is a little poorer perhaps due to degradation of front-end optics or detector's response or even if the field-of-view were less than the desired value, the weapon can still home on to the intended target provided the PRF code programmed in the seeker head matched with the one of received scatter. On the other hand, even if all these parameters were as per specified values, the weapon will have no chance of hitting the target and will go completely haywire if the seeker PRF code does not match with that of the laser radiation produced by laser target designator. A PRF code compatibility check therefore could easily be singled out for performing a serviceability check. Immunity to false PRF codes and capability to lock-on to and track the desired radiation in the presence of radiation operating on false PRF code could be the other important tests for serviceability checks. Other parameters such as sensitivity, field-of-view, and response linearity could be checked only as a part of comprehensive testing exercise done on a periodic basis.

11.6.1 International Test Systems

Most of the information available on the features and capabilities of test systems intended for testing laser-guided munitions comes from the technical literature in the form of data sheets and application notes that come along with these test systems offered by both manufacturers of laser-guided munitions and manufacturers of optoelectronic test systems. In addition, some information is available from research papers published in this field. The Mission Readiness Test Set (MRTS) type no. TTU 594A/E from the Lockheed–Martin Corporation designed to perform Go/ No-Go verification of the Paveway-II class of laser-guided weapons is an example. The system makes use of a semiconductor diode laser whose output is fibre coupled to the probe used to test the seeker head. It is a man-portable, self-contained test setup that can be used to perform Go/ No-Go type of functionality check of Paveway-II class laser-guided munitions. MRTS type no. TTU 594A/E has now been replaced by an advanced MRTS (Figure 11.33) designed to provide Go/No-Go mission readiness test coverage for nine unique functions. These include Self-Test Code, Code Correlation/Mission Pulse Logic, InterPulse stability, Long and Short Laser Pulse Logic, Pulse Width Discrimination, Squib Fire Delay, Steering/Tracking Accuracy, Voltage/Continuity and Multi/Dual Laser. MRTS is upgradable for smart munitions testing and supports O-, I- and D-level maintenance activities.

There are other companies engaged in manufacture of a wide range of optoelectronic simulators and sensors for a variety of test and evaluation functions. These companies offer test systems suitable for both comprehensive characterization as well as serviceability checks. Geotest-Marvin Inc. and CI Systems are two such companies. Laser Source Simulator (LSS) type MT 1888/1888A from Marvin Test Solutions (Figure 11.34) is a handheld device that

Figure 11.33 Mission Readiness Test Set (MRTS) from Lockheed–Martin Corporation.

Figure 11.34 Laser Source Simulator type MT 1888A from Marvin Test Solutions.

simulates the return signatures from a laser designated target. The device operates at 1064 nm and produces a maximum power of 0.6 mW spread over the entire laser beam having a divergence of 500 mrad. The output power is low enough (class-I classification) to be used without protective safety goggles. The device is field programmable and the desired PRF code can be selected from the membrane key pad provided for the purpose. The device under test (DUT) to simulator distance is recommended to be in the range of 0.6–3.0 m. Divergence specification of 500 mrad ensures that the seeker head optics is always overspilled. Power density levels produced by the simulator at the seeker plane are typical of the sensitivity figures of the state-of-the-art laser seekers.

The simulator cannot, however, be used to check the seeker for its full dynamic range. A laser seeker typically receives a power density in the order of tens of milliwatts/cm^2 close to its blind range. A test device that simulates the return signatures of the laser designated target at two different power density values, one simulating the maximum terminal guidance range and the other simulating the blind range would be far more useful. Also, if the device generated a collimated beam with a spot size to overspill the seeker head rather than a highly diverging laser beam as is the case with MT 1888/1888A, it would simulate the real battlefield scenario. Yet another improvement that can possibly be incorporated is the inclusion of a visible laser pointing beam aligned to be at the centre of the IR beam. It will allow the test IR beam to be properly aligned to the seeker front-end, particularly during night time and also when the target seeker is a few metres away from the simulator. Going a step further, using two independent laser beams that are combined to produce the output beam and where the PRF codes of the two beams can be independently chosen would allow the user to perform the mixed code test too to test the response of the seeker in the presence of false code.

To sum up, if the simple test device is modified to incorporate (1) a variable power density feature either by changing the divergence or power level or both to offer at least two power density levels, (2) a visible laser beam at the centre of IR beam to facilitate easy alignment, (3) two IR laser beams with independent PRF code programmability and optically combined to produce the test beam and (4) a collimated laser beam rather than the diverging beam, the device would simulate not only the real battlefield conditions, it would also allow comprehensive and pre-flight testing of this important class of precision-guided weapon.

11.7 Infrared-Guided Weapons

IR-guided weapons are no less important than the laser-guided munitions considering their usage in tactical warfare. While laser-guided munitions are predominantly used in air-to-surface and surface-to-surface roles, IR-guided weapons are largely surface-to-air MANPADS (Man-Portable Air-Defence Systems) and air-to-air missiles. The IR-guided air-to-air missile has been a feature of fighter armament since the 1950s and is likely to remain a key weapon for decades to come. In the following paragraphs different aspects of this class of electro-optically guided weapon are covered, such as operational concept, performance parameters, capabilities and limitations, deployment configurations, international systems and emerging trends.

11.7.1 Introduction

While laser and radar-guided weapons use semi-active or active homing for guidance and therefore need an external laser source or radar for the purpose, IR-guided weapons make use of passive homing guidance in which the weapon homes on to the IR signatures due to hot areas of the target. For this reason, these are popularly known as heat-seeking weapons. Of course, for the weapon to be reliable, the guidance system processor has to perform the complex task of detecting and identifying the genuine signal in the presence of noise produced by unwanted IR emissions from background due to reflection of solar radiation from the Earth's surface, clouds, mountains and so on and IR countermeasures such as chaff, flares and so on deployed by the target.

Initial development of IR-guided (or heat-seeking) missiles began in late 1950s. AIM-4 Falcons and AIM-9 Sidewinders were important air-to-air heat-seeking missiles of that time. These weapons were not very successful initially due to certain issues that were mainly related to reliability. Modern IR-guided missiles are a vast improvement over the ones developed in years past. They not only have increased operational ranges and hit accuracy, they are also far less vulnerable to countermeasures and are capable of discriminating the intended target from passive and active countermeasures.

11.7.2 Infrared Homing Guidance

There are two broad categories of IR-guided weapons, namely those employing *non-imaging type IR seeker heads* and those employing *imaging infrared (IIR) seekers*. The non-imaging IR-guided missiles home on to IR signatures produced by hot areas of the target. Figure 11.35 shows hot spots on the target aircraft as seen by seeker head of an IR-guide missile. The IR sensor used in this case is a single detector. In the case of an IIR seeker, it is an imaging IR sensor, which is a focal plane array of IR/UV detectors. An imaging sensor array sees in IR in a manner similar to the functioning of a CCD sensor in a webcam or a digital camera. IIR sensor output requires far more complex signal processing. In addition to being more resistant to IR countermeasures such as flares and decoys, imaging seekers are also less likely to be fooled into locking onto the Sun, another common trick for avoiding heat-seeking missiles. By using the advanced image-processing techniques, the target shape can be used to find its most vulnerable part. This information can then be used to steer the weapon towards the target.

Non-IIR-guided missiles make use of IR emission corresponding to the thermal signatures of the exhaust and the mainframe of the target aircraft to home on it. Emission in the 3–5 and 8–12 μm bands is characteristic of electromagnetic emission from jet exhaust and mainframe of the aircraft. Figure 11.36 shows a spectral profile of IR emission from different parts of a typical target aircraft. Spectral content of IR emission as received by the seeker head is the superposition of spectral emission of the aircraft on the transmission characteristics of the atmosphere (Figure 11.37). Figure 11.38 shows a typical IR emission spectrum that would be seen by the non-imaging seeker head. This wavelength signature is judiciously used in guidance of air-to-air and surface-to-air IR-guided missiles.

Modern IR seekers also operate in the 8–12 μm wavelength range, which is absorbed least by the atmosphere. Such seekers are called two-colour systems. Two-colour seekers are harder to defeat with countermeasures such as flares and jammers. IR-guided missiles developed in 1970s and 1980s used single colour IR seekers employing the 3–5 μm band. IR seekers used in these missiles were most effective in detecting IR radiation of shorter wavelengths such as a 4.2 μm emission of carbon dioxide efflux from a jet engine. The seekers responding to the 3–5 μm

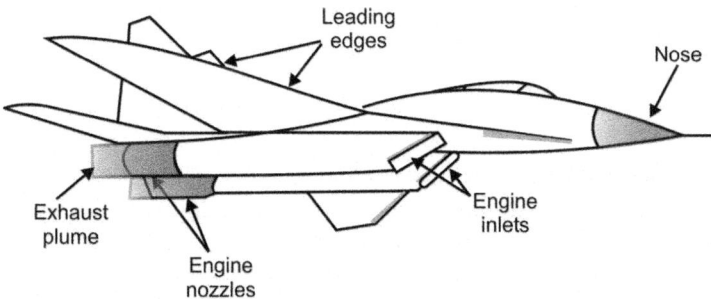

Figure 11.35 Hot spots of the target aircraft.

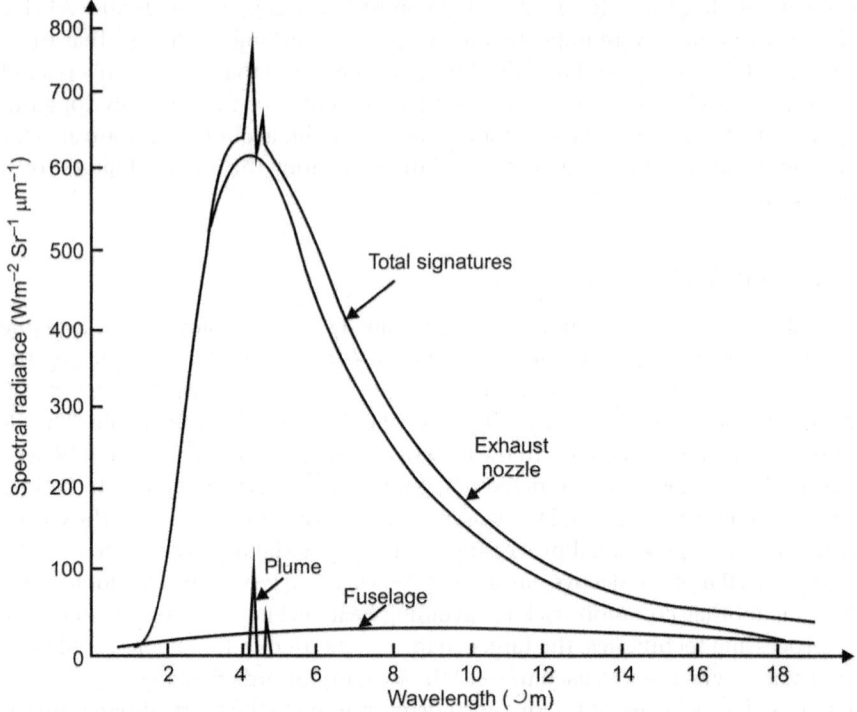

Figure 11.36 Spectral profile of infrared emission from a typical target aircraft.

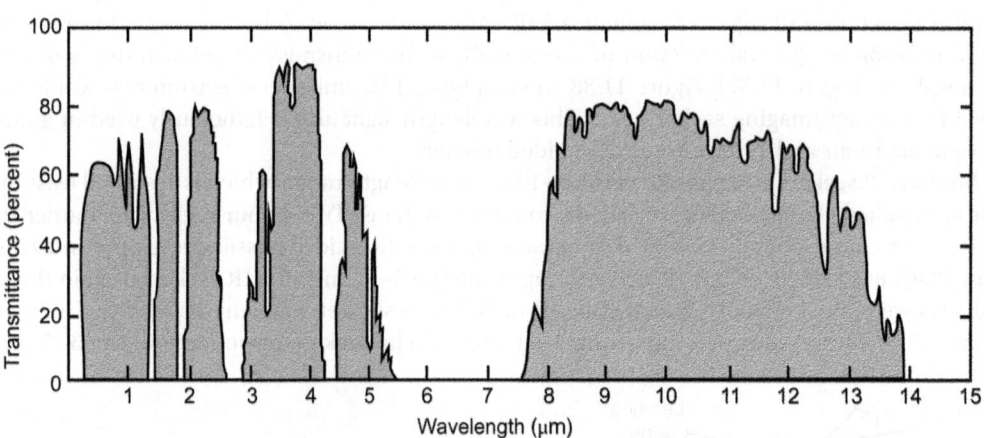

Figure 11.37 Transmission characteristics of atmosphere.

band are called single-colour seekers and missiles single-colour missiles. The MAGIC series air-to-air missiles from France and R-73 air-to-air missiles from Russia are some examples. State-of-the-art IR-guided missiles use seekers that respond to both the 3–5 and 8–12 μm bands to offer improved false alarm rejection and immunity to deception by flares. These seekers are called two-colour seekers and the missiles two-colour missiles. The PYTHON from Israel and RVVAE from Russia are examples of missiles using two-colour seeker heads.

Figure 11.38 Infrared spectrum as seen by the seeker head of an IR-guided missile.

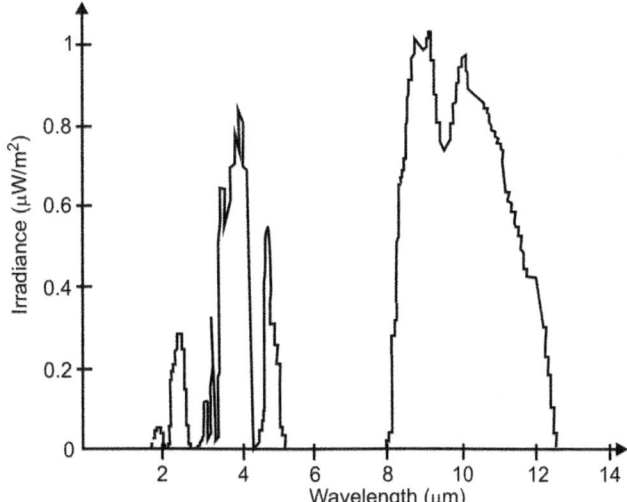

Also, both surface-to-air and air-to-air IR-guided missiles receive a target's IR signatures in the presence of background radiation from the sky and also IR signatures of flares, if any, deployed by target aircraft platform. The seeker head should be able to discriminate between IR signatures of the background and flares from those of the target.

11.7.3 Capabilities and Limitations

IR-guided missiles and radar-guided missiles are deployed in similar roles predominantly as surface-to-air and air-to-air-guided weapons. It would therefore be nothing but logical to compare the two types of guided weapons. IR-guided missiles offer several advantages over radar-guided missiles. (1) IR-guided missiles are far more *immune to electronic countermeasures* than their radar counterparts. (2) IR-guided missiles offer *greater safety of pilots* of aircraft carrying these weapons. Due to the inherent fire-and-forget capabilities of IR-guided missiles, after releasing the missile, the pilot can leave the area while the missile guides itself to the target. (3) IR-guided missiles are almost *impossible to detect* during launch preparation. On the other hand, radar-guided missiles employ either beam rider or semi-active or active guidance. In all forms, it employs radar, which can be detected through radar emission during launch preparation or flight. (4) IR-guided missiles are *manoeuvring missiles* and are particularly suitable for close engagement. (5) IR sensors *perform well during day and night conditions*. (6) IR sensors employed by heat-seeking missiles *cost less per unit*.

While radar-guided missiles are all-weather weapons and have relatively much longer operational ranges, *heat-seeking missiles perform better in close-in ranges*. Also, heat-seeking missiles are *vulnerable to use of passive and active IR countermeasures*. Use of chaff, decoys and flares to deceive heat-seeking missiles is a common occurrence. Heat-seeking missiles employing IIR seekers though are far less vulnerable to these countermeasures. In the early days of the development of heat-seeking missiles, these missiles could only lock-on to intense sources of heat such as jet exhaust pipes and therefore could be fired only from behind the aircraft. Present day missiles using advances IR sensors could even lock-on to friction heated air streaming back from the aircraft's nose. These missiles can be fired from any angle.

11.7.4 Infrared-Guided Weapon Delivery Parameters

This section briefly discusses the factors affecting the IR signatures of the target, the background and the flare. An estimation of IR signatures of the typical target aircraft in terms of amplitude and spectral content as seen by the seeker unit of the IR-guided missile in the presence of static background noise and IR flare will be presented in the paragraphs to follow.

11.7.4.1 IR Signature Model of Background

Major sources of background IR noise include the Earth, the sky, thermal emissions from clouds and solar reflections. The background sources are generally modelled as a grey body whose radiance is dependent upon its temperature and emissivity. The temperature of the background can be considered the ground-level temperature or the ambient aircraft temperature. Though background signatures are dynamic in nature and vary with weather condition, path length and altitude, for the purpose of IR-guided missile guidance, static background noise is a reasonably good approximation.

11.7.4.1.1 Sky Radiance

The sky radiance depends on the weather conditions, path length and altitude. Based on the sky radiance data generated through studies carried out by researchers around the world, the following observations can be made.

1) Radiance is a strong function of altitude both in terms of spectral content as well as magnitude. The magnitude of radiance decreases with altitude.
2) Radiance spectrum is spread over 5–20 µm at sea level and remains more or less the same up to an altitude of a little higher than 1 km. As the altitude increases further, there is a decline in the magnitude of the 8–12 µm band except for a narrow peak around 9.5 µms.
3) The spectrum shifts towards the higher wavelength region with an increase in altitude.
4) In 3–5 and 8–12 µm bands, which are the bands of interest for IR seeker units, the magnitude of radiance is small. It is in the range of 100–800 $\mu W/cm^2/Sr/\mu m$ at an altitude of 1–10 km above sea level.

11.7.4.1.2 Background Radiance Due to Other Objects

The spectral distribution and magnitude of IR radiation from an object are primarily a function of an object's temperature and emissivity. Different objects on the Earth's surface have temperature and emissivity in the range of 250–350 K and 0.8–0.95, respectively. In addition to emissivity and temperature, the area of the background and its range from the missile seeker is needed to model the background. The background radiance can be calculated using eqn. 11.19.

$$N_B = \frac{1}{\pi} \int_{\lambda_1}^{\lambda_2} \frac{\varepsilon_B c_1}{\lambda^5 \left[e^{(c_2/\lambda T_B)} - 1 \right]} d\lambda \qquad (11.19)$$

Where,

N_B = Radiance of the background
T_B = Temperature of the background
ε_B = Total emissivity of the background material
λ_1 = Detection wavelength band start wavelength
λ_2 = Detection wavelength band stop wavelength
c_1, c_2 = Radiation constants.

11.7.4.1.3 Total Background Radiance

The total background radiance is calculated over the complete detection wavelength band for multiple backgrounds. The radiance of each sub-background is calculated separately. The radiance of different sub-backgrounds is then added to determine the total background radiance. In the case of IR-guided missiles, the sky radiance is the primary source of noise and the other sources of noise can be neglected. Therefore, the total background noise radiation spectrum will more or less resemble the sky radiance background noise spectrum. Based on the data generated through various studies, background noise in 3–5 μm and 8–12 μm bands can be considered to be in the range of 100–800 μW/cm^2/Sr/μm for air-to-air missiles flying within an altitude zone of 1–10 km. For a surface-to-air missile flying at an altitude of 5 km, the background radiance would be in the range of 100–1000 μW/cm^2/Sr/μm. As is evident from analysis in subsequent sections, the background signal is more than two orders of magnitude smaller than signal from the target in the wavelength bands of interest.

11.7.4.2 IR Signature Model of Target

The target for the IR-guided missiles is the military aircraft. The military aircraft is a source of IR signatures. Different portions of the aircraft are at different temperatures and hence they emit in different wavelength bands. Different sources of radiation in an aircraft can be classified as internal and external sources. Internal sources include hot engine parts, the airframe skin heated by the engine and aerodynamic heating and combustion of the hot gases in the plume and the skin of the airframe. External sources include reflected ambient radiation from the Sun, sky and ground. The various sources responsible for emission of IR radiation from aircraft are shown in Figure 11.39. The hotter surfaces like the hot engine parts, tail-pipe region and plume mainly contribute in the lower wavelength bands whereas the fuselage region and the airframe and the reflected radiation are the dominant factors in the longer wavelength bands. The target can be modelled as an object with different temperature zones or sub-targets. The IR signature model of target can be determined from known values of emissivity, temperature, range from the guided weapon and area of the sub-target. The data set for a general military aircraft is presented in Table 11.1.

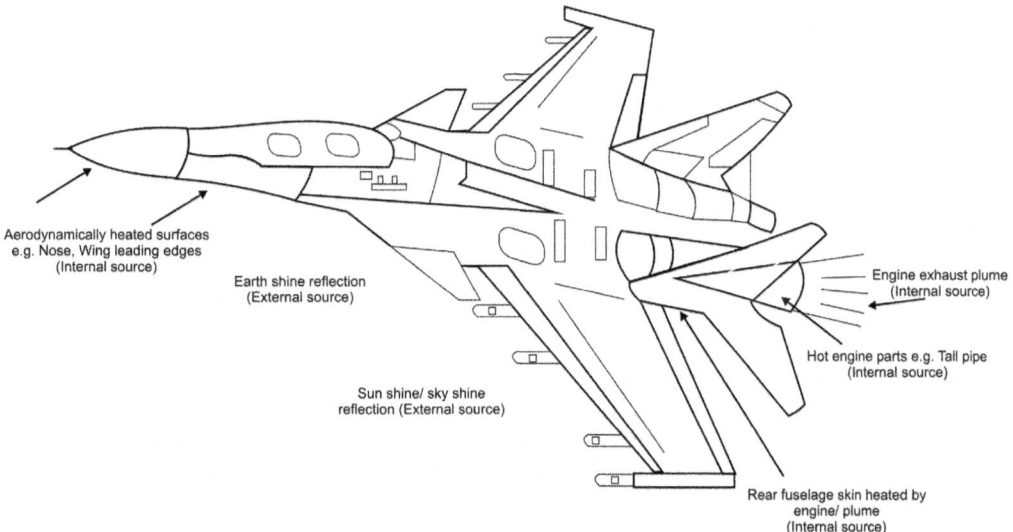

Figure 11.39 Different sources responsible for aircraft IR signatures.

Table 11.1 Temperature of different regions of a typical aircraft.

Sub-target	Area (m^2)	Emissivity	Temperature (K)
Body	30–50	0.9	320–450
Nose	2–5	0.9	350–500
Leading edge	1–2	0.9	350–500
Tail-pipe	2–4	0.9	380–550
Canopy	4–6	0.5	250–400
Plume inner	1–1.5	0.5	1000–1200
Plume outer	2–4	0.5	500–700

The radiance of each temperature zone or a sub-target is calculated independently, assuming that each zone is a grey body radiating uniformly in all directions. The radiance is calculated using Planck's law given in eqn. 11.20.

$$N_{tgt} = \frac{1}{\pi} \int_{\lambda_1}^{\lambda_2} \frac{\varepsilon_{tgt} c_1}{\lambda^5 \left[e^{(c_2/\lambda T_{tgt})} - 1 \right]} d\lambda \tag{11.20}$$

Where,

N_{tgt} = Radiance of the target in a small wavelength interval [$(\lambda_1 - \lambda_2)$ is the wavelength interval and is in the order of 0.01 μm to 0.001 μm]
T_{tgt} = Temperature of the subtarget in Kelvin
ε_{tgt} = Emissivity of the subtarget
λ_1 and λ_2 = Initial and the final wavelengths for every 0.01–0.001 μm interval
c_1, c_2 = Radiation constants.

The radiance values calculated at different wavelength intervals are multiplied by the atmospheric transmittance in those wavelength bands. Total radiance over the wavelength band of the detector is calculated using eqn. 11.21.

$$T_{SR} = \sum_{i=1}^{n} N_{tgti} \tau_{ai} \tag{11.21}$$

Where,

T_{SR} = Total radiance over the wavelength band of the detector
n = Number of intervals in the detector waveband
N_{tgti} = Target radiance at every 0.01 μm interval
τ_{ai} = Atmospheric transmittance at every 0.01 μm interval.

Other than the emissive properties of the targets, their reflective characteristics also contribute to the overall radiance received at the detector. The sources of radiation reflected by the aircraft include the Sun, Earth, sky, atmosphere and clouds. The radiance due to reflections depends upon the relative positions of the sources, the viewers and the angle that they make with the reflecting surface. The total radiance due to the reflections from other sources can be calculated by using eqn. 11.22.

$$N_{reflected} = N_{ss} + N_{es} + N_{sky} + N_{intref} \tag{11.22}$$

Where,

$N_{reflected}$ = Total spectral radiance due to reflection
N_{ss} = Sun shine reflection
N_{es} = Earth shine reflection
N_{sky} = Sky shine reflection
N_{intref} = Reflection from internal portions.

The reflected radiance has two components, one due to specular reflectivity and the other due to diffused reflectivity because of scattering. It may be mentioned here that the body of the target aircraft is opaque in nature. Therefore, it emits thermal radiation and reflects what it receives from the environment.

The turbine engine exhaust gas plume comprises of the residue leftover from the combustion process and is non-opaque in nature. Therefore, the plume emits thermal radiations and transmits the energy through a non-opaque volume. It is usually the dominant source of thermal radiation in the 3–5 μm band. The total radiance of the plume can be calculated using eqn. 11.23.

$$N_{totplume} = N_{plumeemis} + \tau_{plume} \cdot N_{behindplume} \qquad (11.23)$$

Where,

$N_{totplume}$ = Total radiance of the plume
$N_{plumeemis}$ = Radiance due to emissive properties of plume
$N_{behindplume}$ = Radiance of the object behind plume
τ_{plume} = Transmittance of the plume.

The radiance behind the plume may be that of the target itself, sky or background depending upon the target aspect or seeker's viewing angle. Calculation of the signatures of the exhaust gas plume is a complicated task as the shape and volume of the plume is not constant due to the fast movement of the aircraft and the presence of gases (CO and CO_2) in the atmosphere. In fact, the plume gradually gets diluted in the surrounding atmosphere and it is difficult to draw a boundary between the plume and the atmosphere.

For calculating the signatures, the exhaust gas plume can be modelled as co-centric cylindrical regions with the same temperature T, pressure P and molecular density $\aleph$ as shown in Figure 11.40. The bulk of the radiation of the jet engine exhaust plume comes from the vicinity of the exit plane in a region called the inviscid cone (Figure 11.41). As much as three-quarters of the total plume radiance is generated within the inviscid cone. The region beyond the inviscid cone emits more weakly and is more easily absorbed by the cool atmospheric gases.

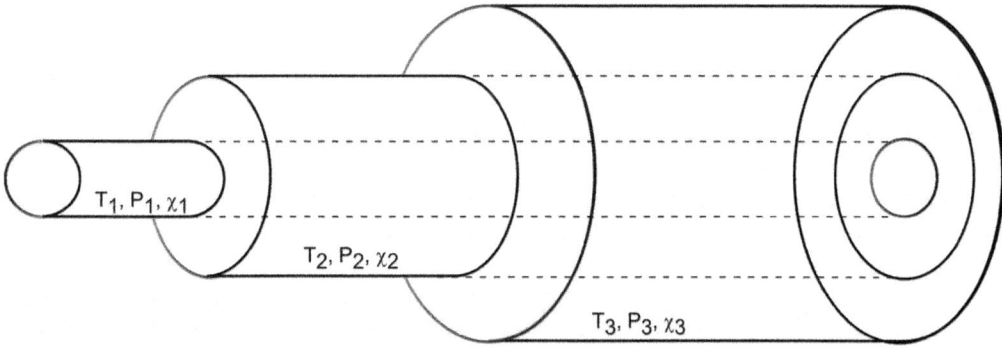

Figure 11.40 Exhaust plume modelled as co-centric cylindrical regions.

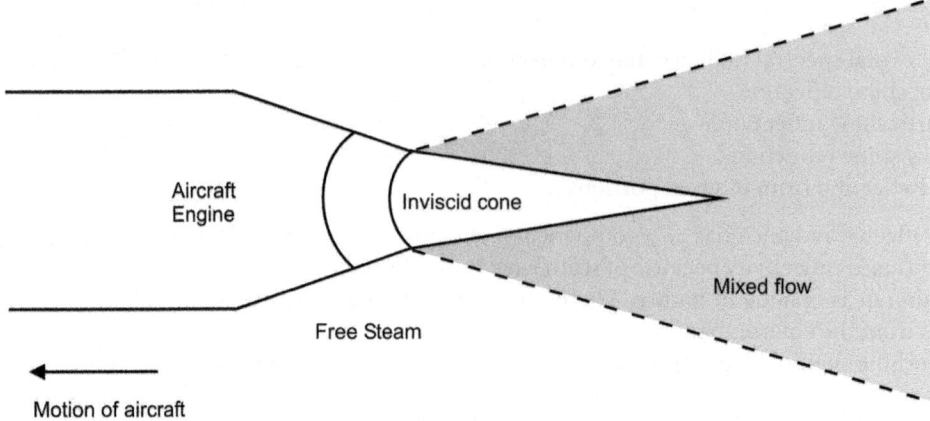

Figure 11.41 Simplified aircraft jet engine exhaust plume.

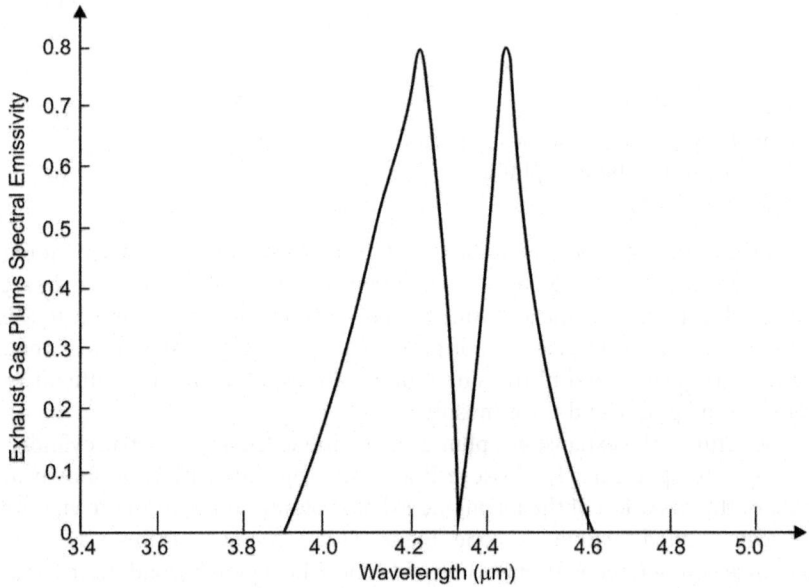

Figure 11.42 Spectral emissivity of the exhaust gas plume.

The spectral emissivity of the plume is calculated in short wavelength intervals using Planck's law and the spectral radiance reaching the detector is calculated by multiplying it with the atmospheric transmittance and then added to calculate the total spectral radiance of the plume due to emissivity. The spectral emissivity of the exhaust gas plume is shown in Figure 11.42.

The IR signatures generated from different portions of the aircraft as well as from the exhaust plume were shown earlier in Figure 11.36. The actual signatures produced by the aircraft will be the addition of the thermal signatures produced by different portions as seen by the approaching missile.

The thermal signatures superimposed on transmission characteristics of the atmosphere are actually seen by an IR-guided missile seeker. Also, the irradiance across the seeker cross-section depends upon approach angle of the missile. The missile sees different parts of the target for different approach angles. Figure 11.43 and Figure 11.44 show typical irradiance

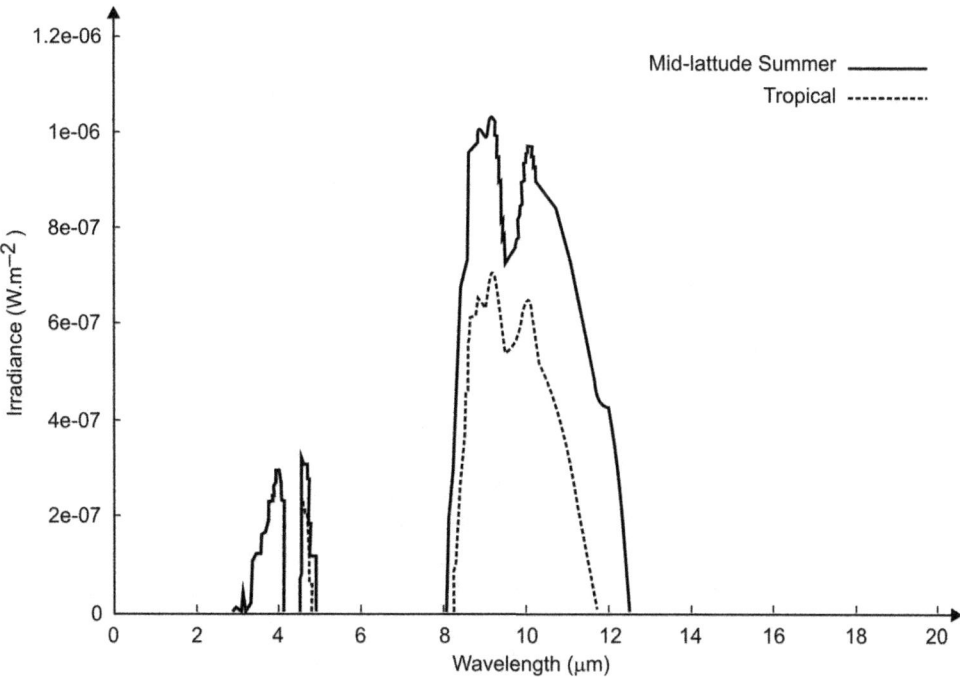

Figure 11.43 Irradiance on a missile seeker with the target at the zenith.

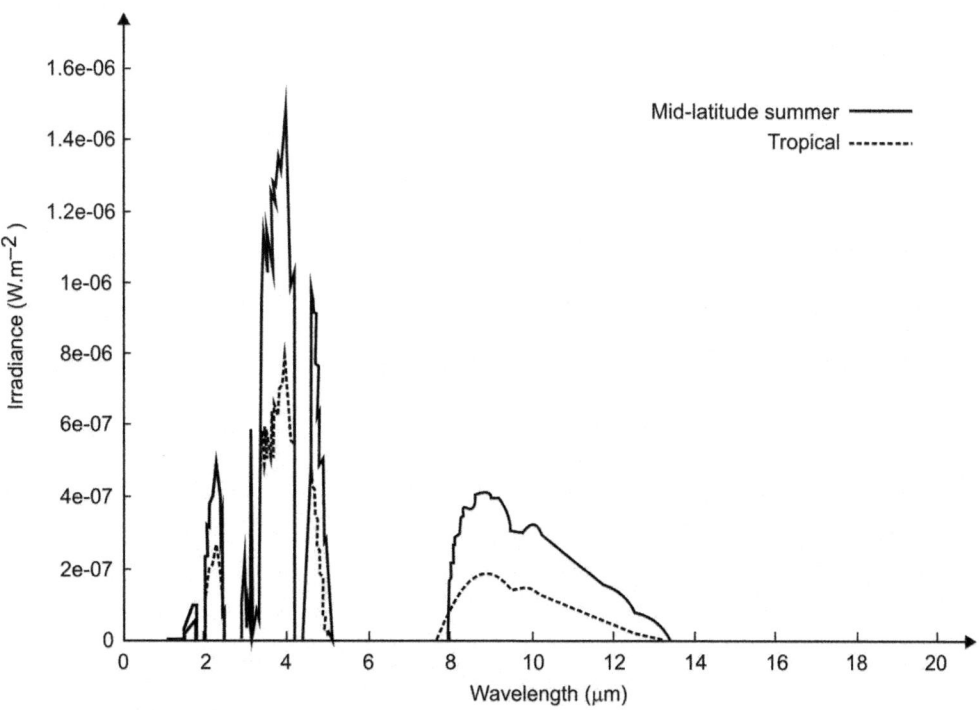

Figure 11.44 Irradiance on a missile seeker with the target at 120° from the horizon.

as seen by a surface-to-air missile seeker for a target cruising at an altitude of 5 km with aircraft at 90° (zenith) and 120° from the horizon, respectively. In the case of the former, IR signatures are more predominant in 8–12 µm band while in the case of latter, IR signatures in the 3–5 µm band are more predominant. Irradiance on a missile seeker can be computed from known values of spectral radiance, atmospheric transmittance, field-of-view of the seeker and detector bandwidth. It has also been observed that irradiance on a missile seeker is significant in five bands: 1.95–2.50 µm, 2.92–3.20 µm, 3.24–4.18 µm, 4.5–4.93 µm and 8.2–11.8 µm. The bands at 1.95–2.50 µm and 2.92–3.20 µm are due to tail-pipe emission only.

11.7.4.3 IR Signature Model of Flare

Flares are chemically heated sources that radiate in accordance with blackbody or grey body characteristics. They are much smaller, lighter and cheaper than the protected target. The smaller physical size of the flare is overcome by heating it to a much higher temperature so as to match its spectral characteristics with that of the target.

The IR signatures of a flare depend upon the spectral size of the flare, the dispensation sequence, ballistic trajectory and the irradiance properties of the flare material. Ideally, the IR signatures of a flare should be spectrally and spatially matched with that of the platform to be protected but with a higher radiant intensity. However, conventional flares act like a hot point source with temperatures as high as 1100–1800 K. For generation of IR signatures of flares, we need to know the important parameters of a flare, which includes flare rise time, burn time, burn time constant, rise time constant, flare temperature and emissivity of the flare.

The full flare plume could be modelled as several co-centric cones of different temperature zones or several discs of different temperatures as shown in Figure 11.45. The inner cones will have higher temperatures than the outer cones. Spectral radiance of each cone can be calculated using Planck's radiation law, which when multiplied by the atmospheric transmittance gives the spectral radiance incident on the detector.

$$N_{flare} \cdot \tau_a = \sum_{i=1}^{n} N_{flarei} \cdot \tau_{ai} \tag{11.24}$$

Where,

N_{flare} = Flare radiance
τ_a = Atmospheric transmittance
n = Number of intervals in the detector waveband
N_{flarei} = Flare radiance at every 0.01 µm interval
τ_{ai} = Atmospheric transmittance at every 0.01 µm interval.

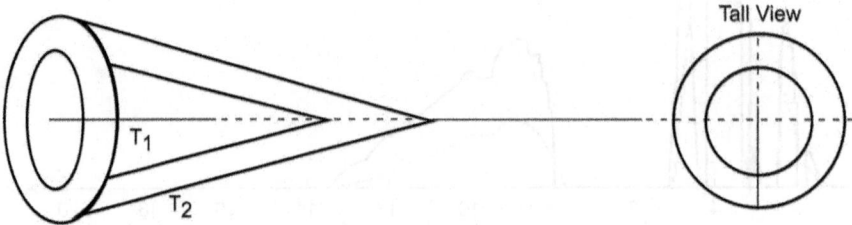

Figure 11.45 IR Flares modelled as co-centric cones.

The outer layers are considered to be semi-transparent so that the inner cone signatures can propagate through the outer cone. The flare's source of radiation is not permanent as the flare's radiant intensity initially rises and then decays with time.

The temporal response of the flare may be modelled as an exponential rise and decay with time. The total flare radiation calculated previously after taking into account the atmospheric effects is multiplied by an exponential function as given in eqn. 11.25 to model the temporal response of the flare.

$$N_{flare}(t) = N_{flare}.\tau_a.f(t) \tag{11.25}$$

Where,

$N_{flare}(t)$ = Flare radiance as a function of time
N_{flare} = Flare radiance
τ_a = Atmospheric transmittance
$f(t)$ = Exponential function.
If $f(t)$ is the rising curve, then $f(t) = 1 - e^{-t/\tau}$ for $0 \leq t \leq t_r$
If $f(t)$ is the decaying curve, then $f(t) = e^{-(t-t_r)/\tau_b}$ for $t_r \leq t \leq t_b$

Where,

t_r = Rise time of the flare (0.2–2 s)
t_b = Burn time of the flare (3–4 s)
τ_r = Rise time constant (0.1–0.2 s)
τ_b = Burn time constant (5–6 s).

Figure 11.46 shows the typical IR signatures of a flare. As can be seen from the figure, the signature of the flare has significant amplitude in 1.5–5 µm band. As a result of this, they are very effective against single colour IR missile seekers and not very effective against dual colour IR missile seekers operating in the 3–5 and 8–12 µm bands. Figure 11.47 shows thermal image of a target aircraft and a flare as seen by a single colour IR-guided missile seeker operating in the 3–5 µm band.

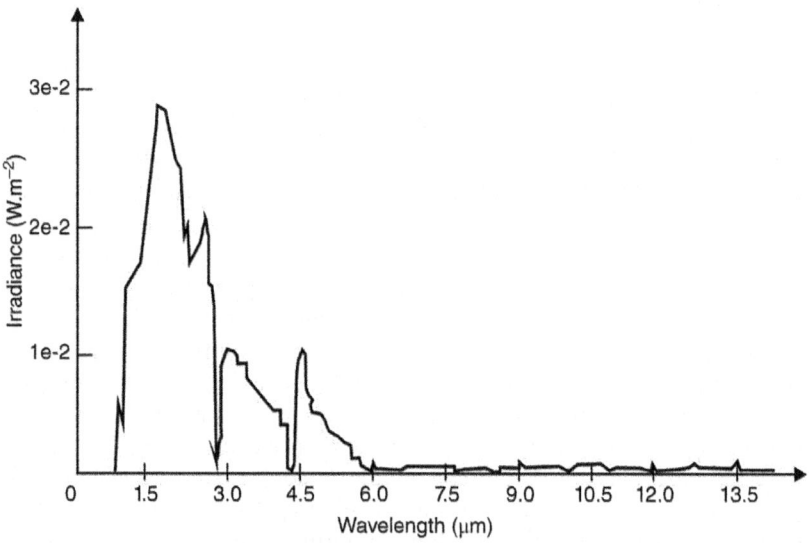

Figure 11.46 Typical IR signatures of a flare.

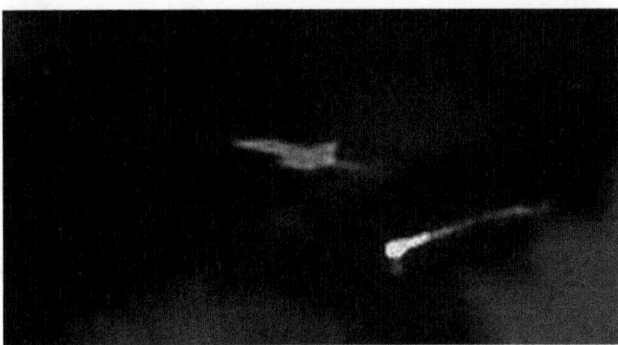

In view of this analysis, an IR-guided missile while operating in the presence of background and IR flare would typically encounter the following IR signatures.

1) Power density (3–5 μm and 8–12 μm bands) across seeker cross-section in the range of 1–1000 μW/m^2 to simulate the target signature for both the maximum and minimum operational range.
2) Power density to simulate static background noise in the range of 10–100 nW/m^2 in the 3–12 μm band.
3) Power density in the range of 10–1000 μW/m^2 to simulate a typical IR flare signature in the 1.5–5 μm band.

11.7.5   Infrared-Guided Missile Seekers

Two broad categories of seekers are used in heat-seeking missiles. These are *reticle seekers* using single IR detector and *imaging seekers* that use a focal plane array of IR/UV detectors. Both types have different variants.

In the most basic reticle seeker (Figure 11.48), IR energy emitted by the target is collected by the optical system and focused on the detector through a rotating reticle. A reticle is nothing but an optical modulator made up of a circular element with sequentially arranged transparent and opaque spokes on it (Figure 11.49). The reticle chops the scene represented by IR energy. The output of the detector is a sequence of pulses whose amplitude is proportional to the magnitude of IR energy and whose frequency is equal to the product of spin rate of reticle and number of transparent/opaque spoke pairs. Also, modern reticle seeker based heat-seeking missiles use spinning mirrors and a fixed reticle. The phase of detector signal with respect to a reference phase is used to determine angular position of seeker axis with respect to the target. The angular error is then used to keep the seeker axis pointed towards the target within its field-of-view and also control the flight path of the missile to intercept the target. We have the spin-scan seekers that suffer from the problem of centre null and the conical scan seekers that eliminate this problem. We shall not go into details of these techniques. Lead sulfide (PbS) with peak sensitivity in the wavelength region of 2 μm when uncooled, indium antimonide (InSb) with peak sensitivity in rgw 4–5 μm band when cooled to liquid nitrogen temperature and mercury cadmium telluride (HgCdTe) with peak sensitivity in the 8–12 μm band when cooled to liquid nitrogen temperature are the commonly used detector materials. Lead sulfide detectors were used in older missile seekers, which forced missiles to look at stern engagements as the missile had to look at the hot turbine in the tail-pipe region to get sufficient signal to track the target. Modern missiles almost invariably use indium antimonide or mercury cadmium telluride detectors.

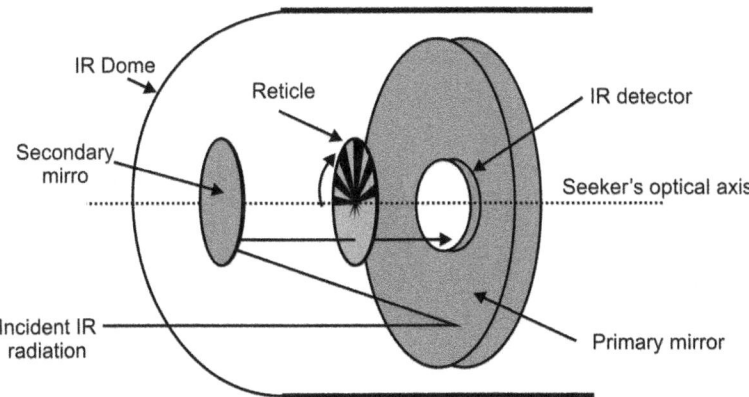

Figure 11.48 Basic reticle seeker.

Figure 11.49 Representative reticles.

Missile seekers would find it very easy and convenient to detect and track a hot target against a uniform benign cool background than doing so in the presence of clouds and extended IR sources. Sunlight reflected from the edge of a cloud would be as attractive a target if not more for the missile seeker than a jet aircraft. Reticle seekers handle all these issues by having very small instantaneous field-of-view (IFOV) and/or advanced signal processing.

Navigation to the target is the next important step after target acquisition and tracking. In one method called *pure pursuit* also called *direct pursuit* the navigation technique ensures that the seeker is looking at the target continuously throughout the engagement duration till it eventually hits the target. In the case of pure pursuit engagement, the flight path is not the most direct one as shown in Figure 11.50 and the trajectory has an ever-decreasing radius turn towards the end of the engagement. This poses a problem as the missile may not be left with

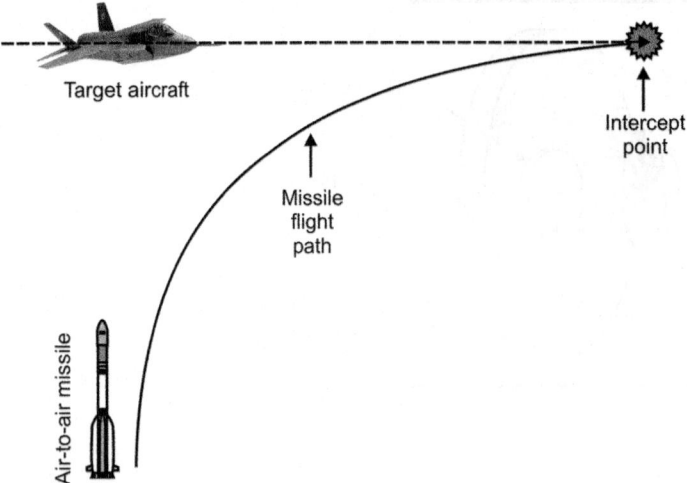

Figure 11.50 Pure pursuit navigation.

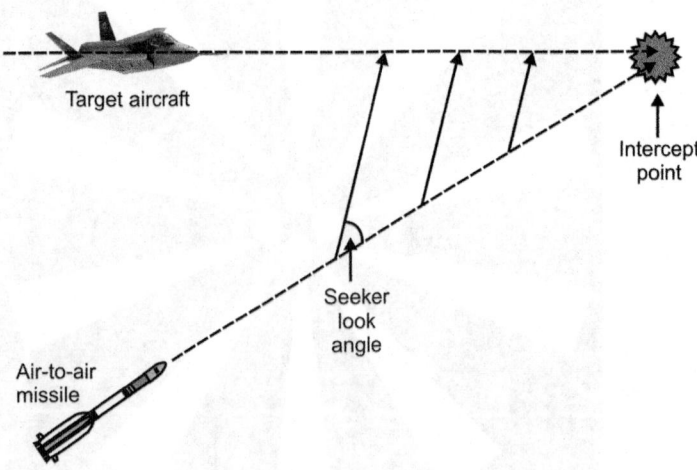

Figure 11.51 Proportional navigation.

enough energy to complete the turn in the close-n range allowing the target to escape. *Proportional navigation* or *proportional pursuit*, which is invariably followed in air-to-air missiles, allows the missile to have the shortest flight path towards the intercept and therefore eliminates the need for a high-g manoeuvre towards the end. The proportional navigation flight path is established by a constant look angle as shown in Figure 11.51 for a given constant missile velocity and assuming that target doesn't manoeuvre. In practice, the target does manoeuver and the missile velocity is also not constant through engagement duration. Therefore, look angle is updated as and when required.

A variant of the reticle seeker is the *pseudo imaging seeker*. Pseudo imaging seeker uses one or more detectors enabling both spatial and temporal information from reticle seekers. A small IFOV ($\cong 2$ mrad) is scanned in a preset pattern and spatial information is used to determine the time instant of appearance of the target within the field-of-view. Detectors are therefore

Figure 11.52 Imaging focal plane array.

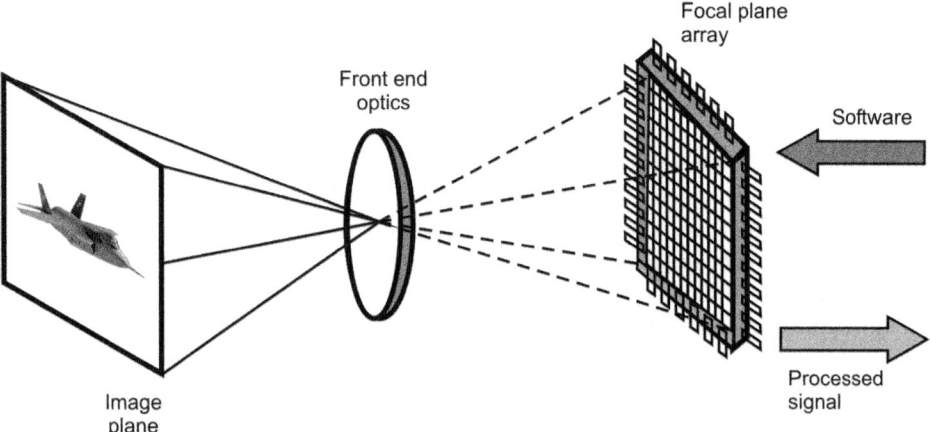

Figure 11.53 Imaging infrared seeker concept.

activated only within the time gate around the predicted time. This allows the missile to avoid large clutter and false targets outside the time gate. It also makes seeker highly immune to IR countermeasures.

Imaging seekers use an array of detectors called an imaging focal plane array (Figure 11.52) instead of a reticle to build an image of the scene in front (Figure 11.53). The image may be created by scanning the scene and using a linear array or a two-dimensional staring array. Imaging seekers are very expensive require huge processing power and complex tracking algorithms. These are therefore employed only under demanding operational requirements. Reticle seekers, on the other hand, are less expensive, easy to manufacture and operate and have a proven reliability and accuracy.

11.8 Major Infrared-Guided Weapon Systems

Some of the better-known IR-guided weapon systems are briefly described in the following paragraphs. The weapon systems are grouped into three categories, namely *Anti-Tank-Guided Missiles* (ATGM), *Surface-to-Air Missiles* (SAM) and *Air-to-Air Missiles* (AAM).

11.8.1 Anti-Tank-Guided Missiles

Anti-tank-guided missiles (ATGMs), also known by other names such as anti-armour missiles or anti-tank missiles, are designed to precisely hit and destroy heavily armoured military vehicle including all combat and transportation vehicles. Different variants of ATGM systems include shoulder fired weapons, large tripod-mounted weapons and vehicle-mounted weapons and systems launched from airborne launchers. United States' *Javelin*, German *PARS-3 LR*, Israeli *Spike* and Indian *Nag* are some examples.

The US *FGM-148 Javelin* (Figure 11.54) is a man-portable third-generation fire-and-forget ATGM jointly developed by Raytheon and Lockheed–Martin. It has an effective firing range of 75–2500 m with 4750 m being the maximum of the range. It uses a tandem shaped charge warhead that can penetrate reactive armour. The missile can be used in both top attack mode to usually hit the thin top armour of the target vehicle and direct mode to hit buildings and airborne targets. It uses an imaging IR seeker and onboard tracker to make it a fire-and-forget missile. It was introduced into service in 1996 and is still in service to day. It was successfully used in Operation Enduring Freedom (War in Afghanistan) and Operation Iraqi Freedom.

The *PARS 3 LR* (Figure 11.55) is an autonomous fire-and-forget missile intended for long-range applications and designed to destroy ground (tanks, armoured vehicles etc.), air (helicopters) and other individual targets. Manufactured by Parsys GmbH, MBDA Deutschland GmbH and Diehl BGT Defence, PARS 3 LR is also known as *TRIGAT-LR* and *AC-3G*. The missile can be launched from a ground vehicle or a helicopter and can be fired in salvos of up to four missiles in 8 s. The missile has a specified operational range of 500–5000 m, which is extendible up to 7000 m. It can be used in both top (terminal dive) mode as well as direct mode. Figure 11.56(a) shows the flight trajectory for top attack mode employed for an anti-armour role and launched from either land vehicle or helicopter. It also shows the flight trajectory for direct attack mode in Figure 11.56(b) employed for an anti-helicopter role and launched from either land vehicle or helicopter. The missile uses a passive IR seeker that locks on to target before the missile is fired. It uses a tandem shaped charge for maximum lethality against modern

Figure 11.54 FGM-148 Javelin. (*Source:* Courtesy of the US Army.)

Figure 11.55 PARS-3 LR.

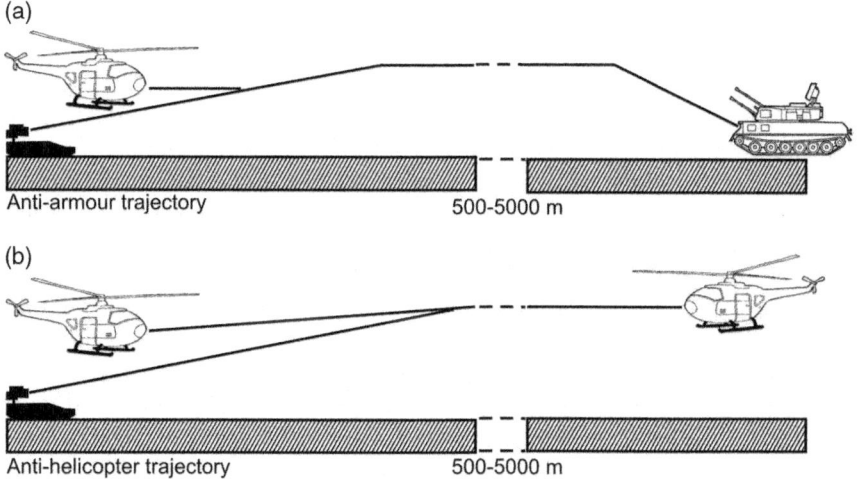

Figure 11.56 Flight trajectory in top attack (a) and direct attack (b) modes.

reactive armour. The German Army authorized series production of the missile system in 2013. The delivery began in 2016 and the delivery of the series production shall end in 2018.

Spike is a fourth-generation man-portable fire-and-forget anti-tank and anti-personnel missile designed and developed by Rafael Advanced Defence Systems. The missile can be launched in fire-and-forget mode destroying targets within the line-of-sight of the launcher and also in fire, observe and update guidance mode while following a top attack flight trajectory. In fire-and-forget mode, the tracker is locked on to the target. The missile is launched and it automatically propels itself towards the target. The missile uses a tandem charged HEAT (High-Explosive Anti-Tank) warhead that can penetrate explosive reactive armour. The guidance system of the Spike missile comprises a charge coupled device (CCD) and IIR seeker. The IIR sensor, in addition to providing higher sensitivity, offers improved thermal background rejection characteristics for all-weather day and night operation. Different variants of the Spike missile system include Spike-SR (Spike-Short-Range) with a maximum operational range of 800 m, Spike-MR (Spike-Medium-Range)

Figure 11.57 NAG ATGM. (*Source:* Courtesy of the US Army.)

with maximum operational range of 2500 m, Spike-LR (Spike-Long-Range) with maximum operational range of 4000 m, Spike-ER (Spike-Extended Range) with maximum operational range of 8000 m, Spike NLOS (Spike-Non-Line-of-Sight) with maximum operational range of 25 km and Mini-Spike with engagement range of 1300 m. Mini-Spike is an anti-personnel guided weapon. The Spike missile system is currently in service with Dutch, Chilean, Colombian, Finnish, German, Polish, Italian, Peruvian, and Spanish and Singaporean Armed Forces. The Spike missile system was successfully used during Lebanon war in 1982, Second Intifada beginning in the year 2000, the Afghanistan War from 2001 until today and the Iraq War in 2006.

Nag (Figure 11.57) is a third-generation fire-and-forget anti-tank-guided missile from India designed and developed by the Defence Research and Development Organization (DRDO) and manufactured by Bharat Dynamics Limited. It has two variants employing an active IIR seeker and miliimetric wave seeker. The operational range for the land version is 500–4000 m and 7–10 km for the air-launched version. It is likely to be inducted into service in 2015.

11.8.2 Surface-to-Air Missiles (SAM)

Short-to-medium range missiles used in the air-defence role against attack helicopters and aircraft are the most common surface-to-air missiles that use IR homing guidance. A large proportion of these missiles belong to the category of man-portable air-defence systems (MANPADS). Some of the more common and better known IR-guided surface-to-air missile systems are Stinger, Igla (Igla and Igla-1) and Strela (Strela-2 and Strela-3).

FIM-92 Stinger is a man-portable IR-guided surface-to-air missile designed by General Dynamics and manufactured by Raytheon Missile Systems in the USA, in Germany by EADS (European Aeronautic Defence and Space Company) and in Turkey by ROCKETSAN. It entered into service in 1981 and continues in service today. It is adaptable to be shoulder fired [Figure 11.58(a)] or from land vehicles [Figure 11.58(b)] as a surface-to-air missile and helicopters as an air-to-air missile. THe Stinger missile has evolved over the years and has undergone significant technological improvements. Three main variants of Stinger include FIM-92A,

(a)

(b)

Figure 11.58 Stinger Missile System. (a) Shoulder fired or (b) fired from land vehicles. (*Source:* Courtesy of the US Airforce.)

FIM-92B and FIM-92C. These are known by the names of Stinger Basic (FIM-92A), Stinger-Passive Optical Seeker Technique or Stinger-POST (FIM-92B) and the Stinger-Reprogrammable Microprocessor or Stinger-RMP (FIM-92C). Stinger is intended to fulfil a Short-Range Air-Defence (SHORAD) role until 2018. Stinger consists of a Stinger round encased in a launch tube and separate a grip stock assembly. Stinger Basic employs an IR seeker. Stinger-POST-uses a dual IR and UV seeker, thereby providing higher immunity to countermeasures compared to

Figure 11.59 9K 338 Igla-S (SA-24 Grinch) Missile System. (*Source:* Vitaly V. Kuzmin, https://commons.wikimedia.org/wiki/File:9K338_Igla-S_(NATO-Code_-_SA-24_Grinch).jpg. CC BY-SA 4.0.)

Stinger. The Stinger-RMP is so-called because of its ability to load a new set of software via ROM chip inserted in the grip at the depot. The missile has maximum effective firing range of 4.8 km and maximum speed of 2.2 Mach (750 m/s). Stinger made its debut in warfare in 1982 during the Falklands War between the UK and Argentina. Subsequently, it was used in the Soviet War in Afghanistan, the Angolan Civil War, the Libyan invasion of Chad, the Chechen War, Sri Lankan Civil War and Syrian Civil War.

Russian *Igla* is man-portable IR homing air-defence system (MANPADS) manufactured by KBM. The missile has maximum operational range of 5.2 km and a peak speed of 800 m/s. It has three variants, namely 9K310 Igla-1E (NATO designation SA-16 Gimlet), 9K38 Igla (NATO designation SA-18 Grouse) and 9K 338 Igla-S (NATO SA-24). The Igla missile system was inducted into service in 1981 and its different variants developed over the years continue to be in service to date. The 9K310 Igla-1 with its 9M313 missile uses a liquid nitrogen cooled indium antimonide IR seeker head. The 9K38 Igla with its 9M39 missile was inducted into service in 1983. It used liquid nitrogen cooled indium antimonide and an uncooled lead sulfide IR seeker head that has higher sensitivity and improved resistance to countermeasures and jamming. The Igla-S (SA-24 Grinch) shown in Figure 11.59 is the latest generation of portable air-defence missile systems designed to target visible aerial platforms such as helicopters, tactical aircraft, unmanned aerial vehicles and cruise missiles. It is an improvement over earlier versions SA-16 and SA-18. It employs a dual band IR seeker and has a maximum engagement range of 6 km compared to 5.2 km in the case of SA-16 and SA-18. It also uses a heavier warhead, which allows it to destroy the target even if it misses the target by 1.5 m.

The *Strela* family of missiles are man-portable surface-to-air missiles that use passive IR homing guidance and a high-explosive warhead. Different members of the family are 9K31 Strela-1, 9K32 Strela-2, 9K34 Strela-3 and 9K37 Strela-10. Strela-1 is commonly known by its NATO designation SA-9 Gaskin and is a short-range, low-altitude self-propelled SAM-carrying system based on the BRDM-2 chassis, an amphibious patrol car mounting two pairs of ready-to-fire 9M31 missiles. The missile has a maximum operational range of 4.2 km and speed of 1.8 Mach. It uses a lead sulfide IR seeker. The 9K32 Strela-2 (NATO designation: SA-7 Grail) is a man-portable, shoulder fired, low-altitude surface-to-air missile system with maximum firing range of 3700 m (Strela-2) and 4200 m (Strela-2M). It was inducted into service in 1968. SA-7 Grail is a tail-chase missile system and its efficacy depends on its ability to lock onto the heat source of low-flying fixed and rotary-wing aircraft. The simple IR seeker mechanism of the missile is easily prone to simple countermeasures and environmental effects. The 9K34 Strela-3 (NATO designation: SA-14 Gremlin) was developed to overcome the shortcomings of its predecessor Strela-2. The 9M36–1 missile from the SA-14 Gremlin used a new IR homing seeker that was less vulnerable to jamming and decoy flares as compared to SA-7 Grail. It has a maximum operational range of 4.5 km and average supersonic speed of 410 m/s. 9K35 Strela-10 (NATO designation: SA-13 Gopher) was designed to replace Strela-1. It was an improvement on Strela-1 and had an effective firing range of 5 km. The 9M37 missiles of Strela-10 used a higher quality IR seeker than was used in 9M31 missiles of Strela-1 system.

11.8.3 Air-to-Air Missiles

Air-to-air missiles are launched from an aircraft with the intention of destroying the aircraft of the adversary. These are broadly grouped as *short-range air-to-air missiles* (SRAAM) also sometimes known as *within visual range air-to-air missiles* (WVRAAM) or dogfight missiles, *medium-range air-to-air missiles* (MRAAM) and *long-range air-to-air missiles* (LRAAM). The second group of missiles is also known as *beyond visual range air-to-air missiles* (BVRAAM). While the missiles in the first group with engagement ranges up to 30 km or so are usually heat-seeking missiles, the missiles in the second group largely employ radar guidance. In the case of long-range missiles, the IR signatures of the target aircraft would be too weak for the detector to be able to track the target. The short-range IR-guided air-to-air missiles have seen five generations of development. These developments have mainly been in the IR seeker technologies and to some extent in digital signal processing.

The first generation of these missiles used an IR seeker that had a field-of-view of 30° and the attack aircraft needed to position itself behind the target aircraft during attack. The target in that case could easily move out of seeker's field-of-view with a simple manoeuvre. Second generation missiles used IR seekers with a field-of-view of 45°. The third-generation missiles were 'all aspect' missiles, which meant that the attack aircraft needed not to position itself behind target aircraft. The fourth-generation missiles used advanced seekers that had higher resistance to IR countermeasures and increased field-of-view of 120° giving them higher off-bore sight capability of 60°. The fifth-generation missiles used IIR seekers and more powerful digital signal processing, which gave them higher immunity to IR countermeasures like flares, greater sensitivity and ability to hit vulnerable points on the target. Some of the well-known contemporary air-to-air missiles include IRIS-T of Germany, Vympel R-73 of Russia, MBDA MICA-IR of France, British AIM-132 ASRAAM, AIM-9X Sidewinder and PYTHON-5 of Israel.

IRIS-T is a short-range air-to-air missile manufactured by Diehl BGT Defence. It employs an IR imaging seeker. It has a maximum speed of 3 Mach and operational range of approximately 25 km. It was developed to replace the AIM-9 Sidewinder missile. IRIS-T has a higher resistance to IR countermeasures such as flares. Extreme close-in agility of IRIS-T with capability to make 60 g turns at 60°/s allows the missile to engage targets even behind the launching aircraft. It was inducted into service in 2005.

Vympel R-73 (NATO designation: AA-11 Archer) manufactured by Tbilisi Aircraft Manufacturing is also a short-range air-to-air missile with maximum speed of 2.5 Mach and maximum operational range of 20 km (R-73E), 30 km (R-73M1) and 40 km (R-73M2). It employs a cryogenically cooled all aspect IR homing seeker with a high off-bore sight capability allowing the missile to see 40° off the missile's centre line. It was inducted into service in 1982. R-73 is also on the inventory of Indian Air Force.

MICA-IR (Figure 11.60) manufactured by MBDA is a short and medium-range air-to-air missile with a maximum operational range of 50 km and a maximum speed of Mach 3. It uses an IIR seeker that gives the missile high resistance to countermeasures such as chaff and decoy flares. It can lock-on after launch, which means that it can engage targets outside missile's acquisition range at the time of launch. It has been in service since 2000. The Indian Air Force has ordered Mica-IR missiles for its Mirage upgrade 2000H multirole fighters.

AIM-132 ASRAAM is a short-range air-to-air missile manufactured by MBDA. It uses an imaging IR seeker with lock-on after launch capability, has a maximum speed of Mach 3+ and maximum operational range of 50 km. It is in use in the Royal Air Force and Royal Australian Air Force, having replaced the AIM-9 Sidewinder. The Indian Air Force is also acquiring ASRAAM to replace the aging *Matra Magic* missiles. The missiles will be integrated on Jaguar strike aircraft.

AIM-9X (Figure 11.61) is the latest addition to the Sidewinder family of short-range air-to-air missiles developed by the Raytheon Company. It features an IR imaging seeker focal plane array seeker with off-bore sight capability of 90°. The imaging IR seeker gives it higher

Figure 11.60 MICA-IR air-to-air missile.

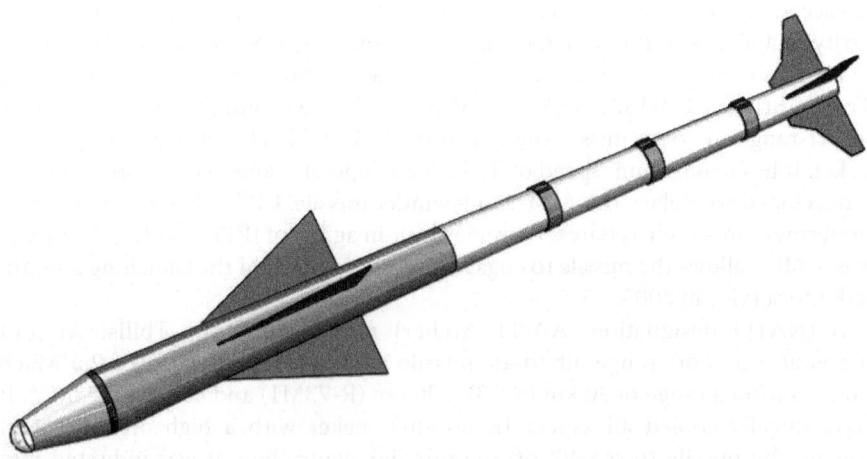

Figure 11.61 AIM-9X air-to-air missile.

resistance to IR countermeasures. The first Sidewinder missile was developed in the 1950s. AIM-9X is the fifth generation Sidewinder and is now in production. AIM-9X uses passive IR energy emitted by target aircraft for acquisition and tracking, which provides a launch-and-leave air combat missile capability. The AIM-9X Sidewinder is characterized by operational range of about 35 km and a speed of 2.5 Mach. AIM-9X Block I was the first in the family of these missiles. Currently, AIM-9X Block II has entered full-scale production. Block II missiles are the upgraded version of Block I missiles with 'lock-on after launch' being the main added feature. The development work has commenced on AIM-9X Block-III missiles. Block-III missiles will be designed to have 60% longer range and use insensitive munitions warhead for increased ground crew safety in addition to replacing old components with state-of-the-art ones. Block-III Sidewinder missiles are expected to achieve operational capability by 2022.

PYTHON-5 is one of the most advanced air-to-air missiles in the world. Different variants of PYTHON family include Shafrir-1, Shafrir-2, PYTHON-3, PYTHON-4 and PYTHON-5. PYTHON-5 is the latest addition to the family and is fifth generation air-to-air missile. Manufactured by Rafael Advanced Defence Systems in Israel, it has many advanced features such as an IIR seeker to give it high immunity to IR countermeasures; target lock-on before and after launch capability to engage targets beyond visual range; higher kill probability and a revolutionary full sphere launch envelope from very short to beyond visual ranges. It can lock-on to target after launch even when the target is 100° off the bore sight. The missile has an operational range of greater than 20 km and a speed of 4 Mach.

IR-guided missile technologies have continued to advance over the years. One of the focal points has been the improvements in seeker technology. Use of IIR seekers has provided much higher immunity to countermeasures such as chaff, flares and decoys. Modern IR-guided missiles using IIR seekers with advanced digital signal processing techniques have much wider detection angles giving them the capability to launch missiles from large off-bore sight angles, approaching 100° in some cases. Helmet-mounted sights with the pilots of the launch aircraft will allow them to distinguish between the target aircraft and a point source of intense heat such as a flare. These missiles almost invariably have a lock-on after launch feature enabling them to engage targets from a very small range to beyond the visual range. Another recent advancement in missile guidance is the use of electro-optical imaging. The electro-optical seeker scans the designated area for targets via optical imaging. Once a target is acquired, the missile locks on to it for the kill. Electro-optical seekers can be programmed to hit the designated spot on the target aircraft. The designated spot could be the most vulnerable point of the target. Since electro-optical imaging does not depend on the target aircraft's heat signature, it can be used against low-heat targets such as unmanned aerial vehicles and cruise missiles. There have been advances in control systems for better manoeuvrability of the flight path. Crew safety on the ground is another concern and has led to development of insensitive munitions warheads that don't detonate accidently.

11.9   Testing Infrared-Guided Weapons

IR-guided missiles developed in 1970s and 1980s used single-colour IR seekers employing the 3–5 μm band. The MAGIC series of IR-guided air-to-air missiles from France and R-73 series of IR-guided air-to-air missiles from Russia are some examples. State-of-the-art IR-guided missiles use two-colour seekers that employ both the 3–5 and 8–12 μm bands to offer improved false alarm rejection and immunity to deception by flares. PYTHON from Israel and RVVAE from Russia are examples of IR-guided missiles using two-colour seeker heads. Another emerging trend is the use of imaging IR seekers even in surface-to-air and air-to-air IR-guided missiles though their use is primarily in short-range anti-tank IR-guided missiles.

Also, both surface-to-air and air-to-air IR-guided missiles receive a target's IR signatures in the presence of background radiation from the sky and also IR signatures of flares, if any, deployed by target aircraft platform. The seeker head should be able to discriminate between IR signatures of the background and flares from those of the target.

When it comes to testing IR-guided missiles, be it a serviceability check or comprehensive characterization, the parameters that stand out include spectral matching of received IR signatures with those of the target as known to the seeker, response of the seeker head to target signatures in the presence of static IR background noise, immunity to deception by flares and field-of-view. Out of these four parameters, spectral matching can be singled out as the one to be used for serviceability checks while all four should be evaluated in the case of comprehensive characterization.

The device designed to carry out serviceability checks would need to have option of generating either 3–5 μm band with desired relative amplitudes of the two sub-bands within the 3–5 μm band or 3–5 and 8–12 μm bands simultaneously. In essence, the test system checks the lock-on sensitivity of the weapon and its ability to perform satisfactorily in the presence of static background noise. In some cases, it may only check the lock-on sensitivity.

To summarize, the spectral output modes of the test system for serviceability checks should include the following.

1) 3–5 μm band target signatures with static background signatures
2) 8–12 μm band target signatures with static background signatures
3) 3–5 and 8–12 μm bands target signatures with static background signatures

It may also be mentioned here that the relative amplitudes of 3–5 μm band and 8–12 μm band signals as seen by the seeker head would also depend upon the direction in which the missile approaches the target. This is an important consideration when it comes to testing IR-guided missiles employing two-colour seeker heads. Approaching the target from rear with seeker head facing the tail region of the target aircraft would lead to IR signatures received by seeker head being predominantly in the 3–5 μm band. On the other hand, if the missile approaches the target aircraft in a direction perpendicular to the mainframe, the received IR signatures would be predominantly in 8–12 μm band. A test system that offers flexibility of generating the two spectra with different relative amplitudes would always be an added advantage. However, this is a desirable feature when it comes to serviceability checks.

Comprehensive characterization is also very important for the following two reasons. (1) It provides required inputs during design and development phase of seeker units of IR-guided missiles. (2) Complete understanding of seeker unit of IR-guided missiles in terms of its response to target's IR signatures against different backgrounds and IR flares of different temporal and spectral profiles provides vital design inputs for development of efficient flares to be used as countermeasures against IR-guided missiles.

Comprehensive characterization of IR-guided missiles requires assessment of their performance in the presence of countermeasures such as IR flares. The missile should also be tested for different target speeds and the target being approached from different directions. Spectral output modes of the test system for comprehensive characterization include the following.

1) 3–5 μm band target signatures with static background signatures
2) 3–5 μm band target signatures with static background and dynamic flare signatures
3) 3–5 and 8–12 μm band target signatures with static background signatures
4) 3–5 and 8–12 μm band target signatures with static background and dynamic flare signatures.

Modes at S. Nos. 3 and 4 have any of the following three user selectable options.

a) Equal amplitudes of 3–5 μm band and 8–12 μm band target signatures
b) 3–5 μm band target signatures stronger than 8–12 μm band target signatures
c) 8–12 μm band target signatures stronger than 3–5 μm band target signatures.

11.9.1 International Test Systems

Target simulators designed to characterize IR-guided missiles are available from international manufacturers such as CI Systems, Israel, Geotest-Marvin Test Systems Inc., USA and SBIR, USA. The test systems offered by these companies range from simple IR sources for checking lock-on sensitivity of IR-guided missiles employing single- and two-colour seekers to more elaborate systems that generate an IR scene comprising of static background, target and flare signatures.

One such system is the *Target Infrared Simulator (TIRS)* from CI Systems. It is a computer-controlled table top test system that simulates an IR scene, which includes static background, moving/growing target and moving/growing flare signatures. The system is designed for characterization of seeker units of single-colour IR-guided missiles. The system has all the necessary features to carry out extensive testing of seeker characteristics. The system has two major limitations; it can only be used to test single-colour IR-guided missiles and it is not portable enough to be used for performing serviceability checks.

Another similar system from CI Systems is the *IR Target Generator (IRTG)* (Figure 11.62) used for hardware in the loop testing of missile guidance subsystems of single- and two-colour IR-guided missiles for target acquisition and tracking capability, and immunity to countermeasures such as flares. The IR scene generated by the system includes a static background, a dynamic target and a dynamic flare. The system is mounted on a flight motion simulator during the characterization process. Again, the system, though capable of generating all the required parameters as seen by the seeker in a real deployment scenario, is not portable enough to be used for serviceability checks of IR-guided missiles in the strap-on condition.

Yet another test system designed for carrying out serviceability checks of IR-guided missiles is the MTS-916 Target Simulator (Figure 11.63) from Geotest-Marvin Test Systems Inc. This is

Figure 11.62 IR target generator from CI Systems, Israel.

Figure 11.63 Target simulator (MTS-916) from Geotest-Marvin Test Systems Inc.

a modular target simulator that tests TV/CCD, IR and laser seekers used in AGM-65 and TGM-65 Maverick missiles. The test system is offered in multiple configurations to support missiles with only TV/CCD seekers (MTS-916–1), only IR seekers (MTS-916–2), both TV/CCD and IR seekers (MTS-916–3) and only laser seekers (MTS-916–4).

The system is primarily intended for performing readiness checks and does not simulate target signatures in the presence of a static IR background, nor does it test the immunity of the missile to dynamic flares. Also, the system is not portable enough or configured the way it should have been to be conducive to performing serviceability checks with the missile in the strap-on condition.

11.10   Radar-Guided Weapons

Radar guidance, like IR homing guidance, is predominantly used in surface-to-air and air-to-air-guided weapons. While IR homing guidance is largely employed in short and medium operational ranges, radar guidance is particularly attractive for long-range air-to-air-guided missiles. State-of-the-art radar-guided air-to-air missiles using active radar homing guidance technology have operational ranges in excess of 150 km. This section focuses on operational aspects of different types of radar-guided weapons. Features and capabilities of major international radar-guided weapon systems are also discussed.

11.10.1   Introduction

Radar-guided weapons are surface-to-air and air-to-air anti-aircraft-guided missiles in which a radar is used to designate the intended target with electromagnetic energy and the missile makes use of the radar energy reflected from the target to steer itself to intercept and destroy the target. The basic concept is same as the one used in the case of laser-guided munitions except for the fact that laser-guided munitions are largely surface-to-surface and air-to-surface weapons and radar-guided weapons are mainly air-to-air and, to a limited extent, surface-to-air missiles. Another difference is in their operational ranges, which are much larger in the case of radar-guided missiles approaching 150–200 km in the case of state-of-the-art air-to-air missiles than they are in the case of laser-guided munitions that seldom exceed 20 km. *AIM-54 Phoenix* and *AIM-120D AMRAAM* have maximum operational ranges of 190 and 180 km, respectively. Also, the laser is either on a land-based platform or aircraft-mounted in semi-active homing guidance. It is an integral part of the weapon hardware in the case of active laser homing. In the case of radar-guided weapons, radar is also mounted either on a land vehicle or an aircraft in semi-active radar homing guidance. Radar is an integral part of the guided missile hardware in the case of active radar homing guidance.

A variation of semi-active radar guidance is TVM guidance. Like semi-active homing missiles, the ground-based radar designates the target with radar energy, which is then reflected off the target and detected by the radar receiver on board the missile. However, unlike a semi-active radar homing missile, the missile doesn't have the hardware and software to compute track and intercept information. Instead, data from the reflected radar energy is relayed back to the ground station via a data link. The ground station computes the track information and communicates it back to the missile. Radar beam riding guidance is another technique used in earlier days for short-range surface-to-surface and surface-to-air missiles. But because of inherent shortcomings of beam riding concept, both laser and radar beam riding weapons are more or less an extinct entity.

11.10.2 Semi-Active Radar Guidance

In a *semi-active radar guidance* system, the target aircraft is illuminated by electromagnetic energy emitted by the fire-control radar located either on the launch aircraft or on an appropriate ground location. The fire-control radar acquires the target and tracks it. A small radar transmitter generating a very narrow beam then selectively illuminates the target using tracking information generated by the fire-control radar. A radar receiver on board the missile receives the radar energy reflected off the target and locks on to the target. Once the missile is locked on to target, it may be launched. The radar seeker unit provides the angular error information to the flight path control system, which steers the missile to intercept the target through tail fins. Semi-active radar homing guidance uses a continuous-wave Doppler radar in the bistatic configuration. It is bistatic because transmit and receive antennas are at different locations with the former being collocated with fire-control radar either on launch aircraft or on ground and the latter on the nose cone of the missile. Doppler shifted receive signal frequency is computed from the closing velocity determined using flight path geometry. Information on the receive signal frequency is used by the missile to acquire the target. Figure 11.64 shows the semi-active radar homing basic concept when both fire-control radar and missile are on board the launch aircraft.

Semi-active radar-guided weapons have the advantages of being low cost and less complex. Due to the fact that the fire-control radar needs to remain committed throughout the duration of the engagement until the target intercept has been achieved, semi-active radar guidance has certain disadvantages; more so when the radar is located on the launch aircraft. It renders the launch aircraft vulnerable to counter attack from the adversary using anti-radiation missiles as the radar itself acts as a beacon for the counter attacking missile. Despite this shortcoming, semi-active radar guidance is likely to stay because of better radar resolution and higher power possible with ground-based or aircraft-mounted radars. It may be mentioned here that radar resolution becomes better with increase in transmit antenna size and radar's size and weight have a direct bearing on power output capability of the radar. A large size transmit antenna and large size radar cannot be accommodated within the missile as would be the requirement in active radar guidance. Raytheon's *AIM-7 Sparrow* with an operational range of 30–50 km for different variants uses semi-active radar homing guidance.

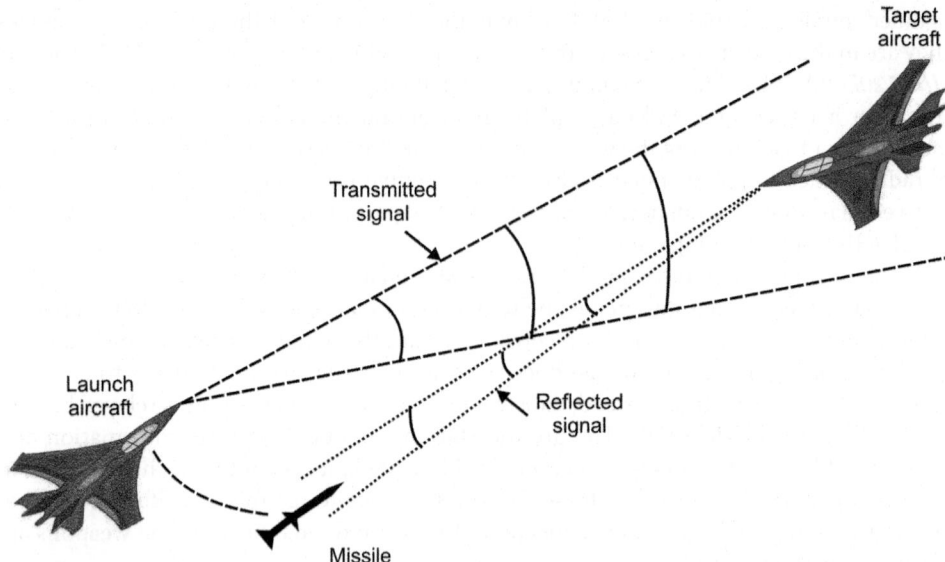

Figure 11.64 Semi-active radar homing guidance concept.

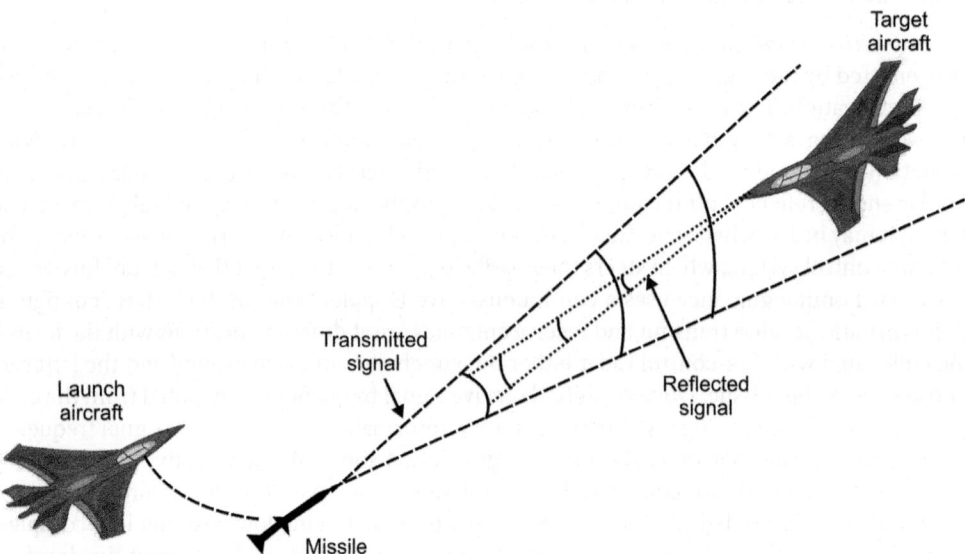

Figure 11.65 Active radar guidance.

11.10.3 Active Radar Guidance

In the case of *active radar guidance*, the missile tracks its target by means of emissions that it generates itself with the radar transceiver located on board the missile. The missile has the required electronics hardware and software to find and track the target autonomously and doesn't depend on the services of an external radar transmitter as is the case in semi-active radar guidance. Active radar guidance is commonly used for terminal homing in anti-ship, surface-to-air and air-to-air missiles. Figure 11.65 shows the basic concept of active radar homing guidance

One of the advantages of active radar homing guidance is higher tracking accuracy and better kill probability compared to semi-active radar guidance. This is due to the missile being much closer to the intended target in the terminal phase than the launch platform would be. This also gives the missile higher resistance to electronic countermeasures. Another major advantage is much reduced vulnerability of launch platform to counter attack. Since the missile is autonomous in the terminal phase, the launch platform, ground-based or aircraft, does not need to keep its radar enabled during this phase. In fact, the aircraft in the case of being the launch platform can exit the scene or undertake other tasks while the missile homes on to the target providing fire-and-forget capability.

Looking at the disadvantages, active radar-guided missiles are likely to have increased size and weight compared to missiles employing semi-active radar guidance as, in the case of former, the missile has to accommodate radar transceiver and electronics for autonomous target tracking. Also, these missiles are more expensive as the sophisticated electronics can only be used once. Another shortcoming is that active radar-guided missiles due to radar emission are more than likely to be detected by the sophisticated radar warning sensors equipping modern fighter aircraft enabling them take an evasive action or deploy countermeasures in plenty of time. Given the fact that even modern aircraft cannot match the manoeuvrability and agility of the state-of-the-art missiles, there is not much an aircraft can do to avoid interception.

Due to restriction on size of the radar transmitter that is linked to the power available at radar output and size of transmitting antenna that is linked to beam width, active radar guidance alone cannot provide guidance to long-range missiles. A practical solution can be to use semi-active radar guidance or inertial guidance for midcourse correction and active guidance for terminal homing. Raytheon's *RIM-174 ERAM*, also called Standard Missile-6 (SM-6), is state-of-the-art surface-to-missile in this category with an operational range of 240 km and using a combination of inertial guidance, semi-active radar guidance and active radar homing. The *AIM-54 Phoenix* air-to-air missile manufactured by the Hughes Aircraft Company and Raytheon Corporation with an operational range of 190 km also uses a combination of semi-active and active radar homing. It was in service from 1974 to 2004. The *MBDA EXOCET* anti-ship sea skimming missile with an operational range of 70–180 km also uses inertial and active radar homing guidance.

11.10.4 Track-via-Missile Radar Guidance

TVM, also called *retransmission guidance*, is a combination of semi-active radar homing and radio command guidance. In this case, the target is illuminated by external radar and the reflected radar energy is intercepted by a receiver on board the missile as is done in semi-active radar homing. However, the missile has no onboard computer to process these signals to generate track information. The signals are instead transmitted back to the radar hosting launch platform over a down-data link for processing. The missile is then commanded over the up-data link to adjust its flight trajectory to intercept the target. Figure 11.66 illustrates the concept of TVM guidance. The MIM-104 Patriot missile system is an example of TVM guidance.

Unlike active radar guidance that alerts the target aircraft about impending danger, in the case of TVM guidance, it may know that it is being illuminated but is never sure whether it is being engaged. Modern phased array radars, by virtue of their thin beams and low sidelobe levels, make detection by the aircraft even more difficult. TVM missiles can be made more accurate by using more sophisticated algorithms for calculating interception than would be possible in the limited processor that could be accommodated in a missile. In addition, operators have the option of adjusting the missile's flight path throughout the engagement. A radar hosting ground station may use radar energy reflected from the target directly and combine this

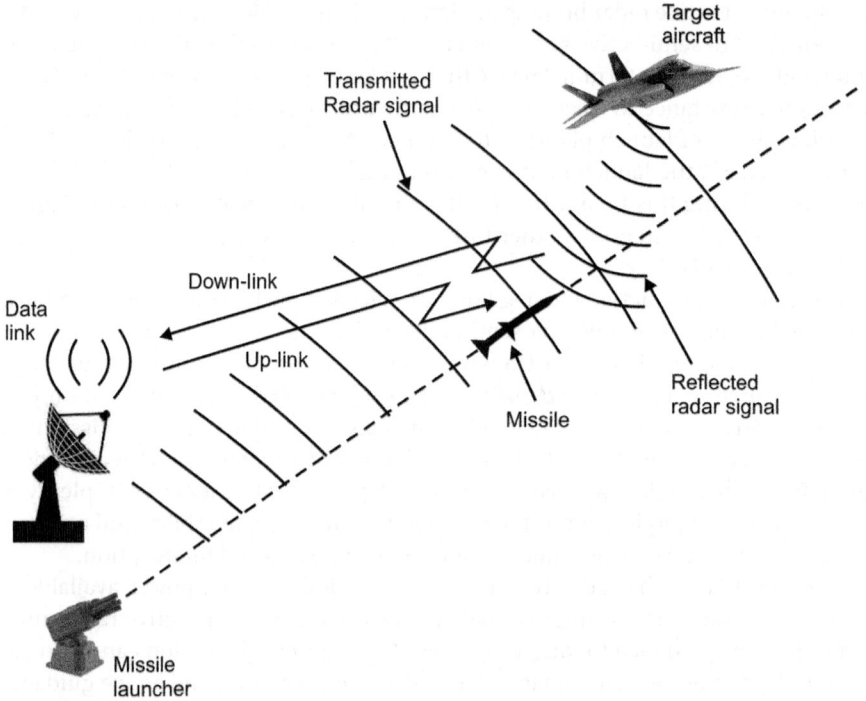

Figure 11.66 Track-via-missile guidance concept.

information with the downloaded information from the missile to generate the interception course. This improves resistance to electronic countermeasures.

One of the disadvantages is the vulnerability of the ground radar station to anti-radiation missiles as the radar has to remain active throughout the engagement. The possibility of data link jamming is another shortcoming. Another potential disadvantage is that the missile will not be able to continue engagement if the target aircraft could manage to put an obstacle such as a hillock between the radar beam and itself, or if it could manage to go out of the radar's tracking envelope.

11.10.5 Missile Guidance and Control

A missile's flight path is generally divided into three separate phases; namely the *launching or initial phase, midcourse phase* and *terminal phase*. During the launch phase, flight controls are locked in neutral position as the missile does not have the aerodynamic stability during this phase. The guidance system takes charge immediately after the initial phase is completed and controls are unlocked. The initial phase lasts for a very short duration.

The main task of the midcourse guidance is to place the missile near the target enabling the terminal phase guidance system to successfully take over. This phase is the longest in both distance travelled by the missile as well as time duration. During the midcourse phase, control may need to be exercised to ensure that missile follows the desired course and also stays on that course. In some cases, midcourse guidance may also additionally perform the task of terminal phase guidance.

The terminal phase is the most crucial phase of the missile's flight path as it is this phase that leads to a target-hit or target-miss. The terminal phase guidance system needs to have high accuracy and fast response to guidance signals. Towards the end of missile's journey to

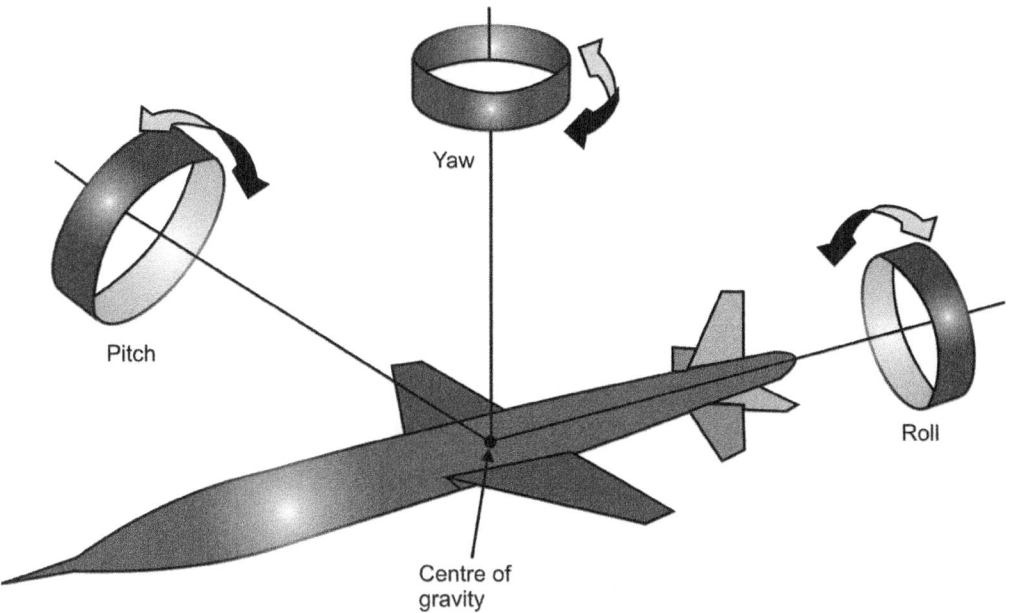

Figure 11.67 Missile stabilization.

intercept the target, when the missile is very close to it, it may not be left with sufficient energy to execute any high-g manoeuvre. Proportional guidance helps address this issue.

In the case of radar guidance, semi-active, active or track-via-missile guidance, the guidance signal is generated based on the radar energy reflected off the target. Every missile guidance system consists of an attitude control system and a flight path control system. The attitude control system controls the attitude of the missile in pitch, yaw and roll parameters on the designated flight path (Figure 11.67). The flight path control system guides the missile to its designated target. This is achieved by determining the flight path errors, generating the necessary commands to correct these errors and sending these commands to the missile's control subsystem that works on the servo principle. The control units make corrective adjustments to the missile control surfaces on receiving an error signal. The control units also adjust the wings or fins to stabilize the missile in roll, pitch and yaw. Guidance and stabilization are two separate processes that occur concurrently.

11.11   Major Radar-Guided Weapon Systems

In this section, we shall discuss some prominent surface-to-air, air-to-air and anti-ship radar-guided missile systems. Missile systems described here include the AIM-7 Sparrow, MIM-104 Patriot missile system (both semi-active radar-guided missile systems), AIM-54 Phoenix, RIM-174 ERAM, AIM 120D AMRAAM, MBDA EXOCET (all active radar-guided missile systems) and AKASH surface-to-air missile.

11.11.1   Surface-to-Air Radar-Guided Missiles

Some representative examples of surface-to-air radar-guided missile systems include the RIM-174 ERAM, AKASH and MIM-104 Patriot missile systems.

Figure 11.68 RIM-174 Standard ERAM (SM-6) in flight.

The *RIM-174 Standard Extended-Range Active Missile (ERAM)*, also called Standard Missile 6 (SM-6), is a long-range anti-air warfare missile designed for use by navies against a range of aerial targets including fixed and rotary-wing aircraft, unmanned aerial vehicles, low attack cruise missiles and anti-ship cruise missiles in flight to provide area and ship defence (Figure 11.68). The RIM-174 Standard ERAM meets the need for a vertically launched, extended-range missile compatible with the AEGIS Weapon System to be used against extended-range threats. The missile is also capable of terminal ballistic defence as a supplement to RIM-161 (SM-3). The missile design is adapted from RIM-156A (also called SM-2ER), an earlier missile of the Standard Missile family. It is a two-stage missile with a booster stage and a second stage. A new addition to the hardware is the active radar homing seeker derived from the seeker used in the AIM-120C AMRAAM missile. An active radar seeker gives it the capability to engage highly agile targets and also the targets that are beyond the effective range of the launch vessels' target illumination radars. The missile has maximum operational range of 240 km with a maximum cruise speed of greater than 3.5 Mach. The system is a combination of inertial guidance, semi-active and active radar homing guidance. The missile may be employed in a number of modes; inertial guided to target with terminal acquisition using an active radar seeker, semi-active radar homing all the way or an-over-the-horizon shot with Cooperative Engagement Capability. The missile was inducted into service in 2013 and is in use by the US Navy, Royal Australian Navy and Republic of Korea Navy.

The *MIM-104 Patriot* manufactured by Raytheon is a long-range, all-altitude, all-weather air-defence system used to counter tactical ballistic missiles, cruise missiles and advanced aircraft. The missile system derives its name from the radar component of the missile system. PATRIOT stands for *Phased Array Tracking Radar to Intercept On Target*. It is now given the name Anti-Ballistic Missile (ABM) system, which is its primary mission. The Patriot system has four major operational functions: communications, command and control, radar surveillance and missile guidance. The four functions combine to provide a coordinated, secure, integrated, mobile air-defence system. The Patriot missile is equipped with a TVM guidance system. The target is acquired in the terminal phase of flight by the target acquisition system in the missile. It transmits the data using the TVM down-link via the ground radar to the engagement control station for computation of final course correction. The course correction commands are transmitted to the missile via the missile track command up-link. The missile has a range of 70 km and maximum altitude is greater than 24 km. The minimum flight time,

Figure 11.69 MIM-104 Patriot missile.

which is the time to arm the missile, is less than 9 s and the maximum flight time is less than 3.5 min. It is a supersonic missile with a maximum speed of Mach 5. Different variants of Patriot missile system include MIM-104A, MIM-104B (PAC-1), MIM-104C (PAC-2), MIM-104D (PAC-2/GEM), MIM-104F (PAC-3) and Patriot Advanced Affordable Capability-4 (PAAC-4).

The missile system was inducted into service in 1981 and is in service today. Other than the USA, a large number of other allied nations, which includes Egypt, Germany, Greece, Israel, Japan, Kuwait, the Netherlands, Saudi Arabia and Taiwan, have the Patriot missile system in their arsenal. Patriot missile systems were first used during the Gulf War of 1991. There have been controversies about the success rate of the missile system in engaging ballistic missiles during the Gulf War. The system was again deployed in Iraq in 2003 during Operation Iraqi Freedom. The systems were stationed in Kuwait and were reportedly used with success against hostile ballistic missiles. Recently, during the Israel-Gaza conflict of 2014 called Operation Protective Edge, the Patriot missile system was successfully used to bring down two unmanned aerial vehicle drones of Hamas. Figure 11.69 shows a Patriot missile.

AKASH (Figure 11.70), developed by India's state-owned Defence Research and Development Organization (DRDO) and manufactured by Ordnance Factories Board, Bharat Dynamics and Bharat Electronics, is an all-weather, medium-range surface-to-air missile system with target intercept range of 30–35 km at altitudes up to 18 km. The missile can be launched from static as well as mobile platforms including wheeled trucks and battle tanks. It is capable of handling multiple targets and destroying manoeuvring targets such as unmanned aerial vehicles, cruise missiles and fighter aircraft. It can carry conventional and nuclear warheads weighing up to 60 kg. The complete system comprises of a launcher, a missile, a control centre, an integral mission guidance system, a multifunctional fire-control radar, a system arming and explosion mechanism, a digital autopilot, C^4I (command, control communication and intelligence) centres and supporting ground equipment. Each AKASH launcher has three missiles. The missile was inducted into service in 2009 and is in use by the Indian Air Force and Indian Army.

Figure 11.70 Akash medium range surface-to-air missile. (*Source:* Frontier India Defense and Strategic News Service, https://commons.wikimedia.org/wiki/File:Akash_SAM.jpg. CC BY-SA 2.5 IN.)

11.11.2 Air-to-Air Radar-Guided Missiles

Representative air-to-air radar-guided missiles discussed in this section include AIM-7 Sparrow, AIM-54 Phoenix and Aim-120 AMRAAM. *AIM-7 Sparrow* is a medium-range, supersonic speed semi-active radar-guided air-to-air missile manufactured by Raytheon. A derivative of this missile is the ship-based version called RIM-7 Sea Sparrow used for an air-defence role. The missile had been in active service from the 1950s to the 1990s. It is being gradually phased out in favour of AIM-120 AMRAAM, all-weather, day/night operation beyond visual range air-to-air missile. AIM-7 Sparrow has a number of variants designated as AIM-7A, AIM-7B, AIM-7C, AIM-7D, AIM-7E, AIM-7F, AIM-7G, AIM-7H, AIM-7M, AIM-7N, AIM-7P and AIM-7R. Development of AIM-7B, AIM-7N and AIM-7R was discontinued after initial interest. Table 11.2 shows a comparison of different variants of AIM-7 Sparrow. Many AIM-7 variants had ship-based version. These included RIM-7E, RIM-7F, RIM-7H, RIM-7M, RIM-7P and RIM-7R. AIM/RIM-7R programmed was abandoned in 1996 due to high costs.

The first operational use of the AIM-7 Sparrow was during the Vietnam War in June 1965 when the US Navy shot down two North Vietnamese MiG-17s. Due to absence of an IFF system on launch aircraft, long-range capability of these missiles could not be used. The first missiles had a kill probability of only 10%. This problem was subsequently overcome in AIM-7E. The last use of AIM-7 Sparrow was in Operation Desert Storm in 1991 where it was extensively used on F-15 and F-16 fighter aircraft with lot of success. The missile has been on the inventory of Armed Forces of a large number of countries including Australia, Canada, Israel, Italy, Japan, South Korea, Spain, Taiwan, Turkey, the UK and the USA. Figure 11.71 shows image of an AIM-7 Sparrow missile.

The *AIM-54 Phoenix*, manufactured by the Hughes Aircraft Company and Raytheon Corporation and capable of attacking more than one aircraft with multiple launches, is a supersonic, radar-guided, long-range air-to-air missile. The missile is carried in clusters of up to six

Table 11.2 Comparison of AIM-7 sparrow variants.

Sparrow	Length (mm)	Wingspan (mm)	Launch Weight (kg)	Speed (mach)	Range (km)	Warhead (kg)
AIM-7A	3740	940	65	2.5	10	20
AIM-7C	3660	1020	78	4	11	30
AIM-7D	3660	1020	78	4	11	30
AIM-7E	3660	1020	89.5	4	30	30
AIM-7F	3660	1020	105	4	70	39
AIM-7H	3660	1020	89.5	4	30	30
AIM-7M	3660	1020	105	4	30	40
AIM-7P	3660	1020	105	4	30	40

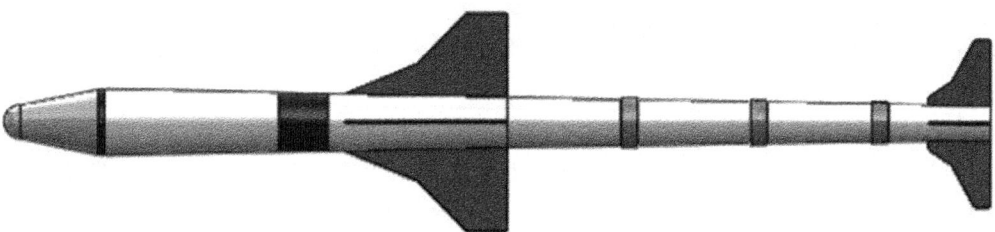

Figure 11.71 AIM-7 Sparrow missile.

missiles and while in service from 1974 to 2004, was used by the United States Navy and the Air Force of Islamic Republic of Iran. F-14 Tomcat was the only launch platform capable of carrying AIM-54 Phoenix. The missile system retired from service in 2004 in favour of AIM-120 AMRAAM. This Mach 5 missile has maximum operational range of 190 km and employs a combination of semi-active (midcourse correction) and active radar homing guidance (terminal guidance).

AIM-120 AMRAAM (Advanced Medium-Range Air-to-Air Missile) is a high-supersonic, day/night/ all-weather beyond visual range (BVR), fire-and-forget air-to-air missile. Manufactured by Hughes from 1991 to 1997 and Raytheon from 1997 until today, It has different variants, which include AIM-120A, AIM-120B, AIM-120C and AIM-120D with the AIM-120D with an operational range in excess of 180 km. The operational ranges for other variants are 55–75 km (AIM-120 A/B) and greater than 105 km (AIM-120C). It employs an inertial navigation system (INS) for midcourse guidance and active radar homing for guidance in terminal phase. Once the missile closes to self-homing distance, the active radar guides it towards the target. This feature provides fire-and-forget capability to the missile and allows the pilot to fire number of missiles simultaneously at multiple targets. Its capabilities include look-down, shoot-down, multiple launches against multiple targets and intercepts at very short-range in dogfight situations. It was inducted into service in 1991 and is in service today with the US Air Force and Navy and more than 25 US allies.

11.11.3 Anti-Ship Radar-Guided Missiles

The *EXOCET family of missiles* currently manufactured by MBDA, France are anti-ship, sea skimming missiles. Different variants of the EXOCET family have operational ranges of 70–180 km and can be launched from surface vessels, submarines, helicopters and fixed-wing aircraft.

Figure 11.72 MBDA EXOCET MM-40 Block-3 missile.

The EXOCET MM 40 series is the latest in the family of EXOCET missiles with EXOCET MM-40 Block-3 being the most recent missile in the series (Figure 11.72). Salient features of the EXOCET MM-40 Block-3 missile include an effective operational range of 180 km, anti-ship as well as littoral operations and land attack capability, automatic computation of engagement plans to support firing decisions and a sophisticated navigation package comprising a hybrid INS/GPS, a radar altimeter and an advanced J band active radar seeker for terminal guidance. The EXOCET missile was inducted in service in 1973 beginning with the air-launched version. It was inducted in the US Navy in 1979. The USA's *Harpoon*, Swedish *RBS-15* and Chinese *Yingji* are its main competitors. It is in use by a large number of countries including India, France, Germany, Greece, Iran, South Africa, South Korea, Pakistan, Indonesia, Turkey and Malaysia. The Indian Navy uses it on the Scorpene class of submarines.

11.11.4 Future Trends

The radar-guided missiles of the future will have enhanced capabilities particularly in terms of their guidance system so as to improve kill probability and make them practically a single shot kill weapon. One such futuristic missile that will incorporate these features is the K-77M from Russia that aims to frustrate any missile evasion manoeuvres enabling a target to escape. The missile would get this feature with a major innovation of the K-77M guidance system. The active radar guidance system in the nose of the missile will have its own Active Phased Array Antenna (APAA) due to which the missile will have zero reaction time to unexpected manoeuvres of target aircraft. The missile is expected to be delivered by 2017.

11.12 GPS/INS-Guided Weapons

As outlined earlier in Section 11.2.7, GPS/INS guidance is used in guided weapons for a variety of roles including guiding weapons to hit static or relocateable targets, enabling weapons adjust their free fall to hit a predefined point programmed into the weapon prior to launch and

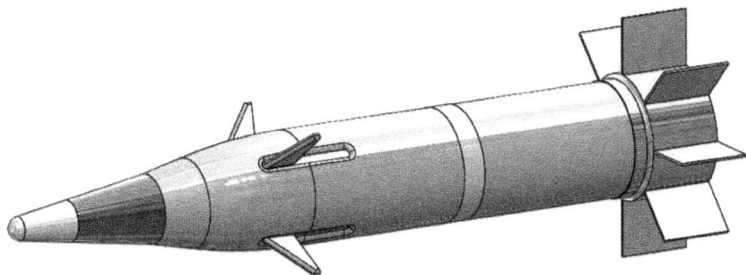

Figure 11.73 XM-982 Excalibur artillery projectile.

providing midcourse guidance to cruise missiles. Enhanced Paveway-III Dual-Mode Laser-Guided Bomb (DMLGB), XM-982 Excalibur extended-range GPS-guided artillery shell and AGM-88 HARM air-to-surface tactical missile are some examples of use of GPS/INS guidance in precision-guided weapons.

Raytheon's *Enhanced Paveway-III Dual-Mode Laser-Guided Bomb* (DMLGB) combines the precision of laser technology together with the resistance to bad weather feature of GPS and INS to give it all-weather capability. The INS transforms raw data into average velocity and distance travelled and does so without any external means. GPS, on the other hand, uses satellite signals for measuring a platform's position and velocity references. GPS and INS are complementary technologies. While INS technology is more resistant to jamming compared to GPS because it does not require external signals for operation, its accuracy degrades with time due to cumulative errors of the inertial sensors.

The *XM-982 Excalibur* 155 mm Extended-Range Artillery Projectile (Figure 11.73) is a family of precision-guided, extended-range modular projectiles that employs GPS-aided inertial guidance and navigation, free spinning base fins, four-axis canard airframe control, base bleed technology and a trajectory glide to achieve increased accuracy and extended ranges beyond 30 km. The XM982 Extended-Range Projectile was developed jointly by Raytheon TI Systems' (RTIS) who were responsible for development of guidance and navigation system, Primex who were responsible for projectile design and manufacturing and KDI who were responsible for the fusing.

The XM-982 smart munitions is configured as three blocks; namely Block I, Block II and Block-III weapons designed to attack three different categories of payload. Block I consists of a high-explosive, unitary penetrating warhead to enhance traditional fire support operations against stationary targets with increased range, improved accuracy and reduced collateral damage against personnel, light materiel and structure targets. Block II munitions are designed to search, detect, acquire and engage moving and time-sensitive or short-dwell targets encountered in open-terrain battlefields. Block-III consists of discriminating munitions designed to selectively identify on the basis of specific target characteristics and engage individual vehicular targets. Salient features of XM-982 include a modular concept that allows multiple warhead payload capability, extended range in excess of 37 km when fired from 39-calibre howitzers and 47 km when fired from the 52-calibre ordnance fitted to the XM2001 Crusader, fire-and-Forget GPS/INS guidance, increased survivability made possible by greater standoff from threats and faster defeat of potential threats and low cost per kill.

The AGM-88 HARM (High-speed Anti-Radiation Missile), shown earlier in Figure 11.1, is a supersonic air-to-surface anti-radiation missile designed by Texas Instruments and manufactured originally by them and currently by the Raytheon Corporation. The primary objective of missile is the destruction of surface-to-air missile radar and radar directed air-defence artillery systems. The guidance system of AGM-88E, an advanced version of AGM-88, is an advanced multi-sensor system comprising a MilliMetre-Wave (MMW) terminal seeker, advanced Anti-Radiation Homing (ARH) receiver and GPS/INS. The guidance system enables the missile to quickly engage both conventional and advanced enemy air-defence targets as well as non-radar time-sensitive strike targets.

Illustrated Glossary

Active Homing Guidance In the case of active homing guidance, the source of target designation is also on board the weapon with the result that this methodology does not require an external source. These features put it in the category of fire-and-forget missiles as the launch platform does not need to continue to illuminate the target after the missile has been launched. Active homing guidance weapons are usually radar-guided.

Active Radar Guidance See *active homing guidance*.

Active Radar Homing See *active homing guidance*.

AGM-88 High-speed anti-radiation air-to-surface tactical missile with the primary objective of destroying surface-to-air missile radars and radar directed defence artillery systems.

AGM-114 Hellfire-II The AGM-114 Hellfire-II by Lockheed–Martin is a combat-proven tactical surface-to-surface and air-to-surface missile system that uses semi-active laser homing. AGM-114L uses millimetre-wave radar guidance.

AIM-120 AMRAAM The AIM-120 AMRAAM (Advanced Medium-Range Air-to-Air Missile) manufactured by Hughes is a high-supersonic, day/night/ all-weather beyond visual range (BVR), fire-and-forget air-to-air missile.

AIM-132 ASRAAM This is a short-range air-to-air missile manufactured by MBDA. It employs a cryogenically cooled all aspect IR homing seeker with high off-bore sight capability.

AIM-54 Phoenix The AIM-54 Phoenix, manufactured by the Hughes Aircraft Company and Raytheon Corporation and capable of attacking more than one aircraft with multiple launches, is a supersonic, radar-guided, long-range air-to-air missile.

AIM-7 Sparrow This is a medium-range, supersonic speed semi-active radar-guided air-to-air missile manufactured by Raytheon.

AIM-9X This is the latest addition to the Sidewinder family of short-range air-to-air missiles developed by the Raytheon Company. It features an IIR seeker.

AKASH AKASH, developed by India's state owned Defence Research and Development Organization (DRDO) and manufactured by Ordnance Factories Board, Bharat Dynamics and Bharat Electronics, is an all-weather, medium-range surface-to-air missile system.

Anti-Radiation Weapons (ARWs) ARWs are mainly designed to target ground-based radars. They do so by detecting radio emission from these radars and then homing onto the source of radio emission. ARWs can also be used to target jammers and radios used for communication.

Automatic Command to Line-of-Sight (ACLOS) Guidance System This is a type of command to line-of-sight guidance system in which all the three functions of target tracking, missile tracking and control are automatic.

Beam Riding Guidance See *beam riding weapons*.

Beam Riding Weapons These make use of radar or laser beam riding guidance. A narrow radar or laser beam is directed at the target and the weapon is launched in the direction of the target. Sometimes after it is launched, it flies into the radar or laser beam. With the help of sensors and a computer on board the missile it keeps itself within the beam. The aiming station keeps the beam pointed at the target until it hits it.

Celestial Navigation In celestial navigation, the missile compares the positions of the stars to an image stored in its memory to determine its flight path.

Command Guidance Here, the missile is commanded on an intercept course with the target. Conventionally, this is achieved by using two separate radars to continuously track the target and the missile. Tracking data from these radars is fed to a computer that computes the trajectories of the two vehicles. The computer in turn sends appropriate command signals over a radio link to the missile. A sensor on board the missile decodes the commands and operates the control surfaces of the missile to adjust its course so as to intercept the target in flight.

Command Line-of-Sight Guidance (CLOS) In the CLOS system, the missile is missile is always to commanded lie on the line-of-sight between the tracking unit and the target.

Command Off Line-of-Sight (COLOS) Guidance The COLOS system ensures missile interception of the target by locating both missile and target in space for which a distance coordinate is needed. Unlike the CLOS system, it does not depend on angular coordinates of the missile and the target. This can be possible only if both missile tracker and target tracker were active. Also, in COLOS guidance missile and target tracker can be oriented in different directions.

Copperhead This is a cannon-launched fin-stabilized terminally laser-guided explosive projectile designed to engage small hard point ground targets such as tanks, armoured vehicles, self-propelled artillery systems and other high-value targets such as bridges, defensive fortifications, C^4I (command, control, communications, computers and intelligence) centres and so on.

Direct Pursuit Navigation In the case of direct pursuit navigation, also called pure pursuit navigation, the seeker continuously looks at the target throughout the engagement duration until it eventually hits the target. In the case of direct pursuit engagement, the flight path is not the most direct one and the trajectory has an ever-decreasing radius turn towards the end of the engagement.

Enhanced Paveway Series of Laser-Guided Bombs This series of dual-mode laser-guided bomb (DMLGB) combines the precision of laser technology together with the resistance to bad weather features of GPS and INS to give it all-weather capability.

EXOCET The EXOCET family of missiles currently manufactured by MBDA, France are anti-ship, sea skimming missiles. EXOCET MM 40 series is the latest in the family of EXOCET missiles with EXOCET MM-40 Block-3 being the most recent missile in EXOCET MM-40 series.

Field-of-View (Laser Seeker) This is maximum value of angle-of-arrival of incoming laser radiation around its optical axis to which it can satisfactorily respond to. It determines the probability of the weapon finding itself with in the laser basket at the maximum guidance range. Laser-guided munitions using a seeker head with a larger field-of-view would have a higher probability of finding themselves in the laser basket and subsequently precisely hit the intended target.

Geophysical Navigation Guidance This depends on the measurements made on the surface of Earth for operation. It uses compasses and magnetometers to measure the Earth's magnetic field as well as gravitometers to measure the Earth's gravitational field. This technique has not found much application in missile guidance.

GPS/INS-guided Weapons These make use of a multichannel GPS receiver to provide information on weapon's location and an inertial measurement unit (IMU) to monitor its attitude to adjust its flight path to precisely hit the intended target.

Griffin This is a laser-guided bomb kit manufactured by Israel Aerospace Industries.

Heat-Seeking Missiles See *infrared-guided missiles.*

Igla This is man-portable IR homing air-defence system (MANPADS) manufactured by KBM.

Imaging Seekers These use an array of detectors called an imaging focal plane array instead of a reticle to build an image of the scene in front. The image may be created by scanning the scene and using a linear array or a two-dimensional staring array.

Inertial Navigation Guidance This system basically works by telling the vehicle where it is at the time of launch and the vehicle's computer uses the signals from the inertial measurement unit to ensure that the vehicle travels along the programmed path. In the case of inertial navigation guidance, the vehicle uses onboard sensors to determine its motion and acceleration with the help of Inertial Measurement Unit (IMU) or Inertial Navigation Sensor (INS).

Infrared-Guided Missiles These make use of electromagnetic radiation emitted from the target predominantly in IR part of the spectrum to track the target and then home onto it. These IR spectrum seeking missiles are also referred to as heat-seeking missiles.

IRIS-T This is a short-range air-to-air missile manufactured by Diehl BGT Defence. It employs imaging IR seeker.

Javelin This is a man-portable third-generation fire-and-forget anti-tank guide missile jointly developed by Raytheon and Lockheed–Martin.

LJDAM JDAM (Joint Direct Attack Munition) is a low-cost guidance kit used to convert existing unguided free-fall bombs into near-precision-guided weapons. LJDAM (Laser Joint Direct Attack Munition) combines laser guidance with GPS/INS to improve performance in adverse weather conditions.

Krasnopol This is a cannon-launched fin-stabilized terminally laser-guided explosive projectile designed to engage small hard point ground targets such as tanks, armoured vehicles, self-propelled artillery systems and other high-value targets such as bridges, defensive fortifications, C^4I (command, control, communications, computers and intelligence) centres and so on. Krasnopol projectile is produced in two variants, namely Krasnopol and Krasnopol-M.

LAHAT LAHAT (a short-range anti-tank missile, manufactured by Israel Aerospace Industries) is an advanced laser homing attack missile that makes use of semi-active laser guidance. The target in this case can be designated either directly from the launching platform or by another land-based or aerial platform.

Laser-Guided Weapons These make use of a laser beam to guide the weapon (bomb, projectile or a missile) to precisely hit the target. Most laser guide munitions with the exception of laser beam riding operate on the principle of semi-active laser homing similar to semi-active radar homing. A laser beam irradiates the intended target, a process called laser designation. The laser beam bounces off the target and gets scattered in all directions. The laser seeker in the munitions detects the direction of arrival of laser energy and guides it to the target.

Manual Command to Line-of-Sight (MCLOS) Guidance This is a type of command to line-of-sight guidance system. In the case of manual command to line-of-sight (MCLOS), target tracking, missile tracking and control functions are all performed manually.

MICA-IR The MICA-IR manufactured by MBDA is a short and medium-range air-to-air missile that uses an imaging IR seeker.

MILAN This is a portable medium-range wire-guided anti-tank missile from France with later versions equipped with tandem heat warheads making it more effective against reactive armour.

MIM-104 Patriot The MIM-104 Patriot (Phased Array Tracking Radar to Intercept on Target), manufactured by Raytheon is a long-range, all-altitude, all-weather air-defence system to counter tactical ballistic missiles, cruise missiles and advanced aircraft.

MM40 EXOCET A surface-to-surface anti-radiation missile that employs an active radar seeker whose receiver component is used to home onto the radar.

NAG This is a third-generation fire-and-forget anti-tank-guided missile from India, designed and developed by the Defence Research and Development Organization (DRDO) and manufactured by Bharat Dynamics Limited.

Non-Imaging Infrared-Guided Missiles These make use of IR emission corresponding to the thermal signatures of the exhaust and the mainframe of the target aircraft to home on it. Emission in the 3–5 and 8–12 μm bands is characteristic of electromagnetic emission from jet exhaust and the mainframe of aircraft.

PARS 3 LR PARS 3 LR, manufactured by Parsys GmbH, MBDA Deutschland GmbH and Diehl BGT Defence, is an autonomous fire-and-forget missile intended for long-range applications and designed to destroy ground (tanks, armoured vehicles etc.), air (helicopters) and other individual targets. Spike is a fourth-generation man-portable fire-and-forget anti-tank and anti-personnel missile designed and developed by Rafael Advanced Defence Systems.

Passive Homing Guidance This makes use of some form of energy emitted by the target. This energy is intercepted by the missile seeker, which is processed to extract guidance information to guide the missile to home on to the target. This energy could be in the form of heat energy generated by the target, which is made use of by the seeker in an IR-guided missile.

Paveway This is a family of laser-guided bomb kits. The Paveway family of laser-guided bombs has revolutionized tactical air-to-ground warfare by converting dumb bombs into smart precision-guided munitions. It has seen four generations; namely Paveway-I, -II, -II Plus, -III and -IV.

Precision-Guided Munitions Precision-guided munitions, also called smart munitions, belong to the group of advanced fire power munitions employing one or more guidance techniques to hit the target more precisely with minimized collateral damage than would be possible with conventional unguided weapons also called dumb weapons. These mainly include projectiles fired from land or ship-based military platforms, surface-to-air and air-to-air missiles and aerially delivered bombs.

PRF Code Compatibility (Laser Seeker) This means the PRF code of the received laser radiation would be the same as the PRF code programmed in the guidance unit before the start of the mission. PRF code is usually expressed as the time interval between two successive laser pulses in milliseconds usually up to the third decimal place. The two codes are considered compatible if the difference in time periods of the two codes is less than a certain specified value.

Proportional Pursuit Navigation This allows the missile to have the shortest flight path towards the intercept. The proportional navigation flight path is established by a constant look angle for a given constant missile velocity and assuming that target doesn't manoeuver.

Pseudo Imaging Seeker This uses one or more detectors enabling both spatial and temporal information from reticle seekers.

PYTHON PYTHON-5, manufactured by Rafael Advanced Defence Systems, is one of the most advanced air-to-air missiles in the world. Different variants of PYTHON family include Shafrir-1, Shafrir-2, PYTHON-3, PYTHON-4 and PYTHON-5. PYTHON-5 is the fifth generation air-to-air missile. It employs an IIR seeker.

Radar Guidance This includes semi-active radar guidance and active radar homing and is commonly used in long-range surface-to-air and air-to-air missiles.

Ranging Navigation This depends on external signals usually provided by radio beacons for guidance. Based on the direction and strength of the signals received by the aircraft, it navigates its way along the desired trajectory.

RBS-70 RBS-70 (a short-range anti-tank-guided missile manufactured by SAAB Bofors Dynamics) is a supersonic laser beam riding missile that is not susceptible to any deception by countermeasures employed by the target aircraft in the form of chaff or flares.

Reticle Seeker This uses a single IR detector with a spinning reticle placed between the focusing optics and the detector. A reticle is nothing but an optical modulator made up of a circular element having sequentially arranged transparent and opaque spokes on it. In the most basic reticle seeker, IR energy emitted by the target is collected by the optical system and focused on the detector through the reticle. The output of the detector is a sequence of pulses whose amplitude is proportional to magnitude of IR energy and whose frequency is equal to the product of spin rate of reticle and number of transparent/opaque spoke pairs. The phase of detector signal with respect to a reference phase is used to determine angular position of seeker axis with respect to the target.

Retransmission Homing Guidance This is similar to semi-active radar homing guidance except that in this case, the missile does not have an onboard computer to process the sensor signal and generate guidance command. Instead, the sensor signal is transmitted back to the launch platform for processing. The command signals generated at the launch platform are retransmitted back to the missile for use by the missile's control surfaces to guide the missile to home on to the target.

RIM-174 ERAM The RIM-174 Standard Extended-Range Active Missile (ERAM) also called Standard Missile 6 (SM-6) is a long-range anti-air warfare missile designed for use by navies against a range of aerial targets including fixed and rotary-wing aircraft, unmanned aerial vehicles, low attack cruise missiles and anti-ship cruise missiles in flight to provide area and ship defence.

Semi-Active Homing Guidance This is a form of homing guidance in which the target is illuminated by an external source, which could be radar or laser. The electromagnetic energy reflected by the target is intercepted by the seeker head of the guided weapon and an onboard computer processes the intercepted signal to determine the target's relative trajectory. It sends appropriate command signals to the control surfaces of the weapon to make it intercept the target.

Semi-Active Radar Guidance In a semi-active radar guidance system, the target aircraft is illuminated by electromagnetic energy emitted by the fire-control radar located either on the launch aircraft or on an appropriate ground location. The fire-control radar acquires the target and tracks it. A small radar transmitter generating a very narrow beam then selectively illuminates the target using tracking information generated by the fire-control radar. A radar receiver on board the missile receives the radar energy reflected off the target and locks on to the target. Once the missile is locked on to target, it may be launched. The radar seeker unit provides the angular error information to the flight path control system, which steers the missile to intercept the target through tail fins.

Semi-Automatic Command to Line-of-Sight (SACLOS) Guidance This is a type of command to line-of-sight guidance system. In a SACLOS guidance system, the target tracking is manual and the missile tracking and control functions are automatic. SACLOS is the most common form of guidance in use against ground targets such as bunkers and tanks.

Semi-Manual Command to Line-of-Sight (SMCLOS) Guidance This is a type of command to line-of-sight guidance system. In the case of SMCLOS, target tracking is automatic but missile tracking and control functions are performed manually.

Sensitivity (Laser Seeker) This is the minimum value of laser power density impinging on the laser seeker cross-section in a plane orthogonal to the optical axis of the seeker head to which it can respond satisfactorily.

Serviceability check (PGMs) These are functional checks that perform Go/No-Go testing of the weapon by confining the evaluation process to one or two vital parameters of the weapon. These tests are usually performed with weapon strapped on to launch platform.

Single Colour Infrared Seekers These are non-imaging IR seekers that respond to the 3–5 µm band in the thermal signatures of the target as seen by the seeker head.

Smart Munitions See *precision-guided munitions*.

Starstreak Starstreak (Short-Range Air-Defence System manufactured by Thales Air-Defence) is a man-portable/vehicle-mounted high velocity beam riding missile designed to counter threats from conventional air threats and fast 'pop up' strikes by helicopter attacks.

Stinger This is a man-portable IR-guided surface-to-air missile designed by General Dynamics and manufactured by Raytheon Missile Systems in the USA, in Germany by EADS (European Aeronautic Defence and Space Company) and in Turkey by ROCKETSAN.

Strela The Strela family of missiles are man-portable surface-to-air missiles that use passive IR homing guidance and a high-explosive warhead.

Sudarshan This is a laser-guided bomb kit developed by the Aeronautical Development Establishment of DRDO and manufactured by Bharat Electronics.

TOW Missile TOW (Tube-launched, optically tracked and wire-guided) from the USA is a family of wire-guided missiles. The TOW family, including TOW 2A, TOW 2B Aero and TOW Bunker Buster missiles, is the premier long-range, heavy assault-precision anti-armour, anti-fortification and anti-amphibious landing weapon system.

Track-via-Missile See *retransmission homing guidance*.

Track-via-Missile Radar Guidance See *retransmission homing guidance*.

Two-Colour Seeker These are non-imaging IR seekers that respond to both 3–5 and 8–12 µm bands in the thermal signatures of the target as seen by seeker head.

Vympel R-27P A Russian anti-radiation air-to-air missile.

Vympel R-73 Vympel R-73 (NATO designation AA-11 Archer) manufactured by Tbilisi Aircraft Manufacturing is a short-range air-to-air missile. It employs a cryogenically cooled all aspect IR homing seeker with high off-bore sight capability.

Wire-Guided Weapons These are guided by electrical signals sent to them through a bundle of wires connected between the weapon and the guidance mechanism located near the launch site. The wires reel out behind the weapon as it flies. Wire guidance is commonly used in anti-tank missiles where its suitability in limited line-of-sight availability is particularly advantageous.

XM-982 Excalibur The XM-982 Excalibur artillery projectile is a family of precision-guided, extended-range modular projectiles that employs GPS-aided inertial guidance and navigation, free spinning base fins, four-axis canard airframe control, base bleed technology and a trajectory glide to achieve increased accuracy and extended ranges.

Bibliography

1 Zarchan, P. (2007), *Tactical and Strategic Missile Guidance (Fifth Edition)*, AIAA.
2 Siouris, G.M. (2004), *Missile Guidance and Control Systems*, Springer.
3 Yanushevsky, R. (2007), *Modern Missile Guidance*, CRC Press.
4 East, D.J. (1977), *Guided Weapon Control Systems*, Pergamon Press.
5 Brooker, G. (2008), *Introduction to Sensors for Ranging and Imaging*, SciTech Publishing.
6 Gillespie, P.G. (2006), *Weapons of Choice – The Development of Precision Guided Munitions*, The University of Alabama Press.
7 Kaufman, R.L. (2012), *Precision Guided Munitions – History and Lessons for the Future*, Biblioscholar.
8 Rouse, J. (2008), *Guided Weapons* (Brassey's Land Warfare into 21st Century), BRASSEYS.

9 Boyne, W.J. (2003), *Operation Iraqi Freedom: What Went Right, What Went Wrong and Why*, Forge Books.

10 Balakrishnan, S.N., Tsourdos, A. and White, B.A. (2017), *Advances in Missile Guidance, Control and Estimation*, CRC Press.

11 Skolnik, M. (2008), *Radar Handbook (Third Edition)*, McGraw-Hill Professional.

12

Directed Energy Weapons

One of the major areas of global interest for scientists and engineers today is that of Directed-Energy Weapons (DEWs). These weapons, with the exception of particle beam and laser-induced plasma channel (LIPC) weapons, generate streams of electromagnetic energy that can be precisely directed over long distances to disable or destroy the intended targets. After decades of research and development, DEWs are becoming an operational reality. This has been possible due to their unique characteristics that potentially enable new concepts of military operation and also because there has been considerable progress over the past two decades in developing relevant technologies such as power sources, beam-control concepts and pointing and tracking techniques. After a brief introduction, the chapter focuses on different categories of DEWs including high-energy laser weapons, high power microwaves, particle beam weapons and laser-induced plasma channel (LIPC) weapons. Particular emphasis is laid on laser-based DEWs and high-power microwaves as these are the most widely exploited and practically realizable technologies for building DEWs. The topics covered in the chapter include operational advantages and limitations, system anatomy, involved technologies, international status and emerging trends.

12.1 Directed-Energy Weapons (DEWs)

A DEW system, with the exception of laser-induced plasma channel (LIPC) weapons, primarily uses directed-energy in the form of concentrated beam of electromagnetic energy or atomic or subatomic particles in the targeted direction to cause intended damage to the enemy's equipment, facilities and personnel. The intended damage could be lethal or nonlethal. Ever since H.G. Wells published *The War of the Worlds* in 1898, DEWs have been a recurring theme in science fiction literature. The idea of a 'Death Ray', which can instantly destroy or burn a target at a distance, in fact dates back to a belief that Archimedes used a 'Burning Glass' to set fire to Roman ships during the siege of Syracuse in 212 B.C. Although many of the images of the death ray depict Archimedes with a parabolic mirror, use of a set of individual flat mirrors appropriately positioned seemed to be a more practical implementation of the concept (Figure 12.1). Though the story has long been dismissed as a myth, the interest generated by it has led to a number of experiments being conducted to verify the technical feasibility of such an event. The experiments conducted by Comte de Buffon and Dr Ioannis Sakkas and more recently by students of MIT have established the feasibility of such an occurrence.

Buffon assembled 168 mirrors, each 8 × 10 inches, adjusted to produce the smallest image 150 feet away. The array turned out to be a formidable weapon. With this elaborate

Handbook of Defence Electronics and Optronics: Fundamentals, Technologies and Systems, First Edition. Anil K. Maini.
© 2018 John Wiley & Sons Ltd. Published 2018 by John Wiley & Sons Ltd.

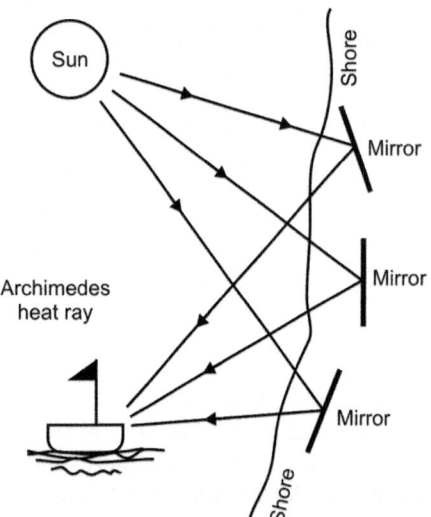

Figure 12.1 Archimedes death ray concept.

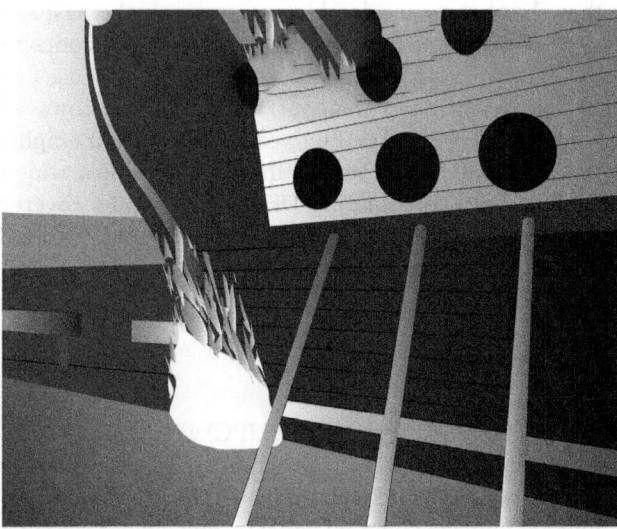

Figure 12.2 MIT experiment.

setup, he performed several experiments. He demonstrated ignition of a creosoted plank at a distance of 66 feet using only 40 mirrors. Just 128 mirrors could ignite a pine plank instantly and, in another experiment, 45 mirrors melted 6 lbs of tin at 20 feet. In another effort, Sakkas lined up almost 60 Greek sailors, each holding an oblong mirror tipped to catch the Sun's rays and direct them at a wooden ship 160 feet away. The ship caught fire at once. As recently as 2009, MIT students carried out an experiment with 127 mirrors of 1 square foot focusing solar radiation on to a boat 100 feet away causing a sustained flame and confirming the technical feasibility of what Archimedes might have achieved with his death ray (Figure 12.2).

At the most fundamental level, DEWs share the concept of delivering a large amount of stored energy from the weapon to the target to produce structural and incendiary damage effects. The difference lies in kinetic energy weapons delivering this effect at subsonic or supersonic speeds while DEWs do so at the speed of light. Both kinetic energy weapons and DEWs need to address two fundamental issues. The first major concern is related to travel or propagation through the atmosphere and hitting the target. In the case of kinetic energy weapons, it is getting the projectile to successfully travel through the atmosphere and hit the target. In the case of DEWs, it is the propagation of a high-energy beam such as high power electromagnetic radiation or a high-energy particle beam through the atmosphere and directing it to hit the target. The second major concern is to produce useful damage effects on the intended target. This is where the interaction of high-energy with matter comes into play. This implies that having a high-power laser or a high-power microwave emitter alone doesn't make a DEW. Three important constituents of a DEW are therefore the high-energy source influencing the operational range, target tracking and beam pointing technology determining probability of target hit and interaction of high-energy beam with the matter that determines the lethality. These aspects are discussed at length in subsequent sections particularly with reference to high-energy laser weapons and high power microwaves.

12.2 Types of DEWs

The four major categories of DEWs include the following.

1) Microwave-based DEWs
2) Laser-based DEWs
3) Particle beam weapons
4) Laser-Induced Plasma Channel (LIPC) weapons.</NL>

The *microwave-based DEW* system is designed to produce the equivalent of electromagnetic interference to damage the enemy's electronics systems. Due to concerns regarding unintended side effects on the host platform, it is usually preferable to put such weapons only on unmanned combat air vehicles (UCAV). Also under consideration is use of high-power microwaves as a weapon to attack underground and deeply buried targets that are resistant to high explosives.

At the core of *laser-based DEW* is a high-power laser that has enough power in the case of CW laser or sufficient pulse energy in the case of pulsed laser to inflict a physical damage to the target. Though the lasers intended for already established applications such as range finding, target designation for munitions guidance and so on will continue to improve as newer technologies evolve and develop, it is the use of lasers as weapon that is going to rewrite the military balance in the next 15–20 years.

Introduction of *laser-based DEWs* is set to dramatically alter the war fighting capabilities of nations by making possible the execution of missions that would be extremely complex if not impossible to realize with conventional kinetic energy weapons. These include ground-based laser systems for disabling low earth orbit satellites and destroying missiles, airborne laser systems for destroying ballistic missiles and space-based laser systems for neutralizing theatre and intercontinental ballistic missiles. Large number of experiments with laser-based DEWs to demonstrate these or similar capabilities have been carried out in different parts of the world. Reliability of these weapons has been established beyond doubt and these weapons have been projected by strategists as the weapons of the twenty-first century.

A *particle beam weapon* uses a high-energy beam of atomic or subatomic particles to inflict the intended damage to the target by disrupting its atomic and/or molecular structure. Particle beam weapons are the least mature of the four DEW technologies and receive by far the least amount of research effort. Particle beam weapons are not true DEWs. Unlike high-energy laser weapons and high power microwaves that direct electromagnetic energy towards the target, particle beam weapons deliver kinetic energy into the target's atomic structure and are solely hard-kill weapons.

The *laser-induced plasma channel (LIPC) weapons* are hybrid weapons that use a laser to ionize a path of molecules to the target via which an electric charge can be delivered into the target to cause damage effects. The LIPC can be used to destroy anything that conducts electricity better than the air or ground surrounding it. It works as follows. A high intensity train of picosecond laser pulses is used to create a powerful electromagnetic pulse around itself that strips electrons from air molecules thereby creating a plasma channel through the air. Since the air is composed of neutral particles that act as insulators, the laser-induced plasma channel is relatively a good conductor. A high voltage current discharge is sent down this conducting filament to the target rather than arcing unpredictably through the air; a phenomenon similar to lightening finding its way from clouds to ground via the path of least resistance.

Of these four categories, high-energy laser weapons (HEL) have the greatest potential in the near term to become worthy of a potent weapon system. High-power microwave (HPM) technology too has a similar potential, but has not been funded as generously as the high-energy laser weapon development programmes. LIPC has significant potential especially as a nonlethal weapon. Particle beam weapons at this time are apt to remain in the science fiction domain, as the weight and cost as yet do not justify the achievable military effect.

The next 10 years will see the emergence of high-energy lasers as an operational capability. These weapons will have the unique capability to attack targets at the speed of light and are likely to significantly impair the effectiveness of many weapon types, especially ballistic weapons. Constrained by the problems associated with propagation of high-energy laser beam through the atmosphere, these weapons will not provide all-weather capabilities and shall largely remain clear-weather systems.

12.3 Particle Beam Weapons

The particle beam weapon (PBW) is a form of the directed-energy weapon that uses atomic or subatomic particles accelerated to speed of light or near speed of light with the help of powerful electric and magnetic fields in a particle accelerator and then directed to deliver fraction of their kinetic energy to the intended target thereby causing severe damage due to disruption of its atomic structure. A PBW is characterized by beam energy in electron-volts (eV), beam current in amperes and beam power in watts.

PBWs come in two primary types: *charged-particle weapons* and *neutral-particle weapons*. When it comes to military application of these different types of PBWs, charged-particle weapons are endoatmospheric, while neutral-particle weapons are exoatmospheric. At the core of a particle beam weapon is the *particle accelerator*. It is also the most complex part of the beam weapon and is built using a linear electric field to accelerate the charged particles similar to the Gauss or coil gun or an induction linear accelerator (linac) system. The induction linac consists of a simple non-resonant structure where the drive voltage is applied to an axially symmetrical gap that encloses a toroidal ferromagnetic material. The change in flux in the magnetic core induces an axial electric field that provides particle acceleration.

12.3.1 Operational Principle

A particle beam consists of protons, electrons or neutral atoms flowing with a real or imaginary current. It is characterized by beam energy, current and power. Beam energy is expressed in mega electron-volts (MeV). One electron-volt is the kinetic energy of an electron that has been accelerated by an electric potential of one volt. Particle beam energy is characterized by energy of a typical particle of the beam as all the particles in a beam will have been accelerated to the same velocity. A typical PBW capable of inflicting serious damage to a target 1000 km away in space would typically require beam energy of 1 GeV.

An estimate of number of charged particles in the beam can be made from the magnitude of beam current. It is possible to assign a current to the particle beam assuming that each particle has a charge quantum equal to that of an electron even if the charged particle were the neutral atom. The beam current for the possible beam weapon described previously would typically be 1000 A.

The power of a particle beam is the rate at which the beam energy is transported, which is also indicative of the rate at which it can deposit energy into a target. As an analogy to the electric circuits, the particle beam in watts is equal to the product of energy in electron-volts and the beam current in amperes.

12.3.2 Types of Particle Beam Weapons

There are two broad categories of particle beam weapons, namely the *charged particle* and *neutral-particle* beam weapons. The charged-particle beam weapons have a set of technological characteristics that are entirely different from the neutral-particle beam weapons. While the characteristics of the former make them suitable for use within the atmosphere; those of neutral beam weapons are better suited to use in space.

Both endoatmospheric and exoatmospheric beam weapons have their own technological hurdles to overcome. A particle beam propagating through atmosphere requires extremely high power and precisely defined beam characteristics. The technologies required for development of a suitable power supply and particle accelerators with sufficient power and appropriately shaped pulses for endoatmospheric weapons are very complex and involve high risk. On the other hand, the greatest challenge in the case of exoatmospheric beam weapons is in the area of beam control. The PBW should not only be able to produce a high intensity low divergence particle beam at the exit of accelerator; it should also have the necessary beam-control mechanism for aiming and beam at the target and ability to detect pointing errors in the beam for applying correction if required. Because of these two different sets of demands, the endo- and exoatmospheric devices represent two different types of weapon systems in appearance and operation. Nevertheless, there are certain fundamental areas of development that are common to both types of PBW. Each of these classes of beam weapons and the associated issues are briefly described in the following paragraphs.

12.3.2.1 Charged-Particle Beam Weapons

A charged-particle beam consists of electrons accelerated to the required energy level in a particle accelerator using a combination of electric and magnetic fields. To be able to destroy the target, the particle energy should be high and so should be the beam current. As an example, a practical electron beam weapon would need to hit a target 1000 km away with a 1000 A beam with energy of 1 GeV for 0.1 ms to destroy it. The particles in the beam have kinetic energies equal to their rest-mass energies with the result that they would travel with nearly speed of light. The particle accelerators researched for high-energy physics have high energies and pulsing rates but low beam currents. On the other hand, particle accelerators related to fusion research generate high beam currents but at low energies and pulsing rates. Particle accelerators suitable for producing beam weapons need to generate high intensity and high-energy particle beams.

Charged-particle beams are of little use in space. The combined effects of emittance and Coulomb's force of repulsion between the like-charged particles broaden the beam. As an example, a 1 GeV, 1000 A charged-particle beam would spread from the initial 1 cm to 5 m over 1000 km. Furthermore, the beam is deflected by the Earth's magnetic field. By the same study, the 1 GeV, 1000 A beam would deflect by 1000 km over 1000 km distance due to Earth's magnetic field. Charged-particle beams though can be made to propagate satisfactorily over a few km through an air channel evacuated by heating air in a straight line. Thus, a charged-particle beam weapon could be employed for ballistic missile defence. The system could be installed in a few ground-based sites in conjunction with either Earth-borne or space-borne radar systems to identify and track incoming ballistic missile warheads. The charged-particle beam weapon could be rapidly pointed at the incoming missile and destroy it. For an interception in air at 10 km, an electron beam weapon would typically require 500 MeV of beam energy and 10 000 A of beam current. However, large fixed installations required for charged-particle beam weapons as per current status of technology may render them vulnerable to sabotage or other forms of attack by an adversary.

12.3.2.2 Neutral-Particle Beam (NPB)

A neutral-particle beam (NPB) weapon consists of neutral atomic particles accelerated to high kinetic energy level in a particle accelerator. The process of generation of a high-energy NPB is as follows. Hydrogen or deuterium gas is subjected to an enormous electrical charge. The electrical charge produces negatively charged ions that are accelerated through a long vacuum tunnel by an electrical potential in the hundreds-of-megavolts range. After the negatively charged ions have been accelerated, at the end of the tunnel, electrons are stripped from the negative ions thereby forming the high-speed NPB.

Weapons-class NPB weapons also require energies in the hundreds of MeV and beam powers in the tens of megawatts. Modern devices have not yet reached this level. Given the state-of-the-art in accelerator technology, achieving the required beam energy and power levels would require hundreds of tonnes of accelerator hardware and enormous power sources to operate. Due to size, weight, power constraints and inherent complexity, it does not appear feasible that a NPB weapon would see the light of the day before 2025.

NPBs travel in a straight line once they have been accelerated and magnetically pointed just before neutralization in the accelerator. Also, a NPB is strongly affected by passage through the atmosphere. They get attenuated and diffused as they passes through dense gas or suspended aerosols, which makes them far more suitable compared to charged-particle beam weapons for applications in space against high flying airborne or space-borne targets. Damage assessment of the target could be possible. When the beam penetrates a target, the target's atomic and subatomic structure produce characteristic emissions that could be used to determine the target's mass or assess the extent of damage to the target. The major disadvantage of a NPB weapon, even in space, is that it is extremely difficult to sense, which complicates the problem of beam control and direction.

12.3.3 Involved Technology Areas

Different technology disciplines related to development of PBWs include *particle accelerator*, *power source* driving the accelerator, *target tracking and beam pointing* system, *beam propagation* through medium separating the source and the target and *lethality*. Different technology disciplines are represented by the block-schematic in Figure 12.3.

12.3.3.1 Particle Accelerator Technology

The particle accelerator is the source of high-energy particle beam and is therefore the heart of the particle beam weapon system. Three main constituent parts of the particle accelerator include the source of particles such as electrons, protons or charged atoms, the device that injects these particles into the accelerating section and finally the accelerating section itself.

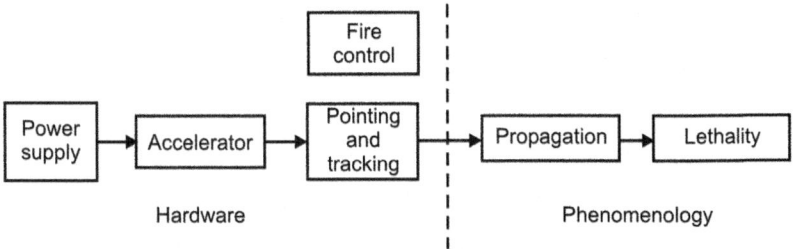

Figure 12.3 Particle beam weapon technology areas.

The particle accelerator accelerates particles to extremely high energies. These particles, as outlined earlier, are elementary particles like electrons or protons or even whole atoms. In modern accelerators, the particles very quickly attain nearly the speed of light, which is about 3×10^5 km/s. According to Einstein's theory of relativity, this speed can only be reached by massless particles and can never be exceeded by any particle. Though the speed of a particle cannot be increased any further after it has attained near speed of light, its kinetic energy can still be increased. All particle accelerators accelerate the particles roughly to the speed of light; the difference between more or less powerful accelerators is the kinetic energy of the accelerated particles.

There are two types of accelerators namely the *linear accelerators* and the *circular accelerators* also called *ring accelerators*. While linear accelerators accelerate particles over a long, straight line where the particle beam travels from one end to the other, in the case of circular accelerators, powerful magnets are used to bend the particle's path into a circle and the beam of particles travels repeatedly round the loop. Technologically, both linear and circular accelerators are in a highly advanced state of development. The preferred accelerator technology depends on the application. In earlier accelerators, strong electric fields were used to accelerate the particles but the maximum usable electric field strength had an upper limit due to spontaneous discharge. The accelerating section of all conventional linear accelerators is therefore made up of a cascade arrangement of a number of accelerating segments that sequentially apply an accelerating electric field to the charged particles. While the voltage in each segment may be relatively low, the repeated application of an accelerating voltage by a large number of modules ultimately produces substantially high particle energies. Modern accelerators use focusing magnets to ensure that particles travelling at the speed of light are not lost and always remain on their reference trajectory. SLAC's 3.22 km long particle accelerator (Figure 12.4) is an example of a linear accelerator. *Circular accelerators* employ strong bending magnets to keep the particles on their circular path. The smaller the diameter of the circular path, the larger and more powerful the focusing magnet needs to be. The Large Hadron Collider (LHC) is an example of a ring accelerator (Figure 12.5). It is the world's largest and most powerful particle accelerator. The LHC consists of a 27-km ring of superconducting magnets with a number of accelerating structures to boost the energy of the particles along the way. Inside the accelerator, two high-energy particle beams travel in opposite directions in separate beam pipes at ultra-high vacuum at close to the speed of light before they are made to collide.

Two common types of linear accelerator technologies used for development of particle weapons include *RF linear accelerators* (RF LINACS) and *induction linear accelerators* (Induction LINACS). Although linear accelerators are capable of accelerating particles to energy levels high enough for use as a weapon, their current carrying ability is severely limited. As a consequence of this they are not suitable for building accelerators for endoatmospheric particle beam weapons. An RF LINAC can be a suitable candidate for building accelerators for

Figure 12.4 Stanford linear accelerator. (*Source:* US Department of Energy.)

Figure 12.5 Large Hadron Collider. (*Source:* Julian Herzog, https://commons.wikimedia.org/wiki/File:CERN_LHC_Tunnel1.jpg. CC BY-SA 3.0.)

exoatmospheric particle beam weapons as space weapons don't call for very high beam power. Both RF LINAC and induction LINAC use successively high voltages across a series of accelerating segments. Induction LINAC differs from RF LINAC in the mechanism used for generating the electric voltage within the segments of the two types of LINACS. Compared to the RF LINAC, the induction LINAC produces a more stable beam at high beam currents. The induction LINAC is therefore a more likely candidate for an endoatmospheric beam weapon.

12.3.3.2 Power Source

Development of power source for an endoatmospheric particle beam weapon is a big technological challenge. The power supply in this case is required to supply high-energy over short time periods, which translates into very high pulsed power levels. Advanced pulsed power technology is therefore required to develop power supply for a particle beam weapon. A pulsed power device comprises of three main parts namely the *prime power source* that provides the required electrical energy over the full operating time of the weapon, an *energy storage device* needed for intermediate storage of the electrical energy as it is generated and the *pulse forming network* that generates power bursts or pulses of desired magnitude and time duration. Each of these three areas represents a major technological challenge.

The *prime power source* of a particle beam weapon is required to deliver power levels in the range of megawatts to gigawatts and at the same time be as lightweight and compact as possible. Also, in many applications, the prime power source needs to be mobile. Though a conventional power station could provide the needed power levels, but it would be neither small nor lightweight and would definitely not meet the mobility requirement. Some of the potential candidates for building the prime power source include advanced technology batteries, turbo-generators and advanced magneto-hydrodynamic (MHD) generators using superconducting circuitry.

A typical *energy storage method* involves charging a bank of capacitors, or spinning a huge mechanical flywheel or simply storing the energy in the form of a high-energy explosive that is released in a contained explosion. There are a number of other energy storage and release mechanisms each having its own set of advantages and disadvantages. The preferred mechanism depends on the particle accelerator and also whether the beam weapon is endo- or exoatmospheric.

The *pulse forming network* is used to shape the power pulse. In the case of an endoatmospheric particle beam weapon, one shot would comprise of a burst of very short duration pulses with burst frequency in the range of kHz. The prime power source may deliver power for a series of bursts of pulsed power followed by the beam weapon going to quiescent state while the energy storage device gets recharged for another series of bursts of pulsed power.

12.3.3.3 Target Tracking and Beam Pointing

Unlike some other areas of particle beam weapon development, such as the generation of high-energy particle beam, which need to address certain basic issues of science and technology, target acquisition and tracking and beam pointing technologies required for a particle beam weapon are not unique to this class of DEWs. In this respect, particle beam weapon programme has immensely benefited from high-energy laser weapon programme. Target acquisition and tracking and beam pointing technologies are discussed under laser-based DEWs. Notwithstanding the commonality of technologies between the particle beam weapons and high-energy laser weapons, there do remain some tracking and pointing problems that need to be solved. Many of them arise out of propagation of the charged-particle beam through the atmosphere and the neutral-particle beam through space.

12.3.3.4   Beam Propagation

Other than the high-energy particle beam source and target tracking and beam pointing technologies, a third important element determining the success of a PBW is the propagation of the charged-particle beam through the atmosphere in the case of endoatmospheric weapons and NPB through space in the case of exoatmospheric weapons. The particle beam must have an extremely precise path of propagation as it traverses the required distance to the target. The propagation problem is more severe in the case of a charged-particle beam through atmosphere as compared to that of an NPB through space. Propagation of a NPB does not suffer from beam instability problems possibly encountered by a charged-particle beam while propagating through the atmosphere.

One of the contributing factors is the beam spreading, which causes increase in beam diameter and consequent decrease in energy density as it travels towards the target. Even for a small amount of beam divergence, the beam diameter may become appreciable enough to be unacceptable for longer ranges. Two factors contribute to beam spreading. One of the factors is the beam divergence imparted by the accelerator itself and present at the exit of the accelerator. The other is due to mutual repulsion of beam particles. While in the case of charged-particle beams both factors are responsible for beam spreading, they would strictly originate from the accelerator in the case of NPBs. Even in the case of a NPB propagating through atmosphere, the surrounding air molecules strip the neutral particles of the electrons and transform it into a charged-particle beam. This would further lead to undesirable beam divergence.

In the atmosphere, however, even if the beam particles were neutral, air molecules would strip the surrounding electrons quickly from the beam's neutral atoms, turning the beam into a charged-particle beam. The charged particles within the beam would then tend to repel one another, producing undesirable beam divergence. The unwanted effect is countered to some extent as the propagating beam may knock off electrons from air molecules, which would in turn intermingle with the beam and neutralize the charged particles. The magnetic field created by the charged-particle beam current also prevents beam spreading by producing a type of conduit. For a charged-particle beam to propagate satisfactorily through the atmosphere, the particle beam needs to have certain threshold values of beam parameters including beam current, particle energy and pulse duration. The problem is still looking for a solution as currently no particle beam accelerator is capable of producing a beam with the required parameters. Two important experimental programmes using particle accelerators are exploring the phenomenon of particle beam propagation through atmosphere. One of them relates to the experiments with Advanced Test Accelerator (ATA) at Lawrence Livermore National Laboratory and the other is a joint Air Force/Sandia National Laboratories programme through the use of a RADial-pulse-Line Accelerator (RADLAC).

12.3.3.5   Lethality

Lethality studies play an important role in understanding the precise effect a particle beam would have on interaction with different types of target materials. The subject becomes more complex as the particle beam-matter interaction would depend on the type of particles in the beam, particle energy and beam power. Parameters determining lethality and hence the efficacy of the beam weapon include beam velocity, dwell time, rapid retargeting capability, beam penetration, ancillary kill mechanisms and all-weather capability.

A *beam velocity* that is much higher than the target speed simplifies the task of fixing the aim point even in the case of an evasive target. As an example, a target at a distance of 50 km and moving at a supersonic speed of 6 Mach would have moved only a metre or so from the time the particle beam weapon is fired till it hits the target. This is made possible by the particle beam travelling at nearly the speed of light of 3×10^5 km/s. The *beam dwell time,* that is the time period for which the particle beam is needed to remain on the aim point on the target to inflict intended damage, is in the order of a few microseconds in the case of endoatmospheric weapons. In this case, beam power is sufficient enough to cause almost instantaneous damage. Short dwell times may be required in the case of exoatmospheric beam weapons due to comparatively lesser power of these weapons. *Rapid retargeting* feature of the particle beam weapons gives them the capability of engaging multiple targets. This is made possible as the charged-particle beam can be deflected in the desired direction within certain limits with the help of a changing magnetic field. Unlike high-energy laser weapons where the laser-matter interaction effects are mainly confined to the surface, particle beams have much *greater penetration*. The structural damage in this case is far more severe sometimes leading even to a catastrophic damage. Also, in the case of laser-induced damage, the material blowing off from the target surface tends to envelop or hide the target thereby acting as a protective shield. PBWs with their penetrating nature do not suffer from this problem. Even target hardening through shielding materials selection proves to be ineffective.

PBWs offer many *ancillary kill mechanisms*, which are capable of inflicting sufficient damage to the target even if the main beam missed the aim point. One of the ancillary kill mechanisms is the presence of a secondary cone of radiation symmetrical about the main particle beam. This radiation cone comprises X-rays, neutrons, α- and β-particles and so on and is created by collisions of beam particles with atoms of the air. Another ancillary mechanism is the electromagnetic pulse (EMP) generated by the presence of beam current pulse. The EMP could play havoc with the electronic components and devices of the target. As these ancillary mechanisms are present due to interaction of charged particles, they are not present in the case of neutral-particle beam based exoatmospheric weapons. Poor weather conditions such as presence of clouds, fog rain and so on severely limit the capabilities of high-energy laser weapons. Particle beam weapons have *all-weather capability*.

12.3.4 Capabilities and Limitations

The major advantages of particle beam weapons include the following.

1) Speed-of-light delivery
2) Rapid retargeting
3) All-weather capability
4) Better target interaction
5) Shorter dwell time

Speed-of-light delivery of the PBWs allows them to precisely target even fast moving and evasive targets, which is a big advantage in anti-ballistic missile (ABM) defence. It also enables precise aim point selection on fast moving targets at longer ranges. *Rapid retargeting* is enabled by use of changing magnetic field to deflect the charged-particle beam within certain limits without any requirement of physical movement. This feature is not available in NPB weapons. PBWs have *all-weather capability* and, unlike high-energy laser weapons, their use is not limited by atmospheric factors such as rain, fog, clouds, dust and so on. PBWs, due to their high particle energy coupled with secondary and tertiary kill mechanisms in the form of cone of radiation of X-rays, α- and β-particles and EMP, interact much better with the target. Particle beams have far more impact damage on the target than the massless photons of the high-energy laser weapon and the penetration increases with increase in particle energy. Presence of ancillary kill mechanisms greatly increases the probability of target damage. This makes it possible to inflict damage to the target even if the main particle beam missed the target. *Dwell-time* requirement of PBWs is much less compared to that of high-energy laser weapons. In the case of charged-particle beam based endoatmospheric weapons, due to high particle energy levels, the dwell time is almost negligible being in the order of few microseconds. In the case of neutral-particle beam weapons for space usage, short dwell times would be required.

Major limitations of particle beam weapons include the following.

1) Line-of-sight weapon
2) Huge size and weight
3) Massive power source requirement
4) Thermal and electrostatic blooming
5) Complicated beam control

PBWs such as high-energy laser weapons are essentially *line-of-sight weapons*. Their efficacy is neutralized or reduced due to the presence of an object obscuring the target. Indirect fire used in artillery warfare is not feasible with line-of-sight weapons, though one might think of reflectors on airborne or space-based platforms in the case of high-energy laser weapons enabling indirect fire. Looking at the state-of-the-art in accelerator technology, the kind of *size and weight* requirement to achieve particle energy levels in the range of hundreds of MeV to GeV is so gigantic making it impractical as a weapon. The accelerators capable of generating weapon grade particle energies need to be driven by electrical power in the range of tens to hundreds of megawatts, almost equivalent to the capacity of a power sub-station. Another limitation of a particle beam weapon arises out of the beam spreading called *blooming*. There is *thermal* blooming and *electrostatic* blooming. Due to blooming, particle energy that would otherwise be focused on the target gets spread out. *Thermal blooming* is present in both charged-particle and neutral-particle beams. It occurs when particles bump into one another under the effects of thermal vibration and also when they bump into air molecules. *Electrostatic blooming* occurs only in charged-particle beams and is due to mutual repulsion of the charged particles. In the case of neutral-particle beams therefore, the beam spreading is due to thermal blooming only. Blooming leads to an increasing beam size as it travels the distance to the target thereby decreasing the beam intensity and consequently the ability to inflict damage. In other words, it reduces operational range. It may be mentioned here that electrostatic blooming is present in charged-particle beams. Neutral-particle beams are not affected by electrostatic blooming as absence of charge means no mutual repulsion and consequent beam spreading. Yet another problem with particle beam weapons is their *complicated beam control*, more so in the case of neutral-particle beam weapons for exoatmospheric applications. One of the reasons for this is deflection caused by Earth's magnetic field. The other reason, applicable to neutral-particle beams, is the difficulty in sensing deviation of the beam from intended path to apply correction if required.

12.3.5 Effects of Particle Beam Weapons

Due to the form of energy propagated by a PBW, the mechanism by which it destroys a target is different from those of other types of DEWs including high-energy laser weapons and high-power microwaves. Both types of PBW including charged-particle and NPB weapons generate their destructive power by accelerating subatomic particles or atoms to near speed of light and focusing these high-energy particles in to a beam whose energy is the aggregate kinetic energy of the particles. When concentrated into a beam, it can melt or fracture the material on interaction with the target. Target destruction in the case of a particle beam weapon occurs by three mechanisms. The first damage mechanism is kinetic penetration due to subatomic particles moving at speed of light, the second is thermal damage and third is the disruption of atomic bonds of the target.

PBWs, like other forms of DEW, affect their targets through either a soft or hard kill depending on some conditions such as distance to target, power generated by the weapon and target hardening level. In the case of soft-kill damage, the effect of attack is to deny or degrade the operation of the target platform or even inflict partial damage. Some examples include disrupting electronics of a guided missile forcing it to miss the intended target or damaging visible, infrared and microwave sensors on board the target platform. Though the soft-kill damage causes temporary loss of function, it can seriously compromise mission success. Hard-kill relates to permanent physical damage to the structure of the target.

12.4   High-Power Microwave (HPM) Weapons

In the era dominated by electronic warfare and smart weapons systems facilitated by advances in electronics technologies, military planners and strategists have always been on look out to acquire improved capabilities in countering artillery fire, providing self-protection to aircraft and credible ship defence against cruise missiles, destruction or disruption of command and control assets, SEAD (suppression of enemy air-defence) systems, space control, security and so on. High-power microwave (HPM) weapons systems with their speed-of-light delivery, all-weather capability to destroy adversary's electronics systems, area coverage of multiple targets, minimum collateral damage, simplified tracking and beam pointing and a deep magazine address most of the requirements of military commanders of the day. HPM weapons generate an intense blast of microwave energy strong enough to overload electrical circuitry, inducing currents large enough to even melt circuitry in some cases. Less intense bursts of microwave energy can temporarily disrupt electronic systems or permanently damage integrated circuits, causing them to fail minutes, days, or even weeks later. Humans exposed to the blast of the HPM weapon remain unharmed and might not even know they have been hit.

Of course, to come to the level of a deployable HPM weapon possessing all of these features, a number of technological challenges need to be addressed. These include development of compact high peak power or average power HPM sources, compact high gain ultra-wideband antennas, compact and efficient pulse power drivers, predictive models for HPM effects and lethality to create a comprehensive database on effects of such a radiation on a wide range of land, sea, air and space-based military assets and system integration to meet requirements of a variety of military platforms such as fixed-wing and rotary aircraft, land vehicle, aircraft pods, naval combatants, unmanned aerial vehicles and so on.

An HPM weapon essentially comprises a *pulse power source* that drives the microwave source, *high power microwave source* and a *transmitting antenna* that directs the microwaves towards the target and acts as an interface between the microwave source and the atmosphere. Besides these three components, are HPM weapons' tracking, aiming and control systems. The microwave power generated by the weapon, beam characteristics and target vulnerabilities together determine the effective operational range of the HPM weapon system. The weapon is designed to disrupt, degrade or destroy electronics of the target by radiating electromagnetic energy in the microwave frequency band. Microwave frequencies ranges from 300 MHz to 300 GHz. HPM weapons under development occupy a frequency range of 500 MHz–3 GHz. Frequencies around 100 GHz have been exploited to build nonlethal weapons due to the significant effects this frequency has on human beings. This is primarily due to coupling effects and generic electronic system vulnerabilities. HPM weapons are broadly categorized into narrowband and ultra-wideband. While narrowband HPM weapons radiate all their energy within a few percent of the centre frequency, which is in the range of tens to hundreds of MHz, an ultra-wideband HPM radiates microwave energy over a bandwidth in the range of hundreds of MHz to several GHz.

Ideally, HPM weapon systems could be used to replace precision-guided munitions to disable or destroy high-value targets or installations located in populated areas with minimized risk of human casualties. Other application under consideration is the suppression of enemy air defences (SEAD). Most air-defence systems are highly susceptible to high power microwaves. Surface-to-air missiles contain highly sensitive guidance systems and the fire-control is usually radar controlled. Environmentally dangerous targets such as chemical and biological weapon production facilities could be targeted more effectively by HPM without fear of releasing deadly toxins into the atmosphere. Different aspects of HPM weapons including types of HPM weapons, technological challenges, capabilities and limitations and effects on targets are discussed in the following paragraphs. Some representative HPM systems are briefly discussed towards the end of the section.

12.4.1 Types of HPM Weapons

HPM weapons are generally categorized into:

1) Narrowband HPM and
2) Ultra-wideband HPM.

Narrowband HPM weapons produce a high-power microwave output in a narrow band of frequencies with the bandwidth equal to only a few percent of the centre frequency. These weapons are capable of generating relatively higher output power levels as compared to ultra-wideband HPM weapons. Narrowband HPM systems are effective only on a given class of targets that would absorb the frequency emitted by the system. Therefore, in the case of a narrowband HPM, knowledge of the absorption by the target material as a function of frequency and aspect angle is an advantage. The frequency absorption data may be generated by scanning the target with a tunable low-energy microwave source and evaluating the reflected signals for missing frequencies. The missing frequencies in this case are those that are heavily absorbed by the target material. Narrowband HPM weapons have the advantages of better transmission characteristics and fewer problems with fratricide. The limitations include their susceptibility to countermeasures such as target hardening and prior knowledge of the threat required to choose the optimum microwave frequency.

Ultra-wideband HPM weapons on the other hand radiate over a broad frequency range, but deliver comparatively less microwave energy at specific frequencies. While narrowband HPM weapons are capable of defeating only well-defined target or class of targets, ultra-wideband weapons are intended for use against a wide range of targets. Ultra-wideband HPM weapons have the advantage that they do not require any prior information on the target's absorption characteristics and provide a broad capability range. Due to comparatively lower radiated microwave power and poorer transmission characteristics than narrowband HPM weapons, they have shorter operational ranges.

12.4.2 Components of HPM Weapon Systems

As outlined earlier, an HPM weapon system essentially comprises a pulsed power source, a source of microwave energy and a transmitting antenna. The pulsed power source further comprises a power supply and a pulse generator. Figure 12.6 shows a simplified block-schematic arrangement of an HPM weapon system.

Figure 12.6 Block-schematic arrangement of an HPM weapon system.

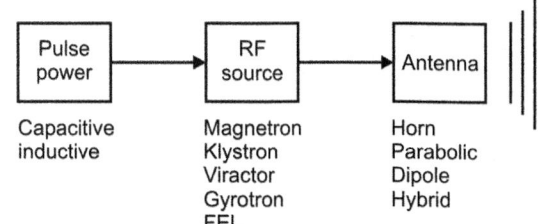

Figure 12.7 Marx bank power supply arrangement.

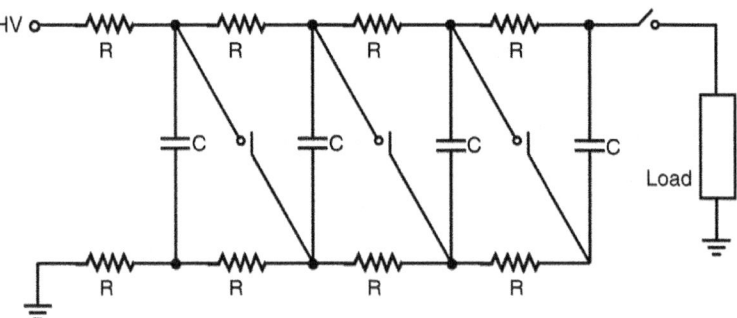

12.4.2.1 Pulse Power Generators

The *pulse power generators* that drive high power microwave sources are generally required to deliver short, intense electrical pulses of 1 MV or more with pulse duration up to 1 μs. One way to generate the required pulses is by using capacitor banks that transform a slowly rising low-voltage signal into a fast rising high voltage signal. A common capacitor bank configuration is the Marx bank where the capacitors in the bank are connected in parallel during the charging process and switched to a series connection during discharge. The series connection multiplies the voltage by the number of capacitors in the Marx bank. Figure 12.7 shows a typical Marx bank. Resistive charging is slow and therefore limits the repetition rate. Inductors could also be used in place of resistors. Inductor charging is preferred for higher repetition rates of few Hz or higher. If the Marx bank had *n* capacitors with each charged to a voltage, V, from a DC power supply; the voltage delivered to the load during discharge would theoretically be equal to nV. Spark gaps are used as switches and the breakdown voltage of spark gaps is kept higher than the voltage, V, across each capacitor. Initially all capacitors are in parallel and are charged to a voltage, V. The spark gaps are in open state. In order to initiate the discharge, the first spark gap is externally triggered to the breakdown state connecting the first two capacitors in series and thereby raising the voltage across the second spark gap to (2 V). The second gap also goes to breakdown state and the process continues till a voltage pulse with amplitude equal to nV is applied across the load. One such Marx generator designed to drive HPM sources is the APELC MG20–22C-2000PF from Applied Physical Electronics LC (APELC). This megavolt-class Marx generator with its 18 Ω source impedance is specifically designed to drive low-impedance loads. With a 50 kV charge voltage, it delivers a 500 kV, 1.1 kJ pulse into a matched load with a peak power of more than 12 GW. This generator uses low-impedance, parallel-switched topology, which makes it well suited for a wide variety of high power microwave applications. Another Marx generator from the same company is the MG30–3C-100nF (Figure 12.8) that is capable of storing a maximum of 1.8 kJ and can deliver 300 kV to a matched load. This Marx

Figure 12.8 Marx generator type MG30-3C-100NF. (*Source:* Courtesy: APELC.)

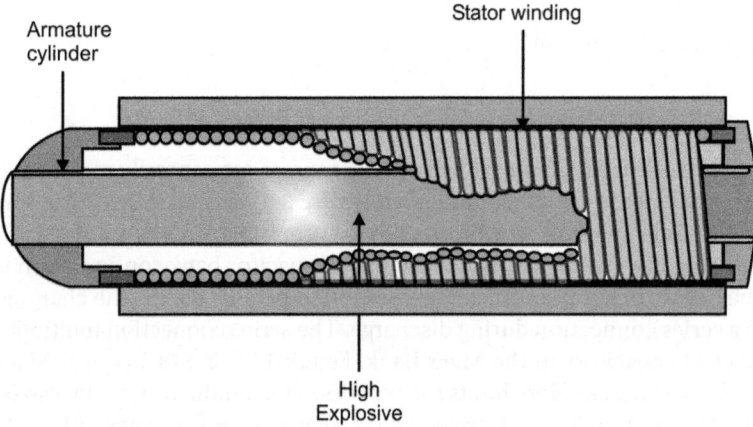

Figure 12.9 Flux compression generator.

generator has low impedance of 33 Ω and is axially-compact in order to drive HPM antennas on remote platforms. Maximum peak power and repetition rate specifications are 5 GW and 10 Hz, respectively.

The other method is to use a flux compression generator. In a flux compression generator, a magnetic coil is compressed either by explosive or magnetic forces leading to rapidly rising current pulse. In an explosively driven flux compression generator, the chemical energy of the explosives to is partially converted into the energy of an intense magnetic field surrounded by a correspondingly large electric current. Explosively driven flux compression generators are for one-time use only as the equipment gets destroyed during operation. Such a flux compression generator constitutes an important component of an E-bomb. Figure 12.9 shows the basic construction of an E-bomb configured around explosive driven flux compression generator (E-bombs are described in Section 12.4.6). It consists of a metal cylinder called the armature, which is surrounded by a coil of wire called the stator winding. The armature cylinder is filled

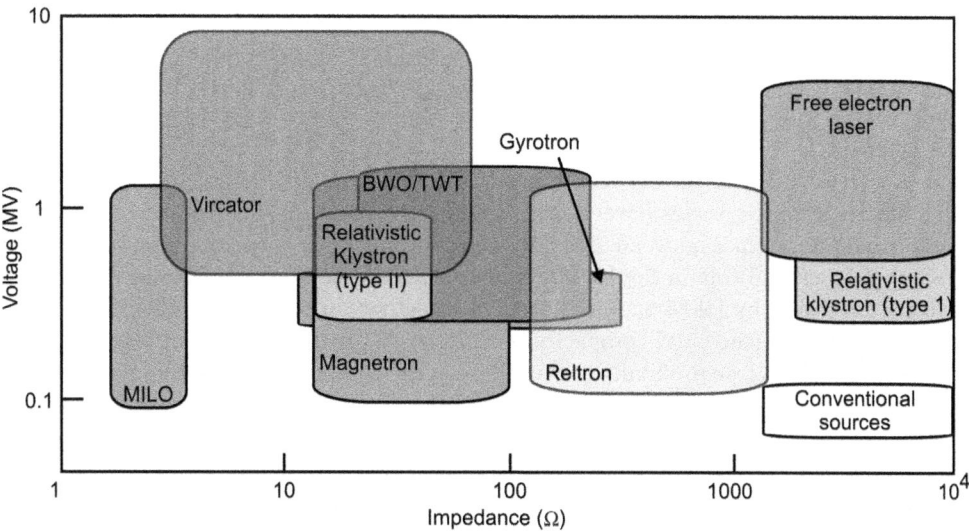

Figure 12.10 Comparison of conventional and HPM sources.

with high explosive, and a sturdy jacket surrounds the entire device. The stator winding and the armature cylinder are separated by empty space. A Marx bank type power source used to drive the stator winding is also a part of the bomb. An electrical current through the stator winding generates an intense magnetic field. The explosive when triggered travels as a wave through the middle of the armature cylinder forcing it to come in contact with the stator winding thereby creating a short circuit. The moving short circuit compresses the magnetic field thereby generating an intense electromagnetic burst. Another common option is to replace the flux compression generator by a pulse forming network.

Caution may be exercised to ensure that the pulsed power source delivers a well-matched signal with regards to voltage and impedance. An impedance mismatch between the driver signal and the HPM source leads to poor energy transfer to the HPM source, more so if the impedance of the source were lower than the impedance of the driver. The choice of pulsed power source is therefore largely governed by the type of HPM source to be used. It may be mentioned here that conventional microwave sources used in radar and so on and HPM sources operate at different voltage and impedance levels. While conventional microwave sources typically operate on relatively low voltage and high impedance, the HPM sources operate at relatively higher voltage and lower impedance levels. Figure 12.10 illustrates a comparison between conventional and HPM sources on the basis of operating ranges of voltage and impedance levels.

12.4.2.2 HPM Sources

HPM sources are broadly classified as *impulsive sources* and *linear beam sources*. In the case of impulsive sources, pulsed microwave energy is generated by charging the antenna, a transmission line, or a tuned circuit directly and making them to ring for one or several cycles by closing a switch. In the case of linear beam sources, microwave energy is generated by converting kinetic energy of an electron beam into electromagnetic energy of the microwave beam. While various ultra-wideband sources are examples of impulsive sources, examples of linear beam sources include klystrons, travelling wave tubes, backward wave oscillators, magnetrons, cross field amplifiers, split-cavity oscillators, virtual cathode oscillators (VIRCATOR), gyrotrons, free electron lasers and orbitron microwave masers. Impulsive and major linear beam sources are briefly described in the next section. Free electron lasers are discussed in Section 12.4.2.2.

12.4.2.3 Antennas

An antenna acts as an interface between the transmitter output and the medium in which the radiated electromagnetic waves have to propagate through. In the case of an HPM weapon system too, it is an interface between the HPM source output and the surrounding atmosphere. Antennas play a crucial role in the HPM system design. Some of the important factors that need to be addressed by HPM antennas include directivity, ultra-wide bandwidth, feed-to-antenna coupling efficiency and compactness. *Directivity* plays an important role. Since the HPM equipment is in close proximity to the antenna, the equipment may suffer damage if the radiated energy were not radiated effectively. HPM signal due to its very short pulse duration inherently has a very wide bandwidth. The antenna therefore has to meet the stringent requirements of handling this ultra-wide bandwidth. Another important requirement is that of *proper matching* between the feed and the radiating element lest the resultant standing waves will cause voltage breakdown. HPM antennas tend to be massive in order to avoid voltage breakdown at the operating electric field levels. To meet the increasing requirements of having HPM weapon systems on smaller platforms, antenna size can play an important role. Antenna shape is also an issue as it influences to a great extent whether air breakdown phenomenon is an issue or not at high power levels. Horn antennas and antenna arrays are the promising candidates for HPM sources with the former being presently the most commonly used type. Different types of horn antennas including conical, circular, rectangular, corrugated, half-oval and TEM horn antennas are used with a particular design preferred depending upon the intended application.

12.4.3 HPM Sources

In the following paragraphs impulsive HPM sources and different types of linear beam sources are discussed such as klystrons, travelling wave tube amplifiers, backward wave oscillators, magnetrons, split-cavity oscillators, virtual cathode oscillators, gyrotrons and free electron lasers.

12.4.3.1 Impulsive HPM Sources

Impulsive sources convert energy from a power supply into a short pulse of electromagnetic radiation by storing electrical energy slowly, typically in a capacitance, and discharging it rapidly. Unlike the linear beam sources discussed in this section, they do not utilize electron beams to generate their electromagnetic energy. These sources are sometimes known as wideband or ultra-wideband (UWB) sources, since their bandwidth is typically a large fraction of their centre frequency. One such impulsive high-power microwave source is the *SNIPER (Sub-Nanosecond ImPulsE Radiator)* developed at Sandia labs. The source runs at 290 kV at greater than 1 kHz pulse repetition rate generating 1.25 GW peak power in the 3.5 ns wide pulse having a rise time specification of 140 ps. The source produces a good spectral content from 100 MHz to 1.2 GHz and field strength of 120 kV/m using a TEM horn antenna. *EMBL (EnantioMorphic BLumlein* meaning *mirror-image Blumlein)* runs at 750 kV and a repetition rate of 700 Hz, which is limited by the capability of the high voltage power supply. The peak power, pulse width and pulse rise time specifications respectively are 11 GW, 3.5 ns and 200 ps. EMBL radiates a 285 ps impulse with a differentiating TEM horn antenna with its spectral content extending from 200 MHz to 1.2 GHz with peak emission at 800 MHz and field strength normalized at a 1 m distance of 350 kV/m. The other type of impulsive HPM sources is built by charging a transmission line or an LC oscillator. A bipolar TLO (Transmission Line Oscillator) built at Sandia labs and operating at 1.4 MV produces a single shot output pulse of 20 GW peak power.

12.4.3.2 Linear Beam Tubes

One type of HPM linear beam sources includes linear beam tubes such as the Relativistic Klystron Amplifier (RKA), Travelling Wave Tube (TWT) and Backward Wave Oscillator. HPM versions of these tubes are very similar in concept to their conventional counterparts.

12.4.3.2.1 Klystron

A *klystron amplifier* converts kinetic energy in the DC electron beam into RF power. It does so by a phenomenon called velocity modulation causing formation of electron bunches. Velocity modulation takes place as the electron beam passes through the buncher cavity where input RF excitation is applied. The polarity of the electric field in the buncher cavity changes every half cycle of the applied RF signal, which alternately accelerates and decelerates the electrons in the beam as it passes through the grids in the buncher cavity. Velocity modulation produces electron bunches in the drift space when the accelerated electrons tend to overtake the decelerated electrons. The electron bunches transfer their kinetic energy to the RF wave in the catcher cavity from where the amplified output is extracted. The catcher grids are placed along the beam at a point where electron bunches are fully formed with the location determined by the transit time of bunches at the natural resonant frequency of buncher and catcher cavities. The two cavities have the same resonant frequency. Figure 12.11 illustrates the process of RF signal amplification in a two-cavity klystron amplifier. Multi-cavity klystrons having intermediate cavities between buncher and catcher cavities yield greatly improved power output and efficiency.

12.4.3.2.2 Travelling Wave Tube

The *travelling wave tube amplifier* (TWT) also employs velocity modulation phenomenon. While in the case of klystron amplifier the RF field is stationary and the electron beam is moving, a TWT amplifier makes use of the distributed interaction between the electron beam and a travelling wave. For this to happen, it is important that both RF field and electron beam travel at same speed. Since an electron beam travels at a speed typically 2–10% of velocity of

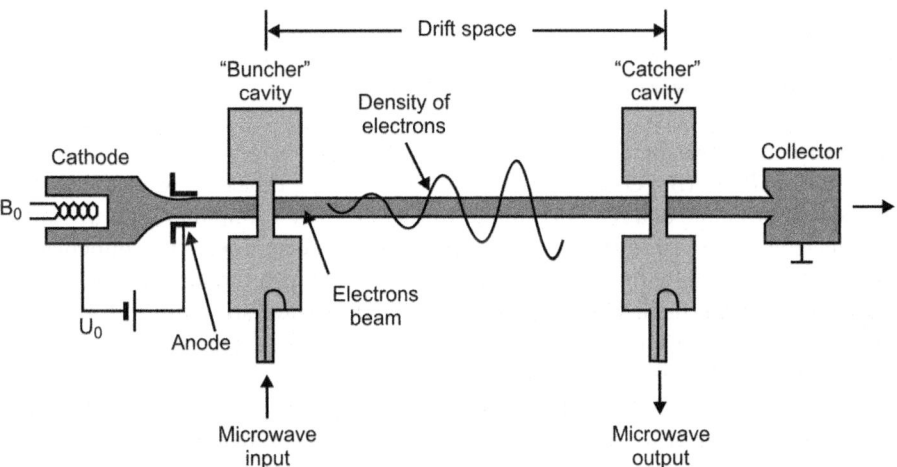

Figure 12.11 Two-cavity klystron.

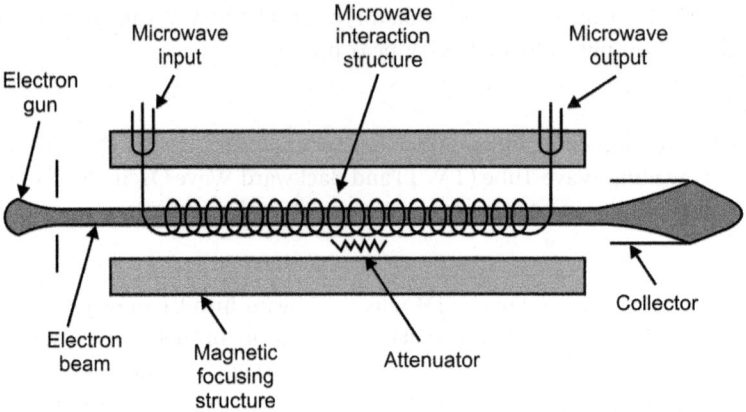

Figure 12.12 TWT amplifier.

electromagnetic waves in free space, the velocity of the travelling RF wave is retarded by using some kind of a slow wave structure such as a helix. Figure 12.12 shows the operation in a TWT amplifier. Ring loop, ring bar and coupled-cavity are the other common slow wave structures. The structure is chosen to achieve desired gain bandwidth and power characteristics. Velocity modulation is produced by the interaction between the travelling wave fields and the electron beam. Bunching would cause the electrons to give up energy to the travelling wave if the fields were of the correct polarity to slow down the electron bunches. This energy transfer from the electron bunches to the travelling RF wave increases the amplitude of the travelling wave in a progressive action. The travelling RF wave therefore builds up as it moves along the length of the TWT.

A *ring loop TWT* uses loops as slow wave structure to tie the rings together. This slow wave structure has high coupling impedance and low harmonic wave components. Ring loop TWTs are capable of higher power levels than conventional helix TWTs. However, they are characterized by a relatively lower cut-off frequency and significantly less bandwidth. *Ring bar TWT* has characteristics similar to those of ring loop TWT. The advantage lies in the easier fabrication of this slow wave structure. The *coupled-cavity TWT* slow wave structure consists of a series of staggered cavities coupled to one another with a transmission line. Since cavities have bandwidth limitations; coupled-cavity TWT too offers much narrower bandwidth of about 10–20% as compared to almost 100% achievable in the case of a helix TWT. The coupled-cavity TWT, though, has higher output power capability than a helix TWT.

12.4.3.2.3 Backwards Wave Oscillator

A *backwards wave oscillator* (BWO) is the oscillator version of the travelling wave tube amplifier. The operational principle of a BWO is similar to that of a TWT amplifier with the difference that the electron beam transfers its energy, present in the form of electron bunches and acquired due to velocity modulation caused by RF field propagating in the slow wave structure, to the backward moving RF field. The output end of the slow wave structure in this case is attached to a suitable termination and the amplified output signal is extracted from the input side of the slow wave structure (Figure 12.13).

The main differences between the conventional linear beam tubes briefly described previously and the HPM version of these tubes lie in the techniques of beam formation, use of pulsed large amplitude axial magnetic fields for beam transport and application of relativistic

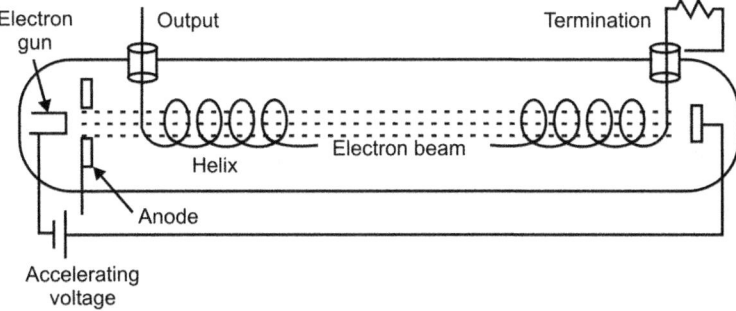

Figure 12.13 Backwards wave oscillator.

voltages and high beam currents. In the case of HPM klystron amplifiers, two categories of devices have been built by following two different approaches. In one type of device, the focus is on building high average power, low beam current tubes that can be operated on continuous, repetitively pulsed basis. The Stanford Linear Accelerator (SLAC) is an example. SLAC klystron produces 67 MW of RF power at 2.856 GHz. The tube operating at 350 kV produces 3.5 μs wide pulses at a repetition rate is 180 Hz. The tube can also operate at 415 kV to produce 100 MW, 1 μs pulses at an efficiency of 48%. The tube has beam current in the order of several hundred amperes. In the second approach, the beam current is high and the space-charge effects become dominant in bunching process. These klystron tubes use electron beams with energy levels in the range of 500–1000 keV guided by axial magnetic fields of about 10 kilo-Gauss. Power levels of 10 GW at 50% efficiency have been realized. Both types of relativistic klystron amplifiers are narrowband devices. While in the case of klystron amplifiers cavities are used for electron beam bunching and power extraction as outlined earlier, TWT amplifiers and BWOs employ continuous slow wave structures. BWOs have demonstrated efficiency levels as high as 35% at moderate power levels. Large axial magnetic fields in the range of 25–50 kilo-Gauss are used to maintain high beam quality at higher power levels. TWT amplifiers are similar to BWOs with the difference that, in the case of former, the interaction of the electron beam is with a forward wave.

12.4.3.3 Magnetrons

The *magnetron* is a type of high power microwave oscillator in which the potential energy of an electron cloud near the cathode is converted into RF energy in a series of cavity resonators whose resonant frequency is determined by the physical dimension of the resonator together with the reactive effect of any perturbations to the inductive or capacitive portion of its equivalent parallel L-C circuit. Magnetrons are capable of producing megawatts of peak power in the centimetre wavelength range and may be operated at wavelengths extending down to millimetre range. The source of electrons in a magnetron is a heated cathode located on the axis of an anode structure containing a number of microwave resonators. Figure 12.14 shows the constructional features of a magnetron. It mainly comprises a cathode, an anode block with cavity resonators, a permanent magnet to produce the axial DC magnetic field and a mechanism to couple out the microwave signal such as a coupling loop and cooling fins.

Electrons emitted by the cathode are accelerated towards the anode structure by a radial DC electric field between the anode and the cathode created by an applied DC voltage. In addition to the DC electric field, there is a DC magnetic field in the region between the anode and cathode in a direction along the axis of the cathode. The magnetic field is perpendicular to

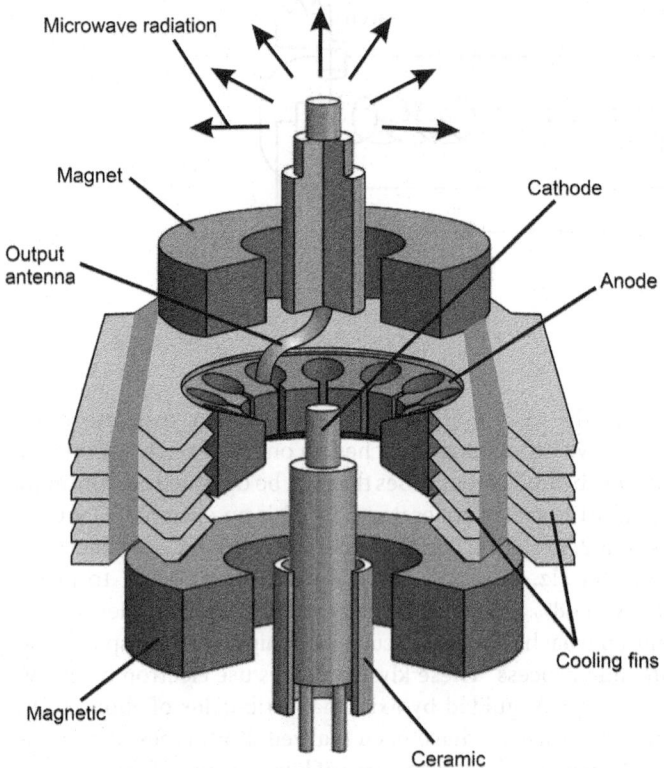

Figure 12.14 Construction of a magnetron.

the radial electric field. The electrons spiral away from the cathode due to the force exerted on the electrons due to presence of a strong DC magnetic field produced by permanent magnet. This force is mutually perpendicular to the DC magnetic field and electron velocity vector. The electron cloud, as it moves away from the cathode, comes under the influence of the RF field in the cavity resonators. The electrons are either retarded or accelerated depending upon whether they encountered an opposing or aiding RF field. Retarded electrons experience reduced curling force and drift towards the anode while accelerated electrons experience increased curling force and curl back away from the anode. As the electron cloud nears the anode, there is a collection of electron spokes with each spoke located at a resonator with an opposing RF field. In the next half cycle of RF field, the spoke pattern rotates to maintain its location in an opposing RF field (Figure 12.15). While the accelerating electrons take energy from the RF field; retarding electrons give energy to the RF field as they slow down. The electron cloud spokes, also called the space-charge wheel, rotate about the cathode at an angular velocity of two poles or anode segments per cycle of the RF field. This phase relationship enables the concentration of electrons to continuously deliver energy to sustain the RF oscillations. The RF energy at microwave frequency is coupled out of the cavity resonator by means of a coaxial line or a waveguide structure.

HPM magnetrons are the cold cathode versions of conventional magnetrons described previously and are basically relativistic. They are characterized by relatively high efficiency, low power densities internal to the tube, robust operation and compact in size. Relativistic magnetrons have achieved efficiencies in the range of 10–30% over 0.5–10 GHz band at power levels

Figure 12.15 Formation of a spoke pattern in a magnetron. (*Source:* Courtesy: Radartutorial.)

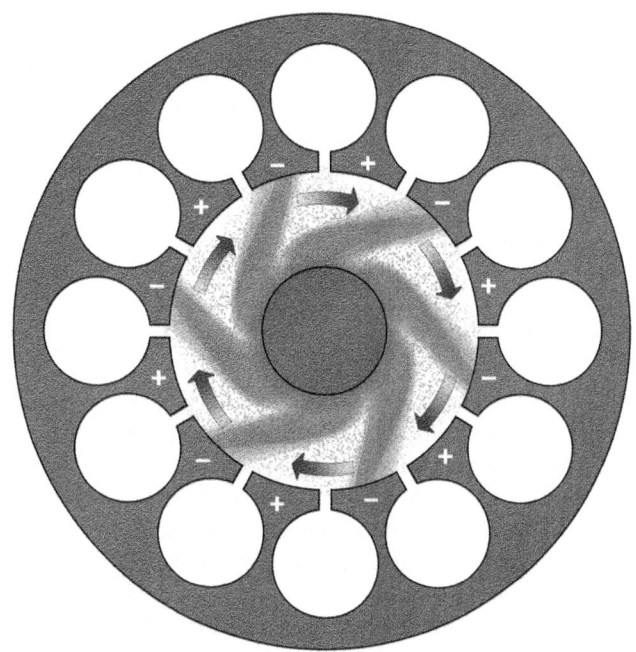

of about 5 GW. Pulse widths are limited by plasma closure of the anode-cathode gap and are typically in the order of 100 ns or less. Magnetrons may be phase locked for higher output power. A type of magnetron known as a Magnetically Insulated Line Oscillator (MILO) has also been developed. It makes use of the magnetic field of the beam current itself to provide the required magnetic insulation between the cathode and anode. Though it has relatively lower efficiency, MILO eliminates the need for pulsed magnets and their power supplies.

12.4.3.4 Split-Cavity Oscillators

The Split-Cavity Oscillator (SCO) utilizes transit-time bunching of an electron beam to generate microwave energy. Since there is no requirement of a magnetic field and low beam quality consistent with simple pulse power sources is adequate, the SCO makes up a very compact high-power microwave source. Pulses in the range of 100 MW for durations on the order of 100 ns have been achieved. An entire system including power supply and mode converting antenna has been built on a roll-around laboratory cart. Multiple SCOs have been injection locked.

12.4.3.5 Virtual Cathode Oscillator

The virtual cathode oscillator (or vircator) is capable of generating short pulses of tunable, narrowband microwaves at very high power levels. The vircator belongs to a class of HPM sources that generate microwave radiation by the phenomena of virtual cathode or/and reflex electron oscillations accompanying injection of an electron beam into a waveguide or cavity in which beam current exceeds the local space-charge limiting current. A cold cathode emits a pulse of highly accelerated beam of electrons by field emission mechanism. The electron beam is attracted towards a thin anode such as aluminized PET film or stainless-steel mesh. A large proportion of electrons pass through the anode forming a bubble of space-charge behind the

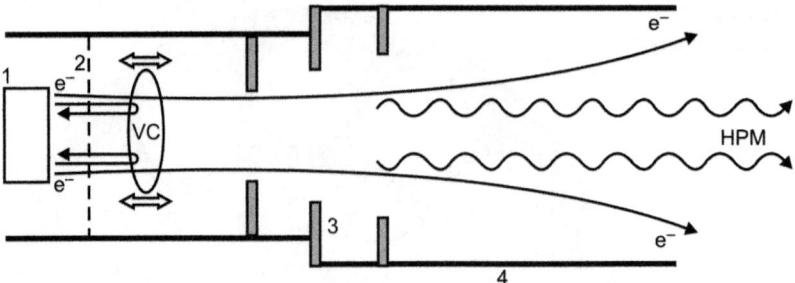

Figure 12.16 Axial vircator.

anode that acts as *virtual cathode*. Due to mutual repulsion of electrons and attraction by the anode, the electrons reverse their direction to pass through the anode again. They are repulsed by real cathode and attracted by anode again. The rapidly accelerating and decelerating electrons, moving back and forth between the real and virtual cathode through the mesh anode, generate oscillations at microwave frequencies. An axial vircator typically comprising a cylindrical waveguide structure and transitioned into a conical horn structure to extract output power is one of the two commonly described configurations (Figure 12.16). The transverse vircator is the other. While an axial vircator typically oscillates in transverse magnetic (TM) modes, a transverse vircator injects cathode current from the side of the cavity and typically oscillates in a transverse electric (TE) mode.

Vircators are capable of generating gigawatt level output in the 1–10 GHz frequency range. Key advantages of vircators are as follows: they are simple to build and in many cases no magnetic field is required; they operate at relatively low impedances, which allows them generate power at relatively low voltages and couple efficiently to low-impedance power sources; they are tunable as frequency depends only on charge density and not on any resonance condition. A single device is capable of tunability of up to two octaves. The disadvantages are that they have relatively low efficiency and are sensitive to the problem of gap closure.

12.4.3.6 Gyrotrons
Gyrotrons are capable of generating high power, high-frequency electromagnetic radiation in the terahertz (THz) range. In the case of conventional microwave tubes described in earlier paragraphs, power versus frequency curve varies as $P \propto (1/f^{2.5})$. These tubes become almost unusable at 30 GHz and beyond. It is because of this that at frequencies of 30 GHz, it is impractical to make the interaction structures that would allow the microwaves to travel at the electron beam velocity. Even if it were feasible, the small structure wouldn't be able to handle high average power due to difficulty of het removal. The operation of a gyrotron is based on the interaction of an axial electric field and an electron beam moving in a spiralling trajectory under the influence of a magnetic field. The spiralling frequency called the cyclotron frequency depends upon the strength of the magnetic field, mass of the electron and the velocity component perpendicular to the magnetic field. The theory of relativity suggests that a fast-moving particle, which has higher energy and therefore larger mass, will result in a decrease in cyclotron frequency. Also, energy coupling causes electrons lose energy, which reduces its mass. This results in increase in cyclotron frequency. All this leads to electron bunching similar to that experienced in the case of klystrons and TWTs. Figure 12.17 shows the constructional features of a typical gyrotron tube. The electron gun and the magnet are so designed as to have a gyrating path for the moving electrons. The cavities are designed to have a beam opening that is 10 times larger than the beam opening in an equivalent klystron cavity in order to accommodate the large diameter hollow beam. A gyrotron amplifier tube becomes an oscillator when

Figure 12.17 Constructional features of a gyrotron

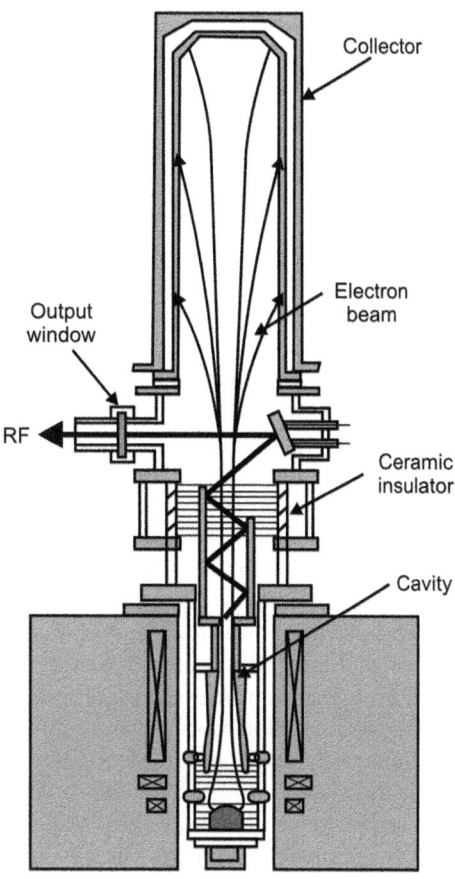

internal feedback is added. Gyrotron oscillators are capable of generating hundreds of kilowatts of CW power at 100 GHz and tens of kilowatts of power at frequencies beyond 100 GHz.

Magnicons and *ubitrons* are the other microwave tubes belonging to the family of gyrotrons. In the case of a magnicon, the interaction between the RF field and the electron beam utilizes TM rather than TE modes. An ubitron is also similar to a gyrotron making use of interaction between a periodic electron trajectory and a linear circuit rather than utilizing a periodic circuit and a linear electron beam as is the case in conventional microwave tubes. The interaction area can be as large as 100 times; for example, in the case of a travelling wave tube at the same frequency. This makes much greater output powers possible in the case of gyrotrons and ubitrons.

12.4.4 HPM Weaponization

A number of technological and operational issues need to be addressed if HPMs were to move out of the confines of research and application laboratories to become serious contenders as operational weapons for defensive and offensive roles. These include issues of weapon systems' compactness and efficiency to suit a range of military platforms, issues related to antenna aperture to get desired operational ranges and peak versus average powers. Target acquisition, tracking and beam pointing technologies and lethality assessment are the other important factors.

Compactness is an important requirement of all military systems, more so when it is to be integrated on a military platform. Size and weight of the system to be integrated has to be commensurate with the capabilities and limitations of the platform. Size and weight often become a major constraint for integration with airborne platforms. The size and weight of an HPM weapon system is not governed by HPM technology used alone. The prime power source, power conditioning, microwave energy transport to the antenna and the antenna contribute significantly in deciding system size and weight. Also, peak or/and average power output capability of the system directly influences the level of compactness of these subsystems. The *efficiency* with which input energy to the HPM source is converted into usable microwave energy output also impacts on a weapon system's compactness as lower efficiency means increased prime power source requirements and, consequently, increased size and weight. Further, efficiency of most HPM sources is strongly dependent on beam quality. The gains in compactness achieved through increased efficiency may be counterbalanced by an associated higher beam quality requiring fairly complex systems that increase size and weight.

In order to achieve desired levels of power density on the target to inflict intended damage at the required operational ranges, the antenna must handle high powers and have high directivity. Both these factors have a bearing on the *antenna size*. High directivity requires an antenna many wavelengths in size. Increase in power level through a given antenna aperture increases the probability of atmospheric breakdown, more so at high altitudes and in dusty environments. A large antenna size is often not able to meets the requirements of manoeuvrability, survivability under fire, drag, visual signature and so on for want of space on practical military platforms.

Peak/average power level and operating frequency are the other factors influencing HPM weaponization. HPM sources have typically been designed to provide high peak power, but not high average power. Average power handling capability of the HPM source is limited by power conditioning performance, vacuum under repetitively pulsed conditions, cooling requirements and materials. Wave structures used in microwave tubes to match beam and wave parameters limit the average power handling capability due to breakdown, heat dissipation and spurious mode generation. The active heat load for these devices is limited to several kW/cm^2. This creates, of course, an average power/volume constraint. Also, cavity dimensions are inversely related to the operating frequency. Higher operating frequency means smaller cavity dimensions and lower breakdown levels even for single shot operation. Magnetrons, linear beam tubes and vircators therefore operate at high peak powers below about 10 GHz. The problems arising from smaller wave structures are not much of an issue in gyrotrons and free electron lasers as the dimensions of their interaction regions are much larger than the wavelength being produced. The high peak powers of many of the HPM sources require high field gradients in the order of megavolts per centimetre, which cause surface emission and the formation of surface plasmas. These plasmas cause gap closure that detunes the tubes and truncates individual pulses. High current densities lead to bipolar space-charge limited flow, where the anode emits positive ions. At still higher current densities, they create anode spots that produce micro-particles leading to gap closure and breakdown. Operation at high repetition rates causes cathode material burnout. Also, at high repetition rates, vacuum systems may not be able to re-establish the base vacuum levels for the next shot thereby limiting the achievable repetition rate and average power.

To ensure that the HPM weapon system meets the projected operational and technical specifications, it is important that the design is comprehensively evaluated by software iterations rather than time consuming and expensive hardware iterations. This task is achieved through *modelling and simulation*. Target acquisition and tracking, beam pointing and lethality assessment are some other important supporting technologies for HPM weaponization. *Target acquisition* and *tracking* is relatively easy in the case of HPM weapons as compared to high-energy laser weapons due to much larger involved beam widths and spot sizes. Since microwaves penetrate through smoke, clouds and fog, target tracking may have to depend on radar imaging techniques. *Beam pointing* can be difficult with large size mechanically steered antennas. Electronic beam steering would depend on accurate phase control of multiple HPM sources. *Lethality assessment* is perhaps the most difficult of all supporting technologies for reasons described in an earlier paragraph.

12.4.5   Capabilities and Limitations

In the following are the key advantages of high power microwave-based DEWs.

1) High power microwaves are *speed-of-light weapons* with essentially no time-of-flight as compared to finite travel time of conventional kinetic energy weapons. A supersonic missile travelling at a 6 Mach speed would take a flight time of about 50 s to hit a target 300 km away. On the other hand, a high-power microwave weapon could hit the same target in a millisecond or so. This feature is particularly suitable for attacking fast moving targets.
2) Unlike high-energy laser weapons, to be described in a later part of the chapter, that are clear-weather weapons, HPM weapons are not adversely affected by weather and atmospheric conditions. For all practical purposes, they are *all-weather weapons.*
3) HPM weapons are *area weapons* unlike high-energy laser weapons that are point weapons. The footprint depends on the weapon's power output, distance from the target and beam divergence. They affect all vulnerable equipment that comes within their lethal footprint (Figure 12.18).
4) HPM weapons are ideally suited to targeting in urban environment due to *minimized collateral damage.* Some collateral damage may occur if friendly or civilian electrical systems are also in the footprint.

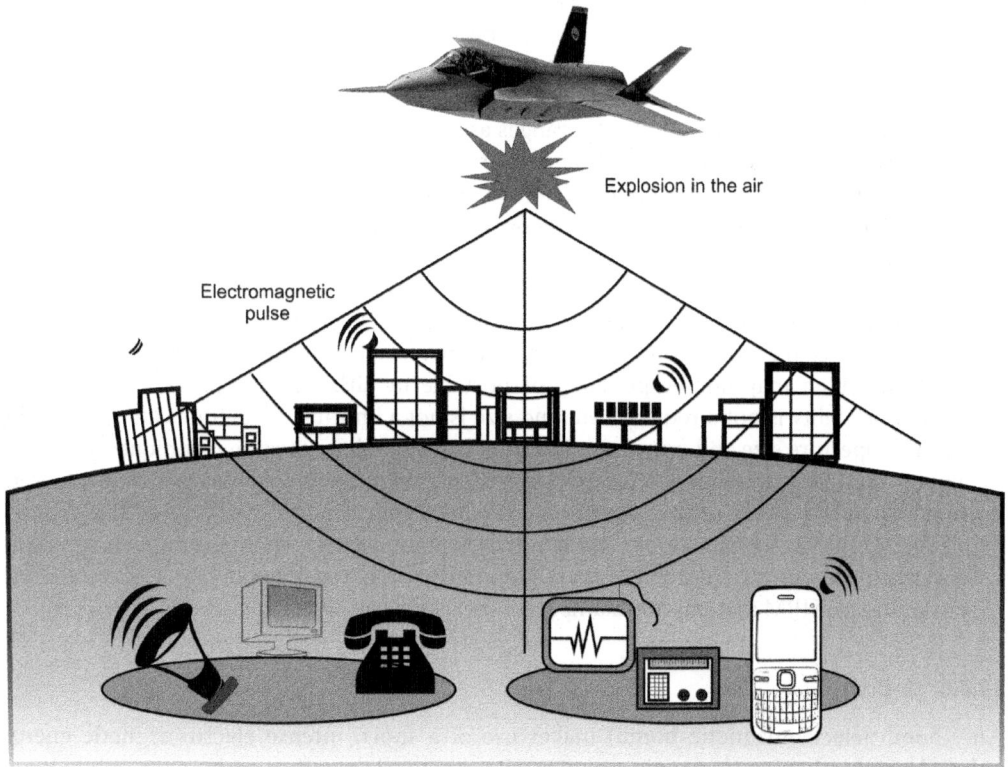

Figure 12.18 Lethal footprint – HPM weapons.

5) HPM weapons have a *deep magazine*. Unlike conventional kinetic energy weapons that are single shot weapons, HPM weapons can operate any number of times as long as there is power source. This also implies reduced logistics cost and cost per shot compared to kinetic energy weapons.

6) HPM weapons' *effects are scalable*. Depending upon the magnitude of energy deposited on the target, the effects may be nonlethal or lethal. The Active Denial System of the USA is a nonlethal HPM system designed for area denial, perimeter security and crowd control. It operates by heating the target's surface such as the skin of targeted human subjects. Also, HPM weapons have the potential of jamming the targeted system without any knowledge about the RF output of the system.

7) In the case of an HPM attack, targeted system recovery is extremely difficult as it may require component or subsystem level troubleshooting.

8) HPM weapons are highly effective against deeply buried bunkers by targeting vulnerable electronic systems such as communications, power and air ventilation systems that support these bunkers.

Major limitations of HPM weapons include the following.

1) HPM weapons are area weapons and therefore adversely affect all unprotected electronic systems within their lethal footprint. These would include civilian and friendly systems. They could be protected by using proper shielding. Proper planning before an HPM attack could also prevent unintended damage to friendly assets.

2) In the case of an HPM weapon attack, target damage assessment is very difficult as there are signs of any physical destruction. Just because a system has stopped operating or emitting does not mean it has been affected by the HPM attack. Both absence and presence of emission from the targeted system doesn't reliably establish the efficacy of attack. However, advanced techniques may be used to measure second and third order effects to corroborate the results and enhance attack assessment.

3) HPM weapons have relatively shorter ranges as compared to high-energy laser weapons. In the case of an HPM weapon system, range is proportional to both power output and antenna size. However, at sufficiently high microwave power levels, the atmosphere at the antenna aperture becomes plasma, a phenomenon called atmospheric breakdown. The plasma density increases with time duration of high power microwave pulses eventually reaching the point where the plasma reflects and absorbs the RF energy. This not only renders the beam ineffective but also creates a shielding problem for the delivery platform. Once atmospheric breakdown occurs, operational range can be increased only by increasing the antenna aperture, which may not be possible for portable or airborne designs. For applications where portability is not a requirement and size is not a limiting factor, one can increase the effective aperture size and power on the target by using phase locking techniques and combining multiple transmitters. Another approach may be adopting designs that integrate antennas into the skin of the system, which allows aperture to be as large as the delivery platform. One such application scenario is that of point defence where the target is travelling towards the weapon system rather than weapon travelling towards the target. Ranets-E and Vigilant Eagle HPM systems are examples of these types of systems.

12.4.6 E-Bomb

An e-bomb (electromagnetic bomb) makes use of a short, intense electromagnetic energy pulse adversely affecting electronic circuitry with practically no effect on humans or buildings. The damage may range from temporarily disabling electronics systems functioning at low levels of EMP (electromagnetic pulse) energy to corrupting computer data at mid-range levels and further to complete destruction of electronic circuitry at high levels. The EMP effect first observed during the early testing of high-altitude airburst nuclear weapons, is characterized by production of a very short, typically hundreds of nanoseconds wide, intense electromagnetic shock wave travelling away from the source. This short and intense magnetic field is capable of producing short-lived transient voltages of kV level on exposed electrical conductors including wires or conducting tracks on printed circuit boards. This makes it militarily significant as it can inflict irreversible damage to a wide range of electronics and computer systems in command, control, communications hardware and radar systems. Radar and electronic warfare equipment, satellite, microwave, UHF, VHF, HF and low-band communications equipment and television equipment are all potentially vulnerable to the EMP effect. The inflicted damage is similar to that experienced through exposure to close proximity lightning strikes and may require complete replacement of the equipment, or at least substantial portions thereof.

Commercial computer equipment is particularly vulnerable to EMP effects due to large-scale use of high density MOS (Metal Oxide Semiconductor) devices that are highly sensitive to high voltage transient exposure. Shielding provided by equipment chassis in this case gives only limited protection as the high voltage transients find their way into the equipment through input/output cables. Computers used in data processing systems and communications systems and those embedded in military equipment are all potentially vulnerable to the EMP effect.

The key technologies involved in generation of high power electromagnetic pulse, which mainly include explosively pumped Flux Compression Generator (FCG), explosive or propellant-driven Magneto-Hydrodynamic Generator (MHD) and Virtual Cathode Oscillator (Vircator), have matured to an extent that realizability of battlefield deployable E-bombs has become feasible. This has opened up many a new application in both tactical and strategic information warfare. The development of conventional E-bomb devices has also allowed their use in nonnuclear confrontations. In the following paragraphs are discussed operational principle of E-bombs, the lethality of E-bombs and their delivery mechanisms. A brief reference is made to some of the experiments carried out with E-bombs.

12.4.6.1 Operational Principle

The explosively pumped flux compression generator (FCG) capable of producing tens of megajoules of electrical energy in a time frame of tens to hundreds of microseconds with corresponding peak power levels of tens of terawatts is the most mature technology applicable to design of E-bombs. An E-bomb configured around an FCG is what is known as the low frequency E-bomb with and its output due to nature of its basic physics constrained to a frequency band below 1 MHz. FCGs may also be used as one-shot pulse power supplies to drive high-power microwave tubes such as vircators to build an HPM E-bomb.

The basic idea behind the construction of FCG-based E-bombs is that of using a fast explosive to rapidly compress a magnetic field, transferring most of the explosive energy into an intense magnetic field. The initial magnetic field in the FCG prior to explosive initiation is produced by a start current, which is supplied by an external source typically a Marx bank. Several types of geometrical configurations have been experimented for building FCGs with coaxial FCG being the most commonly used of all. Its essentially cylindrical form factor lends itself well to packaging into munitions. The basic operation of an FCG was described in an earlier paragraph under the heading of pulse power generators. The current multiplication, ratio of output current to start current, is design dependent. In a munition application, where space and weight are at a premium, the smallest possible start current source is desirable. A current amplification of 60 has been achieved. A cascade arrangement of FCGs may be used in this case. In this case, a small FCG is used to drive a larger FCG with a start current. Figure 12.19 shows component parts of a low frequency E-bomb that employs a two-stage FCG.

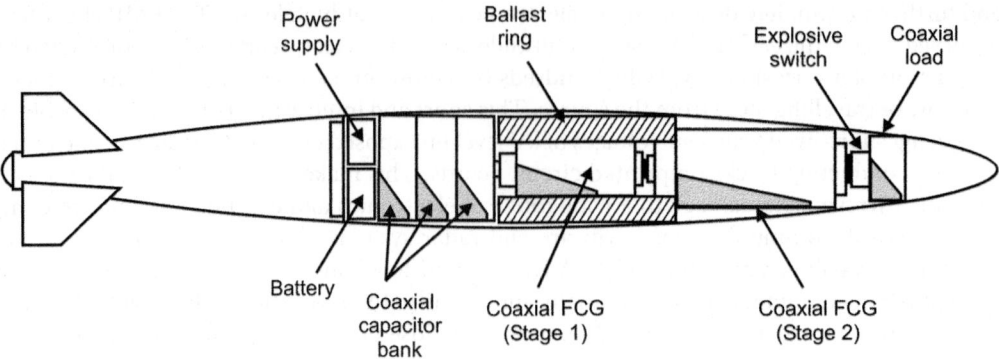

Figure 12.19 FCG-based low-frequency E-bomb.

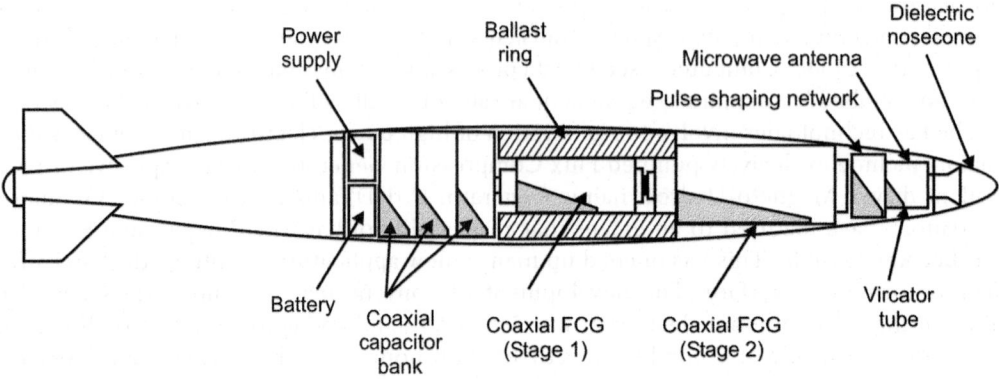

Figure 12.20 HPM E-bomb.

Explosive and propellant-driven *Magnetic-Hydrodynamic (MHD) Generator* technology is another candidate used for generating electromagnetic pulse. MHD generator depends for its operation on the movement of a conductor through a magnetic field producing an electrical current transverse to the direction of the magnetic field and conductor motion. The technology of MHD generator is not mature yet enough to build practical devices and there are issues related to the size and weight of magnetic field generating devices required for their operation. MHD devices may be used for generating start current for FCG devices. The conductor in an MHD generator is a plasma of ionized explosive or propellant gas, which travels through the magnetic field. Current is collected by electrodes that are in contact with the plasma jet. The electrical properties of the plasma are optimized by seeding the explosive or propellant with suitable additives.

Low frequency E-bombs configured around FCGs or MHD generators are far less lethal than HPM E-bombs as a large cross-section of targets are difficult to attack even with very high power levels at their output frequencies. Focusing the output energy from these devices is also a huge problem. On the other hand, microwave weapons not only couple more readily to a variety of targets than low frequency weapons; their output can be tightly focused. Though there are a host of microwave sources including relativistic klystrons, travelling wave tubes, magnetrons, spark gap devices and vircators, from the perspective of a microwave E-bomb's designer, a vircator by virtue of being a one-shot device capable of producing an extremely powerful single pulse of radiation is the preferred choice as of now. In addition, it is mechanically simple, small and robust. Spark gap devices may be used in near future. Figure 12.20 shows the component parts of an HPM E-bomb.

12.4.6.2 Lethality of E-Bombs

Unlike conventional kinetic energy weapons, lethality analysis in the case of HPM weapons is a far more complex proposition. While it would be a simple task to calculate electromagnetic field strength achievable at a given source-to-target distance for a known HPM device, the task of determining lethality or the kill probability for a given class of targets under such conditions is certainly not an easy one. This is because of the simple reason that vulnerability of a given target to HPMs would depend upon the extent of hardening built into it against electromagnetic attack. Equipment that is electromagnetically hardened will withstand orders of magnitude greater field strengths than standard commercially rated equipment. Secondly, the inflicted damage would also depend upon coupling efficiency that ultimately determines how much power is actually transferred from the electromagnetic field generated by the weapon into the target.

There are two principal coupling modes by which electromagnetic energy released by the weapon is coupled into the target. These are the *front door coupling* and the *back door coupling*. In the case of front door coupling, energy is coupled to the targeted equipment such as radar or communications equipment through the antenna. Back door coupling occurs through large transient currents produced in the case of low frequency weapons or electrical standing waves on fixed electrical wiring and cables interconnecting equipment, or providing connections to mains power in the case of HPM weapons.

A low frequency weapon will couple well into a typical wiring infrastructure. Typically, a cable run will comprise multiple linear segments joined at approximately right angles. Irrespective of the orientation of the weapons' electromagnetic field, more than one linear segment of the cable run is likely to be oriented such that a good coupling efficiency can be achieved. HPM weapons operating offer an additional coupling mechanism to back door coupling due to their ability to directly couple into equipment through ventilation holes, gaps between panels and poorly shielded interfaces. Because microwave weapons can couple more readily than low frequency weapons, they have the potential to be significantly more lethal than low frequency weapons.

To maximize the lethality of an electromagnetic bomb it is necessary to maximize the power coupled into the target set. This can be achieved by maximizing the peak power and time duration of the electromagnetic radiation and the coupling efficiency. HPM weapons can further maximize lethality by sweeping the frequency or chirping the vircator, which improves coupling efficiency in comparison with a single frequency weapon enabling them to exploit larger number of coupling opportunities. One could also use polarization of the emitted radiation to further enhance coupling efficiency. Assuming that orientations of possible coupling apertures and resonances in the intended target are of random nature, a circularly polarized output will provide better coupling than a linearly polarized one though designing a high power, wideband, compact circularly polarized antenna would be a technological challenge. One could also enhance the probability of inflicting damage to targets of relatively greater hardness by doing a trade-off between the size of the lethal footprint and intensity of electromagnetic field in that footprint. Lower detonation altitude reduces the lethal footprint but increases the field intensity. Figure 12.21 illustrates how the lethal footprint of an HPM E-bomb reduces with decrease in detonation altitude.

12.4.6.3 Delivery Mechanisms

Like explosive warheads, E-bombs may be fitted to a range of delivery vehicles, such as aircraft in the case of aerially delivered bombs or on board a missile in the fully autonomous configuration. The two modes have their own pros and cons.

In the case of a cruise missile carrying an electromagnetic warhead, the missile hardware will comprise the electromagnetic device such as a flux compression generator, an electrical energy converter such as a Marx capacitor bank and an onboard storage device such as a battery for charging the capacitor bank. The onboard fusing system of the missile is used to detonate the

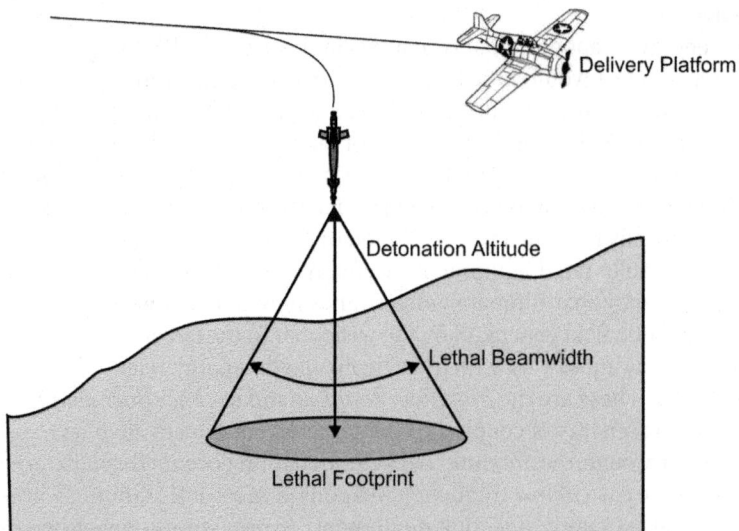

Figure 12.21 Detonation altitude and lethal footprint of an HPM E-bomb.

electromagnetic device. The fusing system will be tied to the navigation system in the case of a cruise missile; in an anti-shipping missile, the radar seeker and in an air-to-air missile, the proximity fusing system. A disadvantage of cruise missile as a delivery vehicle for an E-bomb is that the warhead fraction, which is the ratio of payload mass to the launch mass, may be anywhere between 15 and 30%. An E-bomb to be delivered from an aircraft mainly comprises an electromagnetic device and an electrical energy converter. In this case, the energy storage capacitor bank could be charged from launch aircraft's power supply en-route to the target. A much smaller power supply onboard the weapon could then be used to maintain the charge prior to warhead initiation. Fusing in this case could be provided by a radar altimeter fuse, a barometric fuse or the navigation system as is the case with GPS/INS guided bombs. The warhead fraction in this case could be as high as 85%. To summarize, for a given electromagnetic device design and delivery accuracy, an E-bomb with a mass equal to that of an electromagnetic warhead equipped missile can have a much greater lethality.

Since the E-bombs have a potentially large lethal radius compared to an explosive device of same mass, it is always prudent to deliver the former class of weapons from a standoff distance. Cruise missiles inherently provide this capability. In the case of electromagnetic warheads delivered as glide bombs or onboard anti-ship or air-to-air missiles, one would require to employ some kind of fire-and-forget guidance technology to make sure that the launch aircraft attains adequate separation to be at a safe distance from the weapon at the time of warhead detonation. Glide bomb delivery has the advantages that it can be released from outside the effective radius of target air defences and also that the bomb's autopilot may be programmed to shape the terminal trajectory of the weapon for effective engagement of the intended target.

12.4.7 Representative HPM Weapons

Based on the state-of-the-art in HPM weapons of various categories and emerging trends, these devices will be available for deployment in defensive and offensive roles during the 2030s. These weapons would be mounted on fixed and mobile land and ship-based platforms as part of an integrated air or point defence. There will also be airborne defensive systems to include self-protection suites to target enemy radars and missiles and HPM warheads on air-to-air missiles to target attacking enemy aircraft and missiles. Offensive HPM weapons will include precision-guided e-bombs, cruise missiles, unmanned or manned aircraft with integrated HPM weapons. These offensive weapons will be used to target radars, satellites and command, control and communications infrastructure. In the following paragraphs major HPM systems developed or under development are discussed. These include the *Active Denial System* (ADS) and *Vigilant Eagle* from Raytheon, *HPM Blackout* built by BAE Systems, *RANETS-E* of Russian origin, *Shortstop Electronic Protection System (SEPS)* by the Whittaker Corporation, *Warlock Green* and *Warlock Red* by EDO Communications and Countermeasures, *DS-110* by Diehl, Germany and *CHAMP* developed jointly by the US Air Force Research Laboratory (AFRL) and Boeing Phantom Works.

12.4.7.1 Active Denial System

The *Active Denial System* (ADS) is a nonlethal microwave directed-energy system that can be used in counter-personnel role against hostile human targets at distances beyond the effective range of small arms. Active denial technology is a breakthrough nonlethal technology that uses millimetre-wave electromagnetic energy for area denial, perimeter security and riot control applications from relatively longer ranges without the application of any lethal force. With a range of about 500 m and beam that is little less than 2 m across, an ADS would be capable of dispersing riots from a safe distance. It has been developed by Raytheon for the US Air Force Research Labs and the Department of Defense Joint Nonlethal Weapons Directorate. It operates as follows. Microwave energy at 95 GHz generated by a gyrotron is focused and directed at the targeted personnel by a directional planar antenna. The microwave energy almost instantly produces an intolerable heating sensation within seconds forcing the subject to flee from the scene to avoid exposure to the microwave energy beam (Figure 12.22). The millimetre-wave energy in fact heats up the water molecules just under the subject's skin heating them enough to cause extraordinary pain. The pain sensation immediately ceases when the individual moves out of the beam or when the beam is turned off. The millimetre-wave beam does not cause any injury because of the low-energy levels used and the shallow penetration depth of about 0.5 mm at the operating wavelength. In addition, the pain induced as a consequence of subject's natural defence mechanism serves as a warning to help protect him from any injury. A simple control console on ADS enables the operator to view the scene and precisely aim the millimetre-wave beam to affect only the intended target.

The ADS has been comprehensively evaluated on a large number of volunteers to establish its non-lethality. Based on tests carried out on individuals, Penn State Human Effects Advisory Panel (HEAP) concluded that ADS is a nonlethal weapon with a low probability of injury. Other observations include the following: (1) There are no significant effects for wearers of eyewear

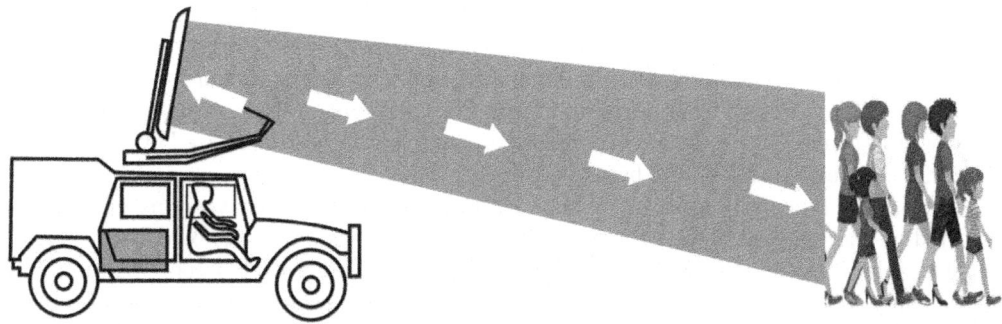

Figure 12.22 Use of millimetre energy for crowd control.

Figure 12.23 Active Denial System. (Courtesy: Raytheon)

including contact lenses, night vision goggles and soon. (2) There is no difference in response of volunteers in different age groups to millimetre-wave exposure from ADS. (3) Exposure to millimetre-wave energy from ADS doesn't affect the male reproduction system. (4) Normal skin applications such as cosmetics have little effect on the interaction of millimetre-wave energy with skin. (5) Less than 0.1% of 10 000 exposures experienced pea-sized blisters.

Two variants of ADS integrated on land vehicles include an ADS mounted on a wheeled vehicular platform and another integrated on a HMMWV (High Mobility Multipurpose Wheeled Vehicle) platform, more commonly known as a Humvee (Figure 12.23). These platforms would be equipped with required power source to drive the weapon system. A mid-range version of the ADS called Silent GUARDIAN with an operational range of about 250 m has also been developed by Raytheon (Figure 12.24). Silent GUARDIAN is primarily developed for use by law enforcement agencies. The system is easily transportable on standard military tactical vehicles and can also be integrated into combat vehicles.

It is also being considered for operation from airborne platforms such as the AC-130 gunship. The airborne version will use more powerful and lightweight version of its land-based counterpart. The US reportedly deployed ADS as a crowd control weapon in Afghanistan but it was withdrawn in 2010 without ever using it due to serious questions raised by critics about the ethics of using a pain beam to break up riots.

According to reports, Russia and China are developing their own versions of ADS. As reported by the Interfax News Agency, the Russian military is testing its own beam weapon under development at the 12th Central Military Research and Development Institute. The device reportedly has a range of about 270 m and is intended to be used for riot control applications. The Chinese system called the Poly WB-1 pain beam weapon also uses millimetre-wave energy to cause intolerable pain in human subjects up to a kilometre away from the source.

Figure 12.24 Silent Guardian protection system.

12.4.7.2 Vigilant Eagle

Vigilant Eagle, an HPM weapon system from Raytheon, is designed to protect both commercial flights as well as military and private aircraft from the threat of surface-to-air missiles including MANPADS (Man-Portable Air-Defence Systems) (Figure 12.25). The system is installed at the airports or airfields and is primarily intended to provide protection to the aircraft while taking off from or landing at the airports or airfields by directing a precisely steered beam of electromagnetic energy, a set of high-frequency microwaves, at the missile threatening the aircraft in proximity of the airport and diverting the threat away from the targeted aircraft.

Major subsystems of Vigilant Eagle include a distributed missile detect and track (MDT) system, command and control system and an active electronically scanned array (AESA) of antennas driven by solid-state RF amplifiers. The MDT consists of a pre-positioned grid of passive infrared sensors mounted on nearby cell phone towers or buildings with communication lines to the command and control system. These sensors detect, identify and track missile threats, forming a dome of security over the airport. The command and control system located on airport premises provides pointing commands to the AESA and rejects false alarms. It also connects to the airport's security interface. The command and control system has the added advantage of being able to determine the missile's launch point, which helps security forces apprehend the enemy. On receiving commands from command and control system, the AESA radiates a precisely aimed small-diameter beam on the missile. This beam of electromagnetic energy is tailored to generate electromagnetic waveform designed to interfere with the missile's guidance circuitry. This forces the missile to abort its intercept course thereby deflecting it away from the aircraft. *Vigilant Eagle* has the design capability of defeating man-portable

Figure 12.25 Vigilant Eagle protection system concept.

missiles in seconds without any alteration to or involvement by the aircraft using the airport. Electromagnetic fields generated by the radiated beam are well within the OSHA (Occupational Safety and Health Administration) standards for personnel exposure limits. Figure 12.25 illustrates the operational concept of the *Vigilant Eagle* airport protection system.

12.4.7.3 RANETS-E

RANETS-E, described as a radio frequency cannon, is a high-power microwave (HPM) weapon system designed to provide terminal by disrupting guidance systems of precision-guided munitions and avionics systems of aircraft. The system is configured around an X band HPM source generating a peak power output of 500 MW. Pulsed output of 10–20 ns pulses at a repetition rate of 500 Hz produces average power of 2.5–5 kW. The roof-top steerable parabolic antenna has gain of 45–50 dB. Figure 12.26 shows RANETS-E deployed on an MAZ-7910 chassis using the 54K6 command post-cabin.

The RANETS-E weapon system reportedly has a lethal range of about 30 km against the electronic guidance systems of PGMs and aircraft avionic systems assuming a vulnerability threshold field strength of about 1.0 kV/m. This makes RANETS-E a credible lethal weapon against electronics systems at ranges typical of a terminal point defence weapon. The lethal footprint would reduce with increase in target hardness. The *Emitter Locating System* (ELS) used as targeting element is located on a separate platform and not on the RANETS-E vehicle. The two platforms are linked together via cables. This is done to avoid the risk of fratricide as even side lobes of the emission from RANETS-E could be electrically lethal at short ranges. SAM (surface-to-air missile) engagement radar with desired angular accuracy could also be used as targeting element. The concept of operations (CONOPS) for the system involves attaching one or more RANETS-E systems to a battery of SAMs and integrating them with the battery fire-control system enabling remote cueing, aiming and firing of RANET-E systems.

Figure 12.26 RANETS-E HPM system.

Figure 12.27 BAE's Bofors HPM Blackout.

12.4.7.4 HPM Blackout

The BAE Systems *Bofors HPM Blackout* is a mobile gigawatt level microwave source operating in the L band. The system comprises a high-power microwave source, an integrated pulsed power unit and an exchangeable conical horn antenna with support systems including compact battery-powered vacuum system for the microwave tube and a gas supply system for the pulsed power unit. The integrated battery unit renders the system operational in all terrains. The *Bofors HPM Blackout*, originally designed as research and evaluation tool for generating data on threats from electromagnetic effects of HPM weapons, could have operational capabilities with its proven destructive effects at considerable distances against a wide range of commercially-off-the-shelf (COTS) equipment. Figure 12.27 shows a photograph of the Bofors HPM Blackout without its support systems.

12.4.7.5 Shortstop Electronic Protection System (SEPS)

SEPS, designed by *Whittaker Corporation* in Simi Valley, California is a portable radio frequency jammer that can be programmed to jam a specific range of frequencies. The system offers effective electronic countermeasures against electronic proximity fuses used in incoming artillery and mortar rounds by countering the threat of RF fused munitions by initiating premature detonation. The proximity fuses of artillery and mortar shells use a radar signal to compute the distance from target and are programmed to generate fuse command at the pre-programmed distance. The *Shortstop system* receives the radar signal and then retransmits a modified signal to make the shell believe it has reached programmed distance while it actually hasn't. Premature detonation offers protection to friendly ground troops, vehicles, structures and other equipment under fire. The system is also capable of effectively defeating remotely controlled improvised explosive devices or suicide bombers. Packaged in a case similar in size to a suitcase, Shortstop's passive electronics and operational features make it immune to detection by the enemy's sensors. While the SEPS defeats proximity fused munitions, weapons with impact fuses go unchallenged. Based on the extensive live fire testing the Shortstop system has been subjected to, the system has demonstrated the ability to significantly enhance survivability of troops and high-value assets from indirect fire proximity fused munitions.

12.4.7.6 Warlock Green and Warlock Red

Warlock Green and *Warlock Red*, designed by *EDO communications and Countermeasures* (now ITT Electronic Systems, EWS) too are electronic countermeasure systems and are modified versions of battle-proven the SEPS from the Whittaker Corporation. Warlock systems have been found to be effective in countering the threat of radio controlled improvised explosive devices (IEDs). The RF signal emitted by Warlock systems jams the communications signals used to detonate the IEDs. It does so by intercepting the signal sent from a remote station to initiate detonation and disallowing to make contact with the IED. To be effective, one needs to transmit the right frequency, which may not be easy. Another application of Warlock systems is to disrupt communication devices used by the enemy. Warlock Red is a less sophisticated low-cost version of Warlock Green and is designed to counter certain specific threats that are used in large numbers. Multiple Warlock systems may be used to provide area coverage without causing any mutual interference. Warlock systems have undergone many revisions to add more frequencies and better software.

12.4.7.7 DS-110 High Power Electromagnetic (HPEM) System

DS-110 developed by Diehl Munitions Systems of Germany is a high-power RF system that generates a wideband electromagnetic pulse capable of destroying process-driven electronic modules in any system by generating resets or inducing power latch-ups thereby neutralizing the targeted devices. Another possible military application of the DS-110 system includes disruption of electronic fusing mechanism of projectiles, artillery shells and improvised explosive devices. The basic system is omnidirectional but it can be configured to provide directional coverage. The system packaged in a suitcase weighs 28 kg. It is powered by a built-in battery capable of powering the system for 30 minutes on a continuous basis or three hours in controlled bursts. Figure 12.28 shows a photograph of the system.

12.4.7.8 CHAMP

CHAMP, an abbreviation for Counter-electronics High-powered microwave Advanced Missile Project, is a technology demonstration programme of the US Air Force Research Laboratory (AFRL) jointly with Boeing Phantom Works for the development and field testing of counter-electronics HPM aerial demonstrator intended for integrating HPM payloads on airborne platforms such as cruise missiles and unmanned combat aerial vehicles (UCAV). CHAMP represents a class of directed-energy HPM weapon in which the weapon travels towards the target rather than the target travelling towards the HPM weapon. Boeing is the prime contractor; an airborne platform provider and system integrator. The HPM source and the pulsed power system, respectively, come from the Ktech Corporation in Albuquerque (the directed-energy capability of Ktech Corporation has now been acquired by Raytheon) and Sandia National Laboratories under a separate contract with AFRL.

Figure 12.28 DS-110 HPEM system.

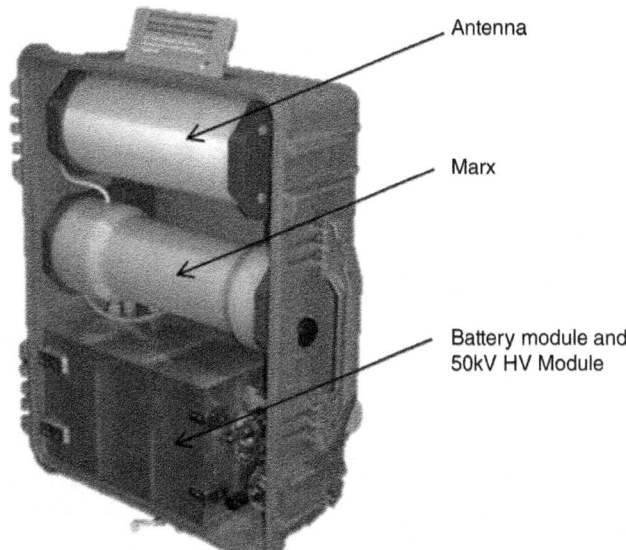

CHAMP brings together directed-energy technology capabilities of AFRL and Boeing's missile design capabilities for development of a new breed of nonlethal, but highly effective weapon systems.

Military use of kinetic energy weapons to counter an adversary's command, control, communications and computer assets have an associated collateral damage. Collateral damage to civilian electronic equipment could be huge depending upon location of the targeted electronic system, targeted spectrum bands and employment techniques. CHAMP that makes use of HPMs is a nonlethal alternative to conventional kinetic weapons to disrupt and destroy electronics assets. While current kinetic energy weapons cannot penetrate hidden, underground targets, the electromagnetic pulses from an HPM weapon can penetrate through metal elements leading into underground command centres, to damage and even destroy sensitive components in command and control systems, communications systems and computers associated with the targeted systems. CHAMP provides the warfighter a nonlethal, low collateral damage capability that can be used against targets that are currently on the kinetic restricted target list.

The HPM payload of CHAMP has been successfully tested including ground testing and flight demonstrations. During the flight testing of CHAMP, the missile was pointed at a set of simulated targets. The test confirmed that the missile could be controlled and timed while using HPM system against multiple targets and locations. Figure 12.29 illustrates the concept of HPM attack by CHAMP mounted on an aerial platform such as X-45 UCAV. In future, development efforts of the CHAMP programme would lead to HPM payloads on cruise missiles or loitering platforms. These payloads would be capable of delivering multiple HPM bursts to a series of targets for applications such as suppression of enemy air defences, disruption and destruction of command and control networks and national infrastructure with minimized collateral damage.

12.5 Directed-Energy Laser Systems

Lasers have been used in various military applications since the early days of development that followed the invention of this magical device. There was a large-scale proliferation of lasers and optoelectronic devices and systems for applications like range finding, target designation, target acquisition and tracking, precision-guided munitions and so on during the 1970s and 1980s. These devices continue to improve in performance and find increased acceptance and

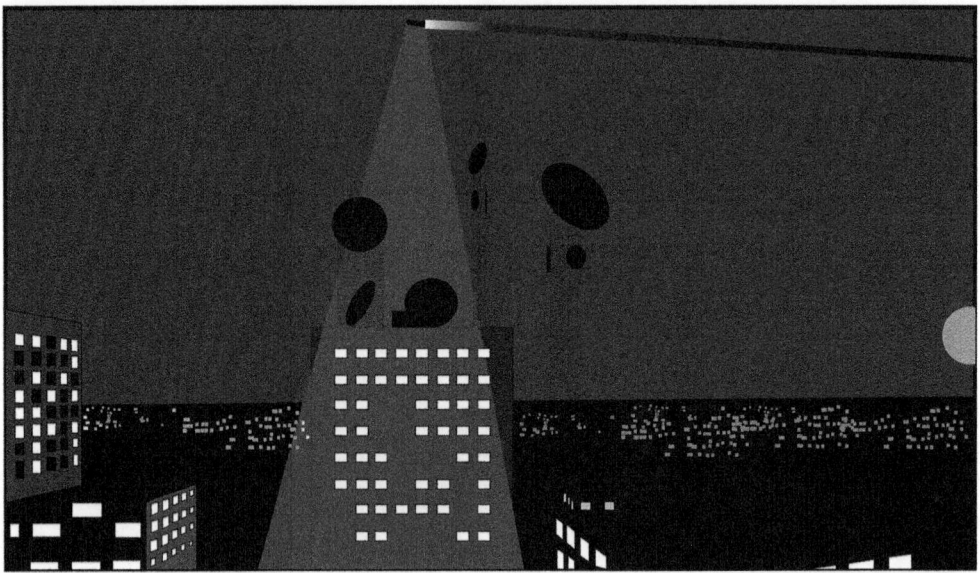

Figure 12.29 Illustration of HPM attack by CHAMP.

usage in the contemporary battle field scenario. Technological advances in optics, optoelectronics and electronics leading to more rugged, reliable, compact and efficient laser devices are largely responsible for making them indispensable in modern warfare. The last decade or so has seen emergence of directed-energy laser applications. A directed-energy laser system primarily uses directed-energy in the form of a concentrated beam of electromagnetic energy in the targeted direction to cause intended damage to the enemy's equipment, facilities and personnel. The intended damage could be lethal or nonlethal.

There are two broad categories of directed-energy laser systems, namely nonlethal (or more appropriately, less-lethal) directed-energy laser system dazzlers and lethal high-energy laser systems. Less-lethal directed-energy laser systems mainly include *laser dazzlers* used in anti-personnel applications for anti-terrorist and counter-insurgency operations and *anti-sensor electro-optic countermeasure systems*. The second category of directed-energy laser systems includes high-energy laser systems capable of inflicting structural damage to the intended targets at tactical and strategic operational ranges. Another important application of high-energy directed-energy laser systems when operated at relatively lower power levels is in safe neutralization of unexploded ordnances with minimized collateral damage. These systems generally configured around kilowatt-class bulk solid-state lasers or high-power fibre lasers have amply demonstrated their efficacy in disposal of unexploded ordnances including improvised explosive devices from a safe distance.

In the following sections, we shall discuss anti-personnel laser dazzlers, laser ordnance disposal systems and high-energy directed-energy laser weapons. Anti-sensor electro-optic countermeasure systems are discussed in Chapter 6 along with electronic warfare systems.

12.6 Less-Lethal Laser Dazzlers

A low intensity conflict is the most common form of warfare today and is likely to be so for the foreseeable future. Data suggests that more than 75% of the armed conflicts since World War II have been of low intensity variety. Low intensity conflict operation is a military term used for deployment and use of troops and/or assets in situations other than conventional war. Compared to a conventional war, in the case of low intensity conflict operations Armed Forces engaged in the conflict operate at a greatly reduced tempo, perhaps with fewer soldiers, reduced range of tactical equipment and limited scope to operate in military manner. Also use of artillery is avoided in the case of conflicts in urban territories and use of air power is often restricted to surveillance and transportation of personnel and equipment. Low intensity conflicts pose an alarming threat to national security is an area of concern for the whole of international community today. Its scope extends from emergency preparedness and response to domestic intelligence activities to riot and mob control, from combating illegal drug trafficking to protection of critical infrastructure, from handling counter-insurgency and anti-terrorist operations to detection of concealed weapons, from detection and identification of chemical and biological warfare agents to detection of explosive materials. Laser and optoelectronics technologies play an important role in handling low intensity conflict situations. The key advantages of use of laser technology in such applications are near-zero collateral damage, speed of light delivery and potential for building nonlethal weapons. Some of the well-established laser devices in low intensity conflict (LIC) applications include laser dazzlers for close combat operations, mob/riot control and protection of critical infrastructures from aerial threats.

12.6.1 Operational Parameters

A *laser dazzler* emits a high intensity laser beam in the visible band, usually in the blue-green region, to temporarily impair the vision of the adversary without causing any permanent or lasting injury or adverse effect to the subject's eyes. Depending upon intended application, laser dazzlers come in a variety of package styles, mounting configurations and performance specifications. These devices can be handheld or weapon mountable for versatility, convenience and ease of use. Some laser dazzlers have adjustable beam divergence that allows them to vary the spot size. While using a broader beam allows them a larger swathe path needed to effectively produce a tactical advantage, tighter beams allow longer range and increased efficacy at longer distances. Choice of operating parameters such as laser power, spot size at the target, laser power density and so on are driven by nature of deployment. The beam shaping and directing optics is so designed to achieve desired value of nominal ocular hazard distance (NOHD) and a laser power density that does not exceed the maximum permissible exposure (MPE) figure dictated by ANSI (American National Standards Institute) standards for eye safety. MPE expressed as power density is dependent on wavelength and exposure time. At 532 nm, the maximum permissible power density equals $2.5\,mW/cm^2$ for a 0.25 s exposure and $1.0\,mW/cm^2$ for an exposure time of 10 s. The blue-green region is the chosen wavelength band as the human eye's response is the highest in this band as shown in Figure 12.30. The most commonly employed wavelength for the purpose is 532 nm usually generated by using either laser diodes or frequency double Nd-YAG laser modules. These devices usually produce a randomly pulsed output in the range of 10–20 Hz riding a DC level for a better overall effect. DC level is usually kept at 30–50% of the peak intensity level. Night time maximum operational range is typically 3–4 times the maximum day time range.

12.6.2 Potential Applications

Nonlethal weapons act as a force multiplier enabling friendly forces to discourage, delay or prevent hostile action. Laser dazzler is a nonlethal weapon specifically designed for applications where subject vision impairment is to be achieved at a distance ranging from few tens of meters to several kilometres in bright ambient conditions. They are particularly effective in situations where use of lethal force is not preferred; examples being limiting escalation and

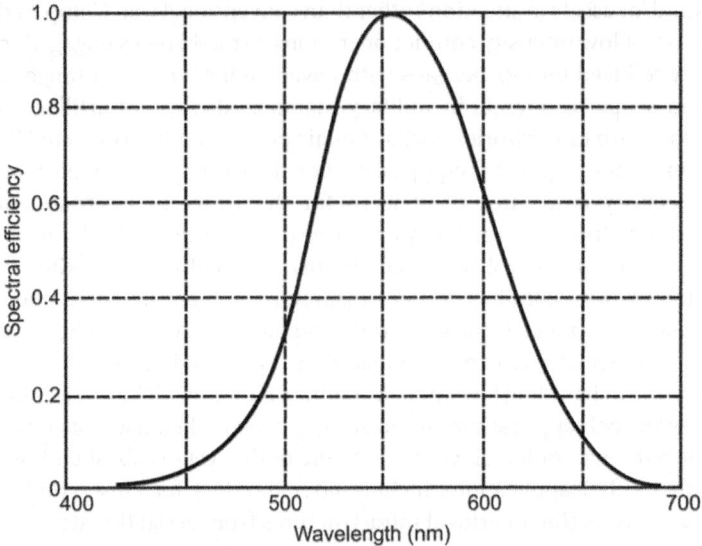

Figure 12.30 Response of the human eye to different wavelengths.

Figure 12.31 Laser dazzler for close combat operation.

temporarily disabling facilities and equipment. Laser-based less-lethal weapons such as laser dazzlers can be used for counter-insurgency, anti-terrorism, counter-sniper, self-defence, crowd control and infrastructure protection applications. Laser dazzlers are emerging internationally as a new nonlethal alternative to lethal force for law enforcement, homeland security, border patrol, coastal protection, infrastructure protection and host of other low intensity conflict scenarios. While Figure 12.31 illustrates the use of a laser dazzler in a close combat scenario, Figure 12.32 shows a photograph of a vehicle-mounted laser dazzler system for unruly crowd/mob control applications. Such vehicles are usually equipped with remotely controlled weapon stations for a wide range of applications. The technology of the short-range laser dazzler and crowd control laser dazzler is more or less similar except that the latter system employs a larger power to be able to produce the desired power density in a larger beam spot. Some means of scanning the laser beam spot may also be necessary.

Figure 12.32 Vehicle-mounted laser dazzler on CROWS-II platform for crowd control application.

Laser dazzlers are also being considered as a potential candidate for warning the crew of commercial airliners or military aircraft who tend to violate intentionally or unintentionally the no fly zone. Such systems in a networked configuration of multiple laser dazzler stations and radars could be effectively used for protection of critical infrastructure or assets. In a system like this, radar provides the initial cue about the rogue or suspect aircraft when it is still more than 100 km away from the actual asset to be protected. The radar keeps a continuous vigil on the suspect aircraft till it comes within the tracking range of the electro-optic tracker, which is usually an integral part of the laser dazzler station. The electro-optic tracker station takes the cue from the radar and tracks the target with much higher accuracy needed for the dazzling action. Exposure to a strong laser light source results in flash blindness and after-images. In flash blindness, exposure to a very bright light source deprives the pilot of vision for a period of time ranging from a few seconds to a few minutes. Also the laser illumination filled the flight deck with a bright light, thus makes it difficult to concentrate on the flight instruments as well and adversely affects pilots intended actions. Before the aircraft cockpit is flooded with dazzling laser light, much lower power levels are employed to send a kind of warning signal to the crew to know their intent. This helps in discriminating rogue aircraft from those that might have gone astray unintentionally.

12.6.3 Representative Systems

A large number of companies are offering short-to-medium range laser dazzler systems for various application scenarios. One such short-range laser dazzler is the *Compact High-Power (CHP) laser dazzler* from LE Systems, USA (Figure 12.33). The dazzler emits a 500 mW flashing green dazzling laser beam. The CHP with its higher power creates a credible glare effect in a larger spot size for use on moving vehicles or individuals. This feature is particularly important for

Figure 12.33 CHP laser dazzler mounted on a tripod.

Figure 12.34 GLARE GBD-IIIC.

protection of entry control points and convoys, at distance and in bright ambient conditions. The CHP laser dazzler has been comprehensively evaluated to demonstrate no lasting eye damage from repeated direct exposures beyond the specified 20-m standoff distance.

The *GLARE MOUT* is a nonlethal visual disruption laser with an effective range of 20 m–2 km. The device is ideally suited for small arms integration as well as mobile crew-served applications. When a suspect individual approaches a restricted or controlled area, such a device can be used to give warning to that person before shots have to be fired. It is configured around 125 mW green laser emitting at 532 nm. It has a nominal ocular hazard distance of 18 m. Its shockingly bright green beam sends an irrefutable, multi-lingual, cross-cultural, unambiguous message that 'you mean business'. Reports confirm that GLARE MOUT has saved numerous lives of both soldiers and non-combatants in Iraq and Afghanistan.

The *GLARE GBD-IIIC* is a long-range variant of GLARE MOUT and is a visual deterrent laser device for hail and warning applications. The GLARE GBD-IIIC is an effective dazzler ideally suiting ship-to-ship signalling or airborne overwatch. With twice the power and a more concentrated beam than the GLAREMOUT, the GLARE GBD-IIIC can effectively hail and warn as far away as 4 km. Like the GLARE MOUT, the GLARE GBD-IIIC produces an overtly bright green diffused laser spot to impair the visual ability of a suspect. Figure 12.34 shows a photograph of the device. The device is configured around a 250 mW green laser operating at 532. It has operational range of 72–4000 m with the lower limit being the nominal ocular hazard distance (NOHD) for unaided eye. A fixed beam divergence of 4–5 mrad produces a spot size of about 30 cm at the minimum operating range. With the optional eye-safety module, the device automatically shuts off when the targeted personnel are within the NOHD.

Figure 12.35 Glare LA-9/P laser dazzler.

The *GLARE LA-9/P*also built around a 250 mW green laser is yet another long-range visual deterrent laser device for hail and warning applications and is intended to be effective out to a range of 0.3–4 km for ship-to-ship signalling or airborne overwatch applications. The GLARE LA-9/P has an additional feature of automatically shutting off the device if the subject target were within the nominal ocular hazard distance. The device is equipped to rapidly detect if a bystander is inside the NOHD, and shuts off the glaring laser output to prevent unintended eye injury. Figure 12.35 shows a photograph of the system.

Saber 203 Grenade Shell Laser Intruder Countermeasure System is a type of laser dazzler that uses a 250 mW red laser diode mounted in a hard plastic capsule in the shape of a standard 40 mm grenade. It is suitable for being loaded into a M203 grenade launcher. It has an effective range of 300 m. It is controlled via a box snapped under the launcher, with the batteries and firing switch housed in this box. In emergency it can be quickly ejected and replaced with a grenade. Saber 203 dazzlers were used in Somalia in 1995 during Operation United Shield. Figure 12.36 shows a photograph of the system.

The Chinese *JD-3 laser dazzler* and *ZM-87 portable laser disruptor* are the other established systems. JD-3 laser dazzler is reported to be mounted on the Chinese Type 98 main battle tank and is coupled with a laser radiation detector, and automatically aims for the enemy's illuminating laser designator, attempting to overwhelm its optical systems or blind the operator. The ZM-87 portable laser disturber is an electro-optic countermeasure laser device. It can blind enemy troops up to a range of 2–3 km and temporarily blind them at up to 10 km range. ZM-87 has reported been widely deployed, which includes their use on naval vessels.

Figure 12.36 Saber-203 laser dazzler system.

Figure 12.37 Dazer Laser GUARDIAN.

The Photonic Disruptor, classified as *TALI (Threat Assessment Laser Illuminator)* is yet another nonlethal high-power green laser developed by Wicked Lasers, USA in cooperation with Xtreme Alternative Defence Systems. This tactical laser is equipped with a versatile focus-adjustable collimating lens to compensate for range and power intensity when used to either incapacitate an attacker in close range or safely identify threats from a distance. TALI-series devices are configured around a 100 mW, 532 nm laser producing a laser beam with 1.5–7.5 mrad adjustable beam divergence.

The *Dazer Laser* from Laser Energetics Inc., USA is yet another very popular device. It comes in two variants named *GUARDIAN* that has a range from 1–300 m (model dependent) and the *Defender* that has a range from 1–2400 m (model dependent). Both variants of the Dazer Laser temporarily impair the vision of the target adversary and succeed in eliminating the threat's ability to see, engage or effectively target the user. This provides the user with a significant advantage over the threat at longer and safer standoff distances. Both variants of the Dazer are designed to be eye-safe at all ranges beyond 1 m and meet the current American National Standard Institute (ANSI) safety standard Z136.1. Figure 12.37 shows the photograph of the Dazer Laser GUARDIAN.

A vehicle-mounted laser dazzler system designed to control violent crowds and unruly mobs was shown in Figure 12.32. The Laser Science & Technology Centre (LASTEC) of Defence Research & Development Organization (DRDO), India has also developed short-range

Figure 12.38 Hand-held laser dazzler developed by LASTEC DRDO. (*Source:* Biswarup Ganguly, https://commons. wikimedia.org/wiki/File:Laser_ Dazzler_-_DRDO_-_Pride_of_India_-_ Exhibition_-_100th_Indian_Science_ Congress_-_Kolkata_2013-01-03_2560. JPG. CC BY 3.0.)

Figure 12.39 Crowd control laser dazzler mounted on a Mahindra Marksman Light Armoured Vehicle.

handheld laser dazzlers with an operational range of up to 50 m (Figure 12.38) and a light armoured vehicle-mounted laser dazzler with an operational range of 50–250 m for crowd control applications (Figure 12.39). Prototypes of these systems have been evaluated.

12.6.4 Emerging Trends

Laser dazzlers have shown lot of promise and potential for a variety of low intensity conflict applications including close quarter battle scenarios, border patrol and coastal surveillance, controlling unruly and violent crowds and mobs, countering asymmetric threats particularly encountered in the naval environment, protection of critical infrastructure from terrorist attacks and so on. Different variants of these devices are being developed to suit different application requirements. While, on one hand, handheld portable devices are being offered to meet short-range requirements of paramilitary and law enforcement agencies, tripod-mounted and vehicle-mounted versions equipped with electro-optic sighting support, target acquisition and tracking devices are also being developed to meet the requirements of emerging scenarios.

While short and medium-range handheld and weapon integrated versions of the laser dazzler are in widespread use for anti-terrorist and counter-insurgency applications, vehicle-mounted systems designed for operations against unruly and violent crowds are beginning to appear on the scene. Ship-borne medium to long-range laser dazzlers mounted on stabilized platforms are also catching the attention of security forces to defeat asymmetric threats. Another interesting development in the recent past has been a keen interest in long-range laser dazzler systems mounted on gimbal platforms and equipped with electro-optic tracker and also integrated with a network of radars to provide 24/7 protection to strategic assets from airspace violation by aerial platforms.

RGB laser dazzlers that emit an intense flashing pattern of red, green and blue lights at a frequency of few kHz are also reportedly under development though not much information is currently available about the existence of any commercial devices. These devices may be configured around a combination of semiconductor diode lasers and diode-pumped solid-state laser modules emitting red, green and blue wavelengths. Arrays of red, green and blue LEDs could also be used instead. RGB laser dazzlers have the potential of being more effective as compared to green laser dazzlers particularly if the targeted personnel were using a protective eyewear. Also, RGB dazzlers are reported to cause nausea in the exposed personnel, which makes these devices particularly effective in a sea environment. One possible application could be against sea pirates.

In the not too distant future, one would see deployment of laser dazzlers with global coverage. These systems are proposed to use remotely controlled membrane reflector to receive the dazzling laser beam from the source station and guide it to intended target location. Figure 12.40 illustrates the concept.

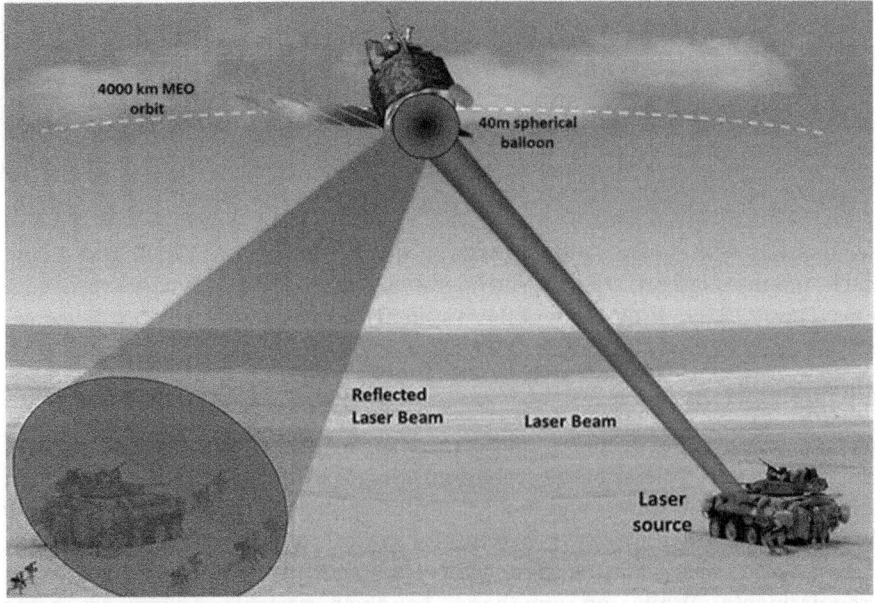

Figure 12.40 Laser dazzler for global coverage.

12.7   High-Power Lasers for Ordnance Disposal

Disposal of unexploded ordnances including surface-laid mines, improvised explosive devices (IED), grenade shells, artillery/mortar rounds, cluster bombs and so on from safe standoff ranges using a high-power laser beam is an emerging application of directed-energy laser systems in homeland security-related applications. Ordnance is disposed of by focusing high-power laser beam on the ordnance casing, thereby heating it until the temperature of backplane of the casing exceeds ignition temperature of explosive filler. The explosive filler ignites and begins to burn and the process is independent of type of fusing used by target explosive. This leads to a low-level detonation or deflagration rather than full power detonation. The advantages of use of laser energy for ordnance disposal include large magazine, high precision, controllable effects with reduced collateral damage and assured and fast disposal from safe standoff ranges.

12.7.1   Application Scenarios

The laser ordnance neutralization system is being considered for a range of application scenarios. One possible application relevant to homeland security is disposal of improvised explosive devices also called roadside bombs. The laser-based disposal system offers a quick, safe and reliable method of neutralizing these roadside bombs from a safe standoff distance in excess of 150 m. Another possible application is in disposal of friendly explosives that have outlived their shelf life. Since full-scale conventional wars are not there to be fought every now and then, there is a large inventory of ammunition stored in the ammunition depots having exceeded their shelf lives and waiting to be neutralized. Neutralization of such large quantities of ammunition is not only a cumbersome exercise, but it also poses a safety hazard. Laser-driven disposal of unexploded ordnances allows not only safe but also rapid neutralization of large quantities of explosive materials.

12.7.2   Representative Laser Ordnance Neutralization Systems

There can be various possible variants of the ordnance neutralization system. One configuration is the standalone high-power laser system mounted on a vehicular platform. ZEUS-HLONS and Laser AVENGER are examples of this category. In another possible variant, the system may be mounted on a remotely controlled vehicle with a robotic arm. While the laser system along with its power and thermal management systems could be mounted on the platform, the laser delivery system could be mounted on the robotic arm. In addition to these configurations, a high-power laser system may be integrated with other means such as ground penetration radar for detection of buried explosives. The integrated system allows detecting and neutralizing both surface-laid as well as buried explosives. In some cases, laser ordnance disposal system is mounted on a remotely controlled weapon station to get both defensive as well as offensive capability. Rafael's high-energy laser weapon system called THOR, developed in Israel to defeat improvised explosive charges (IEDs), roadside bombs, unexploded ordnance (UXO) and other potentially hazardous explosives is an example of this type of system.

The *ZEUS-HLONS* (HUMMWV Laser Ordnance Neutralization System) from the US (Figure 12.41) is a vehicle-mounted laser ordnance neutralization system. The concept of neutralization of live ordnances using laser energy was first demonstrated in the field in 1994 with development and field testing of *Mobile Ordnance Disrupter System* (MODS) that employed a 1.1 kW arc lamp driven solid-state laser mounted on an M113 A2 Armoured

Figure 12.41 Zeus-HLONS laser ordnance neutralization system. (*Source:* Courtesy of the US Military.)

Personnel Carrier. The ZEUS system initially employed a 500 W laser. The latest version of ZEUS system is called ZEUS-II and consists of a multi-kilowatt-class laser, beam director, operator's station and all support subsystems on a single vehicle. The system can be integrated with M1114 HMMWV, M-ATV or other MRAP vehicles. ZEUS has been extensively field evaluated against more than 40 different types of ordinance, which include landmines, improved conventional munitions, mortar rounds, rifle grenades, rockets and artillery projectiles, ranging from small, plastic landmines to large, thick-walled 155-mm projectiles and 500-lb general-purpose bombs. The system has demonstrated its effectiveness in a variety of missions including counter-IED operations, clearing land mines, clearing unexploded ordnances (UXO) from battlefields or during peacekeeping missions, clearing active and formerly used defence sites and clearing active, test and training ranges of exposed UXO.

Another laser system designed for disposal of unexploded ordnances is the *Laser Avenger* system from Boeing combat systems (Figure 12.42). Boeing successfully tested a 1 kW solid-state laser weapon mounted on a converted anti-aircraft vehicle in Redstone Arsenal in Huntsville Alabama in September 2009 by neutralizing multiple types of improvised explosive devices, including large calibre artillery munitions, smaller bomblets and mortar rounds. The system was operated at safe distances from the targets under a variety of conditions including different angles and ranges. The test follows earlier field tests in which the Laser Avenger was used to neutralize five targets representing unexploded ordnance and IEDs and subsequent to that used to shoot down UAVs. The ordnance neutralization test was carried out in 2007 with an earlier relatively lower power version of the system. The UAV shoot-down tests were carried out in 2008 and 2009 at the White Sands Missile Range, NM in which *Laser Avenger* shot down a small unmanned aerial vehicle in each event. *Laser Avenger* used a 1-kW solid-state laser system mounted on a military Humvee that is usually equipped with Stinger anti-aircraft missiles. During the tests, the Laser Avenger's advanced targeting system acquired and tracked three small UAVs flying against a complex background of mountains and desert, shooting down one of the UAVs.

Figure 12.42 Laser Avenger.

Figure 12.43 THOR laser IED neutralization system.

Rafael's *THOR* (Figure 12.43) is a laser ordnance neutralization system with an integrated remotely controlled weapon station. The system comprises of a high-energy laser along with its beam director and a coaxial 12.7 mm M2 machine gun. THOR uses an air-cooled 700 W solid-state laser that offers continuous laser engagement with no cool-down time requirement. The upgraded version of THOR employs a 2 kW water-cooled laser with a continuous duty cycle. The M2 machine gun is used to perform twin functions; as a standoff disrupter, destroying fusing, thick-cased munitions and booby traps and to provide accurate, direct fire upon enemy forces and targets in either an offensive or defensive role. This dual capability enables THOR to be used for offensive and defensive purposes, as well as for safe standoff removal of explosive obstacles by laser directed-energy or projectile kinetic energy. While use of directed, high-energy laser gives the operator ability to neutralize the IED's content by means of burning, deflagrating, or detonating the explosive, the machine gun may be used to neutralize the IED by targeting the operating device by cutting a wire or detonating cord. The IED can be subsequently retrieved by a robot for further neutralization and investigation.

The system is modular and can be installed on a variety of vehicles and weapon stations as an add-on system. THOR, like other laser-based ordnance disposal systems, avoids collateral damage usually associated with other neutralization procedures. Laser-based systems have highly minimized associated collateral damage due to explosive burnout or low-order detonation. In all laser-based ordnance disposal systems, the directed-energy from the laser rapidly clears unexploded ordnance and defeats IEDs by inducing a low-order burning or deflagration reaction in the explosive fill at safe standoff ranges.

Yet another laser-based ordnance neutralization system developed by the Laser Science and Technology Centre (LASTEC) of the Defence Research and Development Organization (DRDO), India is the *Laser Ordnance Disposal System (LORDS)*. The system employs a 1.0 kW fibre-laser and is mounted along with its support systems on a Light Security Vehicle (LSV) made by TATA Motors. The support systems mainly include the thermal management system for waste heat removal, electrical generator to enable standalone operation and command/control system assisted by a CCD camera and a laser range finder for acquisition and beam pointing. The system with its effective range of 30–250 m has been successfully field tested for remote disposal of surface-laid unexploded ordnances, mines, directional mines and IEDs. The system's effective range is 30–250 m. Figure 12.44 shows a photograph of the system.

Figure 12.44 Laser Ordnance Disposal System (LORDS).

12.8   High-Power Directed-Energy Laser Weapons

Conventional weapon systems including bullets, artillery shells, mortar rounds, rockets, missiles and so on propelled by chemical reactions have dominated the tactical battlefield for centuries. Advances in laser technology and the state-of-the-art in high-energy laser development and testing allow the application of directed-energy property of lasers in a new class of weapons. Development of high-energy lasers with military potential is leading to the production of beam weapons that transfer coherent light energy to the intended target to cause structural damage. In the coming years, the monopoly of chemically propelled projectiles may give way to dynamic co-existence and competition with directed-energy weapons as the next-generation weapons for both tactical and strategic missions. Introduction of high-power lasers have made possible the execution of many a new mission hitherto extremely complex to realize with conventional kinetic energy weapons. These include ground-based laser (GBL) for disabling low earth orbit satellites, airborne laser (ABL) for destroying ballistic missiles and space-based laser (SBL) for negation of theatre and intercontinental ballistic missiles. The capability demonstrated by high-pwer lasers, particularly in the last 10–15 years, to destroy various types of fast moving aerial targets such as unmanned aerial vehicles has established high-energy laser weapons as weapons for the twenty-first century.

DEWs are described at length in this section in terms of salient features, application potential, and constituents of a laser-based DEW system. Major subsystems including high-pwer laser sources, beam-control technologies, high-pwer laser propagation effects and lethality considerations are discussed in detail in subsequent sections. Representative laser DEW systems are brief described towards the end.

12.8.1   Operational Advantages and Limitations

Primary advantages of laser-based DEWs include speed-of-light delivery, near-zero collateral damage, multiple target engagement and rapid retargeting capability, immunity to electromagnetic interference and no influence of gravity. Deep magazine and low cost per shot are the other advantages.

Speed-of-light Delivery: Laser weapons engage targets at *speed of light* with essentially no time of flight as compared to conventional kinetic energy weapons that require a finite travel time. As an example, world's one of the fastest cruise missiles BrahMos with a supersonic speed of 2.8–3 Mach would take about 5 min to reach its target located at its maximum operational range of 300 km. On the other hand, the same target when targeted by a laser DEW would be hit in a millisecond.

Near-Zero Collateral Damage: Due to their pin-point accuracy, laser-based directed-energy weapons have near-zero collateral damage.

Multiple Target Engagement and Rapid Retargeting: While kinetic energy weapons such as missiles get destroyed during the mission, the laser weapon is reusable. This feature makes them particularly suitable for engaging fast-moving targets. Multiple target engagement and rapid retargeting feature of laser-based DEWs is attributed to their being powered by rechargeable chemical 9in the case of chemical lasers) or electrical energy (in the case of solid-state and fibre lasers) stores and that shifting from one target to another involves only re-pointing and re-focusing of the beam directing optical system.

Immunity to Electromagnetic Interference: The processes of generation and transfer of lethal laser power to the target are purely in the optical spectrum and hence are immune to any electromagnetic interference and jamming.

No Effect of Gravity: Laser pointing is practically without any inertia and a light bullet has no mass so therefore is not influenced by gravity. As a result, it doesn't require any midcourse correction.

Deep Magazine: A laser-based directed-energy weapon has a practically unlimited magazine. The total number of shots a laser can fire is only limited either by the amount of chemical fuel as is the case for chemical lasers or electrical power for solid-state and fibre lasers.

Low Cost per Shot: Cost per shot in the case of a laser-based DEW is much lower than in the case of conventional kinetic energy weapons. Projectile weapon systems, guided missiles in particular, expend lot of expensive hardware such as rocket motors, guidance systems, avionics, seekers, airframes and so on every time they are fired. In the case of laser weapons, the cost of each laser firing is essentially the cost of the chemical fuel or the electrical power consumed. As an example, a fourth-generation shoulder fired surface-to-air missile MANPADS of the type FIM-92 Stinger series costs about US $40 000. This missile can be used against an aircraft to carry out one such mission. A similar mission from a land-based laser DEW system can be carried out with 50–100 kW solid-state laser by firing the laser beam for a dwell time of about 5 s. Going by present day technology levels, this laser system would draw about 400–500 kW of electrical power for a period of 5 s, which is same as electrical power consumed by a 100 W bulb in 7 h.

Laser-based DEWs also have some limitations. Some of these include their line-of-sight dependence, requirement of finite dwell time, problems due to atmospheric attenuation and turbulence and ineffectiveness against hardened structures.

Line-of-Sight Dependence: Laser weapons require direct line-of-sight to engage a target. Their effectiveness is reduced or neutralized by the presence of any object or structure in front of the target that cannot be burned through.

Finite Swell Time: Unlike projectile weapons that instantly destroy the target upon impact, laser weapons require a minimum dwell time in the order of 3–5 s to deposit sufficient energy for target destruction.

Atmospheric Attenuation and Turbulence: The effectiveness of the laser weapon is adversely affected by the atmospheric conditions. The laser beam suffers attenuation due to absorption and scattering by airborne particles and gas molecules and deterioration of beam quality in the form of deformation of the laser beam wave front and increase in the laser beam spot size at the target.

Minimal Effects on Hardened Structures: Laser weapons are not very effective against hardened structures. However, equipment such as antennas, sensors, external fuel stores and so on mounted on these structures can be targeted effectively.

12.8.2 Application Potential

The operational scenario of directed-energy laser weapons is broadly categorized as short and medium-range tactical missions and long-range strategic missions. Some of the important application areas of tactical class laser weapons include standoff neutralization of ordnances such as mines, unexploded ordnances and IEDs, ground-based defence against rockets, artillery and mortars (RAM), ground-based capability to destroy unmanned aerial vehicles (UAVs) of the adversary, airborne Defence of aircraft against MANPADS such as shoulder fired surface-to-air missiles, ship defence against manoeuvring cruise missiles and tactical ballistic missiles.

Standoff neutralization of ordnances requires laser power in the range of 1–2 kW for operational ranges up to 300 m. Solid-state or fibre-laser sources operating around 1.0 μm are used for the purpose. AVENGER and ZEUS of the US and Israeli THOR are laser

Figure 12.45 Ground-based air defence against aerial targets.

ordnance systems with comparable specifications. Anti-UAV operations up to a range of 8–10 km require about 100 kW of laser power. Operational ranges of 5–6 km are possible with 50 kW laser systems. Again, the preferred lasers are solid-state and fibre lasers. COIL may also be used. For applications such as air-defence against rocket, artillery, mortar (RAM) targets, rocket propelled grenades (RPG), battlefield missiles, laser-guided munitions and so on, typically 100 kW of laser power is needed for operational ranges of 5–10 km. Figures 12.45, 12.46 and 12.47 illustrate operational scenarios of important tactical class directed-energy laser weapons including ground-based air-defence capability against aerial targets (Figure 12.45), ship defence against manoeuvring cruise missiles (Figure 12.46) and airborne defence against MANPADS (Figure 12.47).

Long-range *strategic applications* of laser-based DEW systems mainly include ballistic missile defence (Figure 12.48), space control such as space-based lasers and anti-satellite applications. In all these applications, operational ranges are generally in hundreds to thousands of km and required power levels are in the order of 1–20 MW depending upon the actual mission. Space control applications such as anti-satellite applications require relatively much higher power than the power level needed for ballistic missile defence. Table 12.1 outlines laser weapons' mission capabilities for relevant operational ranges and laser power levels.

12.8.3 Components of a Laser-Based Directed-Energy Weapon System

Unlike conventional military applications of laser systems such as laser range finders and target designators, which primarily comprise of a laser source of desired specifications, laser-based DEWs are much more than a high-pwer laser source. Figure 12.49 shows the block-schematic

Figure 12.46 Laser weapons for ship defence.

Figure 12.47 Airborne laser weapons for defence against MANPADS.

arrangement representative of different components of a laser DEW system and their interconnections. The subsystems other than the high-pwer laser source are required for the purpose of directing the laser beam to the intended point on the target and keeping it there for the desired dwell time and producing the desired value of fluence on the target. A laser-based DEW essentially comprises of two major subsystems, namely a high-power laser source and the beam-control system.

A *high-power laser source* generates a laser beam of desired power level, beam quality and time duration to be able to inflict intended damage to the enemy target. Potential candidates for building high-pwer laser sources include gas dynamic CO_2 lasers, chemical lasers including

Figure 12.48 Airborne laser for ballistic missile defence. (*Source:* Courtesy of the US Airforce.)

Table 12.1 Laser mission capabilities.

S. No.	Mission	Typical Ranges (km)	Laser Power (kW)
1.	Stand-off neutralization of unexploded ordnances	0.2–0.3	1–2
2.	Counter UAV/RPV/Drone	1–3	10
3.	Ship surface threat defence	1–3	10
4.	Short range tactical applications-counter RAM, Missiles	1–3	50
5.	Airborne precision strike of ground targets	5–10	100
6.	Ground based air and missile defence-counter RAM and MANPADS	5–10	100
7.	Anti-tank missiles, Rocket propelled grenades (RPG) and Cruise missiles	5–10	100
8.	Area denial to aircraft, helicopters and UAVs	5–10	100
9.	Ballistic missile defence	hundreds of km	1–2 MW
10.	Space based laser/ASAT capability	thousands of km	10–20 MW

HF/DF and chemical oxyiodine lasers, solid-state lasers and fibre lasers. The ultimate quality desired of a potential high-pwer laser source is its ability to produce as high as possible a power density at the target for a given aperture size. Power density achievable for a given source-beam director combination is inversely proportional to square of wavelength and beam quality parameter for given beam director aperture size and operational range. Therefore, lower wavelength

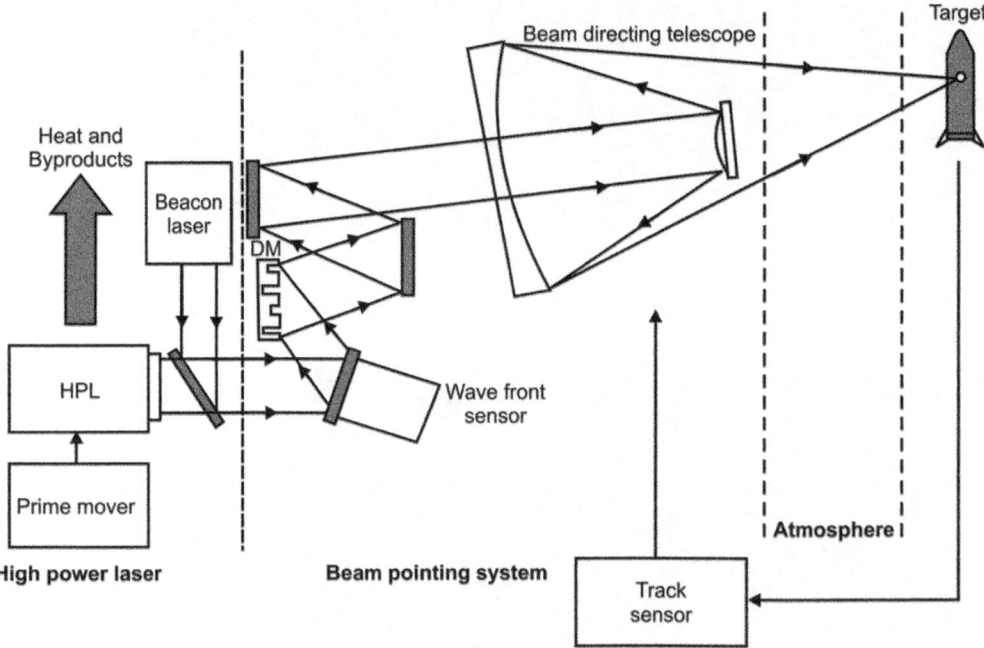

Figure 12.49 Components of a laser-based DEW system.

and smaller value of beam quality parameter signifies its greater suitability as a high-pwer laser source. A smaller value of beam quality parameter indicates higher beam quality with unity being the ideal value. Shorter wavelength also has better target coupling efficiency. In addition, power density at the target also depends upon atmospheric losses. Other important factors include power scalability, operator safety and ease of mounting.

The primary function of the *beam-control system* is to precisely point and focus laser energy at the designated point on the target and keep the same there for sufficient duration to cause intended damage to the target. In addition to target tracking and beam pointing and focusing, the beam-control system also has many other subsystems to perform different tasks, which include beam shaping and beam stabilization, wave front control for atmospheric correction and so on. In addition to laser source and beam pointing system, propagation of high-pwer laser beam through the atmosphere and its coupling with the target also called target lethality are the other important constituents determining the efficacy of laser weapon.

The laser beam has to propagate over long atmospheric paths and the resulting propagation effects control the overall effective range or lethality range of the weapon system. Propagation effects mainly include diffraction, absorption and turbulence. The *diffraction* effect tends to spread out the laser energy as the laser beam propagates and the amount of spreading is proportional to wavelength to beam aperture diameter ratio. This limits the smallest spot diameter to which the laser beam can be focused. *Scattering* is due to atmospheric constituents and results in loss of laser energy from the beam path in other directions. Longer wavelengths have relatively lower scattering losses. *Absorption* in atmospheric constituents also causes loss of energy consequently decreasing the laser power reaching the target end. There is a very strong dependence of absorption parameter on location, time of year and weather pattern. *Turbulence* causes beam to spread resulting in increase in laser spot size and reduction in power density at the target end. Turbulence is measured by the $C_n{}^2$ parameter, which is a strong function of altitude and wind speed and other atmospheric parameters. *Thermal blooming* is another parameter that causes laser beam aberration. This arises out of heating effects produced by absorption leading to refractive index gradient in the laser beam path. This further leads to formation of negative lens and causes the laser beam to spread. Thermal blooming effects can be ignored for rapidly moving targets such as airborne targets.

12.9 High-Power Laser Sources

The high-power laser source is an important constituent of a laser-based DEW system. Different types of laser sources have been exploited and experimented with for DEW application. Gas dynamic lasers and chemical lasers, HF/DF lasers in particular, were used to generate the high-pwer levels required for a DEW application in the 1970s and 1980s as the technology of these lasers was well-established. Due to advances in solid-state laser and fibre-laser technologies, they are considered today as the most potent laser source candidates for deployable laser weapon systems. In the following paragraphs, we shall discuss various potential high-pwer laser sources. A Chemical Oxy-Iodine Laser (COIL) is also very much in contention due to its shorter wavelength of operation, particularly for generating power in megawatts.

12.9.1 Critical Requirements

Laser power and the beam quality are the two most important laser source parameters that decide the high-energy laser weapon's lethal capability. The beam quality determines the smallest achievable focused laser spot size. This together with laser power available at the target site determines the power density and hence the lethality. The basic function of the high-pwer laser source is therefore to produce the required lethal laser power in a near diffraction-limited beam quality. The laser power density at the target is given by the following equation.

$$P_d = \frac{P_0}{A} = \frac{0.21}{\lambda^2 B^2}\left(\frac{D}{R}\right)^2 P_0 \times e^{-\alpha R}$$

λ = Operating wavelength
B = Beam quality
D = Beam director aperture
R = Operational range
α = Atmospheric attenuation coefficient.

It is evident from this equation that for maximizing power density at the target, one would prefer a shorter wavelength, high beam quality and larger beam director aperture size and low atmospheric transmission losses.

Shorter wavelength is preferred as it not only produces a higher power density; target coupling efficiency is also higher at shorter wavelengths. For the same laser power and transmitting telescope aperture, 1 μm high-pwer laser (COIL, Solid-State Lasers) would have an operational range that is approximately 10 times that of a 10 μm high-pwer laser (CO_2 laser) for a given power density requirement at the target. *Beam quality* is essentially a measure of how tightly the laser beam can be focused to form a small and intense spot of light on a distant target. A beam quality of 1 signifies the laser spot size at target that is limited by the laws of diffraction.

For a real laser beam, B is greater than 1 and hence the focused spot sizes are larger than the diffraction-limited values. Maintaining good laser beam quality as the power is scaled up is one of the most complex and challenging tasks in the laser system design. Large telescope *aperture size* is preferred as it produces tighter focusing and hence increased laser power densities at the target. Low *atmospheric transmission losses* are important as the laser beam has to propagate over long atmospheric paths. For this to happen, one of the most important requirements is that the laser wavelength falls within the available transmission windows of the atmosphere in 0.4–1.7 µm (Visible – NIR), 3–5 µm (MWIR), and 8–14 µm (FWIR) bands. In addition to these requirements, another important criterion for high-pwer laser design is the need to adopt *power scalable laser design architecture*. The basic physics and technology of laser system design should allow power scaling to megawatt power level needed for long-range high-energy laser weapons.

Not all laser sources in general meet the above requirements. The elite categories of laser sources for DEW applications are *gas dynamic CO_2 laser, chemical lasers* including HF/DF laser and COIL, *free electron lasers, solid-state lasers* and *fibre lasers.*

12.9.2 Gas Dynamic Lasers

The gas dynamic laser was invented in 1966 by Edward Gerry and Arthur Kantrowitz at the Avco Everett Research Laboratory. Gas dynamic lasers usually use a combustion chamber, supersonic expansion nozzle, and CO_2, in a mixture with nitrogen or helium, as the laser medium. Any hot and compressed gas with appropriate vibrational structure could be utilized. The gas dynamic laser derives its energy from the combustion of a suitable fuel-oxidizer mixture, which means that it does not require any electrical energy for its operation. A gas dynamic laser achieves population inversion by rapid expansion of high-temperature, high-pressure laser gas mixture produced during combustion to a near vacuum in an adiabatic process through an integrated supersonic nozzle bank. Though the expansion reduces the gas temperature, a large number of excited molecules are still in the upper laser level. Population inversion is created if the reduction in pressure and temperature downstream of the nozzle bank takes place in a time that is much shorter than the vibrational relaxation time of the upper laser level corresponding to asymmetric stretching mode of carbon dioxide coupled with nitrogen. The cavity axis is transverse to the direction of gas flow. Since the high-temperature, high-pressure gas mixture is created in a combustion reaction, the laser is called *combustion driven gas dynamic laser* (CD-GDL). Figure 12.50 shows the functional schematic arrangement of a combustion driven gas dynamic laser showing its various constituent

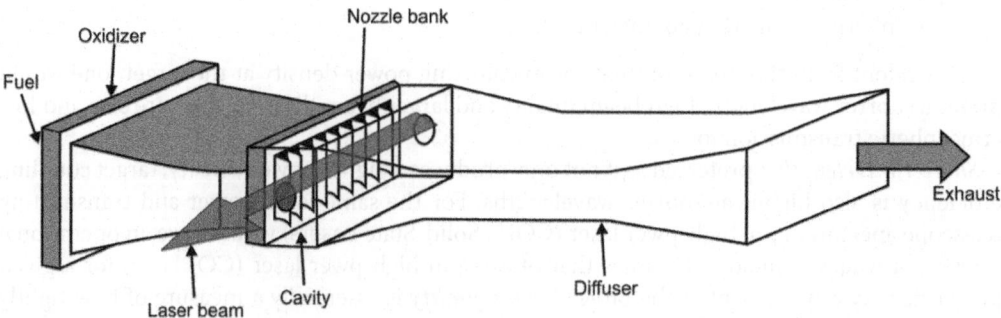

Figure 12.50 Functional schematic of a combustion driven gas dynamic laser.

subsystems. This type of laser, which is also capable of producing hundreds of kW of CW power at 10.6 μm initially assumed importance as a potential HPL weapon but it seems to have lost out in the race in favour of HF/DF and COIL systems due to heavy absorption of 10.6 μm wavelength by water vapour in the atmosphere.

Salient features of a gas dynamic laser include the following *long lasing wavelength* of 10.6 μm, *larger beam divergence* (Beam Divergence $\cong \lambda/D$), *larger telescope aperture size* requirements (Aperture size ~ λ/Beam Divergence), *proven technology* and *non-toxic, power scale-up proven* up to megawatt power level, relatively higher cavity pressure (30 torr) compared to that of other similar HPL sources, well-established technology for *direct atmospheric exhaust* of lasing gases, well suited for *platform mounting*.

The *Airborne Laser Lab* (ALL) programme of the United States launched in 1976 used a gas dynamic CO_2 laser as the high-pwer laser source. The gas dynamic laser used CO_2–N_2–H_2O gas mixture and operated at the 10.6 μm wavelength. The aim of the programme was to develop the technology demonstrator mounted on NKC-135A aircraft to track and destroy airborne targets. Reportedly, this laser produced a raw output power of 456 kW and output power from the aiming system of 380 kilowatts in an 8 s sustained run. The system delivered a power density of 100 W/cm^2 at a distance of 1 km.

12.9.3 Chemical Lasers

A chemical laser derives the energy required to produce population inversion and consequent laser emission from a chemical reaction. Importance of chemical lasers lies in their ability to generate continuous-wave output power level reaching several megawatts. This coupled with the fact that the range of wavelengths generated by these lasers is well absorbed by metals, which makes them suitable for cutting and drilling operations in industry and more so as high-pwer laser sources for directed-energy weapon applications.

There are two categories of chemical lasers. One category has hydrogen fluoride (HF) and deuterium fluoride (DF) and the other has the two iodine lasers including the COIL and an all gas-phase iodine laser (AGIL). Both HF/DF and COIL systems have demonstrated their capability to be operated at megawatt-class CW output power. Lasing action has been demonstrated in AGIL and efforts are on to scale the output to higher power levels.

In general, in a chemical laser, a suitable chemical reaction produces a stream of gas rich in excited atoms or molecules. Another gas is then injected into the stream. This injected gas then reacts with excited particles in gas stream to produce the excited species for laser action. In the case of a hydrogen fluoride (HF) or deuterium fluoride (DF) chemical laser, this stream of gas is abundant in fluorine atoms. Excited fluorine atoms then react with hydrogen or its heavier isotope deuterium to produce vibrationally excited hydrogen fluoride (HF*) or deuterium fluoride (DF*) molecules. The asterisk here denotes the excited state of the molecules. The laser emission is then due to transition of excited HF or DF molecules. In the case of COIL, the initial chemical reaction produces singlet-delta oxygen. It transfers its energy to iodine to generate excited atomic iodine, which subsequently produces laser emission at the wavelength of transition of atomic iodine. Both types of chemical lasers are briefly described in the following paragraphs.

12.9.3.1 Hydrogen Fluoride/Deuterium Fluoride (HF/DF) Lasers

In the HF laser, fluorine atoms are produced by pre-mixing of fluorine gas with helium used as a buffer gas and some other gases. Helium is added to stabilize the reaction and control the temperature. Some other gases may be added to control production of fluorine atoms. Since fluorine gas could be very nasty to handle, it is usually supplied in the form of some other

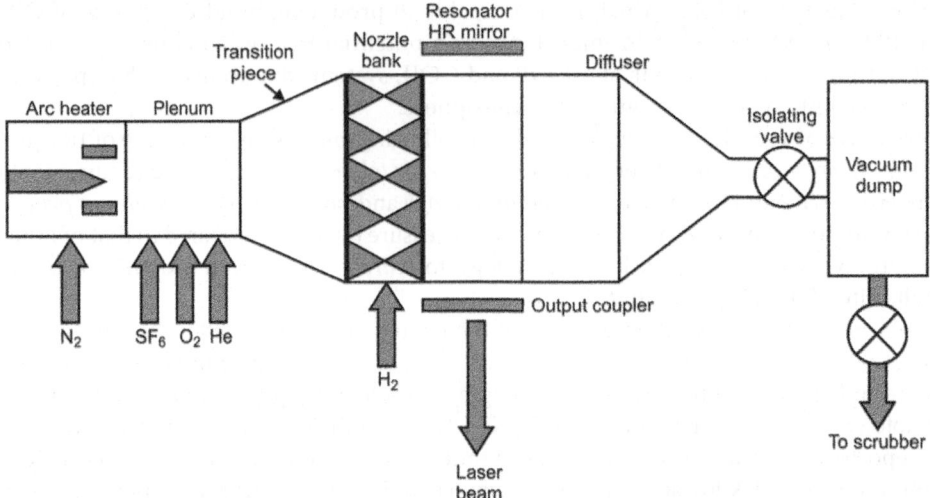

Figure 12.51 Block-schematic of an HF laser.

molecule such as sulfur hexafluoride (SF_6) or nitrogen trifluoride (NF_3). The free excited fluorine atoms produced in the combustor are accelerated through supersonic nozzles into laser cavity. In the cavity, under low temperature, low-pressure conditions, fluorine atoms mix and react with hydrogen to form excited hydrogen fluoride (HF*) molecules. A suitable resonator configuration produces stimulated emission of photons from these excited HF molecules. The output wavelength is in the region of 2.6–3.0 μm. Figure 12.51 shows a typical block-schematic arrangement of an HF laser.

Another laser similar to HF laser is the deuterium fluoride (DF) laser. In this laser, deuterium replaces hydrogen and reacts with fluorine atoms to produce excited DF (DF*) molecules. Deuterium being heavier than hydrogen leads to DF laser producing a longer wavelength. It produces an output in the wavelength range of 3.6–4.0 μm. Between HF and DF laser systems, while HF laser has the advantage of lower wavelength, DF laser propagates relatively much better through atmosphere. As a result, HF is better suited to space-borne platforms. Some of the better known HF/DF laser systems include ALPHA laser, MIRACL (Mid-Infrared Advanced Chemical Laser) and THEL (Tactical High-Energy Laser). While ALPHA is a HF laser, the MIRACL and THEL systems are configured around DF lasers. Figure 12.52 shows a photograph of the THEL system.

MIRACL was the first megawatt-class CW chemical laser. In the MIRACL system, free excited fluorine atoms are generated by a combustion reaction of a fuel (ethylene, C_2H_4) and an oxidizer (nitrogen trifluoride, NF_3). Deuterium and helium are injected into the exhaust downstream of the nozzle to produce excited DF (DF*) molecules. Excited DF molecules produce stimulated emission and laser action in the resonator cavity and the output is spread over several lasing lines in the wavelength range of 3.6–4.2 μms.

Salient features of HF/DF lasers are summarized as follows. Lasing wavelength of a DF laser equal to 3.8 μm has good atmospheric transmission compared to that of a HF laser emitting at a 2.7 μm wavelength. The HF laser is, however, better suited to space-based applications. HF/DF lasers have high specific energy and relatively smaller size as compared to gas dynamic lasers. DF laser of MIRACL system has demonstrated power level of 2.2 MW. However, HF/DF lasers produce highly toxic and explosive gases during operation and also require complex logistics.

Figure 12.52 The THEL system. (*Source:* Courtesy of the US Military.)

12.9.3.2 Chemical Oxygen Iodine Laser (COIL)

Three main parts of a COIL system are the singlet oxygen generator (SOG), the supersonic nozzle and the laser cavity. Singlet-delta oxygen is produced through a reaction of gaseous chlorine and basic hydrogen peroxide (BHP), which is a mixture of hydrogen peroxide and potassium hydroxide. This is a highly exothermic reaction releasing most of the energy as heat into the BHP solution. Rest of the energy goes to generate singlet-delta oxygen, which is an electronically excited state of oxygen. Potassium chloride is produced as a byproduct of this reaction. Molecular iodine is injected into the gas flow in the plenum just upstream of the supersonic nozzle. The molecular iodine dissociates into atoms through a series of energy transfer reactions with singlet-delta oxygen. Singlet-delta oxygen transfers its energy to the iodine molecules injected into the gas stream in a collisional process. Transfer of energy from singlet-delta oxygen to iodine is nearly resonant and thus very rapid. The excited iodine undergoes stimulated emission in the resonator cavity producing a laser output at 1.315 μm. Figure 12.53 shows the block-schematic arrangement of a COIL system.

COIL works at a relatively lower gas pressure. The pressure downstream of the nozzle is in the order of a few torr, which makes its discharge to the atmosphere a difficult proposition due to the large pressure differential involved therein. But it is not so if the laser is to work on an airborne platform. COIL therefore is the laser of choice for a laser-based DEW on an airborne platform.

Salient features of COIL are summarized as follows. COIL is a low temperature, low-pressure device and uses non-toxic fuels. It has high specific energy and a power level of 1.2 MW demonstrated in ABL programme. It operates at shorter lasing wavelength of 1.3 μm that has good atmospheric transmission. It requires a smaller aperture laser beam director and offers a good target coupling efficiency. Low pressure operation makes it better suited to airborne applications. A complex pressure recovery system is required for ground-based applications.

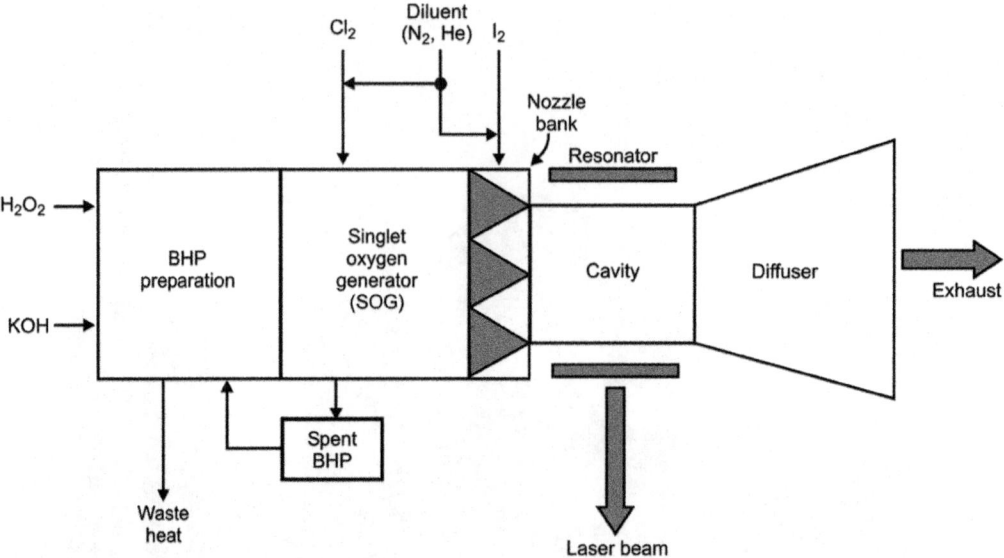

Figure 12.53 Block-schematic of a COIL system.

Figure 12.54 ABL concept.

The Airborne Laser (ABL) programme of US Air Force uses a high-pwer COIL system. ABL is one part of a layered ballistic missile defence system that addresses the ever increasing ballistic missile threat. ABL is designed to destroy the hostile missile while it is still in the vulnerable boost phase of its flight (Figure 12.54). ABL uses an onboard surveillance system to detect and track the missile after launch. The beam-control system locks on to the target and then fires the high-energy laser. The entire system, that is, the high-energy laser, the surveillance system and the beam-control/fire-control system, is configured on a Boeing 747–400 fighter aircraft (Figure 12.55). ABL made its maiden flight in July 2002. In January 2010, the

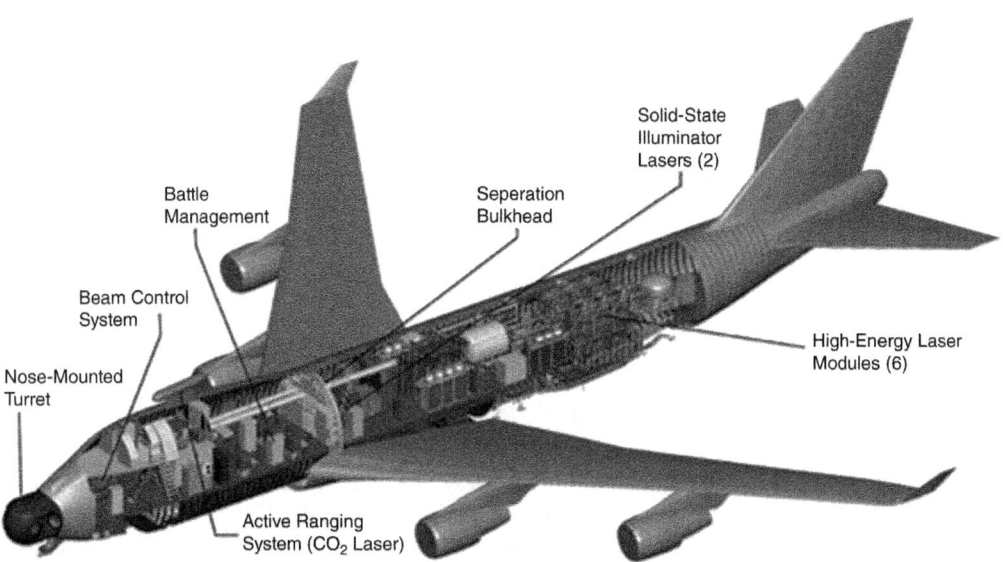

Figure 12.55 ABL configured on a Boeing 747–400 fighter aircraft.

high-energy laser was used in flight, to intercept, although not destroy, a test missile in the boost phase of flight. In February 2010, the system successfully destroyed a liquid-fuel boosting ballistic missile. In December 2011, the ABL programme was reportedly shelved as it was considered to be operationally not viable.

12.9.3.3 All Gas-Phase Iodine Laser (AGIL)

In AGIL, chlorine gas and gaseous hydrogen azide (HN_3) are mixed to produce excited nitrogen chloride (NCl) molecules. Nitrogen chloride then transfers energy to atomic iodine in the same manner as done by a singlet-delta oxygen in COIL. This all gas-phase laser overcomes the disadvantage of heavy, aqueous based chemistry of a COIL device and the undesirable wavelength range of HF laser that is strongly absorbed by the atmosphere. Once scaled to higher power levels, it would be a potent laser technology for the space-borne laser programme.

12.9.4 Free Electron Lasers

The free electron laser is a unique laser where the process of light amplification is a bit unconventional with respect to what we have seen in respect to lasers discussed earlier in this chapter. Unlike conventional lasers that rely on bound atomic or molecular states, free electron lasers use a relativistic electron beam as the active medium. In the case of a free electron laser the lasing medium is a beam of free electrons completely unattached to any atoms.

It may be mentioned here that a relativistic particle is a particle moving at a speed close to speed of light such that effects of special relativity are important for its behaviour. Mass less particles such as photons always move with speed of light and are therefore always relativistic. Particles with some mass are considered relativistic when their kinetic energy is comparable to or greater than the energy mc^2 corresponding to their rest mass. This implies that their speed is close to that of light. Such particles are generated in particle accelerators. In the context of a free electron laser, a beam of electrons is accelerated to relativistic speeds and then made to pass through a periodic or more precisely alternating transverse magnetic field. The transverse

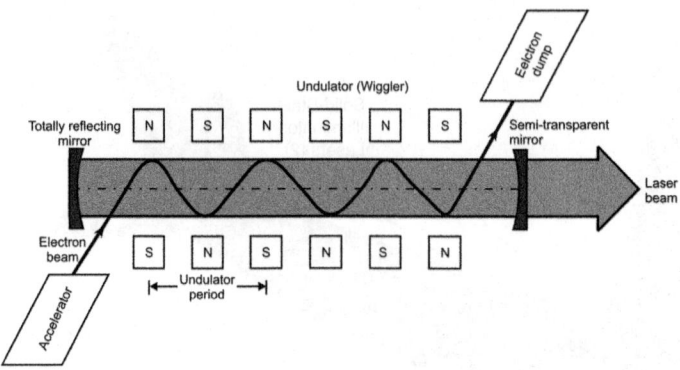

Figure 12.56 Free electron laser. (*Source:* Courtesy: LLNL.)

magnetic field is produced with the help of an array of magnets with alternating poles placed along the beam path. The array of magnets is sometimes called the Wiggler and the magnetic field produced by it as the Wiggler field. It is so-called as it forces the electron beam to assume a sinusoidal path. As the beam travels through the magnetic field, it releases some of its energy as light before it exits from the other end of the field. The emitted wavelength is given by eqn. 12.1.

$$\lambda = \frac{p}{\left[2\left(1 - v^2/c^2\right)\right]} \tag{12.1}$$

Where,

p = Period of Wiggler field
v = Velocity of electrons

The same relationship may also be written in the form of eqn. 12.2.

$$\lambda = \frac{(0.131p)}{(0.511 + E)^2} \tag{12.2}$$

Where,
E = Electron energy in MeV.

The free electron laser is tunable over a wide range of wavelengths in the region from X-rays to microwaves. The tunability is achieved by varying E. The accelerated beam of electrons having desired energy can be obtained from well-established technology of charged-particle accelerators. Figure 12.56 shows the simplistic arrangement of various components in a free electron laser.

Free electron lasers though thought of as potential sources for DEW applications, these lasers have their problems when it comes to generating higher powers at shorter wavelengths. Smaller free electron lasers with power output up to 100 W are particularly important for research and medical applications.

12.9.5 Solid-State Lasers

Solid-state lasers as described earlier in Chapter 7 on *Laser Fundamentals* are electrically driven devices using a solid-state laser gain medium. The gain medium is pumped by either flash lamps or laser diode bar arrays with laser diode pumping being far more widely used in the state-of-the-art solid-state lasers. The all solid-state configuration offers unmatched advantages in terms of compactness, robustness, reliability and logistic simplicity. However, with the present technology status, it can't match the output laser power of megawatt level achievable in chemical lasers. Power level of tens of kilowatts have already been realized in bulk solid-state lasers and the present technology thrust is to realize 100 kW-class solid-state laser source that would be a good enough power level for a short-range tactical directed-energy laser weapon for counter UAV and counter RAM (Rocket Artillery Mortar) applications.

There is a growing need for technologies suitable for light weight, compact and robust high-pwer laser sources suitable for mounting and deployment on mobile platforms for short-range tactical battlefield applications. Chemical lasers are presently the most favoured choice because of their proven scalability to higher power levels with good beam quality. The drawbacks of these lasers are their unusually large size due to the requirements of large quantity of chemical fuels thereby making them logistically complex to engineer for mobile platforms. Technology of electrically driven solid-state lasers is well-established with the advantages of high electrical-to-optical conversion efficiency, robust and compact size for several decades. However, power scaling of solid-state lasers is limited by the thermo-mechanical distortions caused by waste heat deposited in to the gain medium by optical pumping. In the case of chemical lasers, the waste heat is removed and ejected out with the gas mixture at high flow rate thereby allowing power scale-up to very high levels with good laser beam quality. While power scale-up is not much of an issue in chemical lasers including gas dynamic lasers, HF/DF and chemical oxyiodine lasers, there are definitely issues related to size and weight, adaptability to platform mounting, use of toxic fuels, operation on a continuous basis and so on. In the recent years, driven by Armed Forces' requirements to possess compact, mobile laser weapon, there has been a thrust on development of power scalable solid-state lasers. Solid-state lasers are more compact and environmentally friendly. If scalable architectures without the need for complex beam combining were the goal, the output should be generated in a single aperture. This, in fact, is the objective of major high-pwer solid-state laser programmes internationally. Apart from the laser technology, the other critical areas for solid-state lasers are realization of a *compact thermal management system* for efficiently removing the heat generated in the laser system and the *compact electrical power source* for powering the total laser system.

Salient features of solid-state lasers of relevance to directed-energy laser weapon development are summarized as follows. A shorter wavelength of 1 μm yields lower divergence for a given transmitting aperture size and hence longer operating ranges. Also, shorter wavelength offers better laser-matter interaction. All solid-state electrically operated configuration leads to lightweight, compact and rugged system without any chemical hazards and requirement of any special support logistics. High efficiency means low operating cost and a deep magazine.

Research and development efforts in the recent years at *Lawrence Livermore National Laboratory* and *Northrop–Grumman* has generated a lot of interest in high-pwer solid-state lasers internationally for futuristic laser-based DEWs. High-pwer solid-state laser programmes being pursued at the two centres follow two different design architectures. Both programmes have the objective of developing a 100 kilowatt-class single aperture solid-state laser. While *Lawrence Livermore National Laboratory* is developing heat capacity disc laser technologies, Northrop–Grumman's laser is based on oscillator-amplifier configuration. The two programmes are briefly described in the following paragraphs after an introduction to the concept of heat capacity laser.

As outlined earlier, the key obstacle to increasing the output laser power involves efficient management of waste heat generated during laser operation. In a solid-state laser, this waste heat is deposited inside the lasing medium, which could be a crystal or a glass. The heat, if not removed, can damage the optics. Most solid-state laser systems are continuously cooled during operation to avoid such damage. Waste heat is conducted from inside the crystal or glass to the surface from where it can be carried away by the coolant. It leads to creation of a temperature gradient between the heated interior and cooled exterior as the cooling process occurs at the same time as the lasing. These large temperature gradients lead to mechanical stress, physical deformation, optical distortion and consequently fracture of the optics.

In the case of a heat capacity laser, the cooling cycle is separated from the lasing cycle. During lasing operation, the waste heat accumulates evenly throughout the lasing material. At the end of the lasing cycle, which may last for few seconds to few tens of seconds, the laser is shut off and the optical material is aggressively cooled over a period of typically ranging from 30s to several minutes to be ready again for another lasing cycle. By operating in this manner, one can eliminate creation of significant thermal stresses on the optical material during lasing thereby removing the limit on the average power output delivery capability of solid-state laser. The average power emitted by the laser is now limited only by the heat capacity of the lasing material and the output power of the pump source. This allows scaling up the output power to tens of kilowatts.

As a part of the development programme of Lawrence Livermore National Laboratory to develop a 100 kW heat capacity solid-state laser, a 10 kW heat capacity laser was developed in the year 2001. Figure 12.57 shows a photograph of the laser head, which comprises of nine discs of neodymium-doped glass pumped by flash lamps. The laser with average output of 10 kW of power generated 500 J pulses at a repetition rate of 20 Hz in 10s bursts. The laser was tested at the White Sands Missile Range in New Mexico. The laser drew electrical input power of 1 MW working at about 1% efficiency typical of a flash-pumped solid-state laser. The laser could burn a through hole in a 2 cm thick stack of steel samples in 6s.

The ultimate objective of the heat capacity solid-state laser programme was, however, to build a next-generation system with enough electrical efficiency to produce a 100-kW average power laser beam from the same 1 MW of input power. The laser would produce 500 J laser pulses at repetition rate of 200 Hz. The technological challenges that were needed to be overcome included growing large crystals of neodymium-doped gadolinium gallium garnet (Nd-GGG) for building amplifier discs, developing cost-effective technologies to make high-pwer laser diode bar arrays and developing a laser architecture to produce high-quality laser beam. Nd-GGG was chosen as the lasing material due to its higher mechanical strength, thermal conductivity and optical-to-optical

Figure 12.57 A 10 kW heat capacity solid state laser of LLNL. (*Source:* Courtesy: LLNL.)

efficiency and feasibility to grow large high opticalquality crystals. Higher thermal conductivity would allow rapid cooling of discs between runs and therefore a higher repetition rate. Northrop–Grumman and Poly-Scientific are the commercial partners responsible for growing Nd-GGG crystals of the required 20 cm diameter for the 100 kW class laser.

Polycrystalline laser ceramics has been another preferred lasing material. Ceramics have the advantage that they demand no tedious cycles of single crystal growth; can be custom-fabricated with spatially tailored doping concentrations and index profiles; can be easily scaled up to large aperture size; are tougher than single-seed crystal slabs and much less prone to catastrophic fracture; can accommodate higher dopant concentrations that can be precisely controlled and offer the possibility of composite structures.

The challenge with respect to pump diodes is not only in manufacturing large and powerful laser diode arrays but also in developing a cooling system that works in the field. The optical scheme of the resonator would also play an important role in determining the quality of the laser beam produced at the output. One effort in this direction at LLNL has been to develop an adaptive resonator system that would use a wave front sensor to sense distortions in the output wave front during operation and a deformable mirror that would correct the distortions.

In 2005, the Lawrence Livermore National Laboratory reported development of 25 kW average power (125 J @200 Hz) HCSSL that used four ceramic-YAG slabs. The output laser pulse train produced 0.5 millisecond pulses with a duty cycle of 10% in a burst lasting 10 s. The pulse length is about 0.5 ms, giving a duty factor of 10%. This 25 kW average power laser could penetrate a 2.5 cm thick piece of steel in 2–7 s depending on the beam size at the target. Recently, the system has been upgraded to 67 kW average powerwith five ceramic slabs for short fire durations. Based on the success of ceramic slabs, efforts are on to upgrade the system up to 100 kW (500 joule @200 Hz) by designing a megawatt-class, heat capacity solid-state laser using 16 ceramic laser slabs, each measuring 20 × 20 × 4 cm. Figure 12.58 shows a life-size model of a mobile 100 kW heat capacity solid-state laser developed by General Atomics and

Figure 12.58 Life size model of a 100 kW HCSSL. (Courtesy: LLNL)

Figure 12.59 A 105 kW SSL developed under the JHPSSL programme.

PEI Electronics and built on a prototype of a Humvee illustrating the level of compactness achievable in a short-range tactical laser weapon system.

The other important high-pwer solid-state laser programme with the objective of building a 100 kW-class laser is the US military's *Joint High-pwer Solid-State Laser (JHPSSL) programme*. Under this programme, the Northrop–Grumman Corporation adopted a scalable building block approach to build the laser of required output power capability. The basic building block produced 15 kW output power. Seven such lasers were coherently combined to produce 105 kW of output power. The 105 kW laser (Figure 12.59) was demonstrated in 2009. The laser had a turn-on time of less than one second and a continuous operating time of 5 min producing a good quality laser beam. The laser was integrated with an existing beam-control and command control systems of Tactical High-Energy Laser (THEL), another Northrop–Grumman built system. The integrated system called Solid-State Laser Test-bed Experiment (SSLTE) was used to evaluate the capability of a 100 kW-class solid-state laser to accomplish a variety of missions (Figure 12.60).

Another high-power solid-state laser based on the slab architecture previously used by Northrop–Grumman to builda 105 kW laser is the *Gamma laser*. The company announced Gamma in 2012 after completing an extensive series of initial tests. The laser operated at 13.3 kW for a number of shots over a total of 1.5 h with stable performance and a beam quality that exceeded design goals, completing the initial phase of trials. A gamma laser is the basic building block that is designed in such a way that more than one such laser could be combined to produce higher power levels. The laser has also been ruggedized for survivability in real-world operational environment by subjecting key portions of the laser to vibration, shock and thermal testing.

12.9.6   Fibre Lasers

The technology of fibre lasers is the most advanced solid-state laser technology available today. The basic fibre laser comprises of a gain medium in the form of a long optical fibre of suitable material doped with lasing ions. For high-pwer operation, ytterbium-doped glass fibre is typically used. The entire fibre length is pumped with large number of single emitter fibre coupled

Figure 12.60 Solid State Laser Test-bed Experiment (SSLTE). (*Source:* Courtesy: SPI Lasers UK Ltd.)

laser diode arrays. The laser resonator cavity is formed by embedding Bragg grating reflectors at the two ends of the fibre. Due to the small aperture of fibre (a few µms), the output laser beam is emitted with diffraction-limited beam quality resulting in output laser intensity (brightness) that is nearly two orders of magnitude higher as compared to that produced by conventional solid-state lasers. Also, since the resonator is formed within the fibre, there is no free-space optics thereby making the fibre laser extremely robust and reliable compared to other lasers. Fibre lasers were discussed in detail in Chapter 6.

Salient features of fibre lasers are summarized as follows. A shorter wavelength of 1.0 µm offers lower beam divergence and therefore longer operating ranges. Shorter wavelength also means better laser-matter interaction. Fibre lasers produce diffraction-limited beam quality. They are more relatively more efficient with wall plug efficiency of about 25% compared 10% in the case of diode-pumped bulk solid-state lasers. They are also more reliable and robust due to an absence of free-space optics.

The technology of the fibre laser is extremely complex and confined to only couple of international companies. Single-mode fibre lasers with an output power of 400–600 W are commercially available; the technology of kilowatt-class fibre laser is available from only a few companies including SPI lasers UK Ltd and IPG Photonics, USASPI lasers offer single-mode fibre lasers up to 1 kW output power. One such OEM laser is the redPOWER 500/1000 W fibre laser emitting at 1070 nm (Figure 12.61). The laser produces high-quality laser beam with M^2-value and Beam Parameter Product (BPP) specifications of 1.1 and 0.37 mm-mrad, respectively. At IPG Photonics, single-mode fibre lasers up to 10.0 kW power level and a beam quality characterized by M^2 value better than 10 are well-established (Figure 12.62). Multimode fibre lasers up to 50 kW power are also available. Higher single-mode output power levels are realizable by using coherent combination of multiple fibre-laser beams.

Figure 12.61 redPOWER fibre laser. (Courtesy: SPI lasers UK Ltd.)

Figure 12.62 A 10 kW single-mode fibre laser from IPG Photonics. (Courtesy: IPG Photonics.)

12.9.7 Beam Combination of Multiple Lasers

Power scale-up to tens to hundreds of kW in a single laser maintaining the desired beam quality has many a technological challenge to overcome. Development in the field of beam combination techniques has opened new avenues of building higher power lasers than would be possible in a single laser. Combining laser output from multiple fibre lasers is an important area of relevance to high-pwer directed-energy laser weapon systems as it allows achieving higher output power from relatively low power individual lasers. Direct diode lasers, bulk solid-state lasers and fibre lasers have been experimented with in recent years to generate higher powers by combining outputs of multiple lasers of a given type. Power enhancement and beam quality issues have been studied over propagation distances in km range. Three common laser beam combining techniques include *spectral beam combining, coherent beam combining* and *incoherent beam combining*. Each of these techniques is briefly described in the following paragraphs.

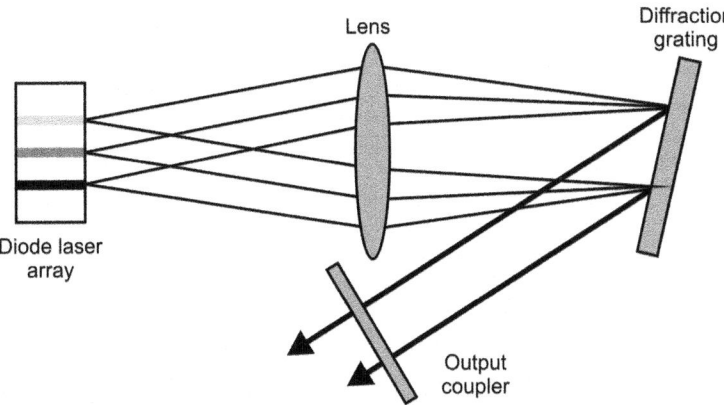

Figure 12.63 Spectral beam combining.

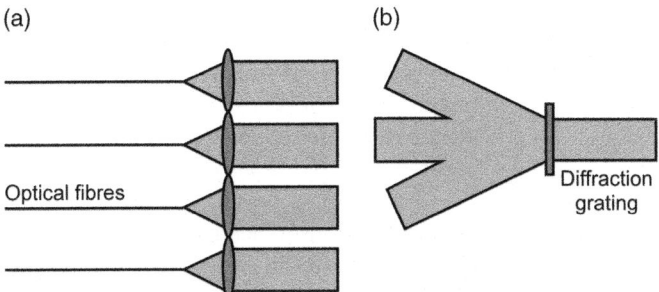

Figure 12.64 Coherent beam combining: (a) tiled aperture technique and (b) filled aperture technique.

In the case of *spectral beam combining*, multiple laser beams with non-overlapping optical spectra are combined by using a wavelength sensitive beam combiner such as a diffraction grating and prisms. Optical components with wavelength sensitive transmission characteristics such as volume Bragg gratings and dichroic mirrors can also be used. The wavelength sensitive optical component ensures that all beams though incident at different angles, subsequently propagate in the same direction. Figure 12.63 illustrates the concept. For spectral beam combination, each laser should have a sufficiently stable output wavelength and an emission bandwidth that is only a small fraction of the gain bandwidth. Beam quality reduces with increase in emission bandwidth.

In the case of *coherent beam combining*, multiple lasers are combined to generate higher output powers with more or less the same beam quality as that of the individual lasers. Coherent combining also preserves the spectral bandwidth. Two commonly used techniques used for coherent beam combining include the *tiled aperture technique* and the *filled aperture technique*. The tiled aperture technique, also called the side-by-side technique, is illustrated in Figure 12.64(a). It is similar in concept to a phased array antenna in microwave spectrum. In the optical domain, the realization is more difficult due to the much smaller wavelength, which introduces correspondingly tighter mechanical tolerances. As is evident from Figure 12.64(a), the tiled aperture approach leads to a larger beam size but reduced divergence. In the case of a filled aperture coherent beam combination, multiple beams are combined into a single beam using some kind of N × 1 grating splitter as shown in Figure 12.64(b). The beam size and divergence of combined beam are equal to those of individual beams.

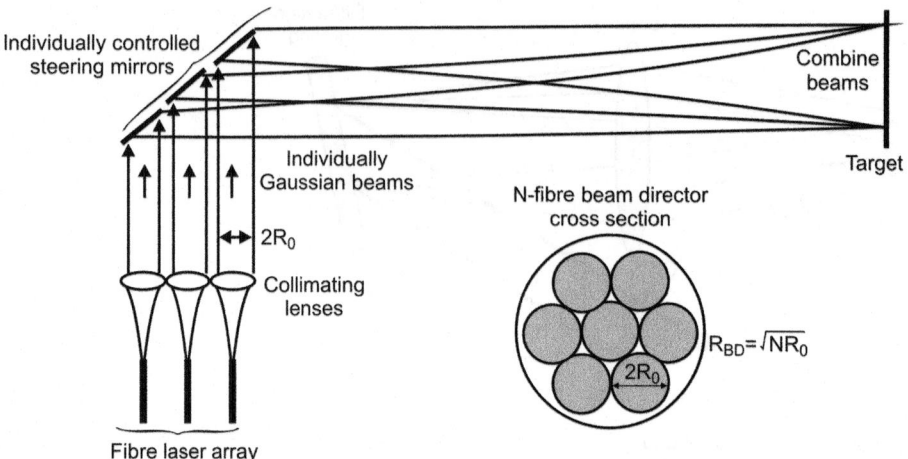

Figure 12.65 Incoherent beam combination.

One variant of filled aperture coherent beam combining is the *polarization beam combing*. When two mutually coherent beams are polarization combined, it is possible to obtain a linear polarization state for the output. Though, polarization-based coherent beam combining can be done with two inputs only; the process can be repeated multiple times due to linear output polarization.

In one method of *incoherent beam combination*, multiple laser beams are combined by overlapping the individual laser beams on the target with a beam director consisting of independently controlled steering mirrors. Figure 12.65 illustrates incoherent beam combination of multiple fibre lasers. Adaptive optics may be used to compensate for the distortions caused by atmospheric turbulence. This technique is relative much simpler than other beam combining techniques including spectral and coherent beam combining as it doesn't require phase locking or polarization locking of individual lasers and can be readily power scaled up for DEW applications.

For N incoherently combined fibre lasers, the total transmitted power equals $N \times P$ and the spot size of the combined beam equals $\sqrt{N} \times R$. Values P and R are, respectively, the power output and laser spot radius of the individual lasers. If nine fibre lasers, each with 5 kW power and 4 cm spot radius, were incoherently combined, the result would be a 45 kW laser system producing a laser spot radius of only 12 cm.

12.10 Beam-Control Technologies

The function of the beam-control system is to precisely point and focus laser energy at the designated point on the intended target and keep it there for sufficient duration to inflict damage to the target. A beam-control system comprises the following subsystems.

1) Beam transport optical system
2) Beam directing telescope system
3) Target acquisition and tracking system
4) Adaptive optical system.

A *beam transport optical system* transfers the laser beam from the stationary high-pwer laser source to the gimbal mounted beam directing telescope system that is used to direct the high-pwer laser beam in the direction of the target. The beam transport system used to couple the laser beam to the telescope system is nothing but a cascade arrangement of a number of gimbal follower mirrors that retain the alignment of laser beam axis with the telescope axis irrespective of its orientation.

The laser is fixed and only the relatively low mass telescope system mounted on a two-axis gimbal platform continuously tracks and points to the target. The *telescope system* is responsible for precisely focusing the laser beam onto the target. The telescope focusing action is controlled in a closed loop with a boresighted laser range finder system that keeps the laser beam focused onto the target in the entire operating range. The telescope aperture size controls the laser spot size and hence the lethal range of the weapon system.

A *target acquisition* video camera either bore sighted or shared with the telescope system acquires the target and tracks it by controlling the movement of gimbal platform. For a directed-energy laser weapon system designed to engage fast moving and manoeuvring aerial targets such as rockets, artillery shells, mortar rounds, unmanned aerial vehicles, missiles and aircraft, it is essential to have a *beam pointing system* with pointing accuracy in the order of few micro-radians. The critical requirement is to aim and maintain the laser beam on the vulnerable spot on the target until a kill has been achieved.

Adaptive optics (AO) is a critical part of the beam-control system for a tactical laser weapon. As the high-energy laser (HEL) propagates through the atmosphere to the target, atmospheric turbulence degrades the laser beam quality and reduces its effectiveness in terms of inflicting damage to the intended target. Adaptive optics senses the wave front aberrations caused by atmospheric turbulence and pre-compensates the outgoing high-energy laser beam to restore its mission performance capability. The adaptive optical system comprises a wavefront sensor, a deformable mirror and a control processor. The compensated beam puts almost more than eight times the peak intensity on the target than an uncompensated beam.

To summarize, the beam-control system performs a number of critical operations, which include target acquisition up to several kilometres, threat analysis and target assignment, course tracking, hit point identification and selection, high precision fast tracking of the hit point, sensing wave front distortion and applying compensation correction, focusing the laser beam for several seconds on the vulnerable spot and damage assessment.

12.11   Laser Propagation Effects

Laser propagation effects have a significant bearing on the effective operational range of directed-energy laser weapon system. Various propagation phenomena, briefly described in the following paragraphs, cause attenuation and spreading of the laser beam as it propagates towards the target. Key atmospheric effects on propagating laser beam include diffraction, scattering, absorption, turbulence and thermal blooming.

The phenomenon of *diffraction* of light is inherent to the nature of light and comes from the wave nature of light. It causes laser beam spreading as it propagates through atmosphere or even through vacuum. This puts a fundamental limit on the smallest achievable focused spot size. The amount of beam spreading is proportional to the ratio λ/D, where D is beam aperture diameter and λ is the wavelength of light. Thus for larger apertures or smaller wavelengths, the effect of diffraction is reduced. Larger transmitting aperture for a given operating wavelength or smaller wavelength for a given transmitting aperture produce smaller focused spot size and hence a higher power density at a given target distance. In other words, for a given desired value of power density at the target that would cause intended damage, a smaller wavelength and larger aperture would mean an increased operational range. Diffraction-limited spot size

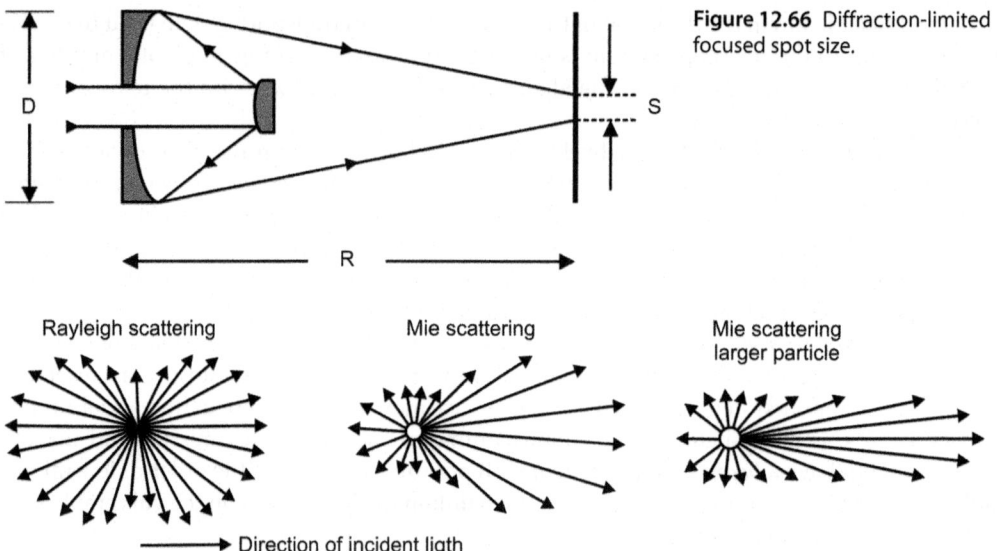

Figure 12.66 Diffraction-limited focused spot size.

Figure 12.67 Rayleigh and Mie scattering.

varies linearly as the ratio λ/D and focal length f. Focal length in this case equals the target distance. Refer to Figure 12.66. Focused spot size, *S*, in the case of high-pwer laser beam may be computed from eqn. 12.3.

$$S = 2.44 \times B \times R \times \left(\frac{\lambda}{D} \right) \tag{12.3}$$

Where *B* is the beam quality parameter.

Atmospheric scattering is another important energy loss mechanism in the propagation of a laser beam that causes reduction in laser power reaching the target. It therefore reduces power density available at the target location or for a given power density value reduces the operational range. Electromagnetic radiation interacts with atmospheric constituents such as water, sea salt, organic matter, dust, soot and urban pollutants, which act as scattering centres. The electromagnetic energy is redistributed in directions that do not contribute to the intended use of the laser. There are three different kinds of scattering; namely Rayleigh scattering, Mie scattering and non-selective scattering.

Rayleigh scattering refers to the scattering of light off of molecules in the air. The radiation from Rayleigh scattering is emitted in all directions, causing a loss of energy as it propagates through the atmosphere. Rayleigh scattering is predominant when the molecular size is smaller than the incident wavelength. Rayleigh scattering is inversely proportional to a fourth power of wavelength $(\propto 1/\lambda^4)$. *Mie scattering* is the scattering type when the scattering centres are roughly comparable to the size of the incident wavelength. These types of Mie scatterers are typically suspended aerosol particles or very small droplets of water. Like Rayleigh scattering, the radiation of incident energy is scattered in all directions with the difference that Mie scattering has predominant scattering in the same direction as the incident photon. When the scattering centres become larger than the incident wavelength, the forward scattering lobe becomes more predominant. Mie scattering is not strongly dependent upon wavelength. Rayleigh and Mie scattering phenomena are illustrated in Figure 12.67. *Non-selective scattering* is the predominant scattering mechanism when the scattering particles are much larger than the incident

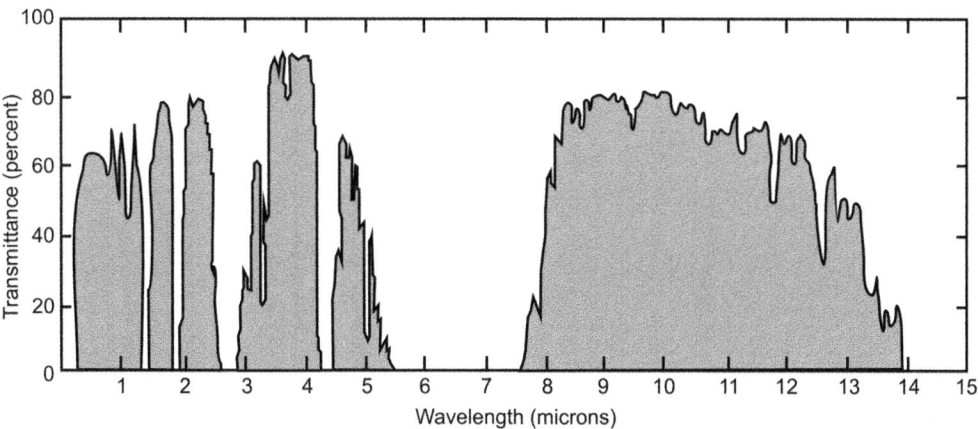

Figure 12.68 Atmospheric transmission windows.

wavelength. This happens when the laser beam passes through fog, haze and clouds. Non-selective scattering is independent of wavelength and loss of energy occurs due to the collisions with the large mass.

Absorption is the process where the incident radiation is absorbed by the medium of propagation. For a given value of fluence or power density at the target location to cause the intended damage, absorption reduces the operational range. The primary atmospheric components that contribute to absorption are water, carbon dioxide, diatomic oxygen (O_2) and ozone. These molecules absorb the electromagnetic radiation and convert it to molecular vibration and rotation. Atmospheric absorption has a strong dependence on location, time of year and weather patterns. Since the composition of atmosphere cannot be controlled, the choice of an appropriate wavelength helps in minimizing absorption. Atmosphere has certain transmission windows of bands of wavelengths where absorption is significantly less. Figure 12.68 shows the graph of transmittance as a function of wavelength illustrating the existence of minimum absorption transmission windows. An operating wavelength within any of these transmission windows will be suitable from the viewpoint of minimizing the effect of absorption.

Turbulence causes laser beam to spread resulting in increase in laser spot size and reduction of power density at the target. A quantitative measure of the optical turbulence in atmosphere is the *refractive index structure parameter* C_n^2 expressed in $m^{-2/3}$. C_n^2 is strongly dependent upon altitude as well as wind speed and other atmospheric parameters. In the atmospheric surface layer, C_n^2 generally varied in the range of 10^{-12} to 10^{-16} $m^{-2/3}$. C_n^2 of $10^{-13} m^{-2/3}$ or greater usually indicates a highly turbulent atmosphere. Such high values of turbulence are generally experienced close to ground on clear days under unstable conditions. Over oceans, high value of C_n^2 may be in the order of 10^{-14} $m^{-2/3}$. Low values of C_n^2 in the range of $10^{-16}–10^{-15}$ $m^{-2/3}$ indicate more adiabatic conditions occurring generally in windy or cloudy conditions and night time, and also during transitional periods after sunrise and before sunset. Figures 12.69 and 12.70, respectively, show graphs of dependence of the C_n^2 parameter on altitude and time of the day. Though the primary effect of turbulence is the beam spread, it also induces a random beam walk around the aim point called *beam wander*. Turbulence is mainly associated with temperature and density fluctuations in the atmosphere. These fluctuations lead to variation in the refractive index of the air thereby causing fluctuations in the direction of the propagating light. If the regions of temperature fluctuations were smaller than the beam itself; different portions of beam front would bend in diverse directions causing beam spread.

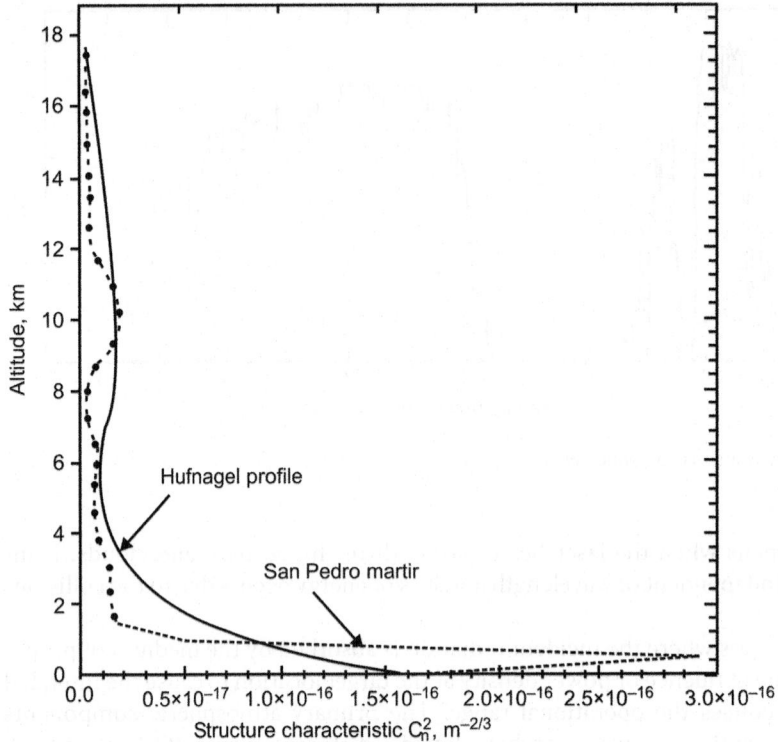

Figure 12.69 $C_n{}^2$ dependence on altitude.

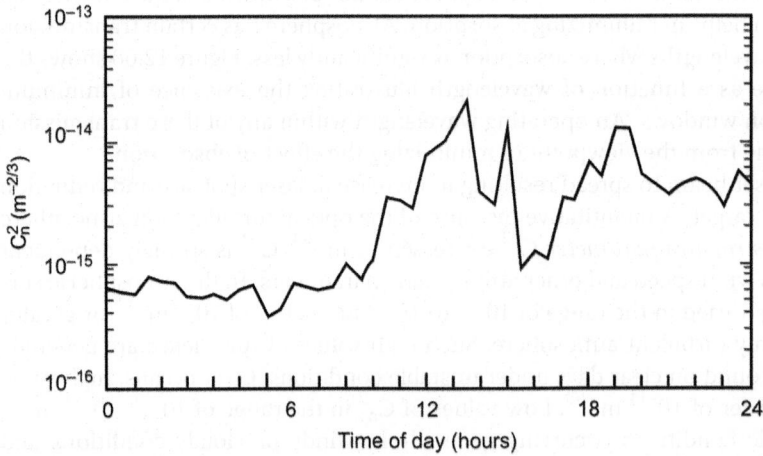

Figure 12.70 $C_n{}^2$ dependence on time of day.

Thermal blooming is distortion of a propagating high-pwer laser beam caused by the heating of the air around it due to the high-energy density of the laser beam. Absorption of laser energy causes the temperature of the air the laser beam is passing through to increase. This creates a refractive index gradient in the laser beam path leading to formation of a negative lens. Formation of negative lens is then the cause of increase in beam spreading. Thermal blooming is far more predominant in the case of high-pwer laser beams aimed at stationary or slow moving targets. In the case of fast moving targets such as aerial targets, the air column through which the laser is passing continuously gets changed, thereby eliminating the possibility of occurrence of temperature rise and thermal blooming. Since thermal blooming is caused by absorption of laser energy in the atmosphere, thermal blooming reduces with increase in altitude of laser platform. The reason is that aerosol content of the atmosphere, which causes absorption, is larger at lower altitudes. Also, crosswinds blowing perpendicular to the beam mitigate thermal blooming by blowing the heated, expanding gases out of the way.

12.12 Lethality

Lethality is the capability of a weapon system to render a target non-functional. Targets may be destroyed, denied, degraded or delayed. The laser-material interaction and the resulting damage mechanism not only depend upon the laser power density but also upon the construction and operational conditions of the target. The basic damage mechanism of DEW is thermal. As the laser impinges on the target, the heat buildup is very quick and consequently the target melts and then vaporizes. It is not necessary that the material must vaporize for the target to be destroyed. Even before melting takes place, the material becomes very soft, thus weakening the structure considerably. The potential targets for DEW applications can be categorized into following categories based on the type of laser-material interaction and the resulting lethality effect on target.

1) Landmines/Improvised explosive devices (IEDs)
2) Rockets, artillery shells and mortars
3) Battlefield/ballistic missiles
4) Drones, RPVs, UAVs
5) Electro-optical devices.

High-power lasers can be employed for neutralization of these improvised explosive devices and landmines by focusing energy on the munitions casing, thereby heating it until the explosive filler ignites and starts to burn. In the case of metal-cased munitions, heat is conducted through the metal case until the side wall in thermal contact with the explosive filler exceeds its combustion temperature. In the case of plastic-cased munitions the laser radiation burns and penetrates the casing material and the explosive filler is ignited either directly from the laser radiation or by the flames burning the plastic case. The combustion created by the laser leads to low-level detonation or deflagration rather than activating the explosive power designed into surface mines/IEDs. The advantage of using a laser to neutralize munitions is its large magazine, ultra precision, assured neutralization, safe standoff range for personnel and controllable effects with reduced collateral damage. The laser power densities required to achieve these function ranges from 50–300 W/cm^2. Such power density values are achievable with kilowatt-class solid-state and fibre lasers. Some representative laser ordnance neutralization systems have already been described in the earlier part of the chapter.

In the case of *RAM* (*Rocker Artillery Mortar*) targets, the damage mechanism is more or less similar to that described earlier in the case of ordnance neutralization. The laser radiation heats up the casing material. Heat is conducted through the casing material until the side wall in thermal contact with the explosive filler exceeds its combustion temperature. The resulting combustion pressure eventually ruptures the heated region of the casing material thereby damaging the device in flight. The laser power density required to neutralize such targets in flight lies in the range of 1–3 kW/cm^2.

Ballistic missiles are highly vulnerable to damage by directed-energy laser weapons. The reason is that missiles have thin load bearing skins, which are heavily stressed during the boost phase. The missile boosters are largely filled with pressurized high-energy propellants and even slight damage to the booster skin can cause a catastrophic failure. For the missiles moving at high speeds, the outer surface of the fuel tank is under severe stress due to low-pressure zone created by the high-speed of the structure. The laser heating of the structure leads to reduction of tensile fracture strength of material that eventually leads to rupture of the structure due to presence of aerodynamic loads. Required laser power density is in the order of few $100\,W/cm^2$.

In the case of *drones*, *RPVs* (*Remotely Piloted Vehicles*) and *UAVs* (*Unmanned Aerial Vehicles*), the high-pwer laser heats and subsequently melts structures leading thereby disabling and grounding of targets. The laser power requirement depends upon the target structure material that ranges from few $100\,W/cm^2$ for plastic structures to few kW/cm^2 for metallic and composite structures.

Directed-energy laser weapons capable of inflicting structural damage to military targets candefinitely destroy *optical* and *electro-optical devices* at much longer distances. The laser power density required to perform this function depends upon the material, sensor type, working wavelength and so on of the device and also upon the wavelength of laser device.

12.13   Representative Directed-Energy Laser Weapon Systems

Presently, a large number of directed-energy laser weapon systems are reported to be under development and upgradation. Some of them are experimental, some are technology demonstrators while others are being upgraded and ruggedized to become realistic battlefield weapon systems in the near future. Some of the more talked about laser weapon systems include the *High-Energy Laser* (HEL) *system* developed by Diehl and LFK of Germany, the*General Area Defence Integrated Anti-missile Laser System* (GARDIAN) from TRW, the *Mid-Infrared Chemical Laser* (MIRACL) from TRW, the *High-Energy Laser Weapon System* (HELWEPS) from TRW, Boeing's *Airborne Laser* (ABL) and *Advanced Tactical Laser* (ATL), *Tactical High-Energy Laser* (THEL), *Mobile-THEL* (M-THEL), Raytheon's *Laser Phalanx* and *Laser Weapon System* (LaWS). These are briefly described in the following paragraphs.

The HEL, jointly developed by two German companies Diehl and LFK, is a short-range tracked vehicle-mounted system for use as an air-defence system against low-flying, high performance aircraft, missiles and attack helicopters. The system is configured around a gas dynamic CO_2 laser emitting at $10.6\,\mu m$ and has associated target acquisition and tracking sensors. The beam director uses a 1 m diameter focusing mirror mounted on an extendible arm for delivery of high-pwer laser beam to the target. Target damage is achieved by focusing the laser radiation into a small-diameter spot producing a very high-energy density causing the target material to get successively heated, melted and vaporized. The system has an operational range of 8 km. The system reportedly has been successfully tested.

Another air-defence laser system is TRW's *General Area Defence Integrated Anti-Missile Laser System* (GARDIAN). This system too is a short-range complement to surface-to-air missile defence designed to engage discrete ballistic threats at longer ranges. The system generates a 400 kW laser beam, which is delivered through a 0.7 m beam pointer/tracker. Depending upon weather conditions, the laser can destroy targets at altitudes from ground-hugging heights up to 15 km. The system has a target acquisition and tracking system, and laser once locked stays locked to the target until it is destroyed. The system can carry fuel for firing 60 shots.

Figure 12.71 Airborne laser. (*Source:* Courtesy of US Airforce.)

One of the very early directed-energy laser systems, a technology demonstrator, is TRW's *Mid-Infrared Advanced Chemical Laser* (MIRACL). It is a 2.2 MW CW deuterium fluoride laser with a maximum lasing duration of 70 s. It uses a 1.8 m Sealite pointing and tracking device. The system was reported to carry out trials against different types of targets at the White Sands Missile Range, NM. In one of the tests conducted in 1996, a small fraction of laser power was used to destroy a 122 mm short-range artillery rocket in flight. The laser beam was locked to the target for 15 s. The laser is also reported to have been used to test against sea skimming missiles.

Yet another recent laser weapon system from TRW is the *High-Energy Laser Weapon System* (HELWEPS), which is again a chemical laser using Ethylene, Hydrogen and Fluorinated Nitrogen as the active medium. The system is based on their earlier experience gained from building MIRACL. The system has an integral electro-optic tracker.

Another laser-based directed-energy weapon system in serious contention until the recent past for use as a laser weapon, particularly for operation from an aerial platform, is the famous Airborne Laser (ABL). The ABL uses a 1.2 MW COIL emitting at 1.315 μm generated by six COIL modules of 200 kW each. It is configured on Boeing 747–400 freighter aircraft as shown in Figure 12.71. The main deck forward of the wings is separated from the aft main deck by a full height bulkhead. The forward fuselage portion houses the battlefield management and the beam-control systems, the aft main deck houses the laser. The battlefield management system comprises computers to manage the weapon system, operator consoles of the weapon system and supporting communications. The beam-control system comprises target acquisition and tracking equipment and adaptive optical system. The adaptive optical system comprises mainly the wavefront sensor and control system for beam distortion control. In addition to the high-pwer COIL located in the aft main deck, there are two supporting lasers. These are the *Tracking Illuminator Laser* (TIL) and *Beacon Illuminator Laser* (BIL). Both the lasers are diode-pumped solid-state devices. While the former is used to illuminate the target to facilitate fine tracking, the latter is part of the setup used to measure atmospheric distortion with the wavefront sensor.

Figure 12.72 Advanced Tactical Laser (ATL). (*Source:* Courtesy of the US Army.)

ABL is capable of destroying a ballistic missile in boost phase. In operation, the aircraft patrols the friendly air space. If an enemy missile launch is detected by a variety of sensors, this information is relayed to the aircraft configured as high-pwer laser system. A notable feature of ABL is the nose-mounted optical turret for the laser's primary mirror. The turret has a +/-120° field of regard in azimuth and is used to point the 1.6 m primary laser mirror. When the laser is not in use, the 1.8 m window is rotated into a stowed position to protect the optical surface from abrasion by atmospheric dust particles and birdstrike damage. The 1.6 m beam director focuses the high-power laser radiation from COIL onto the missile once it rises above the cloud cover. The intercept range of ABL has been put at 200 km from its standoff position. Performance evaluation of ABL began in July 2002. In February 2010, the system was successfully used to destroy ballistic missile. Subsequently, the system was considered to be operationally not viable and the ABL programme has been reportedly shelved since December 2011.

Advance Tactical Laser (ATL) uses 80 kW COIL and is mounted on a modified Boeing C-130H Hercules aircraft with the most obvious visual difference being a rotating turret protruding from the aircraft's underside through a hole (Figure 12.72). This chemical laser is similar to the one developed for the Airborne Laser (ABL) programme with much lower output power. While ABL COIL vents its exhaust into the atmosphere, the Advanced Tactical Laser traps its exhaust, which allows the laser to operate at any altitude. The exhaust can be processed later to get more laser firings. While the Airborne Laser is designed to target ballistic missiles at long ranges; the Advanced Tactical Laser would be striking at nearby tactical targets. A laser beam can be focused and directed to a target through this turret. It is intended to be used for covert activities such as setting fires to vehicles, disabling communication antennas, satellite and radar dishes, break electrical power lines. The ATL is envisioned to offer the mobility of a small aircraft, high-resolution imagery for target identification, and the ability to localize damage to a small area of less than a foot in diameter from a range of 8–10 km.

Another well-known laser-based DEW system is Northrop–Grumman's *Tactical High-Energy Laser* (THEL). The laser is built in two configurations; the baseline static HEL as shown in Figure 12.73 and the relocateable mobile version *Mobile-THEL* (M-THEL) as shown in Figure 12.74. THEL systems are point defence weapon systems designed to engage and destroy artillery rockets, artillery shells, mortar rounds and low-flying aircraft. The system uses a DF laser operating at 3.8 μm. The THEL demonstrator was successfully tested repeatedly between

Figure 12.73 Tactica High Energy Laser – THEL. (*Source:* Courtesy of the U.S. Military.)

Figure 12.74 Mobile-THEL. (*Source:* Courtesy of the US Military.)

2000 and 2004, destroying number of 122 mm and 160 mm Katyusha rockets, multiple artillery shells and mortar rounds, including a salvo attack by mortar.

The *space-based Laser* (SBL) is the ultimate objective of the US directed-energy laser program. The SBL is proposed to be configured around a 20 MW HF laser operating at 2.7 μm. It is designed to intercept ICBMs and other strategic and tactical missiles. It is proposed to deploy the system in a 800–1300 km orbit and the expected target engagement range is 4000–12 000 km.

The SBL programme derives its strength from a number of advanced technologies developed by the *Strategic Defence Initiative Organization* (SDIO) in the 1980s. One of the programmes was the *Large Optics Demonstration Experiment* (LODE) to provide the means to control high-pwer laser beams. In another programme called *Large Advanced Mirror Programme* (LAMP), a 4 m diameter mirror having the required optical figure and surface quality was designed and fabricated. The ALPHA laser developed by the SDIO achieved megawatt power at the requisite operating level in a low-pressure environment similar to space. Also, a number of acquisition, tracking and beam pointing experiments were carried out. *Space Pointing Integrated Controls Experiment* offered near weapons-level results during testing. The technology of pointing and controlling the large structures of the space-based laser was validated by the *Rapid Retargeting and Precision Pointing* (R2P2) programme. Most recently, integrated high-energy ground testing of the laser and beam expander were performed under the ALPHA LAMP Integration (ALI) programme to demonstrate the critical system elements. An integrated space vehicle ground test with a space demonstration to conclusively prove the feasibility of deploying an operational SBL system would be the next step. SBL programme is proposed to be based on a constellation of 20 satellites that would provide nearly full threat negation. Figure 12.75 depicts the operational scenario of an SBL.

Several directed-energy laser systems based on solid-state and fibre lasers are being developed and tested for tactical mission needs ranging from ordnance neutralization to anti-missile and anti-RAM applications. Raytheon has developed and successfully tested a directed-energy laser system called the Laser Phalanx (Figure 12.76) employing a 20 kW industrial fibre-laser. The system has been successfully demonstrated against a static mortar from a distance of 0.5 km. Raytheon has also successfully tested a ship mounted solid-state laser weapon called Laser Weapon System (LaWS) (Figure 12.77) to shoot four drones.

Figure 12.75 Operational scenario of a space-based laser.

Figure 12.76 Laser Phalanx. (*Source:* Courtesy of the US Navy.)

Figure 12.77 Laser weapon system – LaWS. (Courtesy: Raytheon.)

12.14 Laser-Induced Plasma Channel (LIPC) Weapons

The *LIPC weapon* makes use of a laser generating ultra-short laser pulses that create a highly conducting plasma channel between the laser and the intended target and is designed to take out targets that conduct electricity better than air or ground surrounding them. This conducting plasma follows the path of the laser and therefore can be directed to different targets by steering the laser beam. When the plasma comes in contact with a high voltage source, a high voltage current discharge travels down the plasma channel and then through the target to ground thereby causing severe damage to the target similar to what would have been caused had there been a lightning strike.

The concept behind formation of plasma channel is as follows. A pulsed laser of even a modest energy producing ultra-short laser pulses is capable of generating gigantic peak power. For example, 100 mJ, 2 ps laser pulses would generate peak power level of 50 GW. The electromagnetic field produced due to the intense laser beam rips off electrons from the air molecules thereby ionizing the surrounding air and creating plasma. For high intensity laser pulses, the air can act like a lens and the laser focuses on itself in air confining the light to a small-diameter filament. LIPC weapons behave like a lightning strike that always tries to follow the path of least resistance while travelling from cloud to ground. The plasma channel conducts electricity much better than un-ionized air. If the plasma channel comes near a high voltage source, the electrical energy will travel down the ionized conduit. When an LIPC weapon is used on a target such as a vehicular platform, a person or unexploded ordnance, high voltage current discharge flows through the path of least resistance to ground potentially disabling the vehicle or person and initiating ordnance detonation.

LIPC weapon hardware mainly comprises a laser capable of generating ultra-short pulses in the order of few picoseconds and a power source to drive both laser and high voltage discharge. There are many a technological challenge to be overcome, which included synchronizing the laser with the high voltage, ruggedizing the device to survive under the extreme environmental conditions of an operational environment and powering the system for extended periods of time.

Laser-induced plasma channel devices can have a variety of applications. They can be used to kill or incapacitate human targets through electric shock. They can also damage, disable and destroy electronic devices. Laser-induced plasma channels may also be used for a variety experiments including study of lightning discharges, forcing lightning discharges to occur at safe time and place during thunderstorms, inducing thunderheads deliver a precise lightning strike on to a ground target triggered by an airborne laser and harvesting lightning energy for power generation by directing it to a terrestrial collection station.

Illustrated Glossary

Absorption This is the process where the incident radiation is absorbed by the medium of propagation. For a given value of fluence or power density at the target location to cause the intended damage, absorption reduces the operational range.

Active Denial System (ADS) ADS is a nonlethal microwave directed-energy system that can be used in counter-personnel role against hostile human targets at distances beyond the effective range of small arms.

Adaptive Optical System This senses the wave front aberrations caused by atmospheric turbulence and pre-compensates the outgoing high-energy laser beam to restore its mission performance capability. The adaptive optical system comprises a wave front sensor, a deformable mirror and a control processor.

Advanced Tactical Laser (ATL) ATL uses a 80 kW COIL and is mounted on a modified Boeing C-130H Hercules aircraft with the most obvious visual difference being a rotating turret protruding from the aircraft's underside through a hole (Figure 12.71). This chemical laser is similar to the one developed for the Airborne Laser (ABL) programme with much lower output power. While the ABL COIL vents its exhaust into the atmosphere, the Advanced Tactical Laser traps its exhaust, which allows the laser to operate at any altitude. The exhaust can be processed later to get more laser firings.

Airborne Laser (ABL) The ABL programme of the US Air Force uses a high-pwer COIL source. ABL is one part of a layered ballistic missile defence system that addresses the ever increasing ballistic missile threat and is designed to destroy the hostile missile while it is still in the vulnerable boost phase of its flight. The ABL uses a 1.2 MW chemical oxy-iodine laser (COIL) emitting at 1.315 μm generated by six COIL modules of 200 kW each. It is configured on a Boeing 747–400 freighter aircraft.

Airborne Laser Lab (ALL) The ALL programme of the United States was launched in 1976 and used gas dynamic CO_2 laser as the high-pwer laser source. The aim of the programme was to develop the technology demonstrator mounted on NKC-135A aircraft to track and destroy airborne targets.

Back Door Coupling This occurs through large transient currents produced in the case of low frequency weapons or electrical standing waves on fixed electrical wiring and cables interconnecting equipment, or providing connections to mains power in the case of HPM weapons.

Backward Wave Oscillator (BWO) A BWO is the oscillator version of the TWT amplifier. The operational principle of a BWO is similar to that of a TWT amplifier with the difference that the electron beam transfers its energy, present in the form of electron bunches and acquired due to velocity modulation caused by RF field propagating in the slow wave structure, to the backwards moving RF field.

Beam Directing Telescope This system is responsible for precisely focusing the laser beam onto the target. The telescope focusing action is controlled in a closed loop with a bore sighted laser range finder system that keeps the laser beam focused onto the target in the entire operating range.

Beam Dwell Time (Directed-Energy Weapon) This is the time period for which the high-energy beam is needed to remain on the aim point on the target to inflict intended damage.

Beam Transport System A beam transport optical system transfers the laser beam from the stationary high-pwer laser source to the gimbal mounted beam directing telescope system that is used to direct the high-pwer laser beam in the direction of the target.

Bofors HPM Blackout The BAE Systems Bofors HPM Blackout is a mobile gigawatt level microwave source operating in the L band. The system comprises a high-pwer microwave source, an integrated pulsed power unit and an exchangeable conical horn antenna with support systems including compact battery-powered vacuum system for the microwave tube and a gas supply system for the pulsed power unit. The Bofors HPM Blackout, originally designed as research and evaluation tool for generating data on threats from electromagnetic effects of HPM weapons, could have operational capabilities with its proven destructive effects at considerable distances against a wide range of commercially-off-the-shelf (COTS) equipment.

CHAMP CHAMP, an abbreviation for Counter-electronics High-powered microwave Advanced Missile Project, is a technology demonstration programme of the US Air Force Research Laboratory (AFRL) jointly with Boeing Phantom Works for the development and field testing of a counter-electronics HPM aerial demonstrator intended for integrating HPM payloads on airborne platforms such as cruise missiles and unmanned combat aerial vehicles (UCAV).

Charged-Particle Beam Weapons These use a high-energy beam of charged particles such as protons and electrons. These are mainly endoatmospheric weapons.

Chemical Laser This derives the energy required to produce population inversion and consequent laser emission from a chemical reaction. Importance of chemical lasers lies in their ability to generate continuous-wave output power level reaching several megawatts.

CHP Laser Dazzler The Compact High-pwer (CHP) is a short-range laser dazzler from LE systems, USA. The dazzler emits a 500 mW flashing green dazzling laser beam. The CHP with its higher power creates a credible glare effect in a larger spot size for use on moving vehicles or individuals.

Circular Accelerator In the case of circular accelerators, powerful magnets are used to bend the particle's path into a circle and the beam of particles travels repeatedly round the loop.

Coherent Beam Combining In the case of coherent beam combining, multiple lasers are combined to generate higher output powers with more or less the same beam quality as that of the individual lasers. Coherent combining also preserves the spectral bandwidth.

Dazer The Dazer Laser from Laser Energetics Inc., USA is a nonlethal laser dazzler. It comes in two variants named the GUARDIAN that has a range of 1–300 m (model dependent) and the Defender that has a range of 1–2400 m (model dependent). Both variants of the Dazer Laser temporarily impair the vision of the target adversary and succeed in eliminating the threat's ability to see, engage or effectively target the user.

Directed-Energy Laser Weapons A directed-energy laser weapon uses a high-pwer laser that has enough power in the case of CW laser or sufficient pulse energy in the case of pulsed laser to inflict a physical damage to the target.

Directed-Energy Weapon (DEW) A DEW system, with the exception of laser-induced plasma channel (LIPC) weapons, primarily uses directed-energy in the form of concentrated beam of electromagnetic energy or atomic or subatomic particles in the targeted direction to cause intended damage to the enemy's equipment, facilities and personnel.

DS-110 The DS-110, developed by Diehl Munitions Systems of Germany, is a high-pwer RF system that generates wideband electromagnetic pulse capable of destroying process driven electronic modules in any system by generating resets or inducing power latch-ups thereby neutralizing the targeted devices. The DS-110 system can also be used for disruption of electronic fusing mechanism of projectiles, artillery shells and improvised explosive devices.

E-bomb An electromagnetic bomb makes use of a short, intense electromagnetic energy pulse adversely affecting electronics circuitry with practically no effect to humans or buildings. The damage may range from temporarily disabling electronics systems functioning at low levels of EMP (Electromagnetic Pulse) energy to corrupting computer data at mid-range levels and further to complete destruction of electronic circuitry at high levels.

Electrostatic Blooming This is due to mutual repulsion of the charged particles and causes beam spreading. It occurs only in charged-particle beams.

Explosively Driven Flux Compression Generator An explosively pumped flux compression generator (EPFCG) is a device used to generate highly intense burst of electromagnetic energy by compressing magnetic flux using high explosive. The explosively pumped flux compression generator (FCG) capable of producing tens of megajoules of electrical energy in a time frame of tens to hundreds of microseconds with corresponding peak power levels of tens of terawatts is the most mature technology applicable to the design of E-bombs.

Front Door Coupling In the case of front door coupling, energy is coupled to the targeted equipment such as radar or communications equipment through the antenna.

Gas Dynamic Laser This derives its energy from the combustion of a suitable fuel-oxidizer mixture, which means that it does not require any electrical energy for its operation. A gas dynamic laser achieves population inversion by rapid expansion of high-temperature high-pressure laser gas mixture produced during combustion to a near vacuum in an adiabatic process through an integrated supersonic nozzle bank.

General Area Defence Integrated Anti-Missile Laser System (GARDIAN) GARDIAN is TRW's short-range complement to surface-to-air missile defence designed to engage discrete ballistic threats at longer ranges. The system generates a 400 kW laser beam, which is delivered through a 0.7 m beam pointer/tracker.

GLARE GBD-IIIC This is a long-range variant of GLARE MOUT and is a visual deterrent laser device for hail and warning applications.

GLARE LA-9/P The GLARE LA-9/P built around a 250 mW green laser is another long-range visual deterrent laser device for hail and warning applications and is intended to be effective out to a range of 0.3–4 km for ship-to-ship signalling or airborne overwatch applications.

GLARE MOUT This is a nonlethal visual disruption laser with an effective range of 20 m–2 km. The device is ideally suited for small arms integration as well as mobile crew-served applications.

Gyrotron The operation of a gyrotron is based on the interaction of an axial electric field and an electron beam moving in a spiralling trajectory under the influence of a magnetic field. The spiralling frequency called the cyclotron frequency depends upon the strength of the magnetic field, mass of the electron and the velocity component perpendicular to the magnetic field. Gyrotrons are capable of generating high-pwer, high-frequency electromagnetic radiation in terahertz (THz) range.

Hard-Kill Damage Hard kill relates to a permanent physical damage to the structure of the target. High-energy laser weapons and particle beam weapons are usually designed to inflict hard-kill damage.

Heat Capacity Solid-State Laser In a heat capacity solid-state laser, the cooling cycle is separated from the lasing cycle. During lasing operation, the waste heat accumulates evenly throughout the lasing material. At the end of the lasing cycle, which may last for few seconds to few tens of seconds, the laser is shut off and the optical material is aggressively cooled over a period of typically ranging from 30 s to several minutes to be ready again for another lasing cycle. By operating in this manner, one can eliminate creation of significant thermal stresses on the optical material during lasing thereby removing the limit on the average power output delivery capability of solid-state laser.

High-Energy Laser (HEL) HEL, jointly developed by two German companies Diehl and LFK, is a short-range tracked vehicle-mounted system for use as an air-defence system against low-flying, high performance aircraft, missiles and attack helicopters. The system is configured around a gas dynamic $CO2$ laser emitting at 10.6 μm and has associated target acquisition and tracking sensors.

High-Energy Laser Weapon System (HELWEPS) HELWEPS is a high-energy laser weapon system using a combination of ethylene, hydrogen and fluorinated nitrogen as the active medium. The system is based on their earlier experience gained from building MIRACL. The system has an integral electro-optic tracker.

High-Power Microwave Weapons (HPM) HPM weapons use high-power microwaves to produce the equivalent of electromagnetic interference to damage the enemy's electronics devices and systems.

Impulsive sources In the case of impulsive sources, pulsed microwave energy is generated by charging the antenna, a transmission line, or a tuned circuit directly and making them to ring for one or several cycles by closing a switch. Ultra-wideband sources are examples of impulsive sources.

Incoherent Beam Combining In one of the methods of incoherent beam combination, multiple laser beams are combined by overlapping the individual laser beams on the target with a beam director consisting of independently controlled steering mirrors.

JD-3 Laser Dazzler The Chinese JD-3 laser dazzler is a portable laser dazzler system and is reported to be mounted on the Chinese Type 98 main battle tank and is coupled with a laser radiation detector, and automatically aims for the enemy's illuminating laser designator, attempting to overwhelm its optical systems or blind the operator.

Klystron A klystron amplifier is type of source of microwave energy. It converts kinetic energy in the DC electron beam into RF power. It does so by a phenomenon called velocity modulation.

Large Advanced Mirror Programme (LAMP) LAMP was part of space-based laser programme and under this effort, a 4 m diameter mirror having the required optical figure and surface quality was designed and fabricated.

Large Optics Demonstration Experiment (LODE) The LODE programme was part of a space-based laser programme effort. It provided the means to control high-pwer laser beams.

Laser Dazzler This emits a high intensity laser beam in the visible band, usually in the blue-green region, to temporarily impair the vision of the adversary without causing any permanent or lasting injury or adverse effect to the subject's eyes.

Laser-Induced Plasma Channel Weapons (LIPC) LIPC weapons are hybrid weapons, which use a laser to ionize a path of molecules to the target, via which an electric charge can be delivered into the target to cause damage effects. The conducting plasma follows the path of the laser and therefore can be directed to different targets by steering the laser beam.

Laser Phalanx This is a directed-energy laser system developed by Raytheon and configured around a 20 kW industrial fibre-laser. The system has been successfully demonstrated against a static mortar from a distance of 0.5 km.

Laser Weapon System (LaWS) LaWS is a ship mounted solid-state laser weapon by Raytheon that has been successfully tested to shoot drones.

Lethality This is the capability of a weapon system to render a target non-functional. Targets may be destroyed, denied, degraded or delayed. The laser-material interaction and the resulting damage mechanism not only depend upon the laser power density but also upon the construction and operational conditions of the target.

Linear Accelerator These accelerate particles over a long, straight line where the particle beam travels from one end to the other.

Linear Beam Sources In the case of linear beam sources, microwave energy is generated by converting kinetic energy of an electron beam into electromagnetic energy of the microwave beam. Examples of linear beam sources include klystrons, travelling wave tubes, backward wave oscillators, magnetrons, cross field amplifiers, split-cavity oscillators, virtual cathode oscillators (VIRCATOR), gyrotrons, free electron lasers and orbitron microwave masers.

LORDS (Laser Ordnance Disposal System) LORDS is laser-based ordnance neutralization system developed by Laser Science and Technology Centre (LASTEC) of Defence Research and Development Organization (DRDO), India. The system employs a 1.0 kW fibre-laser and is mounted along with its support systems on a Light Security Vehicle (LSV) made by TATA Motors.

Magnetron This is a type of high-pwer microwave oscillator in which the potential energy of an electron cloud near the cathode is converted into RF energy in a series of cavity resonators whose resonant frequency is determined by the physical dimension of the resonator together with the reactive effect of any perturbations to the inductive or capacitive portion of its equivalent parallel L-C circuit. Magnetrons are capable of producing megawatts of peak power in the centimetre wavelength range and may be operated at wavelengths extending down to millimetre range.

Maximum Permissible Exposure (MPE) The MPE is the highest power or energy density (in W/cm^2 or J/cm^2) of a light source that has a negligible probability for creating damage.

It is usually about 10% of the dose that has a 50% chance of creating damage under worst-case conditions. The MPE is measured at the cornea of the human eye or at the skin, for a given wavelength and exposure time.

Mid-Infrared Advanced Chemical Laser (MIRACL) MIRACL is a DEW technology demonstrator configured around a 2.2 MW CW DF laser with a maximum lasing duration of 70 s. It uses a 1.8 m Sealite pointing and tracking device.

Mie Scattering This is the scattering type when the scattering centres are roughly comparable to the size of the incident wavelength. These types of Mie scatterers are typically suspended aerosol particles or very small droplets of water. Like Rayleigh scattering, the radiation of incident energy is scattered in all directions with the difference that Mie scattering has predominant scattering in the same direction as the incident photon.

Narrowband HPM Weapons These produce a high-pwer microwave output in a narrow band of frequencies with the bandwidth equal to only a few percent of the centre frequency. These weapons are capable of generating relatively higher output power levels as compared to ultra-wideband HPM weapons. Narrowband HPM systems are effective only on a given class of targets that would absorb the frequency emitted by the system. Narrowband HPM weapons are capable of defeating only well-defined target or class of targets.

Nominal Ocular Hazard Distance (NOHD) This is a distance within which the irradiance of beam is greater than the MPE.

Neutral-Particle Beam Weapons These use a high-energy beam of neutral particles. These are mainly exoatmospheric weapons.

Non-Selective Scattering This is the predominant scattering mechanism when the scattering particles are much larger than the incident wavelength.

Particle Accelerator This is the source of high-energy particle beam. The particle accelerator accelerates particles to extremely high energies.

Particle Beam Weapons This uses a high-energy beam of atomic or subatomic particles to inflict the intended damage to the target by disrupting its atomic and/or molecular structure.

Photonic Disruptor The Photonic Disruptor, classified as a TALI (Threat Assessment Laser Illuminator) is a nonlethal high-power green laser developed by Wicked Lasers, USA in cooperation with Xtreme Alternative Defence Systems. This tactical laser is equipped with a versatile focus-adjustable collimating lens to compensate for range and power intensity when used to either incapacitate an attacker in close range or safely identify threats from a distance. TALI-series devices are configured around a 100 mW, 532 nm laser producing a laser beam with 1.5–7.5 mrad adjustable beam divergence.

RANETS-E RANETS-E, described as a Radio Frequency Cannon, is an HPM weapon system designed to provide terminal by disrupting guidance systems of precision-guided munitions and avionics systems of aircraft.

Rayleigh Scattering This refers to the scattering of light off of molecules in the air. Rayleigh scattering is predominant when the molecular size is smaller than the incident wavelength. The radiation from Rayleigh scattering is emitted in all directions, causing a loss of energy as it propagates through the atmosphere. It is inversely proportional to fourth power of wavelength.

Saber-203 The Saber-203 Grenade Shell Laser Intruder Countermeasure System is a type of laser dazzler that uses a 250 mW red laser diode mounted in a hard plastic capsule in the shape of a standard 40 mm grenade. It is suitable for being loaded into a M203 grenade launcher.

Shortstop Electronic Protection System (SEPS) The SEPS, designed by the Whittaker Corporation in Simi Valley, California, is a portable RF jammer that can be programmed to jam a specific range of frequencies. The system offers effective electronic countermeasures against electronic proximity fuses used in incoming artillery and mortar rounds by countering the threat of RF fused munitions by initiating premature detonation.

Silent GUARDIAN This is a mid-range version of the active denial system with operational range of about 250 m.

Soft-Kill Damage In the case of soft-kill damage, the effect of attack is to deny or degrade the operation of the target platform or even inflict partial damage. Some examples include disrupting electronics of a guided missile forcing it to miss the intended target or damaging visible, infrared and microwave sensors on board the target platform. Though soft-kill damage causes temporary loss of function, it can seriously compromise mission success. HPMs inflict soft-kill damage.

Space-Based Laser (SBL) An SBL is proposed to be configured around 20 MW HF laser operating at 2.7 μm. It is designed to intercept ICBMs and other strategic and tactical missiles. It is proposed to deploy the system in a 800–1300 km orbit and the expected target engagement range is 4000–12 000 km.

Spectral Beam Combining In the case of spectral beam combining, multiple laser beams with non-overlapping optical spectra are combined by using a wavelength sensitive beam combiner such as a diffraction grating and prisms. Optical components with wavelength sensitive transmission characteristics such as volume Bragg gratings and dichroic mirrors can also be used. The wavelength sensitive optical component ensures that all beams though incident at different angles, subsequently propagate in the same direction.

Split-Cavity Oscillator (SCO) The SCO utilizes transit-time bunching of electron beam generate microwave energy. Since there is no requirement for a magnetic field and a low beam quality consistent with simple pulse power sources is adequate, the SCO makes a very compact HPM source.

Tactical High-Energy Laser (THEL) THEL systems by Northrop–Grumman are point defence weapon systems designed to engage and destroy artillery rockets, artillery shells, mortar rounds and low-flying aircraft. The laser is built in two configurations; baseline static HEL and the relocateable mobile version Mobile-THEL (M-THEL). The system uses a DF laser operating at 3.8 μm.

Thermal Blooming (Particle Beam Weapon) This is distortion of a propagating high-pwer laser beam caused by the heating of the air around it due to the high-energy density of the laser beam. Absorption of laser energy causes the temperature of the air the laser beam is passing through, to increase. This creates a refractive index gradient in the laser beam path leading to formation of a negative lens. Formation of negative lens is then the cause of increase in beam spreading. Thermal blooming is far more predominant in the case of high-pwer laser beams aimed at stationary or slow moving targets.

THOR This is a laser ordnance neutralization system from Rafael with an integrated remotely controlled weapon station.

Travelling Wave Tube Amplifier This, like a klystron, is a source of microwave energy. It employs velocity modulation phenomenon. While in the case of klystron amplifier the RF field is stationary and the electron beam is moving, a TWT amplifier makes use of distributed interaction between the electron beam and a travelling wave.

Turbulence This causes a laser beam to spread resulting in increase in laser spot size and reduction of power density at the target. A quantitative measure of the optical turbulence in atmosphere is the refractive index structure parameter $C_n^{\,2}$ expressed in m$^{-2/3}$. $C_n^{\,2}$ is strongly dependent on altitude as well as wind speed and other atmospheric parameters.

Ultra-Wideband HPM Weapons These radiate over a broad frequency range, but deliver comparatively less microwave energy at specific frequencies. Ultra-wideband weapons are intended for use against a wide range of targets.

Vigilant Eagle This, an HPM weapon system from Raytheon, is designed to protect both commercial flights as well as military and private aircraft from the threat of surface-to-air missiles including MANPADS (Man-Portable Air-Defence Systems). The system is installed at the airports or airfields and is primarily intended to provide protection to the aircraft while taking off from or landing at the airports or airfields by directing a precisely steered beam of electromagnetic energy, a set of high-frequency microwaves, at the missile threatening the aircraft in proximity of the airport and diverting the threat away from the targeted aircraft.

Virtual Cathode Oscillator (Vircator) This is capable of generating short pulses of tunable, narrowband microwaves at very high power levels. A vircator belongs to a class of HPM sources that generate microwave radiation by the phenomena of virtual cathode or/and reflex electron oscillations accompanying injection of an electron beam into a waveguide or cavity in which beam current exceeds the local space-charge limiting current.

Warlock Green and Warlock Red Warlock Green and Warlock Red, designed by EDO communications and Countermeasures (now ITT Electronic Systems, EWS), are electronic countermeasure systems and are modified versions of battle-proven Shortstop Electronic Protection System designed by the Whittaker Corporation. Warlock systems have been found to be effective in countering the threat of radio controlled improvised explosive devices (IEDs).

ZEUS-HLONS (HUMMWV Laser Ordnance Neutralization System) This is a vehicle-mounted laser ordnance neutralization system.

ZM-87 Portable Laser Disturber This is an electro-optic countermeasure laser device. It can temporarily blind enemy troops up to 10 km range. The ZM-87 has reportedly been widely deployed, which includes use on naval vessels.

Bibliography

1 Injeyan, H. and Goodno, G. (2011), *High Power Laser Handbook*, McGraw-Hill, Inc.
2 Zohuri, B. (2012), *Directed Energy Weapon Technologies*, CRC Press.
3 McAulay, A.D. (2011), *Military Laser Technology for Defence*, John Wiley & Sons, Inc.
4 Perram, G.P. (2009), *An Introduction to Laser Weapon Systems*, Directed Energy Laser Society.
5 Accetta, J.S. and Shumaker, D.L. (1993), *The Infrared and Electro-Optic Systems Handbook Volume-7*, SPIE International Society for Optical Engineering.
6 Preston, R. (2002), *Space Weapons Earth Wars*, Rand Corporation.
7 Lele, A. (2009), *Strategic Technologies for the Military*, Sage Publications Pvt Ltd.
8 Motes, R.A. (2009), *Introduction to High Power Fiber Lasers*, Directed Energy Professional.
9 Tyson, R. (2015), *Principles of Adaptive Optics*, CRC Press.
10 Benford, J., Swegle, J.A. and Schamiloglu, E. (2015), *High Power Microwaves (Third Edition)*, CRC Press.
11 Barker, R.J. and Schamiloglu, E. (2001), *High Power Microwave Sources and Technologies*, Wiley-IEEE Press.
12 Kartikeyan, M.V., Borie, E. and Thumm, M. (2004), *Gyrotrons: High Power Microwave and Millimeter Wave Technology*, Springer.
13 Hecht, J. (2000), *Beam Weapons: The Next Arms Race*, iUniverse.
14 Beason, J. D. (2009), *The E-Bomb*, Da Capo Press Inc.
15 Duffner, R.W. (1997), *Airborne Laser: Bullets of Light*, Perseus Books.
16 U.S. DoD, *Laser Weapons* (2010), Progressive Management (Kindle Edition).
17 Brignon, A. (2013), *Coherent Laser Beam Combining*, Wiley-VCH.

Index

Handbook of Defence Electronics and Optronics: Fundamentals, Technologies and Systems, First Edition. Anil K. Maini.
© 2018 John Wiley & Sons Ltd. Published 2018 by John Wiley & Sons Ltd.